PURIFICATION OF LABORATORY CHEMICALS

Fifth Edition

PURIFICATION OF LABORATORY CHEMICALS
Fifth Edition

Wilfred L. F. Armarego
Visiting Fellow
Division of Biomolecular Science
The John Curtin School of Medical Research
Australian National University, Canberra
A. C. T., Australia

Christina Li Lin Chai
Reader in Chemistry
Department of Chemistry
Australian National University, Canberra
A. C. T., Australia

An Imprint of Elsevier Science
www.bh.com

Amsterdam Boston London New York Oxford Paris
San Diego San Francisco Singapore Sydney Tokyo

Butterworth-Heinemann is an imprint of Elsevier Science.

Copyright © 2003, Elsevier Science (USA). All rights reserved.

 Recognizing the importance of preserving what has been written, Elsevier-Science prints its books on acid-free paper whenever possible.

Library of Congress Cataloging-in-Publication Data
A catalog record for this book is available from the Library of Congress.

ISBN: 0-7506-7571-3

British Library Cataloguing-in-Publication Data
A catalogue record for this book is available from the British Library.

The publisher offers special discounts on bulk orders of this book.
For information, please contact:

Manager of Special Sales
Elsevier Science
200 Wheeler Road
Burlington, MA 01803
Tel: 781-313-4700
Fax: 781-313-4882

For information on all Butterworth-Heinemann publications available, contact our World Wide Web home page at: http://www.bh.com

10 9 8 7 6 5 4 3 2 1

Printed and bound in Great Britain by MPG Books Ltd, Bodmin, Cornwall

CONTENTS

vi

Preface to the Fifth Edition

THE DEMAND for **Purification of Laboratory Chemicals** has not abated since the publication of the fourth edition as evidenced by the number of printings and the sales. The request by the Editor for a fifth edition offered an opportunity to increase the usefulness of this book for laboratory purposes. It is with deep regret that mention should be made that Dr Douglas D. Perrin had passed away soon after the fourth edition was published. His input in the first three editions was considerable and his presence has been greatly missed. A fresh, new and young outlook was required in order to increase the utility of this book and it is with great pleasure that Dr Christina L.L. Chai, a Reader in Chemistry and leader of a research group in organic and bio-organic chemistry, has agreed to coauthor this edition. The new features of the fifth edition have been detailed below.

Chapters 1 and 2 have been reorganised and updated in line with recent developments. A new chapter on the 'Future of Purification' has been added. It outlines developments in syntheses on solid supports, combinatorial chemistry as well as the use of ionic liquids for chemical reactions and reactions in fluorous media. These technologies are becoming increasingly useful and popular so much so that many future commercially available substances will most probably be prepared using these procedures. Consequently, a knowledge of their basic principles will be helpful in many purification methods of the future.

Chapters 4, 5 and 6 (3, 4 and 5 in the 4th edn) form the bulk of the book. The number of entries has been increased to include the purification of many recent commercially available reagents that have become more and more popular in the syntheses of organic, inorganic and bio-organic compounds. Several purification procedures for commonly used liquids, e.g. solvents, had been entered with excessive thoroughness, but in many cases the laboratory worker only requires a simple, rapid but effective purification procedure for immediate use. In such cases a **Rapid purification** procedure has been inserted at the end of the respective entry, and should be satisfactory for most purposes. With the increased use of solid phase synthesis, even for small molecules, and the use of reagents on solid support (e.g. on polystyrene) for reactions in liquid media, compounds on solid support have become increasingly commercially available. These have been inserted at the end of the respective entry and have been listed in the General Index together with the above rapid purification entries.

A large number of substances are ionisable in aqueous solutions and a knowledge of their ionisation constants, stated as pK (pKa) values, can be of importance not only in their purification but also in their reactivity. Literature values of the pK's have been inserted for ionisable substances, and where values could not be found they were estimated (pK_{Est}). The estimates are usually so close to the true values as not to affect the purification process or the reactivity seriously. The book will thus be a good compilation of pK values for ionisable substances.

Almost all the entries in Chapters 4, 5 and 6 have CAS (Chemical Abstract Service) Registry Numbers to identify them, and these have been entered for each substance. Unlike chemical names which may have more than one synonymous name, there is only one CAS Registry Number for each substance (with only a few exceptions, e.g. where a substance may have another number before purification, or before determination of absolute configuration). To simplify the method for locating the purification of a substance, a CAS Registry Number Index with the respective page numbers has been included after the General Index at the end of the book. This will also provide the reader with a rapid way to see if the purification of a particular substance has been reported in the book. The brief General Index includes page references to procedures and equipment, page references to abbreviations of compounds, e.g. TRIS, as well as the names of substances for which a Registry Number was not found.

Website references for distributors of substances or/and of equipment have been included in the text. However, since these may be changed in the future we must rely on the suppliers to inform users of their change in website references.

We wish to thank readers who have provided advice, constructive criticism and new information for inclusion in this book. We should be grateful to our readers for any further comments, suggestions, amendments and criticisms which could, perhaps, be inserted in a second printing of this edition. In particular, we thank Professor Ken-chi Sugiura (Graduate School of Science, Tokyo Metropolitan University, Japan) who has provided us with information on the purification of several organic compounds from his own experiences, and Joe Papa BS MS (EXAXOL in Clearwater, Florida, USA) who has provided us not only with his experiences in the purification of many inorganic substances in this book, but also gave us his analytical results on the amounts of other metal impurities at various stages of purification of several salts. We thank them graciously for permission to include their reports in this work. We express our gratitude to Dr William B. Cowden for his generous advice on computer hardware and software over many years and for providing an Apple LaserWriter (16/600PS) which we used to produce the master copy of this book. We also extend our sincere thanks to Dr Bart Eschler for advice on computer hardware and software and for assistance in setting up the computers (iMac and eMac) used to produce this book.

We thank Dr Pauline M. Armarego for assistance in the painstaking task of entering data into respective files, for many hours of proofreading, correcting typographical errors and checking CAS Registry Numbers against their respective entries.

One of us (W.L.F.A) owes a debt of gratitude to Dr Desmond (Des) J. Brown of the Research School of Chemistry, ANU, for unfailing support and advice over several decades and for providing data that was difficult to acquire not only for this edition but also for the previous four editions of this book.

One of us (C.L.L.C) would specially like to thank her many research students (past and present) for their unwavering support, friendship and loyalty, which enabled her to achieve what she now has. She wishes also to thank her family for their love, and would particularly like to dedicate her contribution towards this book to the memory of her brother Andrew who had said that he should have been a scientist.

We thank Mrs Joan Smith, librarian of the Research School of Chemistry, ANU, for her generous help in many library matters which have made the tedious task of checking references more enduring.

W.L.F. Armarego & C.L.L. Chai
November 2002

Preface to the First Edition

WE BELIEVE that a need exists for a book to help the chemist or biochemist who wishes to purify the reagents she or he uses. This need is emphasised by the previous lack of any satisfactory central source of references dealing with individual substances. Such a lack must undoubtedly have been a great deterrent to many busy research workers who have been left to decide whether to purify at all, to improvise possible methods, or to take a chance on finding, somewhere in the chemical literature, methods used by some previous investigators.

Although commercially available laboratory chemicals are usually satisfactory, as supplied, for most purposes in scientific and technological work, it is also true that for many applications further purification is essential.

With this thought in mind, the present volume sets out, firstly, to tabulate methods, taken from the literature, for purifying some thousands of individual commercially available chemicals. To help in applying this information, two chapters describe the more common processes currently used for purification in chemical laboratories and give fuller details of new methods which appear likely to find increasing application for the same purpose. Finally, for dealing with substances not separately listed, a chapter is included setting out the usual methods for purifying specific classes of compounds.

To keep this book to a convenient size, and bearing in mind that its most likely users will be laboratory-trained, we have omitted manipulative details with which they can be assumed to be familiar, and also detailed theoretical discussion. Both are readily available elsewhere, for example in Vogel's very useful book **Practical Organic Chemistry** (Longmans, London, 3rd ed., 1956), or Fieser's **Experiments in Organic Chemistry** (Heath, Boston, 3rd ed, 1957).

For the same reason, only limited mention is made of the kinds of impurities likely to be present, and of the tests for detecting them. In many cases, this information can be obtained readily from existing monographs.

By its nature, the present treatment is not exhaustive, nor do we claim that any of the methods taken from the literature are the best possible. Nevertheless, we feel that the information contained in this book is likely to be helpful to a wide range of laboratory workers, including physical and inorganic chemists, research students, biochemists, and biologists. We hope that it will also be of use, although perhaps to only a limited extent, to experienced organic chemists.

We are grateful to Professor A. Albert and Dr D.J. Brown for helpful comments on the manuscript.

D.D.P., W.L.F.A. & D.R.P.
1966

Preface to the Second Edition

SINCE the publication of the first edition of this book there have been major advances in purification procedures. Sensitive methods have been developed for the detection and elimination of progressively lower levels of impurities. Increasingly stringent requirements for reagent purity have gone hand-in-hand with developments in semiconductor technology, in the preparation of special alloys and in the isolation of highly biologically active substances. The need to eliminate trace impurities at the micro- and nanogram levels has placed greater emphasis on ultrapurification technique. To meet these demands the range of purities of laboratory chemicals has become correspondingly extended. Purification of individual chemicals thus depends more and more critically on the answers to two questions - Purification from what, and to what permissible level of contamination. Where these questions can be specifically answered, suitable methods of purification can usually be devised.

Several periodicals devoted to ultrapurification and separations have been started. These include "Progress in Separation and Purification" Ed. (vol. 1) E.S. Perry, Wiley-Interscience, New York, vols. 1-4, 1968-1971, and **Separation and Purification Methods** Ed. E.S.Perry and C.J.van Oss, Marcel Dekker, New York, vol. 1- , 1973-. Nevertheless, there still remains a broad area in which a general improvement in the level of purity of many compounds can be achieved by applying more or less conventional procedures. The need for a convenient source of information on methods of purifying available laboratory chemicals was indicated by the continuing demand for copies of this book even though it had been out of print for several years.

We have sought to revise and update this volume, deleting sections that have become more familiar or less important, and incorporating more topical material. The number of compounds in Chapters 3 and 4 have been increased appreciably. Also, further details in purification and physical constants are given for many compounds that were listed in the first edition.

We take this opportunity to thank users of the first edition who pointed out errors and omissions, or otherwise suggested improvements or additional material that should be included. We are indebted to Mrs S.Schenk who emerged from retirement to type this manuscript.

D.D.P., W.L.F.A. & D.R.P.
1980

Preface to the Third Edition

THE CONTINUING demand for this monograph and the publisher's request that we prepare a new edition, are an indication that **Purification of Laboratory Chemicals** fills a gap in many chemists' reference libraries and laboratory shelves. The present volume is an updated edition which contains significantly more detail than the previous editions, as well as an increase in the number of individual entries and a new chapter.

Additions have been made to Chapters 1 and 2 in order to include more recent developments in techniques (e.g. Schlenk-type, *cf* p. 10), and chromatographic methods and materials. Chapter 3 still remains the core of the book, and lists in alphabetical order relevant information on *ca* 4000 organic compounds. Chapter 4 gives a smaller listing of *ca* 750 inorganic and metal-organic substances, and makes a total increase of ca 13% of individual entries in these two chapters. Some additions have also been made to Chapter 5.

We are currently witnessing a major development in the use of physical methods for purifying large molecules and macromolecules, especially of biological origin. Considerable developments in molecular biology are apparent in techniques for the isolation and purification of key biochemicals and substances of high molecular weight. In many cases something approaching homogeneity has been achieved, as evidenced by electrophoresis, immunological and other independent criteria. We have consequently included a new section, Chapter 6, where we list upwards of 100 biological substances to illustrate their current methods of purification. In this chapter the details have been kept to a minimum, but the relevant references have been included.

The lists of individual entries in Chapters 3 and 4 range in length from single line entries to *ca* one page or more for solvents such as acetonitrile, benzene, ethanol and methanol. Some entries include information such as likely contaminants and storage conditions. More data referring to physical properties have been inserted for most entries [i.e. melting and boiling points, refractive indexes, densities, specific optical rotations (where applicable) and UV absorption data]. Inclusion of molecular weights should be useful when deciding on the quantities of reagents needed to carry out relevant synthetic reactions, or preparing analytical solutions. The Chemical Abstracts registry numbers have also been inserted for almost all entries, and should assist in the precise identification of the substances.

In the past ten years laboratory workers have become increasingly conscious of safety in the laboratory environment. We have therefore in three places in Chapter 1 (pp. 3 and 33, and bibliography p. 52) stressed more strongly the importance of safety in the laboratory. Also, where possible, in Chapters 3 and 4 we draw attention to the dangers involved with the manipulation of some hazardous substances.

The world wide facilities for retrieving chemical information provided by the Chemical Abstract Service (CAS on-line) have made it a relatively easy matter to obtain CAS registry numbers of substances, and most of the numbers in this monograph were obtained *via* CAS on-line. We should point out that two other available useful files are CSCHEM and CSCORP which provide, respectively, information on chemicals (and chemical products) and addresses and telephone numbers of the main branch offices of chemical suppliers.

The present edition has been produced on an IBM PC and a Laser Jet printer using the **Microsoft Word (4.0)** word-processing program with a set stylesheet. This has allowed the use of a variety of fonts and font sizes which has made the presentation more attractive than in the previous edition. Also, by altering the format and increasing slightly the sizes of the pages, the length of the monograph has been reduced from 568 to 391 pages. The reduction in the number of pages has been achieved in spite of the increase of *ca* 15% of total text.

We extend our gratitude to the readers whose suggestions have helped to improve the monograph, and to those who have told us of their experiences with some of the purifications stated in the previous editions, and in particular with the hazards that they have encountered. We are deeply indebted to Dr M.D. Fenn for the several hours that he has spent on the terminal to provide us with a large number of CAS registry numbers.

This monograph could not have been produced without the expert assistance of Mr David Clarke who has spent many hours to load the necessary fonts in the computer, and for advising one of the authors (W.L.F.A.) on how to use them together with the idiosyncrasies of Microsoft Word.

D.D.P. & W.L.F.A.
1988

Preface to the Fourth Edition

THE AIMS of the first three editions, to provide purification procedures of commercially available chemicals and biochemicals from published literature data, are continued in this fourth edition. Since the third edition in 1988 the number of new chemicals and biochemicals which have been added to most chemical and biochemical catalogues have increased enormously. Accordingly there is a need to increase the number of entries with more recent useful reagents and chemical and biochemical intermediates. With this in mind, together with the need to reorganise and update general purification procedures, particularly in the area of biological macromolecules, as well as the time lapse since the previous publication, this fourth edition of **Purification of Laboratory Chemicals** has been produced. Chapter 1 has been reorganised with some updating, and by using a smaller font it was kept to a reasonable number of pages. Chapters 2 and 5 were similarly altered and have been combined into one chapter. Eight hundred and three hundred and fifty entries have been added to Chapters 3 (25% increase) and 4 (44% increase) respectively, and four hundred entries (310% increase) were added to Chapter 5 (Chapter 6 in the Third Edition), making a total of 5700 entries; all resulting in an increase from 391 to 529 pages, i.e. by *ca* 35%.

Many references to the original literature have been included remembering that some of the best references happened to be in the older literature. Every effort has been made to provide the best references but this may not have been achieved in all cases. Standard abbreviations, listed on page 1, have been used throughout this edition to optimise space, except where no space advantage was achieved, in which cases the complete words have been written down to improve the flow of the sentences.

With the increasing facilities for information exchange, chemical, biochemical and equipment suppliers are making their catalogue information available on the Internet , e.g. Aldrich-Fluka-Sigma catalogue information is available on the World Wide Web by using the address http://www.sigma.sial.com, and GIBCO BRL catalogue information from http://www.lifetech.com, as well as on CD-ROMS which are regularly updated. Facility for enquiring about, ordering and paying for items is available *via* the Internet. CAS on-line can be accessed on the Internet, and CAS data is available now on CD-ROM. Also biosafety bill boards can similarly be obtained by sending SUBSCRIBE SAFETY John Doe at the address "listserv@uvmvm.uvm.edu", SUBSCRIBE BIOSAFETY at the address "listserv@mitvma.mit.edu", and SUBSCRIBE RADSAF at the address "listserv@romulus.ehs.uiuc.edu"; and the Occupational, Health and Safety information (Australia) is available at the address "http://www.worksafe.gov.au/~wsa1". Sigma-Aldrich provide Material Safety data sheets on CD-ROMs.

It is with much sadness that Dr Douglas D. Perrin was unable to participate in the preparation of the present edition due to illness. His contributions towards the previous editions have been substantial, and his drive and tenacity have been greatly missed.

The Third Edition was prepared on an IBM-PC and the previous IBM files were converted into Macintosh files. These have now been reformatted on a Macintosh LC575 computer and all further data to complete the Fourth Edition were added to these files. The text was printed with a Hewlett-Packard 4MV -600dpi Laser Jet printer which gives a clearer resolution.

I thank my wife Dr Pauline M. Armarego, also an organic chemist, for the arduous and painstaking task of entering the new data into the respective files, and for the numerous hours of proofreading as well as the corrections of typographic errors in the files. I should be grateful to my readers for any comments, suggestions, amendments and criticisms which could, perhaps, be inserted in the second printing of this edition.

W.L.F. Armarego
30 June 1996

CHAPTER 1

COMMON PHYSICAL TECHNIQUES
USED IN PURIFICATION

INTRODUCTION

Purity is a matter of degree. Other than adventitious contaminants such as dust, paper fibres, wax, cork, etc., that may have been incorporated into the sample during manufacture, all commercially available chemical substances are in some measure impure. Any amounts of unreacted starting material, intermediates, by-products, isomers and related compounds may be present depending on the synthetic or isolation procedures used for preparing the substances. Inorganic reagents may deteriorate because of defective packaging (glued liners affected by sulfuric acid, zinc extracted from white rubber stoppers by ammonia), corrosion or prolonged storage. Organic molecules may undergo changes on storage. In extreme cases the container may be incorrectly labelled or, where compositions are given, they may be misleading or inaccurate for the proposed use. Where any doubt exists it is usual to check for impurities by appropriate spot tests, or by recourse to tables of physical or spectral properties such as the extensive infrared and NMR libraries published by the Sigma Aldrich Chemical Co.

The important question, then, is not whether a substance is pure but whether a given sample is sufficiently pure for some intended purpose. That is, are the contaminants likely to interfere in the process or measurement that is to be studied. By suitable manipulation it is often possible to reduce levels of impurities to acceptable limits, but absolute purity is an ideal which, no matter how closely approached, can never be attained. A *negative* physical or chemical test indicates only that the amount of an impurity in a substance lies below a certain sensitivity level; no test can demonstrate that a specified impurity is entirely absent.

When setting out to purify a laboratory chemical, it is desirable that the starting material is of the best grade commercially available. Particularly among organic solvents there is a range of qualities varying from *laboratory chemical* to *spectroscopic* and *chromatographic* grades. Many of these are suitable for use as received. With many of the more common reagents it is possible to obtain from the current literature some indications of likely impurities, their probable concentrations and methods for detecting them. However, in many cases complete analyses are not given so that significant concentrations of unspecified impurities may be present.

THE QUESTION OF PURITY

Solvents and substances that are specified as *pure* for a particular purpose may, in fact, be quite impure for other uses. Absolute ethanol may contain traces of benzene, which makes it unsuitable for ultraviolet spectroscopy, or plasticizers which make it unsuitable for use in solvent extraction.

Irrespective of the grade of material to be purified, it is essential that some criteria exist for assessing the degree of purity of the final product. The more common of these include:

1. Examination of physical properties such as:
    (a) Melting point, freezing point, boiling point, and the freezing curve (i.e. the variation, with time, in the freezing point of a substance that is being slowly and continuously frozen).
    (b) Density.
    (c) Refractive index at a specified temperature and wavelength. The sodium D line at 589.26 nm (weighted mean of D_1 and D_2 lines) is the usual standard of wavelength but results from other wavelengths can often be interpolated from a plot of refractive index versus $1/(\text{wavelength})^2$.

(d) Specific conductivity. (This can be used to detect, for example, water, salts, inorganic and organic acids and bases, in non-electrolytes).
(e) Optical rotation, optical rotatory dispersion and circular dichroism.

2. Empirical analysis, for C, H, N, ash, etc.

3. Chemical tests for particular types of impurities, e.g. for peroxides in aliphatic ethers (with acidified KI), or for water in solvents (quantitatively by the Karl Fischer method, see Fieser and Fieser, Reagents for Organic Synthesis J. Wiley & Sons, NY, Vol 1 pp. 353, **528**, *1967*, Library of Congress Catalog Card No 66-27894).

4. Physical tests for particular types of impurities:
    (a) Emission and atomic absorption spectroscopy for detecting organic impurities and determining metal ions.
    (b) Chromatography, including paper, thin layer, liquid (high, medium and normal pressure) and vapour phase.
    (c) Electron spin resonance for detecting free radicals.
    (d) X-ray spectroscopy.
    (e) Mass spectroscopy.
    (f) Fluorimetry.

5. Examination of spectroscopic properties
    (a) Nuclear Magnetic Resonance (^{1}H, ^{13}C, ^{31}P, ^{19}F NMR etc)
    (b) Infrared spectroscopy (IR)
    (c) Ultraviolet spectroscopy (UV)
    (d) Mass spectroscopy [electron ionisation (EI), electron ionisation (CI), electrospray ionisation (ESI), fast atom bombardment (FAB), matrix-associated laser desorption ionisation (MALDI), etc]

6. Electrochemical methods (see Chapter 6 for macromolecules).

7. Nuclear methods which include a variety of radioactive elements as in organic reagents, complexes or salts.

A substance is usually taken to be of an acceptable purity when the measured property is unchanged by further treatment (especially if it agrees with a recorded value). In general, at least two different methods, such as recrystallisation and distillation, should be used in order to ensure maximum purity. Crystallisation may be repeated (from the same solvent or better from different solvents) until the substance has a constant melting point or absorption spectrum, and until it distils repeatedly within a narrow, specified temperature range.

With liquids, the refractive index at a specified temperature and wavelength is a sensitive test of purity. Note however that this is sensitive to dissolved gases such as O_2, N_2 or CO_2. Under favourable conditions, freezing curve studies are sensitive to impurity levels of as little as 0.001 moles per cent. Analogous fusion curves or heat capacity measurements can be up to ten times as sensitive as this. With these exceptions, most of the above methods are rather insensitive, especially if the impurities and the substances in which they occur are chemically similar. In some cases, even an impurity comprising many parts per million of a sample may escape detection.

The common methods of purification, discussed below, comprise distillation (including fractional distillation, distillation under reduced pressure, sublimation and steam distillation), crystallisation, extraction, chromatographic and other methods. In some cases, volatile and other impurities can be removed simply by heating. Impurities can also sometimes be eliminated by the formation of derivatives from which the purified material is regenerated (see Chapter 2).

SOURCES OF IMPURITIES

Some of the more obvious sources of contamination of solvents arise from storage in metal drums and plastic containers, and from contact with grease and screw caps. Many solvents contain water. Others have traces of acidic materials such as hydrochloric acid in chloroform. In both cases this leads to corrosion of the drum and contamination of the solvent by traces of metal ions, especially Fe^{3+}. Grease, for example on stopcocks of separating funnels and other apparatus, e.g. greased ground joints, is also likely to contaminate solvents during extractions and chemical manipulation.

A much more general source of contamination that has not received the consideration it merits comes from the use of plastics for tubing and containers. Plasticisers can readily be extracted by organic solvents from PVC and other plastics, so that most solvents, irrespective of their grade (including spectrograde and ultrapure) have been reported to contain 0.1 to 5ppm of plasticiser [de Zeeuw, Jonkman and van Mansvelt *Anal Biochem* **67** 339 *1975*]. Where large quantities of solvent are used for extraction (particularly of small amounts of compounds), followed by evaporation, this can introduce significant amounts of impurity, even exceeding the weight of the genuine extract and giving rise to spurious peaks in gas chromatography (for example of fatty acid methyl esters [Pascaud, *Anal Biochem* **18** 570 *1967*]. Likely contaminants are di(2-ethylhexyl)phthalate and dibutyl phthalate, but upwards of 20 different phthalate esters are listed as plasticisers as well as adipates, azelates, phosphates, epoxides, polyesters and various heterocyclic compounds. These plasticisers would enter the solvent during passage through plastic tubing or from storage in containers or from plastic coatings used in cap liners for bottles. Such contamination could arise at any point in the manufacture or distribution of a solvent. The problem with cap liners is avoidable by using corks wrapped in aluminium foil, although even in this case care should be taken because aluminium foil can dissolve in some liquids e.g. benzylamine and propionic acid.

Solutions in contact with polyvinyl chloride can become contaminated with trace amounts of lead, titanium, tin, zinc, iron, magnesium or cadmium from additives used in the manufacture and moulding of PVC.

N-Phenyl-2-naphthylamine is a contaminant of solvents and biological materials that have been in contact with black rubber or neoprene (in which it is used as an antioxidant). Although it was only an artefact of the separation procedure it has been isolated as an apparent component of vitamin K preparations, extracts of plant lipids, algae, livers, butter, eye tissue and kidney tissue [Brown *Chem Br* **3** 524 *1967*].

Most of the above impurities can be removed by prior distillation of the solvent, but care should be taken to avoid plastic or black rubber as much as possible.

PRACTICES TO AVOID IMPURITIES
Cleaning practices

Laboratory glassware and Teflon equipment can be cleaned satisfactorily for most purposes by careful immersion into a solution of sodium dichromate in concentrated sulfuric acid, followed by draining, and rinsing copiously with distilled water. This is an exothermic reaction and should be carried out **very** cautiously in an efficient fume cupboard. [To prepare the chromic acid bath, dissolve 5 g of sodium dichromate (CARE: cancer suspect agent) in 5 mL of water. The dichromate solution is then cooled and stirred while 100 mL of concentrated sulfuric acid is added slowly. Store in a glass bottle.] Where traces of chromium (adsorbed on the glass) must be avoided, a 1:1 mixture of concentrated sulfuric and nitric acid is a useful alternative. (*Use in a fumehood to remove vapour and with adequate face protection.*) Acid washing is also suitable for polyethylene ware but prolonged contact (some weeks) leads to severe deterioration of the plastic. Alternatively an alcoholic solution of sodium hydroxide (alkaline base bath) can be used. This strongly corrosive solution (CAUTION: Alkali causes serious burns) can be made by dissolving 120g of NaOH in 120 mL water, followed by dilution to 1 L with 95% ethanol. This solution is conveniently stored in suitable alkali resistant containers (e.g. Nalgene heavy duty rectangular tanks) with lids. Glassware can be soaked overnight in the base bath and rinsed thoroughly after soaking. For much glassware, washing with hot detergent solution, using tap water, followed by rinsing with distilled water and acetone, and heating to 200-300° overnight, is adequate. (Volumetric apparatus should not be heated: after washing it is rinsed with acetone, then hexane, and air-dried. Prior to use, equipment can be rinsed with acetone, then with petroleum ether or hexane, to remove the last traces of contaminants.) Teflon equipment should be soaked, first in acetone, then in petroleum ether or hexane for ten minutes prior to use.

For trace metal analyses, prolonged soaking of equipment in 1M nitric acid may be needed to remove adsorbed metal ions.

Soxhlet thimbles and filter papers may contain traces of lipid-like materials. For manipulations with highly pure materials, as in trace-pesticide analysis, thimbles and filter papers should be thoroughly extracted with hexane before use.

Trace impurities in silica gel for TLC can be removed by heating at 300° for 16h or by Soxhlet extraction for 3h with distilled chloroform, followed by 4h extraction with distilled hexane.

Silylation of glassware and plasticware

Silylation of apparatus makes it repellant to water and hydrophilic materials. It minimises loss of solute by adsorption onto the walls of the container. The glassware is placed in a desiccator containing dichloromethyl silane (1mL) in a small beaker and evacuated for 5min. The vacuum is turned off and air is introduced into the desiccator which allows the silylating agent to coat the glassware uniformly. The desiccator is then evacuated, closed and set aside for 2h. The glassware is removed from the desiccator and baked at 180° for 2h before use.

Plasticware is treated similarly except that it is rinsed well with water before use instead of baking. Note that dichloromethyl silane is highly **TOXIC** and **VOLATILE**, and the whole operation should be carried out in an efficient fume cupboard.

An alternative procedure used for large apparatus is to rinse the apparatus with a 5% solution of dichloromethyl silane in chloroform, followed by several rinses with water before baking the apparatus at 180°/2h (for glass) or drying in air (for plasticware).

Plus One REPEL-SILANE ES (a solution of 2% w/v of dichloromethyl silane in octamethyl cyclooctasilane) is used to inhibit the sticking of polyacrylamide gels, agarose gels and nucleic acids to glass surfaces and is available commercially (Amersham Biosciences).

SAFETY PRECAUTIONS ASSOCIATED WITH THE PURIFICATION OF LABORATORY CHEMICALS

Although most of the manipulations involved in purifying laboratory chemicals are inherently safe, care is necessary if hazards are to be avoided in the chemical laboratory. In particular there are dangers inherent in the inhalation of vapours and absorption of liquids and low melting solids through the skin. In addition to the toxicity of solvents there is also the risk of their flammability and the possibility of eye damage. Chemicals, particularly in admixture, may be explosive. Compounds may be carcinogenic or otherwise deleterious to health. Present day chemical catalogues specifically indicate the particular dangerous properties of the individual chemicals they list and these should be consulted whenever the use of commercially available chemicals is contemplated. Radioisotopic labelled compounds pose special problems of human exposure and of disposal of laboratory waste. Hazardous purchased chemicals are accompanied by detailed MSDS (Material Safety Data Sheets), which contain information regarding their toxicity, safety handling procedures and the necessary precautions to be taken. These should be read carefully and filed for future reference. In addition, chemical management systems such as ChemWatch which include information on hazards, handling and storage are commercially available. There are a number of websites which provide selected safety information: they include the Sigma-Aldrich website (www.sigmaaldrich.com) and other chemical websites e.g. www.ilpi.com/msds).

The most common hazards are:
(1) Explosions due to the presence of peroxides formed by aerial oxidation of ethers and tetrahydrofuran, decahydronaphthalene, acrylonitrile, styrene and related compounds.
(2) Compounds with low flash points (below room temperature). Examples are acetaldehyde, acetone, acetonitrile, benzene, carbon disulfide, cyclohexane, diethyl ether, ethyl acetate and n-hexane.
(3) Contact of oxidising agents ($KMnO_4$, $HClO_4$, chromic acid) with organic liquids.
(4) Toxic reactions with tissues.

The laboratory should at least be well ventilated and safety glasses should be worn, particularly during distillation and manipulations carried out under reduced pressure or elevated temperatures. With this in mind we have endeavoured to warn users of this book whenever greater than usual care is needed in handling chemicals. As a general rule, however, **all chemicals which users are unfamiliar with should be treated with extreme care and assumed to be highly flammable and toxic.** The safety of others in a laboratory should always be foremost in mind, with ample warning whenever a potentially hazardous operation is in progress. Also, unwanted solutions or solvents should never be disposed of *via* the laboratory sink. The operator should be aware of the usual means for disposal of chemicals in her/his laboratories and she/he should remove unwanted chemicals accordingly. **Organic liquids for disposal should be temporarily stored, as is practically possible, in respective containers. Avoid placing all organic liquids in the same container particularly if they contain small amounts of reagents which could react with each other. Halogenated waste solvents should be kept separate from other organic liquids.**

SOME HAZARDS OF CHEMICAL MANIPULATION IN PURIFICATION AND RECOVERY OF RESIDUES

Performing chemical manipulations calls for some practical knowledge if danger is to be avoided. However, with care, hazards can be kept to an acceptable minimum. A good general approach is to consider every operation as potentially perilous and then to adjust one's attitude as the operation proceeds. A few of the most common dangers are set out below. For a larger coverage of the following sections, and of the literature, the bibliography at the end of this chapter should be consulted.

Perchlorates and perchloric acid. At 160° perchloric acid is an exceedingly strong oxidising acid and a strong dehydrating agent. Organic perchlorates, such as methyl and ethyl perchlorates, are unstable and are violently explosive compounds. A number of heavy-metal perchlorates are extremely prone to explode. The use of anhydrous magnesium perchlorate, *Anhydrone, Dehydrite*, as a drying agent for organic vapours is **not** recommended. Desiccators which contain this drying agent should be adequately shielded at all times and kept in a cool place, i.e. **never** on a window sill where sunlight can fall on it.

No attempt should be made to purify perchlorates, except for ammonium, alkali metal and alkaline earth salts which, in water or aqueous alcoholic solutions are insensitive to heat or shock. Note that perchlorates react relatively slowly in aqueous organic solvents, but as the water is removed there is an increased possibility of an explosion. Perchlorates, often used in non-aqueous solvents, are explosive in the presence of even small amounts of organic compounds when heated. Hence stringent care should be taken when purifying perchlorates, and direct flame and infrared lamps should be avoided. Tetra-alkylammonium perchlorates should be dried below 50° under vacuum (and protection). Only very small amounts of such materials should be prepared, and stored, at any one time.

Peroxides. These are formed by aerial oxidation or by autoxidation of a wide range of organic compounds, including diethyl ether, allyl ethyl ether, allyl phenyl ether, dibenzyl ether, benzyl butyl ether, *n*-butyl ether, *iso*-butyl ether, *t*-butyl ether, dioxane, tetrahydrofuran, olefins, and aromatic and saturated aliphatic hydrocarbons. They accumulate during distillation and can detonate violently on evaporation or distillation when their concentration becomes high. If peroxides are likely to be present materials should be tested for peroxides before distillation (for tests see entry under "Ethers", in Chapter 2). Also, distillation should be discontinued when at least one quarter of the residue is left in the distilling flask.

Heavy-metal-containing-explosives. Ammoniacal silver nitrate, on storage or treatment, will eventually deposit the highly explosive silver nitride *"fulminating silver"*. Silver nitrate and ethanol may give silver fulminate (see Chapter 5), and in contact with azides or hydrazine and hydrazides may form silver azide. Mercury can also form such compounds. Similarly, ammonia or ammonium ions can react with gold salts to form *"fulminating gold"*. Metal fulminates of cadmium, copper, mercury and thallium are powerfully explosive, and some are detonators [Luchs, *Photog Sci Eng* **10** 334 *1966*]. Heavy metal containing solutions, particularly when organic material is present should be treated with great respect and precautions towards possible explosion should be taken.

Strong acids. In addition to perchloric acid (see above), extra care should be taken when using strong mineral acids. Although the effects of concentrated sulfuric acid are well known these cannot be stressed strongly enough. Contact with tissues will leave irreparable damage. **Always dilute the concentrated acid by carefully adding the acid down the side of the flask which contains water, and the process should be carried out under cooling. This solution is not safe to handle until the acid has been thoroughly mixed with the water. Protective face, and body coverage should be used at all times.** Fuming sulfuric acid and chlorosulfonic acid are even more dangerous than concentrated sulfuric acid and adequate precautions should be taken. Chromic acid cleaning mixture contains strong sulfuric acid and should be treated in the same way; and in addition the mixture is potentially *carcinogenic*.
Concentrated and fuming nitric acids are also dangerous because of their severe deleterious effects on tissues.

Reactive halides and anhydrides. Substances like acid chlorides, low molecular weight anhydrides and some inorganic halides (e.g. PCl_3) can be **highly toxic and lachrymatory affecting mucous membranes and lung tissues. Utmost care should be taken when working with these materials. Work should be carried out in a very efficient fume cupboard.**

Solvents. The flammability of low-boiling organic liquids cannot be emphasised strongly enough. These invariably have very low flash points and can ignite spontaneously. Special precautions against explosive flammability should be taken when recovering such liquids. Care should be taken with small volumes (ca 250mL) as well as large volumes (> 1L), and the location of all the fire extinguishers, and fire blankets, in the immediate vicinity of the apparatus should be checked. The fire extinguisher should be operational. The following flammable liquids (in alphabetical order) are common fire hazards in the laboratory: acetaldehyde, acetone, acrylonitrile, acetonitrile, benzene, carbon disulfide, cyclohexane, diethyl ether, ethyl acetate, hexane, low-boiling petroleum ether, tetrahydrofuran and toluene. Toluene should always be used in place of benzene wherever possible due to the potential *carcinogenic* effects of the liquid and vapour of the latter.
The drying of flammable solvents with sodium or potassium metal and metal hydrides poses serious potential fire hazards and adequate precautions should be stressed.

Salts. In addition to the dangers of perchlorate salts, other salts such as nitrates, azides and diazo salts can be hazardous and due care should be taken when these are dried. Large quantities should never be prepared or stored for long periods.

SAFETY   DISCLAIMER

Experimental chemistry is a very dangerous occupation and extreme care and adequate safety precautions should be taken at all times. Although we have stated the safety measures that have to be taken under specific entries these are by no means exhaustive and some may have been unknowingly or accidentally omitted. The experimenter without prior knowledge or experience must seek further safety advice on reagents and procedures from experts in the field before undertaking the purification of any material, We take no responsibility whatsoever if any mishaps occur when using any of the procedures described in this book.

METHODS OF PURIFICATION OF REAGENTS AND SOLVENTS

Many methods exist for the purification of reagents and solvents. A number of these methods are routinely used in synthetic as well as analytical chemistry and biochemistry. These techniques, outlined below, will be discussed in greater detail in the respective sections in this Chapter. It is important to note that more than one method of purification may need to be implemented in order to obtain compounds of highest purity.
Common methods of purification are:
(a)   Solvent Extraction and Distribution
(b)   Distillation
(c)   Recrystallisation
(d)   Sublimation
(e)   Chromatography

For substances contaminated with water or solvents, drying with appropriate absorbents and desiccants may be sufficient.

SOLVENT EXTRACTION AND DISTRIBUTION

Extraction of a substance from suspension or solution into another solvent can sometimes be used as a purification process. Thus, organic substances can often be separated from inorganic impurities by shaking an aqueous solution or suspension with suitable immiscible solvents such as benzene, carbon tetrachloride, chloroform, diethyl ether, diisopropyl ether or petroleum ether. After several such extractions the combined organic phase is dried and the solvent is evaporated. Grease from the glass taps of conventional separating funnels is invariably soluble in the solvents used. Contamination with grease can be very troublesome particularly when the amounts of material to be extracted are very small. Instead, the glass taps should be lubricated with the extraction solvent; or better, the taps of the extraction funnels should be made of the more expensive material *Teflon*. Immiscible solvents suitable for extractions are given in Table 1. Addition of electrolytes (such as ammonium sulfate, calcium chloride or sodium chloride) to the aqueous phase helps to ensure that the organic layer separates cleanly and also decreases the extent of extraction into the latter. Emulsions can also be broken up by filtration (with suction) through Celite, or by adding a little octyl alcohol or some other paraffinic alcohol. The main factor in selecting a suitable immiscible solvent is to find one in which the material to be extracted is readily soluble, whereas the substance from which it is being extracted is not. The same considerations apply irrespective of whether it is the substance being purified, or one of its contaminants, that is taken into the new phase. (The second of these processes is described as washing.)

Common examples of washing with aqueous solutions include the following:
   Removal of acids from water-immiscible solvents by washing with aqueous alkali, sodium carbonate or sodium bicarbonate.
   Removal of phenols from similar solutions by washing with aqueous alkali.
   Removal of organic bases by washing with dilute hydrochloric or sulfuric acids.
   Removal of unsaturated hydrocarbons, of alcohols and of ethers from saturated hydrocarbons or alkyl halides by washing with cold concentrated sulfuric acid.
This process can also be applied to purification of the substance if it is an acid, a phenol or a base, by extracting into the appropriate aqueous solution to form the salt which, after washing with pure solvent, is again converted to

the free species and re-extracted. Paraffin hydrocarbons can be purified by extracting them with phenol (in which aromatic hydrocarbons are highly soluble) prior to fractional distillation.

For extraction of solid materials with a solvent, a *Soxhlet* extractor is commonly used. This technique is applied, for example, in the alcohol extraction of dyes to free them from insoluble contaminants such as sodium chloride or sodium sulfate.

Acids, bases and amphoteric substances can be purified by taking advantage of their ionisation constants.

Ionisation constants and pK.

When substances ionise their neutral species produce positive and negative species. The ionisation constants are those constant values (equilibrium constants) for the equilibria between the charged species and the neutral species, or species with a larger number of charges (e.g. between mono and dications). These ionisation constants are given as **pK** values where **pK = -log K** and **K** is the dissociation constant for the equilibrium between the species [Albert and Serjeant *The Determination of Ionisation Constants*, A Laboratory Manual, 3rd Edition, Chapman & Hall, New York, London, 1984, ISBN 0412242907].

The advantage of using pK values (instead of K values) is that theory (and practice) states that the pK values of ionisable substances are numerically equal to the pH of the solution at which the concentrations of ionised and neutral species are equal. For example acetic acid has a pK^{25} value of 4.76 at 25° in H_2O, then at pH 4.76 the aqueous solution contains equal amounts of acetic acid [AcOH] and acetate anion [AcO⁻], i.e. [AcOH]/[AcO⁻] of 50/50. At pH 5.76 (pK + 1) the solution contains [AcOH]/[AcO⁻] of 10/90, at pH 6.76 (pK + 2) the solution contains [AcOH]/[AcO⁻] of 1/99 etc; conversely at pH 3.76 (pK - 1) the solution contains [AcOH]/[AcO⁻] of 90/10, and at pH 2.76 (pK - 2) the solution contains [AcOH]/[AcO⁻] of 99/1.

One can readily appreciate the usefulness of pK value in purification procedures, e.g. as when purifying acetic acid. If acetic acid is placed in aqueous solution and the pH adjusted to 7.76 {[AcOH]/[AcO⁻] with a ratio of 0.1/99.9}, and extracted with say diethyl ether, neutral impurities will be extracted into diethyl ether leaving almost all the acetic acid in the form of AcO⁻ in the aqueous solution. If then the pH of the solution is adjusted to 1.67 where the acid is almost all in the form AcOH, almost all of it will be extracted into diethyl ether.

Aniline will be used as a second example. It has a pK^{25} of 4.60 at 25° in H_2O. If it is placed in aqueous solution at pH 1.60 it will exist almost completely (99.9%) as the anilinium cation. This solution can then be extracted with solvents e.g. diethyl ether to remove neutral impurities. The pH of the solution is then adjusted to 7.60 whereby aniline will exist as the free base (99.9%) and can be extracted into diethyl ether in order to give purer aniline.

See Table 2 for the pH values of selected buffers.

A knowledge of the pK allows the adjustment of the pH without the need of large excesses of acids or base. In the case of inorganic compounds a knowledge of the pK is useful for adjusting the ionic species for making metal complexes which could be masked or extracted into organic solvents [Perrin and Dempsey *Buffers for pH and Metal ion Control*, Chapman & Hall, New York, London, 1974, ISBN 0412117002], or for obtaining specific anionic species in solution e.g. $H_2PO_4^-$, HPO_4^{2-} or PO_4^{3-}.

The **pK** values that have been entered in Chapters 4, 5 and 6 have been collected directly from the literature or from compilations of literature values for organic bases [Perrin *Dissociation Constants of Organic Bases in Aqueous Solution*, Butterworths, London, 1965, Supplement 1972, ISBN 040870408X; Albert and Serjeant *The Determination of Ionisation Constants*, A Laboratory Manual, 3rd Edition, Chapman & Hall, London, New York, 1984, ISBN 0412242907]; organic acids [Kortum, Vogel and Andrussow, *Dissociation Constants of Organic Acids in Aqueous Solution*, Butterworth, London, 1961; Serjeant and Dempsey, *Dissociation Constants of Organic Acids in Aqueous Solution*, Pergamon Press, Oxford, New York, 1979, ISBN 0080223397; and inorganic acids and bases [Perrin, *Ionisation Constants of Inorganic Acids and Bases in Aqueous Solution*, Second Edition, Pergamon Press, Oxford, New York, 1982, ISBN 0080292143]. Where literature values were not available, values have been predicted and assigned **pK_{Est} ~**. Most predictions should be so close to true values as to make very small difference for the purposes intended in this book. The success of the predictions, i.e. how close to the true value, depends on the availability of pK values for closely related compounds because the effect of substituents or changes in structures are generally additive [Perrin, Dempsey and Serjeant, *pKa Prediction for Organic Acids and Bases*, Chapman & Hall, London, New York, 1981, ISBN 041222190X].

All the pK values in this book are pKa values, the acidic pK, i.e. dissociation of H^+ from an acid (AH) or from a conjugate base (BH^+). Occasionally pKb values are reported in the literature but these can be converted using the equation **pKa + pKb = 14**. For strong acids e.g. sulfuric acid, and strong bases, e.g. sodium hydroxide, the pK values lie beyond the 1 to 11 scale and have to be measured in strong acidic and basic media. In these cases appropriate scales e.g. the H_o (for acids) and H_- (for bases) have been used [see Katritzky and Waring *J Chem Soc* 1540 *1962*]. These values will be less than 1 (and negative) for acids and >11 for bases. They are a rough guides to the strengths of acids and bases. Errors in the stated pK and pK_{Est} ~ values can be judged from the numerical values given. Thus pK values of 4.55, 4.5 and 4 mean that the respective errors are better than ± 0.05, ± 0.3 and ± 0.5. Values taken from the literature are written as **pK**, and all the values that were estimated because they were not found in the literature are written as pK_{Est}.

pK and Temperature.

The temperatures at which the literature measurements were made are given as superscripts, e.g. pK^{25}. Where no temperature is given, it is assumed that the measurements were carried out at room temperature, e.g. 15—25°. No temperature is given for estimated values (pK_{Est} ~) and these have been calculated from data at room temperature. The variation of pK with temperature is given by the equation:

$$- d(pK)/dT = (pK + 0.052 \Delta S^o)/T$$

where T is in degrees Kelvin and ΔS^o is in Joules deg^{-1} mol^{-1}. The -d(pK)/dT in the range of temperatures between 5 to 70° is generally small (e.g. between ~0.0024 and ~0.04), and for chemical purification purposes is not a seriously deterring factor. It does however, vary with the compound under study because ΔS^o varies from compound to compound. The following are examples of the effect of temperature on pK values: for imidazole the pK values are 7.57 (0°), 7.33 (10°), 7.10 (20°), 6.99 (25°), 6.89 (30°), 6.58 (40°) and 6.49 (50°), and for 3,5-dinitrobenzoic acid they are 2.60 (10°), 2.73 (20°), 2.85 (30°), 2.96 (40°) and 3.07 (40°), and for *N*-acetyl-β-alanine they are 4.4788 (5°), 4.4652 (10°), 4.4564 (15°), 4.4488 (20°), 4.4452 (25°), 4.4444 (30°), 4.4434 (35°) and 4.4412 (40°).

pK and solvent.

All stated pK values in this book are for data in dilute aqueous solutions unless otherwise stated, although the dielectric constants, ionic strengths of the solutions and the method of measurement, e.g. potentiometric, spectrophotometric etc, are not given. Estimated values are also for dilute aqueous solutions whether or not the material is soluble enough in water. Generally the more dilute the solution the closer is the pK to the real thermodynamic value. The pK in mixed aqueous solvents can vary considerably with the relative concentrations and with the nature of the solvents. For example the pK^{25} values for *N*-benzylpenicillin are 2.76 and 4.84 in H_2O and H_2O/EtOH (20:80) respectively; the pK^{25} values for (-)-ephedrine are 9.58 and 8.84 in H_2O and H_2O/MeOCH$_2$CH$_2$OH (20:80) respectively; and for cyclopentylamine the pK^{25} values are 10.65 and 4.05 in H_2O and H_2O/EtOH (50:50) respectively. pK values in acetic acid or aqueous acetic acid are generally lower than in H_2O.

The dielectric constant of the medium affects the equilibria where charges are generated in the dissociations e.g. AH $\rightleftharpoons$ A^- + H^+ and therefore affects the pK values. However, its effect on dissociations where there are no changes in total charge such as BH^+ $\rightleftharpoons$ B + H^+ is considerably less, with a slight decrease in pK with decreasing dielectric constant.

DISTILLATION

One of the most widely applicable and most commonly used methods of purification of liquids or low melting solids (especially of organic chemicals) is fractional distillation at atmospheric, or some lower, pressure. Almost without exception, this method can be assumed to be suitable for all organic liquids and most of the low-melting organic solids. For this reason it has been possible in Chapter 4 to omit many procedures for purification of organic chemicals when only a simple fractional distillation is involved - the suitability of such a procedure is implied from the boiling point.

The boiling point of a liquid varies with the 'atmospheric' pressure to which it is exposed. A liquid boils when its vapour pressure is the same as the external pressure on its surface, its normal boiling point being the temperature at which its vapour pressure is equal to that of a standard atmosphere (760mm Hg). Lowering the external pressure lowers the boiling point. For most substances, boiling point and vapour pressure are related by an equation of the form,

$$\log p = A + B/(t + 273),$$

where p is the pressure, t is in °C, and A and B are constants. Hence, if the boiling points at two different pressures are known the boiling point at another pressure can be calculated from a simple plot of log p *versus* $1/(t + 273)$. For organic molecules that are not strongly associated, this equation can be written in the form,

$$\log p = 8.586 - 5.703\ (T + 273)/(t + 273)$$

where T is the boiling point in °C at 760mm Hg. Tables 3A and 3B give computed boiling points over a range of pressures. Some examples illustrate its application. Ethyl acetoacetate, **b** 180° (with decomposition) at 760mm Hg has a predicted **b** of 79° at 16mm; the experimental value is 78°. Similarly 2,4-diaminotoluene, **b** 292° at 760mm, has a predicted **b** of 147° at 8mm; the experimental value is 148-150°. For self-associated molecules the predicted **b** are lower than the experimental values. Thus, glycerol, **b** 290° at 760mm, has a predicted **b** of 146° at 8mm: the experimental value is 182°.

Similarly an estimate of the boiling points of liquids at reduced pressure can be obtained using a nomogram (see Figure 1).

For pressures near 760mm, the change in boiling point is given approximately by,

$$Ît = a(760 - p)(t + 273)$$

where $a = 0.00012$ for most substances, but $a = 0.00010$ for water, alcohols, carboxylic acids and other associated liquids, and $a = 0.00014$ for very low-boiling substances such as nitrogen or ammonia [Crafts *Chem Ber* **20** 709 *1887*]. When all the impurities are non-volatile, simple distillation is adequate purification. The observed boiling point remains almost constant and approximately equal to that of the pure material. Usually, however, some of the impurities are appreciably volatile, so that the boiling point progressively rises during the distillation because of the progressive enrichment of the higher-boiling components in the distillation flask. In such cases, separation is effected by fractional distillation using an efficient column.

Techniques.
The distillation apparatus consists basically of a distillation flask, usually fitted with a vertical fractionating column (which may be empty or packed with suitable materials such as glass helices or stainless-steel wool) to which is attached a condenser leading to a receiving flask. The bulb of a thermometer projects into the vapour phase just below the region where the condenser joins the column. The distilling flask is heated so that its contents are steadily vaporised by boiling. The vapour passes up into the column where, initially, it condenses and runs back into the flask. The resulting heat transfer gradually warms the column so that there is a progressive movement of the vapour phase-liquid boundary up the column, with increasing enrichment of the more volatile component. Because of this fractionation, the vapour finally passing into the condenser (where it condenses and flows into the receiver) is commonly that of the lowest-boiling components in the system. The conditions apply until all of the low-boiling material has been distilled, whereupon distillation ceases until the column temperature is high enough to permit the next component to distil. This usually results in a temporary fall in the temperature indicated by the thermometer.

Distillation of liquid mixtures.
The principles involved in fractional distillation of liquid mixtures are complex but can be seen by considering a system which approximately obeys *Raoult's law*. (This law states that the vapour pressure of a solution at any given temperature is the sum of the vapour pressures of each component multiplied by its mole fraction in the solution.) If two substances, A and B, having vapour pressures of 600mm Hg and 360mm Hg, respectively, were mixed in a molar ratio of 2:1 (i.e. 0.666:0.333 mole ratio), the mixture would have (ideally) a vapour pressure of 520mm Hg (i.e. 600 x 0.666 + 360 x 0.333, or 399.6 + 119.88 mm Hg) and the vapour phase would contain 77% (399.6 x 100/520) of A and 23% (119.88 x 100/520) of B. If this phase was now condensed, the new liquid phase would, therefore, be richer in the volatile component A. Similarly, the vapour in equilibrium with this phase is still further enriched in A. Each such liquid-vapour equilibrium constitutes a "theoretical plate". The efficiency of a fractionating column is commonly expressed as the number of such plates to which it corresponds in operation. Alternatively, this information may be given in the form of the height equivalent to a theoretical plate, or HETP. The number of theoretical plates and equilibria between liquids and vapours are affected by the factors listed to achieve maximum separation by fractional distillation in the section below on techniques.

In most cases, systems deviate to a greater or lesser extent from Raoult's law, and vapour pressures may be greater or less than the values calculated. In extreme cases (e.g. azeotropes), vapour pressure-composition curves pass through maxima or minima, so that attempts at fractional distillation lead finally to the separation of a constant-boiling (azeotropic) mixture and one (but not both) of the pure species if either of the latter is present in excess.

Elevation of the boiling point by dissolved solids. Organic substances dissolved in organic solvents cause a rise in boiling point which is proportional to the concentration of the substance, and the extent of rise in temperature is characteristic of the solvent. The following equation applies for dilute solutions and non-associating substances:

$$\frac{M\,Dt}{c} = K$$

Where M is the molecular weight of the solute, Dt is the elevation of boiling point in °C, c is the concentration of solute in grams for 1000gm of solvent, and K is the *Ebullioscopic Constant* (molecular elevation of the boiling point) for the solvent. K is a fixed property (constant) for the particular solvent. This has been very useful for the determination of the molecular weights of organic substances in solution.

The efficiency of a distillation apparatus used for purification of liquids depends on the difference in boiling points of the pure material and its impurities. For example, if two components of an ideal mixture have vapour pressures in the ratio 2:1, it would be necessary to have a still with an efficiency of at least seven plates (giving an enrichment of $2^7 = 128$) if the concentration of the higher-boiling component in the distillate was to be reduced to less than 1% of its initial value. For a vapour pressure ratio of 5:1, three plates would achieve as much separation.

In a fractional distillation, it is usual to reject the initial and final fractions, which are likely to be richer in the lower-boiling and higher-boiling impurities respectively. The centre fraction can be further purified by repeated fractional distillation.

To achieve maximum separation by fractional distillation:

1. The column must be flooded initially to wet the packing. For this reason it is customary to operate a still at reflux for some time before beginning the distillation.

2. The reflux ratio should be high (i.e. the ratio of drops of liquid which return to the distilling flask and the drops which distil over), so that the distillation proceeds slowly and with minimum disturbance of the equilibria in the column.

3. The hold-up of the column should not exceed one-tenth of the volume of any one component to be separated.

4. Heat loss from the column should be prevented but, if the column is heated to offset this, its temperature must not exceed that of the distillate in the column.

5. Heat input to the still-pot should remain constant.

6. For distillation under reduced pressure there must be careful control of the pressure to avoid flooding or cessation of reflux.

Types of distillation

The distilling flask. To minimise superheating of the liquid (due to the absence of minute air bubbles or other suitable nuclei for forming bubbles of vapour), and to prevent bumping, one or more of the following precautions should be taken:

(a) The flask is heated uniformly over a large part of its surface, either by using an electrical heating mantle or, by partial immersion in a bath above the boiling point of the liquid to be distilled.

(b) Before heating begins, small pieces of unglazed fireclay or porcelain (porous pot, boiling chips), pumice, diatomaceous earth, or platinum wire are added to the flask. These act as sources of air bubbles.

(c) The flask may contain glass siphons or boiling tubes. The former are inverted J-shaped tubes, the end of the shorter arm being just above the surface of the liquid. The latter comprise long capillary tubes sealed above the lower end.

(d) A steady slow stream of inert gas (e.g. N_2, Ar or He) is passed through the liquid.

(e) The liquid in the flask is stirred mechanically. This is especially necessary when suspended insoluble material is present.

For simple distillations a Claisen flask is often used. This flask is, essentially, a round-bottomed flask to the neck of which is joined another neck carrying a side arm. This second neck is sometimes extended so as to form a

Vigreux column [a glass tube in which have been made a number of pairs of indentations which almost touch each other and which slope slightly downwards. The pairs of indentations are arranged to form a spiral of glass inside the tube].

For heating baths, see Table 4. For distillation apparatus on a micro or semi-micro scale see Aldrich and other glassware catalogues. Alternatively, some useful websites for suppliers of laboratory glassware are www.wheatonsci.com, www.sigmaaldrich.com and www.kimble-kontes.com.

Types of columns and packings. A slow distillation rate is necessary to ensure that equilibrium conditions operate and also that the vapour does not become superheated so that the temperature rises above the boiling point. Efficiency is improved if the column is heat insulated (either by vacuum jacketing or by lagging) and, if necessary, heated to just below the boiling point of the most volatile component. Efficiency of separation also improves with increase in the heat of vaporisation of the liquids concerned (because fractionation depends on heat equilibration at multiple liquid-gas boundaries). Water and alcohols are more easily purified by distillation for this reason.

Columns used in distillation vary in their shapes and types of packing. Packed columns are intended to give efficient separation by maintaining a large surface of contact between liquid and vapour. Efficiency of separation is further increased by operation under conditions approaching total reflux, i.e. under a high reflux ratio. However, great care must be taken to avoid flooding of the column during distillation. The minimum number of theoretical plates for satisfactory separation of two liquids differing in boiling point by $Ît$ is approximately $(273 + t)/3Ît$, where t is the average boiling point in ^oC.

The packing of a column greatly increases the surface of liquid films in contact with the vapour phase, thereby increasing the efficiency of the column, but reducing its capacity (the quantities of vapour and liquid able to flow in opposite directions in a column without causing flooding). Material for packing should be of uniform size, symmetrical shape, and have a unit diameter less than one eighth that of the column. (Rectification efficiency increases sharply as the size of the packing is reduced but so, also, does the hold-up in the column.) It should also be capable of uniform, reproducible packing.

The usual *packings* are:
(a) Rings. These may be hollow glass or porcelain (Raschig rings), of stainless steel gauze (Dixon rings), or hollow rings with a central partition (Lessing rings) which may be of porcelain, aluminium, copper or nickel.
(b) Helices. These may be of metal or glass (Fenske rings), the latter being used where resistance to chemical attack is important (e.g. in distilling acids, organic halides, some sulfur compounds, and phenols). Metal single-turn helices are available in aluminium, nickel or stainless steel. Glass helices are less efficient, because they cannot be tamped to ensure uniform packing.
(c) Balls or beads. These are usually made of glass.

Condensers. Some of the more commonly used condensers are:
Air condenser. A glass tube such as the inner part of a Liebig condenser (see below). Used for liquids with boiling points above 90°. Can be of any length.
Coil condenser. An open tube, into which is sealed a glass coil or spiral through which water circulates. The tube is sometimes also surrounded by an outer cooling jacket. A double coil condenser has two inner coils with circulating water.
Double surface condenser. A tube in which the vapour is condensed between an outer and inner water-cooled jacket after impinging on the latter. Very useful for liquids boiling below 40°.
Friedrichs condenser. A "cold-finger" type of condenser sealed into a glass jacket open at the bottom and near the top. The cold finger is formed into glass screw threads.
Liebig condenser. An inner glass tube surrounded by a glass jacket through which water is circulated.

Vacuum distillation. This expression is commonly used to denote a distillation under reduced pressure lower than that of the normal atmosphere. Because the boiling point of a substance depends on the pressure, it is often possible by sufficiently lowering the pressure to distil materials at a temperature low enough to avoid partial or complete decomposition, even if they are unstable when boiled at atmospheric pressure.

Sensitive or high-boiling liquids should invariably be distilled or fractionally distilled under reduced pressure. The apparatus is essentially as described for distillation except that ground joints connecting the different parts of the apparatus should be air tight by using grease, or better Teflon sleaves. For low, moderately high, and very high temperatures Apiezon L, M and T greases respectively, are very satisfactory. Alternatively, it is often preferable to avoid grease and to use thin Teflon sleeves in the joints. The distilling flask, must be supplied with a capillary bleed (which allows a fine stream of air, nitrogen or argon into the flask), and the receiver should be of the fraction collector type. When distilling under vacuum it is very important to place a loose packing of glass wool above the liquid to buffer sudden boiling of the liquid. The flask should be not more than two-thirds full of liquid. The vacuum must have attained a steady state, i.e. the liquid has been completely degassed, before the heat source is applied, and the temperature of the heat source must be raised *very slowly* until boiling is achieved.

If the pump is a filter pump off a high-pressure water supply, its performance will be limited by the temperature of the water because the vapour pressure of water at $10°$, $15°$, $20°$ and $25°$ is 9.2, 12.8, 17.5 and 23.8 mm Hg respectively. The pressure can be measured with an ordinary manometer. For vacuums in the range 10^{-2} mm Hg to 10 mm Hg, rotary mechanical pumps (oil pumps) are used and the pressure can be measured with a Vacustat McLeod type gauge. If still higher vacuums are required, for example for high vacuum sublimations, a mercury diffusion pump is suitable. Such a pump can provide a vacuum up to 10^{-6} mm Hg. For better efficiencies, the pump can be backed up by a mechanical pump. In all cases, the mercury pump is connected to the distillation apparatus through several traps to remove mercury vapours. These traps may operate by chemical action, for example the use of sodium hydroxide pellets to react with acids, or by condensation, in which case empty tubes cooled in solid carbon dioxide-ethanol or liquid nitrogen (contained in wide-mouthed Dewar flasks) are used.

Special oil or mercury traps are available commercially and a liquid-nitrogen (**b** -209.9°C) trap is the most satisfactory one to use between these and the apparatus. It has an advantage over liquid air or oxygen in that it is non-explosive if it becomes contaminated with organic matter. Air should not be sucked through the apparatus before starting a distillation because this will cause liquid oxygen (**b** –183°C) to condense in the liquid nitrogen trap and this is potentially explosive (especially in mixtures with organic materials). Due to the potential lethal consequences of liquid oxygen/organic material mixtures, care must be exercised when handling liquid nitrogen. Hence, it is advisable to degas the system for a short period before the trap is immersed into the liquid nitrogen (which is kept in a Dewar flask).

Spinning-band distillation. Factors which limit the performance of distillation columns include the tendency to flood (which occurs when the returning liquid blocks the pathway taken by the vapour through the column) and the increased hold-up (which decreases the attainable efficiency) in the column that should, theoretically, be highly efficient. To overcome these difficulties, especially for distillation under high vacuum of heat sensitive or high-boiling highly viscous fluids, spinning band columns are commercially available. In such units, the distillation columns contain a rapidly rotating, motor-driven, spiral band, which may be of polymer-coated metal, stainless steel or platinum. The rapid rotation of the band in contact with the walls of the still gives intimate mixing of descending liquid and ascending vapour while the screw-like motion of the band drives the liquid towards the still-pot, helping to reduce hold-up. There is very little pressure drop in such a system, and very high throughputs are possible, with high efficiency. For example, a 765-mm long 10-mm diameter commercial spinning-band column is reported to have an efficiency of 28 plates and a pressure drop of 0.2 mm Hg for a throughput of 330mL/h. The columns may be either vacuum jacketed or heated externally. The stills can be operated down to 10^{-5} mm Hg. The principle, which was first used commercially in the Podbielniak Centrifugal Superfractionator, has also been embodied in descending-film molecular distillation apparatus.

Steam distillation. When two immmiscible liquids distil, the sum of their (independent) partial pressures is equal to the atmospheric pressure. Hence in steam distillation, the distillate has the composition

$$\frac{\text{Moles of substance}}{\text{Moles of water}} = \frac{P_{substance}}{P_{water}} = \frac{760 - P_{water}}{P_{water}}$$

where the P's are vapour pressures (in mm Hg) in the boiling mixture.

The customary technique consists of heating the substance and water in a flask (to boiling), usually with the passage of steam, followed by condensation and separation of the aqueous and non-aqueous phases in the distillate. Its advantages are those of selectivity (because only some water-insoluble substances, such as naphthalene, nitrobenzene, phenol and aniline are volatile in steam) and of ability to distil certain high-boiling substances well below their boiling point. It also facilitates the recovery of a non-steam-volatile solid at a relatively low temperature from a high-boiling solvent such as nitrobenzene. The efficiency of steam distillation is increased if superheated steam is used (because the vapour pressure of the organic component is increased relative to water). In this case the flask containing the material is heated (without water) in an oil bath and the steam passing through it is superheated by prior passage through a suitable heating device (such as a copper coil heated electrically or an oil bath).

Azeotropic distillation. In some cases two or more liquids form constant-boiling mixtures, or azeotropes. Azeotropic mixtures are most likely to be found with components which readily form hydrogen bonds or are otherwise highly associated, especially when the components are dissimilar, for example an alcohol and an aromatic hydrocarbon, but have similar boiling points.

Examples where the boiling point of the distillate is a minimum (less than either pure component) include:
Water with ethanol, *n*-propanol and isopropanol, *tert*-butanol, propionic acid, butyric acid, pyridine,
methanol with methyl iodide, methyl acetate, chloroform,
ethanol with ethyl iodide, ethyl acetate, chloroform, benzene, toluene, methyl ethyl ketone,
benzene with cyclohexane,
acetic acid with toluene.
Although less common, azeotropic mixtures are known which have higher boiling points than their components. These include water with most of the mineral acids (hydrofluoric, hydrochloric, hydrobromic, perchloric, nitric and sulfuric) and formic acid. Other examples are acetic acid-pyridine, acetone-chloroform, aniline-phenol, and chloroform-methyl acetate.

The following azeotropes are important commercially for drying ethanol:

ethanol 95.5% (by weight) - water 4.5%	**b** 78.1°
ethanol 32.4% - benzene 67.6%	**b** 68.2°
ethanol 18.5% - benzene 74.1% - water 7.4%	**b** 64.9°

Materials are sometimes added to form an azeotropic mixture with the substance to be purified. Because the azeotrope boils at a different temperature, this facilitates separation from substances distilling in the same range as the pure material. (Conversely, the impurity might form the azeotrope and be removed in this way). This method is often convenient, especially where the impurities are isomers or are otherwise closely related to the desired substance. Formation of low-boiling azeotropes also facilitates distillation.

One or more of the following methods can generally be used for separating the components of an azeotropic mixture:
1. By using a chemical method to remove most of one species prior to distillation. (For example, water can be removed by suitable drying agents; aromatic and unsaturated hydrocarbons can be removed by sulfonation).
2. By redistillation with an additional substance which can form a ternary azeotropic mixture (as in ethanol-water-benzene example given above).
3. By selective adsorption of one of the components. (For example, of water on to silica gel or molecular sieves, or of unsaturated hydrocarbons onto alumina).
4. By fractional crystallisation of the mixture, either by direct freezing or by dissolving in a suitable solvent.

Kügelrohr distillation. The apparatus (Büchi, see www.buchi.com) is made up of small glass bulbs (*ca* 4-5cm diameter) which are joined together *via* Quickfit joints at each pole of the bulbs. The liquid (or low melting solid) to be purified is placed in the first bulb of a series of bulbs joined end to end, and the system can be evacuated. The first bulb is heated in a furnace at a high temperature whereby most of the material distils into the second bulb (which is outside of the furnace). The second bulb is then moved into the furnace and the furnace temperature is reduced by *ca* 5° whereby the liquid in the second bulb distils into the third bulb (at this stage the first bulb is now out at the back of the furnace and the third and subsequent bulbs are outside the front of the furnace). The furnace temperature is lowered by a further *ca* 5° and the third bulb is moved into the furnace. The lower boiling material will distil into the fourth bulb. The process is continued until no more material distils into the subsequent bulb. The vacuum (if applied) and the furnace are removed, the bulbs are separated and the various fractions of distillates are collected from the individual bulbs. For volatile liquids, it may be necessary to cool the receiving bulb with solid CO_2 held in a suitable container (Kügelrohr distillation apparatus with an integrated cooling system is available). This procedure is used for preliminary purification and the distillates are then redistilled or recrystallised.

Isopiestic or isothermal distillation. This technique can be useful for the preparation of metal-free solutions of volatile acids and bases for use in trace metal studies. The procedure involves placing two beakers, one of distilled water and the other of a solution of the material to be purified, in a desiccator. The desiccator is sealed and left to stand at room temperature for several days. The volatile components distribute themselves between the two beakers whereas the non-volatile contaminants remain in the original beaker. This technique has afforded metal-free pure solutions of ammonia, hydrochloric acid and hydrogen fluoride.

RECRYSTALLISATION
Techniques
The most commonly used procedure for the purification of a solid material by recrystallisation from a solution involves the following steps:

(a) The impure material is dissolved in a suitable solvent, by shaking or vigorous stirring, at or near the boiling point, to form a near-saturated solution.

(b) The hot solution is filtered to remove any insoluble particles. To prevent crystallisation during this filtration, a heated filter funnel can be used or the solution can be diluted with more of the solvent.

(c) The solution is then allowed to cool so that the dissolved substance crystallises out.

(d) The crystals are separated from the mother liquor, either by centrifuging or by filtering, under suction, through a sintered glass, a Hirsch or a Büchner, funnel. Usually, centrifugation is preferred because of the greater ease and efficiency of separating crystals and mother liquor, and also because of the saving of time and effort, particularly when very small crystals are formed or when there is entrainment of solvent.

(e) The crystals are washed free from mother liquor with a little fresh cold solvent, then dried.

If the solution contains extraneous coloured material likely to contaminate the crystals, this can often be removed by adding some activated charcoal (decolorising carbon) to the hot, but not boiling, solution which is then shaken frequently for several minutes before being filtered. (The large active surface of the carbon makes it a good adsorbent for this purpose.) In general, the cooling and crystallisation steps should be rapid so as to give small crystals which occlude less of the mother liquor. This is usually satisfactory with inorganic material, so that commonly the filtrate is cooled in an ice-water bath while being vigorously stirred. In many cases, however, organic molecules crystallise much more slowly, so that the filtrate must be set aside to cool to room temperature or left in the refrigerator. It is often desirable to subject material that is very impure to preliminary purification, such as steam distillation, Soxhlet extraction, or sublimation, before recrystallising it. A greater degree of purity is also to be expected if the crystallisation process is repeated several times, especially if different solvents are used. The advantage of several crystallisations from different solvents lies in the fact that the material sought, and its impurities, are unlikely to have similar solubilities as solvents and temperatures are varied.

For the final separation of solid material, sintered-glass discs are preferable to filter paper. Sintered glass is unaffected by strongly acid solutions or by oxidising agents. Also, with filter paper, cellulose fibres are likely to become included in the sample. The sintered-glass discs or funnels can be readily cleaned by washing in freshly prepared *chromic acid cleaning mixture*. This mixture is made by adding 100mL of concentrated sulfuric acid slowly with stirring to a solution of 5g of sodium dichromate (CARE: cancer suspect) in 5mL of water. (The mixture warms to about 70°, see p 3).

For materials with very low melting points it is sometimes convenient to use dilute solutions in acetone, methanol, pentane, diethyl ether or $CHCl_3$-CCl_4. The solutions are cooled to -78° in a dry-ice/acetone bath, to give a slurry which is filtered off through a precooled Büchner funnel. Experimental details, as applied to the purification of nitromethane, are given by Parrett and Sun [*J Chem Educ* **54** 448 *1977*].

Where substances vary little in solubility with temperature, *isothermal crystallisation* may sometimes be employed. This usually takes the form of a partial evaporation of a saturated solution at room temperature by leaving it under reduced pressure in a desiccator.

However, in rare cases, crystallisation is not a satisfactory method of purification, especially if the impurity forms crystals that are isomorphous with the material being purified. In fact, the impurity content may even be greater in such recrystallised material. For this reason, it still remains necessary to test for impurities and to remove or adequately lower their concentrations by suitable chemical manipulation prior to recrystallisation.

Filtration. Filtration removes particulate impurities rapidly from liquids and is also used to collect insoluble or crystalline solids which separate or crystallise from solution. The usual technique is to pass the solution, cold or hot, through a fluted filter paper in a conical glass funnel.

If a solution is hot and needs to be filtered rapidly a Büchner funnel and flask are used and filtration is performed under a slight vacuum (water pump), the filter medium being a circular cellulose filter paper wet with solvent. If filtration is slow, even under high vacuum, a pile of about twenty filter papers, wet as before, are placed in the Büchner funnel and, as the flow of solution slows down, the upper layers of the filter paper are progressively removed. Alternatively, a filter aid, e.g. Celite, Florisil or Hyflo-supercel, is placed on top of a filter paper in the funnel. When the flow of the solution (under suction) slows down, the upper surface of the filter aid is scratched gently. Filter papers with various pore sizes are available covering a range of filtration rates. Hardened filter papers are slow filtering but they can withstand acidic and alkaline solutions without appreciable hydrolysis of the cellulose (see Table 5). When using strong acids it is preferable to use glass micro fibre filters which are commercially available (see Table 5 and 6).

Freeing a solution from extremely small particles [e.g. for optical rotatory dispersion (ORD) or circular dichroism (CD) measurements] requires filters with very small pore size. Commercially available (Millipore, Gelman, Nucleopore) filters other than cellulose or glass include nylon, Teflon, and polyvinyl chloride, and the pore diameter may be as small as 0.01micron (see Table 6). Special containers are used to hold the filters, through which the solution is pressed by applying pressure, e.g. from a syringe. Some of these filters can be used to clear strong sulfuric acid solutions.

As an alternative to the Büchner funnel for collecting crystalline solids, a funnel with a sintered glass-plate under suction may be used. Sintered-glass funnels with various porosities are commercially available and can be easily cleaned with warm chromic or nitric acid (see above).

When the solid particles are too fine to be collected on a filter funnel because filtration is extremely slow, separation by **centrifugation** should be used. Bench type centrifuges are most convenient for this purpose. The solid is placed in the centrifuge tube, the tubes containing the solutions on opposite sides of the rotor should be balanced accurately (at least within 0.05 to 0.1g), and the solutions are spun at maximum speed for as long as it takes to settle the solid (usually *ca* 3-5 minutes). The solid is washed with cold solvent by centrifugation, and finally twice with a pure volatile solvent in which the solid is insoluble, also by centrifugation. After decanting the supernatant, the residue is dried in a vacuum, at elevated temperatures if necessary. In order to avoid "spitting" and contamination with dust while the solid in the centrifuge tube is dried, the mouth of the tube is covered with aluminium foil and held fast with a tight rubber band near the lip. The flat surface of the aluminium foil is then perforated in several places with a pin and the tube and contents are dried in a vacuum desiccator over a desiccant.

Choice of solvents. The best solvents for recrystallisation have the following properties:
(a) The material is much more soluble at higher temperatures than it is at room temperature or below.
(b) Well-formed (but not large) crystals are produced.
(c) Impurities are either very soluble or only sparingly soluble.
(d) The solvent must be readily removed from the purified material.
(e) There must be no reaction between the solvent and the substance being purified.
(f) The solvent must not be inconveniently volatile or too highly flammable. (These are reasons why diethyl ether and carbon disulfide are not commonly used in this way.)

The following generalisations provide a rough guide to the selection of a suitable solvent:
(a) Substances usually dissolve best in solvents to which they are most closely related in chemical and physical characteristics. Thus, hydroxylic compounds are likely to be most soluble in water, methanol, ethanol, acetic acid or acetone. Similarly, petroleum ether might be used with water-insoluble substances. However, if the resemblance is too close, solubilities may become excessive.
(b) Higher members of homologous series approximate more and more closely to their parent hydrocarbon.
(c) Polar substances are more soluble in polar, than in non-polar, solvents.

Although Chapters 4, 5 and 6 provide details of the solvents used for recrystallising a large portion of commercially available laboratory chemicals, they cannot hope to be exhaustive, nor need they necessarily be the best choice. In other cases where it is desirable to use this process, it is necessary to establish whether a given solvent is suitable. This is usually done by taking only a small amount of material in a small test-tube and adding enough solvent to cover it. If it dissolves readily in the cold or on gentle warming, the solvent is unsuitable. Conversely, if it remains insoluble when the solvent is heated to boiling (adding more solvent if necessary), the solvent is again unsuitable. If the material dissolves in the hot solvent but does not crystallise readily within several minutes of cooling in an ice-salt mixture, another solvent should be tried.

Petroleum ethers are commercially available fractions of refined petroleum and are sold in fractions with about 20° boiling ranges. This ensures that little of the hydrocarbon ingredients boiling below the range is lost during standing or boiling when recrystallising a substance. Petroleum ethers with boiling ranges (at 760mm pressure) of 35—60°, 40—60°, 60—80°, 80—100°, and 100—120° are generally free from unsaturated and aromatic hydrocarbons. The lowest boiling petroleum ether commercially available has **b** 30-40°/760mm and is mostly *n*-pentane. The purer spectroscopic grades are almost completely free from olefinic and aromatic hydrocarbons. **Petroleum spirit** (which is sometimes used synonymously with petroleum ether or light petroleum) is usually less refined petroleum, and *ligroin* is used for fractions boiling above 100°. The lower boiling fractions consist of mixtures of *n*-pentane (**b** 36°), *n*-hexane (**b** 68.5°) and *n*-heptane (**b** 98°), and some of their isomers in varying proportions. For purification of petroleum ether b 35-60° see p. 324.

Solvents commonly used for recrystallisation, and their boiling points, are given in Table 7.
For comments on the toxicity and use of **benzene** see the first pages of Chapters 4, 5 and 6.

Mixed Solvents. Where a substance is too soluble in one solvent and too insoluble in another, for either to be used for recrystallisation, it is often possible (provided they are miscible) to use them as a mixed solvent. (In general, however, it is preferable to use a single solvent if this is practicable.) Table 8 contains many of the common pairs of miscible solvents.
The technique of recrystallisation from mixed solvents is as follows:
The material is dissolved in the solvent in which it is the more soluble, then the other solvent (heated to near boiling) is added cautiously to the hot solution until a slight turbidity persists or crystallisation begins. This is cleared by adding several drops of the first solvent, and the solution is allowed to cool and crystallise in the usual way.
A variation of this procedure is simply to precipitate the material in a microcrystalline form from solution in one solvent at room temperature, by adding a little more of the second solvent, filtering off the crystals, adding a little more of the second solvent and repeating the process. This ensures, at least in the first or last precipitation, a material which contains as little as possible of the impurities, which may also be precipitated in this way. With salts, the first solvent is commonly water, and the second solvent is alcohol or acetone.

Recrystallisation from the melt. A crystalline solid melts when its temperature is raised sufficiently for the thermal agitation of its molecules or ions to overcome the restraints imposed by the crystal lattice. Usually, impurities weaken crystal structures, and hence lower the melting points of solids (or the freezing points of liquids). If an impure material is melted and cooled slowly (with the addition, if necessary, of a trace of solid material near the freezing point to avoid supercooling), the first crystals that form will usually contain less of the impurity, so that fractional solidification by partial freezing can be used as a purification process for solids with melting points lying in a convenient temperature range (or for more readily frozen liquids). Some examples of cooling baths that are useful in recrystallisation are summarised in Table 9. In some cases, impurities form higher melting eutectics with substances to be purified, so that the first material to solidify is less pure than the melt. For this reason, it is often desirable to discard the first crystals and also the final portions of the melt. Substances having similar boiling points often differ much more in melting points, so that fractional solidification can offer real advantages, especially where ultrapurity is sought. For further information on this method of recrystallisation, consult the earlier editions of this book as well as references by Schwab and Wichers (*J Res Nat Bur Stand* **25** 747 *1940*). This method works best if the material is already nearly pure, and hence tends to be a final purification step.

Zone refining. Zone refining (or zone melting) is a particular development for fractional solidification and is applicable to all crystalline substances that show differences in the concentrations of impurities in liquid and solid states at solidification. The apparatus used in this technique consists essentially of a device in which the crystalline solid to be purified is placed in a glass tube (set vertically) which is made to move slowly upwards while it passes through a fixed coil (one or two turns) of heated wire. A narrow zone of molten crystals is formed when the tube is close to the heated coil. As the zone moves away from the coil the liquid crystallises, and a fresh molten zone is formed below it at the coil position. The machine can be set to recycle repeatedly. At its advancing side, the zone has a melting interface with the impure material whereas on the upper surface of the zone there is a constantly growing face of higher-melting, resolidified material. This leads to a progressive increase in impurity in the liquid phase which, at the end of the run, is discarded from the bottom of the tube. Also, because of the progressive increase in impurity in the liquid phase, the resolidified material contains correspondingly more of the impurites. For this reason, it is usually necessary to make several zone-melting runs before a sample is satisfactorily purified. This is also why the method works most successfully if the material is already fairly pure. In all these operations the zone must travel slowly enough to enable impurities to diffuse or be convected away from the area where resolidification is occurring.
The technique finds commercial application in the production of metals of extremely high purity (impurities down to 10^{-9} ppm), in purifying refractory oxides, and in purifying organic compounds, using commercially available equipment. Criteria for indicating that definite purification is achieved include elevation of melting point, removal of colour, fluorescence or smell, and a lowering of electrical conductivity. Difficulties likely to be found with organic compounds, especially those of low melting points and low rates of crystallisation, are supercooling and, because of surface tension and contraction, the tendency of the molten zone to seep back into the recrystallised areas. The method is likely to be useful in cases where fractional distillation is not practicable, either because of

unfavourable vapour pressures or ease of decomposition, or where super-pure materials are required. It has been used for the latter purpose for purifying anthracene, benzoic acid, chrysene, morphine, 1,8-naphthyridine and pyrene to name a few. [See E.F.G.Herington, *Zone Melting of Organic Compound*s, Wiley & Sons, NY, 1963; W.Pfann, *Zone Melting*, 2nd edn, Wiley, NY, 1966; H.Schildknecht, *Zonenschmelzen*, Verlag Chemie, Weinheim, 1964; W.R.Wilcox, R.Friedenberg et al. *Chem Rev* **64** 187 *1964*; M.Zief and W.R.Wilcox (Eds), *Fractional Solidification*, Vol I, M Dekker Inc. NY, 1967.]

SUBLIMATION

Sublimation differs from ordinary distillation because the vapour condenses to a solid instead of a liquid. Usually, the pressure in the heated system is diminished by pumping, and the vapour is condensed (after travelling a relatively short distance) onto a cold finger or some other cooled surface. This technique, which is applicable to many organic solids, can also be used with inorganic solids such as aluminium chloride, ammonium chloride, arsenious oxide and iodine. In some cases, passage of a stream of inert gas over the heated substance secures adequate vaporisation. This procedure has the added advantage of removing occluded solvent used in recrystallising the solid.

CHROMATOGRAPHY

Chromatography is often used with advantage for the purification of small amounts of complex organic mixtures. Chromatography techniques all rely on the differential distribution of the various components in a mixture between the mobile phase and the stationary phase. The mobile phase can either be a gas or a liquid whereas the stationary phase can either be a solid or a liquid.

The major chromatographic techniques can also be categorised according to the nature of the mobile phase used - vapour phase chromatography for when a gas is the mobile phase and liquid chromatography for when a liquid is the mobile phase.

A very useful catalog for chromatographic products and information relating to chromatography (from gas chromatography to biochromatography) is that produced by Merck, called the ChromBook and the associated compact disk, ChromCircle.

Vapour phase chromatography (GC or gas-liquid chromatography)

The mobile phase in vapour phase chromatography is a gas (e.g. hydrogen, helium, nitrogen or argon) and the stationary phase is a non-volatile liquid impregnated onto a porous material. The mixture to be purified is injected into a heated inlet whereby it is vaporised and taken into the column by the carrier gas. It is separated into its components by partition between the liquid on the porous support and the gas. For this reason vapour-phase chromatography is sometimes referred to as gas-liquid chromatography (g.l.c.). Vapour phase chromatography is very useful in the resolution of a mixture of volatile compounds. This type of chromatography uses either packed or capillary columns. Packed columns have internal diameters of 3-5 mm with lengths of 2-6 m. These columns can be packed with a range of materials including firebrick derived materials (chromasorb P, for separation of non polar hydrocarbons) or diatomaceous earth (chromasorb W, for separation of more polar molecules such as acids, amines). Capillary columns have stationary phase bonded to the walls of long capillary tubes. The diameters in capillary columns are less than 0.5 mm and the lengths of these columns can go up to 50 m! These columns have much superior separating powers than the packed columns. Elution times for equivalent resolutions with packed columns can be up to ten times shorter. It is believed that almost any mixture of compounds can be separated using one of the four stationary phases, OV-101, SE-30, OV-17 and Carbowax-20M. The use of capillary columns in gas chromatography for analysis is now routinely carried out. An extensive range of packed and capillary columns is available from chromatographic specialists such as Supelco, Alltech, Hewlett-Packard, Phenomenex etc.

Table 10 shows some typical liquids used for stationary phases in gas chromatography.

Although vapour gas chromatography is routinely used for the analysis of mixtures, this form of chromatography can also be used for separation/purification of substances. This is known as preparative GC. In preparative GC, suitable packed columns are used and as substances emerge from the column, they are collected by condensing the vapour of these separated substances in suitable traps. The carrier gas blows the vapour through these traps hence these traps have to be very efficient. Improved collection of the effluent vaporised fractions in preparative work is attained by strong cooling, increasing the surface of the traps by packing them with glass wool, and by applying an electrical potential which neutralises the charged vapour and causes it to condense.

When the gas chromatograph is attached to a mass spectrometer, a very powerful analytical tool (*gas chromatography-mass spectrometry*; **GC-MS**) is produced. Vapour gas chromatography allows the analyses of mixtures but does not allow the definitive identification of unknown substances whereas mass spectrometry is good for the identification of a single compound but is less than ideal for the identification of mixtures of compounds. This means that with GC-MS, both separation *and* identification of substances in mixtures can be achieved. Because of the relatively small amounts of material required for mass spectrometry, a splitting system is inserted between the column and the mass spectrometer. This enables only a small fraction of the effluent to enter the spectrometer, the rest of the effluent is usually collected or vented to the air.

Liquid chromatography

In contrast to vapour phase chromatography, the mobile phase in liquid chromatography is a liquid. In general, there are four main types of liquid chromatography: *adsorption, partition, ion-chromatography*, and *gel filtration*.

Adsorption chromatography is based on the difference in the extent to which substances in solution are adsorbed onto a suitable surface. The main techniques in adsorption chromatography are TLC (Thin Layer Chromatography), paper and column chromatography.

Thin layer chromatography (TLC). In thin layer chromatography, the mobile phase i.e. the solvent, creeps up the stationary phase (the absorbent) by capillary action. The adsorbent (e.g. silica, alumina, cellulose) is spread on a rectangular glass plate (or solid inert plastic sheet or aluminium foil). Some adsorbents (e.g. silica) are mixed with a setting material (e.g. $CaSO_4$) by the manufacturers which causes the film to set hard on drying. The adsorbent can be activated by heating at 100-110° for a few hours. Other adsorbents (e.g. celluloses) adhere on glass plates without a setting agent. Thus some grades of absorbents have prefixes e.g. prefix G means that the absorbent can cling to a glass plate and is used for TLC (e.g. silica gel GF_{254} is for TLC plates which have a dye that fluoresces under 254nm UV light). Those lacking this binder have the letter H after any coding and is suitable for column chromatography e.g. silica gel 60H. The materials to be purified or separated are spotted in a solvent close to the lower end of the plate and allowed to dry. The spots will need to be placed at such a distance so as to ensure that when the lower end of the plate is immersed in the solvent, the spots are a few mm above the eluting solvent. The plate is placed upright in a tank containing the eluting solvent. Elution is carried out in a closed tank to ensure equilibrium. Good separations can be achieved with square plates if a second elution is performed at right angles to the first using a second solvent system. For rapid work, plates of the size of microscopic slides or even smaller are used which can decrease the elution time and cost without loss of resolution. The advantage of plastic backed and aluminium foil backed plates is that the size of the plate can be made as required by cutting the sheet with scissors or a sharp guillotine. Visualisation of substances on TLC can be carried out using UV light if they are UV absorbing or fluorescing substances or by spraying or dipping the plate with a reagent that gives coloured products with the substance (e.g. iodine solution or vapour gives brown colours with amines), or with dilute sulfuric acid (organic compounds become coloured or black when the plates are heated at 100° if the plates are of alumina or silica, but not cellulose). (see Table 11 for some methods of visualisation.) Some alumina and silica powders are available with fluorescent materials in them, in which case the whole plate fluoresces under UV light. Non-fluorescing spots are thus clearly visible, and fluorescent spots invariably fluoresce with a different colour. The colour of the spots can be different under UV light at 254nm and at 365nm. Another useful way of showing up non-UV absorbing spots is to spray the plate with a 1-2% solution of Rhodamine 6G in acetone. Under UV light the dye fluoresces and reveals the non-fluorescing spots. For preparative work, if the material in the spot or fraction is soluble in ether or petroleum ether, the desired substance can be extracted from the absorbent with these solvents which leave the water soluble dye behind.
TLC can be used as an analytical technique, or as a guide to establishing conditions for column chromatography or as a preparative technique in its own right.
The thickness of the absorbent on the TLC plates could be between 0.2mm to 2mm or more. In preparative work, the thicker plates are used and hundreds of milligrams of mixtures can be purified conveniently and quickly. The spots or areas are easily scraped off the plates and the desired substances extracted from the absorbent with the required solvent. For preparative TLC, non destructive methods for visualising spots and fractions are required. As such, the use of UV light is very useful. If substances are not UV active, then a small section of the plate (usually the right or left edge of the plate) is sprayed with a visualising agent while the remainder of the plate is kept covered.
Thin layer chromatography has been used successfully with ion-exchange celluloses as stationary phases and various aqueous buffers as mobile phases. Also, gels (e.g. Sephadex G-50 to G-200 superfine) have been adsorbed on glass plates and are good for fractionating substances of high molecular weights (1500 to 250,000). With this technique, which is called *thin layer gel filtration* (**TLG**), molecular weights of proteins can be determined when suitable markers of known molecular weights are run alongside (see Chapter 6).
Commercially available pre-coated plates with a variety of adsorbents are generally very good for quantitative work because they are of a standard quality. Plates of a standardised silica gel 60 (as medium porosity silica gel with a mean porosity of 6mm) released by Merck have a specific surface of 500 m^2/g and a specific pore volume of 0.75 mL/g. They are so efficient that they have been called *high performance thin layer chromatography* (**HPTLC**) plates (Ropphahn and Halpap *J Chromatogr* **112** 81 *1975*). In another variant of thin layer chromatography the

adsorbent is coated with an oil as in gas chromatography thus producing *reverse-phase thin layer chromatography*. Reversed-phase TLC plates e.g. silica gel RP-18 are available from Fluka and Merck.

A very efficient form of chromatography makes use of a circular glass plate (rotor) coated with an adsorbent (silica, alumina or cellulose). As binding to a rotor is needed, the sorbents used may be of a special quality and/or binders are added to the sorbent mixtures. For example when silica gel is required as the absorbent, silica gel 60 PF-254 with calcium sulfate (Merck catalog 7749) is used. The thickness of the absorbent (1, 2 or 4 mm) can vary depending on the amount of material to be separated. The apparatus is called a **Chromatotron** (available from Harrison Research, USA). The glass plate is rotated by a motor, and the sample followed by the eluting solvent is allowed to drip onto a central position on the plate. As the plate rotates the solvent elutes the mixture, centrifugally, while separating the components in the form of circular bands radiating from the central point. The separated bands are usually visualised conveniently by UV and as the bands approach the edge of the plate, the eluent is collected. The plate with the adsorbent can be re-used many times if care is employed in the usage, and hence this form of chromatography utilises less absorbents as well as solvents.

Recipes and instructions for coating the rotors are available from the Harrison website (http://pw1.netcom.com/~ithres/harrisonresearch.html). In addition, information on how to regenerate the sorbents and binders are also included.

Paper chromatography. This is the technique from which thin layer chromatography developed. It uses cellulose paper (filter paper) instead of the TLC adsorbent and does not require a backing like the plastic sheet in TLC. It is used in the **ascending procedure** (like in TLC) whereby a sheet of paper is hung in a jar, the materials to be separated are spotted (after dissolving in a suitable solvent and drying) near the bottom of the sheet which dips into the eluting solvent just below the spot. As the solvent rises up the paper the spots are separated according to their adsorption properties. A variety of solvents can be used, the sheet is then dried in air (fume cupboard), and can then be run again with the solvent running at right angles to the first run to give a two dimensional separation. The spots can then be visualised as in TLC or can be cut out and analysed as required. A **descending procedure** had also been developed where the material to be separated is spotted near the top of the paper and the top end is made to dip into a tray containing the eluting solvent. The whole paper is placed in a glass jar and the solvent then runs down the paper causing the materials in the spots to separate also according to their adsorption properties and to the eluting ability of the solvent. This technique is much cheaper than TLC and is still used (albeit with thicker cellulose paper) with considerable success for the separation of protein hydrolysates for sequencing analysis and/or protein identification.

Column Chromatography. The substances to be purified are usually placed on the top of the column and the solvent is run down the column. Fractions are collected and checked for compounds using TLC (UV and/or other means of visualisation). The adsorbent for chromatography can be packed dry and solvents to be used for chromatography are used to equilibrate the adsorbent by flushing the column several times until equilibration is achieved. Alternatively, the column containing the adsorbent is packed wet (slurry method) and pressure is applied at the top of the column until the column is well packed (i.e. the adsorbent is settled).

Graded Adsorbents and Solvents. Materials used in columns for adsorption chromatography are grouped in Table 12 in an approximate order of effectiveness. Other adsorbents sometimes used include barium carbonate, calcium sulfate, calcium phosphate, charcoal (usually mixed with Kieselguhr or other form of diatomaceous earth, for example, the filter aid Celite) and cellulose. The alumina can be prepared in several grades of activity (see below).

In most cases, adsorption takes place most readily from non-polar solvents, such as petroleum ether and least readily from polar solvents such as alcohols, esters, and acetic acid. Common solvents, arranged in approximate order of increasing eluting ability are also given in Table 12. Eluting power roughly parallels the dielectric constants of solvents. The series also reflects the extent to which the solvent binds to the column material, thereby displacing the substances that are already adsorbed. This preference of alumina and silica gel for polar molecules explains, for example, the use of percolation through a column of silica gel for the following purposes-drying of ethylbenzene, removal of aromatics from 2,4-dimethylpentane and of ultraviolet absorbing substances from cyclohexane.

Mixed solvents are intermediate in strength, and so provide a finely graded series. In choosing a solvent for use as an eluent it is necessary to consider the solubility of the substance in it, and the ease with which it can subsequently be removed.

Preparation and Standardisation of Alumina. The activity of alumina depends inversely on its water content, and a sample of poorly active material can be rendered more active by leaving for some time in a round bottomed flask heated up to about 200° in an oil bath or a heating mantle while a slow stream of a dry inert gas is passed through it. Alternatively, it is heated to red heat (380-400°) in an open vessel for 4-6h with occasional stirring and then cooled in a vacuum desiccator: this material is then of grade I activity. Conversely, alumina can be rendered less active by adding small amounts of water and thoroughly mixing for several hours. Addition of about 3% (w/w) of water converts grade I alumina to grade II.

Used alumina can be regenerated by repeated extraction, first with boiling methanol, then with boiling water, followed by drying and heating. The degree of activity of the material can be expressed conveniently in terms of the scale due to Brockmann and Schodder (*Chem Ber* B **74** 73 *1941*).

Alumina is normally slightly alkaline. A (less strongly adsorbing) neutral alumina can be prepared by making a slurry in water and adding 2M hydrochloric acid until the solution is acid to Congo red. The alumina is then filtered off, washed with distilled water until the wash water gives only a weak violet colour with Congo red paper, and dried.

Alumina used in TLC can be recovered by washing in ethanol for 48h with occasional stirring, to remove binder material and then washed with successive portions of ethyl acetate, acetone and finally with distilled water. Fine particles are removed by siphoning. The alumina is first suspended in 0.04M acetic acid, then in distilled water, siphoning off 30 minutes after each wash. The process is repeated 7-8 times. It is then dried and activated at 200° [Vogh and Thomson *Anal Chem* **53** 1365 *1981*].

Preparation of other adsorbents

Silica gel can be prepared from commercial water-glass by diluting it with water to a density of 1.19 and, while keeping it cooled to 5°, adding concentrated hydrochloric acid with stirring until the solution is acid to thymol blue. After standing for 3h, the precipitate is filtered off, washed on a Büchner funnel with distilled water, then suspended in 0.2M hydrochloric acid. The suspension is set aside for 2-3 days, with occasional stirring, then filtered, washed well with water and dried at 110°. It can be activated by heating up to about 200° as described for alumina.

Powdered commercial silica gel can be purified by suspending and standing overnight in concentrated hydrochloric acid (6mL/g), decanting the supernatant and repeating with fresh acid until the latter remains colourless. After filtering with suction on a sintered-glass funnel, the residue is suspended in water and washed by decantation until free of chloride ions. It is then filtered, suspended in 95% ethanol, filtered again and washed on the filter with 95% ethanol. The process is repeated with anhydrous diethyl ether before the gel is heated for 24h at 100° and stored for another 24h in a vacuum desiccator over phosphorus pentoxide.

To buffer silica gel for flash chromatography (see later), 200g of silica is stirred in 1L of 0.2M NaH_2PO_4 for 30 minutes. The slurry is then filtered with suction using a sintered glass funnel. The silica gel is then activated at 110°C for 16 hours. The pH of the resulting silica gel is ~4. Similar procedures can be utilized to buffer the pH of the silica gel at various pHs (up to pH ~8: pH higher than this causes degradation of silica) using appropriate phosphate buffers.

Commercial silica gel has also been purified by suspension of 200g in 2L of 0.04M ammonia, allowed to stand for 5min before siphoning off the supernatant. The procedure was repeated 3-4 times, before rinsing with distilled water and drying, and activating the silica gel in an oven at 110° [Vogh and Thomson, *Anal Chem* **53** 1345 *1981*].

Although silica gel is not routinely recycled after use (due to fear of contamination as well as the possibility of reduced activity), the costs of using new silica gel for purification may be prohibitive. In these cases, recycling may be achieved by stirring the used silica gel (1 kg) in a mixture of methanol and water (2L MeOH/4L water) for 30-40 mins. The silica gel is filtered (as described above) and reactivated at 110°C for 16 hours.

Diatomaceous earth (Celite 535 or 545, Hyflo Super-cel, Dicalite, Kieselguhr) is purified before use by washing with 3M hydrochloric acid, then water, or it is made into a slurry with hot water, filtered at the pump and washed with water at 50° until the filtrate is no longer alkaline to litmus. Organic materials can be removed by repeated extraction at 50° with methanol or chloroform, followed by washing with methanol, filtering and drying at 90-100°.

Charcoal is generally satisfactorily activated by heating gently to red heat in a crucible or quartz beaker in a muffle furnace, finally allowing to cool under an inert atmosphere in a desiccator. Good commercial activated charcoal is made from wood, e.g. *Norit* (from Birch wood), *Darco* and *Nuchar*. If the cost is important then the cheaper *animal charcoal* (bone charcoal) can be used. However, this charcoal contains calcium phosphate and other calcium salts and cannot be used with acidic materials. In this case the charcoal is boiled with dilute hydrochloric acid (1:1 by volume) for 2-3h, diluted with distilled water and filtered through a fine grade paper on a Büchner flask, washed with distilled water until the filtrate is almost neutral, and dried first in air then in a vacuum, and activated as above. To improve the porosity, charcoal columns are usually prepared in admixture with diatomaceous earth.

Cellulose for chromatography is purified by sequential washing with chloroform, ethanol, water, ethanol, chloroform and acetone. More extensive purification uses aqueous ammonia, water, hydrochloric acid, water, acetone and diethyl ether, followed by drying in a vacuum. Trace metals can be removed from filter paper by washing for several hours with 0.1M oxalic or citric acid, followed by repeated washing with distilled water.

Flash Chromatography

A faster method of separating components of a mixture is *flash chromatography* (see Still et al. *J Org Chem* **43** 2923 *1978*). In flash chromatography the eluent flows through the column under a pressure of *ca* 1 to 4 atmospheres. The lower end of the chromatographic column has a relatively long taper closed with a tap. The upper end of the column is connected through a ball joint to a tap. Alternatively a specially designed chromatographic column with a solvent reservoir can also be used (for an example, see the Aldrich Chemical Catalog-glassware section). The tapered portion is plugged with cotton, or quartz, wool and *ca* 1 cm of fine washed sand (the latter is optional). The adsorbent is then placed in the column as a dry powder or as a slurry in a solvent and allowed to fill to about one third of the column. A fine grade of adsorbent is required in order to slow the flow rate at the higher pressure, e.g. Silica 60, 230 to 400 mesh with particle size 0.040-0.063mm (from Merck). The top of the adsorbent is layered with *ca* 1 cm of fine washed sand. The mixture in the smallest volume of solvent is applied at the top of the column and allowed to flow into the adsorbent under gravity by opening the lower tap momentarily. The top of the column is filled with eluent, the upper tap is connected by a tube to a nitrogen supply from a cylinder, or to compressed air, and turned on to the desired pressure (monitor with a gauge). The lower tap is turned on and fractions are collected rapidly until the level of eluent has reached the top of the adsorbent (do not allow the column to run dry). If further elution is desired then both taps are turned off, the column is filled with more eluting solvent and the process repeated. The top of the column can be modified so that gradient elution can be performed. Alternatively, an apparatus for producing the gradient is connected to the upper tap by a long tube and placed high above the column in order to produce the required hydrostatic pressure. Flash chromatography is more efficient and gives higher resolution than conventional chromatography at atmospheric pressure and is completed in a relatively shorter time. A successful separation of components of a mixture by TLC using the same adsorbent is a good indication that flash chromatography will give the desired separation on a larger scale.

Paired-ion Chromatography (PIC)

Mixtures containing ionic compounds (e.g. acids and/or bases), non-ionisable compounds, and zwitterions, can be separated successfully by paired-ion chromatography (PIC). It utilises the 'reverse-phase' technique (Eksberg and Schill *Anal Chem* **45** 2092 *1973*). The stationary phase is lipophilic, such as μ-BONDAPAK C_{18} or any other adsorbent that is compatible with water. The mobile phase is water or aqueous methanol containing the acidic or basic counter ion. Thus the mobile phase consists of dilute solutions of strong acids (e.g. 5mM 1-heptanesulfonic acid) or strong bases (e.g. 5 mM tetrabutylammonium phosphate) that are completely ionised at the operating pH values which are usually between 2 and 8. An equilibrium is set up between the neutral species of a mixture in the stationary phase and the respective ionised (anion or cation) species which dissolve in the mobile phase containing the counter ions. The extent of the equilibrium will depend on the ionisation constants of the respective components of the mixture, and the solubility of the unionised species in the stationary phase. Since the ionisation constants and the solubility in the stationary phase will vary with the water-methanol ratio of the mobile phase, the separation may be improved by altering this ratio gradually (gradient elution) or stepwise. If the compounds are eluted too rapidly the water content of the mobile phase should be increased, e.g. by steps of 10%. Conversely, if components do not move, or move slowly, the methanol content of the mobile phase should be increased by steps of 10%.

The application of pressure to the liquid phase in liquid chromatography generally increases the separation (see HPLC). Also in PIC improved efficiency of the column is observed if pressure is applied to the mobile phase (Wittmer, Nuessle and Haney *Anal Chem* **47** 1422 *1975*).

Ion-exchange Chromatography

Ion-exchange chromatography involves an electrostatic process which depends on the relative affinities of various types of ions for an immobilised assembly of ions of opposite charge. The stationary phase is an aqueous buffer with a fixed pH or an aqueous mixture of buffers in which the pH is continuously increased or decreased as the separation may require. This form of liquid chromatography can also be performed at high inlet pressures of liquid with increased column performances.

Ion-exchange Resins. An ion-exchange resin is made up of particles of an insoluble elastic hydrocarbon network to which is attached a large number of ionisable groups. Materials commonly used comprise synthetic ion-exchange resins made, for example, by crosslinking polystyrene to which has been attached non-diffusible ionised or ionisable groups. Resins with relatively high crosslinkage (8-12%) are suitable for the chromatography of small ions, whereas those with low cross linkage (2-4%) are suitable for larger molecules. Applications to hydrophobic systems are possible using aqueous gels with phenyl groups bound to the rigid matrix (Phenyl-Superose/Sepharose, Pharmacia-Amersham Biosciences) or neopentyl chains (Alkyl-Superose, Pharmacia-Amersham Biosciences). (Superose is a cross-linked agarose-based medium with an almost uniform bead size.) These groups are further distinguishable as strong [$-SO_2OH$, $-NR_3^+$] or weak [$-OH$, $-CO_2H$, $-PO(OH)_2$, $-NH_2$]. Their charges are counterbalanced by diffusible ions, and the operation of a column depends on its ability and selectivity to replace these ions. The exchange that takes place is primarily an electrostatic process but adsorptive forces and hydrogen bonding can also be important. A typical sequence for the relative affinities of some common anions (and hence the inverse order in which they pass through such a column), is the following, obtained using a quaternary ammonium (strong base) anion-exchange column:

Fluoride < acetate < bicarbonate < hydroxide < formate < chloride < bromate < nitrite < cyanide < bromide < chromate < nitrate < iodide < thiocyanate < oxalate < sulfate < citrate.

For an amine (weak base) anion-exchange column in its chloride form, the following order has been observed:

Fluoride < chloride < bromide = iodide = acetate < molybdate < phosphate < arsenate < nitrate < tartrate < citrate < chromate < sulfate < hydroxide.

With strong cation-exchangers (e.g. with SO_3H groups), the usual sequence is that polyvalent ions bind more firmly than mono- or di- valent ones, a typical series being as follows:

$Th^{4+} > Fe^{3+} > Al^{3+} > Ba^{2+} > Pb^{2+} > Sr^{2+} > Ca^{2+} > Co^{2+} > Ni^{2+} = Cu^{2+} > Zn^{2+} = Mg^{2+} > UO_2^+ = Mn^{2+} > Ag^+ > Tl^+ > Cs^+ > Rb^+ > NH_4^+ = K^+ > Na^+ > H^+ > Li^+$.

Thus, if an aqueous solution of a sodium salt contaminated with heavy metals is passed through the sodium form of such a column, the heavy metal ions will be removed from the solution and will be replaced by sodium ions from the column. This effect is greatest in dilute solution. Passage of sufficiently strong solutions of alkali metal salts or mineral acids readily displaces all other cations from ion-exchange columns. (The regeneration of columns depends on this property.) However, when the cations lie well to the left in the above series it is often advantageous to use a complex-forming species to facilitate removal. For example, iron can be displaced from ion-exchange columns by passage of sodium citrate or sodium ethylenediaminetetraacetate.

Some of the more common commercially available resins are listed in Table 13.

Ion-exchange resins swell in water to an extent which depends on the amount of crosslinking in the polymer, so that columns should be prepared from the wet material by adding it as a suspension in water to a tube already partially filled with water. (This also avoids trapping air bubbles.) The exchange capacity of a resin is commonly expressed as mg equiv./mL of wet resin. This quantity is pH-dependent for weak-acid or weak-base resins but is constant at about 0.6-2 for most strong-acid or strong-base types.

Apart from their obvious applications to inorganic species, sulfonic acid resins have been used in purifying amino acids, aminosugars, organic acids, peptides, purines, pyrimidines, nucleosides, nucleotides and polynucleotides. Thus, organic bases can be applied to the H^+ form of such resins by adsorbing them from neutral solution and, after washing with water, they are eluted sequentially with suitable buffer solutions or dilute acids. Alternatively, by passing alkali solution through the column, the bases will be displaced in an order that is governed by their pK values. Similarly, strong-base anion exchangers have been used for aldehydes and ketones (as bisulfite addition compounds), carbohydrates (as their borate complexes), nucleosides, nucleotides, organic acids, phosphate esters and uronic acids. Weakly acidic and weakly basic exchange resins have also found extensive applications, mainly in resolving weakly basic and acidic species. For demineralisation of solutions without large changes in pH, mixed-bed resins can be prepared by mixing a cation-exchange resin in its H^+ form with an anion-exchange resin in its OH^- form. Commercial examples include Amberlite MB-1 (IR-120 + IRA-400) and Bio-Deminrolit (Zeo-Karb 225 and Zerolit FF). The latter is also available in a self-indicating form.

Ion-exchange Celluloses and Sephadex. A different type of ion-exchange column that finds extensive application in biochemistry for the purification of proteins, nucleic acids and acidic polysaccharides derives from cellulose by incorporating acidic and basic groups to give ion-exchangers of controlled acid and basic strengths. Commercially available cellulose-type resins are given in Tables 14 and 15. AG 501 x 8 (Bio-Rad) is a mixed-bed resin containing equivalents of AG 50W-x8 H^+ form and AG 1-x8 HO^- form, and Bio-Rex MSZ 501 resin. A dye marker indicates when the resin is exhausted. Removal of unwanted cations, particularly of the transition metals, from amino acids and buffer can be achieved by passage of the solution through a column of Chelex 20 or Chelex 100. The metal-chelating abilities of the resin reside in the bonded iminodiacetate groups.

Chelex can be regenerated by washing in two bed volumes of 1M HCl, two bed volumes of 1M NaOH and five bed volumes of water.

Ion-exchange celluloses are available in different particle sizes. It is important that the amounts of 'fines' are kept to a minimum otherwise the flow of liquid through the column can be extremely slow to the point of no liquid flow. Celluloses with a large range of particle sizes should be freed from 'fines' before use. This is done by suspending the powder in the required buffer and allowing it to settle for one hour and then decanting the 'fines'. This separation appears to be wasteful but it is necessary for reasonable flow rates without applying high pressures at the top of the column. Good flow rates can be obtained if the cellulose column is packed dry whereby the 'fines' are evenly distributed throughout the column. Wet packing causes the 'fines' to rise to the top of the column, which thus becomes clogged.

Several ion-exchange celluloses require recycling before use, a process which must be applied for recovered celluloses. Recycling is done by stirring the cellulose with 0.1M aqueous sodium hydroxide, washing with water until neutral, then suspending in 0.1M hydrochloric acid and finally washing with water until neutral. When regenerating a column it is advisable to wash with a salt solution (containing the required counter ions) of increasing ionic strength up to 2M. The cellulose is then washed with water and recycled if necessary. Recycling can be carried out more than once if there are doubts about the purity of the cellulose and when the cellulose had been used previously for a different purification procedure than the one to be used. The basic matrix of these ion-exchangers is cellulose and it is important not to subject them to strong acid (> 1M) and strongly basic (> 1M) solutions.

When storing ion-exchange celluloses, or during prolonged usage, it is important to avoid growth of microorganisms or moulds which slowly destroy the cellulose. Good inhibitors of microorganisms are phenyl mercuric salts (0.001%, effective in weakly alkaline solutions), chlorohexidine (Hibitane at 0.002% for anion exchangers), 0.02% aqueous sodium azide or 0.005% of ethyl mercuric thiosalicylate (Merthiolate) are most effective in weakly acidic solutions for cation exchangers. Trichlorobutanol (Chloretone, at 0.05% is only effective in weakly acidic solutions) can be used for both anion and cation exchangers. Most organic solvents (e.g. methanol) are effective antimicrobial agents but only at high concentrations. These inhibitors must be removed by washing the columns thoroughly before use because they may have adverse effects on the material to be purified (e.g. inactivation of enzymes or other active preparations).

Sephadex. Other carbohydrate matrices such as *Sephadex* (based on dextran) have more uniform particle sizes. Their advantages over the celluloses include faster and more reproducible flow rates and they can be used directly without removal of 'fines'. *Sephadex*, which can also be obtained in a variety of ion-exchange forms (see Table 15) consists of beads of a cross-linked dextran gel which swells in water and aqueous salt solutions. The smaller the bead size, the higher the resolution that is possible but the slower the flow rate. Typical applications of Sephadex gels are the fractionation of mixtures of polypeptides, proteins, nucleic acids, polysaccharides and for desalting solutions.

Sephadex is a bead form of cross-linked dextran gel. *Sepharose CL* and *Bio-Gel A* are derived from agarose (see below). Sephadex ion-exchangers, unlike celluloses, are available in narrow ranges of particle sizes. These are of two medium types, the G-25 and G-50, and their dry bead diameter sizes are *ca* 50 to 150 microns. They are available as cation and anion exchange Sephadex. One of the disadvantages of using Sephadex ion-exchangers is that the bed volume can change considerably with alteration of pH. *Ultragels* also suffer from this disadvantage to a varying extent, but ion-exchangers of the bead type have been developed e.g. *Fractogels, Toyopearl*, which do not suffer from this disadvantage.

Sepharose (e.g. *Sepharose CL* and *Bio-Gel A*) is a bead form of agarose gel which is useful for the fractionation of high molecular weight substances, for molecular weight determinations of large molecules (molecular weight > 5000), and for the immobilisation of enzymes, antibodies, hormones and receptors usually for affinity chromatography applications.

In preparing any of the above for use in columns, the dry powder is evacuated, then mixed under reduced pressure with water or the appropriate buffer solution. Alternatively it is stirred gently with the solution until all air bubbles are removed. Because some of the wet powders change volumes reversibly with alteration of pH or ionic strength (see above), it is imperative to make allowances when packing columns (see above) in order to avoid overflowing of packing when the pH or salt concentrations are altered.

Cellex CM ion-exchange cellulose can be purified by treatment of 30-40g (dry weight) with 500mL of 1mM cysteine hydrochloride. It is then filtered through a Büchner funnel and the filter cake is suspended in 500mL of 0.05M NaCl/0.5M NaOH. This is filtered and the filter cake is resuspended in 500ml of distilled water and filtered again. The process is repeated until the washings are free from chloride ions. The filter cake is again suspended in 500mL of 0.01M buffer at the desired pH for chromatography, filtered, and the last step repeated several times.

Cellex D and other anionic celluloses are washed with 0.25M NaCl/0.25M NaOH solution, then twice with deionised water. This is followed with 0.25M NaCl and then washed with water until chloride-free. The Cellex is then equilibrated with the desired buffer as above.

Crystalline Hydroxylapatite is a structurally organised, highly polar material which, in aqueous solution (in buffers) strongly adsorbs macromolecules such as proteins and nucleic acids, permitting their separation by virtue of the interaction with charged phosphate groups and calcium ions, as well by physical adsorption. The procedure therefore is not entirely ion-exchange in nature. Chromatographic separations of singly and doubly stranded DNA are readily achievable whereas there is negligible adsorption of low molecular weight species.

Gel Filtration

The gel-like, bead nature of wet Sephadex enables small molecules such as inorganic salts to diffuse freely into it while, at the same time, protein molecules are unable to do so. Hence, passage through a Sephadex column can be used for complete removal of salts from protein solutions. Polysaccharides can be freed from monosaccharides and other small molecules because of their differential retardation. Similarly, amino acids can be separated from proteins and large peptides.

Gel filtration using Sephadex G-types (50 to 200) is essentially useful for fractionation of large molecules with molecular weights above 1000. For Superose, the range is given as 5000 to 5×10^6. Fractionation of lower molecular weight solutes (e,g, ethylene glycols, benzyl alcohols) can now be achieved with Sephadex G-10 (up to Mol.Wt 700) and G-25 (up to Mol.Wt 1500). These dextrans are used only in aqueous solutions. In contrast, Sephadex LH-20 and LH-60 (prepared by hydroxypropylation of Sephadex) are used for the separation of small molecules (Mol.Wt less than 500) using most of the common organic solvents as well as water.

Sephasorb HP (ultrafine, prepared by hydroxypropylation of crossed-linked dextran) can also be used for the separation of small molecules in organic solvents and water, and in addition it can withstand pressures up to 1400 psi making it useful in HPLC. These gels are best operated at pH values between 2 and 12, because solutions with high and low pH values slowly decompose them (see further in Chapter 6).

High Performance Liquid Chromatography (HPLC)

When pressure is applied at the inlet of a liquid chromatographic column the performance of the column can be increased by several orders of magnitude. This is partly because of the increased speed at which the liquid flows through the column and partly because fine column packings which have larger surface areas can be used. Because of the improved efficiency of the columns, this technique has been referred to as high performance, high pressure, or high speed liquid chromatography and has found great importance in chemistry and biochemistry.

The equipment consists of a hydraulic system to provide the pressure at the inlet of the column, a column, a detector, data storage and output, usually in the form of a computer. The pressures used in HPLC vary from a few psi to 4000-5000 psi. The most convenient pressures are, however, between 500 and 1800psi. The plumbing is made of stainless steel or non-corrosive metal tubing to withstand high pressures. Plastic tubing and connectors are used for low pressures, e.g. up to ~500psi. Increase of temperature has a very small effect on the performance of a column in liquid chromatography. Small variations in temperatures, however, do upset the equilibrium of the column, hence it is advisable to place the column in an oven at ambient temperature in order to achieve reproducibility. The packing (stationary phase) is specially prepared for withstanding high pressures. It may be an adsorbent (for adsorption or solid-liquid HPLC), a material impregnated with a high boiling liquid (e.g. octadecyl sulfate, in *reverse-phase* or *liquid-liquid or paired-ion* HPLC), an ion-exchange material (in *ion-exchange* HPLC), or a highly porous non-ionic gel (for high performance *gel filtration*). The mobile phase is water, aqueous buffers, salt solutions, organic solvents or mixtures of these. The more commonly used detectors have UV, visible, diode array or fluorescence monitoring for light absorbing substances, and refractive index monitoring and evaporative light scattering for transparent compounds. UV detection is not useful when molecules do not have UV absorbing chromophores and solvents for elution should be carefully selected when UV monitoring is used so as to ensure the lack of interference in detection. The sensitivity of the refractive index monitoring is usually lower than the light absorbing monitoring by a factor of ten or more. It is also difficult to use a refractive index monitoring system with gradient elution of solvents. When substances have readily oxidised and reduced forms, e.g. phenols, nitro compounds, heterocyclic compounds etc, then electrochemical detectors are useful. These detectors oxidise and reduce these substances and make use of this process to provide a peak on the recorder.

The cells of the monitoring devices are very small (*ca* 5 µl) and the detection is very good. The volumes of the analytical columns are quite small (*ca* 2mL for a 1 metre column) hence the result of an analysis is achieved very quickly. Larger columns have been used for preparative work and can be used with the same equipment. Most machines have solvent mixing chambers for solvent gradient or ion gradient elution. The solvent gradient (for two solvents) or pH or ion gradient can be adjusted in a linear, increasing or decreasing exponential manner.

In general two different types of HPLC columns are available. Prepacked columns are those with metal casings with threads at both ends onto which capillary connections are attached. The cartridge HPLC columns are cheaper and are used with cartridge holders. As the cartridge is fitted with a groove for the holding device, no threads are necessary and the connection pieces can be reused. A large range of HPLC columns (including guard columns, i.e. small pre-columns) are available from Alltech, Supelco (see www.sigmaaldrich.com), Waters (www.waters.com), Agilent Technologies (www.chem.agilent.com), Phenomenex (www.phenomenex.com), YMC (www.ymc.co.jp/en/), Merck (www. merck.de), SGE (www.sge.com) and other leading companies. Included in this range of columns are also columns with chiral bonded phases capable of separating enantiomeric mixtures, such as Chiralpak AS and Chirex™ columns (e.g. from Restek-www.restekcorp.com, Daicel-www.daicel.co.jp/indexe.html).

HPLC systems coupled to mass spectrometers (LC-MS) are extremely important methods for the separation and identification of substances. If not for the costs involved in LC-MS, these systems would be more commonly found in research laboratories.

Other Types of Liquid Chromatography

New stationary phases for specific purposes in chromatographic separation are being continually proposed. *Charge transfer adsorption chromatography* makes use of a stationary phase which contains immobilised aromatic compounds and permits the separation of aromatic compounds by virtue of the ability to form charge transfer complexes (sometimes coloured) with the stationary phase. The separation is caused by the differences in stability of these complexes (Porath and Dahlgren-Caldwell *J Chromatogr* **133** 180 *1977*).

In *metal chelate adsorption chromatography* a metal is immobilised by partial chelation on a column which contains bi- or tri- dentate ligands. Its application is in the separation of substances which can complex with the bound metals and depends on the stability constants of the various ligands (Porath, Carlsson, Olsson and Belfrage *Nature* **258** 598 *1975*; Loennerdal, Carlsson and Porath *FEBS Lett* **75** 89 *1977*).

An application of chromatography which has found extensive use in biochemistry and has brought a new dimension in the purification of enzymes is *affinity chromatography*. A specific enzyme inhibitor is attached by covalent bonding to a stationary phase (e.g. AH-Sepharose 4B for acidic inhibitors and CH-Sepharose 4B for basic inhibitors), and will strongly bind only the specific enzyme which is inhibited, allowing all other proteins to flow through the column. The enzyme is then eluted with a solution of high ionic strength (e.g. 1M sodium chloride) or a solution containing a substrate or reversible inhibitor of the specific enzyme. (The ionic medium can be removed by gel filtration using a mixed-bed gel.) Similarly, an immobilised lectin may interact with the carbohydrate moiety of a glycoprotein. The most frequently used matrixes are cross-linked (4-6%) agarose and polyacrylamide gel. Many adsorbents are commercially available for nucleotides, coenzymes and vitamins, amino acids, peptides, lectins and related macromolecules and immunoglobulins. Considerable purification can be achieved by one passage through the column and the column can be reused several times.

The affinity method may be *biospecific*, for example as an antibody-antigen interaction, or chemical as in the chelation of boronate by *cis*-diols, or of unknown origin as in the binding of certain dyes to albumin and other proteins.

Hydrophobic adsorption chromatography takes advantage of the hydrophobic properties of substances to be separated and has also found use in biochemistry (Hoftsee *Biochem Biophys Res Commun* **50** 751 *1973*; Jennissen and Heilmayer Jr *Biochemistry* **14** 754 *1975*). Specific covalent binding with the stationary phase, a procedure that was called *covalent chromatography*, has been used for separation of compounds and for immobilising enzymes on a support: the column was then used to carry out specific bioorganic reactions (Mosbach *Method Enzymol* **44** 1976; A.Rosevear, J.F.Kennedy and J.M.S.Cabral, *Immobilised Enzymes and Cells: A Laboratory Manual*, Adam Hilger, Bristol, 1987, ISBN 085274515X).

DRYING
Removal of Solvents

Where substances are sufficiently stable, removal of solvent from recrystallised materials presents no problems. The crystals, after filtering at the pump (and perhaps air-drying by suction), are heated in an oven above the boiling point of the solvent (but below this melting point of the crystals), followed by cooling in a desiccator. Where this treatment is inadvisable, it is still often possible to heat to a lower temperature under reduced pressure, for example in an Abderhalden pistol. This device consists of a small chamber which is heated externally by the vapour of a boiling solvent. Inside this chamber, which can be evacuated by a water pump or some other vacuum pump, is placed a small boat containing the sample to be dried and also a receptacle with a suitable drying agent. Convenient liquids for use as boiling liquids in an Abderhalden pistol, and their temperatures, are given in Table 16. Alternatively an electrically heated drying pistol can also be used. In cases where heating above room temperature cannot be used, drying must be carried out in a vacuum desiccator containing suitable absorbents. For example, hydrocarbons, such as cyclohexane and petroleum ether, can be removed by using shredded paraffin wax, and acetic acid and other acids can be absorbed by pellets of sodium or potassium hydroxide. However, in general, solvent removal is less of a problem than ensuring that the water content of solids and liquids is reduced below an acceptable level.

Removal of Water

Methods for removing water from solids depends on the thermal stability of the solids or the time available. The safest way is to dry in a vacuum desiccator over concentrated sulfuric acid, phosphorus pentoxide, silica gel, calcium chloride, or some other desiccant. Where substances are stable in air and melt above 100°, drying in an air oven may be adequate. In other cases, use of an Abderhalden pistol may be satisfactory.

Often, in drying inorganic salts, the final material that is required is a hydrate. In such cases, the purified substance is left in a desiccator to equilibrate above an aqueous solution having a suitable water-vapour pressure. A convenient range of solutions used in this way is given in Table 17.

The choice of desiccants for drying liquids is more restricted because of the need to avoid all substances likely to react with the liquids themselves. In some cases, direct distillation of an organic liquid is a suitable method for drying both solids and liquids, especially if low-boiling azeotropes are formed. Examples include acetone, aniline, benzene, chloroform, carbon tetrachloride, heptane, hexane, methanol, nitrobenzene, petroleum ether, toluene and xylene. Addition of benzene can be used for drying ethanol by distillation. In carrying out distillations intended to yield anhydrous products, the apparatus should be fitted with guard-tubes containing calcium chloride or silica gel to prevent entry of moist air into the system. (Many anhydrous organic liquids are appreciably hygroscopic).

Traces of water can be removed from solvents such as benzene, 1,2-dimethoxyethane, diethyl ether, pentane, toluene and tetrahydrofuran by refluxing under nitrogen a solution containing sodium wire and benzophenone, and fractionally distilling. Drying with, and distilling from CaH_2 is applicable to a number of solvents including aniline, benzene, *tert*-butylamine, *tert*-butanol, 2,4,6-collidine, diisopropylamine, dimethylformamide, hexamethylphosphoramide, dichloromethane, ethyl acetate, pyridine, tetramethylethylenediamine, toluene, triethylamine.

Removal of water from gases may be by physical or chemical means, and is commonly by adsorption on to a drying agent in a low-temperature trap. The effectiveness of drying agents depends on the vapour pressure of the hydrated compound - the lower the vapour pressure the less the remaining moisture in the gas.

The most usually applicable of the specific methods for detecting and determining water in organic liquids is due to Karl Fischer. (See J.Mitchell and D.M.Smith, *Aquametry*, 2nd Ed, J Wiley & Sons, New York, *1977-1984*, ISBN 0471022640; Fieser and Fieser *Reagents for Organic Synthesis*, J.Wiley & Sons, NY, Vol 1, 528 *1967*, ISBN 0271616X). Other techniques include electrical conductivity measurements and observation of the temperature at which the first cloudiness appears as the liquid is cooled (applicable to liquids in which water is only slightly soluble). Addition of anhydrous cobalt (II) iodide (blue) provides a convenient method (colour change to pink on hydration) for detecting water in alcohols, ketones, nitriles and some esters. Infrared absorption measurements of the broad band for water near 3500 cm^{-1} can also sometimes be used for detecting water in non-hydroxylic substances.

For further useful information on mineral adsorbents and drying agents, go to the SigmaAldrich website, under technical library (Aldrich) for technical bulletin AL-143.

Intensity and Capacity of Common Desiccants

Drying agents are conveniently grouped into three classes, depending on whether they combine with water reversibly, they react chemically (irreversibly) with water, or they are molecular sieves. The first group vary in their drying intensity with the temperature at which they are used, depending on the vapour pressure of the hydrate that is formed. This is why, for example, drying agents such as anhydrous sodium sulfate, magnesium sulfate or calcium chloride should be filtered off from the liquids before the latter are heated. The intensities of drying agents belonging to this group fall in the sequence:

$P_2O_5 \gg BaO > Mg(ClO_4)_2$, CaO, MgO, KOH (fused), conc H_2SO_4, $CaSO_4$, Al_2O_3 > KOH (pellets), silica gel, $Mg(ClO_4)_2.3H_2O$ > NaOH (fused), 95% H_2SO_4, $CaBr_2$, $CaCl_2$ (fused) > NaOH (pellets), $Ba(ClO_4)_2$, $ZnCl_2$, $ZnBr_2$ > $CaCl_2$ (technical) > $CuSO_4$ > Na_2SO_4, K_2CO_3.

Where large amounts of water are to be removed, a preliminary drying of liquids is often possible by shaking with concentrated solutions of sodium sulfate or potassium carbonate, or by adding sodium chloride to salt out the organic phase (for example, in the drying of lower alcohols), as long as the drying agent does not react (e.g. $CaCl_2$ with alcohols and amines, see below).

Drying agents that combine irreversibly with water include the alkali metals, the metal hydrides (discussed in Chapter 2), and calcium carbide.

Suitability of Individual Desiccants

Alumina. (Preheated to 175° for about 7h). Mainly as a drying agent in a desiccator or as a column through which liquid is percolated.

Aluminium amalgam. Mainly used for removing traces of water from alcohols *via* refluxing followed by distillation.

Barium oxide. Suitable for drying organic bases.

Barium perchlorate. Expensive. Used in desiccators (*covered with a metal guard*). Unsuitable for drying solvents or organic material where contact is necessary, because of the danger of **EXPLOSION**

Boric anhydride. (Prepared by melting boric acid in an air oven at a high temperature, cooling in a desiccator, and powdering.) Mainly used for drying formic acid.

Calcium chloride (anhydrous). Cheap. Large capacity for absorption of water, giving the hexahydrate below 30°, but is fairly slow in action and not very efficient. Its main use is for preliminary drying of alkyl and aryl halides, most esters, saturated and aromatic hydrocarbons and ethers. Unsuitable for drying alcohols and amines (which form addition compounds), fatty acids, amides, amino acids, ketones, phenols, or some aldehydes and esters. Calcium chloride is suitable for drying the following gases: hydrogen, hydrogen chloride, carbon monoxide, carbon dioxide, sulfur dioxide, nitrogen, methane, oxygen, also paraffins, ethers, olefins and alkyl chlorides.

Calcium hydride. See Chapter 2.

Calcium oxide. (Preheated to 700-900° before use.) Suitable for alcohols and amines (but does not dry them completely). Need not be removed before distillation, but in that case the head of the distillation column should be packed with glass wool to trap any calcium oxide powder that might be carried over. Unsuitable for acidic compounds and esters. Suitable for drying gaseous amines and ammonia.

Calcium sulfate (anhydrous). (Prepared by heating the dihydrate or the hemihydrate in an oven at 235° for 2-3h; it can be regenerated.) Available commercially as Drierite. It forms the hemihydrate, $2CaSO_4.H_2O$, so that its capacity is fairly low (6.6% of its weight of water), and hence is best used on partially dried substances. It is very efficient (being comparable with phosphorus pentoxide and concentrated sulfuric acid). Suitable for most organic compounds. Solvents boiling below 100° can be dried by direct distillation from calcium sulfate.

Copper (II) sulfate (anhydrous). Suitable for esters and alcohols. Preferable to sodium sulfate in cases where solvents are sparingly soluble in water (for example, benzene or toluene).

Lithium aluminium hydride. See Chapter 2.

Magnesium amalgam. Mainly used for removing traces of water from alcohols by refluxing the alcohol in the presence of the Mg amalgam followed by distillation.

Magnesium perchlorate (anhydrous). (Available commercially as Dehydrite. Expensive.) Used in desiccators. Unsuitable for drying solvents or any organic material where contact is necessary, because of the danger of **EXPLOSION**.

Magnesium sulfate (anhydrous). (Prepared from the heptahydrate by drying at 300° under reduced pressure.) More rapid and effective than sodium sulfate but is slightly acidic. It has a large capacity, forming $MgSO_4.7H_2O$ below 48°. Suitable for the preliminary drying of most organic compounds.

Molecular sieves. See later.

Phosphorus pentoxide. Very rapid and efficient, but difficult to handle and should only be used after the organic material has been partially dried, for example with magnesium sulfate. Suitable for anhydrides, alkyl and aryl halides, ethers, esters, hydrocarbons and nitriles, and for use in desiccators. Not suitable with acids, alcohols, amines or ketones, or with organic molecules from which a molecule of water can be eliminated. Suitable for drying the following gases: hydrogen, oxygen, carbon dioxide, carbon monoxide, sulfur dioxide, nitrogen, methane, ethene and paraffins. It is available on a solid support with an indicator under the name *Sicapent* (from Merck). The colour changes in Sicapent depend on the percentage of water present (e.g. in the absence of water, Sicapent is colorless but becomes green with 20% water and blue with 33% w/w water). When the quantity of water in the desiccator is high a crust of phosphoric acid forms a layer over the phosphorus pentoxide powder and decreases its efficiency. The crust can be removed with a spatula to expose the dry pwder and restore the desiccant property.

Potassium (metal). Properties and applications are similar to those for sodium but as the reactivity is greater than that of sodium, the hazards are greater than that of sodium. **Handle with extreme care.**

Potassium carbonate (anhydrous). Has a moderate efficiency and capacity, forming the dihydrate. Suitable for an initial drying of alcohols, bases, esters, ketones and nitriles by shaking with them, then filtering off. Also suitable for salting out water-soluble alcohols, amines and ketones. Unsuitable for acids, phenols, thiols and other acidic substances.

Potassium carbonate. Solid potassium hydroxide is very rapid and efficient. Its use is limited almost entirely to the initial drying of organic bases. Alternatively, sometimes the base is shaken first with a concentrated solution of potassium hydroxide to remove most of the water present. Unsuitable for acids, aldehydes, ketones, phenols, thiols, amides and esters. Also used for drying gaseous amines and ammonia.

Silica gel. Granulated silica gel is a commercially available drying agent for use with gases, in desiccators, and (because of its chemical inertness) in physical instruments (pH meters, spectrometers, balances). Its drying action depends on physical adsorption, so that silica gel must be used at room temperature or below. By incorporating cobalt chloride into the material it can be made self indicating (blue when dry, pink when wet), re-drying in an oven at 110° being necessary when the colour changes from blue to pink.

Sodium (metal). Used as a fine wire or as chips, for more completely drying ethers, saturated hydrocarbons and aromatic hydrocarbons which have been partially dried (for example with calcium chloride or magnesium sulfate). Unsuitable for acids, alcohols, alkyl halides, aldehydes, ketones, amines and esters. Reacts violently if water is present and can cause a fire with highly flammable liquids.

Sodium hydroxide. Properties and applications are similar to those for potassium hydroxide.

Sodium-potassium alloy. Used as lumps. Lower melting than sodium, so that its surface is readily renewed by shaking. Properties and applications are similar to those for sodium.

Sodium sulfate (anhydrous). Has a large capacity for absorption of water, forming the decahydrate below 33°, but drying is slow and inefficient, especially for solvents that are sparingly soluble in water. It is suitable for the preliminary drying of most types of organic compounds.

Sulfuric acid (concentrated). Widely used in desiccators. Suitable for drying bromine, saturated hydrocarbons, alkyl and aryl halides. Also suitable for drying the following gases: hydrogen, nitrogen, carbon dioxide, carbon monoxide, chlorine, methane and paraffins. Unsuitable for alcohols, bases, ketones or phenols. Also available on a solid support with an indicator under the name *Sicacide* (from Merck) for desiccators. The colour changes in Sicacide depends on the percentage of water present (e.g. when dry Sicacide is red-violet but becomes pale violet with 27% water and pale yellow to colorless with 33% w/w water).

For convenience, many of the above drying agents are listed in Table 18 under the classes of organic compounds for which they are commonly used.

Molecular sieves

Molecular sieves are types of adsorbents composed of crystalline zeolites (sodium and calcium aluminosilicates). By heating them, water of hydration is removed, leaving holes of molecular dimensions in the crystal lattices. These holes are of uniform size and allow the passage into the crystals of small molecules, but not of large ones. This *sieving* action explains their use as very efficient drying agents for gases and liquids. The pore size of these sieves can be modified (within limits) by varying the cations built into the lattices. The four types of molecular sieves currently available are:

Type 3A sieves. A crystalline potassium aluminosilicate with a pore size of about 3 Angstroms. This type of molecular sieves is suitable for drying liquids such as acetone, acetonitrile, methanol, ethanol and 2-propanol, and drying gases such as acetylene, carbon dioxide, ammonia, propylene and butadiene. The material is supplied as beads or pellets.

Type 4A sieves. A crystalline sodium aluminosilicate with a pore size of about 4 Angstroms, so that, besides water, ethane molecules (but not butane) can be adsorbed. This type of molecular sieves is suitable for drying chloroform, dichloromethane, diethyl ether, dimethylformamide, ethyl acetate, cyclohexane, benzene, toluene, xylene, pyridine and diisopropyl ether. It is also useful for low pressure air drying. The material is supplied as beads, pellets or powder.

Type 5A sieves. A crystalline calcium aluminosilicate with a pore size of about 5 Angstroms, these sieves adsorb larger molecules than type 4A. For example, as well as the substances listed above, propane, butane, hexane, butene, higher *n*-olefins, *n*-butyl alcohol and higher *n*-alcohols, and cyclopropane can be adsorbed, but not branched-chain C_6 hydrocarbons, cyclic hydrocarbons such as benzene and cyclohexane, or secondary and tertiary alcohols, carbon tetrachloride or boron trifluoride. This is the type generally used for drying gases, though organic liquids such as THF and dioxane can be dried with this type of molecular sieves.

Type 13X sieves. A crystalline sodium aluminosilicate with a pore size of about 10 Angstroms which enables many branched-chain and cyclic compounds to be adsorbed, in addition to all the substances removed by type 5A sieves.

They are unsuitable for use with strong acids but are stable over the pH range 5-11.
Because of their selectivity, molecular sieves offer advantages over silica gel, alumina or activated charcoal, especially in their very high affinity for water, polar molecules and unsaturated organic compounds. Their relative efficiency is greatest when the impurity to be removed is present at low concentrations. Thus, at 25° and a relative humidity of 2%, type 5A molecular sieves adsorb 18% by weight of water, whereas for silica gel and alumina the figures are 3.5 and 2.5% respectively. Even at 100° and a relative humidity of 1.3% molecular sieves adsorb about 15% by weight of water.
The greater preference of molecular sieves for combining with water molecules explains why this material can be used for drying ethanol and why molecular sieves are probably the most universally useful and efficient drying agents. Percolation of ethanol with an initial water content of 0.5% through a 144 cm long column of type 4A molecular sieves reduced the water content to 10ppm. Similar results have been obtained with pyridine.

The main applications of molecular sieves to purification comprise:
1. Drying of gases and liquids containing traces of water.
2. Drying of gases at elevated temperatures.
3. Selective removal of impurities (including water) from gas streams.
(For example, carbon dioxide from air or ethene; nitrogen oxides from nitrogen; methanol from diethyl ether. In general, carbon dioxide, carbon monoxide, ammonia, hydrogen sulfide, mercaptans, ethane, ethene, acetylene (ethyne), propane and propylene are readily removed at 25°. In mixtures of gases, the more polar ones are preferentially adsorbed).

The following applications include the removal of straight-chain from branched-chain or cyclic molecules. For example, type 5A sieves will adsorb *n*-butyl alcohol but not its branched-chain isomers. Similarly, it separates *n*-tetradecane from benzene, or *n*-heptane from methylcyclohexane.
The following liquids have been dried with molecular sieves: acetone, acetonitrile, acrylonitrile, allyl chloride, amyl acetate, benzene, butadiene, *n*-butane, butene, butyl acetate, *n*-butylamine, *n*-butyl chloride, carbon tetrachloride, chloroethane, 1-chloro-2-ethylhexane, cyclohexane, dichloromethane, dichloroethane, 1,2-dichloropropane, 1,1-dimethoxyethane, dimethyl ether, 2-ethylhexanol, 2-ethylhexylamine, *n*-heptane, *n*-hexane, isoprene, isopropyl alcohol, diisopropyl ether, methanol, methyl ethyl ketone, oxygen, *n*-pentane, phenol, propane, *n*-propyl alcohol, propylene, pyridine, styrene, tetrachloroethylene, toluene, trichloroethylene and xylene. In addition, the following gases have been dried: acetylene, air, argon, carbon dioxide, chlorine, ethene, helium, hydrogen, hydrogen chloride, hydrogen sulfide, nitrogen, oxygen and sulfur hexafluoride.

After use, molecular sieves can be regenerated by heating at between 300°–350° for several hours, preferably in a stream of dry inert gas such as nitrogen or preferably under vacumm, then cooling in a desiccator. Special precautions must be taken before regeneration of molecular sieves used in the drying of flammable solvents.
However, care must be exercised in using molecular sieves for drying organic liquids. Appreciable amounts of impurities were *formed* when samples of acetone, 1,1,1-trichloroethane and methyl-*t*-butyl ether were dried in the liquid phase by contact with molecular sieves 4A (Connett *Lab Pract* **21** 545 *1972*). Other, less reactive types of sieves may be more suitable but, in general, it seems desirable to make a preliminary test to establish that no unwanted reaction takes place. Useful comparative data for Type 4A and 5A sieves are in Table 19.

MISCELLANEOUS TECHNIQUES
Freeze-pump-thaw and purging
Volatile contaminants, e.g. traces of low boiling solvent residue or oxygen, in liquid samples or solutions can be very deleterious to the samples on storage. These contaminants can be removed by repeated freeze-pump-thaw cycles. This involves freezing the liquid material under high vacuum in an appropriate vessel (which should be large enough to avoid contaminating the vacuum line with liquid that has bumped) connected to the vacuum line *via* efficient liquid nitrogen traps. The frozen sample is then thawed until it liquefies, kept in this form for some time (*ca* 10-15min), refreezing the sample and the cycle repeated several times without interrupting the vacuum. This procedure applies equally well to solutions, as well as purified liquids, e.g. as a means of removing oxygen from solutions for NMR and other measurements. If the presence of nitrogen, helium or argon, is not a serious contaminant then solutions can be freed from gases, e.g. oxygen, carbon dioxide, and volatile impurities by purging with N_2, He or Ar at room, or slightly elevated, temperature. The gases used for purging are then removed by freeze-pump-thaw cycles or simply by keeping in a vacuum for several hours. Special NMR tubes

with a screw cap thread and a PTFE valve (Wilmad) are convenient for freeze thawing of NMR samples. Alternatively NMR tubes with "J Young" valves (Wilmad) can also be used.

Vacuum-lines, Schlenk and glovebox techniques

Manipulations involving materials sensitive to air or water vapour can be carried out by these procedures. Vacuum-line methods make use of quantitative transfers, and **P**(pressure)-**V**(volume)-**T**(temperature) measurements, of gases, and trap-to-trap separations of volatile substances.

It is usually more convenient to work under an inert-gas atmosphere using **Schlenk** type apparatus. The *principle* of Schlenk methods involve a flask/vessel which has a standard ground-glass joint and a sidearm with a tap. The system can be purged by evacuating and flushing with an inert gas (usually nitrogen, or in some cases, argon), repeating the process until the contaminants in the vapour phases have been diminished to acceptable limits. A large range of Schlenk glassware is commercially available (e.g. see Aldrich Chemical Catalog and the associated technical bulletin AL-166). With these, and tailor-made pieces of glassware, inert atmospheres can be maintained during crystallisation, filtration, sublimation and transfer.

Syringe techniques have been developed for small volumes, while for large volumes or where much manipulation is required, dryboxes (*glove boxes*) or dry chambers should be used.

ABBREVIATIONS

Titles of periodicals are defined as in the Chemical Abstracts Service Source Index (CASSI), except that full stops have been omitted after each abbreviated word. Abbreviations of words in the texts of Chapters 4, 5 and 6 are those in common use and are self evident, e.g. distn, filtd, conc and vac are used for distillation, filtered, concentrated and vacuum.

TABLES

TABLE 1. SOME COMMON IMMISCIBLE OR SLIGHTLY MISCIBLE PAIRS OF SOLVENTS

Carbon tetrachloride with ethanolamine, ethylene glycol, formamide or water.
Dimethyl formamide with cyclohexane or petroleum ether.
Dimethyl sulfoxide with cyclohexane or petroleum ether.
Ethyl ether with ethanolamine, ethylene glycol or water.
Methanol with carbon disulfide, cyclohexane or petroleum ether.
Petroleum ether with aniline, benzyl alcohol, dimethyl formamide, dimethyl sulfoxide, formamide, furfuryl alcohol, phenol or water.
Water with aniline, benzene, benzyl alcohol, carbon disulfide, carbon tetrachloride, chloroform, cyclohexane, cyclohexanol, cyclohexanone, diethyl ether, ethyl acetate, isoamyl alcohol, methyl ethyl ketone, nitromethane, tributyl phosphate or toluene.

TABLE 2.	AQUEOUS BUFFERS
Approx. pH	Composition

0	2N sulfuric acid or N hydrochloric acid
1	0.1N hydrochloric acid or 0.18N sulfuric acid
2	**Either** 0.01N hydrochloric acid or 0.013N sulfuric acid **Or** 50 mL of 0.1M glycine (also 0.1M NaCl) + 50 mL of 0.1N hydrochloric acid
3	**Either** 20 mL of the 0.2M Na_2HPO_4 + 80 mL of 0.1M citric acid **Or** 50 mL of 0.1M glycine + 22.8 mL of 0.1N hydrochloric acid in 100 mL
4	**Either** 38.5 mL of 0.2M Na_2HPO_4 + 61.5 mL of 0.1M citric acid **Or** 18 mL of 0.2M NaOAc + 82 mL of 0.2M acetic acid
5	**Either** 70 mL of 0.2M NaOAc + 30 mL of 0.2M acetic acid **Or** 51.5 mL of 0.2M Na_2HPO_4 + 48.5 mL of 0.1M citric acid
6	63 mL of 0.2M Na_2HPO_4 + 37 mL of 0.1M citric acid
7	82 mL of M Na_2HPO_4 + 18 mL of 0.1M citric acid
8	**Either** 50 mL of 0.1M Tris buffer + 29 mL of 0.1N hydrochloric acid, in 100 mL **Or** 30 mL of 0.05M borax + 70 mL of 0.2M boric acid
9	80 mL of 0.05M borax + 20 mL of 0.2M boric acid
10	**Either** 25 mL of 0.05M borax + 43 mL of 0.1N NaOH, in 100 mL **Or** 50 mL of 0.1M glycine + 32 mL of 0.1N NaOH, in 100 mL
11	50 mL of 0.15M Na_2HPO_4 + 15 mL of 0.1N NaOH
12	50 mL of 0.15M Na_2HPO_4 + 75 mL of 0.1N NaOH
13	0.1N NaOH or KOH
14	N NaOH or KOH

These buffers are suitable for use in obtaining ultraviolet spectra. Alternatively, for a set of accurate buffers of low, but constant, ionic strength (I = 0.01) covering a pH range 2.2 to 11.6 at 20°, see Perrin *Aust J Chem* **16** 572 *1963.* "In 100 mL" means that the solution is made up to 100 mL with pure water.

TABLE 3A.   PREDICTED EFFECT OF PRESSURE ON BOILING POINT[*]

Temperature in degrees Centigrade

760 mmHg	0	20	40	60	80	100	120	140	160	180
0.1	-111	-99	-87	-75	-63	-51	-39	-27	-15	-4
0.2	-105	-93	-81	-69	-56	-44	-32	-19	-7	5
0.4	-100	-87	-74	-62	-49	-36	-24	-11	2	15
0.6	-96	-83	-70	-57	-44	-32	-19	-6	7	20
0.8	-94	-81	-67	-54	-41	-28	-15	-2	11	24
1.0	-92	-78	-65	-52	-39	-25	-12	1	15	28
2.0	-85	-71	-58	-44	-30	-16	-3	11	25	39
4.0	-78	-64	-49	-35	-21	-7	8	22	36	51
6.0	-74	-59	-44	-30	-15	-1	14	29	43	58
8.0	-70	-56	-41	-26	-11	4	19	34	48	63
10.0	-68	-53	-38	-23	-8	7	22	37	53	68
14.0	-64	-48	-33	-23	-2	13	28	44	59	74
16.0	-61	-45	-29	-14	2	17	33	48	64	79
20.0	-59	-44	-28	-12	3	19	35	50	66	82
30.0	-54	-38	-22	-6	10	26	42	58	74	90
40.0	-50	-34	-17	-1	15	32	48	64	81	97
50.0	-47	-30	-14	3	19	36	52	69	86	102
60.0	-44	-28	-11	6	23	40	56	73	86	107
80.0	-40	-23	-6	11	28	45	62	79	97	114
100.0	-37	-19	-2	15	33	50	67	85	102	119
150.0	-30	-12	6	23	41	59	77	95	112	130
200.0	-25	-7	11	29	47	66	84	102	120	138
300.0	-18	1	19	38	57	75	94	113	131	150
400.0	-13	6	25	44	64	83	102	121	140	159
500.0	-8	11	30	50	69	88	108	127	147	166
600.0	-5	15	34	54	74	93	113	133	152	172
700.0	-2	18	38	58	78	98	118	137	157	177
750.0	0	20	40	60	80	100	120	140	160	180
770.0	0	20	40	60	80	100	120	140	160	180
800.0	1	21	41	61	81	101	122	142	162	182

[*] *How to use the Table*: Take as an example a liquid with a boiling point of 80°C at 760mm Hg. The Table gives values of the boiling points of this liquid at pressures from 0.1 to 800mm Hg. Thus at 50mm Hg this liquid has a boiling point of 19°C, and at 2mm Hg its boiling point would be -30°C.

TABLE 3B. PREDICTED EFFECT OF PRESSURE ON BOILING POINT[*]

Temperature in degrees Centigrade

760mmHg	200	220	240	260	280	300	320	340	360	380	400
0.1	8	20	32	44	56	68	80	92	104	115	127
0.2	17	30	42	54	67	79	91	103	116	128	140
0.4	27	40	53	65	78	91	103	116	129	141	154
0.6	33	40	59	72	85	98	111	124	137	150	163
0.8	38	51	64	77	90	103	116	130	143	156	169
1.0	41	54	68	81	94	108	121	134	147	161	174
2.0	53	66	80	94	108	121	135	149	163	176	190
4.0	65	79	93	108	122	136	151	156	179	193	208
6.0	72	87	102	116	131	146	160	175	189	204	219
8.0	78	93	108	123	137	152	167	182	197	212	227
10.0	83	98	113	128	143	158	173	188	203	218	233
14.0	90	105	120	136	151	166	182	197	212	228	243
18.0	95	111	126	142	157	173	188	204	219	235	251
20.0	97	113	129	144	160	176	191	207	223	238	254
30.0	106	123	139	155	171	187	203	219	235	251	267
40.0	113	130	146	162	179	195	211	228	244	260	277
50.0	119	135	152	168	185	202	218	235	251	268	284
60.0	123	140	157	174	190	207	224	241	257	274	291
80.0	131	148	165	182	199	216	233	250	267	284	301
100.0	137	154	171	189	206	223	241	258	275	293	310
150.0	148	166	184	201	219	237	255	273	290	308	326
200.0	156	174	193	211	229	247	265	283	302	320	338
300.0	169	187	206	225	243	262	281	299	318	337	355
400.0	178	197	216	235	254	273	292	311	330	350	369
500.0	185	205	224	244	263	282	302	321	340	360	379
600.0	192	211	231	251	270	290	310	329	349	368	388
700.0	197	217	237	257	277	296	316	336	356	376	396
750.0	200	220	239	259	279	299	319	339	359	279	399
770.0	200	220	241	261	281	301	321	341	361	381	401
800.0	202	222	242	262	282	302	322	342	262	382	403

[*] *How to use the Table*: Taking as an example a liquid with a boiling point of 340°C at 760mm Hg, the column headed 340°C gives values of the boiling points of this liquid at each value of pressures from 0.1 to 800mm Hg. Thus, at 100mm Hg its boiling point is 258°C, and at 0.8mm Hg its boiling point will be 130°C.

FIGURE 1: NOMOGRAM

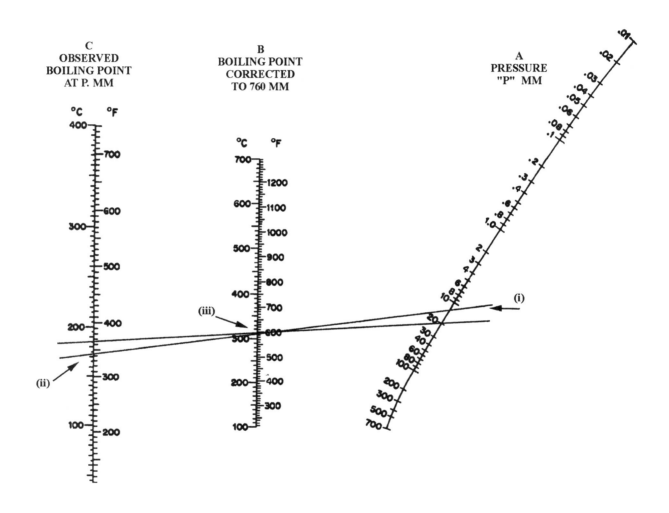

How to use Figure 1:
You can use a nomogram to estimate the boiling points of a substance at a particular pressure. For example, the boiling point of 4-methoxybenzenesulfonyl chloride is 173°C/14mm. Thus to find out what the boiling point of this compound will be at 760mm (atmospheric), draw a point on curve A (pressure) at 14mm (this is shown in (i). Then draw a point on curve C (observed boiling point) corresponding to 173° (or as close as possible). This is shown in (ii). Using a ruler, find the point of intersection on curve B, drawing a line between points (i) and (ii). This is the point (iii) and is the boiling point of 4-methoxybenzenesulfonyl chloride (i.e. approx. 310°C) at atmospheric pressure. If you want to distil 4-methoxybenzenesulfonyl chloride at 20mm, then you will need to draw a point on curve A (at 20mm). Using a ruler, find the point of intersection on curve C drawing through the line intersecting (iii, curve B, i.e. 310°C) and the point in curve A corresponding to 20mm. You should have a value of 185°C, that is, the boiling point of 4-methoxybenzenesulfonyl chloride is estimated to be at 185°C at 20mm.

TABLE 4.	HEATING BATHS
Up to 100°	Water baths
-20 to 200°	Glycerol or di-*n*-butyl phthalate
Up to about 200°	Medicinal paraffin
Up to about 250°	Hard hydrogenated cotton-seed oil (**m** 40-60°) or a 1:1 mixture of cotton-seed oil and castor oil containing about 1% of hydroquinone.
-40 to 250° (to 400° under N_2)	D.C. 550 silicone fluid
Up to about 260°	A mixture of 85% orthophosphoric acid (4 parts) and metaphosphoric acid (1 part)
Up to 340°	A mixture of 85% orthophosphoric acid (2 parts) and metaphosphoric acid (1 part)
60 to 500°	Fisher bath wax (highly unsaturated)
73 to 350°	Wood's Metal*
250 to 800°	Solder*
350 to 800°	Lead*

* In using metal baths, the container (usually a metal crucible) should be removed while the metal is still molten.

TABLE 5.	WHATMAN FILTER PAPERS

Grade No.	1	2	3	4	5	6	113
Particle size retained (in microns)	11	8	5	12	2.4	2.8	28
Filtration speed*(sec/100mL)	40	55	155	20	<300	125	9

Routine ashless filters

Grade No.	40	41	42	43	44
Particle size retained (in microns)	7.5	12	3	12	4
Filtration speed* (sec/100mL)	68	19	200	38	125

	Hardened			**Hardened ashless**		
Grade No.	50	52	54	540	541	542
Particle size retained(in microns)	3	8	20	9	20	3
Filtration speed* (sec/100mL)	250	55	10	55	12	250

Glass microfilters

Grade No	GF/A	GF/B	GF/C	GF/D	GF/F
Particle size retained (in microns)	1.6	1.0	1.1	2.2	0.8
Filtration speed (sec/100mL)*	8.3	20.0	8.7	5.5	17.2

Filtration speeds are rough estimates of initial flow rates and should be considered on a relative basis.

TABLE 6.   MICRO FILTERS*

Nucleopore (polycarbonate) Filters

Mean Pore Size (microns)	8.0	2.0	1.0	0.1	0.03	0.015
Av. pores/cm^2	10^5	2×10^6	2×10^7	3×10^8	6×10^8	$1\text{-}6 \times 10^9$
Water flow rate(mL/min/cm^2)	2000	2000	300	8	0.03	0.1-0.5

Millipore Filters

Type	——Cellulose ester——		——Teflon——		–Microweb#–	
	MF/SC	MF/VF	LC	LS	WS	WH
Mean Pore Size (microns)	8	0.01	10	5	3	0.45
Water flow rate (mL/min/cm^2)	850	0.2	170	70	155	55

Gelman Membranes

Type	——Cellulose ester——				——Copolymer——	
	GA-1	TCM-450	VM-1	DM-800	AN-200	Tuffryn-450
Mean Pore Size (microns)	5	0.45	5	0.8	0.2	0.45
Water flow rate (mL/min/cm^2)	320	50	700	200	17	50

Sartorius Membrane Filters (SM)

Application	Gravi-metric	Biological clarificatn	Sterili-sation	Particle count in H$_2$O	For acids & bases
Type No.	11003	11004	11006	11011	12801
Mean PoreSize (microns)	1.2	0.6	0.45	0.01	8
Water flow rate (mL/min/cm^2)	300	150	65	0.6	1100

* Only a few representative filters are tabulated (available ranges are more extensive).   # Reinforced nylon.

TABLE 7. COMMON SOLVENTS USED IN RECRYSTALLISATION

Acetic acid (118°)	*Cyclohexane (81°)	*Methanol (64.5°)
*Acetone (56°)	Dichloromethane (41°)	*Methyl ethyl ketone (80°)
Acetylacetone (139°)	*Diethyl ether (34.5°)	Methyl isobutyl ketone (116°)
Acetonitrile (82°C)	Dimethyl formamide (76°/39mm)	Nitrobenzene (210°)
*Benzene (80°)	*Dioxane (101°)	Nitromethane (101°)
Benzyl alcohol (93°/10mm)	*Ethanol (78°)	*Petroleum ether (various)
n-Butanol (118°)	2-Ethoxyethanol (cellosolve 135°)	Pyridine (115.5°)
Butyl acetate (126.5°)	*Ethyl acetate (78°)	Pyridine trihydrate (93°)
n-Butyl ether (142°)	Ethyl benzoate (98°/19mm)	*Tetrahydrofuran (64-66°)
γ-Butyrolactone (206°)	Ethylene glycol (68°/4mm)	Toluene (110°)
Carbon tetrachloride (77°)	Formamide (110°/10mm)	Trimethylene glycol (59°/11mm)
Chlorobenzene (132°)	Glycerol (126°/11mm)	Water (100°)
Chloroform (61°)	Isoamyl alcohol (131°)	Xylenes (o 143-145°, m 138-139°, p 138°)

*Highly flammable, should be heated or evaporated on steam or electrically heated water baths only (preferably under nitrogen). None of these solvents should be heated over a naked flame.

TABLE 8. PAIRS OF MISCIBLE SOLVENTS

Acetic acid: with chloroform, ethanol, ethyl acetate, acetonitrile, petroleum ether, or water.

Acetone: with benzene, butyl acetate, butyl alcohol, carbon tetrachloride, chloroform, cyclohexane, ethanol, ethyl acetate, methyl acetate, acetonitrile, petroleum ether or water.

Ammonia: with ethanol, methanol, pyridine.

Aniline: with acetone, benzene, carbon tetrachloride, ethyl ether, n-heptane, methanol, acetonitrile or nitrobenzene.

Benzene: with acetone, butyl alcohol, carbon tetrachloride, chloroform, cyclohexane, ethanol, acetonitrile, petroleum ether or pyridine.

Butyl alcohol: with acetone or ethyl acetate.

Carbon disulfide: with petroleum ether.

Carbon tetrachloride: with cyclohexane.

Chloroform: with acetic acid, acetone, benzene, ethanol, ethyl acetate, hexane, methanol or pyridine.

Cyclohexane: with acetone, benzene, carbon tetrachloride, ethanol or diethyl ether.

Diethyl ether: with acetone, cyclohexane, ethanol, methanol, methylal (dimethoxymethane), acetonitrile, pentane or petroleum ether.

Dimethyl formamide: with benzene, ethanol or ether.

Dimethyl sulfoxide: with acetone, benzene, chloroform, ethanol, diethyl ether or water.

Dioxane: with benzene, carbon tetrachloride, chloroform, ethanol, diethyl ether, petroleum ether, pyridine or water.

Ethanol: with acetic acid, acetone, benzene, chloroform, cyclohexane, dioxane, ethyl ether, pentane, toluene, water or xylene.

Ethyl acetate: with acetic acid, acetone, butyl alcohol, chloroform, or methanol.

Glycerol: with ethanol, methanol or water.

Hexane: with benzene, chloroform or ethanol.

Methanol: with chloroform, diethyl ether, glycerol or water.

Methylal: with diethyl ether.

Methyl ethyl ketone: with acetic acid, benzene, ethanol or methanol.

Nitrobenzene: with aniline, methanol or acetonitrile.

Pentane: with ethanol or diethyl ether.

Petroleum ether: with acetic acid, acetone, benzene, carbon disulfide or diethyl ether.

Phenol: with carbon tetrachloride, ethanol, diethyl ether or xylene.

Pyridine: with acetone, ammonia, benzene, chloroform, dioxane, petroleum ether, toluene or water.

Toluene: with ethanol, diethyl ether or pyridine.

Water: with acetic acid, acetone, ethanol, methanol, or pyridine.

Xylene: with ethanol or phenol.

TABLE 9.   MATERIALS FOR COOLING BATHS

Temperature	Composition	Temperature	Composition
0°	Crushed ice		
-5° to -20°	Ice-salt mixtures	-77°	Solid CO_2 with chloroform or acetone
Up to -20°	Ice-MeOH mixtures	-78°	Solid CO_2 (powdered; CO_2 snow)
-33°	Liquid ammonia		
-40° to -50°	Ice (3.5-4 parts) - $CaCl_2$ $6H_2O$ (5 parts)	-100°	Solid CO_2 with diethyl ether
-72°	Solid CO_2 with ethanol	-196°	liquid nitrogen (see footnote*)

Alternatively, the following liquids can be used, partially frozen, as cryostats, by adding solid CO_2 from time to time to the material in a Dewar-type container and stirring to make a slush:

13°	p-Xylene	-55°	Diacetone
12°	Dioxane	-56°	n-Octane
6°	Cyclohexane	-60°	Di-isopropyl ether
5°	Benzene	-73°	Trichloroethylene or isopropyl acetate
2°	Formamide	-74°	o-Cymene or p-cymene
-8.6°	Methyl salicylate	-77°	Butyl acetate
-9°	Hexane-2,5-dione	-79°	Isoamyl acetate
-10.5°	Ethylene glycol	-83°	Propylamine
-11.9°	tert-Amyl alcohol	-83.6°	Ethyl acetate
-12°	Cycloheptane or methyl benzoate	-86°	Methyl ethyl ketone
-15°	Benzyl alcohol	-89°	n-Butanol
-16.3°	n-Octanol	-90°	Nitroethane
-18°	1,2-Dichlorobenzene	-91°	Heptane
-22°	Tetrachloroethylene	-92°	n-Propyl acetate
-22.4°	Butyl benzoate	-93°	2-Nitropropane or cyclopentane
-22.8°	Carbon tetrachloride	-94°	Ethyl benzene or hexane
-24.5°	Diethyl sulfate	-94.6°	Acetone
-25°	1,3-Dichlorobenzene	-95.1°	Toluene
-29°	o-Xylene or pentachloroethane	-97°	Cumene
-30°	Bromobenzene	-98°	Methanol or methyl acetate
-32°	m-Toluidine	-99°	Isobutyl acetate
-32.6°	Dipropyl ketone	-104°	Cyclohexene
-38°	Thiophene	-107°	Isooctane
-41°	Acetonitrile	-108°	1-Nitropropane
-42°	Pyridine or diethyl ketone	-116°	Ethanol or diethyl ether
-44°	Cyclohexyl chloride	-117°	Isoamyl alcohol
-45°	Chlorobenzene	-126°	Methylcyclohexane
-47°	m-Xylene	-131°	n-Pentane
-50°	Ethyl malonate or n-butylamine	-160°	Isopentane
-52°	Benzyl acetate or diethylcarbitol		

For other organic materials used in low temperature slush-baths with liquid nitrogen see R.E.Rondeau [*J Chem Eng Data* 11 124 1966]. **NOTE:** Use high quality pure nitrogen, do not use liquid air or liquid nitrogen that has been in contact with air for a long period (due to the dissolution of oxygen in it) as this could EXPLODE in contact with organic matter.

TABLE 10. LIQUIDS FOR STATIONARY PHASES IN GAS CHROMATOGRAPHY

Material	Temp.	Retards
Dimethylsulfolane	0-40°	Olefins and aromatic hydrocarbons
Di-*n*-butyl phthalate	0-40°	General purposes
Squalane	0-150°	Volatile hydrocarbons and polar molecules
Silicone oil or grease	0-250°	General purposes
Diglycerol	20-120°	Water, alcohols, amines, esters, and aromatic hydrocarbons
Dinonyl phthalate	20-130°	General purposes
Polydiethylene glycol succinate	50-200°	Aromatic hydrocarbons, alcohols, ketones, esters.
Polyethylene glycol	50-200°	Water, alcohols, amines, esters and aromatic hydrocarbons
Apiezon grease	50-200°	Volatile hydrocarbons and polar molecules
Tricresyl phosphate	50-250°	General purposes

TABLE 11. METHODS OF VISUALISATION OF TLC SPOTS

Reagent	Compound	Preparation	Observations
Iodine	General	Iodine crystals in a closed chamber or spray 1% methanol solution of Iodine	Brown spots which may disappear upon standing. Limited sensitivity.
H_2SO_4	General	50% solution, followed by heating to 150°C	Black or coloured spots
Molybdate	General	5% $(NH_4)_6Mo_7O_{24}$ + 0.2% $Ce(SO_4)_2$ in 5% H_2SO_4, followed by heating to 150°C.	Deep blue spots
Vanillin	General	0.5g vanillin, 0.5 mL H_2SO_4, 9 mL ethanol	various coloured spots
Ammonia	phenols	Ammonia vapour in a closed chamber	various coloured spots
$FeCl_3$	phenols, enolic compounds	1% aqueous $FeCl_3$	various coloured spots
2,4-DNP	aldehydes, ketones	0.5% 2,4-dinitrophenylhydrazine/2M HCl	red to yellow spots
HCl	aromatic acids and amines	HCl vapour in a closed chamber	various coloured spots
Ninhydrin	amino acids, and amines	0.3% ninhydrin in *n*-BuOH with 3% AcOH, followed by heating to 125°C/10 min	blue spots
$PdCl_2$	S and Se compds	0.5% aq. $PdCl_2$ + few drops of conc. HCl	red and yellow spots
Anisaldehyde	carbohydrates	0.5 mL anisaldehyde in 0.5 mL conc H_2SO_4 + 95% EtOH + a few drops of AcOH Heat at 100-110°C for 20-30 minutes	various blue spots

TABLE 12.   **GRADED ADSORBENTS AND SOLVENTS**

Adsorbents (decreasing effectiveness)	**Solvents** (increasing eluting ability)
Fuller's earth (hydrated aluminosilicate)	Petroleum ether, b. 40-60°.
Magnesium oxide	Petroleum ether, b. 60-80°.
Charcoal	Carbon tetrachloride.
Alumina	Cyclohexane.
Magnesium trisilicate	Benzene.
Silica gel	Diethyl ether.
Calcium hydroxide	Chloroform.
Magnesium carbonate	Ethyl acetate.
Calcium phosphate	Acetone.
Calcium carbonate	Ethanol.
Sodium carbonate	Methanol.
Talc	Pyridine.
Inulin	Acetic acid.
Sucrose = starch	

TABLE 13.   **REPRESENTATIVE ION-EXCHANGE RESINS**

Sulfonated polystyrene **Strong-acid cation exchanger**	**Aliphatic amine-type** **weak base anion exchangers**
AG 50W-x8	Amberlites IR-45 and IRA-67
Amberlite IR-120	Dowex 3-x4A
Dowex 50W-x8	Permutit E
Duolite 225	Permutit A 240A
Permutit RS	
Permutite C50D	
Carboxylic acid-type **Weak acid cation exchangers**	**Strong Base, anion exchangers**
Amberlite IRC-50	AG 2x8
Bio-Rex 70	Amberlite IRA-400
Chelex 100	Dowex 2-x8
Duolite 436	Duolite 113
Permutit C	Permutit ESB
Permutits H and H-70	Permutite 330D

TABLE 14.   **MODIFIED FIBROUS CELLULOSES FOR ION-EXCHANGE**

Cation exchange	**Anion exchange**
CM cellulose (carboxymethyl)	DEAE cellulose (diethylaminoethyl)
CM 22, 23 cellulose	DE 22, 23 cellulose
P cellulose (phosphate)	PAB cellulose (p-aminobenzyl)
SE cellulose (sulfoethyl)	TEAE cellulose (triethylaminoethyl)
SM cellulose (sulfomethyl)	ECTEOLA cellulose

SE and SM are much stronger acids than CM, whereas P has two ionisable groups (pK 2-3, 6-7), one of which is stronger, the other weaker, than for CM (3.5-4.5). For basic strengths, the sequence is: TEAE $\gg$ DEAE (pK 8-95) > ECTEOLA (pK 5.5-7) > PAB.   **Their exchange capacities lie in the range 0.3 to 1.0 mg equiv/g.**

TABLE   15.   **BEAD   FORM   ION-EXCHANGE   PACKAGINGS**[1]

Cation exchange	Capacity (meq/g)	Anion exchange	Capacity (meq/g)
CM-Sephadex C-25, C-50.[2](weak acid)	4.5±0.5	DEAE-Sephadex A-25, A-50.[7] (weak base)	3.5±0.5
SP-Sephadex C-25, C-50.[3](strong acid)	2.3±0.3	QAE-Sephadex A-25, A-50.[8] (strong base)	3.0±0.4
CM-Sepharose CL-6B.[4]	0.12±0.02	DEAE-Sepharose CL-6B.[4]	0.13±0.02
		DEAE-Sephacel.[9]	1.4±0.1
Fractogel EMD,CO_2^- (pK ~4.5) , SO_3^{2-} (pK ~<1) .[5]		Fractogel EMD, DMAE (pK ~9), DEAE (pK ~10.8), TMAE (pK >13).[5]	
CM-32 Cellulose.		DE-32 Cellulose.	
CM-52 Cellulose.[6]		DE-52 Cellulose	

[1] May be sterilised by autoclaving at pH 7 and below 120°. [2] Carboxymethyl. [3] Sulfopropyl. [4] Crosslinked agarose gel, no pre-cycling required, pH range 3-10. [5] Hydrophilic methacrylate polymer with very little volume change on change of pH (equivalent to *Toyopearl,* Sigma), available in superfine 650S, and medium 650M particle sizes. [6] Microgranular, pre-swollen, no pre-cycling required. [7] Diethylaminoethyl. [8] Diethyl(2-hydroxy-propyl)aminoethyl. [9] Bead form cellulose, pH range 2-12, no pre-cycling required. Sephadex and Sepharose from Pharmacia-Amersham Biosciences, Fractogel from Merck, Cellulose from Whatman.

TABLE 16.   **LIQUIDS FOR DRYING PISTOLS**

Boiling points (760mm)		Boiling points (760mm)	
Ethyl chloride	12.2°	Toluene	110.5°
Dichloromethane	39.8°	Tetrachloroethylene	121.2°
Acetone	56.1°	Chlorobenzene	132.0°
Chloroform	62.0°	*m*-Xylene	139.3°
Methanol	64.5°	Isoamyl acetate	142.5°
Carbon tetrachloride	76.5°	Tetrachloroethane	146.3°
Ethanol	78.3°	Bromobenzene	155.0°
Benzene	79.8°	*p*-Cymene	176.0°
Trichloroethylene	86.0°	Tetralin	207.0°
Water	100.0°		

TABLE 17.	VAPOUR PRESSURES (mm Hg) OF SATURATED AQUEOUS SOLUTIONS IN EQUILIBRIUM WITH SOLID SALTS					
Salt	Temperature					% Humidity at 20°
	10°	15°	20°	25°	30°	
$LiCl.H_2O$			2.6			15
$CaBr_2.6H_2O$	2.1	2.7	3.3	4.0	4.8	19
KOAc			3.5			20
$CaCl_2.6H_2O$	3.5	4.5	5.6	6.9	8.3	20
CrO_3			6.1			32
$Zn(NO_3)_2.6H_2O$			7.4			42
$K_2CO_3.2H_2O$			7.7	10.7		44
KCNS			8.2			47
$Na_2Cr_2O_7.2H_2O$			9.1			52
$Ca(NO_3)_2.4H_2O$	6.0	7.7	9.6	11.9	14.2	55
$Mg(NO_3)_2.6H_2O$			9.8			56
$NaBr.2H_2O$	5.8	7.8	10.3	13.5	17.5	58
$NaNO_2$			11.6			66
$NaClO_3$			13.1			75
NaCl	6.9	9.6	13.2	17.8	21.4	75
NaOAc			13.3			76
NH_4Cl			13.8			79
$(NH_4)_2SO_4$			14.2			81
KBr			14.7			84
$KHSO_4$			15.1			86
KCl			15.1	20.2	27.0	86
K_2CrO_4			15.4			88
$ZnSO_4.7H_2O$			15.8			90
$NH_4.H_2PO_4$			16.3			93
KNO_3			16.7	22.3	29.8	95
$Pb(NO_3)_2$			17.2			98
H_2O	9.21	12.79	17.53	23.76	31.82	100

TABLE 18. DRYING AGENTS FOR CLASSES OF COMPOUNDS
Class Dried with

Class	Dried with
Acetals	Potassium carbonate.
Acids (organic)	Calcium sulfate, magnesium sulfate, sodium sulfate.
Acyl halides	Magnesium sulfate, sodium sulfate.
Alcohols	Calcium oxide, calcium sulfate, magnesium sulfate, potassium carbonate, followed by magnesium and iodine.
Aldehydes	Calcium sulfate, magnesium sulfate, sodium sulfate.
Alkyl halides	Calcium chloride, calcium sulfate, magnesium sulfate, phosphorus pentoxide, sodium sulfate.
Amines	Barium oxide, calcium oxide, potassium hydroxide, sodium carbonate, sodium hydroxide.
Aryl halides	Calcium chloride, calcium sulfate, magnesium sulfate, phosphorus pentoxide, sodium sulfate.
Esters	Magnesium sulfate, potassium carbonate, sodium sulfate.
Ethers	Calcium chloride, calcium sulfate, magnesium sulfate, sodium, lithium aluminium hydride.
Heterocyclic bases	Magnesium sulfate, potassium carbonate, sodium hydroxide.
Hydrocarbons	Calcium chloride, calcium sulfate, magnesium sulfate, phosphorus pentoxide, sodium (not for olefins).
Ketones	Calcium sulfate, magnesium sulfate, potassium carbonate, sodium sulfate.
Mercaptans	Magnesium sulfate, sodium sulfate.
Nitro compounds and Nitriles	Calcium chloride, magnesium sulfate, sodium sulfate.
Sulfides	Calcium chloride, calcium sulfate.

TABLE 19.	STATIC DRYING FOR SELECTED LIQUIDS (25°C)				
Liquid	Water	Linde Type 4 A	Linde Type 5 A	Activated Alumina	Silicic Acid Gel
MeOH	Residual H_2O %	0.54	0.55	—	0.60
	Wt % absorbed	2.50	1.50	—	—
EtOH	Residual H_2O %	0.25	0.25	0.45	0.68
	Wt % absorbed	7.00	6.80	1.50	—
1-Butylamine	Residual H_2O %	1.65	1.31	1.93	2.07
	Wt % absorbed	10.40	18.20	3.40	—
2-Ethyl-hexylamine	Residual H_2O %	0.25	0.08	0.43	0.53
	Wt % absorbed	15.10	21.10	6.10	1.70
Diethyl ether	Residual H_2O %	0.001	0.013	0.16	0.27
	Wt % absorbed	9.50	9.20	6.20	4.30
Amyl acetate	Residual H_2O %	0.002	—	0.33	0.38
	Wt % absorbed	9.30	—	7.40	1.80

TABLE 20.   BOILING POINTS OF SOME USEFUL GASES AT 760 mm

Argon	-185.6°	Krypton	-152.3°
Carbon dioxide	-78.5°	Methane	-164.0°
(sublimes)		Neon	-246.0°
Carbon monoxide	-191.3°	Nitrogen	-209.9°
Ethane	-88.6°	Nitrous oxide	-88.5°
Helium	-268.6°	Nitric oxide	-195.8°
Hydrogen	-252.6°	Oxygen	-182.96°

TABLE 21. SOLUBILITIES OF HCl AND NH$_3$ AT 760mm (g/100g OF SOLUTION)

Gas	Temperature °C	MeOH	EtOH	Et$_2$O
Hydrogen Chloride*	-10	54.6	—	37.5 (-9.2°)
	0	51.3	45.4	35.6
	20	47.0 (18°)	41.0	24.9
	30	43.0	38.1	19.47
Ammonia	15	21.6 (27.6g/100g MeOH)	13.2 (9.2g/100mL soln)	—
	25	16.5 (19.8g/100g MeOH)	10.0 (6.0g/100mL soln)	—

* Saturated EtOH with HCl is ~ 5.7M at 25°C, i.e. 21.5g/100mL of solution.

TABLE 22.   PREFIXES FOR QUANTITIES

Fractional	deci (d) = 10^{-1}	centi (c) = 10^{-2}	milli (m) = 10^{-3}	micro (μ) = 10^{-6}	nano (n) = 10^{-9}	pico (p) = 10^{-12}	femto (f) = 10^{-15}	atto (atto) = 10^{-18}
Multiple	deca (d) = 10^{1}	hecto (h) = 10^{2}	kilo (k) = 10^{3}	mega (M) = 10^{6}	giga (G) = 10^{9}	tera (T) = 10^{12}	penta (P) = 10^{15}	eka (E) = 10^{18}

BIBLIOGRAPHY

The following books and reviews provide fuller details of the topics indicated in this chapter. The authors' recommendations for excellent introductory and/or reference texts to the topics are indicated with.* [For earlier bibliographies see *Purification of Laboratory Chemicals*, 4th Edn, ISBN 0750628391 (1996, hardback) and 0750637617 (1997, paperback).

AFFINITY CHROMATOGRAPHY

P. Bailon, G.K. Ehrlich, W. Fung and W. Berthold (Eds), *Affinity Chromatography: Methods and Protocols*, Humana Press, Totowa, 2000. ISBN 0896036944.

I. A. Chaiken, *Analytical Affinity Chromatography*, CRC Press Inc, Florida, 1987. ISBN 084935658X.

* P.D.G. Dean, W.S. Johnson and F. Middle (Eds), *Affinity Chromatography : A Practical Approach*, IRL Press, Oxford, 1985. ISBN 0896036944.

P. Matejtschuk (Ed.), *Affinity Separations: A Practical Approach*, Oxford University Press, Oxford, 1997. ISBN 019963551X, 0199635501.

T.M. Phillips and B.F. Dickens, *Affinity and Immunoaffinity Purification Techniques*, Eaton Publishing, Natick MA, 2000. ISBN 1881299228.

CHIRAL CHROMATOGRAPHY

S. Ahuja (Ed.), *Chiral Separations: Applications and Technology*, ACS, Washington DC, 1997. ISBN 0841234078.

* S. Allenmark, *Chromatographic Enantioseparations (Methods and Applications)*, 2nd Edn, Eliss Horwood Publ. New York, 1991. ISBN 0131329782.

T.E. Beesley and R.P.W. Scott, *Chiral Chromatography*, J.Wiley & Son, Chichester,1999. ISBN 0471974277.

* W.J. Lough, *Chiral Liquid Chromatography*, Blackie Academic & Professional, Glasgow, 1989. ISBN 0412017415.

D. Stevenson and I.D. Wilson (Eds), *Recent Advances in Chiral Separations* (Chromatographic Society Symposium Series), Plenum Pub Corp., New York, 1991. ISBN 0306438364.

D. Stevenson and I.D. Wilson (Eds), *Chiral Separations*, Plenum Press, New York, 1988. ISBN 0306432528.

G. Subramanian (Ed.), *Chiral Separation Techniques : A Practical Approach*, VCH Verlagsgesellschaft, Weinheim, 2000. ISBN 3527298754.

M. Zief and L.J. Crane (Eds), *Chromatographic Chiral Separations*, Marcel Dekker, New York, 1988. ISBN 0824777867.

CHROMATOGRAPHY

S. Ahuja, *Chromatography and Separation Chemistry: Advances and Developments*, ACS, Washington DC, 1986. ISBN 0841209537.

* K. Blau and J.M. Halket (Eds), *Handbook of Derivatives for Chromatography*, J. Wiley & Sons, New York, 1994. ISBN 047192699X.

P.R. Brown and A. Grushka (Eds), *Advances in Chromatography*, Marcel Dekker, New York, Vol **34** 1994, vol **35** 1995.

* J. Cazes, *Encyclopedia of Chromatography,* Marcel Dekker, New York, 2001. ISBN 0824705114.

J.C. Giddings, E. Grushka and P.R. Brown (Eds), *Advances in Chromatography*, Marcel Dekker, New York, Vols **1-36**, 1955-1995.

E. Heftmann (Ed.), *Chromatography, 5*[th] Edn, Elsevier Science, Amsterdam,1992. ISBN 0444882375.

E. Heftmann (Ed.), *Chromatography : Fundamentals and Applications of Chromatographic and Electrophoretic Methods*, 2 volume set, Elsevier Science, New York,1983. ISBN 0444420452 (set).

K. Hostettmann, A. Marston and M. Hostettmann, *Preparative Chromatography Techniques: Applications in Natural Product Isolation*, 2nd Edn, Springer-Verlag, Berlin, 1998. ISBN 3540624597.

P. Miller (Ed.), *High Resolution Chromatography: A Practical Approach*, Oxford University Press, Oxford, 1999. ISBN 0199636486.

C.F. Poole and S.A. Schuette, *Contemporary Practice of Chromatography*, Elsevier, Amsterdam, 1984. ISBN 0444424105.

C.F. Poole and S.K. Poole, *Chromatography Today*, Elsevier Science, Amsterdam, 4th reprint 1997. ISBN 0444891617.

* G. Zweig and J. Sherma (Eds), *CRC Handbook of Chromatography*, 2 Volumes, CRC Press, Cleveland, 1972. ISBN 0878195602.

CRYSTALLISATION

T.M. Bergfors (Ed.), *Protein Crystallisation*, International University Line, 1999. ISBN 0963681753.

A. Ducruix and R. Giege (Eds), *Crystallisation of Nucleic Acids and Proteins*, IRL Press, Oxford, 1992. ISBN 0199632456. 2nd Edn, Oxford University Press, Oxford, 1999. ISBN 0199636788.

A. Mersmann (Ed.), *Crystallisation Technology Handbook*, Marcel Dekker, New York, 1994. ISBN 0824792335.

H. Michel (Ed.), *Crystallisation of Membrane Proteins*, CRC Press, Boca Raton, 1991. ISBN 0849348161.

J.M. Mullin, *Crystallisation*, 4th Edn, Butterworth-Heinemann, Oxford, 2001. ISBN 0750648333.

A.S. Myerson and K. Toyokura, *Crystallisation as a Separation Process*, ACS, Washington DC, 1990. ISBN 0841218641.

J. Nyvit and J. Ulrich, *Admixtures in Crystallisation*, Wiley-VCH, New York 1995. ISBN 3527287396.

R.S. Tipson (Chapter 3), in A.Weissberger (Ed.), *Techniques of Organic Chemistry*, Vol III, pt I, 2nd Edn, Interscience, New York, 1956.

DISTILLATION

E.S. Perry and A. Weissberger (Eds), *Techniques of Organic Chemistry*, Vol IV, *Distillation*, Interscience, New York, 1965.

J. Stichlmair and J.R. Fair, *Distillation: Principles and Practices*, J. Wiley and Sons, New York, 1998. ISBN 0471252417.

DRYING

G. Broughton (Chapter VI), in A. Weissberger's (Ed.) *Techniques of Organic Chemistry*, Vol III, pt I, 2nd Edn, Interscience, New York, 1956.

J.A. Riddick and W.B. Bunger, *Organic Solvents: Physical Properties and Methods of Purification, Techniques of Chemistry*, Vol II, Wiley-Interscience, New York, 1970.

J.F. Coetzee, *Recommended Methods for the Purification of Solvents and Tests for Impurities,* Pergamon Press, Oxford, 1982. ISBN 0080223702.

GAS CHROMATOGRAPHY

* P.J. Baugh (Ed.), *Gas Chromatography: A Practical Approach*, IRL Press, Oxford, 1994. ISBN 0199632715.

R.E. Clement (Ed.), *Gas Chromatography: Biochemical, Biomedical and Clinical Applications*, J. Wiley & Sons, New York, 1990. ISBN 0471010480.

* I.A. Fowlis, *Gas Chromatography: Analytical Chemistry by Open Learning*, 2nd Edn, J. Wiley & Sons, Chichester,1995. ISBN 0471954683.

W. Jennings, E. Mittlefehldt and P. Stremple, *Analytical Gas Chromatography*, 2nd Edn, Academic Press, San Diego, 1997. ISBN 012384357X.

G. Guichon and C.L. Guillemin, *Quantitative Gas Chromatography for Laboratory and On-Line Process Control*, Elsevier, Amsterdam, 1988. ISBN 0444428577.

H.H. Hill and D.G. McMinn (Eds), *Detectors for Capillary Chromatography*, J. Wiley & Sons, New York, 1992. ISBN 0471506451.

H.-J. Hubschmann, *Handbook of GC/MS: Fundamentals and Applications*, VCH Verlagsgessellschaft Mbh, Weinheim, 2001. ISBN 3527301704.

F.W. Karasek and R.E. Clement, *Basic Gas-Chromatography-mass spectrometry: Principles and Techniques*, Elsevier, Amsterdam, 1988. ISBN 0444427600.

* M. McMaster, C. McMaster, *GC/MS: A Practical User's Guide*, Wiley-VCH, New York, 1998. ISBN 0471248266.

* H.M. McNair and J.M. Miller, *Basic Gas Chromatography: Techniques in Analytical Chemistry*, J. Wiley & Sons, New York, 1997. ISBN 047117260X.

* D. Rood, *A Practical Guide to the Care, Maintenance, and Troubleshooting of Capillary Gas Chromatographic Systems*, 3rd Edn, Wiley-VCH, New York; 1999. ISBN 3527297502.

R.D. Sacks, J-M.D. Dimandja and DG. Patterson, *High Speed Gas Chromatography* , J. Wiley & Sons, New York, 2001. ISBN 0471415561.

P. Sandra (Ed.), *Sample Introduction in Capillary Gas Chromatography*, Vol 1, Chromatographic Method Series, Huethig, Basel, 1985.

GEL FILTRATION

P.L. Dubin, *Aqueous Size-Exclusion Chromatography*, Elsevier, Amsterdam, 1988. ISBN 0444429573.

L. Fischer, *Gel Filtration*, 2nd Edn, Elsevier, North-Holland, Amsterdam, 1980. ISBN 0444802231.

M. Potschka and P.L. Dubin (Eds), *Strategies in Size-exclusion Chromatography*, ACS, Washington DC, 1996. ISBN 0841234140.

W.W. Yau, D.D. By and J.J. Kirkland, *Modern Size Exclusion Liquid Chromatography: Practice of Gel Permeation and Gel Filtration Chromatography*, J. Wiley and Sons, New York, 1979. ISBN 0471033871.

C-S. Wu (Ed.), *Handbook of Size Exclusion Chromatography*, Marcel Dekker, NY, 1995. ISBN 0824792882

C-S. Wu (Ed.), *Column Handbook for Size Exclusion Chromatography*, Academic Press, NY, 1998. ISBN 0127655557

HIGH PERFORMANCE LIQUID CHROMATOGRAPHY

T.J. Baker, Chemists, *HPLC Solvents Reference Manual*, T.J.Baker Chemical Co., 1985.

* K.D. Bartle and P. Meyers, *Capillary Electrochromatography*, Royal Society of Chemistry Chromatography Monographs, London, 2000. ISBN 0854045309.

A. Berthold (Ed.), *Countercurrent Chromatography (CCC). The Support-Free Liquid Stationary Phase*, Elsevier, Amsterdam, 2002. ISBN 044450737X.

B.A. Bidlingmeyer, *Preparative Liquid Chromatography*, 2nd Reprint, Elsevier, Amsterdam, 1991. ISBN 0444428321.

R. Freitag, *Modern Advances in Chromatography*, Springer-Verlag, NY, 2002. ISBN 3540430423. [Electronic version available at: http://link.springer.de/series/abe/].

W.S. Hancock, J.T. Sparrow, *HPLC Analysis of Biological Compounds: A Laboratory Guide*, (Chromatographic Science Series, Vol 26), Marcel Dekker, New York, 1984. ISBN 0824771400.

T. Hanai and R.M. Smith (Eds), *HPLC: A Practical Guide*, Royal Society of Chemistry, Thomas Graham House, Cambridge, U.K, 1999. ISBN 0854045155.

E.D. Katz (Ed.), *High Performance Liquid Chromatography: Principles and Methods in Biotechnology*, J. Wiley & Sons, Chichester, 1996. ISBN 0471934445.

A.M. Krstulovic and P.R. Brown, *Reversed Phase HPLC. Theory, Practical and Biomedical Applications*, J. Wiley & Sons, New York, 1982. ISBN 0471053694.

C.K. Lim (Ed.), *HPLC of Small Molecules*, IRL Press, Oxford, 1986. ISBN 0947946780, 0947946772.

* W.J. Lough and I.W. Wainer (Eds), *High Performance Liquid Chromatography: Fundamental Principles and Practice*, Blackie Academic and Professional, London, 1996. ISBN 0751400769.

* M.C. McMaster, *HPLC, A Practical User's Guide*, J. Wiley and Sons, New York, 1994. ISBN 0471185868.

* V. Meyer, *Practical High-Performance Liquid Chromatography*, 2nd Edn., J. Wiley & Sons, New York, 1994. ISBN 0471941328, 0471941298.

U.D. Neue, *HPLC Columns: Theory, Technology, and Practice*, Wiley-VCH, New York, 1997. ISBN 0471190373.

R.W.A. Oliver (Ed.), *HPLC of Macromolecules*, 2nd Edn, IRL Press, Oxford, 1998. ISBN 0199635714, 0199635706.

D. Parriott, *A Practical Guide to HPLC Detection*, Academic Press, San Diego, 1997. ISBN 0125456808.

* P.C. Sadek, *The HPLC Solvent Guide*, J. Wiley & Sons, New York, 1996. ISBN 0471118559.

* P.C. Sadek, *Troubleshooting HPLC Systems: A Bench Manual*, Wiley-Interscience, New York, 1999. ISBN 0471178349.

* L.R. Snyder, J. Kirkland and J. Glaich, *Practical HPLC Method Development*, 2nd Edn., Wiley-Interscience, New York, 1997. ISBN 047100703X.

J. Swadesh (Ed.), *HPLC: Practical and Industrial Applications*, CRC Press, Boca Raton, 1997. ISBN 0849326826.

G. Szepesi, *How to Use Reverse Phase HPLC*, VCH, New York, 1992. ISBN 0895737663.

IONIC EQUILIBRIA

G. Kortüm, W. Vogel and K. Andrussow, *Dissociation Constants of Organic Acids in Aqueous Solution*, Butterworths, London, 1961.

D.D. Perrin, *Dissociation Constants of Organic Bases in Aqueous Solution*, Butterworths, London, 1965; and Supplement 1972.

D.D. Perrin, *Dissociation Constants of Inorganic Acids and Bases in Aqueous Solution*, Butterworths, London, 1969.

D.D. Perrin and B. Dempsey, *Buffers for pH and Metal Ion Control*, Chapman and Hall, London, 1974.

E.P. Serjeant and B. Dempsey, *Ionisation Constants of Organic Acids in Aqueous Solution*, Pergamon Press, Oxford, 1979.

ION EXCHANGE

A. Dyer, M.J. Hudson and P.A. Williams (Eds), *Ion Exchange Processes: Advances and Applications*, Royal Society of Chemistry, London, 1993. ISBN 0851864457.

J.S. Fritz and D.T. Gjerde, *Ion Chromatography*, 3rd Edn, VCH Verlagsgessellschaft Mbh, Weinheim, 2000. ISBN 3527299149.

D.T. Gjerde and J.S. Fritz, *Ion Chromatography*, 2nd Edn, J. Wiley and Sons, New York, 1998. ISBN 3527296964.

P.R. Haddad and P.E. Jackson, *Ion Chromatography: Principles and Applications*, Elsevier, New York, 1990. ISBN 0444882324.

F.G. Helfferich, *Ion Exchange*, Dover Publications, New York, 1989. ISBN 0486687848.

J. Korkisch, *Handbook of Ion Exchange Resins: Their Applications to Inorganic Analytical Chemistry*, 6 Vol. Set, CRC Press, Boca Raton, 1989. ISBN 0849331943.

R.E. Smith (Ed.), *Ion Chromatography Applications*, CRC Press, Boca Raton, 1988. ISBN 0849349672.

E.M. Thurman and M.S. Mills, *Solid-phase Extraction: Principles and Practice*, Chemical Analysis Series Vol. 147, J. Wiley & Sons, New York, 1998. ISBN 047161422X.

J. Weiss, *Ion Chromatography*, 2nd Edn, Wiley-VCH, New York, 1994. ISBN 3527286985.

S. Yamamoto, K. Nakanishi and R. Matsuno, *Ion-exchange Chromatography of Proteins*, Chromatographic Science Series Vol. 43, Marcel Dekker, New York, 1988. ISBN 0824779037.

LABORATORY TECHNIQUE AND THEORETICAL DISCUSSION

M. Casey, *Advanced Practical Organic Chemistry*, Chapman and Hall, New York, 1990. ISBN 0412024616, 021692796.

G.S. Coyne, *The Laboratory Companion: A Practical Guide to Materials, Equipment and Technique*, Wiley-Interscience, New York, 1997. ISBN 0471184225.

* L.M. Harwood and C.J. Moody, *Experimental Organic Chemistry: Principles and Practice*, Blackwell Scientific Publications, Oxford, 1989. ISBN 0632020164, 0632020172.

H.J.E. Loewenthal and E. Zass, *A Guide for the Perplexed Organic Experimentalist*, 2nd Edn, J. Wiley & Sons, Chichester, 1992. ISBN 0471935336.

D.W. Mayo, S.S. Butcher, P.K. Trumper, R.M. Pike and S.S. Butcher, *Microscale Techniques for the Organic Laboratory*, J. Wiley & Sons, New York, 1991. ISBN 0471621927.

C.E. Meloan, *Chemical Separations, Principles, Techniques and Experiments*, J. Wiley & Sons, New York, 2000, ISBN 0471351970.

J. Mendham, R.C. Denney, J.D. Barnes and M.J.K. Thomas, *Vogel's Quantitative Chemical Analysis*, 6th Edn, Prentice Hall, Harlow, 2000. ISBN 0582226287.

A.M. Schoffstall, B.A. Gaddis, M.L. Druelinger and M. Druelinger, *Organic Microscale and Miniscale Laboratory Experiments*, McGraw Hill, Boston, 2000. ISBN 0072375493.

J.T. Sharp, I. Gosney and A.G. Rowley, *Practical Organic Chemistry: A Student Handbook of Techniques*, Chapman and Hall, London, 1989. ISBN 0412282305.

* A.I. Vogel and B.S. Furniss, *Vogel's Textbook of Practical Organic Chemistry*, 5th Edn, J. Wiley & Sons, New York, 1989. ISBN 0582462363.

J.W. Zubrick, *The Organic Chemistry Lab Survival Guide*, 5th Edn, J. Wiley & Sons, New York, 2000. ISBN 0471387320.

MOLECULAR SIEVES

A. Dyer, *An Introduction to Zeolite Molecular Sieves*, J. Wiley & Sons, New York, 1988. ISBN 0471919810

R. Szostak, *Molecular Sieves: Principles of Synthesis and Identification*, Van Nostrand Reinhold, New York, 1989. ISBN 0442280238.

REAGENTS AND PURITY

D.J. Bucknell, *Analar Standards for Laboratory Chemicals*, 8th Edn, Analar Standards Ltd., Dorset, 1984. ISBN 095004394X.

K.L. Cheng, K. Ueno, T. Imamura (Eds), *CRC Handbook of Organic Analytical Reagents*, CRC Press, Boca Raton, 1982.

J.A. Dean (Ed.), *Lange's Handbook of Chemistry*, McGraw-Hill, New York, 1999. ISBN 0070163847.

*L.A. Paquette (Ed.), *The Encyclopedia of Reagents for Organic Synthesis*, 8 Volume set, J Wiley & Sons, New York, 1995. ISBN 0471936235.

*A.J. Pearson and W.R. Rouse, *Handbook of Reagents for Organic Synthesis*, 4 Vol. Set, J. Wiley & Sons, Chichester, 2001. ISBN 04719878911.

Reagents Chemicals: American Chemical Society Specifications, Official from January 1 2000 (Reagent Chemicals American Chemical Society Specification) 9th Edn, Washington DC, (Sept. 1999). ISBN 0841236712.

* L.E. Smart, *Separation, Purification and Identification,* The Royal Society of Chemistry, London, 2002 (in paperback and CD-ROM). ISBN 0854046852.

United States Pharmacopeia (USP # 24 NF19) 24-19 edition (January 1, 2000).

G. Wypych (Ed.), *Solvents Database on CD Rom*, William Andrew Publishing, LLC 2001. ISBN 0815514638.

SAFETY IN THE CHEMICAL LABORATORY

* L. Bretherick, P.G. Urben and M.J. Pitt, *Bretherick's Handbook of Reactive Chemical Hazards: An Indexed Guide to Published Data*, 6th Edn, Butterworth-Heinemann, Oxford,1999. ISBN 075063605X.

N.P. Cheremisinoff, *Handbook of Hazardous Chemical Properties*, Butterworth-Heinemann, Oxford, 2000. ISBN 0750672099.

P. Patniak, *A Comprehensive Guide to the Hazardous Properties of Chemical Substances*, 2nd Edn, J. Wiley & Sons, 1999. ISBN 0471291757.

G.J. Hathaway, *Chemical Hazards of the Workplace*, 4th Edn, J. Wiley & Sons,1996. ISBN 0471287024.

M.-A. Armour, *Hazardous Laboratory Chemicals: Disposal Guide*, 2nd Edn, CRC Press, Boca Raton, 1996. ISBN 1566701082.

S.G. Luxon (Ed.), *Hazards in the Chemical Laboratory*, 5th Edn, Royal Society of Chemistry, 1992. ISBN 0851862292.

R.E. Lenga (Ed.), *The Sigma-Aldrich Library of Chemical Safety Data*, 3 Volume set, Aldrich Chemical Co., Milwaukee, 1988. ISBN 0941633160.

R.E. Lenga, K.L. Votoupal (Eds), *The Sigma-Aldrich Library of Regulatory and Safety Data*, 3 Volume set, Aldrich Chemical Co., Milwaukee, 1993. ISBN 0941633357.

R.J. Lewis Sr., *Sax's Dangerous Properties of Industrial Materials*, 10th Edn, 3 Vol. Set., J. Wiley & Sons, New York, 1999. ISBN 0471354074.

R.J. Lewis Sr., *Hazardous Chemicals Desk Reference*, 5th Edn, J. Wiley & Sons, 2002. ISBN 0471441651.

SOLVENTS, SOLVENT EXTRACTION AND DISTRIBUTION

L.C. Craig, D. Craig and E.G. Scheibel (Chapter 2), in A. Weissberger's (Ed.) *Techniques of Organic Chemistry*, vol III, pt I, 2nd Edn, Interscience, New York, 1956.

D.R. Lide, *Handbook of Organic Solvents*, CRC Press, Florida, 1995. ISBN 0849389305.

R.P. Pohanish and S.A. Greene, *Rapid Guide to Chemical Incompatibilities*, J. Wiley & Sons, New York, 1997. ISBN 0471288020.

SPECTROSCOPY

E. Breitmaier, *Structure Elucidation by NMR in Organic Chemistry: A Practical Guide*, J. Wiley & Sons, New York, 1993. ISBN 0471937452, 0471933813.

K. Nakamoto, *Infrared and Raman Spectra of Inorganic and Coordination Compounds*, 2 Vol. Set, 5th Edn, J. Wiley & Sons, New York, 1997. ISBN 0471194069.

* C.J. Pouchert and J. Behnke, *The Aldrich Library of ^{13}C and ^{1}H FT NMR Spectra*, 3 Volume set, Aldrich Chemical Co., Milwaukee, 1993. ISBN 0941633349.

* C.J. Pouchert, *The Aldrich Library of Infrared Spectra*, 3rd Edn, Aldrich Chemical Co., Milwaukee, 1981.

THIN LAYER CHROMATOGRAPHY

B. Fried and J. Sherma, *Thin Layer Chromatography*, Marcel Dekker, New York, 1999. ISBN 0824702220.

B. Fried, J. Sherma (Eds), *Practical Thin-Layer Chromatography: A MultiDisciplinary Approach*, CRC Press, Boca Raton, 1996. ISBN 0849326605.

F. Geiss, *Fundamentals of Thin Layer Chromatography*, Huthig, Heidelberg, 1987. ISBN 3778508547.

E. Hahn-Dienstrop, *Applied Thin Layer Chromatography: Best Practice and Avoidance of Mistakes*, VCH Verlagsgessellschaft Mbh, Weinheim, 2000. ISBN 3527298398.

* H. Jork (Ed.), *Thin Layer Chromatography: Reagents and Detection Methods*, Vol. 1a, 1b, J. Wiley & Sons, 1989 (Vol. 1a) and 1994 (Vol. 1b). ISBN 3527278346 (Vol. 1a) and 352728205X (Vol. 1b).

CHAPTER 2

CHEMICAL METHODS

USED

IN PURIFICATION

GENERAL REMARKS

Greater selectivity in purification can often be achieved by making use of differences in chemical properties between the substance to be purified and the contaminants. Unwanted metal ions may be removed by precipitation in the presence of a *collector* (see p. 54). Sodium-borohydride and other metal hydrides transform organic peroxides and carbonyl-containing impurities such as aldehydes and ketones in alcohols and ethers. Many classes of organic chemicals can be purified by conversion into suitable derivatives, followed by regeneration. This chapter describes relevant procedures.

REMOVAL OF TRACES OF METALS FROM REAGENTS

METAL IMPURITIES

The presence of metal contaminants in reagents may sometimes affect the chemical or biochemical outcomes of an experiment. In these cases, it is necessary to purify the reagents used.

Metal impurities can be determined qualitatively and quantitatively by atomic absorption spectroscopy and the required purification procedures can be formulated. Metal impurities in organic compounds are usually in the form of ionic salts or complexes with organic compounds and very rarely in the form of free metal. If they are present in the latter form then they can be removed by crystallising the organic compound (whereby the insoluble metal can be removed by filtration), or by distillation in which case the metal remains behind with the residue in the distilling flask. If the impurities are in the ionic or complex forms, then extraction of the organic compound in a suitable organic solvent with aqueous acidic or alkaline solutions will reduce their concentration to acceptable levels.

When the metal impurities are present in inorganic compounds as in metals or metal salts, then advantage of the differences in chemical properties should be taken. Properties of the impurities like the solubility, the solubility product (product of the metal ion and the counter-ion concentrations), the stability constants of the metal complexes with organic complexing agents and their solubilities in organic solvents should be considered. Alternatively the impurities can be masked by the addition of complexing agents which could lower the concentration of the metal ion impurities to such low levels that they would not interfere with the main compound (see **complexation** below). Specific procedures and examples are provided below.

DISTILLATION

Reagents such as water, ammonia, hydrochloric acid, nitric acid, perchloric acid, and sulfuric acid can be purified *via* distillation (preferably under reduced pressure and particularly with perchloric acid) using an all-glass still. Isothermal distillation is convenient for ammonia: a beaker containing concentrated ammonia is placed alongside a beaker of distilled water for several days in an empty desiccator so that some of the ammonia distils over into the water. The redistilled ammonia should be kept in polyethylene or paraffin-waxed bottles. Hydrochloric acid can be purified in the same way. To ensure the absence of metal contaminants from some salts (e.g. ammonium acetate), it may be more expedient to synthesise the salts using distilled components rather than to attempt to purify the salts themselves.

USE OF ION EXCHANGE RESINS

Application of ion-exchange columns has greatly facilitated the removal of heavy metal ions such as Cu^{2+}, Zn^{2+} and Pb^{2+} from aqueous solutions of many reagents. Thus, sodium salts and sodium hydroxide can be purified by passage through a column of a cation-exchange resin in its sodium form, prepared by washing the resin with 0.1M aqueous NaOH then washing with water until the pH of the effluent is ~7. Similarly, for acids, a resin in its H^+ form [prepared by washing the column with 0.1M aqueous mineral acid (HCl, H_2SO_4) followed by thorough washing with water until the effluent has pH ~7 is used. In some cases, where metals form anionic complexes, they can be removed by passage through an anion-exchange resin. Iron in hydrochloric acid solution can be removed in this way.

Ion exchange resins are also useful for demineralising biochemical preparations such as proteins. Removal of metal ions from protein solutions using polystyrene-based resins, however, may lead to protein denaturation. This difficulty may be avoided by using a weakly acidic cation exchanger such as Bio-Rex 70.

Heavy metal contamination of pH buffers can be removed by passage of the solutions through a Chelex X-100 column. For example when a solution of 0.02M HEPES [4-(2-HydroxyEthyl)Piperazine-1-Ethanesulfonic acid] containing 0.2M KCl (1L, pH 7.5) alone or with calmodulin, is passed through a column of Chelex X-100 (60g) in the K^+ form, the level of Ca^{2+} ions falls to less than 2×10^{-7} M as shown by atomic absorption spectroscopy. Such solutions should be stored in polyethylene containers that have been washed with boiling deionised water (5min) and rinsed several times with deionised water. TES [N,N,N',N'-Tetraethylsulfamide] and TRIS [Tris-(hydroxymethyl)aminomethane] have been similarly decontaminated from metal ions.

Water, with very low concentrations of ionic impurities (and approaching conductivity standards), is very readily obtained by percolation through alternate columns of cation- and anion-exchange resins, or through a mixed-bed resin, and many commercial devices are available for this purpose. For some applications, this method is unsatisfactory because the final deionised water may contain traces of organic material after passage through the columns. However, organic matter can also be removed by using yet another special column in series for this purpose (see Milli Q water preparation, Millipore Corpn., www.millipore.com).

PRECIPITATION

In removing traces of impurities by precipitation it is necessary to include a material to act as a *collector* of the precipitated substance so as to facilitate its removal by filtration or decantation. The following are a few examples:

Removal of lead contaminants

Aqueous hydrofluoric acid can be freed from lead by adding 1mL of 10% strontium chloride per 100mL of acid, lead being co-precipitated as lead fluoride with the strontium fluoride. If the hydrofluoric acid is decanted from the precipitate and the process repeated, the final lead content in the acid is less than 0.003 ppm. Similarly, lead can be precipitated from a nearly saturated sodium carbonate solution by adding 10% strontium chloride dropwise (1-2mL per 100mL) followed by filtration. (If the sodium carbonate is required as a solid, the solution can be evaporated to dryness in a platinum dish.) Removal of lead from potassium chloride uses precipitation as lead sulfide by bubbling H_2S, followed, after filtration, by evaporation and recrystallisation of the potassium chloride.

Removal of iron contaminants

Iron contaminants have been removed from potassium thiocyanate solutions by adding a slight excess of an aluminium salt, then precipitating aluminum and iron as their hydroxides by adding a few drops of ammonia. Iron is also carried down on the hydrated manganese dioxide precipitate formed in cadmium chloride or cadmium sulfate solutions by adding 0.5% aqueous potassium permanganate (0.5mL per 100mL of solution), sufficient ammonia to give a slight precipitate, and 1mL of ethanol. The solution is heated to boiling to coagulate the precipitate, then filtered. Ferrous ion can be removed from copper solutions by adding some hydrogen peroxide to the solution to oxidise the iron, followed by precipitation of ferric hydroxide by adding a small amount of sodium hydroxide.

Removal of other metal contaminants

Traces of calcium can be removed from solutions of sodium salts by precipitation at pH 9.5-10 as the 8-hydroxyquinolinate. The excess 8-hydroxyquinoline acts as a *collector*.

EXTRACTION

In some cases, a simple solvent extraction is sufficient to remove a particular impurity. For example, traces of gallium can be removed from titanous chloride in hydrochloric acid by extraction with diisopropyl ether.

Similarly, ferric chloride can be removed from aluminium chloride solutions containing hydrochloric acid by extraction with diethyl ether. Usually, however, it is necessary to extract an undesired metal with an organic solvent in the presence of a suitable complexing agent such as dithizone (diphenylthiocarbazone) or sodium diethyl dithiocarbamate. When the former is used, weakly alkaline solutions of the substance containing the metal impurity are extracted with dithizone in chloroform (at about 25mg/L of chloroform) or carbon tetrachloride until the colour of some fresh dithizone solution remains unchanged after shaking. Dithizone complexes metals more strongly in weakly alkaline solutions. Excess dithizone in the aqueous medium is removed by extracting with the pure solvent (chloroform or carbon tetrachloride), the last traces of which, in turn, are removed by aeration. This method has been used to remove metal impurities from aqueous solutions of ammonium hydrogen citrate, potassium bromide, potassium cyanide, sodium acetate and sodium citrate. The advantage of dithizone for such a purpose lies in the wide range of metals with which it combines under these conditions. 8-Hydroxyquinoline (oxine) can also be used in this way. Sodium diethyl dithiocarbamate has been used to remove metals from aqueous hydroxylamine hydrochloride (made just alkaline to thymol blue by adding ammonia) from copper and other heavy metals by repeated extraction with chloroform until no more diethyl dithiocarbamate remained in the solution (which was then acidified to thymol blue by adding hydrochloric acid).

COMPLEXATION

Although not strictly a removal of an impurity, addition of a suitable complexing agent such as ethylenediaminetetraacetic acid often overcomes the undesirable effects of contaminating metal ions by reducing the concentrations of the free metal species to very low levels, i.e. sequestering metal ions by complexation. For a detailed discussion of this *masking*, see *Masking and Demasking of Chemical Reactions*, D.D.Perrin, Wiley-Interscience, New York, 1970.

USE OF METAL HYDRIDES

This group of reagents is commercially available in large quantities; some of its members - notably lithium aluminium hydride ($LiAlH_4$), calcium hydride (CaH_2), sodium borohydride ($NaBH_4$) and potassium borohydride (KBH_4) - have found widespread use in the purification of chemicals.

LITHIUM ALUMINIUM HYDRIDE

This solid is stable at room temperature, and is soluble in ether-type solvents. It reacts violently with water, liberating hydrogen, and is a powerful drying and reducing agent for organic compounds. It reduces aldehydes, ketones, esters, carboxylic acids, peroxides, acid anhydrides and acid chlorides to the corresponding alcohols. Similarly, amides, nitriles, aldimines and aliphatic nitro compounds yield amines, while aromatic nitro compounds are converted to azo compounds. For this reason it finds extensive application in purifying organic chemical substances by the removal of water and carbonyl containing impurities as well as peroxides formed by autoxidation. Reactions can generally be carried out at room temperature, or in refluxing diethyl ether, at atmospheric pressure. *When drying organic liquids with this reagent it is important that the concentration of water in the liquid is below 0.1% - otherwise a violent reaction or* **EXPLOSION** *may occur. $LiAlH_4$ should be added cautiously to a cooled solution of organic liquid in a flask equipped with a reflux condenser.*

CALCIUM HYDRIDE

This powerful drying agent is suitable for use with hydrogen, argon, helium, nitrogen, hydrocarbons, chlorinated hydrocarbons, esters and higher alcohols.

SODIUM BOROHYDRIDE

This solid which is stable in dry air up to 300°, is a less powerful reducing agent than lithium aluminium hydride, from which it differs also by being soluble in hydroxylic solvents and to a lesser extent in ether-type solvents. Sodium borohydride forms a dihydrate melting at $36-37^o$, and its aqueous solutions decompose slowly unless stabilised to above pH 9 by alkali. (For example, a useful sodium borohydride solution is one that is nearly saturated at $30-40^o$ and containing 0.2% sodium hydroxide.) Its solubility in water is 25, 55 and 88g per 100mL of water at 0°, 25° and 60°, respectively. Boiling or acidification rapidly decomposes aqueous sodium borohydride solutions. The reagent, available either as a hygroscopic solid or as an aqueous sodium hydroxide solution, is useful as a water soluble reducing agent for aldehydes, ketones and organic peroxides. This explains its use for the removal of carbonyl-containing impurities and peroxides from alcohols, polyols, esters, polyesters, amino alcohols, olefins, chlorinated hydrocarbons, ethers, polyethers, amines (including aniline), polyamines and aliphatic sulfonates.

Purifications using sodium borohydride can be carried out conveniently using alkaline aqueous or methanolic solutions of sodium borohydride, allowing the reaction mixture to stand at room temperature for several hours. Other solvents that can be used with this reagent include isopropyl alcohol (without alkali), amines (including liquid ammonia, in which its solubility is 104g per 100g of ammonia at 25°, and ethylenediamine), diglyme, formamide, dimethylformamide and tetrahydrofurfuryl alcohol. Alternatively, the material to be purified can be percolated through a column of the borohydride. In the absence of water, sodium borohydride solutions in organic solvents such as dioxane or amines decompose only very slowly at room temperature. Treatment of ethers with sodium borohydride appears to inhibit peroxide formation.

POTASSIUM BOROHYDRIDE
Potassium borohydride is similar in properties and reactions to sodium borohydride, and can similarly be used as a reducing agent for removing aldehydes, ketones and organic peroxides. It is non-hygroscopic and can be used in water, ethanol, methanol or water-alcohol mixtures, provided some alkali is added to minimise decomposition, but it is somewhat less soluble than sodium borohydride in most solvents. For example, the solubility of potassium borohydride in water at 25° is 19g per 100mL of water (as compared to sodium borohydride, 55g).

PURIFICATION *via* DERIVATIVES

Relatively few derivatives of organic substances are suitable for use as aids to purification. This is because of the difficulty in regenerating the starting material. For this reason, we list below, the common methods of preparation of derivatives that can be used in this way.

Whether or not any of these derivatives is likely to be satisfactory for the use of any particular case will depend on the degree of difference in properties, such as solubility, volatility or melting point, between the starting material, its derivative and likely impurities, as well as on the ease with which the substance can be recovered. Purification *via* a derivative is likely to be of most use when the quantity of pure material that is required is not too large. Where large quantities (for example, more than 50g) are available, it is usually more economical to purify the material directly (for example, in distillations and recrystallisations).

The most generally useful purifications *via* derivatives are as follows:

ALCOHOLS
Aliphatic or aromatic alcohols are converted to solid esters. *p*-Nitrobenzoates are examples of convenient esters to form because of their sharp melting points, and the ease with which they can be recrystallised as well as the ease with which the parent alcohol can be recovered. The *p*-nitrobenzoyl chloride used in the esterification is prepared by refluxing dry *p*-nitrobenzoic acid with a 3 molar excess of thionyl chloride for 30min on a steam bath (*in a fume cupboard*). The solution is cooled slightly and the excess thionyl chloride is distilled off under vacuum, keeping the temperature below 40°. Dry toluene is added to the residue in the flask, then distilled off under vacuum, the process being repeated two or three times to ensure complete removal of thionyl chloride, hydrogen chloride and sulfur dioxide. (This freshly prepared *p*-nitrobenzoyl chloride cannot be stored without decomposition; it should be used directly.) A solution of the acid chloride (1mol) in dry toluene or alcohol-free chloroform (distilled from P_2O_5 or by passage through an activated Al_2O_3 column) under a reflux condenser is cooled in an ice bath while the alcohol (1mol), with or without a solvent (preferably miscible with toluene or alcohol-free chloroform), is added dropwise to it. When addition is over and the reaction subsides, the mixture is refluxed for 30min and the solvent is removed under reduced pressure. The solid ester is then recrystallised to constant melting point from toluene, acetone, low boiling point petroleum ether or mixtures of these, but not from alcohols.
Hydrolysis of the ester is achieved by refluxing in aqueous N or 2N NaOH solution until the insoluble ester dissolves. The solution is then cooled, and the alcohol is extracted into a suitable solvent, e.g. ether, toluene or alcohol-free chloroform. The extract is dried ($CaSO_4$, $MgSO_4$) and distilled, then fractionally distilled if liquid or recrystallised if solid. (The *p*-nitrobenzoic acid can be recovered by acidification of the aqueous layer.) In most cases where the alcohol to be purified can be readily extracted from ethanol, the hydrolysis of the ester is best achieved with N or 2N ethanolic NaOH or 85% aqueous ethanolic N NaOH. The former is prepared by dissolving the necessary alkali in a minimum volume of water and diluting with absolute alcohol. The ethanolic solution is refluxed for one to two hours and hydrolysis is complete when an aliquot gives a clear solution on dilution with four or five times its volume of water. The bulk of the ethanol is distilled off and the residue is

extracted as above. Alternatively, use can be made of ester formation with benzoic acid, toluic acid or 3,5-dinitrobenzoic acid, by the above method.

Other derivatives can be prepared by reaction of the alcohol with an acid anhydride. For example, phthalic or 3-nitrophthalic anhydride (1 mol) and the alcohol (1mol) are refluxed for half to one hour in a non-hydroxylic solvent, e.g. toluene or alcohol-free chloroform, and then cooled. The phthalate ester crystallises out, is precipitated by the addition of low boiling petroleum ether or is isolated by evaporation of the solvent. It is recrystallised from water, 50% aqueous ethanol, toluene or low boiling petroleum ether. Such an ester has a characteristic melting point and the alcohol can be recovered by acid or alkaline hydrolysis.

ALDEHYDES

The best derivative from which an aldehyde can be recovered readily is its bisulfite addition compound, the main disadvantage being the lack of a sharp melting point. The aldehyde (sometimes in ethanol) is shaken with a cold saturated solution of sodium bisulfite until no more solid adduct separates. The adduct is filtered off, washed with a little water, followed by alcohol. A better reagent to use is a freshly prepared saturated aqueous sodium bisulfite solution to which 75% ethanol is added to near-saturation. (Water may have to be added dropwise to render this solution clear.) With this reagent the aldehyde need not be dissolved separately in alcohol and the adduct is finally washed with alcohol. The aldehyde is recovered by dissolving the adduct in the least volume of water and adding an equivalent quantity of sodium carbonate (not sodium hydroxide) or concentrated hydrochloric acid to react with the bisulfite, followed by steam distillation or solvent extraction.

Other derivatives that can be prepared are the Schiff bases and semicarbazones. Condensation of the aldehyde with an equivalent of primary aromatic amine yields the Schiff base, for example aniline at 100° for 10-30min. Semicarbazones are prepared by dissolving semicarbazide hydrochloride (*ca* 1g) and sodium acetate (*ca* 1.5g) in water (8-10mL) and adding the aldehyde or ketone (0.5-1g) with stirring. The semicarbazone crystallises out and is recrystallised from ethanol or aqueous ethanol. These are hydrolysed by steam distillation in the presence of oxalic acid or better by exchange with pyruvic acid (Hershberg *J Org Chem* 13 542 *1948*) [see entry under Ketones].

AMINES

Picrates

The most versatile derivative from which the free base can be readily recovered is the picrate. This is very satisfactory for primary and secondary aliphatic amines and aromatic amines and is particularly so for heterocyclic bases. The amine, dissolved in water or alcohol, is treated with excess of a saturated solution of picric acid in water or alcohol, respectively, until separation of the picrate is complete. If separation does not occur, the solution is stirred vigorously and warmed for a few minutes, or diluted with a solvent in which the picrate is insoluble. Thus, a solution of the amine and picric acid in ethanol can be treated with petroleum ether to precipitate the picrate. Alternatively, the amine can be dissolved in alcohol and aqueous picric acid added. The picrate is filtered off, washed with water or ethanol and recrystallised from boiling water, ethanol, methanol, aqueous ethanol, methanol or chloroform. The solubility of picric acid in water and ethanol is 1.4 and 6.23 % respectively at 20°.

It is not advisable to store large quantities of picrates for long periods, *particularly when they are dry due to their potential* **EXPLOSIVE** *nature*. The free base should be recovered as soon as possible. The picrate is suspended in an excess of 2N aqueous NaOH and warmed a little. Because of the limited solubility of sodium picrate, excess hot water must be added. Alternatively, because of the greater solubility of lithium picrate, aqueous 10% lithium hydroxide solution can be used. The solution is cooled, the amine is extracted with a suitable solvent such as diethyl ether or toluene, washed with 5N NaOH until the alkaline solution remains colourless, then with water, and the extract is dried with anhydrous sodium carbonate. The solvent is distilled off and the amine is fractionally distilled (under reduced pressure if necessary) or recrystallised.

If the amines are required as their hydrochlorides, picrates can often be decomposed by suspending them in acetone and adding two equivalents of 10N HCl. The hydrochloride of the base is filtered off, leaving the picric acid in the acetone. Dowex No 1 anion-exchange resin in the chloride form is useful for changing solutions of the more soluble picrates (for example, of adenosine) into solutions of their hydrochlorides, from which sodium hydroxide precipitates the free base.

Salts

Amines can also be purified *via* their salts, e.g. hydrochlorides. A solution of the amine in dry toluene, diethyl ether, dichloromethane or chloroform is saturated with dry hydrogen chloride (generated by addition of concentrated sulfuric acid to dry sodium chloride, or to concentrated HCl followed by drying the gas through sulfuric acid, or from a hydrogen chloride cylinder) and the insoluble hydrochloride is filtered off and dissolved in water. The solution is made alkaline and the amine is extracted, as above. Hydrochlorides can also be prepared by dissolving the amine in ethanolic HCl and adding diethyl ether or petroleum ether. Where hydrochlorides are too hygroscopic or too soluble for satisfactory isolation, other salts, e.g. nitrate, sulfate, bisulfate or oxalate, can be used.

Double salts

The amine (1mol) is added to a solution of anhydrous zinc chloride (1mol) in concentrated hydrochloric acid (42mL) in ethanol (200mL, or less depending on the solubility of the double salt). The solution is stirred for 1h and the precipitated salt is filtered off and recrystallised from ethanol. The free base is recovered by adding excess of 5-10N NaOH (to dissolve the zinc hydroxide that separates) and is steam distilled. Mercuric chloride in hot water can be used instead of zinc chloride and the salt is crystallised from 1% hydrochloric acid. Other double salts have been used, e.g. cuprous salts, but are not as convenient as the above salts.

N-Acetyl derivatives

Purification as their N-acetyl derivatives is satisfactory for primary, and to a limited extent secondary, amines. The base is refluxed with slightly more than one equivalent of acetic anhydride for half to one hour, cooled and poured into ice-cold water. The insoluble derivative is filtered off, dried, and recrystallised from water, ethanol, aqueous ethanol or benzene (CAUTION **toxic**!). The derivative can be hydrolysed to the parent amine by refluxing with 70% sulfuric acid for a half to one hour. The solution is cooled, poured onto ice, and made alkaline. The amine is steam distilled or extracted as above. Alkaline hydrolysis is very slow.

N-Tosyl derivatives

Primary and secondary amines are converted into their tosyl derivatives by mixing equimolar amounts of amine and p-toluenesulfonyl chloride in dry pyridine (ca 5-10mols) and allowing to stand at room temperature overnight. The solution is poured into ice-water and the pH adjusted to 2 with HCl. The solid derivative is filtered off, washed with water, dried (vac. desiccator) and recrystallised from an alcohol or aqueous alcohol solution to a sharp melting point. The derivative is decomposed by dissolving in liquid ammonia (*fume cupboard*) and adding sodium metal (in small pieces with stirring) until the blue colour persists for 10-15min. Ammonia is allowed to evaporate (*fume cupboard*), the residue treated with water and the solution checked that the pH is above 10. If the pH is below 10 then the solution has to be basified with 2N NaOH. The mixture is extracted with diethyl ether or toluene, the extract is dried (K_2CO_3), evaporated and the residual amine recrystallised if solid or distilled if liquid.

AROMATIC HYDROCARBONS

Adducts

Aromatic hydrocarbons can be purified as their picrates using the procedures described for amines. Instead of picric acid, 1,3,5-trinitrobenzene or 2,4,7-trinitrofluorenone can also be used. In all these cases, following recrystallisation, the hydrocarbon can be isolated either as described for amines or by passing a solution of the adduct through an activated alumina column and eluting with toluene or petroleum ether. The picric acid and nitro compounds are more strongly adsorbed on the column.

Sulfonation

Naphthalene, xylenes and alkyl benzenes can be purified by sulfonation with concentrated sulfuric acid and crystallisation of the sodium sulfonates. The hydrocarbon is distilled out of the mixture with superheated steam.

CARBOXYLIC ACIDS

4-Bromophenacyl esters

A solution of the sodium salt of the acid is prepared. If the salt is not available, the acid is dissolved in an equivalent of aqueous NaOH and the pH adjusted to 8-9 with this base. A solution of one equivalent of 4-bromophenacyl bromide (for a monobasic acid, two equivalents for a dibasic acid, etc) in ten times its volume of ethanol is then added. The mixture is heated to boiling, and, if necessary, enough ethanol is added to clarify the solution which is then refluxed for half an hour to three hours depending on the number of carboxylic groups that have to be esterified. (One hour is generally sufficient for monocarboxylic acids.) On cooling, the ester should crystallise out. If it does not, then the solution is heated to boiling, and enough water is added to produce a slight turbidity. The solution is again cooled. The ester is collected, and recrystallised or fractionally distilled.

The ester is hydrolysed by refluxing for 1-2h with 1-5% of barium carbonate suspended in water or with aqueous sodium carbonate solution. The solution is cooled and extracted with diethyl ether, toluene or chloroform. It is then acidified and the acid is collected by filtration or extraction, and recrystallised or fractionally distilled.

p-Nitrobenzyl esters can be prepared in an analogous manner using the sodium salt of the acid and p-nitrobenzyl bromide. They are readily hydrolysed.

Alkyl esters

Of the alkyl esters, methyl esters are the most useful because of their rapid hydrolysis. The acid is refluxed with one or two equivalents of methanol in excess alcohol-free chloroform (or dichloromethane) containing about 0.1g of p-toluenesulfonic acid (as catalyst), using a Dean-Stark apparatus. (The water formed by the esterification is carried away into the trap.) When the theoretical amount of water is collected in the trap, esterification is complete. The chloroform solution in the flask is washed with 5% aqueous sodium carbonate solution, then water, and dried over anhydrous sodium sulfate or magnesium sulfate. The chloroform is distilled off and the ester is fractionally distilled through an efficient column. The ester is hydrolysed by refluxing with 5-10% aqueous NaOH solution until the insoluble ester has completely dissolved. The aqueous solution is concentrated a little by distillation to remove all of the methanol. It is then cooled and acidified. The acid is either extracted with diethyl ether, toluene or chloroform, or filtered off and isolated as above. Other methods for preparing esters are available.

Salts

The most useful salt derivatives for carboxylic acids are the isothiouronium salts. These are prepared by mixing almost saturated solutions containing the acid (carefully neutralised with N NaOH using phenolphthalein indicator) then adding two drops of N HCl and an equimolar amount of S-benzylisothiouronium chloride in ethanol and filtering off the salt that crystallises out. After recrystallisation from water, alcohol or aqueous alcohol the salt is decomposed by suspending or dissolving in 2N HCl and extracting the carboxylic acid from aqueous solution into diethyl ether, chloroform or toluene.

HYDROPEROXIDES

These can be converted to their sodium salts by precipitation below 30° with aqueous 25% NaOH. The salt is then decomposed by addition of solid (powdered) carbon dioxide and extracted with low-boiling petroleum ether. The solvent should be removed under reduced pressure below 20°. **The manipulation should be adequately shielded at all times to guard against EXPLOSIONS for the safety of the operator.**

KETONES

Bisulfite adduct

The adduct can be prepared and decomposed as described for aldehydes. Alternatively, because no Cannizzaro reaction is possible, it can also be decomposed with 0.5N NaOH.

Semicarbazones

A powdered mixture of semicarbazide hydrochloride (1mol) and anhydrous sodium acetate (1.3mol) is dissolved in water by gentle warming. A solution of the ketone (1mol) in the minimum volume of ethanol needed to dissolve it is then added. The mixture is warmed on a water bath until separation of the semicarbazone is complete. The solution is cooled, and the solid is filtered off. After washing with a little ethanol followed by water, it is recrystallised from ethanol or dilute aqueous ethanol. The derivative should have a characteristic melting point. The semicarbazone is decomposed by refluxing with excess of oxalic acid or with aqueous sodium carbonate solution. The ketone (which steam distils) is distilled off. It is extracted or separated from the distillate (after saturating with NaCl), dried with $CaSO_4$ or $MgSO_4$ and fractionally distilled using an efficient column (under vacuum if necessary). [See entry under Aldehydes.]

PHENOLS

The most satisfactory derivatives for phenols that are of low molecular weight or monohydric are the benzoate esters. (Their acetate esters are generally liquids or low-melting solids.) Acetates are more useful for high molecular weight and polyhydric phenols.

Benzoates

The phenol (1mol) in 5% aqueous NaOH is treated (while cooling) with benzoyl chloride (1mol) and the mixture is stirred in an ice bath until separation of the solid benzoyl derivative is complete. The derivative is filtered off, washed with alkali, then water, and dried (in a vacuum desiccator over NaOH). It is recrystallised from ethanol or dilute aqueous ethanol. The benzoylation can also be carried out in dry pyridine at low temperature (ca 0°) instead of in NaOH solution, finally pouring the mixture into water and collecting the solid as above. The ester is hydrolysed by refluxing in an alcohol (for example, ethanol, n-butanol) containing two or three equivalents of the alkoxide of the corresponding alcohol (for example sodium ethoxide or sodium n-butoxide) and a few (ca 5-10) millilitres of water, for half an hour to three hours. When hydrolysis is complete, an aliquot will remain clear on dilution with four to five times its volume of water. Most of the solvent is distilled off. The residue is diluted with cold water and acidified, and the phenol is steam distilled. The latter is collected from the distillate, dried and either fractionally distilled or recrystallised.

Acetates

These can be prepared as for the benzoates using either acetic anhydride with 3N NaOH or acetyl chloride in pyridine. They are hydrolysed as described for the benzoates. This hydrolysis can also be carried out with aqueous 10% NaOH solution, completion of hydrolysis being indicated by the complete dissolution of the acetate in the aqueous alkaline solution. On steam distillation, acetic acid also distils off but in these cases the phenols (see above) are invariably solids which can be filtered off and recrystallised.

PHOSPHATE AND PHOSPHONATE ESTERS

These can be converted to their uranyl nitrate addition compounds. The crude or partially purified ester is saturated with uranyl nitrate solution and the adduct filtered off. It is recrystallised from *n*-hexane, toluene or ethanol. For the more soluble members crystallisation from hexane using low temperatures (-40°) has been successful. The adduct is decomposed by shaking with sodium carbonate solution and water, the solvent is steam distilled (if hexane or toluene is used) and the ester is collected by filtration. Alternatively, after decomposition, the organic layer is separated, dried with $CaCl_2$ or BaO, filtered, and fractionally distilled under high vacuum.

MISCELLANEOUS

Impurities can sometimes be removed by conversion to derivatives under conditions where the major component does not react or reacts much more slowly. For example, normal (straight-chain) paraffins can be freed from unsaturated and branched-chain components by taking advantage of the greater reactivity of the latter with chlorosulfonic acid or bromine. Similarly, the preferential nitration of aromatic hydrocarbons can be used to remove e.g. benzene or toluene from cyclohexane by shaking for several hours with a mixture of concentrated nitric acid (25%), sulfuric acid (58%), and water (17%).

GENERAL METHODS FOR THE PURIFICATION OF CLASSES OF COMPOUNDS

Chapters 4, 5 and 6 list a large number of individual compounds, with a brief statement of how each one may be purified. For substances that are not included in these chapters the following procedures may prove helpful.

PROCEDURES

If the laboratory worker does not know of a reference to the preparation of a commercially available substance, he may be able to make a reasonable guess at the synthetic method used from published laboratory syntheses. This information, in turn, can simplify the necessary purification steps by suggesting probable contaminants. However, for other than macromolecules it is important that *at least* the ^1H NMR spectrum and/or the mass spectrum of the substance should be measured. These measurements require no more than two to three milligrams of material and provide a considerable amount of information about the substance. From the bibliography at the end of this chapter, references to NMR, IR and mass spectral data for a large number of the compounds in the Aldrich catalogue are available and are extremely useful for identifying compounds and impurities. If the material appears to have several impurities these spectra should be followed by examination of their chromatographic properties and spot tests. Purification methods can then be devised to remove these impurities, and a monitoring method will have already been established.

Physical methods of purification depend largely on the melting and boiling points of the materials. For gases and low-boiling liquids use is commonly made of the *freeze-pump-thaw* procedure (see Chapter 1). Gas chromatography is also useful, especially for low-boiling point liquids. Liquids are usually purified by refluxing with drying agents, acids or bases, reducing agents, charcoal, etc., followed by fractional distillation under reduced pressure. For solids, general methods include fractional freezing of the melted material, taking the middle fraction. Another procedure is sublimation of the solid under reduced pressure. The other commonly used method for purifying solids is by recrystallisation from a solution in a suitable solvent, by cooling with or without the prior addition of a solvent in which the solute is not very soluble.

The nature of the procedure will depend to a large extent on the quantity of purified material that is required. For example, for small quantities (50-250mg) of a pure volatile liquid, preparative gas chromatography is probably the best method. Two passes through a suitable column may well be sufficient. Similarly, for small amounts (100-500mg) of an organic solid, column chromatography is likely to be very satisfactory, the eluate being collected as a number of separate fractions (*ca* 5-10mL) which are examined by FT-IR, NMR or UV spectroscopy, TLC or by some other appropriate analytical technique. (For information on suitable adsorbents and eluents the texts referred to in the bibliography at the end of Chapters 1 and 2 should be consulted.) Preparative thin layer chromatography or HPLC can also be used successfully for purifying up to 500mg of solid. HPLC is the more and more commonly used procedure for the purification of small molecules as well as large molecules such as polypeptides and DNA.

Where larger quantities (upwards of 1g) are required, most of the impurities should be removed by preliminary treatments, such as solvent extraction, liquid-liquid partition, or conversion to a derivative (*vide supra*) which can be purified by crystallisation or fractional distillation before being reconverted to the starting material. The substance is then crystallised or distilled. If the final amounts must be in excess of 25g, preparation of a derivative is sometimes omitted because of the cost involved. In all of the above cases, purification is likely to be more laborious if the impurity is an isomer or a derivative with closely similar physical properties.

CRITERIA OF PURITY

Purification becomes meaningful only insofar as adequate tests of purity are applied: the higher the degree of purity that is sought, the more stringent must these tests be. If the material is an organic solid, its melting point should first be taken and compared with the recorded value. *Note that the melting points of most salts, organic or inorganic, are generally decomposition points and are not reliable criteria of purity.* As part of this preliminary examination, the sample might be examined by thin layer chromatography in several different solvent systems and in high enough concentrations to facilitate the detection of minor components. On the other hand, if the substance is a liquid, its boiling point should be measured. If, further, it is a high boiling liquid, its chromatographic behaviour should be examined. Liquids, especially volatile ones, can be studied very satisfactorily by gas chromatography, preferably using at least two different stationary and/or mobile phases. Atomic absorption spectroscopy is a useful and sensitive method for detecting metal impurities and the concentrations of metals and metal salts or complexes.

Application of these tests at successive steps will give a good indication of whether or not the purification is satisfactory and will also show when adequate purification has been achieved. Finally elemental analyses, e.g. of carbon, hydrogen, nitrogen, sulfur, metals etc. are very sensitive to impurities (other than with isomers), and are good criteria of purity.

GENERAL PROCEDURES FOR THE PURIFICATION OF SOME CLASSES OF ORGANIC COMPOUNDS

In the general methods of purification described below, it is assumed that the impurities belong essentially to a class of compounds different from the one being purified. They are suggested for use in cases where substances are not listed in Chapters 4, 5 and the low molecular weight compounds in Chapter 6. In such cases, the experimenter is advised to employ them in conjunction with information given in these chapters for the purification of suitable analogues. Also, for a wider range of drying agents, solvents for extraction and solvents for recrystallisation, the reader is referred to Chapter 1. See Chapter 6 for general purification procedures used for macromolecules.

ACETALS

These are generally diethyl or dimethyl acetal derivatives of aldehydes. They are more stable to alkali than to acids. Their common impurities are the corresponding alcohol, aldehyde and water. Drying with sodium wire removes alcohols and water, and polymerizes aldehydes so that, after decantation, the acetal can be fractionally distilled. In cases where the use of sodium is too drastic, aldehydes can be removed by shaking with alkaline hydrogen peroxide solution and the acetal is dried with sodium carbonate or potassium carbonate. Residual water and alcohols (up to *n*-propyl) can be removed with Linde type 4A molecular sieves. The acetal is then filtered and fractionally distilled. Solid acetals (i.e. acetals of high molecular weight aldehydes) are generally low-melting and can be recrystallised from low-boiling petroleum ether, toluene or a mixture of both.

ACIDS
Carboxylic acids

Liquid carboxylic acids are first freed from neutral and basic impurities by dissolving them in aqueous alkali and extracting with diethyl ether. (The pH of the solution should be at least three units above the pK_a of the acid, see pK in Chapter 1). The aqueous phase is then acidified to a pH at least three units below the pK_a of the acid and again extracted with ether. The extract is dried with magnesium sulfate or sodium sulfate and the ether is distilled off. The acid is fractionally distilled through an efficient column. It can be further purified by conversion to its methyl or ethyl ester (*vide supra*) which is then fractionally distilled. Hydrolysis yields the original acid which is again purified as above.

Acids that are solids can be purified in this way, except that distillation is replaced by repeated crystallisation (preferable from at least two different solvents such as water, alcohol or aqueous alcohol, toluene, toluene/petroleum ether or acetic acid.) Water-insoluble acids can be partially purified by dissolution in N sodium hydroxide solution and precipitation with dilute mineral acid. If the acid is required to be free from sodium ions, then it is better to dissolve the acid in hot N ammonia, heat to *ca* 80°, adding slightly more than an equal volume of N formic acid and allowing to cool slowly for crystallisation. Any ammonia, formic acid or ammonium formate that adhere to the acid are removed when the acid is dried in a vacuum — they are volatile. The separation and purification of naturally occurring fatty acids, based on distillation, salt solubility and low temperature crystallisation, are described by K.S.Markley (Ed.), *Fatty Acids*, 2nd Edn, part 3, Chap. 20, Interscience, New York, 1964.

Aromatic carboxylic acids can be purified by conversion to their sodium salts, recrystallisation from hot water, and reconversion to the free acids.

Sulfonic acids

The low solubility of sulfonic acids in organic solvents and their high solubility in water makes necessary a treatment different from that for carboxylic acids. Sulfonic acids are strong acids, they have the tendency to hydrate, and many of them contain water of crystallisation. The lower-melting and liquid acids can generally be purified with only slight decomposition by fractional distillation, preferably under reduced pressure. A common impurity is sulfuric acid, but this can be removed by recrystallisation from concentrated aqueous solutions. The wet acid can be dried by azeotropic removal of water with toluene, followed by distillation. The higher-melting acids, or acids that melt with decomposition, can be recrystallised from water or, occasionally, from ethanol. For a typical purification of aromatic sulfonic acids using their barium salts refer to benzenesulfonic acid in Chapter 4.

Sulfinic acids

These acids are less stable, less soluble and less acidic than the corresponding sulfonic acids. The common impurities are the respective sulfonyl chlorides from which they have been prepared, and the thiolsulfonates (neutral) and sulfonic acids into which they decompose. The first two of these can be removed by solvent extraction from an alkaline solution of the acid. On acidification of an alkaline solution, the sulfinic acid crystallises out leaving the sulfonic acid behind. The lower molecular weight members are isolated as their metal (e.g. ferric) salts, but the higher members can be crystallised from water (made slightly acidic), or alcohol.

ACID CHLORIDES

The corresponding acid and hydrogen chloride are the most likely impurities. Usually these can be removed by efficient fractional distillation. Where acid chlorides are not readily hydrolysed (e.g. aryl sulfonyl chlorides) the compound can be freed from contaminants by dissolving in a suitable solvent such as alcohol-free chloroform, dry toluene or petroleum ether and shaking with dilute sodium bicarbonate solution. The organic phase is then washed with water, dried with anhydrous sodium sulfate or magnesium sulfate, and distilled or recrystallised. This procedure is *hazardous* with readily hydrolysable acid chlorides such as acetyl chloride and benzoyl chloride. Solid acid chlorides should be thoroughly dried *in vacuo* over strong drying agents and are satisfactorily recrystallised from toluene, toluene-petroleum ether, petroleum ethers, alcohol-free chloroform/toluene, and, occasionally, from dry diethyl ether. Hydroxylic or basic solvents should be strictly avoided. *All operations should be carried out in a fume cupboard because of the **irritant** nature of these compounds which also attack the skin.*

ALCOHOLS
Monohydric alcohols

The common impurities in alcohols are aldehydes or ketones, and water. [*Ethanol* in Chapter 4 is typical.] Aldehydes and ketones can be removed by adding a small amount of sodium metal and refluxing for 2 hours, followed by distillation. Water can be removed in a similar way but it is preferable to use magnesium metal instead of sodium because it forms a more insoluble hydroxide, thereby shifting the equilibrium more completely from metal alkoxide to metal hydroxide. The magnesium should be activated with iodine (or a small amount of methyl iodide), and the water content should be low, otherwise the magnesium will be deactivated. If the amount of water is large it should be removed by azeotropic distillation (see below), or by drying over anhydrous $MgSO_4$ (not $CaCl_2$ which combines with alcohols). Acidic materials can be removed by treatment with anhydrous Na_2CO_3, followed by a suitable drying agent, such as calcium hydride, and fractional distillation, using gas chromatography to establish the purity of the product [Ballinger and Long, *J Am Chem Soc* **82** 795 *1960*]. Alternatively, the alcohol can be refluxed with freshly ignited CaO for 4 hours and then fractionally distilled [McCurdy and Laidler, *Can J Chem* **41** 1867 *1963*].

With higher-boiling alcohols it is advantageous to add some freshly prepared magnesium ethoxide solution (only slightly more than required to remove the water), followed by fractional distillation. Alternatively, in such cases, water can be removed by azeotropic distillation with toluene. Higher-melting alcohols can be purified by crystallisation from methanol or ethanol, toluene/petroleum ether or petroleum ether. Sublimation in vacuum, molecular distillation and gas chromatography are also useful means of purification. For purification *via* derivatives, *vide supra*.

Polyhydric alcohols

These alcohols are more soluble in water than are monohydric alcohols. Liquids can be freed from water by shaking with type 4A Linde molecular sieves and can safely be distilled only under high vacuum. Carbohydrate alcohols can be crystallised from strong aqueous solution or, preferably, from mixed solvents such as ethanol/petroleum ether or dimethyl formamide/toluene. Crystallisation usually requires seeding and is extremely slow. Further purification can be effected by conversion to the acetyl or benzoyl derivatives which are much less soluble in water and which can readily be recrystallised, e.g. from ethanol. Hydrolysis of the acetyl derivatives, followed by removal of acetate or benzoate and metal ions by ion-exchange chromatography, gives the purified material. On no account should solutions of carbohydrates be concentrated above 40° because of darkening and formation of *caramel*. Ion exchange, charcoal or cellulose column chromatography has been used for the purification and separation of carbohydrates.

ALDEHYDES

Common impurities found in aldehydes are the corresponding alcohols, aldols and water from self-condensation, and the corresponding acids formed by autoxidation. Acids can be removed by shaking with aqueous 10% sodium bicarbonate solution. The organic liquid is then washed with water. It is dried with anhydrous sodium sulfate or magnesium sulfate and then fractionally distilled. Water soluble aldehydes must be dissolved in a suitable solvent such as diethyl ether before being washed in this way. Further purification can be effected *via* the bisulfite derivative (see pp. 57 and 59) or the Schiff base formed with aniline or benzidine. Solid aldehydes can be dissolved in diethyl ether and purified as above. Alternatively, they can be steam distilled, then sublimed and crystallised from toluene or petroleum ether.

AMIDES

Amides are stable compounds. The lower-melting members (such as acetamide) can be readily purified by fractional distillation. Most amides are solids which have low solubilities in water. They can be recrystallised from large quantities of water, ethanol, ethanol/ether, aqueous ethanol, chloroform/toluene, chloroform or acetic acid. The likely impurities are the parent acids or the alkyl esters from which they have been made. The former can be removed by thorough washing with aqueous ammonia followed by recrystallisation, whereas elimination of the latter is by trituration or recrystallisation from an organic solvent. Amides can be freed from solvent or water by drying below their melting points. These purifications can also be used for sulfonamides and acid hydrazides.

AMINES

The common impurities found in amines are nitro compounds (if prepared by reduction), the corresponding halides (if prepared from them) and the corresponding carbamate salts. Amines are dissolved in aqueous acid, the pH of the solution being at least three units below the pK_a value of the base to ensure almost complete formation of the cation. They are extracted with diethyl ether to remove neutral impurities and to decompose the carbamate salts. The solution is then made strongly alkaline and the amines that separate are extracted into a suitable solvent (ether or toluene) or steam distilled. The latter process removes coloured impurities. Note that chloroform cannot be used as a solvent for primary amines because, in the presence of alkali, poisonous carbylamines (isocyanides) are formed. However, chloroform is a useful solvent for the extraction of heterocyclic bases. In this case it has the added advantage that while the extract is being freed from the chloroform most of the moisture is removed with the solvent.

Alternatively, the amine may be dissolved in a suitable solvent (e.g. toluene) and dry HCl gas is passed through the solution to precipitate the amine hydrochloride. This is purified by recrystallisation from a suitable solvent mixture (e.g. ethanol/diethyl ether). The free amine can be regenerated by adding sodium hydroxide and isolated as above.

Liquid amines can be further purified *via* their acetyl or benzoyl derivatives (*vide supra*). Solid amines can be recrystallised from water, alcohol, toluene or toluene-petroleum ether. *Care should be taken in handling large quantities of amines because their vapours are* **harmful (possibly carcinogenic)** *and they are readily absorbed through the skin.*

AMINO ACIDS
Because of their zwitterionic nature, amino acids are generally soluble in water. Their solubility in organic solvents rises as the fat-soluble portion of the molecule increases. The likeliest impurities are traces of salts, heavy metal ions, proteins and other amino acids. Purification of these is usually easy, by recrystallisation from water or ethanol/water mixtures. The amino acid is dissolved in the boiling solvent, decolorised if necessary by boiling with 1g of acid-washed charcoal/100g amino acid, then filtered hot, chilled, and set aside for several hours to crystallise. The crystals are filtered off, washed with ethanol, then ether, and dried.
Amino acids have high melting or decomposition points and are best examined for purity by paper or thin layer chromatography. The spots are developed with ninhydrin. Customary methods for the purification of small quantities of amino acids obtained from natural sources (i.e. 1-5g) are ion-exchange chromatography (see Chapter 1). For general treatment of amino acids see Greenstein and Winitz [*The Amino Acids*, Vols 1-3, J.Wiley & Sons, New York 1961] and individual amino acids in Chapters 4 and 6.
A useful source of details such as likely impurities, stability and tests for homogeneity of amino acids is *Specifications and Criteria for Biochemical Compounds*, 3rd edn, National Academy of Sciences, USA, 1972.

ANHYDRIDES
The corresponding acids, resulting from hydrolysis, are the most likely impurities. Distillation from phosphorus pentoxide, followed by fractional distillation, is usually satisfactory. With high boiling or solid anhydrides, another method involves boiling under reflux for 0.5-1h with acetic anhydride, followed by fractional distillation. Acetic acid distils first, then acetic anhydride and finally the desired anhydride. Where the anhydride is a solid, removal of acetic acid and acetic anhydride at atmospheric pressure is followed by heating under vacuum. The solid anhydride is then either crystallised as for acid chlorides or (in some cases) sublimed in a vacuum. A preliminary purification when large quantities of acid are present in a solid anhydride (such as phthalic anhydride) is by preferential solvent extraction of the (usually) more soluble anhydride from the acid (e.g. with $CHCl_3$ in the case of phthalic anhydride). *All operations with liquid anhydrides should be carried out in a fume cupboard because of their* **LACHRYMATORY** *properties. Almost all anhydrides attack skin.*

CAROTENOIDS
These usually are decomposed by light, air and solvents, so that degradation products are probable impurities. Chromatography and adsorption spectra permit the ready detection of coloured impurities, and separations are possible using solvent distribution, chromatography or crystallisation. Thus, in partition between immiscible solvents, xanthophyll remains in 90% methanol while carotenes pass into the petroleum ether phase. For small amounts of material, thin-layer or paper chromatography may be used, while column chromatography is suitable for larger amounts. Colorless impurities may be detected by IR, NMR or mass spectrometry. The more common separation procedures are described by P. Karrer and E. Jucker in *Carotenoids*, E.A. Braude (translator), Elsevier, NY, 1950.
Purity can be checked by chromatography (on thin-layer plates, Kieselguhr, paper or columns), by UV or NMR procedures.

ESTERS
The most common impurities are the corresponding acid and hydroxy compound (i.e. alcohol or phenol), and water. A liquid ester from a carboxylic acid is washed with 2N sodium carbonate or sodium hydroxide to remove acid material, then shaken with calcium chloride to remove ethyl or methyl alcohols (if it is a methyl or ethyl ester). It is dried with potassium carbonate or magnesium sulfate, and distilled. Fractional distillation then removes residual traces of hydroxy compounds. This method does not apply to esters of inorganic acids (e.g. dimethyl sulfate) which are more readily hydrolysed in aqueous solution when heat is generated in the neutralisation of the excess acid. In such cases, several fractional distillations, preferably under vacuum, are usually sufficient.
Solid esters are easily crystallisable materials. It is important to note that esters of alcohols must be recrystallised either from non-hydroxylic solvents (e.g. toluene) or from the alcohol from which the ester is derived. Thus methyl esters should be crystallised from methanol or methanol/toluene, but not from ethanol, *n*-butanol or other alcohols, in order to avoid alcohol exchange and contamination of the ester with a second ester. Useful solvents for crystallisation are the corresponding alcohols or aqueous alcohols, toluene, toluene/petroleum ether, and chloroform (ethanol-free)/toluene. Esters of carboxylic acid derived from phenols

are more difficult to hydrolyse and exchange, hence any alcoholic solvent can be used freely. Sulfonic acid esters of phenols are even more resistant to hydrolysis: they can safely be crystallised not only from the above solvents but also from acetic acid, aqueous acetic acid or boiling *n*-butanol.

Fully esterified phosphoric acid and phosphonic acids differ only in detail from the above mentioned esters. Their major contaminants are alcohols or phenols, phosphoric or phosphonic acids (from hydrolysis), and (occasionally) basic material, such as pyridine, which is used in their preparation. Water-insoluble esters are washed thoroughly and successively with dilute acid (e.g. 0.2N sulfuric acid), water, 0.2N sodium hydroxide and water. After drying with calcium chloride they are fractionally distilled. Water-soluble esters should first be dissolved in a suitable organic solvent and, in the washing process, water should be replaced by saturated aqueous sodium chloride. Some esters (e.g. phosphate and phosphonate esters) can be further purified through their uranyl adducts (*vide supra*). Traces of water or hydroxy compounds can be removed by percolation through, or shaking with, activated alumina (about 100g/L of liquid solution), followed by filtration and fractional distillation in a vacuum. For high molecular weight esters (which cannot be distilled without some decomposition) it is advisable to carry out distillation at as low a pressure as possible. Solid esters can be crystallised from toluene or petroleum ether. Alcohols can be used for recrystallising phosphoric or phosphonic esters of phenols.

ETHERS

The purification of diethyl ether (see Chapter 4) is typical of liquid ethers. The most common contaminants are the alcohols or hydroxy compounds from which the ethers are prepared, their oxidation products (e.g. aldehydes), peroxides and water. Peroxides, aldehydes and alcohols can be removed by shaking with alkaline potassium permanganate solution for several hours, followed by washing with water, concentrated sulfuric acid [CARE], then water. After drying with calcium chloride, the ether is distilled. It is then dried with sodium or with lithium aluminium hydride, redistilled and given a final fractional distillation. The drying process should be repeated if necessary.

Alternatively, methods for removing peroxides include leaving the ether to stand in contact with iron filings or copper powder, shaking with a solution of ferrous sulfate acidified with N sulfuric acid, shaking with a copper-zinc couple, passage through a column of activated alumina, and refluxing with phenothiazine. Cerium(III) hydroxide has also been used .

A simple test for ether peroxides is to add 10mL of the ether to a stoppered cylinder containing 1mL of freshly prepared 10% solution of potassium iodide containing a drop of starch indicator. No colour should develop during one minute if free from peroxides. Alternatively, a 1% solution of ferrous ammonium sulfate, 0.1M in sulfuric acid and 0.01M in potassium thiocyanate should not increase appreciably in red colour when shaken with two volumes of the ether.

As a safety precaution against **EXPLOSION** (in case the purification has been insufficiently thorough) at least a quarter of the total volume of ether should remain in the distilling flask when the distillation is discontinued as peroxides are generally higher boiling. To minimize peroxide formation, ethers should be stored in dark bottles and, if they are liquids, they should be left in contact with type 4A Linde molecular sieves, in a cold place, over sodium amalgam. The rate of formation of peroxides depends on storage conditions and is accelerated by heat, light, air and moisture. The formation of peroxides is inhibited in the presence of diphenylamine, di-*tert*-butylphenol, or other antioxidants as stabiliser.

Ethers that are solids (e.g. phenyl ethers) can be steam distilled from an alkaline solution which will hold back any phenolic impurity. After the distillate is made alkaline with sodium carbonate, the insoluble ether is collected either by extraction (e.g. with chloroform, diethyl ether or toluene) or by filtration. It is then crystallised from alcohols, alcohol/petroleum ether, petroleum ether, toluene or mixtures of these solvents, sublimed in a vacuum and recrystallised if necessary.

HALIDES

Aliphatic halides are likely to be contaminated with halogen acids and the alcohols from which they have been prepared, whereas in aromatic halides the impurities are usually aromatic hydrocarbons, amines or phenols. In both groups the halogen is less reactive than it is in acid halides. Purification is by shaking with concentrated hydrochloric acid, followed by washing successively with water, 5% sodium carbonate or bicarbonate, and water. After drying with calcium chloride, the halide is distilled and then fractionally distilled using an efficient column. For a solid halide the above purification is carried out by dissolving it in a suitable solvent such as toluene. Solid halides can also be purified by chromatography using an alumina column and eluting with toluene or petroleum ether. They can be crystallised from toluene, petroleum ethers, toluene/petroleum ether or toluene/chloroform/petroleum ether. Care should be taken when handling organic halogen compounds because of their **TOXICITY**. It should be noted that methyl iodide is a cancer suspect.

Liquid aliphatic halides are obtained alcohol-free by distillation from phosphorus pentoxide. They are stored in dark bottles to prevent oxidation and, in some cases, the formation of phosgene.

A general method for purifying *chlorohydrocarbons* uses repeated shaking with concentrated sulfuric acid [CARE] until no further colour develops in the acid, then washing with water then a solution of sodium bicarbonate, followed by water again. After drying with calcium chloride, the chlorohydrocarbon is fractionally redistilled to constant boiling point.

HYDROCARBONS
Gaseous hydrocarbons are best freed from water and gaseous impurities by passage through suitable adsorbents and (if olefinic material is to be removed) oxidants such as alkaline potassium permanganate solution, followed by fractional cooling (see Chapter 1 for cooling baths) and fractional distillation at low temperature. To effect these purifications and also to store the gaseous sample, a vacuum line is necessary.

Impurities in hydrocarbons can be characterised and evaluated by gas chromatography and mass spectrometry. The total amount of impurities present can be estimated from the thermometric freezing curve.

Liquid aliphatic hydrocarbons are freed from aromatic impurities by shaking with concentrated sulfuric acid [CARE] whereby the aromatic compounds are sulfonated. Shaking is carried out until the sulfuric acid layer remains colourless for several hours. The hydrocarbon is then freed from the sulfuric acid and the sulfonic acids by separating the two phases and washing the organic layer successively with water, 2N sodium hydroxide, and water. It is dried with $CaCl_2$ or Na_2SO_4, and then distilled. The distillate is dried with sodium wire, P_2O_5, or metallic hydrides, or passage through a dry silica gel column, or preferably, and more safely, with molecular sieves (see Chapter 1) before being finally fractionally distilled through an efficient column. If the hydrocarbon is contaminated with olefinic impurities, shaking with aqueous alkaline permanganate is necessary prior to the above purification. Alicyclic and paraffinic hydrocarbons can be freed from water, non-hydrocarbon and aromatic impurities by passage through a silica gel column before the final fractional distillation. This may also remove isomers. (For the use of chromatographic methods to separate mixtures of aromatic, paraffinic and alicyclic hydrocarbons see references in the bibliography in Chapter 1 under *Chromatography, Gas Chromatography and High Performance Liquid Chromatography*). Another method of removing branched-chain and unsaturated hydrocarbons from straight-chain hydrocarbons depends on the much faster reaction of the former with chlorosulfonic acid.

Isomeric materials which have closely similar physical properties can be serious contaminants in hydrocarbons. With aromatic hydrocarbons, e.g. xylenes and alkyl benzenes, advantage is taken of differences in ease of sulfonation. If the required compound is sulfonated more readily, the sulfonic acid is isolated, crystallised (e.g. from water), and decomposed by passing superheated steam through the flask containing the acid. The sulfonic acid undergoes hydrolysis and the liberated hydrocarbon distils with the steam. It is separated from the distillate, dried, distilled and then fractionally distilled. For small quantities (10-100mg), vapour phase chromatography is the most satisfactory method for obtaining a pure sample (for column materials for packings see Chapter 1).

Azeotropic distillation with methanol or 2-ethoxyethanol (cellosolve) has been used to obtain highly purified saturated hydrocarbons and aromatic hydrocarbons such as xylenes and isopropylbenzenes.

Carbonyl-containing impurities can be removed from hydrocarbons (and other oxygen-lacking solvents such as $CHCl_3$ and CCl_4) by passage through a column of Celite 545 (100g) mixed with concentrated sulfuric acid (60mL). After first adding some solvent and about 10g of granular Na_2SO_4, the column is packed with the mixture and a final 7-8cm of Na_2SO_4 is added at the top [Hornstein and Crowe, *Anal Chem* **34** 1037 *1962*]. Alternatively, Celite impregnated with 2,4-dinitrophenylhydrazine can be used.

With solid hydrocarbons such as naphthalene and polycyclic hydrocarbons, preliminary purification by sublimation in vacuum (or high vacuum if the substance is high melting), is followed by zone refining and finally by chromatography (e.g. on alumina) using low-boiling liquid hydrocarbon eluents. These solids can be recrystallised from alcohols, alcohol/petroleum ether or from liquid hydrocarbons (e.g. toluene) and dried below their melting points. Aromatic hydrocarbons that have been purified by zone melting include anthracene, biphenyl, fluoranthrene, naphthalene, perylene, phenanthrene, pyrene and terphenyl, among others. Some polycyclic hydrocarbons, e.g. benzpyrene, are CARCINOGENIC.

Olefinic hydrocarbons have a very strong tendency to polymerise and commercially available materials are generally stabilised, e.g. with hydroquinone. When distilling compounds such as vinylpyridine or styrene, the stabiliser remains behind and the purified olefinic material is more prone to polymerisation. The most common impurities are higher-boiling dimeric or polymeric compounds. Vacuum distillation in a nitrogen atmosphere not only separates monomeric from polymeric materials but in some cases also depolymerises the impurities. The distillation flask should be charged with a polymerisation inhibitor and the purified material should be used immediately or stored in the dark and mixed with a small amount of stabiliser (e.g. 0.1% of hydroquinone or di-*tert*-butylcatechol). It is also advisable to add to the flask a small amount (*ca* 5-10% by volume of liquid in the flask) of a ground mixture of Kieselguhr and NaCl which will provide nuclei for facilitating boiling and finally for cleaning the flask from insoluble polymeric residue (due to the presence of the water soluble NaCl).

IMIDES

Imides (e.g. phthalimide) can be purified by conversion to their potassium salts by reaction in ethanol with ethanolic potassium hydroxide. The imides are regenerated when the salts are hydrolysed with dilute acid. Like amides, imides readily crystallise from alcohols and, in some cases (e.g. quinolinic imide), from glacial acetic acid.

IMINO COMPOUNDS

These substances contain the -C=NH group and, because they are strong, unstable bases, they are kept as their more stable salts, such as the hydrochlorides. (The free base usually hydrolyses to the corresponding oxo compound and ammonia.) Like amine hydrochlorides, the salts are purified by solution in alcohol containing a few drops of hydrochloric acid. After treatment with charcoal, and filtering, dry diethyl ether (or petroleum ether if ethanol is used) is added until crystallisation sets in. The salts are dried and kept in a vacuum desiccator.

KETONES

Ketones are more stable to oxidation than aldehydes and can be purified from oxidisable impurities by refluxing with potassium permanganate until the colour persists, followed by shaking with sodium carbonate (to remove acidic impurities) and distilling. Traces of water can be removed with type 4A Linde molecular sieves. Ketones which are solids can be purified by crystallisation from alcohol, toluene, or petroleum ether, and are usually sufficiently volatile for sublimation in vacuum. Ketones can be further purified *via* their bisulfite, semicarbazone or oxime derivatives (*vide supra*). The bisulfite addition compounds are formed only by aldehydes and methyl ketones but they are readily hydrolysed in dilute acid or alkali.

MACROMOLECULES   See Chapter 6.

NITRILES

All purifications should be carried out in an efficient fume cupboard because of the **TOXIC** *nature of these compounds.*

Nitriles are usually prepared either by reacting the corresponding halide or diazonium salts with a cyanide salt or by dehydrating an amide. Hence, possible contaminants are the respective halide or alcohol (from hydrolysis), phenolic compounds, amines or amides. Small quantities of phenols can be removed by chromatography on alumina. More commonly, purification of liquid nitriles or solutions of solid nitriles in a solvent such as diethyl ether is by shaking with dilute aqueous sodium hydroxide, followed by washing successively with water, dilute acid and water. After drying with sodium sulfate, the solvent is distilled off. Liquid nitriles are best distilled from a small amount of P_2O_5 which, besides removing water, dehydrates any amide to the nitrile. About one fifth of the nitrile should remain in the distilling flask at the end of the distillation (*the residue may contain some inorganic cyanide*). This purification also removes alcohols and phenols. Solid nitriles can be recrystallised from ethanol, toluene or petroleum ether, or a mixture of these solvents. They can also be sublimed under vacuum. Preliminary purification by steam distillation is usually possible.

Strong alkali or heating with dilute acids may lead to hydrolysis of the nitrile, and should be avoided.

NITRO COMPOUNDS

Aliphatic nitro compounds are generally acidic. They are freed from alcohols or alkyl halides by standing for a day with concentrated sulfuric acid, then washed with water, dried with magnesium sulfate followed by calcium sulfate and distilled. The principal impurities are isomeric or homologous nitro compounds. In cases where the nitro compound was originally prepared by vapour phase nitration of the aliphatic hydrocarbon, fractional distillation should separate the nitro compound from the corresponding hydrocarbon. Fractional crystallisation is more effective than fractional distillation if the melting point of the compound is not too low.

The impurities present in aromatic nitro compounds depend on the aromatic portion of the molecule. Thus, benzene, phenols or anilines are probable impurities in nitrobenzene, nitrophenols and nitroanilines, respectively. Purification should be carried out accordingly. Isomeric compounds are likely to remain as impurities after the preliminary purifications to remove basic and acidic contaminants. For example, *o*-nitrophenol may be found in samples of *p*-nitrophenol. Usually, the *o*-nitro compounds are more steam volatile than the *p*-nitro isomers, and can be separated in this way. Polynitro impurities in mononitro compounds can be readily removed because of their relatively lower solubilities in solvents. With acidic or basic nitro compounds which cannot be separated in the above manner, advantage may be taken of their differences in pK values (see Chapter 1). The compounds can thus be purified by preliminary extractions with several sets of aqueous buffers of known pH (see for example Table 19, Chapter 1) from a solution of the substance in a suitable solvent such as diethyl ether. This method is more satisfactory and less laborious the larger the difference between the pK value of the impurity and the desired compound. Heterocyclic nitro compounds require similar treatment to the nitroanilines. Neutral nitro compounds can be steam distilled.

NUCLEIC ACIDS   See Chapter 6.

PHENOLS

Because phenols are weak acids, they can be freed from neutral impurities by dissolution in aqueous N sodium hydroxide and extraction with a solvent such as diethyl ether, or by steam distillation to remove the non-acidic material. The phenol is recovered by acidification of the aqueous phase with 2N sulfuric acid, and either extracted with ether or steam distilled. In the second case the phenol is extracted from the steam distillate after saturating it with sodium chloride (salting out). A solvent is necessary when large quantities of liquid phenols are purified. The phenol is fractionated by distillation under reduced pressure, preferably in an atmosphere of nitrogen to minimise oxidation. Solid phenols can be crystallised from toluene, petroleum ether or a mixture of these solvents, and can be sublimed under vacuum. Purification can also be effected by fractional crystallisation or zone refining. For further purification of phenols *via* their acetyl or benzoyl derivatives (*vide supra*).

POLYPEPTIDES AND PROTEINS   See Chapter 6.

QUINONES

These are neutral compounds which are usually coloured. They can be separated from acidic or basic impurities by extraction of their solutions in organic solvents with aqueous basic or acidic solutions, respectively. Their colour is a useful property in their purification by chromatography through an alumina column with, e.g. toluene, as eluent. They are volatile enough for vacuum sublimation, although with high-melting quinones a very high vacuum is necessary. *p*-Quinones are stable compounds and can be recrystallised from water, ethanol, aqueous ethanol, toluene, petroleum ether or glacial acetic acid. *o*-Quinones, on the other hand, are readily oxidised. They should be handled in an inert atmosphere, preferably in the absence of light.

SALTS (ORGANIC)
With metal ions

Water-soluble salts are best purified by preparing a concentrated aqueous solution to which, after decolorising with charcoal and filtering, ethanol or acetone is added so that the salts crystallise. They are collected, washed with aqueous ethanol or aqueous acetone, and dried. In some cases, water-soluble salts can be recrystallised satisfactorily from alcohols. Water-insoluble salts are purified by Soxhlet extraction, first with organic solvents and then with water, to remove soluble contaminants. The purified salt is recovered from the thimble.

With organic cations

Organic salts (e.g. trimethylammonium benzoate) are usually purified by recrystallisation from polar solvents (e.g. water, ethanol or dimethyl formamide). If the salt is too soluble in a polar solvent, its concentrated solution should be treated dropwise with a miscible nonpolar, or less polar, solvent (see Table 8, Chapter 1) until crystallisation begins.

With sodium alkane sulfonates

Purified from sulfites by boiling with aqueous HBr. Purified from sulfates by adding $BaBr_2$. Sodium alkane disulfonates are finally pptd by addition of MeOH. [Pethybridge and Taba *J Chem Soc Faraday Trans 1* **78** 1331 *1982*].

SULFUR COMPOUNDS
Disulfides

These can be purified by extracting acidic and basic impurities with aqueous base or acid, respectively. However, they are somewhat sensitive to strong alkali which slowly cleaves the disulfide bond. The lower-melting members can be fractionally distilled under vacuum. The high members can be recrystallised from alcohol, toluene or glacial acetic acid.

Sulfones

Sulfones are neutral and very stable compounds that can be distilled without decomposition. They are freed from acidic and basic impurities in the same way as disulfides. The low molecular weight members are quite soluble in water but the higher members can be recrystallised from water, ethanol, aqueous ethanol or glacial acetic acid.

Sulfoxides

These compounds are odourless, rather unstable compounds, and should be distilled under vacuum in an inert atmosphere. They are water-soluble but can be extracted from aqueous solution with a solvent such as diethyl ether.

Thioethers

Thioethers are neutral stable compounds that can be freed from acidic and basic impurities as described for disulfides. They can be recrystallised from organic solvents and distilled without decomposition. They have sulfurous odours.

Thiols

Thiols, or mercaptans, are stronger acids than the corresponding aliphatic hydroxy or phenolic compounds, but can be purified in a similar manner. However, care must be exercised in handling thiols to avoid their oxidation to disulfides. For this reason, purification is best carried out in an inert atmosphere in the absence of oxidising agents. Similarly, thiols should be stored out of contact with air. They can be distilled without change, and the higher-melting thiols (which are usually more stable) can be crystallised, e.g. from water or dilute alcohol. They oxidise readily in alkaline solution but can be separated from the disulfide which is insoluble in this medium. They should be stored in the dark below 0°. *All operations with thiols should be carried out in an efficient fume cupboard because of their very unpleasant odour and their* **TOXICITY**.

Thiolsulfonates (disulfoxides)

Thiolsulfonates are neutral and are somewhat light-sensitive compounds. Their most common impurities are sulfonyl chlorides (neutral) or the sulfinic acid or disulfide from which they are usually derived. The first can be removed by partial freezing or crystallisation, the second by shaking with dilute alkali, and the third by recrystallisation because of the higher solubility of the disulfide in solvents. Thiolsulfonates decompose slowly in dilute, or rapidly in strong, alkali to form disulfides and sulfonic acids. Thiolsulfonates also decompose on distillation but they can be steam distilled. The solid members can be recrystallised from water, alcohols or glacial acetic acid.

BIBLIOGRAPHY

[For earlier bibliographies see *Purification of Laboratory Chemicals*, 4th Edn, ISBN 0750628391 (1996, hardback) and 0750637617 (1997, paperback)].

CHARACTERISATION OF ORGANIC AND INORGANIC COMPOUNDS

M. Fieser and L. Fieser, *Reagents for Organic Synthesis*, J.Wiley & Sons, Inc., New York, Vol 1 *1967* to Vol **15**, 1990. ISSN 0271 616X.

F. Fiegl, *Spot Tests in Organic Analysis*, 7th Edn, Elsevier Science, New York, 1989. ISBN 0444402098.

A.J. Gordon and R.A. Ford, *The Chemist's Companion: A Handbook of Practical Data, Techniques, and References*, J. Wiley & Sons, New York, 1973. ISBN 0471315907.

W.L. Jolly, *The Synthesis and Characterisation of Inorganic Compounds*, Waveland Press, 1991. ISBN 0881335789.

R.B. King (Ed.), *Encyclopedia of Inorganic Chemistry*, (8 volumes), J. Wiley & Sons, New York, 1994. ISBN 0471936200.

D.R. Lide, *CRC Handbook of Chemistry and Physics*, 82nd Edn, CRC Press, Boca Raton, 2001. ISBN 0849304822.

D.R. Lide, *CRC Handbook of Chemistry and Physics on CD-ROM 2002 Version*, CRC Press, Boca Raton, 2002. ISBN 0849308798. CD-ROM (Windows PCs only) edition.

* J. Mendham, R.C. Denney, J.D. Barnes and M.J.K. Thomas, *Vogel's Quantitative Chemical Analysis*, 6th Edn, Prentice Hall, Harlow, 2000. ISBN 0582226287.

* L.A. Paquette (Ed.), *The Encyclopedia of Reagents for Organic Synthesis*, 8 Volume set, J. Wiley & Sons, New York, 1995. ISBN 0471936235.

R.M. Silverstein, F.X. Webster, and R. M. Silverstein, *Spectrometric Identification of Organic Compounds*, 6th Edn, J. Wiley & Sons, New York, 1998. ISBN 0471134570.

R.L. Shriner, C.K.F. Hermann, T.C. Morrill, D.Y. Curtin, and R.C. Fuson, *The Systematic Identification of Organic Compounds*, 7th Edn, J. Wiley & Sons, New York, 1997. ISBN 0471 59748.

METAL HYDRIDES

A. Dedieu, (Ed.), *Transition Metal Hydrides*, J. Wiley & Sons, New York, 1992. ISBN 0471187682

M.V.C. Sastri, B. Viswanathan, and S.S. Murthy, *Metal Hydrides*, Narosa Publishing House, New Delhi, 1998. ISBN 8173192294.

J. Seyden-Penne, *Reductions by the Alumino- and Borohydrides in Organic Synthesis*, 2nd Edn, J. Wiley & Sons, New York, 1997. ISBN 1560819391.

SPECTROSCOPY

F.W. McLafferty and D.B. Stauffer, *The Wiley/NBS Registry of Mass Spectral Data*, (7 volumes), J. Wiley & Sons, New York, 1989. ISBN 0471628867.

* C.J. Pouchert and J. Behnke, *The Aldrich Library of ^{13}C and ^{1}H FT NMR Spectra*, 3 Volume set, Aldrich Chemical Co., Milwaukee, 1993. ISBN 0941633349.

* C.J. Pouchert, *The Aldrich Library of FT-IR Spectra*, Aldrich Chemical Co., Milwaukee, 3 Volume set, 2nd Edn, 1997. ISBN 0941633349X.

H. Gunzler and M.H. Heise, *IR Spectroscopy,* Wiley-VCH, New York, 2000. ISBN 3527288961.

TRACE METAL ANALYSIS

I.S. Krull (Ed.), *Trace Metal Analysis and Speciation*, Elsevier Science, New York, 1991. ISBN 044488209X.

A.Varma, *Handbook of Atomic Absorption Analysis*, (2 volumes), CRC Press, Boca Raton, Florida, 1984. ISBN 084932985X, 0849329868.

B. Welz (translated by C. Skegg), *Atomic Absorption Spectrometry*, VCH, Weinheim, 1985. ISBN 0895734184.

CHAPTER 3

THE FUTURE OF PURIFICATION

INTRODUCTION

The essence of research is to seek answers wherever there are questions. Regardless of what the answers are the experiments to be conducted must be carried out with utmost care. For this, one must ensure that the quality of the reactants used and the products obtained are of the highest possible purity. In general terms, one can broadly categorise experimental chemistry and biological chemistry into the following areas:

Isolation and identification of substances (natural products from nature, protein purification and characterisation, etc).

Synthesis of substances (organic, or inorganic in nature; these substances may be known substances or new compounds).

Analysis of substances (this is a key process in the identification of new or known chemical and biological substances. Methods of analysis include spectroscopic methods, derivatisation and sequencing methods).

Measurements of particular properties of a compound or substance (enzyme kinetics, reaction kinetics, FACS, fluorescence-activated cell sorting, assay).

Impressive and sophisticated strategies, in the form of new reagents, catalysts and chemical transformations, are currently available for the syntheses of molecules. In recent years there is a deviation in focus from developing new synthetic routes and reactions to improving methods for carrying out reactions. In particular, traditional reactions can be carried out in new ways such that those efficiencies of reactions are greatly improved. The efficiencies of reactions can be measured in terms of the yields of the desired product(s), or in terms of the time taken to obtain the desired product(s). Some of the 'new' lateral ways of thinking to improve efficiencies of reactions recognise the importance of purification of products in the planning of a synthetic sequence. Thus methods such as solid phase synthesis, fluorous chemistry as well as the use of ionic liquids minimise purification procedures and thus improve the ability to rapidly access pure compounds. These techniques also contribute to the efficiencies of reactions in terms of yields. In looking ahead to synthesis in the 21st century, a brief outline of the key aspects of these techniques are presented. In time many commercially available chemicals will be prepared using methods described in this chapter, and knowledge of these now should be useful to the experimenter. Some of these compounds (e.g. peptides) have already been synthesised by such methods (e.g. SPPS, see below).

SOLID PHASE SYNTHESIS

Solid phase synthesis (SPS) has emerged as an important methodology for the rapid and efficient synthesis of molecules. The ease of work-up and purification procedures in solid phase as compared to solution phase chemistry, as well as the scope for combinatorial chemistry provides impetus for further development in this field. The earliest studies on solid phase chemistry were focused on solid phase peptide synthesis (SPPS). The concept of carrying out reactions on a polymer support as distinct to reactants in solution, was conceived by R.B. Merrifield who received the Nobel Prize in Chemistry in 1984 for his pioneering work. However since the mid 1990's, advances in solid phase chemistry have moved beyond the routine (often robotic) synthesis of small to medium peptides and oligonucleotides. SPOS (solid phase organic synthesis) has gained much prominence due to the wealth of compounds (combinatorial libraries) that can be synthesised rapidly. This is especially important for pharmaceutical companies, screening for compounds with certain biological profiles or for chemical companies, screening for new catalysts or reagents. In SPOS, it is envisaged that difficult reactions can be driven to completion by using a large excess of reagents, which are easily removed by filtration. Furthermore, expensive reagents in the form of catalysts or chiral auxiliaries may be recycled easily if supported on a polymer and hence solid phase reactions provide economy in terms of costs and labour. Another strength of SPS is the ease in purification procedures which generally involves filtration of polymer supported products (solid) from soluble reaction components (liquid) in what is effectively a solid-liquid extraction. In the final step of the synthetic sequence, the desired product is then cleaved from the polymer support.

Despite the relative infancy in the development of solid phase reactions, a wide range of functionalised resins are commercially available. The main uses of these functionalised resins can be roughly classified as follows:

SOLID PHASE PEPTIDE SYNTHESIS (SPPS)

Extensive studies on the synthesis of peptides on solid phase have been carried out, so much so that the technique of SPPS can be reliably and routinely used for the synthesis of short peptides by novices in the field. A large number of resins and reagents have been developed specifically for this purpose, and much is known on problems and avoidance of racemisation, difficult couplings, compatibility of reagents and solvents. Methods for monitoring the success of coupling reactions are available. Automated synthesisers are available commercially (e.g. from Protein Technologies, Rainin Inst Inc, Tuscon AZ; protan@dakotacom.net) which can carry out as many as a dozen polypeptide syntheses simultaneously. The most satisfactory chemistry currently used is Fmoc (9-fluorenylmethoxycarbonyl) chemistry whereby the amino group of the individual amino acid residues is protected as the Fmoc. A large number of Fmoc-amino acids are commercially available as well as polymer resins to which the specific Fmoc-amino acid (which will eventually become the carboxy terminal residue of the peptide) is attached. With automated synthesisers, the solvent used is N-methylpyrrolidone and washings are carried out with dimethylformamide. Deprotection of the polypeptide is carried out with anhydrous trifluoroacetic acid (TFA). A cycle for one residue varies with the residue but can take an hour or more. This means that 70-80 mer polypeptides could take more than a week to prepare. This is not a serious drawback because several different polypeptides can be synthesised simultaneously. The success of the synthesis is dependent on the amino acid sequence since there are some twenty or more different amino acids and the facility of forming a peptide bond varies with the pair of residues invloved. However, generally 70 to 80 mers are routinely prepared, and if the sequence is favourable, up to 120 mer polypeptides can be synthesised. After deprotection with TFA the polypeptide is usually purified by HPLC using a C18 column with reverse phase chromatography. There are many commercial firms that will supply custom made polypeptides at a price depending on the degree of purity required.

SOLID PHASE DEOXYRIBONUCLEOTIDE SYNTHESIS

The need for oligodeoxyribonucleotides mainly as primers for the preparation of deoxyribonucleic acids (DNA) and for DNA sequencing has resulted in considerable developments in oligodeoxyribonucleotide synthesis. The solid phase procedure is the method commonly used. Automated synthesisers are commercially avalable, but with the increase in the number of firms which will provide custom made oligodeoxyribonucleotides, it is often not economical to purchase a synthesiser to make one's own oligodeoxyribonucleotides. Unlike in polypeptide synthesis where there are some twenty different residues to "string" together, in DNA synthesis there are only four deoxyribonucleotides, consequently there is usually little difficulty is synthesising 100 mers in quantities from 10 μg to 10 milligrams of material. The deprotected deoxyribonucleic acid which is separated from the solid support is purified on an anion exchange column followed by reverse phase HPLC using C8 to C18 columns for desalting. As for the polypeptides, the cost of DNA will depend on the purification level required.

SOLID PHASE OLIGOSACCHARIDE SYNTHESIS

Although automated solid phase peptide and oligonucleotide synthetic procedures are well established, automated solid phase oligosaccharide synthesis is considerably more difficult. The current awareness of the importance of polysaccharides as surface recognition molecules and in glycoproteins and glycolipids has prompted much interest in oligosaccharide synthesis and some progress has been made (see Kochetkov *Russ Chem Rev* **69** 795 *2000;* Ito and Manabe *Curr Opin Chem Biol* **2** 701 *1998;* Seeberger and Danishefsky *Acc Chem Res* **31** 685 *1998*). A general method for automated oligosaccharide synthesis is not as yet availble. An example of an automated synthesis of *specific* glycosides has been reported by Seeberger (*Science* **291** 1523 *2001;* see also Houlton *Chem Br* **38** (**4**) 46 *2002*).

SOLID PHASE ORGANIC SYNTHESIS (SPOS)

At the time of writing this book, SPOS is in an area of relative infancy but has considerable potential. One of the main difficulties in SPOS lies in the lack of techniques available to monitor reactions carried out on polymer supports. Unlike reactions in solution phase, reactions on solid support cannot be monitored with relative ease and this has hindered the progress as well as the efficacy of solid supported synthesis of small non-peptidic molecules. Despite these difficulties, a large body of studies is available for SPOS. Recent reviews incorporate information on the types of reactions that can be carried out, as well as outline the difficulties and differences with SP (solid phase) reactions as compared with their solution phase counterparts (see bibliography). An interesting application of such procedures is the synthesis of polymeric esters (e.g. polycaprolactones, polyhydroxybutyrates, polylactates) and starch- and cellulose- like polymers using a plasticised starch support. These have been useful for making biodegradable trays and containers for foodstuffs (BenBrahim *Chem Br* **38(4)** 40 *2002*).

POLYMER SUPPORTED REACTANTS

These have become of increasing importance in synthesis and a broad classification of polymer supported reactants are as follows: Polymer bound bases (e.g. dimethylaminopyridine, morpholine, piperidine); Polymer supported catalysts (e.g. Grubbs catalyst for metathesis reactions, palladium for hydrogenation reactions, tributylmethylammonium chloride for phase transfer reactions); Polymer supported condensation reagents {e.g. DEAD (diethyl azodicarboxylate) for Mitsonobu reactions, DEC [1-(3-dimethylamino-propyl)-3-ethylcarbodiimide hydrochloride] {or EDCI [1-ethyl-3-(3-di-methylaminopropyl) carbodiimide HCl]} for peptide synthesis, HOBt (1-hydroxybenzotriazole) for peptide synthesis; Polymer supported oxidizing agents (e.g. osmium tetroxide, perruthenate, pyridinium chlorochromate); Polymer supported reducing agents (e.g. borohydride, tributyltin fluoride); Polymer supported phosphines (for miscellaneous applications depending on the structure) and so on. Commercially available polymer supported reactants are identified in Chapters 4 and 5 of this book.

SCAVENGER RESINS

Though not as extensively utilised as polymer supported reactants, the use of resins to clean up reactions is gaining favour. The type of commercially available scavenger resins are electrophilic scavenger resins (e.g. benzaldehyde derivatised resins to scavenge amines; isocyanate resins to scavenge amines, anilines and hydrazines; tosyl chloride resins to scavenge nucleophiles) and nucleophilic scavenger resins (e.g. diethylenetriamine resins to scavenge acids, acid chlorides, anhydrides; sulfonyl amide resins to scavenge acids, acid chlorides, aldehydes, isocyanates and chloroformates).

RESIN SUPPORT

The common resin matrixes comprise of polystyrene crosslinked with divinylbenzene, graft polymers of polystyrene-polyethylene glycol (PS-PEG) and polyethyleneglycol acrylamide (PEGA) composite resins. For each type of resin matrixes, a range of functionalised polymer supports are available. In addition, a number of these resins are available with different percentage of crosslinking as well as a range of loadings of the reactive functionality. Polystyrene based resins are the most extensively used. Unfortunately these resins do not swell, i.e. do not imbibe water, in polar solvents such as water and methanol and thus cannot be used for carrying out reactions in these solvents. In contrast grafted PS-PEG resins swell in a range of solvents from toluene to water. Examples of grafted PS-PEG resins are NovaSyn® TG and NovaGel® resins. As the success of transformations to be carried out on SPOS depends in part on the swelling properties as well as the robustness of the resin, the choice of resin matrix to be used must be carefully considered. The swelling properties of a number of resin types in a variety of solvents have been documented (see NovaBiochem catalog and also Santini, Griffith and Qi *Tetrahedron Lett* **39** 8951 *1998*). For example, the swelling of a polystyrene resin in DMF is 3 mL/g of resin as compared to that in dichloromethane which is 7 mL/g of resin. It is thought that swelling of resins in the order of greater than 4 mL/g constitutes a good solvent, between 2-4 mL/g a moderate solvent and that less than 2 mL/g a poor solvent choice for carrying out solid phase reactions.

Lightly crosslinked resins are less robust but have greater ability to swell in appropriate solvents. Typically a 1-2% crosslinked divinyl benzene polystyrene resin is employed in organic synthesis.

An extensive list of the commercially available resins is available from Sigma-Aldrich (www.sigmaaldrich.com), Novabiochem (www.novabiochem.com), Fluka and other chemical companies. Sigma-Aldrich and Novabiochem have excellent catalogs. In addition, the Novabiochem catalog and website are a rich source of useful technical information.

CHOICE OF RESIN FOR SPOS

There is a large range of resins available for SPOS. These resins are derivatised polymer supports with a range of linkers. The roles of linkers are (i) to provide point(s) of attachment for the tethered molecule, akin to a solid supported protecting group(s), (ii) to provide distance from the polymeric backbone in order to minimise interactions with the backbone, (iii) to enable cleavage of product molecules under conditions compatible with the stability of the molecules and the reaction conditions employed for chemical transformations. Hence in order to choose an appropriate resin for use in SPOS, one would need to consider the nature of the attachment of the reactant molecule onto the solid support (e.g. in order to tether the carboxy group in a reactant as an ester linkage on a solid support, one may choose to use a hydroxy functionalised linker), the stability of the resin under conditions employed in the chemical transformations (e.g. issues of orthogonality - will the conditions utilised cause premature cleavage of the linker or premature cleavage of the products?), the solvents and reactants needed in the transformations (e.g. will the solvents swell the resin?), conditions of cleavage of products (e.g. will this cause racemisation or rearrangement of the product?), the functionality of the resultant product after cleavage (e.g. will cleavage of the product result in a residual functionality in the molecule?) and so on. Linkers which leave no residual functionalities in the products upon cleavage are known as *traceless* linkers and those which need to be activated in order to be cleaved are known as *safety catch* linkers. A fascinating array of linkers (commercial or otherwise) is available and some excellent reviews are cited in the bibliography at the end of this chapter.

COMBINATORIAL CHEMISTRY
The major impetus for the development of solid phase synthesis centers around applications in combinatorial chemistry. The notion that new drug leads and catalysts can be discovered in a high throughput fashion has been demonstrated many times over as is evidenced from the number of publications that have arisen (see references at the end of this chapter). A number of approaches to combinatorial chemistry exist. These include the split-mix method, serial techniques and parallel methods to generate libraries of compounds. The advances in combinatorial chemistry are also accompanied by sophisticated methods in deconvolution and identification of compounds from libraries. In a number of cases, innovative hardware and software has been developed for these purposes.
Depending on the size of the combinatorial library to be generated as well as the scale of the reactions to be carried out, a wide range of specialised glassware and equipment are commercially available. For example, in order to carry out parallel combinatorial synthesis, reaction stations equipped with temperature and stirring control are available from a number of sources (e.g. www.fisher.co.uk; www.radleys.com; www.sigmaaldrich.com). These reaction stations are readily adapted, using appropriate modules, for conditions under reflux or under inert atmosphere. For automated synthesis of large libraries of compounds, reactions can be carried out using reaction blocks on microtiter plates.
Ready to use CombiKits™ which contain a variety of pre-weighed building blocks are available from Aldrich Chemical Company.

MONITORING SOLID PHASE REACTIONS
This remains the bane of solid phase reactions. Unlike solution phase reactions, where the progress of reactions can be monitored rapidly *via* TLC, GC or HPLC methods, procedures for the rapid monitoring of progress in solid phase reactions are limited. Although a number of spectroscopic methods have been developed for direct monitoring of reactions on solid supports, these methods usually require specialised equipment, not routinely available in chemical laboratories. These methods include on-bead IR analysis (e.g. Huber et al. *Anal Chim Acta* **393** 213 *1999*; Yan and Gremlich *J Chromatogr, B.* **725** 91 *1999*; Yan et al. *J Org Chem* **60** 5736 *1995*) and solid state magic angle spinning NMR techniques (e.g. Warrass and Lippens *J Org Chem* **65** 2946 *2000*; Rousselot-Pailley, Ede and Lippens *J. Comb. Chem.* **3** 559 *2001*).
The most common methods for monitoring solid phase reactions utilized in normal research laboratories are:

Infrared analysis of resin
This is a destructive method in which the resin is ground and pelleted as a KBr disc and analysed by FT-IR analysis. This method works best for systems where distinct functional group transformations (C=O, C-OH, C=C, etc) are expected. No special equipment is needed.

Qualitative and quantitative analyses
There are a number of colour or UV tests which are available for monitoring the presence or absence of certain functional groups. Although some of these tests are routinely used for the quantitative analysis of functional groups in solution phase, the quantification on solid phase is less than reliable. An exception to this is the Fmoc (9-fluorenylmethoxycarbonyl) assay, which is routinely used for quantification of coupling in SPPS using Fmoc amino acids. It should also be noted that the generality of some of these colour tests on a variety of solid phase resins is not known and hence these tests serve only as a *guide* to functional group identification. Some (not an exhaustive list) of the reported methods of analyses are outlined below.

DETECTION OF REACTIVE GROUPS ON RESINS

Detection of hydroxy groups on resin

A method in which the resin is treated with cyanuric chloride (trichlorotriazine, TCT) in DMF followed by a nucleophilic dye (AliR or Mordant Orange 1, beads appear red in the presence of hydroxyl groups, or with fuschin, beads appear fuschia, or with fluorescein, they become fluorescent) has been reported (Attardi and Taddei *Tetrahedron Lett* **42** 2927 *2001*; Attardi, Falchi and Taddei *Tetrahedron Lett* **41** 7395 *2001*). Another colorimetric test for the detection of polymer supported tertiary alcohols utilizes the conversion of the alcohols to the polymer supported diphenylsilylchloride ether, followed by subsequent treatment with methyl red. The beads form a readily distinguishable orange/red colour. The test is also positive for the hydroxy Wang resin and the aminomethylpolystyrene resin [Burkett, Brown and Meloni *Tetrahedron Lett* **42** 5773 *2001*]. Alternatively the conversion of polymer supported alcohols to the tosylate followed by displacement by *p*-nitrobenzylpyridine (PNBP) gives a strongly coloured salt upon treatment with bases such as piperidine, followed by gentle heating [Kuisle et al. *Tetrahedron* **55** 14807 *1999*].

Detection of aldehyde groups on resin

The use of an acidic solution of *p*-anisaldehyde in ethanol to detect aldehyde functionalities on polystyrene polymer supports has been reported (beads are treated with a freshly made solution of *p*-anisaldehyde (2.55 mL), ethanol (88 mL), sulfuric acid (9 mL), acetic acid (1 mL) and heated at 110°C for 4 min). The colour of the beads depends on the percentage of CHO content such that at 0% of CHO groups, the beads are colourless, ~50% CHO content, the beads appear red and at 98% CHO the beads appear burgundy [Vázquez and Albericio *Tetrahedron Lett* **42** 6691 *2001*]. A different approach utilises 4-amino-3-hydrazino-5-mercapto-1,2,4-triazole (Purpald) as the visualizing agent for CHO groups. Resins containing aldehyde functionalities turn dark brown to purple after a 5 min reaction followed by a 10 minute air oxidation [Cournoyer et al. *J Comb Chem* **4** 120 *2002*].

Detection of carboxy groups on resin

The presence of a COOH functionality on a polystyrene resin can be detected using a 0.25% solution of malachite green-oxalate in ethanol in the presence of a drop of triethylamine. Beads with COOH functionalities are coloured dark green or appear as clear gel beads [Attardi, Porcu and Taddei *Tetrahedron Lett* **41** 7391 *2000*].

Detection of amino groups on resin

The methods for the detection of amine functional groups are well established. For example the Kaiser test can be used to detect the presence of amine groups on resins (blue colour is observed). In the Kaiser test, two reagents are prepared. Reagent 1 comprises of a mixture of two solutions: A and B. A is a solution of phenol in absolute ethanol (40g of phenol in 10 mL of absolute ethanol, followed by treatment of this clear solution with 4g of Amberlite mixed bed resin MB-3 for 45 mins. The solution is then filtered.). Solution B is made up of 65 mg of KCN in 100 mL water; 2 mL of this solution is diluted to 100 mL of freshly distilled pyridine. The solution is then stirred with 4 g of Amberlite mixed-bed resin MB-3 and filtered. Solutions A and B are then mixed. Reagent 2 is a solution of ninhydrin (2.5g) in absolute ethanol (50 mL). For a qualitative Kaiser test, 6 drops of reagent 1 and 2 drops of reagent 2 are added to the well washed dried resin (2-5 mg) and mixed, followed by heating to 100°C for 4-6 min. A method for the quantitative determination of amino groups using this test has also been reported [Sarin et al. *Anal Biochem* **117** 147 *1981*]. It is however known that the Kaiser test does not give a positive test with a secondary amino acid such as proline or some 'unnatural' amino acids. In addition some deprotected amino acids (Ser, Asn, Asp) do not show the expected intense blue colour typical of free primary amino groups.

A test for secondary amines (e.g. proline) is the Chloranil test (1 drop of a 2% acetaldehyde solution in DMF, followed by one drop of a 2% solution of *p*-chloranil in DMF, leave for 5 mins). A positive test gives blue stained beads.

Other tests for the detection of amino functionalities on solid supports include the TNBS (2,4,6-trinitrobenzenesulfonic acid, picrylsulfonic acid) [Hancock and Battersby *Anal Biochem* **71** 260 *1976*], the DABITC [Shah et al. *Anal. Commun.* **34** 325 *1997*] and the NF31 [Madder et al. *Eur J Org Chem* 2787 *1999*] tests.

Detection of thiol groups on resin

For quantitative analysis of solid supported thiol residues on free macroporous or PEG grafts, Ellman's reagent has been used [5,5'-dithio-*bis*-(2-nitrobenzoic acid]. However only qualitative information can be gained using lightly crosslinked polystyrene resins [Badyal et al. *Tetrahedron Lett* **42** 8531 *2001*].

Fmoc assay

This is a very important and well tested method for the quantitative determination of loading of Fmoc protected compounds particularly that of Fmoc (fluorenylmethoxycarbonyl) amino acids on solid support. Fmoc groups can be readily deprotected in the presence of base. Generally, in the deprotection and quantification procedures, a known amount of resin is treated with 20% piperidine in DMF at room temperature for 30 mins. The resin is washed with more DMF and the pooled filtrates are combined in a volumetric flask and made up to an accurate volume (e.g. to 10 mL) with more DMF. The UV absorbance at 301 nm of the piperidine-dibenzofulvene adduct which is formed can then be measured against a blank solution of piperidine in DMF [Meienhofer et al. *Int J Pept Protein Res* **13** 35 *1979*]. The loading **L** is then determined using the equation :

$$\mathbf{L} = \frac{\text{(Absorbance value) x (Solution volume in litres)}}{7.8 \text{ x (Weight of resin in mg)}}$$

IONIC LIQUIDS

Ionic liquids are organic or inorganic salts that are liquid at room or reaction temperatures. Although ionic liquids are themselves not new discoveries (e.g. the ionic liquid [$EtNH_3$] [NO_3] was described in 1914), the use of ionic liquids in synthesis is only recent. In particular, the potential applications of ionic liquids as solvents in synthesis and in catalysis have recently been realised. The physical properties of ionic liquids make them unique solvents for synthesis. For example, ionic liquids are good solvents for both organic and inorganic substances and hence can be used to bring reagents into the same phase for reaction. Ionic liquids are also immiscible with a number of organic solvents and thus provide a non-aqueous, polar alternative for two-phase extraction systems. As ionic liquids are non-volatile, they can be used in high vacuum systems without the possibility of loss or contaminants. In addition, this also facilitates the isolation of products as the products can be distilled from the ionic liquid or alternatively extracted with an organic solvent that is immiscible with the ionic liquid. Although ionic liquids are frequently composed of poorly coordinating ions, they are highly polar which are important characteristics in the activation of catalysts.

Commonly used ionic liquids are *N*-alkylpyridinium, *N,N'*-dialkylimidazolium, alkylammonium and alkylphosphonium salts.

To date a number of reactions have been carried out in ionic liquids [for examples, see Dell'Anna et al. *J Chem Soc, Chem Commun* 434 *2002*; Nara, Harjani and Salunkhe *Tetrahedron Lett* **43** 1127 *2002*; Semeril et al. *J Chem Soc Chem Commun* 146 *2002*; Buijsman, van Vuuren and Sterrenburg *Org Lett* **3** 3785 *2001*]. These include Diels-Alder reactions, transition-metal mediated catalysis, e.g. Heck and Suzuki coupling reactions, and olefin metathesis reactions. An example of ionic liquid acceleration of reactions carried out on solid phase is given by Revell and Ganesan [*Org Lett* **4** 3071 *2002*].

FLUOROUS CHEMISTRY

This new approach to synthesis was introduced by Curran early in 1997 and involves the attachment of fluorous phase labels to substrates such that the subsequent fluorinated products can be extracted into the fluorous phase. For example in liquid-liquid extractions (typical work-up procedures), a three-phase extraction is now possible (organic, fluorous and aqueous phases). As organic and inorganic compounds have little or no tendency to dissolve in highly fluorinated solvents and compounds, phase labeling a compound as fluorous will enable successful extraction into the fluorous phase. However in order to carry out homogenous reactions with these fluorinated compounds, organic solvents with a good dissolving power for fluorous compounds or miscible organic and fluorous solvents can be used. Alternatively organic solvents with a few fluorine atoms e.g. trifluoroethanol, benzotrifluoride ('hybrid solvents') will dissolve both organic and fluorous compounds. A number of synthetic applications utilising fluorous chemistry have been reported in the literature. [For examples, see Schneider and Bannwarth *Helv Chim Acta* **84** 735 *2001*; Galante, Lhoste and Sinou *Tetrahedron Lett* **42** 5425 *2001*; Studer and Curran *Tetrahedron* **53** 6681 *1997*; Studer et al. *J Org Chem* **62** 2917 *1997*; Crich and Neelamkavil *Tetrahedron* **58** 3865 *2002*].

BIBLIOGRAPHY

A SAMPLE OF REVIEWS ON SOLID PHASE SYNTHESIS

General

C. Blackburn, Polymer supports for solid-phase organic synthesis, *Biopolymers* **47** 311-351 *1998.*

R.J. Booth and J.C. Hodges, Solid-supported reagent strategies for rapid purification of combinatorial synthesis products, *Acc Chem Res* **32** 18-26 *1999.*

J.A. Ellman, Design, Synthesis, and Evaluation of Small-Molecule Libraries, *Acc Chem. Res* **29** 132-143 *1996.*

J.S. Fruchtel and G. Jung, Organic Chemistry On Solid Supports, *Angew Chem, Int Ed Engl* **35** 17-42 *1996.*

K. Gordon and S. Balasubramanian, Solid phase synthesis - designer linkers for combinatorial chemistry: A review *J Chem Technol Biotechnol* **74** 835-851 *1999.*

F. Guillier, D. Orain and M. Bradley, Linkers and Cleavage Strategies in Solid-phase Organic Synthesis and Combinatorial Chemistry, *Chem Rev* **100** 2091-2157 *2000.*

P. Hodge, Polymer-supported Organic Reactions: What Takes Place In The Beads? *Chem Soc Rev* **26** 417-424 *1997.*

I.W. James, Linkers for solid phase organic synthesis, *Tetrahedron* **55** 4855-4946 *1999.*

J.W. Labadie, Polymeric Supports for Solid Phase Synthesis, *Curr Opin Chem Biol* **2** 346-352 *1998.*

S.V. Ley, I.R. Baxendale, G. Brusotti, M. Caldarelli, A. Massi and M. Nesi, Solid-supported Reagents for Multi-step Organic Synthesis: Preparation and Application, *Il Farmaco* **57** 321-330 *2002.*

P. Seneci, *Solid Phase and Combinatorial Technologies*, J. Wiley & Sons, New York, 2000, ISBN 0471331953.

A.R. Vaino and K.D. Janda, Solid-Phase Organic Synthesis: A Critical Understanding of the Resin, *J Comb Chem* **2** 579-596 *2000.*

Applications to SPOS, SPPS

P. Blaney, R. Grigg and V. Sridharan, Traceless Solid-Phase Organic Synthesis, *Chem Rev* **102** 2607-2624 *2002.*

B. Clapham, T.S. Reger and K.D. Janda, Polymer-supported Catalysis in Synthetic Organic Chemistry, *Tetrahedron* **57** 4637-4662 *2001.*

D.H. Drewry, D.M. Coe and S. Poon, Solid Supported Reagents in Organic Synthesis, *Med Res Rev* **19** 97-148 *1999 .*

S.J. Shuttleworth, S.M. Allin and P.K. Sharma, Functionalised Polymers - Recent Developments and New Applications in Synthetic Organic Chemistry, *Synthesis* 1217-1239 *1997.*

S.J. Shuttleworth, S.M. Allin, R.D. Wilson and D. Nasturica, Functionalised polymers in organic chemistry; Part 2, *Synthesis* **8** 1035-1074 *2000.*

Monitoring SP reactions

Y. Bing, Monitoring the Progress and the Yield of Solid Phase Organic Reactions Directly On Resin Supports, *Acc Chem Res* **31** 621-630 *1998.*

M. Dal Cin, S. Davalli, C. Marchioro, M. Passarini, O. Perini, S. Provera and A. Zaramella, Analytical Methods for Monitoring of Solid Phase Organic Synthesis, *Il Farmaco* **57** 497-510 *2002*.

S.W. Gerritz, Quantitative Techniques for the Comparison of Solid Supports, *Curr Opin Chem Biol*, **5** 264-268 *2001*.

M. Irving, J. Cournoyer, R.S. Li, C. Santos and B. Yan, Qualitative and quantitative analyses of resin-bound organic compounds, *Comb Chem High Throughput Screening* **4** 353-362 *2001*.

COMBINATORIAL CHEMISTRY

F. Balkenhohl, C. Vondembusschehunnefeld, A. Lansky and C. Zechel, Combinatorial Synthesis of Small Organic Molecules, *Angew Chem, Int Ed Engl* **35** 2289-2337 *1996*.

S. Bhattacharyya, Polymer-supported reagents and catalysts: Recent advances in synthetic applications, *Comb Chem High Throughput Screening* **3** 65-92 *2000*.

A. Ganesan, Recent developments in combinatorial organic synthesis, *Drug Discovery Today* **7** 47-55 *2002*.

G. Lowe, Combinatorial Chemistry, *Chem Soc Rev* **24** 309-317 *1995*.

H. Wennemers, Combinatorial chemistry: A tool for the discovery of new catalysts, *Comb Chem High Throughput Screening* **4** 273-285 *2001*.

IONIC LIQUIDS

T. Welton, Room temperature ionic liquids. Solvents for synthesis and catalysis, *Chem Rev* **99** 2071-2083 *1999*.

C.M. Gordon, New developments in catalysis using ionic liquids, *Appl. Catal A: General* **222** 101-117 *2001*.

R. Sheldon, Catalytic reactions in ionic liquids, *J Chem Soc, Chem Commun* 2399-2407 *2001*.

FLUOROUS CHEMISTRY

D.P. Curran, Strategy-level separations in organic synthesis: from planning to practice, *Angew Chem, Int Ed Engl* **37** 1174-1196 *1998*.

D.P. Curran, Combinatorial Organic Synthesis and Phase Separation: Back to the Future, *Chemtracts: Org Chem* **9** 75-87 *1996*.

D.L. Flynn, Phase-trafficking reagents and phase-switching strategies for parallel synthesis, *Med Res Rev* **19** 408-431 *1999*.

A. Studer, S. Hadida, R. Ferritto, S.Y. Kim, P. Jeyer, P. Wipf, D.P. Curran, Fluorous synthesis — A fluorous-phase strategy for improving separation efficiency in organic synthesis, *Science* **275** 823-826 *1997*.

Tetrahedron Symposium **58** Issue 20, pp 3823-4131 *2002*.

CHAPTER 4

PURIFICATION OF ORGANIC CHEMICALS

The general principles, techniques and methods of purification in Chapters 1 and 2 are applicable to this chapter. Most organic liquids and a number of solids can readily be purified by fractional distillation, usually at atmospheric pressure. Sometimes, particularly with high boiling or sensitive liquids, or when in doubt about stability, distillation or fractionation under reduced pressure should be carried out. To save space, the present chapter omits many substances for which the published purification methods involve simple distillation. Where boiling points are given, purification by distillation is another means of removing impurities. Literature references are omitted for methods which require simple recrystallisation from solution if the correct solvent can be guessed readily, and where no further information is given, e.g. spectra. Substances are listed alphabetically, usually with some criteria of purity, giving brief details of how they can be purified. Also noted are the molecular weights (to the first decimal place), melting points and/or boiling points together with the respective densities and refractive indexes for liquids, and optical rotations when the compounds are chiral. When the temperatures and/or the wavelengths are not given for the last three named properties then they should be assumed to be 20ºC and the average of the wavelengths of the sodium D lines repectively; and densities are relative to water at 4º.

The present chapter includes commercially available organic chemicals. Most of the inorganic, metalorganic, organo- bismuth, boron, phosphorus, selenium, silicon and alkali metal compounds and metal ion salts of organic acids are in Chapter 5. Naturally occurring commercially available organic compounds for use in biochemistry, molecular biology and biology are in Chapter 6. Commercially available polymer supported reagents are indicated with § under the appropriate reagent.

Rapid purification procedures are noted for commonly used solvents and reagents which make them suitable for general use in synthetic chamistry.

Abbreviations of titles of periodicals are defined as in the Chemical Abstracts Service Source Index (CASSI). Other abbreviations are self evident (see Chapter 1, p. 30).

Ionisation constants of ionisable compounds are give as **pK** values (published from the literature) and refer to the **pKa** values at room temperature (~ 15ºC to 25ºC). The values at other temperatures are given as superscripts, e.g. pK^{25} for 25ºC. Estimated values are entered as pK_{Est}. (see Chapter 1, p. 7 for further information).

As a good general rule, all low boiling (<100º) organic liquids should be treated as highly flammable and toxic (because they can be inhaled in large quantities) and the necessary precautions should be taken.

Benzene, which has been used as a solvent successfully and extensively in the past for reactions and purification by chromatography and crystallisation is now considered a **very dangerous substance** so it hasto be used with extreme care. We emphasise that an alternative solvent system to benzene (e.g. toluene, toluene-petroleum ether, or a petroleum ether to name a few) should be used first. However, if no other solvent system can be found then all operations involving benzene have to be performed in an efficient fumehood and precautions must be taken to avoid inhalation and contact with skin and eyes. Whenever benzene is mentioned in the text an asterisk e.g. $^*C_6H_6$ or *benzene, is inserted to remind the user that special precaution should be adopted.

Abietic acid *[514-10-3]* M 302.5, m 172-175°, $[\alpha]_D^{25}$ -116° (-106°)(c 1, EtOH), pK
5.27. Crystd by dissolving 100g of acid in 95% EtOH (700mL), adding to H_2O (600mL) and cooling. Filter, dry in a vacuum (over KOH or $CaSO_4$) store in an O_2-free atmosphere. λ in EtOH nm(log ε): 2343(4.3), 241(4.4), 2505(4.2), 235(4.34) and 240(4.36). [*Org Synth* **23** 1 *1952* ; *J Am Chem Soc* **35** 3736 *1949*; *Monatsh Chem* **116** 1345 *1985*.]

S-Abscisic acid *[21293-29-8]* M 264.3, m 160-161°, 161-163° (sublimation), $[\alpha]_{287}$ + 24,000°, $[\alpha]_{245}$ -69,000° (c 1-50µg/mL in acidified MeOH or EtOH), pK_{Est} ~3.9. Crystd from CCl_4-pet.ether, EtOH + hexane and sublimes at 120°.

Acenaphthalene *[208-96-8]* M 152.2, m 92-93°. Dissolved in warm redistd MeOH, filtered through a sintered glass funnel and cooled to -78° to ppte the material as yellow plates [Dainton, Ivin and Walmsley *Trans Faraday Soc* **56** 1784 *1960*]. Alternatively can be sublimed *in vacuo*.

Acenaphthaquinone *[82-86-0]* M 182.2, m 260-261°. Extracted with, then recrystd twice from *C_6H_6. [LeFevre, Sundaram and Sundaram *J Chem Soc* 974 *1963*].

Acenaphthene *[83-32-9]* M 154.2, m 94.0°. Crystd from EtOH. Purified by chromatography from CCl_4 on alumina with *benzene as eluent [McLaughlin and Zainal *J Chem Soc* 2485 *1960*].

RS-Acenaphthenol *[6306-07-6]* M 170.2, m 144.5-145.5°, 146°, 148°. If highly coloured (yellow), dissolve in boiling *benzene (14g in 200mL), add charcoal (0.5g), filter through a heated funnel, concentrate to 100mL and cool to give almost colourless needles. *Benzene vapour is* **TOXIC** *use an efficient fume cupboard*. The *acetate* has **b** 166-168°/5mm (bath temp 180-185°). [*Org Synth* Col.Vol. III 3 *1955*.] It can also be recrystd from *C_6H_6 or EtOH [Fieser and Cason *J Am Chem Soc* **62** 432 *1940*]. It forms a brick-red crystalline ***complex*** with 2,4,5,7-tetranitrofluoren-9-one which is recrystd from AcOH and dried in a vacuum over KOH and P_2O_5 at room temp, **m** 170-172° [Newman and Lutz *J Am Chem Soc* **78** 2469 *1956*].

Acetal (acetaldehyde diethylacetal) *[105-57-7]* M 118.2, b 103.7-104°, d 0.831, n 1.38054, n^{25} 1.3682. Dried over Na to remove alcohols and water, and to polymerise aldehydes, then fractionally distd. Or, treat with alkaline H_2O_2 soln at 40-45° to remove aldehydes, then the soln is saturated with NaCl, separated, dried with K_2CO_3 and distd from Na [Vogel *J Chem Soc* 616 *1948*].

Acetaldehyde *[75-07-0]* M 44.1, b 20.2°, d 0.788, n 1.33113, pK^{25} 13.57 (hydrate). Usually purified by fractional distn in a glass helices-packed column under dry N_2, discarding the first portion of distillate. Or, shaken for 30min with $NaHCO_3$, dried with $CaSO_4$ and fractionally distd at 760mm through a 70cm Vigreux column. The middle fraction was taken and further purified by standing for 2h at 0° with a small amount of hydroquinone, followed by distn [Longfield and Walters *J Am Chem Soc* **77** 810 *1955*].

Acetaldehyde ammonia trimer (hexahydro-2,4,6-trimethyl-1,3,5-triazine trihydrate) *[76231-37-3]* M 183.3, m 94-96°, 95-97°, 97°, b 110°(partly dec). Crystd from EtOH-Et_2O. When prepared it separates as the *trihydrate* which can be dried in a vacuum over $CaCl_2$ at room temp to give the anhydrous compound with the same melting point. The *dihydrate* melts at 25-28° then resolidifies and melts again at 94-95°. *IRRITATES THE EYES AND MUCOUS MEMBRANES*. [*J Org Chem* **38** 3288 *1973*.]

Acetaldehyde dimethyl acetal *[534-15-6]* M 90.1, b 63-65°, d_4^{20} 0.852, n_D^{25} 1.36678. Distd through a fractionating column and fraction boiling at 63.8°/751mm is collected. It forms an azeotrope with MeOH. It has been purified by GLC.

Acetamide *[60-35-5]* M 59.1, m 81°, pK_1^{25} -1.4, pK_2^{25} +0.37. Crystd by soln in hot MeOH (0.8mL/g), diltd with Et_2O and allowed to stand [Wagner *J Chem Educ* **7** 1135 *1930*]. Alternate crystns are from acetone, *benzene, chloroform, dioxane, methyl acetate or from *benzene-ethyl acetate mixture (3:1 and 1:1). It has also been recrystd from hot water after treating with HCl-washed activated charcoal (which had been repeatedly washed with water until free from chloride ions), then crystd again from hot 50% aq. EtOH and finally twice from hot 95% EtOH [Christoffers and Kegeles *J Am Chem Soc* **85** 2562 *1963*]. Final drying is in a vacuum desiccator over P_2O_5. Acetamide is also purified by distn (**b** 221-223°) or by sublimation *in vacuo*. Also purified by recrystn twice from cyclohexane containing 5% (v/v) of *benzene. Needle-like crystals separated by filtn, washed with a small volume of distd H_2O and dried with a flow of dry N_2. [Slebocka-Tilk et al. *J Am Chem Soc* **109** 4620 *1987*.]

Acetamidine hydrochloride *[124-42-5]* **M 94.5, m 164-166°, 165-170° (dec), 174°, pK^{25} 12.40.** It can be recrystd from small volumes of EtOH. Alternatively dissolve in EtOH, filter, add Et_2O, filter the crystalline salt off under N_2, dry in a vacuum desiccator over H_2SO_4. The salt is deliquescent and should be stored in a tightly stoppered container. Solubility in H_2O is 10% at room temperature, soluble in Me_2CO. The *free base* reacts strongly alkaline in H_2O. It has λ_{max} 224nm (ϵ 4000) in H_2O. The *picrate* has **m** 252° (sintering at ~245°). [Dox *Org Synth* Coll Vol I 5 *1941*; Davies and Parsons *Chem Ind (London)* 628 *1958*; Barnes et al. *J Am Chem Soc* **62** 1286 *1940* give **m** 177-178°.]

***N*-(2-Acetamido)-2-aminoethanesulfonic acid (ACES)** *[7365-82-4]* **M 182.2, m > 220°(dec), pK$_{Est}$ ~1.5, pK$_2$ 6.9.** Recrystd from hot aqueous EtOH.

4-Acetamidobenzaldehyde *[122-85-0]* **M 163.2, m 156°.** Recrystd from water.

***p*-Acetamidobenzenesulfonyl chloride (*N*-acetylsulfanilyl chloride)** *[121-60-8]* **M 233.7, m 149°(dec).** Crystd from toluene, $CHCl_3$, or ethylene dichloride.

α-Acetamidocinnamic acid *[5469-45-4]* **M 205.2, m 185-186° (2H$_2$O), 190-191°(anhydr), 193-195°, pK$_{Est}$ ~3.2.** Recrystd from H_2O as the dihydrate and on drying at 100° it forms the anhydrous compound which is *hygroscopic*. Alkaline hydrolysis yields NH_3 and phenylpyruvic acid. [Erlenmeyer and Früstück *Justus Liebigs Ann Chem* **284** 47 *1895*.]

Z-*O*-(2-Acetamido-2-deoxy-D-glycopyranosylideneamino)*N*-phenylcarbamate (PUGNAC) *[132063-05-9]* **M 335.3, m 171-174°(dec), 174-180°(dec), [α]$_D^{20}$+67.5° (c 0.2, MeOH).** Purified by flash chromatography (silica gel and eluted with AcOEt-hexane 3:2) evaporated, and the foam recrystallised from AcOEt-MeOH. TLC on Merck SiO_2 gel 60 F_{254} and detected by spraying with 0.025M I_2 in 10% aqueous H_2SO_4 and heat at 200° gave R_F 0.21. The acetate is hydrolysed with NH_3-MeOH. [*Helv Chim Acta* **68** 2254 *1985*; **73** 1918 *1990*.]

2-Acetamidofluorene *[53-96-3]* **M 223.3, m 194°, 196-198°.** Recrystd from toluene (1.3mg in 100mL). Solubility in H_2O is 1.3mg/L; UV λ_{max} nm(log ϵ) : 288(4.43), 313(4.13). [*J Org Chem* **21** 271 *1956*.] It can also be recrystd from 50% AcOH and sol in H_2O is 1.3mg/100mL at 25° [*Chem Ber* **35** 3285 *1902*]. 9-^{14}C and ω-[14]C 2-acetamidofluorene were recrystd from aqueous EtOH and had **m** 194-195° and 194° respectively. *Potent* **CARCINOGEN.** [*Cancer Res* **10** 616 *1950*; Sadin et al. *J Am Chem Soc* **74** 5073 *1952*.]

***N*-(2-Acetamido)iminodiacetic acid (ADA)** *[26239-55-4]* **M 190.2, m 219° (dec), pK$_1$ ~2.3, pK$_2$ 6.6.** Dissolved in water by adding one equivalent of NaOH soln (to final pH of 8-9), then acidified with HCl to ppte the free acid. Filtered and washed with water.

Acetamidomethanol *[625-51-4]* **M 89.1, m 47-50°, 54-56°, 55°.** Recryst from freshly distd Me_2CO, wash the crystals with dry Et_2O and dry in a vacuum desiccator over P_2O_5. R_F 0.4 on paper chromatography with $CHCl_3$/EtOH (2:8) as solvent and developed with ammoniacal $AgNO_3$. Also crystallises in needles from EtOAc containing a few drops of Me_2CO. It is *hygroscopic* and should be stored under dry conditions. [*J Am Chem Soc* **73** 2775 *1951*; *Chem Ber* **99** 3204 *1966*; *Justus Liebigs Ann Chem* **343** 265 *1905*.]

2-Acetamido-5-nitrothiazole *[140-40-9]* **M 187.2, m 264-265°.** Recrystd from EtOH or glacial acetic acid.

2-Acetamidophenol *[614-80-2]* **M 151.2, m. 209°, pK$_{Est}$ ~9.4.** Recrystd from water or aqueous EtOH.

3-Acetamidophenol *[621-42-1]* **M 151.2, m 148-149°, pK^{25} ~9.59.** Recrystd from water.

4-Acetamidophenol *[103-90-2]* **M 151.2, m 169-170.5°, pK$_{Est}$ ~10.0.** Recrystd from water or EtOH.

4-Acetamido-2,2,6,6-tetramethylpiperidine-1-oxyl **(acetamidoTEMPO)** *[14691-89-5]* **M 213.3, m 144-146°, 146-147°.** Dissolve in CH_2Cl_2, wash with saturated K_2CO_3, then saturated aqueous NaCl, dry (Na_2SO_4), filter and evaporate. The red solid is recrystd from aqueous MeOH, **m 147.5°.** [*J Org Chem* **56** 6110 *1991*; *Bull Acad Sci USSR, Div Chem Sci* **15** 1422 *1966*.]

5-Acetamido-1,3,4-thiadiazole-2-sulfonamide *[59-66-5]* **M 222.3, m 256-259° (dec).** Recrystd from water.

Acetanilide *[103-84-4]* **M 135.2, m 114°, pK^{25} 0.5.** Recrystd from water, aqueous EtOH, *benzene or toluene.

Acetic acid (glacial) *[64-19-7]* **M 60.1, m 16.6°, b 118°, d 1.049, n 1.37171, n^{25} 1.36995, pK^{25} 4.76.** Usual impurities are traces of acetaldehyde and other oxidisable substances and water. (Glacial acetic acid is very *hygroscopic*. The presence of 0.1% water lowers its **m** by 0.2°.) Purified by adding some acetic anhydride to react with water present, heating for 1h to just below boiling in the presence of 2g CrO_3 per 100mL and then fractionally distilling [Orton and Bradfield *J Chem Soc* 960 *1924*, 983 *1927*]. Instead of CrO_3, 2-5% (w/w) of $KMnO_4$, with boiling under reflux for 2-6h, has been used.
Traces of water have been removed by refluxing with tetraacetyl diborate (prepared by warming 1 part of boric acid with 5 parts (w/w) of acetic anhydride at 60°, cooling, and filtering off), followed by distn [Eichelberger and La Mer *J Am Chem Soc* **55** 3633 *1933*].
Refluxing with acetic anhydride in the presence of 0.2g % of 2-naphthalenesulfonic acid as catalyst has also been used [Orton and Bradfield *J Chem Soc* 983 *1927*]. Other suitable drying agents include $CuSO_4$ and chromium triacetate: P_2O_5 converts some acetic acid to the anhydride. Azeotropic removal of water by distn with thiophene-free *benzene or with butyl acetate has been used [Birdwhistell and Griswold *J Am Chem Soc* **77** 873 *1955*]. An alternative purification uses fractional freezing.
Rapid procedure: Add 5% acetic anhydride, and 2% of CrO_3. Reflux and fractionally distil.

Acetic anhydride *[108-24-7]* **M 102.1, b 138°, d 1.082, n 1.3904.** Adequate purification can usually be obtained by fractional distn through an efficient column. Acetic acid can be removed by prior refluxing with CaC_2 or with coarse Mg filings at 80-90° for 5days, or by distn from a large excess of quinoline (1% AcOH in quinoline) at 75mm pressure. Acetic anhydride can also be dried by standing with Na wire for up to a week, removing the Na and distilling from it under vacuum. (Na reacts vigorously with acetic anhydride at 65-70°). Dippy and Evans [*J Org Chem* **15** 451 *1950*] let the anhydride (500g) stand over P_2O_5 (50g) for 3h, then decanted it and stood it with ignited K_2CO_3 for a further 3h. The supernatant liquid was distd and the fraction **b** 136-138°, was further dried with P_2O_5 for 12h, followed by shaking with ignited K_2CO_3, before two further distns through a five-section Young and Thomas fractionating column. The final material distd at 137.8-138.0°. Can also be purified by azeotropic distn with toluene: the azeotrope boils at 100.6°. After removal of the remaining toluene, the anhydride is distd [sample had a specific conductivity of 5 x 10^{-9} ohm^{-1}cm^{-1}].
Rapid procedure: Shake with P_2O_5, separate, shake with dry K_2CO_3 and fractionally distil.

Acetic hydrazide *[1068-57-1]* **M 74.1, m 67°, b 127°/18mm.** Cryst as needles from EtOH. Reduces $NH_3/AgNO_3$.

Acetoacetamide *[5977-14-0]* **M 101.1, m 54-55°, 54-56°.** Recrystallise from $CHCl_3$, or Me_2CO/pet ether. Crystallises from pyridine with 4mol of solvent. Slightly soluble in H_2O, EtOH and AcOH but insoluble in Et$_2$O. *Phenylhydrazone* has **m** 128°. [*Beilstein* **3, 4th Suppl**, p 1545; Kato *Chem Pharm Bull Jpn* **15** 921,923 *1967*; *Chem Ber* **35** 583 *1902*.]

Acetoacetanilide *[102-01-2]* **M 177.2, m 86°, pK 10.68**. Crystd from H$_2$O, aqueous EtOH or pet ether (b 60-80°).

Acetoacetylpiperidide *[1128-87-6]* **M 169.2, b 88.9°/0.1mm, n^{52} 1.4983**. Dissolved in *benzene, extracted with 0.5M HCl to remove basic impurities, washed with water, dried, and distd at 0.1mm [Wilson *J Org Chem* **28** 314 *1963*].

α-Acetobromoglucose (2,3,4,6-tetraacetyl-α-D-glucopyranosyl bromide) *[572-09-8]* **M 411.2, m 88-89°, [α]$_D^{25}$ +199.3° (c 3, CHCl$_3$)**. Crystd from isopropyl ether or pet ether (b 40-60°) [*Org Synth* **65** 236 *1897*].

Acetone *[67-64-1]* **M 58.1, b 56.2°, d 0.791, n 1.35880, pK$_1^{25}$ -6.1 (basic, mono-protonated), pK$_2^{25}$ 20.0 (acidic)** The commercial preparation of acetone by catalytic dehydrogenation of isopropyl alcohol gives relatively pure material. Analytical reagent quality generally contains less than 1% organic impurities but may have up to about 1% H$_2$O. Dry acetone is appreciably *hygroscopic*. The main organic impurity in acetone is mesityl oxide, formed by the aldol condensation. It can be dried with anhydrous CaSO$_4$, K$_2$CO$_3$ or type 4A Linde molecular sieves, and then distd. Silica gel and alumina, or mildly acidic or basic desiccants cause acetone to undergo the aldol condensation, so that its water content is increased by passage through these reagents. This also occurs to some extent when P$_2$O$_5$ or sodium amalgam is used. Anhydrous MgSO$_4$ is an inefficient drying agent, and CaCl$_2$ forms an addition compound. Drierite (anhydrous CaSO$_4$) offers the minimum acid and base catalysis of aldol formation and is the recommended drying agent for this solvent [Coetzee and Siao *Inorg Chem* **14v** 2 *1987*; Riddick and Bunger *Organic Solvents* Wiley-Interscience, N.Y., 3rd edn, 1970]. Acetone was shaken with Drierite (25g/L) for several hours before it was decanted and distd from fresh Drierite (10g/L) through an efficient column, maintaining atmospheric contact through a Drierite drying tube. The equilibrium water content is about 10^{-2}M. Anhydrous Mg(ClO$_4$)$_2$ **should not be used as drying agent because of the risk of EXPLOSION with acetone vapour.**
Organic impurities have been removed from acetone by adding 4g of AgNO$_3$ in 30mL of water to 1L of acetone, followed by 10mL of M NaOH, shaking for 10min, filtering, drying with anhydrous CaSO$_4$ and distilling [Werner *Analyst (London)* **58** 335 *1933*]. Alternatively, successive small portions of KMnO$_4$ have been added to acetone at reflux, until the violet colour persists, followed by drying and distn. Refluxing with chromium trioxide (CrO$_3$) has also been used. Methanol has been removed from acetone by azeotropic distn (at 35°) with methyl bromide, and treatment with acetyl chloride.
Small amounts of acetone can be purified as the NaI addition compound, by dissolving 100g of finely powdered NaI in 400g of boiling acetone, then cooling in ice and salt to -8°. Crystals of NaI.3Me$_2$CO are filtered off and, on warming in a flask, acetone distils off readily. [This method is more convenient than the one using the bisulfite addition compound.] Also purified by gas chromatography on a 20% free fatty acid phthalate (on Chromosorb P) column at 100°.
For efficiency of desiccants in drying acetone see Burfield and Smithers [*J Org Chem* **43** 3966 *1978*]. The water content of acetone can be determined by a modified Karl Fischer titration [Koupparis and Malmstadt *Anal Chem* **54** 1914 *1982*].
Rapid procedure: Dry over anhydrous CaSO$_4$ and distil.

Acetone cyanohydrin *[75-86-5]* **M 85.1, b 48°/2.5mm, 68-70°/11mm, 78-82°/15mm, d$_4^{20}$ 0.93**. Dry with Na$_2$SO$_4$ and distil as rapidly as possible under vacuum to avoid decomposition. Discard fractions boiling below 78-82°/15mm. Store in the dark. **USE AN EFFICIENT FUME HOOD as HCN (POISONOUS) is always present.** [*Org Synth* Col.Vol. II 7 *1940*.]

Acetonedicarboxylic acid *[542-05-2]* **M 146.1, m 138° (dec), pK^{25} 3.10**. Crystd from ethyl acetate and stored over P$_2$O$_5$. Decarboxylates in hot water.

Acetone semicarbazone *[110-20-3]* **M 115.1, m 187°, pK^{25} 1.33**. Crystd from water or from aqueous EtOH.

Acetonitrile (methyl cyanide) *[75-05-8]* M 41.1, b 81.6°, d^{25} 0.77683, n 1.3441, n^{25} 1.34163. Commercial acetonitrile is a byproduct of the reaction of propylene and ammonia to acrylonitrile. The following procedure that significantly reduces the levels of acrylonitrile, allyl alcohol, acetone and *benzene was used by Kiesel [*Anal Chem* **52** 2230 *1988*]. Methanol (300mL) is added to 3L of acetonitrile fractionated at high reflux ratio until the boiling temperature rises from 64° to 80°, and the distillate becomes optically clear down to λ = 240nm. Add sodium hydride (1g) free from paraffin, to the liquid, reflux for 10min, and then distil rapidly until about 100mL of residue remains. Immediately pass the distillate through a column of acidic alumina, discarding the first 150mL of percolate. Add 5g of CaH_2 and distil the first 50mL at a high reflux ratio. Discard this fraction, and collect the following main fraction. The best way of detecting impurities is by gas chromatography.

Usual contaminants in commercial acetonitrile include H_2O, acetamide, NH_4OAc and NH_3. Anhydrous $CaSO_4$ and $CaCl_2$ are inefficient drying agents. Preliminary treatment of acetonitrile with cold, satd aq KOH is undesirable because of base-catalysed hydrolysis and the introduction of water. Drying by shaking with silica gel or Linde 4A molecular sieves removes most of the water in acetonitrile. Subsequent stirring with CaH_2 until no further hydrogen is evolved leaves only traces of water and removes acetic acid. The acetonitrile is then fractionally distd at high reflux, taking precaution to exclude moisture by refluxing over CaH_2 [Coetzee *Pure Appl Chem* **13** 429 *1966*]. Alternatively, 0.5-1% (w/v) P_2O_5 is often added to the distilling flask to remove most of the remaining water. Excess P_2O_5 should be avoided because it leads to the formation of an orange polymer. Traces of P_2O_5 can be removed by distilling from anhydrous K_2CO_3.

Kolthoff, Bruckenstein and Chantooni [*J Am Chem Soc* **83** 3297 *1961*] removed acetic acid from 3L of acetonitrile by shaking for 24h with 200g of freshly activated alumina (which had been reactivated by heating at 250° for 4h). The decanted solvent was again shaken with activated alumina, followed by five batches of 100-150g of anhydrous $CaCl_2$. (Water content of the solvent was then less than 0.2%). It was shaken for 1h with 10g of P_2O_5, twice, and distd in a 1m x 2cm column, packed with stainless steel wool and protected from atmospheric moisture by $CaCl_2$ tubes. The middle fraction had a water content of 0.7 to 2mM.

Traces of unsaturated nitriles can be removed by an initial refluxing with a small amount of aq KOH (1mL of 1% solution per L). Acetonitrile can be dried by azeotropic distn with dichloromethane, *benzene or trichloroethylene. Isonitrile impurities can be removed by treatment with conc HCl until the odour of isonitrile has gone, followed by drying with K_2CO_3 and distn.

Acetonitrile was refluxed with, and distd from alkaline $KMnO_4$ and $KHSO_4$, followed by fractional distn from CaH_2. (This was better than fractionation from molecular sieves or passage through a type H activated alumina column, or refluxing with KBH_4 for 24h and fractional distn)[Bell, Rodgers and Burrows *J Chem Soc, Faraday Trans 1* **73** 315 *1977*; Moore et al. *J Am Chem Soc* **108** 2257 *1986*].

Material suitable for polarography was obtained by refluxing over anhydrous $AlCl_3$ (15g/L) for 1h, distilling, refluxing over Li_2CO_3 (10g/L) for 1h and redistg. It was then refluxed over CaH_2 (2g/L) for 1h and fractionally distd, retaining the middle portion. The product was not suitable for UV spectroscopy use. A better purification procedure used refluxing over anhydrous $AlCl_3$ (15g/L) for 1h, distg, refluxing over alkaline $KMnO_4$ (10g $KMnO_4$, 10g Li_2CO_3/L) for 15min, and distg. A further reflux for 1h over $KHSO_4$ (15g/L), then distn, was followed by refluxing over CaH_2 (2g/L) for 1h, and fractional distn. The product was protected from atmospheric moisture and stored under nitrogen [Walter and Ramalay *Anal Chem* **45** 165 *1973*]. Purificaton of "General Purity Reagent" for this purpose is not usually satisfactory because very large losses occur at the $KMnO_4$, $LiCO_3$ step. For electrochemical work involving high oxidation fluorides, further reflux over P_2O_5 (1g/mL for 0.5h) and distilling (discarding 3% of first and last fractions) and repeating this step is necessary. The distillate is kept over molecular sieves in *vac* after degassing, for 24h and vac distd onto freshly activated 3A molecular sieves. The MeCN should have absorption at 200nm of <0.05 (H_2O reference) and UV cutoff at *ca* 175nm. Also the working potential range of purified Et_4N^+ BF_4^- (0.1mol.dcm^{-3} in the MeCN) should be +3.0 to -2.7V *vs* Ag^+/Ag^o. If these criteria are not realised then further impurities can be removed by treatment with activated neutral alumina (60 mesh) *in vacuo* before final molecular sieves treatment [Winfield *J Fluorine Chem* **25** 91 *1984*].

Acetonitrile has been distd from $AgNO_3$, collecting the middle fraction over freshly activated Al_2O_3. After standing for two days, the liquid was distd from the activated Al_2O_3. Specific conductivity 0.8-1.0 x 10^{-8} mhos [Harkness and Daggett *Can J Chem* **43** 1215 *1965*]. Acetonitrile ^{14}C was purified by gas chromatography and is water free and distd at 81°. [*J Mol Biol* **87** 541 *1974*.]

Rapid procedure: Dry over anhydrous K_2CO_3 for 24h, followed by further drying for 24h over 3A molecular sieves or boric anhydride, followed by distn. Alternatively, stir over P_2O_5 (5% w/v) for 24h then distil. However this last method is not suitable for use in reactions with very acid sensitive compounds.

Acetonylacetone   **(2,5-hexanedione)**   *[110-13-4]*   **M 114.2, m -9°, b 76-78°/13mm, 88°/25mm, 137°/150mm, 188°/atm, d_4^{20} 0.9440, n_D^{20} 1.423, pK 18.7.** Purified by dissolving in Et_2O, stirred with K_2CO_3 (a quarter of the wt of dione), filtered, dried over anhydrous Na_2SO_4 (**not $CaCl_2$**), filtered, evapd and distd in a vacuum. It is then redistd through a 30cm Vigreux column (oil bath temp 150°). It is miscible with H_2O and EtOH. The *dioxime* has **m** 137° (plates from *C_6H_6), *mono-oxime* has **b** 130°/11mm, and the *2,4-dinitrophenylhydrazone* has **m** 210-212° (red needles from EtOH). [*Chem Ber* **22** 2100 *1989*; for enol content see *J Org Chem* **19** 1960 *1954*.]

4-Acetophenetidine (phenacetin) *[62-44-2]* **M 179.2, m 136°.** Crystd from H_2O or purified by soln in cold dilute alkali and reppted by addn of acid to neutralisation point. Air-dried.

Acetophenone *[98-86-2]* **M 120.2, m 19.6°, b 54°/2.5mm, 202°/760mm, d^{25} 1.0238, n^{25} 1.5322, pK 19.2.** Dried by fractional distn or by standing with anhydrous $CaSO_4$ or $CaCl_2$ for several days, followed by fractional distn under reduced pressure (from P_2O_5, optional), and careful, slow and repeated partial crystns from the liquid at 0° excluding light and moisture. It can also be crystd at low temperatures from isopentane. Distn can be followed by purification using gas-liquid chromatography [Earls and Jones *J Chem Soc, Faraday Trans 1* **71** 2186 *1975*.]
§ A commercial polystyrene supported version is available — scavanger resin (for diol substrate).

Aceto-*o*-toluidide *[120-66-1]* **M 149.2, m 110°, b 296°/760mm.** Crystd from H_2O, EtOH or aqueous EtOH.

Aceto-*m*-toluidide *[537-92-8]* **M 149.2, m 65.5°, b 182-183°/14mm, 303°/760mm.** Crystd from H_2O, EtOH or aqueous EtOH.

Aceto-*p*-toluidide *[103-89-9]* **M 149.2, m 146°, b 307°/760mm.** Crystd from aqueous EtOH.

Acetoxime (acetone oxime) *[127-06-0]* **M 73.1, m 63°, b 135°/760mm, pK^{40} 0.99.** Crystd from pet ether (b 40-60°). Can be sublimed.

Acetoxyacetone (acetol acetone) *[592-20-1]* **M 116.1, b 65°/11mm, 73-75°/17mm, 174-176°/atm, d_4^{20} 1.0757, n_D^{20} 1.4141.** Distil under reduced pressure, then redistil at atm pressure. It is miscible with H_2O but is slowly decomposed by it. Store in dry atmosphere. The *2,4-dinitrophenylhydrazone* has **m** 115-115.5° (from $CHCl_3$/hexane). [*J Chem Soc* **59** 789 *1891*; *J Org Chem* **21** 68 *1956*; *Justus Liebigs Ann Chem* **335** 260 *1904*.]

4-Acetoxy-2-azetidinone *[28562-53-0]* **M 129.1, m 38-41°.** Dissolve in $CHCl_3$, dry ($MgSO_4$) concentrate at 40°/70mm, or better at room temperature to avoid decomposition. Wash and stir the residual oil with hexane by decantation and discard wash. Dry the oil at high vacuum when it should solidify, **m** 34°. It can be distd at high vacuum, 80-82°/10^{-3}mm, but this results in extensive losses. The purity can be checked by TLC using Merck Silica Gel F_{254} and eluting with EtOAc. The azetidinone has R_F 0.38 (typical impurities have R_F 0.67). The spots can be detected by the TDM spray. This is prepared from (A) 2.5g 4,4'-tetramethyldiaminodiphenylmethane (TDM) in 10mL AcOH and diluted with 50mL of H_2O, (B) 5g KI in 100mL of H_2O and (C) 0.3g ninhydrin in 10mL of AcOH and 90mL of H_2O. The spray is prepared by mixing (A) and (B) with 1.5mL of (C) and stored in a brown bottle. [*Justus Liebigs Ann Chem* 539 *1974*; *Org Synth* **65** 135 *1987*.]

1-Acetoxy-1,3-butadiene (1,3-butadienyl acetate) *cis-trans* mixture *[1515-76-0]* **M 112.1, b 42-43°/16mm, 51-52°/20mm, 60-61°/40mm, d_4^{20} 0.9466, n_D^{20} 1.4622.** The commercial sample is stabilised with 0.1% of *p-tert*-butylcatechol. If the material contains crotonaldehyde (by IR, used in its synthesis) it should be dissolved in Et_2O, shaken with 40% aqueous sodium bisulfite, then 5% aqueous

Na_2CO_3, water, dried (Na_2SO_4) and distilled several times in a vac through a Widmer (*Helv Chim Acta* **7** 59 *1924*) or Vigreux column [Wicterle and Hudlicky *Collect Czech Chem Commun* **12** 564 *1947*; Hagemeyer and Hull *Ind Eng Chem* **41** 2920 *1949*].

1-Acetoxy-2-butoxyethane *[112-07-2]* **M 160.2, b 61-62°/0.2mm, 75-76°/12mm, 185.5°/ 740mm, 188-192°/atm, d_4^{20} 0.9425, n_D^{20} 1.4121.** Shake with anhydrous Na_2CO_3, filter and distil in a vacuum. Redistn can be then be carried out at atmospheric pressure. [*J Org Chem* **21** 1041 *1956*.]

3R,4R,1'R-4-Acetoxy-3-[1-(*tert*-butylmethylsilyloxy)ethyl]-2-azetinone see Chapter 5.

2-Acetoxyethanol *[542-59-6]* **M 104.1, b 187°/761mm, 187-189°/atm, d_4^{20} 1.108, n_D^{20} 1.42.** Dry over K_2CO_3 (**not CaCl$_2$**), and distil. [*J Chem Soc* 3061 1950; rate of hydrolysis: *J Chem Soc* 2706 *1951*.]

1-Acetoxy-2-ethoxyethane *[111-15-9]* **M 132.2, b 156-159°, d_4^{20} 0.97, n_D^{20} 1.406.** Shake with anhydr Na_2CO_3, filter and distil in vac. Redistn can then be carried out at atm pressure. [*J Org Chem* **21** 1041 *1956*.]

1-Acetoxy-2-methoxyethane *[110-49-6]* **M 118.1, b 141°/732mm, 140144°/atm, d_4^{20} 1.009, n_D^{20} 1.4011.** Shake with anhydrous Na_2CO_3, filter and distil in a vacuum. Redistn can be then be carried out at atmospheric pressure. [*J Org Chem* **21** 1041 *1956*.]

R-(-)-α-Acetoxyphenylacetic (acetyl mandelic) acid *[51019-43-3]* **M 194.2, m 96-98°, $[\alpha]_D^{20}$ -153.7° (c 2.06, Me_2CO), $[\alpha]_{546}^{20}$ -194° (c 2.4, Me_2CO), pK$_{Est}$ ~2.9** Recrysts from H_2O with 1mol of solvent which is removed on drying, or from solvents as for the *S*-isomer. [*J Chem Soc* 227 *1943*.]

S-(+)-α-Acetoxyphenylacetic (acetyl mandelic) acid *[7322-88-5]* **M 194.2, m 80-81°, 95-97.5°, $[\alpha]_D^{27}$ +158° (c 1.78, Me_2CO), $[\alpha]_{546}^{20}$ +186° (c 2, Me_2CO).** Recryst from *benzene-hexane or toluene and has characteristic NMR and IR spectra. [*Justus Liebigs Ann Chem* **622** 10 *1959*; *J Org Chem* **39** 1311 *1974*.]

21-Acetoxypregnenolone *[566-78-9]* **M 374.5, m 184-185°.** Crystd from Me_2CO.

S-(-)-2-Acetoxypropionyl chloride *[36394-75-9]* **M 150.6, b 51-53°/11mm, d_4^{20} 1.19, n_D^{20} 1.423, $[\alpha]_D^{27}$ -33°, (c 4, $CHCl_3$), $[\alpha]_{546}^{20}$ -38° (c 4, $CHCl_3$).** It is moisture sensitive and is hydrolysed to the corresponding acid. Check the IR spectrum. If the OH band above 3000cm^{-1} is too large and broad then the mixture should be refluxed with pure acetyl chloride for 1h, evapd and distd under reduced pressure.

S-Acetoxysuccinic anhydride *[59025-03-5]* **M 158.1, m 58° (RS 81.5-82.5°, 86-87°), $[\alpha]_D^{20}$ -26.0° (c 19, Me_2CO), $[\alpha]_D^{20}$ -28.4° (c 13, Ac_2O).** Recrystd from Ac_2O and dry in a vacuum over KOH, or by washing with dry Et_2O due to its deliquescent nature. [*J Chem Soc* 788 *1933; Synth Commun* **16** 183 *1986*; *J Org Chem* **52** 1040 *1988*; *RS* : *J Am Chem Soc* **88** 5306 *1966*.]

Acetylacetone (2,4-pentanedione) *[123-54-6]* **M 100.1, b 45°/30mm, $d^{30.2}$ 0.9630, $n^{18.5}$ 1.45178, pK_1^{25} -5.0 (enol), -6.6 (keto), pK_2^{25} 8.95** Small amounts of acetic acid were removed by shaking with small portions of 2M NaOH until the aqueous phase remained faintly alkaline. The sample, after washing with water, was dried with anhydrous Na_2SO_4, and distd through a modified Vigreux column [Cartledge *J Am Chem Soc* **73** 4416 *1951*]. An additional purification step is fractional crystn from the liquid. Alternatively, there is less loss of acetylacetone if it is dissolved in four volumes of *benzene and the soln is shaken three times with an equal volume of distd water (to extract acetic acid): the *benzene is then removed by distn at 43-53° and 20-30mm through a helices-packed column. It is then refluxed over P_2O_5 (10g/L) and fractionally distd under reduced pressure. The distillate (sp conductivity 4 x 10^{-8} ohm^{-1}cm^{-1}) was suitable for polarography [Fujinaga and Lee *Talanta* **24** 395 *1977*]. To recover used acetylacetone, metal ions were stripped from the soln at pH 1 (using 100mL 0.1M H_2SO_4/L of acetylacetone). The acetylacetone was washed with

(1:10) ammonia soln (100mL/L) and with distd water (100mL/L, twice), then treated as above. It complexes with Al, Be, Ca, Cd, Ce , Cu, Fe^{2+}, Fe^{3+}, Mn, Mg, Ni, Pb and Zn.

N-Acetyl-L-alaninamide *[15962-47-7]* **M 130.2, m 162°.** Crystd repeatedly from EtOH-diethyl ether.

N-Acetyl-ß-alanine *[3025-95-4]* **M 127.2, m 78.3-80.3°, pK^{25} 4.45.** Crystd from acetone.

N-Acetyl-L-alanyl-L-alaninamide *[30802-37-0]* **M 201.2, m 250-251°.** Crystd repeatedly from EtOH/diethyl ether.

N-Acetyl-L-alanyl-L-alanyl-L-alaninamide *[29428-34-0]* **M 272.3, m 295-300°.** Crystd from MeOH/diethyl ether.

N-Acetyl-L-alanylglycinamide *[76571-64-7]* **M 187.2, m 148-149°.** Crystd repeatedly from EtOH/diethyl ether.

Acetyl-α-amino-n-butyric acid *[34271-24-4]* **M 145.2, pK^{25} 3.72.** Crystd twice from water (charcoal) and air dried [King and King *J Am Chem Soc* **78** 1089 *1956*].

9-Acetylanthracene *[784-04-3]* **M 220.3, m 75-76°.** Crystd from EtOH. [Masnori et al. *J Am Chem Soc* **108** 1126 *1986*.]

N-Acetylanthranilic acid *[89-52-1]* **M 179.1, m 182-184°, 185-186°, 190°(dec), pK^{20} 3.61.** Wash with distilled H_2O and recrystallise from aqueous AcOH, dry and recrystallise again from EtOAc. Also recryst from water or EtOH. [*J Chem Soc* 2495 *1931*; *J Am Chem Soc* **77** 6698 *1955*.]

2-Acetylbenzoic acid *[577-56-0]* **M 164.2, m 115-116°, 116-118°, pK^{25} 4.10.** Recrystallises from *C_6H_6 and H_2O (15g/100mL). The *oxime* has **m** 156-157°, and the *2,4-dinitrophenylhydrazone* has **m** 185-186°(needles from EtOH). [*J Am Chem Soc* **69** 1547 *1947*.]

4-Acetylbenzoic acid *[586-89-0]* **M 164.2, m 207.5-209.5°, 208.6-209.4°, pK^{25} 3.70, 5.10 (EtOH).** Dissolve in 5% aqueous NaOH, extract with Et_2O, and acidify the aqueous soln. Collect the ppte, and recrystallise from boiling H_2O (100 parts) using decolorising charcoal [*J Org Chem* **24** 504 *1959*; *J Chem Soc* 265 *1957*; *J Am Chem Soc* **72** 2882 *1050*, **74** 1058 *1952*].

Acetylbenzonitrile *[1443-80-7]* **M 145.2, m 57-58°.** Recrystd from EtOH [Wagner et al. *J Am Chem Soc* **108** 7727 *1986*].

4-Acetylbiphenyl *[92-91-1]* **M 196.3, m 120-121°, b 325-327°/760mm.** See 4'-phenyl-acetophenone on p. 327.

Acetyl-5-bromosalicylic acid *[1503-53-3]* **M 259.1, m 168-169°, pK_{Est} ~3.0.** Crystd from EtOH.

2-Acetylbutyrolactone *[517-23-7]* **M 128.1, b 105°/5mm, 120-123°/11mm, 142-143°/30mm, d_4^{20} 1.1846, n_D^{20} 1.459.** Purified by distillation, which will convert any free acid to the lactone, alternatively dissolve in Et_2O, wash well with 0.5N HCl, dry the organic layer and distil. The solubility in H_2O is 20% v/v. The *2,4-dinitrophenylhydrazone* forms orange needles from MeOH, **m** 146°. The lactone hydrolyses in mineral acid to 2-acetyl-4-hydroxybutyric acid which can be converted to the *di-n-propylamine salt* with **m** 68-70°. The lactone is a **SKIN IRRITANT**. [*Yakugaku Zasshi (J Pharm Soc Japan)* **62** 417(439) *1942*; *Helv Chim Acta* **35** 2401 *1952*.]

Acetyl chloride *[75-36-5]* **M 78.5, b 52°, d 1.1051, n 1.38976.** Refluxed with PCl_5 for several hours to remove traces of acetic acid, then distd. Redistd from one-tenth volume of dimethylaniline or quinoline to remove free HCl. A.R. quality is freed from HCl by pumping it for 1h at -78° and distg into a trap at -196°.

Acetyl bromide *[506-96-7]* **M 123.0, b 76-77°, d 1.65.** Boiled with PBr_3/Ac_2O for 1h then distd off and redistd. Store dry. [Burton and Degering *J Am Chem Soc* **62** 227 *1940*.] **LACHRYMATORY.**

Acetylcyclohexane (cyclohexyl methylketone) *[823-76-7]* **M 126.2, b 64°/11mm, 76.2-77°/25mm, d_4^{20} 0.9178, n_D^{20} 1.4519.** Dissolve in Et_2O, shake with H_2O, dry, evaporate and fractionate under reduced pressure. [UV: *J Am Chem Soc* **74** 518 *1952*; enol content: *J Org Chem* **19** 1960 *1954*.] The *semicarbazone* has **m** 174°; the *2,4-dinitrophenylhydrazone* has **m** 139-140° [*Helv Chim Acta* **39** 1290 *1956*].

2-Acetylcyclohexanone *[874-23-7]* **M 140.2, m -11°, b 62-64°/2.5mm, 95-98°/10mm, 111-112°/18mm, d_4^{20} 1.08, n_D^{20} 1.51.** Dissolve in ligroin (b 30-60°), wash with saturated aqueous $NaHCO_3$ dry over Drierite and fractionate in a vacuum. [*J Am Chem Soc* **75** 626, 5030 *1953*; *Chem Ber* **87** 108 *1954*.] It forms a *Cu salt* which crystallises in green leaflets from EtOH, **m** 162-163° [UV: *J Chem Soc* 4419 *1957*].

2-Acetylcyclopentanone *[1670-46-8]* **M 126.2, b. 72-75°/8mm, 82-86°/12mm, 88°/18mm, d_4^{20} 1.043, n_D^{20} 1.490.** Dissolve in pet ether (b 30-60°), wash with satd aq $NaHCO_3$, dry over Drierite and fractionate in a vacuum. It gives a violet colour with ethanolic $FeCl_3$ and is only slowly hydrolysed by 10% aq KOH but rapidly on boiling to yield 6-oxoheptanoic acid. [*J Am Chem Soc* **75** 5030 *1953*; *J Chem Soc* 4232 *1956*; UV: *J Am Chem Soc* **81** 2342 *1959* .] It gives a gray green *Cu salt* from Et_2O-pentane, **m** 237-238° [*J Am Chem Soc* **79** 1488 *1957*].

Acetyldigitoxin-α *[25395-32-8]* **M 807.0, m 217-221°, $[\alpha]_D^{20}$+5.0 (c 0.7, pyridine).** Crystd from MeOH as plates.

Acetylene *[74-86-2]* **M 26.0, m -80.8°, b -84°, pK ~25.** If very impure it should be purified by successive passage through spiral wash bottles containing, in this order, satd aq $NaHSO_4$, H_2O, 0.2M iodine in aq KI (two bottles), sodium thiosulfate soln (two bottles), alkaline sodium hydrosulfite with sodium anthraquinone-2-sulfonate as indicator (two bottles), and 10% aqueous KOH soln (two bottles). The gas was then passed through a Dry-ice trap and two drying tubes, the first containing $CaCl_2$, and the second, Dehydrite $[Mg(ClO_4)_2]$ [Conn, Kistiakowsky and Smith *J Am Chem Soc* **61** 1868 *1939*]. Acetone vapour can be removed from acetylene by passage through H_2O, then concd H_2SO_4, or by passage through two gas traps at -65° and -80°, concd H_2SO_4 and a soda lime tower, a tower of 1-mesh Al_2O_3 then into H_2SO_4 [*Org Synth* Coll Vol 1 229 *1941*, 3 853 *1955*; 4 793 *1963*].. Sometimes it contains acetone and air. These can be removed by a series of bulb-to-bulb distns, e.g. a train consisting of a conc H_2SO_4 trap and a cold EtOH trap (-73°), or passage through H_2O and H_2SO_4, then over KOH and $CaCl_2$. [See Brandsma *Preparative Acetylenic Chemistry,* 1st Edn Elsevier 1971, for pK p15, ISBN 0444409475; 2nd Edn Elsevier 1988, ISBN 0444429603, and Chapter 5 for sodium acetylide.] It is also available commercially as 10ppm in helium, and several concentrations in N_2 for instrument calibration.
Sodium acetylide *[1066-26-8]* M 48.0, was prepd by dissolving Na (23g) in liquid NH_3 (1L) and bubbling acetylene until the blue color was discharged (ca 30min) and evapd to dryness [Saunders *Org Synth* Coll Vol III 416 *1955*]; and is available commercially as a suspension in xylene/light mineral oil. [See entry in Chapter 5.]

Acetylenedicarboxamide *[543-21-5]* M 112.1, m 294°(dec). Crystd from MeOH.

Acetylenedicarboxylic acid *[142-45-0]* M 114.1, m 179°(anhydrous), pK_1^{19} 1.04, pK_2^{19} 2.50. Crystd from aqueous ether as dipicrate. For mono K salt see entry in Chapter 5.

N-**Acetylethylenediamine** *[1001-53-2]* **M 102.1, m 50-51°, 51°, b 128°/3mm, 125-130°/5mm, 133-139°/27mm, pK^{25} 9.28.** It has been fractionated under reduced pressure and fraction **b** 125-130°/5mm was refractionated; fraction **b** 132-135°/4mm was collected and solidified. It is a low melting *hygroscopic* solid which can be recrystd from dioxane-Et_2O. It is soluble in H_2O, Et_2O and $*C_6H_6$. The *p-toluenesulfonate salt* can be recrystd from EtOH-EtOAc 1:8, has **m** 125-126° but the free base cannot be recovered from it by basifying and extracting with CH_2Cl_2. The *picrate* has **m** 175° (from EtOH) [*J Am Chem Soc* **63** 853 *1941*, **78** 2570 *1956*].

2-Acetylfluorene *[781-73-7]* M 208.3, m 132°. Crystd from EtOH.

Acetyl fluoride *[557-99-3]* **M 62.0, b 20.5°/760mm, d 1.032.**   Purified by fractional distn.

N-**Acetyl-D-galactosamine** *[14215-68-0]* **M 221.2, m 160-161°, [α]$_{546}$ +102° (c 1, H$_2$O).** Crystd from MeOH/Et$_2$O.

N-**Acetyl-D-glucosamine** *[7512-17-6]* **M 221.2, m ca 215°, [α]$_{546}$ +49° after 2h (c 2, H$_2$O).** Crystd from MeOH/Et$_2$O.

N-**Acetylglutamic acid** *[1188-37-0]* **M 189.2, m 185° (RS); 201° (S), [α]^{25} -16.6° (in H$_2$O), pK$_{Est (1)}$ ~3.4, pK$_{Est(2)}$ ~4.3.**   Likely impurity is glutamic acid. Crystd from boiling water.

N-**Acetylglycinamide** *[2620-63-5]* **M 116.1, m 139-139.5°.**   Repeated crystn from EtOH/Et$_2$O. Dried in a vacuum desiccator over KOH.

N-**Acetylglycine** *[543-24-8]* **M 117.1, m 206-208°, pK$_1^{25}$ -1.92, pK$_2^{25}$ 3.69.**   Treated with acid-washed charcoal and recrystd three times from water or EtOH/Et$_2$O and dried *in vacuo* over KOH [King and King *J Am Chem Soc* **78** 1089 *1956*].

N-**Acetylglycyl-L-alaninamide** *[34017-20-4]* **M 175.2.**   Repeated crystn from EtOH/Et$_2$O. Dried in a vacuum desiccator over KOH.

N-**Acetylglycylglycinamide** *[27440-00-2]* **M 173.2, m 207-208°.**   Repeated crystn from EtOH/Et$_2$O. Dried in a vacuum desiccator over KOH.

N-**Acetylglycylglycylglycinamide** *[35455-24-4]* **M 230.2, m 253-255°.**   Repeated crystn from EtOH/Et$_2$O. Dried in a vacuum desiccator over KOH.

N-**Acetylhistidine (H$_2$O)** *[39145-52-3]* **M 171.2, m 148° (RS); 169° (S) [α]^{25} +46.2° (H$_2$O).** Likely impurity is histidine. Crystd from water, then 4:1 acetone:water.

N-**Acetyl-RS-homocysteine thiolactone (Citiolone)** *[1195-16-0]* *[17896-21-8 for ±]* **M 159.2, m 110°, 109-111°, 111.5-112.5°.**   Dry in a vacuum desiccator and recrystallise from toluene as needles. It is a ninhydrin -ve substance which gives a "slow" nitroprusside test. λ$_{max}$ 238nm (ε 4,400 M^{-1}cm^{-1}); ν (nujol) 1789s and 851ms cm^{-1}. [*J Am Chem Soc* **78** 1597 *1956*; *J Chem Soc* 2758 *1963*.]

N-**Acetylimidazole** *[2466-76-4]* **M 110.1, m 101.5-102.5°, pK^{25} 3.6.**   Crystd from isopropenyl acetate. Dried in a vacuum over P$_2$O$_5$.

3-Acetylindole *[703-80-0]* **M 159.2, m 188-190°, 191-193°, 194°, pK^{25} 12.99 (acidic).** Recrystd from MeOH or *C$_6$H$_6$ containing a little EtOH. The *phenylureido* derivative has m 154°. [*J Chem Soc* 461 *1946*.]

Acetyl iodide *[507-02-8]* **M 170.0, b 108°/760mm.**   Purified by fractional distn.

N-**Acetyl-L-leucinamide** *[28529-34-2]* **M 177.2, m 133-134°.**   Recrystd from CHCl$_3$ and pet ether (b 40-60°).

3-(S-Acetylmercapto)isobutyric acid *[RS 33325-40-5]* **M 162.2, m 40-40.5°, b ca 120°/1.25mm, pK$_{Est}$ ~4.0.**   Distil under vacuum and recrystd from *C$_6$H$_6$. [*Chem Abstr* **38** 3616 *1944*.]

Acetyl methanesulfonate *[5539-53-7]* **M 170.2, b <120°/<0.01mm.**   The main impurity is methanesulfonic acid. Reflux with redistd acetyl chloride for 6-10h, i.e. until no further HCl is absorbed in a trap, and exclude moisture. Dist off excess of AcCl and carefully dist below 0.001mm with the bath temp below 120° to give the anhydride as a pale yellow oil which solidifies below 0°. Below ~130° it decomp to the disulfonic anhydride and above ~130° polymers are formed. It is used for cleaving ethers [Prep, IR, NMR: Karger and Mazur *J Org Chem* **36** 528, 532 *1971*].

N-Acetyl-L-methionine *[65-82-7]* **M 191.3, m 104°, $[\alpha]_{546}$ -24.5° (c 1, in H$_2$O), pK$_{Est}$ ~3.4.** Crystd from water or ethyl acetate. Dried in a vacuum over P$_2$O$_5$.

Acetylmethionine nitrile *[538-14-7]* **M 172.3, m 44-46°·** Crystd from diethyl ether.

5-Acetyl-2-methoxybenzaldehyde *[531-99-7]* **M 166.2 , m 144°.** Crystd from EtOH or Et$_2$O.

N-Acetyl-N'-methyl-L-alanimide *[19701-83-8]* **M 144.2.** Crystd from EtOAc/Et$_2$O, then from EtOH and Et$_2$O.

4-Acetyl-1-methyl-1-cyclohexene *[6090-09-1]* **M 138.2, b 73-75°/7.5mm, 85-86°/13mm, 94-94.7°/20mm, 204.5-206°/747mm, d$_4^{20}$ 1.0238, n$_D^{20}$ 1.469.** Purified by fractionation under reduced pressure *in vacuo*, and when almost pure it can be fractionated at atmospheric pressure, preferably in an inert atm. Forms two *semicarbazones* one of which is more soluble in *C$_6$H$_6$, and both can be recryst from EtOH, more soluble has **m** 149°(151°), and the less soluble has **m** 172-175°(191°). *4-Nitrophenylhydrazone* has **m** 166-167° and the *2,4-dinitrophenylhydrazone* has **m** 114-115°. [*Helv Chim Acta* **17** 129, 140 *1934*; *Justus Liebigs Ann Chem* **564** 109 *1949*.]

N-Acetyl-6N'-methylglycinamide *[7606-79-3]* **M 130.2.** Recrystd from EtOH/Et$_2$O mixture.

N-Acetyl-6N'-methyl-L-leucine amide *[32483-15-1]* **M 186.3.** Recrystd from EtOH/hexane mixture.

4-Acetylmorpholine *[1696-20-4]* **M 129.2, m 13.8-14°, 14°, 14.5°, b 96-97°/6mm, 113-128°/22mm, 242-247°/760mm, d$_4^{20}$ 1.0963, n$_D^{20}$ 1.4830.** Distd through an 8inch Fenske (glass helices packing) column with a manual take-off head. Purified by fractional distn. The *hydrobromide* has **m** 172-175°. [*J Am Chem Soc* **75** 357 *1953, J Org Chem* **21** 1072 *1956*.]

1-Acetylnaphthalene *[941-98-0]* **M 170.1, m 10.5°, b 93-95°/0.1mm, 167°/12mm, 302°/atm, d$_4^{20}$ 1.12, pK -6.22 (H$_0$ scale, aq H$_2$SO$_4$).** If the NMR spectrum indicates the presence of impurities, probably 2-acetylnaphthalene, convert the substance to its picrate by dissolving in *benzene or EtOH and adding excess of satd·picric acid in these solvents until separation of picrates is complete. Recryst the picrate till **m** is 118°. Decompose the picrate with dil NaOH and extract with Et$_2$O. Dry the extract (Na$_2$SO$_4$), filter, evap and dist. The *2,4-dinitrophenylhydrazone* crysts from EtOH and has **m** 259°. [*Justus Liebigs Ann Chem* **380** 95 *1911*; *J Am Chem Soc* **61** 3438 *1939*.]

2-Acetylnaphthalene (2-acetonaphthenone, ß-Acetonaphthone, 2-acetonaphthalene, methyl-2-naphthylketone) *[93-08-3]* **M 170.2, m 52-53°, 55°, 55.8°, b 164-166°/8mm, 171-173°/17mm, 301-303°/atm, pK -6.16 (H$_0$ scale, aq H$_2$SO$_4$).** Separated from the 1-isomer by fractional crystn of the picrate in EtOH (see entry for the 1-isomer above) **m** 82°. Decomposition of the picrate with dil NaOH and extraction with Et$_2$O then evaporation gives purer 2-acetylnaphthalene. If this residue solidifies it can be recrystd from pet ether, EtOH or acetic acid; otherwise it should be distild in a vac and the solid distillate is recrystd [Gorman and Rodgers *J Am Chem Soc* **108** 5074 *1986*; Levanon et al. *J Phys Chem* **91** 14 *1987*]. Purity should be checked by high field NMR spectroscopy. *Oxime* has **m** 145° decomp, and the *semicarbazone* has **m** 235°. [*Justus Liebigs Ann Chem* **380** 95 *1911*; *J Am Chem Soc* **72** 753 and 5626 *1950, J Org Chem* **5** 512 *1940*.]

N-Acetyl-D-penicillamine *[15537-71-0]* **M 191.3, m 189-190° (dec), $[\alpha]_D$ +18° (c 1, in 50% EtOH).** See *N*-acetyl penicillamine on p. 507 in Chapter 6.

N-Acetyl-L-phenylalanine *[2018-61-3]* **M 207.2, m 170-171°, $[\alpha]_D$ +41° (c 1, EtOH), (DL) m 152.5-153°, pK$_{Est}$ 3.5.** Crystd from CHCl$_3$ and stored at 4°. (DL)-isomer crystd from water or acetone.

N-Acetyl-L-phenylalanine ethyl ester *[2361-96-8]* **M 235.3, m 93-94°.** Crystd from aq EtOH or H$_2$O. [Izumiya and Fruton *J Biol Chem* **218** 59 *1956*.]

1-Acetyl-2-phenylhydrazine *[114-83-0]* **M 150.2, m 128.5°, pK^{25} 1.3.** Crystd from aq EtOH.

1-Acetylpiperazine *[13889-98-0]* **M 128.2, m 32-34°, 52°, pK^{25} 7.94.** Purified by recrystn from 40% aqueous EtOH or from EtOH-Et$_2$O. It is an **irritant,** and is *hygroscopic.* The *hydrochloride* has **m** 191° (from EtOH), and the *tosylate* has **m** 148-149° (from EtOH-EtOAc, 1:16). The free base, however, cannot be isolated by basifying the tosylate salt and extractn with CH$_2$Cl$_2$. [*Chem Ber* **66** 113 *1933*; *J Am Chem Soc* **75** 4949 *1953*, 2570 **78** *1956*.]

1-Acetyl-4-piperidone *[32161-06-1]* **M 141.2, b 124-128°/0.2mm, 218°/760mm, d$_4^{25}$ 1.1444, n$_D^{25}$ 1.5023.** Purified by fractional distn through a short Vigreux column (15mm). The *2,4-dinitrophenylhydrazone* has **m** 212-213° (from EtOH). It is freely soluble in H$_2$O but insoluble in Et$_2$O. [*J Am Chem Soc* 901 **71** *1949*.]

3-Acetylpyridine *[350-03-8]* **M 121.1, m 13-14°, b 65-66°/1mm, 92-95°/8-9mm, 105°(113°)/16mm, 219-221°/760mm, d$_4^{20}$ 1.1065, n$_D^{20}$ 1.1065, pK^{25} 3.18.** It is purified by dissolving in HCl, extracting with Et$_2$O to remove the possible impurity of nicotinic acid, basified with NaOH and extracted with Et$_2$O. The dried extract is filtered, evaporated and the residual oil distd. If the NMR spectrum indicates further impurities then convert to the *phenylhydrazone* (**m** 137°, yellow needles from EtOH). This is hydrolysed with HCl [*Chem Ber* **22** 597 *1889*], the phenylhydrazine HCl is removed by filtration, NaNO$_2$ is added, the soln is basified with aq NaOH and extracted with Et$_2$O as before and distd at atmospheric pressure to give 3-acetylpyridine as a colourless oil. Purification can be achieved by shaking with 50% aq KOH, extracting with Et$_2$O, drying the extract and distilling at atmospheric pressure or in a vacuum. [*J Am Chem Soc* **79** 4226 *1957*]. The *hydrochloride* has **m** 180-181° (from MeOH-EtOH), the *picrate* has **m** 133.8-134.8° (from H$_2$O), and the *phenylhydrazone* has **m** 137° (129-130)° (from EtOH) [*J Am Chem Soc* **71** 2285 1949]. The *ketoxime* has **m** 112° (from EtOH or *C$_6$H$_6$. [*J Am Chem Soc* **55** 816 *1933*, **63** 490 *1941*, **67** 1468 *1945*, **79** 4226 *1957*.]

Acetylsalicylic acid (Aspirin) *[50-78-2]* **M 180.2, m 133.5-135°, pK^{25}3.38, (pK^{17}4.56).** Crystd twice from toluene, washed with cyclohexane and dried at 60° under vacuum for several hours [Davis and Hetzer *J Res Nat Bur Stand* **60** 569 *1958*]. Has also been recrystd from isopropanol and from diethyl ether/pet ether (b 40-60°).

O-Acetylsalicyloyl chloride *[5538-51-2]* **M 198.6, m 45°, 46-49°, 48-52°, b 107-110°/0.1mm, 135°/12mm, n$_D^{20}$ 1.536.** Check first the IR to see if an OH frequency is present. If so then some free acid is present. Then reflux with acetyl chloride for 2-3h and fractionate at high vac. The distillate should crystallise. It can be recryst from hexane. [*J Chem Soc* **89** 1318 *1906*.]

O-Acetylsalicylsalicylic acid *[530-75-6]* **M 300.3, m 159°.** Crystd from dilute acetic acid.

N-(4)-Acetylsulfanilamide *[144-80-9]* **M 214.2, m 216°.** Crystd from aqueous EtOH.

2-Acetylthiazole *[24295-03-2]* **M 127.2, b 89-91° (90-95°)/12mm, 95-105°/15mm, d$_4^{20}$ 1.23, n$_D^{20}$ 1.55.** Check NMR spectrum, if not too bad, distil through an efficient column in a vacuum. The *oxime* sublimes at 140-145°, **m** 159° (cryst from H$_2$O) has **m** 163-165.5° [*Helv Chim Acta* **31** 1142 *1948*; *J Am Chem Soc* **79** 4524 *1957*; *Helv Chim Acta* **40** 554 *1957*].

2-Acetylthiophene (methyl 2-thienyl ketone) *[88-15-3]* **M 126.2, m 9.2-10.5°, 10.45°, 10-11°, b 77°/4mm, 89-91°/9mm, 94.5-96.5°/13mm, 213-214°/atm, d$_4^{20}$ 1.17, n$_D^{20}$ 1.5666.** Fractionally distd through a 12 plate column and fraction **b** 77°/4mm was collected. Also wet the acetylthiophene in order to remove and free thiophene which forms an azeotrope with H$_2$O, **b** 68°, Store in a brown bottle and the clear colourless liquid remains thus for extended periods. [*Org Synth* **28** 1 *1948*; *J Am Chem Soc* **69** 3093 *1947*.] The red *4-nitrophenylhydrazone* crysts from EtOH, **m** 181-182°.

3-Acetylthiophene **(methyl 3-thienyl ketone)** *[1468-83-3]* **M 126.2, m 57°, 60-63°, b 106-107°/25mm, 208-210°/748mm.** Recrystd from pet ether (b 30-60°) or EtOH. *2,4-dinitrophenylhydrazone* crystallises from $CHCl_3$, **m** 265°, and the *semicarbazone* crystallises from EtOH, **m** 174-175°. [*J Am Chem Soc* **70** 1555 *1948.*]

N-**Acetylthiourea** *[591-08-2]* **M 118.2, m 164-165°, 165-168°.** Recrystd from AcOH, the solid is washed with Et_2O and dried in air then at 100°. [*Collect Czech Chem Commun* **24** 3678 *1959.*]

Acetyl *p*-toluenesulfonate *[26908-82-7]* **M 214.2, m 54-56°.** The most likely impurity is *p*-toluenesulfonic acid (could be up to 10%). This can be removed by dissolving in dry Et_2O and cooling until the anhydride crystallises out. It decomp on heating; below ~130° it gives the disulfonic anhydride and above ~130° polymers are formed. It is used for cleaving ethers [Prep, IR, NMR: Karger and Mazur *J Org Chem* **36** 528, 532 *1971*].

1-*O*-Acetyl-2,3,5-tri-*O*-benzoyl-β-D-ribofuranose *[6974-32-9]* **M 504.5, m 128-130°, 130-131°, 131-132°, $[\alpha]_D^{20}$ +44.2° (c 1, $CHCl_3$).** Recrystd from EtOH or isoPrOH. [*Helv Chim Acta* **42** 1171 *1959*; NMR: *J Org Chem* **33** 1799 *1968*; IR: *Chem Pharm Bull Jpn* **11** 188 *1963.*]

N-**Acetyltryptophan** **M 246.3,** *[87-32-1]* **m 206° (RS), pK_{Est} ~3.8;** *[1218-34-4]* **m 188° (S), $[\alpha]^{25}$ +30.1° (aq NaOH).** Likely impurity is tryptophan. Crystd from EtOH by adding water.

N-**Acetyl-L-valine amide** *[37933-88-3]* **M 158.2, m 275°.** Recrystd from CH_3OH/Et_2O.

cis-**Aconitic acid** *[585-84-2]* **M 174.1, m 126-129°(dec).** Crystd from water by cooling (sol: 1g in 2mL of water at 25°). Dried in a vacuum desiccator.

trans-**Aconitic acid** **(1,2,3-propenetriscarboxylic acid)** *[4023-65-8]* **M 174.1, m 195°(dec), m 198-199°(dec), 204-205°(dec), pK_1^{25} 2.81, pK_2^{25} 4.46.** Purified by dissolving in AcOH (77g/150mL), filtering and cooling. The acid separates (55g) as colourless needles. A further quantity (10g) can be obtained by reducing the vol of the filtrate. The acid is dried in air then in a vacuum desiccator over NaOH. The acid can be recrystd from Me_2CO-$CHCl_3$. The highest **m** is obtained with the very dry acid. The **m** (209°) is obtained on a Dennis bar [*J Am Chem Soc* **52** 3128 *1930*, *Org Synth* Coll Vol II 12 *1943*].

cis-**Aconitic anhydride** *[6318-55-4]* **M 156.1, m 75°, 76-78°, 78-78.5°.** Reflux in xylene (7.5 parts) for 1h, then evaporate and recrystallise the residue from *C_6H_6. Alternatively, reflux in Ac_2O, evaporate and recrystallise from *C_6H_6. It is sensitive to moisture. [IR: *Acta Chem Scand* **21** 291 *1967*, *Chem Ber* **61** 2523 *1928*; NMR: *Biochemistry* **5** 2335 *1966.*]

Aconitine *[302-27-2]* **M 645.8, m 204°, $[\alpha]_{546}$ +20° (c 1, $CHCl_3$), pK¹⁵ 8.35.** Crystd from EtOH, $CHCl_3$ or toluene.

Aconitine hydrobromide *[6034-57-7]* **M 726.7, m 207°.** Crystd from water or EtOH/ether.

Acridine **(2,3-benzoquinoline)** *[260-94-6]* **M 179.2, m 111° (sublimes), b 346°, pK 5.58 (pK²⁵ of excited state 10.65).** Crystd twice from *benzene/cyclohexane, or from aqueous EtOH, then sublimed, removing and discarding the first 25% of the sublimate. The remainder was again crystd and sublimed, discarding the first 10-15% [Wolf and Anderson *J Am Chem Soc* **77** 1608 *1955*].
Acridine can also be purified by crystn from *n*-heptane and then from ethanol/water after pre-treatment with activated charcoal, or by chromatography on alumina with pet ether in a darkened room. Alternatively, acridine can be ppted as the hydrochloride from *benzene soln by adding HCl, after which the base is regenerated, dried at 110°/50mm, and crystd to constant melting point from pet ether [Cumper, Ginman and Vogel *J Chem Soc* 4518 *1962*]. The regenerated free base may be recrystd, chromatographed on basic alumina, then vac-sublimed and zone-refined. [Williams and Clarke, *J Chem Soc, Faraday Trans 1* **73** 514 *1977*; Albert, *The Acridines*

Arnold Press *1966*]. It can exists in five crystalline forms and is steam volatile. It is a strong IRRITANT to skin and mucous membranes and can become a chronic irritant— handle with CARE.

Acridine Orange *[494-38-2]* **M 349.94, m 181-182° (free base).** The double salt with $ZnCl_2$ (6g) was dissolved in water (200mL) and stirred with four successive portions (12g each) of Dowex-50 ion-exchange resin (K^+ form) to remove the zinc. The soln was then concentrated in vacuum to 20mL, and 100mL of ethanol was added to ppte KCl which was removed. Ether (160mL) was added to the soln from which, on chilling, the dye crystallises as its chloride. It was separated by centrifuging, washed with chilled ethanol and ether, and dried under vac, before being recryst from ethanol (100mL) by adding ether (50mL), and chilling. Yield 1g. [Pal and Schubert *J Am Chem Soc* **84** 4384 *1962*].
It was recrystd twice as the free base from ethanol or methanol/water by dropwise addition of NaOH (less than 0.1M). The ppte was washed with water and dried under vacuum. It was dissolved in $CHCl_3$ and chromatographed on alumina: the main sharp band was collected, concentrated and cooled to -20°. The ppte was filtered, dried in air, then dried for 2h under vacuum at 70°· [Stone and Bradley *J Am Chem Soc* **83** 3627 *1961*; Blauer and Linschitz *J Phys Chem* **66** 453 *1962*.]

Acridine Yellow G *[135-49-9]* **M 273.8, m 325°, CI 46025.** Crystd from 1:1 *benzene/methanol.

Acridone *[578-95-0]* **M 195.2, m >300°, pK$_1$ -0.32 (basic), pK$_2$ 14 (acidic).** Dissolve ~1g in *ca* 1% NaOH (100mL), add 3M HCl to pH 4 when acridone separates as a pale yellow solid with **m** just above 350° (sharp). It can be recrystd from large vols of H_2O to give a few mg. It is soluble in 160 parts of boiling EtOH (540 parts at 22°) [*J Chem Soc* 1294 *1956*]. A few decigms are best crystallised as the *hydrochloride* from 400 parts of 10N HCl (90% recovery) from which the free base is obtained by washing the salt with H_2O. A small quantity can be recrystd (as the neutral species) from boiling AcOH. Larger quantities are best recrystallised from a mixture of 5 parts of freshly distd aniline and 12.5 parts of glacial acetic acid. Acridone distils unchanged at atmospheric pressure, but the boiling point was not recorded, and some sublimation occurs below 350°. It has UV: λ_{max} 399nm. [see Albert, *The Acridines* Arnold Press pp. 201, 372 *1966*.]

***N*-(9-Acridinyl)maleimide (NAM)** *[49759-20-8]* **M 274.3, m 248°, 255-258°.** Purified by chromatography on silica gel using CH_2Cl_2 as eluant. Evaporation of pooled fractions that gave the correct NMR spectra gave a solid which was recrystd from Me_2CO as pale yellow prisms. IR ν (nujol): 1710 (imide); UV (MeOH): λ_{max} (nm), (ϵ $M^{-1}cm^{-1}$): 251 (159 500), 343 shoulder (7 700), 360 (12 400) and 382shoulder (47 000). [*Chem Pharm Bull Jpn* **26** 596 *1978*; *Eur J Biochem* **25** 64 *1972*.]

Acriflavine *[8048-52-0]* **M 196.2, pK >12.** Treated twice with freshly ppted AgOH to remove proflavine, then recrystd from absolute methanol [Wen and Hsu *J Phys Chem* **66** 1353 *1962*].

Acriflavin Mixture (Euflavin, 3,6-diamino-10-methylacridinium chloride) *[8063-24-9]* **M 259.7, m 179-181°.** Purified by dissolving in 50 parts of H_2O, shake with a small excess of freshly ppted and washed Ag_2O. The mixture is set aside overnight at 0° and filtered. The cake is not washed. The pH of the filtrate is adjusted to 7.0 with HCl and evaporated to dryness. The residue is then crystd twice from MeOH, twice from H_2O and dried at 120°· λ_{max} at 452nm has a logε value of 4.67. It is a red powder which readily absorbs H_2O. The solubility is increased in the presence of proflavin. The *dihydrochloride* is a deep red crystn powder. It is available as a mixture of 3,6-diaminoacridinium chloride (35%) and its 10-metho-chloride (65%). [see Albert, *The Acridines* Arnold Press p. 346 *1966*; *Chem Ber* **45** 1787 *1912*].

Acrolein (acraldehyde) *[107-02-8]* **M 56.1, b 52.1°, n 1.3992, d 0.839.** Purified by fractional distn. under nitrogen, drying with anhydrous $CaSO_4$ and then distilling under vac. Blacet, Young and Roof [*J Am Chem Soc* **59** 608 *1937*] distd under nitrogen through a 90cm column packed with glass rings. To avoid formation of diacryl, the vapour was passed through an ice-cooled condenser into a receiver cooled in an ice-salt mixture and containing 0.5g catechol. The acrolein was then distd twice from anhydrous $CuSO_4$ at low pressure, catechol being placed in the distilling flask and the receiver to avoid polymerization. [Alternatively, hydroquinone (1% of the final soln) can be used.]

Acrolein diacetyl acetal **(1,1-diacetoxy-2-propene).** *[869-29-4]* **M 158.2, b 75°/10mm, 184°/atm, d_4^{20} 1.08, n_D^{20} 1.4203.** Check the NMR spectrum. If it is not satisfactory then add Ac_2O and a drop of conc H_2SO_4 and heat at 50° for 10min. Then add anhydrous NaOAc (*ca* 3g/ 100g of liquid) and fractionate. Note that it forms an azeotrope with H_2O, so do not add H_2O at any time. It is a **highly flammable and TOXIC** liquid, keep away from the skin. [*J Am Chem Soc* 73 5282 *1951.*]

Acrolein diethyl acetal *[3054-95-3]* **M 130.2, b 120-125°/atm, n_4^{20} 1.398-1.407.** Add Na_2CO_3 (*ca* 3.5%) and distil using an efficient column, or better a spinning band column. [*Org Synth* 25 1 *1945.*]

Acrolein dimethyl acetal **(1,1-dimethoxy-2-propene)** *[6044-68-4]* **M 102.1, b 87.5-88°/750mm, 89-90°/760mm, d_4^{20} 0.86, n_D^{20} 1.3962.** Fractionally distil (after adding 0.5g of hydroquinone) under reduced press through an all glass column (40cm x 2.5 cm) packed with glass helices and provided with a heated jacket and a total reflux variable take-off head. Stainless steel Lessing rings (1/8 x 1/8 in) or gauze have been used as packing. It is a **highly flammable and TOXIC** liquid, keep away from the skin. [*J Chem Soc* 2657 *1955.*]

Acrolein semicarbazone *[6055-71-6]* **M 113.1, m 171°·** Crystd from water.

Acrylamide *[79-06-1]* **M 71.1, m 84°, b 125°/25mm.** Crystd from acetone, chloroform, ethyl acetate, methanol or *benzene/chloroform mixture, then vac dried and kept in the dark under vac. Recryst from $CHCl_3$ (200g dissolved in 1L heated to boiling and filtered without suction in a warmed funnel through Whatman 541 filter paper. Allowed to cool to room temp and kept at -15° overnight). Crystals were collected with suction in a cooled funnel and washed with 300mL of cold MeOH. Crystals were air-dried in a warm oven. [Dawson et al. *Data for Biochemical Research*, Oxford Press 1986 p. 449.]
CAUTION: *Acrylamide is extremely* **TOXIC** *and precautions must be taken to avoid skin contact or inhalation. Use gloves and handle in a well ventilated fume cupboard.*

Acrylic acid *[79-10-7]* **M 72.1, m 13°, b 30°/3mm, d 1.051, pK^{25} 4.25.** Can be purified by steam distn, or vacuum distn through a column packed with copper gauze to inhibit polymerisation. (This treatment also removes inhibitors such as methylene blue that may be present.) Azeotropic distn of the water with *benzene converts aqueous acrylic acid to the anhydrous material.

Acrylonitrile *[107-13-1]* **M 53.1, b 78°, d 0.806, n^{25} 1.3886.** Washed with dilute H_2SO_4 or dilute H_3PO_4, then with dilute Na_2CO_3 and water. Dried with Na_2SO_4, $CaCl_2$ or (better) by shaking with molecular sieves. Fractionally distd under nitrogen. Can be stabilised by adding 10ppm *tert*-butyl catechol. Immediately before use, the stabilizer can be removed by passage through a column of activated alumina (or by washing with 1% NaOH soln if traces of water are permissible in the final material), followed by distn. Alternatively, shaken with 10% (w/v) NaOH to extract inhibitor, and then washed in turn with 10% H_2SO_4, 20% Na_2CO_3 and distd water. Dried for 24h over $CaCl_2$ and fractionally distd under N_2 taking the fraction boiling at 75.0 to 75.5°C (at 734mm Hg). Stored with 10ppm *tert*-butyl catechol. Acrylonitrile is distilled off as required. [Burton *et al*, *J Chem Soc, Faraday Trans 1* 75 1050 *1979.*]

Acryloyl chloride *[814-68-6]* **M 90.5, b 72-74°/740mm, 74°/760mm, d_4^{20} 1.1127, n_D^{20} 1.4337.** Distil rapidly through an efficient 25cm column after adding 0.5g of hydroquinone/200g of chloride, and then redistil carefully at atmospheric pressure preferably in a stream of dry N_2. [*J Am Chem Soc* 72 72, 2299 *1950.*] **The liquid is an irritant and is TOXIC.**

Actarit (*p*-acetamidophenylacetic acid) *[18699-02-0]* **M 193.2, m 174-175°.** Crystd from MeOH + Me_2CO or aq EtOH.

Adamantane *[281-23-2]* **M 136.2, m 269.6-270.8°** (sublimes). Crystd from acetone or cyclohexane, sublimed in a vacuum below its melting point. [Butler et al. *J Chem Soc, Faraday Trans 1* 82 535 *1986.*] Adamantane was also purified by dissolving in *n*-heptane (*ca* 10mL/g of adamantane) on a hot plate, adding activated charcoal (2g/100g of adamantane), and boiling for 30min, filtering the hot soln through a filter paper, concentrating the filtrate until crystn just starts, adding one quarter of the original volume *n*-heptane and allowing to cool slowly over a period of hours. The supernatant was decanted off and the crystals were dried on a vacuum line at room temperature. [Walter et al. *J Am Chem Soc* **107** 793 *1985*.]

1-Adamantane acetic acid *[4942-47-6]* **M 194.3, m 136°, pK$_{Est}$ ~4.8.** Dissolve in hot N NaOH, treat with charcoal, filter and acidify. Collect solid, wash with H$_2$O, dry and recryst from MeOH. [*Chem Ber* **92** 1629 *1959*.]

1-Adamantane carboxylic acid *[828-51-3]* **M 180.3, m 175-176.5°, 177°, pK$_{Est}$ ~4.9.** Possible impurities are trimethylacetic acid and C9 and C13 acids. Dissolve 15g of acid in CCl$_4$ (300mL) and shake with 110mL of 15N aqueous NH$_3$ and the ammonium salt separates and is collected. Acid impurities form soluble ammonium salts. The salt is washed with cold Me$_2$CO (20mL) and suspended in H$_2$O (250mL). This is treated with 12N HCl and extracted with CHCl$_3$ (100mL). The dried (Na$_2$SO$_4$) is evaporated and the residue recrystd from a mixture of MeOH (30mL) and H$_2$O (*ca* 10mL) to give the pure acid (10-11g). [*Org Synth* Coll Vol.V 20 *1973*.] Also recrystd from absolute EtOH and dried under vacuum at 100°.
Alternatively, the acid (5g) is refluxed for 2h with 15mL of MeOH and 2mL of 98% H$_2$SO$_4$ (cool when mixing this soln). Pour into 10 volumes of H$_2$O and extract with the minimum volume of CHCl$_3$ to give clear separation of phases. The extract is washed with H$_2$O and dried (CaCl$_2$) and distd. The methyl ester is collected at 77-79°/1mm, **m** 38-39°. The ester is hydrolysed with the calculated amount of N KOH and refluxed until clear. Acidification with HCl provides the pure acid with 90% recovery. [*Org Synth* **4** 1 *1964*.] The *amide* crysts from cyclohexane, **m** 189°. [*Chem Ber* **62** 1629 *1959*.]

1,3-Adamantane diamine dihydrochloride *[26562-81-2]* **M 239.2, m >310°, pK$_{Est(1)}$ ~8.1, pK$_{Est(2)}$ ~10.1.** Dissolve in boiling conc HCl (400mg in 15mL) and evaporate to dryness. Dissolve in absolute EtOH and add dry Et$_2$O to crystallise the *dihydrochloride*. [*Chem Ber* **93** 1366 *1960*.]

1,3-Adamantane dicarboxylic acid *[39269-10-8]* **M 224.3, m 276°, 276-278°, 279°, pK$_{Est(1)}$ ~4.9. pK$_{Est(2)}$ 5.9.** Dissolve in aq NaOH, treat with charcoal, filter and acidify with dilute HCl. Recryst from MeOH. [*Chem Ber* **93** 1366 *1960*.]

1-Adamantane methylamine *[17768-41-1]* **M 165.3, b 83-85°/0.3mm, d$_4^{20}$ 0.93, pK$_{Est}$ ~10.2.** Dissolve in Et$_2$O, dry over KOH and distil. The *N-Tosyl* derivative has **m** 134-135° (from EtOH). [*Chem Ber* **96** 550 *1963*.]

1-Adamantanol (1-hydroxyadamantane) *[768-95-6]* **M 152.4, m 288.5-290°.** If 2-adamantanol is a suspected impurity then dissolve substance (10g) in acetone (100mL) and Jones's reagent {CrO$_3$ (10.3g) in H$_2$O (30mL)} and conc H$_2$SO$_4$ (8.7mL) is added dropwise (turns green in colour) until excess reagent is present (slight red colour). Allow to stir overnight, decant the acetone soln from the Cr salts and adamantan-2-one, and dry (Na$_2$SO$_4$) and evaporate to dryness. The residue (*ca* 7g) is chromatographed through Al$_2$O$_3$ (250g) and washed with 50% *benzene-pet ether (b 40-60°), then 100% Et$_2$O (to remove any adamantan-2-one present) and the 1-adamantanol is then eluted with 5% MeOH in Et$_2$O. The eluate is evaporated, and the residue is recrystd from pet ether (b 30-60°) at -70°, **m** 287.2-288.5°. It has characteristic IR, ν 3640, 1114, 1086, 982 and 930cm^{-1}. [*J Am Chem Soc* **83** 182 *1961*.]
Alternatively, if free from the 2-isomer, dissolve in tetrahydrofuran, dilute with H$_2$O to ppte the alcohol. Collect, dry and sublime in a vacuum at 130°. [*Chem Ber* **92** 1629 *1959*.]

2-Adamantanol (2-hydroxyadamantane) *[700-57-2]* **M 152.4, m 296.2-297.7°.** Can be purified by chromatography as for the 1-isomer. It crystallises from cyclohexane and has characteristic IR, ν 3600, 1053, 1029 and 992cm^{-1} [*J Am Chem Soc* **8** 182 *1961*].

2-Adamantanone *[700-58-3]* **M 150.2, m 256-258°(sublimes).** Purified by repeated sublimation *in vacuo*. [Butler et al. *J Chem Soc, Faraday Trans 1* **82** 535 *1986*.]

N-(1-Adamantyl)acetamide *[880-52-4]* **M 193.3, m 149°.** Wash well with H$_2$O, dry and recrystallise from cyclohexane. *It is an* **irritant.** [*Chem Ber* **92** 1629 *1959*.]

1-Adamantylamine *[768-94-5]* **M 151.2, m 160-190°, 208-210°, pK²⁵ 10.58.** Dissolve in Et_2O, dry over KOH, evaporate and sublime in a vacuum. [*Chem Ber* **93** 226 *1960.*]

1-Adamantylamine hydrochloride *[665-66-7]* **M 187.7, m 360° (dec).** Dissolve in dry EtOH, add a few drops of dry EtOH saturated with HCl gas, followed by dry Et_2O to crystallise the hydrochloride. Dry the salt in vacuum. [*Chem Ber* **93** 226 *1960.*]

2-Adamantylamine hydrochloride *[10523-68-9]* **M 187.7, m >300°, pK$_{Est}$ ~10.4.** The free amine in Et_2O, liberated by the action of alkali in H_2O, is dried over KOH, filtered, evap and sublimed at 110°/12Torr, **m** 230-236°. The base is dissolved in EtOH and crystd by the addition of Et_2O, and dried in vac. [*Justus Liebigs Ann Chem* **658** 151 *1962*].

1-Adamantyl bromide *[768-90-1]* **M 215.1, m 117-119°, 118°, 119.5-120°.** If coloured, dissolve in CCl_4, wash with H_2O, treat with charcoal, dry ($CaCl_2$), filter, evap to dryness. Dissolve in a small volume of MeOH and cool in a CO_2/trichloroethylene bath and collect the crystals. Sublime at 90-100°/water pump vacuum. [*Chem Ber* **92** 1629 *1959*; *J Am Chem Soc* **83** 2700 *1961.*]

1-Adamantyl bromomethylketone *[5122-82-7]* **M 257.2, m 76-79°, 78-79°.** Dissolve in Et_2O, wash with H_2O, dry ($MgSO_4$), evaporate and crystallise residue from small volumes of MeOH. **LACHRYMATORY.** [*Chem Ber* **93** 2054 *1960.*]

1-Adamantyl chloride *[935-56-8]* **M 170.7, m 164.3-165.6°.** Crystd from aqueous MeOH and sublimed at 100°/12Torr. Also crystd from MeOH at -70°. [*Chem Ber* **92** 1629 *1959*; *J Am Chem Soc* **83** 2700 *1961.*]

1-Adamantyl fluoride (1-fluoroadamantane) *[768-92-3]* **M 154.2, m 210-212° (dec), 259-260° (dec).** Dissolve in Et_2O, dry over Na_2SO_4, evaporate to dryness and sublime the residue at 90-100°/12mm. Recryst sublimate from MeOH, **m** 259-260°. [*Zh Org Khim* **30** 1609 *1965.*] To remove 1-hydroxyadamantane impurity, dissolve in cyclohexane cool for many hours, filter off the hydroxyadamantane, and evaporate to dryness. Recrystallise the residue from pet ether at -77° and sublime in vacuum, **m** 210-212° dec (sealed tube). [*J Org Chem* **30** 789 *1965.*]

1-Adamantyl fluoroformate *[62087-82-5]* **M 198.2, m 31-32°.** Dissolve in *n*-hexane (*ca* 10g in 150 mL) and keep at 0° for 24h. Any 1-adamantanol present will separate. Filter and evaporate to dryness. Crystalline residue has **m** 31-32° (v 1242, 1824 and 2340 cm^{-1}). There should be no OH str band above 2500 cm^{-1}. [*Z Phys Chem* **357** 1647 *1976*; Haas et al. *J Am Chem Soc* **88** 1988 *1966.*]

1-Adamantyl iodide (1-iodoadamantane) *[768-93-4]* **M 262.1, m 75.3-76.4°.** Dissolve in Et_2O, shake with aqueous $NaHSO_3$, aqueous K_2CO_3, and H_2O, dry (Na_2SO_4), evaporate and recrystallise from MeOH at -70° (to avoid alcoholysis) giving white crystals. [*J Am Chem Soc* **83** 2700 *1961*; lit **m** of 151-152.5° is incorrect.] Also purified by recrystn from pet ether (40-60°C) followed by rigorous drying and repeated sublimation.

1-Adamantyl isocyanate *[4411-25-0]* **M 177.3, m 144-145°.** Recryst from *n*-hexane and sublime. **Irritant.** [*Chem Ber* **95** 2302 *1962.*]

1-Adamantyl isothiocyanate *[4411-26-1]* **M 193.3, m 168-169°.** Dissolve in Et_2O, wash with H_2O, dry (Na_2SO_4), evaporate and sublime the residue in a vacuum at 140°, and recryst from MeOH. **Irritant.** [*Chem Ber* **95** 2302 *1962.*]

N-(1-Adamantyl)urea *[13072-69-0]* **M 194.2, m >250° (dec), 268-272° (dec).** Wash with H_2O and dioxane and recryst from EtOH. [*Chem Ber* **95** 2302 *1962.*]

Adenine *[73-24-5]* **M 135.1, m 360-365° (dec rapid heating), pK$_1^{25}$ 4.12, pK$_2^{25}$ 9.83.** Crystd from distd water.

Adenosine *[58-61-7]* M 267.3, m 234-236°, $[\alpha]_{546}$ -85° (c 2, 5% NaOH), pK_1^{25} 3.48, pK_2^{25} 12.5. Crystd from distilled water.

Adipic acid *[124-04-9]* M 146.1, m 154°, pK_1^{25} 4.44, pK_2^{25} 5.45. For use as a volumetric standard, adipic acid was crystd once from hot water with the addition of a little animal charcoal, dried at 120° for 2h, then recrystd from acetone and again dried at 120° for 2h. Other purification procedures include crystn from ethyl acetate and from acetone/petroleum ether, fusion followed by filtration and crystn from the melt, and preliminary distn under vac.

Adiponitrile (1,4-dicyanobutane) *[111-69-3]* M 108.14, m 2.4°, b 123°/0.5mm, 153°/6mm, 175°/26mm, 184°/30mm, 295°/atm, d_4^{20} 0.9396, n_D^{20} 1.4371. Reflux over P_2O_5 and $POCl_3$, and fractionally distil, then fractionate through an efficient column. **The liquid is TOXIC and is an IRRITANT.** [*Chem Ber* **67** 1770 *1934*; *Justus Liebigs Ann Chem* **596** 127 *1955*; *Can J Chem* **34** 1662 *1956*; *J Am Chem Soc* **62** 228 *1940*.]

Adonitol (Ribitol) *[488-81-3]* M 152.2, m 102°. Crystallise from EtOH by addition of diethyl ether.

Adrenalin see **epinephrine.**

Adrenochrome *[54-06-8]* M 179.2, m 125-130°. Crystd from MeOH/formic acid, as hemihydrate, and stored in a vacuum desiccator.

Adrenosterone (Reichstein's G) *[382-45-6]* M 300.4, m 220-224°. Crystd from EtOH. Can be sublimed under high vacuum.

Agaricic acid [1-(*n*-hexadecyl)citric acid] *[666-99-9]* M 416.6, m 142°(dec), $[\alpha]_D$ -9.8° (in NaOH), $pK_{Est(1)}$ ~2.7, $pK_{Est(2)}$ ~4.2, $pK_{Est(3)}$ ~5.5. Crystd from EtOH.

Agmatine sulfate [5-guanidinopent-1-ylamine sulfate] *[2482-00-0]* M 228.3, m 231°, $pK_{Est(1)}$ ~9.1, $pK_{Est(2)}$ ~13.0. Crystd from aqueous MeOH.

Agroclavin *[548-42-5]* M 238.3, m 198-203°(dec), 205-206°, $[\alpha]_D^{20}$ -155° (c 1, $CHCl_3$), pK_{Est} ~8.0. Crystd from diethyl ether.

Ajmalicine *[483-04-5]* M 352.4, m 250-252°(dec), $[\alpha]_{546}$ -76° (c 0.5, $CHCl_3$), pK_{Est} ~7.4. Crystd from MeOH.

Ajmalicine hydrochloride *[4373-34-6]* M 388.9, m 290°(dec), $[\alpha]_D$ -17° (c 0.5, MeOH). Crystd from EtOH.

Ajmaline [γ-yohimbine] *[4360-12-7]* M 326.4, m 160° (MeOH), 205-206° (anhyd), $[\alpha]_D^{20}$ +144° (c 0.8, $CHCl_3$), pK_{Est} ~7.5. Crystd from MeOH.

Ajmaline hydrochloride *[4410-48-4]* M 388.9, m 140°. Crystd from water.

Alanine *(RS)* *[302-72-7]* M 89.1, m 295-296°, *(S)* *[56-41-7]* m 297°(dec), $[\alpha]_D^{15}$ +14.7° (in 1M HCl), pK_1^{25} 2.34, pK_2^{25} 9.87. Crystd from water or aqueous EtOH, e.g. crystd from 25% EtOH in water, recrystd from 62.5% EtOH, washed with EtOH and dried to constant weight in a vacuum desiccator over P_2O_5. [Gutter and Kegeles *J Am Chem Soc* **75** 3893 *1953*.] 2,2'-Iminodipropionic acid is a likely impurity.

ß-Alanine *[107-95-9]* M 89.1, m 205°(dec), pK_1^{25} 3.55, pK_2^{25} 10.24. Crystd from filtered hot saturated aqueous soln by adding four volumes of absolute EtOH and cooling in an ice-bath. Recrystd in the same way and then finally, crystd from a warm saturated soln in 50% EtOH by adding four volumes of absolute

EtOH cooled in an ice bath. Crystals were dried in a vacuum desiccator over P_2O_5. [Donovan and Kegeles *J Am Chem Soc* **83** 255 *1961*.]

S-Alaninol [*S*-2-Aminopropan-1-ol]   *[2749-11-3]* M 75.1, b 167-169°/760mm, d_4^{20} 0.961, n_D^{20} 1.456, $[\alpha]_{546}$ +26.0° (c 2, EtOH), pK^{25} 9.43. Purification as for *S*-2-amino-3-methylbutan-1-ol

Aldol (3-hydroxybutanal) *[107-89-1]* **M 88.1, b 80-81°/20mm.** An ethereal soln was washed with a saturated aqueous soln of $NaHCO_3$, then with water. The non-aqueous layer was dried with anhydrous $CaCl_2$ and distd immediately before use. The fraction, **b** 80-81°/20mm, was collected, [Mason, Wade and Pouncy *J Am Chem Soc* **76** 2255 *1954*].

Aldosterone *[52-39-1]* **M 360.5, m 108-112°(hydrate), 164°(anhydr),** $[\alpha]_D^{25}$ **+161°** (c 1, $CHCl_3$). Crystd from aqueous acetone. *Acetate*, cryst from Me_2CO + Et_2O, has **m** 198-199°, $[\alpha]_D^{24}$ +121.7° (c 0.7, $CHCl_3$)

Aldrin *[309-00-2]* **M 354.9, m 103-104.5°.** Crystd from MeOH. **POISONOUS**

Aleuritic acid [*RS*-erythro-9,10,16-trihydroxyhexadecanoic acid] *[533-87-9]* **M 304.4, m 100-101°.** Crystd from aqueous EtOH. *Hydrazide* cryst from EtOH has **m** 139-140°.

Alginic acid *[9005-32-7]* **M 48,000-186000.** To 5g in 550mL water containing 2.8g $KHCO_3$, were added 0.3mL acetic acid and 5g potassium acetate. EtOH to make the soln 25% (v/v) in EtOH was added and any insoluble material was discarded. Further addition of EtOH, to 37% (v/v), ppted alginic acid. [Pal and Schubert *J Am Chem Soc* **84** 4384 *1962*.]

Aliquat 336 (methyltricaprylylammonium chloride, tri-*n*-octylmethylammonium chloride) *[5137-55-3]* **M 404.2, d 0.884.** A 30% (v/v) soln in *benzene was washed twice with an equal volume of 1.5M HBr. [Petrow and Allen, *Anal Chem* **33** 1303 *1961*.] Purified by dissolving 50g in $CHCl_3$ (100mL) and shaking with 20% NaOH soln (200mL) for 10min, and then with 20% NaCl (200mL) for 10min. Washed with small amount of H_2O and filtered through a dry filter paper [Adam and Pribil *Talanta* **18** 733 *1971*].

Alizarin (1,2-dihydroxyanthraquinone) *[72-48-0]* **M 240.2, m 290°, d 0.884, pK_1^{25} 7.45, pK_2^{25} 11.80.** Crystd from glacial acetic acid or 95% EtOH. Can also be sublimed at 110°/2mm.

Alizarin-3-methyliminodiacetic acid (Alizarin Complexone) ($2H_2O$) *[3952-78-1]* **M 421.4, m 189°(dec), $pK_{Est(1)}$~4.9, $pK_{Est(2)}$~7.5.** Purified by suspending in 0.1M NaOH (1g in 50mL), filtering the solution and extracting alizarin with 5 successive portions of CH_2Cl_2. Then add HCl dropwise to precipitate the reagent, stirring the solution in a bath. Filter ppte on glass filter, wash with cold water and dry in a vacuum desiccator over KOH [Ingman *Talanta* **20** 135 *1973*].

Alizarin Yellow R [5-(4-nitrophenylazosalicylic acid), Mordant Orange I] *[2243-76-7]* **M 287-2, m 253-254°(dec), >300°, pK^{25} 11.17.** The free acid is ppted by adding HCl to an aq soln of the Na salt. After 2 recrystns from aq AcOH, it has **m** 255°(dec); [m 253-254° dec was reported *J Chem Soc* **79** 49 *1901*]. The free acid can be recrystd from dilute AcOH as orange brown needles. The Na salt changes colour from yellow to red when the pH is increased from 10.2 to 12.0. [*J Am Chem Soc* **75** 5838 *1953*.]

n-**Alkylammonium chloride** **n=2,4,6.** Recrystd from EtOH or an EtOH/Et_2O mixture. [Hashimoto and Thomas *J Am Chem Soc* **107** 4655 *1985;* Chu and Thomas *J Am Chem Soc* **108** 6270 *1986*.]

n-**Alkyltrimethylammonium bromide** **n=10,12,16.** Recrystd from an EtOH/Et_2O mixture. [Hashimoto and Thomas *J Am Chem Soc* **107** 4655 *1985*.]

Allantoin *[97-59-6]* **M 158.1, m 238°(dec).** Crystd from water or EtOH.

Allene (prodiene) *[463-49-0]* **M 40.1, m -146°, b -32°.** Frozen in liquid nitrogen, evacuated, then thawed out. This cycle was repeated several times, then the allene was frozen in a methyl cyclohexane-liquid nitrogen bath and pumped for some time. Also purified by HPLC. [Cripps and Kiefer *Org Synth* **42** 12 *1962*.]

(-)-Alloaromadendrene *[25246-27-9]* **M 204.4, b 96°/2mm, 265-267°/atm, $[\alpha]_D^{25}$ -22° (neat), d_4^{20} 0.923, n_D^{23} 1.501.** Fractionally distd from Na. IR has bands at 6.06 and 11.27μ due to $C=CH_2$. [*J Chem Soc* 715 *1953*; *cf J Am Chem Soc* **91** 6473 *1969*.]

***neo*-Allocimene (*tc*-2,6-dimethyl-2,4,6-octatriene)** *[7216-56-0]* **M 136.2, b 80°/13mm,196-198°/atm, d_4^{20} 0.8161, n_D^{20} 1.5437.** Fractionally distd through an efficient column and stabilised with ca 0.1% of hydroquinone. UV: λ_{max} nm(ε $M^{-1}cm^{-1}$) 290 (32 500), 279 (41 900) and 270 (32 600). [*Justus Liebigs Ann Chem* **609** 1 *1957*; *Anal Chem* **26** 1726 *1954*.]

5α-Allopregnane-3α,20α-diol *[566-58-5]* **M 320.5, m 248-248.5°, $[\alpha]_D$+17° (c 0.15, EtOH).** Crystd from EtOH.

D-Allothreonine *[2R,3R(-)-isomer]* *[24830-94-2]* **M 119.1, m 272-273°(dec), 276°(dec), $[\alpha]_D^{25}$ -9.1° (c 3.9, H_2O), pK_1^{25} 2.11, pK_2^{25} 9.10.** Recrystd from aqueous EtOH or 50% EtOH. [*J Chem Soc* 62 *1950*; *J Am Chem Soc* **194** 455 *1952*; IR: Greenstein & Winitz *The Chemistry of the Amino Acids* J. Wiley, **Vol 3** *1961*.]

Alloxan [2,4,5,6(1H,3H)pyrimidine, tetrone] *[50-71-5]* **M 142.0, m ~170°(dec), pK^{25} 6.64.** Crystn from water gives the tetrahydrate. Anhydrous crystals are obtained by crystn from acetone, glacial acetic acid or by sublimation *in vacuo*.

Alloxan monohydrate *[2244-11-3]* **M 160.1, m 255°(dec), pK 6.64.** Recryst from H_2O as the *tetrahydrate* in large prisms or rhombs. On heating at 100°, or on exposure to air, this is converted to the *monohydrate*. Dissolve it in its own weight of boiling H_2O and cool for several days below 0° [the *tetrahydrate* crystallises from soln much more slowly when free from HNO_3. It is less sol in HCO_3 solns than in H_2O]. Drying the solid over H_2SO_4 yields the *monohydrate*. The *anhydrous* crystals can be obtained by recrystn from dry Me_2CO or AcOH followed by washing with dry Et_2O or by sublimation in a vacuum. On heating it turns pink at 230° and decomposes at *ca* 256°. It is acidic to litmus. [*Org Synth* Coll Vol III 37 *1955*.] It forms a compound with urea which crystallises from H_2O in yellow needles that become red at 170° and dec at 185-186°.

Alloxantin *[76-24-4]* **M 286.2, m 253-255°(dec) (yellow at 225°).** Crystd from water or EtOH and kept under nitrogen. Turns red in air.

Allyl acetate *[591-87-7]* **M 100.1, b 103°, d 0.928, n_4 1.40488, n_D^{27} 1.4004.** Freed from peroxides by standing with crystalline ferrous ammonium sulfate, then washed with 5% $NaHCO_3$, followed by saturated $CaCl_2$ soln. Dried with Na_2SO_4 and fractionally distd in an all-glass apparatus.

Allylacetic acid (pent-4-enoic acid) *[591-80-0]* **M 100.1, m -22.5°, b 83-84°/12mm, 90°/15mm, d_4^{20} 0.9877, n_D^{20} 1.4280, pK^{25} 4.68.** Distil through an efficient column (allyl alcohol has **b** 95-97°). It is characterised as the *S-benzyl isothiouronium salt* **m** 155-158° (96% EtOH, aq EtOH) [*Acta Chem Scand* **9** 1425 *1955*], *4-bromophenacyl ester* **m** 59.5-60.5° (from 90% EtOH). Solubility at 18°: in pyridine (57%), AcOH (7.3%), MeOH (5.4%), Me_2CO (3.2%), MeOAc (2.8%), EtOH (5.4%), H_2O (1.8%), PrOH (1.6%), isoPrOH (0.27%). [*J Am Chem Soc* **74** 1894 *1952*.]

Allyl alcohol *[107-18-6]* **M 58.1, b 98°, d_4 0.857, n_D 1.4134.** Can be dried with K_2CO_3 or $CaSO_4$, or by azeotropic distn with *benzene followed by distn under nitrogen. It is difficult to obtain peroxide free. Also reflux with magnesium and fractionally distd [Hands and Norman *Ind Chem* **21** 307 *1945*].

Allylamine *[107-11-9]* **M 57.1, b 52.9°, d 0.761, n 1.42051, pK^{25} 9.49.** Purified by fractional distn from calcium chloride. Causes sneezing and tears.

1-Allyl-6-amino-3-ethyluracil *[642-44-4]* **M 195.2, m 143-144° (anhydr).** Crystd from water (as monohydrate).

Allyl bromide *[106-95-6]* **M 121, b 70°, d 1.398, n 1.46924.** Washed with NaHCO$_3$, soln then distd water, dried (CaCl$_2$ or MgSO$_4$), and fractionally distd. Protect from strong light. **LACHRYMATOR, HIGHLY TOXIC and FLAMMABLE.**

Allyl butyl ether *[3739-64-8]* **M 114.2, b 64-65°/120mm, 117.8-118°/763mm, d$_4^{20}$ 1.4057, n$_D^{20}$ 0.7829.** Check the IR for the presence of OH str vibrations, if so then wash well with H$_2$O, dry with CaCl$_2$ and distil through a good fractionating column. **The liquid is an irritant.** [*J Org Chem* **23** 1666 *1958*; *J Am Chem Soc* **73** 3528 *1951*.]

Allyl chloride *[107-05-1]* **M 76.5, b 45.1°, d 0.939, n 1.4130.** Likely impurities include 2-chloropropene, propyl chloride, *i*-propyl chloride, 3,3-dichloropropane, 1,2-dichloropropane and 1,3-dichloropropane. Purified by washing with conc HCl, then with Na$_2$CO$_3$ soln, drying with CaCl$_2$, and distn through an efficient column [Oae and Vanderwerf *J Am Chem Soc* **75** 2724 *1953*]. **LACHRYMATOR, TOXIC.**

Allyl chloroformate *[2937-50-0]* **M 120.5, b 56°/97mm, 109-110°/atm, d$_4^{20}$ 1.14, n$_D^{20}$ 1.4223.** Wash several times with cold H$_2$O to remove alcohol and HCl and dry over CaCl$_2$. It is **important** to dry well before distilling *in vacuo*. Note that the receiver should be cooled in ice to avoid loss of distillate into the trap and vacuum pump. The liquid is **highly TOXIC and flammable.** [*J Am Chem Soc* **72** 1254 *1950*.]

Allyl cyanide (3-butene nitrile) *[109-75-1]* **M 67.1, b -19.6°/1.0mm, 2.9°/5 mm, 14.1°/5mm, 26.6°/20mm, 48.8°/60mm, 60.2°/100mm, 98°/400mm, 119°/760mm, d$_4^{20}$ 0.8341, n$_D^{20}$ 1.406.** It should be redistd at atmospheric pressure then distilled under a vacuum to remove final traces of HCN from the residue. Note that the residue from the first distiln may be difficult to remove from the flask and should be treated with conc HNO$_3$ then H$_2$O and finally hot EtOH (**CARE**). Allyl cyanide has an onion-like odour and is stable to heat.. It forms a complex with AlCl$_3$ (2:2) **m 41°**, and (3:2) **m 120°**. **All operations should be done in an efficient fume hood as the liquid is flammable and HIGHLY TOXIC.** [*Org Synth* Coll Vol I 46 *1941*.]

Allyl disulfide (diallyl disulfide) *[2179-57-9]* **M 146.3, b 58-59°/5mm, 79-81°/20mm, 138-139°/atm, d$_4^{20}$ 1.01, n$_D^{20}$ 1.541.** Purified by fractional distn until their molar refractivities are in uniformLy good agreement with the calculated values [*J Am Chem Soc* **69** 1710 *1947*]. Also purified by gas chromatography [retention times: *J Org Chem* **24** 175 *1959*; UV: *J Chem Soc* 395 *1949*].

RS-α-**Allylglycine (2-aminopent-4-enoic acid).** *[7685-44-1]* **M 115.1, m 250-255°(dec), pK$_{Est(1)}$ ~2.3, pK$_{Est(2)}$ ~9.6.** Dissolve in absolute EtOH and ppte with pyridine, then recrystallise from aqueous EtOH [R$_F$ in BuOH:EtOH:NH$_3$:H$_2$O (4:4:1:1:) 0.37]. The *hydrobromide* has **m** 136-140° (from EtOAc) and the *phenylureido* derivative has **m** 159-161°. [*Monatsh Chem* **89** 377 *1958*.]

1-*N*-Allyl-3-hydroxymorphinan *[152-02-3]* **M 283.4, m 180-182°.** Crystd from aqueous EtOH.

Allyl iodide (3-iodopropene) *[556-56-9]* **M 167.7, b 103°, d^{12} 1.848.** Purified in a dark room by washing with aq Na$_2$SO$_3$ to remove free iodine, then drying with MgSO$_4$ and distilling at 21mm pressure, to give a very pale yellow liquid. (This material, dissolved in hexane, was stored in a light-tight container at -5° for up to three months before free iodine could be detected, by its colour in the soln) [Sibbett and Noyes *J Am Chem Soc* **75** 761 *1953*].

5-Allyl-5-*iso***butylbarbituric acid** *[77-26-9]* **M 224.3, m 139°, 139-140°, 140-142°, pK^{38} 12.36.** It can be recrystallised from H_2O or dilute EtOH, and sublimes at 100-120°/8-12mm. It is soluble in *C_6H_6, cyclohexane, tetralin and pet ether at 20°. [*J Am Chem Soc* **77** 1486 *1955*.]

Allylisocyanate *[1476-23-9]* **M 83.1, b 84°/atm, 87-89°/atm, d_4^{20} 0.94, n_D^{20} 1.417.** Purify as for allylisothiocyanate.

Allylisothiocyanate *[57-06-7]* **M 99.2, m -80°, b 84-85°/80mm, 150°/760mm, 151°/atm, d_4^{20} 1.017, n_D^{20} 1.5268.** Fractionate using an efficient column, preferably in a vacuum. It is a yellow **pungent irritating and TOXIC (suspected CARCINOGEN) liquid.** Store in a sealed tube under N_2. The *N'-benzylthiourea* derivative has **m** 94.5° (from aq EtOH) [*J Am Chem Soc* **74** 1104 *1952*].

Allyl Phenyl sulfide *[5296-64-0]* **M 150.2, b 59-60°/1.5mm, 79-80°/3mm, 114-114.3°/23.5mm, 225-226°/740mm, 215-218°/750mm, d_4^{20} 1.0275, n_D^{20} 1.5760.** Dissolve in Et_2O, wash with alkali, H_2O, dry over $CaCl_2$, evaporate and fractionally distil, preferably under vacuum. It should not give a ppte with an alcoholic soln of $Pb(OAc)_2$. [*J Am Chem Soc* **52** 3356 *1930*, **74** 48 *1952*.]

*N***-Allylthiourea (thiosinamine)** *[109-57-9]* **M 116.2, m 70-73°, 78°.** Recrystd from H_2O. Soluble in 30 parts of cold H_2O, soluble in EtOH but insoluble in *C_6H_6. Also recrystd from acetone, EtOH or ethyl acetate, after decolorising with charcoal. The white crystals have a bitter taste with a slight garlic odour and are **TOXIC.** [*Anal Chem* **21** 421 *1949*.]

*N***-Allylurea** *[557-11-9]* **M 100.1, m 85°.** Crystd from EtOH, EtOH/ether, EtOH/chloroform or EtOH/toluene.

Aloin **[10-glucopyranosyl-1,8-dihydroxy-3-(hydroxymethyl)-9(10***H***)anthracenone, Barbaloin]** *[8015-61-0]* **M 418.4, m 148-148.5°, 148-150°.** Lemon yellow crystals from H_2O (450g/1.5L) as the *monohydrate* which has a lower **m** (70-80°). [*J Chem Soc* 2573 *1932*, 3141 *1956*.]

D-Altrose *[1990-29-0]* **M 180.2, m 103-105°, [α]$_{546}$ +35° (c 7.6, H_2O).** Crystd fom aq EtOH.

Amberlite IRA-904 Anion exchange resin (Rohm and Haas) *[9050-98-0]*. Washed with 1M HCl, CH_3OH (1:10) and then rinsed with distilled water until the washings were neutral to litmus paper. Finally extracted successively for 24h in a Soxhlet apparatus with MeOH, *benzene and cyclohexane [Shue and Yan *Anal Chem* **53** 2081 *1981*]. Strongly basic resin also used for base catalysis [Fieser & Fieser *Reagents for Org Synth* **1** 511, Wiley *1967*].

Aminoacetaldehyde dimethyl acetal (2,2-dimethoxyethylamine) *[22483-09-6]* **M 105.1, m <-78°, b 139.5°/768mm, 137-139°/atm, d_4^{20} 0.9676 n_D^{20} 1.4144.** Dry over KOH pellets and distil through a 30cm vac jacketed Vigreux column. [*J Am Chem Soc* **75** 3398 *1953*, **77** 6640 *1955*.]

*p***-Aminoacetanilide** *[122-80-5]* **M 150.2, m 162-163°.** Crystd from water.

Aminoacetic acid (Glycine) *[56-40-6]* **M 75.1, m 262° (dec, goes brown at 226°, sublimes at 200°/0.1mm), pK$_1^{25}$ 2.35, pK$_2^{25}$ 9.78.** Crystd from distilled water by dissolving at 90-95°, filtering, cooling to about -5°, and draining the crystals centrifugally. Alternatively, crystd from distilled water by addition of MeOH or EtOH (e.g. 50g dissolved in 100mL of warm water, and 400mL of MeOH added). The crystals can be washed with MeOH or EtOH, then with diethyl ether. Likely impurities are ammonium glycinate, iminodiacetic acid, nitrilotriacetic acid, ammonium chloride.

Aminoacetonitrile bisulfate *[151-63-3]* **M 154.1, m 188°(dec)** Crystd from aqueous EtOH.

Aminoacetonitrile hydrochloride *[6011-14-9]* **M 92.5, m 166-167°, 172-174°, pK^{25} 5.34.** Recrystd from dil EtOH *hygroscopic* leaflets. Best to crystallise from absolute EtOH-Et_2O (1:1) and then recryst from absolute EtOH. The **m** recorded range from 144° to 174°. The free base has **b** 58°/15mm with partial decomposition. [*J Prakt Chem* **[2] 65** 189 *1902*; *J Am Chem Soc* **56** 2197 *1934*; *J Chem Soc* 1371 *1947*.]

2-Aminoacetophenone hydrochloride *[5468-37-1]* **M 171.6, m 188°(dec), 194°(dec), pK^{25} 5.34.** Crystd from acetone/EtOH or 2-propanol [Castro *J Am Chem Soc* **108** 4179 *1986*].

m-**Aminoacetophenone** *[99-03-6]* **M 135.2, m 98-99°, pK^{25} 3.56.** Recrystd from EtOH.

p-**Aminoacetophenone** *[99-92-3]* **M 135.2, m 104-106°, 105-107°, b 293°/atm, pK^{25} 2.19** Recryst from CHCl$_3$, *C$_6$H$_6$ or H$_2$O. Soluble in hot H$_2$O. UV (EtOH) has λ_{max} 403nm (log ε 4.42) [*J Am Chem Soc* **75** 2720 *1953*]. [*Anal Chem* **26** 726 *1954*.] The *2,4-dinitrophenylhydrazone* has **m** 266-267° (from CHCl$_3$ or EtOH), and the *semicarbazone* has **m** 193-194°(dec)(from MeOH) and the *hydrochloride* has **m** 98°(dec)(from H$_2$O).

α-**Amino acids** see Chapter 6.

9-Aminoacridine [**9-acridineamine**] *[90-45-9]* **M 194.2, m 241°, pK^{20} 9.95.** Crystd from EtOH or acetone and sublimes at 170-180°/0.04mm [Albert and Ritchie *Org Synth* Coll Vol III 53 *1955*; for hydrochloride see Chapter 6.]

dl-α-**Aminoadipic acid (hydrate)** *[542-32-5]* **M 161.2, m 196-198°, pK$_{Est(1)}$ ~2.0, pK$_{Est(2)}$ ~4.5, pK$_{Est(3)}$ ~9.8.** Crystd from water.

2-Amino-4-anilino-*s*-triazine *[537-17-7]* **M 168.2, m 235-236°, pK$_{Est}$ ~5.5.** Crystd from dioxane or 50% aqueous EtOH.

1-Aminoanthraquinone-2-carboxylic acid *[82-24-6]* **M 276.2, m 295-296°.** Crystd from nitrobenzene.

4-Aminoantipyrine *[83-07-8]* **M 203.3, m 109°.** Crystd from EtOH or EtOH/ether.

p-**Aminoazobenzene (*p*-phenylazoaniline)** *[60-09-3]* **M 197.2, m 126°, pK^{25} ~2.82.** Crystd from EtOH, CCl$_4$, pet ether/*benzene, or a MeOH/water mixture.

o-**Aminoazotoluene (Fast Garnet GBC base, 4'-amino-2,3'dimethylazobenzene)** *[97-56-3]* **M 225.3, m 101.4-102.6°, CI 11160, pK$_{Est}$ ~2.8.** Crystd twice from EtOH, once from *benzene, then dried in an Abderhalden drying apparatus [Cilento *J Am Chem Soc* **74** 968 *1952*]. **CARCINOGENIC.**

2-Aminobenzaldehyde *[529-23-7]* **M 121.1, m 39-40°, pK^{20} 1.36.** Distd in steam and crystd from water or EtOH/ether.

2-Aminobenzaldehyde phenylhydrazone (Nitrin) *[63363-93-9]* **M 211.3, m 227-229°.** Crystd from acetone. [Knöpfer *Monatsh Chem* **31** 97 *1910*.]

3-Aminobenzaldehyde *[29159-23-7]* **M 121.1, m 28-30°, pK$_{Est}$ ~2.0.** Crystd from ethyl acetate.

4-Aminobenzamide hydrochloride *[59855-11-7]* **M 199.6, m 284-285°, pK$_{Est}$ ~1.7.** Recrystd from EtOH.

p-**Aminobenzeneazodimethylaniline** *[539-17-3]* **M 240.3, m 182-183°.** Crystd from aqueous EtOH.

o-**Aminobenzoic acid (anthranilic acid)** *[118-92-3]* **M 137.1, m 145°, pK$_1^{25}$ 2.94, pK$_2^{25}$ 4.72.** Crystd from water (charcoal). Has also been crystd from 50% aqueous acetic acid. Can be vacuum sublimed.

m-Aminobenzoic acid *[99-05-8]* M 137.1, m 174°, pK$_1^{25}$ 3.29, pK$_2^{25}$ 5.10. Crystd from water.

p-Aminobenzoic acid *[150-13-0]* M 137.1, m 187-188°, pK$_1^{25}$ 2.45, pK$_2^{25}$ 4.85. Purified by dissolving in 4-5% aqueous HCl at 50-60°, decolorising with charcoal and carefully precipitating with 30% Na$_2$CO$_3$ to pH 3.5-4 in the presence of ascorbic acid. It can be crystd from water, EtOH or EtOH/water mixtures.

p-Aminobenzonitrile *[873-74-5]* M 118.1, m 86-86.5°, 85-87°, pK^{25} 1.74. Crystd from water, 5% aqueous EtOH or EtOH and dried over P$_2$O$_5$ or dried *in vacuo* for 6h at 40°. [Moore et al. *J Am Chem Soc* **108** 2257 *1986*; Edidin et al. *J Am Chem Soc* **109** 3945 *1987*.]

4-Aminobenzophenone *[1137-41-3]* M 197.2, m 123-124°, pK^{25} 2.17. Dissolved in aq acetic acid, filtered and ppted with ammonia. Process repeated several times, then recrystd from aqueous EtOH.

2-Aminobenzothiazole *[136-95-8]* M 150.2, m 132°, pK^{20} 4.48. Crystd from aqueous EtOH.

6-Aminobenzothiazole *[533-30-2]* M 150.2, m 87°, pK$_{Est}$ ~3. Crystd from aqueous EtOH.

N-(*p*-Aminobenzoyl)-L-glutamic acid *[4271-30-1]* M 266.3, m 173° (L-form), [α]$_{546}$ -17.5° (c 2, 0.1m HCl); 197° (DL), pK$_{Est(1)}$ ~1.7, pK$_{Est(2)}$ ~3.4, pK$_{Est(3)}$ ~4.3. Crystd from H$_2$O.

3-*o*-Aminobenzyl-4-methylthiazolium chloride hydrochloride *[534-94-1]* M 277.4, m 213°(dec). Crystd from aqueous EtOH.

4-Amino-1-benzylpiperidine *[50541-93-0]* M 190.3, b ~180°/20mm, d 0.933, n 1.543, pK$_{Est(1)}$~ 8.3 pK$_{Est(2)}$~ 10.4. Purified by distn *in vacuo*, and stored under N$_2$, because it absorbs CO$_2$. The *dihydrochloride salt* *[1205-72-7]* has m 270-273° (255°) after recrystn from MeOH + EtOAc or EtOH. [*J Chem Soc* 3165, 3172 *1957*.] The *4-methylamino-1-benzylpiperidine* derivative has b 168-172°/17mm, n 1.5367 [*J Am Chem Soc* **70** 4009 *1948*]. The *1-(1-benzyl-4-piperidinyl)-3-cyano-2-methylisothiourea* derivative has m 160° from CHCl$_3$/Et$_2$O [Prepn, IR, NMR: Ried et al. *Chem Ber* **116** 1547 *1983*].

o-Aminobiphenyl *[90-41-5]* M 169.2, m 49.0°, pK^{18} 3.83. Crystd from aqueous EtOH (charcoal).

p-Aminobiphenyl *[92-67-1]* M 169.2, m 53°, b 191°/16mm, pK^{18} 4.38. Crystd from water or EtOH. **CARCINOGENIC.**

2-Amino-5-bromotoluene *[583-75-5]* M 186.1, m 59°, pK^{25} 3.58. Steam distd, and crystd from EtOH.

RS-2-Aminobutyric acid *[2835-81-6]* M 103.1, m 303°(dec), pK$_1^{25}$ 2.29, pK$_2^{25}$ 9.83. Crystd from water.

S-2-Aminobutyric acid *[1492-24-6]* M 103.1, m 292°(dec), [α]$_D$ + 20.4° (c 2, 2.5N HCl). Crystd from aqueous EtOH.

3-Aminobutyric acid *[2835-82-7]* M 103.1, m 193-194°, pK$_{Est(1)}$ ~3.5, pK$_{Est(2)}$ ~10.3. Crystd form aqueous EtOH or MeOH + Et$_2$O.

4-Aminobutyric acid (GABA) *[56-12-2]* M 103.1, m 202°(dec), pK$_1^{25}$ 4.14, pK$_2^{25}$ 10.55. Crystd form aqueous EtOH or MeOH + Et$_2$O.

2-Amino-5-chlorobenzoic acid *[635-21-1]* M 171.6, m 100°, pK$_1^{25}$ 1.69, pK$_2^{25}$ 4.35. Crystd from water, EtOH or chloroform.

3-Amino-4-chlorobenzoic acid *[2840-28-0]* M 171.6, m 216-217°, pK$_{Est(1)}$ ~2.7, pK$_{Est(2)}$ ~2.9. Crystd from water.

4-Amino-4'-chlorobiphenyl *[135-68-2]* M 203.5, m 134°, pK$_{Est}$ ~4.0. Crystd from pet ether.

2-Amino-4-chloro-6-methylpyrimidine *[5600-21-5]* M 143.6, m 184-186°, pK$_{Est}$ ~1.0. Crystd from EtOH.

2-Amino-5-chloropyridine *[1072-98-6]* M 128.6, m 135-136°, pK 4.38. Crystd from pet ether, sublimes at 50°/0.5mm.

1-Amino-1-cyclopentanecarboxylic acid (cycloleucine) *[52-52-8]* M 129.2, m 330°(dec), pK$_{Est(1)}$~2.5 pK$_{Est(2)}$~10.3. Crystd from aq EtOH.

2-Amino-3,5-dibromopyridine *[35486-42-1]* M 251.9, m 103-104°, pK$_{Est}$ ~2.4. Steam distd and crystd from aqueous EtOH or pet ether.

2-Amino-4,6-dichlorophenol *[527-62-8]* M 175.0, m 95-96°, pK$_{Est(1)}$~3.1, pK$_{Est(2)}$ ~6.8. Crystd from CS$_2$ or *benzene.

3-Amino-2,6-dichloropyridine *[62476-56-6]* M 164.0, m 119°, b 110°/0.3mm, pK$_{Est}$ ~2.0. Crystd from water.

4-Amino-*N,N*-diethylaniline hydrochloride *[16713-15-8]* M 200.7, m 233.5°, pK^{22} 6.61. Crystd from EtOH.

4-Amino-3,5-diiodobenzoic acid *[2122-61-4]* M 388.9, m >350°, pK$_{Est(1)}$ 0.4, pK$_{Est(2)}$ ~1.6. Purified by soln in dilute NaOH and pptn with dilute HCl. Air dried.

2-Amino-4,6-dimethylpyridine *[5407-87-4]* M 122.2, m 69-70.5°, pK 7.84. Crystd from hexane, ether/pet ether or *benzene. Residual *benzene was removed over paraffin-wax chips in an evacuated desiccator.

2-Amino-4,6-dimethylpyrimidine *[767-15-7]* M 123.2, m 152-153°, pK^{25} 4.95. Crystn from water gives m 197°, and crystn from acetone gives m 153°.

2-Aminodiphenylamine *[534-85-0]* M 184.2, m 79-80°, pK$_{Est(1)}$ ~3.8 (NH$_2$), pK$_{Est(2)}$ <~0. Crystd from H$_2$O.

4-Aminodiphenylamine *[101-54-2]* M 184.2, b 155°/0.026mm, pK^{25} 5.20. Crystn from EtOH gives m 66°, and crystn from ligroin gives m 75°.

2-Amino-1,2-diphenylethanol *[530-36-9]* M 213.3, m 165°, pK$_{Est(1)}$ ~7.5. Crystd from EtOH.

2-Aminodiphenylmethane *[28059-64-5]* M 183.3, m 52°, b 172°/12mm and 190°/22mm, pK$_{Est(1)}$ ~4.2. Crystd from ether.

2-Aminoethanethiol see cysteamine in Chapter 6.

2-Aminoethanol (ethanolamine) *[141-43-5]* M 61.1, f 10.5°, b 72-73°/12mm, 171.1°/760mm, d 1.012, n 1.14539, pK^{25} 9.51. Decomposes slightly when distd at atmospheric pressure, with the formation of conducting impurities. Fractional distn at about 12mm pressure is satisfactory. After distn, 2-aminoethanol was further purified by repeated washing with ether and crystn from EtOH (at low temperature). After fractional distn in the absence of CO$_2$, it was twice crystd by cooling, followed by distn.

Hygroscopic. [Reitmeier, Silvertz and Tartar *J Am Chem Soc* **62** 1943 *1940.*] It can be dried by azeotropic distn with dry *benzene.

2-Aminoethanol hydrochloride *[2002-24-6]* M 97.6, m 75-77°. Crystd from EtOH. It is deliquescent.

2-Aminoethyl hydrogen sulfate (sulfuric acid mono-2-aminoethyl ester) *[926-39-6]* M 141.1, m 285-287° (chars at 275°). Crystd from water or dissolved in water and EtOH added.

S-(2-Aminoethyl)isothiouronium bromide hydrobromide *[56-10-0]* M 281.0, m 194-195°. Crystd from absolute EtOH/ethyl acetate. It is *hygroscopic.*

(2-Aminoethyl)trimethylammonium chloride hydrochloride (chloramine chloride hydrochloride) *[3399-67-5]* M 175.1, m 260°(dec). Crystd from EtOH. (Material is very soluble in H_2O).

2-Aminofluorene *[153-78-6]* M 181.2, m 127.8-128.8°, 132-133°, pK^{25}4.64. Wash well with H_2O and recrystd from Et_2O or 50% aqueous EtOH (25g with 400mL), and dry in a vacuum. Store in the dark. [*Org Synth* Coll Vol II 447 *1943*; Coll Vol V 30 *1973*].

4-Amino hippuric acid *[61-78-9]* M 194.2, m 198-199°, $pK_{Est(1)}$~1.7(NH_2), $pK_{Est(2)}$ ~3.4 (CO_2H). Crystd from H_2O.

4-Amino-3-hydrazino-5-mercapto-1,2,4-triazole (Purpald) *[1750-12-5]* M 146.2, m 228-230°(dec), 234-235°(dec), $pK_{Est(1)}$~2, $pK_{Est(2)}$~3 (NH_2), $pK_{Est(3)}$~8 (SH). Crystd from H_2O (0.6g in 300-400mL). The *benzylidene deriv* has m 245-246°(dec) from *i*-PrOH [Hoggarth *J Chem Soc* 4817 *1952*].

1-Amino-4-hydroxyanthraquinone *[116-85-8]* M 293.2, m 207-208°, $pK_{Est(1)}$ ~2.6 (NH_2), $pK_{Est(2)}$ ~9.0 (OH). Purified by TLC on SiO_2 gel plates using toluene/acetone (9:1) as eluent. The main band was scraped off and extracted with MeOH. The solvent was evaporated and the dye was dried in a drying pistol [Land, McAlpine, Sinclair and Truscott *J Chem Soc, Faraday Trans 1* **72** 2091 *1976*]. Crystd from aq EtOH.

dl-**4-Amino-3-hydroxybutyric acid** *[924-49-2]* M 119.1, m 225°(dec), pK_1^{25}~3.80 (CO_2H), $pK_{Est(2)}$ ~9.3. Crystd from H_2O or aqueous EtOH.

5-Amino-8-hydroxyquinoline hydrochloride *[3881-33-2]* M 196.7, pK_1^{20} 5.67, pK_2^{20} 11.24. Dissolved in minimum of MeOH, then Et_2O was added to initiate pptn. Ppte was filtered off and dried [Lovell et al. *J Phys Chem* **88** 1885 *1984*].

3-Amino-4-hydroxytoluene *[95-84-1]* M 123.2, m 137-138°, $pK_{Est(1)}$~4.7(NH_2), $pK_{Est(2)}$ ~9.6 (OH). Crystd from H_2O or toluene.

4-Amino-5-hydroxytoluene *[2835-98-5]* M 123.2, m 159°, $pK_{Est(1)}$~5.4 (NH_2), $pK_{Est(2)}$ ~10.2 (OH). Crystd from H_2O, 50% EtOH, or toluene.

6-Amino-3-hydroxytoluene *[2835-99-6]* M 123.2, m 162°(dec), $pK_{Est(1)}$~5.4 (NH_2), $pK_{Est(2)}$ ~10.4 (OH). Crystd from 50% EtOH.

4-Aminoimidazole-5-carboxamide hydrochloride (AICAR HCl) *[72-40-2]* M 162.6, m 255-256°(dec), $pK_{Est(1)}$~3.5, $pK_{Est(2)}$ ~9.4. Recrystd from EtOH.

5-Aminoindane *[24425-40-9]* M 133.2, m 37-38°, b 131°/15mm, 146-147°/25mm, 247-249°/745mm, pK^{16} 5.31. Distd and then crystd from pet ether.

6-Aminoindazole *[6967-12-0]* **M 133.2, m 210°, pK^{25} 3.99.** Crystd from H_2O or EtOH and sublimed in a vacuum.

2-Amino-5-iodotoluene *[13194-68-8]* **M 233.0, m 87°, pK$_{Est}$ ~3.6.** Crystd from 50% EtOH.

α-Aminoisobutyric acid *[62-57-7]* **M 103.1, m sublimes at 280°, pK$_1^{25}$ 2.36, pK$_2^{25}$ 10.21.** Crystd from aqueous EtOH and dried at 110°.

RS- **β-Aminoisobutyric acid** (**α-methyl-β-alanine**) *[10569-72-9]* **M 103.1, m 176-178°, 178-180°, 181-182°,** *R* -(-)- *isomer* *[144-90-1]* **m 183°, [α]$_D^{25}$ -21°** (c 0.43, H_2O), **pK$_{Est(1)}$~ 3.7, pK$_{Est(2)}$~ 10.2.** Colorless prisms from hot H_2O, were powdered and dried *in vacuo*. The purity is checked by paper chromatography (Whatman 1) using ninhydrin spray to visualise the amino acid; R_F values in 95% MeOH and *n*-PrOH/5N HCOOH (8:2) are 0.36 and 0.50 respectively. [Kupiecki and Coon *Biochem Prep* **7** 20 *1960*; Pollack *J Am Chem Soc* **65** 1335 *1943*.] The *R-enantiomer*, isolated from iris bulbs or human urine was crystd from H_2O and sublimed *in vacuo* [Asen et al. *J Biol Chem* **234** 343 *1959*]. The *RS-hydrochloride* was recrystd from EtOH/Et$_2$O **m** 128-129°, 130° [Böhme et al. *Chem Ber* **92** 1258, 1260, 1261 *1959*].

5-Aminolaevulinic acid hydrochloride *[5451-09-2]* **M 167.6, m 156-158°(dec), pK$_1^{22}$ 4.05, pK$_2^{22}$ 8.90** Dried in a vacuum desiccator over P_2O_5 overnight then crystd by dissolving in cold EtOH and adding dry Et$_2$O.

Aminomalononitrile toluene-4-sulfonate *[5098-14-6]* **M 253.4, m 168-170°, 172°(dec), pK$_{Est}$ ~ 1.3.** Colourless crysts on recrystn from MeCN (1.8g in 100mL) using activated charcoal. Wash the crystals with dry Et$_2$O and dry at 25°/1mm. Recovery of ~80%. [Ferris et al. *Org Synth* Coll Vol V 32 *1973*.]

3-Amino-5-mercapto-1,2,4-triazole *[16691-43-3]* **M 116.1, m 298°, pK$_{Est(1)}$~ 3.0, pK$_{Est(2)}$~ 9.** Recrystd from H_2O and dried *in vacuo*. The *acetyl derivative* has **m** 325° (dec) after recrystn from H_2O. [*Beilstein* **26, 3rd/4th Suppl** p. 1351.] Also recrystd from EtOH/H_2O (3:1, 1g in 50 mL, 50% recovery), **m** 300-302° dec subject to heating rate (λmax 263nm, log ε 4.12), and *S-Benzyl* derivative when crystd from *C$_6$H$_6$*/EtOH (20:1), or CHCl$_3$/Et$_2$O has **m** 109-111° [Godfrey and Kruzer *J Chem Soc* 3437 *1960*].

2-Amino-4-methoxy-6-methylpyrimidine *[7749-47-5]* **M 139.2, m 157-159°, 158-158.5°, 158-160°, pK$_{Est}$ ~ 6.0.** Recrystd from H_2O. The *picrate* has **m** 220-221°(dec). [Baker et al. *J Am Chem Soc* **69** 3072, 3075 *1947*; Sirakawa et al. *Yakugaku Zasshi* **73** 598 *1953*; Backer and Grevenstuk *Recl Trav Chim, Pays-Bas* **61** 291 *1942*.]

8-Amino-6-methoxyquinoline *[90-52-8]* **M 174.2, m 41-42°, 51°, b 137-138°/1mm, pK^{7} 0.1 3.38.** Distd under N_2 and high vac, then recrystd several times from MeOH (0.4mL/g). It remains colourless for several months when purified in this way [Elderfield and Rubin *J Am Chem Soc* **75** 2963 *1953*].

1-Amino-4-methylaminoanthraquinone *[1220-94-6]* **M 252.3, pK$_{Est(1)}$ ~1, pK$_{Est(2)}$ <~4.** Purified by TLC on silica gel plates using toluene/acetone (3:1) as eluent. The main band was scraped off and extracted with MeOH. The solvent was evaporated and the residue dried in a drying pistol [Land, McAlpine, Sinclair and Truscott *J Chem Soc, Faraday Trans 1* **72** 2091 *1976*].

4-Aminomethylbenzenesulfonamide hydrochloride *[138-37-4]* **M 222.3, m 265-267°, pK$_1^{20}$ 8.18** (NH$_2$), **pK$_2^{20}$ 10.23** (SONH$_2$). Crystd from dilute HCl and dried in a vacuum at 100°.

*S-***2-Amino-3-methyl-1-butanol** (*S-***valinol**) *[2026-48-4]* **M 103.2, m 31-32°, b 88°/11mm, d 0.92, [α]$_{546}$ + 16.5°** (c 6.32, *l* = 2 H_2O), **[α]$_D$ + 15.6°** (EtOH), **pK$_{Est}$ ~10.4.** Purified by vacuum distn using short Vigreux column. Alternatively it is purified by steam distn. The steam distillate is acidified with HCl, the aq layer is collected and evapd. The residue is dissolved in butan-1-ol, filtered and dry Et$_2$O added to cryst the hydrochloride salt (*hygroscopic*), **m** 113°. The free base can be obtained by suspending the salt in Et$_2$O adding small vols of satd K_2CO_3 until effervescence is complete and the mixture is distinctly alkaline. At this stage the aqueous layer should appear as a white sludge. The mixture is heated to boiling and refluxed for 30 min (more Et$_2$O is added if necessary). The Et$_2$O is decanted off from the white sludge, the sludge is extracted twice with Et_2O (by boiling for a few minutes), the combined organic layers are dried (KOH pellets), evapd and the residue distd in a vacuum.

7-Amino-4-methylcoumarin *[26093-31-2]* **M 175.2, m 221-442°(dec), pK_{Est} ~3.2.** Dissolved in 5% HCl, filtered and basified with 2M ammonia. The ppte is dried in a vacuum, and crystd from dilute EtOH. It yields a blue soln and is light sensitive.

4-Amino-2-methyl-1-naphthol hydrochloride *[130-24-5]* **M 209.6, m 283°(dec), p$K_{Est(1)}$~5.6 (NH_2), p$K_{Est(2)}$ ~10.4 (OH).** Crystd from dilute HCl.

2-Amino-2-methyl-1,3-propanediol *[115-69-5]* **M 105.1, m 111°, b 151-152°/10mm, pK^{25} 8.80.** Crystd three times from MeOH, dried in a stream of dry N_2 at room temp, then in a vacuum oven at 55°. Stored over $CaCl_2$ [Hetzer and Bates *J Phys Chem* **66** 308 *1962*].

2-Amino-2-methyl-1-propanol (β-aminoisobutanol) *[124-68-5]* **M 89.4, m 24°, 31°, b 67°/10mm, 164-166°/760mm, d 0.935, n 1.45, pK^{25} 9.71.** Purified by distn and fractional freezing. The *hydrochloride* has **m** 204°-206°.

2-Amino-3-methylpyridine *[1603-40-3]* **M 108.1, m 33.2°, b 221-222°, pK^{25} 7.24.** Crystd three times from *benzene, most of the residual *benzene being removed from the crystals over paraffin wax chips in an evacuated desiccator. The amine, transferred to a separating funnel under N_2, was left in contact with NaOH pellets for 3h with occasional shaking. It was then placed in a vacuum distilling flask where it was refluxed gently in a stream of dry N_2 before being fractionally distd [Mod, Magne and Skau *J Phys Chem* **60** 1651 *1956*].

2-Amino-4-methylpyridine *[695-34-1]* **M 108.1, m 99.2°, b 230°, pK^{25} 7.48.** Crystd from EtOH or a 2:1 *benzene/acetone mixture, and dried under vacuum.

2-Amino-5-methylpyridine *[1603-41-4]* **M 108.1, m 76.5°, b 227°, pK^{25} 7.22.** Crystd from acetone.

2-Amino-6-methylpyridine *[1824-81-3]* **M 108.1, m 44.2°, b 208-209°, pK^{25} 7.41.** Crystd three times from acetone, dried under vacuum at *ca* 45°. After leaving in contact with NaOH pellets for 3h, with occasional shaking, it was decanted and fractionally distd [Mod, Magne and Skau *J Phys Chem* **60** 1651 *1956*]. Also recrystd from CH_2Cl_2 by addition of pet ether. [Marzilli et al. *J Am Chem Soc* **108** 4830 *1986*.]

2-Amino-5-methylpyrimidine *[50840-23-8]* **M 109.1, m 193.5°, pK_{Est} ~4.0.** Crystd from water and *benzene. Sublimes at 50°/0.5mm.

4-Amino-2-methylquinoline *[6628-04-2]* **M 158.2, m 168°, b 333°/760mm, pK^{20} 9.42.** Crystd from *benzene/pet ether.

2-Aminonaphthalene (ß-naphthylamine) *[91-59-8]* **M 143.2, m 111-113°, pK^{25} 4.20.** See entry on p. 306.

3-Amino-2-naphthoic acid *[5959-52-4]* **M 187.2, m 214°(dec), p$K_{Est(1)}$~1.5 p$K_{Est(2)}$ ~4.0.** Crystd from aqueous EtOH.

4-Amino-5-naphthol-2,7-disulfonic acid *[90-20-0]* **M 320.3, pK_1^{25} 3.63, pK_2^{25} 8.83.** Sufficient Na_2CO_3 (*ca* 22g) to make the soln slightly alkaline to litmus was added to a soln of 100g of the dry acid in 750mL of hot distd water, followed by 5g of activated charcoal and 5g of Celite. The suspension was stirred for 10min and filtered by suction. The acid was ppted by adding *ca* 40mL of conc HCl (soln blue to Congo Red), then filtered by suction through sharkskin filter circular sheet (or hardened filter paper) and washed with 100mL of distd water. The purification process was repeated. The acid was dried overnight in an oven at 60° and stored in a dark bottle [Post and Moore *Anal Chem* **31** 1872 *1959*].

1-Amino-2-naphthol hydrochloride *[1198-27-2]* M 195.7, m 250°(dec), $pK_{Est(1)}$~3.7 (NH$_2$), $pK_{Est(2)}$ ~9.9 (OH). Crystd from the minimum volume of hot water containing a few drops of stannous chloride in an equal weight of hydrochloric acid (to reduce atmospheric oxidation).

1-Amino-2-naphthol-4-sulfonic acid *[116-63-2]* M 239.3, m 295°(dec), $pK_{Est(1)}$<0 , $pK_{Est(2)}$ ~2.8 (NH$_2$), $pK_{Est(2)}$ ~8.8. Purified by warming 15g of the acid, 150g of NaHSO$_3$ and 5g of Na$_2$SO$_3$ (anhydrous) with 1L of water to *ca* 90°, shaking until most of the solid had dissolved, then filtering hot. The precipitate obtained by adding 10mL of conc HCl to the cooled filtrate was collected, washed with 95% EtOH until the washings were colourless, and dried under vacuum over CaCl$_2$. It was stored in a dark coloured bottle, in the cold [Chanley, Gindler and Sobotka *J Am Chem Soc* **74** 4347 *1952*].

6-Aminonicotinic acid *[3167-49-5]* M 138.1, m 312°(dec), $pK_{Est(1)}$~2.2 (CO$_2$H), $pK_{Est(2)}$ ~6.5. Crystd from aq acetic acid.

2-Amino-4-nitrobenzoic acid *[619-17-0]* M 182.1, m 269°(dec), pK_1^{25} 0.65, pK_2^{25} 3.70. Crystd from water or aq EtOH.

5-Amino-2-nitrobenzoic acid *[13280-60-9]* M 182.1, m 235°(dec), $pK_{Est(1)}$~1.1, $pK_{Est(2)}$ ~1.2. Crystd from water.

1-Amino-4-nitronaphthalene *[776-34-1]* M 188.2, m 195°, pK^{20} 0.54. Crystd from EtOH or ethyl acetate.

2-Amino-4-nitrophenol *[99-57-0]* M 154.1, m 80-90° (hydrate), 142-143° (anhydr), $pK_{Est(1)}$~3.9 (NH$_2$), $pK_{Est(2)}$ ~9.2. Crystd from water.

2-Amino-5-nitrophenol *[121-88-0]* M 154.1, m 207-208°, $pK_{Est(1)}$~3.8, $pK_{Est(2)}$ ~9.3. Crystd from water.

6-Aminopenicillanic acid *[551-16-6]* M 216.2, m 208-209°, $[\alpha]_{546}$ +327° (in 0.1M HCl), pK_1^{25} 2.30, pK_2^{25} 4.90. Crystd from water..

2-Aminoperimidine hydrobromide *[40835-96-9]* M 264.1, m 299°, pK_{Est} ~7.9 (free base). Purified by boiling a saturated aqueous soln with charcoal, filtering and leaving the salt to crystallise. Stored in a cool, dark place.

2-Aminophenol *[95-55-6]* M 109.1, m 175-176°, pK_1^{25} 4.65, pK_2^{25} 9.75. Purified by soln in hot water, decolorised with activated charcoal, filtered and cooled to induce crystn. Maintain an atmosphere of N$_2$ over the hot phenol soln to prevent its oxidation [Charles and Freiser *J Am Chem Soc* **74** 1385 *1952*]. Can also be crystd from EtOH.

3-Aminophenol *[591-27-5]* M 109.1, m 122-123°, pK_1^{25} 4.25, pK_2^{25} 9.90. Crystd from hot water or toluene.

4-Aminophenol *[123-30-8]* M 109.1, m 190° (under N$_2$), pK_1^{25} 5.38, pK_2^{25} 10.4. Crystd from EtOH, then water, excluding oxygen. Can be sublimed at 110°/0.3mm. Has been purified by chromatography on alumina with a 1:4 (v/v) mixture of absolute EtOH/*benzene as eluent.

4-Aminophenol hydrochloride *[51-78-5]* M 145.6, m 306°(dec). Purified by treating an aqueous soln with saturated Na$_2$S$_2$O$_3$, filtering under an inert atmosphere, then recrystd from 50% EtOH twice and once from absolute EtOH [Livingston and Ke *J Am Chem Soc* **72** 909 *1950*].

4-Aminophenylacetic acid *[1197-55-3]* M 151.2, m 199-200°(dec), pK_1^{20} 3.60, pK_2^{20} 5.26. Crystd from hot water (60-70mL/g).

2-Amino-1-phenylbutan-1-ol [α-(α-aminopropyl)benzyl alcohol] [(±)-threo 5897-76-7] M 165.1, m 79-80°, pK$_{Est}$ ~9.7. Crystd from *benzene/pet ether.

S-(-)-2-Amino-3-phenyl-1-propanol (L-phenylalaninol) [3182-95-4] M 151.2, m 95°. See phenylalaninol on p. 327.

N-Aminophthalimide [1875-48-5] M 162.2, m 200-202°, pK$_{Est}$ ~0. It has been recrystd from 96% EtOH (sol ~2% at boiling temperature) to form a yellow solution. It sublimes in vacuo at ca 150°. Resolidifies after melting, and remelts at 338-341°.

4-Aminopropiophenone [70-69-9] M 163.1, m 140°, pK$_{Est}$ ~2.2. Crystd from water or EtOH.

4-(2-Aminopropyl)phenol [103-86-6] M 151.2, m 125-126°, pK$_{Est(1)}$~9.4 (OH), pK$_{Est(2)}$ ~9.7(NH$_2$). Crystd from *benzene.

1-Aminopyrene [1606-67-3] M 217.3, m 117-118°, pK$_1^{25}$ 2.91 (50% aq EtOH), pK$_2^{25}$ 2.77 (50% aq EtOH). Crystd from hexane.

2-Aminopyridine [504-29-0] M 94.1, m 58°, b 204-210°, pK$_1^{25}$ -7.6, pK$_2^{25}$ 6.71. Crystd from *benzene/pet ether (b 40-60°) or CHCl$_3$ /pet ether.

3-Aminopyridine [462-08-8] M 94.1, m 64°, b 248°, pK$_1^{25}$ -1.5, pK$_2^{25}$ 6.03. Crystd from *benzene, CHCl$_3$/pet ether (b 60-70°), or *benzene/pet ether (4:1).

4-Aminopyridine [504-24-5] M 94.1, m 160°, b 180°/12-13mm, pK$_1^{25}$ -6.55, pK$_2^{25}$ 9.11 (9.18). Crystd from *benzene/EtOH, then recrystd twice from water, crushed and dried for 4h at 105° [Bates and Hetzer J Res Nat Bur Stand **64A** 427 1960]. Has also been crystd from EtOH, *benzene, *benzene/pet ether, toluene and sublimes in vacuum.

2-Aminopyrimidine [109-12-6] M 95.1, m 126-127.5°, pK^{20} 3.45. Crystd from *C$_6$H$_6$, EtOH or H$_2$O.

4-Aminopyrimidine [591-54-8] M 95.1, m 149-151°, 154-156°, pK^{25} 5.69. Recryst 10.6g from hot EtOAc (200mL), 7.4g colorless needles, first crop, evap to 25mL gave 1.7g of second crop. The Hydroiodide has m 180°. Picrate has m 225°. [Brown J Soc Chem Ind (London) **69** 353 1950.]

5-Aminopyrimidine [591-55-9] M 95.1, m 171-172° (with sublimation), pK^{25} 2.52. Purified by conversion to the MgCl$_2$ complex in a small vol of H$_2$O. The complex (~ 5g) is dissolved in the minimum vol of hot H$_2$O, passed through a column of activated Al$_2$O$_3$ (200g) and the column washed with EtOH. Evapn of the EtOH gives a colorless residue of the aminopyrimidine which is recrystd from *C$_6$H$_6$ (toluene could also be used) which forms needles at first then prisms. It melts with sublimation. Acetylation yields 5-acetamidopyrimidine which crysts from *C$_6$H$_6$, m 148-149°. [Whittaker J Chem Soc 1565 1951.]

Aminopyrine (4-dimethylaminoantipyrene) [58-15-1] M 231.3, m 107-109°, pK$_1^{25}$ -2.22, pK$_2^{25}$ 4.94. Crystd from pet ether.

3-Aminoquinoline [580-17-6] M 144.2, m 93.5°, pK^{20} 4.94. Crystd from *C$_6$H$_6$.

4-Aminoquinoline [578-68-7] M 144.2, m 158°, pK$_1^{20}$ -7.11, pK$_2^{20}$ 9.13. Purified by zone refining.

5-Aminoquinoline [611-34-7] M 144.2, m 110°, b 184°/10mm, 310°/760mm, pK$_1^{20}$ 0.97, pK$_2^{20}$ 5.42. Crystd from pentane, then from *benzene or EtOH.

6-Aminoquinoline *[580-15-4]* **M 144.2, m 117-119°, pK$_1^{20}$ 1.63, pK$_2^{20}$ 5.59.** Purified by column chromatography on a SiO$_2$ column using CHCl$_3$/MeOH (4:1) as eluent. It is an **irritant**.

8-Aminoquinoline *[578-66-5]* **M 144.2, m 70°, pK^{20} 3.95.** Crystd from EtOH or ligroin.

4-Aminosalicylic acid *[65-49-6]* **M 153.1, m 150-151°(dec), pK$_1^{25}$ 1.78 (CO$_2$H), pK$_2^{25}$ 3.63 (NH$_2$), pK^{20} 13.74 (OH).** Cryst from EtOH.

5-Aminosalicylic acid (5-amino-2-hydroxybenzoic acid) *[89-57-6]* **M 153.1, m 276-280°, 283° (dec), pK$_1^{25}$ 2.74 (CO$_2$H), pK$_2^{25}$ 5.84 (NH$_2$).** Cryst as needles from H$_2$O containing a little NaHSO$_3$ to avoid aerial oxidation to the quinone-imine. The *Me ester* gives needles from *C$_6$H$_6$, **m 96°**, and the *hydrazide* has **m 180-182°** (From H$_2$O). [Fallab et al. *Helv Chim Acta* **34** 26 *1951*, Shavel *J Amer Pharm Assoc* **42** 402 *1953*.]

2-Amino-5-sulfanilylthiazole *[473-30-3]* **M 238.3, m 219-221°(dec), pK$_{Est}$ ~4.5 (OH).** Crystd from EtOH.

4-Amino-2-sulfobenzoic acid *[527-76-4]* **M 217.1.** Crystd from water.

2-Aminothiazole *[96-50-4]* **M 108.1, m 93°, b 140°/11mm, pK^{20} 5.36.** Crystd from pet ether (b 100-120°), or EtOH.

1-Amino-1,2,4-triazole *[24994-60-3]* **M 84.1, m 91-93°, pK$_{Est}$ ~2.** Crystd from water. [Barszez et al. *J Chem Soc, Dalton Trans* 2025 *1986*.]

3-Amino-1,2,4-triazole *[61-82-5]* **M 84.1, m 159°, pK$_1^{20}$ 4.04, pK$_2^{20}$ 11.08.** Crystd from EtOH (charcoal), then three times from dioxane [Williams, McEwan and Henry *J Phys Chem* **61** 261 *1957*].

4-Amino-1,2,4-triazole *[584-13-4]* **M 84.1, m 80-81°, pK 3.23.** Crystd from water. [Barszez et al. *J Chem Soc, Dalton Trans* 2025 *1986*.]

7-Amino-4-(trifluoromethyl)coumarin, *[53518-15-3]* **M 229.1, m 222°, pK$_{Est}$ ~3.1.** Purified by column chromatography on a C18 column, eluted with acetonitrile/0.01M aq HCl (1:1), and crystd from isopropanol. Alternatively, it is eluted from a silica gel column with CH$_2$Cl$_2$, or by extracting a CH$_2$Cl$_2$ solution (4g/L) with 1M aq NaOH (3 x 0.1L), followed by drying (MgSO$_4$), filtration and evaporation. [Bissell *J Org Chem* **45** 2283 *1980*.]

9-Aminotriptycene *[793-41-9]* **M 269.3, m 223.5-224.5°.** Recrystd from ligroin [Imashiro et al. *J Am Chem Soc* **109** 729 *1987*].

5-Amino-n-valeric acid (5-aminopentanoic acid) *[660-88-8]* **M 117.2, m 157-158°, pK$_1^{25}$ 4.25, pK$_2^{25}$ 10.66.** Crystd by dissolving in H$_2$O and adding EtOH.

5-Amino-n-valeric acid hydrochloride *[627-95-2]* **M 153.6, m 103-104°.** Crystd from CHCl$_3$.

Amodiaquin **[4-(3-dimethylaminomethyl-4-hydroxyanilino)-7-chloroquinoline]** *[86-42-0]* **M 287.5, m 208°.** Crystd from 2-ethoxyethanol.

D-Amygdalin *[29883-15-6]* **M 457.4, m 214-216°, [α]$_D^{22}$ -38° (c 1.2, H$_2$O).** Crystd from water.

n-Amyl acetate *[628-63-7]* **M 130.2, b 149.2°, d 0.876, n 1.40228.** Shaken with saturated NaHCO$_3$ soln until neutral, washed with water, dried with MgSO$_4$ and distd.

n-Amyl alcohol (1-pentanol) *[71-41-0]* **M 88.2, b 138.1°, d^{15} 0.818, n 1.4100.** Dried with anhydrous K$_2$CO$_3$ or CaSO$_4$, filtered and fractionally distd. Has also been treated with 1-2% of sodium and heated at reflux for 15h to remove water and chlorides. Traces of water can be removed from the near-dry alcohol by refluxing with a small amount of sodium in the presence of 2-3% *n*-amyl phthalate or succinate followed by distn (see *ethanol*).

Small amounts of amyl alcohol have been purified by esterifying with *p*-hydroxybenzoic acid, recrystallising the ester from CS_2, saponifying with ethanolic-KOH, drying with $CaSO_4$ and fractionally distilling [Olivier *Recl Trav Chim Pays-Bas* **55** 1027 *1936*].

tert-**Amyl alcohol** *[75-85-4]* **M 88.2, b 102.3°, d^{15} 0.8135, n 1.4058.** Refluxed with anhydrous K_2CO_3, CaH_2, CaO or sodium, then fractionally distd. Near-dry alcohol can be further dried by refluxing with magnesium activated with iodine, as described for *ethanol*. Further purification is possible using fractional crystn, zone refining or preparative gas chromatography.

n-**Amylamine [1-aminopentane]** *[110-58-7]* **M 87.2, b 105°, d 0.752, pK^{25} 10.63.** Dried by prolonged shaking with NaOH pellets, then distd.

n-**Amyl bromide (*n*-pentylbromide)** *[110-53-2]* **M 151.1, b 129.7°, d 1.218, n 1.445.** Washed with conc H_2SO_4, then water, 10% Na_2CO_3 soln, again with water, dried with $CaCl_2$ or K_2CO_3, and fractionally distd just before use.

n-**Amyl chloride** *[543-59-9]* **M 106.6, b 107.8°, d 0.882, n 1.41177.** Same as *sec*-amyl chloride.

sec-**Amyl chloride (1-chloro-2-methylbutane)** *[616-13-7]* **M 106.6, b 96-97°.** Purified by stirring vigorously with 95% H_2SO_4, replacing the acid when it became coloured, until the layer remained colourless after 12h stirring. The amyl chloride was then washed with satd Na_2CO_3 soln, then distd water, and dried with anhydrous $MgSO_4$, followed by filtration, and distn through a 10-in Vigreux column. Alternatively a stream of oxygen containing 5% ozone was passed through the amyl chloride for three times as long as it took to cause the first coloration of starch iodide paper by the exit gas. Washing the liquid with $NaHCO_3$ soln hydrolyzed ozonides and removed organic acids prior to drying and fractional distn [Chien and Willard *J Am Chem Soc* **75** 6160 *1953*].

tert-**Amyl chloride** *[594-36-5]* **M 106.6, b 86°, d 0.866.** Methods of purification commonly used for other alkyl chlorides lead to decomposition. Unsatd materials were removed by chlorination with a small amount of chlorine in bright light, followed by distn [Chien and Willard *J Am Chem Soc* **75** 6160 *1953*].

Amyl ether *[693-65-2]* **M 158.3, b 186.8°, d 0.785, n 1.41195.** Repeatedly refluxed over sodium and distd.

p-*tert*-**Amylphenol** *[80-46-6]* **M 146.3, m 93.5-94.2°, pK$_{Est}$ ~10.2.** Purified *via* its benzoate, as for phenol. After evaporating the solvent from its soln in ether, the material was crystd (from the melt) to constant melting point [Berliner, Berliner and Nelidow *J Am Chem Soc* **76** 507 *1954*].

2-*n*-Amylpyridine *[2294-76-0]* **M 149.2, b 63.0°/2mm, n^{26} 1.4861, pK^{25} 6.00.** Dried with NaOH for several days, then distd from CaO under reduced pressure, taking the middle fraction and redistilling it.

4-*n*-Amylpyridine *[2961-50-4]* **M 149.2, b 78.0°/2.5mm, n 1.4908, pK$_{Est}$ ~6.1.** Dried with NaOH for several days, then distd from CaO under reduced pressure, taking the middle fraction and redistilling it.

α-**Amyrin** *[638-95-9]* **M 426.7, m 186°.** Crystd from EtOH.

ß-**Amyrin** *[508-04-3]* **M 426.7, m 197-197.5°.** Crystd from pet ether or EtOH.

Androstane *[24887-75-0]* **M 260.5, m 50-50.5°.** Crystd from acetone/MeOH.

epi-**Androsterone** *[481-29-8]* **M 290.4, m 172-173°, [α]$_{546}$ +115° (c 1, MeOH).** Crystd from aq EtOH.

cis-**Androsterone** *[53-41-8]* **M 290.4. m 185-185.5°.** Crystd from acetone/Et_2O.

Angelic acid *[565-63-9]* **M 100.1, m 45°, pK^{18} 4.29.** Steam distd, then crystd from H_2O.

Aniline *[62-53-3]* **M 93.1, f -6.0°, b 68.3/10mm, 184.4°/760mm, d 1.0220, n 1.585, n^{25} 1.5832, pK^{25} 4.60.** Aniline is *hygroscopic*. It can be dried with KOH or CaH_2, and distd at reduced pressure. Treatment with stannous chloride removes sulfur-containing impurities, reducing the tendency to become coloured by aerial oxidn. Can be crystd from Et_2O at low temps. More extensive purifications involve preparation of derivatives, such as the double salt of aniline hydrochloride and cuprous chloride or zinc chloride, or *N*-acetylaniline (**m 114°**) which can be recrystd from water.
Recrystd aniline was dropped slowly into an aqueous soln of recrystd oxalic acid. Aniline oxalate was filtered off, washed several times with water and recrystd three times from 95% EtOH. Treatment with satd Na_2CO_3 soln, regenerated aniline which was distd from the soln, dried and redistd under reduced pressure [Knowles *Ind Eng Chem* **12** 881 *1920*].
After refluxing with 10% acetone for 10h, aniline was acidified with HCl (Congo Red as indicator) and extracted with Et_2O until colourless. The hydrochloride was purified by repeated crystn before aniline was liberated by addition of alkali, then dried with solid KOH, and distd. The product was sulfur-free and remained colourless in air [Hantzsch and Freese *Chem Ber* **27** 2529, 2966 *1894*].
Non-basic materials, including nitro compounds were removed from aniline in 40% H_2SO_4 by passing steam through the soln for 1h. Pellets of KOH were added to liberate the aniline which was steam distd, dried with KOH, distd twice from zinc dust at 20mm, dried with freshly prepared BaO, and finally distd from BaO in an all-glass apparatus [Few and Smith *J Chem Soc* 753 *1949*]. Aniline is absorbed by skin and is **TOXIC**

Aniline hydrobromide *[542-11-0]* **M 174.0, m 286°.** Crystd from water or EtOH and dried at 5mm over P_2O_5. Crystd four times from MeOH containing a few drops of conc HCl by addition of pet ether (b 60-70°), then dried to constant weight over paraffin chips, under vacuum [Gutbezahl and Grunwald *J Am Chem Soc* **75** 559 *1953*]. It was ppted from EtOH soln by addition of Et_2O, and the filtered solid was recrystd from EtOH and dried *in vacuo*. [Buchanan et al. *J Am Chem Soc* **108** 1537 *1986*.]

Aniline hydrochloride *[142-04-1]* **M 129.6, m 200.5-201°.** Same as aniline HBr above.

Aniline hydroiodide *[45497-73-2]* **M 220.0, m dec on heating.** Same as aniline HBr, store in thedark.

m-**Anisaldehyde** *[591-31-1]* **M 136.2, b 143°/50mm, d 1.119.** Washed with saturated aq $NaHCO_3$, then H_2O, dried with anhydrous $MgSO_4$ and distd at reduced pressure under N_2. Stored under N_2 in glass sealed ampoules.

p-**Anisaldehyde** (*p*-**methoxybenzaldehyde**) *[123-11-5]* **M 136.2, m -1°, b 249°/atm, 89-90°/2mm, d 1.119, n 1.576.** Washed with saturated aq $NaHCO_3$, then H_2O, steam distd, extracted distillate with Et_2O, dried ($MgSO_4$) and distd under vac and N_2. Store in glass ampules under N_2 in the dark.

o-**Anisidine [2-methoxyaniline]** *[90-04-0]* **M 123.2, m ~5°, b 109°/17mm, 119°/21mm, 225°/atm, d 1.096, n 1.575, pK^{25} 4.52.** It is separated from the *m*- and *p*- isomers by steam distn. It is also separated from its usual synthetic precursor *o*-nitroanisole by dissolving in dil HCl (pH <2.0) extracting the nitro impurity with Et_2O, adjusting the pH to ~8.0 with NaOH extracting the amine in Et_2O or steam distg. Extract the distillate with Et_2O, dry extract (Na_2SO_4), evaporate and fractionate the residual oil. Protect the almost colorless oil from light which turns it yellow in color. [Biggs and Robinson *J Chem Soc* 388*1961*; Nodzu et al. *Yakugaku Zasshi (J Pharm Soc Japan)* **71** 713, 715 *1951*.]

m-**Anisidine [3-methoxyaniline]** *[536-90-3]* **M 123.2, m ~5°, b 79°/1mm, 128°/17mm, 251°/atm, d 1.101, n 1.583, pK^{25} 4.20.** *o*-Isomer impurity can be removed by steam distn. Possible impurity is the precursor 3-nitroanisole which can be removed as for the preceding *o*-isomer and fractionated using an efficient column. Yellow liquid. [Gilman and Kyle *J Am Chem Soc* **74** 3027 *1952*; Bryson *J Am Chem Soc* **82** 4858 *1960*; Kadaba and Massie *J Org Chem* **22** 333 *1957*.]

p-**Anisidine [4-methoxyaniline]**   *[104-94-9]* **M 123.2, m 57°, pK^{25} 5.31.** Crystd from H_2O or aqueous EtOH. Dried in a vacuum oven at 35° for 6h and stored in a dry box. [More et al. *J Am Chem Soc* **108** 2257 *1986*.] Purified by vacuum sublimation [Guarr et al. *J Am Chem Soc* **107** 5104 *1985*].

Anisole *[100-66-3]* **M 108.1, f -37.5°, b 43°/11mm, 153.8°/760mm, d^{15} 0.9988, n^{25} 1.5143, pK^0 -6.61 (aq H_2SO_4).** Shaken with half volume of 2M NaOH, and emulsion allowed to separate. Repeated 3 times, then washed twice with water, dried over $CaCl_2$, filtered, dried over sodium wire and finally distd from fresh sodium under N_2, using a Dean-Stark trap, samples in the trap being rejected until free from turbidity [Caldin, Parbov, Walker and Wilson *J Chem Soc, Faraday Trans 1* **72** 1856 *1976*].
Dried with $CaSO_4$ or $CaCl_2$, or by refluxing with sodium or BaO with crystalline $FeSO_4$ or by passage through an alumina column. Traces of phenols have been removed by prior shaking with 2M NaOH, followed by washing with water. Can be purified by zone refining.

2-*p*-Anisyl-1,3-indanone *[117-37-3]* **M 252.3, m 156-157°, pK^{20} 4.09.** Crystd from acetic acid or EtOH.

Anserine [*N*, β−alanyl-1-methylhistidine]   *[584-85-0]* **M 240.3, m 238-239°, [α]$_D$ +11.3° (H_2O), pK$_1^{25}$ 2.64, pK$_2^{25}$ 7.04, pK$_3^{25}$ 9.49.** Crystd from aqueous EtOH. It is *hygroscopic*.

S-**Anserine nitrate** *[5937-77-9]* **M 303.3, m 225°(dec), [α]$_D^{30}$ +12.2°.** Likely impurities: 1-methylimidazole-5-alanine, histidine. Crystd from aqueous MeOH.

Antheraxanthin *[68831-78-7]* **M 584.8, m 205°, λ$_{max}$ 460.5, 490.5nm, in $CHCl_3$.** Likely impurities: violaxanthin and mutatoxanthin. Purified by chromatography on columns of $Ca(OH)_2$ and of $ZnCO_3$. Crystd from *C_6H_6/MeOH as needles or thin plates. Stored in the dark, in an inert atmosphere, at -20°.

Anthracene *[120-12-7]* **M 178.2, m 218°, pK -7.4 (aq H_2SO_4).** Likely impurities are anthraquinone, anthrone, carbazole, fluorene, 9,10-dihydroanthracene, tetracene and bianthryl. Carbazole is removed by continuous-adsorption chromatography [see Sangster and Irvine *J Phys Chem* **24** 670 *1956*] using a neutral alumina column and passing *n*-hexane. [Sherwood in *Purification of Inorganic and Organic Materials*, Zief (ed), Marcel Dekker, New York, 1969.] The solvent is evaporated and anthracene is sublimed under vacuum, then purified by zone refining, under N_2 in darkness or non-actinic light.
Has been purified by co-distillation with ethylene glycol (boils at 197.5°), from which it can be recovered by additn of water, followed by crystn from 95% EtOH, *benzene, toluene, a mixture of *benzene/xylene (4:1), or Et_2O. It has also been chromatographed on alumina with pet ether in a dark room (to avoid photo-oxidation of adsorbed anthracene to anthraquinone). Other purification methods include sublimation in a N_2 atmosphere (in some cases after refluxing with sodium), and recrystd from toluene [Gorman et al. *J Am Chem Soc* **107** 4404 *1985*].
Anthracene has also been crystd from EtOH, chromatographed through alumina in hot *benzene (*fume hood*) and then vac sublimed in a pyrex tube that has been cleaned and baked at 100°. (For further details see Craig and Rajikan *J Chem Soc, Faraday Trans 1* **74** 292 *1978*; and Williams and Zboinski *J Chem Soc, Faraday Trans 1* **74** 611 *1978*.) It has been chromatographed on alumina, recrystd from *n*-hexane and sublimed under reduced pressure. [Saltiel *J Am Chem Soc* **108** 2674 *1986*; Masnori et al. *J Am Chem Soc* **108** 1126 *1986*.] Alternatively, it was recrystd from cyclohexane, chromatographed on alumina with *n*-hexane as eluent, and recrystd two more times [Saltiel et al. *J Am Chem Soc* **109** 1209 *1987*].

Anthracene-9-carboxylic acid *[723-62-6]* **M 222.2, m 214°(dec), pK^{20} 3.65.** Crystd from EtOH.

9-Anthraldehyde *[642-31-9]* **M 206.2, m 104-105°.** Crystd from acetic acid or EtOH. [Masnori et al. *J Am Chem Soc* **108** 1126 *1986*.]

Anthranol *[529-86-2]* **M 196.2, m 160-170°(dec).** Crystd from glacial acetic acid or aqueous EtOH.

Anthranthrone *[641-13-4]* **M 306.3, m 300°, pK -7.9 (aq H_2SO_4).** Crystd from chlorobenzene or nitrobenzene.

Anthraquinone *[84-65-1]* **M 208.2, m 286°, pK^{25} -8.27 (aq H_2SO_4).** Crystd from $CHCl_3$ (38mL/g), *benzene, or boiling acetic acid, washing with a little EtOH and drying under vacuum over P_2O_5.

Anthrarufin [1,5-dihydroxy-9,10-anthraquinone] *[117-12-4]* **M 240.1, m 280°(dec), pK_1^{25} 9.90, pK_2^{25} 11.05.** Purified by column chromatography on silica gel with $CHCl_3$/Et_2O as eluent, followed by recrystn from acetone. Alternatively recrystd from glacial acetic acid [Flom and Barbara *J Phys Chem* **89** 4489 *1985*].

1,8,9-Anthratriol *[480-22-8]* **M 226.2, m 176-181°, pK_{Est} ~9.5.** Crystd from pet ether.

Anthrimide [1,1'-imino-bis-anthraquinone] *[82-22-4]* **M 429.4, m >250°(dec).** Crystd from chlorobenzene (red needles) or nitrobenzene (red rhombs)

Anthrone *[90-44-8]* **M 194.2, m 155°, pK -5.5 (aq H_2SO_4).** Crystd from a 3:1 mixture of *benzene/pet ether (b 60-80°) (10-12mL/g), or successively from *benzene then EtOH. Dried under vacuum.

Antipyrine [2,3-dihydro-1,5-dimethyl-3-oxo-2-phenylpyrazole] *[60-80-0]* **M 188.2, m 114°, b 319°, pK^{25} 1.45.** Crystd from EtOH/water mixture, *benzene, *benzene/pet ether or hot water (charcoal), and dried under vacuum.

ß-Apo-4'-carotenal, ß-Apo-8'-carotenal, ß-Apo-8'-carotenoic acid ethyl ester, ß-Apo-8'-carotenoic acid methyl ester, Apocodeine, Apomorphine see entries in Chapter 6.

ß-L-Arabinose (natural) *[87-72-9]* **M 150.1, m 158°, $[\alpha]_D$ +104° (c 4, H_2O after 24h).** Crystd slowly twice from 80% aq EtOH, then dried under vacuum over P_2O_5.

D-Arabinose *[10323-20-3, 28697-53-2 (pyranoside)]* **M 150.1, m 164°, $[\alpha]_{546}$ -123° (c 10, H_2O after 24h), pK^{25} 12.54.** Crystd three times from EtOH, vacuum dried at 60° for 24h and stored in a vacuum desiccator.

L-Arabitol *[7643-75-6]* **M 152.2, m 102°, $[\alpha]_{546}$ -16° (c 5, 8% borax soln).** Crystd from 90% EtOH.

DL-Arabitol *[2152-56-9]* **M 152.2, m 105-106°.** Crystd from 90% EtOH.

Arachidic (eicosanoic C_{20}) acid *[506-30-9]* **M 312.5, m 77°, pK_{Est} ~5.0.** Crystd from abs EtOH.

Arachidic alcohol (1-eicosanol) *[629-96-9]* **M 298.6, m 65.5° (71°), b 200°/3mm.** Crystd from *benzene or *benzene/pet ether.

p-Arbutin *[497-76-7]* **M 272.3, m 163-164°.** Crystd from water.

S-Arginine *[74-79-3]* **M 174.2, m 207°(dec), $[\alpha]_D$ +26.5° (c 5, in 5M HCl), $[\alpha]_{546}$ +32° (c 5, in 5M HCl), pK_1^{25} 2.18, pK_2^{25} 9.36, pK_3^{25} 11.5.** Crystd from 66% EtOH.

S-Arginine hydrochloride *[1119-34-2]* **M 210.7, m 217°(dec), $[\alpha]_D^{20}$ +26.9° (c 6, M HCl).** Likely impurity is ornithine. Crystd from water at pH 5-7, by adding EtOH to 80% (v/v).

S-Argininosuccinic acid *[2387-71-5]* **M 290.3, $[\alpha]_D^{24}$ +16.4° (H_2O).** Likely impurity is fumaric acid. In neutral or alkaline soln it readily undergoes ring closure to the 'anhydride'. Crystd from water by adding 1.5 vols of EtOH. Barium salt is stable at 0-5° if dry. [Westfall *Biochem J* **77** 135 *1960*.]

S-Argininosuccinic anhydride *[28643-94-9]* M 272.3, $[\alpha]_D^{23}$ -10° (H$_2$O for anhydride formed at neutral pH). Crystd from water by adding two volumes of EtOH. An isomeric anhydride is formed if the free acid is allowed to stand at acid pH. In soln, the mixture of anhydrides and free acid is formed [see above entry].

L(+)-Ascorbic acid *[50-81-7]* M 176.1, m 193°(dec), $[\alpha]_{546}$ +23° (c 10, H$_2$O), pK$_1^{25}$ 4.04, pK$_2^{25}$ 11.34. Crystd from MeOH/Et$_2$O/pet ether [Herbert et al. *J Chem Soc* 1270 *1933*].

S-Asparagine *[70-47-3]* M 150.1, m 234-235°, (monohydrate) *[5794-13-8]* $[\alpha]_D$ +32.6° (0.1M HCl), pK$_1^{25}$ 1.98, pK$_2^{25}$ 8.84. Likely impurities are aspartic acid and tyrosine. Crystd from H$_2$O or aqueous EtOH. Slowly effloresces in dry air.

Aspartic acid M 133.1, m 338-339° (*RS*, *[617-45-8]*); m 271° (*S*, requires heating in a sealed tube *[56-84-8]*), $[\alpha]_D^{25}$ +25.4° (3M HCl), pK$_1^{25}$ 1.99, pK$_2^{25}$ 3.90. Likely impurities are glutamic acid, cystine and asparagine. Crystd from water by adding 4 volumes of EtOH and dried at 110°.

L-Aspartic acid ß-methyl ester hydrochloride *[16856-13-6]* M 183.6, m 194°, pK^{25} 8.62. Recrystd from MeOH by using anhydrous diethyl ether [Bach et al. *Biochem Prep* 13 20 *1971*].

DL-Aspartic acid dimethyl ester hydrochloride *[14358-33-9]* M 197.7, 116-117°. Crystd from absolute MeOH. [Kovach et al. *J Am Chem Soc* 107 7360 *1985*.] *Diethyl ester* has pK^{25} 6.4.

Aspergillic acid *[490-02-8]* M 224.3, m 97-99°, pK 5.5. Sublimed at 80°/10^{-3}mm. Crystd from MeOH.

Astacin (β,β-carotene-3,3',4,4'-tetraone) *[514-76-1]* M 592.8, m 228°, 240-243°(evacuated tube), $\varepsilon_{1cm}^{1\%}$ 550,000 at 498mm (pyridine). Probable impurity is astaxanthin. Purified by chromatography on alumina/fibrous clay (1:4) or sucrose, or by partition between pet ether and MeOH (alkaline). Crystd from pyridine/water. Stored in the dark under N$_2$ at -20°. [Davis and Weedon *J Chem Soc* 182 *1960*.]

Atrolactic acid (0.5H$_2$O) (2-hydroxy-2-phenylpropionic acid) *[515-30-0]* M 166.2, m 94.5° (anhydr), 88-91° (0.5H$_2$O), pK^{18} 3.53. Crystd from water and dried at 55°/0.5mm.

Atropine *[51-55-8]* M 289.4, m 114-116°, pK^{18} 9.85. Crystd from acetone or hot water.

Auramine O (4,4'-bis-dimethylaminobenzophenone imine hydrochloride) *[2465-27-2]* M 321.9, pK^{25} 10.71 (free base), 9.78 (carbinolamine). Crystd from EtOH as hydrochloride, very slightly soluble in CHCl$_3$, UV: λ_{max} 434 (370) nm. The free base has m 136° after crystn from *benzene. [*J Chem Soc* 1724 *1949*; *Biochemistry* 9 1540 *1970*].

Aurin tricarboxylic acid *[4431-00-9]* M 422.4, m 300°. The acid is dissolved in aqueous NaOH, NaHSO$_3$ solution is added until the colour is discharged and then the tricarboxylic acid is ppted with HCl [*Org Synth* Coll Vol I 54 *1947*]. Do not extract the acid with hot water because it softens forming a viscous mass. Make a solution by dissolving in aqueous NH$_3$. See **Aluminon** for the ammonium salt.

8-Azaadenine *[1123-54-2]* M 136.1, m 345°(dec), pK$_1^{20}$ 2.65, pK$_2^{20}$ 6.29. Crystd from H$_2$O.

2-Azacyclotridecanone (laurolactam) *[947-04-6]* M 197.3, m 152°. Crystd from CHCl$_3$, stored over P$_2$O$_5$ in a vacuum desiccator.

8-Azaguanine *[134-58-7]* M 152.1, m >300°, pK$_1^{20}$ 1.04, pK$_2^{20}$ 6.29. Dissolved in hot M NH$_4$OH, filtered, and cooled; recrystd, and washed with water.

7-Azaindole *[271-63-6]* M 118.1, m 105-106°, pK^{20} 4.57. Repeatedly recrystd from EtOH, then vacuum sublimed [Tokumura et al. *J Am Chem Soc* 109 1346 *1987*].

1-Azaindolizine *[274-76-0]* **M 118.1, b 72-73°/1mm, pK^{20} 1.43.** Purified by distn or gas chromatography.

Azaserine *[115-02-6]* **M 173.1, m 146-162°(dec), $[\alpha]_D^{27.5}$ -0.5° (c 8.5, H$_2$O, pH 5.2), pK$_{Est(1)}$~4.53, pK$_{Est(2)}$ ~5.40.** Crystd from 90% EtOH.

8-Azapurine (1H-1,2,3-triazolo[4,5-d]pyrimidine, 1,2,3,4,6[3H]penta-azaindene) *[273-40-5]* **M 121.1, m 174-175° (effervescence, m depends on heating rate), pK$_1^{20}$ 2.05 (equilib with covalent hydrate), pK$_2^{20}$ 4.84.** Sublimed at 120-130°/0.01mm and recryst from 3 parts of EtOH. [Albert *J Chem Soc(B)* 427 *1966*.]

Azelaic acid *[123-99-9]* **M 188.2, m 105-106°.** Crystd from H$_2$O (charcoal) or thiophene-free *benzene. The material cryst from H$_2$O was dried by azeotropic distn in toluene, the residual toluene soln was cooled and filtered, the ppte being dried in a vacuum oven. Also purified by zone refining or by sublimation onto a cold finger at 10^{-3} torr.

Azetidine (trimethyleneimine) *[503-29-7]* **M 57.1, b 61-62°, d 0.846, n 1.432, pK^{25} 11.29.** It is a flammable, hygroscopic liquid smelling of ammonia, which absorbs CO$_2$ from air and should be kept under Argon. Purified by drying over solid KOH and distd through a short Vigreux column at atm pressure (under Argon) and keeping the pot temp below 210°. [Searles et al. *J Am Chem Soc* **78** 4917 *1956*.]

Aziridine (ethyleneimine) *[151-56-4]* **M 43.1, b 55-56°/756mm, 56°/760mm, d^{24} 0.8321, pK^{25} 8.00.** Redistd in an Ar or N$_2$ atmosphere in a fume hood, and stored over KOH in sealed bottles in a refrigerator. Commercial aziridine has been dried over sodium and distd from the metal through an efficient column before use [Jackson and Edwards *J Am Chem Soc* **83** 355 *1961*; Wenker *J Am Chem Soc* **57** 2328 *1935*]. It is a weaker base than Me$_2$NH (pK 10.87) but is caustic to the skin. It should not be inhaled, causes inflammation of the eyes, nose and throat and one may become sensitised. It is sol in H$_2$O and has an ammoniacal smell and reacts with CO$_2$. Pure aziridine is comparatively stable but polymerises in the presence of traces of H$_2$O and is occasionally explosive in the presence of acids. CO$_2$ is sufficiently acidic to cause polymerisation (forms linear polymers) which is not free radical promoted. It is stable in the presence of bases. The violet *2:1 Cu complex* crystd from EtOH containing a few drops of Aziridine and adding Et$_2$O has **m** 142°(decomp). The *picrate* has **m** 142°. [O'Rourke et al. *J Am Chem Soc* **78** 2159 *1956*.] It has also been dried with BaO, and distd from sodium under nitrogen. **TOXIC.**

Azobenzene *[103-33-3]* **M 182.2, m 68°, pK^{25} 2.48.** Ordinary azobenzene is nearly all in the *trans*-form. It is partly converted into the *cis*-form on exposure to light [for isolation see Hartley *J Chem Soc* 633 *1938*, and for spectra of *cis*- and *trans*-azobenzenes, see Winkel and Siebert *Chem Ber* **74B** 670*1941*]. *trans*-Azobenzene is obtained by chromatography on alumina using 1:4 *benzene/heptane or pet ether, and crystd from EtOH (after refluxing for several hours) or hexane. All operations should be carried out in diffuse red light or in the dark.

1,1'-Azobis(cyclohexane carbonitrile) *[2094-98-6]* **M 244.3, m 114-114.5°, ε_{350nm} 16.0.** Crystd from EtOH.

α,α'-Azobis(isobutyronitrile) (AIBN) *[78-61-1]* **M 164.2, m 103°(dec).** Crystd from acetone, Et$_2$O, CHCl$_3$, aq EtOH or MeOH. Has also been crystd from abs EtOH below 40° in subdued light. Dried under vacuum at room temp over P$_2$O$_5$ and stored under vacuum in the dark at <-10° until used. Also crystd from CHCl$_3$ soln by addn of pet ether (b <40°). [Askham et al. *J Am Chem Soc* **107** 7423 *1985*; Ennis et al. *J Chem Soc, Dalton Trans* 2485 *1986*; Inoue and Anson *J Phys Chem* **91** 1519 *1987*; Tanner *J Org Chem* **52** 2142 *1987*].

Azolitmin B *[1395-18-2]* **M ~3300, m >250°(dec).** Crystd from water as dark violet scales, or ppted from H$_2$O by addtn of EtOH as a red powder. It is an indicator which is red at pH 4.5 and blue at pH 8.3.

Azomethane *[503-28-6]* **M 58.1, m -78°, b 1.5°.** Purified by vacuum distn and stored in the dark at -80°. Can be **EXPLOSIVE.**

p,p'-**Azoxyanisole** (**4,4'-dimethoxyazoxybenzene**) *[1562-94-3]* **M 258.3, transition temps: 118.1-118.8°, 135.6-136.0°, pK^{25} -5.23 (20% aq EtOH + 80% aq H$_2$SO$_4$).** Crystd from absolute or 95% EtOH, or acetone, and dried by heating under vacuum or sublimed in a vac onto a cold finger.

Azoxybenzene *[495-48-7]* **M 198.2, m 36°, pK^{25} -6.16 (20% aq EtOH + 80% aq H$_2$SO$_4$).** Crystd from EtOH or MeOH, and dried for 4h at 25° and 10^{-3}mm. Sublimed before use.

p,p-**Azoxyphenetole** *[4792-83-0]* **M 286.3, m 137-138°** (**turbid liquid clarifies at 167°**). Crystd from toluene or EtOH.

Azulene *[275-51-4]* **M 128.2, m 98.5-99°, pK^{25} -1.65 (aq H$_2$SO$_4$).** Crystd from EtOH.

Azuleno(1,2-b)thiophene *[25043-00-9]* **M 184.2.** Crystd from cyclohexane, then sublimed *in vacuo*.

Azuleno(2,1-b)thiophene *[248-13-5]* **M 184.2.** Crystd from cyclohexane, then sublimed *in vacuo*.

Azure A (**3-amino-7-dimethylaminophenazin-5-ium chloride**) *[531-53-3]* **M 291.8, CI 52005, m > 290°(dec), λ_{max} 633nm, pK 7.2.** Twice recrystd from H$_2$O, and dried at 100°/1h in an oven.

Azure B (**3-dimethylamino-7-methylaminophenazin-5-ium chloride**) *[531-55-5]* **M 305.8, CI 52010, m > 201°(dec), λ_{max} 648nm, pK 7.4.** Twice recrystd from H$_2$O, and dried at 100°/1h in an oven.

Azure C (**3-amino-7-methylaminophenazin-5-ium chloride**) *[531-57-7]* **M 277.8, λ_{max} 616nm, pK 7.0.** Twice recrystd from H$_2$O, and dried at 100°/1h in an oven.

B.A.L. (British Anti-Lewesite) see 1,2-dimercapto-3-propanol.

Barbituric acid [**6-hydroxypyrimidin-2,4-dione**] *[67-52-7]* **M 128.1, m 250°(dec), pK$_1^{25}$ 3.99, pK$_2^{25}$ 12.5.** Crystd twice from H$_2$O, then dried for 2 days at 100°.

Bathophenanthroline (**4,7-diphenyl-1,10-phenanthroline**) *[1662-01-7]* **M 332.4, m 215-216°, 218-220°, pK^{25} 4.67.** Best purified by recrystn from *C$_6$H$_6$ or toluene. Its solubility (per L): H$_2$O (1mg), M HCl (20mg), heptane (110mg), Et$_2$O (530mg), Me$_2$CO (2.3g), dioxane (3.4g), MeOH (6.0g), EtOH (10.5g), isoPrOH (10.0g), *n*-pentanol (18.7g), *C$_6$H$_6$ (12.2g), pyridine (33g), nitrobenzene (44.7g), CHCl$_3$ (78g) and AcOH (450.4g). [UV: *Bull Soc Chim Fr* 371 *1972*.] For di-Na salt 3H$_2$O see entry in Chapter 5.

Batyl alcohol *[544-62-7]* **M 344.6, m 70.5-71°.** Crystd from aq Me$_2$CO, EtOH or pet ether (b 40-60°).

Behenoyl chloride (**docosanoyl chloride**) *[21132-76-3]* **M 359.0, m 40°.** If the IR shows OH bands then it should be dissolved in oxalyl chloride in *C$_6$H$_6$ soln and warmed at 35° for 24h in the absence of moisture, evaporated and distd in a vacuum of 10^{-5}mm. It is sol in *C$_6$H$_6$ and Et$_2$O. It is moisture sensitive and is **LACHRYMATORY.** [*J Chem Soc* 1001 *1937*; *J Biol Chem* **59** 905 *1924*.]

Benzalacetone (*trans*-**4-phenyl-3-buten-2-one**) *[122-57-6]* **M 146.2, m 42°.** Crystd from pet ether (b 40-60°), or distd (**b 137-142° /16mm**).

Benzalacetophenone (Chalcone) *[94-41-7]* **M 208.3, m 56-58°, b 208°/25mm, pK²⁵ -5.73 (aq H₂SO₄).** Crystd from EtOH warmed to 50° (about 5mL/g), iso-octane, or toluene/pet ether, or recrystd from MeOH, and then twice from hexane. SKIN IRRITANT.

Benzaldehyde *[100-52-7]* **M 106.1, f -26°, b 62° (58°)/10mm, 179.0°/760mm, d 1.044, n 1.5455, pK²⁵ -7.1 (aq H₂SO₄).** To diminish its rate of oxidation, benzaldehyde usually contains additives such as hydroquinone or catechol. It can be purified *via* its bisulfite addition compound but usually distn (under nitrogen at reduced pressure) is sufficient. Prior to distn it is washed with NaOH or 10% Na₂CO₃ (until no more CO₂ is evolved), then with satd Na₂SO₃ and H₂O, followed by drying with CaSO₄, MgSO₄ or CaCl₂.

anti-**Benzaldoxime** *[932-90-1]* **M 121.1, m 33-34°.** Crystd from diethyl ether by adding pet ether (b 60-80°). The *syn*-isomer [622-32-2] has **b** 121-124°/12mm, **m** 34-36°.

Benzamide *[55-21-0]* **M 121.1, m 129.5°, pK²⁵ -2.16 (aq H₂SO₄).** Crystd from hot water (about 5mL/g), EtOH or 1,2-dichloroethane, and air dried. Crystd from dilute aqueous ammonia, water, acetone and then *benzene (using a Soxhlet extractor). Dried in an oven at 110° for 8h and stored in a desiccator over 99% H₂SO₄. [Bates and Hobbs *J Am Chem Soc* **73** 2151 *1951*.]

Benzamidine *[618-39-3]* **M 120.2, m 64-66°, pK²⁰ 11.6.** Liberated from chloride by treatment with 5M NaOH. Extracted into diethyl ether. Sublimed *in vacuo*.

Benzanilide *[93-98-1]* **M 197.2, m 164°, pK⁵⁵ 1.26.** Crystd from pet ether (b 70-90°) using a Soxhlet extractor, and dried overnight at 120°. Also crystd from EtOH.

Benz[a]anthracene *[56-55-3]* **M 228.3, m 159-160°.** Crystd from MeOH, EtOH or *benzene (charcoal), then chromatographed on alumina from sodium-dried *benzene (twice), using vacuum distn to remove *benzene. Final purification was by vacuum sublimation.

Benz[a]anthracene-7,12-dione *[2498-66-0]* **M 258.3, m 169.5-170.5°.** Crystd from MeOH (charcoal).

Benzanthrone *[82-05-3]* **M 230.3, m 170°, pK -3.2 (aq H₂SO₄).** Crystd from EtOH or xylene.

***Benzene** *[71-43-2]* **M 78.1, f 5.5°, b 80.1°, d 0.874, n 1.50110, n²⁵ 1.49790.** For most purposes, *benzene can be purified sufficiently by shaking with conc H₂SO₄ until free from thiophene, then with H₂O, dilute NaOH and water, followed by drying (with P₂O₅, sodium, LiAlH₄, CaH₂, 4X Linde molecular sieve, or CaSO₄, or by passage through a column of silica gel, for a preliminary drying, CaCl₂ is suitable), and distn. A further purification step to remove thiophene, acetic acid and propionic acid, is crystn by partial freezing. The usual contaminants in dry thiophene-free *benzene are non-benzenoid hydrocarbons such as cyclohexane, methylcyclohexane, and heptanes, together with naphthenic hydrocarbons and traces of toluene. Carbonyl-containing impurities can be removed by percolation through a Celite column impregnated with 2,4-dinitrophenylhydrazine, phosphoric acid and H₂O. (Prepared by dissolving 0.5g DNPH in 6mL of 85% H₃PO₄ by grinding together, then adding and mixing 4mL of distd H₂O and 10g Celite.) [Schwartz and Parker *Anal Chem* **33** 1396 *1961*.] *Benzene has been freed from thiophene by refluxing with 10% (w/v) of Raney nickel for 15min, after which the nickel was removed by filtration or centrifugation.
Dry *benzene was obtained by doubly distilling high purity *benzene from a soln containing the blue ketyl formed by the reaction of sodium-potassium alloy with a small amount of benzophenone.
Thiophene has been removed from *benzene (absence of bluish-green coloration when 3mL of *benzene is shaken with a soln of 10mg of isatin in 10mL of conc H₂SO₄) by refluxing the *benzene (1Kg) for several hours with 40g HgO (freshly pptd) dissolved in 40mL glacial acetic acid and 300mL of water. The ppte was filtered off, the aq phase was removed and the *benzene was washed twice with H₂O, dried and distd. Alternatively, *benzene dried with CaCl₂ has been shaken vigorously for half an hour with anhydrous AlCl₃ (12g/L) at 25-35°, then decanted, washed with 10% NaOH, and water, dried and distd. The process was repeated, giving thiophene-free *benzene. [Holmes and Beeman *Ind Eng Chem* **26** 172 *1934*.]

After shaking successively for about an hour with conc H_2SO_4, distd water (twice), 6M NaOH, and distd water (twice), *benzene was distd through a 3-ft glass column to remove most of the water. Abs EtOH was added and the *benzene-alcohol azeotrope was distd. (This low-boiling distn leaves any non-azeotrope-forming impurities behind.) The middle fraction was shaken with distd water to remove EtOH, and again redistd. Final slow and very careful fractional distn from sodium, then $LiAlH_4$ under N_2, removed traces of water and peroxides. [Peebles, Clarke and Stockmayer *J Am Chem Soc* **82** 2780 *1960*.] *Benzene liquid and vapour are very* **TOXIC** *and* **HIGHLY FLAMMABLE,** *and all operations should be carried out in an efficient fume cupboard and in the absence of naked flames in the vicinity.*

Rapid purification: To dry benzene, alumina, CaH_2 or 4A molecular sieves (3% w/v) may be used (dry for 6h). Then benzene is distd, discarding the first 5% of distillate, and stored over molecular sieves (3A, 4A) or Na wire.

[$^2H_6$]*Benzene (*benzene-d_6) *[1076-43-3]* **M 84.2, b 80°/773.6mm, 70°/562mm, 60°/399mm, 40°/186.3mm, 20°/77.1mm, 10°/49.9mm, 0°/27.5mm, d 0.9488, d^{40} 0.9257, n 1.4991, n^{40} 1.4865.** Hexadeuteriobenzene of 99.5% purity is refluxed over and distd from CaH_2 onto Linde type 5A sieves under N_2.

Benzeneazodiphenylamine (4-phenylazodiphenylamine) *[28110-26-1]* **M 273.3, m 82°, pK^{22} 1.52.** Purified by chromatography on neutral alumina using anhydrous *C_6H_6 with 1% anhydrous MeOH. The major component, which gave a stationary band, was cut out and eluted with EtOH or MeOH. [Högfeldt and Bigeleisen *J Am Chem Soc* **82** 15 *1960*.] Crystd from pet ether or EtOH. See Sudan I.

1-Benzeneazo-2-naphthol *[842-07-9]* **M 248.3, m 134°, pK$_{Est}$ ~9.5 (OH).** Crystd from EtOH.

1-Benzeneazo-2-naphthylamine (Yellow AB) *[85-84-7]* **M 247.3, m 102-104°, pK$_{Est}$ ~4.1.** Crystd from glacial acetic acid, acetic acid/water or ethanol.

1,2-Benzenedimethanol (1,2-bishydroxymethylbenzene) *[612-14-6]* **M 138.2, m 61-64°, 63-64°, 64-65°, 65-66.5°, b 145°/3mm.** Recrystd from *C_6H_6, H_2O, pet ether or pentane. It has been extracted in a Soxhlet with Et_2O, evaporated and recrystd from hot pet ether. Also dissolve in Et_2O, allow to evaporate till crystals are formed, filter off and wash the colourless crystals with warm pet ether or pentane. The *diacetate* has **m 35°, 35-36°**. [*J Am Chem Soc* **69** 1197 *1947*, IR and UV: *J Am Chem Soc* **74** 441 *1952*.]

m-Benzenedisulfonic acid *[98-48-6]* **M 238.2, pK$_{Est}$ <0.** Freed from H_2SO_4 by conversion to the calcium or barium salts (using $Ca(OH)_2$ or $Ba(OH)_2$, and filtering). The calcium salt was then converted to the potassium salt, using K_2CO_3. Both the potassium and the barium salts were recrystd from H_2O, and the acid was regenerated by passing through the H^+ form of a strong cation exchange resin. The acid was recrystd twice from conductivity water and dried over $CaCl_2$ at 25°. [Atkinson, Yokoi and Hallada *J Am Chem Soc* **83** 1570 *1961*.] It has also been crystd from Et_2O and dried in a vacuum oven.

m-Benzenedisulfonyl chloride *[585-47-7]* **M 275.1, m 63°.** Crystd from $CHCl_3$ (EtOH free, by passing through an alumina column) and dried at 20mm pressure.

Benzene-1,2-dithiol *[17534-15-5]* **M 142.2, m 24-25°, 27-28°, b 110-112°, pK$_{Est(1)}$ ~6.0, pK$_{Est(2)}$~9.4.** Likely impurities are the oxidation products, the disulfides which could be polymeric. Dissolve in aq NaOH until the soln is alkaline. Extract with Et_2O and discard the extract. Acidify with cold HCl (diluted 1:1 by vol with H_2O) to Congo Red paper under N_2 and extract three times with Et_2O. Dry the Et_2O with Na_2SO_4, filter, evaporate and distil residue under reduced press in an atmosphere of N_2. The distillate solidifies on cooling. [UV: *J Chem Soc* 3076 *1958*; *J Am Chem Soc* **81** 4939 *1951*; *Org Synth* Coll Vol V 419 *1973*.]

Benzenesulfinic acid *[618-41-7]* **M 142.2, m 84°, pK^{25} 2.16 (2.74).** The acid is purified by dissolving the Na salt in H_2O, acidifying to Congo Red paper with HCl and adding a concentrated soln of $FeCl_3$ whereby Fe sulfinate ppts. Collect the salt, wash with a little H_2O, drain, suspend in H_2O and add a slight excess of 1.5M aq NaOH. The $Fe(OH)_3$ ppts, it is filtd off, the sulfinic acid in the aq soln is extracted with

Et$_2$O, the extract is dried (Na$_2$SO$_4$) and evapd to give colorless crysts of benzenesulfinic acid **m** 84° which are stored under N$_2$ in the dark, as it slowly oxidises in air to the sulfonic acid [see *Org Synth* **42** 62 *1966*].

Benzenesulfonic acid *[98-11-3]* **M 158.2, m 43-44°, 50-55°** (anhydrous), **65-66°, pK^{25} -2.7, 0.70 (2.53?)** Purified by dissolving in a small volume of distd H$_2$O and stirring with slightly less than the theoretical amount of BaCO$_3$. When effervescence is complete and the solution is still acidic, filter off the insoluble barium benzenesulfonate. The salt is collected and dried to constant weight *in vacuo*, then suspended in H$_2$O and stirred with a little less than the equivalent (half mol.) of sulfuric acid. The insoluble BaSO$_4$ (containing a little barium benzenesulfonate) is filtd off and the filtrate containing the free acid is evapd in a high vacuum. The oily residue will eventually crystallise when completely anhydrous. A 32% commercial acid was caused to fractionally cryst at room temp over P$_2$O$_5$ in a vac desiccator giving finally colorless deliquescent plates **m** 52.5°. The anhydrous crystn acid is deliquescent and should be stored over anhyd Na$_2$SO$_4$ in the dark and should be used in subdued sunlight as it darkens under sunlight. The main impurifty is Fe which readily separates as the Fe salt in the early fractions [Taylor and Vincent *J Chem Soc* 3218 *1952*]. It is an IRRITANT to the skin and eyes. [see *Org Synth* Coll Vol I 84 *1948*; Michael and Adair *Chem Ber* **10** 585 *1877*.]

Benzenesulfonic anhydride *[512-35-6]* **M 298.3, m 88-91°.** Crystd from Et$_2$O.

Benzenesulfonyl chloride *[98-09-9]* **M 176.6, m 14.5°, b 120°/10mm, 251.2°/760mm(dec), d 1.384.** Distd, then treated with 3mole % each of toluene and AlCl$_3$, and allowed to stand overnight. The free benzenesulfonyl chloride was distd off at 1mm pressure, and then carefully fractionally distd at 10mm in an all-glass column. [Jensen and Brown *J Am Chem Soc* **80** 4042 *1958*.]

Benzene-1,2,4,5-tetracarboxylic (pyromellitic) acid *[89-05-4]* **M 254.2, m 281-284°, pK$_1^{25}$ 1.87, pK$_2^{25}$ 2.72, pK$_3^{25}$ 4.30, pK$_4^{25}$ 5.52.** See entry on p. 345.

Benzene-1,2,3-tricarboxylic (hemimellitic) acid (H$_2$O) *[36362-97-7]* **M 210.1, m 190°(dec), pK$_1^{25}$ 2.62, pK$_2^{25}$ 3.82, pK$_3^{25}$ 5.51.** Crystd from water.

Benzene-1,3,5-tricarboxylic (trimesic or trimellitic) acid *[554-95-0]* **M 210.1, m 360°(dec), pK$_1^{25}$ 2.64, pK$_2^{25}$ 3.71, pK$_3^{25}$ 5.01.** Crystd from water.

1,2,4-Benzenetriol *[533-73-3]* **M 126.1, m 141°, pK$_1^{20}$ 9.08, pK$_2^{20}$ 11.82.** Crystd from Et$_2$O.

Benzethonium chloride *[121-54-0]* **M 448.1, m 164-166°.** Crystd from 1:9 MeOH/Et$_2$O mixture.

Benzhydrol (diphenylmethanol) *[91-01-0]* **M 184.2, m 69°, b 297°/748mm, 180°/20mm.** Crystd from hot H$_2$O or pet ether (b 60-70°), pet ether containing a little *benzene, from CCl$_4$, or EtOH (1mL/g). An additional purification step is passage of a *benzene soln through an activated alumina column. Sublimes in a vacuum. Also crystd three times from MeOH/H$_2$O [Naguib *J Am Chem Soc* **108** 128 *1986*]. § A commercial polystyrene supported version is available.

Benzidine (4,4'-diaminobiphenyl) *[92-87-5]* **M 184.2, m 128-129°, pK$_1^{20}$ 3.85, pK$_2^{20}$ 4.95.** Its soln in *benzene was decolorized by percolation through two 2-cm columns of activated alumina, then concentrated until benzidine crystd on cooling. Recrystd alternatively from EtOH and *benzene to constant absorption spectrum [Carlin, Nelb and Odioso *J Am Chem Soc* **73** 1002 *1951*]. Has also been crystd from hot water (charcoal) and from diethyl ether. Dried under vac in an Abderhalden pistol. Stored in the dark in a stoppered container. **CARCINOGENIC.**

Benzidine dihydrochloride *[531-85-1]* **M 257.2, m >250°(dec).** Crystd by soln in hot H$_2$O, with addition of conc HCl to the slightly cooled soln. **CARCINOGENIC.**

Benzil *[134-81-6]* **M 210.2, m 96-96.5°.** Crystd from *benzene after washing with alkali. (Crystn from EtOH did not free benzil from material reacting with alkali.) [Hine and Howarth *J Am Chem Soc* **80** 2274

1958.] Has also been crystd from CCl$_4$, diethyl ether or EtOH [Inoue et al. *J Chem Soc, Faraday Trans 1* **82** 523 *1986*].

Benzilic acid (diphenylglycollic acid) *[76-93-7]* **M 228.3, m 150°, pK^{18} 3.06.** Crystd from *benzene (*ca* 6mL/g), or hot H$_2$O.

Benzil monohydrazone *[5433-88-7]* **M 224.3, m 151°.** Crystd from EtOH.

α-Benzil monoxime *[14090-77-8]*, [E, *574-15-2*], [Z, *574-16-3*] **M 105.1, m 140°.** Crystd from *C$_6$H$_6$ (must not use animal charcoal).

Benzimidazole *[51-17-2]* **M 118.1, m 172-173°, pK$_1^{25}$ 5.53, pK$_2^{25}$ 11.70.** Crystd from water or aqueous EtOH (charcoal), and dried at 100° for 12h.

2-Benzimidazolylacetonitrile *[4414-88-4]* **M 157.2, m 200-205° dec, 209.7-210.7°(corrected), 210°.** Recrystd from aqueous EtOH. It has been recrystd from hot H$_2$O using charcoal, and finally from aqueous EtOH. [*J Am Chem Soc* **65** 1072 *1943*].

Benzo[*b*]biphenylene *[259-56-3]* **M 202.2, m dec >250°.** Purified by sublimation under reduced pressure.

Benzo-15-crown-5 *[14098-44-3]* **M 268.3, m 78-80°.** Recrystd from *n*-heptane. **IRRITANT.**

Benzo-18-crown-6 *[14098-24-9]* **M 312.2, m 42-45°, 43-43.5°.** Purified by passage through a DEAE cellulose column in cyclohexane. Recryst from *n*-hexane. Its complex with thiourea has **m** 127° [5-6 mol of urea to ether, *J Org Chem* **36** 1690 *1971*]. The stability constants of Na$^+$, K$^+$, Rb$^+$, Cs$^+$, Tl$^+$ and Ba^{2+} are in *Inorg Chim Acta* **28** 73 *1978*] [NMR: *J Am Chem Soc* **98** 3769 *1976*]. **IRRITANT.**

Benzo[3,4]cyclobuta[1,2-*b*]quinoxaline *[259-57-4]* **M 204.2, m dec >250°.** Purified by sublimation under reduced pressure.

Benzofuran (coumarone) *[271-89-6]* **M 118.1, b 62-63°/15mm, 97.5-99.0°/80mm, 170-173°/atm, 173-175°(169)/760mm, d$_4^{20}$ 1.0945, n$_D^{20}$ 1.565.** Steam distil, dissolve in Et$_2$O, wash with 5% aqueous NaOH, saturated NaCl, dry (Na$_2$SO$_4$), evaporate and distil. UV: λ$_{max}$ 245, 275, 282nm (log ε 4.08, 3.45, 3.48). The *picrate* has **m** 102-103°. [*Org Synth* Coll Vol V 251 *1973*; NMR: Black and Heffernan *Aust J Chem* **18** 353 *1965*.]

2-Benzofurancarboxylic acid *[496-41-3]* **M 162.1, m 192-193°, pK$_{Est}$ ~3.2.** Crystd from water.

Benzofurazan *[273-09-6]* **M 120.1, m 55°.** Purified by crystn from EtOH and sublimed.

Benzoic acid *[65-85-0]* **M 122.1, m 122.6-123.1°, pK^{25} 4.12.** For use as a volumetric standard, analytical reagent grade benzoic acid should be carefully fused to *ca* 130° (to dry it) in a platinum crucible, and then powdered in an agate mortar. Benzoic acid has been crystd from boiling water (charcoal), aq acetic acid, glacial acetic acid, *C$_6$H$_6$, aq EtOH, pet ether (b 60-80°), and from EtOH soln by adding water. It is readily purified by fractional crystn from its melt and by sublimation in a vacuum at 80°.

***o*-Benzoic acid sulfimide (saccharin, 1,2-benzisothiazol-3(2*H*)-one 1,1-dioxide)** *[81-07-2]* **M 183.2, m 227-229°, 229°, 228.8-229.7°, pK^{25} 1.31, pK^{25} 12.8.** Purified by recrystn from Me$_2$CO [solubility 7.14% at 0°, 14.4% at 50°], or aqueous isoPrOH to give a fluorescent soln. [*Am J Pharm* **41** 17 *1952*.]

Benzoic anhydride *[93-97-0]* **M 226.2, m 42°.** Freed from benzoic acid by washing with NaHCO$_3$, then water, and drying. Crystd from *benzene (0.5mL/g) by adding just enough pet ether (b 40-60°), to cause cloudiness, then cooling in ice. Can be distd at 210-220°/20mm.

(±)-Benzoin   (2-hydroxy-2-phenylacetophenone) *[119-53-9]* **M 212.3, m 137°.** Crystd from CCl$_4$, hot EtOH (8mL/g), or 50% acetic acid. Crystd from high purity *benzene, then twice from high purity MeOH, to remove fluorescent impurities [Elliott and Radley *Anal Chem* **33** 1623 *1961*]. Sublimes.

(±)-α-Benzoinoxime *[441-38-3]* **M 227.3, m 151°.** Crystd from diethyl ether.

Benzonitrile *[100-47-0]* **M 103.1, f -12.9°, b 191.1°, d 1.010, n 1.528.** Dried with CaSO$_4$, CaCl$_2$, MgSO$_4$ or K$_2$CO$_3$, and distd from P$_2$O$_5$ in an all-glass apparatus, under reduced pressure (**b** 69°/10mm), collecting the middle fraction. Distn from CaH$_2$ causes some decomposition of solvent. Isonitriles can be removed by preliminary treatment with conc HCl until the smell of isonitrile has gone, followed by preliminary drying with K$_2$CO$_3$. (This treatment also removes amines).
Steam distd (to remove small quantities of carbylamine). The distillate was extracted into ether, washed with dil Na$_2$CO$_3$, dried overnight with CaCl$_2$, and the ether removed by evaporation. The residue was distd at 40mm (**b** 96°) [Kice, Perham and Simons *J Am Chem Soc* **82** 834 *1960*].
Conductivity grade benzonitrile (specific conductance 2 x 10^{-8} mho) was obtained by treatment with anhydrous AlCl$_3$, followed by rapid distn at 40-50° under vacuum. After washing with alkali and drying with CaCl$_2$, the distillate was vac distd several times at 35° before being fractionally crystd several times by partial freezing. It was dried over finely divided activated alumina from which it was withdrawn as required [Van Dyke and Harrison *J Am Chem Soc* **73** 402 *1951*].

Benzo[ghi]perylene   (1,12-benzoperylene) *[191-24-2]* **M 276.3, m 273°, 277-278.5°, 278-280°.** Purified as light green crystals by recrystn from *C$_6$H$_6$ or xylene and sublimes at 320-340° and 0.05mm [UV *Helv Chim Acta* **42** 2315 *1959*; *Chem Ber* **65** 846 *1932*; Fluoresc. Spectrum: *J Chem Soc* 3875 *1954*]. *1,3,5-Trinitrobenzene complex* **m** 310-313° (deep red crystals from *C$_6$H$_6$); *picrate* **m** 267-270° (dark red crystals from *C$_6$H$_6$); *styphnate (2,4,6-trinitroresorcinol complex)* **m** 234° (wine red crystals from *C$_6$H$_6$). It recrystallises from propan-1-ol [*J Chem Soc* 466 *1959*].

3,4-Benzophenanthrene *[195-19-7]* **M 228.3, m 68°.** Crystd from EtOH, pet ether, or EtOH/Me$_2$CO.

Benzophenone *[119-61-9]* **M 182.2, m 48.5-49°, pK -6.0 (aq H$_2$SO$_4$).** Crystd from MeOH, EtOH, cyclohexane, *benzene or pet ether, then dried in a current of warm air and stored over BaO or P$_2$O$_5$. Also purified by zone melting and by sublimation [Itoh *J Phys Chem* **89** 3949 *1985*; Naguib et al. *J Am Chem Soc* **108** 128 *1986*; Gorman and Rodgers *J Am Chem Soc* **108** 5074 *1986*; Ohamoto and Teranishi *J Am Chem Soc* **108** 6378 *1986*; Naguib et al. *J Phys Chem* **91** 3033 *1987*].

Benzophenone oxime *[574-66-3]* **M 197.2, m 142°, pK 11.18.** Crystd from MeOH (4mL/g).

Benzopinacol *[464-72-2]* **M 366.5, m 170-180° (depends on heating rate).** Crystd from EtOH.

Benzo[a]pyrene   (3,4 benzpyrene) *[50-32-8]* **M 252.3, m 177.5-178°, 179.0-179.5°.** A soln of 250mg in 100mL of *benzene was diluted with an equal volume of hexane, then passed through a column of alumina, Ca(OH)$_2$ and Celite (3:1:1). The adsorbed material was developed with a 2:3 *benzene/hexane mixture. (It showed as an intensely fluorescent zone.) The main zone was eluted with 3:1 acetone/EtOH, and was transferred into 1:1 *benzene-hexane by adding H$_2$O. The soln was washed, dried with Na$_2$SO$_4$, evaporated and crystd from *benzene by the addition of MeOH [Lijinsky and Zechmeister *J Am Chem Soc* **75** 5495 *1953*]. Alternatively it can be chromatographed on activated alumina, eluted with a cyclohexane-*benzene mixture containing up to 8% *benzene, and the solvent evapd under reduced pressure [Cahnmann *Anal Chem* **27** 1235 *1955*], and recrystd from EtOH [Nithipatikom and McGown *Anal Chem* **58** 3145 *1986*]. **CARCINOGENIC.**

Benzo[e]pyrene   (1,2-benzpyrene) *[192-97-2]* **M 252.3, m 178-179°, 178-180°.** Purified by passage through an Al$_2$O$_3$ column (Woelm, basic, activity I) and eluted with *C$_6$H$_6$ and recrystd from 2 volumes of EtOH-*C$_6$H$_6$ (4:1). Forms colourless or light yellow prisms or needles. [*J Chem Soc* 3659 *1954*; *Justus Liebigs Ann Chem* **705** 190 *1967*.] *1,3,5-Trinitrobenzene complex* **m** 253-254° (orange needles from

EtOH); the *picrate* prepared by mixing 20mg in 1mL of *C_6H_6 with 20mg of picric acid in 2mL *C_6H_6, collecting the deep red crystals, and recrystallising from *C_6H_6 **m** 228-229° [Synth *J Chem Soc* 398 *1967*; NMR: *J Chem Phys* **47** 2020 *1967*]. **CARCINOGEN.**

3,4-Benzoquinoline (phenanthridine) *[229-87-8]* **M** 179.2, **m** 108-109°, **b** 350°, **pK**^{20} **4.61.** Chromatographed on activated alumina from *benzene soln, with diethyl ether as eluent. Evapn of ether gave crystalline material which was freed from residual solvent under vacuum, then further purified by fractional crystn under N_2, from its melt . Sublimes in vacuo. See also p. 324.

5,6-Benzoquinoline *[85-02-9]* **M** 179.0, **m** 93°, **b** 350°, **pK**^{20} **5.11.** As 3,4-benzoquinoline above.

7,8-Benzoquinoline *[230-27-3]* **M** 179.0, **m** 52.0-52.5°, **pK**^{20}**4.21.** As 3,4-benzoquinoline above.

p-**Benzoquinone** *[106-51-4]* **M** 108.1, **m** 115.7°. Usually purified in one or more of the following ways: steam distn, followed by filtration and drying (e.g. in a desiccator over $CaCl_2$); crystn from pet ether (b 80-100°), *benzene (with, then without, charcoal), water or 95% EtOH; sublimation under vacuum (e.g. from room temperature to liquid N_2). It slowly decomposes, and should be stored, refrigerated, in an evacuated or sealed glass vessel in the dark. It should be resublimed before use. [Wolfenden et al. *J Am Chem Soc* **109** 463 *1987*.]

1-Benzosuberone **(6,7,8,9-tetrahydrobenzocyclohepten-5-one)** *[826-73-3]* **M** 160.2, **b** 80-85°/0.5mm, 90-93°/1mm, 138-139°/12mm, 154°/15mm, 175-175°/40mm, **d**$^{20}_4$ 1.086, **n**$^{20}_D$ 1.5638. Purified by dissolving in toluene, washing with aqueous 5% NaOH, then brine, dried (MgSO$_4$), and distd. *2,4-Dinitrophenylhydrazone* has **m** 210.5°, 207-208° (from CHCl$_3$ + MeOH). *Z-O-Picryloxime* has **m** 156-157° (from Me$_2$CO+MeOH); the *E-O-picryloxime* has **m** 107°. The *oxime* has **m** 106.5-107.5°. [UV *J Am Chem Soc* **73** 1411 *1951*, **75** 3744 *1953*; *Chem Ber* **90** 1844 *1957*.]

1,2,3-Benzothiadiazole *[273-77-8]* **M** 136.2, **m** 35°, **pK**$_{Est}$ ~<0. Crystd from pet ether.

2,1,3-Benzothiadiazole *[272-13-2]* **M** 136.2, **m** 44°, **b** 206°/760mm, **pK**$_{Est}$ <0. Crystd from pet ether.

1-Benzothiophene **(benzo[b]thiophene, thianaphthene)** *[95-15-8]* **M** 134.2, **m** 29-32°, 30°, 31-32°, 32°, **b** 100°/16mm, 103-105°/20mm, 221-222°/760mm, **d**$^{32.2}_4$ 1.1484, **n**$^{39}_D$1.6306. It has the odour of naphthalene. If the IR spectrum is not very good then suspend in a faintly alkaline aqueous soln and steam distil. Extract the distillate with Et$_2$O, dry the extract with CaCl$_2$, filter, evaporate the solvent and fractionate the residue. Distillate sets solid. The *sulfoxide* has **m** 142°, the *picrate* has **m** 148-149° (yellow crystals from EtOH) and the *styphnate* has **m** 136-137°. [*J Org Chem* **10**, 381 *1945*; *Chem Ber* **52B** 1249 *1919*, **53** 1551 *1920*; *The Chemistry of Heterocyclic Compounds* Hartough and Weisel eds, Interscience Publ, NY, p23, 28, *1954*.]

1,2,3-Benzotriazole *[95-14-7]* **M** 119.1, **m** 96-97°, 98.5°, 100°, **b** 159°/0.2mm, 204°/15mm, **pK**$^{20}_1$**1.6**, **pK**$^{20}_2$**8.64.** Crystd from toluene, CHCl$_3$, Me$_2$NCHO or satd aq soln, and dried at room temperature or in a vacuum oven at 65°. Losses are less if material is distd in a vacuum. **CAUTION: may EXPLODE during vac distn, necessary precautions must be taken.** [*Org Synth* Coll Vol III 106 *1955*.]

O-**Benzotriazol-1-yl-***N,N,N',N'*-**tetramethyluronium hexafluorophosphate (HBTU)** *[94790-37-1]* **M** 379.2, **m** 200° (dec), 250°, 254°(dec). Wash with H$_2$O (3 x), CH$_2$Cl$_2$ (3 x), dry and recryst from MeCN. Dry in a vacuum and store cold in the dark [Dourtoglou et al. *Tetrahedron Lett* 1269 *1978*, NMR: *Synthesis* 572 *1984*].

Benzoylacetone (1-phenyl-1,3-butanedione) *[93-91-4]* **M** 162.2, **m** 58.5-59.0°. Crystd from Et$_2$O or MeOH and dried under vacuum at 40°.

2-Benzoylbenzoic acid *[85-52-9]* **M 226.2, m 126-129°, 129.2, 130°, pK^{25} 3.54.** Recrystd from *C$_6$H$_6$ or cyclohexane, but is best recrystallised by dissolving in a small volume of hot toluene and then adding just enough pet ether to cause pptn and cool. Dry in a low vacuum at 80°. It can be sublimed at 230-240°/0.3mm [*J Chem Soc* 265 *1957*]. The *S-benzylthiouronium salt* has **m** 177-178° (from EtOH). [*J Am Chem Soc* **75** 4087 *1953*; *Chem Ber* **90** 1208 *1957*.]

3-Benzoylbenzoic acid *[579-18-0]* **M 226.2, m 164-166°, pK$_{Est}$~3.5.** Cryst from EtOH; vac subl.

4-Benzoylbenzoic acid *[611-95-0]* **M 226.2, m 196.5-198°, 197-200°, pK$_{Est}$ ~3.7.** Dissolve in hot H$_2$O by adding enough aqueous KOH soln till distinctly alkaline, filter and then acidify with drops of conc HCl. Filter off, wash solid with cold H$_2$O, dry at 100°, and recrystallise from EtOH. [*J Am Chem Soc* **55** 2540 *1933*.]

(S +) and (R -) 1-Benzoyl-2-*tert*-butyl-3-methyl-4-imidazolinone *[R- 101055-57-6] [S-101055-56-5]* **M 260.3, m 142-143°, 145.6-146.6°, 145-147°, [α]$^{20}_{546}$ (+) and (-) 155°, [α]$^{20}_{D}$ (+) and (-) 133° (c 1, CHCl$_3$).** Recrystd from boiling EtOH (sol 1.43g/mL) or better by dissolving in CH$_2$Cl$_2$ and adding pentane, filter and dry for at least 12h at 60°/0.1mm and sublimed at 135°/0.01mm. It has also been purified by flash column chromatography with Merck silica gel at 0.04-0.063mm and using Et$_2$O/pet ether/MeOH (60:35:5) as eluent. It is then recrystd from EtOH/pet ether. [IR, NMR: *Helv Chim Acta* **70** 237 *1987*; *Angew Chem, Int Ed Engl* **25** 345 *1986*.] The *racemate* is purified in a similar manner and has **m** 104-105° [NMR: *Helv Chim Acta* **68** 949 *1985*].

Benzoyl chloride *[98-88-4]* **M 140.6, b 56°/4mm, 196.8°/745mm, d 1.2120, n^{10} 1.5537.** A soln of benzoyl chloride (300mL) in *C$_6$H$_6$ (200mL) was washed with two 100mL portions of cold 5% NaHCO$_3$ soln, separated, dried with CaCl$_2$ and distd [Oakwood and Weisgerber *Org Synth* **III** 113 *1955*]. Repeated fractional distn at 4mm Hg through a glass helices-packed column (avoiding porous porcelain or silicon-carbide boiling chips, and hydrocarbon or silicon greases on the ground joints) gave benzoyl chloride that did not darken on addition of AlCl$_3$. Further purification was achieved by adding 3 mole% each of AlCl$_3$ and toluene, standing overnight, and distilling off the benzoyl chloride at 1-2mm [Brown and Jenzen *J Am Chem Soc* **80** 2291 *1958*]. Refluxing for 2h with an equal weight of thionyl chloride before distn, has also been used. *Strong* **IRRITANT.** *Use in a fume cupboard.*

Benzoylformic acid (phenylglyoxylic acid) *[611-73-4]* **M 150.14, m 62-65°, 64.5-65.5°, 67°, b 84°/0.1mm, 163-167°/15mm, pK^{25} 1.39 (1.79).** If the sample is oily then it may contain H$_2$O. In this case dry in a vacuum desiccator over P$_2$O$_5$ or KOH until crisp. For further purification dissolve 5.5g in hot CCl$_4$ (750mL), add charcoal (2g, this is necessary otherwise the acid may separate as an oil), filter, cool in ice-water until crystallisation is complete. Filter the acid, and the solvent on the crystals is removed by keeping the acid (4.5g) in a vacuum desiccator for 2 days. Slightly yellow crystals are obtained. It can be recrystd also from *C$_6$H$_6$/pet ether, and can be distilled in vacuum. The acid is estimated by titration with standard NaOH. The *phenylhydrazone* is recrystallised form EtOH, **m** 163-164°; the *semicarbazone acid* has **m** 259°(dec) (from EtOH). The *methyl ester* distils at 137°/14mm, 110-111°/2mm, n$^{20}_{D}$ 1.5850. [*J Am Chem Soc* **67** 1482 *1945*; *J Org Chem* **24** 1825 *1959*.]

Benzoyl glycine (hippuric acid) *[495-69-2]* **M 179.2, m 188°, pK^{40} 3.59.** Crystd from boiling H$_2$O. Dried over P$_2$O$_5$.

Benzoyl isothiocyanate *[532-55-8]* **M 163.2, m 25.5-26°, b 72.5-73°/6mm, 88-91°/20mm, 94-96°/21mm, 202.5-204°/724mm, 250-255°/atm, d$^{20}_{4}$ 1.213, n$^{20}_{D}$ 1.637.** Distil over a small amount of P$_2$O$_5$, whereby the distillate crystallises in prisms. It is readily hydrolysed by H$_2$O to give benzamide and benzoylurea, but with NH$_3$ it gives *benzoylurea* **m** 210° which can be recrystd from EtOH. [*J Am Chem Soc* **62** 1595 *1940*, **76** 580 *1954*; *Org Synth* Coll Vol **III** 735 *1955*.]

Benzoyl peroxide *[94-36-0]* **M 242.2, m 95°(dec).** Dissolved in CHCl$_3$ at room temperature and ppted by adding an equal volume of MeOH or pet ether. Similarly ppted from acetone by adding two volumes of distilled water. Has also been crystd from 50% MeOH, and from diethyl ether. Dried under vacuum at room temperature for 24h. Stored in a desiccator in the dark at 0°. When purifying in the absence of water it can be **EXPLOSIVE** and it should be done on a very small scale with adequate protection. Large amounts should be kept moist with water and stored in a refrigerator. [Kim et al. *J Org Chem* **52** 3691 *1987*.]

p-**Benzoylphenol (4-hydroxybenzophenone)** *[1137-42-4]* **M 198.2, m 133.4-134.8°, pK^{25} 7.95.** Dissolved in hot EtOH (charcoal), crystd once from EtOH/H$_2$O and twice from *benzene [Grunwald *J Am Chem Soc* **73** 4934 *1951*; Dryland and Sheppard *J Chem Soc Perkin Trans 1* 125 *1986*].

N-**Benzoyl-*N*-phenylhydroxylamine** *[304-88-1]* **M 213.2, m 121-122°.** Recrystd from hot water, *benzene or acetic acid.

2-**Benzoylpyridine** *[91-02-1]* **M 183.2, m 41-43°, 48-50°, 72°/0.02mm, 104-105°/0.01, n$_D^{24}$ 1.6032, pK$_{Est}$ ~2.4.** Dissolve in Et$_2$O, shake with aqueous NaHCO$_3$, H$_2$O, dry over MgSO$_4$, it solidifies on cooling. The solid can be recrystd from pet ether. Its *hydrochloride* crystallises from Me$_2$CO, **m** 126-127°, and the *2,4-dinitrophenylhydrazone* has **m** 193-195°. [*J Organomet Chem* **24** 623 *1970*.]

Benzoyl sulfide *[644-32-6]* **M 174.4, m 131.2-132.3°.** About 300mL of solvent was blown off from a filtered soln of benzoyl disulfide (25g) in acetone (350mL). The remaining acetone was decanted from the solid which was recrystd first from 300mL of 1:1 (v/v) EtOH/ethyl acetate, then from 300mL of EtOH, and finally from 240mL of 1:1 (v/v) EtOH/ethyl acetate. Yield about 40% [Pryor and Pickering *J Am Chem Soc* **84** 2705 *1962*]. *Handle in a fume cupboard because of* **TOXICITY** *and obnoxious odour.*

2,1-Benzoxathiol-3-one-1,1-dioxide (sulfobenzoic acid anhydride) *[81-08-3]* **M 184.2, m 116-124°, 126-127°, 128°, b 184-186°/18mm.** Purified by distn in a vacuum and readily solidifies to a crystalline mass on cooling. [*J Am Chem Soc* **34** 1594 *1912*.] Alternatively purified by dissolving in the minimum vol of toluene and reflux for 2h using a Dean-Stark trap. Evaporate under reduced pressure and distil the anhydride at 18mm. It can then be recrystd three times from its own weight of dry *C$_6$H$_6$. It is sensitive to moisture and should be stored in the dark in a dry atmosphere. The *O-methyloxime* has **m** 110-112° [*Tetrahedron Lett* 3289 *1972*]. [*Org Synth* Coll Vol I 495 *1941*.] (See also p. 568 in Chapter 6.)

Benzoxazolinone *[59-49-4]* **M 135.1, m 137-139°, 142-143°(corrected), b 121-213°/17mm, 335-337°/760mm.** It can be purified by recrystn from aqueous Me$_2$CO then by distn at atm pressure then in a vacuum. The *methyl mercury salt* recryst from aq EtOH has **m** 156-158°. [*J Am Chem Soc* **67** 905 *1945*.]

N-**Benzoyl-*o*-tolylhydroxylamine** *[1143-74-4]* **M 227.3, m 104°.** Recrystd from aqueous EtOH.

Benzyl-2-acetamido-4,6-*O*-benzylidene-2-deoxy-α-D-glucopyranoside *[13343-63-0]* **M 399.4, m 256-261°, 263-264°, [α]$_D^{26}$ +120° (c 1, pyridine).** Wash with cold isoPrOH and crystallise from dioxane/isoPrOH. [*J Org Chem* **32** 2759 *1967*.]

Benzyl acetate *[140-11-4]* **M 150.2, m -51°, b 92-93°/10mm, 134°/102mm, 214.9°/760mm, d$_4^{20}$ 1.0562, n$_D^{25}$ 1.4994.** Purified by fractional distn, preferably in a good vacuum. Values of n^{25} of 1.5232-1.5242 seem too high and should be 1.4994. [*J Org Chem* **26** 5180 *1961*.]

Benzyl acetoacetate *[5396-89-4]* **M 192.2, b 130°/2mm, 156-157°/10mm, 162-167°/15mm, 275-277°/atm, d$_4^{20}$ 1.114, n$_D^{20}$ 1.514.** Fractionate and collect fractions of expected physical properties. Otherwise add *ca* 10% by weight of benzyl alcohol and heat in an oil bath (160-170°, open vessel) for 30min during which time excess of benzyl alcohol will have distd off, then fractionate. [*J Org Chem* **17** 77 *1952*.]

4'-Benzylacetophenone *[782-92-3]* **M 210.3, m 73°.** Crystd from EtOH (*ca* 1mL/g).

Benzyl alcohol *[100-51-6]* **M 107.2, f -15.3°, b 205.5°, 93°/10mm, d 0.981, n 1.54033, pK^{25} 15.4.** Usually purified by careful fractional distn at reduced pressure in the absence of air. Benzaldehyde, if present, can be detected by UV absorption at 283nm. Also purified by shaking with aq KOH and extracting with peroxide-free diethyl ether. After washing with water, the extract was treated with satd NaHS sol, filtered, washed and dried with CaO and distd under reduced pressure [Mathews *J Am Chem Soc* **48** 562 *1926*]. Peroxy compounds can be removed by shaking with a soln of Fe(II) followed by washing the alcohol layer with distd water and fractionally distd.

Benzylamine *[100-46-9]* **M 107.2, b 178°/742mm, 185°/768mm, d 0.981, n 1.5392, pK^{25} 9.33.** Dried with NaOH or KOH, then distd under N_2, through a column packed with glass helices, taking the middle fraction. Has also been distd from zinc dust under reduced pressure.

Benzylamine hydrochloride *[3287-99-8]* **M 143.6, m 248°** (rapid heating). Crystd from water.

N-**Benzylaniline** (*N*-**phenylbenzylamine**) *[103-32-2]* **M 183.4, m 36°, b 306-307°, d 1.061, pK^{25} 4.04.** Crystd from pet ether (b 60-80°) (*ca* 0.5mL/g).

1-Benzyl-1-aza-12-crown-4 (**10-benzyl-1,4,7-trioxa-10-azacyclododecane**) *[84227-47-4]* **M 265.4, 122-125°/0.03mm, 140-143°/0.05mm, d$_4^{20}$ 1.09, n$_D^{20}$ 1.52, pK$_{Est}$ ~ 7.7.** Dissolve in CH_2Cl_2 or CCl_4 (1g in 30mL) wash with H_2O (30mL), brine (30mL), H_2O (30 mL) again, dry over $MgSO_4$ or Na_2SO_4 and evaporate. The residue in CH_2Cl_2 is chromatographed through Al_2O_3 (eluting with 10% EtOAc in hexane), evaporate, collect the correct fractions and distil (Kügelrohr). Log K_{Na} in dry MeOH at 25° for Na^+ complex is 2.08. [*Tetrahedron Lett* **26** 151 *1985*; *J Org Chem* **53** 5652 *1988*.]

Benzyl bromide *[100-39-0]* **M 171.0, m -4°, b 85°/12mm, 192°/760mm, d 1.438, n 1.575.** Washed with conc H_2SO_4 (CARE), water, 10% Na_2CO_3 or $NaHCO_3$ soln, and again with water. Dried with $CaCl_2$, Na_2CO_3 or $MgSO_4$ and fractionally distd in the dark, under reduced pressure. It has also been thoroughly degassed at 10^{-6} mm and redistd in the dark. This gave material with λ_{max} (MeCN): 226nm (ε 8200) [Mohammed and Kosower *J Am Chem Soc* **93** 2709 *1971*]. *Handle in a fume cupboard, extremely* **LACHRYMATORY.**

Benzyl bromoacetate *[5437-45-6]* **M 229.1, b 96-98°/0.1mm, 146°/12mm, 166-170°/22mm, d$_4^{20}$ 1.444, n$_D^{25}$ 1.5412.** Dilute with Et_2O, wash with 10% aqueous $NaHCO_3$, H_2O, dry ($MgSO_4$) and fractionate using a Fenske (glass helices packing) column. [*J Chem Soc* 1521 *1956*.] **LACHRYMATORY**

N-**Benzyl-*tert*-butylamine** (*N*-*tert*-**butylbenzylamine**) *[3378-72-1]* **M 163.3, b 91°/12mm, 109-110°/25mm, 218-220°/atm, d$_4^{20}$ 0.899, n$_D^{25}$ 1.4942., pK^{25} 10.19.** Dissolve in Et_2O, dry over KOH pellets, filter and fractionate in a N_2 atmosphere to avoid reaction with CO_2 from the air. The *hydrochloride* has **m** 245-246°(dec) (from MeOH + Me_2CO) and the *perchlorate* has **m** 200-201°. [*J Am Chem Soc* **80** 4320 *1958*.]

Benzyl carbamate *[621-84-1]* **M 151.2, m 86°, 86-88°, 90-91°.** If it smells of NH_3 then dry in a vac desiccator and recryst from 2 vols of toluene and dry in a vac desiccator again. It forms glistening plates from toluene, and can be recrystd from H_2O [*J Org Chem* **6** 878 *1941*; *Org Synth* Coll Vol III 168 *1955*].

Benzyl chloride *[100-44-7]* **M 126.6, m 139°, b 63°/8mm, d 1.100, n 1.538.** Dried with $MgSO_4$ or $CaSO_4$, or refluxed with fresh Ca turnings, then fractionally distd under reduced pressure, collecting the middle fraction and storing with CaH_2 or P_2O_5. Has also been purified by passage through a column of alumina. Alternatively it is dried over $MgSO_4$ and distd in a vacuum. The middle fraction is degassed by several freeze-thaw cycles and then fractionated in an 'isolated fractionating column' (which has been evacuated and sealed off at ~10^{-6} mm) over a steam bath. The middle fraction is retained. The final samples were vacuum distd from this sample and again retaining the middle fraction. The purity is >99.9% (no other peaks are visible on GLC and the NMR spectrum is consistent with the structure. [Mohammed and Kosower *J Am Chem Soc* **93** 1709 *1971*.] **IRRITANT** and *strongly* **LACHRYMATORY.**

N-**Benzyl-ß-chloropropionamide** *[24752-66-7]* **M 197.7, m 94°.** Crystd from MeOH.

Benzyl cinnamate *[103-41-3]* **M 238.3, m 34-35°, 39°, b 154-157°/0.5mm, 228-230°/22mm.** Recrystd to constant melting point from 95% EtOH and has the odour of balsam. Alternatively dissolve in

Et$_2$O, wash with 10% aqueous Na$_2$CO$_3$, H$_2$O, dry (Na$_2$SO$_4$), evaporate and fractionate under reduced press using a short Vigreux column. It decomposes when boiled at atm press. [*J Am Chem Soc* **74** 547 *1952*; **84** 2550 *1962*.]

Benzyl cyanide *[140-29-4]* **M 117.1, b 100°/8mm, 233.5°/760mm, d 1.015, n 1.523.** Benzyl isocyanide can be removed by shaking vigorously with an equal volume of 50% H$_2$SO$_4$ at 60°, washing with satd aq NaHCO$_3$, then half-saturated NaCl soln, drying and fractionally distilling under reduced pressure. Distn from CaH$_2$ causes some decomposition of this compound: it is better to use P$_2$O$_5$. Other purification procedures include passage through a column of highly activated alumina, and distn from Raney nickel. *Precautions should be taken because of possible formation of free* **TOXIC** *cyanide; use an efficient fume cupboard.*

N-**Benzyl dimethylamine** *[103-83-3]* **M 135.2, b 66-67°/15mm, 83-84°/30mm, 98-99°/24mm, d$_4^{20}$ 0.898, n$_D^{20}$ 1.516, pK^{25} 8.91.** Dry over KOH pellets and fractionate over Zn dust in a CO$_2$—free atmosphere. It has a pKa^{25} of 8.25 in 45% aq EtOH. Store under N$_2$ or in a vacuum. The *picrate* has **m** 94-95°, and the *picrolonate* has **m** 151° (from EtOH). [*Chem Ber* **63** 34 *1930*; *J Am Chem Soc* **55** 3001 *1933*; *J Chem Soc* 2845 *1957*.] The *tetraphenyl borate salt* has **m** 182-185°. [*Anal Chem* **28** 1794 *1956*.]

Benzyldimethyloctadecylammonium chloride *[122-19-0]* **M 442.2, m 63°.** Crystd from acetone.

2-Benzyl-1,3-dioxolane *[101-49-5]* **M 164.2, b 98-99°/1mm, 110°/5mm, 137-138°/34mm, 240-242°/atm, d$_4^{20}$ 1.087, n$_D^{20}$ 1.532.** Dissolve in CH$_2$Cl$_2$, wash well with 1M NaOH, dry over K$_2$CO$_3$, filter, evaporate and distil through a short path still (Kügelrohr). It has also been purified by preparative gas chromatography. [*Synthesis* 808 *1974*; *J Org Chem* **34** 3949 *1969*.]

Benzyl ether *[103-50-4]* **M 198.3, b 298°, 158-160°/0.1mm, d 1.043, n 1.54057.** Refluxed over sodium, then distd under reduced pressure. Also purified by fractional freezing.

N-**Benzyl-*N*-ethylaniline** *[92-59-1]* **M 221.3, b 212-222°/54mm, 285-286°/710mm, 312-313°/atm (dec), d$_4^{20}$ 1.029, n$_D^{20}$ 1.595, pK$_{Est}$ ~4.6.** Dry over KOH pellets and fractionate. The *picrate* crystallises from *C$_6$H$_6$ as yellow lemon crystals **m** 126-128° (softening at 120°). [*J Chem Soc* 303 *1951*; IR: *J Chem Soc* 760 *1958*.]

Benzyl ethyl ether *[539-30-0]* **M 136.2, b 186°, 65°/10mm, d 0.949, n 14955.** Dried with CaCl$_2$ or NaOH, then fractionally distd. [*J Am Chem Soc* **78** 6079 *1956*.]

Benzyl ethyl ketone (1-phenylbutan-2-one) *[1007-32-5]* **M 148.2, b 49-49.5°/0.01mm, 66-69°/1mm, 83-85°/5mm, 101-102°/10mm, 229-233°/atm, d$_4^{20}$ 0.989, n$_D^{25}$ 1.5015.** Purified by fractionation using an efficient column. It can be converted into the *oxime* and distd, **b** 117-118°/2mm, 145-146°/15mm, d$_{25}^{25}$ 1.036, n$_D^{25}$ 1.5363, decompose oxime and the ketone is redistilled. It can also be purified *via* the *semicarbazone* which has **m** 154 155°. [*J Am Chem Soc* **77** 5655 *1955*; *J Org Chem* **15** 8 *1950*.]

S-(+)- and *R*-(-)- **Benzyl glycidyl ether (1-benzyloxyoxirane)** *[S:14618-80-5]* *[R:16495-13-9]* **M 164.2, b 68°/10^{-4} mm, 105°/0.4mm, d$_4^{20}$ 1.072, n$_D^{20}$ 1.517, [α]$_{546}^{20}$ (+) and (-) 5.5°, [α]$_D^{20}$ (+) and (-) 5.1° (c 5, toluene), [α]$_D^{20}$ (+) and (-) 1.79° (c 5.02, CHCl$_3$), [α]$_D^{21}$ (+) and (-) 15.3° (neat).** The ether in EtOAc is dried (Na$_2$SO$_4$) then purified by flash chromatography using pet ether/EtOAc (5:1) as eluent. The ether is then distd through a short path dist apparatus (Kugelrohr) as a colourless liquid. Alternatively, dissolve in CHCl$_3$, wash with H$_2$O, dry (Na$_2$SO$_4$), evaporate and purify through silica gel chromatography. [*J Chem Soc* 1021 *1967*; *Heterocycles* **16** 381 *1981*; *Org Synth* **69** 82 *1990*; *Synthesis* 539 *1989*; *Chem Pharm Bull Jpn* **39** 1385 *1991*.]

3-Benzyl-5-(2-hydroxyethyl)-4-methylthiazolinium chloride *[4568-71-2]* **M 269.8, m 142-144°, 145-147°.** Purified by recrystn from EtOH or H$_2$O. If placed in a bath at 125° and heated at 2°/min the **m** is 140.5-141.4°. [*J Biol Chem* **167** 699 *1947*, *J Am Chem Soc* **79** 4386 *1957*.]

O-**Benzylhydroxylamine hydrochloride** *[2687-43-6]* **M 159.6, m 234-238°(sublimes), pK$_{Est}$ ~5.9.** Recrystd from H_2O or EtOH.

N-**Benzylideneaniline** *[538-51-2]* **M 181.2, m 48° (54°), b 300°/760mm.** Steam volatile and crystd from *benzene or 85% EtOH.

Benzyl isocyanate *[3173-56-6]* **M 133.2, b 82-84°/10mm, 87°/14mm, 95°/17mm, 101-104°/33mm, d$_4^{20}$ 1.08, n$_D^{20}$ 1.524.** Purified by fractionation through a two-plate column. It is a viscous liquid and is **TOXIC.** [*J Chem Soc* 182 *1947*; *J Am Chem Soc* 81 4838 *1959*; IR: *Monatsh Chem* 88 35 *1957*.]

Benzyl isothiocyanate *[622-78-6]* **M 149.2, b 123-124°/1mm, 138-140°/20mm, 255-260°/atm, d$_4^{20}$ 1.1234, n$_D^{20}$ 1.6039.** Dissolve in Et_2O, filter, if there is any solid, and distil through an efficient column at 11mm with bath temperature at *ca* 150°. Characterise by reacting (0.5mL) in EtOH (1mL) with 50% $NH_2NH_2.H_2O$ (2 mL) to give *4-benzylthiosemicarbazide* as colourless needles which are recrystallised from EtOH, **m** 130°. [*J Chem Soc* 1582 *1950*; *Justus Liebigs Ann Chem* 612 11 *1958*; IR and UV: *Acta Chem Scand* 13 442 *1959*.]

S-**Benzyl-isothiouronium chloride** *[538-28-3]* **M 202.7, two forms, m 150° and 175°, pK$_{Est}$ ~9.8 (free base).** Crystd from 0.2M HCl (2mL/g) or EtOH and dried in air.

Benzylmalonic acid *[616-75-1]* **M 194.2, m 121°, pK$_1^{25}$ 2.91, pK$_2^{25}$ 5.87.** Crystd from *C_6H_6.

Benzylidene malononitrile *[2700-22-3]* **M 154.2, m 83-84°.** Recrystd from EtOH [Bernasconi et al. *J Am Chem Soc* 107 3612 *1985*].

Benzyl mercaptan *[100-53-8]* **M 124.2, b 70.5-70.7°/9.5mm, d 1.058, n 1.5761, pK^{25} 9.43.** Purified *via* the mercury salt [see Kern *J Am Chem Soc* 75 1865 *1953*], which was crystd from *benzene as needles (**m** 121°), and then dissolved in $CHCl_3$. Passage of H_2S gas regenerated the mercaptan. The HgS ppte was filtered off, and washed thoroughly with $CHCl_3$. The filtrate and washings were evaporated to remove $CHCl_3$, then residue was fractionally distd under reduced pressure [Mackle and McClean, *Trans Faraday Soc* 58 895 *1962*].

(-)-*N*-**Benzyl-*N*-methylephedrinium bromide** [benzyl(2-hydroxy-1-methyl-2-phenethyl) dimethylammonium bromide] *[58648-09-2]* **M 350.3, m 209-211°, 212-214°, [α]$_D^{25}$ -3.8° (c 1.45, MeOH), [α]$_D^{20}$ -5.3° (c 1.45, MeOH).** Recrystd from MeOH/Et_2O. [*Justus Liebigs Ann Chem* 710 *1978*.] The *chloride* is recrystd from EtOAc/*n*-hexane, **m** 198-199° [α]$_D^{25}$ -8.67° (c 1.45, MeOH). [*J Chem Soc, Perkin Trans 1* 574 *1981*.]

Benzyl 4-nitrophenyl carbonate *[13795-24-9]* **M 273.2, m 78-80°.** Dissolve in Et_2O, wash with H_2O (3x) and satd aq NaCl, dry (MgSO$_4$), evap in vac and recryst residue from a small vol of MeOH, **m** 78-79°. Alternatively dissolve in Et_2O, wash with N HCl (2x), 0.5N NaHCO$_3$ (4x) then H_2O, dry (Na$_2$SO$_4$), evap Et_2O and recryst residue from *C_6H_6-pet ether, **m** 79-80°. [Khosla et al. *Indian J Chem* 5 279 *1967*; Wolman et al. *J Chem Soc (C)* 596 *1976*.]

Benzyloxyacetyl chloride *[19810-31-2]* **M 184.6, b 81°/0.2mm, 84-87°/0.4mm, 105-107°/5mm, d$_4^{20}$ 1.19, n$_D^{20}$ 1.523.** Check IR to see if there are OH bands. If so then it may be contaminated with free acid formed by hydrolysis. Add oxalyl chloride (amount depends on contamination and needs to be judged, *ca* 3mols) heat at 50° in the absence of moisture for 1h and fractionate twice, **b** 81°/0.2mm (with bath temp at 81°). Excessive heating results in decomposition to give benzyl chloride. The *anilide* is formed by adding aniline in $CHCl_3$ soln, **m** 49°. [*Helv Chim Acta* 16 1130 *1933*.]

Benzyloxybutan-2-one *[6278-91-7]* **M 178.2, b 90-92°/0.1mm, 88-91°/0.5mm, 121-126°/5mm, d$_4^{20}$ 1.0275, n$_D^{20}$ 1.5040.** Dissolve in $CHCl_3$, wash with H_2O, aqueous saturated NaHCO$_3$, H_2O, dry (MgSO$_4$), evaporate the $CHCl_3$, and fractionate. [*J Am Chem Soc* 79 2316 *1957*.]

Benzyloxycarbonyl chloride (Cbz-Cl, benzyl chloroformate) *[501-53-1]* **M 170.6, b 103°/20mm, d 1.195, n 1.5190.** Commercial material is better than 95% pure and may contain some toluene, benzyl alcohol, benzyl chloride and HCl. After long storage (e.g. two years at 4°, Greenstein and Winitz [*The Chemistry of the Amino Acids* **Vol 2** p. 890, J Wiley and Sons NY, *1961*] recommended that the liquid should be flushed with a stream of dry air, filtered and stored over sodium sulfate to remove CO_2 and HCl which are formed by decomposition. It may further be distilled from an oil bath at a temperature below 85° because Thiel and Dent [*Annalen* **301** 257 *1898*] stated that benzyloxycarbonyl chloride decarboxylates to benzyl chloride slowly at 100° and vigorously at 155°. Redistillation at higher vac below 85° yields material which shows no other peaks than those of benzyloxycarbonyl chloride by NMR spectroscopy. **LACHRYMATORY** and **TOXIC**.

N-**Benzyloxycarbonylglycyl-L-alaninamide** *[17331-79-2]* **M 279.3, m dec >200°.** Recrystd from EtOH/Et_2O.

N-**Benzyloxycarbonyl-*N'*-methyl-L-alaninamide** *[33628-84-1]* **M 236.3, m dec >200°.** Recrystd from EtOAc.

5-Benzyloxyindole *[1215-59-4]* **M 223.3, m 96-97°; 100-103°, 104-106°, pK <0.** Recrystd from *C_6H_6-pet ether or pet ether. The *picrate*, red crystals from *C_6H_6, has **m 142-143°.** [*Chem Ind (London)* 1035 *1953*; *J Am Chem Soc* **76** 5579 *1954*; fluorescence: *Biochem J* **107** 225 *1968*.]

p-**(Benzyloxy)phenol** *[103-16-2]* **M 200.2, m 122.5°, pK_{Est} ~10.1.** Crystd from EtOH or water, and dried over P_2O_5 under vacuum. [Walter et al. *J Am Chem Soc* **108** 5210 *1986*.]

S-**(-)-3-Benzyloxypropan-1,2-diol** *[17325-85-8]* **M 182.2, m 24-26°, b 117-118°/10^{-4}mm, 115-116°/0.02mm, 121-123°/0.2mm, d_4^{20} 1.1437, n_D^{22} 1.5295, $[\alpha]_D^{25}$ -5.9° (neat).** Purified by repeated fractional distn. [*J Biol Chem* **193** 835 *1951*, **230** 447 *1958*.]

2-Benzylphenol *[28994-41-4]* **M 184.2, m 54.5°, b 312°/760mm, 175°/18mm, pK_{Est} ~10.0** Crystd from EtOH, stable form has **m 52°** and unstable form has **m 21°**.

4-Benzylphenol (α-Phenyl-*p*-cresol) *[101-53-1]* **M 184.2, m 84°, pK_{Est} ~10.2.** Crystd from water.

1-Benzyl-4-piperidone *[3612-20-2]* **M 189.3, b 107-108°/0.2mm, 114-116°/0.3mm, 143-146°/5mm, 157-158°/11mm, d 1.059, n 1.538.** If physical properties show contamination then dissolve in the minimum volume of H_2O, made strongly alkaline with aqueous KOH, extract with toluene several times, dry the extract with K_2CO_3, filter, evaporate and distil the residue at high vacuum using a bath temp of 160-190°, and redistil. [*J Chem Soc* 3173 *1957*, *J Am Chem Soc* **53** 1030 *1930*.] The *hydrochloride* has **m 159-161°** (from Me_2CO + Et_2O), and the *picrate* has **m 174-182°** (from Me_2CO + Et_2O). [*Helv Chim Acta* **41** 1184 *1958*.]

2-Benzylpyridine *[101-82-6]* **M 169.2, b 98.5°/4mm, d 1.054, n^{26} 1.5771, pK^{25} 5.13.** Dried with NaOH for several days, then distd from CaO under reduced pressure, redistilling the middle fraction.

4-Benzylpyridine *[2116-65-6]* **M 169.2, b 110.0°/6mm, d 1.065, n^{26} 1.5814, pK^{25} 5.59.** Dried with NaOH for several days, then distd from CaO under reduced pressure, redistilling the middle fraction.

4-*N*-Benzylsulfanilamide *[1709-54-2]* **M 262.3, m 175°.** Crystd from dioxane/H_2O.

Benzyl sulfide *[538-74-9]* **M 214.3, m 50°.** See dibenzylsulfide on p. 192.

Benzylthiocyanate *[3012-37-1]* **M 149.2, m 43°, b 256°(dec).** Crystd from EtOH or aqueous EtOH.

Benzyl toluene-*p*-sulfonate *[1024-41-5]* **M 162.3, m 58°.** Crystd from pet ether (b 40-60°).

Benzyltributylammonium bromide *[25316-59-0]* **M 356.4, m 169-171°, 174-175°.** Recrystd from EtOAc/EtOH and EtOH/Et$_2$O. [*J Am Chem Soc* **73** 4122 *1951*, **81** 3264 *1959*.]

Benzyl 2,2,2-trichloroacetimidate *[81927-55-1]* **M 252.5, b 106°/0.5mm, m 3°, d 1.349, n 1.545.** Purify by distn to remove up to 1% of PhCH$_2$OH as stabiliser. A soln in hexane can be stored for up to 2 months without decompn. It is hygroscopic and has to be stored dry. [Wessel et al. *J Chem Soc, Perkin Trans 1* 2247 *1985*.]

Benzyltrimethylammonium chloride *[56-93-9]* **M 185.7, m 238-239°(dec).** A 60% aq soln was evapd to dryness under vac on a steam bath, and then left in a vac desiccator containing a suitable dehydrating agent. The solid residue was dissolved in a small amount of boiling absolute EtOH and pptd by adding an equal volume of diethyl ether and cooling. After washing, the ppte was dried under vac [Karusch *J Am Chem Soc* **73** 1246 *1951*].

Benzyltrimethylammonium hydroxide (Triton B) *[100-85-6]* **M 167.3, d 0.91.** A 38% soln (as supplied) was decolorized (charcoal), then evaporated under reduced pressure to a syrup, with final drying at 75° and 1mm pressure. Prepared anhydrous by prolonged drying over P$_2$O$_5$ in a vacuum desiccator.

Berbamine *[478-61-5]* **M 608.7, m 197-210°, $[\alpha]_D^{20}$ +115° (CHCl$_3$), pK^{20} 7.33.** Crystd from pet ether.

Berberine *[2086-83-1]* **M 336.4, m 145°, pK$_1^{20}$ 2.47, pK$_2^{20}$ 11.73 (pseudobase?).** Crystd from pet ether or ether as yellow needles.

Berberine hydrochloride (2H$_2$O) *[633-65-8]* **M 407.9, m 204-206°(dec), pK 2.47.** Crystn from water gives the dihydrate. The anhydrous salt may be obtained by recrystn from EtOH/Et$_2$O, wash with Et$_2$O and dry in a vacuum. The *iodide* has **m** 250°(dec) (from EtOH). [*J Chem Soc* **113** 503 *1918*; *J Chem Soc* 2036 *1969*.]

Betaine *[107-43-7]* **M 117.1, m 301-305°(dec) (anhydrous), pK^{25} 1.83.** Crystd from aq EtOH.

Betamethasone (9α-fluoro-11β,17α,21-trihydroxy-16β-methylpregna-1,4-diene-3,20-dione) *[378-44-9]* **M 392.5, m 231-136°(dec), 235-237°(dec), $[\alpha]_D^{20}$ +108° (c 1, Me$_2$CO).** Crystd from ethyl acetate, and has λ_{max} 238nm (log ϵ 4.18) in MeOH.

Biacetyl (butan-2,3-dione) *[431-03-8]* **M 86.1, b 88°, d 0.981, n$^{18.5}$1.3933.** Dried with anhydrous CaSO$_4$, CaCl$_2$ or MgSO$_4$, then vacuum distd under nitrogen, taking the middle fraction and storing it at Dry-ice temperature in the dark (to prevent polymerization).

Bibenzyl *[103-29-7]* **M 182.3, m 52.5-53.5°.** Crystd from hexane, MeOH, or 95% EtOH. It has also been sublimed under vacuum, and further purified by percolation through columns of silica gel and activated alumina.

Bicuculline *[485-49-4]* **M 367.4, m 215° (196°, 177°), $[\alpha]_{546}^{20}$ +159° (c 1, CHCl$_3$), pK 4.84.** See bicuculline entry on p. 515 in Chapter 6.

Bicyclohexyl *[92-51-3]* **M 166.3, b 238° (*cis-cis*), 217-219° (*trans-trans*).** Shaken repeatedly with aqueous KMnO$_4$ and with conc H$_2$SO$_4$, washed with water, dried, first from CaCl$_2$ then from sodium, and distd. [Mackenzie *J Am Chem Soc* **77** 2214 *1955*.]

Bicyclo[3.2.1]octane *[6221-55-2]* **M 110.2, m 141°.** Purified by zone melting.

Biguanide *[56-03-1]* **M 101.1, m 130° pK$_1^{25}$ 3.1, pK$_2^{25}$ 12.8.** Crystd from EtOH.

Bilirubin *[635-65-4]* **M 584.7, ε_{450nm} 55,600 in CHCl$_3$, pK$_{Est}$ ~3.0.** An acyclic tetrapyrrole bile pigment with impurities which can be eliminated by successive Soxhlet extraction with diethyl ether and MeOH. It crystallises from CHCl$_3$ as deep red-brown rhombs, plates or prisms, and is dried to constant weight at 80° under vacuum. [Gray et al. *J Chem Soc* 2264, 2276 *1961*.]

Biliverdine *[114-25-0]* **M 582.6, m >300°, pK 3.0.** The precursor of bilirubin (above) and forms dark green plates or prisms, with a violet reflection, from MeOH. [Gray et al. *J Chem Soc* 2264 *1961*; Sheldrick *J Chem Soc, Perkin Trans 2* 1457 *1976*.]

(±)-1,1'-Bi-(2-naphthol) **[1,1'-di-(2-naphthol)]** *[602-09-5; 41024-90-2]* **M 286.3, m 215-217°, 218°, pK$_{Est(1)}$~7.1, pK$_{Est(2)}$ ~11.2.** Crystd from toluene or *benzene (10mL/g). When crystd from chlorobenzene it had **m 238°**. Sol in dioxane is 5%.

1,1'-Bi-(2-naphthol) **[1,1'-di-(2-naphthol)]** *[R-(+)- 18531-94-7], [S-(-)- 18531-99-2]* **M 286.3, m 207.5-208.5°, 209-211°, [α]$_D^{20}$ (+) and (-) 37.4.0° (c 0.5, THF), [α]$_{546}^{25}$ (+) and (-) 51° (c 0.1, THF), pK as above.** Dissolve in cold 2.5N NaOH, extract with CH$_2$Cl$_2$, and acidify with 5% HCl. Collect the white ppte and recryst from aq EtOH and dry in a vacuum [*Tetrahedron* **27** 5999 *1971*]. Optically stable in dioxane-water (100°/24h). *Racemisation*: 72% in 1.2N HCl at 100°/24h and 68% in 0.67M KOH in BuOH at 118°/23h [*J Am Chem Soc* **95** 2693 *1973*]. Cryst from *C$_6$H$_6$ (sol 1%) using Norite or aq EtOH after chromatography through silica gel, eluting with Me$_2$CO-*C$_6$H$_6$. [Kyba et al. *J Org Chem* **42** 4173 *1977*; see also Brussee and Jansen *Tetrahedron Lett* **24** 3261 *1983*; Akimoto and Yamada *Tetrahedron* **27** 5999 *1971*.]

1,1'-Binaphthyl *[± 32507-32-7 and 604-53-5; R(-)- 24161-30-6; S(+)- 734-77-0]* **M 254.3, m 145°, 159° (±, 2 forms), 153-154° (+ and -), [α]$_D^{20}$ (-) and (+) ~220° (*C$_6$H$_6$).** Purified through a silica gel column with Me$_2$CO-*C$_6$H$_6$ [or Al$_2$O$_3$ with 10% *C$_6$H$_6$/pet ether (b 30-60°)] and recrystd from EtOH, pentane, or slow evap of *C$_6$H$_6$, Me$_2$CO or Et$_2$O solns. Half life ~10h at 25° in various solvents. [Wilson and Pincock *J Am Chem Soc* **97** 1474 *1975*; Akimoto and Yamada *Tetrahedron* **27** 5999 *1971*.]

2,2'-Binaphthyl (β, β'-binaphthyl) *[61-78-2]* **M 254.3, m 188°.** Crystd from *benzene.

Biphenyl *[92-52-4]* **M 154.2, m 70-71°, b 255°, d 0.992.** Crystd from EtOH, MeOH, aq MeOH, pet ether (b 40-60°) or glacial acetic acid. Freed from polar impurities by passage through an alumina column in *benzene, followed by evapn. A in CCl$_4$ has been purified by vac distn and by zone refining. Treatment with maleic anhydride removed anthracene-like impurities. Recrystd from EtOH followed by repeated vacuum sublimation and passage through a zone refiner. [Taliani and Bree *J Phys Chem* **88** 2351 *1984*.]

p-**Biphenylamine** *[92-67-1]* **M 169.2, m 53°, b 191°/15mm, pK^{18} 4.38.** See *p*-aminobiphenyl entry on p. 104.

4-Biphenylcarbonyl chloride *[14002-51-8]* **M 216.7, m 114-115°.** Dissolve in a large volume of pet ether (10 x, b 50-70°), filter through a short column of neutral alumina, evaporate to dryness *in vacuo* and recryst from pet ether (b 60-80°). **LACHRYMATORY.**

Biphenyl-2-carboxylic (2-phenylbenzoic) acid *[947-84-2]* **M 198.2, m 114°, b 343-344°, pK 3.46.** Crystd from *C$_6$H$_6$-pet ether or aq EtOH.

Biphenyl-4-carboxylic (4-phenylbenzoic) acid *[92-92-2]* **M 198.2, m 228°, pK^{25} 5.66 (in 50% 2-butoxyethanol).** Crystd from *C$_6$H$_6$-pet ether or aq EtOH.

2,4'-Biphenyldiamine *[492-17-1]* **M 184.2, m 45°, b 363°/760mm, pK$_{Est(1)}$~4.8, pK$_{Est(2)}$ ~3.9.** Crystd from aqueous EtOH.

Biphenylene *[259-79-0]* **M 152.2, m 152°.** Recrystd from cyclohexane then sublimed in vacuum.

α-(4-Biphenylyl)butyric acid *[959-10-4]* M 240.3, m 175-177°, pK$_{Est}$ ~4.5. Crystd from MeOH.

γ-(4-Biphenylyl)butyric acid *[6057-60-9]* M 240.3, m 118°, pK$_{Est}$ ~4.8. Crystd from MeOH.

2,2'-Bipyridyl *[366-18-7]* M 156.2, m 70.5°, b 273°, pK$_1^{25}$ -0.52, pK$_2^{25}$ 4.44. Crystd from hexane, or EtOH, or (after charcoal treatment of a CHCl$_3$ soln) from pet ether. Also ppted from a conc soln in EtOH by addition of H$_2$O. Dried in a vacuum over P$_2$O$_5$. Further purification by chromatography on Al$_2$O$_3$ or by sublimation. [Airoldi et al. *J Chem Soc, Dalton Trans* 1913 *1986.*]

4,4'-Bipyridyl *[553-26-4]* M 156.2, m 73°(hydrate), 114° (171-171°)(anhydrous), b 305°/760mm, 293°/743mm, pK$_1^{20}$ 3.17, pK$_2^{20}$ 4.82. Crystd from water, *benzene/pet ether, ethyl acetate and sublimed *in vacuo* at 70°. Also purified by dissolving in 0.1M H$_2$SO$_4$ and twice ppted by addition of 1M NaOH to pH 8. Recrystd from EtOH. [Man et al. *J Chem Soc, Faraday Trans 1* 82 869 *1986*; Collman et al. *J Am Chem Soc* 109 4606 *1987.*]

2,2'-Bipyridylamine *[1202-34-2]* M 171.2, m 95.1°, pK$_{Est}$ ~5.0. Crystd from Me$_2$CO.

2,2'-Biquinolin-4,4'-dicarboxylic (2,2'-bicinchoninic) acid *[1245-13-2]* M 344.3, m 367°, pK$_{Est(1)}$~1.5, pK$_{Est(2)}$ ~4.0. Dissolve in dilute NaOH and ppte with acetic acid, filter, wash well with H$_2$O and dry at 100° in a vacuum oven. Attempts to form a picrate failed. The *methyl ester* (SOCl$_2$-MeOH) has m 165.6-166°. [*J Am Chem Soc* 64 1897 *1942*; 68 2705 *1946.*] For di-K salt see entry in Chapter 5.

2,2'-Biquinolyl (α,α'-diquinolyl) *[119-91-5]* M 256.3, m 196°, pK$_{Est}$ ~4.2. Decolorised in CHCl$_3$ soln (charcoal), then crystd to constant melting point from EtOH or pet ether [Cumper, Ginman and Vogel *J Chem Soc* 1188 *1962*].

Bis-acrylamide (*N,N'*-methylene bisacrylamide) *[110-26-9]* M 154.2, m >300°. Recrystd from MeOH (100g dissolved in 500mL boiling MeOH) and filtered without suction in a warmed funnel. Allowed to stand at room temperature and then at -15°C overnight. Crystals collected with suction in a cooled funnel and washed with cold MeOH). Crystals air-dried in a warm oven. **TOXIC.**

Bis-(4-aminophenyl)methane *[101-77-9]* M 198.3, m 92-93°, b 232°/9mm, pK$_{Est}$ ~4.9. See 4.4'-diaminodiphenylmethane on p. 189.

2,5-Bis-(4-aminophenyl)-1,3,4-oxadiazole (BAO) *[2425-95-8]* M 252.3, m 252-255°, 254-255°. Recrystd from EtOH using charcoal and under N$_2$ to avoid oxidation.

2,5-Bis(2-benzothiazolyl)hydroquinone *[33450-09-8]* M 440.3, m dec >200°. Purified by repeated crystn from dimethylformamide followed by sublimation in vacuum [Erusting et al. *J Phys Chem* 91 1404 *1987*].

Bis-(*p*-bromophenyl)ether *[53563-56-7]* M 328.0, m 60.1-61.7°. Crystd twice from EtOH, once from *benzene and dried under vac [Purcell and Smith *J Am Chem Soc* 83 1063 *1961*].

Bis-*N-tert*-butyloxycarbonyl-L-cystine, *[10389-65-8]* M 440.5, m 144.5-145°, [α]$_D^{20}$ -133.2° (c 1, MeOH), pK$_{Est}$ ~2.9. Crystd from in EtOAc by adding hexane [Ferraro *Biochem Prep* 13 39 *1971*].

2R,3R-(+)-1,4-Bis-(4-chlorobenzoyl)-2,3-butanediol *[85362-86-3]* and 2S,3S-(-)-1,4-Bis-(4-chlorobenzoyl)-2,3-butanediol *[85362-85-2]* M 371.3, m 76-77°, [α]$_D^{20}$ (+) and (-) 6.4° (c 3.11 CHCl$_3$). Recrystd from toluene-hexane. [*Tetrahedron* 40 4617 *1984*.]

Bis-(β-chloroethyl)amine hydrochloride *[821-48-7]* M 178.5, m 214-215°, pK$_{Est}$ ~5.8 (free base). Crystd from Me$_2$CO.

Bis-(β-chloroethyl) ether *[111-44-4]* **M 143.0, b 94°/33mm, 178.8°, d 1.220, n 1.45750.** Wash with conc H_2SO_4, then Na_2CO_3 soln, dry with anhydrous Na_2CO_3, and finally pass through a 50cm column of activated alumina before distn. Alternatively, wash with 10% ferrous sulfate soln to remove peroxides, then H_2O, dry with $CaSO_4$, and dist in vac. Add 0.2% of catechol to stabilise it. **VERY TOXIC.**

N,N-Bis-(2-chloroethyl)2-naphthylamine (chlornaphthazine) *[494-03-1]* **M 268.3, m 54-56°, b 210°/5mm, pK_{Est} ~5.3.** Crystd from pet ether. **CARCINOGENIC.**

Bis-(chloromethyl)durene *[3022-16-0]* **M 231.2, m 197-198°.** Crystd three times from *benzene, then dried under vacuum in an Abderhalden pistol.

3,3'-Bis-(chloromethyl)oxacyclobutane *[78-71-7]* **M 155.0, m 18.9°.** Shaken with aqueous $NaHCO_3$ or $FeSO_4$ to remove peroxides. Separated, dried with anhydrous Na_2SO_4, then distd under reduced pressure from a little CaH_2 [Dainton, Ivin and Walmsley *Trans Faraday Soc* **65** 17884 *1960*].

2,2-Bis-(p-chlorophenyl)-1,1-dichloroethane (p,p'-DDD) *[72-54-8]* **M 320.1, m 109-111°, 111-112°.** Crystd from EtOH and dried in a vac. Purity checked by TLC. **TOXIC INSECTICIDE.**

2,2-Bis-(p-chlorophenyl)-1,1-dichloroethylene (p,p'-DDE) *[72-55-9]* **M 318.0, m 89-91°.** Crystd from EtOH and dried in a vac. Purity checked by TLC. **POSSIBLE CARCINOGEN.**

2,2-Bis-(4-chlorophenyl)-1,1,1-trichloroethane (p,p'-DDT, 1,1,1-trichloro-2,2-bis(p-chlorophenyl)ethane) *[50-29-3]* **M 354.5, m 108.5-109°, 108°.** Crystd from *n*-propyl alcohol (5mL/g), then dried in air or an air oven at 50-60°. Alternatively crystd from 95% EtOH, and checked by TLC.

2,2'-Bis-[di-(carboxymethyl)-amino]diethyl ether, $(HOOCCH_2)_2NCH_2CH_2OCH_2CH_2N-(CH_2COOH)_2$ *[923-73-9]* **M 336.3, pK_1^{20} 1.8, pK_2^{20} 2.76, pK_3^{20} 8.84, K_4^{20} 9.47.** Crystd from EtOH.

4,4'-Bis-(diethylamino)benzophenone *[90-93-7]* **M 324.5, m 95-96°, $pK_{Est(1)}$~1.8, $pK_{Est(2)}$ ~3.3.** Crystd from EtOH (25mL/g) and dried under vacuum.

Bis-(4-dimethylaminobenzylidene)benzidine *[6001-51-0]* **M 454.5, m 318°, pK_{Est} ~0.** Crystd from nitrobenzene.

1,8-Bis-(dimethylamino)naphthalene (Proton sponge) *[20734-58-1]* **M 214.3, m 47-48°, pK 12.34 (pK_2 -10.5 from half protonation in 86% H_2SO_4).** Crystd from EtOH and dried in a vacuum oven. Stored in the dark. Also see *N,N,N',N'*-Tetramethyl-1,8-naphthalenediamine on p. 364.

Bis-(dimethylthiocarbamyl)disulfide (tetramethylthiuram disulfide, Thiram) *[137-26-8]* **M 240.4, m 155-156°.** See tetramethylthiuram disulfide on p. 365.

Bis-(4-fluoro-3-nitrophenyl) sulfone *[312-30-1]* **M 344.3, m 193-194°.** Recrystd from Me_2CO and H_2O (5:1). It should give a yellow colour in aqueous base. [*Chem Ber* **86** 172 *1953*.]

N,N-Bis-(2-hydroxyethyl)-2-aminoethanesulfonic acid (BES) *[10191-18-1]* **M 213.3, m 150-155°, pK^{20} 7.17.** Crystd from aqueous EtOH.

Bis-(2-hydroxyethyl)amino-tris-(hydroxymethyl)methane (BIS-TRIS) *[6976-37-0]* **M 209.2, m 89°, pK^{20} 6.46.** Crystd from hot 1-butanol. Dried in a vacuum at 25°.

N,N-Bis-(2-hydroxyethyl)glycine *[150-25-4]* **M 163.2, m 191-194°(dec).** See *N,N*-di-(hdyroxyethyl)glycine (BICINE) on p. 208.

3,4-Bis-(4-hydroxyphenyl)hexane *[5635-50-7]* **M 270.4, m 187°.** Freed from diethylstilboestrol by zone refining.

1,4-Bismethylaminoanthraquinone (Disperse Blue 14) *[2475-44-7]* **M 266.3, λmax 640 (594)nm.** Purified by thin-layer chromatography on silica gel plates, using toluene/acetone (3:1) as eluent. The main band was scraped off and extracted with MeOH. The solvent was evapd and the dye was dried in a drying pistol [Land, McAlpine, Sinclair and Truscott *J Chem Soc, Faraday Trans 1* **72** 2091 *1976*].

Bis-(1-naphthylmethyl)amine *[5798-49-2]* **M 329.4, m 62°, pK$_{Est}$ ~8.4.** Crystd from pet ether.

N,N'-Bis-(nicotinic acid) hydrazide *[840-78-8]* **M 227-228°, m dec 200°, pK$_{Est}$ ~3.3.** Crystd from water.

Bis-(4-nitrophenyl) carbonate *[5070-13-3]* **M 304.3, m 142-143°.** Dissolve in CHCl$_3$, wash with 2N NaOH (3 x) and once with conc HCl, dry (Na$_2$SO$_4$), evaporate and crystallise from toluene (authors say 15 vols of *benzene, prisms). [*Helv Chim Acta* **46** 795 *1963*.]

Bis-(2-nitrophenyl) disulfide *[1155-00-6]* **M 308.3, m 192-195°, 195°, 194-197°, 198-199°.** Purified by recrystn from glacial AcOH or from *C$_6$H$_6$ and the yellow needles are dried in an oven at 100° until the odour of the solvent is absent. It is sparingly soluble in EtOH and Me$_2$CO. [Bogert and Stull *Org Synth Coll Vol I* 220 *1941*; Bauer and Cymerman *J Chem Soc* 3434 *1949*.]

Bis-(4-nitrophenyl) ether *[101-63-3]* **M 260.2, m 142-143°.** Crystd twice from *C$_6$H$_6$, and dried under vacuum.

Bis-(4-nitrophenyl) methane *[1817-74-9]* **M 258.2, m 183°.** Crystd twice from *C$_6$H$_6$, and dried under vacuum.

Bisnorcholanic acid (pregnane-20-carboxylic acid) *[28393-20-6]* **M 332.5, m 214° (α-form), 242° (β-form), 210-211° (γ-form), 184° (δ-form), 181° (ε-form), pK$_{Est}$ ~5.0.** Crystd from EtOH (α-form), or acetic acid (all forms).

3,3'-Bis-(phenoxymethyl)oxacyclobutane *[1224-69-7]* **M 270.3, m 67.5-68°.** Crystd from MeOH.

1,4-Bis-(2-pyridyl-2-vinyl)benzene *[20218-87-5]* **M 284.3, pK$_{Est}$ ~5.4.** Recrystd from xylene, then chromatography (in the dark) on basic silica gel (60-80-mesh), using CH$_2$Cl$_2$ as eluent. Vacuum sublimed in the dark to a cold surface at 10^{-3} torr.

Bis-(trichloromethyl) carbonate (triphosgene) *[32315-10-9]* **M 296.8, m 79-83°, 81-83°, b 203-206°(slight decomp).** A good solid substitute for phosgene (using a third mol per mol). Cryst from pet ether and wash with anhydrous cold Et$_2$O, degas at 200mm then dry at 0.1mm (over H$_2$SO$_4$). It is a **lachrymator**, is **TOXIC** and should be handled with gloves and in an efficient fume hood. [Eckert and Forster *Angew Chem, Int Ed Engl* **26** 894 *1987*; *Aldrichimica Acta* **21** 47 *1988*.]

Bistrifluoroacetamide *[407-24-9]* **M 209.1, m 85°, b 135-136°/744mm, 141°/760mm.** Major impurity is trifluoroacetamide. Add trifluoroacetic anhydride, reflux for 2h and fractionate using a Vigreux column at atmospheric pressure. [*J Chromatogr* **78** 273 *1973*.]

Bis-(trifluoroacetoxy)iodobenzene *[2712-78-9]* **M 430.0, m 112-114° (dec), 120-121°, 124-126°.** Cryst from warm trifluoroacetic acid and dry over NaOH pellets. Recrystd from Me$_2$CO/pet ether. Melting point depends on heating rate. [*Synthesis* 445 *1975*.]

Biuret (allophanic acid amide, carbamoylurea) *[108-19-0]* **M 103.1, sinters at 218° and chars at 270°, pK$_1^{25}$ -0.88, pK$_2^{25}$>4.** Crystd from EtOH.

Bixin (6,6'-diapo-ψ,ψ- carotenedioic acid monomethyl ester) *[6983-79-5]* **M 394.5, m 198°, 217°(dec),** λmax **(CHCl₃) 209, 475 and 443nm, pK$_{Est}$ ~4.3.** Crystd from Me₂CO (violet prisms) [Pattenden et al. *J Chem Soc (C)* 235 *1970*].

Blue Tetrazolium *[1871-22-3]* **M 727.7, m 254-255°(dec).** Crystd from 95% EtOH/anhydrous diethyl ether, to constant absorbance at 254nm.

*R-2-endo-***Borneol** *[464-43-7]* **M 154.3, m 208° [α]$_D^{20}$ +15.8° (in EtOH).** Crystd from boiling EtOH (charcoal).

(±)-Borneol *[6627-72-1]* **M 154.3, m 130°(dec).** Crystd from pet ether (b 60-80°).

Brazilin *[474-07-7]* **M 269.3, m 130°(dec), pK$_{Est(1)}$~9.3, pK$_{Est(2)}$ ~10.0, pK$_{Est(3)}$ ~12.5 (all phenolic).** Crystd from EtOH.

Brilliant Cresyl Blue *[4712-70-3]* **M 332.8, pK^{25} 3.2.** Crystd from pet ether.

Brilliant Green (4-dimethylaminotriphenyl carbinol) *[633-03-4]* **M 482.7, m 209-211°(dec), pK^{25} 4.75.** Purified by pptn as the perchlorate from aqueous soln (0.3%) after filtering, heating to 75° and adjustment to pH 1-2. Recrystd from EtOH/water (1:4) [Kerr and Gregory *Analyst (London)* **94** 1036 *1969*].

*N-***Bromoacetamide** *[79-15-2]* **M 138.0, m 102-105°, 107-109°, 108° (anhyd).** Possible contaminant is CH₃CONBr₂. Recrystd from CHCl₃/hexane (1:1, seed if necessary) or water and dried over CaCl₂. [Oliveto and Gerold *Org Synth* Coll Vol IV 104 *1963*).

4-Bromoacetanilide *[103-88-8]* **M 214.1, m 167°.** Crystd from aq MeOH or EtOH. Purified by zone refining.

Bromoacetic acid *[79-08-3]* **M 138.9, m 50°, b 118°/15mm, 208°/760mm, pK^{25} 2.92.** Crystd from pet ether (b 40-60°). Diethyl ether soln passed through an alumina column, and the ether evaporated at room temperature under vacuum. **LACHRYMATORY.**

Bromoacetone *[598-31-2]* **M 137.0, b 31.5°/8mm.** Stood with anhydrous CaCO₃, distd under low vacuum, and stored with CaCO₃ in the dark at 0°. **LACHRYMATORY.**

4-Bromoacetophenone *[99-90-1]* **M 199.1, m 54°.** Crystd from EtOH, MeOH or from pet ether (b 80-100°). [Tanner *J Org Chem* **52** 2142 *1987*.]

ω−**Bromoacetophenone (phenacyl bromide)** *[70-11-1]* **M 199.1, m 57-58°.** Crystd from EtOH, MeOH or from pet ether (b 80-100°). [Tanner *J Org Chem* **52** 2142 *1987*.]

4-Bromoaniline *[106-40-1]* **M 172.0, m 66°, pK^{25} 3.86.** Crystd (with appreciable loss) from aqueous EtOH.

2-Bromoanisole *[578-57-4]* **M 187.0, f 2.5°, b 124°/40mm, d 1.513, n^{25} 1.5717.** Crystd by partial freezing (repeatedly), then distd under reduced pressure.

4-Bromoanisole *[104-92-7]* **M 187.0, f 13.4°, b 124°/40mm, d 1.495, n^{25} 1.5617.** Crystd by partial freezing (repeatedly), then distd under reduced pressure.

9-Bromoanthracene *[1564-64-3]* **M 257.1, m 98-100°.** Crystd from MeOH or EtOH followed by sublimation *in vacuo*. [Masnori et al. *J Am Chem Soc* **108** 126 *1986*.]

4-Bromobenzal diacetate *[55605-27-1]* **M 287.1, m 95°.** Crystd from hot EtOH (3mL/g).

Bromobenzene *[108-86-1]* **M 157.0, b 155.9°, d 1.495, n 1.5588, n^{15} 1.56252.** Washed vigorously with conc H_2SO_4, then 10% NaOH or $NaHCO_3$ solns, and H_2O. Dried with $CaCl_2$ or Na_2SO_4, or passed through activated alumina, before refluxing with, and distilling from, CaH_2, using a glass helix-packed column.

4-Bromobenzene diazonium tetrafluoroborate *[673-40-5]* **M 270.8, m 133° (dec), 135-140° (dec), 135° (dec).** Wash with Et_2O until the wash is colourless and allow to dry by blowing N_2 over it. Store at 0-4° in the dark. [*Chem Ber* **64** 1340 *1931*.]

4-Bromobenzenesulfonyl chloride *[98-58-8]* **M 255.5, m 73-75°, 74.3-75.1, 75-76°, 77°, b 153°/15mm, 150.6°/13mm.** Wash with cold water, dry and recryst from pet ether, or from ethyl ether cooled in powdered Dry-ice after the ether soln had been washed with 10% NaOH until colourless, then dried with anhydrous Na_2SO_4. Alternatively dissolve in $CHCl_3$, wash with H_2O, dry (Na_2SO_4), evaporate and crystallise. [*J Am Chem Soc* **62** 511 *1940*.] Test for the SO_2Cl group by dissolving in EtOH and boiling with NH_4CNS whereby a yellow amorphous ppte forms on cooling [*J Am Chem Soc* **25** 198 *1901*].

o-**Bromobenzoic acid** *[88-65-3]* **M 201.0, m 148.9°, pK^{20} 2.88.** Crystd from *C_6H_6 or MeOH.

m-**Bromobenzoic acid** *[585-76-2]* **M 201.0, m 155°, pK^{25} 3.81.** Crystd from acetone/water, MeOH or acetic acid.

p-**Bromobenzoic acid** *[586-76-5]* **M 201.0, m 251-252°, 254-256°, 257-258°, pK^{25} 3.96.** Crystd from MeOH, or MeOH/water mixture, 90% EtOH and Et_2O. The *methyl ester* has **m** 81° from Et_2O or dilute MeOH. [Male and Thorp *J Am Chem Soc* **35** 269 *1913*; Lamneck *J Am Chem Soc* **76** 406 *1954*, Vandenbelt et al. *Anal Chem* **26** 926 *1954*.]

p-**Bromobenzophenone** *[90-90-4]* **M 261.1, m 81°.** Crystd from EtOH.

p-**Bromobenzoyl chloride** *[586-75-4]* **M 219.5, m 36-39°, 39.8°, 41°, b 62°/0.1mm, 104.5°/6mm, 126.4-127.2°/14mm.** Check IR of a film to see if OH bands are present. If absent then recryst from pet ether and dry in a vacuum. If OH bands are weak then distil *in vacuo* and recryst if necessary. If OH bands are very strong then treat with an equal volume of redistilled $SOCl_2$ reflux for 2h then evaporate excess of $SOCl_2$ and distil residual oil or low melting solid. Store in the dark away from moisture. **LACHRYMATORY.** [Martin and Partington *J Chem Soc* 1175 *1936*.]

p-**Bromobenzyl bromide** *[589-15-1]* **M 249.9, m 60-61°.** Crystd from EtOH. **LACHRYMATORY.**

p-**Bromobenzyl chloride** *[589-17-3]* **M 205.5, m 40-41°, b 105-115°/12mm.** Crystd from EtOH. **LACHRYMATORY.**

p-**Bromobiphenyl** *[92-66-0]* **M 233.1, m 88.8-89.2°.** Crystd from abs EtOH and dried under vacuum.

2-Bromobutane *[78-76-2]* **M 137.0, b 91.2°, d 1.255, n 1.4367, n^{25} 1.4341.** Washed with conc HCl, water, 10% aqueous $NaHSO_3$, and then water. Dried with $CaCl_2$, Na_2SO_4 or anhydrous K_2CO_3, and fractionally distd through a 1m column packed with glass helices.

(+)-3-Bromocamphor-8-sulfonic acid *[5344-58-1]* **M 311.2, m 195-196°(anhydrous), $[\alpha]_D^{20}$ +88.3° (in H_2O), pK ~0.** Crystd from water.

1R(endo,anti)-**3-Bromocamphor-8-sulfonic acid ammonium salt** see entry in Chapter 5.

(+)-3-Bromocamphor-10-sulfonic acid hydrate *[67999-30-8]* **M 329.2, m 119-121°, $[\alpha]_D^{20}$ +98.3° (in H_2O), pK ~0.** Crystd from water.

4-Bromo-4'-chlorobenzophenone   *[27428-57-5]*   **M 295.6, m 150°.** Crystd from EtOH or *$*C_6H_6$ and further purified by zone refining (100 passes) [Grove and Turner *J Chem Soc* 509 *1929*; Lin and Hanson *J Phys Chem* **91** 2279 *1987*].

Bromocresol Green (3',3",5',5"-tetrabromo-*m*-cresolsulfonephthalein) *[76-60-8]* **M 698.0, m 218-219°(dec), 225°(dec), pK 4.51.** Crystd from glacial acetic acid or dissolved in aqueous 5% NaHCO$_3$ soln and ppted from hot soln by dropwise addition of aqueous HCl. Repeated until the extinction did not increase (λ_{max} 423nm). Indicator at pH 3.81 (yellow) and pH 5.4 (blue-green).

Bromocresol Purple (5',5"-dibromo-*o*-cresolsulfonephthalein) *[115-40-2]* **M 540.2, m 241-242°(dec), pK$_1$ -2.15, pK$_2$ 6.3.** Dissolved in aqueous 5% NaHCO$_3$ soln and ppted from hot soln by dropwise addition of aqueous HCl. Repeated until the extinction did not increase (λ_{max} 419nm). Can also be crystd from *benzene. Indicator at pH 5.2 (yellow) and pH 6.8 (purple).

5-Bromocytosine *[2240-25-7]* **M 190.0, m 245-255°(dec), 250°(dec), pK$_1^{25}$ 3.04, pK$_2^{25}$ 10.33.** Recryst from H$_2$O or 50% aq EtOH. Alternatively, dissolve *ca* 3g in conc HCl (10mL) and evaporate to dryness. Dissolve the residual hydrochloride in the minimum volume of warm H$_2$O and make faintly alkaline with aq NH$_3$. Collect the crystals and dry in a vacuum at 100°. [*J Am Chem Soc* **56** 134 *1934*.]

p-**Bromo-*N*,*N*-dimethylaniline** *[586-77-6]* **M 200.1, m 55°, b 264°, pK^{25} 4.23.** Refluxed for 3h with two equivalents of acetic anhydride, then fractionally distd under reduced pressure

1-Bromo-2,4-dinitrobenzene *[584-48-5]* **M 247.0, m 75°.** Crystd from ethyl ether, isopropyl ether, 80% EtOH or absolute EtOH.

2-(2-Bromoethyl)-1,3-dioxane *[33884-43-4]* **M 195.1, b 67-70°/2.8mm, 71-72°/4mm, 95°/15mm, d $_4^{20}$ 1.44, n $_D^{20}$ 1.4219.** Purify by vacuum fractionation. Also dissolve in Et$_2$O, wash with aqueous NaHCO$_3$, dry extract with Na$_2$SO$_4$, filter and fractionate. NMR in CCl$_4$ has δ 1.3 (m, 1H), 2.1 (m, 3H), 3.36 (t, 2H), 3.90 (m, 4H) and 4.57 (t, H). [*J Org Chem* **41** 560 *1976*; NMR, MS: *Tetrahedron* **35** 1969 *1979*; *J Pharm Sci* **60** 1250 *1971*.]

2-(2-Bromoethyl)-1,3-dioxolane *[18742-02-4]* **M 181.1, b 68-80°/8mm, 68-73°/10mm, 78-80°/20mm, d$_4^{20}$ 1.510, n$_D^{20}$ 1.479.** Dissolve in pentane, wash with 5% aqueous NaHCO$_3$, dry (Na$_2$SO$_4$), and evaporate. Distil the residue. [NMR: *J Org Chem* **34** 1122 *1969*; *J Pharm Sci* **60** 1250 *1971*.]

N-**(2-Bromoethyl)phthalimide** *[574-98-1]* **M 254.1, m 81-83°, 82.5-83.5°.** The following is to be carried out in an efficient **FUME HOOD**. Dissolve the compound (180g) in CS$_2$ (500 mL) by refluxing for 15 min (to cause the separation of the most likely impurity, 1,2-diphthalimidoethane), filter and evaporate under reduced pressure. The product forms light tan crystals. (**m** 78-80°). Recryst from EtOH (charcoal) [the compound (50g) is dissolved in hot 75% EtOH (200mL), boiled for *ca* 10 min, carbon added (5g, Norite), filtered and cooled to 0°], as white crystals (40g) which can be recrystd (**m** 80-81°) and further recrystn gave **m** 82-83°. [*Org Synth* Coll Vol I 119 *1932*, *Synthesis* 389 *1976*; NMR: *Bull Soc Chim Fr* II-165 *1979*.]

Bromoform *[75-25-2]* **M 252.8, f 8.1°, 55-56°/35mm, 149.6°/760mm, d^{15} 2.9038, d^{30} 2.86460, n^{15} 1.60053, n 1.5988.** Storage and stability of bromoform and chloroform are similar. Ethanol, added as a stabilizer, is removed by washing with H$_2$O or with saturated CaCl$_2$ soln, and the CHBr$_3$, after drying with CaCl$_2$ or K$_2$CO$_3$, is fractionally distd. Prior to distn, CHBr$_3$ has also been washed with conc H$_2$SO$_4$ until the acid layer no longer became coloured, then dilute NaOH or NaHCO$_3$, and H$_2$O. A further purification step is fractional crystn by partial freezing.

3-Bromofuran *[22037-28-1]* **M 147.0, b 38.5°/40mm, 50°/110mm, 102.5-103°/atm, d 1.661, n 1.4970.** Purified by two steam distillations and dried over fresh CaO. It can be dried over Na metal (no obvious reaction) and fractionated. It is not very soluble in H$_2$O but soluble in organic solvents. Freshly distilled, it is a clear oil, but darkens on standing and eventually resinifies. It can be stored for long periods by covering the oil with an alkaline soln of hydroquinone and redistilled when required. It forms a characteristic *maleic anhydride adduct*, **m** 131.5-132°. [*J Am Chem Soc* **52** 2083 *1930*, **53** 737 *1931*, adduct: **55** 430 *1933*.]

(±)-2-Bromohexadecanoic acid (2-bromopalmitic acid) *[18263-25-7]* **M 335.3, m 51-53°, 52.3-52.5°, 53°, pK$_{Est}$ ~3.2.** Recrystd from pet ether (b 60-80°, charcoal) and finally from EtOH. The *ethyl ester* has **b** 177-178°/2mm, d$_{28}^{28}$ 1.0484, n$_D^{20}$ 1.4560. [IR: *J Org Chem* **21** 1426 *1956*.]

5-Bromoindole *[10075-50-0]* **M 196.1, m 90.5-91°, 90-92°, pK 16.13 (NH).** Purified by steam distn from a faintly alkaline soln. Cool the aqueous distillate, collect the solid, dry in a vacuum desiccator over P$_2$O$_5$ and recryst from aqueous EtOH (35% EtOH) or pet ether-Et$_2$O. λ_{max} in MeOH: 279, 287 and 296 (logε 3.70, 3.69 and 3.53. The *picrate* has **m** 137-138°(dec) (from Et$_2$O-pet ether). [UV: *Chem Ber* **95** 2205 *1962*; UV and NMR: *Bull Soc Chim Fr* 4091 *1970*.]

5-Bromoisatin *[87-48-9]* **M 226.0, m 245°(dec), 251-153°, 255-256°.** Forms red prisms or needles from EtOH. The *N-acetate* crystallises as yellow prisms from *C$_6$H$_6$, **m** 170-172°, and the *N-methyl* derivative form orange-red needles from MeOH, **m** 172-173°. [*Chem Ber* **47** 360 *1914*, **53** 1545 *1920*; *Recl Trav Chim Pays-Bas* **73** 197 *1954*; *Tetrahedron Lett* 215 *1978*.]

6-Bromoisatin *[6326-79-0]* **M 226.0, m 270°, pK 10.35.** Recrystd from AcOH (yellow needles). It is a plant growth substance. [Sadler *J Org Chem* **21** 169 *1956*.]

2-Bromomethylanthraquinone *[7598-10-9]* **M 301.1, m 200-202°.** Recrystd from AcOH, the crystals are washed with a little Et$_2$O, dried in air and then in vac at 100°. It is prepared by bromination of 2-methylanthraquinone with Br$_2$/PhNO$_2$ at 145-150°, or *N*-bromosuccinimide in CCl$_4$ containing a trace of (PhCOO)$_2$.

2-(Bromomethyl)benzonitrile *[22115-41-9]* **M 195.1, m 72-73°, 79°, b 152-155°/15mm.** Purified by steam distn. Extract the distillate with Et$_2$O, dry extract (Na$_2$SO$_4$), evap and distil residue. The solidified distillate can be recrystd from pet ether or cyclohexane. NMR (CDCl$_3$) δ: 7.8-7.2 (m 4H), 4.62 (s, 2H); IR ν: 2238 cm^{-1}. **LACHRYMATORY** [*Chem Ber* **24** 2570 *1891*, **74** 675 *1934*; *Aust J Chem* **22** 577 *1969*.]

S-(+)-**1-Bromo-2-methylbutane** *[534-00-9]* **M 151.1, b 38.2°/39mm, 49°/62mm, 60.8°(57-58°)/100mm, 65-65.6°/140mm, 116-122°/atm, d$_4^{20}$ 1.2232, n$_D^{20}$ 1.4453, [α]$_D^{20}$ +5.1° (neat, +5.8° (c 5, CHCl$_3$).** Wash with ice-cold H$_2$O, dried by freezing, shake twice with an equal vol of H$_2$SO$_4$ at 0°, and twice with an equal volume of H$_2$O at 0°. Freeze dried and kept over freshly heated (and then cooled) K$_2$CO$_3$, and distd through a vacuum jacketed column of broken glass. Alternatively, dissolve in pet ether (b 40-60°), wash with 5% NaOH, conc H$_2$SO$_4$ (at 0°), then H$_2$O, dry (CaCl$_2$), evaporate and distil. [*J Am Chem Soc* **74** 4858 *1952*, **81** 2779 *1959*; *J Chem Soc* 1413 *1959*, 2685 *1950*.]

2-Bromo-3-methylindole (2-bromoskatole) *[1484-28-2]* **M 210.1, m 102-104°, pK$_{Est}$ <0.** Purified by chromatography on silica gel in CHCl$_3$/pet ether (1:2) followed by crystn from aqueous EtOH. [Phillips and Cohen *J Am Chem Soc* **108** 2023 *1986*.]

4-(Bromomethyl)-7-methoxycoumarin *[35231-44-8]* **M 269.1, m 208-209°, 213-215°, 216-218°.** Cryst from boiling AcOH, crystals are washed with AcOH, EtOH and dried in a vacuum, NMR (TFA) δ 3.97s, 4.57s, 6.62s, 6.92-7.19m and 7.80d. [*Biochem Biophys Res Commun* **45** 1262 *1971*.]

2-(Bromomethyl)-naphthalene *[939-26-4]* **M 221.1, m 52-54°, 56°, 56-57°, b 133-136°/0.8mm, 214°/100mm.** Dissolve in toluene, wash with saturated aqueous NaHCO$_3$, dry (Mg SO$_4$), evaporate and fractionally distil the residue and recrystallise the distillate from EtOH. [*J Chem Soc* 5044, *1952*; *Bull Soc Chim Fr* 566 *1953*.]

2-Bromo-2-methylpropane *[507-19-7]* **M 137.0, b 71-73°, d 1.218, n 1.429.** Neutralised with K_2CO_3, distd, and dehydrated using molecular sieves (5A), then vacuum distd and degassed by freeze-pump-thaw technique. Sealed under vacuum.

1-Bromonaphthalene *[90-11-9]* **M 207.1, b 118°/6mm, d 1.489.** Purified by passage through activated alumina, and three vacuum distns.

2-Bromonaphthalene *[580-13-2]* **M 207.1, m 59°.** Purified by fractional elution from a chromatographic column. Crystd from EtOH.

1-Bromo-2-naphthol *[573-97-7]* **M 223.1, m 76-78°, pK$_{Est}$ ~8.0.** Crystd from EtOH.

6-Bromo-2-naphthol *[15231-91-1]* **M 223.1, m 122-126°, pK$_{Est}$ ~9.1.** Crystd from EtOH.

5-Bromonicotinic acid *[20826-04-4]* **M 202.0, m 178-182°, 189-190°, pK$_{Est}$ ~4.4, pK^{25} 4.02 (50% aq EtOH).** Recryst from H_2O and then from EtOH using charcoal. The *amide* has **m** 219-219.5° (from aq EtOH) and the *methyl ester* prepared by addition of ethereal diazomethane can be purified by sublimation in a vacuum and has **m** 98-99°, the *acid chloride* also can be sublimed in *vacuo* and has **m** 74-75° and gives the methyl ester in MeOH. [*J Prakt Chem* **138** 244 *1933*; *J Am Chem Soc* **70** 2381 *1948*; **82** 4430 *1960*; *J Chem Soc* 35 *1978*.]

ω-Bromo-4-nitroacetophenone *[99-81-0]* **M 244.1, m 98°.** Crystd from *C_6H_6-pet ether.

o-**Bromonitrobenzene** *[577-19-5]* **M 202.1, m 43°.** Crystd twice from pet ether, using charcoal before the first crystn.

m-**Bromonitrobenzene** *[585-79-5]* **M 202.1, m 55-56°.** Crystd twice from pet ether, using charcoal before the first crystn.

p-**Bromonitrobenzene** *[586-78-7]* **M 202.1, m 127°.** Crystd twice from pet ether, using charcoal before the first crystn.

1-Bromooctadecane *[112-89-0]* **M 333.4, m 26°, 27.3°, 28-30°, b 178-179°/2mm, 214-218°/15mm, d^{20} 0.976, n$_D^{20}$ 1.461.** Twice recrystd from the melt then distilled under vacuum three times and using the middle cut. Alternatively, wash the oil with aqueous Na_2SO_4, then conc H_2SO_4 (cool) and again with aqueous Na_2SO_4 and then fractionate. [*J Am Chem Soc* **55** 1574 *1933*, **72** 171 *1950*; IR: *Aust J Chem* **12** 743 *1959*; IR: *Bull Soc Chim Fr* 516 *1957*.]

(±)-2-Bromopentane *[107-81-3]* **M 151.1, b 117.2°/753mm, 116-117°/atm, 117.5°/740mm, d$_4^{20}$ 1.2190, n$_D^{20}$ 1.4401.** Dry over K_2CO_3 and distil through a short Vigreux column. [IR: *J Am Chem Soc* **74** 4063 *1952*, **78** 2199 *1956*.]

p-**Bromophenacyl bromide** *[99-73-0]* **M 277.9, m 110-111°.** Crystd from EtOH (*ca* 8mL/g).

o-**Bromophenol** *[95-56-7]* **M 173.0, b 194°, d 1.490, pK^{25} 8.45.** Purified by at least two passes through a chromatographic column.

p-**Bromophenol** *[106-41-2]* **M 173.0, m 64°, pK^{25} 9.36.** Crystd from $CHCl_3$, CCl_4, pet ether (b 40-60°), or water and dried at 70° under vacuum for 2h.

Bromophenol Blue (3,3',5,5'-tetrabromophenolsulfonephthalein) *[115-39-9]* **M 670.0, m 270-271°(dec), 273°(dec), λmax 422max, pK 3.62.** Crystd from *benzene or acetone/glacial acetic acid, and air dried. Indicator at pH 3.0 (yellow) and pH 4.6 (purple).

(4-Bromophenoxy)acetic acid *[1878-91-7]* **M 231.1, m 158°, pK^{25} 3.13.** Crystd from EtOH.

3-(4-Bromophenoxy)propionic acid *[93670-18-9]* **M 247.1, m 146°, pK$_{Est}$ ~4.2.** Crystd from EtOH.

4-Bromophenylacetic acid *[1878-68-8]* **M 215.1, m 112-113°, 113-115°, 114°, pK 4.19.** Recrystd from H$_2$O as needles. The *acid chloride* has **b** 238°/atm, **m** 50°, and the *anilide* has **m** 174-175°. [*J Chem Soc* 161 *1934, 1251* 1948; *J Org Chem* 11 798 *1946*.]

4-Bromophenylhydrazine *[589-21-9]* **M 187.1, m 108-109°, pK^{20}-5.6 (aq H$_2$SO$_4$), pK^{25} 5.05.** Crystd from H$_2$O.

4-Bromophenyl isocyanate *[2492-02-9]* **M 189.0, m 41-42°.** Crystd from pet ether (b 30-40°).

4-Bromophenyl isothiocyanate *[1985-12-2]* **M 214.1, m 56-58°.** Recryst from boiling *n*-hexane. Any insoluble material is most probably the corresponding urea. It can be purified by steam distn, cool the receiver, add NaCl and extract in Et$_2$O, wash extract with N H$_2$SO$_4$; dry (MgSO$_4$), evaporate and recrystallise the residual solid. [*Org Synth* Coll Vol IV 700 *1963*; Coll Vol I 447 *1941*.]

Bromopicrin (tribromonitromethane) *[464-10-8]* **M 297.8, m 10.2-10.3°, b 85-87°/16mm, d 2.788, n 1.579.** Steam distd, dried with anhydrous Na$_2$SO$_4$ and vacuum distd. **TOXIC.**

R-(+)-**2-Bromopropionic acid** *[10009-70-8]* **M 153.0, b 78°/4mm, [α]$_D^{25}$+27.2° (neat), pK^{25}4.07.** Dissolve in Et$_2$O, dry (CaCl$_2$), evap and distil through a short column. Distillation through a Podbielniak column led to decomposition. [**Podbielniak column.** A plain tube containing "Heli-Grid" Nichrome or Inconel wire packing. This packing provides a number of passage-ways for the reflux liquid, while the capillary spaces ensure very even spreading of the liquid, so that there is a very large area of contact between liquid and vapour while, at the same time, channelling and flooding are minimised. A column 1m high has been stated to have an efficiency of 200-400 theoretical plates (for further details, see Podbielniak *Ind Eng Chem (Anal Ed)* 13 639 *1941*; Mitchell and O'Gorman *Anal Chem* 20 315 *1948*)]. Store in the dark under N$_2$, preferably in sealed ampoules. Even at -10° it slowly decomposes. [*J Am Chem Soc* 76 6054 *1954*.]

3-Bromopropionic acid *[590-92-1]* **M 153.0, m 62.5°, 62.5-63.5°, 63-64°.** Crystallises as plates from CCl$_4$. It is soluble in organic solvents and H$_2$O. It has a pKa^{25} in H$_2$O of 4.01, and its *methyl ester* has **b** 65°/18mm and 80°/27mm. The *S-benzylisothiouronium salt* has **m** 136°. [*Org Synth* Coll Vol I 134 *1948*; *Justus Liebigs Ann Chem* 599 140 *1956*.]

N-(**3-Bromopropyl)phthalimide** *[5460-29-7]* **M 268.1, m 72-74°, 74°.** Place in a Soxhlet and extract with Et$_2$O, whereby the bis-phthalimido impurity is not extracted. Evaporate the Et$_2$O and recryst from EtOH or aqueous EtOH or pet ether. [*Chem Ber* 21 2669 1888; *Justus Liebigs Ann Chem* 614 83 *1958*; *Can J Chem* 31 1060 *1953*.]

2-Bromopyridine *[109-04-6]* **M 158.0, b 49.0°/2.7mm, d 1.660, n 1.5713, pK^{25} 0.90.** Dried over KOH for several days, then distd from CaO under reduced pressure, taking the middle fraction.

Bromopyrogallol Red (5,5'-dibromopyrogallolsulfonephthalein) *[16574-43-9]* **M 576.2, m 300°, λmax 538nm (ε 54,500 H$_2$O pH 5.6-7.5), pK$_1$ 2.9, pK$_2$ 4.39, pK$_3$ 9.15, pK$_4$ 11.72.** Crystd from 50% EtOH, or aq alkaline soln by acid [Suk *Collect Czech Chem Commun* 31 3127 *1966*].

Bromopyruvic acid *[1113-59-3]* **M 167.0, m 79-82°, pK$_{Est}$ ~1.6.** Dried by azeotropic distn (toluene), and then recrystd from dry CHCl$_3$. Dried for 48h at 20° (0.5 Torr) over P$_2$O$_5$. Stored at 0°. [Labandiniere et al. *J Org Chem* 52 157 *1987*].

5-Bromosalicyl hydroxamic acid *[5798-94-7]* **M 210.1, m 232°(dec), pK$_{Est(1)}$~ 1.5, pK$_{Est(2)}$~ 7.0, pK$_{Est(3)}$~ 8.7.** Crystd from EtOH.

4-Bromostyrene *[2039-82-9]* **M 183.1, b 49.5-50°/2.5mm, 87-88°/12mm, 102-104°/20mm, d 1.3984, n 1.5925.** It polymerises above 75° in the presence of benzoyl peroxide. To purify, if it has not gone to a solid resin, dissolve in Et₂O, dry (MgSO₄), add *ca* 0.1g of 4-*tert*butylcatechol (polymerisation inhibitor) per 100g of bromostyrene. Filter, evap under reduced press (use as high a vac as possible) and distil. Store in dark bottles in the presence of the inhibitor (concn as above). [*Org Synth* Coll Vol III 204 *1955*.]

N-**Bromosuccinimide** *[128-08-5]* **M 178.0, m 183-184°(dec).** *N*-Bromosuccinimide (30g) was dissolved rapidly in 300mL of boiling water and filtered through a fluted filter paper into a flask immersed in an ice bath, and left for 2h. The crystals were filtered, washed thoroughly with *ca* 100mL of ice-cold water and drained on a Büchner funnel before drying under vac over P₂O₅ or CaCl₂ [Dauben and McCoy *J Am Chem Soc* **81** 4863 *1959*]. Has also been crystd from acetic acid or water (10 parts, washed in water and dried *in vacuo*, [Wilcox et al. *J Am Chem Soc* **108** 7693 *1986;* Shell et al. *J Am Chem Soc* **108** 121 *1986*; Phillips and Cohen *J Am Chem Soc* **108** 2013 *1986*.]

Bromotetronic acid *[21151-51-9]* **M 179.0, m 183°(dec), pK²⁵ 2.23.** Decolorised, and free bromine was removed by charcoal treatment of an ethyl acetate soln, then recrystd from ethyl acetate [Schuler, Bhatia and Schuler *J Phys Chem* **78** 1063 *1974*].

Bromotheophylline *[10381-75-6]* **M 259.1, m 309°, 315-320° (with browning and dec), pK$_{Est(1)}$~5.5, pK$_{Est(2)}$ ~9.2.** It is purified by dissolving in the minimum volume of dilute NaOH (charcoal), filter and acidify to pH *ca* 3.5-4 and the solid that separates is collected, dried *in vacuo* at 100° and stored in a dark container. [*J Prakt Chem* [2] **118** 158 *1928*; *Chem Ber* **28** 3142 *1895*.]

Bromothymol Blue (3',3"-dibromothymolsulfonephthalein) *[76-59-5]* **M 624.4, m 201-203°, pK₁ -0.66, pK₂ 6.99.** Dissolved in aq 5% NaHCO₃ soln and ppted from the hot soln by dropwise addn of aq HCl. Repeated until the extinction did not increase (λ_{max} 420nm). Indicator at pH 6.0 (yellow) and 7.6 (blue).

p-**Bromotoluene** *[106-38-7]* **M 171.0, m 28°, b 184°, d 1.390.** Crystd from EtOH [Taylor and Stewart *J Am Chem Soc* **108** 6977 *1986*].

α-**Bromo-4-toluic acid** *[6232-88-8]* **M 215.1, m 229-230°, pK$_{Est}$ ~3.2.** Crystd from Me₂CO.

Bromotrichloromethane *[75-62-7]* **M 198.5, f -5.6°, m 21°, b 104.1°, d 2.01, n 1.5061.** Washed with aq NaOH soln or dilute Na₂CO₃, then with H₂O, and dried with CaCl₂, BaO, MgSO₄ or P₂O₅ before distilling in diffuse light and storing in the dark. Has also been purified by treatment with charcoal and by fractional crystn by partial freezing. Purified also by vigorous stirring with portions of conc H₂SO₄ until the acid did not discolour during several hours stirring. Washed with Na₂CO₃ and water, dried with CaCl₂ and then illuminated with a 1000W projection lamp at 15cm for 10h, after making it 0.01M in bromine. Passed through a 30 x 1.5cm column of activated alumina and fractionally redistilling through a 12-in Vigreux column. [Firestone and Willard *J Am Chem Soc* **83** 3511 *1961*; see also Cadogan and Duell *J Chem Soc* 4154 *1962*.]

1-Bromo-3,3,3-trifluoroethane *[421-06-7]* **M 163.0, m -94°, b 26-27°, d 1.788, n 1.332.** Washed with water, dried (CaCl₂) and distd.

Bromotrifluoromethane (Freon 13B1) *[75-63-8]* **M 148.9, b -59°, d 1.590.** Passed through a tube containing P₂O₅ on glass wool into a vac system where it was frozen out in a quartz sample tube and degassed by a series of cycles of freezing, evacuating and thawing.

5-Bromouracil *[51-20-7]* **M 191.0, m 293°, 303-305°, 312°(dec), pK₁²⁵ -7.25, pK₂²⁵ 7.83.** Purified by dissolving in 2N NaOH (charcoal), filter and acidify with HCl. The ppte is dried *in vacuo* at 100° and recryst (prisms) twice from H₂O. [*J Am Chem Soc* **56** 134 *1934*, UV: *J Am Chem Soc* **81** 3786 *1959*; *J Org Chem* **23** 1377 *1958*.]

5-Bromovaleric (γ–bromopentanoic) acid *[2067-33-6]* **M 181.0, m 40°, pK$_{Est}$ ~4.6.** Crystd from pet ether.

α-Bromo-*p*-xylene *[104-81-4]* **M 185.1, m 35°, b 218-220°/740mm.** Crystd from EtOH or pentane.

Bromural [*N*-(aminocarbonyl)-2-bromo-3-methylbutanamide, bromisovalum] *[496-67-3]* **M 223.1, m 154-155°.** Crystd from toluene, and air dried.

Bufotenine hydrogen oxalate *[2963-79-3]* **M 294.3, m 96.5°.** Crystd from Et$_2$O.

1,3-Butadiene *[106-99-0]* **M 54.1, b -2.6°.** Dried by condensing with a soln of triethylaluminium in decahydronaphthalene; then flash distd. Also dried by passage over anhydrous CaCl$_2$ or distd from NaBH$_4$. Also purified by passage through a column packed with molecular sieves (4A), followed by cooling in Dry-ice/MeOH bath overnight, filtering off the ice and drying over CaH$_2$ at -78° and distd in a vacuum line.

***n*-Butane** *[106-97-8]* **M 58.1, m -135°, b -0.5°.** Dried by passage over anhydrous Mg(ClO$_4$)$_2$ and molecular sieves type 4A. Air was removed by prolonged and frequent degassing at -107°.

1,4-Butanediol (tetramethylene glycol) *[110-63-4]* **M 90.1, f 20.4°, b 107-108°/4mm, 127°/20mm, d 1.02, n 1.4467.** Distd and stored over Linde type 4A molecular sieves, or crystd twice from anhydrous diethyl ether/acetone, and redistd. Also purified by recrystn from the melt and doubly distd *in vacuo* in the presence of Na$_2$SO$_4$.

***meso*-2,3-Butanediol** *[513-85-9]* **M 90.1, m 25°.** Crystd from isopropyl ether.

***threo*-2,3-butanediol** *[R,R(-): 24347-58-8]* *[S,S (+):19132-06-0]* **M 90.1, m 16-19°, 19.7°, b 77,5-78°/10mm, 179-180°/atm, [α]$_D^{20}$ (-) or (+) 13.1° (neat).** Purified by fractional distn. The *bis-(4-nitrobenzoate)* has **m 141-142°** and **[α]$_D^{25}$± 52°** (c 4 CHCl$_3$). [*J Am Chem Soc* **79** 734 *1957*, **74** 425 *1952*, *Can J Res* **27** 457 *1949*.]

1-Butanesulfonyl chloride *[2386-60-9]* **M 156.6, b 75-76°/7mm, 98°/13mm, d$_4^{20}$1.2078, n$_D^{20}$ 1.4559.** It has a pungent odour and is **LACHRYMATORY.** If IR shows OH bands then dissolve in Et$_2$O, wash with cold saturated aq NaHCO$_3$ (care since CO$_2$ will be generated) then H$_2$O, dry over solid Na$_2$SO$_4$, filter evaporate and distil the residue twice. Characterised by shaking a soln in Et$_2$O or *C$_6$H$_6$ with aq NH$_3$, collect the solid and recryst from CHCl$_3$, CCl$_4$ or Et$_2$O-pet ether, **m 48°**. [*J Am Chem Soc* **60** 1488 *1938*; *J Org Chem* **5** 83 *1940*.]

1-Butanethiol *[109-79-5]* **M 90.2, b 98.4°, d^{25} 0.837, n 1.443, n^{25} 1.440, pK$_{Est}$ ~11.3.** Dried with CaSO$_4$ or Na$_2$SO$_4$, then refluxed from magnesium; or dried with, and distd from CaO, under nitrogen [Roberts and Friend *J Am Chem Soc* **108** 7204 *1986*.] Has been separated from hydrocarbons by extractive distn with aniline.
Dissolved in 20% NaOH, extracted with a small amount of *C$_6$H$_6$, then steam distd, until clear. The soln was then cooled and acidified slightly with 15% H$_2$SO$_4$. The thiol was distd out, dried with CaSO$_4$ or CaCl$_2$, and fractionally distd under N$_2$ [Mathias and Filho *J Phys Chem* **62** 1427 *1958*]. Also purified by pptn as lead mercaptide from alcoholic soln, with regeneration by adding dilute HCl to the residue after steam distn. *All operations should be carried out in a fume cupboard due to the* **TOXICITY** *and obnoxious odour of the thiol.*

2-Butanethiol *[513-53-1]* **M 90.2, b 37.4°/134mm, d^{25} 0.846, n^{25} 1.4338, pK$_{Est}$ ~11.4.** Purified as for 1-butanethiol.

***n*-Butanol** *[71-36-3]* **M 74.1, b 117.7°, d^{25} 0.80572, n 1.39922, n^{15} 1.40118.** Dried with MgSO$_4$, CaO, K$_2$CO$_3$, Ca or solid NaOH, followed by refluxing with, and distn from, calcium, magnesium activated with iodine, aluminium amalgam or sodium. Can also dry with molecular sieves, or by refluxing with *n*-butyl phthalate or succinate. (For method, see *Ethanol.*) *n*-Butanol can also be dried by efficient fractional distn, water passing over in the first fractn as a binary azeotrope (contains about 37% water). An ultraviolet-transparent distillate has been obtained by drying with magnesium amd distilling from sulfanilic acid. To remove bases, aldehydes and ketones, the alcohol has been washed with dil H_2SO_4, then $NaHSO_4$ soln; esters were removed by boiling for 1.5h with 10% NaOH.

Also purified by adding 2g $NaBH_4$ to 1.5L butanol, gently bubbling with argon and refluxing for 1 day at 50°. Then added 2g of freshly cut sodium (washed with butanol) and refluxed for 1 day. Distd and the middle fraction collected [Jou and Freeman *J Phys Chem* **81** 909 *1977*].

2-Butanone (methyl ethyl ketone, MEK) *[78-93-0]* **M 72.1, b 79.6°, d 0.853, n 1.37850, n^{25} 1.37612, pK^{25} -7.2 (aq H_2SO_4).** In general, purification methods are the same as for acetone. Aldehydes can be removed by refluxing with $KMnO_4$ + CaO, until the Schiff aldehyde test is negative, prior to distn. Shaking with satd K_2CO_3, or passage through a small column of activated alumina, removes cyclic impurities. The ketone can be dried by careful distn (an azeotrope containing 11% water boils at 73.4°), or by $CaSO_4$, P_2O_5, Na_2SO_4, or K_2CO_3, followed by fractional distn. Purification as the bisulfite addition compound is achieved by shaking with excess satd Na_2SO_3, cooled to 0°, filtering off the ppte, washing with a little ethyl ether and drying in air; this is followed by decomposition with a slight excess of Na_2CO_3 soln and steam distn, the distillate being satd with K_2CO_3 so that the ketone can be separated, dried with K_2CO_3, filtered, and distd. Purification as the *NaI addition compound* (m 73-74°) is more convenient. (For details, see *Acetone*.) Small quantities of 2-butanone can be purified by conversion to the semicarbazone, recrystn to constant melting point, drying under vac over $CaCl_2$ and paraffin wax, refluxing for 30min with excess oxalic acid, followed by steam distn, salting out, drying and distilling [Cowan, Jeffery and Vogel *J Chem Soc* 171 *1940*].

***cis*-2-Butene** *[590-18-1]* **M 56.1, b 2.95-3.05°/746mm.** The gas is dried with CaH_2. Purified by gas chromatography. **HIGHLY FLAMMABLE.**

***trans*-2-Butene** *[624-64-6]* **M 56.1, b 0.3-0.4°/744mm.** The gas is dried with CaH_2. Purified by gas chromatography. **HIGHLY FLAMMABLE.**

2-Butene-1,4-dicarboxylic acid (*trans*-3-hexenedioic acid, *trans*-ß-hydromuconic acid) *[4436-74-2]* **M 144.1, m 194-197°, 195-196°, $pK_{Est(1)}$~4.2, $pK_{Est(2)}$ ~5.00.** Crystd from boiling water, then dried at 50-60° in a vacuum oven.

But-3-en-2-one (methyl vinyl ketone) *[78-94-4]* **M 70.1, b 79-80°/760mm, d 0.842.** See entry on p.302.

2-*tert*-Butoxycarbonyloxyimino-2-phenylacetonitrile (BOC-ON) *[58632-95-4]* **M 246.3, m 87-89°.** Triturate solid with 90% aq MeOH, filter, wash with 90% aq MeOH and dry in a vac. Recryst from MeOH (needles or plates), but use warm MeOH and cool to cryst, *do not boil as it decomposes slowly*. IR has v 1785 (C=O) cm^{-1} and NMR ($CDCl_3$) usually shows two *tert*-butyl singlets for *syn* and *anti* isomers. Store in a brown bottle (fridge). It evolves CO_2 at room temp (stoppered bottle can explode!), but can be stored over silica gel which can extend its useful life to more than a year. [Itoh et al. *Org Synth* **59** 95 *1980*.]

2-Butoxyethanol (butyl cellosolve) *[111-76-2]* **M 118.2, b 171°/745mm, d 0.903, n 1.4191.** Peroxides can be removed by refluxing with anhydrous $SnCl_2$ or by passage under slight pressure through a column of activated alumina. Dried with anhydrous K_2CO_3 and $CaSO_4$, filtered and distd, or refluxed with, and distd from NaOH.

4-Butoxyphenylacetic acid *[4547-57-3]* **M 208.3, m 86-87°, 88.5°, pK_{Est} ~4.4.** Purified by recrystn from pet ether (b 40-60°). [*J Am Chem Soc* **68** 2592 *1946*.]

***n*-Butyl acetate** *[123-86-4]* **M 116.2, b 126.1°, d 0.882, n 1.394.** Distd, refluxed with successive small portions of $KMnO_4$ until the colour persisted, dried with anhydrous $CaSO_4$, filtered and redistd.

***tert*-Butyl acetate** *[540-88-5]* **M 116.2, b 97-98°, d 0.72.** Washed with 5% Na_2CO_3 soln, then saturated aqueous $CaCl_2$, dried with $CaSO_4$ and distd.

tert-**Butyl acetoacetate** *[1694-31-1]* M 158.2, b 71°/10mm, 85°/20mm, d_4^{20} 0.954, n_4^{20} 1.42. Dist under reduced press through a short column. [*Org Synth* **42** 28 *1962*.] **HARMFUL VAPOUR.**

tert-**Butylacetylchloride** *[7065-46-5]* M 134.6, b 68-71°/100mm, 81°/180mm, 128-132°/atm, d_4^{20} 0.964, n_D^{20} 1.423. Distil under vacuum. If IR shows OH group then treat with thionyl chloride or oxalyl chloride at *ca* 50° for 30min, evap and fractionate using a short column. Strongly **LACHRYMATORY**, use a good fume hood. [*J Am Chem Soc* **72** 222 *1950*; *J Org Chem* **22** 1551 *1957*.]

Butyl acrylate *[141-32-2]* M 128.2, b 59°/25mm, d 0.894, n^{12} 1.4254. Washed repeatedly with aqueous NaOH to remove inhibitors such as hydroquinone, then with distilled water. Dried with $CaCl_2$. Fractionally distd under reduced pressure in an all-glass apparatus. The middle fraction was sealed under nitrogen and stored at 0° in the dark until used [Mallik and Das *J Am Chem Soc* **82** 4269 *1960*].

(±)-*sec*-Butyl alcohol *[15892-23-6]* M 74.1, b 99.4°, d 0.808. Purification methods are the same as for *n*-Butanol. These include drying with K_2CO_3 or $CaSO_4$, followed by filtration and fractional distn, refluxing with CaO, distn, then refluxing with magnesium and redistn; and refluxing with, then distn from CaH_2. Calcium carbide has also been used as a drying agent. Anhydrous alcohol is obtained by refluxing with *sec*-butyl phthalate or succinate. (For method see *Ethanol*.) Small amounts of alcohol can be purified by conversion to the alkyl hydrogen phthalate and recrystn [Hargreaves, *J Chem Soc* 3679 *1956*]. For purification of optical isomers, see Timmermans and Martin [*J Chem Phys* **25** 411 *1928*].

tert-**Butyl alcohol** *[75-65-0]* M 74.1, m 23-25°, 25.7°, b 28.3°/60mm, 43.3°/123.8mm, 61.8°/315mm, 72.5°/507mm, 82.45°/760mm, d_4^{20} 0.7858, n_D^{20} 1.3878. Synthesised commercially by the hydration of 2-methylpropene in dilute H_2SO_4. Dried with CaO, K_2CO_3, $CaSO_4$ or $MgSO_4$, filtered and fractionally distd. Dried further by refluxing with, and distilling from, either magnesium activated with iodine, or small amounts of calcium, sodium or potassium, under nitrogen. Passage through a column of type 4A molecular sieve is another effective method of drying. So, also, refluxing with *tert*-butyl phthalate or succinate. (For method see *Ethanol*.) Other methods include refluxing with excess aluminium *tert*-butylate, or standing with CaH_2, and distilling as needed. Further purification is achieved by fractional crystn by partial freezing, taking care to exclude moisture. *tert*-Butyl alcohol samples containing much water can be dried by adding *benzene, so that the water distils off as a tertiary azeotrope, b 67.3°. Traces of isobutylene have been removed from dry *tert*-butyl alcohol by bubbling dry pre-purified nitrogen through for several hours at 40-50° before using. It form azeotropic mixtures with a large number of compounds. It has also been purified by distn from CaH_2 into Linde 4A molecular sieves which had been activated at 350° for 24h [Jaeger et al. *J Am Chem Soc* **101** 717 *1979*].
Rapid purification: Dry *tert*-butanol with CaH_2 (5% w/v), distil and store over 3A molecular sieves.

n-**Butylamine** *[109-73-9]* M 73.1, b 77.8°, d 0.740, n 1.4009, n^{25} 1.399, pK^{25} 10.66. Dried with solid KOH, K_2CO_3, $LiAlH_4$, CaH_2 or $MgSO_4$, then refluxed with, and fractionally distd from P_2O_5, CaH_2, CaO or BaO. Further purified by pptn as the *hydrochloride*, m 213-213.5°, from ether soln by bubbling HCl gas into it. Re-ppted three times from EtOH by adding ether, followed by liberation of the free amine using excess strong base. The amine was extracted into ether, which was separated, dried with solid KOH, the ether removed by evapn and then the amine was distd. It was stored in a desiccator over solid NaOH [Bunnett and Davis *J Am Chem Soc* **82** 665 *1960*; Lycan et al. *Org Synth* Coll Vol II 319 *1943*]. **SKIN IRRITANT.**

R-(-)-*sec*-**Butylamine** *[13250-12-9]* M 73.1, b 61-63°/atm, 62.5°/atm, d_4^{20} 0.731, n_D^{20} 1.393, $[\alpha]_D^{20}$ +7.5° (neat), pK 10.56. Dry over solid NaOH overnight and fractionate through a short helices packed column. The L-*hydrogen tartrate salt* has m 139-140° (from H_2O), the *$1H_2O$* has m 96° $[\alpha]_D^{21}$ +18.1° (c 11, H_2O); the *hydrochloride* has m 152° $[\alpha]_D^{21}$ -1.1° (c 13, H_2O) and the *benzoyl derivative* crystallises from EtOH as needles m 97°, $[\alpha]_D^{21}$ -34.9° (c 11, H_2O). [*J Chem Soc* 921 *1956*; *Acta Chem Scand* **11** 898 *1957*.]

tert-**Butylamine** *[75-64-9]* M 73.1, b 42°, d 0.696, pK 10.68. Dried with KOH or $LiAlH_4$. Distd from CaH_2 or BaO.

n-Butyl *p*-aminobenzoate *[94-25-7]* **M 193.2, m 57-59°, pK$_{Est}$ ~2.5.** Crystd from EtOH.

tert-**Butylammonium bromide** *[60469-70-7]* **M 154.1, m >250°(dec).** Recrystd several times from absolute EtOH and thoroughly dried at 105°.

4-*tert*-Butylaniline *[769-92-6]* **M 149.2, m 14.5-15°, 15-16°, b 98.5-99°/3mm, 122°/20mm, d$_4^{20}$ 0.945, n$_D^{20}$ 1.538, pK^{25} 4.95.** Isolate as sulfate salt then liberate the free base with 10% aqueous NaOH, separate layers, dry over solid KOH and dist twice from Zn dust in a vacuum and store in brown containers. It has pKa^{25} (H$_2$O) 4.95 and (50% aq EtOH) 4.62. [*J Am Chem Soc* **76** 2349 *1954*.] The *anilide* has **m** 171.5-172.3°, and the *hydrochloride* has **m** 270-274°. [*J Chem Soc* 680 *1952*; *J Am Chem Soc* **76** 6179 *1954*.]

2-*tert*-Butylanthracene *[13719-97-6]* **M 234.3, m 148-149°.** Recrystd from EtOH and finally purified by TLC.

n-**Butylbenzene** *[104-51-8]* **M 134.2, b 183.3°, d 0.860, n 1.4897, n^{25} 1.487.** Distd from sodium. Washed with small portions of conc H$_2$SO$_4$ until the acid was no longer coloured, then with water and aqueous Na$_2$CO$_3$. Dried with anhydrous MgSO$_4$, and distd twice from Na, collecting the middle fraction [Vogel *J Chem Soc* 607 *1948*].

tert-**Butylbenzene** *[98-06-6]* **M 134.2, b 169.1°, d 0.867, n 1.493, n^{25} 1.490.** Washed with cold conc H$_2$SO$_4$ until a fresh portion of acid was no longer coloured, then with 10% aqueous NaOH, followed by distd water until neutral. Dried with CaSO$_4$ and distd in a glass helices-packed column, taking the middle fraction.

4-*tert*-Butyl benzoyl chloride *[1710-98-1]* **M 196.7, b 135°/10mm, 149.9-150.5°/14mm, 266-268°(dec), d$_4^{20}$ 1.082, n$_D^{20}$ 1.536.** Distil under vac. If IR shows OH group then treat with thionyl chloride or oxalyl chloride at *ca* 50° for 30min, evap and fractionate in a vac using a short column. Strongly **LACHRYMATORY**, use a good fume hood. [*Bull Chem Soc Jpn* **32** 960 *1959*; *J Am Chem Soc* **72** 5433 *1950*.]

n-**Butyl bromide** *[109-65-9]* **M 137.0, b 101-102°, d^{25} 1.2678, n 1.4399, n^{25} 1.4374.** Washed with conc H$_2$SO$_4$, water, 10% Na$_2$CO$_3$ and again with H$_2$O. Dried with CaCl$_2$, CaSO$_4$ or K$_2$CO$_3$, and distd. Redistd after drying with P$_2$O$_5$, or passed through two columns containing 5:1 silica gel/Celite mixture and stored with freshly activated alumina.

tert-**Butyl bromoacetate** *[5292-43-3]* **M 195.1, b 52°/10mm, 74-76°/25mm, d$_4^{20}$ 1.324, n$_D^{25}$ 1.4162.** Dissolve in Et$_2$O, wash well with ice cold 10% aqueous K$_2$CO$_3$, dry over CaCl$_2$, filter and evaporate the Et$_2$O then fractionate through a Vigreux column in a vacuum. **LACHRYMATORY** [*Org Synth* **34** 28 *1954*, Coll Vol III 144 *1955*; *J Am Chem Soc* **64** 2274 *1942*, **65** 986 *1943*.]

4-*tert*-Butylcalix[4]arene *[60705-62-6]* **M 648.9, m >300° (dec), 380° (dec), 344-346°.** Recrystd from CHCl$_3$ in large solvated prisms (**m** 380° dec) effloresces on drying in air; *tetra-acetate* crysts from Ac$_2$O in colourless prisms **m** 332-333° dec. Crysts from CCl$_4$ or chlorobenzene + EtOH (**m** >300°) and *tetra-acetate* cryst from CHCl$_3$ + EtOH **m** >290° dec. Crysts from toluene in white plates with toluene of crystallisation **m** 344-346° (330-332°); the *tetra-acetate* crystallises with 1AcOH of crystallisation **m** 383-386° (softening at 330-340°, also **m** 283-286°), but acetylation with Ac$_2$O-NaOAc gives *triacetate* which recrysts from AcOH with 1AcOH of crystn **m** 278-281°. 4-*tert*-Butylcalix[4]arene (100mg) is unchanged after boiling for 4h with 10N KOH (0.04mL) in xylene (4mL). [*Br J Pharmacol* **10** 73 *1955*; *Monatsh Chem* **109** 767 *1978*; *J Am Chem Soc* **103** 3782 *1981*; see also J.Vicens and V.Böhner eds,*Calixarenes*, Kluawer Academic Publ., Boston, 1991.]

4-*tert*-Butylcalix[6]arene *[78092-53-2]* **M 972.3, m >300°, 380-381°.** Recryst from CHCl$_3$ or CHCl$_3$ - MeOH as a white solid from the mother liquors of the calix[8]arene preparation. The *hexa-acetate* (Ac$_2$O-H$_2$SO$_4$) crystallises from CHCl$_3$-MeOH **m** 360-362° dec, and the *(SiMe$_3$)$_6$* derivative crystallises from

CHCl$_3$-MeOH **m** 410-412°. Stability in KOH-xylene is same as for the 4-*tert*-butylcalix[4]arene. [*J Am Chem Soc* **103** 3782 *1981*; see also J.Vicens and V.Böhner eds,*Calixarenes*, Kluawer Academic Publ., Boston, 1991.]

4-*tert*-Butylcalix[8]arene *[68971-82-4]* **M 1297.8, m 411-412°.** Recryst from CHCl$_3$ in fine colourless glistening needles. It melts sharply between 400-401° and 411-412° depending on the sample and is sensitive to traces of metal ions. TLC on silica gel (250μm thick) and elution with CHCl$_3$-hexane (3:4); it has R$_F$ 0.75. The *octa-acetate* is prepared from 8g in Ac$_2$O (50mL) and 2 drops of conc H$_2$SO$_4$ refluxed for 2h. On cooling a colourless ppte separates and is recrystd from Ac$_2$O (1.2g 48%) **m** 353-354°. The *(SiMe$_3$)$_8$* is prepared from 4-*tert*-butylcalix[8]arene (0.65g) in pyridine (4mL) with excess of hexamethyldisilazane (1mL) and trimethylchlorosilane (0.5mL) and refluxed under N$_2$ for 2h. Cool, evaporate the pyridine, triturate gummy residue with MeOH. Chromatography on silica gel using hexane-CH$_2$Cl$_2$ gave 0.5g (61%) with one spot on TLC. Crystallises from hexane-Me$_2$CO as colourless needles **m** 358-360°. [*J Am Chem Soc* **103** 3782 *1981*; *J Org Chem* **43** 4905 1978; **44** 3962 *1979*; *J Chem Soc, Chem Commun* 533 *1981*; see also J.Vicens and V.Böhner eds,*Calixarenes*, Kluawer Academic Publ., Boston, 1991.]

***tert*-Butyl carbazate** *[870-46-2]* **M 132.2, m 41-42°, b 64°/0.01mm, 55-57°/0.4mm.** Dist in a Claisen flask with a water or oil bath at *ca* 80°. After a couple of drops have distd the carbazate is collected as an oil which solidifies to a snow white solid. It can be crystd with 90% recovery from a 1:1 mixt of pet ether (b 30-60°) and pet ether (b 60-70°). [*Org Synth* **44** 20 *1964*.]

4-*tert*-Butylcatchol *[98-29-3]* **M 166.22, m 47-48°, 55-56°, 75°, b 265°/atm, pK$_{Est(1)}$~9.5, p K$_{Est(2)}$ ~13.0.** Vacuum distd and recrystd from pentane or pet ether (or *C$_6$H$_6$).

***n*-Butyl chloride** *[109-69-3]* **M 92.6, b 78°, d 0.886, n 1.4021.** Shaken repeatedly with conc H$_2$SO$_4$ (until no further colour developed in the acid), then washed with water, aq NaHCO$_3$ or Na$_2$CO$_3$, and more water. Dried with CaCl$_2$, or MgSO$_4$ (then with P$_2$O$_5$ if desired), decanted and fractionally distd. Alternatively, a stream of oxygen continuing *ca* three times as long as was necessary to obtain the first coloration of starch iodide paper by the exit gas. After washing with NaHCO$_3$ soln to hydrolyze ozonides and to remove the resulting organic acid, the liquid was dried and distd [Chien and Willard *J Am Chem Soc* **75** 6160 *1953*].

***tert*-Butyl chloride** *[507-20-0]* **M 92.6, f -24.6°, b 50.4°, d 0.851, n 1.38564.** Purification methods commonly used for other alkyl halides lead to decomposition. Some impurities can be removed by photochlorination with a small amount of chlorine prior to use. The liquid can be washed with ice water, dried with CaCl$_2$ or CaCl$_2$ + CaO and fractionally distd. It has been further purified by repeated fractional crystn by partial freezing.

***tert*-Butyl chloroacetate** *[107-59-5]* **M 150.6, b 48-49°/11mm, 60.2°/15mm, 155°/atm (dec), d$_4^{25}$ 1.4204, n$_D^{20}$ 1.4259.** Check the NMR spectrum, if satisfactory then dist in a vac, if not then dissolve in Et$_2$O, wash with H$_2$O, 10% H$_2$SO$_4$ until the acid extract does not become cloudy when made alkaline with NaOH. Wash the organic layer again with H$_2$O, then satd aq NaHCO$_3$, dry over Na$_2$SO$_4$, evap and fractionate through a carborundum-packed column or a 6-inch Widmer column (*see tert-butyl ethyl malonate for precautions to avoid decomposition during distn*). [*J Chem Soc* 940 *1940*; *J Am Chem Soc* **75** 4995 *1953*; *Org Synth* Coll Vol 144 *1944*.]

6-*tert*-Butyl-1-chloro-2-naphthol *[525-27-9]* **M 232.7, m 76°, b 185°/15mm, pK$_{Est}$ ~8.0.** Crystd from pet ether.

***tert*-Butyl cyanide (trimethylacetonitrile)** *[630-18-2]* **M 83.1, m 16-18°, d 0.765, b 104-106°.** Purified by a two stage vac distn and degassed by freeze-pump-thaw technique. Stored under vac at 0°. **TOXIC,** use efficient fume hood.

***tert*-Butyl cyanoacetate** *[1116-98-9]* **M 141.2, b 40-42°/0.1mm, 54-56°/0.3mm, 90°/10mm, 107-108°/23mm, d$_4^{20}$ 0.989, n$_D^{20}$ 1.4198.** The IR spectrum of a film should have bands at 1742 (ester

CO) and 2273 (C≡N) but not OH band (*ca* 3500 broad) cm^{-1}. If it does not have the last named band then fractionally dist, otherwise dissolve in Et$_2$O, wash with satd aq NaHCO$_3$, dry over K$_2$CO$_3$, evap Et$_2$O, and dist residue under a vacuum (*see tert-butyl ethyl malonate for precautions to avoid decomposition during distn*). [*J Chem Soc* 423 *1955*; *Helv Chim Acta* **42** 1214 *1959*].

4-*tert*-Butyl-1-cyclohexanone [98-53-3] **M 154.3, m 49-50°.** Crystd from pentane.

***n*-Butyl disulfide** [629-45-8] **M 178.4, b 110-113°/15mm, d 0.938, n^{22} 1.494.** Shaken with lead peroxide, filtered and distd in vacuum under N$_2$.

***n*-Butyl ether (di-*n*-butyl ether)** [142-96-1] **M 130.2, b 52-53°/26mm, 142.0°/760mm, d 0.764, n 1.39925, n^{25} 1.39685, pK^{25} -5.40 (aq H$_2$SO$_4$).** Peroxides (detected by the liberation of iodine from weakly acid (HCl) solns of 2% KI) can be removed by shaking 1L of ether with 5-10mL of a soln comprising 6.0g of ferrous sulfate and 6mL conc H$_2$SO$_4$ and 110mL of water, with aq Na$_2$SO$_3$, or with acidified NaI, water, then Na$_2$S$_2$O$_3$. After washing with dil NaOH, KOH, or Na$_2$CO$_3$, then water, the ether is dried with CaCl$_2$ and distd. It can be further dried by distn from CaH$_2$ or Na (after drying with P$_2$O$_5$), and stored in the dark with Na or NaH. The ether can also be purified by treating with CS$_2$ and NaOH, expelling the excess sulfide by heating. The ether is then washed with water, dried with NaOH and distd [Kusama and Koike *J Chem Soc Japan, Pure Chem Sect* **72** 229 *1951*]. Other purification procedures include passage through an activated alumina column to remove peroxides, or through a column of silica gel, and distn after adding about 3% (v/v) of a 1M soln of MeMgI in *n*-butyl ether.

***n*-Butyl ethyl ether** [628-81-9] **M 102.2, b 92.7°, d 0.751, n 1.38175, n^{25} 1.3800.** Purified by drying with CaSO$_4$, by passage through a column of activated alumina (to remove peroxides), followed by prolonged refluxing with Na and then fractional distn.

***tert*-Butyl ethyl ether** [637-92-3] **M 102.2, b 71-72°, d 0.741.** Dried with CaSO$_4$, passed through an alumina column, and fractionally distd.

***tert*-Butyl ethyl malonate** [32864-38-3] **M 188.2, b 83-85°/8mm, 93-95°/17mm, 107-109°/24mm, d$_4^{25}$ 0.994, n$_D^{24}$ 1.4150.** Likely impurity is monoethyl malonate, check IR for OH bands at 3330 br. To *ca* 50g of ester add ice cold NaOH (50g in 200mL of H$_2$O and 200g of ice). Swirl a few times (filter off ice if necessary), place in a separating funnel and extract with 2 x 75mL of Et$_2$O. Dry extract (MgSO$_4$) (since traces of acid decompose the *t*-Bu group of the ester, the distillation flask has to be washed with aq NaOH, rinsed with H$_2$O and allowed to dry). Addition of some K$_2$CO$_3$ or MgO before distilling is recommended to inhibit decomposition. Distil under reduced press through a 10 cm Vigreux column. *Decomposition is evidenced by severe foaming due to autocatalytic decomposition and cannot be prevented from accelerating except by stopping the distillation and rewashing the distillation flask with alkali again.* [*J Am Chem Soc* **66** 1287 *1944*, **64** 2714 *1942*; *Org Synth* Coll Vol IV 417 *1963*; *Org Synth* **37** 35 *1957*.]

***n*-Butyl formate** [592-84-7] **M 102.1, b 106.6°, d 0.891, n 1.3890.** Washed with satd NaHCO$_3$ soln in the presence of satd NaCl, until no further reaction occurred, then with saturated NaCl soln, dried (MgSO$_4$) and fractionally distd.

Butyl glycolate [7397-62-8] **M 132.2, b 191-192°/755mm, 187-190°/atm, d$_4^{20}$ 1.019, n$_D^{20}$ 1.4263.** Dissolve in CHCl$_3$ (EtOH-free), wash with 5% KHCO$_3$ until effervescence ceases (if free acid is present), dry over CaCl$_2$, filter, evaporate and distil through a short column. [Bøhme and Opfer *Z Anal Chem* **139** 255 *1953*; *cf J Am Chem Soc* **73** 5265 *1951*.]

***tert*-Butyl hydroperoxide (TBHP)** [75-91-2] **M 90.1, f 5.4°, m 0.5-2.0°, b 38°/18mm, d 0.900, n 1.4013, pK^{20} 12.8.** *Care should be taken when handling this peroxide because of the possibility of* **EXPLOSION. It explodes when heating over an open flame.** Alcoholic and volatile impurities can be removed by prolonged refluxing at 40° under reduced pressure, or by steam distn. For example, Bartlett, Benzing and Pincock [*J Am Chem Soc* **82** 1762 *1960*] refluxed at 30mm pressure in an azeotropic separation apparatus until two phases no longer separated, and then distilled at 41°/23mm. Pure material is stored under N_2, in the dark at 0°. Crude commercial material has been added to 25% NaOH below 30°, and the crystals of the sodium salt have been collected, washed twice with *benzene and dissolved in distd water. After adjusting the pH of the soln to 7.5 by adding solid CO_2, the peroxide was extracted into pet ether, from which, after drying with K_2CO_3, it was recovered by distilling off the solvent under reduced pressure at room temperature [O'Brien, Beringer and Mesrobian *J Am Chem Soc* **79** 6238 *1957*]. **The temperatures should be kept below 75°**. It has also been distilled through a helices packed column (*ca* 15 plates) and material collected had b 34-35°/20 mm. Similarly, a soln in pet ether has been extracted with cold aq NaOH, and the hydroperoxide has been regenerated by adding at 0°, $KHSO_4$ at a pH not higher than 4.5, then extracted into diethyl ether, dried with $MgSO_4$, filtered and the ether evapd in a rotary evaporator under reduced pressure [Milac and Djokic *J Am Chem Soc* **84** 3098 *1962*].

A 3M soln of TBHP in CH_2Cl_2 is prepared by swirling 85mL (0.61mol) of commercial TBHP (70% TBHP-30% H_2O, **d** 0.935 *ca* 7.2mmol/mL) with 140mL of CH_2Cl_2 in a separating funnel. The milky mixture is allowed to stand until the phases separate (*ca* 30min). The organic (lower) layer (*ca* 200mL) containing 0.60mole of TBHP was separated from the aqueous layer (*ca* 21mL) and used without further drying. TBHP is assayed by iodometric titration. With 90% grade TBHP (w/w, d 0.90, *ca* 9.0mmole/mL) no separation of layers occurs; i.e. when TBHP (66.67mL, 0.60mole) is added to CH_2Cl_2 (140mL) the resulting soln (*ca* 200mL) is clear. [*J Am Chem Soc* **77** 60032 *1955*, **74** 4742 *1952*; Akashi, Palermo and Sharpless *J Org Chem* **43** 2063 *1978* states quality of available grades, handling and compatibility for reactions.]

2-*tert*-Butyl hydroquinone *[1948-33-0]* **M 166.2, m 125-127°, 127-128°, 129°, $pK_{Est(1)}$ ~10.5. $pK_{Est(2)}$ ~11.6.** Recrysts from H_2O or MeOH and dried in a vacuum at 70°. Store in a dark container. [*Angew Chem* **69** 699 *1957*.]

***n*-Butyl iodide (1-iodobutane)** *[542-69-8]* **M 184.0, b 130.4°, d 1.616, n^{25} 1.44967.** Dried with $MgSO_4$ or P_2O_5, fractionally distd through a column packed with glass helices, taking the middle fraction and storing with calcium or mercury in the dark. Also purified by prior passage through activated alumina or by shaking with conc H_2SO_4 then washing with Na_2SO_3 soln. It has also been treated carefully with sodium to remove free HI and H_2O, before distilling in a column containing copper turnings at the top. Another purification consisted of treatment with bromine, followed by extraction of free halogen with $Na_2S_2O_3$, washing with H_2O, drying and fractional distn.

***tert*-Butyl iodide** *[558-17-8]* **M 184.0, b 100°(dec), d 1.544.** Vacuum distn has been used to obtain a distillate which remained colourless for several weeks at -5°. More extensive treatment has been used by Boggs, Thompson and Crain [*J Phys Chem* **61** 625 *1957*] who washed with aq $NaHSO_3$ soln to remove free iodine, dried for 1h with Na_2SO_3 at 0°, and purified by four or five successive partial freezings of the liquid to obtain colourless material which was stored at -78°.

***tert*-Butyl isocyanate** *[1609-86-5]* **M 99.1, m 10.5-11.5°, b 30.5-32°/10mm, 64°/52mm, d^{25}_{25} 0.9079, n^{25}_D 1.470.** It is **LACHRYMATORY** and **TOXIC**, and should have IR with 2251 ($C\equiv N$) cm^{-1} and no OH bands. The NMR should have one band at 1.37 ppm from TMS. Purified by fractional distn under reduced pressure. [*J Org Chem* **36** 3056 *1971*; *J Prakt Chem.* **125** 152 *1930*.]

***tert*-Butyl isocyanide** *[7188-38-7]* **M 83.1, b 91-92°/730mm, 90°/758mm, d^{20} 0.735.** Dissolve in pet ether (b 40-60°) wash with H_2O, dry (Na_2SO_4), remove pet ether under slight vacuum, dist using a vacuum-jacketed Vigreux column at atmospheric pressure, IR: ν 2134 cm^{-1}. [*Chem Ber* **93** 239 *1960*.]

***tert*-butyl isocyanoacetate** *[2769-72-4]* **M 141.2, b 50°/0.1mm, 49-50°/10mm, 63-65°/15mm, d^{20}_4 0.970, n^{20}_D 1.420.** If it contains some free acid (OH bands in IR) then dissolve in Et_2O, shake with 20% Na_2CO_3, dry over anhydrous K_2CO_3, evaporate and distil. [*Chem Ber* **94** 2814 *1961*.]

***n*-Butyl methacrylate** *[97-88-1]* **M 142.2, b 49-52°/0.1mm, d 0.896, n 1.424.** Purified as for butyl acrylate.

tert-**Butyl methacrylate** *[585-07-9]* **M 142.2, f -48°, b 135-136°/760mm, d 0.878, n 1.415.** Purified as for butyl acrylate.

2-*tert*-Butyl-4-methoxyphenol (**2-*tert*-butyl-4-hydroxyanisole**) *[121-00-6]* **M 180.3, m 64.1°, pK$_{Est}$ ~10.8.** Fractionally distd *in vacuo*, then passed as a soln in CHCl$_3$ through alumina, and the solvent evaporated from the eluate. Recrystd from pet ether.

n-**Butyl methyl ether** *[628-28-4]* **M 88.2, b 70°, d 0.744, pK -3.50 (aq H$_2$SO$_4$).** Dried with CaSO$_4$, passed through an alumina column to remove peroxides, and fractionally distd.

tert-**Butyl methyl ether** *[1634-04-4]* **M 88.2, b 56°, n 1.369.** Same as for *n*-butyl methyl ether.

tert-**Butyl methyl ketone** *[75-97-8]* **M 100.2, b 105°/746mm, 106°/760mm, d 0.814, n 1.401.** Refluxed with a little KMnO$_4$. Dried with CaSO$_4$ and distd.

8-*sec*-Butylmetrazole *[25717-83-3]* **M 194.3, m 70°.** Crystd from pet ether, and dried for 2 days under vacuum over P$_2$O$_5$.

tert-**Butyl nitrite** *[540-80-7]* **M 103.1, b 34°/250mm, 61-63°/atm, d$_4^{20}$ 0.8671, n$_D^{25}$ 1.3660.** If it is free from OH bands (IR) then distil through a 12inch helices packed column under reduced pressure, otherwise wash with aq 5% NaHCO$_3$ (**effervescence**), then H$_2$O, dry (Na$_2$SO$_4$) and fractionate through a 10 theoretical plates column at *ca* 10mm pressure. [*J Chem Soc* 1968 *1954*, *J Am Chem Soc* **70** 1516 *1948*; UV: *J Org Chem* **21** 993 *1956*; IR: *Bull Soc Chim Belg* **60** 240 *1951*.]

p-*tert*-**Butylnitrobenzene** *[3282-56-2]* **M 179.2, m 28.4°.** Fractionally crystd three times by partially freezing a mixture of the mono-nitro isomers, then recryst from MeOH twice and dried under vacuum [Brown *J Am Chem Soc* **81** 3232 *1959*].

N-(*n*-**Butyl**)-**5-nitro-2-furamide** *[14121-89-2]* **M 212.2, m 89-90°.** Recrystd twice from EtOH/water mixture.

Butyloxirane (**1-hexene oxide**) *[1436-34-6]* **M 100.2, b 116-117°/atm, 116-119°/atm, d$_4^{20}$ 0.833, n$_D^{20}$ 1.44051.** Purified by fractional distn through a 2ft helices packed column at atmospheric pressure in a N$_2$ atm. [*J Org Chem* **30** 1271 *1965*; *J Chem Soc* 2433 *1927* ; ^{13}C NMR *J Chem Soc Perkin Trans 2* 861 *1975*.]

tert-**Butyl peracetate** *[107-71-1]* **M 132.2, b 23-24°/0.5mm, n^{25} 1.4030.** Washed with NaHCO$_3$ from a *benzene soln, then redistd to remove *benzene [Kochi *J Am Chem Soc* **84** 774 *1962*]. *Handle with adequate protection due to possible* **EXPLOSIVE** *nature.*

tert-**Butylperoxy isobutyrate** *[109-13-7]* **M 160.2, f -45.6°.** After diluting 90mL of the material with 120mL of pet ether, the mixture was cooled to 5° and shaken twice with 90mL portions of 5% NaOH soln (also at 5°). The non-aqueous layer, after washing once with cold water, was dried at 0° with a mixture of anhydrous MgSO$_4$ and MgCO$_3$ containing *ca* 40% MgO. After filtering, this material was passed, twice, through a column of silica gel at 0° (to remove *tert*-butyl hydroperoxide). The soln was evapd at 0°/0.5-1mm to remove the solvent, and the residue was recrystd several times from pet ether at -60°, then subjected to high vac to remove traces of solvent [Milos and Golubovic *J Am Chem Soc* **80** 5994 *1958*]. *Handle with adequate protection due to possible* **EXPLOSIVE** *nature.*

tert-**Butyl perphthalic acid** *[15042-77-0]* **M 238.2, pK$_{Est}$ ~6.2.** Crystd from Et$_2$O and dried over H$_2$SO$_4$. *Possibly* **EXPLOSIVE**.

p-*tert*-**Butylphenol** *[98-54-4]* **M 150.2, m 99°, pK^{25} 10.39.** Crystd to constant melting point from pet ether (b 60-80°). It sublimes. Also purified *via* its benzoate, as for phenol.

p-tert-**Butylphenoxyacetic acid** *[1798-04-5]* **M 208.3, m 88-89°, pK$_{Est}$ ~2.9.** Crystd from pet ether/*C$_6$H$_6$ mixture.

tert-**Butyl phenyl carbonate** *[6627-89-0]* **M 194.2, b 74-78°/0.5mm, 83°/0.6mm, d$_4^{20}$ 1.05, n$_D^{20}$ 1.480.** If IR is free from OH then purify by redistillation, otherwise, dissolve in Et$_2$O, wash with 5% HCl, then H$_2$O, dry over MgSO$_4$, evap and distil through a Claisen head under vacuum. Care should be taken in the distillation as distn of large quantities can lead to decomposition with liberation of CO$_2$ and isobutylene, **use the necessary precautions.** [*J Am Chem Soc* **79** 98 *1957*.]

n-**Butyl phenyl ether** *[1126-79-0]* **M 150.2, b 210.5°, d 0.935.** Dissolved in diethyl ether, washed first with 10% aq NaOH to remove traces of phenol, then repeatedly with distilled water, followed by evaporation of the solvent and distn under reduced pressure [Arnett and Wu *J Am Chem Soc* **82** 5660 *1960*].

N-tert-**Butyl α-phenyl nitrone** *[3376-24-7]* **M 177.2, m 73-74°.** Crystd from hexane.

Butyl phthalate *[84-74-2]* **M 278.4, f -35°, b 340°/760mm, d 1.043.** Freed from alcohol by washing with H$_2$O, or from acids and butyl hydrogen phthalate by washing with dilute NaOH. Distd at 10torr or less. (See also p. 195.)

4-*tert*-Butyl pyridine *[3978-81-2]* **M 135.2, f -44.4°, b 194-197°atm, 197°/765mm, d$_4^{20}$ 0.923, n$_D^{20}$ 1.495, pK^{25} 5.82.** It is dried over solid KOH and is purified by fractional distn through an efficient column under dry N$_2$. Its *picrate* has **m** 153.9-154°, and the *hydrochloride* has **m** 151.7-154.8° (from Me$_2$CO). [*J Am Chem Soc* **73** 3308, 3310 *1951*, IR: *J Am Chem Soc* **100** 214 *1978*; *J Chem Soc* 4454 *1960*.]

Butyl stearate *[123-95-5]* **M 340.6, m 26.3°, d 0.861.** Acidic impurities removed by shaking with 0.05M NaOH or a 2% NaHCO$_3$ soln, followed by several water washes, then purified by fractional freezing of the melt and fractional crystn from solvents with boiling points below 100°.

S-*tert*-Butyl thioacetate *[999-90-6]* **M 132.2, b 31-32°/11mm, 38°/14mm, 44-45°/28mm, 67°/54mm, 135.6-135.9°/773mm, d$_4^{25}$ 0.9207, n$_D^{20}$ 1.4532.** Dissolve in CHCl$_3$ (EtOH-free), wash with H$_2$O, 10% H$_2$SO$_4$, saturated aqueous NaHCO$_3$ (care CO$_2$ liberated), H$_2$O again, dried over Drierite and anhydrous K$_2$CO$_3$, and fractionate under reduced pressure. [*J Am Chem Soc* **72** 3021 *1950*.]

p-tert-**Butyltoluene** *[98-51-1]* **M 148.3, f -53.2°, b 91°/28mm, d 0.854, n 1.4920.** A sample containing 5% of the *meta*-isomer was purified by selective mercuration. Fractional distn of the solid arylmercuric acetate, after removal from the residual hydrocarbon, gave pure *p-tert*-butyltoluene [Stock and Brown *J Am Chem Soc* **81** 5615 *1959*].

tert-**Butyl 2,4,6-trichlorophenyl carbonate** *[16965-08-5]* **M 297.6, m 64-66°.** Crystd from a mixture of MeOH (90mL) and water (6mL) using charcoal [Broadbent et al. *J Chem Soc (C)* 2632 *1967*].

N-tert-**Butyl urea** *[1118-12-3]* **M 116.2, m 182°, 185°(dec).** Possible impurity is *N,N'*-di-*tert*-butyl urea which is quite insol in H$_2$O. Recrystd from hot H$_2$O, filter off insol material, and cool to 0° to -5° with stirring. Dry in vac at room temp over KOH or H$_2$SO$_4$. If dried at higher temperatures it sublimes slowly. It can be recrystd from EtOH as long white needles or from 95% aq EtOH as plates. During melting point determination the bath temp has to be raised rapidly as the urea sublimes slowly above 100° at 760mm. [*Org Synth* Coll Vol III 151 *1955*.]

n-**Butyl vinyl ether** *[111-34-2]* **M 100.2, b 93.3°, d 0.775.** After five washings with equal volumes of water to remove alcohols (made slightly alkaline with KOH), the ether was dried with sodium and distd under vacuum, taking the middle fraction [Coombes and Eley *J Chem Soc* 3700 *1957*]. Stored over KOH.

2-Butyne *[503-17-3]* **M 54.1, b 0°/253mm, d 0.693.** Stood with sodium for 24h, then fractionally distd under reduced pressure.

2-Butyne-1,4-diol *[110-65-6]* **M 86.1, m 54-57°, 56-58°, b 238°.** Crystd from EtOAc.

n-Butyraldehyde *[123-72-8]* **M 72.1, b 74.8°, d 0.810, n 1.37911, n^{15} 1.38164.** Dried with $CaCl_2$ or $CaSO_4$, then fractionally distd under N_2. Lin and Day [*J Am Chem Soc* **74** 5133 *1952*] shook with batches of $CaSO_4$ for 10min intervals until a 5mL sample, on mixing with 2.5mL of CCl_4 containing 0.5g of aluminium isopropoxide, gave no ppte and caused the soln to boil within 2min. Water can be removed from *n*-butyraldehyde by careful distn as an azeotrope distilling at 68°. The aldehyde has also been purified through its bisulfite compound which, after decomposing with excess $NaHCO_3$ soln, was steam distd, extracted under N_2 into ether and, after drying, the extract was fractionally distd [Kyte, Jeffery and Vogel *J Chem Soc* 4454 *1960*].

Butyramide *[514-35-5]* **M 87.1, m 115°, b 230°.** Crystd from acetone, *benzene, CCl_4-pet ether, 20% EtOH or water. Dried under vacuum over P_2O_5, $CaCl_2$ or 99% H_2SO_4.

n-Butyric acid *[107-92-6]* **M 88.1, f -5.3°, b 163.3°, d 0.961, n^{25} 1.396, pK^{25} 2.82.** Distd, mixed with $KMnO_4$ (20g/L), and fractionally redistd, discarding the first third [Vogel *J Chem Soc* 1814 *1948*].

n-Butyric anhydride *[106-31-0]* **M 158.2, b 198°, d 0.968.** Dried by shaking with P_2O_5, then distd.

γ-Butyrolactone *[96-48-0]* **M 86.1, b 83.8°/12mm, d 1.124.** Dried with anhydrous $CaSO_4$, then fractionally distd. *Handle in a fume cupboard due to* **TOXICITY.**

Butyronitrile *[109-74-0]* **M 69.1, b 117.9°, d 0.793, n 1.3846, n^{30} 1.37954.** Treated with conc HCl until the smell of the isonitrile had gone, then dried with K_2CO_3 and fractionally distd [Turner *J Chem Soc* 1681 *1956*]. Alternatively it was twice heated at 75° and stirred for several hours with a mixture of 7.7g Na_2CO_3 and 11.5g $KMnO_4$ per L of butyronitrile. The mixture was cooled, then distd. The middle fraction was dried over activated alumina. [Schoeller and Wiemann *J Am Chem Soc* **108** 22 *1986*.]

Butyryl chloride (butanoyl chloride) *[141-75-3]* **M 106.6, f -89°, b 101-102°/atm, d_4^{20} 1.026, n_D^{20} 1.412.** Check IR to see if there is a significant peak at 3000-3500 cm^{-1} (br) for OH. If OH is present then reflux with less than one mol equiv of $SOCl_2$ for 1h and distil directly. The fraction boiling between 85-100° is then refractionated at atm pressure. Keep all apparatus free from moisture and store the product in sealed glass ampoules under N_2. **LACHRYMATORY -** *handle in a good fume hood.* [*Org Synth* Coll Vol I 147 *1941*.]

Cacotheline   (2,3-dihydro-4-nitro-2,3-dioxo-9,10-*seco*strychnidin-10-oic acid) *[561-20-6]* **M 508.4, $pK_{Est(1)}$~4.4 (CO_2H), $pK_{Est(2)}$ ~10.2 (N).** Yellow crystals from H_2O. It is then dried over H_2SO_4 which gives the *dihydrate*, and in a vacuum over H_2SO_4 at 105° to give the anhydrous compound. The *hydrochloride* separates as the hydrate (on heating in vacuum at 80°) in orange-yellow prisms or plates, **m** 250°(dec), and forms a *resorcinol complex* which gives brown crystals from EtOH, **m** 325°, and a *hydroquinone complex* as dark red crystals from EtOH, **m** 319°. [*Chem Ber* **43** 1042 *1910*, **86** 232, UV: 242 *1953*; complexes: Gatto *Gazz Chim Ital* **85** 1441 *1955*.] Used in the titrimetric estimation of Sn^{2+} ions [Szrvas and Lantos *Talanta* **10** 477 *1963*].

Caffeic (3,4-dihydroxycinnamic) acid *[331-39-5]* **M 180.2, m 195°, 223-225°, pK_1^{25} 4.62, pK_2^{25} 9.07** Crystd from water.

Caffeine *[58-08-2]* **M 194.2, m 237°, pK_1^{40} -0.10, pK_2^{55} 1.22.** Crystd from water or absolute EtOH.

(+)-Calarene (+ β-gurjunen, 1,3,3,11-tetramethyltricyclo[5.4.0.0^{2,4}]undecan-7-ane, (1a*R*)-1,1,7c,7ac-tetramethyl-1a,2,3,5,6,7,7a,7b-octahydro-1*H*-cyclopropa[α]naphthalene, new name 1(10)aristolene) *[17334-55-3]* **M 204.35, b 45-47°/0.008-0.01mm, 255-258°/atm, d_4^{20} 0.9340, n_D^{20} 1.55051, $[α]_D^{20}$ +58° (EtOH), +81.8° (neat).** Purified by gas chromatography (7%

propylene glycol adipate on unglazed tile particles of size 0.2-0.3mm, 400 cm column length and 0.6 cm diameter, at 184°, with N_2 carrier gas at a flow rate of 0.54 mL/sec using a thermal detector). Also purified by chromatography on alumina (200 times the weight of calarene) and eluted with pet ether. UV: λ_{max} 200 and 210 nm (ϵ 9560, 5480) in EtOH. [IR: Sorm *Collect Czech Chem Commun* **18** 512 *1953*, **29** 795 *1964*; *Tetrahedron Lett* 827 *1962*, 225 *1963*.]

Calcon carboxylic acid [3-hydroxy-4-(2-hydroxy-4-sulfo-1-naphthylazo)naphthalene-2-carboxylic acid] *[3737-95-9]* **M 428.4, m 300°, λmax 560nm, pK$_1$ 1.2, pK$_2$ 3.8, pK$_3$ 9.26, pK$_4$ 13.14.** Purified through its *p*-toluidinium salt. The dye was dissolved in warm 20% aq MeOH and treated with *p*-toluidine to ppte the salt after cooling. Finally recrystd from hot water. [Itoh and Ueno *Analyst (London)* **95** 583 *1970*.] Patton and Reeder (*Anal Chem* **28** 1026 *1956*) indicator and complexes with Ca in presence of Mg and other metal ions.

Calmagite *[3147-14-6]* **M 358.4, m 300°, pK$_1$ 8.1, pK$_2$ 12.4.** Crude sample was extracted with anhydrous diethyl ether [Lindstrom and Diehl *Anal Chem* **32** 1123 *1960*]. Complexes with Ca, Mg and Th.

Campesterol (24R-24-methylcholest-5-en-3β-ol) *[474-62-4]* **M 400.7, m 156-159°, 157-158°, [α]$_D^{24}$ -35.1°** (c 1.2, $CHCl_3$). Recryst twice from hexane and once from Me_2CO. The *benzoyl derivative* has **m** 158-160° [α]$_D^{23}$ -8.6° ($CHCl_3$), the *acetyl derivative* has **m** *137-138°* (from EtOH) and [α]$_D^{23}$ -35.1° (c 2.9, $CHCl_3$) [*J Am Chem Soc* **63** 1155 *1941*].

1R,4S-(-)-Camphanic acid *[13429-83-9]* **M 198.2, m 190-192°, 198-200°, [α]$_{548}^{20}$ -22.5°** (c 1, dioxane), **-4.4°** (c 8, EtOH), **pK$_{Est}$ ~3.8.** Dissolve in CH_2Cl_2, dry (MgSO$_4$), filter, evaporate and residue is sublimed at 120°/0.5mm or 140°/1mm. [*Helv Chim Acta* **61** 2773 *1978*.]

1R,4S-(-)-Camphanic acid chloride *[39637-74-6]* **M 216.7, m 65-66.5°, 70.5-71°, [α]$_{548}$ -23°** (c 2, CCl_4), **-7.5°** (c 0.67, *benzene*). Soluble in toluene (50g/100mL at 0°) and crysts from pet ether (b 40-60°). It sublimes at 70°/5mm, Store dry at 0°, ν (CCl_4) 1805s and 1780m cm^{-1}. [*J Chem Soc, Dalton Trans* 2229 *1976*.]

RS-Camphene *[565-00-4]* **M 136.2, m 51-52°, b 40-70°/10mm.** Crystd twice from EtOH, then repeatedly melted and frozen at 30mm pressure. [Williams and Smyth *J Am Chem Soc* **84** 1808 *1962*.] Alternatively it is dissolved in Et_2O, dried over $CaCl_2$ and Na, evaporated and the residue sublimed in a vacuum [NMR: *Chem Ber* **111** 2527 *1978*].

(-)-Camphene (1S-2,2-dimethyl-3-methylene norbornane) *[5794-04-7]* **M 136.2, m 49.2-49.6°, 49-50°, b 79-80°/58mm, 91.5°/100mm, d$_4^{54}$ 0.8412, n$_D^{54}$ 1.4564, [α]$_D^{25}$ -106.2°** (c 40, *C_6H_6*), **-117.5°** (c 19, toluene), **-113.5°** (c 9.7, Et_2O). Purified by fractionation through a Stedman column (see p. 441) at 100mm in a N_2 atmosphere, crystallised from EtOH and sublimed in a vacuum below its melting point. It is characterised by its *camphenilone semicarbazone*, **m** 217-218.5°, or *camphor semicarbazone*, **m** 236-238°. [NMR: *Chem Ber* **111** 2527 *1978*; *Justus Liebigs Ann Chem* **623** 217 *1959*; Bain et al. *J Am Chem Soc* **72** 3124 *1950*]

Camphor (1R-bornan-2-one) *[R-(+)- 464-49-3]*; *S-(-)- 464-48-2]* **M 136.2, m 178.8°, 179.97°(open capillary), b 204°/atm, [α]$_{546}^{35}$ (+) and (-) 59.6° (in EtOH), [α]$_D^{20}$ (+) and (-) 44.3°** (c 10, EtOH), **[α]$_{579}^{179}$ (+) and (-) 70.85°** (melt). Crystd from EtOH, 50% EtOH/water, MeOH, or pet ether or from glacial acetic acid by addition of water. It can be sublimed (50°/14mm) and also fractionally crystd from its own melt. It is steam volatile. It should be stored in tight containers as it is appreciably volatile at room temperature. The solubility is 0.1% (H_2O), 100% (EtOH), 173% (Et_2O) and 300% ($CHCl_3$). The *R-oxime* (from Et_2O, $CHCl_3$, or dil EtOH) **m** 119° [α]$_D^{20}$ -42.4° (c 3, EtOH); the ± *oxime* has **m** 118-119°. [*Chem Ber* **67** 1432 *1934*; Allan and Rodgers *J Chem Soc (B)* 632 *1971*; UV, NMR: Fairley et al. *J Chem Soc, Perkin Trans 1* 2109 *1973*; *J Am Chem Soc* **62** 8 *1940*.]

Camphoric acid (1,2,2-trimethyl-cyclopentan-1r,3c-dicarboxylic acid) *[1R,2S)-(+)- 124-83-4*; *1S,2R)-(-)- 560-09-8]* **M 200.2, m 186-188°, 187°, 186.5-189°, [α]$_{546}^{20}$ (+) and (-) 57°** (c 1,

EtOH), $[\alpha]_D^{20}$ (+) and (-) 47.7° (c 4, EtOH), pK_1^{25} 4.71, pK_2^{25} 5.83 (for + isomer). Purified by repptn from an alkaline soln by HCl, filtered, and recrystd from water several times, rejecting the first crop. It forms leaflets from EtOH and Me$_2$CO and H$_2$O and is insol in CHCl$_3$. Sol in H$_2$O is 0.8% at 25° and 10% at 100°; 50% (EtOH) and 5% in ethylene glycol. The (±)-acid has **m** 202-203°. The (+)-1-methyl ester had **m** 86° (from pet ether) $[\alpha]_D^{20}$ +45° (c 4, EtOH), and the (+)-3-methyl ester has **m** 77° (from pet ether) $[\alpha]_D^{17.5}$ +53.9° (c 3, EtOH). [*J Am Chem Soc* **53** 1661 *1931*; *Helv Chim Acta* **30** 933 *1947*; *Acta Chem Scand* **2** 597 *1948*; *J Am Chem Soc* **80** 6316 *1958*.]

(±)-Camphoric anhydride [595-30-2] **M 182.2, transitn temp 135°, m 223.5°.** Crystd from EtOH.

Camphorquinone (borna-2,3-dione) *[1R-(-)- 10334-26-6; 1S-(+)- 2767-84-2]* **M 166.2, m 198.7°, 198-199°, 197-201°, $[\alpha]_D^{25}$ (-) and (+) 101.1° (c 2, EtOH).** It can be purified by steam distillation, recrystn (yellow prisms) from EtOH, *C$_6$H$_6$ or Et$_2$O-pet ether and can be sublimed in a vacuum. The (±)-quinone forms needles from EtOH, **m** 197-198°, 203°. [*Helv Chim Acta* **13** 1026; *Chem Ber* **67** 1432 *1934*.]

***RS*-Camphorquinone** *[10373-78-1]* **M 166.2, m 199-202°.** Purification is same as for above isomers.

(1*R*)-(-)Camphor-10-sulfonic acid *[35963-20-3]* **M 232.3, m 197.4-198°(dec), 197-198°, $[\alpha]_D^{20}$ -20.7° (c 5.4, H$_2$O), pK$_{Est}$ ~-1.** Forms prisms from AcOH or EtOAc, and is deliquescent in moist air. Store in tightly stoppered bottles. The *NH$_4$ salt* forms needles from H$_2$O $[\alpha]_D^{16}$ ±20.5° (c 5, H$_2$O). [*J Chem Soc* **127** 279 *1925*; *J Am Chem Soc* **78** 3063 *1956*.]

(1*S*)-(+)Camphor-10-sulfonic acid *[3144-16-9]* **M 232.3, m 193°(dec), 197-198°, $[\alpha]_{546}^{20}$ +27.5° (c 10, H$_2$O), $[\alpha]_D^{20}$ +43.5° (c 4.3, EtOH), pK$_{Est}$ ~-1.** Crystd from ethyl acetate and dried under vacuum.

Camphor-10-sulfonyl chloride *[1S-(+)- 21286-54-4; 1R-(-)- 39262-22-1]* **M 250.7, m 67-68°, 70°, $[\alpha]_D^{20}$ (+) and (-) 32.2° (c 3, CHCl$_3$).** If free from OH bands in the IR then recryst from Et$_2$O or pet ether, otherwise treat with SOCl$_2$ at 50° for 30min, evaporate, dry residue over KOH in a vacuum and recrystallise. The (±)-acid chloride has **m** 85°. Characterised as the *amide* (prisms from EtOH) **m** 132°, $[\alpha]_D^{17}$ (+) and (-) 1.5° (EtOH). [Read and Storey *J Chem Soc* 2761 *1930*; *J Am Chem Soc* **58** 62 *1936*.]

2,10-Camphorsultam *[1R-(+)- 108448-77-7; 1S-(-)- 94594-90-8]* **M 215.3, m 181-183°, 185-187°, $[\alpha]_D^{20}$ (+) and (-) 32° (c 5, CHCl$_3$).** The (-)-enantiomer is recrystd from 95% EtOH. It dissolves in dil aq NaOH and can be pptd without hydrolysis by acidifying. It forms the *N*-Na salt in EtOH (by addition of Na to the EtOH soln) and the salt can be methylated with MeI to give the (-)-*N*-*Me lactam* with **m** 80° after recrystn from hot H$_2$O and has $[\alpha]_D^{25}$ -59.6° (c 5, CHCl$_3$) [Shriner et al. *J Am Chem Soc* **60** 2794 *1938*].

***S*-Canavanine** *[543-38-4]* **M 176.2, m 184°, $[\alpha]_D^{17}$ +19.4° (c 2, H$_2$O), pK_1^{25} 2.43, pK_2^{25} 9.41.** Crystd from aqueous EtOH.

***S*(L)-Canavanine sulfate** *[2219-31-0]* **M 274.3, m 172°(dec).** See L-canavanine sulfate on p. 518 in Chapter 6.

Cannabinol *[521-35-7]* **M 310.4, m 76-77°, b 185°/0.05mm.** Crystd from pet ether. Sublimed.

Canthaxanthin (*trans*) *[514-78-3]* **M 564.9, m 211-212°, $A_{1cm}^{1\%}$ 2200 (470nm) in cyclohexane.** Purified by chromatography on a column of deactivated alumina or magnesium oxide, or on a thin layer of silica gel G (Merck), using dichloromethane/diethyl ether (9:1) to develop the chromatogram. Stored in the dark and in an inert atmosphere at -20°.

Capric acid (decanoic acid) *[334-48-5]* **M 172.3, m 31.5°, b 148°/11mm, d 0.886, n^{25} 1.424, pK$_{Est}$ ~4.9.** Purified by conversion to its *methyl ester*, **b** 114.0°/15mm (using excess MeOH, in the presence of H$_2$SO$_4$). After removal of the H$_2$SO$_4$ and excess MeOH, the ester was distd under vacuum through a 3ft column packed with glass helices. The acid was then obtained from the ester by saponification. [Trachtman and Miller *J Am Chem Soc* **84** 4828 *1962*].

n-**Caproamide** (*n*-**hexanamide**) *[628-02-4]* M 115.2, m 100°. Crystd from hot water.

Caproic acid (hexanoic acid) *[142-62-1]* M 116.2, b 205.4°, d 0.925, n 1.417, pK^{25} 4.85. Dried with $MgSO_4$ and fractionally distilled from $CaSO_4$.

ε-**Caprolactam (azepan-2-one, aza-2-cycloheptanone, 2-oxohexamethyleneimine)** *[105-60-2]* M 113.2, m 70°, 70.5-71.5°, 70-71°, 262.5°/760mm. Distd at reduced pressure, crystd from acetone or pet ether and redistd. Purified by zone melting. Very *hygroscopic*. Discolours in contact with air unless small amounts (0.2g/L) of NaOH, Na_2CO_3 or $NaBO_2$ are present. Crystd from a mixture of pet ether (185mL of b 70°) and 2-methyl-2-propanol (30mL), from acetone, or pet ether. Distd under reduced pressure and stored under nitrogen. [*Synthesis* 614 *1978*.]

Capronitrile (hexanenitrile) *[124-12-9]* M 125.2, b 163.7°, n 1.4069, n^{25} 1.4048. Washed twice with half-volumes of conc HCl, then with saturated aqueous $NaHCO_3$, dried with $MgSO_4$, and distilled.

Caprylolactam (azanon-2-one, azacyclononan-2-one, 8-aminooctanoic acid lactam) *[935-30-8]* M 141.2, m 72°, 73°, 74-76°, 75°, 76-77°, b 119-122°/0.7mm, 150-151°/7-8mm, 164°/14mm, d_4^{73} 1.009, n_D^{73} 1.489, pK^{25} 0.55 (AcOH). Dissolve in $CHCl_3$, decolorise with charcoal, evaporate to dryness and recrystallise from $CHCl_3$-hexane. Sublime at high vacuum. The *oxime* has **m** 117° (from $*C_6H_6$ or pet ether). [*J Med Chem* **14** 501 *1971*; *Justus Liebigs Ann Chem* **607** 67 *1957*.]

Capsaicin (E-N-[(4-hydroxy-3-methoxyphenyl)-methyl]-8-methyl-6-nonenamide) *[404-86-4]* M 305.4, m 64-66°, 65°, 66.1°, b 210-220°/0.01mm. Recrystd from pet ether (b 40-60°), or pet ether-Et_2O (9:1). Also purified by chromatography on neutral Al_2O_3 (grade V) and eluted successively with $*C_6H_6$, $*C_6H_6$-EtOAc (17:3) and $*C_6H_6$-EtOAc (7:3), and distilled at 120°/10^{-5}mm, and repeatedly recrystd from isopropanol (charcoal), needles. [*J Chem Soc* 11025 *1955*, *J Chem Soc(C)* 442 *1968*.]

Capsorubin (3,3'-dihydroxy-κ,κ-carotene-6,6'dione) *[470-38-2]* M 604.9, m 218°, λ$_{max}$ 443, 468, 503 nm, in hexane. Possible impurities: zeaxanthin and capsanthin. Purified by chromatography on a column of $CaCO_3$ or MgO. Crystd from *benzene/pet ether or CS_2.

Captan (N-trichloromethylmercapto-cyclohex-4-ene-1,2-dicarboxamide) *[133-06-2]* M 300.5, m 172-173°. Crystd from CCl_4. Large quantities internally cause diarrhoea and vomiting.

Captopril (S-1-[3-mercapto-2-methyl-1-oxopropyl]-L-proline) *[62571-86-2]* M 217.3, m 103-104°(polymorphic unstable form m 86°, melts at 87-88° solidifies and then melts again at 104-105°), [α]$_D^{22}$ -131° (c 1.7, EtOH), pK$_1$ 3.7, pK$_2$ 9.8. Purified by recrystn from EtOAc-hexane. Also purified by dissolving in EtOAc and chromatographed on a column of Wakogel C200 using a linear gradient of MeOH in EtOAc (0-100°) and fractions which give a positive nitroprusside test (for SH) are combined, evap and recrystd from EtOAc-hexane (1:1), white crystals [α]$_D^{20}$ -128.2° (c 2.0, EtOH).[Nam *J Pharm Sci* **73** 1843 *1984*]. Alternatively, dissolve in H_2O, apply to a column of AG-50Wx2 (BioRad) and eluted with H_2O. The free acid is converted to the dicyclohexylamine salt in MeCN by addition until the pH is 8-9 (moist filter paper). The salt is converted to the free acid by shaking with EtOAc and 10% aq $KHSO_4$ or passage through an AG50Wx2 column. The EtOAc soln is dried ($MgSO_4$) and recrystd as above from EtOAc-hexane [*Biochem J* **16** 5484 *1977*; NMR and IR: Horii and Watanabe *Yakugaku Zasshi* (*J Pharm Soc Japan*) **81** 1786 *1961*].

4-(Carbamoylmethoxy)acetanilide *[14260-41-4]* M 208.2, m 208°. Crystd from water.

3-Carbamoyl-1-methylpyridinium chloride (1-methylnicotinamide chloride, Trigonellamide) *[1005-24-9]* M 172.6, m 240°(dec). Crystd from MeOH.

Carbanilide (sym-diphenylurea) *[102-07-8]* **M 212.3, m 242°.** Crystd from EtOH or a large volume (40mL/g) of hot water.

9-Carbazolacetic acid *[524-80-1]* **M 225.2, m 215°, pK$_{Est}$ ~3.5.** Crystd from ethyl acetate.

Carbazole *[86-74-8]* **M 167.2, m 240-243°, pK <0.** Dissolved (60g) in conc H$_2$SO$_4$ (300mL), extracted with three 200mL portions of *benzene, then stirred into 1600mL of an ice-water mixture. The ppte was filtered off, washed with a little water, dried, crystd from *benzene and then from pyridine/*benzene. [Feldman, Pantages and Orchin *J Am Chem Soc* **73** 4341 *1951*]. Has also been crystd from EtOH or toluene, sublimed in vacuum, zone-refined, and purified by TLC.

Carbazole-9-carbonyl chloride *[73500-82-0]* **M 300.0, m 100-103°, 103.5-104.5°.** Recrystd from *C$_6$H$_6$. If it is not very pure (presence of OH or NH bands in the IR) dissolve in pyridine, shake with phosgene in toluene, evaporate and recrystallise the residue. Carry out this experiment in a good fume cupboard as COCl$_2$ is very **TOXIC**, and store the product in the dark. It is moisture sensitive. The *amide* has **m** 246.5-247°, and the *dimethylaminoethylamide hydrochloride* has **m** 197-198°. [Weston et al. *J Am Chem Soc* **75** 4006 *1953*.]

4-Carboethoxy-3-methyl-2-cyclohexen-1-one (Hagemann's ester) *[487-51-4]* **M 182 , b 79-80°/0.2mm, 121-123°/4mm, 142-144°/15mm, d$_4^{20}$ 1.038.** Dissolve in ether, shake with solid K$_2$CO$_3$, aqueous saturated NaHCO$_3$, dry (MgSO$_4$) and distil. *Semicarbazone* has **m** 165-167° (169°). [*J Am Chem Soc* **65**, 631, *1943*.]

1-Carbethoxy-4-methylpiperazine hydrochloride *[532-78-5]* **M 204.7, m 168.5-169°, pK 7.31.** Crystd from absolute EtOH.

N-Carboethoxyphthalimide (N-ethoxycarbonylphthalimide) *[22509-74-6]* **M 219.2, m 87-89°, 90-92°.** Crystd from toluene-pet ether (or *benzene-pet ether). Partly soluble in Et$_2$O, *benzene and CHCl$_3$. [*Chem Ber* **54** 1112 *1921*.]

γ–Carboline {9H-pyrido[3,4-b]indole} *[244-69-9]* **M 168.2, m 225°, pK~0.** Crystd from water.

Carbon Black Leached for 24h with 1:1 HCl to remove oil contamination, then washed repeatedly with distd water. Dried in air, and eluted for one day each with *benzene and acetone. Again dried in air at room temp, then heated in a vacuum for 24h at 600° to remove adsorbed gases. [Tamamushi and Tamaki *Trans Faraday Soc* **55** 1007 *1959*.]

Carbon disulfide *[75-15-0]* **M 76.1, b 46.3°, d 1.264, n 1.627.** Shaken for 3h with three portions of KMnO$_4$ soln (5g/L), twice for 6h with mercury (to remove sulfide impurities) until no further darkening of the interface occurred, and finally with a soln of HgSO$_4$ (2.5g/L) or cold, satd HgCl$_2$. Dried with CaCl$_2$, MgSO$_4$, or CaH$_2$ (with further drying by refluxing with P$_2$O$_5$), followed by fractional distn in diffuse light. **Alkali metals cannot be used as drying agents.** Has also been purified by standing with bromine (0.5mL/L) for 3-4h, shaking with KOH soln, then copper turnings (to remove unreacted bromine), and drying with CaCl$_2$. CS$_2$ is highly **TOXIC** and highly **FLAMMABLE.** *Work in a good fumehood.*
Small quantities of CS$_2$ have been purified (including removal of hydrocarbons) by mechanical agitation of a 45-50g sample with a soln of 130g of sodium sulfide in 150mL of H$_2$O for 24h at 35-40°. The aqueous sodium thiocarbonate soln was separated from unreacted CS$_2$, then ppted with 140g of copper sulfate in 350g of water, with cooling. After filtering off the copper thiocarbonate, it was decomposed by passing steam into it. The distillate was separated from H$_2$O and distd from P$_2$O$_5$. [Ruff and Golla *Z Anorg Chem* **138** 17 *1924*.]

Carbon tetrabromide *[558-13-4]* **M 331.7, m 92.5°.** Reactive bromide was removed by refluxing with dilute aqueous Na$_2$CO$_3$, then steam distd, crystd from EtOH, and dried in the dark under vacuum. [Sharpe and Walker *J Chem Soc* 157 *1962*.] Can be sublimed at 70° at low pressure.

Carbon tetrachloride *[56-23-5]* **M 153.8, b 76.8°, d^{25} 1.5842.** For many purposes, careful fractional distn gives adequate purification. Carbon disulfide can be removed by shaking vigorously for several hours with saturated KOH, separating, and washing with water: this treatment is repeated. The CCl$_4$ is shaken with conc H$_2$SO$_4$ until there is no further coloration, then washed with water, dried with CaCl$_2$ or MgSO$_4$ and distd (from P$_2$O$_5$ if desired). **It must not be dried with sodium.** An initial refluxing with mercury for 2h removes sulfides. Other purification steps include passage of dry CCl$_4$ through activated alumina, and distn from KMnO$_4$. Carbonyl containing impurities can be removed by percolation through a Celite column impregnated with 2,4-dinitrophenylhydrazine (DNPH), H$_3$PO$_4$ and water. (Prepared by dissolving 0.5g DNPH in 6mL of 85% H$_3$PO$_4$ by grinding together, then mixing with 4mL of distd water and 10g Celite.) [Schwartz and Parks *Anal Chem* **33** 1396 *1961*]. Photochlorination of CCl$_4$ has also been used: CCl$_4$ to which a small amount of chlorine has been added is illuminated in a glass bottle (e.g. for 24h with a 200W tungsten lamp near it), and, after washing out the excess chlorine with 0.02M Na$_2$SO$_3$, the CCl$_4$ is washed with distd water and distd from P$_2$O$_5$. It can be dried by passing through 4A molecular sieves and distd. Another purification procedure is to wash CCl$_4$ with aq NaOH, then repeatedly with water and N$_2$ gas bubbled through the liquid for several hours. After drying over CaCl$_2$ it is percolated through silica gel and distd under dry N$_2$ before use [Klassen and Ross *J Phys Chem* **91** 3664 *1987*].
Rapid purification: Distil, discarding the first 10% of distillate or until the distillate is clear. The distilled CCl$_4$ is then stored over 5A molecular sieves.

Carbon tetrafluoride *[75-73-0]* **M 88.0, b -15°.** Purified by repeated passage over activated charcoal at solid-CO$_2$ temperatures. Traces of air were removed by evacuating while alternately freezing and melting. Alternatively, liquefied by cooling in liquid air and then fractionally distilled under vacuum. (The chief impurity originally present was probably CF$_3$Cl).

Carbon tetraiodide *[507-25-5]* **M 519.6, m 168°(dec).** Sublimed *in vacuo*.

N,N'-**Carbonyldiimidazole** *[530-62-1]* **M 162.2, m 115.5-116°.** Crystd from *benzene or tetrahydrofuran, in a dry-box.

1,1'-Carbonyldi(1,2,4-triazole) *[41864-22-6]* **M 164.1, m 134-136°, 145-150°.** Dissolve in tetrahydrofuran and evaporate at 10mm until it crystallises. Wash crystals with cold tetrahydrofuran and dry in a vacuum desiccator over P$_2$O$_5$ in which it can be stored for months. [*Recl Trav Chim Pays-Bas* **80** 1372 *1961*; Potts *J Org Chem* **27** 2631 *1962*; Staab *Justus Liebigs Ann Chem* **106** 75 *1957*.]

Carbonyl sulfide *[463-58-1]* **M 60.1.** See carbonyl sulfide entry on p. 409 in Chapter 5.

o-**Carboxyphenylacetonitrile** *[6627-91-4]* **M 161.2, m 114-115°.** Crystd (with considerable loss) from *benzene or glacial acetic acid.

(-)-Caryophyllene oxide **(1-*S*-5*c*-6*t*-epoxy-6*c*,10,10-trimethyl-2-methylene-1*r*,9*t*-bicyclo-[7.2.0]undecane)** *[1139-30-6]* **M 220.4, m 62-63°, 63.5-64°, 64°, b 114-117°/1.8mm, 141-142°/11mm, d$_4^{20}$ 0.9666, n$_D^{20}$ 1.49564, [α]$_D^{20}$ -79° (c 2,CHCl$_3$), [α]$_D^{20}$ -68° (supercooled melt).** Purified by TLC on silica gel with EtOAc-pet ether (b 60-80°) (15:85), and recrystallised from MeOH or *C$_6$H$_6$. [NMR: Warnhoff *Can J Chem* **42** 1664 *1964*, Ramage and Whitehead *J Chem Soc* 4336 *1954*.]

(±)-Catechin *[7295-85-4]* **M 272.3, m 177° (anhyd).** Crystd from hot water. Dried at 100°.

Catechol (1,2-dihydroxybenzene, pyrocatechol) *[120-80-9]* **M 110.1, m 105°, pK$_1^{25}$ 9.45, pK$_2^{25}$ 12.8.** Crystd from *benzene or toluene. Sublimed under vacuum. [Rozo et al. *Anal Chem* **58** 2988 *1986*.]

Cation exchange resin. Conditioned before use by successive washing with water, EtOH and water, and taken through two H$^+$-Na$^+$-H$^+$ cycles by successive treatment with M NaOH, water and M HCl then washed with water until neutral.

(+)-Cedrol   (octahydro-3,6,8,8-tetramethyl-1-3a,7-methanoazulen-6-ol,   8aS-6c-hydroxy-3c,6t,8,8-tetramethyl[8ar-H]-octahydro-3H,3at,7t-methanoazulene), *[77-53-2]* **m 82-86°, 86-87°,** $[\alpha]_D^{28}$ **+10.5°** **(c 5,CHCl$_3$),** $[\alpha]_D^{18}$**+13.1°** **(c 5.5, EtOH),** $[\alpha]_D^{18}$**+14.3°** **(c 10, dioxane).**
Purified by recrystn from aqueous MeOH.   It is estimated colorimetrically with H$_3$PO$_4$ in EtOH followed by
vanillin and HCl [Hayward and Seymour *Anal Chem* **20** 572 *1948*].   The *3,5-dinitrobenzoyl derivative* has **m**
92-93°. [*J Am Chem Soc* **83** 3114 *1961.*]

β-**Cellobiose** *[528-50-7]* **M 342.3, m 228-229°(dec),** $[\alpha]_D^{25}$**+33.3°** **(c 2, water).**   Crystd from
75% aqueous EtOH.

Cellulose triacetate *[9012-09-3]* **M 72,000-74,000.**   Extracted with cold EtOH, dried in air, washed
with hot distd water, again dried in air, then dried at 50° for 30min. [Madorsky, Hart and Straus *J Res Nat Bur
Stand* **60** 343 *1958.*]

Cerulenin (helicocerin, 2R,3S-2,3-epoxy-4-oxo-7E,10E-dodecadienamide)   *[17397-89-6]* **M
223.3, m 93-94°, 93-95°, b 120°/10^{-8}mm,** $[\alpha]_D^{16}$**+63°** **(c 2, MeOH).**   White needles from *C$_6$H$_6$.
Also purified by repeated chromatography from Florisil and silica gel.   It is soluble in EtOH, MeOH, *C$_6$H$_6$,
slightly soluble in H$_2$O and pet ether.   The dl-form has **m** 40-42° (from *C$_6$H$_6$-hexane), and the 2R,3S-
tetrahydrocerulenin has **m** 86-87°, $[\alpha]_D^{20}$ +44.4 (c 0.25, MeOH after 24h). [*Tetrahedron Lett* 2095 *1978*, 2039
1979; *J Am Chem Soc* **99** 2805 *1977*; *J Org Chem* **47** 1221 *1982.*]

Cetyl acetate *[629-70-9]* **M 284.5, m 18.3°.**   Vacuum distd twice, then crystd several times from
diethyl ether/MeOH.

Cetyl alcohol (1-hexadecanol) *[36653-82-4]* **M 242.5, m 49.3°.**   Crystd from aqueous EtOH or from
cyclohexane.   Purified by zone refining.   Purity checked by gas chromatography.

Cetylamide *[629-54-9]* **M 255.4, m 106-107°, b 235-236°/12mm.**   Crystd from thiophene-free
*benzene and dried under vacuum over P$_2$O$_5$.

Cetylamine (1-hexadecylamine)   *[143-27-1]* **M 241.5, m 48°, b 162-165°/5.2mm, pK^{25}
10.60.**   Crystd from thiophene-free *benzene and dried under vacuum over P$_2$O$_5$.

Cetylammonium chloride *[1602-97-7]* **M 278.0, m 178°.**   Crystd from MeOH.

Cetyl bromide (1-bromohexadecane) *[112-82-3]* **M 305.4, m 15°, b 193-196°/14mm.**   Shaken
with H$_2$SO$_4$, washed with water, dried with K$_2$CO$_3$ and fractionally distd.

Cetyl ether *[4113-12-6]* **M 466.9, m 54°.**   Vacuum distd then crystd several times from
MeOH/*benzene.

Cetylpyridinium chloride (H$_2$O) *[6004-24-6]* **M 358.0, m 80-83°.**   Crystd from MeOH or
EtOH/diethyl ether and dried *in vacuo.* [Moss et al. *J Am Chem Soc* **108** 788 *1986*; Lennox and McClelland *J
Am Chem Soc* **108** 3771 *1986.*]

Cetyltrimethylammonium bromide (cetrimonium bromide, CTAB)   *[57-09-0]* **M 364.5, m
227-235°(dec).**   Crystd from EtOH, EtOH/*benzene or from wet acetone after extracting twice with pet ether.
Shaken with anhydrous diethyl ether, filtered and dissolved in a little hot MeOH.   After cooling in the
refrigerator, the ppte was filtered at room temperature and redissolved in MeOH.   Anhydrous ether was added and,
after warming to obtain a clear soln, it was cooled and crystalline material was filtered. [Dearden and Wooley *J
Phys Chem* **91** 2404 *1987*; Hakemi et al. *J Am Chem Soc* **91** 120 *1987.*]

Cetyltrimethylammonium chloride *[112-02-7]* **M 320.0.**   Crystd from acetone/ether mixture,
EtOH/ether, or from MeOH. [Moss et al. *J Am Chem Soc* **109** 4363 *1987.*]

Charcoal. Charcoal (50g) was added to 1L of 6M HCl and boiled for 45min. The supernatant was discarded, and the charcoal was boiled with two more lots of HCl, then with distilled water until the supernatant no longer gave a test for chloride ion. The charcoal (which was now phosphate-free) was filtered on a sintered-glass funnel and air dried at 120° for 24h. [Lippin, Talbert and Cohn *J Am Chem Soc* **76** 2871 *1954*.] The purification can be carried out using a Soxhlet extractor (without cartridge), allowing longer extraction times. Treatment with conc H_2SO_4 instead of HCl has been used to remove reducing substances.

Chaulmoogric acid [(13-cyclopent-2-enylyl)tridecanoic acid] *[29106-32-9]* **M 280.4, m 68.5°, b 247-248°/20mm, $[\alpha]_D^{20}$ +60° (c 4, $CHCl_3$), pK_{Est} ~5.0.** Crystd from pet ether or EtOH. The *Me ester [24828-59-9]* has **m** 22°, **b** 227°/20mm and $[\alpha]_D^{15}$ +50° (c 5, $CHCl_3$).

Chelerythrine *[34316-15-9]* **M 389.4, m 207°.** Crystd from $CHCl_3$ by addition of MeOH.

Chelex 100 *[11139-85-8]*. Washed successively with 2M ammonia, water, 2M nitric acid and water. Chelex 100 may develop an odour on long standing. This can be removed by heating to 80° for 2h in 3M ammonia, then washing with water. [Ashbrook *J Chromatogr* **105** 151 *1975*.]

Chelidonic acid (4-oxopyran-2,6-dicarboxylic acid) *[99-32-1]* **M 184.1, m 262°, pK_2^{25} 2.36.** Crystd from aqueous EtOH.

Chenodesoxycholic acid *[474-25-9]* **M 392.6, m 143°.** See 3α,7α-dihydroxycholanic acid on p. 207.

Chimyl alcohol (1-*O*-*n*-hexadecylglycerol) *[(±) 506-03-6; 10550-58-0 (chimyl alcohol)]* **M 316.5, m 64°.** Crystd from hexane.

Chloral *[75-87-6; 302-17-0 (hydrate)]* **M 147.4, b 98°, pK^{25} 10.04.** Distd, then dried by distilling through a heated column of $CaSO_4$.

Chloralacetone chloroform *[512-47-0]* **M 324.9, m 65°.** Crystd from *benzene.

α−**Chloralose (*R*-1,2-*O*-[2,2,2-trichloroethylidene]-α-D-glucofuranose)** *[15879-93-3]* **M 309.5, m 180-182°, 187°, 186-188°, $[\alpha]_D^{26}$ +19.5° (c 11, pyridine).** Recrystd from EtOH, 38% aqueous EtOH, Et_2O, H_2O or $CHCl_3$ The solubility is 0.44% in H_2O at 15°, 0.83% in H_2O at 37°, 6.7% in EtOH at 25°. [Whiton and Hixon *J Am Chem Soc* **55** 2438 *1933*; *Helv Chim Acta* **6** 621 *1923*.] The β-*isomer* is less soluble in H_2O, EtOH or Et_2O and has **m** 237.5-238° [*J Am Chem Soc* **59** 1955 *1937*; *Acta Chem Scand* **19** 359 *1965*].

2-Chloroacetophenone *[532-27-4]* **M 154.6, m 54-56°.** Crystd from MeOH [Tanner *J Org Chem* **52** 2142 *1987*].

p-**Chloranil (2,3,5,6-tetrachloro-1,4-benzoquinone)** *[118-75-2]* **M 245.9, m 290°, 294.2-294.6° (sealed tube).** Crystd from acetic acid, acetone, *benzene, EtOH or toluene, drying under vac over P_2O_5, or from acetic acid, drying over NaOH in a vacuum desiccator. It can be sublimed under vacuum at 290°. Sample may contain significant amounts of the *o*-chloranil isomer as impurity. Purified by triple sublimation under vacuum. Recrystd before use. **It is a skin and mucous membrane irritant.** [UV: *Rec Trav Chim Pays Bas* **276** 684 *1924*; Brook *J Chem Soc* 5040 *1952*.]

Chloranilic acid (2,5-dichloro-3,6-dihydroxy-1,4-benzoquinone) *[87-88-7]* **M 209.0, m 283-284° pK_1^{25} 1.22, pK_2^{25} 3.01.** A soln of 8g in 1L of boiling water was filtered while hot, then extracted twice at about 50° with 200mL portions of *benzene. The aq phase was cooled in ice-water. The crystals were filtered off, washed with three 10mL portions of water, and dried at 115°. It can be sublimed in vacuum. [*J Phys Chem* **61** 765 *1957*.] The *diacetate* has **m** 182-185° [*J Am Chem Soc* **46** 1866 *1924*; Thamer and Voight *J Phys Chem* **56** 225 *1952*.]

Chlorendic anhydride **(1,4,5,6,7,7,-hexachloro-5-norbornene-2,3-dicarboxylic anhydride)** *[115-27-5]* **M 370.9, m 234-236°. 235-237°, 238°.** Steam distn or recrystn from H_2O yields the diacid. The purified diacid yields the anhydride with Ac_2O. [Prill *J Am Chem Soc* **69** 62 *1947.*]

Chloroacetaldehyde dimethyl acetal *[97-97-2]* **M 124.6, m -34.4°, b 64°/23mm, 71-72°/35mm, d_4^{20} 1.0172, n_D^{20} 1.4175.** Purified by fractional distillation. [Melhotra *J Indian Chem Soc* **36** 4405 *1959*; *Bull Soc Chim Belg* **61** 393 *1952.*]

α-Chloroacetamide *[79-07-2]* **M 93.5, m 121°, b 224-225°/743mm.** Crystd from acetone and dried under vacuum over P_2O_5.

p-Chloroacetanilide *[539-03-7]* **M 169.6, m 179°.** Crystd from EtOH or aqueous EtOH.

Chloroacetic acid *[79-11-8]* **M 94.5, m 62.8°, b 189°, pK^{25} 2.87.** Crystd from $CHCl_3$, CCl_4, *benzene or water. Dried over P_2O_5 or conc H_2SO_4 in a vacuum desiccator. Further purification by distn from $MgSO_4$, and by fractional crystn from the melt. Stored under vac or under dry N_2. [Bernasconi et al. *J Am Chem Soc* **107** 3621 *1985.*]

Chloroacetic anhydride *[541-88-8]* **M 171.0, m 46°, d 1.5494.** Crystd from *benzene.

Chloroacetone *[78-95-5]* **M 92.5, b 119°/763mm, d 1.15.** Dissolved in water and shaken repeatedly with small amounts of diethyl ether which extracts, preferentially, 1,1-dichloroacetone present as an impurity. The chloroacetone was then extracted from the aqueous phase using a large amount of diethyl ether, and distd at slightly reduced pressure. It was dried with $CaCl_2$ and stored at Dry-ice temperature. Alternatively, it was stood with $CaSO_4$, distd and stored over $CaSO_4$. **LACHRYMATORY.**

Chloroacetonitrile *[107-14-2]* **M 75.5, b 125°.** Refluxed with P_2O_5 for one day, then distd through a helices-packed column. Also purified by gas chromatography. **LACHRYMATOR, HIGHLY TOXIC.**

o-Chloroaniline *[95-51-2]* **M 127.6, m -1.9°, b 208.8°, d 1.213, n 1.588, pK^{25} 2.66.** Freed from small amounts of the *p*-isomer by dissolving in one equivalent of H_2SO_4 and steam distilling. The *p*-isomer remains behind as the sulfate. [Sidgwick and Rubie *J Chem Soc* 1013 *1921.*] An alternative method is to dissolve in warm 10% HCl (11mL/g of amine) and on cooling, the hydrochloride of *o*-chloroaniline separates out. The latter can be recrystd until the acetyl derivative has a constant melting point. (In this way, yields are better than for the recrystn of the picrate from EtOH or of the acetyl derivative from pet ether.) [King and Orton *J Chem Soc* 1377 *1911*].

p-Chloroaniline *[106-47-8]* **M 127.6, m 70-71°, pK^{25} 3.98.** Crystd from MeOH, pet ether (b 30-60°), or 50% aq EtOH, then *benzene/pet ether (b 60-70°), then dried in a vacuum desiccator. Can be distd under vacuum (**b 75-77°/33mm**).

p-Chloroanisole *[623-12-1]* **M 142.6, b 79°/11.5mm, 196.6°/760mm, d 1.164, $n^{25.5}$ 1.5326.** Washed with 10% (vol) aqueous H_2SO_4 (three times), 10% aqueous KOH (three times), and then with water until neutral. Dried with $MgSO_4$ and fractionally distd from CaH_2 through a glass helices-packed column under reduced pressure.

9-Chloroanthracene *[716-53-0]* **M 212.9, m 105-107°.** Crystd from EtOH. [Masnori *J Am Chem Soc* **108** 1126 *1986.*]

10-Chloro-9-anthraldehyde *[10527-16-9]* **M 240.7, m 217-219°.** Crystd from EtOH.

o-Chlorobenzaldehyde *[89-98-5]* **M 140.6, m 11°, b 213-214°, d 1.248, n 1.566.** Washed with 10% Na_2CO_3 soln, then fractionally distd in the presence of a small amount of catechol.

3-Chlorobenzaldehyde *[587-04-2]* **M 140.6, m 18°, b 213-214°, d 1.241, n 1.564.** Purified by low temperature crystn from pet ether (b 40-60°).

4-Chlorobenzaldehyde *[104-88-1]* **M 140.6, m 47°.** Crystd from EtOH/water (3:1), then sublimed twice at 2mm pressure at a temperature slightly above the melting point.

Chlorobenzene *[108-90-7]* **M 112.6, b 131.7°, d 1.107, n 1.52480.** The main impurities are likely to be chlorinated impurities originally present in the *benzene used in the synthesis of chlorobenzene, and also unchlorinated hydrocarbons. A common purification procedure is to wash several times with conc H_2SO_4 then with aq $NaHCO_3$ or Na_2CO_3, and water, followed by drying with $CaCl_2$, K_2CO_3 or $CaSO_4$, then with P_2O_5, and distn. It can also be dried with Linde 4A molecular sieve. Passage through, and storage over, activated alumina has been used to obtain low conductance material. [Flaherty and Stern *J Am Chem Soc* **80** 1034 *1958*.]

4-Chlorobenzenesulfonyl chloride *[98-60-2]* **M 211.1, m 53°, b 141°/15mm.** Crystd from ether in powdered Dry-ice, after soln had been washed with 10% NaOH until colourless and dried with Na_2SO_4.

4-Chlorobenzhydrazide *[536-40-3]* **M 170.6, m 164°.** Crystd from water.

2-Chlorobenzoic acid *[118-91-2]* **M 156.6, m 139-140°, pK^{25} 2.91.** Crystd successively from glacial acetic acid, aq EtOH, and pet ether (b 60-80°). Other solvents include hot water or toluene (*ca* 4mL/g). Crude material can be given an initial purification by dissolving 30g in 100mL of hot water containing 10g of Na_2CO_3, boiling with 5g of charcoal for 15min, then filtering and adding 31mL of 1:1 aq HCl: the ppte is washed with a little water and dried at $100°$.

3-Chlorobenzoic acid *[535-80-8]* **M 156.6, m 154-156°, 158°, d$_4^{25}$ 1.496, pK^{25} 3.82 (5.25 in 50% dimethylacetamide).** Crystd successively from glacial acetic acid, aqueous EtOH and pet ether (b 60-80°). It also recrysts from *C_6H_6 or Et_2O-hexane, and sublimes at 55° in a vacuum. [*Anal Chem* **26** 726 *1954*] The *methyl ester* has **m** 21°, **b** 231°/atm. The *S-benzyl thiouronium salt* has **m** 164-165° (from EtOH) [*Acta Chem Scand* **9** 1425 *1955*; *J Chem Soc* 1318 *1960*].

4-Chlorobenzoic acid *[74-11-3]* **M 156.6, m 238-239°, pK^{25} 3.99.** Same as for *m*-chlorobenzoic acid. Has also been crystd from hot water, and from EtOH.

2-Chlorobenzonitrile *[873-32-5]* **M 137.6, m 45-46°.** Crystd to constant melting point from *benzene/pet ether (b 40-60°).

4-Chlorobenzophenone *[134-85-0]* **M 216.7, m 75-76°.** Recrystd for EtOH. [Wagner et al. *J Am Chem Soc* **108** 7727 *1986*.]

2-Chlorobenzothiazole *[615-20-3]* **M 169.6, m 21°, 90-91.4°/4mm, 135-136°/28mm, d$_4^{20}$ 1.303, n$_D^{20}$ 1.6398.** It is purified by fractional distn *in vacuo*. The *2-chloro-3-methylbenzothiazolinium 2,4-dinitrobenzenesulfonate* crystallises from Ac_2O, **m** 162-163° (dec). [*J Am Chem Soc* **73** 4773 *1951*; *J Org Chem* **19** 1830 *1954*; *J Chem Soc* 2190 *1930*.]

o-**Chlorobenzotrifluoride** *[88-16-4]* **M 180.6, b 152.3°.** Dried with $CaSO_4$, and distd at high reflux ratio through a silvered vacuum-jacketed glass column packed with one-eight inch glass helices [Potter and Saylor *J Am Chem Soc* **73** 90 *1951*].

m-**Chlorobenzotrifluoride** *[98-15-7]* **M 180.6, b 137.6°.** Same as for *o*-chlorobenzotrifluoride above.

p-**Chlorobenzotrifluoride** *[98-56-6]* **M 180.6, b 138.6°.** Same as for *o*-chlorobenzotrifluoride above.

2-Chlorobenzoxazole *[615-18-9]* **M 153.6, b 95-96°/20mm, 198-202°/atm, d_4^{20} 1.331, n_D^{20} 1.570.** Purified by fractional distn, preferably in a vacuum. [Siedel *J Prakt Chem* (2) **42** 456 *1890*; *J Am Chem Soc* **75** 712 *1953*.]

p-**Chlorobenzyl chloride** *[104-83-6]* **M 161.0, m 28-29°, b 96°/15mm.** Dried with $CaSO_4$, then fractionally distd under reduced pressure. Crystd from heptane or dry diethyl ether. **LACHRYMATORY.**

p-**Chlorobenzylisothiuronium chloride** *[544-47-8]* **M 237.1, m 197°, pK_{Est} ~9.6 (free base).** Crystd from conc HCl by addition of water.

2-Chlorobutane *[78-86-4]* **M 92.6, b 68.5°, d 0.873, n^{25} 1.3945.** Purified in the same way as *n*-butyl chloride.

2-(4-Chlorobutyl)-1,3-dioxolane *[118336-86-0]* **M 164.6, b 56-58°/0.1mm, d_4^{20} 1.106, n_D^{20}1.457.** If the IR has a CHO band then just distil in vacuum. If it is present then dissolve in Et_2O, wash with H_2O, then saturated $NaHCO_3$, dry over $MgSO_4$, evaporate and distil. [*J Am Chem Soc* **73** 1365 *1951*.]

N-**Chlorocarbonyl isocyanate** *[27738-96-1]* **M 105.5, m -68°, b 63.6°/atm, d_4^{20} 1.310.** Fractionally distd at atmospheric pressure using a 40cm column. **TOXIC** *vapour use a good fume hood.* Store dry, ν 2260 (NCO), 1818 (CO) and 1420 (NCO sym) cm^{-1}. [*Chem Ber* **106** 1752 *1975*.]

trans-**4-Chlorocinnamic acid** *[1615-02-7]* **M 182.6, m 243°, 248-250°, 249-251°, pK^{25} 4.41.** Recrystd from EtOH or aq EtOH (charcoal). [*Org Synth* Coll Vol IV 731 *1963*, Walling and Wolfstin *J Am Chem Soc* **69** 852 *1947*.]

Chlorocyclohexane *[542-18-7]* **M 118.6, b 142.5°, d 1.00, n^{25} 1.46265.** Washed several times with dilute $NaHCO_3$, then repeatedly with distilled water. Dried with $CaCl_2$ and fractionally distd.

4-Chloro-2,6-diaminopyrimidine (**2,4-diamino-6-chloropyrimidine**) *[156-83-2]* **M 144.6, m 198°, 199-202°, pK^{25} 3.57.** Purified by recrystn from boiling H_2O (charcoal) as needles; also crystallises from Me_2CO. [Büttner *Chem Ber* **36** 2232 *1903*; Roth *J Am Chem Soc* **72** 1914 *1950*; UV: *J Chem Soc* 3172 *1962*.]

4-Chloro-3,5-dimethylphenol *[88-04-0]* **M 156.6, m 115.5°, pK^{25} 9.70.** Crystd from *benzene or toluene.

1-Chloro-2,4-dinitrobenzene *[97-00-7]* **M 202.6, m 48-50°, 51°, 52-54°, 54°, b 315°/atm, d_4^{22} 1.697.** Usually crystd from EtOH or MeOH. Has also been crystd from Et_2O, *C_6H_6, *C_6H_6-pet ether or isopropyl alcohol. A preliminary purification step has been to pass its soln in *benzene through an alumina column. Also purified by zone refining. It exists in three forms: one stable and two unstable. The stable form crysts as yellow needles from Et_2O, **m** 51°, **b** 315°/atm with some dec, and is sol in EtOH. The labile forms also crystallises from Et_2O, **m** 43°, and is more soluble in organic solvents. The second labile form has **m** 27°. [Hoffman and Dame, *J Am Chem Soc* **41** 1015 *1919*, Welsh *J Am Chem Soc* **63** 3276 *1941*; *J Chem Soc* 2476 *1957*.]

4-Chloro-3,5-dinitrobenzoic acid *[118-97-8]* **M 246.6, m 159-161°, pK_{Est} ~2.5.** Crystd from EtOH/water, EtOH or *benzene.

2-Chloro-3,5-dinitropyridine *[2578-45-2]* **M 203.5, m 62-65°, 63-65°, 64°, pK_{Est} <-5.** Dissolve in $CHCl_3$, shake with saturated $NaHCO_3$, dry ($MgSO_4$), evaporate and apply to an Al_2O_3 column, elute with pet ether (b 60-80°), evaporate and recryst from *C_6H_6 or pet ether. [*Chem Pharm Bull Jpn* **8** 28 *1960*; *Recl Trav Chim Pays-Bas* **72** 573 *1953*.]

2-Chloroethanol (ethylene chlorohydrin) *[107-07-3]* **M 80.5, b 51.0°/31mm, 128.6°/760mm, d 1.201, n^{15} 1.444.** Dried with, then distd from, CaSO$_4$ in the presence of a little Na$_2$CO$_3$ to remove traces of acid.

2-Chloroethyl bromide (1-bromo-2-chloroethane) *[107-04-0]* **M 143.4, b 106-108°.** Washed with conc H$_2$SO$_4$, water, 10% Na$_2$CO$_3$ soln, and again with water, then dried with CaCl$_2$ and fractionally distd before use.

2-Chloroethyl chloroformate *[627-11-2]* **M 143.0, b 52-54°/12mm, 153°/760mm, d$_4^{18}$ 1.3760, n$_D^{20}$ 1.446.** Purified by fractional distn, preferably in a vacuum and stored in dry atmosphere. [*J Chem Soc* 2735 *1957*.]

1-(2-Chloroethyl)pyrrolidine hydrochloride *[7250-67-1]* **M 170.1, m 167-170°, 173.5-174°, pK$_{Est}$ ~8.5 (free base).** Purified by recrystn from isopropanol-di-isopropyl ether (charcoal) and recrystallised twice more. The *free base*, **b 55-56°/11mm, 60-63°/23mm and 90°/56mm,** is relatively unstable and should be converted to the hydrochloride immediately, by dissolving in isopropanol and bubbling dry HCl through the soln at 0°, and filtering off the hydrochloride and recrystallising it. The *picrate* has **m 107.3-107.8°** (from EtOH) [Cason *J Org Chem* **24** 247 *1959*; *J Am Chem Soc* **70** 3098 *1948*].

2-Chloroethyl vinyl ether *[110-75-8]* **M 106.6, b 109°/760mm, d 1.048, n 1.437.** Washed repeatedly with equal volumes of water made slightly alkaline with KOH, dried with sodium, and distd under vacuum. **TOXIC.**

Chloroform *[67-66-3]* **M 119.4, b 61.2°, d^{15} 1.49845, d^{10} 1.47060, n^{15} 1.44858.** Reacts slowly with oxygen or oxidising agents, when exposed to air and light, giving, mainly, phosgene, Cl$_2$ and HCl. Commercial CHCl$_3$ is usually stabilized by addn of up to 1% EtOH or of dimethylaminoazobenzene. Simplest purifications involve washing with water to remove the EtOH, drying with K$_2$CO$_3$ or CaCl$_2$, refluxing with P$_2$O$_5$, CaCl$_2$, CaSO$_4$ or Na$_2$SO$_4$, and distilling. **It must not be dried with sodium.** The distd CHCl$_3$ should be stored in the dark to avoid photochemical formation of phosgene. As an alternative purification, CHCl$_3$ can be shaken with several small portions of conc H$_2$SO$_4$, washed thoroughly with water, and dried with CaCl$_2$ or K$_2$CO$_3$ before filtering and distilling. EtOH can be removed from CHCl$_3$ by passage through a column of activated alumina, or through a column of silica gel 4-ft long by 1.75-in diameter at a flow rate of 3mL/min. (The column, which can hold about 8% of its weight of EtOH, is regenerated by air drying and then heating at 600° for 6h. It is pre-purified by washing with CHCl$_3$, then EtOH, leaving in conc H$_2$SO$_4$ for about 8hr, washing with water until the washings are neutral, then air drying, followed by activation at 600° for 6h. Just before use it is reheated for 2h to 154°.) [McLaughlin, Kaniecki and Gray *Anal Chem* **30** 1517 *1958*]. Carbonyl-containing impurities can be removed from CHCl$_3$ by percolation through a Celite column impregnated with 2,4-dinitrophenylhydrazine, phosphoric acid and water. (Prepared by dissolving 0.5g DNPH in 6mL of 85% H$_3$PO$_4$ by grinding together, then mixing with 4mL of distilled water and 10g of Celite.) [Schwartz and Parks *Anal Chem* **33** 1396 *1961*]. Chloroform can be dried by distn from powdered type 4A Linde molecular sieves. For use as a solvent in IR spectroscopy, chloroform is washed with water (to remove EtOH), then dried for several hours over anhydrous CaCl$_2$ and fractionally distd. This treatment removes material absorbing near 1600 cm^{-1}. (Percolation through activated alumina increases this absorbing impurity). [Goodspeed and Millson *Chem Ind (London)* 1594 *1967*].
Rapid purification: Pass through a column of basic alumina (Grade I, 10g/mL of CHCl$_3$), and either dry by standing over 4A molecular sieves, or alternatively, distil from P$_2$O$_5$ (3% w/v). Use immediately.

Chlorogenic [3-(3,4-dihydroxycinnamoyl)quinic] acid *[327-97-9]* **M 354.3, m 208°, [α]$_D^{25}$ -36°** (c 1, H$_2$O), **pK$_1^{25}$ 3.59, pK$_2^{25}$ 8.59.** Crystd from water. Dried at 110°.

Chlorohydroquinone (2-chloro-1,4-dihydroxybenzene) *[615-67-8]* **M 144.6, m 106°, b 263°, pK$_1^{25}$ 8.81, pK$_2^{25}$ 10.78.** Crystd from CHCl$_3$ or toluene.

5-Chloro-8-hydroxy-7-iodoquinoline (vioform) *[130-26-7]* **M 305.5, m 178-179°, pK$_{Est(1)}$~2.6, pK$_{Est(2)}$ ~7.0.** Crystd from abs EtOH.

5-Chloroindole   *[17422-32-1]* M **151.6, m 69-71°, 72-73°, b 120-130°/0.4mm, pK$_{Est}$ <0** It is distd at high vacuum and recrystallises from pet ether (b 40-60°) or (b 80-100°) as glistening plates. The *picrate* has **m** 147° (146.5-147.5°)(from *C$_6$H$_6$). [*J Chem Soc* 3493 *1955*; *J Org Chem* **44** 578 *1979*].

4-Chloroiodobenzene *[637-87-6]* M **238.5, m 53-54°.** Crystd from EtOH.

2,3-Chloromaleic anhydride   *[1122-17-4]*   M **167.0, m 121-121.5°.** See 2,3-dichloromaleic anhydride on p. 198.

5-Chloro-2-methoxyaniline   (**2-amino-4-chloroanisole**)   *[95-03-4]* M **157.6, m 81-83°, 82-84°, 84°, pK^{25} 3.56.** Purified by steam distn and recrystn from H$_2$O or 40% aqueous EtOH. The *N-acetate* forms needles from hot H$_2$O **m** 104°; the *N-benzoyl* derivative forms needles from aq EtOH **m** 77-78°; the *picrate* has **m** 194° dec. [*J Am Chem Soc* **48** 2657 *1926*.]

9-Chloromethyl anthracene   *[24463-19-2]* M **226.7, m 141-142° dec, 141-142.5°.** If it is free from OH in the IR then recryst from hexane-*C$_6$H$_6$ or *C$_6$H$_6$ as needles. If OH is present then some solvolysis has occurred. In this case treat 8.5g with SOCl$_2$ (4.8g) in dioxane (60mL) and reflux for 5h, then evaporate to dryness and wash the residue with cold *C$_6$H$_6$ and recrystallise. With KI/Me$_2$CO it forms the *iodomethyl* derivative. [Martin et al. *Helv Chim Acta* **38** 2009 *1955*; *J Org Chem* **21** 1512 *1956*.]

2-Chloro-3-methylindole (**2-chloroskatole**) *[51206-73-6]*   M **165.6, m 114.5-115.5°, pK$_{Est}$ <0.** Purified by chromatography on silica gel in CH$_2$Cl$_2$/pet ether (1:2), followed by recrystn from aqueous EtOH or aqueous acetic acid. [Phillips and Cohen *J Am Chem Soc* **108** 2023 *1986*.]

Chloromethyl methyl ether (MOMCl)   *[107-30-2]* M **80.5, b 55-57°, d 1.060, n 1.396.** If suspect (check IR), shake with satd aq CaCl$_2$ soln, dry over CaCl$_2$ and fractionally distil taking middle fraction. [Marvel and Porter *Org Synth* Coll Vol I 377 *1941*.] **VERY TOXIC** and **CARCINOGENIC**.

4-Chloro-2-methylphenol *[1570-64-5]* M **142.6, m 49°, pK^{25} 9.71.** Purified by zone melting.

4-Chloro-3-methylphenol *[59-50-7]* M **142.6, m 66°, pK^{25} 9.55.** Crystd from pet ether.

4-Chloro-2-methylphenoxyacetic acid (**MCPA**) *[94-74-6]* M **200.6, m 113-117°, 120°, 122-123°, pK^{20} 3.62 (3.05).** It is insoluble in H$_2$O (sol 0.55g/L at 20°), and recrystallises from *C$_6$H$_6$ or chlorobenzene as plates [*Acta Chem Scand* **6** 993 *1952*]. The *S-benzylthiouronium salt* has **m** 164-165°, and the Cu^{2+} salt has **m** 247-249°dec [Armarego et al. *Nature* **183** 1176 *1959*; UV: Duvaux and Grabe *Acta Chem Scand* **4** 806 *1950*; IR: Jöberg *Acta Chem Scand* **4** 798 *1950*].

Chloromethyl phenyl sulfide   *[7205-91-6]* M **158.7, b 63°/0.1mm, 98°/12mm, 113-115°/20mm.** Dissolve in CH$_2$Cl$_2$ or CCl$_4$ and dry over CaCl$_2$, or pass through a tube of CaCl$_2$ and fractionally distil using a fractionating column. ***Harmful* vapours.** It gives the *sulfone* (b 130°/1mm and **m** 53° from EtOH) on oxidation with permonophthalic acid. [*Justus Liebigs Ann Chem* **563** 54 64 *1949*.]

***N*-(Chloromethyl)phthalimide** *[17564-64-6]* M **195.6, m 131-135°, 134-135°, 136.5°.** Purified by recrystn from EtOAc or CCl$_4$ [*J Am Chem Soc* **70** 2822 *1948*; Böhme et al. *Chem Ber* **92** 1258 *1959*].

4-(Chloromethyl)pyridine hydrochloride *[1822-51-1]*   M **164.0, m 170-175°, 172-173°, pK$_{Est}$ ~5.6.** Purified by recrystn from EtOH or EtOH-dry Et$_2$O. It melts between 171° and 175° and the clear melt resolidifies on further heating at 190° and turns red to black at 280° but does not melt again. The *picrate-hydrochloride* (prepared in EtOH) has **m** 146-147°. The free base is an oil, [Mosher and Tessieri *J Am Chem Soc* **73** 4925 *1951*.]

2-Chloro-1-methylpyridinium iodide *[14338-32-0]* M **255.5, m 203-205°, 205-206°(dec), 207°.** Purified by dissolving in EtOH and adding dry Et$_2$O. The solid is washed with Me$_2$CO and dried at

20°/0.35mm. Store in the dark. Attempted recrystn from Me$_2$CO-EtOH-pet ether (b 40-60°) caused some exchange of the Cl substituent by I. The *picrate* has **m** 106-107°, and the *perchlorate* has **m** 212-213°. [UV and solvolysis: Barlin and Benbow *J Chem Soc, Perkin Trans 2* 790 *1974*.]

1-Chloronaphthalene *[90-13-1]* **M 162.6, f -2.3°, b 136-136.5°/20mm, 259.3°/760mm, d 1.194, n 1.6326.** Washed with dilute NaHCO$_3$, then dried with Na$_2$SO$_4$ and fractionally distd under reduced pressure. Alternatively, before distn, it was passed through a column of activated alumina, or dried with CaCl$_2$, then distd from sodium. It can be further purified by fractional crystn by partial freezing or by crystn of its picrate to constant melting point (132-133°) from EtOH, and recovering from the picrate.

2-Chloronaphthalene *[91-58-7]* **M 162.6, m 61°, b 264-266°.** Crystd from 25% EtOH/water and dried under vacuum.

1-Chloro-2-naphthol *[633-99-8]* **M 178.6, m 70°, pK$_{Est}$ ~8.3.** Cryst from pet ether. *Acetate* has **m** 42-43°.

2-Chloro-1-naphthol *[606-40-6]* **M 178.6, m 64-65°, pK$_{Est}$ ~7.9.** Crystd from pet ether.

4-Chloro-1-naphthol *[604-44-4]* **M 178.6, m 116-117°, 120-121°, pK^{25} 8.86.** Crystd from EtOH or chloroform.

6-Chloronicotinic acid *[5326-23-8]* **M 157.6, m 190-193°, 198-199°(dec), pK^{25} 4.22 (50% aq EtOH).** Purified by recrystn from hot H$_2$O and is sublimed in a vacuum. [Pechmann and Welsch *Chem Ber* **17** 2384 *1884*; Herz and Murty *J Org Chem* **26** 122 *1961*.]

4-Chloro-2-nitroaniline *[89-63-4]* **M 172.6, m 116-116.5°, pK^{25} -0.99.** Crystd from hot water or EtOH/water and dried for 10h at 60° under vacuum.

2-Chloro-4-nitrobenzamide *[3011-89-0]* **M 200.6, m 172°.** Crystd from EtOH.

2-Chloro-1-nitrobenzene *[88-73-3]* **M 157.6, m 32.8-33.2°.** Crystd from EtOH, MeOH or pentane (charcoal).

3-Chloro-1-nitrobenzene *[121-73-3]* **M 157.6, m 45.3-45.8°.** Crystd from MeOH or 95% EtOH (charcoal), then pentane.

4-Chloro-1-nitrobenzene *[100-00-5]* **M 157.6, m 80-83°, 83.5-84°, b 113°/8mm, 242°/atm, d $^{100.5}_4$ 1.2914.** Crystd from 95% EtOH (charcoal) and sublimes in a vacuum. [Emmons *J Am Chem Soc* **76** 3470 *1954*; Newman and Forres *J Am Chem Soc* **69** 1221 *1947*.]

4-Chloro-7-nitrobenzofurazane **(7-chloro-4-nitrobenzoxadiazole)** *[10199-89-0]* **M 199.6, m 96.5-97°, 97°, 99-100°.** Wash the solid with H$_2$O and recrystallise from aqueous EtOH (1:1) as pale yellow needles. It sublimes in a vacuum [UV, NMR: Bolton, Gosh and Katritzky *J Chem Soc* 1004 *1966*].

1-Chloro-2-nitroethane *[625-47-8]* **M 109.5, b 37-38°/20mm, n 1.4224, n^{25} 1.4235.** Dissolved in alkali, extracted with ether (discarded), then the aqueous phase was acidified with hydroxylamine hydrochloride, and the nitro compound fractionally distd under reduced pressure. [Pearson and Dillon *J Am Chem Soc* **75** 2439 *1953*.]

2-Chloro-3-nitropyridine *[5470-18-8]* **M 158.5, m 100-103°, 101-102°, 103-104° (sublimes), pK^{20} -2.6.** Forms needles from H$_2$O. Purified by continuous sublimation over a period of 2 weeks at 50-60°/0.1mm [Barlin *J Chem Soc* 2150 *1964*]. The *N-oxide* has **m** 99-100°(from CH$_2$Cl$_2$-Et$_2$O). [Taylor and Driscoll *J Org Chem* **25** 1716 *1960*; Ochiai and Kaneko *Chem Pharm Bull Jpn* **8** 28 *1960*.]

2-Chloro-5-nitropyridine *[4548-45-2]* **M 158.5, m 108°, pK$_{Est}$ ~-2.6.** Crystd from *benzene or *benzene/pet ether.

3-Chloroperbenzoic acid *[937-14-4]* **M 172.6, m 92-94°(dec), pK^{25} 7.57.** Recrystd from CH$_2$Cl$_2$ [Traylor and Mikztal *J Am Chem Soc* **109** 2770 *1987*]. Peracid of 99+% purity can be obtained by washing commercial 85% material with phosphate buffer pH 7.5 and drying the residue under reduced pressure. Alternatively the peracid can be freed from *m*-chlorobenzoic acid by dissolving 50g/L of *benzene and washing with an aq soln buffered at pH 7.4 (NaH$_2$PO$_4$/NaOH) (5 x 100mL). The organic layer was dried over MgSO$_4$ and carefully evaporated under vacuum. *Necessary care should be taken in case of* **EXPLOSION**. The solid was recrystd twice from CH$_2$Cl$_2$/Et$_2$O and stored at 0° in a plastic container as glass catalyses the decomposition of the peracid. The acid is assayed iodometrically. [*J Org Chem* **29** 1976 *1964*; Bortolini et al. *J Org Chem* **52** 5093 *1987*.]

2-Chlorophenol *[95-57-8]* **M 128.6, m 8.8°, b 61-62°/10mm, 176°/atm, pK^{25} 8.34.** Passed at least twice through a gas chromatograph column. Also purified by fractional distn.

3-Chlorophenol *[108-43-0]* **M 128.6, m 33°, b 44.2°/1mm, 214°/atm, pK^{25} 9.13.** Could not be obtained solid by crystn from pet ether. Purified by distn under reduced pressure.

4-Chlorophenol *[106-48-9]* **M 128.6, m 43°, 100-101°/10mm, pK^{25} 9.38.** Distd, then crystd from pet ether (b 40-60°) or hexane, and dried under vacuum over P$_2$O$_5$ at room temp. It has pKa 9.38 at 20° in water. [Bernasconi and Paschalis *J Am Chem Soc* **108** 2969 *1986*.]

Chlorophenol Red (3,3'-dichlorophenolsulfonephthalein) *[4430-20-0]* **M 423.3, m dec on heating, λ$_{max}$ 573nm, pK^{25} 5.96.** Crystd from glacial acetic acid. It is yellow at pH 4.8 and violet at pH 6.7.

4-Chlorophenoxyacetic acid *[122-88-3]* **M 186.6, m 157°, pK^{20} 3.00.** Crystd from EtOH.

α-4-Chlorophenoxypropionic acid *[3307-39-9]* **M 200.6, m 116°, pK$_{Est}$ ~3.2.** Crystd from EtOH.

ß-4-Chlorophenoxypropionic acid *[3284-79-5]* **M 200.6, m 138°, pK$_{Est}$ ~4.2.** Crystd from EtOH.

3-Chlorophenylacetic acid *[1878-65-5]* **M 170.6, m 74°, pK^{25} 4.11.** Crystd from EtOH/water, or as needles from *C$_6$H$_6$ or H$_2$O (charcoal). The *acid chloride* (prepared by boiling with SOCl$_2$) has **b** 127-129°/15mm. [Dippy and Williams *J Chem Soc* 161 *1934*; Misra and Shukla *J Indian Chem Soc* **28** 480 *1951*.]

4-Chlorophenylacetic acid *[1878-66-6]* **M 170.6, m 105°, 106°, pK^{25} 4.12.** Same as for 3-chlorophenylacetic acid.

4-Chloro-1-phenylbutan-1-one *[939-52-6]* **M 182.7, m 19-20°, b 134-137°/5mm, d$_4^{20}$ 1.149, n$_D^{20}$ 1.55413.** Fractionate several times using a short column. It can be recrystd from anhydrous pet ether at -20° as glistening white rosettes and filtered at 0° and dried in a vacuum desiccator over H$_2$SO$_4$. The *semicarbazone* has **m** 136-137°. [*J Am Chem Soc* **46** 1882 *1924*, **51** 1174 *1929*, Hart and Curtis *J Am Chem Soc* **79** 931 *1957*.]

1-(2-Chlorophenyl)-1-(4-chlorophenyl)-2,2-dichloroethane (Mitotane, *o,p'*-DDD) *[53-19-0]* **M 320.1, m 75.8-76.8°, 76-78°.** Purified by recrystallisation from pentane and from MeOH or EtOH. It is sol in isooctane and CCl$_4$. [Haller et al. *J Am Chem Soc* **67** 1600 *1945*.]

3-(4-Chlorophenyl)-1,1-dimethylurea (monuron) *[150-68-5]* **M 198.7, m 171°.** Crystd from MeOH.

4-Chloro-1,2-phenylenediamine *[95-83-0]* **M 142.6, m 69-70°, pK_1^{25} -0.27 (aq H_2SO_4), pK_2^{25} 3.35 (3.67).** Recrystd from pet. ether.

4-Chlorophenyl isocyanate *[104-12-1]* **M 153.6, m 28-31°, 31-32°, 32°, 32.5°, b 80.6-80.9°/9.5mm, 115-117°/45mm.** Purified by recrystn from pet ether (b 30-40°) or better by fractional distn. **TOXIC irritant.**

4-Chlorophenyl isothiocyanate *[2131-55-7]* **M 169.6, m 44°, 43-45°, 45°, 46°, 47°, b 110-115°/4mm, 135-136°/24mm.** Check the IR first. Triturate with pet ether (b 30-60°) and decant the solvent. Repeat 5 times. The combined extracts are evap under reduced press to give almost pure compound as a readily crystallisable oil with a pleasant anise odour. It can be recrystd from the minimum vol of EtOH at 50° (do not boil too long in case it reacts). It can be purified by vac distn. **IRRITANT.** [*Org Synth* Coll Vol V 223 *1973*.]

4-Chlorophenyl 2-nitrobenzyl ether *[109669-56-9]* **M 263.7, m 69°.** Crystd from EtOH.

4-Chlorophenyl 4-nitrobenzyl ether *[5442-44-4]* **M 263.7, m 102°.** Crystd from EtOH.

9-Chloro-9-phenylxanthene (Pixyl chloride) *[42506-03-6]* **M 292.8, m 105-106°.** Possible impurity is 9-hydroxy-9-phenylxanthene. If material contains a lot of the hydroxy product then boil 10g in $CHCl_3$ (50mL) with redistd acetyl chloride (1mL) until liberation of HCl is complete. Evapn leaves the chlorophenylxanthene as the hydrochloride which on heating with *benzene loses HCl; and on adding pet ether prisms of chlorophenylxanthene separate and contain 0.5mol of *benzene. The *benzene-free compound is obtained on drying and melts to a colourless liquid. [*Justus Liebigs Ann Chem* **370** 142 *1909*.] The 9-phenylxanthyl group is called pixyl. [*J Chem Soc, Chem Commun* 639 *1978*.]

Chlorophyll *a* *[479-61-8]* **M 983.5, m 117-120°, 150-153°, 178-180° (sinters at ~150°), $[\alpha]_D^{20}$ -262° (Me_2CO).** Forms green crystals from Me_2CO, Et_2O + H_2O, Et_2O + hexane + H_2O or Et_2O + pentane + H_2O. It is sparingly soluble in MeOH and insol in pet ether. In alkaline soln it gives a blue-green colour with deep red fluorescence. A very crude chlorophyll mixture has been purified by chromatography on low melting polyethylene (MI 0.044; 'Dow' melting index MI <2) and developed with 70% aq Me_2CO. The order of effluent from the bottom of the column is: xanthophylls, chlorophyll *b*, chlorophyll *a*, phaeophytins and carotenes. A mixture of chlorophylls *a* and *b* is best separated by chromatography on sugar and the order is chlorophyll *b* elutes first followed by chlorophyll *a*. To an Me_2CO-H_2O soln of chlorophylls 200mL of iso-octane is added and the mixt shaken in a separating funnel and the H_2O is carefully removed. The iso-octane layer is dried (Na_2SO_4) and applied to a glass column (5cm diameter) dry packed with 1000mL of powdered sucrose which has been washed with 250mL of iso-octane. Elution with 0.5% of isopropanol in iso-octane gives chlorophyll *a*. Keeping the eluate overnight at 0° yields micro crystals which are collected by filtration or centrifugation (Yield 40mg). UV_{EtOH} has λ_{max} **660**, 613, 577, 531, 498, **429** and 409 nm. [Anderson and Calvin *Nature* **194** 285 *1962*; Stoll and Weidemann *Helv Chim Acta* **16** 739 757 *1933*; NMR: Katz et al. *J Am Chem Soc* **90** 6841 *1968*, **85** 3809 *1963* for *a* and *b*; ORD: Inhoffen et al. *Justus Liebigs Ann Chem* **704** 208 *1967*; Willstätter and Isler *Justus Liebigs Ann Chem* **390** 269, 233 *1912*.]

Chlorophyll *b* *[519-62-0]* **M 907.52, sinters at 86-92°, sinters at 170°, dec at 160-170°, m 183-185°, 190-195°, $[\alpha]_D^{20}$ -267° (Me_2CO + MeOH), $[\alpha]_{720}^{25}$ -133° (MeOH + Pyridine 95:5).** See purification of chlorophyll *a*, and is separated from "*a*" by chromatography on sucrose [UV, IR: Stoll and Weidemann *Helv Chim Acta* **42** 679, 681 *1959*]. It forms red-black hexagonal bipyramids or four sided plates from dilute EtOH and has been recrystd from $CHCl_3$-MeOH. It is soluble in MeOH, EtOH, EtOAc and insoluble in pet ether. [*J Am Chem Soc* **88** 5037 *1966*.]

Chloropicrin (trichloronitromethane) *[76-06-2]* **M 164.5, b 112°.** Dried with $MgSO_4$ and fractionally distd. **EXTREMELY NEUROTOXIC, use appropriate precautions.**

***RS*-2-Chloropropionic acid** *[598-78-7]* **M 108.5, b 98°/3mm, d 1.182, n 1.453 pK^{25} 2.89.** Dried with P_2O_5 and fractionally distd under vacuum.

S-(-)-2-Chloropropionic acid *[29617-66-1]* **M 108.5, b 77°/10mm, 80.7°/10mm, 185-188°/atm, d_4^{25} 1.2485, n_D^{25} 1.436, $[\alpha]_D^{25}$ -14.6° (neat).** Purified by twice fractionating through a 115cm Podbielniak column (calcd 50 theoretical plates at atm pressure, see p. 141) using a take-off ratio of 1:5. Ths *acid chloride* is prepared by dissolving the acid in $SOCl_2$ adding a few drops of PCl_3, refluxing and then distilling through a 30 cm column, **b 53°/100mm, $[\alpha]_D^{25}$ - 4.6° (neat), d_4^{25} 1.2689, n_D^{25} 1.4368.** [Fu et al. *J Am Chem Soc* **76** 6954 *1954*].

3-Chloropropionic acid *[107-94-8]* **M 108.5, m 41°, pK^{25} 4.08.** Crystd from pet ether or *benzene.

3-Chloropropyl bromide (1-bromo-3-chloropropane) *[109-70-6]* **M 157.5, b 142-145°, n^{25} 1.4732.** Washed with conc H_2SO_4, water, 10% Na_2CO_3 soln, water again and then dried with $CaCl_2$ and fractionally distd just before use [Akagi, Oae and Murakami *J Am Chem Soc* **78** 4034 *1956*].

6-Chloropurine *[87-42-3]* **M 154.6, m 179°(dec), pK_1^{20} 0.45, pK_2^{20} 7.88.** Crystd from water.

2-Chloropyrazine *[14508-49-7]* **M 114.5, b 62-63°/31mm, 153-154°/atm, d_4^{20} 1.302, n_D^{26} 1.535, pK_{Est} <0.** Fractionally distil through a short column packed with glass helices. It has a penetrating mildly pungent odour with a high vapour pressure at room temperature. [Erickson and Spoerri *J Am Chem Soc* **68** 400 *1946*; *J Org Chem* **28** 1682 *1963*.]

2-Chloropyridine *[109-09-1]* **M 113.6, b 49.0°/7mm, d 1.20, n 1.532, pK^{20} 0.49 (0.72).** Dried with NaOH for several days, then distd from CaO under reduced pressure.

3-Chloropyridine *[626-60-8]* **M 113.6, b 148°, d 1.194, n 1.5304, pK^{25} 2.84.** Distd from KOH pellets.

4-Chloropyridine *[626-61-9]* **M 113.6, b 85-86°/100mm, 147-148°/760mm, pK^{20} 3.84.** Dissolved in distilled water and excess of 6M NaOH was added to give pH 12. The organic phase was separated and extracted with four volumes of diethyl ether. The combined extracts were filtered through paper to remove water and the solvent evaporated. The dark brown residual liquid was kept under high vacuum [Vaidya and Mathias *J Am Chem Soc* **108** 5514 *1986*]. It can be distd but readily darkens and is best kept as the *hydrochloride* *[7379-35-3]* **M 150.1, m 163-165°(dec).**

2-Chloropyrimidine *[1722-12-9]* **M 114.5, m 63-65°, 66°, b 91°/26mm, pK^{20} -1.90.** It has been recrystd from *C_6H_6, pet ether or a mixture of both. It sublimes at 50°/18mm and can be distd in a vacuum. [IR: Short and Thompson *J Chem Soc* 168 *1952*; Boarland and McOmie *J Chem Soc* 1218 *1951*.]

2-Chloroquinoline *[612-62-4]* **M 163.6, m 34°, b 147-148°/15mm, d^{35} 1.235, n^{25} 1.629, pK_{Est} ~0.3.** Purified by crystn of its picrate to constant melting point (123-124°) from *benzene, regenerating the base and distilling under vacuum [Cumper, Redford and Vogel *J Chem Soc* 1183 *1962*]. 2-Chloroquinoline can be crystd from EtOH. Its *picrate* has **m 122°** (from EtOH).

4-Chloroquinoline *[611-35-8]* **M 163.6, m 29-32°, 31°, b 130°/15mm, 261°/744mm, pK 3.72.** Possible impurities include the 2-isomer. Best purified by converting to the *picrate* (**m 212-213° dec**) in EtOH and recryst from EtOH (where the picrate of the 2-chloroquinoline stays in soln) or EtOAc . The picrate is decomposed with 5% aqueous NaOH, extracted in $CHCl_3$, washed with H_2O, dried ($MgSO_4$), evapd and distd in a vacuum. It can be steam distd from slightly alkaline aqueous solns, the aqueous distillate is extracted with Et_2O, evaporated and distd. The distillate solidifies on cooling. [Bobránski *Chem Ber* **71** 578 *1938*.]

8-Chloroquinoline *[611-33-6]* **M 163.6, b 171-171.5°/24mm, d 1.278, n 1.644, pK^{25} 3.12.** Purified by crystn of its $ZnCl_2$ complex (**m 228°**) from aqueous EtOH.

4-Chlororesorcinol *[95-88-5]* **M 144.6, m 105°, $pK_{Est(1)}$~9.2, $pK_{Est(2)}$ ~10.1.** Crystd from boiling CCl_4 (10g/L, charcoal) and air dried.

5-Chlorosalicaldehyde *[635-93-8]* **M 156.6, m 98.5-99°.** Steam distd, then crystd from aq EtOH.

N-Chlorosuccinimide *[128-09-6]* **M 133.5, m 149-150°.** Rapidly crystd from *benzene, or glacial acetic acid and washed well with water then dried *in vacuo.* [Phillips and Cohen *J Am Chem Soc* **108** 2023 *1986.*]

2-Chlorothiophene (2-thienyl chloride) *[96-43-5]* **M 118.6, b 126-128°, d 1.285, n 1.551.** Purified by fractional distn at atmospheric pressure or by gas chromatography.

8-Chlorotheophylline *[85-18-7]* **M 214.6, m 311°(dec), pK$_{Est(1)}$~5.4, pK$_{Est(2)}$ ~9.1.** Crystd from H$_2$O.

4-Chlorothiophenol *[106-54-7]* **M 144.6, m 51-52°, pK^{25} 6.14.** Recrystd from aqueous EtOH [D'Sousa et al. *J Org Chem* **52** 1720 *1987*].

2-Chlorotoluene *[95-49-8]* **M 126.6, b 159°, d 1.083, n 1.5255.** Dried for several days with CaCl$_2$, then distd from Na using a glass helices-packed column.

3-Chlorotoluene *[108-41-8]* **M 126.6, m -48°, b 161-163°, d 1.072, n 1.522.** Purified as for 2-chlorotoluene above.

4-Chlorotoluene *[106-43-4]* **M 126.6, f 7.2°, b 162.4°, d 1.07, n 1.521.** Dried with BaO, fractionally distd, then fractionally crystd by partial freezing.

2-Chlorotriethylamine hydrochloride *[869-24-9]* **M 172.1, m 208-210°, pK$_{Est}$ ~8.6 (free base).** Crystd from absolute MeOH (to remove highly coloured impurities).

Chlorotrifluoroethylene *[79-38-9]* **M 116.5, b -26 to -24°.** Scrubbed with 10% KOH soln, then 10% H$_2$SO$_4$ soln to remove inhibitors, and dried. Passed through silica gel.

Chlorotrifluoromethane *[75-72-9]* **M 104.5, m -180°, b -81.5°.** Main impurities were CO$_2$, O$_2$, and N$_2$. The CO$_2$ was removed by passage through saturated aqueous KOH, followed by conc H$_2$SO$_4$. The O$_2$ was removed using a tower packed with activated copper on Kieselguhr at 200°, and the gas dried over P$_2$O$_5$.

Chlorotriphenylmethane see **triphenylmethyl chloride (trityl chloride).**

5-Chlorouracil (5-chloro-2,4(6)-dihydroxypyrimidine) *[1820-81-1]* **M 146.5, m 314-418° dec, 324-325° dec, pK$_1^{25}$ 7.95, pK$_2^{25}$ >13.** Recrystallised from hot H$_2$O (4g/500mL) using charcoal. [McOmie et al. *J Chem Soc* 3478 *1955;* West and Barrett *J Am Chem Soc* **76** 3146 *1954.*]

5β-Cholanic acid *[25312-65-6]* **M 360.6, m 164-165°, [α]$_D^{14}$ +21.7°** (CHCl$_3$), **pK$_{Est}$ ~4.9.** Crystd from EtOH. The *Ethyl ester* has **m** 93-94° (from 80% EtOH), **b** 273°/12mm, [α]$_D^{20}$ +21° (CHCl$_3$).

Cholanthrene (1,2,-dihydrobenz[j]aceanthrylene) *[479-23-2]* **M 254.3, m 173°.** Crystd from *benzene/diethyl ether.

5α-Cholestane *[481-21-0]* **M 372.7, m 80°, [α]$_{546}^{20}$ +29.5° (c 2, CHCl$_3$).** Crystd from diethyl ether/EtOH.

5α-Cholestan-3β-ol *[80-97-7]* **M 388.7, m 142-143°(monohydrate), [α]$_{546}^{20}$ +28° (c 1, CHCl$_3$), [α]$_D$ +27.4° (in CHCl$_3$).** Crystd from EtOH or slightly aqueous EtOH, or MeOH. [Mizutani and Whitten *J Am Chem Soc* **107** 3621 *1985.*]

Cholest-2-ene *[15910-23-3]* **M 370.6, m 75-76°, [α]$_D^{24}$ +64°.** Recrystd from MeOH or diethyl ether/acetone. [Berzbrester and Chandran *J Am Chem Soc* **109** 174 *1987.*]

Cholesterol *[57-88-5]* **M 386.7, m 148.9-149.4°, [α]$_D^{25}$ -35° (hexane).** Crystd from ethyl acetate, EtOH or isopropyl ether/MeOH. [Hiromitsu and Kevan *J Am Chem Soc* **109** 4501 *1987*.] For extensive details of purification through the dibromide, see Fieser [*J Am Chem Soc* **75** 5421 *1953*] and Schwenk and Werthessen [*Arch Biochem Biophys* **40** 334 *1952*], and by repeated crystn from acetic acid, see Fieser [*J Am Chem Soc* **75** 4395 *1953*].

Cholesteryl acetate *[604-35-3]* **M 428.7, m 112-115°, [α]$_{546}^{20}$ -51° (c 5, CHCl$_3$).** Crystd from *n*-pentanol.

Cholesteryl myristate *[1989-52-2]* **M 597.0.** Crystd from *n*-pentanol. Purified by column chromatography with MeOH and evaporated to dryness. Recrystd and finally, dried in vacuum over P$_2$O$_5$. [Malanik and Malat *Anal Chim Acta* **76** 464 *1975*].

Cholesteryl oleate *[303-43-5]* **M 651.1, m 48.8-49.4°.** Purified by chromatography on silica gel.

Cholic acid *[81-25-4]* **M 408.6, m 198-200°, [α]$_{546}$ +41° (c 0.6, EtOH), pK^{20} 4.98.** Crystd from EtOH. Dried under vacuum at 94°.

Choline chloride *[67-48-1]* **M 139.6, m 302-305°(dec).** *Extremely deliquescent.* Purity checked by AgNO$_3$ titration or by titration of free base after passage through an anion-exchange column. Crystd from absolute EtOH, or EtOH-diethyl ether, dried under vacuum and stored in a vacuum desiccator over P$_2$O$_5$ or Mg(ClO$_4$)$_2$.

4-Chromanone *[491-37-2]* **M 148.2, m 35-37°, 39°, 41°, b 92-93°/3mm, 130-132°/15mm, 160°/50mm.** It has been recryst from pet ether, or purified by dissolving in *C$_6$H$_6$ washing with H$_2$O, drying (MgSO$_4$), evaporate and dist in a vacuum, then recryst the residue. The liquid has a pleasant lemon-like odour. The *semicarbazone* has **m** 227°. [Loudon and Razdan *J Chem Soc* 4299 *1954*.] The *oxime* is prepared from 3g of chromanone, 3g NH$_2$OH.HCl in EtOH (50mL), 6g K$_2$CO$_3$ and refluxed on a water bath for 6h. The soln is poured into H$_2$O, the solid is filtered off, dried and dissolved in hot *C$_6$H$_6$ which on addition of pet ether yields the oxime as glisteneing needles **m** 140°. Decomposition of this gives very pure chromanone. The *benzal derivative* is prepared from 3g of chromanone, 4g PhCHO in 50mL EtOH, heated to boiling, 10mL of conc HCl are added dropwise and set aside for several days. The derivative separates and is recrystd from EtOH to give yellow needles, **m** 112° [*J Am Chem Soc* **45** 2711 *1923*]. Reaction with Pb(OAc)$_4$ yields the *3-acetoxy derivative* **m** 74° (from pet ether + trace of EtOAc) [Cavill et al. *J Chem Soc* 4573 *1954*].

Chrysene *[218-01-9]* **M 228.3, m 255-256°.** Purified by chromatography on alumina from pet ether in a darkened room. Its soln in *C$_6$H$_6$ was passed through a column of decolorising charcoal, then crystd by concentration of the eluate. Also purified by crystn from *C$_6$H$_6$ or *C$_6$H$_6$-pet ether , and by zone refining. [Gorman et al. *J Am Chem Soc* **107** 4404 *1985*]. It was freed from 5*H*-benzo[*b*]carbazole by dissolving in *N,N*-dimethylformamide and successively adding small portions of alkali and iodomethane until the fluorescent colour of the carbazole anion no longer appeared when alkali was added. The chrysene (and alkylated 5*H*-benzo[*b*]carbazole) separated on addition of water. Final purification was by crystn from ethylcyclohexane and from 2-methoxyethanol [Bender, Sawicki and Wilson *Anal Chem* **36** 1011 *1964*]. It can be sublimed in a vacuum.

Chrysoidine G (4-phenylazo-1,3-benzenediamine monohydrochloride) *[532-82-1]* **M 248.7, m 118-118.5°, pK$_1$ 3.32, pK$_2$ 5.21.** Red-brown powder which is recrystd from H$_2$O. It gives a yellow soln in conc H$_2$SO$_4$ which turns orange on dilution. Its solubility at 15° is 5.5% (H$_2$O), 4.75% (EtOH), 6.0% (cellosolve), 9.5% (ethylene glycol), 0.005% (xylene) and insol in *C$_6$H$_6$. The *hydroiodide* has **m** 184° (from EtOH) and the *picrate* forms red needles **m** 196°. [*Bull Chem Soc Jpn* **31** 864 *1958*; *Chem Ber* **10** 213 *1877*.]

1,8-Cineole (1,8-epoxy-*p*-menthane) *[470-82-6]* **M 154.2, f 1.3°, b 176.0°, d 0.9251.** See eucaliptol on p. 242.

trans-Cinnamaldehyde   *[14271-10-9]* **M** 132.2, **m** -4°, -7.5°, -9°, **b** 80°/0.4mm, 85.8°/1.1mm, 125-128°/11mm, 152.2°/40mm, 163.7°/60mm, 199.3°/200mm, 246°/760mm dec, d_4^{20} 1.0510, n_D^{20} 1.623. Purified by steam distn (sol 1 in 700 parts H_2O) followed by distn *in vacuo*. The *cis*-isomer has **b** 67-69°/40mm and d_4^{20} 1.0436 and n_D^{20} 1.5937. The *trans-semicarbazone* has **m** 210° dec from $CHCl_3$-MeOH (*cis-semicarbazone* has **m** 196°); the *trans-phenylsemicarbazone* has **m** 177° from $CHCl_3$-MeOH (the *cis-phenylsemicarbazone* has **m** 146°); the *trans-2,4-dinitrophenylhydrazone* has **m** 250° dec from MeOH as the *cis*-isomer [Gamboni et al. *Helv Chim Acta* **38** 255 *1955*; Peine *Chem Ber* **17** 2117 *1884*; *J Org Chem* **26** 4814 *1961*; *J Am Chem Soc* **86** 198 *1964*].

cis-Cinnamic acid (*Z*-3-phenyl-2-propenoic acid) *[102-94-3]* **M** 148.2, **m** 68° (for *allo*-form), pK^{25} 3.93. The *cis*-acid is prepared by catalytic reduction of phenylpropiolic acid and after distn in high vacuum at ~95° gives the most stable *allo*-isomer **m** 68°. Recrystn from pet ether yields **Liebermann's iso-cinnamic acid m 58°**. When the *allo*-acid (**m** 68°) is heated at 20° above its melting point in a sealed capillary for 0.5h and allowed to cool slowly **Erlenmyer's iso-cinnamic acid m 42°** is formed. This form can also be obtained in larger amounts by heating the *allo*-acid at 80° for 3h and on cooling it remains liquid for several weeks but gives the 42° acid on innoculation with the crystals from the capillary tube. This form is unchanged in 6 weeks when kept in a dark cupboard. All three forms have the same pK values and the same rate of bromination. There is also a very labile form with **m** 32°. [Liebermann, *Chem Ber* **26** 1572 *1893*; Claisen and Crismer *Justus Liebigs Ann Chem* **218** 135 *1883*; Robinson and James *J Chem Soc* 1453 *1933*; Berthoud and Urech *Helv Chim Acta* **13** 437 *1930*; McCoy and McCoy *J Org Chem* **33** 2354 *1968*.]

trans-Cinnamic (*E*-3-phenyl-2-propenoic) acid *[140-10-3; 621-82-9 for E-Z mixture]* **M** 148.2, **m** 134.5-135°, pK^{25} 4.42 (4.50). Crystd from *benzene, CCl_4, hot water, water/EtOH (3:1), or 20% aqueous EtOH. Dried at 60° *in vacuo*. Steam volatile.

Cinnamic anhydride *[538-56-7]* **M** 278.4, **m** 136°. Crystd from *C_6H_6 or toluene/pet ether (b 60-80°).

N-Cinnamoyl-*N*-phenylhydroxylamine   *[7369-44-0]* **M** 239.3, **m** 158-163°. Recrystd from EtOH.

Cinnamyl   alcohol   *[104-54-1]* **M** 134.2, **m** 33°, **b** 143.5°/14mm, λ_{max} 251nm (ϵ 18,180M^{-1} cm^{-1}). Crystd from diethyl ether/pentane.

Cinnoline *[253-66-7]* **M** 130.2, **m** 38°, pK^{20} 2.37. Crystd from pet ether. Kept under N_2 in sealed tubes in the dark at 0°.

Citraconic acid *[498-23-7]* **M** 130.1, **m** 91°, pK_1^{25} 2.2, pK_2^{25} 5.60 (cis). Steam distd and crystd from EtOH/ligroin.

Citraconic anhydride *[616-02-4]* **M** 112.1, **m** 8-9°, **b** 47°/0.03mm, 213°/760mm, d_4^{20} 1.245, n_D^{20} 1.472. Possible contamination is from the acid formed by hydrolysis. If the IR has OH bands then reflux with Ac_2O for 30 min, evaporate then distil the residue in a vacuum; otherwise distil in a vacuum. Store in a dry atmosphere. [*Biochem J* **191** 269 *1980*.]

Citrazinic acid (2,6-dihydroxyisonicotinic acid) *[99-11-6]* **M** 155.1, **m** >300°, pK_1 3.0, pK_2 4.76. Yellow powder with a greenish shade, but is white when ultra pure and turns blue on long standing. It is insoluble in H_2O but slightly soluble in hot HCl and soluble in alkali or carbonate solutions. Purified by precipitation from alkaline solutions with dilute HCl, and dry in a vacuum over P_2O_5. The *ethyl ester* has **m** 232° (evacuated tube) and a pKa of 4.81 in $MeOCH_2CH_2OH$ [IR: Pitha *Coll Czech Chem Comm* **28** 1408 *1963*].

Citric acid (H_2O) *[5949-29-1; 77-92-9 (anhydr)]* **M** 210.1, **m** 156-157°, 153° (anhyd), pK_1^{25} 2.96, pK_2^{25} 4.38, pK_3^{25} 5.68. Crystd from water.

Citronellal **(3,7-dimethyloctan-6-al)** *[R(+): 2385-77-5; S(-): 5949-05-3]* **M 154.3, b 67°/4mm, 89°/14mm, 104-105°/21mm, 207°/760mm,** $[\alpha]_{546}^{20}$ **(+) and (-) 20°,** $[\alpha]_D^{20}$ **(+) and (-) 16.5° (neat).** Fractionally distd. Alternatively extracted with $NaHSO_3$ solution, washed with Et_2O then acidified to decompose the bisulfite adduct and extracted with Et_2O, dried (Na_2SO_4), evaporated and distd. Check for purity by hydroxylamine titration. The ORD in MeOH (c 0.167) is: $[\alpha]_{700}$ +9°, $[\alpha]_{589}$ +11°, $[\alpha]_{275}$ +12° and $[\alpha]_{260}$ 12°. The *semicarbazone* has **m** 85°, and the *2,4-dinitrophenylhydrazone* has **m** 79-80°. [IR: *J Chem Soc* 3457 *1950*; ORD: Djerassi and Krakower *J Am Chem Soc* **81** 237 *1959*.]

β−Citronellene **(2,6-dimethylocta-2,7-diene)** *[S-(+): 2436-90-0; R-(-): 10281-56-8]* **M 138.3, b 153-154°/730mm, 155°/atm,** d_4^2 **0.757,** n_D^{22} **1.431,** $[\alpha]_{546}^{22}$ **(+) and (-) 13°,** $[\alpha]_{546}^{20}$ **(+) and (-) 10° (neat).** Purified by distillation over Na three times and fractionation. [(-) Arigoni and Jeager *Helv Chim Acta* **37** 881 *12954*; (+) Eschenmoser and Schinz *Helv Chim Acta* **33** 171 *1950*.]

β−Citronellol **(3,7-dimethyloctan-6-ol)** *[R-(+): 11171-61-9; S-(-): 106-22-9]* **M 156.3, b 47°/1mm, 102-104(110°)°/10mm, 112-113°/12mm, 221-224°/atm, 225-226°/atm,** d_4^{24} **0.8551,** n_D^{24} **1.4562,** $[\alpha]_{546}^{20}$ **(+) and (-) 6.3°,** $[\alpha]_D^{20}$ **(+) and (-) 5.4° (neat).** Purified by distn through a cannon packed (Ni) column and the main cut collected at 84°/14mm and redistd. Also purified *via* the benzoate. [IR: Eschenazi *J Org Chem* **26** 3072 *1961*; *Bull Soc Chim Fr* 505 *1951*.]

S-**Citrulline** **(2-amino-5-ureidopentanoic acid)** *[372-75-8]* **M 175.2, m 222°,** $[\alpha]_D^{20}$ **+24.2° (in 5M HCl), pK^{25} 9.71.** Likely impurities are arginine, and ornithine. Crystd from water by adding 5 volumes of EtOH. Also crystd from water by addn of MeOH.

Clofazimine **[2-(4-chloroanilino)-3-isopropylimino-5-(4-chlorophenyl)-3,4-dihydrophen-azine]** *[2030-63-9]* **M 473.5, m 210-212°.** Recrystd from acetone.

Coenzyme Q$_0$ **(2,3-Dimethoxy-5-methyl-1,4-benzoquinone, 3,4-dimethoxy-2,5-tolu-quinone, fumigatin methyl ether), colchicine** and **colchicoside** see entries in Chapter 6.

Conessine *[546-06-5]* **M 356.6, m 125°, 127-128.5°,** $[\alpha]_D^{20}$ **-1.9° (in CHCl$_3$) and +25.3° (in EtOH), pK$_{Est(1)}$~10.4 , pK$_{Est(2)}$~10.7,.** Crystd from acetone. The *dihydrochloride* has **m** >340° and $[\alpha]_D^{20}$ +9.3° (c 0.9, H_2O).

Coniferyl alcohol **[4-hydroxy-3-methoxy-cinnamyl alcohol, 3-(4-hydroxy-3-methoxy-phenyl)-2-propen-1-ol]** *[458-35-5]* **M 180.2, m 73-75°, b 163-165°/3mm, pK^{25} 9.54.** It is soluble in EtOH and insoluble in H_2O. It can be recrystd from EtOH and distd in a vacuum. It polymerises in dilute acid. The *benzoyl derivative* has **m** 95-96° (from pet ether), and the *tosylate* has **m** 66°. [Derivatives: Freudenberg and Achtzehn *Chem Ber* **88** 10 *1955*; UV: Herzog and Hillmer *Chem Ber* **64** 1288 *1931*.]

Congo Red *[573-58-0]* **M 696.7,** λ_{max} **497nm, pK$_2^{28}$ 4.19.** Crystd from aq EtOH (1:3). Dried in air.

(-)-α-Copaene **{1R,2S,6S,7S,8S-8-isopropyl-1,3-dimethyltricyclo[4.4.0.0^{2,7}]dec-3-ene}** *[3856-25-5]* **M 204.4, b 119-120°/10mm, 246-251°, d 0.908, n 1.489,** $[\alpha]_D^{20}$ **-6.3° (c 1.2, CHCl$_3$).** Purified by distn, preferably under vacuum.

4,5-Coprosten-3-ol **(cholest-4-ene-3ß-ol)** *[517-10-2]* **M 386.7, m 132°.** Crystd from MeOH/diethyl ether.

Coprosterol (5α-cholestan-3ß-ol, dihydrocholesterol) *[80-97-7]* **M 388.7, m 101°, 139-140°,** $[\alpha]_D^{20}$ **+24° (c 1, CHCl$_3$).** See entry on p. 169.

Coronene *[191-07-1]* **M 300.4, m 438-440°,** λ_{max} **345nm (log ε 4.07).** Crystd from *benzene or toluene, then sublimed in vacuum.

Cortisol, corticosterone, cortisone and **cortisone-21-acetate** see entries in Chapter 6.

Coumalic acid (2-pyrone-5-carboxylic acid) *[500-05-0]* **M 140.1, m 205-210°(dec), pK_{Est} ~0.** Crystd from MeOH. *Methyl ester* , from pet ether, has **m 74-74°** and **b 178-180°/60 mm.**

Coumarin *[91-64-5]* **M 146.2, m 68-69°, b 298°, pK -4.97 (aq H_2SO_4)** . Crystd from ethanol or water and sublimed *in vacuo* at 43° [Srinivasan and deLevie *J Phys Chem* **91** 2904 *1987*].

Coumarin-3-carboxylic acid *[531-81-7]* **M 190.2, m 188°(dec), pK_{Est} ~1.5.** Crystd from water.

Creatine (H_2O) and **creatinine** see entries in Chapter 6.

o-**Cresol** *[95-48-7]* **M 108.1, m 30.9°, b 191°, n^{41} 1.536, n^{46} 1.534, pK^{25} 10.22.** Can be freed from *m*- and *p*-isomers by repeated fractional distn, Crystd from *benzene by addition of pet ether. Fractional crystd by partial freezing of its melt.

m-**Cresol** *[108-39-4]* **M 108.1, f 12.0°, b 202.7°, d 1.034, n 1.544, pK^{25} 10.09.** Separation of the *m*- and *p*-cresols requires chemical methods, such as conversion to their sulfonates [Brüchner *Anal Chem* **75** 289 *1928*]. An equal volume of H_2SO_4 is added to *m*-cresol, stirred with a glass rod until soln is complete. Heat for 3h at 103-105°. Dilute carefully with 1-1.5 vols of water, heat to boiling point and steam distil until all unsulfonated cresol has been removed. Cool and extract residue with ether. Evaporate the soln until the boiling point reaches 134° and steam distil off the *m*-cresol. Another purification involves distn, fractional crystn from the melt, then redistn. Freed from *p*-cresol by soln in glacial acetic acid and bromination by about half of an equivalent amount of bromine in glacial acetic acid. The acetic acid was distd off, then fractional distn of the residue under vac gave bromocresols from which 4-bromo-*m*-cresol was obtained by crystn from hexane. Addn of the bromocresol in glacial acetic acid slowly to a reaction mixture of HI and red phosphorus or (more smoothly) of HI and hypophosphorus acid, in glacial acetic acid, at reflux, removed the bromine. After an hour, the soln was distd at atmospheric pressure until layers were formed. Then it was cooled and diluted with water. The cresol was extracted with ether, washed with water, $NaHCO_3$ soln and again with water, dried with a little $CaCl_2$ and distd [Baltzly, Ide and Phillips *J Am Chem Soc* **77** 2522 *1955*].

p-**Cresol** *[106-44-5]* **M 108.1, m 34.8°, b 201.9°, n^{41} 1.531, n^{46} 1.529, pK^{25} 10.27.** Can be separated from *m*-cresol by fractional crystn of its melt. Purified by distn, by pptn from *benzene soln with pet ether, and *via* its benzoate, as for phenol. Dried under vacuum over P_2O_5. Has also been crystd from pet ether (b 40-60°) and by conversion to sodium *p*-cresoxyacetate which, after crystn from water was decomposed by heating with HCl in an autoclave [Savard *Ann Chim (Paris)* **11** 287 *1929*].

o-**Cresolphthalein complexon** *[2411-89-4]* **M 636.6, m 186°(dec), λmax 575nm, pK_1 2.2, pK_2 2.9, pK_3 7.0, pK_4 7.8, pK_5 11.4, pK_6 12.0.** *o*-Cresolphthalein (a complexone precursor without the two bis-carboxymethylamino groups) is a contaminant and is one of the starting materials. It can be removed by dissolving the reagent in water and adding a 3-fold excess of sodium acetate and fractionally precipitating it by dropwise addition of HCl to the clear filtrate. Wash the ppte with cold H_2O and dry the monohydrate at 30° in a vacuum. The pure material gives a single spot on paper chromatography (eluting solvent EtOH/water/phenol, 6:3:1; and developing with NaOH). [Anderegg et al. *Helv Chim Acta* **37** 113 *1954*.] Complexes with Ba, Ca, Cd, Mg and Sr.

o-**Cresol Red** *[1733-12-6]* **M 382.4, m 290°(dec), pK^{25} 1.26.** Crystd from glacial acetic acid. Air dried. Dissolved in aqueous 5% $NaHCO_3$ soln and ppted from hot soln by dropwise addition of aqueous HCl. Repeated until extinction coefficients did not increase.

o-**Cresotic acid (methylsalicylic acid)** *[83-40-9]* **M 152.2, m 163-164°, pK_1^{25} 3.32.** Crystd from water.

m-**Cresotic acid** *[50-85-1]* **M 152.2, m 177°, pK_1^{25} 3.15, pK_2^{25} 13.35.** Crystd from water.

p-**Cresotic acid** *[89-56-5]* **M 152.2, m 151°, pK_1^{25} 3.40, pK_2^{25} 13.45.** Crystd from water.

Crocetin diethyl ester *[5056-14-4]* **M 384.5, m 218-219°, 222.5°, $A_{1cm}^{1\%}$ (λmax) 2340 (400nm), 3820 (422nm), 3850 (450nm) in pet ether.** Purified by chromatography on a column of silica gel G. Crystd from *benzene. Stored in the dark, under an inert atmosphere, at 0°.

Crotonaldehyde (2-butenal) *[123-73-9]* **M 70.1, b 104-105°, d 0.851, n 1.437.** Fractionally distd under N_2, through a short Vigreux column. Stored in sealed ampoules. Stabilised with 0.01% of 2,6-di-tert-butyl-*p*-cresol

trans-**Crotonic acid** *[107-93-7]* **M 86.1, m 72-72.5°, pK_1^{25}-6.17 (aq H_2SO_4), pK_2^{18} 4.71.** Distd under reduced pressure. Crystd from pet ether (b 60-80°) or water, or by partial freezing of the melt.

E- **and** *Z*-**Crotonitrile (mixture)** *[4786-20-3]* **M 67.1, b 120-121°, d 1.091, n 1.4595.** Separated by preparative GLC on a column using 5% FFAP on Chromosorb G. [Lewis et al. *J Am Chem Soc* **108** 2818 *1986*.]

γ-**Crotonolactone [2(5H)-furanone]** *[497-23-4]* **M 84.1, m 3-4°, 76-77°/3.5mm, 90.5-91°/11.5mm, 92-93°/14mm, 107-109°/24mm, 212-214°/760mm, d_4^{20} 1.197, n_D^{20} 1.470.** Fractionally distd under reduced pressure. IR: (CCl_4) 1784 and 1742 cm^{-1}, UV no max above 205nm (ε 1160 $cm^{-1} M^{-1}$) and ^1H NMR: (CCl_3) τ: 2.15 (pair of triplets 1H), 3.85 (pair of triplets 1H) and 5.03 (triplet 2H). [*Org Synth* Coll Vol V 255 *1973*; Smith and Jones *Can J Chem* **37** 2007, 2092 *1959*].

Crotyl bromide *[29576-14-5]* **M 135.0, b 103-105°/740mm, n^{25} 1.4792.** Dried with $MgSO_4$, $CaCO_3$ mixture. Fractionally distd through an all-glass Todd column. [**Todd column.** A column (which may be a Dufton type, fitted with a Monel metal rod and spiral, or a Hempel type, fitted with glass helices) which is surrounded by an open heating jacket so that the temperature can be adjusted to be close to the distillation temperature (Todd *Ind Eng Chem (Anal Ed)* **17** 175 *1945*)].

15-Crown-5 *[33100-27-5]* **M 220.3, b 93-96°/0.1mm, d 1.113, n 1.465.** Dried over 3A molecular sieves.

18-Crown-6 *[17455-13-9]* **M 264.3, m 37-39°.** Recrystd from acetonitrile and vacuum dried. Purified by pptn of 18-crown-6/nitromethane 1:2 complex with Et_2O/nitromethane (10:1 mixture). The complex is decomposed in vacuum and distilled under reduced pressure. Also recrystd from acetonitrile and vacuum dried.

Cryptopine *[482-74-6]* **M 369.4, m 220-221°.** Crystd from *benzene.

Cryptoxanthin *[472-70-8]* **M 552.9, $A_{1cm}^{1\%}$ 2370 (452nm), 2080 (480nm) in pet ether.** Purified by chromatography on MgO, $CaCO_3$ or deactivated alumina, using EtOH or diethyl ether to develop the column. Crystd from $CHCl_3$/EtOH. Stored in the dark, under inert atmosphere, at -20°.

Crystal Violet Chloride {Gentian violet, *N*-4[bis[4-(dimethylaminophenyl)methylene]-2,5-cyclohexadien-1-ylidene]-*N*-methylmethaninium chloride} *[548-62-9]* **M 408.0, pK 9.36.** Crystd from water (20mL/g), the crystals being separated from the chilled soln by centrifugation, then washed with chilled EtOH (sol 1g in 10 mL of hot EtOH) and diethyl ether and dried under vac. It is sol in $CHCl_3$ but insol in Et_2O. The carbinol was ppted from an aqueous soln of the HCl dye, using excess NaOH, then dissolved in HCl and recrystd from water as the chloride [UV and kinetics: Turgeon and La Mer *J Am Chem Soc* **74** 5988 *1952*]. The *carbinol base* has **m** 195° (needles from EtOH). The *diphthalate* (blue and turns red in H_2O) crystallises from H_2O, **m** 153-154° (dec 185-187°)[Chamberlain and Dull *J Am Chem Soc* **50** 3089 *1928*].

Cumene (isopropyl benzene) *[98-82-8]* **M 120.2, b 69-70°/41mm, 152.4°/760mm, d 0.864, n 1.49146, n^{25} 1.48892.** Usual purification is by washing with several small portions of conc H_2SO_4 (until the acid layer is no longer coloured), then with water, 10% aq Na_2CO_3, again with water, and drying with $MgSO_4$, $MgCO_3$ or Na_2SO_4, followed by fractional distn. It can then be dried with, and distd from, Na, NaH or CaH_2. Passage through columns of alumina or silica gel removes oxidation products. Has also been steam distd from 3% NaOH, and azeotropically distd with 2-ethoxyethanol (which was subsequently removed by washing out with water).

Cumene hydroperoxide *[80-15-9]* **M 152.2, b 60°/0.2mm, d 1.028, n^{24} 1.5232.** Purified by adding 100mL of 70% material slowly and with agitation to 300mL of 25% NaOH in water, keeping the temperature below 30°. The resulting crystals of the sodium salt were filtered off, washed twice with 25 mL portions of *benzene, then stirred with 100mL of *benzene for 20min. After filtering off the crystals and repeating the washing, they were suspended in 100mL of distilled water and the pH was adjusted to 7.5 by addn of 4M HCl. The free hydroperoxide was extracted into two 20mL portions of *n*-hexane, and the solvent was evaporated under vacuum at room temperature, the last traces being removed at 40-50° and 1mm [Fordham and Williams *Canad J Res* **27B** 943 *1949*]. Petroleum ether, **but not diethyl ether**, can be used instead of *benzene, and powdered solid CO_2 can replace the 4M HCl. *The material is potentially* **EXPLOSIVE.**

Cuminaldehyde (4-isopropylbenzaldehyde) *[122-03-2]* **M 148.2, b 82-84°/3.5mm, 120°/23mm, 131-135°/35mm, 235-236°/760mm, d^{20} 0.978, n$_D^{20}$ 1.5301.** Likely impurity is the benzoic acid. Check the IR for the presence of OH from CO_2H and the CO frequencies. If acid is present then dissolve in Et_2O, wash with 10% $NaHCO_3$ until effervescence ceases, then with brine, dry over $CaCl_2$, evap and distil the residual oil, preferably under vacuum. It is almost insoluble in H_2O, but soluble in EtOH and Et_2O. The *thiosemicarbazone* has **m** 147° after recrystn from aqueous EtOH, or MeOH or *C_6H_6. [Crounse *J Am Chem Soc* **71** 1263 *1949*; Bernstein et al. *J Am Chem Soc* **73** *906* 1951; Gensler and Berman *J Am Chem Soc* **80** 4949 *1958*.]

Cuprein (6'-hydroxycinchonidine) *[524-63-0]* **M 310.4, m 202°, [α]$_D^7$ -176° (in MeOH), pK^{15} 7.63.** Crystd from EtOH.

Curcumin [bis-(4-hydroxy-3-methoxycinnamoyl)methane] *[458-37-7]* **M 368.4, m 183°.** Crystd from EtOH or acetic acid.

Cyanamide *[420-04-2]* **M 42.0, m 41°, pK$_1^{25}$-0.36, pK$_2^{25}$10.27.** See cyanamide on p. 416 in Chapter 5.

Cyanoacetamide *[107-91-5]* **M 84.1, m 119.4°.** Crystd from MeOH/dioxane (6:4), then water. Dried over P_2O_5 under vacuum.

Cyanoacetic acid *[372-09-8]* **M 85.1, m 70.9-71.1°, pK^{25} 2.47.** Crystd to constant melting point from *benzene/acetone (2:3), and dried over silica gel.

Cyanoacetic acid hydrazide *[140-87-4]* **M 99.1, m 114.5-115°.** Crystd from EtOH.

p-**Cyanoaniline** *[873-74-5]* **M 118.1, m 85-87°.** See *p*-aminobenzonitrile on p. 104.

9-Cyanoanthracene (anthracene-9-carbonitrile) *[1210-12-4]* **M 203.2, m 134-137°.** Purified by crystn from EtOH or toluene, and vacuum sublimed in the dark and in an inert atmosphere [Ebied et al. *J Chem Soc, Faraday Trans 1* **76** 2170 *1980*; Kikuchi et al. *J Phys Chem* **91** 574 *1987*].

9-Cyanoanthracene photodimer *[33998-38-8]* **M 406.4, dec to monomer above ~147°.** Purified by dissolving in the minimum amount of $CHCl_3$ followed by addition of EtOH at 5° [Ebied et al. *J Chem Soc, Faraday Trans 1* **75** 1111 *1979*; **76** 2170 *1980*].

p-**Cyanobenzoic acid** *[619-65-8]* **M 147.1, m 219°, pK^{25} 3.55.** Crystd from water and dried in vacuum desiccator over Sicapent.

4-Cyanobenzoyl chloride *[6068-72-0]* **M 165.6, m 68-70°,69-70°, 73-74°, b 132°/8mm, 150-151°/25mm.** If the IR shows presence of OH then treat with $SOCl_2$ boil for 1h, evaporate and distil in vacuum. The distillate solidifies and can be recrystallised from pet ether. It is moisture sensitive and is an **IRRITANT**. [Ashley et al. *J Chem Soc* 103 *1942*; Fison et al. *J Org Chem* 16 648 *1951*.]

Cyanoguanidine (dicyanodiamide) *[461-58-5]* **M 84.1, m 209.5°, pK -0.4.** Crystd from water or EtOH.

5-Cyanoindole *[15861-24-2]* **M 142.2, m 106-108°, 107-108°, pK <0.** Dissolve in 95% EtOH boil in the presence of charcoal, filter, evaporate to a small volume and add enough H_2O to cause crystallisation and cool. Recrystd directly from aqueous EtOH and dried in a vacuum. UV: λ_{max} 276 nm (log ε 3.6) in MeOH. [Lindwall and Mantell *J Org Chem* 18 345 *1953*, 20 1458 *1955*; Thesing at al. *Chem Ber* 95 2205 *1962*; NMR: Lallemend and Bernath *Bull Soc Chim Fr* 4091 *1970*.]

p-**Cyanophenol** *[767-00-0]* **M 119.1, m 113°, pK^{25} 7.97.** Crystd from pet ether, *benzene or water and kept under vacuum over P_2O_5. [Bernasconi and Paschelis *J Am Chem Soc* 108 2969 *1986*.]

3-Cyanopyridine *[100-54-9]* **M 104.1, m 50°, pK^{25} 1.38.** Crystd to constant melting point from *o*-xylene/hexane.

4-Cyanopyridine *[100-48-1]* **M 104.1, m 76-79°, pK^{25} 1.86.** Crystd from dichloromethane/diethyl ether mixture.

Cyanuric acid (2,4,6-trichloro-1,3,5-triazine) *[108-80-5]* **M 120.1, m >300°, pK 6.78.** Crystd from water. Dried at room temperature in a desiccator under vacuum.

Cyanuric chloride (TCT, 2,4,6-trichloro-1,3,5-triazine) *[108-77-0]* **M 184.4, m 146-149°, 154°, b 190°.** Crystd from CCl_4 or pet ether (b 90-100°), and dried under vacuum. Recrystd twice from anhydrous *benzene immediately before use [Abuchowski et al. *J Biol Chem* 252 3582 *1977*].

Cyclobutane carboxylic acid *[3721-95-7]* **M 100.1, m 3-4°, -5.4°, b 84-84.5°/10mm, 110°/25mm, 135-138°/110mm, 194°/760mm, d_4^{20} 1.061, n_D^{20} 1.453, pK^{25} 4.79.** Dissolve in aqueous HCO_3^- and acidify with HCl and extract into Et_2O, wash with H_2O, dry (Na_2SO_4), concentrate to a small volume, then distil through a glass helices packed column. The *S-benzylthiouronium salt* has **m 176°** (from EtOH), and the *anilide* has **m** 112.5-113°, and the *p-toluide* has **m** 123°. [Payne and Smith *J Org Chem* 22 1680 *1957*; Kantaro and Gunning *J Am Chem Soc* 73 480 *1951*.]

trans-**Cyclobutane-1,2-dicarboxylic acid** *[1124-13-6]* **M 144.1, m 131°, pK$_1^{25}$ 4.11, pK$_2^{25}$ 5.15.** Crystd from *benzene.

Cyclobutanone *[1191-95-3]* **M 70.1, b 96-97°, d 0.931, n^{52} 1.4189.** Treated with dilute aqueous $KMnO_4$, dried with molecular sieves and fractionally distd. Purified *via* the semicarbazone, then regenerated, dried with $CaSO_4$, and distd in a spinning-band column. Alternatively, purified by preparative gas chromatography using a Carbowax 20-M column at 80°. (This treatment removes acetone).

Cyclobutylamine *[2516-34-9]* **M 71.1, b 82-83°/atm, 83.2-84.2°/760mm, d_4^{20} 0.839, n_D^{20} 1.437, pK^{25} 10.04 (9.34 in 50% aq EtOH).** It has been purified by steam distn. The aqueous distillate (e.g. 2L) is acidified with 3N HCl (90mL) and evapd to dryness in a vacuum. The *hydrochloride* is treated with a few mL of H_2O, cooled in ice and a slush of KOH pellets ground in a little H_2O is added slowly in portions and keeping the soln very cold. The amine separates as an oil from the strongly alkaline soln. The oil is collected dried over solid KOH and distd using a vac jacketed Vigreux column and protected from CO_2 using a soda lime tube. The fraction boiling at 79-83° is collected, dried over solid KOH for 2 days and redistd over a few pellets of KOH (b 80.5-81.5°). Best distil in a dry N_2 atmosphere. The purity can be checked by GLC using a polyethylene glycol on Teflon column at 72°, 15 psi, flow rate of 102 mL/min of He. The sample can appear homogeneous but because of tailing it is not possible to tell if H_2O is present. The NMR in CCl_4 should show no signals less than 1 ppm from TMS. The *hydrochloride* has a multiplet at *ca* 1.5-2.6ppm (H 2,2,4,3,3,4,4), a quintet at 3.8 ppm (H 1) and a singlet at 4.75 for NH_2 [Roberts and Chambers *J Am Chem*

Soc **73** 2509 *1951*]. The *benzenesulfonamide* has **m** 85-86° (from aq MeOH) and the *benzoyl derivative* has **m** 120.6-121.6° [Roberts and Mazur *J Am Chem Soc* **73** 2509 *1951*; Iffland et al. *J Am Chem Soc* **75** 4044 *1953*; *Org* Synth Coll Vol V 273 *1973*.]

Cyclodecanone *[1502-06-3]* **M 154.2, m 21-24°, b 100-102°/12mm.** Purified by sublimation in a vac.

cis-**Cyclodecene** *[935-31-9]* **M 138.3, m -3°, -1°, b 73°/15mm, 90.3°/33mm, 194-195°/740mm, 197-199°/atm, d_4^{20} 0.8770, n_D^{20} 1.4854.** Purified by fractional distn. It forms an AgNO$_3$ complex which crystallises from MeOH, **m** 167-187° [Cope et al. *J Am Chem Soc* **77** 1628 *1955*; IR: Blomqvist et al. *J Am Chem Soc* **74** 3636 *1952*; Prelog et al. *Helv Chim Acta* **35** 1598 *1952*].

α-**Cyclodextrin** (H$_2$O) *[10016-20-3]* **M 972.9, m >280°(dec),** $[\alpha]_{546}^{20}$ **+175°** (c 10, H$_2$O). See entry on p. 524 in Chapter 6.

β-**Cyclodextrin** (H$_2$O) *[7585-39-9, 68168-23-0]* **M 1135.0, m >300°(dec),** $[\alpha]_{546}^{20}$ **+170°** (c 10, H$_2$O). See entry on p. 524 in Chapter 6.

trans-cis-cis-**1,5,9-Cyclododecatriene** (**cyclododec-1c,5c,9t-triene**) *[2765-29-9]* **M 162.3, m -9°, -8°, b 117.5°/2mm, 237-239°/atm, 244°/760mm, d_4^{20} 0.907, n_D^{20} 1.5129.** Purified by fractional distn, preferably in a vacuum under N$_2$, and forms an insoluble AgNO$_3$ complex. [Breil et al. *Makromol Chemie* **69** 28 *1963*.]

Cyclododecylamine *[1502-03-0]* **M 183.3, m 27-29°, b 140-150°/ca 18mm, 280°/atm, pK 9.62 (in 80% methyl cellosolve).** It can be purified *via* the *hydrochloride salt* **m** 274-275° (from EtOH) or the *picrate* **m** 232-234°, and the free base is distd at water-pump vacuum [Prelog et al. *Helv Chim Acta* **33** 365 *1950*].

1,3-Cycloheptadiene *[4054-38-0]* **M 94.2, b 55°/75mm, 71.5°/150mm, 120-121°/atm, d_4^{20} 0.868, n_D^{20} 1.4972.** It was purified by dissolving in Et$_2$O, wash with 5% HCl, H$_2$O, dry (MgSO$_4$), evap and the residue is distd under dry N$_2$ through a semi-micro column (some foaming occurs) [Cope et al. *J Am Chem Soc* **79** 6287 *1957*; UV: Pesch and Friess *J Am Chem Soc* **72** 5756 *1950*].

Cycloheptane *[291-64-5]* **M 98.2, b 114.4°, d 0.812, n 1.4588.** Distd from sodium, under nitrogen.

Cycloheptanol *[502-41-0]* **M 114.2, b 77-81°/11mm, 185°/atm, d 0.951, n 1.477.** Purified as described for cyclohexanol.

Cycloheptanone *[502-42-1]* **M 112.2, b 105°/80mm, 172.5°/760, d 0.952, n^{24} 1.4607.** Shaken with aq KMnO$_4$ to remove material absorbing around 230-240nm, then dried with Linde type 13X molecular sieves and fractionally distd.

Cycloheptatriene *[544-25-2]* **M 92.1, b 114-115°, d 0.895, n 1.522.** Washed with alkali, then fractionally distd.

Cycloheptylamine *[5452-35-7]* **M 113.2, b 50-52°/11mm, 60°/18mm, d_4^{20} 0.887, n_D^{20} 1.472, pK$_{Est}$ ~10.5 (H$_2$O), pK^{24} 9.99 (in 50% aq methyl cellosolve).** It can be purified by conversion to the *hydrochloride* **m** 242-246°, and the free base is distd under dry N$_2$ in a vacuum [Cope et al. *J Am Chem Soc* **75** 3212 *1953*; Prelog et al. *Helv Chim Acta* **33** 365 *1950*].

1,3-Cyclohexadiene *[592-57-4]* **M 80.1, b 83-84°/atm, d_4^{20} 0.840, n_D^{20} 1.471.** Distd from NaBH$_4$.

1,4-Cyclohexadiene   *[628-41-1]* **M 80.1, b 83-86°/714mm, 88.3°/741mm, 86-88°/atm, 88.7-89°/760mm, d_4^{20} 0.8573, n_4^{20} 1.4725.** Dry over $CaCl_2$ and distil in a vacuum under N_2. [Hückel and Wörffel *Chem Ber* **88** 338 *1955*; Giovannini and Wegmüller *Helv Chim Acta* **42** 1142 *1959*.]

Cyclohexane   *[110-82-7]* **M 84.2, f 6.6°, b 80.7°, d^{24} 0.77410, n 1.42623, n^{25} 1.42354.** Commonly, washed with conc H_2SO_4 until the washings are colourless, followed by water, aq Na_2CO_3 or 5% NaOH, and again water until neutral. It is next dried with P_2O_5, Linde type 4A molecular sieves, $CaCl_2$, or $MgSO_4$ then Na and distd. Cyclohexane has been refluxed with, and distd from Na, CaH_2, $LiAlH_4$ (which also removes peroxides), sodium/potassium alloy, or P_2O_5. Traces of *benzene can be removed by passage through a column of silica gel that has been freshly heated: this gives material suitable for ultraviolet and infrared spectroscopy. If there is much *benzene in the cyclohexane, most of it can be removed by a preliminary treatment with nitrating acid (a cold mixture of 30mL conc HNO_3 and 70mL of conc H_2SO_4) which converts *benzene into nitrobenzene. The impure cyclohexane and the nitrating acid are placed in an ice bath and stirred vigorously for 15min, after which the mixture is allowed to warm to 25° during 1h. The cyclohexane is decanted, washed several times with 25% NaOH, then water dried with $CaCl_2$, and distd from sodium. Carbonyl-containing impurities can be removed as described for chloroform. Other purification procedures include passage through columns of activated alumina and repeated crystn by partial freezing. Small quantities may be purified by chromatography on a Dowex 710-Chromosorb W gas-liquid chromatographic column.
Rapid purification: Distil, discarding the forerun. Stand distillate over Grade I alumina (5% w/v) or 4A molecular sieves.

Cyclohexane butyric acid   *[4441-63-8]* **M 170.3, m 31°, 26.5-28.5°, b 136-139°/4mm. 169°/20mm, 188.8°/46mm, pK^{25} 4.95.** Distil through a Vigreux column, and the crystalline distillate is recrystd from pet ether. The *S-benzylthiouronium salt* has **m** 154-155° (from EtOH) [*Acta Chem Scand* **9** 1425 *1955*; English and Dayan *J Am Chem Soc* **72** 4187 *1950*].

Cyclohexane-1,2-diaminetetraacetic acid (H_2O; CDTA)   *[13291-61-7]* **M 364.4, pK_1 1.34, pK_2 3.20, pK_3 5.75 (6.12), pK_4 9.26 (12.35).** Dissolved in aq NaOH as its disodium salt, then pptd by adding HCl. The free acid was filtered off and boiled with distd water to remove traces of HCl [Bond and Jones *Trans Faraday Soc* **55** 1310 *1959*]. Recrystd from water and dried under vacuum.

trans-**Cyclohexane-1,2-dicarboxylic acid**   *[2305-32-0]* **M 172.2, m 227.5-228°, 228-230.5°, pK_1^{25} 4.30, pK_2^{25} 6.06 [cis , pK_1^{25} 4.25, pK_2^{25} 6.74].** It is purified by recrystn from EtOH or H_2O. The *dimethyl ester* has **m** 95-96° (from *C_6H_6-pet ether). [Abell *J Org Chem* **22** 769 *1957*; Smith and Byrne *J Am Chem Soc* **72** 4406 *1950*; Linstead et al. *J Am Chem Soc* **64** 2093 *1942*.]

(±)-*trans*-1,2-Cyclohexanediol   *[1460-57-7]* **M 116.2, m 104°, 105°, 120°/14mm.** Crystd from Me_2CO and dried at 50° for several days. It can also be recrystd from CCl_4 or EtOAc and can be distilled. The *2,4-dinitrobenzoyl derivative* has **m** 179°. [Winstein and Buckles *J Am Chem Soc* **64** 2780 *1942*.]

trans-**1,2-Cyclohexanediol**   *[1R,2R-(-)- 1072-86-2; 1S,2S-(+)- 57794-08-8]* **M 116.2, m 107-109°, 109-110.5°, 109-111°, 111-112°, 113-114°, $[\alpha]_D^{22}$ (-) and (+) 46.5° (c 1, H_2O).** The enantiomers have been recrystd from *C_6H_6 or EtOAc. The (±) diol has been resolved as the distrychnine salt of the hemisulfate [Hayward, Overton and Whitham *J Chem Soc Perkin Trans 1* 2413 *1976*]; or the *1-menthoxy acetates.* {l-trans- diastereoisomer has **m** 64°, $[\alpha]_D$ -91.7° (c 1.4 EtOH) from pet ether or aqueous EtOH and yields the (-)-*trans*-diol } and {d-trans-diastereoisomer has **m** 126-127°, $[\alpha]_D$ -32.7° (c 0.8 EtOH) from pet ether or aq EtOH and yields the (+)-trans diol}. The bis-*4-nitrobenzoate* has **m** 126.5° $[\alpha]_D$ (-) and (+) 25.5° (c 1.1 $CHCl_3$), and the bis-*3,5-dinitrobenzoate* has **m** 160° $[\alpha]_D$ ± -83.0° (c 1.8 $CHCl_3$) [Wilson and Read *J Chem Soc* 1269 *1935*].

cis-**1,3-Cyclohexanediol**   *[823-18-7]* **M 116.2, m 86°.** Crystd from ethyl acetate and acetone.

trans-**1,3-Cyclohexanediol**   *[5515-64-0]* **M 116.2, m 117°.** Crystd from ethyl acetate.

cis-1,4-Cyclohexanediol *[556-58-9]* M 116.2, m 102.5°. Crystd from acetone (charcoal), then dried and sublimed under vacuum.

Cyclohexane-1,3-dione *[504-02-9]* M 112.1, m 107-108°. Crystd from *benzene.

Cyclohexane-1,4-dione *[637-88-7]* M 112.1, m 76-77°, 78°, 79.5°, 79-80°, b 130-133°/20mm, d_4^{91} 1.0861, n_D^{102} 1.4576. Crystd from water, then *benzene. It can also be recrystd from CHCl$_3$/pet ether or Et$_2$O. It has been purified by distn in a vacuum and the pale yellow distillate which solidified is then recrystd from CCl$_4$ (14.3 g/100 mL) and has m 77-79°. The *di-semicarbazone* has m 231°, the *dioxime HCl* has m 150° (from MeOH-*C$_6$H$_6$) and the bis-*2,4-dinitrophenylhydrazone* m 240° (from PhNO$_2$). [*Org Synth* Coll Vol V 288 *1973*; IR: LeFevre and LeFevre *J Chem Soc* 3549 *1956*.]

Cyclohexane-1,2-dione dioxime (Nioxime) *[492-99-9]* M 142.2, m 189-190°, pK_1^{25} 10.68, pK_2^{25} 11.92. Crystd from alcohol/water and dried in a vacuum at 40°.

1,4-Cyclohexanedione monoethylene acetal (**1,4-dioxa-spiro[4.5]decan-8-one**) *[4746-97-8]* M 156.2, m 70-73°, 73.5-74.5°. Recrystd from pet ether and sublimes slowly on attempted distillation. Also purified by dissolving in Et$_2$O and adding pet ether (b 60-80°) until turbid and cool. [Gardner et al. *J Am Chem Soc* 22 1206 *1957*; Britten and Lockwood *J Chem Soc Perkin Trans 1* 1824 *1974*.]

cis,cis-1,3,5-Cyclohexane tricarboxylic acid *[16526-68-4]* M 216.2, m 216-218°, $pK_{Est(1)}$~4.1, $pK_{Est(2)}$ ~5.4, $pK_{Est(3)}$ ~6.8. Purified by recrystn from toluene + EtOH or H$_2$O. It forms a *1.5 hydrate* with m 216-218°, and a *dihydrate* at 110°. Purified also by conversion to the *triethyl ester* b 217-218°/10mm, 151°/1mm and distillate solidifies on cooling, m 36-37° and is hydrolysed by boiling in aq HCl. The *trimethyl ester* can be distd and recrystd from Et$_2$O, m 48-49°. [Newman and Lawrie *J Am Chem Soc* 76 4598 *1954*, Lukes and Galik *Coll Czech Chem Comm* 19 712 *1954*.]

Cyclohexanol *[108-93-0]* M 100.2, m 25.2°, b 161.1°, d 0.946, n 1.466, n^{25} 1.437, n^{30} 1.462. Refluxed with freshly ignited CaO, or dried with Na$_2$CO$_3$, then fractionally distd. Redistd from Na. Further purified by fractional crystn from the melt in dry air. Peroxides and aldehydes can be removed by prior washing with ferrous sulfate and water, followed by distillation under nitrogen from 2,4-dinitrophenylhydrazine, using a short fractionating column: water distils as the azeotrope. Dry cyclohexanol is *very hygroscopic*.

Cyclohexanone *[108-94-1]* M 98.2, f -16.4°, b 155.7°, d 0.947, n^{15} 1.452. n 1.451, pK^{25} -6.8 (aq H$_2$SO$_4$), pK^{25} 11.3 (enol), 16.6 (keto). Dried with MgSO$_4$, CaSO$_4$, Na$_2$SO$_4$ or Linde type 13X molecular sieves, then distd. Cyclohexanol and other oxidisable impurities can be removed by treatment with chromic acid or dil KMnO$_4$. More thorough purification is possible by conversion to the bisulfite addition compound, or the semicarbazone, followed by decompn with Na$_2$CO$_3$ and steam distn. [For example, equal weights of the bisulfite adduct (crystd from water) and Na$_2$CO$_3$ are dissolved in hot water and, after steam distn, the distillate is saturated with NaCl and extracted with *benzene which is then dried and the solvent evaporated prior to further distn.]

Cyclohexanone oxime *[100-64-1]* M 113.2, m 90°. Crystd from water or pet ether (b 60-80°).

Cyclohexanone phenylhydrazone *[946-82-7]* M 173.3, m 77°. Crystd from EtOH.

Cyclohexene *[110-83-8]* M 82.2, b 83°, d 0.810, n 1.4464, n^{25} 1.4437. Freed from peroxides by washing with successive portions of dil acidified ferrous sulfate, or with NaHSO$_3$ soln then with distd water, dried with CaCl$_2$ or CaSO$_4$, and distd under N$_2$. Alternative methods of removing peroxides include passage through a column of alumina, refluxing with sodium wire or cupric stearate (then distilling from sodium). Diene is removed by refluxing with maleic anhydride before distg under vac. Treatment with 0.1moles of MeMgI in 40mL of diethyl ether removes traces of oxygenated impurities. Other purification procedures include washing with aq NaOH, drying and distg under N$_2$ through a spinning band column; redistg from CaH$_2$; storage with sodium wire; and passage through a column of alumina, under N$_2$, immediately before use. Stored in a refrigerator under argon. [Woon et al. *J Am Chem Soc* **108** 7990 *1986*; Wong et al. *J Am Chem Soc* **109** 3428 *1987*.]

(±)-2-Cyclohexen-1-ol   (3-hydroxycyclohex-1-ene)   *[822-67-3]* **M 242.2, b 63-65°/12mm, 65-66°/13mm, 67°/15mm, 74°/25mm, 85°/35mm, 166°/atm, d_4^{20} 0.9865, n_D^{20} 1.4720.** Purified by distillation through a short Vigreux column. The *2,4-dinitrobenzoyl derivative* has **m** 120.5°, and the *phenylurethane* has **m** 107°. [*Org Synth* **48** 18 *1968*, Cook *J Chem Soc* 1774 *1938*; Deiding and Hartman *J Am Chem Soc* **75** 3725 *1953*.]

Cyclohexene oxide *[286-20-4]* **M 98.2, b 131-133°/atm, d_4^{20} 0.971, n_D^{20} 1.452.** Fractionated through an efficient column. The main impurity is probably H_2O. Dry over $MgSO_4$, filter and distil several times (**b** 129-134°/atm). The residue is sometimes hard to remove from the distilling flask. To avoid this difficulty, add a small amount of a mixture of ground NaCl and Celite (1:1) to help break the residue particularly if H_2O is added. [*Org Synth* Coll Vol I 185 *1948*.]

Cycloheximide *[68-81-9]* **M 281.4, m 119.5-121°, $[\alpha]_{546}^{20}$ +9.5° (c 2, H_2O).** Crystd from water/MeOH (4:1), amyl acetate, isopropyl acetate/isopropyl ether or water.

Cyclohexylamine *[108-91-8]* **M 99.2, b 134.5°, d 0.866, d^{25} 0.863, n 1.4593, n^{25} 1.456, pK^{25} 10.63.** Dried with $CaCl_2$ or $LiAlH_4$, then distd from BaO, KOH or Na, under N_2. Also purified by conversion to the hydrochloride, several crystns from water, then liberation of the amine with alkali and fractional distn under N_2.

Cyclohexylbenzene *[827-52-1]* **M 160.3, f 6.8°, b 237-239°, d 0.950, n 1.5258.** Purified by fractional distn, and fractional freezing.

Cyclohexyl bromide *[108-85-0]* **M 156.3, b 72°/29mm, d 0.902, n^{25} 1.4935.** Shaken with 60% aqueous HBr to remove the free alcohol. After separation from excess HBr, the sample was dried and fractionally distd.

Cyclohexyl chloride *[542-18-7]* **M 118.6, b 142-142.5°, d 1.000, n 1.462.** See chlorocyclohexane on p. 162.

1-Cyclohexylethylamine   *[S-(+): 17430-98-7; R-(-): 5913-13-3]* **M 127.2, b 177-178°/atm, d_4^{20} 0.866, n_D^{20} 1.446, $[\alpha]_D^{15}$ (-) and (+) 3.2° (neat), pK_{Est} ~10.6.** Purified by conversion to the *bitartrate salt* (**m** 172°), then decomposing with strong alkali and extracting into Et_2O, drying (KOH), filtering, evaporating and distilling. The *hydrochloride salt* has **m** 242° (from EtOH-Et_2O), $[\alpha]_D^{15}$ -5.0° (c 10 H_2O; from (+) amine). The *oxalate salt* has **m** 132° (from H_2O). The *(±)-base* has **b** 176-178°/760mm, and *HCl* has **m** 237-238°. [Reihlen, Knöpfle and Sapper *Justus Liebigs Ann Chem* **532** 247 *1938*; *Chem Ber* **65** 660 *1932*.]

Cyclohexylidene fulvene *[3141-04-6]* **M 134.2.** Purified by column chromatography and eluted with *n*-hexane [Abboud et al. *J Am Chem Soc* **109** 1334 *1987*].

Cyclohexyl mercaptan   (cyclohexane thiol)   *[1569-69-3]* **M 116.2, b 38-39°/12mm, 57°/23mm, 90°/100mm, 157°/763mm, d_4^{20} 0.949, n_D^{20} 1.493, pK_{Est} ~10.8.** Possible impurities are the sulfide and the disulfide. Purified by conversion to the Na salt by dissolving in 10% aq NaOH, extract the sulfide and disulfide with Et_2O, and then acidify the aq soln (with cooling and under N_2) with HCl, extract with Et_2O, dry $MgSO_4$, evaporate and distil in a vacuum (**b** 41°/12mm). The *sulfide* has **b** 74°/0.2mm, $n_D^{18.5}$ 1.5162 and the *disulfide* has **b** 110-112°/0.2mm, $n_D^{18.5}$ 1.5557. The *Hg-mercaptide* has **m** 77-78° (needles from EtOH). [Naylor *J Chem Soc* 1532 *1947*.]

Cyclohexyl methacrylate *[101-43-9]* **M 168.2, b 81-86°/0.1mm, d 0.964, n 1.458.** Purification as for methyl methacrylate.

1-Cyclohexyl-5-methyltetrazole *[7707-57-5]* **M 166.2, m 124-124.5°.** Crystd from absolute EtOH, then sublimed at 115°/3mm.

Cyclononanone *[3350-30-9]* **M 140.2, m 142.0-142.8°, b 220-222°.** Repeatedly sublimed at 0.05-0.1mm pressure.

***cis,cis*-1,3-Cyclooctadiene** *[29965-97-7]* **M 108.2, m -5°, -49°, b 55°/34mm, 142-144°/760mm, d_4^{20} 0.8690, n_D^{20} 1.48921.** Purified by GLC. Fractionally distd through a Widmer column [as a mobile liquid and redistilled with a Claisen flask or through a semi-micro column [Gould, Holzman and Neiman *Anal Chem* **20** 361 *1948*]. NB: It has a strong characteristic disagreeable odour detectable at low concentrations and causes headaches on prolonged exposure. [IR: Cope and Estes *J Am Chem Soc* **72** 1128 *1950*; UV: Cope and Baumgardner *J Am Chem Soc* **78** 2812 *1956*.] [**Widmer column.** A Dufton column, modified by enclosing within two concentric tubes the portion containing the glass spiral. Vapour passes up the outer tube and down the inner tube before entering the centre portion. In this way flooding of the column, especially at high temperatures, is greatly reduced (Widmer *Helv Chim Acta* **7** 59 *1924*).]

***cis-cis*-1,5-cyclooctadiene,** *[1552-12-1]* **M 108.2, m -69.5°, -70°, b 51-52°/25mm, 97°/144mm, 150.8°/757mm, d_4^{20} 0.880, n_D^{20} 1.4935, .** Purified by GLC. It has been purified *via* the AgNO$_3$ salt. This is prepared by shaking with a soln of 50% aq AgNO$_3$ w/w several times (e.g. 3 x 50 mL and 4 x 50 mL) at 70° for *ca* 20min to get a good separation of layers. The upper layers are combined and further extracted with AgNO$_3$ at 40° (2 x 20 mL). The upper layer (19 mL) of original hydrocarbon mixture gives colourless needles AgNO$_3$ complex on cooling. The adduct is recrystd from MeOH (and cooling to 0°). The hydrocarbon is recovered by steam distilling the salt. The distillate is extracted with Et$_2$O, dried (MgSO$_4$), evap and distd. [Jones *J Chem Soc* 312 *1954*.]

Cyclooctanone *[502-49-8]* **M 126.2, m 42°.** Purified by sublimation after drying with Linde type 13X molecular sieves.

1,3,5,7-Cyclooctatetraene *[629-20-9]* **M 104.2, b 141-141.5°, d 1.537, n^{25} 1.5350.** Purified by shaking 3mL with 20mL of 10% aqueous AgNO$_3$ for 15min, then filtering off the silver nitrate complex as a ppte. The ppte was dissolved in water and added to cold conc ammonia to regenerate the cyclooctatetraene which was fractionally distd under vacuum onto molecular sieves and stored at 0°. It was passed through a dry alumina column before use [Broadley et al. *J Chem Soc, Dalton Trans* 373 *1986*].

***cis*-Cyclooctene** *[931-88-4]* **M 110.2, b 32-34°/12mm, 66.5-67°/60mm, 88°/141mm, 140°/170mm, 143°/760mm, d_4^{20} 0.84843, n_D^{20} 1.4702, .** The *cis*-isomer was freed from the *trans*-isomer by fractional distn through a spinning-band column, followed by preparative gas chromatography on a Dowex 710-Chromosorb W GLC column. It was passed through a short alumina column immediately before use [Collman et al. *J Am Chem Soc* **108** 2588 *1986*]. It has also been distd in a dry nitrogen glove box from powdered fused NaOH through a Vigreux column and then passed through activated neutral alumina before use [Wong et al. *J Am Chem Soc* **109** 4328 *1987*]. Alternatively it can be purified *via* the AgNO$_3$ salt. This salt is obtained from crude cyclooctene (40 mL) which is shaken at 70-80° with 50% w/w AgNO$_3$ (2 x 15 mL) to remove cyclooctadienes (aq layer). Extraction is repeated at 40° (4 x 20 mL, of 50% AgNO$_3$). Three layers are formed each time. The middle layer contains the AgNO$_3$ adduct of cyclooctene which crystallises on cooling the layer to room temperature. The adduct (complex 2:1) is highly soluble in MeOH (at least 1g/mL) from which it crystallises in large flat needles when cooled at 0°. It is dried under slight vacuum for 1 week in the presence of CaCl$_2$ and paraffin wax soaked in the cyclooctene. It has **m** 51° and loses hydrocarbon on exposure to air. *cis*-Cyclooctene can be recovered by steam distn of the salt, collected, dried (CaCl$_2$) and distilled in vacuum. [Braude et al. *J Chem Soc* 4711 *1957*; AgNO$_3$: Jones *J Chem Soc* 1808 *1954*; Cope and Estes *J Am Chem Soc* **72** 1128 *1950*.]

***cis*-Cyclooctene oxide {(1r,8c)-9-oxabicyclo[6.1.0]nonane}** *[286-62-4]* **M 126.7, m 56-57°, 57.5-57.8°,50-60°, b 85-88°/17mm, 82.5°/22mm, 90-93°/37mm, 189-190°/atm.** It can be distd in vacuum and the solidified distillate can be sublimed in vacuum below 50°. It has a characteristic odour.

[IR: Cope et al. *J Am Chem Soc* **74** 5884 *1952,* **79** 3905 *1957*; Reppe et al. *Justus Liebigs Ann Chem* **560** 1 *1948*].

Cyclopentadecanone *[502-72-7]* **M 224.4, m 63°.** Sublimation is better than crystn from aq EtOH.

Cyclopentadiene *[542-92-7]* **M 66.1, b 41-42°.** Dried with $Mg(ClO_4)_2$ and distd.

Cyclopentane *[287-92-3]* **M 70.1, b 49.3°, d 0.745, n 1.40645, n^{25} 1.4340.** Freed from cyclopentene by two passages through a column of carefully dried and degassed activated silica gel.

Cyclopentane carbonitrile *[4254-02-8]* **M 95.2, m -75.2°, -76°, b 43-44°/7mm, 50-62°/10mm, 67-68°/14mm, 74.5-75°/30mm, d_4^{20} 0.912, n_D^{20} 1.441.** Dissolve in Et_2O, wash thoroughly with saturated aqueous K_2CO_3, dry ($MgSO_4$) and distil through a 10 cm Vigreux column. [McElvain and Stern *J Am Chem Soc* **77** 457 *1955*, Bailey and Daly *J Am Chem Soc* **81** 5397 *1959*.]

Cyclopentane-1,1-dicarboxylic acid *[5802-65-3]* **M 158.1, m 184°, pK₁ 3.23, pK₂ 4.08.** Recrystd from water.

1,3-Cyclopentanedione *[3859-41-4]* **M 98.1, m 149-150°, 151-152.5°, 151-154°, 151-153°.** Purified by Soxhlet extraction with $CHCl_3$. The $CHCl_3$ is evaporated and the residue is recrystd from EtOAc and/or sublimed at 120°/4mm. It has an acidic pKa of 4.5 in H_2O. [IR: Boothe et al. *J Am Chem Soc* **75** 1732 *1953*; DePuy and Zaweski *J Am Chem Soc* **81** 4920 *1959*.]

Cyclopentanone *[120-92-3]* **M 84.1, b 130-130.5°, d 0.947, n 1.4370, n^{25} 1.4340.** Shaken with aq $KMnO_4$ to remove materials absorbing around 230 to 240nm. Dried with Linde type 13X molecular sieves and fractionally distd. Has also been purified by conversion to the $NaHSO_3$ adduct which, after crystallising four times from EtOH/water (4:1), was decomposed by adding to an equal weight of Na_2CO_3 in hot H_2O. The free cyclopentanone was steam distd from the soln. The distillate was saturated with NaCl and extracted with *benzene which was then dried and evaporated; the residue was distd [Allen, Ellington and Meakins *J Chem Soc* 1909 *1960*].

Cyclopentene *[142-29-0]* **M 68.1, b 45-46°, d 0.772, n 1.4228.** Freed from hydroperoxide by refluxing with cupric stearate. Fractionally distd from Na. Chromatographed on a Dowex 710-Chromosorb W GLC column. Methods for **cyclohexene** should be applicable here. Also washed with 1M NaOH soln followed by water. It was dried over anhydrous Na_2SO_4, distd over powdered NaOH under nitrogen, and passed through neutral alumina before use [Woon et al. *J Am Chem Soc* **108** 7990 *1986*]. It was distd in a dry nitrogen atmosphere from powdered fused NaOH through a Vigreux column, and then passed through activated neutral alumina before use [Wong et al. *J Am Chem Soc* **109** 3428 *1987*].

1-Cyclopentene-1,2-dicarboxylic anhydride *[3205-94-5]* **M 138.1, m 42-54°, 46-47°, b 130°/5mm, 133-135°/5mm, n_D^{20} 1.497.** If IR has OH peaks then some hydrolysis to the diacid (**m** 178°) must have occurred. In this case reflux with an appropriate volume of Ac_2O for 30min, evaporate the Ac_2O and distil *in vacuo*. The distillate solidifies and can be recrystd from EtOAc-hexane (1:1). The diacid distils without dec due to formation of the anhydride. The *dimethyl ester* has **m** 120-125°/11mm. [Askain *Chem Ber* **98** 2322 *1965*.]

Cyclopentylamine *[1003-03-8]* **M 85.2, m -85.7°, b 106-108°/760mm, 108.5°/760mm, d_4^{20} 0.869, n_D^{20} 1.452, pK^{25} 10.65, (pK^{25} 4.05 in 50% aq EtOH).** May contain H_2O or CO_2 in the form of carbamate salt. Dry over KOH pellets and then distil from a few pellets of KOH. Store in a dark, dry CO_2-free atmosphere. It is characterised as the *thiocyanate salt* **m** 94.5°. The *benzenesulfonyl derivative* has **m** 68.5-69.5°. [Roberts and Chambers *J Am Chem Soc* **73** 5030 *1951*; Bollinger et al. *J Am Chem Soc* **75** 1729 *1953*.]

Cyclopropane *[75-19-4]* **M 42.1, b -34°.** Washed with a soln of $HgSO_4$, and dried with $CaCl_2$, then $Mg(ClO_4)_2$.

Cyclopropanecarbonyl chloride *[4023-34-1]* **M 104.5, b 117.9-118.0°/723mm, 119.5-119.6[°/760mm, d$_4^{20}$ 1.142, n$_4^{20}$ 1.453.** If the IR shows OH bands then some hydrolysis to the free acid must have occurred. In this case heat with oxalyl chloride at 50° for 2h or SOCl$_2$ for 30min, then evap and distil three times using a Dufton column. Store in an inert atm, preferably in sealed tubes. Strong **IRRITANT.** If it is free from OH bands then just distil *in vacuo* and store as before. [Jeffrey and Vogel *J Chem Soc* 1804 *1948.*]

Cyclopropane-1,1-dicarboxylic acid *[598-10-7]* **M 130.1, m 140°, pK$_1^{25}$ 1.8, pK$_2^{25}$ 5.42.** Recrystd from CHCl$_3$.

Cyclopropylamine *[765-30-0]* **M 57.1, b 49-49.5°/760mm, 48-50°/atm, 49-50°/750mm, d$_4^{20}$ 0.816, n$_D^{20}$ 1.421, pK^{25} 9.10 (pK^{25} 5.33 in 40% aq EtOH).** It has been isolated as the *benzamide* **m** 100.6-101.0° (from aqueous EtOH). It forms a *picrate* **m** 149° (from EtOH-pet ether) from which the free base can be recovered using a basic ion exchange resin and can then be distd through a Todd column (see p. 174) using an automatic still head which only collects products boiling below 51°/atm. Polymeric materials if present will boil above this temperature. The *hydrochloride* has **m** 85-86° [Roberts and Chambers *J Am Chem Soc* **73** 5030 *1951*; Jones *J Org Chem* **9** 484 *1944*; Emmons *J Am Chem Soc* **79** 6522 *1957*].

Cyclopropyldiphenylcarbinol *[5785-66-0]* **M 224.3, m 86-87°.** Crystd from *n*-heptane.

Cyclopropyl methyl ketone *[765-43-5]* **M 84.1, b 111.6-111.8°/752mm, d 0.850, n 1.4242.** Stored with anhydrous CaSO$_4$, distd under nitrogen. Redistd under vacuum.

Cyclotetradecane *[295-17-0]* **M 192.3, m 56°.** Recrystd twice from aq EtOH then sublimed *in vacuo* [Dretloff et al. *J Am Chem Soc* **109** 7797 *1987*].

Cyclotetradecanone *[3603-99-4]* **M 206.3, m 25°, b 145°/10mm, d 0.926, n 1.480.** It was converted to the semicarbazone which was recrystd from EtOH and reconverted to the free cyclotetradecanone by hydrolysis [Dretloff et al. *J Am Chem Soc* **109** 7797 *1987*].

Cyclotrimethylenetrinitramine (RDX, 1,3,5-trinitrohexahydro-1,3,5-triazine) *[121-82-4]* **M 222.2, m 203.8°(dec).** Crystd from acetone. **EXPLOSIVE.**

p-**Cymene** *[99-87-6]* **M 134.2, b 177.1°, d 0.8569, n 1.4909, n^{25} 1.4885.** Washed with cold, conc H$_2$SO$_4$ until there is no further colour change, then repeatedly with H$_2$O, 10% aqueous Na$_2$CO$_3$ and H$_2$O again. Dried with Na$_2$SO$_4$, CaCl$_2$ or MgSO$_4$, and distd. Further purification steps include steam distn from 3% NaOH, percolation through silica gel or activated alumina, and a preliminary refluxing for several days over powdered sulfur. Stored over CaH$_2$.

Cystamine dihydrochloride, *S,S*-(L,L)-Cystathionine, Cysteamine, Cysteamine hydrochloride, (±)-Cysteic acid, *S*-Cysteic acid (H$_2$O), L-Cysteine hydrochloride (H$_2$O), (±)-Cysteine hydrochloride and **L-Cystine, Cytidine,** see entries in Chapter 6.

Cytisine (7*R*,9*S*-7,9,10,11,12,13-hexahydro-7,9-methano-12*H*-pyrido[1,2-*a*][1,5]diazocin-8-one, Laburnine, Ulexine) *[485-35-8]* **M 190.3, m 152-153°, 155°, b 218°/2mm, [α]$_D^{17}$ -120° (H$_2$O), [α]$_D^{25}$ -115° (c 1, H$_2$O), pK$_1^{15}$ 1.20, pK$_2^{15}$ 8.12 [also stated are pK$_1$ 6.11, pK$_2$ 13.08].** Crystd from acetone and sublimed in a vacuum. Its solubilities are: 77% (H$_2$O), 7.7% (Me$_2$CO), 28.6% (EtOH), 3.3% (*C$_6$H$_6$), 50% (CHCl$_3$) but is insoluble in pet ether. The *tartrate* has **m** 206-207° [α]$_D^{24}$ +45.9°, the *N-tosylate* has **m** 206-207°, and the *N-acetate* has **m** 208°. [Bohlmann et al. *Angew Chem* **67** 708 *1955*; van Tamelen and Baran *J Am Chem Soc* **77** 4944 *1955*; Isolation: Ing *J Chem Soc* 2200 *1931*; Govindachari et al. *J Chem Soc* 3839 *1957*; Abs config: Okuda et al. *Chem Ind (London)* 1751 *1961*.] **TOXIC.**

Cytosine see entry in Chapter 6.

cis-Decahydroisoquinoline *[2744-08-3]* M 139.2, b 97-98°/15mm, 208-209°/730mm, pK²⁰ 11.32.

cis-**Decahydroisoquinoline** *[2744-08-3]* **M 139.2, b 97-98°/15mm, 208-209°/730mm, pK²⁰ 11.32.** The free base is treated with satd aq picric acid, allowed to stand for 12h, filtd, washed with MeOH to remove the more soluble *trans* isomer and recrystd from MeOH to give pure cis-*picrate* **m 149-150°**. The picrate (~5g) is shaken with 5M aq NaOH (50mL) and Et₂O (150mL) while H₂O is added to the aq phase to dissolve insoluble Na picrate. The Et₂O extract is dried over solid NaOH and then shaken with Al₂O₃ (Merck for chromatography) until the yellow color of traces of picric acid disappears (this color cannot be removed by repeated shaking with 5-10 M aq NaOH). The extract is concentrated to 50mL and dry HCl is bubbled through until separatn of the white crysts of the cis-*HCl* is complete. These are washed with Et₂O, dried at 100° and recrystd from EtOH + EtOAc to yield pure cis-*Hydrochloride* **m 182-183°** (dried in a vac desiccator over KOH) with IR (KBr) ν_{max} 2920, 2820, 1582, 1470, 1445, 1410, 1395, 1313, 1135, 1080, 990, 870 cm⁻¹. The pure *free base* is prepared by dissolving the *hydrochloride* in 10 M aq NaOH, extracted with Et₂O, dried over solid KOH, filtd and distd in vac. It has IR (film) ν_{max} 2920, 2820, 2720, 2560, 1584, 1470, 1445, 1415, 1395, 1315, 1300, 1135, 1080, 1020, 990, 873 cm⁻¹. The ¹H NMR in CDCl₃ is characteristically different from that of the *trans*-isomer. [Armarego *J Chem Soc (C)* 377 *1967*; Gray and Heitmeier *J Am Chem Soc* **80** 6274 *1958*; Witkop *J Am Chem Soc* **70** 2617 *1948*; Skita *Chem Ber* **57** 1982 *1924*; Helfer *Helv Chim Acta* **6** 799*1923*.]

trans-**Decahydroisoquinoline** *[2744-09-4]* **M 139.2, b 106°/15mm, pK²⁰ 11.32.** This is purified as the *cis*-isomer above. The trans-*picrate* has **m 175-176°**, and the trans-*hydrochloride* has **m 221-222°** and has IR (KBr) ν_{max} 2930, 3800, 1589, 1450, 1400, 1070, 952, 837 cm⁻¹. The pure *free base* was prepared as above and had IR (film) ν_{max} 2920, 2820, 2720, 2560, 1584, 1470, 1445, 1415, 1395, 1315, 1300, 1135, 1080, 1020, 990, 873 cm⁻¹. The ¹H NMR in CDCl₃ is characteristically different from that of the *cis*-isomer. (references as above and Helfer *Helv Chim Acta* **9** 818 *1926*).

Decahydronaphthalene (decalin, mixed isomers) *[91-17-8]* **M 138.2, b 191.7°, d 0.886, n 1.476.** Stirred with conc H₂SO₄ for some hours. Then the organic phase was separated, washed with water, saturated aqueous Na₂CO₃, again with water, dried with CaSO₄ or CaH₂ (and perhaps dried further with Na), filtered and distd under reduced pressure (**b 63-70°/10mm**). Also purified by repeated passage through long columns of silica gel previously activated at 200-250°, followed by distn from LiAlH₄ and storage under N₂. Type 4A molecular sieves can be used as a drying agent. Storage over silica gel removes water and other polar substances.

cis-**Decahydronaphthalene** *[493-01-6]* **M 138.2, f -43.2°, b 195.7°, d 0.897, n 1.48113.** Purification methods described for the mixed isomers are applicable. The individual isomers can be separated by very efficient fractional distn, followed by fractional crystn by partial freezing. The *cis*-isomer reacts preferentially with AlCl₃ and can be removed from the *trans*-isomer by stirring the mixture with a limited amount of AlCl₃ for 48h at room temperature, filtering and distilling.

trans-**Decahydronaphthalene** *[493-02-7]* **M 138.2, f -30.6°, b 187.3°, d 0.870, n 1.46968.** See purification of *cis*-isomer above.

cis-**Decahydroquinoline** *[10343-99-4]* **M 139.2, b 207-208°/708mm , pK²⁰ 11.29.** It is available as a *cis-trans*-mixture (**b 70-73°/10mm**, Aldrich, ~ 18% *cis*-isomer *[2051-28-7]*), but the isomers can be fractionated in a spinning band column (1~1.5 metre, type E) at atmospheric pressure and collecting 2mL fractions with a distillation rate of 1 drop in 8-10sec. The lower boiling fraction solidifies and contains the *trans*-isomer (see below, **m 48°**). The higher boiling fraction **b 207-208°/708mm**, remains liquid and is mostly the *cis*-isomer. This is reacted with PhCOCl and M aq NaOH to yield the *N-benzoyl derivative* **m 96°** after recrystn from pet ether (b 80-100°). It is hydrolysed with 20% aq HCl by refluxing overnight. PhCO₂H is filtd off, the filtrate is basified with 5M aq NaOH and extracted with Et₂O. The dried extract (Na₂SO₄) is satd with dry HCl gas and the *cis-decahydroquinoline hydrochloride* which separates has **m 222-224°** after washing with Et₂O and drying at 100°; and has IR (KBr) ν_{max} 2900, 2780, 2560, 1580, 1445, 1432, 1403, 1165, 1080,

1036, 990, 867 cm^{-1}. The *free base* is obtained by dissolving the *hydrchloride* salt in 5M aq NaOH, extracting with Et$_2$O and drying the extrtact (Na$_2$SO$_4$), evaporating and distilling the residue; it has IR (film) v_{max} 2900, 2840, 2770, 1445, 1357, 1330, 1305, 1140, 1125, 1109, 1068, 844 cm^{-1}. The ^{1}H NMR in CDCl$_3$ is characteristically different from that of the *trans*-isomer. [Armarego *J Chem Soc (C)* 377 *1967*; Hückel and Stepf *Justus Liebigs Ann Chem* **453** 163 *1927*; Bailey and McElvain *J Am Chem Soc* **52** 4013 *1930*.]

***trans*-Decahydroquinoline** *[767-92-0]* **M 139.2, m 48°, b 205-206°/708mm , pK^{20} 11.29.** The lower boiling fraction from the preceding spinning band column fractionation of the commercial *cis-trans*-mixture (~ 20:60; see the *cis*-isomer above) solidifies readily (**m** 48°) and the receiver has to be kept hot with warm water. It is further purified by conversion to the *Hydrochloride* **m** 285-286° after recrystn from EtOH/AcOEt. This has IR (KBr) v_{max} 2920, 2760, 2578, 2520, 1580, 1455, 1070, 1050, 975, 950, 833 cm^{-1}. The *free base* is prepared as for the *cis*-isomer above and distd; and has IR (film, at ca 50°) v_{max} 2905, 2840, 2780, 1447, 1335, 1305, 1240, 1177, 1125, 987, 900, 835 cm^{-1}. The ^{1}H NMR in CDCl$_3$ is characteristically different from that of the *cis*-isomer. [Armarego *J Chem Soc (C)* 377 *1967*; Hückel and Stepf *Justus Liebigs Ann Chem* **453** 163 *1927*; Bailey and McElvain *J Am Chem Soc* **52** 4013 *1930*; Prelog and Szpilfogel *Helv Chim Acta* **28** 1684 *1945*.]

***n*-Decane** *[124-18-5]* **M 142.3, b 174.1°, d 0.770, n 1.41189, n^{25} 1.40967.** It can be purified by shaking with conc H$_2$SO$_4$, washing with water, aqueous NaHCO$_3$, and more water, then drying with MgSO$_4$, refluxing with Na and distilling. Passed through a column of silica gel or alumina. It can also be purified by azeotropic distn with 2-butoxyethanol, the alcohol being washed out of the distillate, using water; the decane is next dried and redistilled. It can be stored with NaH. Further purification can be achieved by preparative gas chromatography on a column packed with 30% SE-30 (General Electric methyl-silicone rubber) on 42/60 Chromosorb P at 150° and 40psig, using helium [Chu *J Chem Phys* **41** 226 *1964*]. Also purified by zone refining.

Decan-1,10-diol *[112-47-0]* **M 174.3, m 72.5-74°.** Crystd from dry ethylene dichloride.

n*-Decanol** (n*-decyl alcohol**) *[112-30-1]* **M 158.3, f 6.0°, b 110-119°/0.1mm, d 0.823, n 1.434.** Fractionally distd in an all-glass unit at 10mm pressure (**b** 110°), then fractionally crystd by partial freezing. Also purified by preparative GLC, and by passage through alumina before use.

***n*-Decyl bromide** *[112-29-8]* **M 221.2, b 117-118°/15.5mm, d 1.066.** Shaken with H$_2$SO$_4$, washed with water, dried with K$_2$CO$_3$, and fractionally distd.

Decyltrimethylammonium bromide *[2082-84-0]* **M 280.3, m 239-242°.** Crystd from 50% (v/v) EtOH/diethyl ether, or from acetone and washed with ether. Dried under vacuum at 60°. Also recrystd from EtOH and dried over silica gel. [McDonnell and Kraus *J Am Chem Soc* **73** 2170 *1952*; Dearden and Wooley *J Phys Chem* **91** 2404 *1987*.]

(+)-Dehydroabietylamine (**abieta-8,11,13-triene-18-ylamine**) *[1446-61-3]* **M 285.5, m 41°, 42.5-45°, b 192-193°/1mm, 250°/12mm, n$_D^{40}$ 1.546, [α]$_{546}^{20}$ +51° (c 1, EtOH), pK$_{Est}$ ~10.3.** The crude base is purified by converting 2g of base in toluene (3.3mL) into the acetate salt by heating at 65-70° with 0.46g of AcOH and the crystals are collected and dried (0.96g from two crops; **m** 141-143°). The acetate salt is dissolved in warm H$_2$O, basified with aqueous NaOH and extracted with *C$_6$H$_6$. The dried extract (MgSO$_4$) is evaporated in vacuum leaving a viscous oil which crystallises and can be distd. [Gottstein and Cheney *J Org Chem* **30** 2072 *1965*.] The *picrate* has **m** 234-236° (from aq MeOH), and the *formate* has **m** 147-148° (from heptane).

Dehydro-L(+)-ascorbic acid *[490-83-5]* **M 174.1, m 196°(dec), [α]$_{546}^{20}$ +42.5° (c 1, H$_2$O), pK 3.90.** Crystd from MeOH

7-Dehydrocholesterol *[434-16-2]* **M 384.7, m 142-143°, [α]$_D^{20}$ -122° (c 1, CHCl$_3$).** Crystd from MeOH.

Dehydrocholic acid *[81-23-2]* **M 402.5, m 237°, $[\alpha]^{20}_{546}$ -159° (c 1, in CHCl$_3$), pK$_{Est}$ ~4.9.** Crystd from acetone.

Dehydroepiandrosterone *[54-43-0]* **M 288.4, m 140-141° and 152-153° (dimorphic), $[\alpha]^{20}_{D}$ +13° (c 3, EtOH).** Crystd from MeOH and sublimed in vacuum.

Delphinine *[561-07-9]* **M 559.7, m 197-199°, $[\alpha]^{20}_{D}$ +26° (EtOH).** Crystd from EtOH.

3-Deoxy-D-allose *[6605-21-6]* **M 164.2, $[\alpha]^{20}_{D}$ +8° (c 0.25 in H$_2$O).** Obtained from diethyl ether as a colourless syrup.

Deoxybenzoin *[451-40-1]* **M 196.3, m 60°, b 177°/12mm, 320°/760mm.** Crystd from EtOH.

Deoxycholic acid *[83-44-3]* **M 392.6, m 171-174°, 176°, 176-178°, $[\alpha]^{20}_{546}$ +64° (c 1, EtOH), $[\alpha]^{20}_{D}$ +55° (c 2.5, EtOH), pK 6.58.** Refluxed with CCl$_4$ (50mL/g), filtered, evaporated under vacuum at 25°, recrystd from acetone and dried under vacuum at 155° [Trenner et al. *J Am Chem Soc* **76** 1196 *1954*]. A soln of (cholic acid-free) material (100mL) in 500mL of hot EtOH was filtered, evaporated to less than 500mL on a hot plate, and poured into 1500mL of cold diethyl ether. The ppte, filtered by suction, was crystd twice from 1-2 parts of absolute EtOH, to give an alcoholate, m 118-120°, which was dissolved in EtOH (100mL for 60g) and poured into boiling water. After boiling for several hours the ppte was filtered off, dried, ground and dried to constant weight [Sobotka and Goldberg *Biochem J* **26** 555 *1932*]. Deoxycholic acid was also freed from fatty acids and cholic acid by silica gel chromatography by elution with 0.5% acetic acid in ethyl acetate [Tang et al. *J Am Chem Soc* **107** 4058 *1985*]. It can also be recrystd from butanone. Its solubility in H$_2$O at 15° is 0.24g/L but in EtOH it is 22.07g/L.

11-Deoxycorticosterone **(21-hydroxyprogesterone)** *[64-85-7]* **M 330.5, m 141-142°, $[\alpha]^{20}_{546}$ +178° and $[\alpha]_{546}$ +223° (c 1, EtOH).** Crystd from diethyl ether.

2-Deoxy-β-D-galactose *[1949-89-9]* **M 164.2, m 126-128°, $[\alpha]^{20}_{D}$ +60° (c 4, H$_2$O).** Crystd from diethyl ether.

2-Deoxy-α-D-glucose *[154-17-6]* **M 164.2, m 146°, $[\alpha]^{20}_{D}$ +46° (c 0.5, H$_2$O after 45h).** Crystd from MeOH/acetone.

6-Deoxy-D-glucose **(D-quinovose)** *[7658-08-4]* **M 164.2, m 146°, $[\alpha]^{20}_{D}$ +73° (after 5 min) and +30° (final, after 3h) (c 8.3, H$_2$O).** It is purified by recrystn from EtOAc and is soluble in H$_2$O, EtOH but almost insoluble in Et$_2$O and Me$_2$CO. [Srivastava and Lerner *Carbohydr Res* **64** 263 *1978*; NMR: Angyal and Pickles *Aust J Chem* **25** 1711 *1972*.]

2-Deoxy-β-L-ribose *[18546-37-7]* **M 134.1, m 77°, 80°, $[\alpha]^{25}_{D}$ +91.7° (c 7, pyridine, 40° final).** Crystd from diethyl ether.

2-Deoxy-β-D-ribose *[533-67-5]* **M 134.1, m 86-87°, 87-90°, $[\alpha]^{20}_{D}$ -56° (c 1, H$_2$O after 24h).** Crystd from diethyl ether.

Desyl bromide **(α-bromo-desoxybenzoin, ω-bromo-ω−phenyl acetophenone)** *[484-50-0]* **M 275.2, m 57.1-57.5°.** Crystd from 95% EtOH.

Desyl chloride **(α-chloro-desoxybenzoin, ω-chloro-ω−phenyl acetophenone)** *[447-31-4]* **M 230.7, m 62-64°, 66-67°,67.5°, 68°.** For the purification of small quantities recrystallise from pet ether (b 40-60°), but use MeOH or EtOH for larger quantities. For the latter solvent, dissolve 12.5g of chloride in 45mL of boiling EtOH (95%), filter and the filtrate yields colourless crystals (7.5g) on cooling. A further crop (0.9g) can be obtained by cooling in an ice-salt bath. It turns brown on exposure to sunlight but it is stable in sealed dark containers. [Henley and Turner *J Chem Soc* 1182 *1931*; *Org Synth* Coll Vol II 159 *1943*.]

Dexamethasone **(9-α−fluoro-16-α-methylprednisolone)** *[50-02-2]* **M 392.5, m 262-264°, 268-271°, [α]$_D^{25}$ +77.5° (c 1, dioxane).** It has been recrystallised from Et$_2$O or small volumes of EtOAc. Its solubility in H$_2$O in 10 mg/100mL at 25°; and is freely soluble in Me$_2$CO, EtOH and CHCl$_3$. [Arth et al. *J Am Chem Soc* **80** 3161 *1958*; for the β-methyl isomer see Taub et al. *J Am Chem Soc* **82** 4025 *1960*.]

Dexamethasone 21-acetate **(9-α−fluoro-16-α-methylprednisolone-21-acetate)** *[1177-87-3]* **M 434.5, m 215-225°, 229-231°, [α]$_D^{25}$ +77.6° (c 1, dioxane), +73° (c 1, CHCl$_3$).** Purified on neutral Al$_2$O$_3$ using CHCl$_3$ as eluent, fraction evaporated, and recrystd from CHCl$_3$. UV has λ$_{max}$ at 239nm. [Oliveto et al. *J Am Chem Soc* **80** 4431 *1958*].

Diacetamide *[625-77-4]* **M 101.1, m 75.5-76.5°, b 222-223°.** Purified by crystn from MeOH [Arnett and Harrelson *J Am Chem Soc* **109** 809 *1987*].

Diacetoxyiodobenzene **(iodobenzenediacetate)** *[3240-34-4]* **M 322.1, m 163-165°.** Purity can be checked by treatment with H$_2$SO$_4$ then KI and the liberated I$_2$ estimated with standard thiosulfate. It has been recrystd from 5M acetic acid and dried overnight in a vac desiccator over CaCl$_2$. The surface of the crystals may become slightly yellow but this does not affect its usefulness. [Sharefkin and Saltzman *Org Synth* Coll Vol V 600 *1973*.]

1,2-Diacetyl benzene *[704-00-7]* **M 162.2, m 39-41°, 41-42°, b 110°/0.1mm, 148°/20mm.** Purified by distn and by recrystn from pet ether. The *bis-2,4-dinitrophenyl hydrazone* has **m** 221° dec. [Halford and Weissmann *J Org Chem* **17** 1646 *1952*; Riemschneider and Kassahn *Chem Ber* **92** 1705 *1959*.]

1,4-Diacetyl benzene *[1009-61-6]* **M 162.2, m 113-5-114.2°.** Crystd from *benzene and vacuum dried over CaCl$_2$. Also dissolved in acetone, treated with Norit, evapd and recrystd from MeOH [Wagner et al. *J Am Chem Soc* **108** 7727 *1986*].

(+)-Di-*O*-acetyl-L-tartaric anhydride **[(*R,R*)-2,3-diacetoxysuccinic anhydride]** *[6283-74-5]* **M 216.2, m 129-132°, 133-134°, 135°, 137.5°, [α]$_D^{20}$ +97.2° (c 0.5, dry CHCl$_3$).** If the IR is good, i.e. no OH bands, then keep in a vacuum desiccator overnight (over P$_2$O$_5$/paraffin) before use. If OH bands are present then reflux 4g in Ac$_2$O (12.6mL) containing a few drops of conc H$_2$SO$_4$ for 10min (use a relatively large flask), pour onto ice, collect the crystals, wash with dry *C$_6$H$_6$ (2 x 2mL), stir with 17mL of cold Et$_2$O, filter and dry in a vacuum desiccator as above, and store in dark evacuated ampoules under N$_2$ in small aliquots. It is not very stable in air; the melting point of the crystals drop one degree in the first four days then remains constant (132-134°). If placed in a stoppered bottle it becomes gummy and the **m** falls 100° in three days. Recrystn leads to decomposition. If good quality anhydride is required it should be prepared fresh from tartaric acid. It sublimes in a CO$_2$ atmosphere. [*Org Synth* Coll Vol IV 242 *1963*.]

Diallyl amine **(*N*-2-propenyl-2-propen-1-amine)** *[124-02-7]* **M 97.2, b 107-111°/760mm, 112°/760mm, d$_4^{20}$ 0.789, n$_D^{20}$ 1.440, pK^{20} 9.42.** Keep over KOH pellets overnight, decant and distil from a few pellets of KOH at atm pressure (**b** 108-111°), then fractionate through a Vigreux column. [Vliet *J Am Chem Soc* **46** 1307 *1924*; *Org Synth* Coll Vol I 201 *1941*.] The *hydrochloride* has **m** 164-165° (from Me$_2$CO + EtOH). [Butler and Angels *J Am Chem Soc* **79** 3128 *1957*.]

(+)-*N,N*'-Diallyl tartrimide **(DATD)** *[58477-85-3]* **M 228.3, m 184°, [α]$_{546}$ +141° (c 3, MeOH).** Wash with Et$_2$O containing 10% EtOH until the washings are clear and colourless, and dry *in vacuo*. [*FEBS Lett* **7** 293 *1970*.]

Diamantane *[2292-79-7]* **M 188.3, m 234-235°.** Purified by repeated crystn from MeOH or pentane. Also dissolved in CH$_2$Cl$_2$, washed with 5% aq NaOH and water, and dried (MgSO$_4$). The soln was concentrated to a small volume, an equal weight of alumina was added, and the solvent evaporated. The residue was placed on an activated alumina column (*ca* 4 x weight of diamantane) and eluted with pet ether (b 40-60°). Eight sublimations and twenty zone refining experiments gave material **m** 251° of 99.99% purity by differential analysis [*Tetrahedron Lett* 3877 *1970*; *J Chem Soc (C)* 2691 *1972*].

3,6-Diaminoacridine hydrochloride *[952-23-8]* **M** 245.7, **m** 270°(dec), ε_{456} 4.3 x 10^4, pK_1^{20} **1.5, pK_2^{20} 9.60 (9.65 free base).** First purified by pptn of the free base by adding aq NH_3 soln to an aq soln of the hydrochloride or hydrogen sulfate, drying the ppte and subliming at 0.01mm Hg [Müller and Crothers *Eur J Biochem,* **54** 267 *1975*].

3,6-Diaminoacridine sulfate (proflavin sulfate) *[1811-28-5]* **M** 516.6, **m** >300°(dec), λ_{max} **456nm.** An aqueous soln, after treatment with charcoal, was concentrated, chilled overnight, filtered and the ppte was rinsed with a little diethyl ether. The ppte was dried in air, then overnight in a vacuum oven at 70°.

1.3-Diaminoadamantane *[10303-95-4]* **M** 164.3, **m** 52°, $pK_{Est(1)}$~8.6, $pK_{Est(2)}$ ~10.6. Purified by zone refining.

1,4-Diaminoanthraquinone *[128-95-0]* **M** 238.3, **m** 268°. Purified by thin-layer chromatography on silica gel using toluene/acetone (9:1) as eluent. The main band was scraped off and extracted with MeOH. The solvent was evaporated and the quinone was dried in a drying pistol [Land, McAlpine, Sinclair and Truscott *J Chem Soc, Faraday Trans 1* **72** 2091 *1976*]. Crystd from EtOH in dark violet crystals.

1,5-Diaminoanthraquinone *[129-44-2]* **M** 238.3, **m** 319°. Recrystd from EtOH or acetic acid [Flom and Barbara *J Phys Chem* **89** 4481 *1985*].

2,6-Diaminoanthraquinone *[131-14-6]* **M** 238.3, **m** 310-320°. Crystd from pyridine. Column-chromatographed on Al_2O_3 / toluene to remove a fluorescent impurity, then recrystd from EtOH.

3,3'-Diaminobenzidine tetrahydrochloride (2H$_2$O) *[7411-49-6]* **M** 396.1, **m** >300°(dec), $pK_{Est(1))}$ ~3.3, $pK_{Est(2)}$~4.7 **(free base).** Dissolved in water and ppted by adding conc HCl, then dried over solid NaOH.

3,4-Diaminobenzoic acid *[619-05-6]* **M** 152.2, **m** 213°(dec), 228-229°, pK_1^{25} 2.57 **(4-NH$_2$),** pK_2^{25} **3.39 (3-NH$_2$),** $pK_{Est(3)}$~5.1 **(CO$_2$H).** Crystd from H_2O or toluene.

3,5-Diaminobenzoic acid *[535-87-5]* **M** 152.2, **m** 235-240°(dec), pK^{25} 5.13 **(CO$_2$H).** Crystd from water. *Dihydrochloride* has **m** 226-228°(dec).

4,4'-Diaminobenzophenone *[611-98-3]* **M** 212.3, **m** 242-244°, 243-245°, 246.5-247.5° **(after sublimation at 0.0006 mm)** , pK_1^{25} **1.37,** pK_2^{25} **2.92.** Purified by recrystn from EtOH and by sublimation in high vacuum. The *dihydrochloride* has **m** 260° dec (from EtOH) and the *thiosemicarbazone* has **m** 207-207.5° dec (from aq EtOH). [Kuhn et al. *Chem Ber* **75** 711 *1942*.]

1,4-Diaminobutane dihydrochloride (putrescine 2HCl) *[333-93-7]* **M** 161.1, **m** >290°, pK_1^{25} **9.63,** pK_2^{25} **10.80.** Crystd from EtOH/water.

1,2-Diamino-4,5-dichlorobenzene *[5348-42-5]* **M** 177.0, **m** 163°, $pK_{Est(1))}$ ~0, $pK_{Est(2)}$~2.9. Refluxed with activated charcoal in CH_2Cl_2, followed by recrystn from diethyl ether/pet ether or pet ether [Koolar and Kochi *J Org Chem* **52** 4545 *1987*].

2,2'-Diaminodiethylamine (diethylenetriamine) *[111-40-0]* **M** 103.2, **b** 208°, **d** 0.95, **n** **1.483, pK_1^{25} 4.34, pK_2^{25} 9.13, pK_3^{25} 9.94.** Dried with Na and distd, preferably under reduced pressure, or in a stream of N_2. § Polymer-bound diethylenetriamine is commercially available.

4,5-Diamino-2,6-dihydroxypyrimidine (diamino uracil) sulfate *[32014-70-3]* **M** 382.3, **m** **>300°, pK_1^{20} 1.7, pK_2^{20} 3.20, pK_3^{20} 4.56.** The salt is quite insoluble in H_2O but can be converted to the free base which is recrystd from H_2O and converted to the sulfate by addition of the required amount of H_2SO_4. The *hydrochloride* has **m** 300-305° dec and can be used to prepare the sulfate by addition of H_2SO_4; It is more soluble than the sulfate. The *perchlorate* has **m** 252-254°. The free base has λ_{max} 260nm (log ε 4.24) in 0.1M HCl. [Bogert and Davidson *J Am Chem Soc* **86** 1668 *1933*; Bredereck et al. *Chem Ber* **86** 850 *1953*; *Org Synth* Coll Vol IV 247 *1963*; Barlin and Pfleiderer *J Chem Soc (B)* 1425 *1971*.]

5,6-Diamino-1,3-dimethyluracil hydrate **(5,6-diamino-1,3-dimethyl-2-pyrimidine-2,4-dione hydrate)** *[5440-00-6]* **M 188.2, m 205-208° dec, 209° dec, 210°dec, pK$_1$ 1.7, pK$_2$ 4.6.** Recryst from EtOH. The *hydrochloride* has **m** 310° (from MeOH) and the *perchlorate* has **m** 246-248°. [UV: Bredereck et al. *Chem Ber* **92** 583 *1959*; Taylor et al. *J Am Chem Soc* **77** 2243 *1955*.]

4,4'-Diamino-3,3'-dinitrobiphenyl *[6271-79-0]* **M 274.2, m 275°, pK$_{Est}$ ~-0.2.** Crystd from aqueous EtOH.

4,4'-Diaminodiphenylamine *[537-65-5]* **M 199.3, m 158°, pK$_{Est}$ ~5.0.** Crystd from water.

4,4'-Diaminodiphenylmethane *[101-77-9]* **M 198.3, m 91.6-92°, pK$_{Est}$ ~4.9.** Crystd from water, 95% EtOH or *benzene.

3,3'-Diaminodipropylamine *[56-18-8]* **M 131.2, b 152°/50mm, d 0.938, n 1.481, pK$_1^{25}$ 7.72, pK$_2^{25}$ 9,57, pK$_3^{25}$ 10.65.** Dried with Na and distd under vacuum.

6,9-Diamino-2-ethoxyacridine **(Ethacridine)** *[442-16-0]* **M 257.3, m 226°, pK$_{Est}$ ~11.5.** Crystd from 50% EtOH.

2,7-Diaminofluorene *[524-64-4]* **M 196.3, m 165°, pK$_{Est}$ ~4.6.** Recrystd from H_2O.

2,4-Diamino-6-hydroxypyrimidine *[56-06-4]* **M 126.1, m 260-270°(dec), pK$_1^{25}$ 1.34, pK$_2^{25}$ 3.27, pK$_3^{25}$ 10.83.** Recrystd from H_2O.

4,5-Diamino-6-hydroxypyrimidine hemisulfate *[102783-18-6]* **M 350.3, m 268°, 270°, pK$_1^{25}$ 1.34, pK$_2^{25}$ 3.57, pK$_3^{25}$ 9.86.** Recrystd from H_2O. The free base also recrystallises from H_2O (**m** 239°). [UV: Mason *J Chem Soc* 2071 *1954*; Elion et al. *J Am Chem Soc* **74** 411 *1952*.]

1,5-Diaminonaphthalene *[2243-62-1]* **M 158.2, m 190°, pK^{25} 4.12.** Rerystd from boiling H_2O, but is wasteful due to poor solubility. Boil in chlorobenzene (charcoal), filter hot and cool the filtrate. This gives colourless crystals. Dry in a vac till free from chlorobenzene (odour), and store away from light.

1,8-Diaminonaphthalene *[479-27-6]* **M 158.2, m 66.5°, b 205°/12mm, pK 4.44.** Crystd from water or aqueous EtOH, and sublimed in a vacuum. *N,N'-DiMe* deriv [20734-56-9] has **m** 103-104° and **pK** 5.61; *N,N,N'-TriMe* deriv *[20734-57-0]* has **m** 29-30° and **pK** 6.43. [Hodgson et al. *J Chem Soc* 202 *1945*.]

2,3-Diaminonaphthalene *[771-97-1]* **M 158.2, m 199°, pK^{21} 3.54 (in 50% aq EtOH).** Crystd from water, or dissolved in 0.1M HCl, heated to 50°. After cooling, the soln was extracted with decalin to remove fluorescent impurities and centrifuged.

1,8-Diaminooctane *[373-44-4]* **M 144.3, m 50-52°, 51-52°, 52-53°, b 121°/18mm, 120°/24mm, pK$_1^{20}$ 10.1, pK$_2^{20}$ 11.0.** Distil under vacuum in an inert atmosphere(N_2, Ar), cool and store distillate in an inert atmosphere in the dark. The *dihydrochloride* has **m** 273-274°. [Nae and Le *Helv Chim Acta* **15** 55 *1955*.]

2,4-Diamino-5-phenylthiazole **(Amiphenazole)** *[490-55-1]* **M 191.3, m 163-164°(dec).** Crystd from aqueous EtOH or water. Stored in the dark under N_2.

1,5-Diaminopentane *[462-94-2]* **M 102.2, m 14-16°, b 78-80°/12mm, 101-103°/35mm, 178-180°/750mm, d_4^{20} 0.869, n_D^{20} 1.458, pK_1^{20} 10.02, pK_2^{20} 10.96.** Purified by distn, after standing over KOH pellets (at room temp; i.e. liquid form). It has pKa^{20} values of 10.02 and 10.96 in H_2O. Its *dihydrochloride* has **m** 275° (sublimes in vac), and its *tetraphenyl boronate* has **m** 164°. [Schwarzenbach et al. *Helv Chim Acta* **35** 2333 *1952.*]

***d,l*-2,6-Diaminopimelic acid** *[583-93-7]* **M 190.2, m 313-315°(dec) $pK_{Est(1)}$ ~2.2, $pK_{Est(2)}$~9.7.** Crystd from water.

1,3-Diaminopropane dihydrochloride *[10517-44-9]* **M 147.1, m 243°, pK_1^{25} 8.29, pK_2^{25} 10.30.** Crystd from EtOH/water.

1,3-Diaminopropan-2-ol *[616-29-5]* **M 90.1, m 38-40°, pK_1^{25} 7.94, pK_2^{25} 9.57.** Dissolved in an equal amount of water, shaken with charcoal and vacuum distd at 68°/0.1mm. It is too viscous to be distd through a packed column.

L(*S*)-2,3-Diaminopropionic acid monohydrochloride   (3-amino-L-alanine hydrochloride) *[1482-97-9]* **M 140.6, m 132-133°dec, 237°dec, $[\alpha]_D^{25}$ +26.1° (c 5.8, M HCl), pK_1^{25} 1.30, pK_2^{25} 6.79, pK_3^{25} 9.51.** Forms needles from H_2O and can be recrystd from aqueous EtOH. [Gmelin et al. *Z Physiol Chem.* **314** 28 *1959*; IR: Koegel et al. *J Am Chem Soc* **77** 5708 *1977.*]

2,3-Diaminopyridine *[452-58-4]* **M 109.1, m 116°, pK_1^{25} -0.50, pK_2^{25} 6.92.** Crystd from *benzene and sublimed *in vacuo*.

2,6-Diaminopyridine *[141-86-6]* **M 109.1, m 121.5° $pK_{Est(1)}$ <-6.0, $pK_{Est(2)}$ ~7.3.** Crystd from *benzene and sublimed *in vacuo*.

3,4-Diaminopyridine *[54-96-6]* **M 109.1, m 218-219°, pK_1^{20} 0.49, pK_2^{20} 9.14.** Crystd from *benzene and stored under H_2 because it is deliquescent and absorbs CO_2.

***meso*-2,3-Diaminosuccinic acid** *[23220-52-2]* **M 148.1, m 305-306°(dec, and sublimes), $pK_{Est(1)}$ ~3.6, $pK_{Est(2)}$~9.8.** Crystd from water.

Diaminotoluene see **toluenediamine.**

3,5-Diamino-1,2,4-triazole (Guanazole) *[1455-77-2]* **M 99.1, m 206° pK_1^{20} 4.43, pK_2^{20} 12.12.** Crystd from water or EtOH.

2,5-Di-*tert*-amylhydroquinone *[79-74-3]* **M 250.4, m 185.8-186.5°.** Crystd under N_2 from boiling glacial acetic acid (7mL/g) plus boiling water (2.5mL/g) [Stolow and Bonaventura *J Am Chem Soc* **85** 3636 *1963*].

Di-n-amyl phthalate *[131-18-0]* **M 306.4, b 204-206°/11mm, d^{25} 1.023, n 1.489.** Washed with aqueous Na_2CO_3, then distilled water. Dried with $CaCl_2$ and distd under reduced pressure. Stored in a vacuum desiccator over P_2O_5.

1,3-Diazaazulene (cycloheptimidazole) *[275-94-5]* **M 130.1, m 120°.** Recrystd repeatedly from de-aerated cyclohexane in the dark.

1,5-Diazabicyclo[4.3.0]non-5-ene (DBN, 2,3,4,,6,7,8-hexahydropyrrolo[1,2-*a*]-pyrimidine) *[3001-72-7]* **M 124.2, b 96-98°/11mm, 100-102°/12mm, 118-121°/32mm, d_4^{20} 1.040, n_D^{20} 1.520, pK >13.0.** Distd from BaO. It forms a *hydroiodide* by addn of 47% HI, dry and dissolve in MeCN, evaporate and repeat, recrystallise from EtOH then dry at 25°/1mm for 5h, then at 80°/0.03mm for 12h and store and dispense in a dry box, **m** 154-156° [Jaeger et al. *J Am Chem Soc* **101** 717 *1979*]. The *methiodide* is recrystd from $CHCl_3$ + Et_2O, **m** 248-250°, and *hydrogen fumarate* has **m** 159-160° and is crystd from *iso-*

PrOH [Rokach et al. *J Med Chem* **22** 237 *1979*; Oediger et al. *Chem Ber* **99** 2012 *1966*; Reppe et al. *Justus Liebigs Ann Chem* **596** 210 *1955*].

1,4-Diazabicyclo[2.2.2]octane (DABCO, triethylenediamine TED) *[280-57-9]* **M 112.2, m 156-157° (sealed tube), pK_1^{25} 2.97, pK_2^{25} 8.82** Crystd from 95% EtOH, pet ether or MeOH/diethyl ether (1:1). Dried under vacuum over CaCl$_2$ and BaO. It can be sublimed *in vacuo*, and readily at room temperature. Also purified by removal of water during azeotropic distn of a *benzene soln. It was then recrystd twice from anhydrous diethyl ether under argon, and stored under argon [Blackstock et al. *J Org Chem* **52** 1451 *1987*].

1,8-Diazabicyclo[5.4.0]undec-7-ene (DBU, 2,3,4,6,7,8,9,10-octahydropyrimidino[1,2-*a*]-azepine) *[6674-22-2]* **M 152.2, b 115°/11mm, d 1.023, n 1.522, pK_{Est} ~ >13.** Fractionally dist under vac. Also purified by chromatography on Kieselgel and eluting with CHCl$_3$/EtOH/25% aq NH$_3$ (15:5:2) and checked by IR and MS. [Oediger et al. *Chem Ber* **99** 2012 *1962; Angew Chem, Int Ed Engl* **6** 76 *1967*; Guggisberg et al. *Helv Chim Acta* **61** 1057 *1978*.]

1,8-Diazabiphenylene *[259-84-7]* **M 154.2, pK_{Est} ~4.4.** Recrystd from cyclohexane, then sublimed in a vacuum.

2,7-Diazabiphenylene *[31857-42-8]* **M 154.2, pK_{Est} ~4.5.** Recrystd from cyclohexane, then sublimed in a vacuum.

Diazoaminobenzene (1,3-diphenyltriazene) *[136-36-6]* **M 197.2, m 99°.** Crystd from pet ether (b 60-80°), 60% MeOH/water or 50% aqueous EtOH (charcoal) containing a small amount of KOH. Also purified by chromatography on alumina/toluene and toluene-pet ether. Stored in the dark.

6-Diazo-5-oxo-L-norleucine *[157-03-9]* **M 171.2, m 145-155°(dec), $[\alpha]_D^{20}$ +21° (c 5, EtOH), pK_1 2.1, pK_2 8.95.** Crystd from EtOH, aq EtOH or MeOH.

Dibenzalacetone *[538-58-9]* **M 234.3, m 112°.** Crystd from hot ethyl acetate (2.5mL/g) or EtOH.

Dibenz[*a,h*]anthracene *[53-70-3]* **M 278.4, m 266-267°.** The yellow-green colour (due to other pentacyclic impurities) has been removed by crystn from *benzene or by selective oxidation with lead tetraacetate in acetic acid [Moriconi et al. *J Am Chem Soc* **82** 3441 *1960*].

Dibenzo-18-crown-6 *[14187-32-7]* **M 360.4, m 163-164°.** Crystd from *benzene, *n*-heptane or toluene and dried under vacuum at room temperature for several days. [Szezygiel *J Phys Chem* **91** 1252 *1987*.]

Dibenzo-18-crown-8 *[14174-09-5]* **M 448.5, m 103-106°.** Recrystd from EtOH, and vacuum dried at 60° over P$_2$O$_5$ for 16hours. [Delville et al. *J Am Chem Soc* **109** 7293 *1987*.]

Dibenzofuran *[132-64-9]* **M 168.2, m 82.4°.** Dissolved in diethyl ether, then shaken with two portions of aqueous NaOH (2M), washed with water, separated and dried (MgSO$_4$). After evaporating the ether, dibenzofuran was crystd from aq 80% EtOH and dried under vacuum. [Cass et al. *J Chem Soc* 1406 *1958*.] High purity material was obtained by zone refining.

Dibenzopyran (xanthene) *[92-83-1]* **M 182.2, m 100.5°, b 310-312°.** See entry on p. 386.

Dibenzothiophene *[132-65-0]* **M 184.3, m 99°.** Purified by chromatography on alumina with pet ether, in a darkened room. Crystd from water or EtOH.

trans-**1,2-Dibenzoylethylene** *[959-28-4]* **M 236.3, m 109-112°, 111°.** Recrystd from MeOH or EtOH as yellow needles [Koller et al. *Helv Chim Acta* **29** 512 *1946*]. The *dioxime* has **m** 210-211°dec from AcOH. [IR: Kuhn et al. *J Am Chem Soc* **72** 5058 *1950*; Yates *J Am Chem Soc* **74** 5375 *1952*; Erickson et al. *J Am Chem Soc* **73** 5301 *1951*.]

Dibenzoylmethane (1,3-diphenyl-1,3-propanedione) *[120-46-7]* **M 224.3, m 80°.** Crystd from pet ether or MeOH.

Di-*O*-benzoyl-(*R* and *S*)-tartaric acid (H$_2$O) *[R-(+)- 17026-42-5; S-(-)- 2743-38-6]* **M 376.3, m 88-89°** **(hydrate), 173° (anhydrous),** $[\alpha]_{546}^{20}$ **(+) and (-) 136° (c 2, EtOH),** $[\alpha]_D^{20}$ **(+) and (-) 117° (c 5, EtOH), p$K_{Est(1)}$ ~2.9, p$K_{Est(2)}$~4.2.** Crystd from water (18g from 400 mL boiling H$_2$O) and stir vigorously while cooling in order to obtain crystals; otherwise an oil will separate which solidifies on cooling. Dry in a vacuum desiccator over KOH-H$_2$SO$_4$ - yield 16.4g) as monohydrate, **m 88-89°.** It crystallises from xylene as the anhydrous acid, **m 173° (150-153°).** It does not cryst from *C$_6$H$_6$, toluene, *C$_6$H$_6$-pet ether (oil), or CHCl$_3$-pet ether. [Butler and Cretcher *J Am Chem Soc* **55** 2605 *1933*; *Tetrahedron* **41** 2465 *1085*.]

2,3,6,7-Dibenzphenanthrene *[222-93-5]* **M 276.3, m 257°.** Crystd from xylene.

Dibenzyl amine *[103-49-1]* **M 197.3, m -26°, b 113-114°/0.1mm, 174-175°/6mm, 270°/250mm, 300° (partial dec), d$_4^{20}$1.027, n$_D^{20}$1.576, pK^{25} 8.52.** Purified by distn in a vacuum. **It causes burns to the skin.** The *dihydrochloride* has **m 265-266°** after recrystn from MeOH-HCl, and the *tetraphenyl boronate* has **m 129-133°.** [Bradley and Maisey *J Chem Soc* 247 *1954*; Hall *J Phys Chem* **60** 63 *1956*; Donetti and Bellora *J Org Chem* **37** 3352 *1972*.]

Dibenzyl disulfide *[150-60-7]* **M 246.4, m 71-72°.** Crystd from EtOH.

Dibenzylethylenediamine (benzathine, DBED) *[140-28-3]* **M 240.4, m 26°, b 195°/4mm, d 1.02, n 1.563, pK$_{Est(1)}$~ 5.9, pK$_{Est(2)}$~ 8.9.** Dissolve in acid, extract with toluene, basify, extract with Et$_2$O, dry over solid KOH, evap and fractionate *in vacuo*. The *diacetate* cryst from H$_2$O by addn of EtOH, has **m 111°** (sol in H$_2$O is ~25%). [Frost et al., *J Am Chem Soc* **71** 3842 *1949*.]

1,3,4,6-Di-*O*-benzylidene-D-mannitol *[28224-73-9]* **M 358.4, m 192-195°, 193°,** $[\alpha]_D^{20}$ **-11.9° (c 0.7, Me$_2$CO).** Recryst from Et$_2$O in long fine needles. λmax 256nm (ε 435) in 95% EtOH, R$_F$ 0.21 (1:1 CCl$_4$-EtOAc) on TLC Silica Gel G. [Sinclair *Carbohydr Res* **12** 150 *1970*; ORD, CD, NMR, IR, MS: Brecknell et al. *Aust J Chem* **29** 1749 *1976*.]

Dibenzyl ketone (1,3-diphenyl-2-propanone) *[102-04-5]* **M 210.3, m 34.0°.** Fractionally crystd from its melt, then crystd from pet ether. Stored in the dark.

Dibenzyl malonate *[15014-25-2]* **M 284.3, b 188-190°/0.2mm, 193-196°/1mm, d$_4^{20}$ 1.158, n$_D^{20}$ 1.5452.** Dissolve in toluene, wash with aqueous NaHCO$_3$, H$_2$O, dry over MgSO$_4$, filter, evaporate and distil. [Ginsburg and Pappo *J Am Chem Soc* **75** 1094 *1953*; Baker et al. *J Org Chem* **17** 77 *1952*.]

Dibenzyl sulfide *[538-74-9]* **M 214.3, m 48.5°, 50°.** Crystd from EtOH/water (10:1), or repeatedly from Et$_2$O. Also chromatographed on Al$_2$O$_3$ (pentane as eluent), then recrystd from EtOH [Kice and Bowers *J Am Chem Soc* **84** 2390 *1962*]. Vacuum dried at 30° over P$_2$O$_5$, fused under nitrogen and re-dried.

2,4-Dibromoaniline *[615-57-6]* **M 250.9, m 79-80°, pK^{25} 1.87.** Crystd from aqueous EtOH.

9,10-Dibromoanthracene *[523-27-3]* **M 336.0, m 226°.** Recrystd from xylene and vacuum sublimed [Johnston et al. *J Am Chem Soc* **109** 1291 *1987*].

***p*-Dibromobenzene** *[106-37-6]* **M 235.9, m 87.8°.** Steam distd, crystd from EtOH or MeOH and dried in the dark under vacuum. Purified by zone melting.

2,5-Dibromobenzoic acid *[610-71-9]* **M 279.9, m 157°, pK$_{Est}$ ~1.5.** Crystd from water or EtOH.

4,4'-Dibromobiphenyl *[92-86-4]* **M 312.0, m 164°, b 355-360°/760mm.** Crystd from MeOH.

***trans*-1,4-Dibromobut-2-ene** *[821-06-7]* **M 213.9, m 54°, b 85°/10mm.** Crystd from ligroin.

α,α-**Dibromodeoxybenzoin** *[15023-99-1]* **M 354.0, m 111.8-112.7°.** Crystd from acetic acid.

Dibromodichloromethane *[594-18-3]* **M 242.7, m 22°.** Crystd repeatedly from its melt, after washing with aqueous $Na_2S_2O_3$ and drying with BaO.

1,3-Dibromo-5,5-dimethylhydantoin *[77-48-5]* **M 285.9, m 190-192°dec, 190-193°dec.** Recrystd from H_2O. Solubility in CCl_4 is 0.003 mol/L at 25° and 0.024 mol/L at 76.5°.

1,2-Dibromoethane *[106-93-4]* **M 187.9, f 10.0°, b 29.1°/10mm, 131.7°/760mm, d 2.179, n^{15} 1.54160.** Washed with conc HCl or H_2SO_4, then water, aqueous $NaHCO_3$ or Na_2CO_3, more water, and dried with $CaCl_2$. Fractionally distd. Alternatively, kept in daylight with excess bromine for 2hours, then extracted with aqueous Na_2SO_3, washed with water, dried with $CaCl_2$, filtered and distd. It can also be purified by fractional crystn by partial freezing. Stored in the dark.

4',5'-Dibromofluorescein *[596-03-2]* **M 490.1, m 285°.** Crystd from aqueous 30% EtOH.

5,7-Dibromo-8-hydroxyquinoline *[521-74-4]* **M 303.0, m 196°, pK_1^{25} 5.84, pK_2^{25} 9.56.** Crystd from acetone/EtOH. It can be sublimed.

Dibromomaleic acid *[608-37-7]* **M 273.9, m 123.5°, 125°dec, pK_1^{25} 1.45, pK_2^{25} 4.62.** It has been recrystd from Et_2O or Et_2O-$CHCl_3$. It is slightly soluble in H_2O, soluble also in AcOH but insoluble in *C_6H_6 and $CHCl_3$. [Salmony and Simonis *Chem Ber* **38** 2583 *1905*; Ruggli *Helv Chim Acta* **3** 566 *1929*.]

2,5-Dibromonitrobenzene *[3460-18-2]* **M 280.9, m 84°.** Crystd from acetone.

2,6-Dibromo-4-nitrophenol *[99-28-5]* **M 280.9, m 143-144°, pK^{25} 3.39.** Crystd from aq EtOH.

2,4-Dibromophenol *[615-58-7]* **M 251.9, m 37°, 41-42°, b 154°/10mm, 239°/atm, pK^{25} 7.79.** Crystd from $CHCl_3$ at -40°.

2,6-Dibromophenol *[608-33-3]* **M 251.9, m 56-57°, b 138°/10mm, 255-256°/740mm, pK^{25} 6.67.** Vacuum distd (at 18mm), then crystd from cold $CHCl_3$ or from EtOH/water.

1,3-Dibromopropane *[109-64-8]* **M 201.9, f -34.4°, b 63-63.5°/26mm, 76-77°/40mm, 90°/80mm, 165°/atm, d 1.977, n 1.522.** Washed with dilute aqueous Na_2CO_3, then water. Dried and fractionally distd under reduced pressure.

2,6-Dibromopyridine *[626-05-1]* **M 236.9, m 117-119°, 118.5-119°, b 249°/757.5mm, pK_{Est} <0.** Purified by steam distn then twice recrystd from EtOH. Does not form an $HgCl_2$ salt. [den Hertog and Wibaut *Rec Trav Chim Pays Bas* **51** 381 *1932*.]

meso-**2,3-Dibromosuccinic acid** *[526-78-3]* **M 275.9, m 288-290° (sealed tube, dec), pK_1^{20} 1.56, pK_2^{20} 2.71.** Crystd from distilled water, keeping the temperature below 70°.

1,2-Dibromotetrafluoroethane *[124-73-2]* **M 259.8, b 47.3°/760mm.** Washed with water, then with weak alkali. Dried with $CaCl_2$ or H_2SO_4 and distd. [Locke et al. *J Am Chem Soc* **56** 1726 *1934*.] Also purified by gas chromatography on a silicone DC-200 column.

α,α'-**Dibromo**-*o*-**xylene** *[91-13-4]* **M 264.0, m 95°, b 129-130°/4.5mm.** Crystd from $CHCl_3$

α,α'-**Dibromo**-*m*-**xylene** *[626-15-3]* **M 264.0, m 77°, b 156-160°/12mm.** Crystd from acetone.

α,α'-**Dibromo**-*p*-**xylene** *[623-24-5]* **M 264.0, m 145-147°, b 155-158°/12-15mm, 245°/760mm.** Crystd from *benzene or chloroform.

Di-*n*-butylamine *[111-92-2]* **M 129.3, b 159°, n 1.41766, d 0.761, pK²⁵ 11.25.** Dried with LiAlH₄, CaH₂ or KOH pellets, filtered and distd from BaO or CaH₂.

α-Dibutylamino-α-(*p*-methoxyphenyl)acetamide **(Ambucetamide)** *[519-88-0]* **M 292.4, m 134°.** Crystd from EtOH containing 10% diethyl ether.

2,5-Di-*tert*-butyl aniline *[21860-03-7]* **M 205.4, m 103-104°, 103-106°, pK²⁵ 3.34 (50% aq MeOH), 3.58 (90% aq MeOH).** Recrystd from EtOH in fine needles after steam distn. It has a pKa²⁵ of 3.58 (50% aq EtOH) and 3.34 (90% aq MeOH). The *tosylate* has **m** 164° (from AcOH). [Bell and Wilson *J Chem Soc* 2340 *1956*; Carpenter et al. *J Org Chem* 16 586 *1951*; Bartlett et al. *J Am Chem Soc* 76 2349 *1954*.]

Di-*tert*-butylazodicarboxylate *[870-50-8]* **M 230.3, m 90-92°.** Cryst from ligroin. Best purified by covering the dry solid (22g) with pet ether (b 30-60°, 35-40 mL) and adding ligroin (b 60-90°) to the boiling soln until the solid dissolves. On cooling, large lemon yellow crystals of the ester separate (~ 20g), **m** 90.7-92°. Evapn of the filtrate gives a further crop of crystals [Carpino and Crowley *Org Synth* 44 18 *1964*].

***p*-Di-*tert*-butylbenzene** *[1012-72-2]* **M 190.3, m 80°.** Crystd from diethyl ether, EtOH and dried under vacuum over P₂O₅ at 55°. [Tanner et al. *J Org Chem* 52 2142 *1987*.]

2,6-Di-*tert*-butyl-1,4-benzoquinone *[719-22-2]* **M 220.3, m 66-67°.** It can be recrystd from MeOH and sublimed in a vaccum.

3,5-Di-*tert*-butyl-*o*-benzoquinone *[3383-21-9]* **M 220.3, m 112-114°, 113-114°.** It can be recrystd from MeOH or pet ether, and forms fine red plates or rhombs. [Flaig et al. *Justus Liebigs Ann Chem* 597 196 *1955*; IR: Ley and Müller *Chem Ber* 89 1402 *1956*.]

3,5-Di-*tert*-butyl catechol *[1020-31-1]* **M 222.3, m 99°, 99-100°, pK_Est(1) ~11.0, pK_Est(2)~13.1.** Recrystd from pet ether. [Ley and Müller *Chem Ber* 89 1402 *1956*; UV Flaig et al. *Z Naturforschung* 10b 668 *1955*.] Also crystd three times from pentane [Funabiki et al. *J Am Chem Soc* 108 2921 *1986*].

Dibutylcarbitol **[di(ethyleneglycol)dibutyl ether]** *[112-73-2]* **M 218.3, b 125-130°/0.1mm, d 0.883, n 1.424.** Freed from peroxides by slow passage through a column of activated alumina. The eluate was shaken with Na₂CO₃ (to remove any remaining acidic impurities), washed with water, and stored with CaCl₂ in a dark bottle [Tuck *J Chem Soc* 3202 *1957*].

2,6-Di-*tert*-butyl-*p*-cresol **(2,6-di-*tert*-butyl-4-methylphenol,   butylatedhydroxytoluene, BHT)** *[128-37-0]* **M 230.4, m 71.5°, pK²⁵ 12.23.** Dissolved in *n*-hexane at room temperature, then cooled with rapid stirring, to -60°. The ppte was separated, redissolved in hexane, and the process was repeated until the mother liquor was no longer coloured. The final product was stored under N₂ at 0° [Blanchard *J Am Chem Soc* 82 2014 *1960*]. Also crystd from EtOH, MeOH, *benzene, *n*-hexane, methylcyclohexane or pet ether (b 60-80°), and dried under vacuum.

Di-*tert*-butyl dicarbonate **(di-*tert*-butyl pyrocarbonate)** *[24424-99-5]* **M 218.3, m 23° (21-22°), b 55-56°/0.15mm, 62-65°/0.4mm, d 0.950, n 1.409.** Melt by heating at ~35°, and distil in vac. If IR and NMR (ν 1810m 1765 cm⁻¹, δ in CCl₄ 1.50 singlet) suggest very impure then wash with equal vol of H₂O containing citric acid to make the aqueous layer slightly acidic, collect the organic layer and dry over anhyd MgSO₄ and distil in vac. [Pope et al. *Org Synth* 57 45 *1977*.] **FLAMMABLE.**

2,6-Di-*tert*-butyl-4-dimethylaminomethylphenol *[88-27-7]* **M 263.4, m 93-94°, b 172°/30mm, pK_Est ~12.0.** Crystd from *n*-hexane.

Di-*tert*-butyldiperphthalate *[2155-71-7]* **M 310.3, m dec explosively, CARE.** Crystd from diethyl ether. Dried over H₂SO₄.

2,6-Di-*tert*-butyl-4-ethylphenol *[4130-42-1]* **M 234.4, m 42-44°, pK$_{Est}$ ~12.3.** Cryst from aqueous EtOH or *n*-hexane.

N,N-Dibutyl formamide *[761-65-9]* **M 157.3, b 63°/0.1mm, 118-120°/15mm, 244-246°/760mm, d$_4^{20}$ 0.878, n$_D^{20}$ 1.445.** Purified by fractn distn [Mandel and Hill *J Am Chem Soc* **76** 3981 *1954*].

2,5-Di-*tert*-butylhydroquinone *[88-58-4]* **M 222.3, m 222-223°.** Crystd from *C$_6$H$_6$ or AcOH.

2,4-Di-*tert*-butyl-4-isopropylphenol *[5427-03-2]* **M 248.4, m 39-41°, pK$_{Est}$ ~12.3.** Crystd from *n*-hexane or aq EtOH.

2,6-Di-*tert*-butyl-4-methylpyridine *[38222-83-2]* **M 205.4, m 31-32°, 33-36°, b 148-153°/95mm, 223°/760mm, n$_4^{20}$ 1.476, pK$_{Est}$ ~5.7.** Possible impurity is 2,6-di-*tert*-butyl-4-neopentylpyridine. Attempts to remove coloured impurities directly by distn, acid-base extraction or treatment with activated charcoal were unsuccessful. Pure material can be obtained by dissolving 0.3mole of the alkylpyridine in pentane (150mL) and introducing it at the top of a water jacketed chromatographic column (40 x 4.5cm) the cooling is necessary because the base in pentane reacts exothermically with alumina) containing activated and acidic alumina (300g). The column is eluted with pentane using a 1L constant pressure funnel fitted at the top of the column to provide slight press. All the pyridine is obtained in the first two litres of eluent (the progress of elution is monitored by spotting a fluorescent TLC plate and examining under short wave UV light - a dark blue spot is evidence for the presence of the alkylpyridine). Elution is complete in 1h. Pentane is removed on a rotovap with 90-93% recovery yielding a liquid which solidifies on cooling, **m** 31-32°, and the base can be distilled. The *HPtCl$_6$ salt* has **m** 213-314° (dec), and the *CF$_3$SO$_3$H salt* has **m** 202.5-203.5° (from CH$_2$Cl$_2$). [*Org Synth* **60** 34 *1981*.]

Di-*tert*-butyl peroxide (*tert*-butyl peroxide) *[110-05-4]* **M 146.2, d 0.794, n 1.389.** Washed with aqueous AgNO$_3$ to remove olefinic impurities, water and dried (MgSO$_4$). Freed from *tert*-butyl hydroperoxide by passage through an alumina column [Jackson et al. *J Am Chem Soc* **107** 208 *1985*], and if necessary two high vacuum distns from room temp to a liquid-air trap [Offenbach and Tobolsky *J Am Chem Soc* **79** 278 *1957*]. *The necessary protection from* **EXPLOSION** *should be used.*

2,6-Di-*tert*-butylphenol *[128-39-2]* **M 206.3, m 37-38°, pK^{25} 11.70.** Crystd from aqueous EtOH or *n*-hexane.

Dibutyl phthalate *[84-74-2]* **M 278.4, b 206°/20mm, 340°/760mm, d 1.4929, d^5 1.0426, n^{25} 1.490.** Washed with dilute NaOH (to remove any butyl hydrogen phthalate), aqueous NaHCO$_3$ (charcoal), then distd water. Dried with CaCl$_2$, distd under vacuum, and stored in a desiccator over P$_2$O$_5$. (See also p. 151.)

2,6-Di-*tert*-butylpyridine *[585-48-8]* **M 191.3, b 100-101°/23mm, d 0.852, n 1.474, pK^{25} 5.02.** Redistd from KOH pellets.
§ Polystyrene supported version is commercially available.

Di-*n*-butyl sulfide *[544-40-1]* **M 146.3, α-form b 182°, β-form 190-230°(dec).** Washed with aq 5% NaOH, then water. Dried with CaCl$_2$ and distd from sodium.

Di-*n*-butyl sulfone *[598-04-9]* **M 162.3, m 43.5°.** Purified by zone melting.

N,N'-Di-*tert*-butylthiourea *[4041-95-6]* **M 188.3, m 174-175° (evac capillary).** Recrystd from H$_2$O [Bortnick et al. *J Am Chem Soc* **78** 4358 *1956*].

3,5-Dicarbethoxy-1,4-dihydrocollidine *[632-93-9]* **M 267.3, m 131-132°.** Crystd from hot EtOH/water.

Dichloramine-T (*N,N*-dichloro-*p*-toluenesulfonamide) *[473-34-7]* **M 240.1, m 83°.** Crystd from pet ether (b 60-80°) or CHCl₃/pet ether. Dried in air. (see also chloramine-T in Chapter 5).

Dichloroacetic acid *[79-43-6]* **M 128.9, m 13.5°, b 95.0-95.5°/17-18mm, d 1.563, n 1.466, pK²⁵ 1.35.** Crystd from *benzene or pet ether. Dried with MgSO₄ and fractionally distd. [Bernasconi et al. *J Am Chem Soc* **107** 3612 *1985*.]

sym-**Dichloroacetone** (**1,3-dichloropropan-2-one**) *[534-07-6]* **M 127.0, m 41-43°, 45°, b 86-88°/12mm, 75-77°/22mm, 172-172.5°/atm, 170-175° /atm, d 1.383.** Crystd from CCl₄, CHCl₃ and *benzene. Distd under vacuum. [Conant and Quayle *Org Synth* Coll Vol 211 *1941*; Hall and Sirel *J Am Chem Soc* **74** 836 *1952*]. It is dimorphic [Daasch and Kagarise *J Am Chem Soc* **77** 6156 *1955*]. The *oxime* has **m** 130-131°, **b** 106°/25mm [*Arzneimittel-Forsch* **8** 638 *1958*].

Dichloroacetonitrile *[3018-12-0]* **M 110.0, b 110-112°, d 1.369, n 1.440.** Purified by distn and by gas chromatography. **FLAMMABLE.**

2,4-Dichloroaniline *[554-00-7]* **M 162.0, m 63°, pK²⁵ 2.02.** Crystd from EtOH/water. Also crystd from EtOH and dried *in vacuo* for 6h at 40° [Moore et al. *J Am Chem Soc* **108** 2257 *1986*; Edidin et al. *J Am Chem Soc* **109** 3945 *1987*].

3,4-Dichloroaniline *[95-76-1]* **M 162.0, m 71.5°, pK²⁵ 2.97.** Crystd from MeOH.

9,10-Dichloroanthracene *[605-49-1]* **M 247.1, m 214-215°.** Purified by crystn from MeOH or EtOH, followed by sublimation under reduced pressure. [Masnori and Kochi *J Am Chem Soc* **107** 7880 *1985*.]

2,4-Dichlorobenzaldehyde *[874-42-0]* **M 175.0, m 72°.** Crystd from EtOH or ligroin.

2,6-Dichlorobenzaldehyde *[83-38-5]* **M 175.0, m 70.5-71.5°.** Crystd from EtOH/water or pet ether (b 30-60°).

o-**Dichlorobenzene** *[95-50-1]* **M 147.0, b 81-82°/31-32mm, 180.5°/760mm, d 1.306, n 1.551, n²⁵ 1.549.** Contaminants may include the *p*-isomer and trichlorobenzene [Suslick et al. *J Am Chem Soc* **106** 4522 *1984*]. It was shaken with conc or fuming H₂SO₄, washed with water, dried with CaCl₂, and distd from CaH₂ or sodium in a glass-packed column. Low conductivity material (*ca* 10⁻¹⁰ mhos) has been obtained by refluxing with P₂O₅, fractionally distilled and passed through a column packed with silica gel or activated alumina: it was stored in a dry-box under N₂ or with activated alumina.

m-**Dichlorobenzene** *[541-73-1]* **M 147.0, b 173.0°, d 1.289, n 1.54586, n²⁵ 1.54337.** Washed with aqueous 10% NaOH, then with water until neutral, dried and distd. Conductivity material (*ca* 10⁻¹⁰ mhos) has been prepared by refluxing over P₂O₅ for 8h, then fractionally distilling, and storing with activated alumina. *m*-Dichlorobenzene dissolves rubber stoppers.

p-**Dichlorobenzene** *[106-46-7]* **M 147.0, m 53.0°, b 174.1°, d 1.241, n⁶⁰ 1.52849.** *o*-Dichlorobenzene is a common impurity. Has been purified by steam distn, crystn from EtOH or boiling MeOH, air-dried and dried in the dark under vacuum. Also purified by zone refining.

2,2'-Dichlorobenzidine *[84-68-4]* **M 253.1, m 165°, pK_Est(1) ~3.0, pK_Est(2)~4.0.** Crystd from EtOH.

3,3'-Dichlorobenzidine *[91-94-1]* **M 253.1, m 132-133°, pK_Est(1) ~4.8, pK_Est(2)~5.7.** Crystd from EtOH or *benzene. **CARCINOGEN.**

2,4-Dichlorobenzoic acid *[50-84-0]* **M 191.0, m 163-164°, pK²⁵ 2.68.** Crystd from aqueous EtOH (charcoal), then *benzene (charcoal). It can also be recrystd from water.

2,5-Dichlorobenzoic acid *[50-79-3]* **M 191.0, m 154°, b 301°/760mm, pK²⁵ 2,47.** Crystd from water.

2,6-Dichlorobenzoic acid *[50-30-6]* **M 191.0, m 141-142°, pK²⁵ 1.59.** Crystd from EtOH and sublimed *in vacuo*.

3,4-Dichlorobenzoic acid *[51-44-5]* **M 191.0, m 206-207°, pK²⁵ 3.64.** Crystd from aqueous EtOH (charcoal) or acetic acid.

3,5-Dichlorobenzoic acid *[51-36-5]* **M 191.0, m 188°, pK²⁵ 3.54.** Crystd from EtOH and sublimed in a vacuum.

2,6-Dichlorobenzonitrile *[1194-65-6]* **M 172.0, m 145°.** Crystd from acetone.

4,4'-Dichlorobenzophenone *[90-98-2]* **M 251.1, m 145-146°.** Recrystd from EtOH [Wagner et al. *J Am Chem Soc* **108** 7727 *1986*].

2,5-Dichloro-1,4-benzoquinone *[615-93-0]* **M 177.0, m 161-162°.** Recrystd twice from 95% EtOH as yellow needles [Beck et al. *J Am Chem Soc* **108** 4018 *1986*].

2,6-Dichloro-1,4-benzoquinone *[697-91-6]* **M 177.0, m 122-124°.** Recrystd from pet ether (b 60-70°) [Carlson and Miller *J Am Chem Soc* **107** 479 *1985*].

2,6-Dichlorobenzoyl chloride *[4659-45-4]* **M 209.5, m 15-17, b 122-124°/15mm, d 1.464.** Reflux for 2h with excess of acetyl chloride (3 vols), distil off AcCl followed by the benzoyl chloride. Store away from moisture. It is an **IRRITANT**.

3,4-Dichlorobenzyl alcohol *[1805-32-9]* **M 177.0, m 38-39°.** Crystd from water.

2,3-Dichloro-1,3-butadiene *[1653-19-6]* **M 123.0, b 41-43°/85mm, 98°/760mm.** Crystd from pentane to constant melting point about -40°. A mixture of *meso* and *d,l* forms was separated by gas chromatography on an 8m stainless steel column (8mm i.d.) with 20% DEGS (diethyleneglycolsilyl chloride) on Chromosorb W (60-80 mesh) at 60° and 80mL He/min. [Su and Ache *J Phys Chem* **80** 659 *1976*.]

(+) and (-) (8,8-Dichlorocamphorylsulfonyl)oxaziridine *[127184-05-8]* **M 298.2, m 178-180°, 183-186°, [α]²⁰_D (+) and (-) 88.3° (c 1.3, CHCl₃), (+) and (-) 91° (c 5, CHCl₃).** Recrystd from EtOH [Davis and Weismiller *J Org Chem* **55** 3715 *1990*].

cis-**3,4-Dichlorocyclobutene** *[2957-95-1]* **M 123.0, b 70-71°/55mm, 74-76°/55mm, d₄²⁰ 1.297, n_D²⁰ 1.499.** Distd at 55mm through a 36-in platinum spinning band column, a fore-run b 58-62°/55mm is mainly 1,4-dichlorobutadiene. When the temperature reaches 70° the reflux ratio is reduced to 10:1 and the product is collected quickly. It is usually necessary to apply heat frequently with a sun lamp to prevent any dichlorobutadiene from clogging the exit in the early part of the distn [Pettit and Henery *Org Synth* **50** 36 *1970*].

2,3-Dichloro-5,6-dicyano-*p*-benzoquinone (DDQ) *[84-58-2]* **M 227.0, m 203° (dec).** Crystd from CHCl₃, CHCl₃/*benzene (4:1), or *benzene and stored at 0°. [Pataki and Harvey *J Org Chem* **52** 2226 *1987*.]

ß,ß'-Dichlorodiethyl ether *[111-44-4]* **M 143.0, b 79-80°/20mm, 176-177.0°/743mm, n 1.457, d 1.219.** See bis-(β-dichloroethyl)ether on p. 134.

1,2-Dichloro-1,2-difluoroethane *[431-06-1]* **M 134.9, b 59°, n 1.376.** Purified by fract dist [Hazeldine *J Chem Soc* 4258 *1952*]. For purification of diastereoisomeric mixture, with resolution into *meso* and *rac* forms, see Machulla and Stocklin [*J Phys Chem* **78** 658 *1974*].

Dichlorodifluoromethane (Freon 12) *[75-71-8]* **M 120.9, m -158°, b -29.8°/atm, 42.5°/10atm.** Passage through saturated aqueous KOH then conc H_2SO_4, and a tower packed with activated copper on Kielselguhr at 200° removed CO_2 and O_2. A trap cooled to -29° removed a trace of high boiling material. It is a non-flammable propellant.

1,3-Dichloro-5,5'-dimethylhydantoin *[118-52-5]* **M 197.0, m 132-134°, 136°.** Purified by dissolving in conc H_2SO_4 and diluting with ice H_2O, dry and rerystd from $CHCl_3$. It sublimes at 100° in a vacuum. Exhibits time dependent hydrolysis at pH 9. [Petterson and Grzeskowiak *J Org Chem* **24** 1414 *1959*.]

4,5-Dichloro-3H-1,2-dithiol-3-one *[1192-52-5]* **M 187.1, m 52-56°, 61°, b 87°/0.5mm, 125°/11mm.** Distd *in vacuo* and then recrystd from pet ether. IR: ν 1650 cm^{-1} [Boberg *Justus Liebigs Ann Chem* **693** 212 *1966*].

1,1-Dichloroethane (ethylidene dichloride) *[75-34-3]* **M 99.0, b 57.3°, d^{15} 1.18350, d 1.177, n^{15} 1.41975.** Shaken with conc H_2SO_4 or aqueous $KMnO_4$, then washed with water, saturated aqueous $NaHCO_3$, again with water, dried with K_2CO_3 and distd from CaH_2 or $CaSO_4$. Stored over silica gel.

1,2-Dichloroethane *[107-06-2]* **M 99.0, b 83.4°, d 1.256, n^{15} 1.44759.** Usually prepared by chlorinating ethylene, so that likely impurities include higher chloro derivatives and other chloro compounds depending on the impurities originally present in the ethylene. It forms azeotropes with water, MeOH, EtOH, trichloroethylene, CCl_4 and isopropanol. Its azeotrope with water (containing 8.9% water, and b 77°) can be used to remove gross amounts of water prior to final drying. As a preliminary purification step, it can be steam distd, and the lower layer was treated as below.
Shaken with conc H_2SO_4 (to remove alcohol added as an oxidation inhibitor), washed with water, then dilute KOH or aqueous Na_2CO_3 and again with water. After an initial drying with $CaCl_2$, $MgSO_4$ or by distn, it is refluxed with P_2O_5, $CaSO_4$ or CaH_2 and fractionally distd. Carbonyl-containing impurities can be removed as described for chloroform.

1,2-Dichloroethylene *[cis + trans 540-59-0]* **M 96.9, b 60° (cis), d 1.284, b 48° (trans), d 1.257.** Shaken successively with conc H_2SO_4, water, aqueous $NaHCO_3$ and water. Dried with $MgSO_4$ and distn separated the *cis*- and *trans*-isomers.

cis-1,2-Dichloroethylene *[156-59-2]* **M 96.9, b 60.4°, d 1.2830, n^{15} 1.44903, n 1.4495.** Purified by careful fractional distn, followed by passage through neutral activated alumina. Also by shaking with mercury, drying with K_2CO_3 and distn. from $CaSO_4$.

trans-1,2-Dichloroethylene *[156-60-5]* **M 96.9, b 47.7°, n^{15} 1.45189, n 1.4462, d 1.2551.** Dried with $MgSO_4$, and fractionally distd under CO_2. Fractional crystn at low temperatures has also been used.

5,7-Dichloro-8-hydroxyquinoline *[773-76-2]* **M 214.1, m 180-181°, pK$_1$ 1.89, pK$_2$ 7.62.** Crystd from acetone/EtOH.

2,3-Dichloromaleic anhydride *[1122-17-4]* **M 167.0, m 105-115°, 120°, 121-121.5°.** Purified by sublimation *in vacuo* [Katakis et al. *J Chem Soc, Dalton Trans* 1491 *1986*]. It has also been purified by Soxhlet extraction with hexane, recrystd from $CHCl_3$ and sublimed [MS, Relles *J Org Chem* **37** 3630 *1972*].

Dichloromethane (methylene dichloride) *[75-09-2]* **M 84.9, b 40.0°, d 1.325, n 1.42456, n^{25}1.4201.** Shaken with portions of conc H_2SO_4 until the acid layer remained colourless, then washed with water, aqueous 5% Na_2CO_3, $NaHCO_3$ or NaOH, then water again. Pre-dried with $CaCl_2$, and distd from $CaSO_4$, CaH_2 or P_2O_5. Stored away from bright light in a brown bottle with Linde type 4A molecular sieves, in an atmosphere of dry N_2. Other purification steps include washing with aq $Na_2S_2O_3$, passage through a column of silica gel, and removal of carbonyl-containing impurities as described under **Chloroform**. It has also been purified by treatment with basic alumina, distd, and stored over molecular sieves under nitrogen [Puchot et al. *J Am Chem Soc* **108** 2353 *1986*].

Dichloromethane from Japanese sources contained MeOH as stabiliser which is not removed by distn. It can, however, be removed by standing over activated 3A Molecular Sieves (note that 4A Sieves cause the development of pressure in bottles), passed through activated Al_2O_3 and distd [Gao et al. *J Am Chem Soc* **109** 5771 *1987*]. It has been fractionated through a platinum spinning band column, degassed, and distd onto degassed molecular sieves, Linde 4A, heated under high vacuum at over 450° until the pressure readings reached the low values of 10^{-6} mm — ~1-2h [Mohammad and Kosower *J Am Chem Soc* **93** 2713 *1971*].
Rapid purification: Reflux over CaH_2 (5% w/v) and distil. Store over 4A molecular sieves.

3,9-Dichloro-7-methoxyacridine *[86-38-4]* **M 278.1, m 160-161°.** Crystd from *benzene.

5,7-Dichloro-2-methyl-8-hydroxyquinoline **(5,7-dichloro-8-hydroxyquinaldine)** *[72-80-0]* **M 228.1, m 114-115°, $pK_{Est(1)}$ ~2.0, $pK_{Est(2)}$~8.4.** Crystd from EtOH.

2,4-Dichloro-6-methylphenol *[1570-65-6]* **M 177.0, m 55°, b 129-132°/40mm, pK^{20} 8.14.** Crystd from water.

2,4-Dichloro-1-naphthol *[2050-76-2]* **M 213.1, m 106-107°, pK_{Est} ~7.7.** Crystd from MeOH.

2,3-Dichloro-1,4-naphthoquinone *[117-80-6]* **M 227.1, m 193°.** Crystd from EtOH.

2,5-Dichloro-4-nitroaniline *[6627-34-5]* **M 207.0, m 157-158°, pK^{25} -1.74 (aq H_2SO_4)** . Crystd from EtOH, then sublimed.

2,6-Dichloro-4-nitroaniline *[99-30-9]* **M 207.0, m 193°.** Crystd from aq EtOH or *benzene/EtOH.

2,5-Dichloro-1-nitrobenzene *[89-61-2]* **M 192.0, m 56°.** Crystd from absolute EtOH.

3,4-Dichloro-1-nitrobenzene *[99-54-7]* **M 192.0, m 43°.** Crystd from absolute EtOH.

2,4-Dichloro-6-nitrophenol *[609-89-2]* **M 208.0, m 122-123°, pK_{Est} ~5.0.** Crystd from AcOH.

2,6-Dichloro-4-nitrophenol *[618-00-4]* **M 208.0, m 125°, pK^{25} 3.55.** Crystd from EtOH and dried *in vacuo* over anhydrous $MgSO_4$.

4,6-Dichloro-5-nitropyrimidine *[4316-93-2]* **M 194.0, m 100-103°, 101-102°, pK_{Est} <0.** If too impure then dissolve in Et_2O, wash with H_2O, dry over $MgSO_4$, evaporate to dryness and recrystallise from pet ether (b 85-105°) as a light tan solid. It is sol in *ca* 8 parts of MeOH [Boon et al, *J Chem Soc* 96 *1951*; Montgomery et al. in *Synthetic Procedures in Nucleic Acid Chemistry* (Zorbach and Tipson eds) Wiley & Sons, NY, p76 *1968*].

Dichlorophen **[2,2'-methylenebis(4-chlorophenol)]** *[97-23-4]* **M 269.1, b 177-178°, pK_{Est} ~9.7.** Crystd from toluene.

2,3-Dichlorophenol *[576-24-9]* **M 163.0, m 57°, pK^{25} 7.70.** Crystd from ether.

2,4-Dichlorophenol *[120-83-2]* **M 163.0, m 42-43°, pK^{25} 7.89.** Crystd from pet ether (b 30-40°). Purified by repeated zone melting, using a P_2O_5 guard tube to exclude moisture. Very *hygroscopic* when dry.

2,5-Dichlorophenol *[583-78-8]* **M 163.0, m 58°, b 211°/744mm, pK^{25} 7.51.** Crystd from ligroin and sublimed.

3,4-Dichlorophenol *[95-77-2]* **M 163.0, m 68°, b 253.5°/767mm, pK^{25} 8.58.** Crystd from pet ether/*benzene mixture.

3,5-Dichlorophenol *[591-35-5]* **M 163.0, m 68°, b 122-124°/8mm, 233-234°/760mm, pK^{25} 8.81.** Crystd from pet ether/*benzene mixture.

2,4-Dichlorophenoxyacetic acid (2,4-D) *[94-75-7]* **M 221.0, m 146°, pK^{25} 2.90.** Crystd from MeOH. **TOXIC.**

α-(2,4-Dichlorophenoxy)propionic acid (2,4-DP, Dichloroprop) *[120-36-5]* **M 235.1, m 117°, pK^{20} 2,86,** Crystd from MeOH. **TOXIC.**

2,4-Dichlorophenylacetic acid *[19719-28-9]* **M 205.0, m 131°, 132-133°, pK$_{Est}$ ~4.0.** Crystd from aqueous EtOH.

2,6-Dichlorophenylacetic acid *[6575-24-2]* **M 205.0, m 157-158°, pK$_{Est}$ ~3.8.** Crystd from aqueous EtOH.

3-(3,4-Dichlorophenyl)-1,1-dimethyl urea (Diuron) *[330-54-1]* **M 233.1.** Crystd four times from 95% EtOH [Beck et al. *J Am Chem Soc* **108** 4018 *1986*].

4,5-Dichloro-*o*-phenylenediamine *[5348-42-5]* **M 177.1, m 162°, 162-163°, pK$_{Est(1)}$ ~-1.0, pK$_{Est(2)}$~2.9.** Recrystd from hexane, *C$_6$H$_6$, pet ether or H$_2$O (Na$_2$SO$_4$) and sublimed at 150°/15mm.

4,5-Dichlorophthalic acid *[56962-08-4]* **M 235.0, m 200° (dec to anhydride), pK$_{Est(1)}$ 2.2, pK$_{Est(2)}$~4.7.** Crystd from water. Can be purified by converting to the anhydride, reacting with boiling EtOH to form the *monoethyl ester* (**m** 133-134°) and hydrolysing back to the diacid

3,6-Dichlorophthalic anhydride *[4466-59-5]* **M 189-191°, 191-191.5°, b 339°.** Boil in xylene (allowing any vapours which would contain H$_2$O to be removed, e.g. Dean and Stark trap), which causes the acid to dehydrate to the anhydride and cool. Recryst from xylene [Villiger *Chem Ber* **42** 3539 *1909*; Fedoorow *Izv Akad Nauk SSSR Otd Khim Nauk* 397 *1948*, *Chem Abstr* 1585 *1948*].

1,2-Dichloropropane *[78-87-5]* **M 113°, b 95.9-96.2°, d 1.158, n 1.439.** Distd from CaH$_2$.

2,2-Dichloropropane *[594-20-7]* **M 113.0, b 69.3°, d 1.090, n 1.415.** Washed with aqueous Na$_2$CO$_3$ soln, then distilled water, dried over CaCl$_2$ and fractionally distd.

2,6-Dichloropurine *[5451-40-1]* **M 189.0, m 180-181.5°, 181°, 185-195°(dec), 188-189°, pK$_1^{20}$ 1.16 (aq H$_2$SO$_4$) , pK$_2^{20}$ 7.06.** It can be recrystd from 150 parts of boiling H$_2$O and dried at 100° to constant weight. Soluble in EtOAc. The HgCl$_2$ salt separates from EtOH soln. UV: λmax 275nm (ε 8.9K) at pH 1; and 280nm (ε 8.5K) at pH 11 [Elion and Hitchings *J Am Chem Soc* **78** 3508 *1956*; Schaeffer and Thomas *J Am Chem Soc* **80** 3738 *1958*; Beaman and Robins *J Appl Chem (London)* **12** 432 *1962*; Montgomery *J Am Chem Soc* **78** 1928 *1956*].

2,6-Dichloropyridine *[2402-78-0]* **M 148.0, m 87-88°, pK -2.86 (aq H$_2$SO$_4$).** Crystd from EtOH.

3,5-Dichloropyridine *[2457-47-8]* **M 148.0, m 64-65°, pK^{25} 0.67.** Crystd from EtOH.

4,7-Dichloroquinoline *[86-98-6]* **M 198.1, m 86.4-87.4°, b 148°/10mm, pK^{25} 2.80.** Crystd from MeOH or 95% EtOH.

2,3-Dichloroquinoxaline *[2213-63-0]* **M 199.0, m 152-153°, 152-154°, pK$_{Est}$ <0.** Recrystd from *C$_6$H$_6$ and dried in a vacuum [Cheeseman *J Chem Soc* 1804 *1955*].

2,6-Dichlorostyrene *[28469-92-3]* **M 173.0, b 72-73°/2mm, d 1.4045, n 1.5798.** Purified by fractional crystn from the melt and by distn.

2,4-Dichlorotoluene *[95-73-8]* **M 161.1, m -13.5°, b 61-62°/3mm, d 1.250, n 1.5513.** Recrystd from EtOH at low temperature or fractionally distd.

2,6-Dichlorotoluene *[118-69-4]* **M 161.1, b 199-200°/760mm, d 1.254, n 1.548.** Fractionally distd and collecting the middle fraction.

3,4-Dichlorotoluene *[95-75-0]* **M 161.1, m -16°, b 205°/760mm, d 1.2541, n 1.549.** Recrystd from EtOH at very temperature or fractionally distd.

α,α'-Dichloro-*p*-xylene *[623-25-6]* **M 175.1, m 100°.** Crystd from *benzene and dried under vacuum.

Dicinnamalacetone (1,9-diphenyl-1,3,6,8-nonatetraen-5-one) *[622-21-9]* **M 314.4, m 146°.** Crystd from *benzene/isooctane (1:1).

Dicumyl peroxide *[80-43-3]* **M 270.4, m 39-40°.** Crystd from 95% EtOH (charcoal). Stored at 0°. *Potentially* **EXPLOSIVE.**

9,10-Dicyanoanthracene *[1217-45-4]* **M 228.3, m 340°.** Recrystd twice from pyridine [Mattes and Farid *J Am Chem Soc* **108** 7356 *1986*].

1,2-Dicyanobenzene *[91-15-6]* **M 128.1, m 141°.** (See phthalonitrile on p. 334.)

1,4-Dicyanobenzene *[623-26-7]* **M 128.1, m 222°.** Crystd from EtOH.

1,4-Dicyanonaphthalene *[3029-30-9]* **M 178.2, m 206°.** Purified by crystn and sublimed *in vacuo*.

1,3-Dicyclohexylcarbodimide (DCC) *[538-75-0]* **M 206.3, m 34-35°, b 95-97°/0.2mm, 120-121°/0.6mm, 155°/11mm.** It is sampled as a liquid after melting in warm H_2O. It is sensitive to air and *it is a potent skin irritant.* It can be distd in a vacuum and stored in a tightly stoppered flask in a freezer. It is very soluble in CH_2Cl_2 and pyridine where the reaction product with H_2O, after condensation, is dicyclohexyl urea which is insoluble and can be removed by filtration. Alternatively dissolve in CH_2Cl_2 add powdered anhyd $MgSO_4$ shake 4h, filter, evaporate and distil at 0.6 mm press and oil bath temperature 145°. [*Biochem Prep* **10**, 122 *1963*; *Justus Liebigs Ann Chem* **571** 83 *1951*; *Justus Liebigs Ann Chem* **612** 11 *1958*.]

***cis*-Dicyclohexyl-18-crown-6** *[16069-36-6]* **M 372.5, m 47-50°.** Purified by chromatography on neutral alumina and eluting with an ether/hexane mixture [see *Inorg Chem* **14** 3132 *1975*]. Dissolved in ether at *ca* 40°, and spectroscopic grade MeCN was added to the soln which was then chilled. The crown ether ppted and was filtered off. It was dried *in vacuo* at room temperature [Wallace *J Phys Chem* **89** 1357 *1985*]. **SKIN IRRITANT.**

Di-*n*-decylamine *[1120-49-6]* **M 297.6, m 34°. b 153°/1mm, 359°/760mm, pK_{Est} ~11.0.** Dissolved in *benzene and ppted as its bisulfate by shaking with 4M H_2SO_4. Filtered, washed with *benzene, separating by centrifugation, then the free base was liberated by treating with aqueous NaOH [McDowell and Allen *J Phys Chem* **65** 1358 *1961*].

Didodecylamine *[3007-31-6]* **M 353.7, m 51.8°, pK^{25} 11.00.** Crystd from EtOH/*C_6H_6 under N_2.

Didodecyldimethylammonium bromide *[3282-73-3]* **M 463.6, m 157-162°.** Recrystd from acetone, acetone/ether mixture, then from ethyl acetate, washed with ether and dried in a vacuum oven at 60° [Chen et al. *J Phys Chem* **88** 1631 *1984*; Rupert et al. *J Am Chem Soc* **107** 2628 *1985*; Halpern et al. *J Am Chem Soc* **108** 3920 *1986*; Allen et al. *J Phys Chem* **91** 2320 *1987*].

Dienestrol [4,4'-(diethylidene-ethylene)diphenol, Dienol] *[84-17-3]* **M 266.3, m 227-228°, 231-233°, pK$_{Est}$ ~9.8.** Crystd from EtOH or dilute EtOH, sublimes at 130°/1mm. The *diacetate* has **m** 119-120° (from EtOH) [Hobday and Short *J Chem Soc* 609 *1943*].

Diethanolamine (2,2'-iminodiethanol) *[111-42-2]* **M 105.1, m 28°, b 154-155°/10mm, 270°/760mm pK^{25} 8.88.** Fractionally distd twice, then fractionally crystd from its melt.

3,4-Diethoxy-3-cyclobutene-1,2-dione (diethyl squarate) *[5321-87-8]* **M 170.2, b 89-91°/0.4mm, 88-92°/0.4mm, d$_4^{20}$ 1.162, n$_D^{25}$ 1.5000.** Dissolve in Et$_2$O, wash with Na$_2$CO$_3$, H$_2$O and dry (Na$_2$SO$_4$), filter, evaporate and distil using a Kügelrohr or purify by chromatography. Use a Kieselgel column and elute with 20% Et$_2$O-Pet ether (b 40-60°) then with Et$_2$O-pet ether (1:1), evaporate and distil *in vacuo*. [Dehmlow and Schell *Chem Ber* **113** 1 *1980*; Perri and Moore *J Am Chem Soc* **112** 1897 *1990*; IR: Cohen and Cohen *J Am Chem Soc* **88** 1533 *1966*.] **It can cause severe dermatitis** [Foland et al. *J Am Chem Soc* **111** 975 *1989*].

N,N-Diethylacetamide *[685-91-6]* **M 157.2, b 86-88°, n 1.474, d 0.994.** Dissolved in cyclohexane, shaken with anhydrous BaO and then filtered. The procedure was repeated three times, and the cyclohexane was distd off at 1 atmosphere pressure. The crude amide was also fractionally distd three times from anhydrous BaO.

Diethyl acetamidomalonate *[1068-90-2]* **M 217.2, m 96°.** Crystd from *benzene/pet ether.

Diethyl acetylenedicarboxylate *[762-21-0]* **M 170.2, b 60-62°/0.3mm, 107-110°/11mm, 118-120°/20mm, d$_4^{20}$ 1.0735, n$_D^{20}$ 1.4428.** Dissolve in *C$_6$H$_6$, wash with NaHCO$_3$, H$_2$O, dry over Na$_2$SO$_4$, filter, evaporate and distil in a vacuum [IR: Walton and Hughes *J Am Chem Soc* **79** 3985 *1957*; Truce and Kruse *J Am Chem Soc* **81** 5372 *1959*].

Diethylamine *[109-89-7]* **M 73.1, b 55.5°, d 0.707, n 1.38637, pK^{15} 11.38.** Dried with LiAlH$_4$ or KOH pellets. Refluxed with, and distd from, BaO or KOH. Converted to the *p*-toluenesulfonamide and crystd to constant melting point from dry pet ether (b 90-120°), then hydrolysed with HCl, excess NaOH was added, and the amine passed through a tower of activated alumina, redistd and dried with activated alumina before use [Swift *J Am Chem Soc* **64** 115 *1942*].
§ A polystyrene diethylaminomethyl supported version is commercially available.

Diethylamine hydrochloride *[660-68-4]* **M 109.6, m 223.5°.** Crystd from absolute EtOH. Also crystd from dichloroethane/MeOH. *Hygroscopic.*

***trans*-4-(Diethylamino)azobenzene** *[3588-91-8]* **M 320.5, m 171° pK$_{Est(1)}$ ~-5.4, pK$_{Est(2)}$~3.0.** Purified by column chromatography [Flamigni and Monti *J Phys Chem* **89** 3702 *1985*].

N,N-Diethylaniline *[91-66-7]* **M 149.2, b 216.5°, d 0.938, n 1.5409 pK^{25} 6.57.** Refluxed for 4h with half its weight of acetic anhydride, then fractionally distd under reduced pressure (b 92°/10mm).

Diethyl azodicarboxylate (DEAD) *[1972-28-7]* **M 174.2, b 104.5°/12mm, 211-213°/atm, d$_4^{20}$ 1.110, n$_D^{20}$ 1.420.** Dissolve in toluene, wash with 10% NaHCO$_3$ till neutral (may require several washes if too much hydrolysis had occurred (check IR for OH bands), then wash with H$_2$O (2 x), dry over Na$_2$SO$_4$, filter, evaporate the toluene and distil through a short Vigreux column. Main portion boils at 107-111°/15mm [*Org Synth* Coll Vol III 376 *1955*].
§ A polystyrene supported DEAD version is commercially available.

5,5-Diethylbarbituric acid (Barbital) *[57-44-3]* **M 184.2, m 188-192°, pK$_1^{25}$ 8.02, pK$_2^{25}$ 12.7.** Crystd from water or EtOH. Dried in a vacuum over P$_2$O$_5$.

Diethyl bromomalonate *[685-87-0]* **M 239.1, b 116-118°/10mm, 122-123°/20mm, d_4^{20} 1.420, n_D^{20} 1.4507.** Purified by fractional distn in a vacuum. IR: 1800 and 1700cm^{-1} [Abramovitch *Can J Chem* **37** 1146 *1959*; Bretschneider and Karpitschka *Monatsh Chem* **84** 1091 *1053*].

Diethyl *tert*-butylmalonate *[759-24-0]* **M 216.3, b 40-42°/0.03, 102-104°/11mm, 109.5-110.5°/17mm, 205-210°/760mm, d_4^{20} 0.980, n_D^{20} 1.425.** Dissolve in Et_2O, wash with aqueous $NaHCO_3$, H_2O, dry ($MgSO_4$), filter, evaporate and distil residue. Identified by hydrolysis to the acid and determining the neutralisation equiv (theor: 80.0). The *acid* has **m 155-157° efferv** [Hauser, Abramovitch and Adams *J Am Chem Soc* **64** 2715 *1942*; Bush and Beauchamp *J Am Chem Soc* **75** 2949 *1953*].

***N,N'*-Diethylcarbanilide** (*sym*-**Diethyldiphenylurea**) *[85-98-3]* **M 268.4, m 79°.** Crystd from EtOH.

Diethyl carbonate *[105-58-8]* **M 118.1, b 126.8°, d 0.975, n^{25} 1.38287.** It was washed (100mL) with an aqueous 10% Na_2CO_3 (20mL) solution, saturated $CaCl_2$ (20mL), then water (30mL). After drying by standing over solid $CaCl_2$ for 1h (note that prolonged contact should be avoided because slow combination with $CaCl_2$ occurs), it should be fractionally distd. Also dried over $MgSO_4$ and distd.

1,1'-Diethyl-2,2'-cyanine iodide *[977-96-8]* **M 454.4, m 274°(dec).** Crystd from EtOH and dried in a vacuum oven at 80° for 4h.

***N,N*-Diethylcyclohexylamine** *[91-65-6]* **M 155.3, b 193°/760mm, d 0.850, n 1.4562, pK^{25} 10.72.** Dried with BaO and fractionally distd.

Diethylene glycol *[111-46-6]* **M 106.1, f -10.5°, b 244.3°, d 1.118, n^{15} 1.4490, n 1.4475.** Fractionally distd under reduced pressure (**b 133°/14mm**), then fractionally crystd by partial freezing.

Diethylene glycol diethyl ether *[112-36-7]* **M 162.2, b 85-86°/10mm, 188.2-188.3°/751mm, d 0.909, n 1.412.** Dried with $MgSO_4$, then CaH_2 or $LiAlH_4$, under N_2. If sodium is used the ether should be redistd alone to remove any products which may be formed by the action of sodium on the ether. As a preliminary purification, the crude ether (2L) can be refluxed for 12h with 25mL of conc HCl in 200mL of water, under reduced pressure, with slow passage of N_2 to remove aldehydes and other volatile substances. After cooling, addn of sufficient solid KOH pellets (slowly and with shaking until no more dissolves) gives two liquid phases. The upper of these is decanted, dried with fresh KOH pellets, decanted, then refluxed over, and distd from, sodium. Can be passed through (alkaline) alumina prior to purification.

Diethylene glycol ditosylate *[7460-82-4]* **M 414.5, m 86-87°, 87-88°, 88-89°.** Purified by recrystn from Me_2CO and dried in a vacuum.

Diethylene glycol mono-*n*-butyl ether (**butyl carbitol**) *[112-34-5]* **M 162.2, b 69-70°/0.3mm, 230.5°/760mm, d 0.967, n 1.4286.** Dried with anhydrous K_2CO_3 or $CaSO_4$, filtered and fractionally distd. Peroxides can be removed by refluxing with stannous chloride or a mixture of $FeSO_4$ and $KHSO_4$ (or, less completely, by filtration under slight pressure through a column of activated alumina).

Diethylene glycol monoethyl ether *[111-90-0]* **M 134.2, b 201.9°, d 0.999, n 1.4273, n^{25} 1.4254.** Ethylene glycol can be removed by extracting 250g in 750mL of *benzene with 5mL portions of water, allowing for phase separation, until successive aqueous portions show the same volume increase. Dried, and freed from peroxides, as described for diethylene glycol mono-*n*-butyl ether.

Diethylene glycol monomethyl ether *[111-77-3]* **M 120.2, b 194°, d 1.010, n 1.423.** Purified as for diethylene glycol mono-*n*-butyl ether.

Diethylenetriaminepenta-acetic acid (**DTPA**) *[67-43-6]* **M 393.4, m 219-220°, pK_1^{25} 1.79, pK_2^{25} 2.56, pK_3^{25} 4.42, pK_4^{25} 8.76, pK_5^{25} 10.42.** Crystd from water. Dried under vacuum or at 110°. [Bielski and Thomas *J Am Chem Soc* **109** 7761 *1987*].

Diethyl ether (ethyl ether) *[60-29-7]* **M 74.1, b 34.6°/760mm, d 0.714, n^{15} 1.3555, n 1.35272.** Usual impurities are water, EtOH, diethyl peroxide (which is explosive when concentrated), and aldehydes. Peroxides [detected by liberation of iodine from weakly acid (HCl) solutions of KI, or by the blue colour in the ether layer when 1mg of $Na_2Cr_2O_7$ and 1 drop of dil H_2SO_4 in 1mL of water is shaken with 10mL of ether] can be removed in several different ways. The simplest method is to pass dry ether through a column of activated alumina (80g Al_2O_3/700mL of ether). More commonly, 1L of ether is shaken repeatedly with 5-10mL of a soln comprising 6.0g of ferrous sulfate and 6mL of conc H_2SO_4 in 110mL of water. Aqueous 10% Na_2SO_3 or stannous chloride can also be used. The ether is then washed with water, dried for 24h with $CaCl_2$, filtered and dried further by adding sodium wire until it remains bright. The ether is stored in a dark cool place, until distd from sodium before use. Peroxides can also be removed by wetting the ether with a little water, then adding excess $LiAlH_4$ or CaH_2 and leaving to stand for several hours. (This also dried the ether.)
Werner [*Analyst* **58** 335 *1933*] removed peroxides and aldehydes by adding 8g $AgNO_3$ in 60mL of water to 1L of ether, then 100mL of 4% NaOH and shaking for 6min. Fierz-David [*Chimia* **1** 246 *1947*] shook 1L of ether with 10g of a zinc-copper couple. (This reagent was prepared by suspending zinc dust in 50mL of hot water, adding 5mL of 2M HCl and decanting after 20sec, washing twice with water, covering with 50mL of water and 5mL of 5% cuprous sulfate with swirling. The liquid was decanted and discarded, and the residue was washed three times with 20mL of ethanol and twice with 20mL of diethyl ether).
Aldehydes can be removed from diethyl ether by distn from hydrazine hydrogen sulfate, phenyl hydrazine or thiosemicarbazide. Peroxides and oxidisable impurities have also been removed by shaking with strongly alkaline satd $KMnO_4$ (with which the ether was left to stand in contact for 24h), followed by washing with water, conc H_2SO_4, water again, then drying ($CaCl_2$) and distn from sodium, or sodium containing benzophenone to form the ketyl. Other purification procedures include distn from sodium triphenylmethide or butyl magnesium bromide, and drying with solid NaOH or P_2O_5.
Rapid purification: Same as for 1,4-dioxane.

Diethyl ethoxymethylene malonate *[87-13-8]* **M 216.2, b 014°/0.2mm, 109°/0.5mm, 279-283°/atm, d_4^{20} 1.079, n_D^{20} 1.4623.** Likely impurity is diethyl diethoxymethylene malonate which is difficult to separate from diethyl ethoxymethylene malonate by distn and it is necessary to follow the course of the distn by the change in refractive index instead of boiling point. After a low boiling fraction is collected, there is obtained an intermediate fraction (n_D^{20} 1.414—1.458) the size of which depends on the amount of diethoxymethylene compound. This fraction is fractionated through a 5-inch Vigreux column at low pressure avoiding interruption in heating. Fraction **b** 108-110°/0.25mm was *ca* 10° lower than the submitters' (**b** 97.2°/0.25mm (n $_D^{20}$ 1.4612—1.4623) [*Org Synth* Coll Vol III 395 *1955*; Fuson et al. *J Org Chem* **11** 197 *1946*; Duff and Kendal *J Chem Soc* 893 *1948*].

N,N'-Diethylformamide *[617-84-5]* **M 101.2, b 29°/0.5mm, 61-63°/10mm, 178.3-178.5°/760mm, d_4^{20} 0.906, n_D^{25} 1.4313.** Distd under reduced pressure then at atmospheric pressure [Wintcler et al. *Helv Chim Acta* **37** 2370 *1954*; NMR: Hoffmann *Z Anal Chem* **170** 177 *1959*].

Diethyl fumarate *[623-91-6]* **M 172.2, b 218°, d 1.052, n 1.441.** Washed with aqueous 5% Na_2CO_3, then with saturated $CaCl_2$ soln, dried with $CaCl_2$ and distd.

Di-(2-ethylhexyl)phthalate ('di-iso-octyl' phthalate) *[117-81-7]* **M 390.6, b 384°, 256-257°/1mm, d 0.9803, n 1.4863.** Washed with Na_2CO_3 soln, then shaken with water. After the resulting emulsion had been broken by adding ether, the ethereal soln was washed twice with water, dried ($CaCl_2$), and evaporated. The residual liquid was distd several times under reduced pressure, then stored in a vacuum desiccator over P_2O_5 [French and Singer *J Chem Soc* 1424 *1956*]

Diethyl ketone (3-pentanone) *[96-22-0]* **M 86.1, b 102.1°, d 0.8099, n 1.392.** Dried with anhydrous $CaSO_4$ or $CuSO_4$, and distd from P_2O_5 under N_2 or under reduced pressure. Further purification by conversion to the semicarbazone (recrystd to constant **m** 139°, from EtOH) which, after drying under vacuun over $CaCl_2$ and paraffin wax, was refluxed for 30min with excess oxalic acid, then steam distd and salted out with K_2CO_3. Dried with Na_2SO_4 and distd [Cowan, Jeffrey and Vogel *J Chem Soc* 171 *1940*].

Diethyl malonate *[105-53-3]* **M 160.2, b 92°/22mm, 198-199°/760mm, d 1.056, d^{25} 1.0507, n 1.413.** If too impure (IR, NMR) the ester (250g) has been heated on a steam bath for 36h with absolute EtOH (125mL) and conc H_2SO_4 (75mL), then fractionally distd under reduced pressure. Otherwise fractionally distil under reduced pressure and collect the steady boiling middle fraction.

Diethyl phenyl orthoformate (diethoxy phenoxy ethane) *[14444-77-0]* **M 196.3, b 111°/11mm, 122°/13mm, d$_4^{20}$ 1.0099, n$_D^{20}$ 1.4799.** Fractionated through an efficient column under vacuum [Smith *Acta Chem Scand* **10** 1006 *1956*].

Diethyl phthalate *[84-66-2]* **M 222.2. b 172°/12mm, b 295°/760mm, d^{25} 1.1160, n 1.5022.** Washed with aqueous Na_2CO_3, then distilled water, dried ($CaCl_2$), and distd under reduced pressure. Stored in a vacuum desiccator over P_2O_5.

Diethyl phthalimidomalonate *[56680-61-5]* **M 305.3, m 72-74°, 73-74°, pK 9.17.** Dissolve in xylene and when the temperature is 30° add pet ether (b 40-60°) and cool to 20° whereby the malonate separates as a pale brown powder [Booth et al. *J Chem Soc* 666 *1944*]. Alternatively, dissolve in *C_6H_6, dry over $CaCl_2$, filter, evaporate and the residual oil solidifies. This is ground with Et_2O, filter and wash with Et_2O until white in colour, and dry in a vacuum. The anion has λmax 254nm (ε 18.5K) [Clark and Murray *Org Synth* Coll Vol I 271 *1941*; UV of Na salt: Nnadi and Wang *J Am Chem Soc* **92** 4421 *1970*].

2,2-Diethyl-1,3-propanediol *[115-76-4]* **M 132.2, m 61.4-61.8°.** Crystd from pet ether (b 65-70°).

Diethyl pyrocarbonate (DEP) *[1609-47-8]* **M 162.1, b 38-40°/12mm, 160-163°/atm, d$_4^{20}$ 1.119, n$_D^{20}$ 1.398.** Dissolve in Et_2O, wash with dilute HCl, H_2O, dry over Na_2SO_4, filter, evaporate and distil the residue first *in vacuo* then at atmospheric pressure. It is soluble in alcohols, esters, ketones and hydrocarbon solvents. A 50% w/w soln is usually prepared for general use. **Treat with great CAUTION as DEP irritates the eyes, mucous membranes and skin.** [Boehm and Mehta *Chem Ber* **71** 1797 *1938*; Thoma and Rinke *Justus Liebigs Ann Chem* **624** 30 *1959*.]

Diethylstilboesterol *[56-23-1]* **M 268.4, m 169-172°.** Crystd from *benzene.

Diethyl succinate *[123-25-1]* **M 174.2, b 105°/15mm, d 1.047, n 1.4199.** Dried with $MgSO_4$, and distd at 15mm pressure.

Diethyl sulfate *[64-67-5]* **M 154.2, b 96°/15mm, 118°/40mm, d 1.177, n 1.399.** Washed with aqueous 3% Na_2CO_3 (to remove acidic material), then distilled water, dried ($CaCl_2$), filtered and distd. *Causes blisters to the skin.*

Diethyl disulfide *[110-81-6]* **M 122.3, b 154-155°, d 0.993, n 1.506.** Dried with silica gel or $MgSO_4$ and distd under reduced pressure (optionally from $CaCl_2$).

Diethyl sulfide *[352-93-2]* **M 90.2, m 0°/15mm, 90.1°/760mm, d 0.837, n 1.443.** Washed with aq 5% NaOH, then water, dried with $CaCl_2$ and distd from sodium. Can also be dried with $MgSO_4$ or silica gel. Alternative purification is *via* the Hg(II) chloride complex [(Et)$_2$S.2HgCl$_2$] (see dimethyl sulfide).

Diethyl (-)-D- (from the non-natural) *[13811-71-7]* **and (+)-L- (from the natural acid)** *[89-91-2]* **tartrate M 206.2, m 17°, b 80°/0.5mm, 162°/19mm, 278-282°/atm, d$_4^{20}$ 1.204, n$_D^{20}$ 1.4476, [α]$_D^{20}$ (-) and (+) 26.5° (c 1, H_2O) and (-) and (+) 8.5° (neat), [α]$_{546}^{20}$ (-) and (+) 30° (c 1, H_2O).** Distd under high vacuum and stored under vacuum or in an inert atm in a desiccator in round bottomed flasks equiped with a vac stopcock. Have also been dist by Kügelrohr distn and/or by 'wiped-film' molecular distn. Slightly sol in H_2O but miscible in EtOH and Et_2O. [Gao et al. *J Am Chem Soc* **109** 5770 (5771) *1987*; IR: Pristera *Anal Chem* **25** 844 *1953*.]

Diethyl terephthalate *[636-09-0]* **M 222.2, m 44°, 142°/2mm, 302°/760mm.** Crystd from toluene and distd under reduced pressure.

sym-Diethylthiourea *[105-55-5]* M 132.2, m 76-77°. Crystd from *benzene.

Difluoroacetic acid *[381-73-7]* M 96.0, m -0.35°, b 67-70°/20mm, 134°/760mm, d_4^{20} 1.530, n_D^{20} 1.3428, pK^{25} 1.28. Purified by distilling over P_2O_5. The *acid chloride* is a fuming liquid b 25°/atm, and the *amide* has b 108.6°/35mm, m 52° (from *C_6H_6), and the *anilide* has b 90°/1mm, 114°/5mm, m 58° [Henne and Pelley *J Am Chem Soc* 74 1426 *1952*, Coffman et al. *J Org Chem* 14 749 *1949*; NMR: Meyer et al. *J Am Chem Soc* 75 4567 *1953*; pK: Wegscheider *Z Phys Chem* 69 614 1909].

Diglycolic acid (2-oxapentane-1,5-dioic acid) *[110-99-6]* M 134.1, m 148° (monohydrate), pK_1^{25} 2.97, pK_2^{25} 4.37. Crystd from water.

Diglycyl glycine *[556-33-2]* M 189.2, m 246°(dec), pK_1^{25} 3.30, pK_2^{25} 7.96. Crystd from H_2O or H_2O/EtOH and dried at 110°.

Diglyme [bis-(2-methoxyethyl) ether, diethylene glycol dimethyl ether] *[111-96-6]* M 134.2, b 62°/17mm, 75°/35mm, 160°/760mm, d 0.917, n 1.4087. Dried with NaOH pellets or CaH_2, then refluxed with, and distd (under reduced pressure) from Na, CaH_2, $LiAlH_4$, $NaBH_4$ or NaH. These operations were carried out under N_2. The amine-like odour of diglyme has been removed by shaking with a weakly acidic ion-exchange resin (Amberlite IR-120) before drying and distn. Addn of 0.01% $NaBH_4$ to the distillate inhibits peroxidation. Purification as for dioxane. Also passed through a 12-in column of molecular sieves to remove water and peroxides.

Digoxin *[20830-75-5]* M 781.0, m 265°(dec), $[\alpha]_{546}^{20}$ +14.0° (c 10, pyridine). Crystd from aqueous EtOH or aqueous pyridine.

4,4'-Di-*n*-heptyloxyazoxybenzene *[2635-26-9]* M 426.6, m 75°, 95° (smectic → nematic) and 127° (nematic → liquid), pK_{Est} ~-5. Purified by chromatography on Al_2O_3 (*benzene), recrystd from hexane or 95% EtOH and dried by heating under vacuum. The liquid crystals can be sublimed *in vacuo*. [Mellifiori et al. *Spectrochim Acta Part A* 37(A) 605 *1981*; Dewar and Schroeder *J Am Chem Soc* 86 5235 *1964*; Weygand and Glaber *J Prakt Chem* 155 332 *1940*].

9,10-Dihydroanthracene *[613-31-0]* M 180.3, m 110-110.5°. Crystd from EtOH [Rabideau et al. *J Am Chem Soc* 108 8130 *1986*].

2,3-Dihydrobenzofuran (coumaran) *[496-16-2]* M 120.2, m -21.5°, 72-73°/12mm, 84°/17mm, 188°/atm, d_4^{20} 1.065, n_D^{20} 1.5524. Suspend in aqueous NaOH and steam distil. Saturate the distillate with NaCl and extract with Et_2O, dry extract ($MgSO_4$), filter, evap and distil the residue. It gives a strong violet colour with $FeCl_3$ + H_2SO_4 and forms a yellow *picrate*, m 76°, from EtOH or *C_6H_6 which loses coumaran in a desiccator [Bennett and Hafez *J Chem Soc* 287 *1941*; Baddeley et al. *J Chem Soc* 2455 *1956*].

Dihydrochloranil (tetrachloro-1,4-hydroquinone) *[87-87-6]* M 247.9, m 240.5°. Crystd from EtOH or AcOH+EtOH. Sublimes at 77°/0.6x10⁻³mm. The *dibenzoyl derivative* has m 233°. [Conant and Fieser *J Am Chem Soc* 45 2207 *1923*; Rabideau et al. *J Am Chem Soc* 108 8130 *1986*.]

Dihydrocodeine *[125-28-0]* M 301.4, m 112-113°, b 248/14mm°. Crystd from aqueous methanol.

1,4-Dihydro-1,4-epoxynaphthalene *[573-57-9]* M 144.2, m 53-54.5°, 53-56°, 55-56°. Dissolve in Et_2O, wash with H_2O, dry over K_2CO_3, filter, evaporate and dry the residue at 15mm, then recrystallise from pet ether (b 40-60°), dry at 25°/0.005mm and sublime (sublimes slowly at room temp)[Wittig and Pohmer *Chem Ber* 89 1334 *1956*; Gilman and Gorsich *J Am Chem Soc* 79 2625 *1957*].

Dihydropyran (3,4-dihydro-2*H*-pyran) *[110-87-2]* M 84.1, b 84.4°/742mm, 85.4-85.6°/760mm, d_4^{20} 0.9261, n_D^{20} 1.4423, pK_{Est} ~ 4.2. Partially dried with Na_2CO_3, then fractionally distd. The fraction b 84-85°, was refluxed with Na until hydrogen was no longer evolved when fresh Na was added. It was then dried, and distd again through a 60 x 1.2cm column packed with glass rings [Brandon et al. *J Am Chem Soc* **72** 2120 *1950*; UV: Elington et al. *J Chem Soc* 2873 *1952*, NMR: Bushweller and O'Neil *Tetrahedron Lett* 4713 *1969*]. It has been characterised as the *2,3,5-dinitrobenzoyloxy-tetrahydrofuran* derivative, **m** 103° which forms pale yellow crystals from dihydropyran-Et$_2$O [Woods and Kramer *J Am Chem Soc* **69** 2246 *1947*].

3,4-Dihydro-2*H*-pyrido[1,2*a*]-pyrimidin-2-one *[5439-14-5]* **M 148.2, m 185-187°, 187-188°, 191-191.5°.** Dissolve in CHCl$_3$, filter, evaporate then recrystallise the residue from EtOH-Me$_2$CO (needles) which can be washed with Et$_2$O and dried. It can also be recrystd from CHCl$_3$-pet ether or CHCl$_3$-hexane. The *hydrochloride* has **m** 295-295° (dec, from EtOH or MeOH-Et$_2$O), the *hydrobromide* has **m** 299-300°(dec, from MeOH-Et$_2$O) and the *picrate* has **m** 224-226°(corr), **m** 219-220° from EtOH. [Adams and Pachter *J Am Chem Soc* **74** 4906 *1952*; Lappin *J Org Chem* **23** 1358 *1958*; Hurd and Hayao *J Am Chem Soc* **77** 115 *1955*.]

Dihydrotachysterol *[67-96-9]* **M 398.7, m 125-127°, $[\alpha]_D^{20}$ +97° (CHCl$_3$).** Crystd from 90% MeOH.

1,8-Dihydroxyanthraquinone *[117-10-2]* **M 240.1, m 193-197°, pK_1^{25} 8.30, pK_2^{25} 12.46.** Crystd from EtOH and sublimed in a vacuum.

2,4-Dihydroxyazobenzene (Sudan orange G) *[2051-85-6]* **M 214.2, m 228°, pK\S$_{(,Est(1))}$<0, $pK_{Est(2)}$ ~7.3, $pK_{Est(3)}$ ~9.3.** Crystd from hot EtOH (charcoal).

2,3-Dihydroxybenzaldehyde *[24677-78-9]* **M 138.1, m 135-136°, pK_1^{20} 7.73, pK_2^{20} 10.91.** Crystd from water.

2,4-Dihydroxybenzoic acid *[89-86-1]* **M 154.1, m 226-227°(dec), pK_1^{25} 3.30, pK_2^{25} 9.12, pK_3^{25} 15.6.** Crystd from water.

2,5-Dihydroxybenzoic acid *[490-79-9]* **M 154.1, m 204.5-205°, pK^{25} 2.95.** Crystd from hot water or *benzene/acetone. Dried in a vacuum desiccator over silica gel.

2,6-Dihydroxybenzoic acid *[303-07-1]* **M 154.1, m 167°(dec), pK^{25} 1.05.** Dissolved in aqueous NaHCO$_3$ and the soln was washed with ether to remove non-acidic material. The acid was ppted by adding H$_2$SO$_4$, and recrystd from water. Dried under vacuum and stored in the dark [Lowe and Smith *J Chem Soc, Faraday Trans 1* **69** 1934 *1973*].

2,4-Dihydroxybenzophenone *[131-56-6]* **M 214.2, m 145.5-147° $pK_{Est(1)}$ ~7.0, $pK_{Est(2)}$ ~12.0.** Recrystd from MeOH.

2,5-Dihydroxybenzyl alcohol (Gentisyl alcohol) *[495-08-9]* **M 140.1, m 100° $pK_{Est(1)}$ ~9.3, $pK_{Est(2)}$ ~11.3.** Crystd from CHCl$_3$. Sublimes at ~70° under high vacuum.

2,2'-Dihydroxybiphenyl *[1806-29-7]* **M 186.2, m 108.5-109.5°, pK_1^{25} 7.56, pK_2^{25} 11.80.** Repeatedly crystd from toluene, then sublimed at 60°/10^{-4}mm.

3α,7α-Dihydroxycholanic acid (Chenodeoxycholic acid) *[474-25-9]* **M 239.6, m 143°, $[\alpha]_{546}^{20}$ +14° (c 2, EtOH), pK_{Est} ~4.9.** Crystd from ethyl acetate.

7,8-Dihydroxycoumarin (Daphnetin) *[486-35-1]* **M 178.2, m 256°(dec), $pK_{Est(1)}$ ~8.5, $pK_{Est(2)}$ ~12.3.** Crystd from aqueous EtOH. Sublimed.

2,2'-Dihydroxy-6,6'-dinaphthyl disulfide *[6088-51-3]* **M 350.5, m 220-223°.** See 6-hydroxy-2-naphthyl disulfide on p. 264.

trans-2,3-Dihydroxy-1,4-dioxane   *[4845-50-5]* **M** 120.1, **m** 91-95°, 100°. Recryst from Me$_2$CO. With phenylhydrazine it gives *glyoxal phenylhydrazone* **m** 175° (from Me$_2$CO-pet ether). The *diacetyl* derivative has **m** 105-106° [Head *J Chem Soc* 1036 *1955*, Raudnitz *Chem Ind (London)* 166 *1956*].

2,5-Dihydroxy-1,4-dithiane   *[40018-26-6]* **M** 152.2, **m** (142-147° ?) 150-152°, 151°. Recrystd from EtOH. The *2,5-diethoxy-dithiane* has **m** 91° (92-93°) crystallises from pet ether and can be sublimed at 60°/0.001mm [Hormatka and Haber *Monatsh Chem* **85** 1088 *1954*; Thiel et al. *Justus Liebigs Ann Chem* **611** 121 *1958*; Hesse and Jøeder *Chem Ber* **85** 924 *1952*].

(*N,N*-Dihydroxyethyl)glycine   (BICINE)   *[150-25-4]* **M** 163.2, **m** 193°(dec), pK$_1^{25}$1.81, pK$_2^{25}$ 8.27. Dissolved in a small volume of hot water and ppted with EtOH, twice. Repeated once more but with charcoal treatment of the aqueous soln, and filtered before addition of EtOH.

Dihydroxyfumaric (**1,2-dihydroxybut-1-ene-1,2-dioic**) **acid dihydrate**   *[133-38-0]* **M** 184.1, **m** 155°(dec), pK$_1^{25}$1.57, pK$_2^{25}$3.36. Crystd from water.

3,4'-Dihydroxyisoflavone   *[578-86-9]* **M** 256.3, **m** 234-236°. Crystd from aqueous 50% EtOH.

5,7-Dihydroxy-4'-methoxyflavone   *[480-44-4]* **M** 284.3, **m** 261°. Crystd from 95% EtOH.

1,8-Dihydroxy-3-methylanthraquinone (**chrysophanic acid**)   *[481-74-3]* **M** 245.3, **m** 196°, pK$_{Est(1)}$ ~8.2, pK$_{Est(2)}$~12.4. Crystd from EtOH or *benzene and sublimed in a vacuum.

1,5-Dihydroxynaphthalene   *[83-56-7]* **M** 160.2, **m** 260°, 250-261°, pK$_{Est}$ ~9.6. Crystd from nitromethane.

1,6-Dihydroxynaphthalene   *[575-44-0]* **M** 160.2, **m** 138-139° (**with previous softening**), pK$_{Est}$ ~9.4. Crystd from *benzene or *benzene/EtOH after treatment with charcoal.

2,5-Dihydroxyphenylacetic acid (**homogentisic acid**)   *[451-13-8]* **M** 168.2, **m** 152°, 154-152°, pK^{20} 4.14 (COOH). Crystd from EtOH/CHCl$_3$ or H$_2$O (sol 85% at 25°).

3,4-Dihydroxytoluene   *[452-86-8]* **M** 124.1, **m** 65-66°, 68°, **b** 112°/3mm, 241°/760mm, pK$_1^{25}$ 9.44 (9.7), pK$_2^{25}$10.90 (11.9). Crystd from *C$_6$H$_6$. Purity checked by TLC. Crystd from high-boiling pet ether and distd in a vacuum.

1,3-Diiminoisoindoline   *[3468-11-9]*   **M** 145.2, **m** 193-194° (dec), 196° (dec), pK 8.27. It crystallises from H$_2$O, MeOH or MeOH-Et$_2$O (charcoal) in colourless prisms that become green on heating. [Elvidge and Linstead *J Chem Soc* 5000 *1952*]. IR (nujol): 3150 and 690 cm^{-1}, and UV: λmax 251nm (ε 12.5K), 256nm (ε 12.5K) and 303nm (ε 4.6K) [Elvidge and Golden *J Chem Soc* 700 *1957*; Clark et al. *J Chem Soc* 3593 *1953*]. The *thiocyanate* has **m** 250-255° (dec), the *monohydrochloride* has **m** 300-301° (turns green) and the *dihydrochloride* has **m** 326-328° (turns green) and the *picrate* cryst from EtOH has **m** 299° (dec).

1,4-Diiodobenzene   *[624-38-4]* **M** 329.9, **m** 132-133°. Crystd from EtOH or boiling MeOH, then air dried.

1,2-Diiodoethane   *[624-73-7]*   **M** 281.9, **m** 81-84°, **d** 2.134. Dissolved in ether, washed with satd aq Na$_2$S$_2$O$_3$, drying it over MgSO$_4$ and evap the ether *in vacuo* [Molander et al. *J Am Chem Soc* **109** 453 *1987*].

5,7-Diiodo-8-hydroxyquinoline   *[83-73-8]* **M** 397.0, **m** 214-215°(dec) pK$_{Est(1)}$~3.2, pK$_{Est(2)}$~8.2. Crystd from xylene and dried at 70° in a vacuum.

Diiodomethane (**methylene diiodide**)   *[75-11-6]* **M** 267.8, **m** 6.1°, **b** 66-70°/11-12mm, **d** 3.325. Fractionally distd under reduced pressure, then fractionally crystd by partial freezing, and stabilized with silver wool if necessary. It has also been purified by drying over $CaCl_2$ and fractionally distd from Cu powder.

S-3,5-Diiodotyrosine (iodogorgoic acid) *[300-39-0]* **M 469.0, m 204°(dec), [α]$_D$ +1.5° (in 1M HCl) pK$_1$ 2.12, pK$_2$ 6.48, pK$_3$ 7.82.** See 3,5-diiodo-L-tyrosine dihydrate on p. 530 in Chapter 6.

Diisopropanolamine *[110-97-4]* **M 133.2, m 41-44°, d 1.004, pK$_{Est}$ ~10.7.** Repeatedly crystd from dry diethyl ether.

Diisopropylamine *[108-18-9]* **M 101.2, b 83.5°/760mm, n 1.39236, d 0.720, pK^{25} 11.20.** Distd from NaOH, or refluxed over Na wire or NaH for three minutes and distd into a dry receiver under N_2. § A polystyrene supported version of diisopropylamine is commercially available.

Diisopropylethylamine *[7087-68-5]* **M 129.3, b 127°, pK$_{Est}$ ~10.9.** Distd from ninhydrin, then from KOH [Dryland and Sheppard, *J Chem Soc, Faraday Trans 1* 125 *1986*]. It is a strong base and should be stored in the absence of carbon dioxide.

(-)-2,3:4,6-Di-O-isopropylidene-2-keto-L-gulonic acid monohydrate (- DAG) *[18467-77-1]* **M 292.3, m 100-101°, 103°, [α]$_D^{25}$ -21.6° (c 2.3, MeOH).** Dissolve in Et_2O, filter, dry ($MgSO_4$), filter, evaporate to give a yellow oil. Addition of one drop of H_2O induces crystn to the monohydrate, which also forms rhombic crystals by recrystn from 95% EtOH-H_2O at room tempereture. [Flatt et al. *Synthesis* 815 *1979*; Reichstein and Grussner *Helv Chim Acta* 17 311 *1934*; Takagi and Jeffrey *Acta Crystallogr Sect B* 34 2932 *1978*; cf *Org Synth* 55 80 *1976*.]

1,2:5,6-Di-O-isopropylidene-D-mannitol *[1707-77-3]* **M 262.3, m 121-125°, 122°, [α]$_D^{25}$ +1.2° (c 3, H_2O).** Although quite soluble in H_2O it gives a purer product from this solvent, forming needles [Baer *J Am Chem Soc* 67 338 *1945*; NMR: Curtis et al. *J Chem Soc, Perkin Trans 1* 1756 *1977*].

Diisopropyl ketone (2,4-dimethyl-3-pentanone) *[565-80-0]* **M 114.2, b 124°, d 0.801, n 1.400.** Dried with $CaSO_4$, shaken with chromatographic alumina and fractionally distd from P_2O_5 under N_2.

Diketene *[674-82-8]* **M 84.1, m -7°, b 66-68°/90mm, d 1.440, n 1.4376, n^{25} 1.4348.** Diketene polymerizes violently in the presence of alkali. Distd at reduced pressure, then fractionally crystd by partial freezing (using as a cooling bath a 1:1 soln of $Na_2S_2O_3$ in water, cooled with Dry-ice until slushy, and stored in a Dewar flask). Freezing proceeds slowly, and takes about a day for half completion. The crystals are separated and stored in a refrigerator under N_2. See ketene on p. 276.

2,2'-Diketospirilloxanthin *[24009-17-4]* **M 624.9, m 225-227°, ε$_{1cm}^{1\%}$(λmax) 550(349nm), 820(422nm), 2125(488nm), 2725(516nm), 2130(551nm) in hexane.** Purified by chromatography on a column of partially deactivated alumina. Crystd from acetone/pet ether. Stored in the dark, in an inert atmosphere at 0°.

Dilauroyl peroxide *[105-74-8]* **M 398.6, m 53-54°.** See lauryl peroxide (dodecyl peroxide) on p. 278.

Dimedone (5,5-dimethylcyclohexane-1,3-dione) *[126-81-8]* **M 140.2, m 148-149°, pK^{25} 5.27.** Crystd from acetone (*ca* 8mL/g), water or aqueous EtOH. Dried in air.

2,3-Dimercapto-1-propanol (BAL, British Anti-Lewisite) *[59-52-9]* **M 124.2, b 82-84°/0.8mm, d 1.239, n 1.5732, pK$_1^{25}$8.62, pK$_2^{25}$10.75.** Ppted as the Hg mercaptide [see Bjöberg *Chem Ber* 75 13 *1942*], regenerated with H_2S, and distd at 2.7mm [Rosenblatt and Jean *Anal Chem* 951 *1955*].

1,3-Dimercapto-2-propanol *[584-04-3]* **M 124.2, b 82°/1.5mm.** Purified as for 2,3-dimercapto-1-propanol above.

meso-2,3-Dimercaptosuccinic acid　　*[304-55-2]*　　M 182.2, m 191-192° (dec), 210° (dec), 210-211° (dec), pK_1^{25} 2.71, pK_2^{25} 3.48, pK_3^{25} 8.89, pK_4^{25} 10.75. Purified by dissolving in NaOH and precipitating with dilute HCl, dry and recrystallise from MeOH. IR has ν at 2544 (SH) and 1689 (CO_2H) cm^{-1}. The *bis-S-acetyl* deriv has **m** 183-185° (from EtOAc or Me_2CO) and its *Me ester* has **m** 119-120° (from pet ether) [Gerecke et al. *Helv Chim Acta* **44** 957 *1961*; Owen and Sultanbawa *J Chem Soc* 3112 *1949*].

4,4'-Dimethoxyazobenzene *[501-58-6]* **M 242.3, m 162.7-164.7°, pK_{Est} ~0.** Chromatographed on basic alumina, eluted with *benzene. Crystd from 2:2:1 (v/v) methanol/ethanol/*benzene.

4,4'-Dimethoxyazoxybenzene *[1562-94-3]* **M 258.3, transition temp 118-121°.** See *p,p'*-azoxyanisole on p. 118.

1,2-Dimethoxybenzene (veratrole) *[91-16-7]* **M 137.2, m 23°, b 208.5-208.7, d 1.085, n^{25} 1.53232.** Steam distd. Fractionally distd from BaO, CaH_2 or Na. Crystd from *benzene or low-boiling pet ether at 0°. Fractionally crystd from its melt. Stored over anhydrous Na_2SO_4.

1,3-Dimethoxybenzene *[151-10-0]* **M 137.2, b 212-213°, d 1.056, n 1.5215.** Extracted with aqueous NaOH, and water, then dried. Fractionally distd from BaO or Na.

1,4-Dimethoxybenzene *[150-78-7]* **M 137.2, m 57.2-57.8°.** Steam distd. Crystd from hexane or *benzene, and from MeOH or EtOH but these are wasteful due to high solubilities. Dried under vacuum. Also sublimes under vacuum.

2,4-Dimethoxybenzoic acid *[91-52-1]* **M 182.2, m 109°, pK^{25} 4.36.** Crystd from water and dried in a vacuum desiccator over H_2SO_4.

2,6-Dimethoxybenzoic acid *[1466-76-8]* **M 182.2, m 186-187°, pK^{25} 3.44.** Crystd from water.

3,4-Dimethoxybenzoic acid (veratric acid) *[93-07-2]* **M 182.2, m 181-182°, pK^{25} 4.43.** Crystd from water or aq acetic acid.

3,5-Dimethoxybenzoic acid *[1132-21-4]* **M 182.2, m 185-186°, pK^{25} 3.97.** Crystd from water, EtOH or aq acetic acid.

p,p'-**Dimethoxybenzophenone** *[90-96-0]* **M 242.3, m 144.5°.** Crystd from absolute EtOH.

2,6-Dimethoxy-1,4-benzoquinone *[530-55-2]* **M 168.1, m 256°.** Crystd from acetic acid. Sublimes in a vacuum.

1,1-Dimethoxyethane (acetaldehyde dimethyl acetal) *[534-15-6]* **M 90.1, b 212°/760mm, d 0.828, n 1.4140.** Purification as for acetal on p. 81. Also purified by GLC.

1,2-Dimethoxyethane (glycol dimethyl ether, glyme) *[110-71-4]* **M 90.1, b 84°, d 0.867, n 1.380.** Traces of water and acidic materials have been removed by refluxing with Na, K or CaH_2, decanting and distilling from Na, K, CaH_2 or $LiAlH_4$. Reaction has been speeded up by using vigorous high-speed stirring and molten potassium. For virtually complete elimination of water, 1,2-dimethoxyethane has been dried with Na-K alloy until a characteristic blue colour was formed in the solvent at Dry-ice/cellosolve temperatures: the solvent was kept with the alloy until distd for use [Ward *J Am Chem Soc* **83** 1296 *1961*]. Alternatively, glyme, refluxed with benzophenone and Na-K, was dry enough if, on distn, it gave a blue colour of the ketyl immediately on addition to benzophenone and sodium [Ayscough and Wilson *J Chem Soc* 5412 *1963*]. Also purified by distn under N_2 from sodium benzophenone ketyl (see above).

5,6-Dimethoxy-1-indanone *[2107-69-9]* **M 192.2, m 118-120°.** Crystd from MeOH, then sublimed in a vacuum.

Dimethoxymethane (methylal) *[109-87-5]* **M 76.1, b 42-46°/atm, d_4^{20} 0.8608, n_D^{20} 1.35335.** See formaldehyde dimethyl acetal on p. 245.

1,4-Dimethoxynaphthalene *[10075-62-4]* **M 188.2, m 87-88°.** Crystd from EtOH.

1,5-Dimethoxynaphthalene *[10075-63-5]* **M 188.2, m 183-184°.** Crystd from EtOH.

2,6-Dimethoxyphenol *[91-10-1]* **M 154.2, m 54-56°, pK_{Est} ~9.6.** Purified by zone melting or sublimation in a vacuum.

3,5-dimethoxyphenol (phloroglucinol dimethylether) *[500-99-2]* **M 154.2, m 42-43°, b 115°/0.04mm, pK^{25} 9.35.** Purified by distn followed by sublimation in a vacuum.

3,4-Dimethoxyphenyl acetic acid (homoveratric acid) *[93-40-3]* **M 196.2, m 97-99°, pK^{25} 4.33.** Crystd from water or *benzene/ligroin.

3,5-Dimethoxyphenylacetonitrile *[13388-75-5]* **M 177.1, m 53°.** Crystd from MeOH or pet ether (b 90-110°). [Adams et al. *J Am Chem Soc* **70** 664 *1948*; Sankaraman et al. *J Am Chem Soc* **109** 5235 *1987*.]

4,4'-Dimethoxythiobenzophenone *[958-80-5]* **M 258.3, m 120°.** Recrystd from a mixture of cyclohexane/dichloromethane (4:1).

2,6-Dimethoxytoluene *[5673-07-4]* **M 152.2, m 39-41°.** Sublimed *in vacuo* [Sankaraman et al. *J Am Chem Soc* **109** 5235 *1987*].

4,4'-Dimethoxytrityl chloride (DMT) *[40615-36-9]* **M 338.8, m 114°.** Crysts from cyclohexane-acetyl chloride as the hydrochloride and dry over KOH pellets in a desiccator. When dissolved in *C_6H_6 and air is blown through, HCl is removed. It crystallises from Et_2O. [*Justus Liebigs Ann Chem* **370** 142 *1909*; *Chem Ber* **36** 2774 *1903*; Smith et al. *J Am Chem Soc* **84** 430 *1962*; Smith et al. *J Am Chem Soc* **85** 3821 *1963*.] If it had hydrolysed considerably (see OH in IR) then repeat the crystallisation from cyclohexane-acetyl chloride — excess of AcCl is removed in vac over KOH.

***N,N*-Dimethylacetamide** *[127-19-5]* **M 87.1, b 58.0-58.5°/11.4mm, d 0.940, n 1.437.** Shaken with BaO for several days, refluxed with BaO for 1h, then fractionally distd under reduced pressure, and stored over molecular sieves.

β,β-Dimethylacrylic acid (senecioic acid) *[541-47-9]* **M 100.1, m 68°, pK^{25} -5.4 (aq H_2SO_4).** Crystd from hot water or pet ether (b 60-80°).

Dimethyl adipate *[627-93-0]* **M 174.2, m 9-11°, b 109°/10mm, 121-123°/20mm, 235°/760mm, d_4^{20} 1.0642, n_D^{20} 1.4292.** Dissolve in Et_2O, wash with $NaHCO_3$, H_2O, dry over $MgSO_4$, filter, evaporate and distil several times until the IR and NMR are consistent with the structure [Lorette and Brown *J Org Chem* **24** 261 *1959*; Hoffmann and Weiss *J Am Chem Soc* **79** 4759 *1957*].

Dimethyl adipimidate dihydrochloride *[14620-72-5]* **M 245.1, m 218-220°, 222-224°.** If the salt smells of HCl then wash with MeOH and dry Et_2O (1:3) under N_2 until the HCl is completely removed. Recryst from MeOH-Et_2O (it is very important that the solvents are super dry) [Hartman and Wold *Biochemistry* **6** 2439 *1967*; McElvain and Shroeder *J Am Chem Soc* **71** 40 *1949*].

Dimethylamine *[124-40-3]* **M 45.1, f -92.2°, b 0°/563mm, 6.9°/760mm, pK^{25} 10.73.** Dried by passage through a KOH-filled tower, or by standing with sodium pellets at 0° during 18h.
§ A dimethylaminomethyl polystyrene supported version is commercially available.

Dimethylamine hydrochloride *[506-59-2]* **M 81.6, m 171°.** Crystd from hot CHCl$_3$ or abs EtOH. Also recrystd from MeOH/ether soln. Dried in a vacuum desiccator over H$_2$SO$_4$, then P$_2$O$_5$. *Hygroscopic.*

4-*N*,*N*'-Dimethylaminoazo-benzene-4'-isothiocyanate {**DABITC**, **4-[(4-isocyanatophenyl)-azo]-*N*,*N*'-dimethylaniline**} *[7612-98-8]* **M 282.4, m 170-171°, pK$_{Est}$~ 2.5.** Crystd by dissolving 1g in 150mL of boiling Me$_2$CO, filtering hot and allowing to cool at -20° overnight collecting the solid and drying in vac. Solns in pyridine should be used immediately otherwise it dec. Moistue sensitive. [Chang *Methods Enzymol* **91** 79, 455 *1983*.]

***p*-Dimethylaminoazobenzene (Methyl Yellow)** *[60-11-7]* **M 225.3, m 118-119°(dec), pK$_1^{25}$ -5.34 (aq H$_2$SO$_4$), pK$_2^{25}$ 2.96.** Crystd from acetic acid or isooctane, or from 95% EtOH by adding hot water and cooling. Dried over KOH under vacuum at 50°. **CARCINOGEN.**

***p*-Dimethylaminobenzaldehyde (Ehrlich's Reagent)** *[100-10-7]* **M 149.2, m 74-75°, pK$_{Est}$ ~2.6.** Crystd from water, hexane, or from EtOH (2mL/g), after charcoal treatment, by adding excess of water. Also dissolved in aqueous acetic acid, filtered, and ppted with ammonia. Finally recrystd from EtOH.

***p*-Dimethylaminobenzoic acid** *[619-84-1]* **M 165.2, m 242.5-243.5°(dec), pK$_1$ 2.51, pK$_2$ 6.03.** Crystd from EtOH/water.

***p*-Dimethylaminobenzophenone** *[530-44-9]* **M 225.3, m 92-93°, pK$_{Est}$ ~2.7.** Crystd from EtOH.

***N*,*N*-Dimethylamino-*p*-chlorobenzene (*p*-chloro-*N*,*N*-dimethylaniline)** *[698-69-1]* **M 155.6, m 32-33.5°, 35.5°, b 231°/atm.** Purified by vacuum sublimation [Guarr et al. *J Am Chem Soc* **107** 5104 *1985*]. The *picrate* has **m 126-128°** (from methanol).

2*S*,3*R*-(+)-4-Dimethylamino-1,2-diphenyl-3-methyl-2-butanol *[38345-66-3]* **M 283.4, m 55-57°, [α]$_{546}^{20}$ +9.3° (c 9.6, EtOH), [α]$_D^{20}$ +7.7° (c 9.6, EtOH), pK$_{Est}$ ~10.0.** Purification of the *hydrochloride* by dissolving 1.5g in 13.5 mL of 5N HCl heating to boiling and evaporate in a vacuum. Recrystn of the *hydrochloride* three times from MeOH-EtOAc gives **m 189-190°, [α]$_D$ -33.7° (c 1, H$_2$O)** {enantiomer has +34.2°}. The *hydrochloride* in the minimum volume of water is basified with aqueous 5N NaOH and extracted with Et$_2$O. The extract is dried (K$_2$CO$_3$) and evap leaving a residue which is stored in a desiccator over solid KOH as a low melting solid. It can be recovered with these procedures from asymmetric reductions with LAH, and reused. [*J Am Chem Soc* **77** 3400 *1955*; *J Org Chem* **28** 2381 2483 *1963*.]

***dl*-4-Dimethylamino-2,2-diphenylvaleramide** *[60-46-8]* **M 296.4, m 183-184°, pK$_{Est}$ ~9.8.** Crystd from aqueous EtOH.

(-)-L-4-Dimethylamino-2,2-diphenylvaleramide *[6078-64-4]* **M 296.4, m 136.5-137.5°.** Crystd from pet ether or EtOH.

2-Dimethylaminoethanol *[108-01-0]* **M 89.1, b 134.5-135.5°, d 1.430, n 1.4362, pK^{25} 9.23.** Dried with anhydrous K$_2$CO$_3$ or KOH, and fractionally distd.

1-(3-Dimethylamino-propyl)-3-ethylcarbodiimide hydrochloride (EDCI, DEC, 1-ethyl-3-(3-dimethyl-aminopropyl) carbodiimide hydrochloride) *[25952-53-8]* **M 191.7, m 113.5-114.5°, 114-116°, pK$_{Est}$ ~ 10.3.** An excellent H$_2$O-soluble peptide coupling reagent. It is purified by dissolving (*ca* 1g) in CH$_2$Cl$_2$ (10mL) at room temperature and then add dry Et$_2$O (~110mL) dropwise and the crystals that separate are collected, washed with dry Et$_2$O and recrystd from CH$_2$Cl$_2$-Et$_2$O and dried in a vacuum over P$_2$O$_5$. It is important to work in a dry atmosphere or work rapidly and then dry the solid as soon as possible. Material is moderately *hygroscopic* but once it becomes wet it reacts slowly with H$_2$O. Store away from moisture and at -20° to slow down the hydrolysis process. The *free base* has **b 47-48°/0.27mm, 53-54°/0.6mm, n$_D^{25}$ 1.4582.** The *methiodide* is recrytstallised from CHCl$_3$-EtOAc, the crystals are filtered off, washed with dry Et$_2$O and recrystd from CHCl$_3$-Et$_2$O, and dried *in vacuo* over P$_2$O$_5$, **m 93-95°, 94-95°.** [Sheehan et al. *J Am Chem Soc* **87** 2492 *1965*; Sheehan and Cruickshank *Org Synth* Coll Vol V 555 *1973*.]

§ A polymer bound version is commercially avaiable.

6-Dimethylaminopurine *[938-55-6]* **M 163.1, m 257.5-258.5°, 259-262°, 263-264° , pK$_1^{25}$ 3.87, pK$_2^{25}$ 10.5.** It is purified by recrystn from H_2O, EtOH (0.32g in 10mL) or $CHCl_3$. [Albert and Brown *J Chem Soc* 2060 *1954*; UV: Mason *J Chem Soc* 2071 *1954*.] The *monohydrochloride* crystallises from EtOH-Et_2O, **m** 253° (dec) [Elion et al. *J Am Chem Soc* **74** 411 *1952*], the *dihydrochloride* has **m** 225° (dec) and the *picrate* has **m** 245° (235-236.5°) [Fryth et al. *J Am Chem Soc* **80** 2736 *1958*].

4-Dimethylaminopyridine (DMAP) *[1122-58-3]* **M 122.2, m 108-109°, b 191°, pK^{25} 9.61.** Recrystd from toluene [Sadownik et al. *J Am Chem Soc* **108** 7789 *1986*].
§ A polystyrene supported version (PS-DMAP) is commercially available.

***N,N*-Dimethylaniline** *[121-69-7]* **M 121.2, f 2°, b 84°/15mm, 193°/760mm, d 0.956, n^{25} 1.5556, pK^{25} 5.07.** Primary and secondary amines (including aniline and monomethylaniline) can be removed by refluxing for some hours with excess acetic anhydride, and then fractionally distilling. Crocker and Jones (*J Chem Soc* 1808 *1959*) used four volumes of acetic anhydride, then distd off the greater part of it, and took up the residue in ice-cold dil HCl. Non-basic materials were removed by ether extraction, then the dimethylaniline was liberated with ammonia, extracted with ether, dried, and distd under reduced pressure. Metzler and Tobolsky (*J Am Chem Soc* **76** 5178 *1954*) refluxed with only 10% (w/w) of acetic anhydride, then cooled and poured into excess 20% HCl, which, after cooling, was extracted with diethyl ether. (The amine hydrochloride, remains in the aqueous phase.) The HCl soln was cautiously made alkaline to phenolphthalein, and the amine layer was drawn off, dried over KOH and fractionally distd under reduced pressure, under nitrogen. Suitable drying agents for dimethylaniline include NaOH, BaO, $CaSO_4$, and CaH_2.
Other purification procedures include the formation of the picrate, prepared in *benzene soln and crystd to constant melting point, then decomposed with warm 10% NaOH and extracted into ether: the extract was washed with water, and distd under reduced pressure. The oxalate has also been used. The base has been fractionally crystd by partial freezing and also from aq 80% EtOH then from absolute EtOH. It has been distd from zinc dust, under nitrogen.

2,6-Dimethylaniline *[87-62-7]* **M 121.2, m 11°, b 210-211°/736mm, d 0.974, n 1.5604, pK^{25} 3.95.** Converted to its hydrochloride which, after recrystn, was decomposed with alkali to give the free base. Dried over KOH and fractionally distd.

3,4-Dimethylaniline *[95-64-7]* **M 121.2, m 51°, b 116-118°/25mm, b 226°/760mm, pK^{25} 5.17.** Crystd from ligroin and distilled under vacuum.

9,10-Dimethylanthracene *[781-43-1]* **M 206.3, m 180-181°.** Crystd from EtOH, and by recrystn from the melt.

1,3-Dimethylbarbituric acid *[769-42-6]* **M 156.1, m 123°, pK^{25} 4.56.** Crystd from water and sublimed in a vacuum. Also purified by dissolving 10g in 100mL of boiling $CCl_4/CHCl_3$ (8:2) (1g charcoal), filtered and cooled to 25°. Dried *in vacuo* [Kohn et al. *Anal Chem* **58** 3184 *1986*].

7,12-Dimethylbenz[a]anthracene *[57-97-6]* **M 256.4, m 122-123°.** Purified by chromatography on alumina/toluene or *benzene. Crystd from acetone/EtOH.

5,6-Dimethylbenzimidazole *[582-60-5]* **M 146.2, m 205-206°, pK$_1^{25}$ 5.96, pK$_2^{25}$ 12.52.** Crystd from diethyl ether. Sublimed at 140°/3mm.

2,3-Dimethylbenzoic acid *[603-79-2]* **M 150.2, m 146°, pK^{25} 3.72.** Crystd from EtOH and is volatile in steam.

2,4-Dimethylbenzoic acid *[611-01-8]* **M 150.2, m 126-127°, b 267°/727mm, pK^{25} 4.22.** Crystd from EtOH, and sublimed in a vacuum.

2,5-Dimethylbenzoic acid *[610-72-0]* M 150.2, m 134°, b 268°/760mm, pK^{25} 4.00. Steam distd, and crystd from EtOH.

2,6-Dimethylbenzoic acid *[632-46-2]* M 150.2, m 117°, pK^{25} 3.35. Steam distd, and crystd from EtOH.

3,4-Dimethylbenzoic acid *[619-04-5]* M 150.2, m 166°, pK^{25} 4.50. Crystd from EtOH and sublimed *in vacuo*.

3,5-Dimethylbenzoic acid *[499-06-9]* M 150.2, m 170°, pK^{25} 4.30. Distd in steam, crystd from water or EtOH and sublimed in a vacuum.

4,4'-Dimethylbenzophenone *[611-97-2]* M 210.3, m 95°, b 333-334°/725mm. Purified by zone refining.

2,5-Dimethyl-1,4-benzoquinone *[137-18-8]* M 136.1, m 124-125°. Crystd from EtOH.

2,6-Dimethyl-1,4-benzoquinone *[527-61-7]* M 136.1, m 72° (sealed tube). Crystd from water/EtOH (8:1).

2,3-Dimethylbenzothiophene *[31317-17-6]* M 212.3, b 123-124°/10mm, n^{19} 1.6171. Fractionated through a 90cm Monel spiral column, or other efficient fractionating or spinning band column and collecting the middle fraction.

N,N-**Dimethylbenzylamine** *[103-83-3]* M 135.2, b 66-67°/15mm, 181°/760mm, d$_4^{20}$ 0.898, n$_D^{20}$ 1.516, pK^{25} 8.91. See *N*-benzyl dimethylamine on p. 128.

4,4'-Dimethyl-2,2'-bipyridine *[1134-35-6]* M 184.2, m 175-176°, pK$_{Est(1)}$ ~0.2, pK$_{Est(2)}$~4.9. Crystd from ethyl acetate. [Elliott et al. *J Am Chem Soc* **107** 4647 *1985*.]

1,1'-Dimethyl-4,4'-bipyridylium dichloride (3H$_2$O; Methyl Viologen Dichloride, paraquat dichloride) *[1910-42-5]* M 311.2, m >300°(dec). Recrystd from MeOH/acetone mixture. Also crystd three times from absolute EtOH [Bancroft et al. *Anal Chem* **53** 1390 *1981*]. Dried at 80° in a vacuum.

N,N-**Dimethylbiuret** *[7710-35-2]* M 131.1, m 178°. Purified by repeated crystn from the melt, or H$_2$O. [Bredereck and Richter *Chem Ber* **99** 2461 *1968*; Dunning and Close *J Am Chem Soc* **75** 3615 *1953*.]

2,3-Dimethyl-1,3-butadiene *[513-81-5]* M 82.2, m -69-70°, b 68-69°/760mm, d 0.727, n 1.4385. Distd from NaBH$_4$, and purified by zone melting.

1,3-Dimethylbutadiene sulfone (1,3-dimethylsulfolene) *[10033-92-8]* M 145.2, m 40.4-41.0°. Crystd from diethyl ether.

2,2-Dimethylbutane *[75-83-2]* M 86.2, b 49.7°, d 0.649, n^{25} 1.36595. Distd azeotropically with MeOH, then washed with water, dried and distd.

2,3-Dimethylbutane *[79-29-8]* M 86.2, b 58.0°, d 1.375, n^{25} 1.37231. Distd from sodium, passed through a column of silica gel (activated by heating in nitrogen to 350° before use) to remove unsaturated impurities, and again distd from sodium. Also distilled azeotropically with MeOH, then washed with water, dried and redistd.

2,3-Dimethylbut-2-ene *[563-79-1]* M 84.2, b 72-73°/760mm, d 0.708, n 1.41153. Purified by GLC on a column of 20% squalene on chromosorb P at 50° [Flowers and Rabinovitch *J Phys Chem* **89** 563 *1985*]. Also washed with 1M NaOH soln followed by H$_2$O. Dried over Na$_2$SO$_4$, distd over powdered KOH under nitrogen and passed through activated alumina before use. [Woon et al. *J Am Chem Soc* **108** 7990 *1986*; Wong et al. *J Am Chem Soc* **109** 3428 *1987*.]

Dimethylcarbamoyl chloride *[79-44-7]* **M 107.5, m -33°, b 34°/0.1mm, d 1.172, n 1.4511.** Must distil under high vacuum to avoid decomposition.

3,3'-Dimethylcarbanilide *[620-50-8]* **M 240.3, m 225°.** Crystd from ethyl acetate.

Dimethyl carbonate *[616-38-5]* **M 90.1, m 4.65°, b 90-91°, d 1.070, n 1.369.** Contains small amounts of water and alcohol which form azeotropes. Stood for several days in contact with Linde type 4A molecular sieves, then fractionally distd. The middle fraction was frozen slowly at 2°, several times, retaining 80% of the solvent at each cycle.

***cis-* and *trans*-1,4-Dimethylcyclohexane** *[589-90-2]* **M 112.2, b 120°, d 0.788, n 1.427.** Freed from olefins by shaking with conc H_2SO_4, washing with water, drying and fractionally distilling.

1,2-Dimethylcyclohexene *[1674-10-8]* **M 110.2, b 135-136°/760mm, d 0.826, n 1.4591.** Passed through a column of basic alumina and distd.

1,5-Dimethyl-1,5-diazaundecamethylene polymethobromide (Hexadimethrene, Polybrene) *[28728-55-4]* **M 5000—10,000 polymer** Purified by chromatography on Dowex 50 and/or by filtration through alumina before use [Frank *Hoppe-Seyler's Z Physiol Chemie* **360** 997 *1979*]. Hygroscopic, sol in H_2O is 10%.

2,9-Dimethyl-4,7-diphenyl-1,10-phenanthroline *[4733-39-5]* **M 360.5, m >280°, pK$_{Est}$ ~5.6.** Purified by recrystn from *benzene.

Dimethyl disulfide *[624-92-0]* **M 94.2, f -98°, b 40°/12mm, 110°/760mm, d 1.0605, n 1.5260.** Passed through neutral alumina before use.

2,2-Dimethylethyleneimine *[2658-24-4]* **M 71.1, b 70.5-71.0°, pK^{25} 8.64.** Freshly distd from sodium before use.

N,N-Dimethylformamide (DMF) *[68-12-2]* **M 73.1, b 40°/10mm, 61°/30mm, 88°/100mm, 153°/760mm, d 0.948, n^{25} 1.4269, pK -0.3.** Decomposes slightly at its normal boiling point to give small amounts of dimethylamine and carbon monoxide. The decomposition is catalysed by acidic or basic materials, so that even at room temperature DMF is appreciably decomposed if allowed to stand for several hours with solid KOH, NaOH or CaH_2. If these reagents are used as dehydrating agents, therefore, they should not be refluxed with the DMF. Use of $CaSO_4$, $MgSO_4$, silica gel or Linde type 4A molecular sieves is preferable, followed by distn under reduced pressure. This procedure is adequate for most laboratory purposes. Larger amounts of water can be removed by azeotropic distn with *benzene (10% v/v, previously dried over CaH_2), at atmospheric pressure: water and *benzene distil below 80°. The liquid remaining in the distn flask is further dried by adding $MgSO_4$ (previously ignited overnight at 300-400°) to give 25g/L. After shaking for one day, a further quantity of $MgSO_4$ is added, and the DMF distd at 15-20mm pressure through a 3-ft vacuum-jacketed column packed with steel helices. However, $MgSO_4$ is an inefficient drying agent, leaving about 0.01M water in the final DMF. More efficient drying (to around 0.001-0.007M water) is achieved by standing with powdered BaO, followed by decanting before distn, with alumina powder (50g/L; previously heated overnight to 500-600°), and distilling from more of the alumina; or by refluxing at 120-140° for 24h with triphenylchlorosilane (5-10g/L), then distilling at *ca* 5mm pressure [Thomas and Rochow *J Am Chem Soc* **79** 1843 *1957*]. Free amine in DMF can be detected by colour reaction with 1-fluoro-2,4-dinitrobenzene. It has also been purified by drying overnight over KOH pellets and then distd from BaO through a 10 cm Vigreux column [*Exp Cell Res* **100** 213 *1976*]. [For efficiency of desiccants in drying dimethylformamide see Burfield and Smithers [*J Org Chem* **43** 3966 *1978*, and for a review on purification, tests of purity and physical properties, see Juillard *Pure Appl Chem* **49** 885 *1977*].

It has been purified by distilling from K_2CO_3 under high vac and fractionated in an all-glass apparatus. The middle fraction is collected, degassed (seven or eight freeze-thaw cycles) and redistd under as high a vacuum as possible [Mohammad and Kosower *J Am Chem Soc* **93** 2713 *1971*].

Rapid purification: Stir over CaH_2 (5% w/v) overnight, filter, then distil at 20mmHg. Store the distd DMF over 3A or 4A molecular sieves. For solid phase synthesis, the DMF used must be of high quality and free from amines.

d,l-2,4-Dimethylglutaric acid *[2121-67-7]* **M 160.2, m 144-145°** $pK_{Est(1)}$~4.4, $pK_{Est(2)}$~5.4. Distd in steam and crystd from ether/pet ether.

3,3-Dimethylglutaric acid *[4839-46-7]* **M 160.2, m 103-104°, b 89-90°/2mm, 126-127°/4.5mm, pK_1^{25} 3.85, pK_2^{25} 6.45.** Crystd from water, *benzene or ether/pet ether. Dried in a vacuum.

3,3-Dimethylglutarimide *[1123-40-6]* **M 141.2, m 144-146°.** Recrystd from EtOH [Arnett and Harrelson *J Am Chem Soc* **109** 809 *1987*].

N,N-Dimethylglycinehydrazide hydrochloride *[539-64-0]* **M 153.6, m 181°.** Crystd by adding EtOH to a conc aqueous soln.

Dimethylglyoxime *[95-45-4]* **M 116.1, m 240°, pK_1^{25} 10.60, pK_2^{25} 11.85.** Crystd from EtOH (10mL/g) or aqueous EtOH. **TOXIC.**

2,5-Dimethyl-2,4-hexadiene *[764-13-6]* **M 110.2, f 14.5°, b 132-134°, d 0.773, n 1.4796.** Distd, then repeatedly fractionally crystd by partial freezing. Immediately before use, the material was passed through a column containing Woelm silica gel (activity I) and Woelm alumina (neutral) in separate layers.

2,2-Dimethylhexane *[590-73-8]* **M 114.2, m -121.2°, b 107°, d 0.695.** Dried over type 4A molecular sieves and distd.

2,5-Dimethylhexane *[592-13-2]* **M 114.2, m -91.2°, b 109°, d 0.694.** Dried over type 4A molecular sieves and distd.

2,5-Dimethylhexane-2,5-diol *[110-03-2]* **M 146.2, m 88-90°.** Purified by fractional crystn. Then the diol was dissolved in hot acetone, treated with activated charcoal, and filtered while hot. The soln was cooled and the diol was filtered off and washed well with cold acetone. The crystn process was repeated several times and the crystals were dried under a vac in a freeze-drying apparatus [Goates et al. *J Chem Soc, Faraday Trans 1* **78** 3045 *1982*].

5,5-Dimethylhydantoin *[77-71-4]* **M 128.1, m 177-178° pK^{24} 9.19.** Crystd from EtOH and sublimed *in vacuo.*

1,1-Dimethylhydrazine *[57-14-7]* **M 60.1, b 60.1°/702mm, d 0.790, n 1.408 pK^{30} 7.21.** Fractionally distd through a 4-ft column packed with glass helices. Ppted as its oxalate from diethyl ether soln. After crystn from 95% EtOH, the salt was decomposed with aqueous saturated NaOH, and the free base was distd, dried over BaO and redistd [McBride and Kruse *J Am Chem Soc* **79** 572 *1957*]. Distn and storage should be under nitrogen.

4,6-Dimethyl-2-hydroxypyrimidine *[108-79-2]* **M 124.1, m 198-199°, pK_1^{20} 3.77, pK_2^{20} 10.50.** Crystd from absolute EtOH (charcoal).

1,2-Dimethylimidazole *[1739-84-0]* **M 96.1, m 38-40°, b 206°/760mm, d 1.084, pK_{Est} ~8.1.** Crystd from *benzene and stored at 0-4°. [Gorun et al. *J Am Chem Soc* **109** 4244 *1987*.]

1,1-Dimethyl-1H-indene *[18636-55-0]* **M 144.2, b 57°/4.8mm, 115°/20mm.** Purified by gas chromatography or by fractional distn.

Dimethyl itaconate *[617-52-7]* **M 158.2, m 38°, b 208°, d 1.124.** Crystd from MeOH by cooling to -78°.

Dimethylmaleic anhydride *[766-39-2]* **M 126.1, m 96°, b 225°/760mm.** Distd from *benzene/ligroin and sublimed in a vacuum.

Dimethylmalonic acid *[595-46-0]* **M 132.1, m 192-193° pK$_1^{25}$ 3.03, pK$_2^{25}$ 5.73.** Crystd from *benzene/pet ether and sublimed in a vacuum with slight decomposition.

1,5-Dimethylnaphthalene *[571-61-9]* **M 156.2, m 81-82°, b 265-266°.** Crystd from 85% aq EtOH.

2,3-Dimethylnaphthalene *[581-40-8]* **M 156.2, m 104-104.5°.** Steam distd and crystd from EtOH.

2,6-Dimethylnaphthalene *[581-42-0]* **M 156.2, m 110-111°, b 122.5-123.5°/10mm, 261-262°/760mm.** Distd in steam and crystd from EtOH.

3,3'-Dimethylnaphthidine (4,4'-diamino-3,3'-dimethyl-1,1'-binaphthyl) *[13138-48-2]* **M 312.4, m 213°.** Recrystd from EtOH or pet ether (b 60-80°).

N,N-**Dimethyl-*m*-nitroaniline** *[619-31-8]* **M 166.1, m 60°, pK^{25} 2.63.** Crystd from EtOH.

N,N-**Dimethyl-*p*-nitroaniline** *[100-23-2]* **M 166.1, m 164.5-165.2°, pK^{25} 0.61 (0.92).** Crystd from EtOH or aqueous EtOH. Dried under vacuum.

Dimethylnitrosamine (*N*-nitrosodimethylamine) *[62-75-9]* **M 74.0, m -28°, b 149-150°/atm, 153°/774mm, d. 1.006, n 1.4370.** Dry over anhyd K_2CO_3 or dissolve in Et_2O, dry over solid KOH, filter, evap Et_2O and distil yellow oily residue through a 30cm fractionating column discarding the first fraction which may contain Me_2N. Also dried over $CaCl_2$ and distd at atm pressure. All should be done in an efficient fume cupboard as the vapors are **TOXIC** and **CARCINOGENIC**. [Fischer *Chem Ber* **8** 1588 *1875*; Romberg *Recl Trav Chim, Pays-Bas* **5** 248 *1886*; Hatt *Org Synth* Coll Vol II 211 *1961*; Krebs and Mandt *Chem Ber* **108** 1130 *1975*.]

N,N-**Dimethyl-*p*-nitrosoaniline** *[138-89-6]* **M 150.2, m 86-87°.** See 4-nitroso-*N,N*-dimethylaniline on p. 314.

N,N-**Dimethyl-*p*-nitrosoaniline hydrochloride** *[42344-05-8]* **M 186.7, m 177°.** Crystd from hot water in the presence of a little HCl.

2,6-Dimethyl-2,4,6-octatriene *[7216-56-0; cis/trans mixt 673-84-7; trans, trans 3016-19-1]* **M 136.2, b 80-82°/15mm, ε$_{278nm}$ 42,870.** Repeated distn at 15mm through a long column of glass helices, the final distn being from sodium under nitrogen. See *neo*-allocimene on p. 100.

Dimethylolurea *[140-95-4]* **M 120.1, m 137-139°.** Crystd from aqueous 75% EtOH.

Dimethyl oxalate *[553-90-2]* **M 118.1, m 54°, b 163-165°, d 1.148.** Crystd repeatedly from EtOH. Degassed under nitrogen high vacuum and distd.

3,3-Dimethyloxetane *[6921-35-3]* **M 86.1, b 79.2-80.3°/760mm.** Purified by gas chromatography using a 2m silicone oil column.

2,3-Dimethylpentane *[565-59-3]* **M 100.2, b 89.8°, d 0.695, n 1.39197, n^{25} 1.38946.** Purified by azeotropic distn with EtOH, followed by washing out the EtOH with water, drying and distn [Streiff et al. *J Res Nat Bur Stand* **37** 331 *1946*].

2,4-Dimethylpentane *[108-08-7]* **M 100.2, b 80.5°, d 0.763, n 1.3814, n²⁵ 1.37882.** Extracted repeatedly with conc H_2SO_4, washed with water, dried and distd. Percolated through silica gel (previously heated in nitrogen to 350°). Purified by azeotropic distn with EtOH, followed by washing out the EtOH with water, drying and distn.

4,4-Dimethyl-1-pentene *[762-62-9]* **M 98.2, b 72.5°/760mm, d 0.6827, n 1.3918.** Purified by passage through alumina before use [Traylor et al. *J Am Chem Soc* **109** 3625 *1987*].

Dimethyl peroxide *[690-02-8]* **M 62.1, b 13.5°/760mm, d 0.8677, n 1.3503.** Purified by repeated trap-to-trap fractionation until no impurities could be detected by gas IR spectroscopy [Haas and Oberhammer *J Am Chem Soc* **106** 6146 *1984*]. *All necessary precautions should be taken in case of* **EXPLOSION**.

2,9-Dimethyl-1,10-phenanthroline *[484-11-7]* **M 208.3, m 162-164°, pK²⁵ 5.85.** Purified as hemihydrate from water, and as anhydrous from *benzene.

***R*-(+)-*N,N'*-Dimethyl-1-phenethylamine** *[19342-01-9]* **and *S*-(-)-*N,N'*-Dimethyl-1-phenethyl-amine** *[17279-31-1]* **M 149.2, b 81°/16mm, [α] $_D^{20}$ (+) and (-) 50.2° (c 1, MeOH), [α] $_D^{26}$ +61.8° and -64.4° (neat *l* 1), d 0.908, pK$_{Est}$ ~9.0 (for *RS*).** The amine is mixed with aqueous 10N NaOH and extracted with toluene. The extract is washed with saturated aqueous NaCl and dried over K_2CO_3, and transfered to fresh K_2CO_3 until the soln is clear, and filtered. The filtrate is distd. If a short column packed with glass helices is used, the yield is reduced but a purer product is obtained. [*Org Synth* **25** 89 *1945*; *J Am Chem Soc* **71** 291 3929 3931 4165 *1949*.] The (-)-*picrate* has **m** 140-141° (cryst from EtOH). The *racemate* *[1126-71-2]* has **b** 88-89°/16mm, 92-94°/30mm, 194-195°/atm, d$_4^{20}$ 0.908.

2,3-Dimethylphenol *[526-75-0]* **M 122.2, m 75°, b 120°/20mm, 218°/760mm, pK²⁵ 10.54.** Crystd from aqueous EtOH.

2,5-Dimethylphenol *[95-87-1]* **M 122.2, m 73°, b 211.5°/762mm, pK²⁵ 10.41.** Crystd from EtOH/ether.

2,6-Dimethylphenol *[576-26-1]* **M 122.2, m 49°, b 203°/760mm, pK²⁵ 10.61.** Fractionally distd under nitrogen, crystd from *benzene or hexane, and sublimed at 38°/10mm.

3,4-Dimethylphenol *[95-65-8]* **M 122.2, m 65°, b 225°/757mm, pK²⁵ 10.36.** Heated with an equal weight of conc H_2SO_4 at 103-105° for 2-3h, then diluted with four volumes of water, refluxed for 1h, and either steam distd or extracted repeatedly with diethyl ether after cooling to room temperature. The steam distillate was also extracted and evaporated to dryness. (The purification process depends on the much slower sulfonation of 3,5-dimethylphenol than most of its likely contaminants.). It can also be crystd from water, hexane or pet ether, and vacuum sublimed. [Kester *Ind Eng Chem (Anal Ed)* **24** 770 *1932*; Bernasconi and Paschalis *J Am Chem Soc* **108** 2969*1986*.]

3,5-Dimethylphenol *[108-68-9]* **M 122.2, m 68°, b 219°, pK²⁵ 10.19.** Purification as for 3,4-dimethylphenol.

Dimethyl phthalate *[131-11-3]* **M 194.2, b 282°, n 1.5149, d 1.190, d²⁵ 1.1865.** Washed with aqueous Na_2CO_3, then distilled water, dried ($CaCl_2$) and distd under reduced pressure (**b** 151-152°/0.1mm).

2,2-Dimethyl-1,3-propanediol (neopentyl glycol) *[126-30-7]* **M 104.2, m 128.4-129.4°, b 208°/760mm.** Crystd from *benzene or acetone/water (1:1).

2,2-Dimethyl-1-propanol (*neo*-pentyl alcohol) *[75-84-3]* **M 88.2, m 52°, b 113.1°/760mm.** Difficult to distil because it is a solid at ambient temperatures. Purified by fractional crystallisation and sublimation.

N,N-**Dimethylpropionamide** *[758-96-3]* **M 101.2, b 175-178°, d 0.920, n 1.440.** Shaken over BaO for 1-2 days, then distd at reduced pressure.

2,5-Dimethylpyrazine *[123-32-0]* **M 108.1, b 156°, d 0.990, n 1.502, pK_1^{25} -4.6** (aq H_2SO_4), pK_2^{25} **1.85.** Purified *via* its picrate (m 150°) [Wiggins and Wise *J Chem Soc* 4780 *1956*].

3,5-Dimethylpyrazole *[67-51-6]* **M 96.1, m 107-108°, pK^{20} 4.16.** Crystd from cyclohexane or water. [Barszez et al. *J Chem Soc, Dalton Trans* 2025 *1986*.]

2,3-Dimethylquinoxaline *[2379-55-7]* **M 158.2, m 106°, pK^{25} -3.84** (aq H_2SO_4). Crystd from distilled water.

2,4-Dimethylresorcinol *[634-65-1]* **M 138.1, m 149-150° $pK_{Est(1)}$ ~9.8, $pK_{Est(2)}$~11.7.** Crystd from pet ether (b 60-80°).

meso-α,β-**Dimethylsuccinic acid** *[608-40-2]* **M 146.1, m 211°, pK_1^{25} 3.77, pK^{25} 5.36.** Crystd from EtOH/ether or EtOH/chloroform.

2,2-Dimethylsuccinic acid *[597-43-3]* **M 146.1, m 141°, pK_1^{20} 4.15, pK_2^{20} 6.40.** Crystd from EtOH/ether or EtOH/chloroform.

(±)-2,3-Dimethylsuccinic acid *[13545-04-5]* **M 146.1, m 129°, pK_1^{25} 3.82, pK_2^{25} 5.98.** Crystd from water.

Dimethyl sulfide *[75-18-3]* **M 62.1, f -98.27°, b 0°/172mm, 37.5-38°/760mm, d^{21} 0.8458, n^{25} 1.4319.** Purified *via* the Hg(II) chloride complex by dissolving 1 mole of Hg(II)Cl$_2$ in 1250mL of EtOH and slowly adding the boiling alcoholic soln of dimethyl sulfide to give the right ratio for 2(CH$_3$)$_2$S.3HgCl$_2$. After recrystn of the complex to constant melting point, 500g of complex is heated with 250mL conc HCl in 750mL of water. The sulfide is separated, washed with water, and dried with CaCl$_2$ and CaSO$_4$. Finally, it is distd under reduced pressure from sodium. Precautions should be taken (*efficient fume hood*) because of its very **UNPLEASANT ODOUR.**

2,4-Dimethylsulfolane *[1003-78-7]* **M 148.2, b 128°/77mm, d^{25} 1.1314.** Vacuum distd.

Dimethyl sulfone *[67-71-0]* **M 94.1, m 109°.** Crystd from water. Dried over P$_2$O$_5$.

Dimethyl sulfoxide (DMSO) *[67-68-5]* **M 78.1, m 18.0-18.5°, b 75.6-75.8°/12mm, 190°/760mm, d 1.100, n 1.479.** Colourless, odourless, very *hygroscopic* liquid, synthesised from dimethyl sulfide. The main impurity is water, with a trace of dimethyl sulfone. The Karl-Fischer test is applicable. It is dried with Linde types 4A or 13X molecular sieves, by prolonged contact and passage through a column of the material, then distd under reduced pressure. Other drying agents include CaH$_2$, CaO, BaO and CaSO$_4$. It can also be fractionally crystd by partial freezing. More extensive purification is achieved by standing overnight with freshly heated and cooled chromatographic grade alumina. It is then refluxed for 4h over CaO, dried over CaH$_2$, and then fractionally distd at low pressure. For efficiency of desiccants in drying dimethyl sulfoxide see Burfield and Smithers [*J Org Chem* 43 3966 *1978*; Sato et al. *J Chem Soc, Dalton Trans* 1949 *1986*].
Rapid purification: Stand over freshly activated alumina, BaO or CaSO$_4$ overnight. Filter and distil over CaH$_2$ under reduced pressure (~ 12 mm Hg). Store over 4A molecular sieves.

Dimethyl terephthalate *[120-61-6]* **M 194.2, m 150°.** Purified by zone melting.

N,N-**Dimethylthiocarbamoyl chloride** *[16420-13-6]* **M 123.6, m 42-43°, b 64-65°/0.1mm.** Crystd twice from pentane.

N,*N*-Dimethyl-*o*-toluidine *[609-72-3]* **M 135.2, b 68°/10mm, 211-211.5°/760mm, d 0.937, n 1.53664, pK^{25} 5.85.** Isomers and other bases have been removed by heating in a water bath for 100h with two equivalents of 20% HCl and two and a half volumes of 40% aq formaldehyde, then making the soln alkaline and separating the free base. After washing well with water it was distd at 10mm pressure and redistd at ambient pressure [von Braun and Aust *Chem Ber* **47** 260 *1914*]. Other procedures include drying with NaOH, distilling from zinc in an atmosphere of nitrogen under reduced pressure, and refluxing with excess of acetic anhydride in the presence of conc H$_2$SO$_4$ as catalyst, followed by fractional distn under vacuum.

N,*N*-Dimethyl-*m*-toluidine *[121-72-2]* **M 135.2, b 211.5-212.5°, d 0.93, pK^{25} 5.22.** See *m*-methyl-*N*,*N*-dimethylaniline on p. 291.

N,*N*-Dimethyl-*p*-toluidine *[99-97-8]* **M 135.2, b 93-94°/11mm, b 211°, d 0.937, n 1.5469, pK^{25} 4.76.** See *p*-methyl-*N*,*N*-dimethylaniline on p. 291.

1,3-Dimethyluracil *[874-14-6]* **M 140.1, m 121-122°, pK -3.25 (aq H$_2$SO$_4$).** Crystd from EtOH/ether.

sym-**Dimethylurea** *[96-31-1]* **M 88.1, m 106°.** Crystd from acetone/diethyl ether by cooling in an ice bath. Also crystd from EtOH and dried at 50° and 5mm for 24h [Bloemendahl and Somsen *J Am Chem Soc* **107** 3426 *1985*].

2,2'-Dinaphthylamine *[532-18-3]* **M 269.3, m 170.5°, pK$_{Est}$ <0.** Crystd from *benzene.

2,4-Dinitroaniline *[97-02-9]* **M 183.1, m 180°, ε$_{348}$ 12,300 in dil aq HClO$_4$, pK^{25} -4.27 (aq H$_2$SO$_4$).** Crystd from boiling EtOH by adding one-third volume of water and cooling slowly. Dried in a steam oven.

2,6-Dinitroaniline *[606-22-4]* **M 183.1, m 139-140°, pK^{25} -5.37 (aq H$_2$SO$_4$).** Purified by chromatography on alumina, then crystd from *benzene or EtOH.

2,4-Dinitroanisole *[5327-44-6]* **M 198.1, m 94-95°.** Crystd from aq EtOH.

3,5-Dinitroanisole *[119-27-7]* **M 198.1, m 105-106°.** Purified by repeated crystn from water and dried in a vacuum desiccator over P$_2$O$_5$.

1,2-Dinitrobenzene *[528-29-0]* **M 168.1, m 116.5°.** Crystd from EtOH.

1,3-Dinitrobenzene *[99-65-0]* **M 168.1, m 90.5-91°.** Crystd from alkaline EtOH soln (20g in 750mL 95% EtOH at 40°, plus 100mL of 2M NaOH) by cooling and adding 2.5L of water. The ppte, after filtering off, washing with water and sucking dry, was crystd from 120mL, then 80mL of absolute EtOH [Callow, Callow and Emmens *Biochem J* **32** 1312 *1938*]. Has also been crystd from MeOH, CCl$_4$ and ethyl acetate. Can be sublimed in a vacuum. [Tanner *J Org Chem* **52** 2142 *1987*.]

1,4-Dinitrobenzene *[100-25-4]* **M 168.1, m 173°.** Crystd from EtOH or ethyl acetate. Dried under vacuum over P$_2$O$_5$. Can be sublimed in a vacuum.

2,4-Dinitrobenzenesulfenyl chloride *[528-76-7]* **M 234.6, m 96°.** Crystd from CCl$_4$.

2,4-Dinitrobenzenesulfonyl chloride *[1656-44-6]* **M 266.6, m 102°.** Crystd from *benzene or *benzene/pet ether.

2,4-Dinitrobenzoic acid *[610-30-3]* **M 212.1, m 183°, pK^{25} 1.42.** Crystd from aqueous 20% EtOH (10mL/g), dried at 100°.

2,5-Dinitrobenzoic acid *[610-28-6]* **M 212.1, m 179.5-180°, pK^{25} 1.62.** Crystd from distd water. Dried in a vacuum desiccator.

2,6-Dinitrobenzoic acid *[603-12-3]* **M 212.1, m 202-203°, pK^{25} 1.14.** Crystd from water.

3,4-Dinitrobenzoic acid *[528-45-0]* **M 212.1, m 166°, pK^{25} 2.81.** Crystd from EtOH by addition of water.

3,5-Dinitrobenzoic acid *[99-34-3]* **M 212.1, m 205°, pK^{25} 2.73 (2.79).** Crystd from distilled water or 50% EtOH (4mL/g). Dried in a vacuum desiccator or at 70° over BaO under vacuum for 6h.

4,4'-Dinitrobenzoic anhydride *[902-47-6]* **M 406.2, m 189-190°.** Crystd from acetone.

3,5-Dinitrobenzoyl chloride *[99-33-2]* **M 230.6, m 69.5°.** Crystd from CCl$_4$ or pet ether (b 40-60°). It reacts readily with water, and should be kept in sealed tubes or under dry pet ether.

2,2'-Dinitrobiphenyl *[2436-96-6]* **M 244.2, m 123-124°.** Crystd from EtOH.

2,4'-Dinitrobiphenyl *[606-81-5]* **M 244.2, m 92.7-93.7°.** Crystd from EtOH.

4,4'-Dinitrobiphenyl *[1528-74-1]* **M 244.2, m 240.9-241.8°.** Crystd from *benzene, EtOH (charcoal) or acetone. Dried under vacuum over P$_2$O$_5$.

2,6-Dinitro-*p*-cresol (2,6-dinitro-4-methylphenol) *[609-93-3]* **M 198.1, m 78-79°, pK$_{Est}$~3.7.** Recrystd from EtOH and is steam volatile. **TOXIC IRRITANT.**

4,6-Dinitro-*o*-cresol *[534-52-1]* **M 198.1, m 85-86°, 87°, pK^{25}4.70.** Crystd from aqueous EtOH.

2,4-Dinitrodiphenylamine *[961-68-2]* **M 259.2, m 157°, pK$_{Est}$ <0.** Crystd from aqueous EtOH.

4,4'-Dinitrodiphenylurea *[587-90-6]* **M 302.2, m 312°(dec).** Crystd from EtOH. Sublimes in vac.

2,4-Dinitrofluorobenzene (Sanger's reagent) *[70-34-8]* **M 186.1, m 25-27°, b 133°/2mm, 140-141°/5mm, d 1.483.** Crystd from ether or EtOH. Vacuum distd through a Todd Column (see p. 174). If it is to be purified by distn *in vacuo*, the distn unit must be allowed to cool before air is allowed into the apparatus otherwise the residue carbonises spontaneously and an **EXPLOSION** may occur. The material is a **skin irritant** and may cause serious dermatitis.

1,8-Dinitronaphthalene *[602-38-0]* **M 218.2, m 170-171°.** Crystd from *benzene.

2,4-Dinitro-1-naphthol (Martius Yellow) *[605-69-6]* **M 234.2, m 81-82°, pK$_{Est}$ ~3.7.** Crystd from *benzene or aqueous EtOH.

2,4-Dinitrophenetole *[610-54-8]* **M 240.2, m 85-86°.** Crystd from aqueous EtOH.

2,4-Dinitrophenol *[51-28-5]* **M 184.1, m 114°, pK^{25} 4.12.** Crystd from *benzene, EtOH, EtOH/water or water acidified with dil HCl, then recrystd from CCl$_4$. Dried in an oven and stored in a vac desiccator over CaSO$_4$.

2,5-Dinitrophenol *[329-71-5]* **M 184.1, m 108°, pK^{25} 5.20.** Crystd from H$_2$O with a little EtOH.

2,6-Dinitrophenol *[573-56-8]* **M 184.1, m 63.0-63.7°, pK^{25} 3.73.** Crystd from *benzene/cyclohexane, aqueous EtOH, water or *benzene/pet ether (b 60-80°, 1:1).

3,4-Dinitrophenol *[577-71-9]* **M 184.1, m 138°, pK²⁵ 5.42.** Steam distd and crystd from water and air-dried. **CAUTION - EXPLOSIVE** when dry, store with 10% water.

3,5-Dinitrophenol *[586-11-8]* **M 184.1, m 126°, pK²⁵ 6.68.** Crystd from *benzene or CHCl₃/pet ether. Should be stored with 10% water because it is **EXPLOSIVE** when dry.

2,4-Dinitrophenylacetic acid *[643-43-6]* **M 226.2, m 179°(dec), pK²⁵ 3.50.** Crystd from water.

2,4-Dinitrophenylhydrazine (DNPH) *[119-26-6]* **M 198.1, m 200°(dec), pK$_{Est}$ ~2.0.** Crystd from butan-1-ol, dioxane, EtOH, *C₆H₆ or ethyl acetate. *HCl* has **m** 186°(dec).

2,2-Dinitropropane *[595-49-3]* **M 162.1, m 53.5°.** Crystd from EtOH or MeOH. Dried over CaCl₂ or under vacuum for 1h just above the melting point.

2,4-Dinitroresorcinol *[519-44-8]* **M 200.1, m 149°, pK²⁵ 3.05.** Crystd from aq EtOH. **Explosive.**

3,5-Dinitrosalicylic acid *[609-99-4]* **M 228.1, m 173-174°, pK₁²⁵ 0.70, pK₂²⁵ 7.40.** Crystd from water.

2,6-Dinitrothymol *[303-21-9]* **M 240.2, m 53-54°.** Crystd from aq EtOH.

2,3-Dinitrotoluene *[602-01-7]* **M 182.1, m 63°.** Distd in steam and crystd from water or *benzene/pet ether. Stored with 10% water. *Could be* **EXPLOSIVE** *when dry.*

2,4-Dinitrotoluene *[121-14-2]* **M 182.1, m 70.5-71.0°.** Crystd from acetone, isopropanol or MeOH. Dried under vacuum over H₂SO₄. Purified by zone melting. *Could be* **EXPLOSIVE** *when dry.*

2,5-Dinitrotoluene *[619-15-8]* **M 182.1, m 51.2°.** Crystd from *benzene.

2,6-Dinitrotoluene *[606-20-2]* **M 182.1, m 64.3°.** Crystd from acetone.

3,4-Dinitrotoluene *[610-39-9]* **M 182.1, m 61°.** Distil in steam and cryst from *benzene/pet ether. Store with 10% of water to avoid **EXPLOSION.**

3,5-Dinitro-*o*-toluic acid *[28169-46-2]* **M 226.2, m 206°, pK$_{Est}$ ~3.0.** Crystd from H₂O or aq EtOH.

2,4-Dinitro-*m*-xylene *[603-02-1]* **M 196.2, m 83-84°.** Crystd from EtOH.

Dinonyl phthalate (mainly 3,5,5-trimethylhexyl phthalate isomer) *[28553-12-0; 14103-61-8]* **M 418.6, m 26-29°, b 170°/2mm, d 0.9640, n 1.4825.** Washed with aqueous Na₂CO₃, then shaken with water. Ether was added to break the emulsion, and the soln was washed twice with water, and dried (CaCl₂). After evaporating the ether, the residual liquid was distd three times under reduced pressure. It was stored in a vacuum desiccator over P₂O₅

Dioctadecyldimethylammonium bromide *[3700-67-2]* **M 630.9, m 161-163°.** Crystd from acetone then MeOH [Lukac *J Am Chem Soc* **106** 4387 *1984*]. Also purified by chromatography on alumina by washing with *C₆H₆ and eluting with Me₂CO, evap and cryst from MeCN [Swain and Kreevoy *J Am Chem Soc* **77** 1126 *1955*].

Diosgenin *[512-04-9]* **M 294.5, m 204-207°, [α]$_D^{25}$ -129° (in Me₂CO).** Crystd from acetone.

1,3-Dioxane *[505-22-6]* **M 88.1, b 104.5°/751mm, d 1.040, n 1.417.** Dried with sodium and fractionally distd.

1,4-Dioxane *[123-91-1]* **M 88.1, f 11.8°, b 101.3°, d^{25} 1.0292, n^{15} 1.4236, n^{25} 1.42025.**
Prepared commercially either by dehydration of ethylene glycol with H_2SO_4 and heating ethylene oxide or bis(ß-chloroethyl)ether with NaOH. Usual impurities are acetaldehyde, ethylene acetal, acetic acid, water and peroxides. Peroxides can be removed (and the aldehyde content decreased) by percolation through a column of activated alumina (80g per 100-200mL solvent), by refluxing with $NaBH_4$ or anhydrous stannous chloride and distilling, or by acidification with conc HCl, shaking with ferrous sulfate and leaving in contact with it for 24h before filtering and purifying further.

Hess and Frahm [*Chem Ber* **71** 2627 *1938*] refluxed 2L of dioxane with 27mL conc HCl amd 200mL water for 12h with slow passage of nitrogen to remove acetaldehyde. After cooling the soln KOH pellets were added slowly and with shaking until no more would dissolve and a second layer had separated. The dioxane was decanted, treated with fresh KOH pellets to remove any aq phase, then transferred to a clean flask where it was refluxed for 6-12h with sodium, then distd from it. Alternatively, Kraus and Vingee [*J Am Chem Soc* **56** 511 *1934*] heated on a steam bath with solid KOH until fresh addition of KOH gave no more resin (due to acetaldehyde). After filtering through paper, the dioxane was refluxed over sodium until the surface of the metal was not further discoloured during several hours. It was then distd from sodium.

The acetal (**b 82.5°**) is removed during fractional distn. Traces of *benzene, if present, can be removed as the *benzene/MeOH azeotrope by distn in the presence of MeOH. Distn from $LiAlH_4$ removes aldehydes, peroxides and water. Dioxane can be dried using Linde type 4X molecular sieves. Other purification procedures include distn from excess C_2H_5MgBr, refluxing with PbO_2 to remove peroxides, fractional crystn by partial freezing and the addition of KI to dioxane acidified with aq HCl. Dioxane should be stored out of contact with air, preferably under N_2.

A detailed purification procedure is as follows: Dioxane was stood over ferrous sulfate for at least 2 days, under nitrogen. Then water (100mL) and conc HCl (14mL) / litre of dioxane were added (giving a pale yellow colour). After refluxing for 8-12h with vigorous N_2 bubbling, pellets of KOH were added to the warm soln to form two layers and to discharge the colour. The soln was cooled rapidly with more KOH pellets being added (magnetic stirring) until no more dissolved in the cooled soln. After 4-12h, if the lower phase was not black, the upper phase was decanted rapidly into a clean flask containing sodium, and refluxed over sodium (until freshly added sodium remained bright) for 1h. The middle fraction was collected (and checked for minimum absorbency below 250nm). The distillate was fractionally frozen three times by cooling in a refrigerator, with occasional shaking or stirring. This material was stored in a refrigerator. For use it was thawed, refluxed over sodium for 48h, and distilled into a container for use. All joints were clad with Teflon tape.

Coetzee and Chang [*Pure Appl Chem* **57** 633 *1985*] dried the solvent by passing it slowly through a column (20g/L) of 3A molecular sieves activated by heating at 250° for 24h. Impurities (including peroxides) were removed by passing the effluent slowly through a column packed with type NaX zeolite (pellets ground to 0.1mm size) activated by heating at 400° for 24h or chromatographic grade basic Al_2O_3 activated by heating at 250° for 24h. After removal of peroxides the effluent was refluxed several hours over sodium wire, excluding moisture, distilled under nitrogen or argon and stored in the dark.

One of the best tests of purity of dioxane is the formation of the purple disodium benzophenone complex during reflux and its persistence on cooling. (Benzophenone is better than fluorenone for this purpose, and for the storing of the solvent.) [Carter, McClelland and Warhurst *Trans Faraday Soc* **56** 343 *1960*]. **TOXIC.**

Rapid purification: Check for peroxides (see Chapter 1 and Chapter 2 for test under ethers). Pre-dry with $CaCl_2$ or better over Na wire. Then reflux the pre-dried solvent over Na (1% w/v) and benzophenone (0.2% w/v) under an inert atmosphere until the blue colour of the benzophenone ketyl radical anion persists. Distil, and store over 4A molecular sieves in the dark.

1,3-Dioxolane *[646-06-0]* **M 74.1, b 75.0-75.2°, d 1.0600, n^{21} 1.3997.** Dried with solid NaOH, KOH or $CaSO_4$, and distd from sodium or sodium amalgam. Barker et al. [*J Chem Soc* 802 *1959*] heated 34mL of dioxalane under reflux with 3g of PbO_2 for 2h, then cooled and filtered. After adding xylene (40mL) and PbO_2 (2g) to the filtrate, the mixture was fractionally distd. Addition of xylene (20mL) and sodium wire to the main fraction (**b 70-71°**) led to vigorous reaction, following which the mixture was again fractionally distd. Xylene and sodium additions were made to the main fraction (**b 73-74°**) before it was finally distd.

4,4'-Di-*n*-pentyloxyazoxybenzene *[64242-26-8]* **M 370.5.** Crystd from Me_2CO, and dried by heating under vacuum.

Diphenic acid *[482-05-3]* **M 242.2, m 228-229°, pK²⁵ 3.46.** Crystd from water.

Diphenic anhydride *[6050-13-1]* **M 466.3, m 217°.** After removing free acid by extraction with cold aq Na_2CO_3, the residue has been crystd from acetic anhydride and dried at 100°. Acetic anhydride also converts the acid to the anhydride.

***N,N*-Diphenylacetamidine** *[621-09-0]* **M 210.3, m 131°.** Crystd from EtOH, then sublimed under vacuum at *ca* 96° onto a "finger" cooled in solid CO_2/MeOH, with continuous pumping to free it from occluded solvent.

Diphenylacetic acid *[117-34-0]* **M 212.3, m 147.4-148.4°, pK²⁵ 3.94.** Crystd from *benzene, H_2O or aq 50% EtOH.

Diphenylacetonitrile *[86-29-3]* **M 193.3, m 73-75°.** Crystd from EtOH or pet ether (b 90-100°).

Diphenylacetylene (tolan) *[501-65-5]* **M 178.2, m 62.5°, b 90-97°/0.3mm.** Crystd from EtOH.

Diphenylamine *[122-39-4]* **M 169.2, m 62.0-62.5°, pK²⁵ 0.77 (aq H_2SO_4).** Crystd from pet ether, MeOH, or EtOH/water. Dried under vacuum.

Diphenylamine-2-carboxylic acid *[91-40-7]* **M 213.2, m 184°, pK$_1^{25}$ -1.28 (aq H_2SO_4), pK$_2^{25}$ 3.86.** See *N*-phenylanthranilic acid on p. 327.

Diphenylamine-2,2'-dicarboxylic acid **(2,2'-iminodibenzoic acid)** *[579-92-0]* **M 257.2, m 298°(dec), pK$_{Est}$ ~3.7.** Crystd from EtOH.

9,10-Diphenylanthracene *[1499-10-1]* **M 330.4, m 248-249°.** Crystd from acetic acid or xylene [Baumstark et al. *J Org Chem* **52** 3308 *1987*].

***N,N'*-Diphenylbenzidine** *[531-91-9]* **M 336.4, m 245-247°, 251-252°, pK²⁵ 0.30.** Crystd from toluene or ethyl acetate. Stored in the dark.

***trans-trans*-1,4-Diphenylbuta-1,3-diene** *[538-81-8]* **M 206.3, m 153-153.5°.** Its soln in pet ether (b 60-70°) was chromatographed on an alumina-Celite column (4:1) and the column was washed with the same solvent. The main zone was cut out, eluted with ethanol and transferred to pet ether, which was then dried and evaporated [Pinckard, Wille and Zechmesiter *J Am Chem Soc* **70** 1938 *1948*]. Recrystd from hexane.

***sym*-Diphenylcarbazide** *[140-22-7]* **M 242.3, m 172°.** A common impurity is phenylsemicarbazide which can be removed by chromatography [Willems et al. *Anal Chim Acta* **51** 544 *1970*]. Crystd from EtOH or glacial acetic acid.

1,5-Diphenylcarbazone *[538-62-5]* **M 240.3, m 124-127°.** Crystd from EtOH (*ca* 5mL/g), and dried at 50°. A commercial sample, nominally *sym*-diphenylcarbazone but of **m 154-156°**, was a mixture of diphenylcarbazide and diphenylcarbazone. The former was removed by dissolving 5g of the crude material in 75mL of warm EtOH, then adding 25g Na_2CO_3 dissolved in 400mL of distd water. The alkaline soln was cooled and extracted six times with 50mL portions of diethyl ether (discarded). Diphenylcarbazone was then ppted by acidifying the alkaline soln with 3M HNO_3 or glacial acetic acid. It was filtered on a Büchner funnel, air dried, and stored in the dark [Gerlach and Frazier *Anal Chem* **30** 1142 *1958*]. Other impurities were phenylsemicarbazide and diphenylcarbodiazone. Impurities can be detected by chromatography [Willems et al. *Anal Chim Acta* **51** 544 *1970*].

Diphenyl carbonate *[102-09-0]* **M 214.2, m 80°.** Purified by sublimation, and by preparative gas chromatography with 20% Apiezon on Embacel, and crystn from EtOH.

Diphenylcyclopropenone (Diphencyprone) *[886-38-4]* **M 206.2, m 87-90°(hydrate), 119-121°(anhydr).** Crystd from cyclohexane. UV (MeCN): λmax 226, 282, 297nm.

Diphenyl disulfide (phenyl disulfide) *[882-33-7]* **M 218.3, m 60.5°.** Crystd from MeOH. [Alberti et al. *J Am Chem Soc* **108** 3024 *1986*]. Crystd repeatedly from hot diethyl ether, then vac dried at 30° over P_2O_5, fused under nitrogen and re-dried, the whole procedure being repeated, with a final drying under vac for 24h. Also recrystd from hexane/EtOH soln. [Burkey and Griller *J Am Chem Soc* **107** 246 *1985*.]

1,1-Diphenylethanol *[599-67-7]* **M 198.3, m 80-81°.** Crystd from *n*-heptane. [Bromberg et al. *J Am Chem Soc* **107** 83 *1985*.]

1,1-Diphenylethylene *[530-48-3]* **M 180.3, b 268-270°, d 1.024, n 1.6088.** Distd under reduced pressure from KOH. Dried with CaH_2 and redistd.

N,N'-Diphenylethylenediamine (Wanzlick's reagent) *[150-61-8]* **M 212.3, m 67.5°, b 178-182°/2mm** $pK_{Est(1)}$ ~0.5, $pK_{Est(2)}$~3.8. Crystd from aqueous EtOH or MeOH.

N,N'-Diphenylformamidine *[622-15-1]* **M 196.2, m 142°, 137°, 136-139°.** Crystd from absolute EtOH, gives the hydrate with aqueous EtOH.

1,3-Diphenylguanidine *[102-06-7]* **M 211.3, m 148°, pK^{25} 10.12.** Crystd from toluene, aqueous acetone or EtOH, and vacuum dried.

1,6-Diphenyl-1,3,5-hexatriene *[1720-32-7]* **M 232.3, m 200-203°.** Crystd from $CHCl_3$ or EtOH/$CHCl_3$ (1:1).

5,5-Diphenylhydantoin *[57-41-0]* **M 252.3, m 293-295°.** Crystd from EtOH.

1,1-Diphenylhydrazine (hydrazobenzene) *[122-66-7]* **M 184.2, m 34°, 44°, 175°/10mm, 222°/40mm, pK_{Est} ~1.7.** Crystd from hot EtOH containing a little ammonium sulfide or H_2SO_3 (to prevent atmospheric oxidation), preferably under nitrogen. Dried in a vacuum desiccator. Also crystd from pet ether (b 60-100°) to constant absorption spectrum. *HCl*, from EtOH has **m** 163-164°(dec). *Picrate*, from *C_6H_6, has **m** 123°(dec).

1,3-Diphenylisobenzofuran *[5471-63-6]* **M 270.3, m 129-130°.** Recrystd from EtOH or EtOH/$CHCl_3$ (1:1) under red light (as in photographic dark rooms) or from *benzene in the dark.

Diphenylmethane *[101-81-5]* **M 168.2, m 25.4°.** Sublimed under vacuum, or distd at 72-75°/4mm. Crystd from EtOH. Purified by fractional crystn of the melt.

1,1-Diphenylmethylamine *[91-00-9]* **M 183.2, m 34°, pK_{Est} ~9.1.** Crystd from water.

Diphenylmethyl chloride (benzhydryl chloride) *[90-99-3]* **M 202.7, m 17.0°, b 167°/17mm, n 1.5960.** Dried with Na_2SO_4 and fractionally distd under reduced pressure.

***all-trans*-1,8-Diphenyl-1,3,5,8-octatetraene** *[3029-40-1]* **M 258.4, m 235-237°.** Crystd from EtOH.

2,5-Diphenyl-1,3,4-oxadiazole (PPD) *[725-12-2]* **M 222.3, m 70° (hydrate), 139-140° (anhydrous), b 231°/13mm, 248°/16mm.** Crystd from EtOH and sublimed *in vacuo*.

2,5-Diphenyloxazole (PPO) *[92-71-7]* **M 221.3, m 74°, b 360°/760mm.** Distd in steam and crystd from ligroin.

N,N'-**Diphenyl**-*p*-**phenylenediamine** *[74-31-7]* **M 260.3, m 148-149°, b 219-224°/0.7mm, pK$_{Est}$ <0.** Crystd from EtOH, chlorobenzene/pet ether or *benzene. Has also been crystd from aniline, then extracted three times with absolute EtOH.

1,1-Diphenyl-2-picrylhydrazine *[1707-75-1]* **M 395.3, m 174°(dec), 178-179.5°(dec).** Crystd from CHCl$_3$, or *benzene/pet ether (1:1), then degassed at 100° and <10^{-5}mm Hg for *ca* 50h to decompose the 1:1 molar complex formed with *benzene.

2,2-Diphenylpropionic acid *[5558-66-7]* **M 226.3, m 173-174°, pK$_{Est}$ ~3.8.** Crystd from EtOH.

3,3-Diphenylpropionic acid *[606-83-7]* **M 226.3, m 155°, pK$_{Est}$ ~4.5.** Crystd from EtOH.

Diphenyl sulfide *[139-66-2]* **M 186.3, b 145°/8mm, d 1.114, n 1.633.** Washed with aqueous 5% NaOH, then water. Dried with CaCl$_2$, then with sodium. The sodium was filtered off and the diphenyl sulfide was distd under reduced pressure.

Diphenyl sulfone *[127-63-9]* **M 218.3, m 125°, b 378°(dec).** Crystd from diethyl ether. Purified by zone melting.

sym-**Diphenylthiourea (thiocarbanilide)** *[102-08-9]* **M 228.3, m 154°.** Crystd from boiling EtOH by adding hot water and allowing to cool.

1,1-Diphenylurea *[603-54-3]* **M 212.3, m 238-239°.** Crystd from MeOH.

Dipicolinic acid (pyridine-2,6-dicarboxylic acid) *[499-83-2]* **M 167.1, m 255°(dec), λ$_{max}$ 270nm, pK$_1^{20}$ 2.10, pK$_2^{20}$ 4.68.** Crystd from water, and sublimed in a vacuum.

N,N-**Di**-*n*-**propylaniline** *[2217-07-4]* **M 177.3, b 127°/10mm, 238-241°/760mm, pK^{23} 5.68.** Refluxed for 3hr with acetic anhydride, then fractionally distd under reduced pressure.

Dipropylene glycol (octan-4,5-diol) *[110-98-5]* **M 134.2, b 109-110°/8mm, d 1.022, n 1.441.** Fractionally distd below 15mm pressure, using packed column and taking precautions to avoid absorption of water.

Di-*n*-**propyl ketone** *[123-19-3]* **M 114.2, b 143.5°, d 0.8143, n 1.40732.** Dried with CaSO$_4$, then distd from P$_2$O$_5$ under nitrogen.

Di-*n*-**propyl sulfide** *[111-47-7]* **M 118.2, b 141-142°, d 0.870, n 1.449.** Washed with aqueous 5% NaOH, then water. Dried with CaCl$_2$ and distd from Na [Dunstan and Griffiths *J Chem Soc* 1344 *1962*].

Di-(4-pyridoyl)hydrazine *[4329-75-3]* **M 246.2, m 254-255°.** Crystd from water.

2,2'-Dipyridylamine *[1202-34-2]* **M 171.2, m 84° and remelts at 95° after solidifying, b 176-178°/13mm, 307-308°/760mm, pK 6.69 (in 20% EtOH).** Crystd from *benzene or toluene [Blakley and De Armond *J Am Chem Soc* **109** 4895 *1987*].

2,2'-Dipyridyl disulfide (2,2'dithiopyridine) *[2127-03-9]* **M 220.3, m 53°, 56-58°, 57-58°, pK$_1^{25}$ 0.35, pK$_2^{25}$ 2.45.** Recrystd from *C$_6$H$_6$/pet ether (6:7), ligroin or *C$_6$H$_6$. *Picrate* has m 119° (from EtOH). [Walter et al. *Justus Liebigs Ann Chem* **695** 7785 *1966*; Marckwald et al. *Chem Ber* 33 1556 *1900*; Brocklehurst and Little *Biochem J* **133** 67,78 *1973*.]

1,2-Di-(4-pyridyl)-ethane *[4916-57-8]* **M 184.2, pK$_{Est(1)}$ ~3.8, pK$_{Est(2)}$~5.4.** Crystd from cyclohexane/*benzene (5:1).

trans-1,2-Di-(4-pyridyl)-ethylene *[13362-78-2]* M 182.2, m 153-154°, 155.5-156.5°, pK_1^{25} 3.65, pK_2^{25} 5.6. Crystd from water (1.6g/100mL at 100°). *Di-HCl* has m 347°, from EtOH

1,3-Di-(4-pyridyl)-propane *[17252-51-6]* M 198.3, m 60.5-61.5°, $pK_{Est(1)}$ ~4.5, $pK_{Est(2)}$ ~5.5. Crystd from *n*-hexane/*benzene (5:1).

S-1,2-Distearin *[1188-58-5]* M 625.0, m 76-77°, $[\alpha]_D^{20}$ -2.8° (c 6.3, $CHCl_3$), $[\alpha]_{546}^{20}$ +1.4° (c 10, $CHCl_3$/MeOH, 9:1). Crystd from chloroform/pet ether.

2,5-Distyrylpyrazine *[14990-02-4]* M 284.3, m 219°. Recrystd from xylene; chromatographed on basic silica gel (60-80 mesh) using CH_2Cl_2 as eluent, then vac sublimed on to a cold surface at 10^{-3} torr [Ebied *J Chem Soc, Faraday Trans 1* **78** 3213 *1982*]. Operations should be carried out in the dark.

1,3-Dithiane *[505-23-7]* M 120.2, m 54°. Crystd from 1.5 times its weight of MeOH at 0°, and sublimed at 40-50°/0.1mm.

2,2'-Dithiobis(benzothiazole) *[120-78-8]* M 332.2, m 180°. Crystd from *benzene.

4,4'-Dithiodimorpholine *[103-34-4]* M 236.2, m 124-125°. Crystd from hot aq dimethylformamide.

1,4-Dithioerythritol (DTE, *erythro*-2,3-dihydroxy-1,4-dithiobutane) *[6892-68-8]* M 154.3, m 82-84°, pK_1 9.0, pK_2 9.9. Crystd from ether/hexane and stored in the dark at 0°.

Dithiooxamide (rubeanic acid) *[79-40-3]* M 120.2, m >300°. Crystd from EtOH and sublimed in a vacuum.

RS-1,4-Dithiothreitol (DTT, Cleland's reagent) *[27565-41-9]* M 154.3, m 42-43°, pK_1 8.3, pK_2 9.5. Crystd from ether and sublimed at 37°/0.005mm. Should be stored at 0°.

Dithizone (diphenylthiocarbazone) *[60-10-6]* M 256.3, ratio of $\varepsilon_{620nm}/\varepsilon_{450nm}$ should be ≥1.65, ε_{620} 3.4 x 10^4 ($CHCl_3$), pK_2 4.6. The crude material is dissolved in CCl_4 to give a concentrated soln. This is filtered through a sintered glass funnel and shaken with 0.8M aq ammonia to extract dithizonate ion. The aqueous layer is washed with several portions of CCl_4 to remove undesirable materials. The aqueous layer is acidified with dil H_2SO_4 to precipitate pure dithizone. It is dried in a vacuum. When only small amounts of dithizone are required, purification by paper chromatography is convenient. [Cooper and Hibbits *J Am Chem Soc* **75** 5084 *1933*.] Instead of CCl_4, $CHCl_3$ can be used, and the final extract, after washing with water, can be evapd in air at 40-50° and dried in a desiccator. Complexes with Cd, Hg, Ni and Zn.

Di-*p*-tolyl carbonate *[621-02-3]* M 242.3, m 115°. Purified by GLC with 20% Apiezon on Embacel followed by sublimation *in vacuo*.

N,N'-Di-*o*-tolylguanidine *[97-39-2]* M 239.3, m 179° (175-176°), pK_{Est} ~10.3. Crystd from aqueous EtOH.

Di-*p*-tolylphenylamine *[20440-95-3]* M 273.4, m 108.5°, pK_{Est} ~-5.0. Crystd from EtOH.

Di-*p*-tolyl sulfone *[599-66-6]* M 278.3, m 158-159°, b 405°. Crystd repeatedly from diethyl ether. Purified by zone melting.

Djenkolic acid (S,S'-methylene-bis-L-cysteine) *[498-59-9]* M 254.3, m 300-350°(dec), $[\alpha]_D^{20}$ -65° (c 2, HCl) [See pK of S-methyl-L-cysteine]. Crystd from a large volume of water (sol 0.5g%).

cis-4,7,10,13,16,19-Docosahexaenoic acid *[6217-54-5]* M 328.5, m -44/1°, -44.1°, n_D^{20} 1.5017, pK_{Est} ~4.6. Its solubility in $CHCl_3$ is 5%. It has been purified from fish oil by GLC using Ar as mobile phase and EGA as stationary phase with an ionisation detector [UV: Stoffel and Ahrens *J Lipid Res* **1**

139 *1959*], and *via* the ester by evaporative "molecular" distillation using a 'continuous molecular still' at 10^{-4} mm with the highest temperature being 110°, and a total contact time with the hot surface being 60sec [Farmer and van den Heuvel *J Chem Soc* 427 *1938*]. The *methyl ester* has **b** 208-211°/2mm, d_4^{20} 0.9398, n_D^{20} 1.5035. With Br$_2$ it forms a *dodecabromide* **m** *ca* 240° dec. Also the acid was converted to the methyl ester and purified through a three stage molecular still [as described by Sutton *Chem Ind (London)* 11383 *1953*] at 96° with the rate adjusted so that one third of the material was removed each cycle of three distillations. The distillate (numbered 4) (13g) was dissolved in EtOH (100mL containing 8g of KOH) at -70° and set aside for 4h at 30° with occasional shaking under a vac. Water (100mL) is added and the soln is extracted with pentane, washed with HCl, dried (MgSO$_4$), filtered and evapd to give a clear oil (11.5g) **m** -44.5° to -44.1°. In the catalytic hydrogenation of the oil six mols of H$_2$ were absorbed and *docosanoic acid (behenic acid)* was produced with **m** 79.0-79.3° undepressed with an authentic sample (see docosanoic acid below) [Whitcutt *Biochem J* **67** 60 *1957*].

Docosane *[629-97-0]* **M 310.6, m 47°, b 224°/15mm.** Crystd from EtOH or ether.

Docosanoic acid (behenic acid) *[112-85-6]* **M 340.6, m 81-82°, pK$_{Est}$ ~4.9.** Crystd from ligroin. [Francis and Piper *J Am Chem Soc* **61** 577 *1939*].

1-Docosanol (behenyl alcohol) *[661-19-8]* **M 182.3, m 70.8°.** Crystd from ether or chloroform/ether.

n-Dodecane *[112-40-3]* **M 170.3, b 97.5-99.5°/5mm, 216°/760mm, d 0.748, n 1.42156.** Passed through a column of Linde type 13X molecular sieves. Stored in contact with, and distd from, sodium. Passed through a column of activated silica gel. Has been crystd from diethyl ether at -60°. Unsaturated dry material which remained after passage through silica gel has been removed by catalytic hydrogenation (Pt$_2$O) at 45lb/in^2 (3.06 atmospheres), followed by fractional distn under reduced pressure [Zook and Goldey *J Am Chem Soc* **75** 3975 *1953*]. Also purified by partial crystn from the melt.

Dodecane-1,10-dioic acid (decane-1,10-dicarboxylic acid) *[693-23-2]* **M 230.3, m 129°, b 245°/10mm, pK$_{Est}$ ~4.8.** Crystd from water, 75% or 95% EtOH (sol 10%), or glacial acetic acid.

1-Dodecanol (dodecyl alcohol) *[112-53-8]* **M 186.3, m 24°, b 91°/1mm, 135°/10mm, 167°/40mm, 213°/200mm, 259°/atm, d^{24} 0.8309 (liquid).** Crystd from aqueous EtOH, and vacuum distd in a spinning-band column. [Ford and Marvel *Org Synth* **10** 62 *1930*.]

1-Dodecanthiol *[112-55-0]* **M 202.4, b 111-112°/3mm, 153-155°/24mm, d 0.844, n 1.458, pK$_{Est}$ ~10.8.** Dried with CaO for several days, then distd from CaO.

Dodecylammonium butyrate *[17615-97-3]* **M 273.4, m 39-40°, pK^{25} 10.63 (for free base).** Recrystd from *n*-hexane.

Dodecylammonium propionate *[17448-65-6]* **M 259.4, m 55-56°.** Recrystd from hexanol/pet ether (b 60-80°).

Dodecyldimethylamine oxide *[1643-20-5]* **M 229.4, m 102°.** Crystd from acetone or ethyl acetate. [Bunton et al. *J Org Chem* **52** 3832 *1987*].

Dodecyl ether *[4542-57-8]* **M 354.6, m 33°.** Vacuum distd, then crystd from MeOH/*benzene.

1-Dodecylpyridinium chloride *[104-74-5]* **M 301.9, m 68-70°.** Purified by repeated crystn from acetone (charcoal); twice recrystd from EtOH [Chu and Thomas *J Am Chem Soc* **108** 6270 *1986*].

Dodecyltrimethylammonium bromide *[1119-94-4]* **M 308.4, m 246°(dec).** Purified by repeated crystn from acetone. Washed with diethyl ether and dried in a vacuum oven at 60° [Dearden and Wooley *J Phys Chem* **91** 2404 *1987*].

Dodecyltrimethylammonium chloride *[112-00-5]* **M 263.9, m 246°(dec).** Dissolved in MeOH, treated with active charcoal, filtered and dried *in vacuo* [Waldenburg *J Phys Chem* **88** 1655 *1984*], or recrystd several times from 10% EtOH in acetone. Also repeatedly crystd from EtOH/ether or MeOH. [Cella et al. *J Am Chem Soc* **74** 2062 *1952*.]

Dulcitol *[608-66-2]* **M 182.2, m 188-189°, b 276-280°/1.1mm.** Crystd from water by addition of EtOH.

Duroquinone (tetramethylbenzoquinone) *[527-17-3]* **M 164.2, m 110-111°.** Crystd from 95% EtOH. Dried under vacuum.

α-Ecdyson *[3604-87-3]* **M 464.7, m 239-242°, 242°, $[\alpha]_D^{20}$ +72° (c 1, EtOH).** Recrystd from tetrahydrofuran-pet ether, and from H_2O as a hydrate. It has been purified by chromatogaphy on Al_2O_3 and elution with EtOAc-MeOH. It has λmax at 242nm (ε 12.400). Its *acetate* has **m** 214-216° from EtOAc-pet ether, and the *2,4-dinitrophenylhydrazone* has **m** 170-175° (dec) from EtOAc. [Karlson and Hoffmeister *Justus Liebigs Ann Chem* **662** 1 *1963*; Karlson *Pure Appl Chem* **14** 75 *1967*.]

β-Ecdyson (β-echdysterone) *[5289-74-7]* **M 480.7, m 245-247°, $[\alpha]_D^{20}$ +66° (c 1, MeOH).** Crystd from water or tetrahydrofuran/pet ether.

Echinenone *[432-68-8]* **M 550.8, m 178-179°, $A_{1cm}^{\%}$ (λmax) 2160 (458nm) in pet ether.** Purified by chromatography on partially deactivated alumina or magnesia, or by using a thin layer of silica gel G with 4:1 cyclohexane/diethyl ether as the developing solvent. Stored in the dark at -20°.

Eicosane *[112-95-8]* **M 282.6, m 36-37°, b 205°/15mm, d $^{36.7}$ 0.7779, n^{40} 1.43453.** Crystd from EtOH.

Elaidic (*trans*-oleic) acid *[112-79-8]* **M 282.5, m 44.5°, pK^{25} 4.9.** Crystd from acetic acid, then EtOH.

Ellagic acid (2H$_2$O) *[476-66-4]* **M 338.2, m >360°, pK$_{Est(1)}$~8, pK$_{Est(2)}$~11.** Crystd from pyridine.

Elymoclavine (8,9-didehydro-6-methylergoline-8-methanol) *[548-43-6]* **M 254.3, m 249-253°(dec), $[\alpha]_D^{20}$ -59° (c 1, EtOH).** Crystd from MeOH.

Embonic acid (Pamoic acid, 4,4'-methylene bis[3-hydroxy-2-naphthalenecarboxylic acid]) *[130-85-8]* **M 388.4, m 295°, >300°, pK$_{Est(1)}$ ~2.2, pK$_{Est(2)}$~13.2.** Forms crystals from dilute pyridine which decomposition above 280° without melting. It is almost insoluble in H_2O, EtOH, Et_2O, *C_6H_6, CH_3CO_2H, sparingly soluble in $CHCl_3$ but soluble in nitrobenzene, pyridine and alkalis [Barber and Gaimster *J Appl Chem (London)* **2** 565 *1952*]. Used for making salts of organic bases.

Emetine hydrochloride hydrate *[316-42-7]* **M 553.6 + aq, m 235-240°, 235-250°, 240-250°, 248-250° (depending on H$_2$O content), $[\alpha]_D^{20}$ -49.2° (free base, c 4, CHCl$_3$), +18° (c 6, H$_2$O, dry salt), pK$_1$ 5.77, pK$_2$ 6.64.** It crystallises from MeOH-Et$_2$O, MeOH or Et$_2$O-EtOAc. The *free base* has **m** 104-105°, and the *(-)-phenyl thiourea derivative* has **m** 220-221° [from EtOAc-pet ether, $[\alpha]_D^{25}$ - 29.3° (CHCl$_3$)]. IR: 3413 (OH) and 2611 (NH$^+$) cm^{-1}; UV λmax 230nm (ε 16 200) and 282nm (ε 6 890) [Brossi et al. *Helv Chim Acta* **42** 1515 *1959*; Barash et al. *J Chem Soc* 3530 *1959*].

Emodine (1,3,8-trihydroxy-6-methyl-9,10-anthracenedione, archin) *[518-82-1]* **M 270.2, m 253-257°, 255-256°, 256-257°, 262°, 264° (phenolic pKs 7—10).** Forms orange needles from EtOH, Et$_2$O, *C_6H_6, toluene or pyridine. It sublimes above 200° at 12mm. [Tutin and Clewer *J Chem*

Soc **99** 946 *1911*; IR: Bloom et al. *J Chem Soc* 178 *1959*; UV: Birkinshaw *Biochem J* **59** 495 *1955*; Raistrick *Biochem J* **34** 159 *1940*.]

1R,2S-(-)Ephedrine *[299-42-3]* **M 165.2, m 40°, b 225°,** $[\alpha]_{546}^{20}$ **-47°** and $[\alpha]_D^{20}$ **-40°** (**c 5, 2.2M HCl), pK^{25} 9.57.** See (-)-ephedrine (1R,2S-2-methylamino-1-phenylpropanol) on p. 533 in Chapter 6.

(-)Ephedrine hydrochloride *[50-98-6]* **M 201.7, m 218°,** $[\alpha]_{546}^{20}$ **-48°** (**c 5, 2M HCl).** Crystd from water.

Epichlorohydrin *[106-89-8]* **M 92.5, b 115.5°, n 1.438, d 1.180.** Distd at atmospheric pressure, heated on a steam bath with one-quarter its weight of CaO, then decanted and fractionally distd.

R(-)Epinephrine (adrenalin) *[51-43-4]* **M 183.2, m 215°(dec),** $[\alpha]_{546}^{20}$ **-61°** (**c 5, 0.5M HCl), pK$_1^{25}$ 8.75, pK$_2^{25}$ 9.89, pK$_3^{25}$ ~13.** Dissolved in dilute aqueous acid, then ppted by addn of dilute aqueous ammonia or alkali carbonates. (Epinephrine readily oxidises in neutral alkaline soln. This can be diminished if a little sulfite is added).

1,2-Epoxybutane *[106-88-7]* **M 72.1, b 66.4-66.6°, d 0.837, n 1.3841.** Dried with CaSO$_4$, and fractionally distd through a long (126cm) glass helices-packed column. The first fraction contains a water azeotrope.

(+)-Equilenine *[517-09-9]* **M 266.3, m 258-259°,** $[\alpha]_D^{16}$ **+87°** (**c 7.1, H$_2$O).** Crystd from EtOH and dried in a vacuum.

Ergocornine *[564-36-3]* **M 561.7, m 182-184°,** $[\alpha]_D^{20}$ **-176°** (**c 0.5, CHCl$_3$).** Crystd with solvent of crystn from MeOH.

Ergocristine *[511-08-0]* **M 573.7, m 165-170°,** $[\alpha]_D^{20}$ **-183°** (**c 0.5, CHCl$_3$).** Crystd with 2 moles of solvent of crystn, from *benzene.

Ergocryptine *[511-09-1]* **M 575.7, m 212-214°,** $[\alpha]_D^{20}$ **-180°** (**c 0.5, CHCl$_3$).** Crystd with solvent of crystn, from acetone, *benzene or methanol.

Ergosterol *[57-87-4]* **M 396.7, m 165-166°,** $[\alpha]_{546}^{20}$ **-171°** (**in CHCl$_3$).** Crystd from ethyl acetate, then from ethylene dichloride.

Ergotamine *[113-15-5]* **M 581.6, m 212-214°(dec),** $[\alpha]_D^{20}$ **-160°** (**c 0.5, CHCl$_3$), pK^{25} 6.40.** Crystd from *benzene, then dried by prolonged heating in high vacuum. Very *hygroscopic.*

Ergotamine tartrate *[379-79-3]* **M 657.1, m 203°(dec).** Crystd from MeOH.

Erucic acid (cis-13-docosenoic acid) *[112-86-7]* **M 338.6, m 33.8°, b 358°/400mm, pK$_{Est}$ ~4.9.** Crystd from MeOH.

meso-**Erythritol** *[149-32-6]* **M 122.1, m 122°.** Crystd from distd water and dried at 60° in a vac oven.

Erythrityl tetranitrate *[7297-25-8]* **M 302.1, m 61°.** Crystd from EtOH.

β-**Erythroidine** *[466-81-9]* **M 273.3,** $[\alpha]_D^{20}$ **+89°** (**H$_2$O).** Crystd from EtOH.

D-Erythronic acid (3R-3,4-dihydroxyfuran-2-one) *[15667-21-7]* **M 118.1, m 98-100°, 103-104°, 104-105°, 105°,** $[\alpha]_D^{20}$ **-73.2°** (**c 0.5, H$_2$O),** $[\alpha]_{546}^{20}$ **-87.6°** (**c 4, H$_2$O).** Recrystd from EtOAc (20 parts) or isoPrOH (3 parts). [Baker and MacDonald *J Am Chem Soc* **82** 230 *1960*; Glattfeld and

Forbrich *J Am Chem Soc* **56** 1209 *1934*; Weidenhagen and Wegner *Chem Ber* **72** 2010 *1939*, Musich and Rapoport *J Am Chem Soc* **100** 4865 *1978*.]

Esculetin (cichorigenin, 6,7-dihydroxycoumarin) *[305-01-1]* **M 178.2, m 272-275°** (dec), **274°** (dec), $pK_{Est(1)}$ **~8.7,** $pK_{Est(2)}$ **~12.4.** Forms prisms from AcOH or aq EtOH and provides leaflets on sublimation in a vacuum. [Sethna and Shah *Chem Rev*; Merz *Arch Pharm (Weinheim Ger)* **270** 486 *1932*.] *Esculin (the 6-glucoside)* has **m** 215° (dec), $[\alpha]_D^{20}$ -41° (c 5, pyridine).

Eserine (Physostigmine, Physostol, [(3aS-*cis*)-1,2,3,3a,8,8a-hexahydro-1,3a,8-trimethyl-pyrrolo[2,3-b]indol-5-ol methylcarbamate ester] *[57-47-6]* **M 275.4, m 102-104°, 105-106°,** $[\alpha]_D^{17}$ **-67°** (c 1.3, $CHCl_3$), $[\alpha]_D^{25}$ **-120°** (*C_6H_6), pK_1^{15} **1.96,** pK_2^{15} **8.08.** Recrystallises form Et_2O or *C_6H_6 and forms an unstable low melting form **m** 86-87° [Harley-Mason and Jackson *J Chem Soc* 3651 *1954*; Wijnberg and Speckamp *Tetrahedron* **34** 2399 *1978*].

1,3,5-Estratrien-3-ol-17-one (Estrone, Folliculin) *[53-16-7]* **M 270.4, m 260-261°, polymorphic also m 254° and 256°,** $[\alpha]_{546}^{20}$ **+198°** (c 1, dioxane), pK^{25} **10.91.** Crystd from EtOH.

1,3,5-Estratrien-3ß,16a,17ß-triol (Estriol) *[50-27-1]* **M 288.4, m 283°,** $[\alpha]_{546}^{20}$ **+66°** (c 1, dioxane). Crystd from EtOH/ethyl acetate.

Ethane *[74-84-0]* **M 30.1, f -172°, b -88°, d_4^0 1.0493 (air = 1).** Ethylene can be removed by passing the gas through a sintered-glass disc into fuming H_2SO_4 then slowly through a column of charcoal satd with bromine. Bromine and HBr were removed by passage through firebrick coated with *N,N*-dimethyl-*p*-toluidine. The ethane was also passed over KOH pellets (to remove CO_2) and dried with $Mg(ClO_4)_2$. Further purification was by several distns of liquified ethane, using a condensing temperature of -195°. Yang and Gant [*J Phys Chem* **65** 1861 *1961*] treated ethane by standing it for 24h at room temperature in a steel bomb containing activated charcoal treated with bromine. They then immersed the bomb in a Dry-ice/acetone bath and transferred the ethane to an activated charcoal trap cooled in liquid nitrogen. (The charcoal had previously been degassed by pumping for 24h at 450°.) By allowing the trap to warm slowly, the ethane was distd, retaining only the middle third. Removal of methane was achieved using Linde type 13X molecular sieves (previously degassed by pumping for 24h at 450°) in a trap which, after cooling in Dry-ice/acetone, was satd with ethane. After pumping for 10min, the ethane was recovered by warming the trap to room temperature. (The final gas contained less than 10^{-4} mole % of either ethylene or methane).

Ethanesulfonyl chloride *[594-44-5]* **M 128.6, b 55°/9mm, 62°/12mm, 74°/19mm, 76-79°/22mm, 95-98°/50mm, 177°/760mm, d_4^{20} 1.357, n_D^{20} 1.4539.** Purified by repeated distn to remove HCl formed from hydrolysis. **Fuming, corrosive liquid, handle in a good fumehood.** It is hydrolysed by aq N NaOH at room temperature and is best stored in aliquots in sealed ampules under N_2. [Davies and Dick *J Chem Soc* 484 *1932*; Klamann and Drahowzal *Monatsh Chem* **83** 463 *1952*; Saunders et al. *Biochem J* **36** 372 *1942*.]

Ethanethiol (ethyl mercaptan) *[75-08-1]* **M 62.1, b 32.9°/704mm, d^{52} 0.83147, pK^{25} 10.61.** Dissolved in aqueous 20% NaOH, extracted with a small amount of *benzene and then steam distd until clear. After cooling, the alkaline soln was acidified slightly with 15% H_2SO_4 and the thiol was distd off, dried with $CaSO_4$, $CaCl_2$ or 4A molecular sieves, and fractionally distd under nitrogen [Ellis and Reid *J Am Chem Soc* **54** 1674 *1932*].

Ethanol *[64-17-5]* **M 46.1, b 78.3°, d^{15} 0.79360, d^5 0.78506, n 1.36139, pK^{25} 15.93.** Usual impurities of fermentation alcohol are fusel oils (mainly higher alcohols, especially pentanols), aldehydes, esters, ketones and water. With synthetic alcohol, likely impurities are water, aldehydes, aliphatic esters, acetone and diethyl ether. Traces of *benzene are present in ethanol that has been dehydrated by azeotropic distillation with *benzene. Anhydrous ethanol is very *hygroscopic*. Water (down to 0.05%) can be detected by formation of a voluminous ppte when aluminium ethoxide in *benzene is added to a test portion, Rectified spirit (95% ethanol) is converted to *absolute* (99.5%) ethanol by refluxing with freshly ignited CaO (250g/L) for 6h, standing overnight and distilling with precautions to exclude moisture.

Numerous methods are available for further drying of *absolute* ethanol for making "Super dry ethanol". Lund and Bjerrum [*Chem Ber* **64** 210 *1931*] used reaction with magnesium ethoxide, prepared by placing 5g of clean dry magnesium turnings and 0.5g of iodine (or a few drops of CCl_4), to activate the Mg, in a 2L flask, followed by 50-75 mL of *absolute* ethanol, and warming the mixture until a vigorous reaction occurs. When this subsides, heating is continued until all the magnesium is converted to magnesium ethoxide. Up to 1L of ethanol is added and, after an hour's reflux, it is distd off. The water content should be below 0.05%. Walden, Ulich and Laun [*Z Phys Chem* **114** 275 *1925*] used amalgamated aluminium chips, prepared by degreasing aluminium chips (by washing with Et_2O and drying in a vac to remove grease from machining the Al), treating with alkali until hydrogen was vigorously evolved, washing with H_2O until the washings were weakly alkaline and then stirring with 1% $HgCl_2$ soln. After 2min, the chips were washed quickly with H_2O, then alcohol, then ether, and dried with filter paper. (The amalgam became warm.) These chips were added to the ethanol, which was then gently warmed for several hours until evolution of hydrogen ceased. The alcohol was distd and aspirated for some time with pure dry air. Smith [*J Chem Soc* 1288 *1927*] reacted 1L of *absolute* ethanol in a 2L flask with 7g of clean dry sodium, and added 25g of pure ethyl succinate 27g of pure ethyl phthalate was an alternative), and refluxed the mixture for 2h in a system protected from moisture, and then distd the ethanol. A modification used 40g of ethyl formate, instead, so that sodium formate separated out and, during reflux, the excess of ethyl formate decomposed to CO and ethanol.

Drying agents suitable for use with ethanol include Linde type 4A molecular sieves, calcium metal, and CaH_2. The calcium hydride (2g) was crushed to a powder and dissolved in 100mL *absolute* ethanol by gently boiling. About 70mL of the ethanol were distd off to remove any dissolved gases before the remainder was poured into 1L of *ca* 99.9% ethanol in a still, where it was boiled under reflux for 20h, while a slow stream of pure, dry hydrogen (better use nitrogen or Ar) was passed through. It was then distd [Rüber *Z Elektrochem* **29** 334 *1923*]. If calcium was used for drying, about ten times the theoretical amount should be taken, and traces of ammonia (from some calcium nitride in the Ca metal) would be removed by passing dry air into the vapour during reflux.

Ethanol can be freed from traces of basic materials by distn from a little 2,4,6-trinitrobenzoic acid or sulfanilic acid. *Benzene can be removed by fractional distn after adding a little water (the *benzene/water/ethanol azeotrope distils at 64.9°); the alcohol is then redried using one of the methods described above. Alternatively, careful fractional distn can separate *benzene as the *benzene/ethanol azeotrope (**b** 68.2°). Aldehydes can be removed from ethanol by digesting with 8-10g of dissolved KOH and 5-10g of aluminium or zinc per L, followed by distn. Another method is to heat under reflux with KOH (20g/L) and $AgNO_3$ (10g/L) or to add 2.5-3g of lead acetate in 5mL of water to 1L of ethanol, followed (slowly and without stirring) by 5g of KOH in 25mL of ethanol: after 1hr the flask is shaken thoroughly, then set aside overnight before filtering and distilling. The residual water can be removed by standing the distillate over activated aluminium amalgam for 1 week, then filtering and distilling. Distn of ethanol from Raney nickel eliminates catalyst poisons.

Other purification procedures include pre-treatment with conc H_2SO_4 (3mL/L) to eliminate amines, and with $KMnO_4$ to oxidise aldehydes, followed by refluxing with KOH to resinify aldehydes, and distilling to remove traces of H_3PO_4 and other acidic impurities after passage through silica gel, and drying over $CaSO_4$. Water can be removed by azeotropic distn with dichloromethane (azeotrope boils at 38.1° and contains 1.8% water) or 2,2,4-trimethylpentane.

Rapid purification: Place degreased Mg turnings (grease from machining the turnings is removed by washing with dry EtOH then Et_2O, and drying in a vac) (5g) in a dry 2L round bottomed flask fitted with a reflux condenser (protect from air with a drying tube filled with $CaCl_2$ or KOH pellets) and flush with dry N_2. Then add iodine crystals (0.5g) and gently warm the flask until iodine vapour is formed and coats the turnings. Cool, then add EtOH (50mL) and carefully heat to reflux until the iodine disappears. Cool again then add more EtOH (to 1L) and reflux under N_2 for several hours. Distil and store over 3A molecular sieves (pre-heated at 300° –350° for several hours and cooled under dry N_2 or argon).

S-Ethionine [*13073-35-3*] **M 163.2, m 282°(dec), $[\alpha]_D^{25}$ +23.7° (in 5M HCl), pK^{25} 9.02 (for** *RS*). Likely impurities are *N*-acetyl-(*R* and *S*)-ethionine, *S*-methionine, and *R*-ethionine. Crystd from water by adding 4 volumes of EtOH.

Ethoxycarbonyl isocyanate   *[19617-43-7]* **M 115.1, b 51-55°/13mm, 56°/18mm, d_4^{20} 1.15.** Fractionally distilled. [*J Heterocycl Chem* **5** 837 *1968*.]

Ethoxycarbonyl isothiocyanate   *[16182-04-0]* **M 131.5, b 43°/14mm, 51-55°/13mm, 56°/18mm, d_4^{20} 1.12.** Fractionally distd through a short column. It also distils at 83°/30mm with some decomposition liberating CO_2 and sulfurous gases, best distil below 20mm vacuum. [*J Chem Soc* **93** 697 *1908*; 1340, *1948*; *J Heterocycl Chem* **5** 837 *1968*.]

3-Ethoxy-*N,N*-diethylaniline *[1846-92-2]* **M 193.3, b 141-142°/15mm pK$_{Est}$ ~6.1.** Refluxed for 3h with acetic anhydride, then fractionally distilled under reduced pressure.

2-Ethoxyethanol *[110-80-5]* **M 90.1, b 134.8°, d 0.931, n 1.40751.** Dried with $CaSO_4$ or K_2CO_3, filtered and fractionally distd. Peroxides can be removed by refluxing with anhydrous $SnCl_2$ or by filtration under slight pressure through a column of activated alumina.

2-Ethoxy-1-ethoxycarbonyl-1,2-dihydroquinoline (EEDQ) *[16357-59-8]* **M 247.3, m 63.5-65°, 66-67°.** Dissolve ~180g in $CHCl_3$, evap to dryness under vac. Add dry Et_2O (20mL) when a white solid separates on standing. Set aside for a few hours, collect solid, wash thoroughly with cold Et_2O and dry in vac (~140g, **m** 63.5-65°). A further crop of solid (~25g) is obtained from the filtrate on standing overnight. [Fieser and Fieser *Reagents for Organic Synthesis* **2** 191 *1969*; Belleau et al. *J Am Chem Soc* **90** 823 *1968* and **90** 1651 *1968*.]

2-Ethoxyethyl ether [*bis*-(2-ethoxyethyl) ether] *[112-36-7]* **M 162.2, b 76°/32mm, d 0.910, n 1.412.** See diethyleneglycol diethyl ether on p. 203.

2-Ethoxyethyl methacrylate *[2370-63-0]* **M 158.2, b 91-93°/35mm, d 0.965, n 1.429.** Purified as described under methyl methacrylate.

1-Ethoxynaphthalene *[5328-01-8]* **M 172.2, b 136-138°/14mm, 282°/760mm, d 1.061, n 1.604.** Fractionally distd (twice) under a vacuum, then dried with, and distd under a vacuum from, sodium.

2-Ethoxynaphthalene *[93-18-5]* **M 172.2, m 35.6-36.0°, b 142-143°/12mm.** Crystd from pet ether. Dried under vacuum or distd in a vacuum.

Ethyl acetimidate *[1000-84-6]* **M 87.1, b 92-95°/atm, 89.7-90°/765mm, d 0.8671, n 1.4025, pK$_{Est}$ ~5.5.** It is best to prepare it freshly from the *hydrochloride* (see below). Dissolve the hydrochloride (123.5g) by adding it slowly to an ice-cold mixt of H_2O (500mL), K_2CO_3 (276g) and Et_2O (200mL) and stirring rapidly. The Et_2O layer was separated, the aq layer was extd with Et_2O (100mL), the combined Et_2O layers were dried ($MgSO_4$), evapd and the residual oil distd through a glass helices packed column (70x1.2cm). The yield was 19g (22%). [Glickman and Cope *J Am Chem Soc* **67** 1020 *1945*; Chaplin and Hunter *J Chem Soc* 1118 *1937*; *Methods Enzymol* **25** 585 *1972*.]

Ethyl acetimidate hydrochloride *[2208-07-3]* **M 123.6, m 98-100°(dec), 110-115° (dec), 112-113°(dec), m 112-114°(dec), pK$_{Est}$ ~5.5.** Recrystd by dissolving in the minimum volume of super dry EtOH and addition of dry Et_2O or from dry Et_2O. Dry in vacuum and store in a vacuum desiccator with P_2O_5. Alternatively it could be crystd from EtOH (containing a couple of drops of ethanolic HCl) and adding dry Et_2O. Filter and dry in a vac desiccator over H_2SO_4 and NaOH. [Pinner *Chem Ber* **16** 1654 *1883*.] [Glickman and Cope *J Am Chem Soc* **67** 1020 *1945*; Chaplin and Hunter *J Chem Soc* 1118 *1937*; McElvain and Schroeder *J Am Chem Soc* **71** 40 *1949*; McElvain and Tate *J Am Chem Soc* **73** 2233 *1951*; *Methods Enzymol* **25** 585 *1972*.]

Ethyl acetate *[141-78-6]* **M 88.1, b 77.1°, d 0.9003, n 1.37239, n^{25} 1.36979, pK^{25} -6.93 (aq H_2SO_4).** The commonest impurities are water, EtOH and acetic acid. These can be removed by washing with aqueous 5% Na_2CO_3, then with saturated aqueous $CaCl_2$ or NaCl, and drying with K_2CO_3, $CaSO_4$ or $MgSO_4$. More efficient drying is achieved if the solvent is further dried with P_2O_5, CaH_2 or molecular sieves before

distn. CaO has also been used. Alternatively, ethanol can be converted to ethyl acetate by refluxing with acetic anhydride (*ca* 1mL per 10mL of ester); the liquid is then fractionally distd, dried with K_2CO_3 and redistd.
Rapid purification: Distil, dry over K_2CO_3, distil again and store over 4A molecular sieves.

Ethyl acetoacetate *[141-97-9]* **M 130.1, b 71°/12mm, 100°/80mm, d 1.026, n 1.419, pK^{25} 10.68.** Shaken with small amounts of saturated aqueous $NaHCO_3$ (until no further effervescence), then with water. Dried with $MgSO_4$ or $CaCl_2$. Distd under reduced pressure.

Ethyl acrylate *[140-88-5]* **M 100.1, b 20°/40mm, 99.5°/atm, d 0.922, n 1.406.** Washed repeatedly with aqueous NaOH until free from inhibitors such as hydroquinone, then washed with saturated aqueous $CaCl_2$ and distd under reduced pressure. Hydroquinone should be added if the ethyl acrylate is to be stored for extended periods. **LACHRYMATORY.**

Ethylamine *[75-04-7]* **M 45.1, b 16.6°/760mm, d 1.3663, pK^{20} 10.79.** Condensed in an all-glass apparatus cooled by circulating ice-water, and stored with KOH pellets below 0°.

Ethylamine hydrochloride *[557-66-4]* **M 81.5, m 109-110°.** Crystd from absolute EtOH or MeOH/CHCl$_3$.

Ethyl *o*-aminobenzoate *[94-09-7]* **M 165.2, m 92°, pK^{25} 2.39.** Crystd from EtOH/H$_2$O and air dried.

***p*-Ethylaniline** *[589-16-2]* **M 121.2, b 88°/8mm, d 0.975, n 1.554, pK^{25} 5.00.** Dissolved in *benzene, then acetylated. The acetyl derivative was recrystallised from *benzene/pet ether, and hydrolysed by refluxing 50g with 500mL of water and 115mL of conc H_2SO_4 until the soln becomes clear. The amine sulfate was isolated, suspended in water and solid KOH was added to regenerate the free base, which was separated, dried and distd from zinc dust under a vacuum [Berliner and Berliner *J Am Chem Soc* **76** *6179 1954*].

Ethylbenzene *[100-41-6]* **M 106.2, b 136.2°, d 0.867, n 1.49594, n^{25} 1.49330.** Shaken with cold conc H_2SO_4 until a fresh portion of acid remained colourless, then washed with aqueous 10% NaOH or $NaHCO_3$, followed by distilled water until neutral. Dried with $MgSO_4$ or $CaSO_4$, then dried further with, and distd from, sodium, sodium hydride or CaH_2. Can also be dried by passing through silica gel. Sulfur-containing impurities have been removed by prolonged shaking with mercury. Also purified by fractional freezing.

Ethyl benzoate *[93-89-0]* **M 150.2, b 98°/19mm, 212.4°/760mm, d 1.046, n^{15} 1.5074, n^{25} 1.5043, pK -7.37 (aq H_2SO_4).** Washed with aq 5% Na_2CO_3, then satd $CaCl_2$, dried with $CaSO_4$ and distd under reduced pressure.

Ethyl bis-(2,4-dinitrophenyl)acetate *[5833-18-1]* **M 358.3, m 150-153°.** Crystd from toluene as pale yellow crystals.

Ethyl bixin *[6895-43-8]* **M 436.6, m 138°.** Crystd from EtOH.

Ethyl bromide *[74-96-4]* **M 109.0, b 0°/165mm, 38°/745mm, d 1.460, n 1.4241.** The main impurities are usually EtOH and water, with both of which it forms azeotropes. Ethanol and unsaturated compounds can be removed by washing with conc H_2SO_4 until no further coloration is produced. The ethyl bromide is then washed with water, aq Na_2CO_3, and water again, then dried with $CaCl_2$, $MgSO_4$ or CaH_2, and distd. from P_2O_5. Olefinic impurities can also be removed by storing the ethyl bromide in daylight with elementary bromine, later removing the free bromine by extraction with dil aq Na_2SO_3, drying the ethyl bromide with $CaCl_2$ and fractionally distilling. Alternatively, unsaturated compounds can be removed by bubbling oxygen containing *ca* 5% ozone through the liquid for an hour, then washing with aqueous Na_2SO_3 to hydrolyse ozonides and remove hydrolysis products, followed by drying and distn.

Ethyl bromoacetate *[105-36-2]* **M 167.0, b 158-158.5°/758mm, d 1.50, n 1.450.** Washed with saturated aqueous Na_2CO_3 (three times), 50% aq $CaCl_2$ (three times) and saturated aqueous NaCl (twice). Dried with $MgSO_4$, $CaCl_2$ or $CaCO_3$, and distd. **LACHRYMATORY.**

Ethyl 2-(bromomethyl)acrylate *[17435-72-2]* **M 193.1, b 38°/0.8mm, d 1.398, n 1.479.** If it contains some free acid, add H_2O, cool, and neutralise with $NaHCO_3$ until evolution of CO_2 ceases. Extract the mixt with Et_2O (3x) and dry combined extracts (Na_2SO_4, 3h). Evap Et_2O and dist ester collecting fraction **b** 39-40°/0.9mm and check spectra. [Prep and NMR: Ramarajan et al. *Org Synth* Coll Vol VII 211 *1990*.]

Ethyl α-bromopropionate *[535-11-5]* **M 181.0, b 69-70°/25mm, d 1.39, n 1.447.** Washed with saturated aqueous Na_2CO_3 (three times), 50% aq $CaCl_2$ (three times) and saturated aqueous NaCl (twice). Dried with $MgSO_4$, $CaCl_2$ or $CaCO_3$, and distd. **LACHRYMATORY.**

Ethyl bromopyruvate *[70-23-5]* **M 195.0, b 47°/0.5mm, 71-73°/5mm, 87°/9mm, 89-104°/14mm, d_4^{20} 1.561, n_D^{20} 1.464.** Most likely impurity is free carboxylic acid (bromopyruvic or bromoacetic acids). Dissolve in dry Et_2O or dry $CHCl_3$, stir with $CaCO_3$ until effecrvescence ceases, filter, (may wash with a little H_2O rapidly), dry ($MgSO_4$) and distil at least twice. The *2,4-dinitrophenylhydrazone* has **m** 144-145°. [Burros and Holland *J Chem Soc* 672 *1947*; Letsinger and Laco *J Org Chem* **21** 764 *1956*; Kruse et al. *J Am Chem Soc* **76** 5796 *1954*.] **LACHRYMATORY.**

2-Ethyl-1-butanol *[97-95-0]* **M 102.2, b 146.3°, n^{15} 1.4243, n^{25} 1.4205.** Dried with $CaSO_4$ for several weeks, filtered and fractionally distd.

2-Ethylbut-1-ene *[760-21-4]* **M 84.1, b 66.6°, d 0.833, n 1.423.** Washed with saturated aqueous NaOH, then water. Dried with $CaCl_2$, filtered and fractionally distd.

Ethyl *n*-butyrate *[105-54-4]* **M 116.2, b 49°/50mm, 119-120°/760mm, d 0.880, n 1.393.** Dried with anhydrous $CuSO_4$ and distd under dry nitrogen.

Ethyl carbamate (urethane) *[51-79-6]* **M 88.1, m 48.0-48.6°.** Crystd from *benzene.

Ethyl carbazate (*N*-ethoxycarbonyl hydrazine) *[4114-31-2]* **M 104.1, m 44-48°, 51-52°, b 95.5°/10m, 92-95°/12mm, 100-102°/11mm.** Fractionated using a Vigreux column until the distillate crystallises [Allen and Bell *Org Synth* Coll Vol III 404 *1955*.]

***N*-Ethylcarbazole** *[86-28-2]* **M 195.3, m 69-70°.** Recrystd from EtOH, EtOH/water or isopropanol and dried below 55°.

Ethyl carbonate *[105-58-8]* **M 118.1, b 124-125°, d 0.975, n 1.38287.** See diethyl carbonate on p. 203.

Ethyl chloride *[75-00-3]* **M 64.5, b 12.4°, d 0.8978, n 1.3676.** Passed through absorption towers containing, successively, conc H_2SO_4, NaOH pellets, P_2O_5 on glass wool, or soda-lime, $CaCl_2$, P_2O_5. Condensed into a flask containing CaH_2 and fractionally distd. Has also been purified by illumination in the presence of bromine at 0° using a 1000W lamp, followed by washing, drying and distn.

Ethyl chloroacetate *[105-39-5]* **M 122.6, b 143-143.2°, d 1.150, n^{25} 1.4192.** Shaken with satutated aqueous Na_2CO_3 (three times), aqueous 50% $CaCl_2$ (three times) and saturated aqueous NaCl (twice). Dried with Na_2SO_4 or $MgSO_4$ and distd. **LACHRYMATORY.**

Ethyl chloroformate *[541-41-3]* **M 108.5, m -81°, b 94-95°, d 1.135, n 1.3974.** Washed several times with water, redistd using an efficient fractionating column at atmospheric pressure and a $CaCl_2$ guard tube to keep free from moisture [Hamilton and Sly *J Am Chem Soc* **47** 435 *1925*; Saunders, Slocombe and Hardy, *J Am Chem Soc* **73** 3796 *1951*]. **LACHRYMATORY AND TOXIC.**

Ethyl chrysanthemate (ethyl ±2,2-dimethyl-3{c and t}-[2-methylpropenyl]-cyclopropane carboxylate) *[97-41-6]* **M 196.3, b 98-102°/11mm, 117-121°/20mm.** Purified by vacuum distn. The free *trans-acid* has **m** 54° (from, EtOAc) and the free *cis-acid* has **m** 113-116° (from EtOAc). The *4-nitrophenyl ester* has **m** 44-45° (from pet ether) [Campbell and Harper *J Chem Soc* 283 *1945*; IR: Allen et al. *J Org Chem* **22** 1291 *1957*].

Ethyl cinnamate *[103-36-6]* **M 176.2, f 6.7°, b 127°/6mm, 272.7°/768mm, d 1.040, n 1.55983.** Washed with aqueous 10% Na_2CO_3, then water, dried ($MgSO_4$), and distd. The purified ester was saponified with aqueous KOH, and, after acidifying the soln, cinnamic acid was isolated, washed and dried. The ester was reformed by refluxing for 15h the cinnamic acid (25g) with abs EtOH (23g), conc H_2SO_4 (4g) and dry *benzene (100mL), after which it was isolated, washed, dried and distd under reduced pressure [Jeffery and Vogel *J Chem Soc* 658 *1958*].

Ethyl *trans*-crotonate *[623-70-1]* **M 114.2, b 137°, d 0.917, n 1.425.** Washed with aqueous 5% Na_2CO_3, washed with saturated aqueous $CaCl_2$, dried with $CaCl_2$ and distd.

Ethyl cyanoacetate *[105-56-6]* **M 113.1, b 206.0°, d 1.061, n 1.41751.** Shaken several times with aqueous 10% Na_2CO_3, washed well with water, dried with Na_2SO_4 and fractionally distd.

Ethyl cyanoformate *[623-49-4]* **M 99.1, b 113-114°/740mm, 116.5-116.8°/765.5mm, d_4^{20} 1.0112, n_D^{20} 1.3818.** Dissolve in Et_2O, dry over Na_2SO_4, filter, evaporate and distil [Malachowsky et al. *Chem Ber* **70** 1016 *1937*; Adickes et al. *J Prakt Chem* [2] **133** 313 *1932*; Grundmann et al. *Justus Liebigs Ann Chem* **577** 77 *1952*].

Ethylcyclohexane *[1678-91-7]* **M 112.2, b 131.8°, d 0.789, n 1.43304, n^{25} 1.43073.** Purified by azeotropic distn with 2-ethoxyethanol, then the alcohol was washed out with water and, after drying, the ethylcyclohexane was redistd.

Ethyl cyclohexanecarboxylate *[3289-28-9]* **M 156.2, b 76-77°/10mm, 92-93°/34mm, d 0.960, n 1.420.** Washed with M sodium hydroxide solution, then water, dried with Na_2SO_4 and distd.

Ethyl diazoacetate *[623-73-4]* **M 114.1, m -22°, b 42°/5mm, 45°/12mm, 85-86°/88mm, 140-141°/720mm, 140-143°/atm, $d_4^{17.6}$ 1.0852, $n_D^{17.6}$ 1.4588.** A very volatile yellow oil with a strong pungent odour. **EXPLOSIVE [distillation even under reduced pressure is dangerous and may result in an explosion — TAKE ALL THE NECESSARY PRECAUTIONS IF DISTILLATION IS TO BE CARRIED OUT].** It explodes in contact with conc H_2SO_4 - trace acid causes rapid decomp. It is slightly sol in H_2O, but is miscible with EtOH, $*C_6H_6$, pet ether and Et_2O. To purify dissolve in Et_2O [using CH_2Cl_2 instead of Et_2O protects the ester from acid], wash with 10% aq Na_2CO_3, dry ($MgSO_4$), filter and repeat as many times as possible until the Et_2O layer loses its yellow colour, remove the solvent below 20° (vac). Note that prolonged heating may lead to rapid decomp and low yields. It can also be purified by steam distn under reduced pressure but with considerable loss in yield. Place the residual oil in a brown bottle and keep below 10°, and use as soon as possible without distilling. [Womack and Nelson *Org Synth* Coll Vol III 392 *1955*; UV: Miller and White *J Am Chem Soc* **79** 5974 *1957*; Fieser **1** 367 *1967*.]

Ethyl dibromoacetate *[617-33-4]* **M 245.9, b 81-82°/14.5mm, n^{22} 1.4973.** Washed briefly with conc aqueous $NaHCO_3$, then with aqueous $CaCl_2$. Dried with $CaSO_4$ and distd under reduced pressure.

Ethyl α,β-dibromo-β-phenylpropionate *[5464-70-0, erythro 30983-70-1]* **M 336.0, m 75°.** Crystd from pet ether (b 60-80°).

Ethyl dichloroacetate *[535-15-9]* **M 157.0, b 131.0-131.5°/401mm, d 1.28, n 1.438.** Shaken with aqueous 3% $NaHCO_3$ to remove free acid, washed with distd water, dried for 3 days with $CaSO_4$ and distd under reduced pressure.

Ethyl 3,3-diethoxypropionate *[10601-80-6]* **M 190.2, b 58.5°/1.5mm, 65°/2mm, 95-96°/12mm, d_4^{20} 0.78, n_D^{25} 1.4101.** Dissolve in dry Et_2O, and dry with solid $NaHCO_3$, filter and distil and carefully fractionate [Dyer and Johnson *J Am Chem Soc* **56** 223 *1934*].

Ethyl 1,3-dithiane-2-carboxylate *[20462-00-4]* **M 192.3, b 75-77°/0.2mm, 96°/0.4mm, d_4^{20} 1.220, n_D^{25} 1.5379.** Dissolve in $CHCl_3$, wash with aqueous K_2CO_3, 2 x with H_2O, dry over $MgSO_4$, filter, evaporate and distil. [Eliel and Hartman *J Org Chem* **37** 505 *1972*; Seebach *Synthesis* **1** 17 *1969*.]

Ethyl 1,3-dithiolane-2-carboxylate *[20461-99-8]* **M 178.3, b 85°/0.1mm, d_4^{20} 1.250, n_D^{20} 1.538.** Dissolve in $CHCl_3$, wash with aqueous K_2CO_3, 2 x with H_2O, dry over $MgSO_4$, filter, evaporate and distil [Hermann et al *Tetrahedron Lett* 2599 *1973*; Corey and Erickson *J Org Chem* **36** 3553 *1971*].

Ethylene (ethene) *[74-85-1]* **M 28.0, m -169.4°, b -102°/700mm.** Purified by passage through a series of towers containing molecular sieves or anhydrous $CaSO_4$ or a cuprous ammonia soln, then conc H_2SO_4, followed by KOH pellets. Alternatively, ethylene has been condensed in liquid nitrogen, with melting, freezing and pumping to remove air before passage through an activated charcoal trap, followed by a further condensation in liquid air. A sputtered sodium trap has also been used, to remove oxygen.

Ethylene N,N'-bis[(o-hydroxyphenyl)glycine] *[1170-02-1]* **M 360.4, m 249°(dec), $pK_{Est(1)}$~1.8, $pK_{Est(2)}$~4.8, $pK_{Est(3)}$~9.0.** Purified by extensive Soxhlet extraction with acetone. [Bonadies and Carrano *J Am Chem Soc* **108** 4088 *1986*].

Ethylene carbonate (1,3-dioxolan-2-one) *[96-49-1]* **M 88.1, m 37°, d 1.32, n^{40} 1.4199.** Dried over P_2O_5 then fractionally distd at 10mm pressure. Crystd from dry diethyl ether.

Ethylenediamine (1,2-diaminoethane) *[107-15-3]* **M 60.1, f 11.0°, b 117.0°, d 0.897, n 1.45677, n^{30} 1.4513, pK_1^{25} 6.86, pK_2^{25} 9.92.** Forms a constant-boiling (b 118.5°) mixture with water (15%) [*hygroscopic* and miscible with water]. Recommended purification procedure [Asthana and Mukherjee in J.F.Coetzee (ed), *Purification of Solvents,* Pergamon Press, Oxford, 1982 cf p 53]: to 1L of ethylenediamine was added 70g of type 5A Linde molecular sieves and shaken for 12h. The liquid was decanted and shaken for a further 12h with a mixture of CaO (50g) and KOH (15g). The supernatant was fractionally distd (at 20:1 reflux ratio) in contact with freshly activated molecular sieves. The fraction distilling at 117.2° /760mm was collected. Finally it was fractionally distilled from sodium metal. All distns and storage of ethylenediamine should be carried out under nitrogen to prevent reaction with CO_2 and water. Material containing 30% water was dried with solid NaOH (600g/L), heated on a water bath for 10h. Above 60°, separation into two phases took place. The hot ethylenediamine layer was decanted off, refluxed with 40g of sodium for 2h and distd [Putnam and Kobe *Trans Electrochem Soc* **74** 609 *1938*]. Ethylenediamine is usually distd under nitrogen. Type 5A Linde molecular sieves (70g/L), then a mixture of 50g of CaO and 15g of KOH/L, with further dehydration of the supernatant with molecular sieves has also been used for drying this diamine, followed by distn from molecular sieves and, finally, from sodium metal. A spectroscopically improved material was obtained by shaking with freshly baked alumina (20g/L) before distn.

N,N'-Ethylenediaminediacetic acid (EDDA) *[5657-17-0]* **M 176.2, m 222-224°(dec), pK_1^{25} 6.48, pK_2^{25} 9.57 (for NH groups).** Crystd from water.

Ethylenediamine dihydrochloride *[333-18-6]* **M 133.0, pK_1^{25} 6.86, pK_2^{25} 9.92.** Crystd from water.

Ethylenediaminetetraacetic acid (EDTA) *[60-00-4]* **M 292.3, m 253°(dec), pK_1^{25} 0.26 pK_2^{25} 0.96, pK_3^{25} 2.60, pK_4^{25} 2.67, pK_5^{25} 6.16, pK_6^{25} 10.26.** Dissolved in aqueous KOH or ammonium hydroxide, and ppted with dil HCl or HNO_3, twice. Boiled twice with distd water to remove mineral acid, then recrystd from water or dimethylformamide. Dried at 110°. Also recrystd from boiling 1N HCl, wash crystals with distd H_2O and dried *in vacuo*. [Ma and Ray *Biochemistry* **19** 751 *1980*.]

Ethylene dimethacrylate *[97-90-5]* **M 198.2, b 98-100°/5mm, d 1.053, n 1.456.** Distd through a short Vigreux column at about 1mm pressure, in the presence of 3% (w/w) of phenyl-ß-naphthylamine.

Ethylene dimyristate *[627-84-9]* **M 482.8, m 61.7°.** Crystd from *benzene-MeOH or diethyl ether-MeOH, and dried in a vacuum desiccator.

Ethylene dipalmitate *[624-03-3]* **M 538.9, m 69.1°.** Crystd from *benzene-MeOH or diethyl ether-MeOH and dried in a vacuum desiccator.

Ethylene distearate *[627-83-8]* **M 595.0, m 75.3°.** Crystd from *benzene-MeOH or diethyl ether-MeOH and dried in a vacuum desiccator.

Ethylene glycol *[107-21-1]* **M 62.1, b 68°/4mm, 197.9°/760mm, d 1.0986, n^{15} 1.43312, n^{25} 1.43056, pK^{25} 10.6.** Very *hygroscopic*, and also likely to contain higher diols. Dried with CaO, $CaSO_4$, $MgSO_4$ or NaOH and distd under vacuum. Further dried by reaction with sodium under nitrogen, refluxed for several hours and distd. The distillate was then passed through a column of Linde type 4A molecular sieves and finally distd under nitrogen, from more molecular sieves. Fractionally distd.

Ethylene glycol bis(ß-aminoethylether)-N,N'-tetraacetic acid (EGTA) *[67-42-5]* **M 380.4, m >245°(dec), pK_1^{20} 1.15 (2.40), pK_2^{20} 2.40 (2.50), pK_3^{20} 8.40 (8.67), pK_4^{20} 8.94 (9.22).** Dissolved in aq NaOH, pptd by addn of aq HCl, washed with water and dried at 100° *in vacuo*.

Ethylene glycol diacetate *[111-55-7]* **M 146.2, b 190.1°, 79-81°/11mm, d^{25} 1.4188, n 1.4150.** Dried with $CaCl_2$, filtd (excluding moisture) fractionally distd under reduced pressure.

Ethylene glycol dibutyl ether *[112-48-1]* **M 174.3, b 78-80°/0.2mm, d 1.105, n 1.42.** Shaken with aq 5% Na_2CO_3, dried with $MgSO_4$ and stored with chromatographic alumina to prevent peroxide formation.

Ethylene glycol diethyl ether (1,2-diethoxyethane) *[629-14-1]* **M 118.2, b 121.5°, d 0.842, n 1.392.** After refluxing for 12h, a mixture of the ether (2L), conc HCl (27mL) and water (200mL), with slow passage of nitrogen, the soln was cooled, and KOH pellets were added slowly and with shaking until no more dissolved. The organic layer was decanted, treated with some KOH pellets and again decanted. It was refluxed with, and distd from sodium immediately before use. Alternatively, after removal of peroxides by treatment with activated alumina, the ether has been refluxed in the presence of the blue ketyl formed by sodium-potassium alloy with benzophenone, then distd.

Ethylene glycol dimethyl ether (monoglyme) *[110-71-4]* **M 90.1.** See 1,2-dimethoxyethane on p. 210.

Ethylene oxide *[75-21-8]* **M 44.0, b 13.5°/746mm, d^{10} 0.882, n^7 1.3597.** Dried with $CaSO_4$, then distd from crushed NaOH. Has also been purified by its passage, as a gas, through towers containing solid NaOH.

Ethylene thiourea (2-imidazolidinethione) *[96-45-7]* **M 102.2, m 203-204°.** Crystd from EtOH or amyl alcohol.

Ethylene urea (2-imidazolidone) *[120-93-4]* **M 86.1, m 131°.** Crystd from MeOH (charcoal).

Ethylenimine (aziridine) *[151-56-4]* **M 43.1, b 55.5°/760mm, d 0.8321, pK^{25} 8.00.** See aziridine on p. 117.

2-Ethylethylenimine *[2549-67-9]* **M 71.1, b 88.5-89°, pK^{25} 8.31.** Freshly distd from sodium before use. **TOXIC.**

Ethyl formate *[109-94-4]* **M 74.1, b 54.2°, d 0.921, d^{30} 0.909, n 1.35994, n^{25} 1.3565.** Free acid or alcohol is removed by standing with anhydrous K_2CO_3, with occasional shaking, then decanting and distilling from P_2O_5. Alternatively, the ester can be stood wih CaH_2 for several days, then distd from fresh CaH_2. Cannot be dried with $CaCl_2$ because it reacts rapidly with the ester to form a crystalline compound.

Ethyl gallate *[831-61-8]* **M 198.2, m 150-151°, 163-165°.** Recryst from 1,2-dichloroethane, UV: λmax (neutral species) 275nm (ϵ 10 000); (anion) 235nm (ϵ 10 300), 279nm (ϵ 11 400) and 324nm (ϵ 8 500) [Campbell and Coppinger *J Am Chem Soc* **73** 2708 *1951*].

2-Ethyl-1-hexanol *[104-76-7]* **M 130.2, b 184.3°, d 0.833, n 1.431.** Dried with sodium, then fractionally distd.

2-Ethylhexyl vinyl ether *[37769-62-3, 103-44-6]* **M 156.3, b 177-178°/atm.** Usually contains amines as polymerization inhibitors. These are removed by fractional distn.

Ethyl hydrocupreine hydrochloride (Optochin) *[3413-58-9]* **M 376.9, m 249-251°, pK_1^{25} 5.50, pK_2^{25} 9.95.** Recryst from H_2O [UV: Heidt and Forbes *J Am Chem Soc* **55** 2701 *1933*].

Ethyl iodide (iodoethane) *[75-03-6]* **M 156.0, b 72.4°, d 1.933, n^{15} 1.5682, n^{25} 1.5104.** Drying with P_2O_5 is unsatisfactory, and with $CaCl_2$ is incomplete. It is probably best to dry with sodium wire and distil [Hammond et al. *J Am Chem Soc* **82** 704 *1960*]. Exposure of ethyl iodide to light leads to rapid decomposition, with the liberation of iodine. Free iodine can be removed by shaking with several portions of dil aq $Na_2S_2O_3$ (until the colour is discharged), followed by washing with water, drying (with $CaCl_2$, then sodium), and distn. The distd ethyl iodide is stored, over mercury, in a dark bottle away from direct sunlight. Other purification procedures include passage through a 60cm column of silica gel, followed by distn; and treatment with elemental bromine, extraction of free halogen with $Na_2S_2O_3$ soln, followed by washing with water, drying and distn. Free iodine and HI have also been removed by direct distn through a LeBel-Henninger column containing copper turnings. Purification by shaking with alkaline solns, and storage over silver, are reported to be unsatisfactory.

Ethyl isobutyrate *[97-62-1]* **M 116.2, b 110°, d 0.867, n 1.388.** Washed with aqueous 5% Na_2CO_3, then with saturated aqueous $CaCl_2$. Dried with $CaSO_4$ and distd.

Ethyl isocyanate *[109-90-0]* **M 71.1, b 559.8°/759mm, 59-61°/atm, 60-63°/atm, d_4^{20} 0.9031, n_D^{20} 1.3808.** Fractionate through an efficient column preferably in an inert atmosphere and store in aliquots in sealed tubes [Bieber *J Am Chem Soc* **74** 4700 *1952*; Slocombe et al. *J Am Chem Soc* **72** 1888 *1950*].

3-Ethylisothionicotinamide *[10605-12-6]* **M 166.2, m 164-166°(dec).** Crystd from EtOH.

Ethyl isovalerate *[108-64-5]* **M 130.2, b 134.7°, d 0.8664, n 1.39621, n^{25} 1.3975.** Washed with aqueous 5% Na_2CO_3, then saturated aqueous $CaCl_2$. Dried with $CaSO_4$ and distd.

Ethyl levulinate (4-oxopentanoic acid ethyl ester) *[539-88-8]* **M 144.2, m 37.2°, b 106-108°/2mm, 138.8°/8mm, 203-205°/atm, d_4^{20} 1.012, n_D^{20} 1.423.** Stir ester with Na_2CO_3 and charcoal, filter and distil. It is freely soluble in H_2O and EtOH [IR, NMR: Sterk *Monatsh Chem* **99** 1770 *1968*; Thomas and Schuette *J Am Chem Soc* **53** 2328 *1931*; Cox and Dodds *J Am Chem Soc* **55** 3392 *1933*].

Ethyl malonate monoamide *[7597-56-0]* **M 131.1, m 47-50°, 49.5-50°, 50°, b 130-135°/2mm.** Crystallise from Et_2O or by slow evaporation of an aqueous soln as colourless crystals [Snyder and Elston *J Am Chem Soc* **76** 3039 *1954*; McAlvain and Schroeder *J Am Chem Soc* **71** 45 *1949*; Rising et al. *J Biol Chem* **89** 20 *1930*].

Ethyl methacrylate *[97-63-2]* **M 114.2, b 59°/100mm, d 0.915, n 1.515.** Washed successively with 5% aqueous $NaNO_2$, 5% $NaHSO_3$, 5% NaOH, then water. Dried with $MgSO_4$, added 0.2% (w/w) of phenyl-ß-naphthylamine, and distd through a short Vigreux column [Schultz *J Am Chem Soc* **80** 1854 *1958*].

Ethyl methyl ether *[540-67-0]* **M 60.1, b 10.8°, d° 0.725.** Dried with $CaSO_4$, passed through an alumina column (to remove peroxides), then fractionally distd.

3-Ethyl-2-methyl-2-pentene *[19780-67-7]* **M 112.2, b 114.5°/760mm.** Purified by preparative GLC on a column of 20% squalene on Chromosorb P at 70°.

3-Ethyl-4-methylpyridine *[529-21-5]* **M 121.2, b 76°/12mm, 194.5°/750mm, d 0.947, n 1.510, pK_{Est} ~6.3.** Dried with solid NaOH, and fractionally distd.

5-Ethyl-2-methylpyridine *[104-90-5]* **M 121.2, b 178.5°/765mm, d 0.919, n 1.497, pK^{20} 6.51.** Purified by conversion to the picrate, crystn, and regeneration of the free base, then distn.

N-**Ethylmorpholine** *[100-74-3]* **M 115.2, b 138-139°/763mm, d 0.912, n 1.445, pK^{25} 7.67.** Distd twice, then converted by HCl gas into the hydrochloride (extremely deliquescent) which was crystd from anhydrous EtOH-acetone (1:2) [Herries, Mathias and Rabin *Biochem J* **85** 127 *1962*].

Ethyl nitroacetate *[626-35-7]* **M 133.1, b 42-43°/0.2mm, 71-72°/3mm, 93-96°/9mm, 194-195°/atm, d_4^{20} 1.1953, n_D^{20} 1.4260, pK^{25} 5.82.** Purified by repeated distn. IR 1748 (CO_2), 1570 and 1337 (NO_2), and 800cm^{-1} [Hazeldine *J Chem Soc* 2525 *1953*]. The *hydrazine salt* crystallises from 95% EtOH or MeOH as yellow crystals **m** 104-105° [Ungnade and Kissinger *J Org Chem* **22** 1661 *1957*, Emmons and Freeman *J Am Chem Soc* **77** 4391 *1955*].

Ethyl *p*-nitrobenzoate *[99-77-4]* **M 195.2, m 56°.** Dissolved in diethyl ether and washed with aqueous alkali, then the ether was evaporated and the solid recrystd from EtOH.

Ethyl orthoformate *[122-51-0]* **M 148.2, b 144°/760mm, d 0.892, n 1.391.** Shaken with aqueous 2% NaOH, dried with solid KOH andd distd from sodium through a 20cm Vigreux column.

o-**Ethylphenol** *[90-00-6]* **M 122.2, f 45.1°, b 210-212°, d 1.020, n 1.537, pK^{25} 10.20.** Purified as for *p*-ethylphenol below.

p-**Ethylphenol** *[123-07-9]* **M 122.2, m 47-48°, b 218.0°/762mm, n^{25} 1.5239, pK^{25} 10.21.** Non-acidic impurities were removed by passing steam through a boiling soln containing 1 mole of the phenol and 1.75 moles of NaOH (as aq 10% soln). The residue was cooled and acidified with 30% (v/v) H_2SO_4, and the free phenol was extracted into diethyl ether. The extract was washed with water, dried with $CaSO_4$ and the ether was evapd. The phenol was distd at 100mm pressure through a **Stedman** gauze-packed column (see p. 441). It was further purified by fractional crystn by partial freezing, and by zone refining, under nitrogen [Biddiscombe et al. *J Chem Soc* 5764 *1963*]. Alternative purification is *via* the benzoate, as for phenol.

Ethyl phenylacetate *[101-97-3]* **M 164.2, b 99-99.3°/14mm, d 1.030, n 1.499.** Shaken with saturated aqueous Na_2CO_3 (three times), aqueous 50% $CaCl_2$ (twice) and saturated aqueous NaCl (twice). Dried with $CaCl_2$ and distd under reduced pressure.

3-Ethyl-5-phenylhydantoin (Ethotoin) *[86-35-1]* **M 204.2, m 94°.** Crystd from water.

N-**Ethyl-5-phenylisoxazolinium-3'-sulfonate** *[4156-16-5]* **M 253.3, m 220°(dec).** Crystd from diethyl ether or ethyl acetate/pet ether. [Lamas et al. *J Am Chem Soc* **108** 5543 *1986*.]

3-Ethyl-3-phenyl-2,6-piperidinedione (Glutethimide) *[77-21-4]* **M 217.3, m 84°.** Crystd from diethyl ether or ethyl acetate/pet ether.

Ethyl propionate *[105-37-3]* **M 102.1, b 99.1°, d 0.891, n^{15} 1.38643, n 1.38394.** Treated with anhydrous $CuSO_4$ and distd under nitrogen.

2-Ethylpyridine *[100-71-0]* **M 107.2, b 148.6°, d 0.942, pK^{25} 5.89.** Dried with BaO, and fractionally distd. Purified by conversion to the picrate, recrystn and regeneration of the free base followed by distn.

4-Ethylpyridine *[536-75-4]* **M 107.2, b 168.2-168.3°, d 0.942, pK^{25} 6.02.** Dried with BaO, and fractionally distd. Also converted to the picrate, recrystd and the free base regenerated and distd.

4-Ethylpyridine-1-oxide *[14906-55-9]* **M 123.1, m 109-110°, pK$_{Est}$~1.1.** Crystd from acetone/ether.

Ethyl pyruvate *[617-35-6]* **M 116.1, m -50°, b 44-45°/10mm, 56°/20mm, 69-71°/42mm, 63°/23mm, 155.5°/760mm, d$_4^{20}$ 1.047, n$_D^{20}$ 1.4052.** Shake the ester with 10mL portions of satd aq CaCl$_2$ soln (removes ethyl acetate) and the organic layer is removed by centrifugation, decantation and filtration, and is distilled under reduced pressure. Purification of small quantities is carried out *via* the bisulfite adduct: the ester (2.2mL) is shaken with saturated NaHSO$_3$ (3.6mL), chill in a freezing mixture when crystals separate rapidly (particularly if seeded). After 5min EtOH (10mL) is added and the crystals are filtered off, washed with EtOH and Et$_2$O and dried. Yield *ca* 3g of *bisulfite adduct*. Then treat the adduct (16g) with saturated aqueous MgSO$_4$ (32mL) and 40% formaldehyde (5mL) and shake, whereby the ester separates as an oil which is extracted with Et$_2$O, the extract is dried (MgSO$_4$), filtered, evapd and the residue is distd (**b** 56°/20mm), and then redistd (**b** 147.5°/750mm) to give 5.5g of pure ester. [Cornforth *Org Synth* Coll Vol IV 467 *1963*.]

Ethyl Red [2-(4-diethylaminophenylazo)benzoic acid] *[76058-33-8]* **M 197.4, m 150-152°, pK$_1$ 2.5, pK$_2$ 9.5.** Crystd from EtOH/diethyl ether or toluene. Indicator: pH 4.4 (red) and 6.2 (yellow)

Ethyl stearate *[111-61-5]* **M 312.5, m 33°, b 213-215°/15mm.** The solid portion was separated from the partially solid starting material, then crystd twice from EtOH, dried by azeotropic distn with *benzene, and fractionally distd in a spinning-band column at low pressure [Welsh *Trans Faraday Soc* **55** 52 *1959*].

Ethyl thiocyanate (ethyl rhodanide) *[542-90-5]* **M 87.1, b 144-145°, d 1.011, n 1.462.** Fractionally distd at atmospheric pressure. **(CARE LACHRYMATOR.)**

Ethyl thioglycolate *[623-51-8]* **M 120.2, b 50-51°/10mm, 55°/17mm, 62.5-64°/22mm, 67-68°/24mm, 155-158°/atm, d$_4^{20}$ 1.096, n$_D^{20}$ 1.457.** Dissolve in Et$_2$O, wash with H$_2$O, dry over Na$_2$SO$_4$, filter, evaporate and distil the residue under reduced pressure [Bredereck et al. *Chem Ber* **90** 1837 *1957*). The *Ni complex [Ni(SCH$_2$CO$_2$Et)$_2$]* recrystallised twice from EtOH gives crystals which became black when dried in a vacuum over H$_2$SO$_4$, **m** 104-105° [Dranet and Cefola *J Am Chem Soc* **76** 1975 *1954*].

N-Ethyl thiourea *[625-53-6]* **M 104.2, m 110°.** Crystd from EtOH, MeOH or ether.

Ethyl trichloroacetate *[515-84-4]* **M 191.4, b 100-100.5°/30mm, d 1.383.** Shaken with saturated aqueous Na$_2$CO$_3$ (three times), aqueous 50% CaCl$_2$ (three times), saturated aqueous NaCl (twice), then distd with CaCl$_2$ and distd under reduced pressure.

Ethyl trifluoroacetate *[383-63-1]* **M 142.1, b 61.3°/750, 60-62°/atm, 62-64°/755mm, d$_4^{20}$ 1.191, n$_D^{20}$ 1.30738.** Fractionate through a long Vigreux column. IR has ν at 1800 (CO$_2$) and 1000 (OCO) cm^{-1} [Fuson et al. *J Chem Phys* **20** 1627 *1952*; Bergman *J Org Chem* **23** 476 *1958*].

Ethyl trifluoromethanesulfonate *[425-75-2]* **M 178.1, b 115°/atm, 118-120°/atm, d$_4^{20}$ 1.378, n$_D^{20}$ 1.336.** The ester reacts slowly with H$_2$O and aqueous alkali. If its IR has no OH bands (~3000 cm^{-1}) then purify by redistillation. If OH bands are present then dilute with dry Et$_2$O and shake (carefully) with aqueous NaHCO$_3$ until effervescence ceases, then wash with H$_2$O and dry (MgSO$_4$), filter, evaporate and distil the residue under slight vacuum then at atmospheric pressure in a N$_2$ atmosphere. **IT IS A POWERFUL ALKYLATING AGENT, AND THE FUMES ARE VERY TOXIC - CARRY ALL OPERATIONS IN AN EFFICIENT FUME CUPBOARD.** [Gramstad and Hazeldine *J Chem Soc* 173 *1956*; Howells and McCown *Chem Rev* **77** 69 *1977*.]

S-Ethyl trifluorothioacetate *[383-64-2]* **M 158.1, b 88-90°/atm, 90.5°/760mm, d$_4^{20}$ 1.255, n$_D^{20}$ 1.372.** If IR is free of OH bands then fractionate, but if OH bands are present then dilute with dry Et$_2$O, wash with 5% KOH and H$_2$O, dry over MgSO$_4$ and fractionate through an efficient column [Hauptschein et al. *J Am Chem Soc* **74** 4005 *1952*]. *Powerful obnoxious odour.*

Ethyl vinyl ether *[109-92-2]* **M 72.1, b 35.5°, d 0.755.** Contains polymerization inhibitors (usually amines, e.g. triethanolamine) which can be removed by fractional distn. Redistd from sodium. **LACHRYMATORY.**

1-Ethynyl-1-cyclohexanol *[78-27-3]* **M 124.2, m 30-33°, 32-33°, b 74°/12mm, 76-78°/17mm, 171-172°/694mm, 180°/atm, d$_4^{25}$ 0.9734, n$_D^{25}$ 1.4801.** Dissolve in Et$_2$O, wash with H$_2$O, dilute NaHCO$_3$, H$_2$O again, dry (Na$_2$SO$_4$), filter, evaporate and distil the residue. IR (CCl$_4$): 3448 (OH), 2941 (CH), 1449-1123 and 956 cm^{-1}; NMR (CCl$_4$) δ: 3.2 (OH), 2.5 (≡CH), 1.70 (m 10H, CH$_2$) [Hasbrouck and Kiessling *J Org Chem* **38** 2103 *1972*].

Ethynyl p-tolylsulfone *[13894-21-8]* **M 180.2, m 73-74°.** Recrystd from pet ether and dried in vac.

Etiocholane (5ß-androsterone) *[438-23-3]* **M 260.5, m 78-80°.** Crystd from acetone.

Etiocholanic acid *[438-08-4]* **M 304.5, m 228-229°, pK$_{Est}$ ~4.7.** Crystd from glacial acetic acid and sublimes at 160°/0.002mm. The *methyl ester* has **m** 99-101°. [Weiland et al. *Z Physiol Chem* **161** 80 *1926*.]

Etioporphyrin I *[448-71-5]* **M 478.7, m 360-363°.** Crystd from pyridine or CHCl$_3$-pet ether.

Eucaliptol **(1,8-cineol, 1,8-epoxy-p-menthane, 1,3,3-trimethyl-2-oxabicyclo[2.2.2]-octane)** *[470-82-6]* **M 154.2, m 1.3°, 1.5°, b 39-39.3°/4mm, 176-176.4°/760mm, d$_4^{20}$ 0.9232, n$_D^{20}$ 1.4575.** Purified by dilution with an equal volume of pet ether, then saturated with dry HBr. The ppte was filtered off, washed with small vols of pet ether, then cineole was regenerated by stirring the crystals with H$_2$O. It can also be purified *via* its o-cresol or resorcinol addition compds. Stored over Na until required. Purified by fractional distn. Insoluble in H$_2$O but soluble in organic solvents. [IR: Kome et al. *Nippon Kagaku Zasshi* [*J Chem Soc Japan* (Pure Chem Sect)] **80** 66 *1959*; *Chem Abstr* 603 *1961*.]

Eugenol **(4-allyl-2-methoxyphenol)** *[97-53-0]* **M 164.2, b 253°/760mm, 255°/760mm, d$_4^{20}$ 1.066, n$_D^{20}$ 1.540, pK^{25} 10.19.** Fractional distn gives a pale yellow liquid which darkens and thickens on in air. Should store under N$_2$ at -20°. [Waterman and Priedster *Recl Trav Chim Pays-Bas* **48** 1272 *1929*.]

Eugenol methyl ether **(4-allyl-1,2-dimethoxybenzene)** *[93-15-2]* **M 178.2, m -4°, b 127-129°/11mm, 146°/30mm, 154.7°/760mm, d$_4^{20}$ 1.0354, n$_D^{20}$ 11.53411.** Recrystd from hexane at low temp and redistd (preferably *in vacuo*). [Hillmer and Schorning *Z Phys Chem* [A] **167** 407 *1934*; Briner and Fliszár *Helv Chim Acta* **42** 2063 *1959*.]

(+)-α-Fenchol (1R-1,3,3-trimethyl-norbornan-2-ol) *[1632-73-1]* M 154.3, m 40-43°, 47-47.5°, b 201-202°, [α]$_D^{20}$ +12.5° (c 10, EtOH).

It is prepared by reduction of (-)-fenchone and is purified by recrystallisation from *C$_6$H$_6$-pet ether, or distn, or both. The *2-carboxybenzoyl (monophthalate)* derivative has **m** 146.5-147.5° [α]$_D^{20}$ -20.4° (EtOH), and the *2-phenylurethane* has **m** 81°. [Beckmann and Metzger *Chem Ber* **89** 2738 *1956*].

(+)- Fenchone **(1S-1,3,3-trimethyl-norbornan-2-one)** *[4695-62-9]* **M 152.2, m 5-7°, 6.1°, b 63-65°/13mm, 66°/15mm, 122°/10mm, d$_4^{20}$ 0.9434, n$_D^{20}$ 1.4636, [α]$_D^{20}$ +66.9° (neat, or in c 1.5, EtOH), [α]$_{546}^{20}$ +60.4° (neat).** The oily liquid is purified by distn in a vacuum, and is very soluble in EtOH and Et$_2$O. [Boyle et al. *J Chem Soc, Chem Commun* 395 *1971*, Hückel *Justus Liebigs Ann Chem* **549** 186 *1941*; (±)-isomer: Braun and Jacob *Chem Ber* **66** 1461 *1933*.] It forms two *oximes; cis-oxime:* **m**

167° (cryst from pet ether) $[\alpha]_D^{20}$ +46.5° (c 2, EtOH), *O-benzoyloxime* **m** 81° $[\alpha]_D^{20}$ +49° (EtOH) and *oxime-HCl* **m** 136° (dec). The *trans-oxime* has **m** 123° (cryst from pet ether) $[\alpha]_D^{18}$+148° (c 2, EtOH) and the *O-benzoyloxime* has **m** 125° $[\alpha]_D^{20}$ +128.5° (c 2, EtOH) [Hückel *Justus Liebigs Ann Chem* **549** 186 *1941*; Hückel and Sachs *Justus Liebigs Ann Chem* **498** 166 *1932*].

(-)- Fenchone (1R-1,3,3-trimethyl-norbornan-2-one) *[7787-20-4]* **M 152.2, m 5.2°, b 67.2°/10mm, 191-195°/atm, d_4^{20} 0.9484, n_D^{20} 1.4630, $[\alpha]_D^{20}$ -66.8° (neat).** Purification as for the (+)-enantiomer above and should have the same physical properties except for the optical rotations. UV: λmax 285nm (ε 12.29). [Braun and Jacob *Chem Ber* **66** 1461 *1933*; UV: Ohloff et al. *Chem Ber* **90** 106 *1957*.]

Flavone (2-phenyl-4H-1-benzopyran-4-one) *[525-82-6]* **M 222.3, m 100°.** Crystd from pet ether.

Fluoranthene (benzo[j,k]fluorene) *[206-44-0]* **M 202.3, m 110-111°.** Purified by chromatography of CCl_4 solns on alumina, with *benzene as eluent. Crystd from EtOH, MeOH or *benzene. Purified by zone melting. [Gorman et al. *J Am Chem Soc* **107** 4404 *1985*.]

2-Fluorenamine *[153-78-6]* **M 181.2.** See 2-aminofluorene on p. 106.

9-Fluorenamine *[525-03-1]* **M 181.2, m 64-65°, pK_{Est} ~3.5.** Crystd from hexane.

Fluorene *[86-73-7]* **M 166.2, m 114.7-115.1°, b 160°/15mm.** Purified by chromatography of CCl_4 or pet ether (b 40-60°) soln on alumina, with *benzene as eluent. Crystd from 95% EtOH, 90% acetic acid and again from EtOH. Crystn using glacial acetic acid retained an impurity which was removed by partial mercuration and pptn with LiBr [Brown, Dubeck and Goldman *J Am Chem Soc* **84** 1229 *1962*]. Has also been crystd from hexane, or *benzene/EtOH, distd under vacuum and purified by zone refining. [Gorman et al. *J Am Chem Soc* **107** 4404 *1985*.]

9-Fluorenone *[486-25-9]* **M 180.2, m 82.5-83.0°, 85-86°, b 341°/760mm.** Crystd from absolute EtOH, MeOH or *benzene/pentane. [Ikezawa *J Am Chem Soc* **108** 1589 *1986*.] Also twice recrystd from toluene and sublimed in a vac [Saltiel *J Am Chem Soc* **108** 2674 *1986*]. Can be distd under high vacuum.

Fluorene-2,7-diamine *[525-64-4]* **M 196.3, m 165-166°.** Crystd from hot H_2O or aq EtOH, dried in a vac and stored in the dark.

9-Fluorenylmethyl chloroformate (FMOC-Cl) *[28920-43-6]* **M 258.7, m 61-63°, 61.4-63°.** The IR should contain no OH bands (at ~3000 cm^{-1}) due to the hydrolysis product 9-fluorenylmethanol. Purify by recrystn from dry Et_2O. IR ($CHCl_3$) has band at 1770 cm^{-1} (C=O) and the NMR ($CDCl_3$) has δ at 4-4.6 (m 2H, CHCH_2) and 7.1-7.8 (m, 8 aromatic H). The *azide* (*FMOC-N$_3$*) has **m** 89-90° (from hexane) and IR ($CHCl_3$) at 2135 (N_3) and 1730 (C=O) cm^{-1}; and the *carbazate* (*FMOC-NHNH$_2$*) has **m** 171° dec (from nitromethane), IR (KBr) 3310, 3202 (NH) and 1686 (CONH) cm^{-1}. [Caprino and Han *J Org Chem* **37**, 3404 *1972* and *J Am Chem Soc* **92** 5748 *1970*; Koole et al. *J Org Chem* **59** 1657 *1989*; Fürst et al. *J Chromatogr* **499** 537 *1990*.]

9-Fluorenylmethyl succinimidyl carbonate *[82911-69-1]* **M 337.3, m 147-151° (dec), 151° (dec).** Recrystd from $CHCl_3$-Et_2O, or from pet ether (b 40-60°). [Pauet *Can J Chem* **60** 976 *1982*; Lapatsaris et al. *Synthesis* 671 *1983*.]

Fluorescein [9-(o-carboxyphenyl-6-hydroxy-3H-xanthene-3-one] *[2321-07-5]* **M 320.0, ε$_{495nm}$ 7.84 x 10^4 (in 10^{-3}M NaOH), pK_1 2.2, pK_2 4.4, pK_3 6.7.** Dissolved in dilute aqueous NaOH, filtered and ppted by adding dilute (1:1) HCl. The process was repeated twice more and the fluorescein was dried at 100°. Alternatively, it has been crystd from acetone by allowing the soln to evaporate at 37° in an open beaker. Also recrystd from EtOH and dried in a vacuum oven.

Fluoresceinamine (mixture of 5- and 6-aminofluorescein) *[27599-63-9]* **M 347.3, m 314-316° (dec, 5-amino) and m >200° (dec).** Dissolve in EtOH, treat with charcoal, filter, evaporate and dry residue in vacuum at 100° overnight. Also recrystallise from 6% HCl, then dissolve in 0.5% aqueous NaOH and ppte by acidifying with acetic acid. The separate amines are made from the respective nitro compounds which are best separated *via* their acetate salts. They have similar R_F of 0.26 on Silica Gel Merck F_{254} in 5 mL MeOH + 150 mL Et_2O satd with H_2O. IR (Me_2SO) has a band at 1690 cm^{-1} (CO_2^-) and sometimes a weak band at 1750 cm^{-1} due to lactone. UV (EtOH) of 6-isomer λmax 222 (ϵ 60 000) and 5-isomer λmax 222 (ϵ 60 000) and 285 (ϵ 20.600). [IR: McKinney and Churchill *J Chem Soc (C)* 654 *1970*; McKinney et al. *J Org Chem* **27** 3986 *1962*; UV: Verbiscar *J Org Chem* **29** 490 *1964*.]

Fluorescein isothiocyanate isomer I (5-isocyanato isomer) *[3326-32-7; 27072-45-3 mixture of 5- and 6-isomers]* **M 389.4, m >160° (slow dec).** It is made from the pure 5-amino isomer. Purified by dissolving in boiling Me_2CO, filtering and adding pet ether (b 60-70°) until it becomes turbid. If an oil separates then decant and add more pet ether to the supernatant and cool. Orange-yellow crystals separate, collect and dry *in vacuo*. Should give one spot on TLC (silica gel) in EtOAc, pyridine, AcOH (50:1:1) and in Me_2NCHO, $CHCl_3$, 28% N_4OH (10:5:4). IR (Me_2SO): 2110 (NCS) and 1760 (C=O). The NMR spectra in Me_2CO-d_6 of the 5- and 6-isomers are distinctly different for the protons in the *benzene ring; the UV in phosphate buffer pH 8.0 shows a max at ~490nm. [Sinsheimer et al. *Anal Biochem* **57** 227 *1974*; McKinney et al. *Anal Biochem* **7** 74 *1964*.]

Fluoroacetamide *[640-19-7]* **M 77.1, m 108°.** Crystd from chloroform.

Fluorobenzene *[462-06-6]* **M 96.1, b 84.8°, d 1.025, n 1.46573, n^{30} 1.4610.** Dried for several days with P_2O_5, then fractionally distd.

o-**Fluorobenzoic acid** *[445-29-4]* **M 140.1, m 127°, pK^{25} 3.27.** Crystd from 50% aqueous EtOH, then zone melted or vacuum sublimed at 130-140°.

m-**Fluorobenzoic acid** *[445-38-9]* **M 140.1, m 124°, pK^{25} 3.86.** Crystd from 50% aqueous EtOH, then vacuum sublimed at 130-140°.

p-**Fluorobenzoic acid** *[456-22-4]* **M 140.1, m 182°, pK^{25} 4.15.** Crystd from 50% aqueous EtOH, then zone melted or vacuum sublimed at 130-140°.

3-Fluoro-4-hydroxyphenylacetic acid *[458-09-3]* **M 170.1, m 33°, pK$_{Est(1)}$~4.4, pK$_{Est(2)}$~9.4.** Crystd from water.

1-Fluoro-4-nitrobenzene *[350-46-9]* **M 141.1, m 27° (stable form), 21.5° (unstable form), b 205.3°/735mm, 95-97.5°/22mm, 86.6°/14mm.** Crystd from EtOH.

1-Fluoro-4-nitronaphthalene *[341-92-4]* **M 191.2, m 80°.** Recrystd from EtOH as yellow needles [Bunce et al. *J Org Chem* **52** 4214 *1987*].

o-**Fluorophenol** *[367-12-4]* **M 112.1, m 16°, b 53°/14mm, d 1.257, n 1.514, pK^{25} 8.70.** Passed at least twice through a gas chromatographic column for small quantities, or fractionally distd under reduced pressure.

p-**Fluorophenoxyacetic acid** *[405-79-8]* **M 170.1, m 106°, pK^{25} 3.13.** Crystd from EtOH.

4-Fluorophenylacetic acid *[405-50-5]* **M 154.1, m 86°, pK^{25} 4.22.** Crystd from heptane.

4-Fluorophenyl isocyanate *[1195-45-5]* **M 137.1, b 55°/8mm, n$_D^{20}$ 1.514.** Purify by repeated fractionation through an efficient column. If IR indicated that there is too much urea (in the presence of moisture the symmetrical urea is formed) then dissolve in dry EtOH-free $CHCl_3$, filter, evaporate and distil. **It is a pungent LACHRYMATORY liquid.** [see Hardy *J Chem Soc* 2011 *1934*; and Hickinbottom *Reactions of Organic Compounds* Longmans p. 493 *1957*.]

4-Fluorophenyl isothiocyanate *[1544-68-9]* **M 153.2, m 24-26°, 26-27°, b 66°/2mm, 215°/atm, 228°/760mm, n_D^{20} 1.6116.** Likely impurity is the symmetrical thiourea. Dissolve the isothiocyanate in dry $CHCl_3$, filter and distil the residue in a vacuum. It can also be steam distd, the oily layer separated, dried over $CaCl_2$ and distilled *in vacuo*. *Bis-(4-fluorophenyl)thiourea* has **m** 145° (from aq EtOH). [Browne and Dyson *J Chem Soc* 3285 *1931*; Buu Hoi et al. *J Chem Soc* 1573 *1955*; Olander *Org Synth* Coll Vol I 448 *1941*].

p-**Fluorophenyl**-*o*-**nitrophenyl ether** *[448-37-3]* **M 247.2, m 62°.** Crystd from EtOH.

o-**Fluorotoluene** *[95-52-3]* **M 110.1, b 114.4°, d 1.005, n 1.475.** Dried with P_2O_5 or $CaSO_4$ and fractionally distd through a silvered vacuum-jacketed glass column with 1/8th-in glass helices. A high reflux ratio is necessary because of the closeness of the boiling points of the *o-*, *m-* and *p-* isomers [Potter and Saylor *J Am Chem Soc* **37** 90 *1951*].

m-**Fluorotoluene** *[352-70-5]* **M 110.1, b 116.5°, d 1.00, n^{27} 1.46524.** Purification as for *o*-fluorotoluene.

p-**Fluorotoluene** *[352-32-9]* **M 116.0°, d 1.00, n 1.46884.** Purification as for *o*-fluorotoluene.

Formaldehyde *[50-00-0]* **M 30.0, m 92°, b -79.6°/20mm, d^{20} 0.815, pK^{25} 13.27 (hydrate).** Commonly contains added MeOH. Addition of KOH soln (1 mole KOH: 100 moles HCHO) to 40% formaldehyde soln, or evaporation to dryness, gives paraformaldehyde polymer which, after washing with water, is dried in a vacuum desiccator over P_2O_5 or H_2SO_4. Formaldehyde is regenerated by heating the paraformaldehyde to 120° under vacuum, or by decomposing it with barium peroxide. The monomer, a gas, is passed through a glass-wool filter cooled to -48° in $CaCl_2$/ice mixture to remove particles of polymer, then dried by passage over P_2O_5 and either condensed in a bulb immersed in liquid nitrogen or absorbed in ice-cold conductivity water.

Formaldehyde dimethyl acetal (dimethoxymethane, methylal, formal) *[109-87-5]* **M 76.1, m -108°, b 41-42°/736mm, 41-43°/atm, 42-46°/atm, d_4^{20} 0.8608, n_D^{20} 1.35335.** It is a volatile flammable liquid which is soluble in three parts of H_2O, and is readily hydrolysed by acids. Purify by shaking with an equal vol of 20% aq NaOH, stand for 20min, dry over fused $CaCl_2$, filter and fractionally distil through an efficient column, store over molecular sieves. [Buchler et al. *Org Synth* Coll Vol III 469 *1955*; *Ind Eng Chem* **18** 1092 *1926*; Rambaud and Besserre *Bull Soc Chim Fr* 45 *1955*; IR: *Can J Chem* **36** 285 *1958*.]

Formaldehyde dimethyl mercaptal (bis-[methylthio]methane) *[1618-26-4]* **M 108.2, b 44-47°/13mm, 45.5°/18mm, 148-149°/atm, d_4^{20} 1.0594, n_D^{20} 1.5322. Work in an efficient fume cupboard as the substance may contain traces (or more) of methylmercaptan which has a very bad odour.** Dissolve in Et_2O, shake with aqueous alkalis then dry over anhydrous K_2CO_3, filter and distil over K_2CO_3 under a stream of N_2. If the odour is very strong then allow all gas efluents to bubble through 5% aqueous NaOH soln which is then treated with dilute $KMnO_4$ in order to oxidise MeSH to odourless products. UV: λmax 238 nm (log ϵ 2.73) [Fehnel and Carmack *J Am Chem Soc* **71** 90 *1949*; Fehér and Vogelbruch *Chem Ber* **91** 996 *1958*; Bøhme and Marz *Chem Ber* **74** 1672 *1941*]. Oxidation with aq $KMnO_4$ yields *bis-(methylsulfonyl)methane* which has **m** 142-143° [Fiecchi et al. *Tetrahedron Lett* 1681 *1967*].

Formamide *[75-12-7]* **M 45.0, f 2.6°, b 103°/9mm, 210.5°/760mm(dec), d 1.13, n 1.44754, n^{25} 1.44682.** Formamide is easily hydrolysed by acids and bases. It also reacts with peroxides, acid halides, acid anhydrides, esters and (on heating) alcohols; while strong dehydrating agents convert it to a nitrile. It is very *hygroscopic*. Commercial material often contains acids and ammonium formate. Vorhoek [*J Am Chem Soc* **58** 2577 *1956*] added some bromothymol blue to formamide and then neutralised it with NaOH before heating to 80-90° under reduced pressure to distil off ammonia and water. The amide was again neutralised and the process was repeated until the liquid remained neutral on heating. Sodium formate was added, and the formamide was reduced under reduced pressure at 80-90°. The distillate was again neutralised and redistd. It was then fractionally crystd in the absence of CO_2 and water by partial freezing.

Formamide (specific conductance 2 x 10^{-7} ohm^{-1} cm^{-1}) of low water content was dried by passage through a column of 3A molecular sieves, then deionized by treatment with a mixed-bed ion-exchange resin loaded with H$^+$ and HCONH$^-$ ions (using sodium formamide in formamide)[Notley and Spiro *J Chem Soc (B)* 362 *1966*].

Formamidine acetate *[3473-63-0]* **M 104.1, m 159-161°(dec), 164°(dec), pK$_{Est}$ ~ 12.** Unlike the hydrochloride, the acetate salt is not hygroscopic. It is recrystd from a small volume of acetic acid, by addition of EtOH and the crysts are washed with EtOH then Et$_2$O and dried in a vac. [Taylor, Ehrhart and Karanisi *Org Synth* **46** 39 *1966*.]

Formamidine sulfinic acid (thiourea-S-dioxide) *[1758-73-2]* **M 108.1, m 124-126°(dec).** Dissolved in five parts of aq 1:1% NaHSO$_3$ at 60-63° (charcoal), then crystd slowly, with agitation, at 10°. Filtered. Dried immediately at 60° [Koniecki and Linch *Anal Chem* **30** 1134 *1958*].

Formanilide *[103-70-8]* **M 121.1, m 50°, b 166°/14mm, 216°/120mm, d 1.14.** Crystd from ligroin/xylene.

Formic acid *[64-18-6]* **M 46.0 (anhydr), f 8.3°, b 25°/40mm, 100.7°/760mm, d 1.22, n 1.37140, n^{25} 1.36938, pK^{25} 3.74.** Anhydrous formic acid can be obtained by direct fractional distillation under reduced pressure, the receiver being cooled in ice-water. The use of P$_2$O$_5$ or CaCl$_2$ as dehydrating agents is unsatisfactory. Reagent grade 88% formic acid can be satisfactorily dried by refluxing with phthalic anhydride for 6h and then distilling. Alternatively, if it is left in contact with freshly prepared anhydrous CuSO$_4$ for several days about one half of the water is removed from 88% formic acid: distn removes the remainder. Boric anhydride (prepared by melting boric acid in an oven at a high temperature, cooling in a desiccator, and powdering) is a suitable dehydrating agent for 98% formic acid; after prolonged stirring with the anhydride the formic acid is distd under vacuum. Formic acid can be further purified by fractional crystn using partial freezing.

Forskolin (5-[acetyloxy]-3-ethenyldodecahydro-6,10,10b-trihydroxy-3,4a,7,7,10a-penta-methyl-[3R-{3α-4aβ, 5β, 6β, 6aα,10α, 10aβ, 10bα}-1H-naphtho[2,1-b]pyran-1-one) *[66575-29-9]* **M 410.5, m 229-232°, 228-233°.** Recrystd from *C$_6$H$_6$-pet ether. It is antihypertensive, positive ionotropic, platelet aggregation inhibitory and adenylate cyclase activating properties [*Chem Abstr* **89** 1978 244150, de Souza et al. *Med Res Rev* **3** 201 *1983*].

D(-)-Fructose *[57-48-7]* **M 180.2, m 103-106°, [α]$^{20}_{546}$ -190° (after 1h, c 10, H$_2$O), pK^{25} 12.03.** Dissolved in an equal weight of water (charcoal, previously washed with water to remove any soluble material), filtered and evaporated under reduced pressure at 45-50° to give a syrup containing 90% of fructose. After cooling to 40°, the syrup was seeded and kept at this temperature for 20-30h with occasional stirring. The crystals were removed by centrifugation, washed with a small quantity of water and dried to constant weight under a vacuum over conc H$_2$SO$_4$. For higher purity, this material was recrystd from 50% aqueous ethanol [Tsuzuki, Yamazaki and Kagami *J Am Chem Soc* **72** 1071 *1950*].

D(+)-Fucose *[3615-37-0]* **M 164.2, m 144°, [α]$^{20}_{546}$ +89° (after 24h, c 10 in H$_2$O).** Crystd from EtOH.

Fullerene C$_{60}$ (Buckminsterfullerene C$_{60}$, Footballene, Buckyball 60) *[99685-96-8]* **M 720.66 and Fullerene C$_{70}$** *[115383-22-7]* **M 840.77.** Purified from the soluble toluene extract (400mg) of the soot (Fullerite) formed from resistive heating of graphite by adsorption on neutral alumina (100g; Brockmann I; 60 x 8cm). Elution with toluene-hexane (5:95 v/v) gives *ca* 250mg of quite pure C$_{60}$. It has characteristic spectral properties (see below). Further elution with toluene-hexane (20:80 v/v; i.e. increased polarity of solvent) provides 50mg of "pure" C$_{70}$ [*J Am Chem Soc* **113** 1050 *1991*].
Chromatography on alumina can be improved by using conditions which favour adsorption rather than crystn. Thus the residue from toluene extraction (1g) in CS$_2$ (*ca* 300mL) is adsorbed on alumina (375g, standard grade, neutral *ca* 150 mesh, Brockmann I) and loaded as a slurry in toluene-hexanes (5:95 v/v) to a 50 x 8cm column of alumina (1.5Kg) in the same solvent. To avoid crystn of the fullerenes, 10% of toluene in hexanes is added quickly followed by 5% of toluene in hexanes after the fullerenes had left the loading fraction (2-3h). With a flow rate of 15mL/min the purple C$_{60}$ fraction is eluted during a 3-4h period. Evapn of the eluates gives 550-

630mg of product which, after recrystn from CS_2-cyclohexane yields 520-600mg of C_{60} which contains adsorbed solvent. On drying at $275^o/10^{-3}$mm for 48h a 2% weight loss is observed although the C_{60} still contains traces of solvent. Further elution of the column with 20% of toluene in hexanes provides 130mg of C_{70} containing 10-14% of C_{60} (by ^{13}C NMR). This was rechromatographed as above using a half scale column and adsorbing the 130mg in CS_2 (20mL) on alumina (24g) and gave 105mg of recrystd C_{70} (containing 2% of C_{60}). The purity of C_{60} can be improved further by washing the crystalline product with Et_2O and Me_2CO followed by recrystn from $*C_6H_6$ and vacuum drying at high temperatures. [*J Chem Soc, Chem Commun* 956 *1992.*]

Carbon soot from resistive heating of a carbon rod in a partial helium atmosphere (0.3bar) under specified conditions is extracted with boiling $*C_6H_6$ or toluene, filtered and the red-brown soln evapd to give crystalline material in 14% yield which is mainly a mixture of fullerenes C_{60} and C_{70}. Chromatographic filtration of the 'crude' mixture with $*C_6H_6$ allows no separation of components, but some separation was observed on silica gel TLC with n-hexane or n-pentane, but not cyclohexane. Analytical HPLC with hexanes (5μm Econosphere silica) gave satisfactory separation of C_{60} and C_{70} (retention times of 6.64 and 6.93min respectively) at a flow rate of 0.5mL/min and using a detector at 256nm. HPLC indicated the presence of minor (<1.5% of total mass) unidentified C_n with species (retention times of 5.86 and 8.31min. Column chromatography on flash silica gel with hexanes gives a few fractions of C_{60} with ≥95% purity but later fractions contain mixtures of C_{60} and C_{70}. These can be obtained in 99.85 and >99% purity respectively by column chromatography on neutral alumina. [*J Phys Chem* **94** 8630 *1990*].

Separation of C_{60} and C_{70} can be achieved by HPLC on a dinitroanilinopropyl (DNAP) silica (5μm pore size, 300Å pore diameter) column with a gradient from n-hexane to 50% CH_2Cl_2 using a diode array detector at wavelengths 330nm (for C_{60}) and 384nm (for C_{70}). [*J Am Chem Soc* **113**, 2940, *1991.*]

Soxhlet extraction of the "soot" is a good preliminary procedure, or if material of only *ca* 98% purity is required. Soxhlet extraction with toluene is run (20min per cycle) until colourless solvent filled the upper part of the Soxhlet equipment (10h). One third of the toluene remained in the pot. After cooling, the solution was filtered through a glass frit. This solid (purple in toluene) was *ca* 98% C_{60}. This powder was again extracted in a Soxhlet using identical conditions as before and the C_{60} was recrystd from toluene to give 99.5% pure C_{60}. C_{70} has greater affinity than C_{h60} for toluene. [*J Chem Soc, Chem Commun* 1402 *1992.*]

Purification of C_{60} from a C_{60}/C_{70} mixture was achieved by dissolving in an aqueous soln of γ (but not β) cyclodextrin (0.02M) upon refluxing. The rate of dissolution (as can be followed by UV spectra) is quite slow and constant up to $10^{-5}M$ of C_{60}. The highest concn of C_{60} in H_2O obtained was 8 x $10^{-5}M$ and a 2 γ-cyclodextrin:1 C_{60} clathrate is obtained. C_{60} is extracted from this aqueous soln by toluene and C_{60} of >99 purity is obtained by evaporation. With excess of γ-cyclodextrin more C_{60} dissolves and the complex precipitates. The ppte is insol in cold H_2O but sol in boiling H_2O to give a yellow soln. [*J Chem Soc, Chem Commun* 604 *1922.*]

C_{60} and C_{70} can also be readily purified by inclusion complexes with *p-tert*-butylcalix[6] and [8]arenes. Fresh carbon-arc soot (7.5g) is stirred with toluene (250mL) for 1h and filtered. To the filtrate is added *p-tert*-butylcalix[8]arene, refluxed for 10min and filtered. The filtrate is seeded and set aside overnight at 20°. The C_{60} complex separated as yellow-brown plates and recrystd twice from toluene (1g from 80mL), 90% yield. Addition of $CHCl_3$ (5mL) to the complex (0.85g) gave C_{60} (0,28g, 92% from recryst complex).

p-tert-Butylcalix[6]arene-$(C_{60})_2$ complex is prepared by adding to a refluxing soln of C_{60} (5mg) in toluene (5mL), *p-tert*-butylcalix[6]arene (4.4mg). The hot soln was filtered rapidly and cooled overnight to give prisms (5.5mg, 77% yield). Pure C_{60} is obtained by decomposing the complex with $CHCl_3$ as above.

The *p-tert*-butylcalix[6]arene-$(C_{70})_2$ complex is obtained by adding *p-tert*-butylcalix[6]arene (5.8mg) to a refluxing soln of C_{70} (5mg) in toluene (2mL), filtering hot and slowly cooling. to give red-brown needles (2.5mg, 31% yield) of the complex. Pure C_{70} is then obtained by decomposing the complex with $CHCl_3$.

Decomposition of these complexes can also be achieved by boiling a toluene soln over KOH pellets for *ca* 10min. The calixarenes form Na salts which do not complex with the fullerenes. These appear to be the most satisfactory means at present for preparing large quantities of relatively pure fullerene C_{60} and C_{70} and is considerably cheaper than previous methods. [*Nature* **368** 229 *1994.*]

Repeated chromatography on neutral alumina yields minor quantities of solid samples of C_{76}, C_{84}, C_{90} and C_{94} believed to be higher fullerenes. A stable oxide $C_{70}O$ has been identified. Chromatographic procedures for the separation of these compounds are reported. [*Science* **252** 548 *1991.*]

Physical properties of Fullerene C_{60}: It does not melt below 360°, and starts to sublime at 300° *in vacuo*. It is a mustard coloured solid that appears brown or black with increasing film thickness. It is soluble in common organic solvents, particularly aromatic hydrocarbons which give a beautiful magenta colour. Toluene solutions are purple in colour. Sol in *C_6H_6 (5mg/mL), but dissolves slowly. Crysts of C_{60} are both needles and plates. UV-Vis in hexanes: λmax nm(log ε) 211(5.17), 227sh(4.91), 256(5.24), 328(4.71), 357sh(4.08), 368sh(3.91), 376sh(3.75), 390(3.52), 395sh(3.30), 403(3.48), 407(3.78), 492sh(2.72)< 540(2.85), 568(2.78), 590(2.86), 598(2.87) and 620(2.60).
IR (KBr): ν 1429m, 1182m, 724m, 576m and 527s cm^{-1}. ^{13}C NMR: one signal at 142.68ppm.

Physical properties of Fullerene C_{70}: It does not melt below 360°, and starts to sublime at 350° *in vacuo*. A reddish-brown solid, greenish black in thicker films. Solns are port-wine red in colour. Mixtures of C_{60} and C_{70} are red due to C_{70} being more intensely coloured. It is less soluble than C_{60} in *C_6H_6 but also dissolves slowly. C_{70} gives orange coloured soln in toluene. Drying at 200-250° is not sufficient to remove All solvent. Samples need to be sublimed to be free from solvent.
UV-Vis in hexanes: λmax nm(log ε) 214(5.05), 235(5.06), 249sh(4.95), 268sh(4.78), 313(4.23), 330(4.38), 359(4.29), 377(4.45), 468(4.16), 542(3.78), 590sh(3.47), 599sh(3.38), 609(3.32), 623sh(3.09), 635sh(3.13) and 646sh(2.80).
IR (KBr): ν 1430m, 1428m, 1420m, 1413m, 1133mw, 1087w, 795s, 674ms, 642ms, 5778s, 566m, 535ms and 458m cm^{-1}.
^{13}C NMR [run in the presence of Cr(pentan-2,4-dione)$_3$ which induces a *ca* 0.12ppm in the spectrum]: Five signals at 150.07, 147.52, 146.82, 144.77 and 130.28ppm, unaffected by proton decoupling.
[Further reading: Kroto, Fischer and Cox *Fullerenes* Pergamon Press, Oxford 1993 ISBN 0080421520; Kadish and Ruoff (Eds) *Fullerenes: Recent Advances in the Chemistry and Physics of Fullerenes and Related Materials* The Electrochemical Soc. Inc, Pennington, NJ, 1994 ISBN 1566770823]

Fumagillin {2,4,6,8-decatetraene-1,10-dioic acid mono[4-(1,2-epoxy-1,5-dimethyl-4-hexenyl)-5-methoxy-1-oxaspiro[2.5]oct-6-yl] ester} *[23110-15-8]* **M 458.5, m 194-195°,** $[\alpha]_D^{20}$ **-26.2° (in 95% EtOH), pK$_{Est}$ ~4.5.** Forty grams of a commercial sample containing 42% fumagillin, 45% sucrose, 10% antifoam agent and 3% of other impurities were digested with 150mL of CHCl$_3$. The insoluble sucrose was filtered off and washed with CHCl$_3$. The combined CHCl$_3$ extracts were evapd almost to dryness at room temperature under reduced pressure. The residue was triturated with 20mL of MeOH and the fumagillin was filtered off by suction. It was crystd twice from 500mL of hot MeOH by standing overnight in a refrigerator (yellow needles). (The long chain fatty ester used as antifoam agent was still present, but was then removed by repeated digestion, on a steam bath, with 100mL of diethyl ether.) For further purification, the fumagillin (10g) was dissolved in 150mL of 0.2M ammonia, and the insoluble residue was filtered off. The ammonia soln (cooled in running cold water) was then brought to pH 4 by careful addn of M HCl with constant shaking in the presence of 150mL of CHCl$_3$. (Fumagillin is acid-labile and must be removed rapidly from the aq acid soln.) The CHCl$_3$ extract was washed several times with distd water, dried (Na$_2$SO$_4$) and evaporated under reduced pressure. The solid residue was washed with 20mL of MeOH. The fumagillin was filtered by suction, then crystd from 200mL of hot MeOH. [Tarbell et al. *J Am Chem Soc* **77** 5610 *1955*.]
Alternatively, 10g of fumagillin in 100mL CHCl$_3$ was passed through a silica gel (5g) column to remove tarry material, and the CHCl$_3$ was evaporated to leave an oil which gave fumagillin on crystn from amyl acetate. It recrystallises from MeOH (charcoal). The fumagillin was stored in dark bottles in the absence of oxygen and at low temperatures. [Schenk, Hargie and Isarasena *J Am Chem Soc* **77** 5606 *1955*.]

Fumaraldehyde bis-(dimethyl acetal) (*trans*-1,1,4,4-tetramethoxybut-2-ene) *[6068-62-8]* **M 176.2, b 100-103°/15mm, 101-103°/25mm, d_4^{20} 1.011, n_D^{20} 1.425.** Dry over fused CaCl$_2$ and dist *in vacuo*. The maleic (cis) isomer has **b** 112°/11mm, and d^{23} 0.932 and n_D^{25}1.4243. [Zeik and Heusner *Chem Ber* **90** 1869 *1957*; Clauson-Kaas et al. *Acta Chem Scand* **9** 111 *1955;* Clauson-Kaas *Acta Chem Scand* **6** 569 *1952*.]

Fumaric (*trans*-but-2-ene-1,4-dioic) acid *[110-17-8]* **M 116.1, m 289.5-291.5° (sealed tube), pK$_1^{25}$ 3.10, pK$_2^{25}$ 4.60 (4.38).** Crystd from hot M HCl or water. Dried at 100°.

Furan *[110-00-9]* **M 68.1, b 31.3°, d 1.42, n 1.4214.** Shaken with aqueous 5% KOH, dried with $CaSO_4$ or Na_2SO_4, then distd under nitrogen, from KOH or sodium, immediately before use. A trace of hydroquinone could be added as an inhibitor of oxidation.

3-(2-Furanayl)acrylic acid *[539-47-9]* **M 138.1, m 141°, pK_{Est} ~3.8.** Crystd from H_2O or pet ether (b 80-100°)(charcoal).

Furan-2-carboxylic (2-furoic) acid *[88-14-2]* **M 112.1, m 133-134°, b 141-144°/20mm, 230-232°/760mm, pK_1^{25} -7.3 (O-protonation), pK_2^{25} 3.32.** Crystd from hot water (charcoal), dried at 120° for 2h, then recrystd from $CHCl_3$, and again dried at 120° for 2h. For use as a standard in volumetric analysis, good quality commercial acid should be crystd from $CHCl_3$ and dried as above or sublimed at 130-140° at 50-60mm or less.

Furan-3-carboxylic (3-furoic) acid *[488-93-7]* **M 112.1, m 122-123°, pK^{25} 4.03** Crystd from water.

Furan-3,4-dicarboxylic acid *[3387-26-6]* **M 156.1, m 217-218° pK_1^{25} 1.44, pK_2^{25} 7.84.** Crystd from water.

Furfural (2-furfuraldehyde) *[98-01-1]* **M 96.1, b 54-56°/11mm, 59-60°/15mm, 67.8°/20mm, 90°/65mm, 161°/760mm, d_4^{20} 1.159, n_D^{20} 1.52608, pK -6.5 (O-protonation).** Unstable to air, light and acids. Impurities include formic acid, ß-formylacrylic acid and furan-2-carboxylic acid. Distd over an oil bath from 7% (w/w) Na_2CO_3 (added to neutralise acids, especially pyromucic acid). Redistd from 2% (w/w) Na_2CO_3, and then, finally fractionally distd under vacuum. It is stored in the dark. [Evans and Aylesworth *Ind Eng Chem (Anal ed)* **18** 24 *1926*.]
Impurities resulting from storage can be removed by passage through chromatographic grade alumina. Furfural can be separated from impurities other than carbonyl compounds by the bisulfite addition compound. The aldehyde is steam volatile.
It has been purified by distn (using a Claisen head) under reduced pressure. This is essential as is the use of an oil bath with temperatures of no more than 130° are highly recommended. When furfural is distd at atm press (in a stream of N_2), or under reduced pressure with a free flame (caution because the aldehyde is flammable) an almost colourless oil is obtained. After a few days and sometimes a few hours the oil gradually darkens and finally becomes black. This change is accelerated by light but occurs more slowly when kept in a brown bottle. However, when the aldehyde is distd under vacuum and the bath temperature kept below 130° during the distn, the oil develops only a slight colour when exposed to direct sunlight during several days. The distn of very impure material should NOT be attempted at atm pressure otherwise the product darkens rapidly. After one distn under vacuum a distn at atmospheric pressure can be carried out without too much decomposition and darkening. The liquid **irritates mucous membranes.** Store in dark containers under N_2. [Adams and Voorhees *Org Synth* Coll Vol I 280 *1941*.]

Furfuryl alcohol (2-furylmethanol) *[98-00-0]* **M 98.1, b 68-69°/20mm, 170.0°/750mm, d 1.132, n 1.4873, n^{30} 1.4801, pK^{25} 2.61.** Distd under reduced pressure to remove tarry material, shaken with aqueous $NaHCO_3$, dried with Na_2SO_4 and fractionally distd under reduced pressure from Na_2CO_3. Further dried by shaking with Linde 5A molecular sieves.

Furfurylamine (2-aminomethylfuran) *[617-89-0]* **M 97.1, b 142.5-143°/735mm, d 1.059, n 1.489, pK^{30} 8.89.** Distd under nitrogen from KOH through a column packed with glass helices.

Furil *[492-94-4]* **M 190.2, m 165-166°.** Crystd from MeOH or *benzene (charcoal).

Furoin *[552-86-3]* **M 192.2, m 135-136°.** Crystd from MeOH (charcoal).

Galactaric Acid (mucic acid) [526-99-6] M 210.1, m 212-213°(dec) pK$_1^{25}$ 3.09
(3.29), pK$_2^{25}$ 3.63 (4.41). Dissolved in the minimum volume of dil aq NaOH, and ppted by adding dil
HCl. The temperature should be kept below 25°.

D-Galactonic acid [576-36-3] M 196.2, m 148°, pK$_{Est}$ ~3.5. Crystd from EtOH. Cyclises to *D-galactonono-1,4-lactone*, m 134-136°, $[\alpha]_{546}^{30}$ mutarotates in 1h to -92° (c 5,H$_2$O).

D(-)-Galactono-1,4-lactone [2782-07-2] M 178.1, m 134-137°, $[\alpha]_D^{20}$ -78° (in H$_2$O). Crystd
from EtOH.

D(+)-Galactosamine hydrochloride [1772-03-8] M 215.6, m 181-185°, $[\alpha]_D^{25}$ +96.4° (after
24h, c 3.2 in H$_2$O), pK$_{Est}$ ~7.7 (free base). Dissolved in a small volume of H$_2$O. Then added three
volumes of EtOH, followed by acetone until faintly turbid and stood overnight in a refrigerator. [Roseman and
Ludoweig *J Am Chem Soc* **76** 301 *1954*.]

α-D-Galactose [59-23-4] M 180.2, m 167-168°, $[\alpha]_D^{20}$ +80.4° (after 24h, c 4 in H$_2$O), pK^{25}
12.48. Crystd twice from aqueous 80% EtOH at -10°, then dried under vacuum over P$_2$O$_5$.

Gallic acid (H$_2$O) (3,4,5-trihydroxybenzoic acid) [5995-86-8 (H$_2$O), 149-91-7 (anhydr)] M
188.1, m 253°(dec), pK$_1^{25}$ 4.27, pK$_2^{25}$ 8.68. Crystd from water.

Galvinoxyl [2,6-di-*tert*-butyl-α-(3,5-di-*tert*-butyl-4-oxo-2,5-cyclohexadiene-1-ylidene)-*p*-tolyloxy] [2370-18-5] M 421.65, m 153.2-153.6°, 158-159°. A stable free radical scavanger of
short-lived free radicals with odd electrons on C or O. Best prepared freshly by oxidation of 3,3',5,5'-tetra-*tert*-butyl-4,4'-dihydroxydiphenyl-methane [m 154°, 157.1-157.6°; obtained by gently heating for 10-15min 2,6-di-*tert*-butylphenol, formaldehyde and NaOH in EtOH and recryst from EtOH (20g/100mL) as colorless plates,
Karasch and Joshi *J Org Chem* **22** 1435 *1957*; Bartlett et al. *J Am Chem Soc* **82** 1756 *1960* and **84** 2596
1962.] Oxidation is carried out under N$_2$ with PbO$_2$ in Et$_2$O or isooctane [Galvin A. Coppinger *J Am Chem
Soc* **79** 501 *1957* ; Bartlett et al. above] or with alkaline potassium ferricyanide [Karasch and Joshi, above],
whereby Galvinoxyl separates as deep blue crystals, and is recrystd twice under N$_2$ from *C$_6$H$_6$ soln by suction
evaporation at 30°. The VIS spectrum has λmax 407nm (ε 30,000), 431nm (ε 154,000) and weak absorption at
772.5nm, and IR: ν 1577 and 2967cm^{-1}, and is estimated by iodometric titration. It is sensitive to O$_2$ in
presence of OH$^-$ ions and to traces of strong acid in hydroxylic or hydrocarbon solvents. At 62.5° in a 0.62mM
soln in *C$_6$H$_6$ the radical decays with a first order k = 4 x 10^{-8} sec^{-1} (half life 1.7 x 10^{17} sec, ~200 days) as
observed by the change in OD at 550nm [see also Green and Adam *J Org Chem* **28** 3550 *1963*].

Genistein (4',5,7-trihydroxyisoflavone) [446-72-0] M 270.2, m 297-298°, $[\alpha]_D^{20}$ -28° (c
0.6, 20mM NaOH), (phenolic pKs 8-10). Crystd from 60% aqueous EtOH or water.

Genistin (genistein-7-D-glucoside) [529-59-9] M 432.4, m 256°. Crystd from 80% EtOH/water.

α-Gentiobiose (amygdalose, 6-O-α-D-glucopyranosyl-D-glucopyranose) [5995-99-5 (bi-pyranose)] M 342.3, m 86°, $[\alpha]_{546}^{20}$ +11° (after 24h, c 4, H$_2$O). Crystd from MeOH (retains
solvent of crystn).

β-Gentiobiose (see above) [5996-00-9 (bi-pyranose); 554-91-6 (one open ring)] M 342.3, m 190-195°, $[\alpha]_{546}^{20}$ +8° (after 6h, c 3, H$_2$O). Crystd from MeOH or EtOH.

Geraniol [106-24-1] M 154.3, b 230°, d 0.879, n 1.4766. Purified by ascending chromatography
or by thin layer chromatography on plates of kieselguhr G with acetone/water/liquid paraffin (130:70:1) as
solvent system. Hexane/ethyl acetate (1:4) is also suitable. Also purified by GLC on a silicone-treated column
of Carbowax 20M (10%) on Chromosorb W (60-80 mesh). [Porter *Pure Appl Chem* **20** 499 *1969*.] Stored
in full, tightly sealed containers in the cool, protected from light.

Gibberillic acid (GA$_3$) [77-06-5] M 346.4, m 233-235°(dec), $[\alpha]_{546}^{20}$ +92° (c 1, MeOH), pK
4.0. Crystd from ethyl acetate.

Girard Reagent T (2-hydrazino-*N,N,N*-trimethyl-2-oxo-ethanaminium chloride) *[123-46-6]* **M 167.6, m 192°.** Should be crystd from absolute EtOH (slight decomposition) when it has a slight odour, and stored in tightly stoppered containers because it is hygroscopic.

Glucamine *[488-43-7]* **M 181.2, m 127°, $[\alpha]_D^{20}$ -8°** (c 10, H_2O), **pK_{Est} ~9.0.** Crystd from MeOH.

D-Gluconamide *[3118-85-2]* **M 197.2, m 144°, $[\alpha]_D^{23}$ +31°** (c 2, H_2O). Crystd from EtOH.

D-Glucono-δ-lactone *[90-80-2]* **M 178.1, m 152-153°, $[\alpha]_{546}^{20}$ +76°** (c 4, H_2O). Crystd from ethylene glycol monomethyl ether and dried for 1h at 110°.

D-Glucosamine *[3416-24-8]* **M 179.2, m 110°(dec), $[\alpha]_D^{20}$ +28° → +48°** (c 5, H_2O), **pK^{24} 7.71.** Crystd from MeOH. *N-Acetyl deriv,* **m 205°** from MeOH/Et_2O has $[\alpha]_D^{20}$ +64° → +41° (c 5, H_2O).

D-Glucosamine hydrochloride *[66-84-2]* **M 215.6, m >300°, $[\alpha]_D^{25}$ +71.8°** (after 20h, c 4, H_2O). Crystd from 3M HCl, water, and finally water/EtOH/acetone as for galactosamine hydrochloride.

α-D-Glucose *[492-62-6]* **M 180.2, m 146°, $[\alpha]_D^{20}$ +52.5°** (after 24h, c 4, H_2O), **pK^{25} 12.46.** Recrysts slowly from aqueous 80% EtOH, then vacuum dried over P_2O_5. Alternatively, crystd from water at 55°, then dried for 6h in a vacuum oven between 60-70° at 2mm.

β-D-Glucose *[50-99-7]* **M 180.2, m 148-150°.** Crystd from hot glacial acetic acid.

α-D-Glucose pentaacetate *[604-68-2]* **M 390.4, m 110-111°, 112°, $[\alpha]_{546}^{20}$ +119°** (c 5, $CHCl_3$). Crystd from MeOH or EtOH.

β-D-Glucose pentaacetate *[604-69-3]* **M 390.4, m 131-132°, $[\alpha]_{546}^{20}$ +5°** (c 5, $CHCl_3$). Crystd from MeOH or EtOH.

D-Glucose phenylhydrazone *[534-97-4]* **M 358.4, m 208°.** Crystd from aqueous EtOH.

D-Glucuronic acid *[6556-12-3]* **M 194.1, m 165°, $[\alpha]_D^{20}$ +36°** (c 3, H_2O), **pK_2^{20} 3.18.** Crystd from EtOH or ethyl acetate.

D-Glucuronolactone *[32449-92-6]* **M 176.1, m 175-177°, $[\alpha]_{546}^{20}$ +22°** (after 24h, c 10, H_2O). Crystd from water.

L-Glutamic acid *[56-86-0]* **M 147.1, m 224-225°(dec), $[\alpha]_D^{25}$ +31.4°** (c 5, 5M HCl), **pK_1^{20}, pK_2^{20} 2.06,4.35, pK_3^{20} 9.85.** Crystd from H_2O acidified to pH 3.2 by adding 4 volumes of EtOH, and dried at 110°. Likely impurities are aspartic acid and cysteine.

L-Glutamic acid-γ-benzyl ester *[1676-73-9]* **M 237.3, m 179-181°, $[\alpha]_{589}^{20}$ 19.3°** (c 1, HOAc), **pK_1^{25} 2.17, pK_2^{25} 9.00.** Recrystd from H_2O and stored at 0°. [Estrin *Biochem Prep* **13** 25 *1971.*]

L-Glutamine *[56-85-9]* **M 146.2, m 184-185°, $[\alpha]_D^{25}$ +31.8°** (M HCl), **pK_1^{25} 2.17, pK_2^{25} 9.13.** Likely impurities are glutamic acid, ammonium pyroglutamate, tyrosine, asparagine, isoglutamine, arginine. Crystd from water.

Glutaraldehyde *[111-30-8]* **M 100.1, b 71°/10mm, as 50% aq soln.** Likely impurities are oxidation products - acids, semialdehydes and polymers. It can be purified by repeated washing with activated charcoal (Norit) followed by vacuum filtration, using 15-20g charcoal/100mL of glutaraldehyde soln. Vacuum distn at 60-65°/15mm, discarding the first 5-10%, was followed by dilution with an equal volume of freshly distilled water at 70-75°, using magnetic stirring under nitrogen. The soln is stored at low temp (3-4°),

in a tightly stoppered container, and protected from light.   Standardised by titration with hydroxylamine.
[Anderson *J Histochem Cytochem* **15** 652 *1967*.]

Glutaric acid   *[110-94-1]* **M 132.1, m 97.5-98°, pK$_1^{25}$ 4.35, pK$_2^{25}$ 5.40.** Crystd from *benzene,
CHCl$_3$, distilled water or *benzene containing 10% (w/w) of diethyl ether. Dried under vacuum.

*dl–***Glyceraldehyde**   *[56-82-6]* **M 90.1, m 145°.** Crystd from EtOH/diethyl ether.

Glycerol   *[56-81-5]* **M 92.1, m 18.2°, b 182°/20mm, 290°/760mm, d 1.261, n^{25} 1.47352,
pK 14.4.** Glycerol was dissolved in an equal volume of *n*-butanol (or *n*-propanol, amyl alcohol or liquid
ammonia) in a water-tight container, cooled and seeded while slowly revolving in an ice-water slurry. The
crystals were collected by centrifugation, then washed with cold acetone or isopropyl ether. [Hass and Patterson
Ind Eng Chem (Anal Ed) **33** 615 *1941*.] Coloured impurities can be removed from substantially dry glycerol
by extraction with 2,2,4-trimethylpentane. Alternatively, glycerol can be decolorised and dried by treatment
with activated charcoal and alumina, followed by filtering. Glycerol can be distd at 15mm in a stream of dry
nitrogen, and stored in a desiccator over P$_2$O$_5$. Crude glycerol can be purified by digestion with conc H$_2$SO$_4$ and
saponification with a lime paste, then re-acidified with H$_2$SO$_4$, filtered, treated with an anion exchange resin and
fractionally distd under vacuum.

Glycidol (oxirane-2-methanol)   *[RS-(±)- 556-52-5;   R-(+)- 57044-25-4;   S-(-)- 60456-23-7]* **M 74.1, b
61-62°/15mm, d 1.117, n 1.433 (±) , b 49-50°/7mm, 66-67°/19mm, [α] $_D^{20}$ -15° (neat) (*S*-
isomer, § also available on polymer support), b 56-56.5°/11mm, d 1.117, n 1.429, [α]
$_D^{20}$ +15° (neat).** Purified by fractional distn. The *4-nitrobenzoates* have m 56° (±); m 60-62°, [α] $_D^{20}$ -37.9°
(c 3.38 CHCl$_3$) for *R*-(-)-isomer *[106268-95-5]*; m 60-62°, [α]$_D^{20}$ +38° (c 1 CHCl$_3$) for *S*-(+)-isomer m 60-62°,
[α] $_D^{20}$ -38° (c 1 CHCl$_3$) *[115459-65-9]*, and are recrystd from Et$_2$O or Et$_2$O/pet ether (b 40-60°) [*S*-isomer:
Burgos et al. *J Org Chem* **52** 4973 *1987*; Sowden and Fischer *J Am Chem Soc* **64** 1291 *1942*.]

Glycinamide hydrochloride   *[1668-10-6]* **M 110.5, m 186-189° (207-208°), pK$_1^{25}$ -6.10, pK$_2^{25}$
-1.78, pK$_3^{25}$ 7.95.** Crystd from EtOH.

Glycine see **aminoacetic acid.**

Glycine ethyl ester hydrochloride   *[623-33-6]* **M 136.9, m 145-146°, pK^{25} 7.69.**   Crystd
from absolute EtOH.

Glycine hydrochloride   *[6000-43-7]* **M 111.5, m 176-178°.** Crystd from absolute EtOH.

Glycine methyl ester hydrochloride   *[5680-79-5]*   **M 125.6, m 174°(dec), pK^{25} 7.66.** Crystd
from MeOH.

Glycine *p*-nitrophenyl ester hydrobromide.   *[7413-60-7]* **M 277.1, m 214° (dec).** Recryst from
MeOH by adding diethyl ether. [Alners et al. *Biochem Preps* **13** 22 *1971*].

Glycocholic acid (*N*-cholylglycine)   *[475-31-0]* **M 465.6, m 154-155°, 165-168°, [α]$_{546}^{20}$
+37° (c 1, EtOH), pK 4.4.** Crystd from hot water as sesquihydrate. Dried at 100°.

Glycolic (α-hydroxyacetic) acid   *[79-14-1]* **M 76.1, m 81°, pK^{25} 3.62.** Crystd from diethyl
ether.

N-**Glycylanilide**   *[555-48-6]* **M 150.2, m 62°, pK$_{Est}$~8.0.** Crystd from water, sol in Et$_2$O..

Glycylglycine   *[556-50-3]* **M 132.1, m 260-262°(dec), pK^{20} 8.40, pK^{30} 8.04.** Crystd from
aqueos 50% EtOH or water at 50-60° by addition of EtOH. Dried at 110°.

Glycylglycine hydrochloride *[13059-60-4]* M 168.6, m 215-220°, 235-236°, 260-262°, pK_1^{25} 3.12, pK_2^{25} 8.17. Crystd from 95% EtOH.

Glycyl-L-proline *[704-15-4]* M 172.2, m 185°, pK_1^{25} 2.81, pK_2^{25} 8.65. Crystd from water at 50-60° by addition of EtOH.

dl-**Glycylserine** *[687-38-7]* M 162.2, m 207°(dec), pK_1^{25} 2.92, pK_2^{25} 8.10. Crystd from H_2O (charcoal) by addition of EtOH.

Glycyrrhizic acid ammonium salt ($3H_2O$) *[53956-04-0]* M 823.0, m 210°(dec), $[\alpha]_{546}^{20}$ +60° (c 1, 50% aq EtOH), pK_{Est} ~4.0. Crystd from glacial acetic acid, then dissolved in ethanolic ammonia and evaporated.

Glyoxal bis(2-hydroxyanil) *[1149-16-2]* M 240.3, m 210-213°, ε_{294nm} 9880. Crystd from MeOH or EtOH.

Glyoxylic acid *[298-12-4]* M 74.0, m 98°(anhydr), 50-52°(monohydrate), pK^{25} 2.98. Crystd from water as the monohydrate.

Gramine (3-dimethylaminoethylindole) *[87-52-5]* M 174.3, m 134°, pK^{25} 16.00 (NH acidic). Crystd from diethyl ether, ethanol or acetone.

Griseofulvin *[126-07-8]* M 352.8, m 220°, $[\alpha]_D^{22}$ +365° (c 1, acetone). Crystd from *benzene.

Guaiacic acid [4,4'-(2,3-dimethyl-1-butene-(1,4-diyl)-bis-(2-methoxyphenol)] *[500-40-3]* M 328.4, m 99-100.5°, pK_{Est} ~10.0. Crystd from EtOH.

Guaiacol (2-methoxyphenol) *[90-05-1]* M 124.1, m 32°, b 106°/24mm, 205°/746mm, pK^{25} 9.90. Crystd from *benzene/pet ether or distd.

Guaiacol carbonate *[553-17-3]* M 274.3, m 88.1°. Crystd from EtOH.

Guanidine *[113-00-8]* M 59.1, m ~50°, pK^{25} 13.6. Crystd from water/EtOH under nitrogen. Very deliquescent and absorbs CO_2 from the air readily.

Guanidine carbonate *[593-85-1]* M 180.2, m 197°. Crystd from MeOH.

Guanidine hydrochloride *[50-01-1]* M 95.5, m 181-183°. Crystd from hot methanol by chilling to about -10°, with vigorous stirring. The fine crystals were filtered through fritted glass, washed with cold (-10°) methanol, dried at 50° under vacuum for 5h. (The product is more pure than that obtained by crystn at room temperature from methanol by adding large amounts of diethyl ether.) [Kolthoff et al. *J Am Chem Soc* **79** 5102 *1957*].

Guanosine (H_2O) *[118-00-3]* M 283.2, m 240-250°(dec), $[\alpha]_{546}^{20}$ -86° (c 1, 0.1M NaOH), pK_1^{25} 1.9, pK_2^{25} 9.24, pK_3^{25} 12.33. Crystd from water. Dried at 110°.

Guanylic acid (guanosine-5'-monophosphoric acid) *[85-32-5]* M 363.2, m 208°(dec), pK_2^{25} 2.4, pK_3^{25} 6.66 (6.1), pK_4^{25} 9.4. Crystd from water. Dried at 110°.

Harmine *[442-51-3]* M 212.3, m 261°(dec), pK^{20} 7.61. Crystd from MeOH.

Harmine hydrochloride (hydrate) *[343-27-1]* M 248.7, m 280°(dec). Crystd from water.

Hecogenine acetate *[915-35-5]* M 472.7, m 265-268°, $[\alpha]_D^{23}$ -4.5° (c 1, CHCl$_3$). Crystd from MeOH.

Heptadecanoic acid (margaric) *[506-12-7]* M 270.5, m 60-61°, b 227°/100mm, pK$_{Est}$ ~4.9. Crystd from MeOH or pet ether.

1-Heptadecanol *[1454-85-9]* M 256.5, m 54°. Crystd from acetone.

Heptafluoro-2-iodopropane *[677-69-0]* M 295.9, b 41°. Purified by gas chromatography on a triacetin (glyceryl triacetate) column, followed by bulb-to-bulb distn at low temperature. Stored over Cu powder to stabilise it.

n-**Heptaldehyde** *[111-71-7]* M 114.2, b 40.5°/12mm, 152.8°/760mm, d 0.819, n^{25} 1.4130. Dried with CaSO$_4$ or Na$_2$SO$_4$ and fractionally distd under reduced pressure. More extensive purification by pptn as the bisulfite compound (formed by adding the aldehyde to saturated aqueous NaHSO$_3$) which was filtered off and recrystd from hot H$_2$O. The crystals, after being filtered and washed well with H$_2$O, were hydrolysed by adding 700mL of aqueous Na$_2$CO$_3$ (12.5% w/w of anhydrous Na$_2$CO$_3$) per 100g of aldehyde. The aldehyde was then steam distd, separated, dried with CuSO$_4$ and distd under reduced pressure in a slow stream of nitrogen. [McNesby and Davis *J Am Chem Soc* **76** 2148 *1954*].

n-**Heptaldoxime** *[629-31-2]* M 129.2, m 53-55°. Crystd from 60% aqueous EtOH.

n-**Heptane** *[142-18-5]* M 100.2, b 98.4°, d 0.684, n 1.38765, n^{25} 1.38512. Passage through a silica gel column greatly reduces the ultraviolet absorption of *n*-heptane. (The silica gel is previously heated to 350° before use.) For more extensive purification, heptane is shaken with successive small portions of conc H$_2$SO$_4$ until the lower (acid) layer remains colourless. The heptane is then washed successively with water, aq 10% Na$_2$CO$_3$, water (twice), and dried with CaSO$_4$, MgSO$_4$ or CaCl$_2$. It is distd from sodium. *n*-Heptane can be distd azeotropically with methanol, then the methanol can be washed out with water and, after drying, the heptane is redistd. Other purification procedures include passage through activated basic alumina, drying with CaH$_2$, storage with sodium, and stirring with 0.5*N* KMnO$_4$ in 6N H$_2$SO$_4$ for 12h after treatment with conc H$_2$SO$_4$. Carbonyl-containing impurities have been removed by percolation through a column of impregnated Celite made by dissolving 0.5g of 2,4-dinitrophenylhydrazine in 6mL of 85% H$_3$PO$_4$ by grinding together, then adding 4mL of distilled water and 10g Celite. [Schwartz and Parks *Anal Chem* **33** 1396 *1961*].

Hept-1-ene *[592-76-7]* M 98.2, b 93°/771mm, d 0.698, n 1.400. Distd from sodium, then carefully fractionally distd using an 18-in gauze-packed column. Can be purified by azeotropic distn with EtOH. Contained the 2- and 3-isomers as impurities. These can be removed by gas chromatography using a Carbowax column at 70°.

n-**Heptyl alcohol** *[111-70-6]* M 116.2, b 175.6°, d 0.825, n 1.425. Shaken with successive lots of alkaline KMnO$_4$ until the colour persisted for 15min, then dried with K$_2$CO$_3$ or CaO, and fractionally distd.

n-**Heptylamine** *[111-68-2]* M 115.2, b 155°, d 0.775, n 1.434, pK^{25} 10.66. Dried in contact with KOH pellets for 24h, then decanted and fractionally distd.

n-**Heptyl bromide** *[629-04-9]* M 179.1, b 70.6°/19mm, 180°/760mm, d 1.140, n 1.45. Shaken with conc H$_2$SO$_4$, washed with water, dried with K$_2$CO$_3$, and fractionally distd.

Heptyl-β–D-glucopyranoside *[78617-12-6]* M 278.4, m 74-77°, 76-77°, $[\alpha]_D^{20}$ -34.2° (c 5, H$_2$O). Purified by several recrystns from M$_2$CO which is a better solvent than EtOAc. The *acetate* has m 66-68.5°, $[\alpha]_D^{20}$ -20.5° (c 4, CHCl$_3$) [Pigman and Richtmyer *J Am Chem Soc* **64** 369 *1942*].

Heptyl-β–D-1-thioglucopyranoside *[85618-20-8]* M 294.4, m 98-99°. The *tetra-acetyl derivative* is purified by silica gel column chromatography and eluted with a *C$_6$H$_6$-Me$_2$CO (gradient up to 5% of Me$_2$CO) and recrystd from *n*-hexane as colourless needles m 72-74° (Erbing and Lindberg *Acta Chem Scand*

B30 611 *1976* gave **m** 69-70°). Hydrolysis using an equivalent of base in methanol gave the desired glucoside. This is a non-ionic detergent for reconstituting membrane proteins and has a critical micelle concentration of 30 mM. [Shimamoto et al. *J Biochem (Tokyo)* **97** 1807 *1985*; Saito and Tsuchiya *Chem Pharm Bull Jpn* **33** 503 *1985*].

Hesperetin (3',5,7-trihydroxy-4'-methoxyflavanone) *[520-33-2]* **M 302.3, m 227-228°, pK$_{Est}$ ~8.5-10.5 (phenolic).** Crystd from EtOAc or ethanol. The natural (-) form has $[\alpha]_D^{20}$ -38° (c 2, EtOH). Note that C2 is chiral.

Hesperidin (hesperetin 7-rhamnoside) *[520-26-3]* **M 610.6, m 258-262°, $[\alpha]_{546}^{20}$ -82° (c 2, pyridine).** Dissolved in dilute aqueous alkali and ppted by adjusting the pH to 6-7.

Hexachlorobenzene *[118-74-1]* **M 284.8, m 230.2-231.0°.** Crystd repeatedly from *benzene. Dried under vacuum over P_2O_5.

Hexachloro-1,3-butadiene *[87-68-3]* **M 260.8.** See perchlorobutadiene on p. 323.

1,2,3,4,5,6-Hexachlorocyclohexane *[α-319-84-6; γ-58-89-9]* **M 290.8, m 158° (α-), 312° (ß-), 112.5° (γ-isomer).** Crystd from EtOH. Purified by zone melting. **Possible CANCER AGENT, TOXIC.**

Hexachlorocyclopentadiene *[77-47-4]* **M 272.8, b 80°/1mm, d 1.702, n^{25} 1.5628.** Dried with $MgSO_4$. Distd under vacuum in nitrogen.

Hexachloroethane *[67-72-1]* **M 236.7, m 187°.** Steam distd, then crystd from 95% EtOH. Dried in the dark under vacuum.

Hexacosane *[630-01-3]* **M 366.7, m 56.4°, b 169°/0.05mm, 205°/1mm, 262°/15mm.** Distd under vacuum and crystd from diethyl ether.

Hexacosanoic acid (cerotinic acid) *[506-46-7]* **M 396.7, m 86-87°, 88-89°, pK$_{Est}$ ~4.9.** Crystd from EtOH, aq EtOH and pet ether+Me_2CO.

1,14-Hexadecanedioic acid (thaspic acid). *[505-54-4]* **M 286.4, m 126°, pK$_{Est(1)}$~4.5, p K$_{Est(2)}$ ~5.5.** Crystd from EtOH, ethyl acetate or *C_6H_6.

n-**Hexadecane (Cetane)** *[544-76-3]* **M 226.5, m 18.2°, b 105°/0.1mm, d 0.773, n 1.4345, n^{25} 1.4325.** Passed through a column of silica gel and distd under vacuum in a column packed with Pyrex helices. Stored over silica gel. Crystd from acetone, or fractionally crystd by partial freezing.

Hexadecanoic acid (palmitic acid) *[57-10-3]* **M 256.4, m 62-63°, b 215°/15mm, pK^{25} 6.46 (50% aq EtOH), 5.0 (H_2O).** Purified by slow (overnight) recrystn from hexane. Some samples were also crystd from acetone, EtOH or EtOAc. Crystals were stood in air to lose solvent, or were pumped dry of solvent on a vacuum line. [Iwahashi et al. *J Chem Soc, Faraday Trans 1* **81** 973 *1985*; pK: White *J Am Chem Soc* **72** 1858 *1950*].

Hexadecyl 3-hydroxynaphthalene-2-carboxylate *[531-84-0]* **M 412.6, m 73-74°.** Recrystd from hot EtOH and sublimed in a vacuum. [Oshima and Hayashi *J Soc Chem Ind Jpn* **44** 821 *1941*.]

1,5-Hexadiene *[592-42-7]* **M 82.2, b 59.6°, d 0.694, n 1.4039.** Distd from $NaBH_4$.

Hexaethylbenzene *[604-88-6]* **M 246.3, m 128.7-129.5°.** Crystd from *benzene or *benzene/EtOH.

Hexafluoroacetone *[684-16-2, 34202-69-2 (3H$_2$O)]* **M 166.1, m -129°, (trihydrate m 18-21°), b -28°.** Dehydrated by passage of the vapour over P_2O_5. Ethylene was removed by passing the dried vapour through a tube containing Pyrex glass wool moistened with conc H_2SO_4. Further purification was by low temperature distn using Warde-Le Roy stills. Stored in the dark at -78°. [Holmes and Kutschke *Trans Faraday Soc* **58** 333 *1962*].

Hexafluoroacetylacetone   (1,1,1,5,5,5-hexafluoro-2,4-pentanedione)   *[1522-22-1]* **M 208.1, b 68°/736mm, 70-70.2°/760mm, 68-71°/atm, d_4^{20} 1.490, n_D^{20} 1.333.** It forms a dihydrate which has no UV spectrum compared with λmax ($CHCl_3$) 273nm (ε 7,800) for the anhydrous ketone. The dihydrate dec at ~90°. The hydrate (10g) plus anhyd $CaSO_4$ (Drierite, 30g) are heated and distd; the distillate is treated with more $CaSO_4$ and redist.   When the distillate is treated with aqueous NaOH and heated, the dihydrate crystallises on cooling. The Cu complex has **m** 135° (after sublimation). [Gilman et al. *J Am Chem Soc* **78** 2790 *1956*; Belford et al. *J Inorg Nucl Chem* **2** 11 *1956*].

Hexafluorobenzene   *[392-56-3]*   **M 186.1, m 5.1°, b 79-80°, d 1.61, n 1.378.** Main impurities are incompletely fluorinated benzenes. Purified by standing in contact with oleum for 4h at room temperature, repeating until the oleum does not become coloured. Washed several times with water, then dried with P_2O_5. Final purification was by repeated fractional crystn.

Hexafluoroethane   *[76-16-4]*   **M 138.0, b -79°.** Purified for pyrolysis studies by passage through a copper vessel containing CoF_3 at *ca* 270°, and held for 3h in a bottle with a heated (1300°) platinum wire. It was then fractionally distd. [Steunenberg and Cady *J Am Chem Soc* **74** 4165 *1962*.]

1,1,1,3,3,3-Hexafluoropropan-2-ol   *[920-66-1]*   **M 168.1, b 57-58°/760mm, d 1.4563, n^{22} 1.2750.** Distd from 3A molecular sieves, retaining the middle fraction.

Hexahydro-1H-azepine (hexamethyleneimine, Azepane)   *[111-49-9]* **M 99.2, b 70-72°/30mm, 135-138°/atm, d 0.879, n 1.466, pK^{25} 11.10 (pK^0 9.71, pK^{75} 9.71).** Purified by dissolving in Et_2O and adding ethanolic HCl until all the base separates as the white *hydrochloride* , filter, wash with Et_2O and dry (**m** 236°). The salt is dissolved in the minimum vol of H_2O and basified to pH ~ 14 with 10N KOH. The soln is extracted with Et_2O, the extract is dried over KOH, evapd and distd. The base is a **FLAMMABLE** and **TOXIC** liquid, and best kept as the salt. The *nitrate* has **m** 120-123°, *Picrate* has **m** 145-147°, and the *Tosylate* has **m** 76.5° (ligroin). [Müller and Sauerwald *Monatsh Chem* **48** 727 *1027*; Hjelt and Agback *Acta Chem Scand* **18** 194 *1964*].

Hexahydromandelic acid   *[R-(-)- 53585-93-6;   S-(+)- 61475-31-8]* **M 158.2, m 127-129°, 128-129°, 129.7°, $[\alpha]_d^{20}$ (-) and (+) 25.5° (c 1, AcOH) and $[\alpha]_d^{20}$ (-) and (+) 13.6° (c 7.6, EtOH).** For hexagonal clusters by recrystallisation from CCl_4 or Et_2O. [Wood and ComLey *J Chem Soc* 2638 *1924*; Lettré et al. *Chem Ber* **69** 1594 *1936*]. The *racemate* has **m** 137.2-137.6° (134-135°) [Smith et al. *J Am Chem Soc* **71** 3772 *1949*].

Hexamethylbenzene   *[87-85-4]*   **M 162.3, m 165-165.5°.** Sublimed, then crystd from abs EtOH, *benzene, EtOH/*benzene or EtOH/cyclohexane. Also purified by zone melting. Dried under vac over P_2O_5.

Hexamethyl(Dewar)benzene   *[7641-77-2]*   **M 162.3, m 7°, b 60°/20mm, d 0.803, n 1.4480.** Purified by passage through alumina [Traylor and Miksztal *J Am Chem Soc* **109** 2770 *1987*].

Hexamethylenediamine   *[124-09-4]*   **M 116.2, m 42°, b 46-47°/1mm, 84.9°/9mm, 100°/20mm, 204-205°/760mm, pK_1^{25} 10.24, pK_2^{25} 11.02.** Crystd in a stream of nitrogen. Sublimed in a vacuum.

Hexamethylenediamine dihydrochloride   *[6055-52-3]*   **M 189.2, m 248°.** Crystd from water or EtOH.

Hexamethylene glycol (1,6-hexanediol)   *[629-11-8]*   **M 118.2, m 41.6°, 43-45°, b 134°/10mm, 250°, n 1.458.** Fractionally crystd from its melt or from water. Distils *in vacuo*.

Hexamethylenetetramine (Urotropine, hexamine, HMTA) *[100-97-0]* **M 140.1, m 280°** **(subln), 290-292° (sealed tube, CARE), d 1.331, pK²⁵ 4.85 (6.30).** It is soluble in H_2O (67%), $CHCl_3$ (10%), EtOH (8%) and Et_2O (0.3%), and a 0.2M soln has a pH of 8.4. Dissolve in hot abs EtOH (reflux, Norite), filter using a heated funnel, cool at room temp first then in ice. Wash crysts with cold Et_2O, dry in air or under a vacuum. A further crop can be obtained by adding Et_2O to the filtrate. It sublimes above 260° without melting. The *picrate* has **m 179°** (dec). [pK 4.85:Reilley and Schmid *Anal Chem* **30** 947 *1958*; pK 6.30: Pummerer and Hofmann *Chem Ber* **56** 1255 *1923*.]

n-**Hexane** *[110-54-3]* **M 86.2, b 68.7°, d 0.660, n 1.37486, n²⁵ 1.37226.** Purification as for *n*-heptane. Modifications include the use of chlorosulfonic acid or 35% fuming H_2SO_4 instead of conc H_2SO_4 in washing the alkane, and final drying and distn from sodium hydride. Unsatd compounds can be removed by shaking the hexane with nitrating acid (58% H_2SO_4, 25% conc HNO_3, 17% water, or 50% HNO_3, 50% H_2SO_4), then washing the hydrocarbon layer with conc H_2SO_4, followed by H_2O, drying, and distg over sodium or *n*-butyl lithium. Also purified by distn under nitrogen from sodium benzophenone ketyl solubilised with tetraglyme. Also purified by passage through a silica gel column followed by distn [Kajii et al. *J Phys Chem* **91** 2791 *1987*]. FLAMMABLE liquid and possible nerve toxin.
Rapid purification: Distil, discarding the first forerun and stored over 4A molecular sieves.

1,2-Hexanediol *[6920-22-5]* **M 118.2, b 214-215, d 0.951, n 1.442.** Fractionally distd.

1-Hexene *[592-41-6]* **M 84.2, b 63°, d 0.674, n 1.388.** Purified by stirring over Na/K alloy for at least 6h, then fractionally distd from sodium under nitrogen.

cis-**2-Hexene** *[7688-21-3]* **M 84.2, b 68-70°, d 0.699, n 1.399.** Purification as for 1-hexene above.

trans-**2-Hexene** *[4050-45-7]* **M 84.2, b 65-67°, n 1.390.** Purifn as for 1-hexene above.

trans-**3-Hexene** *[13269-52-8]* **M 84.2, b 67-69°, d 0.678, n 1.393.** Purifn as for 1-hexene above.

meso-**Hexoestrol** *[84-16-2]* **M 270.4, m 185-188°.** Crystd from *benzene or aqueous EtOH.

n-**Hexyl alcohol (1-hexanol)** *[111-27-3]* **M 102.2, b 157.5°, d 0.818, n¹⁵ 1.4198, n²⁵ 1.4158.** Commercial material usually contains other alcohols which are difficult to remove. A suitable method is to esterify with hydroxybenzoic acid, recrystallise the ester and saponify. [Olivier *Recl Trav Chim, Pays-Bas* **55** 1027 *1936*.] Drying agents include K_2CO_3 and $CaSO_4$, followed by filtration and distn. (Some decomposition to the olefin occurred when Al amalgam was used as drying agent at room temperature, even though the amalgam was removed prior to distn.) If the alcohol is required anhydrous, the redistd material can be refluxed with the appropriate alkyl phthalate or succinate, as described under *Ethanol*.

n-**Hexylamine** *[111-26-2]* **M 101.2, b 131°, d 0.765, n 1.419, pK²⁵ 10.64.** Dried with, and fractionally distd from, KOH or CaH_2.

n-**Hexyl bromide** *[111-25-1]* **M 165.1, b 87-88°/90mm, 155°/743mm, d 1.176, n 1.448.** Shaken with H_2SO_4, washed with water, dried with K_2CO_3 and fractionally distd.

n-**Hexyl methacrylate** *[142-09-6]* **M 154.2, b 65-66°/4mm, 88-88.5°/14mm, d 0.8849, n 1.4320.** Purified as for *methyl methacrylate*. [IR: Hughes and Walton *J Am Chem Soc* **79** 3985 *1957*.]

Hexyltrimethylammonium bromide *[2650-53-5]* **M 224.3, m 186°.** Recrystd from acetone. Extremely *hygroscopic* salt. [McDowell and Kraus *J Am Chem Soc* **73** 2170 *1951*.]

1-Hexyne *[693-02-7]* **M 82.2, b 12.5°/75mm, 71°/760mm, d 0.7156, n 1.3989.** Distd from $NaBH_4$ to remove peroxides. Stood with sodium for 24h, then fractionally distd under reduced pressure. Also dried by repeated vac transfer into freshly activated 4A molecular sieves, followed by vacuum transfer into Na/K alloy and stirring for 1h before fractionally distilling.

2-Hexyne *[764-35-2]*   **M 82.2, b 83.8°/760mm, d 0.73146, n 1.41382.**   Purification as for 1-hexyne above.

3-Hexyne *[928-49-4]*   **M 82.2, b 81°/760mm, d 0.7231, n 1.4115.**   Purification as for 1-hexyne above.

Histamine *[51-45-6]*   **M 111.2, m 86° (sealed tube), b 167°/0.8mm, 209°/18mm, pK_1^{25} 6.02, pK_2^{25} 9.70.**   Crystd from *benzene or chloroform.

Histamine dihydrochloride *[56-92-8]*   **M 184.1, m 249-252° (244-245°).**   Crystd from aq EtOH.

S-**Histidine** *[71-00-1]*   **M 155.2, m 287°(dec), $[\alpha]_D^{25}$ -39.7° (c 1, H_2O), +13.0° (6M HCl), pK_1^{25} 1.96, pK_2^{25} 6.12, pK_3^{25} 9.17.**   Likely impurity is arginine.   Adsorbed from aqueous soln on to Dowex 50-H$^+$ ion-exchange resin, washed with 1.5M HCl (to remove other amino acids), then eluted with 4M HCl as the dihydrochloride.   Histidine is also purified as the dihydrochloride which is finally dissolved in water, the pH adjusted to 7.0, and the free zwitterionic base crystallises out on addition of EtOH.   Sol in H_2O is 4.2% at 25°.

S-**Histidine dihydrochloride** *[1007-42-7]*   **M 242.1, m 245°, $[\alpha]_D^{20}$ +47.5° (c 2, H_2O).**   Crystd from water or aqueous EtOH, and washed with acetone, then diethyl ether.   Converted to the histidine di-(3,4-dichlorobenzenesulfonate) salt by dissolving 3,4-dichlorobenzenesulfonic acid (1.5g/10mL) in the aqueous histidine soln with warming, and then the soln is cooled in ice.   The resulting crystals (**m** 280° dec) can be recrystd from 5% aqueous 3,4-dichlorobenzenesulfonic acid, then dried over $CaCl_2$ under vacuum, and washed with diethyl ether to remove excess reagent.   The dihydrochloride can be regenerated by passing the soln through a Dowex-1 (Cl$^-$ form) ion-exchange column.   The solid is obtained by evapn of the soln on a steam bath or better in a vacuum.   [Greenstein and Winitz, The Amino Acids **Vol 3** p. 1976 *1961*.]

S-**Histidine monohydrochloride (H$_2$O)** *[5934-29-2 (H$_2$O); 7048-02-4]*   **M 209.6, m 80° monohydrate, 254°(dec, anhyd), $[\alpha]_D^{25}$ +13.0° (6M HCl).**   Crystd from aqueous EtOH.

dl-**Homocysteine** *[6027-13-0]*   **M 135.2, pK_2^{25} 8.70, pK_3^{30} 10.46.**   Crystd from aqueous EtOH.

Homocystine *[462-10-2]*   **M 268.4, m 260-265°(dec), pK_1^{25} 1.59, pK_2^{25} 2.54, pK_3^{25} 8.52, pK_4^{25} 9.44.**   Crystd from water.

Homophthalic acid *[89-51-0]*   **M 180.2, m 182-183°, 189-190°, (depends on heating rate) $pK_{Est(1)}$ ~3.5, $pK_{Est(2)}$ ~4.3.**   Crystd from boiling water (25mL/g).   Dried at 100°.

Homopiperazine (1,4-diazepane) *[505-66-8]* **M 100.2, m 38-40°, 43°, b 60°/10mm, 92°/50mm, 169°/atm, pK_1^{20} 6.70, pK_2^{20} 10.41.**   Purified by fractionation through a column of 10 theoretical plates with a reflux ratio of 3:1.   It boiled at 169° and the cool distillate crystallises in plates **m** 43°.   [Poppelsdorf and Myerly *J Org Chem* **26** 131 *1961*.]   Its pKa values are 6.89 and 10.65 at 40°, and 6.28 and 9.86 at 40° [Pagano et al. *J Phys Chem* **65** 1062 *1961*].   The *1,4-bis(4-bromobenzoyl* derivative has **m** 194-198° (from EtOH); the *hydrochloride* has **m** 270-290° (from EtOH) and the *picrate* has **m** 265° (dec) [Lloyd et al. *J Chem Soc (C)* 780 *1966*].

L-Homoserine (2-amino-4-hydroxybutyric acid) *[672-15-1]*   **M 119.1, m 203°, $[\alpha]_D^{26}$ +18.3° (in 2M HCl), $pK_{Est(1)}$ ~2.1, $pK_{Est(2)}$ ~9.3.**   Likely impurities are *N*-chloroacetyl-L-homoserine, *N*-chloroacetyl-D-homoserine, L-homoserine, homoserine lactone, homoserine anhydride (formed in strong solns of homoserine if slightly acidic).   Cyclises to the lactone in strongly acidic soln.   Crystd from water by adding 9 volumes of EtOH.

Homoveratronitrile (3,4-dimethoxybenzylnitrile) *[93-17-4]* **M 177.2, m 62-64°, 68°, b 184°/20mm, 195-196°/2mm, 208°/atm.**   Its solubility is 10% in MeOH. and has been recrystd from

EtOH or MeOH. Purified by distillation followed by recrystn. [Niederl and Ziering *J Am Chem Soc* **64** 885 *1952*; Julian and Sturgis *J Am Chem Soc* **57** 1126 *1935*.]

Homoveratrylamine (2-[3,4-dimethoxyphenyl]ethylamine) *[120-20-7]* **M 181.2, b 99.3-101.3°/0.5mm, 168-170°/15mm, d_4^{20} 1.091, n_D^{20} 1.5460, pK_{Est} ~9.8.** Purified by fractionation through an efficient column in an inert atmosphere as it is a relatively strong base. [Horner and Sturm *Justus Liebigs Ann Chem* **608** 12819 *1957*; Jung et al. *J Am Chem Soc* **75** 4664 *1953*.] The *hydrochloride* has **m** 152°, 154°, 156° (from EtOH, Me$_2$CO or EtOH/Et$_2$O) and the *picrate* has **m** 165-167° dec, and the *4-nitrobenzoyl* derivative has **m** 147° [Buck *J Am Chem Soc* **55** 2593 *1933*].

Hordenine {4-[(2-dimethylamino)ethyl]phenol} *[539-15-1]* **M 165.2, m 117-118°, pK^{25} 9.46 (OH).** Crystd from EtOH or water.

Humulon *[26472-41-3]* **M 362.5, m 65-66.5°, $[\alpha]_D^{26}$ -212° (95% EtOH).** Crystd from Et$_2$O.

Hyamine 1622 [(diisobutylphenoxyethoxyethyl)dimethylbenzylammonium chloride, benzethionium chloride] *[121-54-0]* **M 448.1, m 164-166° (sinters at 120°, monohydrate).** Crystd from boiling acetone after filtering, or from CHCl$_3$-pet ether. The ppte was filtered off, washed with diethyl ether and dried for 24h in a vacuum desiccator.

Hydantoin (2,4-dihydroxyimidazole) *[461-72-3]* **M 100.1, m 216°, 220°, pK^{25} 9.15.** Crystd from MeOH. The *diacetate* has **m** 104-105°.

Hydrazine N,N'-dicarboxylic acid diamide *[110-21-4]* **M 116.1, m 248°.** Crystd from water and dried in vac over P$_2$O$_5$.

4-Hydrazinobenzoic acid *[619-67-0]* **M 152.2, m 217° (dec), pK^{25} 4.13.** Crystd from water.

1-Hydrazinophthalazine hydrochloride (hydralazine hydrochloride) *[304-20-1]* **M 196.6, m 172-173°, pK^{20} 6.57.** Crystd from MeOH.

2-Hydrazinopyridine *[4930-98-7]* **M 109.1, m 41-44°, 46-47°, 49-50°, b 105°/0.5mm, 128-135°/13mm.** Purified by distn and by recrystn from Et$_2$O-hexane. [Kauffmann et al. *Justus Liebigs Ann Chem* **656** 103 *1962*, Potts and Burton *J Org Chem* **31** 251 *1966*.] The *mono-hydrochloride* has **m** 183° (dec) from aq HCl and the *di-hydrochloride* has **m** 214-215°.

Hydrazobenzene *[122-66-7]* **M 184.2.** See 1,1-diphenylhydrazine on p. 225.

Hydrobenzamide [1-phenyl-N,N'-bis(phenylmethylene)-methanediamine] *[92-29-5]* **M 298.4, m 101-102°, 107-108°.** Crystd from absolute EtOH or cyclohexane/*benzene. Dried under vacuum over P$_2$O$_5$. [Pirrone *Gazz Chim Ital* **67** 534 *1937*.]

dl-**Hydrobenzoin** *[492-70-6]* **M 214.3, m 120°.** Crystd from diethyl ether/pet ether.

meso-**Hydrobenzoin** *[579-43-1]* **M 214.3, m 139°.** Crystd from EtOH or water.

Hydroquinone (1,4-dihydroxybenzene, quinol) *[123-31-9]* **M 110.1, m 175.4, 176.6°, pK$_1^{20}$ 9.91, p K$_2^{20}$ 11.56.** Crystd from acetone, *benzene, EtOH, EtOH/*benzene, water or acetonitrile (25g in 30mL), preferably under nitrogen. Dried under vacuum. [Wolfenden et al. *J Am Chem Soc* **109** 463 *1987*.]

4'-Hydroxyacetanilide *[103-90-2]* See 4-acetamidophenol on p. 83.

p-**Hydroxyacetophenone** *[99-93-4]* **M 136.2, m 109°, pK^{25} 8.01.** Crystd from diethyl ether, aqueous EtOH or *benzene/pet ether.

4-Hydroxyacridine *[18123-20-1]* **M 195.2, m 116.5° pK$_1^{15}$ 5.28, pK$_2^{15}$ 9.75.** Crystd from EtOH.

5-Hydroxyanthranilic acid *[548-93-6]* **M 153.1, m >240°(dec),** λ_{max} **298nm, log ε 3000 (0.1M HCl), pK$_1^{20}$ 2.7, pK$_2^{20}$ 5.37, pK$_3^{20}$ 10.12.** Crystd from water. Sublimes below its melting point in a vacuum.

erythro-**3-Hydroxy-*RS*-aspartic acid** *[6532-76-9]* **M 149.1, pK$_1^{25}$ 1.91, pK$_2^{25}$ 3.51, pK$_3^{25}$ 9.11.** Likely impurities are 3-chloromalic acid, ammonium chloride, *threo*-3-hydroxyaspartic acid. Crystd from water.

m-**Hydroxybenzaldehyde** *[100-83-4]* **M 122.1, m 108° pK$_1^{25}$ 8.98, pK$_2^{25}$ 15.81.** Crystd from water.

p-**Hydroxybenzaldehyde** *[123-08-0]* **M 122.1, m 115-116°, pK^{25} 7.61.** Crystd from water (containing some H$_2$SO$_4$). Dried over P$_2$O$_5$ under vacuum.

m-**Hydroxybenzoic acid** *[99-06-9]* **M 138.1, m 200.8°, pK$_1^{25}$ 4.08, pK$_2^{25}$ 9.98.** Crystd from absolute EtOH.

p-**Hydroxybenzoic acid** *[99-96-7]* **M 138.1, m 213-214°, pK$_1^{25}$ 4.50, pK$_2^{25}$ 9.11.** Crystd from water.

p-**Hydroxybenzonitrile** *[767-00-0]* **M 119.1, m 113-114°.** See *p*-cyanophenol on p. 176.

4-Hydroxybenzophenone *[1137-42-4]* **M 198.2, m 133.4-133.8°, pK^{25} 7.95.** See *p*-benzoylphenol on p. 126.

2-Hydroxybenzothiazole *[934-34-9]* **M 183.1, m 117-118°** Crystd from aqueous EtOH or water. [Dryland and Sheppard *J Chem Soc Perkin Trans 1* 125 *1986*.]

1-Hydroxybenzotriazole hydrate (HOBt) *[2592-95-2]* **M 135.1, m 159-160°.** Crystd from aqueous EtOH or water. [Dryland and Sheppard *J Chem Soc Perkin Trans 1* 125 *1986*.]
§ A polystyrene supported version is available.

2-Hydroxybenzyl alcohol *[90-01-7]* **M 124.1, m 87°, pK^{25} 9.84.** Crystd from water or *benzene.

3-Hydroxybenzyl alcohol *[620-24-6]* **M 124.1, m 71°, pK$_{Est}$ ~9.8.** Crystd from *benzene.

4-Hydroxybenzyl alcohol *[623-05-2]* **M 124.1, m 114-115°, pK^{25} 9.73.** Crystd from water.

2-Hydroxybiphenyl *[90-43-7]* **M 170.2, m 56°, b 145°/14mm, 275°/760mm, pK^{20} 10.01.** Crystd from pet ether.

4-Hydroxybiphenyl (4-phenylphenol) *[92-69-3]* **M 170.2, m 164-165°, b 305-308°/760mm, pK^{23} 9.55.** Crystd from aqueous EtOH, aq EtOH, *C$_6$H$_6$, and vac dried over CaCl$_2$ [Buchanan et al. *J Am Chem Soc* 108 7703 *1986*].

3-Hydroxy-2-butanone (acetoin) *[513-86-0]* **M 88.1, b 144-145°, [m 100-105° dimer].** Washed with EtOH until colourless, then with diethyl ether or acetone to remove biacetyl. Air dried by suction and further dried in a vacuum desiccator.

(±)-α-Hydroxy-γ-butyrolactone *[19444-84-9]* **M 102.1, b 84°/0.2mm, 133°/10mm, d$_4^{20}$ 1.310, n$_D^{20}$ 1.4656.** It has been purified by repeated fractionation, forms a colourless liquid. It has to be distd at high vacuum otherwise it will dehydrate. The *acetoxy* derivative has b 94°/0.2mm, [NMR: Daremon and Rambaud *Bull Soc Chim Fr* 294 *1971*; Schmitz et al. *Chem Ber* 108 1010 *1975*.]

4-Hydroxycinnamic acid (*p*-coumaric acid) *[501-98-4]* M 164.2, m 210-213°, 214-215°, 215° pK_1^{25} 4.64, pK_2^{25} 9.45. Crystd from H_2O (charcoal). Needles from conc aqueous solutions as the *anhydrous acid*, but from hot dilute solutions the *monohydrate acid* separates on slow cooling. The acid (33g) has been recrystd from 2.5L of H_2O (1.5g charcoal) yielding 28.4g of recrystd acid, m 207°. It is insol in *C_6H_6 or pet ether. The UV in 95% EtOH has λ_{max} 223 and 286nm (ϵ 14,450 and 19000 $M^{-1}cm^{-1}$). [UV Wheeler and Covarrubias *J Org Chem* **28** 2015 *1963*; Corti *Helv Chim Acta* **32** 681 *1949*.]

4-Hydroxycoumarin *[1076-38-6]* M 162.1, m 206°, pK_{Est} ~9.0. Crystd from water and dried in a vacuum desiccator over Sicapent.

3-(4-Hydroxy-3,5-dimethoxyphenyl)acrylic acid *[530-59-6]* M 234.1, m 204-205°(dec), $pK_{Est(1)}$~4.6, $pK_{Est(2)}$ ~9.3. Crystd from water.

4-Hydroxydiphenylamine *[122-37-2]* M 185.2, m 72-73°, pK_{Est} ~10.0. Crystd from chlorobenzene/pet ether.

12-Hydroxydodecanoic acid *[505-95-3]* M 216.3, m 86-88°, pK_{Est} ~4.8. Crystd from toluene [Sadowik et al. *J Am Chem Soc* **108** 7789 *1986*].

2-Hydroxy-4-(*n*-dodecyloxy)benzophenone *[2985-59-3]* M 382.5, m 50-52°, pK_{Est} ~7.1. Recryst from *n*-hexane and then 10% (v/v) EtOH in acetonitrile [Valenty et al. *J Am Chem Soc* **106** 6155 *1984*].

***N*-[2-Hydroxyethyl]ethylenediamine** [2-(2-aminoethylamino)ethanol] *[111-41-1]* M 104.1, b 91.2°/5mm, 238-240°/752mm, n 1.485, d 1.030, pK_1^{20} 3.75, pK_2^{20} 9.15. Distilled twice through a Vigreux column. Redistilled from solid NaOH, then from CaH_2. Alternatively, it can be converted to the dihydrochloride and recrystallised from water. It is then dried, mixed with excess of solid NaOH and the free base distilled from the mixture. It is finally redistilled from CaH_2. [Drinkard, Bauer and Bailar *J Am Chem Soc* **82** 2992 *1960*.]

***N*-[2-Hydroxyethyl]ethylenediaminetriacetic acid** (HEDTA) *[150-39-0]* M 278.3, m 212-214°(dec), pK_1^{20} 2.51, pK_2^{20} 5.31, pK_3^{20} 9.86. Crystd from warm H_2O, after filtering, by addition of 95% EtOH and allowing to cool. The crystals, collected on a sintered-glass funnel, were washed three times with cold absolute EtOH, then again crystd from H_2O. After leaching with cold H_2O, the crystals were dried at 100° under vacuum. [Spedding, Powell and Wheelwright *J Am Chem Soc* **78** 34 *1956*.]

***N*-Hydroxyethyliminodiacetic acid** (HIMDA) *[93-62-9]* M 177.2, m 181°(dec), pK_1^{25} 2.16, pK_2^{25} 8.72, pK_3^{25} 13.7 (OH). Crystd from water.

2-Hydroxyethylimino-tris(hydroxymethyl)methane (MONO-TRIS) *[7343-51-3]* M 165.2, m 91°, pK_{Est} ~9.8. Crystd twice from EtOH. Dried under vacuum at 25°.

2-Hydroxyethyl methacrylate *[868-77-9]* M 130.1, b 67°/3.5mm, d 1.071, n 1.452. Dissolved in water and extracted with *n*-heptane to remove ethylene glycol dimethacrylate (checked by gas-liquid chromatography and by NMR) and distilled twice under reduced pressure [Strop, Mikes and Kalal *J Phys Chem* **80** 694 *1976*].

***N*-2-Hydroxyethylpiperazine-*N'*-2-ethanesulfonic acid** (HEPES) *[7365-45-9]* M 238.3, pK^{20} 7.55. Crystd from hot EtOH and water.

3-Hydroxyflavone *[577-85-5]* M 238.2, m 169-170°, 171-172°. Recrystd from MeOH, EtOH or hexane. Also purified by repeated sublimation under high vacuum, and dried by high vacuum pumping for at least one hour [Bruker and Kelly *J Phys Chem* **91** 2856 *1987*].

ß-Hydroxyglutamic acid *[533-62-0]* **M 163.1, m 100°(dec), pK$_1^{25}$ 2.27, pK$_2^{25}$ 4.29, pK$_3^{25}$ 9.66.** Crystd from water.

4-Hydroxyindane *[1641-41-1]* **M 134.2, m 49-50°, b 120°/12mm, pK^{25} 10.32.** Crystd from pet ether. *Acetyl* deriv has **m** 30-32° (from EtOH), **b** 127°/14mm. [Dallacker et al. *Chem Ber* **105** 2568 *1972*.]

5-Hydroxyindane *[1470-94-6]* **M 134.2, m 55°, b 255°/760mm, pK$_{Est}$ ~10.4.** Crystd from pet ether.

5-Hydroxy-L-lysine monohydrochloride *[32685-69-1]* **M 198.7, [α]$_D^{25}$ +17.8° (6M HCl), pK$_2^{25}$ 8.85, pK$_3^{25}$ 9.83.** Likely impurities are 5-*allo*-hydroxy-(D and L)-lysine, histidine, lysine, ornithine. Crystd from water by adding 2-9 volumes of EtOH stepwise.

4-Hydroxy-3-methoxyacetophenone *[498-02-2]* **M 166.2, m 115°, pK$_{Est}$ ~7.9.** Crystd from water, or EtOH/pet ether.

4-Hydroxy-3-methoxycinnamic acid (ferulic acid) *[1135-24-6]* **M 194.2, m 174°, pK$_1^{25}$ 4.58, pK$_2^{25}$ 9.39.** Crystd from H$_2$O.

1-Hydroxymethyladamantane *[770-71-8]* **M 166.3, m 115°.** Dissolve in Et$_2$O, wash with aqueous 0.1N NaOH and H$_2$O, dry over CaCl$_2$, evaporate and recryst residue from aqueous MeOH. [*Chem Ber* **92** 1629 *1959*.]

17β-Hydroxy-17α-methyl-3-androsterone (Mestanolone) *[521-11-9]* **M 304.5°, m 192-193°.** Crystd from ethyl acetate.

4-Hydroxy-2-methylazobenzene *[1435-88-7]* **M 212.2, m 100-101°, pK$_{Est}$ ~9.5.** Crystd from hexane.

4-Hydroxy-3-methylazobenzene *[621-66-9]* **M 212.2, m 125-126°.** Crystd from hexane.

3-Hydroxy-4-methylbenzaldehyde *[57295-30-4]* **M 136.1, m 116-117°, b 179°/15mm, pK$_{Est}$ ~10.2.** Crystd from water.

***dl*-2-Hydroxy-2-methylbutyric acid** *[3739-30-8]* **M 118.1, m 72-73°, pK^{25} 3.73.** Crystd from *benzene, and sublimed at 90°.

***dl*-2-Hydroxy-3-methylbutyric (α-hydroxyisovaleric) acid** *[600-37-3]* **M 118.1, m 86°, pK$_{Est}$ ~3.9.** Crystd from ether/pentane.

***R*-γ-Hydroxymethyl-γ-butyrolactone** *[52813-63-5]* **M 116.1, b 101-102°/0.048mm, d$_4^{20}$ 1.2238, n$_D^{20}$ 1.471, [α]$_{546}^{20}$ -38°, [α]$_D^{20}$ -33° (c 3, EtOH), [α]$_D^{30}$ -53.5° (c 3, EtOH).** Purified by column chromatography in Silica gel 60 (Merck 70-230 mesh) and eluting with 7% EtOH-73% CHCl$_3$. IR (film): 3400 (OH), 1765 (C=O) and 1180 (COC) cm^{-1}. [Eguchi and Kakuta *Bull Chem Soc Jpn* **47** 1704 *1974*; IR and NMR: Ravid et al. *Tetrahedron* **34** 1449 *1978*.]

7-Hydroxy-4-methylcoumarin (4-methylumbelliferone) *[90-33-5]* **M 176.2, m 185-186°, pK$_{Est}$ ~10.0.** Crystd from absolute EtOH. (See also entry on p. 548 in Chapter 6.)

2-Hydroxymethyl-12-crown-4 *[75507-26-5]* **M 206.2, d$_4^{20}$ 1.186, n$_D^{20}$ 1.480.** Purified by chromatography on Al$_2$O$_3$ with EtOAc as eluent to give a *hygroscopic* colourless oil with IR 3418 (OH) and 1103 (COC) cm^{-1}, NMR δ 3.70 (s). [Pugia et al. *J Org Chem* **52** 2617 *1987*.]

***S*-(-)-5-Hydroxymethyl-2(5H)-furanone** *[78508-96-0]* **M 114.1, 39-42°, 40-44°, b 130°/0.3mm, [α]$_{546}^{20}$ -180°, [α]$_D^{20}$ -148° (c 1.4, H$_2$O).** It has been purified by chromatography on

Silica gel using hexane-EtOAc (1:1) to give a colourless oil which was distd using a Kügelrohr apparatus and the distillate crystallises on cooling. It has R_F 0.51 on Whatman No 1 paper using pentan-1-ol and 85% formic acid (1:1) and developing with ammoniacal AgNO$_3$. [Boll *Acta Chem Scand* **22** 3245 *1968*; NMR: Oppolzer et al. *Helv Chim Acta* **68** 2100 *1985*.]

5-(Hydroxymethyl)furfural *[67-47-0]* **M 126.1, m 33.5°, b 114-116°/1mm.** Crystd from diethyl ether/pet ether.

3-Hydroxy-3-methylglutaric acid (Meglutol) *[503-49-1]* **M 162.1, m 99-102°, 108-109°, 100°, pK$_{Est(1)}$ ~4.0, pK$_{Est(2)}$~5.0.** Recrystd from diethyl ether/hexane and dried under vac at 60° for 1h.

dl-**3-Hydroxy-*N*-methylmorphinan** *[297-90-5]* **M 257.4, m 251-253°.** Crystd from anisole + aqueous EtOH.

6-Hydroxy-2-methyl-1,4-naphthaquinone *[633-71-6]* **M 188.2, pK$_{Est}$ ~10.0.** Crystd from aqueous EtOH. Sublimes on heating.

4-Hydroxy-4-methyl-2-pentanone *[123-42-2]* **M 116.2, b 166°, d 0.932, n 1.4235, n^{25} 1.4213.** Loses water when heated. Can be dried with CaSO$_4$, then fractionally distd under reduced pressure.

17α-Hydroxy-6α-methylprogesterone (Medroxyprogesterone) *[520-85-4]* **M 344.5, m 220°, [α]$_D^{25}$ +75°.** Crystd from chloroform.

2-Hydroxy-2-methylpropionic acid (α-hydroxyisobutyric acid, 2-methyllactic acid)) *[594-61-6]* **M 104.1, m 79°, b 114°/12mm, 212°/760mm, pK^{25} 3.78.** Distd in steam, crystd from diethyl ether or *benzene, sublimed at 50° and dried under vacuum.

8-Hydroxy-2-methylquinoline *[826-81-3]* **M 159.2, m 74-75°, b 266-267°, pK$_1^{25}$ 5.61, pK$_2^{25}$ 10.16.** Crystd from EtOH or aqueous EtOH.

2-Hydroxy-1-naphthaldehyde *[708-06-5]* **M 172.2, m 82°, b 192°/27mm, pK$_{Est}$ ~7.8.** Crystd from EtOH (1.5mL/g), ethyl acetate or water.

2-Hydroxy-1-naphthaleneacetic acid *[10441-45-9]* **M 202.2, pK$_{Est(1)}$ ~4.2, pK$_{Est(2)}$~8.3.** Treated with activated charcoal and crystd from EtOH/water (1:9, v/v). Dried under vacuum, over silica gel, in the dark. Stored in the dark at -20° [Gafni, Modlin and Brand *J Phys Chem* **80** 898 *1976*]. Forms a lactone (**m** 107°) readily.

6-Hydroxy-2-naphthalenepropionic acid *[553-39-9]* **M 216.2, m 180-181°, pK$_{Est(1)}$ ~4.6, pK$_{Est(2)}$~9.0** Crystd from aqueous EtOH or aqueous MeOH.

3-Hydroxy-2-naphthalide *[92-77-3]* **M 263.3, m 248.0-248.5°, CI 37505.** Crystd from xylene [Schnopper, Broussard and La Forgia *Anal Chem* **31** 1542 *1959*].

3-Hydroxy-2-naphtho-4'-chloro-*o*-toluidide *[92-76-2]* **M 311.8, m 243.5-244.5°.** Crystd from xylene [Schnopper, Broussard and La Forgia *Anal Chem* **31** 1542 *1959*].

3-Hydroxy-2-naphthoic-1'-naphthylamide *[123-68-3]* **M 314.3, m 217-.5-218.0°.** Crystd from xylene [Schnopper, Broussard and La Forgia *Anal Chem* **31** 1542 *1959*].

3-Hydroxy-2-naphthoic-2'-naphthylamide *[136-64-8]* **M 305.3, m 243.5-244.5°, and other naphthol AS derivatives.** Crystd from xylene [Schnopper, Broussard and La Forgia *Anal Chem* **31** 1542 *1959*].

2-Hydroxy-1,4-naphthaquinone *[83-72-7]* **M 174.2, m 192°(dec), pK$_1^{25}$-5.6 (C=O protonation), pK$_2^{25}$4.00 (phenolic OH).** Crystd from *benzene.

5-Hydroxy-1,4-naphthaquinone (Juglone) *[481-39-0]* **M 174.2, m 155°, 164-165°, pK 8.7.** Crystd from *benzene/pet ether or pet ether.

6-Hydroxy-2-naphthyl disulfide *[6088-51-3]* **M 350.5, m 221-222°, 226-227°, pK$_{Est}$ ~9.0.** Crystallises as leaflets from AcOH and is slightly soluble in EtOH, and AcOH, but is soluble in *C$_6$H$_6$ and in alkalis to give a yellow soln. [Zincke and Dereser *Chem Ber* **51** 352 *1918*.] The *acetoxy* derivative has **m** 198-200° (from AcOH or dioxane-MeOH) and the *diacetyl* derivative has **m** 167-168° (from AcOH). A small amount of impure disulfide can be purified by dissolving in a small volume of Me$_2$CO and adding a large volume of toluene, filter rapidly and concentrate to one third of its volume. The hot toluene soln is filtered rapidly from any tarry residue, and crystals separate on cooling. After recrystn from hot acetic acid gives crystals **m** 220-223° [Barrett and Seligman *Science* **116** 323 *1952*].

2-Hydroxy-5-nitrobenzyl bromide *[772-33-8]* **M 232.0, m 147°, pK$_{Est}$ ~8.0.** Crystd from *benzene or *benzene/ligroin.

4-Hydroxy-2-*n*-nonylquinoline *N*-oxide *[316-66-5]* **M 287.4, m 148-149°, pK$_{Est}$ ~6.0.** Crystd from EtOH.

***N*-Hydroxy-5-norbornene-2,3-dicarboxylic acid imide** *[21715-90-2]* **M 179.2, m 165-166°, 166-169°, pK$_{Est}$~6** Dissolve in CHCl$_3$, filter, evaporate and recrystallise from EtOAc. IR (nujol): 1695, 1710 and 1770 (C=O), and 3100 (OH) cm^{-1}. *O-Acetyl* derivative has **m** 113-114° (from EtOH) with IR bands at 1730, 1770 and 1815 cm^{-1} only, and the *O-benzoyl* derivative has **m** 143-144° (from propan-2-ol or *C$_6$H$_6$). [Bauer and Miarka *J Org Chem* **24** 1293 *1959*; Fujino et al. *Chem Pharm Bull Jpn* **22** 1857 *1974*].

DL-*erythro*-3-Hydroxynorvaline (2-amino-3-hydroxypentanoic acid) *[34042-00-7]* **M 133.2, m 257-259° (dec), 263° (dec), pK$_1^{20}$2.32, pK$_2^{20}$9.12.** Purified by recrystn from aqueous EtOH. The *Cu salt* has **m** 255-256° (dec), the *benzoyl* derivative has **m** 181°, and the *N-phenylcarbamoyl* derivative has **m** 164°. [Buston et al. *J Biol Chem* **204** 665 *1953*].

2-Hydroxyoctanoic acid (2-hydroxycaprylic acid) *[617-73-2]* **M 160.2, m 69.5°, b 160-165°/10mm, pK$_{Est}$ ~3.7.** Crystd from EtOH/pet ether or ether/ligroin.

1-Hydroxyphenazine (Hemipyocyanine) *[528-71-2]* **M 196.2, m 157-158°, pK$_1^{25}$1.61, pK$_2^{15}$8.33.** Chromatographed on acidic alumina with *benzene/ether. Crystd from *benzene/heptane, and sublimed.

2-Hydroxyphenylacetic acid *[614-75-5]* **M 152.2, m 148-149°, b 240-243°/760mm, pK$_{Est(1)}$~4.3, pK$_{Est(2)}$~10.1.** Crystd from ether or chloroform (**m** from latter is always lower).

3-Hydroxyphenylacetic acid *[621-37-4]* **M 152.2, m 137°, pK$_{Ext(1)}$~4.3, pK$_{Ext(2)}$~10.** Crystd from *benzene/ligroin.

4-Hydroxyphenylacetic acid *[156-38-7]* **M 152.2, m 150-151°, 152°, pK$_1$ 4.28, pK$_2$ 10.1.** Crystd from water or Et$_2$O/pet ether.

2-(2-Hydroxyphenyl)benzothiazole *[3411-95-8]* **M 227.2, m 132-133°, b 173-179°/3mm.** Recrystd several times from aqueous EtOH and by sublimation. [Itoh and Fujiwara *J Am Chem Soc* **107** 1561 *1985*.]

2-(2-Hydroxyphenyl)benzoxazole *[835-64-3]* **M 211.2, m 127°, b 338°/760mm.** Recrystd several times from aqueous EtOH and by sublimation. [Itoh and Fujiwara *J Am Chem Soc* **107** 1561 *1985*.]

3-Hydroxy-2-phenylcinchoninic acid *[485-89-2]* **M 265.3, m 206-207°(dec).** Crystd from EtOH.

N-(*p*-Hydroxyphenyl)glycine *[22818-40-2]* **M 167.2, m >240°(dec), pK$_{Est(1)}$~2, pK$_{Est(2)}$~4.5, pK$_{Est(3)}$~10.3.** Crystd from water.

N-(4-Hydroxyphenyl)-3-phenylsalicylamide *[550-57-2]* **M 305.3, m 183-184°, pK$_{Est}$ ~9.5.** Crystd from aqueous MeOH.

L-2-Hydroxy-3-phenylpropionic acid (3-phenyl lactic acid) *[20312-36-1]* **M 166.2, m 125-126°, [α]$_D^{12}$ -18.7° (EtOH), pK see below.** Crystd from water, MeOH, EtOH or *benzene.

dl-2-Hydroxy-3-phenylpropionic acid *[828-01-3]* **M 166.2, m 97-98°, b 148-150°/15mm, pK$_{Est}$ ~3.7.** Crystd from *benzene or chloroform.

3-*p*-Hydroxyphenylpropionic acid (phloretic acid) *[501-97-3]* **M 166.2, m 129-130°, 131-133°, pK$_{Est(1)}$~4.7, pK$_{Est(2)}$~10.1.** Crystd from ether or H$_2$O.

p-Hydroxyphenylpyruvic acid *[156-39-8]* **M 180.2, m 220°(dec), pK$_{Est}$ ~2.3.** Crystd three times from 0.1M HCl/EtOH (4:1, v/v) immediately before use [Rose and Powell *Biochem J* **87** 541 *1963*], or from Et$_2$O. The *3,4-Dinitrophenylhydrazone* has **m** 178°.

N-Hydroxyphthalimide *[524-38-9]* **M 163.1, m 230°, ~235° (dec), 237-240°, pK^{30} 7.0.** Dissolve in H$_2$O by adding Et$_3$N to form the salt and while hot acidify, cool and pour into a large volume of H$_2$O. Filter off the solid, wash with H$_2$O, dry over P$_2$O$_5$ in vacuum. [Nefken And Teser *J Am Chem Soc* **83** 1263 *1961*; Fieser **1** 485 *1976*; Nefkens et al. *Recl Trav Chim Pays-Bas* **81** 683 *1962*] The *O-acetyl* derivative has **m** 178-180° (from EtOH).

3-β-Hydroxy-5-pregnen-20-one (pregnenolone) *[145-13-1]* **M 316.5, m 189-190°, [α]$_D^{20}$ +30° (EtOH), [α]$_{546}$ +34° (c 1, EtOH).** Crystd from MeOH.

17α-Hydroxyprogesterone *[604-09-1]* **M 330.5, m 222-223°, [α]$_{546}^{20}$ +141° (c 2, dioxane), λ$_{max}$ 240nm.** Crystd from acetone or EtOH. *Acetate:* **m** 239-240° and *caproate:* **m** 119-121° crystallised from CHCl$_3$/MeOH.

R-(+)-3-Hydroxyprolidine *[2799-21-5]* **M 87.1, b 215-216°, d$_4^{20}$ 1.078, n$_D^{20}$ 1.490, [α]$_D^{20}$ +6.5° (c 1.5, MeOH), pK$_{Est}$ ~10.1.** Purify by repeated distn. The *hydrochloride* has -ve rotation and the *dimethiodide* has **m** 230° and [α]$_D^{24}$ -8.02°. [Suyama and Kanno *Yakugaku Zasshi (J Pharm Soc Japan)* **85** 531 *1965*; Uno et al. *J Heterocycl Chem* **24** 1025 *1987*; Flanagan and Joullie *Heterocycles* **26** 2247 *1987*.]

trans-L-4-Hydroxyproline *[51-35-4]* **M 131.1, m 274°, [α]$_D^{20}$ -76.0° (c 5, H$_2$O), pK$_1^{25}$ 1.86, pK$_2^{25}$ 9.79.** Crystd from MeOH/EtOH (1:1). Separation from normal *allo*-isomer can be achieved by crystn of the copper salts [see *Biochem Prep* **8** 114 *1961*].

4'-Hydroxypropiophenone *[70-70-2]* **M 150.2, m 149°, pK$_{Est}$ ~10.** Crystd from water.

2-(α-Hydroxypropyl)piperidine (2-piperidinepropanol) *[24448-89-3]* **M 143.2, m 121°, b 226°, pK$_{Est}$ ~10.2.** Crystd from ether.

7-(2-Hydroxypropyl)theophylline (Proxyphylline) *[603-00-9]* **M 238.2, m 135-136°.** Crystd from EtOH.

6-Hydroxypurine (hypoxanthine) *[68-94-0]* **M 136.1, m 150°(dec), pK$_1^{20}$ 8.96, pK$_2^{20}$ 12.18.** Crystd from hot water. Dried at 105°.

2-Hydroxypyridine (2-pyridone) *[142-08-5]* **M 95.1, m 105-107°, b 181-185°/24mm, ε$_{293nm}$ 5900 (H$_2$O) pK$_1^{25}$ 1.25, pK$_2^{25}$ 11.99.** Distd under vacuum to remove coloured impurity, then crystd from *benzene, CCl$_4$, EtOH or CHCl$_3$/diethyl ether. It can be sublimed under high vacuum. [DePue et al. *J Am Chem Soc* **107** 2131 *1985*.]

3-Hydroxypyridine *[109-00-2]* **M 95.1, m 129° pK$_1^{25}$ 5.10, pK$_2^{25}$ 8.6.** Crystd from water or EtOH.

4-Hydroxypyridine (4-pyridone) *[626-64-2]* **M 95.1, m 65°(hydrate), 148.5° (anhydr), b >350°/760mm, pK$_1^{20}$ 3.20, pK$_2^{20}$ 11.12.** Crystd from H$_2$O. Loses H$_2$O on drying *in vacuo* over H$_2$SO$_4$. Stored over KOH because it is *hygroscopic*.

2(6)-Hydroxypyridine-5(3)-carboxylic acid (6-hydroxynicotinic acid) *[5006-66-6]* **M 139.1, m 304°(dec), pK^{20} 3.82.** Crystd from water.

4-Hydroxypyridine-2,6-dicarboxylic acid (chelidamic acid) *[138-60-3]* **M 183.1, m 254°(dec), pK$_1^{22}$ 1.9, pK$_2^{22}$ 3.18, pK$_3^{22}$ 10.85.** Crystd from water.

2-Hydroxypyrimidine *[557-01-7]* **M 96.1, m 179-180°, pK$_1^{20}$ 2.15, pK$_2^{20}$ 9.2.** Crystd from EtOH or ethyl acetate.

4-Hydroxypyrimidine *[4562-27-0]* **M 96.1, m 164-165°, pK$_1^{20}$ 1.66, pK$_2^{20}$ 8.63.** Crystd from *benzene or ethyl acetate.

2-Hydroxypyrimidine hydrochloride *[38353-09-2]* **M 132.5, m 205°(dec).** Crystd from EtOH.

2-Hydroxyquinoline (carbostyril) *[59-31-4]* **M 145.2, m 199-200°, pK$_1^{20}$ -0.31, pK$_2^{20}$ 11.76.** Crystd from MeOH.

8-Hydroxyquinoline (oxine, 8-quinolinol) *[148-24-3]* **M 145.2, m 71-73°, 75-76°, 76°, b ~ 267° pK$_1^{25}$ 4.91, pK$_2^{25}$ 9.81.** Crystd from hot EtOH, acetone, pet ether (b 60-80°) or water. Crude oxine can be purified by pptn of copper oxinate, followed by liberation of free oxine with H$_2$S or by steam distn after acidification with H$_2$SO$_4$. Stored in the dark. Forms metal complexes. [Manske et al. *Can J Research* **27F** 359 *1949;* Phillips *Chem Rev* **56** 271 *1956*.]

8-Hydroxyquinoline-5-sulfonic acid (H$_2$O) *[84-88-8]* **M 243.3, m >310° pK$_1^{25}$ 4.09, pK$_2^{25}$ 8.66.** Crystd from water or dil HCl (*ca* 2% by weight).

5-Hydroxysalicylic acid *[490-79-9]* **M 154.1.** See 2,5-dihydroxybenzoic acid on p. 207.

trans-**4-Hydroxystilbene** *[6554-98-9]* **M 196.3, m 189°.** Crystd from *benzene or acetic acid.

N-**Hydroxysuccinimide** *[6066-82-6]* **M 115.1, m 96-98°, pK 6.0.** Recrystd from EtOH/ethyl acetate [Manesis and Goodmen *J Org Chem* **52** 5331 *1987*].

dl-**2-Hydroxytetradecanoic acid** *[2507-55-3]* **M 244.4, m 81-82°, pK$_{Est}$ ~3.7.** Crystd from chloroform.

R-**2-Hydroxytetradecanoic acid** *[26632-17-7]* **M 244.4, m 88-2-88.5°, [α]$_D^{20}$ -31° (CHCl$_3$).** Crystd from chloroform.

4-Hydroxy-2,2,6,6-tetramethylpiperidine *[2403-88-5]* **M 157.3, m 130-131°, pK 10.05.** Crystd from water as hydrate, and crystd from ether as the anhydrous base.

Hydroxy(tosyloxy)iodobenzene [phenyl(hydroxyl)tosyloxyiodine,   hydroxy(4-methyl-benzenesulfonato-O)phenyliodine, Koser's reagent] *[27126-76-7]* **M 392.2, m 134-136°, 135-138°, 134-136°, 136-138.5°.** Possible impurities are tosic acid (removed by washing with Me$_2$CO) and acetic acid (removed by washing with Et$_2$O). It is purified by dissolving in the minimum vol of MeOH, adding

Et_2O to cloud point and setting aside for the prisms to separate [Koser and Wettach *J Org Chem* **42** 1476 *1977*; NMR: Koser et al. *J Org Chem* **41** 3609 *1976*]. It has also been crystd from CH_2Cl_2 (needles, **m** 140-142°) [Neiland and Karele *J Org Chem, USSR (Engl Transl)* **6** 889 *1970*].

4(6)-Hydroxy-2,5,6(2,4,5)-triaminopyrimidine sulfate *[35011-47-3]* **M 257.22, m >340°, pK$_1$ 2.0, pK$_2$ 5.1, pK$_3$ 10.1.** This salt has very low solubility in H_2O. It is best purified by conversion into the dihydrochloride salt which is then reconverted to the insoluble sulfate salt. The sulfate salt (2.57g, 10mmoles) is suspended in H_2O (20mL) containing $BaCl_2$ (10mmoles) and stirred in a boiling water bath for 15min. After cooling the insoluble $BaSO_4$ is filtered off and washed with boiling H_2O (10mL). The combined filtrate and washings are made acidic with HCl and evaporated to dryness. The residual hydrochloride salt is recrystd from H_2O by adding conc HCl whereby the *dihydrochloride salt* separates as clusters which darken at 260° and dec > 300° [Baugh and Shaw *J Org Chem* **29** 3610 *1964*; King and Spengley *J Chem Soc* 2144 *1952*]. The hydrochloride is then dissolved in H_2O and while hot an equivalent of H_2SO_4 is added when the sulfate separates as a white microcrystalline solid which is filtered off washed liberally with H_2O and dried in vacuum over P_2O_5. [Albert and Wood *J Appl Chem London* **3** 521 *1953*; UV: Cavalieri et al. *J Am Chem Soc* **70** 3875 *1948*; see also Pfleiderer *Chem Ber* **90** 2272 *1957*; Traube *Chem Ber* **33** 1371 *1900*].

9-Hydroxytriptycene *[73597-16-7]* **M 270.3, m 245-246.5°.** Crystd from *benzene/pet ether. Dried at 100° in a vacuum [Imashiro et al. *J Am Chem Soc* **109** 729 *1987*].

5-Hydroxy-L-tryptophan *[4350-09-8]* **M 220.2, m 273°(dec), $[\alpha]_D^{22}$ -32.5°, $[\alpha]_{546}^{20}$ -73.5° (c 1, H_2O), pK$_{Est(1)}$~2.4, pK$_{Est(2)}$~9.0, pK$_{Est(3)}$~9.4, pK$_{Est(4)}$ 16 (NH).** Likely impurities are 5-hydroxy-D-tryptophan and 5-benzyloxytryptophan. Crystd under nitrogen from water by adding EtOH. Stored under nitrogen.

Hydroxyurea *[127-07-1]* **M 76.1, m 70-72° (unstable form), m 135-140°, 141° (stable form).** See hydroxyurea on p. 431 in Chapter 5.

3-Hydroxyxanthone *[3722-51-8]* **M 212.2, m 246°.** Purified by chromatography on SiO_2 gel with pet ether/*benzene). Recrystd from *benzene or EtOH [Itoh et al. *J Am Chem Soc* **107** 4819 *1985*].

α-Hyodeoxycholic acid *[83-49-8]* **M 392.6, m 196-197°, $[\alpha]_{546}^{20}$ +8° (c 2, EtOH), pK$_{Est}$ ~4.9.** Crystd from ethyl acetate.

Hyoscine (scopolamine, atroscine) *[51-34-3]* **M 321.4, m 59°, $[\alpha]_D^{20}$ -18° (c 5, EtOH), -28° (c 2, H_2O), $[\alpha]_{546}^{20}$ -30° (c 5, $CHCl_3$), pK^{25} 7.55.** Crystd from *benzene/pet ether. Racemate has **m** 56-57° (H_2O), 37-38° ($2H_2O$), syrup (anhydr), *l* and *d* isomers can separate as syrups when anhydrous.

Hypericin *[548-04-9]* **M 504.4, m 320°(dec).** Crystd from pyridine by addition of methanolic HCl.

Ibogaine *[83-74-9]* M 300.3, m 152-153°, $[\alpha]_D^{20}$ -54° (EtOH), pK 8.1 (80% aq $MeOCH_2CH_2OH$). Crystd from aqueous EtOH and sublimes at 150°/0.01mm.

Imidazole (glyoxaline) *[288-32-4]* **M 68.1, m 89.5-91°, b 256°, pK$_1^{25}$ 6.99, pK$_2^{25}$ 14.44.** Crystd from *benzene, CCl_4, CH_2Cl_2, EtOH, pet ether, acetone/pet ether and distd deionized water. Dried at 40° under vacuum over P_2O_5. Distd at low pressure. Also purified by sublimation or by zone melting. [Caswell and Spiro *J Am Chem Soc* **108** 6470 *1986*.] ^{15}N-imidazole was crystd from *benzene [Scholes et al. *J Am Chem Soc* **108** 1660 *1986*].

4'-(Imidazol-1-yl)acetophenone *[10041-06-2]* **M 186.2, m 104-107°, pK 4.54.** Twice recrystd from CH_2Cl_2/hexane [Collman et al. *J Am Chem Soc* **108** 2588 *1986*].

Iminodiacetic acid *[142-73-4]* **M 133.1, m 225°(dec), pK$_1^{25}$ 2.50, pK$_2^{25}$ 9.40.** Crystd from water.

1,3-Indandione *[606-23-5]* **M 146.2, m 129-132°, pK^{18} 7.2 (1% aq EtOH).** Recrystd from EtOH [Bernasconi and Paschalis *J Am Chem Soc* **108** 2969 *1986*].

Indane *[496-11-7]* **M 118.1, b 177°, d 0.960, n 1.538.** Shaken with conc H$_2$SO$_4$, then water, dried and fractionally distd.

Indanthrene *[81-77-6]* **M 442.4, m 470-500°.** Crystd repeatedly from 1,2,4-trichlorobenzene.

Indazole *[271-44-3]* **M 118.1, m 147°, pK$_1^{20}$ 1.32, p K$_2^{25}$ 13.80 (acidic NH).** Crystd from water, sublimed under a vacuum, then pet ether (b 60-80°).

Indene *[95-13-6]* **M 116.2, f -1.5°, b 114.5°/100mm, d 0.994, n 1.5763.** Shaken with 6M HCl for 24h (to remove basic nitrogenous material), then refluxed with 40% NaOH for 2h (to remove benzonitrile). Fractionally distd, then fractionally crystd by partial freezing. The higher-melting portion was converted to its sodium salt by adding a quarter of its weight of sodamide under nitrogen and stirring for 3h at 120°. Unreacted organic material was distd off at 120°/1mm. The sodium salts were hydrolysed with water, and the organic fraction was separated by steam distn, followed by fractional distn. Before use, the distillate was passed, under nitrogen, through a column of activated silica gel. [Russell *J Am Chem Soc* **78** 1041 *1956*.]

Indigo *[482-89-3]* **M 262.3, sublimes at ~300°, m 390°(dec), and halogen-substituted indigo dyes.** Reduced in alkaline soln with sodium hydrosulfite, and filtered. The filtrate was then oxidised by air, and the resulting ppte was filtered off, dried at 65-70°, ground to a fine powder, and extracted with CHCl$_3$ in a Soxhlet extractor. Evapn of the CHCl$_3$ gave the purified dye. [Brode, Pearson and Wyman *J Am Chem Soc* **76** 1034 *1954*; spectral characteristics are listed.]

Indole *[120-72-9]* **M 117.2, m 52°, b 124°/5mm, 253-254°/760mm, pK$_1^{25}$ -2.47 (H$_0$ scale), pK$_2^{25}$ 16.97 (acidic NH).** Crystd from *benzene, hexane, water or EtOH/water (1:10). Further purified by sublimation in a vacuum or zone melting.

Indole-3-acetic acid *[87-51-4]* **M 175.2, m 167-169°, pK$_1^{25}$ -6.13 (aq H$_2$SO$_4$), pK$_2^{25}$ 4.54 (CO$_2$H).** Recrystd from EtOH/water [James and Ware *J Phys Chem* **89** 5450 *1985*].

3-Indoleacetonitrile *[771-51-7]* **M 156.2, m 33-36°, 36-38°, b 157°/0.2mm, 158-160°/0.1mm, viscous oil n$_D^{20}$ 1.6097.** Distil in very high vacuum and the viscous distillate crystallises on standing after a few days; the *picrate* has **m** 127-128° (from EtOH) [Coker et al. *J Org Chem* **27** 850 *1962*; Thesing and Schülde *Chem Ber* **85** 324 *1952*]. The *N-acetate* has **m** 118° (from MeOH) and has R$_F$ = 0.8, on Silica Gel F$_{254}$ in CHCl$_2$-MeOH 19:1 [Buzas et al. *Synthesis* 129 *1977*].

Indole-3-butanoic acid *[133-32-4]* **M 203.2, m 124-125°.** See 3-indolylbutyric acid on p. 543 in Chapter 6.

Indole-3-propionic acid *[830-96-6]* **M 189.2, m 134-135°, pK$_{Est}$ ~4.7.** Recrystd from EtOH/water [James and Ware *J Phys Chem* **89** 5450 *1985*].

Indolizine [pyrrocoline, pyrrolo(1,2-a)pyridine] *[274-40-8]* **M 117, m 73-74°, 75°, pK^{20} 3.94 (C-protonation).** Purified through an alumina column in *C$_6$H$_6$ and eluted with *C$_6$H$_6$ (toluene could be used instead). The eluate contained in the fluorescent band (using UV light λ 365mn) was collected, evapd and the cryst residues sublimed twice at 40-50°/0.2-0.5mm. The colourless crystals darkend on standing and should be stored in dark sealed containers. If the original sample is dark in color then is should be covered with water and steam distd. The crysts in the distillate are collected and, dried between filter paper and sublimed. It protonates on C3 in aqueous acid. It should give one fluorescent spot on paper chromatography (Whatman 1) in 3% aq ammonia and in *n*-BuOH, AcOH, H$_2$O (4:1:1). The *picrate* has **m** 101° from EtOH. [Armarego *J Chem Soc* 226 *1944*; Armarego *J Chem Soc (B)* 191 *1966*; Scholtz *Chem Ber* **45** 734 *1912*.]

(-)-Inosine *[58-63-9]* **M 268.2, m 215°, $[\alpha]_{546}^{20}$ -76°** (c 1, 0.1M NaOH), pK_1^{25} **1.06,** pK_2^{25} **8.96,** pK_3^{25} **11.36.** Crystd from aqueous 80% EtOH.

i-**Inositol** (*myo*) *[87-89-8]* **M 180.2, m 228°.** See entry on p. 543 in Chapter 6.

Inositol monophosphate *[15421-51-9]* **M 260.1, m 195-197°(dec).** Crystd from water and EtOH.

Iodinin (1,6-phenazine-5,10-dioxide) *[68-81-5]* **M 244.1, m 236°(dec), pK 12.5.** Crystd from CHCl$_3$.

Inulin *[9005-80-5]* **M (162.14)$_n$.** Crystd from water.

Iodoacetamide *[144-48-9]* **M 185.0, m** *ca* **143°(dec).** Crystd from water or CCl$_4$.

Iodoacetic acid *[64-69-7]* **M 160.6, m 78°, pK^{25} 3.19.** Crystd from pet ether (b 60-80°) or CHCl$_3$/CCl.

2-Iodoaniline *[615-43-0]* **M 219.0, m 60-61°, pK^{25} 2.54.** Distd with steam and crystd from *benzene/pet ether.

4-Iodoaniline *[540-37-4]* **M 219.0, m 62-63°, pK^{25} 3.81.** Crystd from pet ether (b 60-80°) by refluxing, then cooling in an ice-salt bath freezing mixture. Dried in air. Also crystd from EtOH and dried in a vacuum for 6h at 40° [Edidin et al. *J Am Chem Soc* **109** 3945 *1987*].

4-Iodoanisole *[696-62-8]* **M 234.0, m 51-52°, b 139°/35mm, 237°/726mm.** Crystd from aqueous EtOH.

Iodobenzene *[591-50-4]* **M 204.0, b 63-65°/10mm, 188°/atm, d 1.829, n^{25} 1.6169.** Washed with dilute aqueous Na$_2$S$_2$O$_3$, then water. Dried with CaCl$_2$ or CaSO$_4$. Decolorised with charcoal. Distd under reduced pressure and stored with mercury or silver powder to stabilise it.

o-**Iodobenzoic acid** *[88-67-5]* **M 248.4, m 162°, pK^{20} 2.93.** Crystd repeatedly from water and EtOH. Sublimed under vacuum at 100°.

m-**Iodobenzoic acid** *[618-51-9]* **M 248.4, m 186.6-186.8°, pK^{25} 3.85.** Crystd repeatedly from water and EtOH. Sublimed under vacuum at 100°.

p-**Iodobenzoic acid** *[619-58-9]* **M 248.4, m 271-272°, pK^{25} 4.00.** Crystd repeatedly from water and EtOH. Sublimed under vacuum at 100°.

4-Iodobiphenyl *[1591-31-7]* **M 280.1, m 113.7-114.3°.** Crystd from EtOH/*benzene and dried under vacuum over P$_2$O$_5$.

2-Iodobutane *[513-48-4]* **M 184.0, b 120.0, d 1.50, n^{25} 1.4973.** Purified by shaking with conc H$_2$SO$_4$, then washing with water, aq Na$_2$SO$_3$ and again with water. Dried with MgSO$_4$ and distd. Alternatively, passed through a column of activated alumina before distn, or treated with elemental bromine, followed by extraction of the free halogen with aqueous Na$_2$S$_2$O$_3$, thorough washing with water, drying and distilling. It is stored over silver powder and distd before use.

1-Iodo-2,4-dinitrobenzene *[709-49-9]* **M 294.0, m 88°.** Crystd from ethyl acetate.

Iodoform *[75-47-8]* **M 393.7, m 119°.** Crystd from MeOH, EtOH or EtOH/EtOAc. Steam volatile.

1-Iodo-4-nitrobenzene *[636-98-6]* **M 249.0, m 171-172°.** Ppted from acetone by addition of water, then recrystd from EtOH.

o-**Iodophenol** *[533-58-4]*   **M 280.1, m 42°, pK^{25} 8.51.** Crystd from $CHCl_3$ or diethyl ether.

p-**Iodophenol**   *[540-38-5]*   **M 280.1, m 94°, 138-140°/5mm, pK^{25} 9.30.** Crystd from pet ether (b 80-100°) or distd *in vacuo*. If material has a brown or violet color, dissolve in $CHCl_3$, shake with 5% sodium thiosulfate soln until the $CHCl_3$ is colorless. Dry (Na_2SO_4), extract, evap and dist residue *in vacuo*. [Dains and Eberly *Org Synth* Coll Vol II, 355 *1948*.]

5-Iodosalicylic acid (2-hydroxy-5-iodobenzoic acid) *[119-30-2]* **M 264.0, m 197° pK$_1^{25}$ 2.65, pK$_2^{25}$ 13.05.** Crystd from water.

o-**Iodosobenzoic acid** *[304-91-6]* **M 264.0, m >200°, pK$_{Est}$ ~2.6.** Crystd from EtOH.

N-**Iodosuccinimide** *[516-12-1]* **M 225.0, m 200-201°.** Crystd from dioxane/CCl_4.

p-**Iodotoluene** *[624-31-7]* **M 218.0, m 35°, b 211-212°.** Crystd from EtOH.

3-Iodo-L-tyrosine   *[70-78-0]*   **M 307.1, m 205-208°(dec), $[\alpha]_D^{25}$ -4.4° (c 5, 1M HCl), pK$_{Est(2)}$~2.1, pK$_{Est(3)}$~6.4, pK$_4^{25}$ 8.7.** Likely impurities are tyrosine, diiodotyrosine and iodide. Crystd by soln in dilute ammonia, at room temperature, followed by addition of dilute acetic acid to pH 6. Stored at 0°.

α-Ionone *[127-41-3]* **M 192.3, b 131°/13mm, d 0.931, n 1.520, $[\alpha]_D^{23}$ +347° (neat).** Purified on a spinning band fractionating column.

β-Ionone *[79-77-6]* **M 192.3, b 150-151°/24mm, d 0.945, n 1.5211, ε_{296nm} 10,700.** Converted to the *semicarbazone* (m 149°) by adding 50g of semicarbazide hydrochloride and 44g of potassium acetate in 150mL of water to a soln of 85g of ß-ionone in EtOH. (More EtOH was added to redissolve any ß-ionone that ppted.) The semicarbazone crystallised on cooling in an ice-bath and was recrystallised from EtOH or 75% MeOH to constant **m** (148-149°). The semicarbazone (5g) was shaken at room temperature for several days with 20mL of pet ether and 48mL of M H_2SO_4, then the ether layer was washed with water and dilute aqueous $NaHCO_3$, dried and the solvent was evaporated. The ß-ionone was distilled under vacuum. (The customary steam distillation of ß-ionone semicarbazone did not increase the purity.) [Young et al. *J Am Chem Soc* **66** 855 *1944*].

Iproniazid (isonicotinic acid 2-isopropylhydrazide) phosphate *[305-33-9]* **M 277.2, m 178-179°, 180-182°, pK$_{Est}$ ~3.5 (free base).** Crystd from H_2O and Me_2CO. *Free base* has **m** 113-114° from *C_6H_6/pet ether.

(±)-Irone   (6-methyl-ionone, ±-*trans*-(α)-4*t*-[2,5,6,6-tetramethyl-cyclohex-2-yl]but-3*t*-en-2-one) *[79-69-6]* **M 206.3, b 85-86°/0.05mm, 109°/0.7mm, d$_4^{20}$ 0.9340, n$_D^{20}$ 1.4998.** If large amounts are available then fractionate through a Podbielniak column (see p. 141) or an efficient spinning band column, but small amounts are distilled using a Kügelrohr apparatus. The *4-phenyl-semicarbazone* has **m** 174-175° (165-165.5°). [IR: Seidel and Ruzocka *Helv Chim Acta* **35** 1826 *1952*; Naves *Helv Chim Acta* **31** 1280 *1948*; Lecomte and Naves *J Chim Phys* **53** 462 *1956*.]

Isatin (indole-2,3-dione) *[91-56-5]* **M 147.1, m 201-203°, 205°, pK >12 (acidic NH).** Crystd from amyl alcohol and sublimed at 180°/1mm. In aq NaOH the ring opens to yield sodium *o*-aminobenzoylformate.

Isatoic anhydride   (3,1-benzoxazin-2,4[1-*H*]-dione) *[118-48-9]* **M 163.1, m 235-240°, 240-243°, 243°, 243-245°.** Recryst from EtOH or 95% EtOH (30mL/g) or dioxane (10mL/g) and dried in a vacuum. [Wagner and Fegley *Org Synth* Coll Vol III 488 *1955*; Ben-Ishai and Katchalski *J Am Chem Soc* **74** 3688 *1952*; UV: Zentmyer and Wagner *J Org Chem* **14** 967 *1949*.]

Isoamyl acetate (1-butyl-3-methyl acetate) *[123-92-2]* **M 130.2, b 142.0°, d 0.871, n 1.40535.** Dried with finely divided K_2CO_3 and fractionally distd.

Isoamyl alcohol (1-butyl-3-methyl alcohol) *[123-51-3]* **M 88.2, b 132°/760mm, d^{15} 0.8129, n^{15} 1.4085.** See 3-methyl-1-butanol on p. 290.

Isoamyl bromide (1-butyl-3-methyl bromide) *[107-82-4]* **M 151.1, f -112°, b 119.2°/737mm, d 1.208, n 1.444.** Shaken with conc H_2SO_4, washed with water, dried with K_2CO_3 and fractionally distd.

Isoamyl chloride (1-butyl-3-methyl chloride) *[513-36-0]* **M 106.6, b 99°/734mm, d 0.8704, n 1.4084.** Shaken vigorously with 95% H_2SO_4 until the acid layer no longer became coloured during 12h, then washed with water, saturated aq Na_2CO_3, and more water. Dried with $MgSO_4$, filtered and fractionally distd. Alternatively, a stream of oxygen containing 5% of ozone was passed through the chloride for a time, three times longer than was necessary to cause the first coloration of starch iodide paper by the exit gas. Subsequent washing of the liquid with aqueous $NaHCO_3$ hydrolysed the ozonides and removed organic acids. After drying and filtering, the isoamyl chloride was distd. [Chien and Willard *J Am Chem Soc* **75** 6160 *1953*.]

Isoamyl ether [diisopentyl ether, di-(1-butyl-3-methyl) ether] *[544-01-4]* **M 158.3, b 173.4°, d 0.778, n 1.40850.** This is a mixture of 2- and 3-methylbutyl ether. It is purified by refluxing with sodium for 5h, then distilled under reduced pressure, to remove alcohols. Isoamyl ether can also be dried with $CaCl_2$ and fractionally distd from P_2O_5.

D(-)-Isoascorbic acid (araboascorbic acid) *[89-65-6]* **M 176.1, m 174°(dec), [α]$_D^{25}$ -16.8° (c 2, H_2O), pK^{18} 3.99.** Crystd from H_2O or dioxane.

dl-**Isoborneol** *[124-76-5]* **M 154.3, m 212° (sealed tube).** Crystd from EtOH or pet ether (b 60-80°). Sublimes in a vacuum.

Isobutane *[75-28-5]* **M 58.1, b -10.2°, d 0.557.** Olefines and moisture can be removed by passage at 65° through a bed of silica-alumina catalyst which has previously been evacuated at about 400°. Alternatively, water and CO_2 can be taken out by passage through P_2O_5 then asbestos impregnated with NaOH. Treatment with anhydrous $AlBr_3$ at 0° then removes traces of olefins. Inert gases can be separated by freezing the isobutane at -195° and evacuating out the system.

Isobutene *[115-11-7]* **M 56.1, b -6.6°/760mm.** Dried by passage through anhydrous $CaSO_4$ at 0°. Purified by freeze-pump-thaw cycles and trap-to-trap distn.

Isobutyl alcohol (2-methyl-1-propanol) *[78-83-1]* **M 74.1, b 108°/760mm, d 0.801, n 1.396.** Dried with K_2CO_3, $CaSO_4$ or $CaCl_2$, filtered and fractionally distd. For further drying, the redistd alcohol can be refluxed with the appropriate alkyl phthalate or succinate as described under *ethanol* (see also p. 271).

Isobutyl bromide (1-bromo-2-methylpropane) *[78-77-3]* **M 137.0, b 91.2°, d 1.260, n 1.437.** Partially hydrolysed to remove any tertiary alkyl halide, then fractionally distd, washed with conc H_2SO_4, water and aqueous K_2CO_3, then redistd from dry K_2CO_3. [Dunbar and Hammett *J Am Chem Soc* **72** 109 *1950*.]

Isobutyl chloride (1-chloro-2-methylpropane) *[513-36-0]* **M 92.3, b 68.8°/760mm, d 0.877, n 1.398.** Same methods as described under *isoamyl chloride*.

Isobutyl chloroformate *[543-27-1]* **M 136.6, b 123-127°/atm, 128.8°/atm, d 1.053, n 1.4070.** It can be dried over $CaCl_2$ and fractionated at atm press while keeping moisture out. Its purity can be checked by conversion to the *phenyl urethane derivative* with PhNCO [Saunders et al. *J Am Chem Soc* **73** 3796 *1951*.] IR: ν 1780cm^{-1} [Thompson and Jameson *Spectrochim Acta* **13** 236 *1959*; Röse *Justus Liebigs Ann Chem* **205** 227 *1880*].

Isobutyl formate *[542-55-2]* **M 102.1, b 98.4°, d 0.885, n 1.38546.** Washed with saturated aqueous $NaHCO_3$ in the presence of saturatedd NaCl, until no further reaction occurred, then with saturated aqueous NaCl, dried ($MgSO_4$) and fractionally distd.

Isobutyl iodide (1-iodo-2-methylpropane) *[513-38-2]* **M 184.0, b 83°/250mm, 120°/760mm, d 1.60, n 1.495.** Shaken with conc H_2SO_4, and washed with water, aqueous Na_2SO_3, and water, dried with $MgSO_4$ and distd. Alternatively, passed through a column of activated alumina before distn. Stored under nitrogen with mercury in a brown bottle or in the dark.

Isobutyl vinyl ether *[109-53-5]* **M 100.2, b 108-110°, d 0.768, n 1.398.** Washed three times with equal volumes of aqueous 1% NaOH, dried with CaH_2, refluxed with sodium for several hours, then fractionally distd from sodium.

Isobutyraldehyde *[78-84-2]* **M 72.1, b 62.0°, d 0.789, n 1.377.** Dried with $CaSO_4$ and used immediately after distn under nitrogen because of the great difficulty in preventing oxidation. Can be purified through its acid bisulfite derivative.

Isobutyramide *[563-83-7]* **M 87.1, m 128-129°, b 217-221°.** Crystd from acetone, *benzene, $CHCl_3$ or water, then dried under vacuum over P_2O_5 or 99% H_2SO_4. Sublimed under vacuum.

Isobutyric acid *[79-31-2]* **M 88.1, b 154-154.5°, d 0.949, n 1.393, pK^{25} 4.60.** Distd from $KMnO_4$, then redistd from P_2O_5.

Isobutyronitrile (2-methylpropionitrile, isopropyl cyanide) *[78-82-0]* **M 69.1, b 103.6°, d^{25} 0.7650, n 1.378.** Shaken with conc HCl (to remove isonitriles), then with water and aq $NaHCO_3$. After a preliminary drying with silica gel or Linde type 4A molecular sieves, it is shaken or stirred with CaH_2 until hydrogen evolution ceases, then decanted and distd from P_2O_5 (not more than 5g/L, to minimize gel formation). Finally it is refluxed with, and slowly distd from CaH_2 (5g/L), taking precautions to exclude moisture.

(-)-γ-Isocaryophyllene (8-methylene-4,11,11-trimethylbicyclo[7.2.0]undec-4-ene) *[118-65-0]* **M 204.4, b 122-124°/12mm, 131-133°/16mm, 130-131°/24mm, 271-273°/atm, d_4^{20} 0.8959, n_D^{20} 1.496, $[\alpha]_{546}^{20}$ -31°, $[\alpha]_D^{20}$ -27° (neat).** Purified by vac dist or GLC using a nitrile-silicone column [Corey et al. *J Am Chem Soc* **86** 485 *1964*; Ramage and Simonsen *J Chem Soc* 741 *1936*; Kumar et al. *Synthesis* 461 *1976*].

L-Isoleucine *[73-32-5]* **M 131.2, m 285-286°(dec), $[\alpha]_D^{20}$ +40.6° (6M HCl) pK_1^{25} 2.66, pK_2^{25} 9.69.** Crystd from water by addition of 4 volumes of EtOH.

(-)-β-Isolongifolene (1-*R*-(-)- 2,2,7,7-tetramethyltricyclo[6.2.1.0^{1,6}]undec-5-ene) *[1135-66-6]* **M 204.4, b 82-83°/0.4mm, 144-146°/30mm, 255-256°/atm, d_4^{20} 0.930, n_D^{20} 1.4992, $[\alpha]_{546}^{20}$ -166°, $[\alpha]_D^{20}$ -138° (c 1, EtOH).** Refluxed over and distd from Na. [Zeiss and Arakawa *J Am Chem Soc* **76** 1653 *1954*; IR: Reinaecker and Graafe *Angew Chem, Int Ed Engl* **97** 348 *1985*; UV and NMR: Ranganathan et al. *Tetrahedron* **26** 621 *1970*.]

Isolysergic acid *[478-95-5]* **M 268.3, m 218°(dec), $[\alpha]_D^{20}$ +281° (c 1, pyridine) pK_1^{24} 3.33, pK_2^{24} 8.46.** Crystd from water.

Isonicotinamide *[1453-82-3]* **M 122.1, m 155.5-156°, pK_1^{20} -1.0 (protonation of $CONH_2$), pK_2^{20} 3.61, pK_3^{25} 11.47 (acidic $CONH_2$).** Recrystd from hot water.

Isonicotinic acid (pyridine-4-carboxylic acid) *[55-22-1]* **M 123.1, m 320°, pK^{25} 4.90.** Crystd repeatedly from water. Dried under vac at 110°.

Isonicotinic acid hydrazide (isoniazide) *[54-85-3]* **M 137.1, m 172°, pK_1 1.75 ($NHNH_2$), pK_2 3.57 (=N-), pK_3 10.75 (-NH).** Crystd from 95% EtOH.

1-Isonicotinyl-2-isopropylhydrazide *[54-92-2]* **M 179.2, m 112.5-113.5°.** Crystd from *benzene/pet ether.

1-Isonicotinyl-2-salicylidenehydrazide *[495-84-1]* **M 241.2, m 232-233°.** Crystd from EtOH.

Isonitrosoacetone (*anti*-pyruvic aldehyde-1-oxime) *[31915-82-9]* **M 87.1, m 69°.** Crystd from ether/pet ether or CCl_4.

Isonitrosoacetophenone (phenylglyoxaldoxime) *[532-54-7]* **M 149.2, m 126-128°.** Crystd from water.

5-Isonitrosobarbituric acid (violuric acid) *[26851-19-9]* **M 175.1, m 221-223°, 245-250°, pK$_1$ 4.41, pK$_2$ 9.66 (10.1).** Crystd from water or EtOH. *1,1-Dimethylvioluric acid*, **m** 144-147° has pK 4.72 [Taylor and Robinson *Talanta* **8** 518 *1961*].

Isononane *[34464-40-9]* **M 128.3, b 142°/760mm.** Passed through columns of activated silica gel and basic alumina (activity 1). Distd under high vacuum from Na/K alloy.

Isopentyl formate *[110-45-2]* **M 116.2, b 121-123°/atm, 123-123.6°/atm, 123-124°/atm, d_4^{20} 0.8713, n_D^{20} 1.391.** Colourless liquid which is soluble in 300 volumes of H_2O and is soluble in common organic solvents. It is purified by repeated distn using an efficient column at atmospheric pressure.

Isophorone *[78-59-1]* **M 138.2, b 94°/16mm, d 0.921, n^{18} 1.4778.** Washed with aqueous 5% Na_2CO_3 and then distd under reduced pressure, immediately before use. Alternatively, can be purified *via* the semicarbazone. [Erskine and Waight *J Chem Soc* 3425 *1960*.]

Isophthalic acid (benzene-1,3-dicarboxylic acid) *[121-91-5]* **M 166.1, m 345-348°, pK$_1^{25}$ 3.70, pK$_2^{25}$ 4.60.** Crystd from aqueous EtOH.

Isopinocampheol (pinan-3-ol, 2,6,6-trimethylbicyclo[3.1.1]heptan-3-ol) *[1S,2S,3S,5R-(+)-27779-29-9; 1R,2R,3R,5S-(-)- 25465-65-0]* **M 154.25, m 52-55°, 55-56°, 55-57°, b 103°/11mm, n_D^{20} 1.4832, $[\alpha]_{546}^{20}$ (+) and (-) 43°, $[\alpha]_D^{20}$ (+) and (-) 36° (c 20, EtOH).** Dissolve in Et_2O, dry $MgSO_4$, filter, evaporate, then recryst from pet ether. Also recryst from aqueous EtOH and has been distd in a vacuum. [Kergomard and Geneix *Bull Soc Chim Fr* 394 *1958*; Zweifel and Brown *J Am Chem Soc* **86** 393 *1964*.] The *3,4-dinitrobenzoyl* deriv has **m** 100-101°, the *phenylcarbamoyl* derivative has **m** 137-138° and the *acid -phthalate* has **m** 125-126°.

Isoprene (2-methyl-1,3-butadiene) *[78-79-5]* **M 68.1, b 34.5-35°/762mm, d 0.681, n^{25} 1.4225.** Refluxed with sodium. Distd from sodium or $NaBH_4$ under nitrogen, then passed through a column containing KOH, $CaSO_4$ and silica gel. *tert*-Butylcatechol (0.02% w/w) was added, and the isoprene was stored in this way until redistd before use. The inhibitor (*tert*-butylcatechol) in isoprene can be removed by several washings with dil NaOH and water. The isoprene is then dried over CaH_2, distd under nitrogen at atmospheric pressure, and the fraction distilling at 32° is collected. Stored under nitrogen at -15°.

Isopropanol *[67-63-0]* **M 60.1, b 82.5°, d 0.783, $n^{25.8}$ 1.3739, pK^{25} 17.1.** Isopropyl alcohol is prepared commercially by dissolution of propene in H_2SO_4, followed by hydrolysis of the sulfate ester. Major impurities are water, lower alcohols and oxidation products such as aldehydes and ketones. Purification of isopropanol follows substantially the same procedure as for *n*-propyl alcohol.
Isopropanol forms a constant-boiling mixture, **b** 80.3°, with water. Most of the water can be removed from this 91% isopropanol by refluxing with CaO (200g/L) for several hours, then distilling. The distillate can be dried further with CaH_2, magnesium ribbon, BaO, $CaSO_4$, calcium, anhydrous $CuSO_4$ or Linde type 5A molecular sieves. Distn from sulfanilic acid removes ammonia and other basic impurities. Peroxides [indicated by liberation of iodine from weakly acid (HCl) solns of 2% KI] can be removed by refluxing with solid stannous chloride or with $NaBH_4$ then fractionally distilling. To obtain isopropanol containing only 0.002M of water, sodium (8g/L) has been dissolved in material dried by distn from $CaSO_4$, 35mL of isopropyl benzoate has been

added and, after refluxing for 3h, the alcohol has been distd through a 50-cm Vigreux column. [Hine and Tanabe *J Am Chem Soc* **80** 3002 *1958*.] Other purification steps for isopropanol include refluxing with solid aluminium isopropoxide, refluxing with NaBH$_4$ for 24h, and the removal of acetone by treatment with, and distn from 2,4-dinitrophenylhydrazine. Peroxides re-form in isopropanol if it is stood for several days.

Isopropenylcyclobutane *[3019-22-5]* **M 98.1, b 98.7°, d 0.7743, n 1.438.** Purified by preparative chromatography (silicon oil column), or fractionally distd. Dried with molecular sieves.

Isopropyl acetate *[108-22-5]* **M 102.1, b 88.4°, d 0.873, n 1.3773.** Washed with 50% aq K$_2$CO$_3$ (to remove acid), then with saturated aq CaCl$_2$ (to remove any alcohol). Dried with CaCl$_2$ and fractionally distd.

Isopropyl bromide (2-bromopropane) *[75-26-3]* **M 123.0, b 0°/69.2mm, 59.4°/760mm, d 1.31, n^{15} 1.42847, n 1.4251.** Washed with 95% H$_2$SO$_4$ (conc acid partially oxidised it) until a fresh portion of acid did not become coloured after several hours, then with water, aq NaHSO$_3$, aq 10% Na$_2$CO$_3$ and again with water. (The H$_2$SO$_4$ can be replaced by conc HCl.) Prior to this treatment, isopropyl bromide has been purified by bubbling a stream of oxygen containing 5% ozone through it for 1h, followed by shaking with 3% hydrogen peroxide soln, neutralising with aq Na$_2$CO$_3$, washing with distilled water and drying. Alternatively, it has been treated with elemental bromine and stored for 4 weeks, then extracted with aq NaHSO$_3$ and dried with MgSO$_4$. After the acid treatment, isopropyl bromide can be dried with Na$_2$SO$_4$, MgSO$_4$ or CaH$_2$, and fractionally distd.

N-**Isopropylcarbazole** *[1484-09-9]* **M 209.3, m 120°.** Crystd from isopropanol. Sublimed under vacuum. Zone refined. The *picrate* has **m** 143° after recrystn from EtOH.

Isopropyl chloride (2-chloropropane) *[75-29-6]* **M 78.5, b 34.8°, d 0.864, n 1.3779, n^{25} 1.3754.** Purified with 95% H$_2$SO$_4$ as described for *isopropyl bromide*, then dried with MgSO$_4$, P$_2$O$_5$ or CaH$_2$, and fractionally distd from Na$_2$CO$_3$ or CaH$_2$. Alternatively, a stream of oxygen containing *ca* 5% ozone has been passed through the chloride for about three times as long as was necessary to obtain the first coloration of starch iodide paper by the exit gas, and the liquid was then washed with NaHCO$_3$ soln to hydrolyse ozonides and remove organic acids before drying and distilling.

Isopropyl ether (diisopropyl ether) *[108-20-3]* **M 102.2, b 68.3°, d 0.719, n 1.3688, n^{25} 1.36618.** Common impurities are water and peroxides [detected by the liberation of iodine from weakly acid (HCl) solns of 2% KI]. Peroxides can be removed by shaking with aqueous Na$_2$SO$_3$ or with acidified ferrous sulfate (0.6g FeSO$_4$ and 6mL conc H$_2$SO$_4$ in 110mL of water, using 5-10g of soln per L of ether), or aqueous NaBH$_4$ soln. The ether is then washed with water, dried with CaCl$_2$ and distd. Alternatively, refluxing with LiAlH$_4$ or CaH$_2$, or drying with CaSO$_4$, then passage through an activated alumina column, can be used to remove water and peroxides. Other dehydrating agents used with isopropyl ether include P$_2$O$_5$, sodium amalgam and sodium wire. (The ether is often stored in brown bottles, or in the dark, with sodium wire.) Bonner and Goishi (*J Am Chem Soc* **83** 85 *1961*) treated isopropyl ether with dil sodium dichromate/sulfuric acid soln, followed by repeated shaking with a 1:1 mixture of 6M NaOH and saturated KMnO$_4$. The ether was washed several times with water, dilute aqueous HCl, and water, with a final washing with, and storage over, ferrous ammonium sulfate acidified with H$_2$SO$_4$. Blaustein and Gryder (*J Am Chem Soc* **79** 540 *1957*), after washing with alkaline KMnO$_4$, then water, treated the ether with ceric nitrate in nitric acid, and again washed with water. Hydroquinone was added before drying with CaCl$_2$ and MgSO$_4$, and refluxing with sodium amalgam (108g Hg/100g Na) for 2h under nitrogen. The distillate (nitrogen atmosphere) was made 2 x 10^{-5}M in hydroquinone to inhibit peroxide formation (which was negligible if the ether was stored in the dark). Catechol (pyrocatechol) and resorcinol are alternative inhibitors.

4,4'-Isopropylidenediphenol *[80-05-7]* **M 228.3, m 158°, pK$_{Est}$ ~10.3.** Crystd from acetic acid/water (1:1).

Isopropyl iodide (2-iodopropane) *[75-30-9]* **M 170.0, b 88.9°, d 1.70, n 1.4987.** Treated with bromine, followed by extraction of free halogen with aqueous Na$_2$S$_2$O$_3$ or NaHSO$_3$, washing with water, drying (MgSO$_4$ or CaCl$_2$) and distn. (The treatment with bromine is optional.) Other purification methods include passage through activated alumina, or shaking with copper powder or mercury to remove iodine, drying with P_2O_5 and distn. Washing with conc H_2SO_4 or conc HCl (to remove any alcohol), water, aqueous Na_2SO_3, water and aqueous Na_2CO_3 has also been used. Treatment with silica gel causes some liberation of iodine. Distillations should be carried out at slightly reduced pressure. Purified isopropyl iodide is stored in the dark in the presence of a little mercury.

Isopropyl methyl ether *[598-53-8]* **M 74.1, b 32.5°/777mm, d^{15} 0.724, n 1.3576.** Purified by drying with $CaSO_4$, passage through a column of alumina (to remove peroxides) and fractional distn.

Isopropyl *p*-nitrobenzoate *[13756-40-6]* **M 209.2, m 105-106°.** Dissolved in diethyl ether, washed with aqueous alkali, then water and dried. Evapn of the ether and recrystn from EtOH gave pure material.

p-**Isopropyl toluene** (*p*-cymene) *[99-87-6]* **M 134.2, b 176.9°/744mm, d 0.8569, n 1.4902.** See entry on p. 183.

Isoquinoline *[119-65-3]* **M 129.2, m 24°, b 120°/18mm, d 1.0986, n 1.6148, pK^{25} 5.40.** Dried with Linde type 5A molecular sieves or Na_2SO_4 and fractionally distd at reduced pressure. Alternatively, it was refluxed with, and distd from, BaO. Also purified by fractional crystn from the melt and distd from zinc dust. Converted to its *phosphate* (**m** 135°) or *picrate* (**m** 223°), which were purified by crystn and the free base recovered and distd. [Packer, Vaughn and Wong *J Am Chem Soc* **80** 905 *1958*.] The procedure for purifying *via* the picrate comprises the addition of quinoline to picric acid dissolved in the minimum volume of 95% EtOH to yield yellow crystals which are washed with EtOH and air dried before recrystn from acetonitrile. The crystals are dissolved in dimethyl sulfoxide (previously dried over 4A molecular sieves) and passed through a basic alumina column, on which picric acid is adsorbed. The free base in the effluent is extracted with *n*-pentane and distd under vacuum. Traces of solvent are removed by vapour phase chromatography. [Mooman and Anton *J Phys Chem* **80** 2243 *1976*.]

Isovaleric acid *[502-74-2]* **M 102.1, b 176.5°/762mm, d 0.927, n^{15} 1.4064, n 1.40331, pK^{25} 4.77.** Dried with Na_2SO_4, then fractionally distd.

L-Isovaline (2-amino-2-methylbutyric acid) *[595-40-4]* **M 117.2, m *ca* 300° (sublimes in vac), [α]$_D^{25}$ +10° (5M HCl), pK$_{Est(1)}$~2.4, pK$_{Est(2)}$~9.7.** Crystd from aqueous acetone.

Isovanillin (3-hydroxy-4-methoxybenzaldehyde) *[621-59-0]* **M 152.2, m 117°, b 175°/14mm, pK^{25} 8.89.** Cryst from H_2O or *C_6H_6. The *oxime* has **m** 147°.

Isoviolanthrone *[128-64-3]* **M 456.5, m 510-511°(uncorrected).** Dissolved in 98% H_2SO_4 and ppted by adding water to reduce the acid concentration to about 90%. Sublimes *in vacuo*. [Parkyns and Ubblehode *J Chem Soc* 4188 *1960*.]

Itaconic acid (2-propen-1,2-dicarboxylic acid) *[97-65-4]* **M 130.1, m 165-166°, pK$_1^{25}$ 3.63, pK$_2^{25}$ 5.00.** Crystd from EtOH, EtOH/water or EtOH/*benzene.

Itaconic anhydride (2-propen-1,2-dicarboxylic anhydride) *[2170-03-8]* **M 112.1, m 66-68°, 67-68°, 68°, b 139-140°/30mm.** Crystd from $CHCl_3$/pet ether. Can be distd under reduced press. Distn at atm press, or prolonged distn causes rearrangement to citraconic anhydride (2-methylmaleic anhydride). If the material (as seen in the IR spectrum) contains much free acid then heat with acetyl chloride or $SOCl_2$, evaporate and distil at as high a vacuum as possible. The crude anhydride deposits crystals of itaconic acid on standing probably due to hydrolysis by H_2O — store in sealed ampoules under dry N_2. [*Org Synth* Coll Vol II 369 *1943*; IR: Nagai *Bull Chem Soc Jpn* **37** 369 *1964*; Kelly and Segura *J Am Chem Soc* **56** 2497 *1934*.]

Janus Green B (3-dimethylamino-7-[4-dimethylaminoazo]-5-phenylphenazonium chloride) *[2869-83-2]* **M 511.1, m >200°.** Dissolves in H_2O to give a bluish violet soln which

becomes colourless when made 10M in. NaOH. Dissolve in EtOH to give a blue-violet colour, filter from insoluble material then add dry Et_2O whereby the dye separates out leaving a small amount of blue colour in soln. Filter off the solid and dry in vacuum. Store in a dark bottle.

Janus Red B {3-[(2-hydroxy-1-naphtholenyl)azo-2-methylphenylazo]N,N,N-trimethyl-benzenaminium chloride} *[2636-31-9]* **M 460.0.** Crystd from $EtOH/H_2O$ (1:1 v/v) and dry in vacuum. Store in a dark bottle.

Jervine (3β,23β-17,23-epoxy-2-hydroxyvertraman-11-one, a steroidal alkaloid) *[469-59-0]* **M 425.6, m 243-245°, $[\alpha]_D^{20}$ -150° (in EtOH), pK_{Est}~9.4** Crystd from $MeOH/H_2O$. The *hydrochloride* has **m 300-302°.** [Kutney et al. *Can J Chem* **53** 1796 *1975.*]

Julolidine (2,3,6,7-tetrahydro-1H,5H-benzo[ij]quinolizidine) *[479-59-4]* **M 173.3, m 34-36°, 40°, b 105-110°/1mm, 155-156°/17mm, 280° (dec), pK_{Est} ~7.0.** Purified by dissolving in dilute HCl, steam is bubbled through the soln and the residual acidic soln is basified with 10N NaOH, extracted with Et_2O, washed with H_2O, dried (NaOH pellets), filtered, evaporated and distd *in vacuo*. The distillate crystallises on standing (**m 39-40°**). On standing in contact with air for several days it develops a red colour. The colour can be removed by distilling or dissolving in 2-3 parts of hexane, adding charcoal, filtering and cooling in Me_2CO-Dry-ice when julolidine crystallises (85-90% yield). The *hydrochloride [83646-41-7]* has **m** 218° (239-242°), the *picrate* has **m** 165° and the *methiodide* crystallises from MeOH, **m** 186° [Glass and Weisberger *Org Synth* Coll Vol III 304 *1955.*] **Highly TOXIC.**

Kainic acid monohydrate (2S,3S,4S-2-carboxy-4-isoprenyl-3-pyrrolidine-

acetic acid) *[487-79-6]* **M 231.4, m 235-245° (dec), 251° (dec), $[\alpha]_D^{20}$ -14.6° (c 1.46, H_2O), pK_1 2.09, pK_2 4.58, pK_3 10.21.** Purified by adsorbing on to a strongly acidic ion exchange resin (Merck), elution of the diacid with aqueous M NaOH, the eluate is evaporated, H_2O is added, and filtered through a weakly acidic ion exchange resin (Merck). The filtrate is then evaporated and recrystd from EtOH. Its solubility is 0.1g in 1mL of 0.5N HCl. (±)-α-Kainic acid recryst from H_2O, **m 230-260°.** UV (MeOH): λmax 219 (log ε 3.9); ^{1}H NMR (CCl_4, 100MHz, Me_4Si standard) δ: 1.64 (s 1H), 1.70 (s 3H), 3.24 (d J 7.5, 2H), 3.3-4.2 (1H), 3.70 (s 3H), 3.83 (s 3H), 4.35 (dd J 7.5, J 14.5, 1H), 5.21 (t J 7.5, 1H), 7.26 (t J 7.5, 1H). [Oppolzer and Andres *Helv Chim Acta* **62** 2282 *1979.*]

Kerosene *[8008-20-6]* **(mixture of hydrocarbons) b ~175-325°, d 0.75-0.82, n 1.443.** Stirred with conc H_2SO_4 until a fresh portion of acid remains colourless, then washed with water, dried with solid KOH and distd in a Claisen flask. For more complete drying, the kerosene can be refluxed with, and distd from Na.

Ketanserine [3(4-p-fluorobenzoylpiperidinyl-N-ethyl)quinazolin-2,4-dione] *[74050-98-9]* **M 395.4, m 227-235°, pK 7.5.** Solubility is 0.001%in H_2O, 0.038% in EtOH and 2.34 in Me_2NCHO. It has been purified by recrystn from 4-methyl-3-pentanone [Peeters et al. *Cryst Structure Commun* **11** 375 *1982*; Kacprowicz et al. *J Chromatogr* **272** 417 *1983*; Davies et al. *J Chromatogr* **275** 232 *1983*].

Ketene *[463-51-4]* **M 42.0, b 127-130°, d 1.093, n 1.441.** Prepared by pyrolysis of acetic anhydride. Purified by passage through a trap at -75° and collected in a liquid-nitrogen-cooled trap. Ethylene was removed by evacuating the ethylene in an isopentane-liquid-nitrogen slush pack at -160°. Stored at room temperature in a suitable container in the dark. See diketene on p. 209.

2-Keto-L-gulonic acid *[526-98-7]* **M 194.1, m 171°.** Crystd from water and washed with acetone.

Ketone moschus (4-$tert$-butyl-2,6-dimethyl-3,5-dinitroacetophenone) *[81-14-1]* **M 234.1, m 134-137°, 137-138°.** Purified by recryst from MeOH. [Fuson et al. *J Org Chem* **12** 587 *1947.*]

Khellin (4,9-dimethoxy-7-methyl-5-oxofuro[3,2-g]-1,2-chromene) *[82-02-0]* **M 260.3, m 154-155°, b 180-200°/0.05mm.** Crystd from MeOH or diethyl ether.

Kojic acid [(2-hydroxy-5-hydroxymethyl)-4H-pyran-4-one] *[501-30-4]* **M 142.1, m 154-155°, pK$_1^{25}$-1.38, pK$_2^{25}$7.66.** Crystd from MeOH (charcoal) by adding Et$_2$O. Sublimed at 0.1 torr.

Kynurenic acid (4-hydroxyquinoline-2-carboxylic acid) *[492-27-3]* **M 189.1, m 282-283°, pK$_{Est(1)}$~2, pK$_{Est(2)}$~10.** Crystd from absolute EtOH.

L-Kynurenine *[343-65-7]* **M 208.2, m 190°(dec), 210°(dec), [α]$_D^{20}$ -30° (c 0.4, H$_2$O), pK$_{Est(1)}$~2.3, pK$_{Est(2)}$~3.5, pK$_{Est(3)}$~9.2.** Crystd from H$_2$O or aq AcOH. *Picrate* has **m** 188.5-189°(dec) after crystn from H$_2$O.

L-Kynurenine sulfate *[16055-80-4]* **M 306.3, m 194°, monohydrate m 178°, [α]$_D^{25}$ +9.6° (H$_2$O).** Crystd from water by addition of EtOH.

L(+)-Lactic acid *[79-33-4]* **M 90.1, m 52.8°, b 105°/0.1mm, [α]$_D^{20}$ +3.82° (H$_2$O), pK^{31} 3.83.** Purified by fractional distn at 0.1mm pressure, followed by fractional crystn from diethyl ether/isopropyl ether (1:1, dried with sodium). [Borsook, Huffman and Liu *J Biol Chem* **102** 449 *1933*.] The solvent mixture, *benzene/diethyl ether (1:1) containing 5% pet ether (b 60-80°) has also been used.

Lactobionic acid *[96-82-2]* **M 358.3, m 128-130°, [α]$_{546}^{20}$ +28° (c 3, after 24h in H$_2$O), pK$_{Est}$ ~3.6.** Crystd from water by addition of EtOH.

α-Lactose (H$_2$O) *[63-42-3]* **M 360.3, m 220°(dec), [α]$_D^{20}$ +52.3° (c 4.2, H$_2$O), pK 12.2 (OH).** Crystd from water below 93.5°.

Lactulose *[4618-18-2]* **M 342.2, m 167-169°(dec), [α]$_{546}^{20}$ -57° (c 1, H$_2$O).** Crystd from MeOH.

Lanatoside A *[17575-20-1]* **M 969.1, m 245-248°, [α]$_D^{20}$ +32° (EtOH).** Crystd from MeOH.

Lanatoside B *[17575-21-2]* **M 985.1, m 233°(dec), [α]$_D^{20}$ +35° (MeOH).** Crystd from MeOH.

Lanatoside C *[17575-22-3]* **M 297.1, m 246-248°, [α]$_D^{20}$ +34° (EtOH).** Crystd from MeOH.

Lanosterol *[79-63-0]* **M 426.7, m 138-140°, [α]$_D^{20}$+62.0° (c 1, CHCl$_3$).** Recrystd from anhydrous MeOH. Dried *in vacuo* over P$_2$O$_5$ for 3h at 90°. Purity checked by proton magnetic resonance.

Lanthanide shift reagents A variety of these reagents are available commercially and they are generally quite stable and should not deteriorate on long storage in a dry state and in the absence of light. [See G.R.Sullivan in *Top Stereochem* (Eliel and Allinger Eds) J Wiley & Sons Vol 10 287 *1978*; T.C.Morrill Ed. *Lanthanide Shift Reagents* Deerfield Beach Florida *1986*, ISBN 0895731193.]

Lapachol *[84-79-7]* **M 226.3, m 140°.** Crystd from EtOH or diethyl ether.

dl- and *l-***Laudanosine** *[(±) 1699-51-0; (-) 2688-77-9]* **M 357.4, m 114-115°.** Crystd from EtOH. The *(-)*-isomer has **m** 83-85° and [α]$_D^{20}$ -85° (c 0.5, EtOH).

Lauraldehyde (1-dodecanal) *[112-54-9]* **M 184.3, b 99.5-100°/3.5mm, n$^{24.7}$ 1.4328.** Converted to the addition compound by shaking with saturated aqueous NaHSO$_3$ for 1h. The ppte was filtered off, washed with ice cold water, EtOH and ether, then decomposed with aqueous Na$_2$CO$_3$. The aldehyde was extracted into diethyl ether which, after drying and evap, gave an oil which was fractionally distd under vacuum.

Lauric acid (1-dodecanoic acid) *[143-07-7]* **M 200.3, m 44.1°, b 141-142°/0.6-0.7mm, 225°/100mm, pK^{20} 5.3.** Vacuum distd. Crystd from absolute EtOH, or from acetone at -25°.

Alternatively, purified *via* its *methyl ester* (**b** 140.0°/15mm), as described for *capric* acid. Also purified by zone melting.

Lauryl peroxide (dodecyl peroxide)   *[105-74-8]*   **M 398.6, m 53-54°.** Crystd from *n*-hexane or *benzene and stored below 0°. Potentially **EXPLOSIVE**.

L-Leucine   *[61-90-5]*   **M 131.2, m 293-295°(dec), $[\alpha]_D^{25}$ +15.6°** (5M HCl), **pK_1^{25} 2.33, pK_2^{25} 9.74.** Likely impurities are isoleucine, valine, and methionine. Crystd from water by adding 4 volumes of EtOH.

Leucomalachite Green   *[129-73-7]*   **M 330.5, m 92-93°, pK^{25} 6.90 (several pK's).** Crystd from 95% EtOH (10mL/g), then from *benzene/EtOH, and finally from pet ether.

Lithocholic acid   *[434-13-9]*   **M 376.6, m 184-186°, $[\alpha]_D^{23}$ +35°** (c 1, EtOH), **pK_{Est} ~4.8.** Crystd from EtOH or acetic acid.

Lumichrome   *[1086-80-2]*   **M 242.2, m >290°, $pK_{Est(1)}$~-0.1 (basic), $pK_{Est(2)}$~9.9 (acidic),** Recrystd twice from glacial AcOH and dried at 100° in a vacuum.

Luminol (5-aminophthalazin-1,4-dione)   *[521-31-3]*   **M 177.2, m 329-332°, pK_1 3.37, pK_2 6.35.** Dissolved in KOH soln, treated with Norit (charcoal), filtered and ppted with conc HCl. [Hardy, Sietz and Hercules *Talanta* **24** 297 *1977*.] Stored in the dark in an inert atmosphere, because its structure changes during its luminescence. It has been recrystd from 0.1M KOH [Merenyi et al. *J Am Chem Soc* **108** 77716 *1986*].

dl-**Lupinane**   *[10248-30-3]*   **M 169.3, m 98-99°.** Crystd from acetone.

Lupulon   *[468-28-0]*   **M 414.6, m 92-94°.** Crystd from 90% MeOH.

Lutein (α-carotene-3,3'-diol, xanthophyll)   *[127-40-2]*   **M 568.9, m 196°, $\varepsilon_{1cm}^{1\%}$ 1750 (423nm), 2560 (446nm), 2340 (477.5nm) in EtOH; λ_{max} in CS_2 446, 479 and 511nm.** Crystd from MeOH (copper-coloured prisms) or from diethyl ether by adding MeOH. Also purified by chromatography on columns of magnesia or calcium hydroxide, and crystd from CS_2/EtOH. May be purified *via* the dipalmitate ester. Stored in the dark, in an inert atmosphere.

Lutidine (mixture). For the preparation of pure 2,3-, 2,4- and 2,5-lutidine from commercial "2,4- and 2,5-lutidine" see Coulson et al. *J Chem Soc* 1934 *1959*, and Kyte, Jeffery and Vogel *J Chem Soc* 4454 *1960*.

2,3-Lutidine   *[583-61-9]*   **M 107.2, f -14.8°, b 160.6°, d 0.9464, n 1.50857, pK^{25} 6.57.** Steam distd from a soln containing about 1.2 equivalents of 20% H_2SO_4, until *ca* 10% of the base has been carried over with the non-basic impurities. The acid soln was then made alkaline, and the base was separated, dried over NaOH or BaO, and fractionally distd. The distd lutidine was converted to its urea complex by stirring 100g with 40g of urea in 75mL of H_2O, cooling to 5°, filtering at the pump, and washing with 75mL of H_2O. The complex, dissolved in 300mL of H_2O was steam distd until the distillate gave no turbidity with a little solid NaOH. The distillate was then treated with excess solid NaOH, and the upper layer was removed: the aqueous layer was then extracted with diethyl ether. The upper layer and the ether extract were combined, dried (K_2CO_3), and distd through a short column. Final purification was by fractional crystn using partial freezing. [Kyte, Jeffery and Vogel *J Chem Soc* 4454 *1960*].

2,4-Lutidine   *[108-47-4]*   **M 107.2, b 157.8°, d 0.9305, n 1.50087, n^{25} 1.4985, pK^{25} 6.77.** Dried with Linde type 5A molecular sieves, BaO or sodium, and fractionally distd. The distillate (200g) was heated with *benzene (500mL) and conc HCl (150mL) in a Dean and Stark apparatus on a water bath until water no longer separated, and the temperature just below the liquid reached 80°. When cold, the supernatant *benzene was decanted and the 2,4-lutidine hydrochloride, after washing with a little *benzene, was dissolved in water (350mL). After removing any *benzene by steam distn, an aqueous soln of NaOH (80g) was added, and the free lutidine was steam distd. It was isolated by saturating the distillate with solid NaOH, and distd through a short column. The pptn cycle was repeated, then the final distillate was partly frozen in an apparatus at -67.8-68.5° (cooled by acetone/CO_2). The crystals were then melted and distd. [Kyte, Jeffery and Vogel *J Chem Soc* 4454 *1960*.] Alternative purifications are *via* the picrate [Clarke and Rothwell *J Chem Soc* 1885 *1960*], or the hydrobromide [Warnhoff *J Org Chem* **27** 4587 *1962*]. The latter is ppted from a soln of lutidine in *benzene by passing dry HBr gas: the salt is recrystd from $CHCl_3$/methyl ethyl ketone, then decomposed with NaOH, and the free base is extracted into diethyl ether, dried, evaporated and the residue distd.

2,5-Lutidine *[589-93-5]* **M 107.2, m -15.3°, b 156.7°/759mm, d 0.927, n^{25} 1.4982, pK^{25} 6.40.** Steam distd from a soln containing 1-2 equivalents of 20% H_2SO_4 until about 10% of the base had been carried over with the non-basic impurities, then the acid soln was made alkaline, and the base separated, dried with NaOH and fractionally distd twice. Dried with Na and fractionally distd through a Todd column packed with glass helices (see p. 174).

2,6-Lutidine *[108-48-5]* **M 107.2, m -59°, b 144.0°, d 0.92257, n 1.49779, pK^{25} 6.72.** Likely contaminants include 3- and 4-picoline (similar boiling points). However, they are removed by using BF_3, with which they react preferentially, by adding 4mL of BF_3 to 100mL of dry fractionally distd 2,6-lutidine and redistilling. Distn of commercial material from $AlCl_3$ (14g per 100mL) can also be used to remove picolines (and water). Alternatively, lutidine (100mL) can be refluxed with ethyl benzenesulfonate (20g) or ethyl *p*-toluenesulfonate (20g) for 1h, then the upper layer is cooled, separated and distd. The distillate is refluxed with BaO or CaH_2, then fractionally distd, through a glass helices-packed column.
2,6-Lutidine can be dried with KOH or sodium, or by refluxing with (and distilling from) BaO, prior to distn. For purification *via* its picrate, 2,6-lutidine, dissolved in abs EtOH, is treated with an excess of warm ethanolic picric acid. The ppte is filtered off, recrystd from acetone (to give **m** 163-164.5°), and partitioned between ammonia and $CHCl_3$/diethyl ether. The organic soln, after washing with dilute aqueous KOH, is dried with Na_2SO_4 and fractionally distd. [Warnhoff *J Org Chem* **27** 4587 *1962*.] Alternatively, 2,6-lutidine can be purified *via* its urea complex, as described under 2,3-lutidine. Other purification procedures include azeotropic distn with phenol [Coulson et al. *J Appl Chem (London)* **2** 71 *1952*], fractional crystn by partial freezing, and vapour-phase chromatography using a 180-cm column of polyethylene glycol-400 (Shell, 5%) on Embacel (May and Baker) at 100°, with argon as carrier gas [Bamford and Block *J Chem Soc* 4989 *1961*].

3,5-Lutidine *[591-22-0]* **M 107.2, f -6.3°, b 172.0°/767mm, d 0.9419, n 1.50613, n^{25} 1.5035, pK^{25} 6.15.** Dried with sodium and fractionally distd through a Todd column packed with glass helices (see p. 174). Dissolved (100mL) in dil HCl (1:4) and steam distd until 1L of distillate was collected. Excess conc NaOH was added to the residue which was again steam distd. The base was extracted from the distillate, using diethyl ether. The extract was dried with K_2CO_3, and distd. It was then fractionally crystd by partial freezing.

Lycopene *[502-65-8]* **M 536.9, m 172-173°, $\varepsilon_{1cm}^{1\%}$ 2250 (446nm), 3450 (472nm), 3150 (505nm) in pet ether.** Crystd from CS_2/MeOH, diethyl ether/pet ether, or acetone/pet ether, and purified by column chromatography on deactivated alumina, $CaCO_3$, calcium hydroxide or magnesia. Stored in the dark, in an inert atmosphere.

Lycorine *[476-28-8]* **M 552.9, m 275-280°(dec) $[\alpha]_D^{20}$ -130° (c 0.16, EtOH).** Crystd from EtOH.

Lycoxanthin (Ψ,Ψ-carotene-16-ol) *[19891-74-8]* **M 268.3, m 173-174°, $\varepsilon_{1cm}^{1\%}$ 3360 (472.5nm), also λ_{max} 444 and 503nm in pet ether.** Crystd from diethyl ether/light petroleum, *benzene/pet ether or CS_2. Purified by chromatography on columns of $CaCO_3$, $Ca(OH)_2$ or deactivated alumina, washing with *benzene and eluting with 3:1 *benzene/MeOH. Stored in the dark, in an inert atmosphere, at -20°.

Lysergic acid *[82-58-6]* **M 268.3, m 240°(dec), $[\alpha]_D^{20}$ +40° (pyridine), pK_1^{25} 3.32, pK_2^{25} 7.82.** Crystd from water.

L-Lysine *[56-87-1]* **M 146.2, m >210°(dec), pK$_1$ 2.18, pK$_2$ 8.95, pK$_3$ 10.53.** Crystd from aqueous EtOH.

L-Lysine dihydrochloride *[657-26-1]* **M 219.1, m 193°, [α]$_D^{25}$ +25.9° (5M HCl).** Crystd from MeOH, in the presence of excess HCl, by adding diethyl ether.

L-Lysine monohydrochloride *[657-27-2]* **M 182.7, [α] as above.** Likely impurities are arginine, D-lysine, 2,6-diaminoheptanedioic acid and glutamic acid. Crystd from water at pH 4-6 by adding 4 volumes of EtOH. Above 60% relative humidity it forms a dihydrate.

β-D-Lyxose *[1114-34-7]* **M 150.1, m 118-119°, [α]$_D^{20}$ -14° (c 4, H$_2$O).** Crystd from EtOH or aqueous 80% EtOH. Dried under vacuum at 60°, and stored in a vacuum desiccator over P$_2$O$_5$ or CaSO$_4$.

Malachite Green (carbinol) *[510-13-4]* M 346.4, m 112-114°, CI 42000, pK^{24} 6.84.

The oxalate was recrystd from hot water and dried in air. The carbinol was ppted from the oxalate (1g) in distd water (100mL) by adding M NaOH (10mL). The ppte was filtered off, recrystd from 95% EtOH containing a little dissolved KOH, then washed with ether, and crystd from pet ether. Dried in a vacuum at 40°. An acid soln (2 x 10^{-5}M in 6 x 10^{-5}M H$_2$SO$_4$) rapidly reverted to the dye. [Swain and Hedberg *J Am Chem Soc* **72** 3373 *1950*.]

Z-Maleamic acid (*cis*-maleic acid monoamide) *[557-24-4]* **M 115.1, m 158-161°, 172-173°(dec), pK$_{Est}$ ~2.65.** Crystd from EtOH. **IRRITANT.**

Maleic acid *[110-16-7]* **M 116.1, m 143.5°, pK$_1^{25}$ 1.91, pK$_2^{25}$ 6.33.** Crystd from acetone/pet ether (b 60-80°) or hot water. Dried at 100°.

Maleic anhydride *[108-31-6]* **M 98.1, m 54°, b 94-96°/20mm, 199°/760mm.** Crystd from *benzene, CHCl$_3$, CH$_2$Cl$_2$ or CCl$_4$. Sublimed under reduced pressure. [Skell et al. *J Am Chem Soc* **108** 6300 *1986*.]

Maleic hydrazide *[123-33-1]* **M 112.1, m 144°(dec), pK$_1^{25}$ 5.67, pK$_2^{25}$ 13.3.** Crystd from water.

Maleimide (pyrrol-2,5-dione) *[541-59-3]* **M 97.1, m 91-93°, 92.6-93°, d$_D^{105.5}$ 1.2493, n$_D^{110.7}$ 1.49256.** Purified by sublimation in a vacuum. The UV has λ$_{max}$ at 216 and 280nm in EtOH. [de Wolf and van de Straete *Bull Soc Chim Belg* **44** 288 *1935*; UV: Rondestvedt et al. *J Am Chem Soc* **78** 6115 *1956*; IR: Chiorboli and Mirone *Ann Chim (Rome)* **42** 681 *1952*.]

Maleuric acid *[105-61-3]* **M 158.1, m 167-168°(dec).** Crystd from hot water.

***dl*-Malic acid** *[617-48-1 and 6915-15-7]* **M 134.1, m 128-129°.** Crystd from acetone, then from acetone/CCl$_4$, or from ethyl acetate by adding pet ether (b 60-70°). Dried at 35° under 1mm pressure to avoid formation of the anhydride.

L-Malic acid *[97-67-6]* **M 134.1, m 104.5-106°, [α]$_D^{20}$ -2.3° (c 8.5, H$_2$O), pK$_1^{25}$ 3.46, pK$_2^{25}$ 5.10.** Crystd (charcoal) from ethyl acetate/pet ether (b 55-56°), keeping the temperature below 65°. Or, dissolved by refluxing in fifteen parts of anhydrous diethyl ether, decanted, concentrated to one-third volume and crystd at 0°, repeatedly to constant melting point.

Malonamide *[108-13-4]* **M 102.1, m 170°.** Crystd from water.

Malonic acid *[141-82-2]* **M 104.1, m 136°, pK$_1^{25}$ 2.58, pK$_2^{25}$ 5.69.** Crystd from *benzene/diethyl ether (1:1) containing 5% of pet ether (b 60-80°), washed with diethyl ether, then recrystd from H$_2$O or acetone. Dried under vac over conc H$_2$SO$_4$.

Malononitrile *[109-77-3]* **M 66.1, m 32-34°, b 109°/20mm, 113-118°/25mm, 220°/760mm.** Crystd from water, EtOH, *benzene or chloroform. Distd in a vacuum from, and stored over, P_2O_5. [Bernasconi et al. *J Am Chem Soc* **107** 7692 *1985*; Gratenhuis *J Am Chem Soc* **109** 8044 *1987*].

Maltol (3-hydroxy-2-methyl-4-pyrone) *[118-71-8]* **M 126.1, m 161-162°.** Crystd from $CHCl_3$ or aqueous 50% EtOH. Volatile in steam. It can be readily sublimed in a vacuum.

Maltose (H_2O) *[6363-53-7]* **M 360.3, m 118°.** Purified by chromatography from aqueous soln on to a charcoal/Celite (1:1) column, washed with water to remove glucose and other monosaccharides, then eluted with aqueous 75% EtOH. Crystd from water, aqueous EtOH or EtOH containing 1% nitric acid. Dried as the monohydrate at room temperature under vacuum over H_2SO_4 or P_2O_5.

Mandelic acid (α-hydroxyphenylacetic acid) *[S-(+)- 17199-29-0; R-(-)- 611-71-2]* **M 152.2, m 130-133°, 133°, 133.1° (evacuated capillary), 133-133.5°, $[α]_{546}^{20}$ (+) and (-) 188° (c 5, H_2O), $[α]_D^{20}$ (+) and (-) 155° (c 5, H_2O) and (+) and (-) 158° (c 5, Me_2CO), pK^{25} 3.41.** Purified by recrystn from H_2O, *C_6H_6 or $CHCl_3$. [Roger *J Chem Soc* 2168 *1932*; Jamison and Turner *J Chem Soc* 611 *1942*.] They have solubilities in H_2O of *ca* 11% at 25°. [Banks and Davies *J Chem Soc* 73 *1938*.] The *S-benzylisothiuronium salt* has **m** 180° (from H_2O) and $[α]_D^{25}$ (+) and (-) 57° (c 20, EtOH) [El Masri et al. *Biochem J* **68** 199 *1958*].

RS-(±)-**Mandelic acid** *[61-72-3]* **M 152.2, m 118°, 120-121°.** Purified by Soxhlet extraction with *benzene (about 6mL/g), allowing the extract to crystallise. Also crystallises from $CHCl_3$. The *S-benzylisothiuronium salt* has **m** 169° (166°) (from H_2O). Dry at room temperature under vacuum.

D-Mannitol *[69-65-8]* **M 182.2, m 166.1°, $[α]_{546}^{20}$ + 29° (c 10, after 1h in 8% borax soln).** Crystd from EtOH or distilled water and dried at 100°.

Mannitol hexanitrate *[15825-70-4]* **M 452.2, m 112-113°.** Crystd from EtOH. **EXPLOSIVE (on detonation).**

α-D-Mannose *[3458-28-4]* **M 180.2, m 132°, $[α]_D^{20}$ +14.1° (c 4, H_2O).** Crystd repeatedly from EtOH or aq 80% EtOH, then dried under vacuum over P_2O_5 at 60°.

Meconic acid (3-hydroxy-γ-pyrone-2,6-dicarboxylic acid) *[497-59-6]* **M 200.1, m 100° (loses H_2O), pK_1^{25} 1.83, pK_2^{25} 2.3, pK_3^{25} 10.10.** Crystd from water and dried at 100° for 20min.

Melamine (2,4,6-triamino-1,3,5-triazine) *[108-78-1]* **M 126.1, m 353°, pK^{25} 5.00.** Crystd from water or dilute aqueous NaOH.

D(+)-Melezitose (H_2O) *[597-12-6]* **M 540.5, m 153-154°(dec), $2H_2O$ m 160°(dec), $[α]_D^{20}$ +88° (c 4, H_2O).** Crystallises from water as the dihydrate, then dried at 110° (anhydrous).

D(+)-Melibiose ($2H_2O$) *[585-99-9, 66009-10-7]* **M 360.3, m 84-85°, $[α]_D^{20}$+135° (c 5, after 10h H_2O).** Crystallises as a hydrate from water or aqueous EtOH.

(±)-Mellein [(±)-3,4-dihydro-8-hydroxy-3-methyl-2-benzopyran-1-one] *[1200-93-7]* **M 178.2, m 37-39°, 39°, pK_{Est} ~9.5.** Purified by recrystn from aqueous EtOH. It has UV max at 247 and 314nm. [Arakawa et al. *Justus Liebigs Ann Chem* **728** 152 *1969*; Blair and Newbold *Chem Ind (London)* 93 *1955*, *J Chem Soc* 2871 *1955*.] The *methyl ether* has **m** 66-67° and UV: $λ$max 242nm ($ε$ 7,400) and 305nm ($ε$ 4,600).

Melphalan (4-[bis-{2-chloroethyl}amino]-L-phenylalanine) *[148-82-3]* **M 305.2, m 182-183° (dec), 183-185°, $[α]_D^{25}$ +7.5° (c 1.33, 1.0 N HCl), $[α]_D^{20}$-28° (c 0.8, MeOH), pK_{Est} ~6.4.** Purified by recrystn from MeOH and its solubility is 5% in 95% EtOH containing one drop of 6N HCl.

It is soluble in EtOH and propylene glycol but is almost insoluble in H_2O. The *RS-form* has **m** 180-181° and the *R-form* crystallises from MeOH with a **m** 181.5-182° and $[\alpha]_D^{21}$ -7.5° (c 1.26, 1.0 N HCl). [Bergel and Stock *J Chem Soc* 2409 *1954*.]

(-)-Menthol   *[2216-51-5]*   **M 156.3, m 44-46.5°, $[\alpha]_D$ 50° (c 10, EtOH).** Crystd from $CHCl_3$, pet ether or EtOH/water.

1R-(-)-Menthyl chloride (1S,2R,4R-2-chloro-1-isopropyl-1-methylcyclohexane) *[16052-42-9]* **M 174.7, m -20.1° to -16.5°, b 88.5°/12.5mm, 101-105°/21mm, d_D^{20} 0.936, n_D^{20} 1.463(neat).** Dissolve in pet ether (b 40-60°), wash with H_2O, conc H_2SO_4 until no discoloration of the organic layer occurs (care with the use of conc H_2SO_4 during shaking in a separating funnel), again with H_2O and dry over $MgSO_4$. Evaporate the pet ether and dist the residual oil through a Claisen head with a Vigreux neck (head) of *ca* 40 cm length. [Smith and Wright *J Org Chem* 17 1116 *1952*; Barton et al. *J Chem Soc* 453 *1952*.]

Meprobamate [2,2-di(carbamoyloxymethyl)pentane]   *[57-53-4]*   **M 246.3, m 104-106°.** Crystd from hot water. Could be an addictive drug.

2-Mercaptobenzimidazole   *[583-39-1]*   **M 150.2, m 302-304°, 312°, pK^{20} 10.24.** Crystd from aq EtOH, AcOH or aq ammonia.

2-Mercaptobenzothiazole   *[149-30-4]*   **M 167.2, m 182°, pK^{25} 7.5 (50% aq AcOH).** Crystd repeatedly from 95% EtOH, or purified by incomplete pptn by dilute H_2SO_4 from a basic soln, followed by several crystns from acetone/H_2O or *benzene. Complexes with Ag, Au, Bi, Cd, Hg, Ir, Pt, and Tl.

2-Mercaptoethanol *[60-24-2]* **M 78.1, b 44°/4mm, 53.5°/10mm, 58°/12mm, 68°/20mm, 78.5°/40mm, 96-97° (92°)/100mm, 157°/748mm, d_4^{20} 1.114, n_D^{20} 1.500, pK^{25} 9.72 (9.43).** Purified by distn in a vacuum. Distn at atmospheric pressure causes some oxidation and should be done in an inert atmosphere. [Woodward *J Chem Soc* 1892 *1948*.] **It has a foul odour, is irritating to the eyes, nose and skin — should be handled in an efficient fume cupboard.** It is miscible with H_2O, EtOH, Et_2O and *C_6H_6 and has a UV max at 235nm. The *2,4-dinitrophenyl thioether* has **m** 101-102°(from EtOH or aq MeOH) [Grogen et al. *J Org Chem* 20 50 *1955*].

2-Mercaptoethylamine *[60-23-1]* **M 77.2.** See cysteamine p. 525 in Chapter 6.

2-Mercaptoimidazole *[872-35-5]* **M 100.1, m 221-222°, pK_1^{20} -1.6, p K_2^{20} 11.6.** Crystd from H_2O.

2-Mercapto-1-methylimidazole   *[60-56-0]*   **M 114.2, m 145-147°, pK_1^{20} -2.0, p K_2^{20} 11.9.** Crystd from EtOH.

6-Mercaptopurine (H_2O) *[6112-76-1]* **M 170.2, m >315°(dec), pK_1^{20} 0.5, pK_2^{20} 7.77, pK_3^{20} 10.8.** Crystd from pyridine (30mL/g), washed with pyridine, then triturated with water (25mL/g), adjusting to pH 5 by adding M HCl. Recrystd by heating, then cooling, the soln. Filtered, washed with water and dried at 110°. Has also been crystd from water (charcoal).

8-Mercaptoquinoline (2H_2O, thioxine) *[491-33-8]* **M 197.3, m 58-59°, pK_1^{25} 2.0, pK_2^{25} 8.40.** Easily oxidised in air to give diquinolyl-8,8'-disulfide (which is stable). It is more convenient to make 8-mercaptoquinoline by reduction of the material. [Nakamura and Sekido *Talanta* 17 515 *1970*.]

Mesaconic acid (methylfumaric acid)   *[498-24-8]*   **M 130.1, m 204-205°, pK^{18} 4.82.** Crystd from water or EtOH [Katakis et al. *J Chem Soc, Dalton Trans* 1491 *1986*].

Mescaline sulfate [2-(3,4,5-trimethoxyphenyl)ethylamine sulfate]   *[5967-42-0]*   **M 309.3, m 183-184°, pK_{Est} ~9.7.** Crystd from water.

Mesitylene (1,3,5-trimethylbenzene) *[108-67-8]* **M 120.2, m -44.7°, b 99.0-99.8°/100mm, 166.5-167°/760mm, n^{25} 1.4967, d 0.865.** Dried with $CaCl_2$ and distd from Na in a glass helices packed column. Treated with silica gel and redistd. Alternative purifications include vapour-phase chromatography, or fractional distn followed by azeotropic distn with 2-methoxyethanol (which is subsequently washed out with H_2O), drying and fractional distn. More exhaustive purification uses sulfonation by dissolving in two volumes of conc H_2SO_4, precipitating with four volumes of conc HCl at 0°, washing with conc HCl and recrystallising from $CHCl_3$. The mesitylene sulfonic acid is hydrolysed with boiling 20% HCl and steam distd. The separated mesitylene is dried ($MgSO_4$ or $CaSO_4$) and distilled. It can also be fractionally crystd from the melt at low temperatures.

Mesityl oxide *[141-79-7]* **M 98.2, b 112°/760mm, n^{24} 1.4412, d 0.854, pK^{20} -5.36 (H$_0$ scale, aq H$_2$SO$_4$).** Purified *via* the *semicarbazone* (m 165°). [Erskine and Waight *J Chem Soc* 3425 *1960.*]

Metalphthalein (H$_2$O) (o-cresolcomplexon) *[2411-89-4]* **M 636.6, m 186°(dec).** See *o*-cresolphthalein complexone on p. 173.

Metanilic acid (3-aminobenzenesulfonic acid) *[121-47-1]* **M 173.2, m <300°(dec), pK$_1^{25}$ < 1, pK$_2^{25}$ 3.74.** Crystd from water (as the hydrate), under CO_2 in a semi-darkened room. (The soln is photosensitive.) Dried over 90% H_2SO_4 in a vac desiccator.

α-Methacraldehyde *[78-85-3]* **M 68.1, b 68.4°, d 0.849, n 1.416.** Fractionally distd under nitrogen through a short Vigreux column. Stored in sealed ampoules. (Slight polymerisation may occur.)

Methacrylamide *[79-39-0]* **M 85.1, m 111-112°.** Crystd from *benzene or ethyl acetate and dried under vacuum at room temperature.

Methacrylic acid *[79-41-4]* **M 86.1, b 72°/14mm, 160°/760mm, d 1.015, n 1.431, pK 4.65.** Aq methacrylic acid (90%) was satd with NaCl (to remove the bulk of the water), then the organic phase was dried with $CaCl_2$ and distd under vacuum. Polymerisation inhibitors include 0.25% *p*-methoxyphenol, 0.1% hydroquinone, or 0.05% *N,N'*-diphenyl-*p*-phenylenediamine.

Methacrylic anhydride *[760-93-0]* **M 154.2, b 65°/2mm, d 1.040, n 1.454.** Distd at 2mm pressure, immediately before use, in the presence of hydroquinone.

Methacrylonitrile *[126-98-7]* **M 67.1, b 90.3°, d 0.800, n 1.4007, n^{30} 1.3954.** Washed (to remove inhibitors such as *p-tert*-butylcatechol) with satd aq $NaHSO_3$, 1% NaOH in saturated NaCl and then with saturated NaCl. Dried with $CaCl_2$ and fractionally distd under nitrogen to separate from impurities such as methacrolein and acetone.

Methane *[74-82-8]* **M 16.0, m -184°, b -164°/760mm, -130°/6.7atm, d^{-164} 0.466 (air 1).** Dried by passage over $CaCl_2$ and P_2O_5, then passed through a Dry-ice trap and fractionally distd from a liquid-nitrogen trap. Oxygen can be removed by prior passage in a stream of hydrogen over reduced copper oxide at 500°, and higher hydrocarbons can be removed by chlorinating about 10% of the sample: the hydrocarbons, chlorides and HCl are readily separated from the methane by condensing the sample in the liquid-nitrogen trap and fractionally distilling it. Methane has also been washed with conc H_2SO_4, then solid NaOH and then 30% NaOH soln. It was dried with $CaCl_2$, then P_2O_5, and condensed in a trap at liquid air temp, then transferred to another trap cooled in liquid nitrogen. CO_2, O_2, N_2 and higher hydrocarbons can be removed from methane by adsorption on charcoal. [Eiseman and Potter *J Res Nat Bur Stand* **58** 213 *1957.*] **HIGHLY FLAMMABLE.**

Methanesulfonic acid *[76-75-2]* **M 96.1, m 20°, b 134.5-135°/3mm, d 1.483, n 1.432, pK^{25} -1.86 (-1.2).** Dried, either by azeotropic removal of water with *benzene or toluene, or by stirring 20g of P_2O_5 with 500mL of the acid at 100° for 0.5h. Then distd under vacuum and fractionally crystd by partial freezing. Sulfuric acid, if present, can be removed by prior addition of $Ba(OH)_2$ to a dilute soln, filtering off the $BaSO_4$ and concentrating under reduced pressure, and is sufficiently pure for most applications.

Methanesulfonyl chloride *[124-63-0]* **M 114.5. b 55°/11mmm, d 1.474, n 1.452.** Distd from P_2O_5 under vacuum.

Methanol *[67-56-1]* **M 32.0, b 64.5°, d^{15} 0.79609, d^{25} 1.32663, n^{15} 1.33057, n^{25} 1.32663, pK^{25} 15.5.** Almost all methanol is now obtained synthetically. Likely impurities are water, acetone, formaldehyde, ethanol, methyl formate and traces of dimethyl ether, methylal, methyl acetate, acetaldehyde, carbon dioxide and ammonia. Most of the water (down to about 0.01%) can be removed by fractional distn. Drying with CaO is unnecessary and wasteful. Anhydrous methanol can be obtained from "absolute" material by passage through Linde type 4A molecular sieves, or by drying with CaH_2, $CaSO_4$, or with just a little more sodium than required to react with the water present; in all cases the methanol is then distd. Two treatments with sodium reduces the water content to about 5×10^{-5}%. [Friedman, Gill and Doty *J Am Chem Soc* **83** 4050 *1961*.] Lund and Bjerrum [*Chem Ber* **64** 210 *1931*] warmed clean dry magnesium turnings (5g) and iodine (0.5g) with 50-75mL of "absolute" methanol in a flask until the iodine disappeared and all the magnesium was converted to methoxide. Up to 1L of methanol was added and, after refluxing for 2-3h, it was distd off, excluding moisture from the system. Redistn from tribromobenzoic acid removes basic impurities and traces of magnesium oxides, and leaves conductivity-quality material. The method of Hartley and Raikes [*J Chem Soc* **127** 524 *1925*] gives a slightly better product. This consists of an initial fractional distn, followed by distn from aluminium methoxide, and then ammonia and other volatile impurities are removed by refluxing for 6h with freshly dehydrated $CuSO_4$ (2g/L) while dry air is passed through: the methanol is finally distd. (The aluminium methoxide is prepared by warming with aluminium amalgam (3g/L) until all the aluminium has reacted. The amalgam is obtained by warming pieces of sheet aluminium with a soln of $HgCl_2$ in dry methanol). This treatment also removes aldehydes.

If acetone is present in the methanol, it is usually removed prior to drying. Bates, Mullaly and Hartley [*J Chem Soc* 401 *1923*] dissolved 25g of iodine in 1L of methanol and then poured the soln, with constant stirring, into 500mL of M NaOH. Addition of 150mL of water ppted iodoform. The soln was stood overnight, filtered, then boiled under reflux until the odour of iodoform disappeared, and fractionally distd. (This treatment also removes formaldehyde.) Morton and Mark [*Ind Eng Chem (Anal Edn)* **6** 151 *1934*] refluxed methanol (1L) with furfural (50mL) and 10% NaOH soln (120mL) for 6-12h, the refluxing resin carrying down with it the acetone and other carbonyl-containing impurities. The alcohol was then fractionally distd. Evers and Knox [*J Am Chem Soc* **73** 1739 *1951*], after refluxing 4.5L of methanol for 24h with 50g of magnesium, distd off 4L of it, which they then refluxed with $AgNO_3$ for 24h in the absence of moisture or CO_2. The methanol was again distd, shaken for 24h with activated alumina before being filtered through a glass sinter and distd under nitrogen in an all-glass still. Material suitable for conductivity work was obtained.

Variations of the above methods have also been used. For example, a sodium hydroxide soln containing iodine has been added to methanol and, after standing for 1day, the soln has been poured slowly into about a quarter of its volume of 10% $AgNO_3$, shaken for several hours, then distd. Sulfanilic acid has been used instead of tribromobenzoic acid in Lund and Bjerrum's method. A soln of 15g of magnesium in 500mL of methanol has been heated under reflux, under nitrogen, with hydroquinone (30g), before degassing and distilling the methanol, which was subsequently stored with magnesium (2g) and hydroquinone (4g per 100mL). Refluxing for about 12h removes the bulk of the formaldehyde from methanol: further purification has been obtained by subsequent distn, refluxing for 12h with dinitrophenylhydrazine (5g) and H_2SO_4 (2g/L), and again fractionally distilling.

Rapid purification: Methanol purification is the same as for Ethanol.

Another simple purification procedure consists of adding 2g of $NaBH_4$ to 1.5L methanol, gently bubbling with argon and refluxing for a day at 30°, then adding 2g of freshly cut sodium (washed with methanol) and refluxing for 1day before distilling. The middle fraction is taken. [Jou and Freeman *J Phys Chem* **81** 909 *1977*.]

dl-**Methionine** *(RS-α-aminohexanoic acid)* *[59-51-8]* **M 149.2, m 281°(dec).** Crystd from hot water.

L-Methionine *[63-68-3]* **M 149.2, m 283°(dec), $[\alpha]_D^{25}$ +21.2° (0.2M HCl) pK$_1^{25}$ 2.13, pK$_2^{25}$ 9.73.** Crystd from aqueous EtOH.

dl-**Methionine sulfoxide**   *[454-41-1, 62697-73-8]*   **M 165.2, m >240°(dec).**   Likely impurities are *dl*-methionine sulfone and *dl*-methionine.   Crystd from water by adding EtOH in excess.

Methoxyacetic acid   *[625-45-6]*   **M 90.1, b 97°/13-14mm, d 1.175, n 1.417, pK^{25} 3.57.** Fractionally crystd by repeated partial freezing, then fractionally distd under vacuum through a vacuum-jacketed Vigreux column 20cm long.

p-**Methoxyacetophenone**   *[100-06-1]*   **M 150.2, m 39°, b 139°/15mm, 264°/736mm.**   Crystd from diethyl ether/pet ether.

Methoxyamine hydrochloride   *[593-56-6]*   **M 83.5, m 151-152°, pK^{25} 4.60.**   Crystd from absolute EtOH or EtOH by addition of diethyl ether.   [Kovach et al. *J Am Chem Soc* **107** 7360 *1985*.]

p-**Methoxyazobenzene**   *[2396-60-3]*   **M 212.3, m 54-56°.**   Crystd from EtOH.

3-Methoxybenzanthrone   *[3688-79-7]*   **M 274.3, m 173°.**   Crystd from *benzene, EtOH or Me$_2$CO as yellow needles.

m-**Methoxybenzoic acid** (*m*-**anisic acid**)   *[586-38-9]*   **M 152.2, m 110°, pK^{25} 4.09.**   Crystd from EtOH/water.

p-**Methoxybenzoic acid** (*p*-**anisic acid**)   *[100-09-4]*   **M 152.2, m 184.0-184.5°, pK^{25} 4.51.** Crystd from EtOH, water, EtOH/water or toluene.

4-Methoxybenzyl chloride (anisyl chloride)   *[824-94-2]*   **M 156.6, m -1°, b 76°/0.1mm, 95°/5mm, 110°/10mm, 117-117/5°/14mm, 117°/18mm, d$_4^{20}$ 1.15491, n$_D^{20}$ 1.55478.**   Purified by fractional distn under vacuum and the middle fraction is redistd at 10^{-6} mm at room temperature by intermittent cooling of the receiver in liquid N$_2$, and the middle fraction is collected.   [Mohammed and Kosower *J Am Chem Soc* **93** 2709 *1971*.]

3-Methoxycarbonyl-2,5-dihydrothiophen-1,1-dioxide   *[67488-50-0]*   **M 176.1, m 57-58°, 60-62°.**   If IR show CO bands then dissolve in CHCl$_2$, wash with aqueous Na$_2$CO$_3$ and H$_2$O, dry over MgSO$_4$, filter, evaporate and wash the residue with cold Et$_2$O and dry *in vacuo*.   NMR (CDCl$_3$): δ 7.00 (m 1H), 3.98 (bs 4H) and 3.80 (s Me).   [Mcintoch and Sieber *J Org Chem* **43** 4431 *1978*.]

"Methoxychlor", 1,1-Bis(*p*-methoxyphenyl)-2,2,2-trichloroethane (**dimorphic**) *[72-43-5]* **M 345.7, m 78-78.2°, or 86-88°.**   Freed from 1,1-bis(*p*-chlorophenyl)-2,2,2-trichloroethane by crystn from EtOH.

trans-*p*-**Methoxycinnamic acid**   *[830-09-1, 943-89-5 (trans)]*   **M 178.2, m 173.4-174.8°, pK^{25} 4.54.**   Crystd from MeOH to constant melting point and UV spectrum.

2-Methoxyethanol (methylcellosolve)   *[109-86-4]*   **M 76.1, b 124.4°, d 0.964, n 1.4017, pK^{25} 14.8.**   Peroxides can be removed by refluxing with stannous chloride or by filtration under slight pressure through a column of activated alumina.   2-Methoxyethanol can be dried with K$_2$CO$_3$, CaSO$_4$, MgSO$_4$ or silica gel, with a final distn from sodium.   Aliphatic ketones (and water) can be removed by making the solvent 0.1% in 2,4-dinitrophenylhydrazine and allowing to stand overnight with silica gel before fractionally distilling.

2-Methoxyethoxymethylchloride   (**MEMCl**)   *[3970-21-6]*   **M 124.6, b 50-52°/13mm, 140-145°(dec)/atm, d 1.092, n 1.427.**   Possible impurites are methoxyethanol (b 124°/atm), HCHO and HCl which can be removed below the **b** of MEMCl.   Purify by fractional distn in a vacuum.   If too impure, prepare from methoxyethanol (152g) and *s*-trioxane (66g) by bubbling a stream of dry HCl (with stirring) until a clear mixt is obtained.   Dilute with pentane (900mL), dry (3h over 100g MgSO$_4$, at 5°), evaporate and the residue is distd in a vac.   It is MOISTURE SENSITIVE and **TOXIC**.   The *MEM.NEt$_3^+$Cl⁻* salt, prepared by reactn with

1.3 equivs of Et$_3$N (16h/25°) and dried in vac has **m** 58-61°, and is moisture sensitive. [Corey et al. *Tetrahedron Lett* 809 *1976*.]

β-Methoxyethylamine *[109-85-3]* **M 75.1, b 94°, d 0.874, n 1.407, pK^{25} 9.40.** An aqueous 70% soln was dehydrated by azeotropic distn with *benzene or methylene chloride and the amine was distilled twice from zinc dust. Store in a tight container as it absorbs CO$_2$ from the atmosphere.

6-Methoxy-1-indanone *[13623-25-1]* **M 162.2, m 151-153°.** Crystd from MeOH, then sublimed in a high vacuum.

5-Methoxyindole *[1006-94-6]* **M 147.2, m 55°, 57°, b 176-178°/17mm, pK$_{Est}$ ~0.** Crystd from cyclohexane pet ether or pet ether/Et$_2$O.

1-Methoxynaphthalene *[2216-69-5]* **M 158.2, b 268.4-268.5°.** See methyl 1-naphthyl ether on p. 295.

2-Methoxynaphthalene *[93-04-9]* **M 158.2, m 73.0-73.6°, b 273°/760mm.** Fractionally distd under vacuum. Crystd from absolute EtOH, aqueous EtOH, *benzene or *n*-heptane, and dried under vacuum in an Abderhalden pistol or distd *in vacuo*. [Kikuchi et al. *J Phys Chem* **91** 574 *1987*.]

1-Methoxy-4-nitronaphthalene *[4900-63-4]* **M 203.2, m 85°.** Purified by chromatography on silica gel and recrystd from MeOH. [Bunce et al. *J Org Chem* **52** 4214 *1987*.]

***p*-Methoxyphenol** *[150-76-5]* **M 124.1, m 54-55°, b 243°, pK^{25} 10.21.** Crystd from *benzene, pet ether or H$_2$O, and dried under vacuum over P$_2$O$_5$ at room temp. Sublimes *in vacuo*. [Wolfenden et al. *J Am Chem Soc* **109** 463 *1987*.]

α-Methoxyphenylacetic acid (**O-methyl mandelic acid**), *[R-(-)- 3966-32-3; S-(+)- 26164-26-1]* **M 166.2, m 62.9°, 62-65°, 65-66°, [α]$^{20}_{546}$ (-) and (+) 179° (169.8°), [α]$^{20}_D$ (-) and (+) 150.7° (148°) (c 0.5, EtOH), pK$_{Est}$ ~3.1.** Purified by recrystn from *C$_6$H$_6$-pet ether (b 80-100°). [Neilson and Peters *J Chem Soc* 1519 *1962*; Weizmann et al. *J Am Chem Soc* **70** 1153 *1948*; Pirie and Smith *J Chem Soc* 338 *1932*; NMR: Dale and Mosher *J Am Chem Soc* **95** 512 *1973*; for resolution: Roy and Deslongchamps *Can J Chem* **63** 651 *1985;* Trost et al. *J Am Chem Soc* **108** 4974 *1986*.] The *racemic mixture* has **m** 72°, **b** 121-122°/0.4mm, 165°/18mm (from pet ether) [Braun et al. *Chem Ber* **63** 2847 *1930*].

***o*-Methoxyphenylacetic acid** *[93-25-4]* **M 166.2, m 71.0-71.2°, pK$_{Est}$ ~4.4.** Crystd from H$_2$O, EtOH or aq EtOH and dried in a vacuum desiccator over Sicapent.

***m*-Methoxyphenylacetic acid** *[1798-09-0]* **M 166.2, m 71.0-71.2°, pK$_{Est}$ ~4.3.** Crystd from H$_2$O, or aq EtOH.

***p*-Methoxyphenylacetic acid** *[104-01-8]* **M 166.2, m 85-87°, pK^{25} 4.36.** Crystd from EtOH/water.

5-(*p*-Methoxyphenyl)-1,2-dithiole-3-thione *[42766-10-9]* **M 240.2, m 111°.** Crystd from butyl acetate.

***N*-(*p*-Methoxyphenyl)-*p*-phenylenediamine** *[101-64-4]* **M 214.3, m 102°, b 238°/12mm, pK 6.6 (5.9).** Crystd from ligroin.

8-Methoxypsoralen see xanthotoxin p. 577 in Chapter 6.

α-Methoxy-α-trifluoromethylphenylacetic acid (MTPA, Mosher's acid) *[R-(+)- 20445-31-2; S-(-)- 17257-71-5]* **M 234.2, m 43-45°, 90°/0.1mm, 105-107°/1mm, [α]$^{20}_{546}$ (+) and (-) 87°, [α]$^{20}_D$ (+) and (-) 73° (c 2, MeOH), pK$_{Est}$ ~2.5.** A likely impurity is phenylethylamine from the resolution. Dissolve acid in ether-*benzene (3:1), wash with 0.5N H_2SO_4, then water, dry over magnesium sulfate, filter, evaporate and distil. [Dale et al. *J Org Chem* **34** 2543 *1969, J Am Chem Soc* **75** 512 *1973*.]

α-Methoxy-α-trifluoromethylphenylacetyl chloride *[R-(-)- 39637-99-5; S-(+)- 20445-33-4]* **M 252.6, b 54-56°/1mm, 213-214°/atm, d_4^{20} 1.353, n_D^{20} 1.468, $[α]_{546}^{20}$ (-) and (+) 167°, $[α]_D^{20}$ (-) and (+) 137° (c 4, CCl_4), $[α]_D^{24}$ (-) and (+)10.0° (neat).** The most likely impurity is the free acid due to hydrolysis and should be checked by IR. If free from acid then distil taking care to keep moisture out of the apparatus. Otherwise add $SOCl_2$ and reflux for 50h and distil. Note that shorter reflux times resulted in a higher boiling fraction (**b** 130-155°/1mm) which has been identified as the anhydride. [Dale et al. *J Org Chem* **34** 2543 *1969*; for enantiomeric purity see *J Am Chem Soc* **97** 512 *1973*.]

N-Methylacetamide *[79-16-3]* **M 73.1, m 30°, b 70-71°/2.5-3mm, pK_1^{25}-3.70, pK_2^{25}-0.42.** Fractionally distd under vacuum, then fractionally crystd twice from its melt. Impurities include acetic acid, methyl amine and H_2O. For detailed purification procedure, see Knecht and Kolthoff, *Inorg Chem* **1** 195 *1962*. Although *N*-methylacetamide is commercially available it is often extensively contaminated with acetic acid, methylamine, water and an unidentified impurity. The recommended procedure is to synthesise it in the laboratory by direct reaction. The gaseous amine is passed into hot glacial acetic acid, to give a partially aq soln of methylammonium acetate which is heated to *ca* 130° to expel water. Chemical methods of purificatn such as extractn by pet ether, treatment with H_2SO_4, K_2CO_3 or CaO can be used but are more laborious.
Tests for purity include the Karl Fischer titration for water; this can be applied directly. Acetic acid and methylamine can be detected polarographically.
In addition to the above, purification of *N*-methylacetamide can be achieved by fractional freezing, including zone melting, repeated many times, or by chemical treatment with vacuum distn under reduced pressures. For details of zone melting techniques, see Knecht in *Recommended Methods for Purification of Solvents and Tests for Impurities*, Coetzee Ed. Pergamon Press *1982*.

N-Methylacetanilide *[579-10-2]* **M 149.2, m 102-104°.** Crystd from water, ether or pet ether (b 80-100°).

Methyl acetate *[79-20-9]* **M 74.1, b 56.7-57.2°, d 0.934. n 1.36193, n^{25} 1.3538, pK^{20} -7.28 (H_0 scale, aq H_2SO_4).** Methanol in methyl acetate can be detected by measuring solubility in water. At 20°, the solubility of methyl acetate in water is *ca* 35g per 100mL, but 1% MeOH confers miscibility. Methanol can be removed by conversion to methyl acetate, using refluxing for 6h with acetic anhydride (85mL/L), followed by fractional distn. Acidic impurities can be removed by shaking with anhydrous K_2CO_3 and distilling. An alternative treatment is with acetyl chloride, followed by washing with conc NaCl and drying with CaO or $MgSO_4$. (Solid $CaCl_2$ cannot be used because it forms a crystalline addition compound.) Distn from copper stearate destroys peroxides. Free alcohol or acid can be eliminated from methyl acetate by shaking with strong aq Na_2CO_3 or K_2CO_3 (three times), then with aq 50% $CaCl_2$ (three times), satd aq NaCl (twice), drying with K_2CO_3 and distn from P_2O_5.

Methyl acetimidate hydrochloride *[14777-27-6]* **M 109.6, m 93-95°, 105°(dec), pK_{Est} ~ 5.5.** Crystd from methanol by adding dry ether to a ratio of 1:1 and cooled at 0°. Filter off the crystals in a cold room, wash with methanol/ether (1:2), then dry in a vacuum. [Hunter and Ludwig *J Am Chem Soc* **84** 3491 *1962*.] The *free base* has **b** 90-91°/765mm, **d** 0.867, **n** 1.403 [Hunter and Ludwig *Methods Enzymol* **25** 585 *1973*.]

p-Methylacetophenone *[122-00-9]* **M 134.2, m 22-24°, b 93.5°/7mm, 110°/14mm, d 1.000, n 1.5335.** Impurities, including the *o*- and *m*-isomers, were removed by forming the semicarbazone which, after repeated crystn, was hydrolysed to the ketone. [Brown and Marino *J Am Chem Soc* **84** 1236 *1962*.] Also purified by distn under reduced pressure, followed by low temperature crystn from isopentane.

Methyl acrylate *[96-33-3]* **M 86.1, b 80°, d 0.9535, n 1.4040.** Washed repeatedly with aqueous NaOH until free from inhibitors (such as hydroquinone), then washed with distd water, dried ($CaCl_2$) and fractionally distd under reduced pressure in an all-glass apparatus. Sealed under nitrogen and stored at 0° in the dark. [Bamford and Han *J Chem Soc, Faraday Trans 1* **78** 855 *1982*.]

1-Methyladamantane *[768-91-2]* **M 150.2, m 103°.** Purified by zone melting and sublimes at 90-95°/12mm.

2-Methyladamantane *[700-56-1]* **M 150.2.** Purified by zone melting.

Methylamine (gas) *[74-89-5]* **M 31.1, b -7.55°/719mm, pK²⁵ 10.62.** Dried with sodium or BaO.

Methylamine hydrochloride *[593-51-1]* **M 67.5, m 231.8-233.4°, b 225-230°/15mm, pK²⁵ 10.62.** Crystd from *n*-butanol, absolute EtOH or MeOH/CHCl₃. Washed with CHCl₃ to remove traces of dimethylamine hydrochloride. Dried under vacuum first with H₂SO₄ then P₂O₅. Deliquescent, stored in a desiccator over P₂O₅.

1-Methylaminoanthraquinone *[82-38-2]* **M 237.3, m 166.5°, pK$_{Est}$~2.** Crystd to constant melting point from butan-1-ol, then crystd from EtOH. It can be sublimed under vacuum.

***N*-Methyl-*o*-aminobenzoic acid (*N*-methylanthranilic acid)** *[119-68-6]* **M 151.2, m 178.5°, pK$_1^{25}$ 1.97, pK$_2^{25}$ 5.34.** Crystd from water or EtOH.

***p*-Methylaminophenol sulfate** *[55-55-0]* **M 344.4, m 260°(dec), pK²⁵ 5.9.** Crystd from MeOH.

6-Methylaminopurine *[443-72-1]* **M 149.2, m >300°, 312-314° (dec), pK$_1^{20}$<1, pK$_2^{20}$ 4.15, pK$_3^{20}$ 10.02.** Best purified by recrystallising 2g from 50mL of H₂O and 1.2g of charcoal. [UV: Albert and Brown *J Chem Soc* 2060 *1954*; UV: Mason *J Chem Soc* 2071 *1954*; see also Elion et al. *J Am Chem Soc* **74** 411 *1952*.] The *picrate* has **m** 265°(257°) [Bredereck et al. *Chem Ber* **81** 307 *1948*].

Methyl 3-aminopyrazine-2-carboxylate *[16298-03-6]* **M 153.1, m 169-172°, 172°.** Forms yellow needles from H₂O (100 parts using charcoal). If it contains the free acid then dissolve in CH₂Cl₂ wash with saturated aqueous Na₂CO₃, brine, dry over MgSO₄ filter, evaporate and recrystallise the residue. The *free acid* has **m** 203-204° (dec) [UV: Brown and Mason *J Chem Soc* 3443 *1956*] and pK₁ <1 and pK₂ 3.70. The *ammonium salt* has **m** 232° (dec) (from aq Me₂CO) and the *amide* has **m** 239.2° (from H₂O) [Ellingson et al. *J Am Chem Soc* **67** 1711 *1945*].

***N*-Methylaniline** *[100-61-8]* **M 107.2, b 57°/4mm, 81-82°/14mm, d 0.985, n 1.570, pK²⁵ 4.56.** Dried with KOH pellets and fractionally distd under vacuum. Acetylated and the acetyl derivative was recrystd to constant melting point (**m** 101-102°), then hydrolysed with aqueous HCl and distd from zinc dust under reduced pressure. [Hammond and Parks *J Am Chem Soc* **77** 340 *1955*.]

***N*-Methylaniline hydrochloride** *[2739-12-0]* **M 143.7, m 123.0-123.1°.** Crystd from dry *benzene/CHCl₃ and dried under vacuum.

Methyl *p*-anisate *[121-98-2]* **M 166.2, m 48°.** Crystd from EtOH.

4-Methyl anisole *[104-93-8]* **M 122.2, b 175-176°, d$_{15}^{15}$ 0.9757, n 1.512.** Dissolved in diethyl ether, washed with M NaOH, water, dried (Na₂CO₃), evaporated and the residue distd under vacuum.

2-Methylanthracene *[613-12-7]* **M 192.3, m 204-206°** Chromatographed on silica gel with cyclohexane as eluent and recrystd from EtOH [Werst *J Am Chem Soc* **109** 32 *1987*].

4-Methylanthracene *[779-02-2]* **M 192.3, m 77-79°, b 196-197°/12mm, d 1.066.** Chromatographed on silica gel with cyclohexane as eluent and recrystd from EtOH [Werst *J Am Chem Soc* **109** 32 *1987*].

2-Methylanthraquinone *[84-54-8]* **M 222.3, m 176°.** Crystd from EtOH, then sublimed.

Methylarenes (see also pentamethyl- and hexamethyl- benzenes). Recrystd from EtOH and sublimed in vacuum [Schlesener et al. *J Am Chem Soc* **106** 7472 *1984*].

Methyl benzoate *[93-58-3]* M **136.2**, b **104-105°/39mm, 199.5°/760mm, d 1.087, n**^{15} **1.52049, n 1.51701, pK**^{20} **-8.11, -6.51 (H$_0$ scale, aq H$_2$SO$_4$).** Washed with dilute aqueous NaHCO$_3$, then water, dried with Na$_2$SO$_4$ and fractionally distd under reduced pressure.

p-**Methylbenzophenone** *[134-84-9]* M **196.3, m 57°.** Crystd from MeOH and pet ether.

Methyl-1,4-benzoquinone *[553-97-9]* M **122.1, m 68-69°.** Crystd from heptane or EtOH, dried rapidly (vacuum over P$_2$O$_5$) and stored under vacuum.

Methyl benzoylformate *[15206-55-0]* M **164.2, m 246-248°.** Purified by radial chromatography (diethyl ether/hexane, 1:1), and dried at 110-112° at 6mm pressure. [Meyers and Oppenlaender *J Am Chem Soc* **108** 1989 *1986*.]

2-Methyl-3,4-benzphenanthrene *[652-04-0]* M **242.3, m 70°.** Crystd from EtOH.

dl-α-**Methylbenzyl alcohol** *[13323-81-4]* M **122.2, b 60.5-61.0°/3mm.** See *dl*-1-phenylethanol on p. 330.

R-(+)-α-**Methylbenzylamine** *[R(+) 3886-69-9, RS 618-36-0]* M **121.2, b 187-188°/atm, [α]**$^{20}_{546}$ **+35° (c 10, EtOH), [α]**$^{25}_D$ **+39.7° (neat), pK 9.08 (for *RS*).** Dissolve in toluene, dry over NaOH and distd, fraction boiling at 187-188°/atm is collected. Store under N$_2$ to avoid forming the carbamate and urea. Similarly for the *S*-(-) *enantiomer {2627-86-3}*. [*Org Synth* Coll Vol II 503 *1943*.]

p-**Methylbenzyl bromide** *[104-81-4]*. See α-bromo-*p*-xylene on p. 143.

p-**Methylbenzyl chloride** *[104-82-5]* M **140.6, b 80°/2mm, d 1.085, n 1.543.** Dried with CaSO$_4$ and fractionally distd under vacuum.

Methylbixin *[26585-94-4]* M **408.5, m 163°.** Crystd from EtOH/CHCl$_3$.

Methyl bromide *[74-83-9]* M **94.9, b 3.6°.** Purified by bubbling through conc H$_2$SO$_4$, followed by passage through a tube containing glass beads coated with P$_2$O$_5$. Also purified by distn from AlBr$_3$ at -80°, by passage through a tower of KOH pellets and by partial condensation.

Methyl *o*-bromobenzoate *[610-94-6]* M **215.1, b 122°/17mm, 234-244°/760mm.** Soln in ether is washed with 10% aqueous Na$_2$CO$_3$, water, then dried and distd.

Methyl *p*-bromobenzoate *[619-42-1]* M **215.1, m 79.5-80.5°.** Crystd from MeOH.

2-Methylbutane (isopentane) *[78-78-4]* M **72.2, b 27.9°, d 0.621, n 1.35373, n**^{25} **1.35088.** Stirred for several hours in the cold with conc H$_2$SO$_4$ (to remove olefinic impurities), then washed with H$_2$O, aqueous Na$_2$CO$_3$ and H$_2$O again. Dried with MgSO$_4$ and fractionally distd using a Todd column packed with glass helices (see p. 174). Material transparent down to 180nm was obtained by distilling from sodium wire, and passing through a column of silica gel which had previously been dried in place at 350° for 12h before use. [Potts *J Phys Chem* **20** 809 *1952*].

2-Methyl-1-butanol *[137-32-6; RS 34713-94-5; S(-)- 1565-80-6]* M **88.2, b 130°(*RS*), 128.6°(*S*), [α]**$^{25}_D$ **-5.8° (neat), d 0.809, n**^{52} **1.4082.** Refluxed with CaO, distd, refluxed with magnesium and again fractionally distd. A small sample of highly purified material was obtained by fractional crystn after conversion into a suitable ester such as the trinitrophthalate or the 3-nitrophthalate. The latter was converted to the cinchonine salt in acetone and recrystd from CHCl$_3$ by adding pentane. The salt was saponified, extracted with ether, and fractionally distd. [Terry et al. *J Chem Eng Data* **5** 403 *1960*.]

3-Methyl-1-butanol   *[123-51-3]*   **M 88.2, b 128°/750mm, 132°/760mm, d^{15} 0.8129, n^{15} 1.4085, n 1.4075.**   Dried by heating with CaO and fractionally distilling, then heating with BaO and redistilling.   Alternatively, boiled with conc KOH, washed with dilute H_3PO_4, and dried with K_2CO_3, then anhydrous $CuSO_4$, before fractionally distilling.   If very dry alcohol is required, the distillate can be refluxed with the appropriate alkyl phthalate or succinate as described for *ethanol.*   It is separated from 2-methyl-1-butanol by fractional distn, fractional crystn and preparative gas chromatography.

3-Methyl-2-butanol   *[598-75-4]*   **M 88.2, b 111.5°, d 0.807, n 1.4095, n^{25} 1.4076.**   Refluxed with magnesium, then fractionally distd.

3-Methyl-2-butanone (methyl isopropyl ketone)   *[563-80-4]*   **M 86.1, b 93-94°/752mm, d 0.818, n 1.410, pK^{25} -7.1 (aq H$_2$SO$_4$).**   Refluxed with a little $KMnO_4$.   Fractionated on a spinning-band column, dried with $CaSO_4$ and distd.

2-Methyl-2-butene   *[513-35-9]*   **M 70.1, f -133.8°, b 38.4°/760mm, d^{15} 0.66708, d 0.6783, d^{25} 0.65694, n^{15} 1.3908.**   Distd from sodium.

Methyl *n*-butyrate   *[623-42-7]*   **M 102.1, b 102.3°/760mm, d 0.898, n 1.389.**   Treated with anhydrous $CuSO_4$, then distd under dry nitrogen.

***S*-(+)-2-Methylbutyric acid**   *[1730-91-2]*   **M 102.1, b 64°/2mm, 78°/15mm, 90-94°/23mm, 174-175°/atm, d$_4^{20}$ 0.938, n$_D^{20}$ 1.406, [α]$_{546}^{20}$ +23°, [α]$_D^{20}$ +19.8° (neat), [α]$_D^{13}$ +18.3° (c 6, EtOH), pK^{25} 4.76 (for *RS*).**   Purified by distn *in vacuo* [Sax and Bergmann *J Am Chem Soc* 77 1910 *1955*; Doering and Aschner *J Am Chem Soc* 75 393 *1953*].   The *methyl ester* is formed by addition of diazomethane and has **b** 112-115°/atm, [α]$_D^{27}$ +21.1° (c 1.7, MeOH).

Methyl carbamate   *[598-55-0]*   **M 75.1, m 54.4-54.8°.**   Crystd from *benzene.

9-Methylcarbazole   *[1484-12-4]*   **M 181.2, m 89°.**   Purified by zone melting.

4-Methylcatechol   *[452-86-8]*   **M 124.1.**   See 3,4-dihydroxytoluene on p. 208.

Methyl chloride   *[74-87-3]*   **M 50.5, b -24.1°.**   Bubbled through a sintered-glass disc dipping into conc H_2SO_4, then washed with water, condensed at low temperature and fractionally distd.   Has been distd from $AlCl_3$ at -80°.   Alternatively, passed through towers containing $AlCl_3$, soda-lime and P_2O_5, then condensed and fractionally distd.   Stored as a gas.

Methyl chloroacetate   *[96-34-4]*   **M 108.5, b 129-130°, d 1.230, n 1.423.**   Shaken with satd aq Na_2CO_3 (three times), aq 50% $CaCl_2$ (three times), satd aq NaCl (twice), dried (Na_2SO_4) and fractionally distd.

***R*-(+) Methyl 2-chloropropionate**   *[77287-29-7]*   **M 122.6, b 49-50°/35mm, 78-80°/120mm, 132-134°/760mm, d$_4^{20}$ 1.152, n$_D^{20}$ 1.417, [α]$_D^{20}$ +26° (19.0°) (neat).**   Purified by repeated distillation [Walker *J Chem Soc* 67 916 *1895*; Walden *Chem Ber* 28 1293 *1985*; see also Gless *Synth Commun* 16 633 *1986*].

3-Methylcholanthrene   *[56-49-5]*   **M 268.4, m 179-180°.**   Crystd from *benzene and diethyl ether. **CARCINOGEN.**

Methyl cyanoacetate   *[105-34-0]*   **M 99.1, f -13°, b 205°, d 1.128, n 1.420.**   Purified by shaking with 10% Na_2CO_3 soln, washing well with water, drying with anhydrous Na_2SO_4, and distilling.

Methyl cyanoformate   *[17640-15-2]*   **M 85.1, b 81°/47mm, 97°/751mm, 100-101°/760mm, d$_4^{20}$ 1.072, n$_D^{20}$ 1.37378.**   Purified by fractionation through a 45cm glass helices packed column and with a 30cm spinning band column. [Sheppard *J Org Chem* 27 3756 *1962*.] It has been distd through a short Vigreux column, and further purified by recrystn from Et$_2$O at -40° as white crystals which melt at room temperature. NMR: δ 4.0 (CH$_3$), and IR: 2250 (CN) and 1750 (CO) cm^{-1}. [Childes and Weber *J Org Chem* **41** 3486 *1976*.]

Methylcyclohexane *[108-87-2]* **M 98.2, b 100.9°, d^{25} 0.7650, n 1.4231, n^{52} 1.42058.** Passage through a column of activated silica gel gives material transparent down to 220nm. Also purified by passage through a column of activated basic alumina, or by azeotropic distn with MeOH, followed by washing out the MeOH with H$_2$O, drying and distilling. Methylcyclohexane can be dried with CaSO$_4$, CaH$_2$ or sodium. Has also been purified by shaking with a mixture of conc H$_2$SO$_4$ and HNO$_3$ in the cold, washing with H$_2$O, drying with CaSO$_4$ and fractionally distilling from potassium. Percolation through a Celite column impregnated with 2,4-dinitrophenylhydrazine (DNPH), phosphoric acid and H$_2$O (prepared by grinding 0.5g DNPH with 6mL 85% H$_3$PO$_4$, then mixing with 4mL of distilled H$_2$O and 10g of Celite) removes carbonyl-containing impurities.

2-Methylcyclohexanol *[583-59-5]* **M 114.2, b 65°/20mm, 167.6°/760mm, d 0.922, n 1.46085.** Dried with Na$_2$SO$_4$ and distd under vacuum.

cis- and *trans*-**3-Methylcyclohexanol** *[591-23-1]* **M 114.2, b 69°/16mm, 172°/760mm, d 0.930, n 1.45757, n$^{25.5}$ 1.45444.** Dried with Na$_2$SO$_4$ and distd under vacuum.

4-Methylcyclohexanone *[589-92-4]* **M 112.2, b 165.5°/743mm, d 0.914, n 1.44506.** Dried with CaSO$_4$, then fractionally distd.

1-Methylcyclohexene *[591-49-1]* **M 96.2, b 107.4-108°/atm, 110-111°/760mm, d 0.813, n 1.451.** Freed from hydroperoxides by passing through a column containing basic alumina or refluxing with cupric stearate, filtered and fractionally distd from sodium.

Methylcyclopentane *[96-37-7]* **M 84.2, b 71.8°, d 0.749, n 1.40970, n^{25} 1.40700.** Purification procedures include passage through columns of silica gel (prepared by heating in nitrogen to 350° prior to use) and activated basic alumina, distn from sodium-potassium alloy, and azeotropic distn with MeOH, followed by washing out the methanol with water, drying and distilling. It can be stored with CaH$_2$ or sodium.

3'-Methyl-1,2-cyclopentenophenanthrene *[549-38-2]* **M 232.3, m 126-127°.** Crystd from AcOH.

S-Methyl-L-cysteine *[1187-84-4]* **M 135.2, m 207-211°, [α]$_D^{26}$ -32.0° (c 5, H$_2$O), pK$_1^{25}$ 1.94 (COOH), pK$_2^{25}$ 8.73 (NH$_2$, 8.97).** Likely impurities are cysteine and S-methyl-*dl*-cysteine. Crystd from water by adding 4 volumes of EtOH.

5-Methylcytosine [4-amino-5-methylpyrimidin-2(1H)-one] *[554-01-8]* **M 125.1, m 270°(dec), pK$_1$ 4.6, pK$_2$ 12.4.** Crystd from water (sol 3.4%).

Methyl decanoate *[110-42-9]* **M 186.3, b 114°/15mm, 224°/760mm, d 0.874, n 1.426.** Passed through alumina before use.

Methyl 2,4-dichlorophenoxyacetate *[1928-38-7]* **M 235.1, m 43°, b 119°/11mm.** Crystd from MeOH.

m-**Methyl-*N*,*N*-dimethylaniline** *[121-72-2]* **M 135.2, b 72-74°/5mm, 215°/760mm, pK^{25} 5.22.** Refluxed for 3h with 2 molar equivalents of acetic anhydride, then fractionally distd under reduced pressure. Also dried over BaO, distd and stored over KOH. Methods described for *N*,*N*-dimethylaniline are applicable.

p-**Methyl-*N*,*N*-dimethylaniline** *[99-97-8]* **M 135.2, b 76.5-77.5°/4mm, 211°/760mm, pK^{25} 4.76.** Refluxed for 3h with 2 molar equivalents of acetic anhydride, then fractionally distd under reduced pressure. Also dried over BaO, distd and stored over KOH. Methods described for *N*,*N*-dimethylaniline are applicable.

2-Methyl-1,3-dithiane *[6007-26-7]* **M 134.3, b 53-54°/1.1mm, 66°/5mm, 79-80°/8-10mm, 85°/12mm, d_4^{20} 1.121, n_D^{20} 1.560.** Wash with H_2O, 2.5 M aqueous NaOH, H_2O, brine, dried over K_2CO_3 (use toluene as solvent if volume of reagent is small), filter, evaporate and distil the colourless residue. IR film: 1455, 1371 and 1060 (all medium and CH_3), 1451m, 1422s, 1412m, 1275m, 1236m, 1190m, 1171w, 918m and 866w (all dithiane) cm^{-1} [Corey and Erickson *J Org Chem* **36** 3553 *1971*; Seebach and Corey *J Org Chem* **40** 231 *1975*].

Methyl dodecanoate *[111-82-0]* **M 214.4, m 5°, b 141°/15mm, d 0.870, n^{50} 1.4199.** Passed through alumina before use.

***N*-Methyleneaminoacetonitrile** *[109-82-0]* **M 68.1, m 129°.** Crystd from EtOH or acetone.

p,p'-**Methylene-bis-(*N,N*-dimethylaniline)** *[101-61-1]* **M 254.4, m 89.5°.** See *p,p'*-tetramethyldiaminodiphenylmethane on p. 364.

Methylene Blue [3,7-bis-(dimethylamino)phenothiazin-5-ium chloride *[61-73-4]* **M 319.9, CI 52015, ε_{654} 94,000 (EtOH), ε_{664} 81,000 (H_2O), pK^{25} 3.8.** Crystd from 0.1M HCl (16mL/g), the crystals were separated by centrifugation, washed with chilled EtOH and diethyl ether and dried under vacuum. Crystd from 50% aqueous EtOH, washed with absolute EtOH, and dried at 50-55° for 24h. Also crystd from *benzene-MeOH (3:1). Salted out with NaCl from a commercial conc aqueous soln, then crystd from water, dried at 100° in an oven for 8-10h.

3,4-Methylenedioxyaniline *[14268-66-7]* **M 137.1, m 45-46°, b 144°/14mm, pK$_{Est}$ ~3.8.** Crystd from pet ether.

3,4-Methylenedioxycinnamic acid *[2373-80-0]* **M 192.2, m 243-244°(dec), pK$_{Est}$ ~4.6.** Crystd from glacial acetic acid.

5,5'-Methylenedisalicylic acid *[122-25-8]* **M 372.3, m 238°(dec).** Crystd from acetone and *benzene.

Methylene Green [3,7-bis-(dimethylamino)-4-nitrophenothiazin-5-ium chloride] *[2679-01-8]* **M 364.9, m >200°(dec), CI 52020, pK^{25} 3.2.** Crystd three times from water (18mL/g).

***N*-Methylephedrine (2-dimethylamino-1-phenylpropanol)** *[1S,2R-(+)- 42151-56-4; 1R,2S-(-)- 552-79-4]* **M 179.3, m 85-86°, 87-87.5°, 90°, b 115°/2mm, $[\alpha]_{546}^{20}$ (+) and (-) 35°, $[\alpha]_D^{20}$ (+) and (-) 30° (c 4.5, MeOH), pK^{26} 9.22.** It has been recrystd from Et_2O, pet ether, of aq EtOH or aq MeOH and has been distilled under reduced pressure. [Smith *J Chem Soc* 2056 *1927* ; Tanaka and Sugawa *Yakugaku Zasshi (J Pharm Soc Japan)* **72** 1548 *1952* (*Chem Abstr* **47** 8682 *1953*); Takamatsu *Yakugaku Zasshi (J Pharm Soc Japan)* **76** 1227 *1956* ,*Chem Abstr* **51** 4304 *1957*.] The *hydrochloride* has **m 192-193°** and $[\alpha]_D^{20}$ +30° (c 5,H_2O)[Prelog and Hüfliger *Helv Chim Acta* **33** 2021 *1950*].

Methyl ether (dimethyl ether) *[115-10-6]* **M 46.1, b -63.5°/96.5mm.** Dried by passing over alumina and then BaO, or over CaH_2, followed by fractional distn at low temperatures.

***N*-Methyl ethylamine hydrochloride** *[624-60-2]* **M 95.6, m 126-130°, pK 10.9 (free base).** Crystd from absolute EtOH or diethyl ether.

***N*-Methyl formamide** *[123-39-7]* **M 59.1, m -3.5°, b 100.5°/25mm, d 1.005., n^{52} 1.4306** Dried with molecular sieves for 2days, then distd under reduced pressure through a column packed with glass helices. Fractionally crystd by partial freezing and the solid portion was vac distd.

Methyl formate *[107-31-3]* **M 60.1, b 31.5°, d 0.971, n^{15} 1.34648, n 1.34332.** Washed with strong aq Na_2CO_3, dried with solid Na_2CO_3 and distd from P_2O_5. (Procedure removes free alcohol or acid.)

2-Methylfuran *[534-22-5]* **M 82.1, b 62.7-62.8°/731mm, d 0.917, n 1.436.** Washed with acidified satd ferrous sulfate soln (to remove peroxides), separated, dried with $CaSO_4$ or $CaCl_2$, and fractionally distd from KOH immediately before use. To reduce the possibility of spontaneous polymerisation, addition of about one-third of its volume of heavy mineral oil to 2-methylfuran prior to distn has been recommended.

Methyl gallate *[99-24-1]* **M 184.2, m 202°.** Crystd from MeOH.

N-**Methylglucamine** *[6284-40-8]* **M 195.2, m 128-129°, $[\alpha]_{546}^{20}$ -19.5°(c 2, H_2O), pK^{28} 9.62.** Crystd from MeOH.

Methyl α-D-glucosamine *[97-30-3]* **M 194.2, m 165°, $[\alpha]_D^{25}$ +157.8° (c 3.0, H_2O), pK^{30} 7.1.** Crystd from MeOH.

α-**Methylglutaric acid** *[18069-17-5]* **M 146.1, m 79°, pK_1^{25} 4.36, pK_2^{25} 5.37.** Crystd from distd water, then dried under vacuum over conc H_2SO_4.

β-**Methylglutaric acid** *[626-51-7]* **M 146.1, m 87°, pK_1^{25} 4.35, pK_2^{25} 5.44.** Crystd from distd water, then dried under vacuum over conc H_2SO_4.

Methylglyoxal *[78-98-8]* **M 72.1, b ca 72°/760mm.** Commercial 30% (w/v) aqueous soln was diluted to about 10% and distd twice, taking the fraction boiling below 50°/20mm Hg. (This treatment does not remove lactic acid).

Methyl Green *[82-94-0, 7114-03-6 ($ZnCl_2$ salt)]* **M 458.5, m >200°(dec).** Crystd from hot water.

1-Methylguanine *[938-85-2]* **M 165.2, m >300°(dec), pK_1^{20} 3.13, pK_2^{20} 10.54.** Crystd from 50% aqueous acetic acid.

7-Methylguanine *[578-76-7]* **M 165.2, pK_1^{20} 3.50, pK_2^{20} 9.95.** Crystd from water.

2-Methylhexane *[591-76-4]* **M 100.2, b 90.1°, d 0.678, n 1.38485, n^{25} 1.38227.** Purified by azeotropic distn with MeOH, then washed with water (to remove the MeOH), dried over type 4A molecular sieves and distd.

3-Methylhexane *[589-34-4]* **M 100.2, b 91.9°, d 0.687, n 1.38864, n^{25} 1.38609.** Purification as for 2-methylhexene.

Methyl hexanoate *[106-70-7]* **M 130.2, b 52°/15mm, 150°/760mm, d 0.885, n 1.410.** Passed through alumina before use.

Methylhydrazine *[60-34-4]* **M 46.1, b 87°/745mm, d 0.876, n 1.436, pK^{30} 7.87.** Dried with BaO, then vacuum distd. Stored under nitrogen.

Methyl hydrazinocarboxylate *[6294-89-9]* **M 90.1, m 70-73°.** To remove impurities, the material was melted and pumped under vacuum until the vapours were spectroscopically pure [Caminati et al. *J Am Chem Soc* 108 4364 *1986*].

Methyl 4-hydroxybenzoate *[99-76-3]* **M 152.2, m 127.5°, pK_{Est} ~9.3.** Fractionally crystd from its melt, recrystd from *benzene, then from *benzene/MeOH and dried over $CaCl_2$ in a vacuum desiccator.

Methyl 3-hydroxy-2-naphthoate *[883-99-8]* **M 202.2, m 73-74°, pK_{Est} ~9.0.** Crystd from MeOH (charcoal) containing a little water.

N-Methylimidazole   *[616-47-7]*   **M 82.1, b 81-84°/27mm, 197-198°/760mm, d 1.032, n 1.496, pK^{25} 7.25.** Dried with sodium metal and then distd. Stored at 0° under dry argon.

2-Methylimidazole   *[693-98-1]*   **M 82.1, m 140-141°, b 267°/760mm, pK^{25} 7.86.** Recrystd from *benzene or pet ether.

4-Methylimidazole   *[822-36-6]*   **M 82.1, m 47-48°, b 263°/760mm, pK^{25} 7.61.** Recrystd from *benzene or pet ether.

2-Methylindole   *[95-20-5]*   **M 131.2, m 61°, pK^{25} -0.28 (C-3 protonation, aq H$_2$SO$_4$).** Crystd from *benzene. Purified by zone melting.

3-Methylindole (skatole)   *[83-34-1]*   **M 131.2, m 95°, pK^{25} -4.55 (C-3-protonation, aq H$_2$SO$_4$).** Crystd from *benzene. Purified by zone melting.

Methyl iodide   *[74-88-4]*   **M 141.9, b 42.8°, d 2.281, n 1.5315.** Deteriorates rapidly with liberation of iodine if exposed to light. Usually purified by shaking with dilute aqueous Na$_2$S$_2$O$_3$ or NaHSO$_3$ until colourless, then washed with water, dilute aqueous Na$_2$CO$_3$, and more water, dried with CaCl$_2$ and distd. It is stored in a brown bottle away from sunlight in contact with a small amount of mercury, powdered silver or copper. (Prolonged exposure of mercury to methyl iodide forms methylmercuric iodide.) Methyl iodide can be dried further using CaSO$_4$ or P$_2$O$_5$. An alternative purification is by percolation through a column of silica gel or activated alumina, then distn. The soln can be degassed by using a repeated freeze-pump-thaw cycle.

O-Methylisourea hydrogen sulfate   (2-methylpseudourea sulfate) *[29427-58-5]* **M 172.2, m 114-118°, 119°.** Recrystd from MeOH-Et$_2$O (327g of salt dissolved in 1L of MeOH and 2.5L of Et$_2$O is added) [Fearing and Fox *J Am Chem Soc* **76** 4382 *1954*]. The *picrate* has **m 192°** [Odo et al. *J Org Chem* **23** 1319 *1958*].

N-Methyl maleimide   *[930-88-1]*   **M 111.1, m 94-96°.** Crystd three times from diethyl ether.

Methylmalonic acid   *[516-05-2]*   **M 118.1, m 135°(dec), pK$_1^{25}$ 3.05, pK$_2^{25}$ 5.76.** Crystallises as the hydrate from water.

3-Methylmercaptoaniline   *[1783-81-9]* **M 139.2, b 101.5-102.5°/0.3mm, 163-165°/16mm, d$_4^{20}$ 1.147, n$_D^{20}$ 1.641, pK^{25} 4.05.** Purified by fractional distn in an inert atmostphere. It has UV max at 226 and 300. [Bordwell and Cooper *J Am Chem Soc* **74** 1058*1952*.] The *N-acetyl* derivative has **m 78-78.5°** (after recrystn from aq EtOH).

4-Methylmercaptoaniline   *[104-96-1]*   **M 139.2, b 140°/15mm, 151°/25mm, 155°/23mm, d$_4^{20}$ 1.137, n$_D^{20}$ 1.639, pK^{25} 4.40.** Purified by fractional distn in an inert atmosphere. [Lumbroso and Passerini *Bull Soc Chim Fr* 311 *1957*; Mangini and Passerini *J Chem Soc* 4954 *1956*.]

Methyl methacrylate   *[80-62-6]*   **M 100.1, f -50°, b 46°/100mm, d 0.937, n 1.4144.** Washed twice with aqueous 5% NaOH (to remove inhibitors such as hydroquinone) and twice with water. Dried with CaCl$_2$, Na$_2$CO$_3$, Na$_2$SO$_4$ or MgSO$_4$, then with CaH$_2$ under nitrogen at reduced pressure. The distillate is stored at low temperatures and redistd before use. Prior to distn, inhibitors such as ß-naphthylamine (0.2%) or di-β-naphthol are sometimes added. Also purified by boiling aqueous H$_3$PO$_4$ soln and finally with saturated NaCl soln. It was dried for 24h over anhydrous CaSO$_4$, distd at 0.1mm Hg at room temperature and stored at -30° [Albeck et al. *J Chem Soc, Faraday Trans 1* **1** 1488 *1978*].

Methyl methanesulfonate *[66-27-3]* **M 110.3, b 59°/0.6mm, 96-98°/19mm, d 1.300, n 1.4140.** Purified by careful fractionation and collecting the middle fraction. Suspected **CARCINOGEN**. Note that MeSO$_3$H has **b 134.5-135°, 167-167.5°/10mm** and methanesulfonic anhydride has **b 138°/10mm)**—both are possible impurities.

Methyl methanethiolsulfonate *[2949-92-0]* **M 126.2, b 69-71°/0.4mm, 96-97°/4.5mm, 104-105°/10mm, 119°/16mm, d 1.226, n 1.515.** Purified by fractional distn uner reduced pressure, IR: v 1350, 750 cm^{-1}. [Applegate et al. *J Org Chem* **38** 943 *1973*.]

α-Methylmethionine *[562-48-1]* **M 163.0, m 283-284°, pK^{30} 9.45.** Crystd from aqueous EtOH.

S-Methyl-L-methionine chloride (Vitamin U) *[1115-84-0]* **M 199.5, $[\alpha]_D^{23}$ +33° (0.2M HCl), pK$_1$ 1.9, pK$_2$ 7.9.** Likely impurities are methionine, methionine sulfoxide and methionine sulfone. Crystd from water by adding a large excess of EtOH. Stored in a cool, dry place, protected from light.

N-Methylmorpholine *[109-02-4]* **M 101.2, b 116-117°/764mm, d 0.919, n 1.436, pK^{25} 7.38.** Dried by refluxing with BaO or sodium, then fractionally distd through a helices-packed column.

4-Methylmorpholine-4-oxide monohydrate *[7529-22-8]* **M 135.2, m 71-73°.** When dried for 2-3h at high vacuum it dehydrates. Add MeOH to the oxide and distil off the solvent under vacuum until the temp is *ca* 95°. Then add Me$_2$CO at reflux then cool to 20°. The crystals are filtered off washed with Me$_2$CO and dry. The degree of hydration may vary and may be important for the desired reactions. [van Rheenan et al. *Tetrahedron Lett* 1973 1076; Schneider and Hanze *US Pat 2 769 823*; see also Sharpless et al. *Tetrahedron Lett* 2503 *1976*.]

1-Methylnaphthalene *[90-12-0]* **M 142.2, f -30°, b 244.6°, d 1.021, n 1.6108.** Dried for several days with CaCl$_2$ or by prolonged refluxing with BaO. Fractionally distd through a glass helices-packed column from sodium. Purified further by soln in MeOH and pptn of its picrate complex by adding to a saturated soln of picric acid in MeOH. The picrate, after crystn to constant melting point (**m** 140-141°) from MeOH, was dissolved in *benzene and extracted with aqueous 10% LiOH until the extract was colourless. Evaporation of the *benzene under vacuum gave 1-methylnaphthalene [Kloetzel and Herzog *J Am Chem Soc* **72** 1991 *1950*]. However, neither the picrate nor the styphnate complexes satisfactorily separates 1- and 2-methylnaphthalenes. To achieve this, 2-methylnaphthalene (10.7g) in 95% EtOH (50mL) has been ppted with 1,3,5-trinitrobenzene (7.8g) and the complex has been crystd from MeOH to **m** 153-153.5° (**m** of the 2-methyl isomer is 124°). [Alternatively, 2,4,7-trinitrofluorenone in hot glacial acetic acid could be used, and the derivative (**m** 163-164°) recrystd from glacial acetic acid]. The 1-methylnaphthalene was regenerated by passing a soln of the complex in dry *benzene through a 15-in column of activated alumina and washing with *benzene/pet ether (b 35-60°) until the coloured band of the nitro compound had moved down near the end of the column. The complex can also be decomposed using tin and acetic-hydrochloric acids, followed by extraction with diethyl ether and *benzene; the extracts were washed successively with dilute HCl, strongly alkaline sodium hypophosphite, water, dilute HCl and water. [Soffer and Stewart *J Am Chem Soc* **74** 567 *1952*.] It can be purified from anthracene by zone melting.

2-Methylnaphthalene *[91-57-6]* **M 142.2, m 34.7-34.9°, b 129-130°/25mm.** Fractionally crystd repeatedly from its melt, then fractionally distd under reduced pressure. Crystd from *benzene and dried under vacuum in an Abderhalden pistol. Purified *via* its picrate (**m** 114-115°) as described for 1-methylnaphthalene.

6-Methyl-2-naphthol *[17579-79-2]* **M 158.2, m 128-129°, pK$_{Est}$ ~9.8.** Crystd from EtOH or ligroin. Sublimed *in vacuo*.

7-Methyl-2-naphthol *[26593-50-0]* **M 158.2, m 118°, pK$_{Est}$ ~9.7.** Crystd from EtOH or ligroin. Sublimed *in vacuo*.

Methyl 1-naphthyl ether *[2216-69-5]* **M 158.2, b 90-91°/2mm, d 1.095, n^{26} 1.6210.** Steam distd from alkali. The distillate was extracted with diethyl ether. After drying (MgSO$_4$) the extract and evaporating diethyl ether, the methyl naphthyl ether was then fractionated under reduced pressure from CaH$_2$.

Methyl nitrate *[598-58-3]* **M 77.0, b 65°/760mm, d^5 1.2322, d^{15} 1.2167, d^{25} 1.2032.** Distd at -80°. The middle fraction was subjected to several freeze-pump-thaw cycles. **VAPOUR EXPLODES ON HEATING.**

Methyl nitrite *[624-91-9]* **M 61.0, b -12°, d^{15} (liq) 0.991.** Condensed in a liquid nitrogen trap. Distd under vacuum, first trap containing dry Na_2CO_3 to free it from acid impurities then into further Na_2CO_3 traps before collection.

***N*-Methyl-4-nitroaniline** *[100-15-2]* **M 152.2, m 152.2°, pK^{25} 0.55.** Crystd from aqueous EtOH.

2-Methyl-5-nitroaniline *[99-55-8]* **M 152.2, m 109°, pK^{25} 2.35.** Acetylated, and the acetyl derivative crystd to constant melting point, then hydrolysed with 70% H_2SO_4 and the free base regenerated by treatment with ammonia [Bevan, Fayiga and Hirst *J Chem Soc* 4284 *1956*].

4-Methyl-3-nitroaniline *[119-32-4]* **M 152.2, m 81.5°, pK^{25} 3.02.** Crystd from hot water (charcoal), then ethanol and dried in a vacuum desiccator.

Methyl 3-nitrobenzoate *[618-95-1]* **M 181.2, m 78°.** Crystd from MeOH (1g/mL).

Methyl 4-nitrobenzoate *[619-50-1]* **M 181.2, m 95-95.5°.** Dissolved in diethyl ether, then washed with aqueous alkali, the ether was evaporated and the ester was recrystd from EtOH.

2-Methyl-2-nitro-1,3-propanediol *[77-49-6]* **M 135.1, m 145°.** Crystd from *n*-butanol.

2-Methyl-2-nitro-1-propanol *[76-39-1]* **M 119.1, m 87-88°.** Crystd from pet ether.

***N*-Methyl-4-nitrosoaniline** *[10595-51-4]* **M 136.2, m 118°.** Crystd from *benzene.

***N*-Methyl-*N*-nitroso-*p*-toluenesulfonamide** (diazald) *[80-11-5]* **M 214.2, m 62°.** Crystd from *benzene by addition of pet ether, store in a refrigerator.

Methylnorbornene-2,3-dicarboxylic anhydride (5-methylnorborn-5-ene-2-endo-3-endo-dicarboxylic anhydride) *[25134-21-8]* **M 178.2, m 88.5-89°.** Purified by thin layer chromatography on Al_2O_3 (previously boiled in EtOAc) and eluted with hexane-*C_6H_6 (1:2) then recrystd from *C_6H_6-hexane. The free acid has **m** 118.5-119.5°. [Miranov et al. *Tetrahedron* **19** 1939 *1963*.]

3-Methyloctane *[2216-33-3]* **M 128.3, b 142-144°/760mm, d 0.719, n 1.407.** Passed through a column of silica gel [Klassen and Ross *J Phys Chem* **91** 3668 *1987*].

Methyl octanoate (methyl caprylate) *[111-11-5]* **M 158.2, b 83°/15mm, 193-194°/760mm, d 0.877, n 1.419.** Passed through alumina before use.

Methyl oleate *[112-62-9]* **M 296.5, f -19.9°, b 217°/16mm, d 0.874, n 1.4522.** Purified by fractional distn under reduced pressure, and by low temperature crystn from acetone.

3-Methyl-2-oxazolidone *[19836-78-3]* **M 101.1, m 15°, b 88-91°/1mm, d 1.172, n 1.455.** Purified by successive fractional freezing, then dried in a dry-box over 4A molecular sieves for 2 days.

3-Methyl-3-oxetanemethanol (3-hydroxymethyl-3-methyloxetane) *[3143-02-0]* **M 102.1, b 80°/4mm, 92-93°/12mm, d$_4^{20}$ 1.033, n$_D^{25}$ 1.4449.** Purified by fractionation through a glass column [Pattison *J Am Chem Soc* **79** 3455 *1957*].

Methylpentane (mixture of isomers). Passage through a long column of activated silica gel (or alumina) gave material transparent down to 200nm by UV.

2-Methylpentane *[107-83-5]* **M 86.2, b 60.3°, d 0.655, n 1.37145, n^{25} 1.36873.** Purified by azeotropic distn with MeOH, followed by washing out the MeOH with water, drying (CaCl$_2$, then sodium), and distn. [Forziati et al. *J Res Nat Bur Stand* **36** 129 *1946*.]

3-Methylpentane *[96-14-0]* **M 86.2, b 63.3°, d 0.664, n 1.37652, n^{25} 1.37384.** Purified by azeotropic distn with MeOH, as for 2-methylpentane. Purified for ultraviolet spectroscopy by passage through columns of silica gel or alumina activated by heating for 8h at 210° under a stream of nitrogen. Has also been treated with conc (or fuming) H_2SO_4, then washed with water, aqueous 5% NaOH, water again, then dried ($CaCl_2$, then sodium), and distd through a long, glass helices-packed column.

2-Methyl-2,4-pentanediol *[107-41-5]* **M 118.2, b 107.5-108.5°/25mm, d 0.922, n^{25} 1.4265.** Dried with Na_2SO_4, then CaH_2 and fractionally distd under reduced pressure through a packed column, taking precautions to avoid absorption of water.

2-Methyl-1-pentanol *[105-30-6]* **M 102.2, b 65-66°/60mm, 146-147°/760mm, d 0.827, n 1.420.** Dried with Na_2SO_4 and distd.

4-Methyl-2-pentanol *[108-11-2]* **M 102.2, b 131-132°, d 0.810, n 1.413.** Washed with aqueous $NaHCO_3$, dried and distd. Further purified by conversion to the phthalate ester by adding 120mL of dry pyridine and 67g of phthalic anhydride per mole of alcohol, purifying the ester and steam distilling it in the presence of NaOH. The distillate was extracted with ether, and the extract was dried and fractionally distd. [Levine and Walti *J Biol Chem* **94** 367 *1931*].

3-Methyl-3-pentanol carbamate (Emylcamate) *[78-28-4]* **M 145.2, m 56-58.5°.** Crystd from 30% EtOH.

4-Methyl-2-pentanone (methyl isobutyl ketone) *[108-10-1]* **M 100.2, b 115.7°, d 0.801, n 1.3958, n^{25} 1.3938.** Refluxed with a little $KMnO_4$, washed with aqueous $NaHCO_3$, dried with $CaSO_4$ and distd. Acidic impurities were removed by passage through a small column of activated alumina.

2-Methyl-1-pentene *[763-29-1]* **M 84.2, b 61.5-62°, d 0.680, n 1.395.** Water was removed, and peroxide formation prevented by several vacuum distns from sodium, followed by storage with sodium-potassium alloy.

cis-4-Methyl-2-pentene *[691-38-3]* **M 84.2, m -134.4°, b 57.7-58.5°, d 0.672, n 1.388.** Dried with CaH_2, and distd.

trans-4-Methyl-2-pentene *[674-76-0]* **M 84.2, m -140.8°, b 58.5°, d 0.669, n 1.389.** Dried with CaH_2, and distd.

5-Methyl-1,10-phenanthroline *[3002-78-6]* **M 194.2, m 113°(anhydr), pK^{25} 5.28.** Crystd from *benzene/pet ether.

N-Methylphenazonium methosulfate see 5-methylphenazinium methyl sulfate on p. 547 in Chapter 6.

N-Methylphenothiazine *[1207-72-3]* **M 213.2, α-form m 99.3° and b 360-365°, ß-form m 78-79°.** Recrystn (three times) from EtOH gave α-form (prisms). Recrystn from EtOH/*benzene gave the ß-form (needles). Also purified by vacuum sublimation and carefully dried in a vacuum line. Also crystd from toluene and stored in the dark [Guarr et al. *J Am Chem Soc* **107** 5104 *1985;* Olmsted et al. *J Am Chem Soc* **109** 3297 *1987.*]

4-Methylphenylacetic acid *[622-47-9]* **M 150.2, m 94°, pK^{25} 4.37.** Crystd from heptane or water.

1-Methyl-1-phenylhydrazine sulfate *[33008-18-3]* **M 218.2, pK^{25} 4.98 (free base).** Crystd from hot H_2O by addition of hot EtOH.

3-Methyl-1-phenyl-5-pyrazolone *[89-25-8]* **M 174.2, m 127°.** Crystd from hot H_2O, or EtOH/water (1:1).

N-Methylphthalimide   *[550-44-7]*   **M 161.1, m 133.8°.**   Recrystd from absolute EtOH.

2-Methylpiperazine   *[109-07-9]*   **M 100.2, m 61-62°, 66°, b 147-150°/739mm, pK_1^{25} 5.46, pK_2^{25} 9.90.**   Purified by zone melting and by distn.

3-Methylpiperidine   *[626-56-2]*   **M 99.2, b 125°/763mm, d 0.846, n^{25} 1.4448, pK^{25} 11.07.**   Purified *via* the *hydrochloride* (**m** 172°). [Chapman, Isaacs and Parker *J Chem Soc* 1925 *1959*.]

4-Methylpiperidine   *[626-58-4]*   **M 99.2, b 124.4°/755mm, d 0.839, n^{25} 1.4430, pK^{25} 10.78.**   Purified *via* the *hydrochloride* (**m** 189°). Freed from 3-methylpyridine by zone melting.

1-Methyl-4-piperidone   *[1445-73-4]*   **M 113.2, b 53-56°/0.5mm, 54-56°/9mm, 68-71°/17mm, 85-87°/45mm, d_4^{20} 0.972, n_D^{25} 1.4588, pK^{25} 7.9.**   It is best purified by fractional distn. The *hydrochloride* of the hydrate (4-diol) has **m** 94.7-95.5°, but the anhydrous *hydrochloride* which crystallises from CHCl$_3$-Et$_2$O and has **m** 165-168° (164-167°) and can also be obtained by sublimation at 120°/2mm. The *oxime* has **m** 130-132° (from Me$_2$CO). The *methiodide* crystallises from MeOH and the crystals with 1MeOH has **m** 189-190°, and the solvent-free *iodide* has **m** 202-204° dec. [Lyle et al. *J Org Chem* 24 342 *1959*; Bowden and Greeen *J Chem Soc* 1164 *1952*; Tomita *Yakugaku Zasshi (J Pharm Soc Japan)* 71 1053 *1951*.]

2-Methylpropane-1,2-diamine (1,2-diamino-2-methylpropane)   *[811-93-8]*   **M 88.2, b 47-48°/17mm, pK_1^{25} 6.25 (6.18), pK_2^{25} 9.82 (9.42).**   Dried with sodium for 2 days, then distd under reduced pressure from sodium.

2-Methylpropane-1-thiol   *[513-44-0]*   **M 90.2, b 41.2°/142mm, n^{25} 1.43582, pK_{Est} ~10.8.**   Dissolved in EtOH, and added to 0.25M Pb(OAc)$_2$ in 50% aqueous EtOH. The ppted lead mercaptide was filtered off, washed with a little EtOH, and impurities were removed from the molten salt by steam distn. After cooling, dilute HCl was added dropwise to the residue, and the mercaptan was distd directly from the flask. Water was separated from the distillate, and the mercaptan was dried (Na$_2$CO$_3$) and distd under nitrogen. [Mathias *J Am Chem Soc* 72 1897 *1950*.]

2-Methylpropane-2-thiol   *[75-66-1]*   **M 90.2, b 61.6°/701mm, d^{25} 0.79426, n^{25} 1.41984, pK^{25} 11.22.**   Dried for several days with CaO, then distd from CaO. Purified as for *2-methylpropane-1-thiol*.

2-Methyl-1-propanol (isobutanol)   *[78-83-1]*   **M 74.1, b 107.9°, d 0.804, n^{15} 1.39768, n^{25} 1.3939.**   Dried by refluxing with CaO and BaO for several hours, followed by treatment with calcium or aluminium amalgam, then fractional distn from sulfanilic or tartaric acids. More exhaustive purifications involve formation of phthalate or borate esters. Heating with phthalic anhydride gives the *acid phthalate* which, after crystn to constant melting point (**m** 65°) from pet ether, is hydrolysed with aqueous 15% KOH. The alcohol is distd as the water azeotrope and dried with K$_2$CO$_3$, then anhydrous CuSO$_4$, and finally magnesium turnings, followed by fractional distn. [Hückel and Ackermann *J Prakt Chem* 136 15 *1933*.] The borate ester is formed by heating the dried alcohol for 6h in an autoclave at 160-175° with a quarter of its weight of boric acid. After fractional distns under vac the ester is hydrolysed by heating for a short time with aq alkali and the alcohol is dried with CaO and distd. [Michael, Scharf and Voigt *J Am Chem Soc* 38 653 *1916*.] (see p. 271).

Methyl propiolate   *[922-67-8]*   **M 84.1, b 100°/atm, 102°/atm, 103-105°/atm, d 0.945, n 1.4080.**   Purified by fractional distn and collecting the middle fraction; note that propiolic acid has a high **b** [144°(dec)/atm]. **LACHRYMATORY.**

N-**Methylpropionamide**   *[1187-58-2]*   **M 87.1, f -30.9°, b 103°/12-13mm, d 0.934, n^{25} 1.4356.**   A colourless, odourless, neutral liquid at room temperature with a high dielectric constant. The amount of water present can be determined directly by Karl Fischer titration; GLC and NMR have been used to detect unreacted propionic acid. Commercial material of high quality is available, probably from the condensation of anhydrous methylamine with 50% excess of propionic acid. Rapid heating to 120-140° with stirring favours the reaction by removing water either directly or as the ternary xylene azeotrope. The quality of the distillate improves during the distn.

The propionamide can be dried over CaO. H_2O and unreacted propionic acid were removed as their xylene azeotropes. It was vacuum dried. Material used as an electrolyte solvent (specific conductance less than 10^{-6} ohm^{-1} cm^{-1}) was obtained by fractional distn under reduced pressure, and stored over BaO or molecular sieves because it readily absorbs moisture from the atmosphere on prolonged storage. [Hoover *Pure Appl Chem* **37** 581 *1974*; *Recommended Methods for Purification of Solvents and Tests for Impurities,* Coetzee Ed., Pergamon Press, *1982.*]

Methyl propionate *[554-12-1]* **M 88.1, b 79.7°.** Washed with satd aq NaCl, then dried with Na_2CO_3 and distd from P_2O_5. (This removes any free acid and alcohol.) It has also been dried with anhydrous $CuSO_4$.

Methyl *n*-propyl ether *[557-17-5]* **M 74.1, b 39°, d 0.736, n^{14} 1.3602, pK^{25} -3.79 (aq H_2SO_4).** Dried with $CaSO_4$, then passed through a column of alumina (to remove peroxides) and fractionally distd.

Methyl *n*-propyl ketone *[107-87-9]* **M 86.1, b 102.4°, d 0.807, n 1.3903.** Refluxed with a little $KMnO_4$, dried with $CaSO_4$ and distd. It was converted to its bisulfite addition compound by shaking with excess saturated aqueous $NaHSO_3$ at room temperature, cooling to 0°, filtering, washing with diethyl ether and drying. Steam distillation gave a distillate from which the ketone was recovered, washed with aq $NaHCO_3$ and distd water, dried (K_2CO_3) and fractionally distd. [Waring and Garik *J Am Chem Soc* **78** 5198 *1956.*]

3-Methyl-1-propyn-3-ol carbamate *[302-66-9]* **M 141.2, m 55.8-57°.** Crystd from ether/pet ether or cyclohexane.

2-Methylpyrazine *[109-08-0]* **M 94.1, b 136-137°, d 1.025, n 1.505, pK$_1^{25}$ -5.25 (aq H_2SO_4), pK$_2^{25}$ 1.47.** Purified *via* the picrate. [Wiggins and Wise *J Chem Soc* 4780 *1956.*]

2-Methylpyridine (2-picoline) *[109-06-8]* **M 93.1, b 129.4°, d 0.9444, n 1.50102, pK^{25} 5.96.** Biddiscombe and Handley [*J Chem Soc* 1957 *1954*] steam distd a boiling soln of the base in 1.2 equivalents of 20% H_2SO_4 until about 10% of the base had been carried over, along with non-basic impurities. Excess aqueous NaOH was then added to the residue, the free base was separated, dried with solid NaOH and fractionally distd.

2-Methylpyridine can also be dried with BaO, CaO, CaH$_2$, LiAlH$_4$, sodium or Linde type 5A molecular sieves. An alternative purification is *via* the ZnCl$_2$ adduct, which is formed by adding 2-methylpyridine (90mL) to a soln of anhydrous ZnCl$_2$ (168g) and 42mL conc HCl in absolute EtOH (200mL). Crystals of the complex are filtered off, recrystd twice from absolute EtOH (to give **m** 118.5-119.5°), and the free base is liberated by addition of excess aqueous NaOH. It is steam distd, and solid NaOH added to the distillate to form two layers, the upper one of which is then dried with KOH pellets, stored for several days with BaO and fractionally distd. Instead of ZnCl$_2$, HgCl$_2$ (430g in 2.4L of hot water) can be used. The complex, which separates on cooling, can be dried at 110° and recrystd from 1% HCl (to **m** 156-157°).

3-Methylpyridine (3-picoline) *[108-99-6]* **M 93.1, m -18.5°, b 144°/767mm, d 0.957, n 1.5069, pK^{25} 5.70.** In general, the same methods of purification that are described for *2-methylpyridine* can be used. However, 3-methylpyridine often contains 4-methylpyridine and 2,6-lutidine, neither of which can be removed satisfactorily by drying and fractionation, or by using the ZnCl$_2$ complex. Biddiscombe and Handley [*J Chem Soc* 1957 *1954*], after steam distn as for *2-methylpyridine*, treated the residue with urea to remove 2,6-lutidine, then azeotropically distd with acetic acid (the azeotrope had **b** 114.5°/712mm), and recovered the base by adding excess of aqueous 30% NaOH, drying with solid NaOH and carefully fractionally distilling. The distillate was then fractionally crystd by slow partial freezing. An alternative treatment [Reithof et al. *Ind Eng Chem (Anal Edn)* **18** 458 *1946*] is to reflux the crude base (500mL) for 20-24h with a mixture of acetic anhydride (125g) and phthalic anhydride (125g) followed by distn until phthalic anhydride begins to pass over. The distillate was treated with NaOH (250g in 1.5L of water) and then steam distd. Addition of solid NaOH (250g) to this distillate (*ca* 2L) led to the separation of 3-methylpyridine which was removed, dried (K_2CO_3, then BaO) and fractionally distd. (Subsequent fractional freezing would probably be advantageous.)

4-Methylpyridine (4-picoline) *[108-89-4]* **M 93.1, m 4.25°, b 145.0°/765mm, d 0.955, n 1.5058, pK^{25} 4.99.** Can be purified as for *2-methylpyridine*. Biddescombe and Handley's method for 3-methylpyridine is also applicable. Lidstone [*J Chem Soc* 242 *1940*] purified *via* the *oxalate* (**m** 137-138°) by heating 100mL of 4-methylpyridine to 80° and adding slowly 110g of anhydrous oxalic acid, followed by 150mL of boiling EtOH. After cooling and filtering, the ppte was washed with a little EtOH, then recrystd from EtOH, dissolved in the minimum quantity of water and distd with excess 50% KOH. The distillate was dried with solid KOH and again distd. Hydrocarbons can be removed from 4-methylpyridine by converting the latter to its hydrochloride, crystallising from EtOH/diethyl ether, regenerating the free base by adding alkali and distilling. As a final purification step, 4-methylpyridine can be fractionally crystd by partial freezing to effect a separation from 3-methylpyridine. Contamination by 2,6-lutidine is detected by its strong absorption at 270nm.

4-Methylpyridine 1-oxide *[1003-67-4]* **M 109.1, m 182-184°.** See 4-picoline-*N*-oxide on p. 335.

N-**Methylpyrrole** *[96-54-8]* **M 81.1, b 115-116°/756mm, d 0.908, n 1.487, pK -3.4 (-2.90).** Dried with CaSO$_4$, then fractionally distd from KOH immediately before use.

1-Methyl-2-pyrrolidinone *[872-50-4]* **M 99.1, f -24.4, b 65-76°/1mm, 78-79°/12mm, 94-96°/20mm, 202°/760mm, d$_4^{20}$ 1.0328, n$_D^{20}$ 1.4678, pK -0.17 (also -0.92, and 0.2).** Dried by removing water as *benzene azeotrope. Fractionally distd at 10 torr through a 100-cm column packed with glass helices. [Adelman *J Org Chem* 29 1837 *1964*; McElvain and Vozza *J Am Chem Soc* 71 896 *1949*.] The *hydrochloride* has **m** 86-88° (from EtOH or Me$_2$CO-EtOH) [Reppe et al. *Justus Liebigs Ann Chem* 596 1 *1955*].

2-Methylquinoline (quinaldine) *[91-63-4]* **M 143.2, b 86-87°/1mm, 155°/14mm, 246-247°/760mm, d 1.058, n 1.6126, pK^{25} 5.65.** Dried with Na$_2$SO$_4$ or by refluxing with BaO, then fractionally distd under reduced pressure. Redistd from zinc dust. Purified by conversion to its *phosphate* (**m** 220°) or *picrate* (**m** 192°) from which after recrystn, the free base was regenerated. [Packer, Vaughan and Wong *J Am Chem Soc* 80 905 *1958*.] Its ZnCl$_2$ complex can be used for the same purpose.

4-Methylquinoline (lepidine) *[491-35-0]* **M 143.2, b 265.5°, d 1.084, n 1.61995, pK^{25} 5.59.** Refluxed with BaO, then fractionally distd. Purified *via* its recrystd *dichromate salt* (**m** 138°). [Cumper, Redford and Vogel *J Chem Soc* 1176 *1962*.]

6-Methylquinoline *[91-62-3]* **M 143.2, b 258.6°, d 1.067, n 1.61606, pK^{25} 4.92.** Refluxed with BaO, then fractionally distd. Purified *via* its recrystd *ZnCl$_2$ complex* (**m** 190°). [Cumper, Redford and Vogel *J Chem Soc* 1176 *1962*.]

7-Methylquinoline *[612-60-2]* **M 143.2, m 38°, b 255-260°, d 1.052, n 1.61481, pK^{25} 5.29.** Purified *via* its *dichromate complex* (**m** 149°, after five recrystns from water). [Cumper, Redford and Vogel *J Chem Soc* 1176 *1962*.]

8-Methylquinoline *[611-32-5]* **M 143.2, b 122.5°/16mm, 247.8°/760mm, d 1.703, n 1.61631, pK^{25} 4.60.** Purified as for 2-methylquinoline. The *phosphate* and *picrate* have **m** 158° and **m** 201° respectively.

Methyl Red (4-dimethylaminoazobenzene-2'-carboxylic acid) *[493-52-7]* **M 269.3, m 181-182°, CI 13020, pK$_1^{25}$ 2.30, pK$_2^{25}$ 4.82.** The acid is extracted with boiling toluene using a Soxhlet apparatus. The crystals which separated on slow cooling to room temperature are filtered off, washed with a little toluene and recrystd from glacial acetic acid, *benzene or toluene followed by pyridine/water. Alternatively, dissolved in aq 5% NaHCO$_3$ soln, and ppted from hot soln by dropwise addition of aq HCl. Repeated until the extinction coefficients did not increase.

Methyl salicylate (methyl 2-hydroxybenzoate) *[119-36-8]* **M 152.2, m -8.6°, b 79°/6mm, 104-105°/14mm, 223.3°/atm, d$_4^{20}$ 1.1149, n$_D^{20}$ 1.5380, pK^{25} 10.19.** Dilute with Et$_2$O, wash with satd NaHCO$_3$ (it may effervesce due to the presence of free acid), brine, dry MgSO$_4$, filter, evaporate and distil.

Its solubility is 1g/1.5L of H_2O. The *benzoyl* derivative has **m** 92° (b 270-280°/120mm), and the *3,5-dinitrobenzoate* has **m** 107.5°, and the *3,5-dinitrocarbamoyl* derivative has **m** 180-181°. [Hallas *J Chem Soc* 5770 *1965*.]

Methyl stearate *[122-61-8]* **M 298.5, m 41-43°, b 181-182°/4mm.** Crystd from pet ether or distd.

α-Methylstyrene (monomer) *[98-83-9]* **M 118.2, b 57°/15mm, d 0.910, n 1.5368.** Washed three times with aqueous 10% NaOH (to remove inhibitors such as quinol), then six times with distd water, dried with $CaCl_2$ and distd under vacuum. The distillate is kept under nitrogen, in the cold, and redistd if kept for more than 48h before use. It can also be dried with CaH_2.

***trans*-ß-Methylstyrene** *[873-66-5]* **M 118.2, b 176°/760mm, d 0.910, n 1.5496.** Distd under nitrogen from powdered NaOH through a Vigreux column, and passed through activated neutral alumina before use [Wong et al. *J Am Chem Soc* **109** 3428 *1987*].

4-Methylstyrene *[622-97-9]* **M 118.2, b 60°/12mm, 106°/10mm, d_4^{20} 0.9173, n_D^{20} 1.542.** Purified as the above styrenes and add a small amount of antioxidant if it is to be stored, UV in EtOH λmax 285nm (log ε 3.07), and in EtOH + HCl 295nm (log ε 2.84) and 252nm (log ε 4.23). [Schwartzman and Carson *J Am Chem Soc* **78** 322 *1956*; Joy and Orchin *J Am Chem Soc* **81** 305 *1959*; Buck et al. *J Chem Soc* 2377 *1949*.]

Methylsuccinic acid *[498-21-5]* **M 132.1, m 115.0°, pK_1^{25} 3.88, pK_2^{25} 5.35.** Crystd from water.

(±)-3-Methylsulfolane (3-methyl-tetrahydrothiophene-1,1-dioxide) *[872-93-5]* **M 134.2, m 0.5°, b 101°/2mm, 125-130°/12mm, 278-282°/763.5mm, d_4^{20} 1.1885, n_D^{20} 1.4770.** Distil under vacuum and recryst from Et_2O at -60° to -70°. IR film has strong bands at 570 and 500 cm^{-1}. [Eigenberger *J Prakt Chem* [2] **131** 289 *1931*; Freaheller and Katon *Spectrochim Acta* **20** 1099 *1964*.]

17α-Methyltestosterone *[58-18-4]* **M 302.5, m 164-165°, $[α]_{546}^{20}$ +87°** (c 1, dioxane). Crystd from hexane/*benzene.

Methyl tetradecanoate (methyl myristate) *[124-10-7]* **M 382.7, m 18.5°, b 155-157°/7mm.** Passed through alumina before use.

2-Methyltetrahydrofuran *[96-47-9]* **M 86.1, b 80.0°, d_4^{20} 0.856, n_D^{20} 1.4053.** Likely impurities are 2-methylfuran, methyldihydrofurans and hydroquinone (stabiliser, which is removed by distn under reduced pressures). It was washed with 10% aqueous NaOH, dried, vacuum distd from CaH_2, passed through freshly activated alumina under nitrogen, and refluxed over sodium metal under vacuum. Stored over sodium. [Ling and Kevan *J Phys Chem* **80** 592 *1976*.] Vacuum distd from sodium, and stored with sodium-potassium alloy. (Treatment removes water and prevents the formation of peroxides.) Alternatively, it can be freed from peroxides by treatment with ferrous sulfate and sodium bisulfate, then solid KOH, followed by drying with, and distilling from, sodium, or type 4A molecular sieves under argon. It may be difficult to remove *benzene if it is present as an impurity (can be readily detected by its ultraviolet absorption in the 249-268nm region). [Ichikawa and Yoshida *J Phys Chem* **88** 3199 *1984*.] It has also been purifed by percolating through Al_2O_3 and fractionated collecting fraction **b** 79.5-80°. After degassing, the material was distd onto degassed molecular sieves, then distd onto anthracene and a sodium mirror. The solvent was distd from the green soln onto potassium mirror or sodium-potassium alloy, from which it was distilled again. [Mohammad and Kosower *J Am Chem Soc* **93** 2713 *1971*.] It should be stored in the presence of 0.1% of hydroquinone as stabiliser. **HARMFUL VAPOURS.**

N-Methylthioacetamide *[5310-10-1]* **M 89.1, m 59°.** Recrystd from *benzene.

3-Methylthiophene *[616-44-4]* **M 98.2, b 111-113°, d 1.024, n 1.531.** Dried with Na_2SO_4, then distd from sodium.

6(4)-Methyl-2-thiouracil *[56-04-2]* **M 142.2, m 330°(dec), 299-303° (dec), 323-324° (dec), pK 8.1.** Crystd from a large volume of H_2O. Purified by dissolving in base adding charcoal, filtering and acidifying with AcOH. Suspend the wet solid (*ca* 100g) in boiling H_2O (1L), stir and add AcOH (20mL), stir and refrigerate. Collect the product, wash with cold H_2O (4 x 200mL), drain for several hours then place in an oven at 70° to constant weight. [IR: Short and Thompson *J Chem Soc* 168 *1952*; Foster and Snyder *Org Synth* Coll Vol IV 638 *1063*.]

Methyl 4-toluenesulfonate *[80-48-8]* **M 186.2, m 25-28°, 28°, b 144.6-145.2°/5mm, 168-170°/13mm, d_4^{20} 1.23.** It is purified by distn *in vacuo* and could be crystd from pet ether or Et_2O-pet ether at low temperature. It is a powerful methylating agent and is **TOXIC** and a **skin irritant**, so it is better to purify by repeated distn. [IR: Schreiber *Anal Chem* 21 1168 *1949*; Buehler et al. *J Org Chem* 2 167 *1937*; Roos et al. *Org Synth* Coll Vol I 145 *1948*.]

4-Methyl-1,2,4-triazoline-3,5-dione (MTAD) *[13274-43-6]* **M 113.1, m 103-104°, m 107-109°.** Obtained as pink needles by sublimation at 40-50°/0.1mm (see 4-phenyl-1,2,4-triazoline-3,5-dione, PTAD below). [Cookson et al. *Org Synth* 51 121 *1971*; Cheng et al. *J Org Chem* 49 2910 *1984*.]

2-Methyltricycloquinazoline *[2642-52-6]* **M 334.4, m >300°.** Purified by vac sublimation. **CARCINOGEN.**

Methyl trifluoromethanesulfonate (methyl triflate) *[333-27-7]* **M 164.1, b 97-97.5°/736mm, 99°/atm, 100-102°/atm, d_4^{20} 1.496, n_D^{25} 1.3238.** It is a strong methylating agent but is corrosive and **POISONOUS**. Fractionate carefully and collecting the middle fraction (use efficient fume cupboard) and keep away from moisture. It is **POWERFUL ALKYLATING AGENT** and a strong **IRRITANT**. [IR: Gramstad and Haszeldine *J Chem Soc* 173 *1956*, 4069 *1957*.] *Trifluoromethanesulfonic acid* (triflic acid) *[1493-13-6]* **M 151.1**, boils higher (**b** 162°/atm), has a **pKa** of 3.10, and is **TOXIC** and hygroscopic. [Hansen *J Org Chem* 30 4322 *1965*; Kurz and El-Nasr *J Am Chem Soc* 104 5823 *1982*.]

N-Methyltryptophan (L-abrine) *[526-31-8]* **M 218.3, m 295°(dec), $[\alpha]_D^{21}$ +44.4° (c 2.8, 0.5M HCl), $pK_{Est(1)}$~2.3, $pK_{Est(2)}$~9.7.** Crystd from water.

dl-5-Methyltryptophan *[951-55-3]* **M 218.3, m 275°(dec) [pK see tryptophan].** Crystd from aqueous EtOH. *Picrate* has **m** 202° (dec).

6-Methyluracil *[626-48-2]* **M 126.1, m 270-280°(dec), λ_{max} 260$_{nm}$ logε 3.97, pK_1 ~1.1, pK_2 9.8.** Crystd from EtOH or acetic acid.

3-Methyluric acid *[39717-48-1]* **M 182.1, m >350°, pK_1 5.75 (6.2), pK_2 >12.** Crystd from water.

7-Methyluric acid *[30409-21-3]* **M 182.1, m >380°, pK_1 5.6, pK_2 10.3.** Crystd from water.

9-Methyluric acid *[30345-24-5]* **M 182.1, m >400°.** Crystd from water.

Methyl vinyl ketone *[78-94-4]* **M 70.1, b 62-68°/400mm, 79-80°/760mm, d 0.845, n 1.413.** Forms an 85% azeotrope with water. After drying with K_2CO_3 and $CaCl_2$ (with cooling), the ketone is distd at low pressures.

Methyl vinyl sulfone *[3680-02-2]* **M 106.1, b 116-118°/20mm, d 1.215, n 1.461.** Passed through a column of alumina, then degassed and distd on a vacuum line and stored at -190° until required.

Methyl Violet 2B [4,4'-bis-(diethylamino)-4"-methyliminotriphenylmethyl hydrochloride) *[8004-87-3]* **M 394.0, m 137°(dec), CI 42535, max ~580nm.** Crystd from absolute EtOH by pptn with diethyl ether during cooling in an ice-bath. Filtered off and dried at 105°.

1-Methylxanthine *[6136-37-4]* **M 166.1, m >360°** pK_1^{20} **7.90,** pK_2^{20} **12.23.** Crystd from water.

3-Methylxanthine *[1076-22-8]* **M 166.1, m >360°** pK_1^{20} **8.45,** pK_2^{20} **11.92.** Crystd from water.

7-Methylxanthine *[552-62-5]* **M 166.1, m >380°(dec)** pK_1^{20} **8.42,** pK_2^{20} **>13.** Crystd from water.

8-Methylxanthine *[17338-96-4]* **M 166.1, m 292-293°(dec).** Crystd from water.

9-Methylxanthine *[1198-33-0]* **M 166.1, m 384°(dec),** pK_1^{20} **2.0,** pK_2^{20} **6.12,** pK_3^{20} **10.5 (>13).** Crystd from water.

Metrazol (Cardiazol, Leptazol, 3a,4,5,6,7,8-hexahydro-1,2,3,3a-tetraaza-azulene, 1,5-pentamethylene-1,2,3,4-tetrazole) *[54-95-5]* **M 138.2, m 61°, b 194°/12mm,** pK_{Est} **~<0.** Crystd from diethyl ether. Dried under vacuum over P_2O_5.

Michler's ketone [4,4'-bis(dimethylamino)benzophenone] *[90-94-8]* **M 268.4, m 179°,** pK^{25} **9.84.** Dissolved in dilute HCl, filtered and ppted by adding ammonia (to remove water-insoluble impurities such as benzophenone). Then crystd from EtOH or pet ether. [Suppan *J Chem Soc, Faraday Trans1* **71** 539 *1975*.] It was also purified by dissolving in *benzene, then washed with water until the aqueous phase was colourless. The *benzene was evaporated off and the residue recrystd three times from *benzene and EtOH [Hoshino and Kogure *J Phys Chem* **72** 417 *1988*].

Monensin *[17090-79-8]* **M 670.9, m 103-105° (1 H_2O),** $[\alpha]_D$ **+47.7°,** pK_{Est} **~ 4.6, pK 6.6 (66% Me_2NCHO).** Purified by chromatography, stable in aq alkaline soln. Slightly sol in H_2O but sol in EtOH, EtOAc and Et_2O.

N-Monobutyl urea *[592-31-4]* **M 116.2, m 96-98°.** Crystd from EtOH/water, then dried under vacuum at room temperature.

N-Monoethyl urea *[625-52-5]* **M 88.1, m 92-95°.** Crystd from EtOH/water, then dried under vacuum at room temperature.

N-Monomethyl urea *[598-50-5]* **M 74.1, m 93-95°.** Crystd from EtOH/water, then dried under vacuum at room temperature.

Monopropyl urea *[627-06-5]* **M 102.1, m 110°.** Crystd from EtOH.

Morin (hydrate) (2',3,4',5,7-pentahydroxyflavone) *[480-16-0]* **M 302.2, m 289-292°, pK_1 5.3, pK_2 8.74.** Stirred at room temperature with ten times its weight of absolute EtOH, then left overnight to settle. Filtered, and evaporated under a heat lamp to one-tenth its volume. An equal volume of water was added, and the ppted morin was filtered off, dissolved in the minimum amount of EtOH and again ppted with an equal volume of water. The ppte was filtered, washed with water and dried at 110° for 1h. (Yield ca 2.5%.) [Perkins and Kalkwarf *Anal Chem* **28** 1989 *1956*.] Complexes with W and Zr.

Morphine (H_2O) *[57-27-2]* **M 302.2, m 230°(dec),** $[\alpha]_D^{23}$ **-130.9° (MeOH), pK_1 8.31, pK_2 9.51.** Crystd from MeOH.

Morpholine *[110-91-8]* **M 87.1, f -4.9°, b 128.9°, d 1.0007, n 1.4540, n^{25} 1.4533, pK^{25} 8.33.** Dried with KOH, fractionally distd, then refluxed with Na, and again fractionally distd. Dermer and Dermer [*J Am Chem Soc* **59** 1148 *1937*] ppted as the oxalate by adding slowly to slightly more than 1 molar equivalent of oxalic acid in EtOH. The ppte was filtered and recrystd twice from 60% EtOH. Addition of the oxalate to conc aq NaOH regenerated the base, which was separated and dried with solid KOH, then sodium, before being fractionally distd.
§ A polystyrene supported morpholine is commercially available.

2-(N-Morpholino)ethanesulfonic acid (MES)   *[4432-31-9]* M 213.3, m >300°(dec), pK^{20} 6.15. Crystd from hot EtOH containing a little water.

Mucochloric acid (2,3-dichloro-4-oxo-2-butenoic acid)   *[87-56-9]* M 169.0, m 124-126°, pK^{25} 4.20. Crystd twice from water (charcoal).

***trans,trans*-Muconic acid (hexa-2,4-dienedioic acid)**   *[3588-17-8]* M 142.1, m 300°, pK^{25} 4.51, for *cis,cis* pK^{25} 4.49. Cryst from H$_2$O.

Muramic acid (H$_2$O) (3-*O*-α-carboxyethyl-D-glucosamine)   *[1114-41-6]* M 251.2, m 152-154°(dec). See muramic acid on p. 549 in Chapter 6.

Murexide (ammonium purpurate)   *[3051-09-0]* M 284.2, m >300°, λ$_{max}$ 520nm (ε 12,000), pK$_2$ 9.2, pK$_3$ 10.9. The sample may be grossly contaminated with uramil, alloxanthine, etc. Difficult to purify. It is better to synthesise it from pure alloxanthine [Davidson *J Am Chem Soc* **58** 1821 *1936*]. Crystd from water.

Myristic acid (tetradecanoic acid)   *[544-63-8]* M 228.4, m 58°, pK^{20} 6.3 (50% EtOH), pK$_{Est}$ ~4.9 (H$_2$O). Purified *via* the *methyl ester* (b 153-154°/10mm, n^{25} 1.4350), as for capric acid. [Trachtman and Miller *J Am Chem Soc* **84** 4828 *1962*.] Also purified by zone melting. Crystd from pet ether and dried in a vacuum desiccator containing shredded wax.

Naphthacene (benz[b]anthracene, 2,3-benzanthracene, rubene) *[92-24-0]* M 228.3, m >300°, 341° (open capillary), 349°, 357°. Crystd from EtOH or *benzene. Dissolved in sodium-dried *benzene and passed through a column of alumina. The *benzene was evaporated under vacuum, and the chromatography was repeated using fresh *benzene. Finally, the naphthacene was sublimed under vacuum. [Martin and Ubblehode *J Chem Soc* 4948 *1961*.] Also recrysts in orange needles from xylene and sublimes *in vacuo* at 186°. [UV: *Chem Ber* **65** 517 *1932*, **69** 607 *1936*; IR: *Spectrochim Acta* **4** 373 *1951*.]

2-Naphthaldehyde   *[66-99-9]* M 156.2, m 59°, b 260°/19mm, pK^{20} -7.04 (aq H$_2$SO$_4$). Distilled with steam and crystd from water or EtOH.

Naphthalene   *[91-20-3]* M 128.2, m 80.3°, b 87.5°/10mm, 218.0°/atm, d 1.0253, d^{100} 0.9625, n^{85} 1.5590. Crystd one or more times from the following solvents: EtOH, MeOH, CCl$_4$, *benzene, glacial acetic acid, acetone or diethyl ether, followed by drying at 60° in an Abderhalden drying apparatus. Also purified by vacuum sublimation and by fractional crystn from its melt. Other purification procedures include refluxing in EtOH over Raney Ni, and chromatography of a CCl$_4$ soln on alumina with *benzene as eluting solvent. Baly and Tuck [*J Chem Soc* 1902 *1908*] purified naphthalene for spectroscopy by heating with conc H$_2$SO$_4$ and MnO$_2$, followed by steam distn (repeating the process), and formation of the picrate which, after recrystallisation, was decomposed and the naphthalene was steam distd. It was then crystd from dilute EtOH. It can be dried over P$_2$O$_5$ under vacuum. Also purified by sublimation and subsequent crystn from cyclohexane. Alternatively, it has been washed at 85° with 10% NaOH to remove phenols, with 50% NaOH to remove nitriles, with 10% H$_2$SO$_4$ to remove organic bases, and with 0.8g AlCl$_3$ to remove thianaphthalenes and various alkyl derivatives. Then it was treated with 20% H$_2$SO$_4$, 15% Na$_2$CO$_3$ and finally distd. [Gorman et al. *J Am Chem Soc* **107** 4404 *1985*.]
Zone refining purified naphthalene from anthracene, 2,4-dinitrophenylhydrazine, methyl violet, benzoic acid, methyl red, chrysene, pentacene and indoline.

Naphthalene-2,5-disulfonic acid   *[92-41-1]* M 288.2, pK$_{Est}$ <0. Crystd from conc HCl.

Naphthalene-1-sulfonic acid   *[85-47-2]* M 208.2, m (2H$_2$O) 90°, (anhydrous) 139-140°, pK^{20} -0.17. Crystd from conc HCl and twice from water.

Naphthalene-2-sulfonic acid *[120-18-3]* **M 208.2, m 91°, pK$_{Est}$ <1.** Crystd from conc HCl.

Naphthalene-1-sulfonyl chloride *[85-46-1]* **M 226.7, m 64-67°, 68°, b 147.5°/0.9mm, 147.5°/13mm.** If the IR indicates the presence of OH then treat with an equal weight of PCl$_5$ and heat at *ca* 100° for 3h, cool and pour into ice + H$_2$O, stir well and filter off the solid. Wash the solid with cold H$_2$O and dry the solid in a vacuum desiccator over P$_2$O$_5$ + solid KOH. Extract the solid with pet ether (b 40-60°) filter off any insoluble solid and cool. Collect the crystalline sulfonyl chloride and recryst from pet ether or *C$_6$H$_6$ pet ether. If large quantities are available then it can be distd under high vacuum. [Fierz-Davaid and Weissenbach *Helv Chim Acta* **3** 2312 *1920*.] The *sulfonamide* has **m** 150° (from EtOH or H$_2$O).

Naphthalene-2-sulfonyl chloride *[93-11-8]* **M 226.7, m 74-76°, 78°, 79°, b 148°/0.6mm, 201°/13mm.** Crystd (twice) from *benzene/pet ether (1:1 v/v). Purified as the 2-sulfonyl chloride. [Fierz-Davaid and Weissenbach *Helv Chim Acta* **3** 2312 *1920*.] The *sulfonamide* has **m** 217° (from EtOH).

1,8-Naphthalic acid (naphthalene-1,8-dicarboxylic acid) *[518-05-8]* **M 216.9, m 270°, pK$_{Est(1)}$~ 2.1, pK$_{Est(2)}$~ 4.5.** Crystd from EtOH or aq EtOH.

1,8-Naphthalic anhydride *[81-84-5]* **M 198.2, m 274°.** Extracted with cold aqueous Na$_2$CO$_3$ to remove free acid, then crystd from acetic anhydride.

Naphthamide *[2243-82-5]* **M 171.2, m 195°, pK^{20} -2.30 (H$_o$ scale, aq H$_2$SO$_4$).** Crystd from EtOH.

Naphthazarin (5,8-dihydroxy-1,4-naphthoquinone) *[475-38-7]* **M 190.2, m ~ 220-230°(dec), m 225-230°, pK$_{Est(1)}$~9.5, pK$_{Est(2)}$~11.1.** Red-brown needles with a green shine from EtOH. Also recrystd from hexane and purified by vacuum sublimation. [Huppert et al. *J Phys Chem* **89** 5811 *1985*.] It is sparingly soluble in H$_2$O but soluble in alkalis. It sublimes at 2-10mm. The *diacetate* forms golden yellow prisms from CHCl$_3$, **m** 192-193° and the *5,8-dimethoxy* derivative has **m** 157° (155°) (from pet ether) [Bruce and Thompson *J Chem Soc* 1089 *1955*; IR: Schmand and Boldt *J Am Chem Soc* **97** 447 *1975*; NMR: Brockmann and Zeeck *Chem Ber* **101** 4221 *1968*]. The *monothiosemicarbazone* has **m** 168°(dec) from EtOH [Gardner et al. *J Am Chem Soc* 74 2106 *1952*].

Naphthionic acid (4-aminonaphthalene-1-sulfonic acid) *[84-86-6]* **M 223.3, m > 300°(dec), pK^{25} 2.68.** It crystallises from H$_2$O as needles of the 0.5 hydrate . Salt solns fluoresce strongly blue.

1-Naphthoic acid *[86-55-5]* **M 172.2, m 162.5-163.0°, pK^{25} 3.60.** Crystd from toluene (3mL/g) (charcoal), pet ether (b 80-100°), or aqueous 50% EtOH.

2-Naphthoic acid *[93-09-4]* **M 172.2, m 184-185°, pK^{25} 4.14.** Crystd from EtOH (4mL/g), or aqueous 50% EtOH. Dried at 100°.

1-Naphthol *[90-15-3]* **M 144.2, m 95.5-96°, pK^{25} 9.34.** Sublimed, then crystd from aqueous MeOH (charcoal), aq 25% or 50% EtOH, *benzene, cyclohexane, heptane, CCl$_4$ or boiling water. Dried over P$_2$O$_5$ under vacuum. [Shizuka et al. *J Am Chem Soc* **107** 7816 *1985*.]

2-Naphthol *[135-19-3]* **M 144.2, m 122.5-123.5°, pK^{25} 9.57.** Crystd from aqueous 25% EtOH (charcoal), water, *benzene, toluene or CCl$_4$, e.g. by repeated extraction with small amounts of EtOH, followed by dissolution in a minimum amount of EtOH and pptn with distilled water, then drying over P$_2$O$_5$ under vacuum. Has also been dissolved in aqueous NaOH, and ppted by adding acid (repeated several times), then ppted from *benzene by addition of heptane. Final purification can be by zone melting or sublimation *in vacuo*. [Bardez et al. *J Phys Chem* **89** 5031 *1985*; Kikuchi et al. *J Phys Chem* **91** 574 *1987*.]

Naphthol AS-D (3-hydroxy-2-naphthoic-*o*-toluide) *[135-61-5]* **M 277.3, m 1196-198°.** Purified by recrystn from xylene. Gives yellow-green fluorescent solutions at pH 8.2-9.5, [IR: Schnopper et al. *Anal Chem* **31** 1542 *1959*.] With AcCl *naphthol AS-D acetate* is obtained **m** 168-169°, and with chloroacetyl chloride *naphthol AS-D-chloroacetate* is obtained [Moloney et al. *J Histochem Cytochem* **8** 200 *1960*; Burstone *Arch Pathology* **63** 164 *1957*].

α-**Naphtholbenzein** [bis-(α-{4-hydroxynaphth-1-yl})-benzyl alcohol] *[6948-88-5]* **M 392.5, m 122-125°, pK_{Est} ~ 9.3.** Crystd from EtOH, aqueous EtOH or glacial acetic acid.

1-Naphthol-2-carboxylic acid (1-hydroxy-2-naphthoic acid) *[86-48-6]* **M 188.2, m 203-204°, $pK_{Est(1)}$~2.5, $pK_{Est(2)}$~12.** Successively crystd from EtOH/water, diethyl ether and acetonitrile, with filtration through a column of charcoal and Celite. [Tong and Glesmann *J Am Chem Soc* **79** 583 *1957*.]

2-Naphthol-3-carboxylic acid (2-hydroxy-3-naphthoic acid) *[92-70-6]* **M 188.2, m 222-223°, pK_1^{25} 2.79, pK_2^{25} 12.84.** Crystd from water or acetic acid.

1,2-Naphthoquinone *[524-42-5]* **M 158.2, m 140-142°(dec).** Crystd from ether (red needles) or *benzene (orange leaflets).

1,4-Naphthoquinone *[130-15-4]* **M 158.2, m 125-125.5°.** Crystd from diethyl ether (charcoal). Steam distd. Crystd from *benzene or aqueous EtOH. Sublimed in a vacuum.

β-**Naphthoxyacetic acid** *[120-23-0]* **M 202.2, m 156°, pK_{Est} ~3.0.** Crystd from hot water or *benzene.

β-**Naphthoyltrifluoroacetone** (4,4,4-trifluoro-2-naphthylbutan-1,3-dione) *[893-33-4]* **M 266.2, m 70-71°, 74-76°, pK^{20} 6.35.** Crystd from EtOH. The *mono oxime* crystd from H_2O or aq EtOH has **m** 137-138°. [Reid and Calvin *J Am Chem Soc* **72** 2948 *1950*.]

Naphthvalene *[34305-47-0]* **M 104.1, m dec at 175° to benzvalene.** Purified by chromatography on alumina and eluting with pentane It is stable at room temp [Abelt et al. *J Am Chem Soc* **107** 4148 *1985*]. The ^{1}H NMR in CCl$_4$ has τ 3.18 (4H), 6.17 (t *J* 1.5Hz, 2H), 7.60 (t *J* 1.5Hz 2H).

1-Naphthyl acetate *[830-81-9]* **M 186.2, m 45-46°.** Chromatographed on silica gel and crystd as the 2-isomer below.

2-Naphthyl acetate *[1523-11-1]* **M 186.2, m 71°.** Crystd from pet ether (b 60-80°) or dilute aq EtOH.

1-Naphthylacetic acid *[86-87-3]* **M 186.2, m 132°, pK^{25} 4.23.** Crystd from EtOH or water.

2-Naphthylacetic acid *[581-96-4]* **M 186.2, m 143.1-143.4°, pK^{25} 4.30.** Crystd from water or *benzene.

1-**Naphthylamine** *[134-32-7]* **M 143.2, m 50.8-51.2°, b 160°, pK^{25} 3.94.** Sublimed at 120° in a stream of nitrogen, then crystd from pet ether (b 60-80°), or abs EtOH then diethyl ether. Dried under vacuum in an Abderhalden pistol. Has also been purified by crystn of its hydrochloride from water, followed by liberation of the free base and distn; finally purified by zone melting. **CARCINOGEN.**

1-**Naphthylamine hydrochloride** *[552-46-5]* **M 179.7, m sublimes on heating.** Crystd from water (charcoal).

2-**Naphthylamine** *[91-59-8]* **M 143.2, m 113°, pK^{25} 4.20.** Sublimed at 180° in a stream of nitrogen. Crystd from hot water (charcoal) or *benzene. Dried under vacuum in an Abderhalden pistol. **CARCINOGEN.**

1-Naphthylamine-5-sulfonic acid *[84-89-9]* **M 223.3, m >200°(dec), $pK_{Est(1)}$<1, pK_2^{25} 3.69** (NH$_2$) Crystd under nitrogen from boiling water and dried in a steam oven [Bryson *Trans Faraday Soc* **47** 522, 527 *1951*].

2-Naphthylamine-1-sulfonic acid *[81-16-3]* M 223.3, m >200°(dec), pK_1^{25}<1, pK_2^{25} 2.35 (NH₂). Crystd under nitrogen from boiling water and dried in a steam oven [Bryson *Trans Faraday Soc* **47** 522, 527 *1951*].

2-Naphthylamine-6-sulfonic acid *[93-00-5]* M 223.3, m >200°(dec). Crystd from a large volume of hot water.

1-(1-naphthyl) ethanol *[R-(+)- 42177-25-3; S-(-)- 15914-84-8]* M 172.2, m 46°, 45-47.5°, 48°, $[\alpha]_{546}^{20}$ (+) and (-) 94°, $[\alpha]_D^{20}$ (+) and (-) 78° (c 1, MeOH). Purified by recrystn from Et₂O-pet ether, Et₂O, hexane [Balfe et al. *J Chem Soc* 797 *1946*; IR, NMR: Theisen and Heathcock *J Org Chem* **53** 2374 *1988*; see also Fredga et al. *Acta Chem Scand* **11** 1609 *1957*]. The *RS-alcohol* *[57605-95-5]* has **m** 63-65,°, 65-66° from hexane.

1-(1-Naphthyl)ethylamine *[R-(+)- 3886-70-2; S-(-)- 10420-89-0]* M 171.2, b 153°/11mm, 178-181°/20mm, d_4^{20} 1.067, n_D^{20} 1.624, $[\alpha]_{546}^{20}$ (+) and (-) 65°, $[\alpha]_D^{20}$ (+) and (-) 55° (c 2, MeOH); $[\alpha]_D^{17}$ (+) and (-) 82.8° (neat), pK_{Est} ~9.3. Purified by distn in a good vacuum. [Mori et al. *Tetrahedron* **37** 1343 *1981*; cf Wilson in *Top Stereochem* (Allinger and Eliel eds) **vol 6** 135 *1971*; Fredga et al. *Acta Chem Scand* **11** 1609 *1957*.] The *hydrochlorides* crystallises from H₂O $[\alpha]_D^{18}$ ±3.9° (c 3, H₂O) and the *sulfates* recrystallises from H₂O as *tetrahydrates* **m** 230-232°. The *RS-amine* has **b** 153°/11mm, 156°/15mm, 183.5°/41mm [Blicke and Maxwell *J Am Chem Soc* **61** 1780 *1939*].

2-Naphthylethylene (2-vinylnaphthalene) *[827-54-3]* M 154.2, m 66°, b 95-96°/2.1mm, 135-137°/18mm. Crystd from aqueous EtOH.

N-(α-Naphthyl)ethylenediamine dihydrochloride *[1465-25-4]* M 291.2, m 188-190°, $pK_{Est(1)}$~3.8, $pK_{Est(2)}$~9.4. Crystd from water.

1-Naphthyl isocyanate *[86-84-0]* M 169.2, m 3-5°, b 269-270°/atm, d_4^{20} 1.18. Distd at atmospheric pressure or in a vacuum. Can be crystd from pet ether (b 60-70°) at low temperature. *It has a pungent odour, is* **TOXIC** *and is absorbed through the skin.*

1-Naphthyl isothiocyanate *[551-06-4]* M 185.3, m 58-59°. Crystd from hexane (1g in 9 mL). White needles soluble in most organic solvents but is insoluble in H₂O. *It is absorbed through the skin and may cause dermatitis.* [*Org Synth* Coll Vol IV 700 *1963*.]

2-Naphthyl lactate *[93-43-6]* M 216.2. Crystd from EtOH.

2-(2-Naphthyloxy)ethanol *[93-20-9]* M 188.2, m 76.7°. Crystd from *benzene/pet ether.

N-1-Naphthylphthalamic acid *[132-66-1]* M 291.3, m 203°. Crystd from EtOH.

2-Naphthyl salicylate *[613-78-5]* M 264.3, m 95°, pK_{Est} ~10.0. Crystd from EtOH.

1-Naphthyl thiourea (ANTU) *[86-88-4]* M 202.2, m 198°. Crystd from EtOH.

1-Naphthyl urea *[6950-84-1]* M 186.2, m 215-220°. Crystd from EtOH.

2-Naphthyl urea *[13114-62-0]* M 186.2, m 219-220°. Crystd from EtOH.

1,5-Naphthyridine *[254-79-5]* M 130.1, m 75°, b 112°/15mm, pK^{20} 2.84. Purified by repeated sublimation.

Narcein {6-[6-(2-dimethylaminoethyl)]-2-methoxy-3,4-(methylenedioxy)phenylacetyl]-2,3-dimethoxybenzoic acid} *[131-28-2]* **M 445.4, m 176-177° (145° anhydrous), pK$_1^{15}$ 3.5, pK$_2^{15}$ 9.3.** Crystd from water (as trihydrate).

Naringenin (4',5,7-trihydroxyflavanone) *[480-41-1]* **M 272.3, m 251° (phenolic pKs~ 8-11).** Crystd from aqueous EtOH.

Naringin (naringenin 7-rhamnoglucoside) *[10236-47-2]* **M 580.5, m 171° (2H$_2$O), [α]$_D^{19}$ -90° (c 1, EtOH), [α]$_{546}^{20}$ -107° (c 1, EtOH).** Crystd from water. Dried at 110°(to give the dihydrate).

Neopentane (2,2-dimethylpropane) *[463-82-1]* **M 72.2, b 79.3°, d 0.6737, n 1.38273.** Purified from isobutene by passage over conc H$_2$SO$_4$ or P$_2$O$_5$, and through silica gel.

Neostigmine [(3-dimethylcarbamoylphenyl)trimethylammonium] bromide *[114-80-7]* **M 303.2, m 176°(dec).** Crystd from EtOH/diethyl ether. (*Highly* **TOXIC**).

Neostigmine methyl sulfate *[51-60-5]* **M 334.4, m 142-145°.** Crystd from EtOH. (*Highly* **TOXIC**.]

Nerolidol (3,7,11-trimethyl-1,6,10-dodecatrien-3-ol) **M 222.4** *[cis/trans 7212-44-4]* **b 122°/3mm, n 1.477, d 0.73,** *[cis 3790-78-1]* **b 70°/0.1mm,** *[trans 40716-66-3]* **b 78°/0.2mm, 145-146°/2mm.** Purified by TLC on plates of Kieselguhr G [McSweeney *J Chromatogr* **17** 183 *1965*] or silica gel impregnated with AgNO$_3$, using 1,2-CH$_2$Cl$_2$/CHCl$_3$/EtOAc/PrOH (10:10:1:1) as solvent system. Also by GLC on butanediol succinate (20%) on Chromosorb W. Stored under N$_2$ at ~5° in the dark.

Neutral Red (2-amino-8-dimethylamino-3-methylphenazine HCl, Basic Red 5, CI 50040) *[553-24-2]* **M 288.8, m 290°(dec), pK^{25} 6.5.** Crystd from *benzene/MeOH (1:1). In aq sol it is red at pH 6.8 and yellow at pH 8.0.

Nicotinaldehyde thiosemicarbazone *[3608-75-1]* **M 180.2, m 222-223°.** Crystd from water.

Nicotinamide *[98-92-0]* **M 122.1, m 128-131°, pK$_1^{20}$ 0.5, pK$_2^{20}$ 3.33.** Crystd from *benzene.

Nicotinic acid (niacin, 2-yridine-3-carboxylic acid) *[59-67-6]* **M 123.1, m 232-234°, pK$_1^{25}$ 2.00, pK$_2^{25}$ 4.82.** Crystd from *benzene.

Nicotinic acid hydrazide *[553-53-7]* **M 137.1, m 158-159°, pK$_1^{20}$ 3.3, pK$_2^{20}$ 11.49(NH).** Crystd from aqueous EtOH or *benzene.

Nile Blue A (a benzophenoxazinium sulfate dye) *[3625-57-8]* **M 415.5, m >300°(dec), CI 51180, pK^{25} 2.4.** Crystd from pet ether.

Ninhydrin (1,2,3-triketohydrindene hydrate) *[485-47-2]* **M 178.1, m 241-243°(dec), pK^{30} 8.82.** Crystd from hot water (charcoal). Dried under vacuum and stored in a sealed brown container.

Nitrilotriacetic acid [tris(carboxymethyl)amine, NTA, Complexone 1] *[139-13-9]* **M 191.1, m 247°(dec), pK$_1$ 0.8, pK$_2$ 1.71, pK$_3$ 2.47, pK$_4$ 9.71.** Crystd from water. Dried at 110°.

2-Nitroacetanilide *[552-32-9]* **M 180.2, m 93-94°, pK$_{Est}$ <0.** Crystd from water.

4-Nitroacetanilide *[104-04-1]* **M 180.2, m 217°, pK$_{Est}$ <0.** Ppted from 80% H$_2$SO$_4$ by adding ice, then washed with water, and crystd from EtOH. Dried in air.

3-Nitroacetophenone *[121-89-1]* **M 165.2, m 81°, b 167°/18mm, 202°/760mm.** Distilled in steam and crystd from EtOH.

4-Nitroacetophenone *[100-19-6]* **M 165.2, m 80-81°, b 145-152°/760mm.** Crystd from EtOH or aqueous EtOH.

3-Nitroalizarin (1,2-dihydroxy-3-nitro-9,10-anthraquinone, Alizarin Orange) *[568-93-4]* **M 285.2, m 244° (dec), $pK_{Est(1)}$~4.6, $pK_{Est(2)}$~9.6.** Crystd from acetic acid.

o-**Nitroaniline** *[88-74-4]* **M 138.1, m 72.5-73.0°, pK^{25} -0.25 (-0.31).** Crystd from hot water (charcoal), then crystd from water, aqueous 50% EtOH, or EtOH, and dried in a vacuum desiccator. Has also been chromatographed on alumina, then recrystd from *benzene.

m-**Nitroaniline** *[99-09-2]* **M 138.1, m 114°, pK^{25} 2.46.** Purified as for *o*-nitroaniline. **Warning: it is absorbed through the skin.**

p-**Nitroaniline** *[100-01-6]* **M 138.1, m 148-148.5°, pK^{25} 1.02.** Purified as for *o*-nitroaniline. Also crystd from acetone. Freed from *o*- and *m*-isomers by zone melting and sublimation.

o-**Nitroanisole (2-methoxynitrobenzene)** *[91-23-6]* **M 153.1, f 9.4°, b 265°/737mm, d 1.251, n 1.563.** Purified by repeated vacuum distn in the absence of oxygen.

p-**Nitroanisole (4-methoxynitrobenzene)** *[100-17-4]* **M 153.1, m 54°.** Crystd from pet ether or hexane and dried *in vacuo*.

9-Nitroanthracene *[602-60-8]* **M 223.2, m 142-143°.** Purified by recrystn from EtOH or MeOH. Further purified by sublimation or TLC.

5-Nitrobarbituric acid (dilituric acid) *[480-68-2]* **M 173.1, m 176°, 176-183°, pK^{20} 10.25.** Crystd from water.

o-**Nitrobenzaldehyde** *[552-89-6]* **M 151.1, m 44-45°, b 120-144°/3-6mm.** Crystd from toluene (2-2.5mL/g) by addition of 7mL pet ether (b 40-60°) for 1mL of soln. Can also be distd at reduced pressures.

m-**Nitrobenzaldehyde** *[99-61-6]* **M 151.1, m 58°.** Crystd from water or EtOH/water, then sublimed twice at 2mm pressure at a temperature slightly above its melting point.

p-**Nitrobenzaldehyde** *[555-16-8]* **M 151.1, m 106°.** Purification as for *m*-nitrobenzaldehyde above.

Nitrobenzene *[98-95-3]* **M 123.1, f 5.8°, b 84-86.5°/6.5-8mm, 210.8°/760mm, d 1.206, n^{15} 1.55457, n 1.55257, pK^{18} -11.26 (aq H_2SO_4).** Common impurities include nitrotoluene, dinitrothiophene, dinitrobenzene and aniline. Most impurities can be removed by steam distn in the presence of dilute H_2SO_4, followed by drying with $CaCl_2$, and shaking with, then distilling at low pressure from BaO, P_2O_5, $AlCl_3$ or activated alumina. It can also be purified by fractional crystn from absolute EtOH (by refrigeration). Another purification process includes extraction with aqueous 2M NaOH, then water, dilute HCl, and water, followed by drying ($CaCl_2$, $MgSO_4$ or $CaSO_4$) and fractional distn under reduced pressure. The pure material is stored in a brown bottle, in contact with silica gel or CaH_2. It is very *hygroscopic*.

4-Nitrobenzene-azo-resorcinol (magneson II) *[74-39-5]* **M 259.2, m 199-200°.** Crystd from EtOH.

2-Nitrobenzenesulfenyl chloride (NPS-Cl) *[7669-54-7]* **M 189.6, m 73-74.5°, 74.5-75°, 74-76°.** Recrystd from CCl_4 (2mL/g), filter off the soln at 5° (recovery 75%). Also recrystd from pet ether (b 40-60°), dried rapidly at 50° and stored in a brown glass bottle, sealed well and stored away from moisture. [Hubacher *Org Synth* Coll Vol II 455 *1943*; Ito et al. *Chem Pharm Bull (Jpn)* **26** 296 *1978*.]

4-Nitrobenzhydrazide *[606-26-8]* **M 181.1, m 213-214°.** Crystd from EtOH.

4'-Nitrobenzo-15-crown-5 *[60835-69-0]* **M 313.3, m 84-85°, 93-95°.** Recrystd from EtOH, MeOH or *C₆H₆-hexane as for the 18-crown-6 compound below. It complexes with Na⁺, K⁺, NH₄⁺, Ca²⁺, Mg²⁺ and Cd²⁺. NMR (CDCl₃) has δ: 3.6-4.4 (m 16CH₂), 6.8 (d 1H arom), 7.65 (d 1H arom), 7.80 (dd 1H arom J_{ab} 9Hz and J_{bc} 3Hz) [Shmid et al. *J Am Chem Soc* **98** 5198 *1976*; Kikukawa et al. *Bull Chem Soc Jpn* **50** 2207 *1977*; Toke et al. *Justus Liebigs Ann Chem* **349** 349, 761 *1988*; Lindner et al. *Z Anal Chem* **322** 157 *1985*].

4'-Nitrobenzo-18-crown-6 *[53408-96-1]* **M 357.4, m 83-84°, 83-84°.** If impure and discoloured then chromatograph on Al₂O₃ and eluting with *C₆H₆ (1:1) with 1% MeOH added. The fractions are followed by TLC on Al₂O₃ (using detection with Dragendorff's reagent R_F 0.6 in the above solvent system). Recrystallise the residues from the fractions containing the product from *C₆H₆-hexane to give yellowish leaflets. It complexes with Na or K ions with **logK_Na 3.95 and logK_K 4.71.** [Petranek and Ryba *Collect Chem Czech Chem Commun* **39** 2033 *1974*.]

2-Nitrobenzoic acid *[552-16-9]* **M 167.1, m 146-148°, pK²⁵ 2.21.** Crystd from *benzene (twice), *n*-butyl ether (twice), then water (twice). Dried and stored in a vacuum desiccator. [Le Noble and Wheland *J Am Chem Soc* **80** 5397 *1958*.] Has also been crystd from EtOH/water.

3-Nitrobenzoic acid *[121-92-6]* **M 167.1, m 143-143.5°, pK²⁵ 3.46.** Crystd from *benzene, water, EtOH (charcoal), glacial acetic acid or MeOH/water. Dried and stored in a vacuum desiccator.

4-Nitrobenzoic acid *[62-23-7]* **M 167.1, m 241-242°, pK²⁵ 3.43.** Purification as for 3-nitrobenzoic acid above.

4-Nitrobenzoyl chloride *[122-04-3]* **M 185.6, m 75°, b 155°/20mm.** Crystd from dry pet ether (b 60-80°) or CCl₄. Distilled under vacuum. **Irritant**.

4-Nitrobenzyl alcohol *[619-73-8]* **M 153.1, m 93°.** Crystd from EtOH and sublimed in a vacuum. Purity should be at least 99.5%. Sublimed samples should be stored in the dark over anhydrous CaSO₄ (Drierite). It the IR contains OH bands then the sample should be resublimed before use. [Mohammed and Kosower *J Am Chem Soc* **93** 2709 *1979*.]

4-Nitrobenzyl bromide *[100-11-8]* **M 216.0, m 98.5-99.0°.** Recrystd four times from abs EtOH, then twice from cyclohexane/hexane/*benzene (1:1:1), followed by vac sublimation at 0.1mm and a final recrystn from the same solvent mixture. [Lichtin and Rao *J Am Chem Soc* **83** 2417 *1961*.] Has also been crystd from pet ether (b 80-100°, 10mL/g, charcoal). It slowly decomposes even when stored in a desiccator in the dark. **IRRITANT.**

m-**Nitrobenzyl chloride** *[619-23-8]* **M 171.6, m 45°.** Crystd from pet ether (b 90-120°). **IRRITANT**.

p-**Nitrobenzyl chloride** *[100-14-1]* **M 171.6, m 72.5-73°.** Crystd from CCl₄, dry diethyl ether, 95% EtOH or *n*-heptane, and dried under vacuum. **IRRITANT**.

p-**Nitrobenzyl cyanide** *[555-21-5]* **M 162.2, m 117°.** Crystd from EtOH. **TOXIC.**

4-(4-Nitrobenzyl)pyridine (PNBP) *[1083-48-3]* **M 214.2, m 70-71°, pK_Est ~5.0.** Crystd from cyclohexane.

2-Nitrobiphenyl *[86-00-0]* **M 199.2, m 36.7°.** Crystd from EtOH (seeding required). Sublimed under vacuum.

3-Nitrocinnamic acid *[555-68-0]* **M 193.2, m 200-201°, pK²⁵ 2.58 (*trans*).** Crystd from *benzene or EtOH.

4-Nitrocinnamic acid *[882-06-4]* **M 193.2, m 143° (*cis*), 286°(*trans*), pK$_{Est}$ ~2.6 (*trans*).** Crystd from water.

4-Nitrodiphenylamine *[836-30-6]* **M 214.2, m 133-134°, pK^{25} -2.5.** Crystd from EtOH.

2-Nitrodiphenyl ether *[2216-12-8]* **M 215.2, b 106-108°/0.01mm, 137-138°/0.5mm, 161-162°/4mm, 188-189°/12mm, 195-200°/25mm, d$_4^{20}$ 1.241, n$_D^{25}$ 1.600.** Purified by fractional distn. UV (EtOH): 255, 315mm (ϵ 6200 and 2800); IR (CS$_2$): 1350 (NO$_2$) and 1245, 1265 (COC) cm^{-1} [UV, IR: Dahlgard and Brewster *J Am Chem Soc* **80** 5861 *1958*; Tomita and Takase *Yakugaku Zasshi (J Pharm Soc Japan)* **75** 1077 *1955*; Fox and Turner *J Chem Soc* 1115 *1930*, Henley *J Chem Soc* 1222 *1930*].

Nitrodurene (1-nitro-2,3,5,6-tetramethylbenzene) *[3463-36-3]* **M 179.2, m 53-55°, b 143-144°/10mm.** Crystd from EtOH, MeOH, acetic acid, pet ether or chloroform.

Nitroethane *[79-24-3]* **M 75.1, b 115°, d 1.049, n 1.3920, n^{25} 1.39015, pK^{25} 8.60 (8.46, pH equilibrium requires *ca* 5 min).** Purified as described for *nitromethane*. A spectroscopic impurity has been removed by shaking with activated alumina, decanting and rapidly distilling.

2-Nitrofluorene *[607-57-8]* **M 211.2, m 156°.** Crystd from aqueous acetic acid.

Nitroguanidine *[556-88-7]* **M 104.1, m 246-246.5°(dec), 257°, pK$_1^{25}$ -0.55, pK$_2^{25}$ 12.20.** Crystd from water (20mL/g).

5-Nitroindole *[6146-52-7]* **M 162.1, m 141-142°, pK^{25} -7.4 (aq H$_2$SO$_4$).** Decolorised (charcoal) and recrystd twice from aqueous EtOH.

Nitromesitylene (2-nitro-1,3,5-trimethylbenzene) *[603-71-4]* **M 165.2, m 44°, b 255°.** Crystd from EtOH.

Nitromethane *[75-52-5]* **M 61.0, f -28.5°, b 101.3°, d 1.13749, d^{30} 1.12398, n 1.3819, n^{30} 1.37730, pK^{25} 10.21.** Nitromethane is generally manufactured by gas-phase nitration of methane. The usual impurities include aldehydes, nitroethane, water and small amounts of alcohols. Most of these can be removed by drying with CaCl$_2$ or by distn to remove the water/nitromethane azeotrope, followed by drying with CaSO$_4$. Phosphorus pentoxide is not suitable as a drying agent. [Wright et al. *J Chem Soc* 199 *1936*.] The purified material should be stored by dark bottles, away from strong light, in a cool place. Purifications using extraction are commonly used. For example, Van Looy and Hammett [*J Am Chem Soc* **81** 3872 *1959*] mixed about 150mL of conc H$_2$SO$_4$ with 1L of nitromethane and allowed it to stand for 1 or 2days. The solvent was washed with water, aqueous Na$_2$CO$_3$, and again with water, then dried for several days with MgSO$_4$, filtered again with CaSO$_4$. It was fractionally distd before use. Smith, Fainberg and Winstein [*J Am Chem Soc* **83** 618 *1961*] washed successively with aqueous NaHCO$_3$, aqueous NaHSO$_3$, water, 5% H$_2$SO$_4$, water and dilute NaHCO$_3$. The solvent was dried with CaSO$_4$, then percolated through a column of Linde type 4A molecular sieves, followed by distn from some of this material (in powdered form). Buffagni and Dunn [*J Chem Soc* 5105 *1961*] refluxed for 24h with activated charcoal while bubbling a stream of nitrogen through the liquid. The suspension was filtered, dried (Na$_2$SO$_4$) and distd, then passed through an alumina column and redistd. It has also been refluxed over CaH$_2$, distd and kept under argon over 4A molecular sieves.
Can be purified by zone melting or by distn under vacuum at 0°, subjecting the middle fraction to several freeze-pump-thaw cycles. An impure sample containing higher nitroalkanes and traces of cyanoalkanes was purified (on the basis of its NMR spectrum) by crystn from diethyl ether at -60° (cooling in Dry-ice)[Parrett and Sun *J Chem Educ* **54** 448 *1977*].
Fractional crystn was more effective than fractional distn from Drierite in purifying nitromethane for conductivity measurements. [Coetzee and Cunningham *J Am Chem Soc* **87** 2529 *1965*.] Specific conductivities around 5 x 10^{-9} ohm^{-1}cm^{-1} were obtained.

Nitron [1,4-diphenyl-3-phenylamino-(1H)-1,2,4-triazolium (hydroxide) inner salt] *[2218-94-2]* **M 312.4, m 189°(dec).** Crystd from EtOH, chloroform or EtOH/*C$_6$H$_6$.

1-Nitronaphthalene *[86-57-7]* **M 173.2, m 57.3-58.3°, b 30-40°/0.01mm.** Fractionally distd under reduced pressure, then crystd from EtOH, aqueous EtOH or heptane. Chromatographed on alumina from *benzene/pet ether. Sublimes *in vacuo*.

2-Nitronaphthalene *[581-89-5]* **M 173.2, m 79°, b 165°/15mm.** Crystd from aqueous EtOH and sublimed in a vacuum.

1-Nitro-2-naphthol *[550-60-7]* **M 189.2, m 103°, pK^{25} 5.93.** Crystd (repeatedly) from *benzene-pet ether (b 60-80°)(1:1).

2-Nitro-1-naphthol *[607-24-9]* **M 189.2, m 127-128°, pK^{25} 5.89.** Crystd (repeatedly) from EtOH.

5-Nitro-1,10-phenanthroline *[4199-88-6]* **M 225.2, m 197-198°, pK^{25} 3.33.** Crystd from *benzene/pet ether, until anhydrous.

2-Nitrophenol *[88-75-5]* **M 139.1, m 44.5-45.5°, pK^{25} 7.23.** Crystd from EtOH/water, water, EtOH, *benzene or MeOH/pet ether (b 70-90°). Can be steam distd. Petrucci and Weygandt [*Anal Chem* 33 275 *1961*] crystd from hot water (twice), then EtOH (twice), followed by fractional crystn from the melt (twice), drying over CaCl$_2$ in a vacuum desiccator and then in an Abderhalden drying pistol.

3-Nitrophenol *[554-84-7]* **M 139.1, m 96°, b 160-165°/12mm, pK^{25} 8.36.** Crystd from water, CHCl$_3$, CS$_2$, EtOH or pet ether (b 80-100°), and dried under vacuum over P$_2$O$_5$ at room temperature. Can also be distd at low pressure.

4-Nitrophenol *[100-02-7]* **M 139.1, m 113-114°, pK^{25} 7.16.** Crystd from water (which may be acidified, e.g. N H$_2$SO$_4$ or 0.5N HCl), EtOH, aqueous MeOH, CHCl$_3$, *benzene or pet ether, then dried under vacuum over P$_2$O$_5$ at room temperature. Can be sublimed at 60°/10^{-4}mm.

2-Nitrophenoxyacetic acid *[1878-87-1]* **M 197.2, m 150-159°, pK^{25} 2.90.** Crystd from water, and dried over P$_2$O$_5$ under vacuum..

p-**Nitrophenyl acetate** *[830-03-5]* **M 181.2, m 78-79°.** Recrystd from absolute EtOH [Moss et al. *J Am Chem Soc* **108** 5520 *1986*].

2-Nitrophenylacetic acid *[3740-52-1]* **M 181.2, m 120°, pK^{25} 3.95.** Crystd from EtOH/water and dried over P$_2$O$_5$ under vacuum.

4-Nitrophenylacetic acid *[104-03-0]* **M 181.2, m 80.5°, pK^{25} 3.92.** Crystd from EtOH/water (1:1), then from sodium-dried diethyl ether and dried over P$_2$O$_5$ under vacuum.

4-Nitro-1,2-phenylenediamine *[99-56-9]* **M 153.1, m 201°, pK$_1^{25}$ 1.39 (1-NH$_2$), pK$_2^{25}$ 2.61 (2-NH$_2$).** Crystd from water.

1-(4-Nitrophenyl)ethylamine hydrochloride *[R-(+)- 57233-86-0; S-(-)- 132873-57-5]* **M 202.6, m 225°, 240-242° (dec), 243-245° (dec), 248-250°, [α]$_D^{20}$ (+) and (-) 72° (c 1, 0.05 M NaOH), (+) and (-) 0.3° (H$_2$O), pK$_{Est}$ ~8.6.** To ensure dryness the hydrochloride (*ca* 175 g) is extracted with EtOH (3x100mL) and evaporated to dryness (any residual H$_2$O increases the solubility in EtOH and lowers the yield). The hydrochloride residue is triturated with absolute EtOH and dried *in vacuo*. The product is further purified by refluxing with absolute EtOH (200 mL for 83g) for 1h, cool to 10° to give 76.6g of hydrochloride **m** 243-245° (dec). The *free base* is prepd by dissolving in N NaOH, extract with CH$_2$Cl$_2$ (3 x 500mL), dry (Na$_2$CO$_3$), filter, evaporate and distil, **m** 27°, **b** 119-120°/0.5mm (105-107°/0.5mm, 157-159°/9mm, d$_4^{20}$

1.1764, n $_D^{20}$ 1.5688, $[\alpha]_D^{24}$ ±17.7° (neat)[Perry et al. *Synthesis* 492 *1977*; ORD: Nerdel and Liebig *Justus Liebigs Ann Chem* **621** 142 *1959*].

4-Nitrophenylhydrazine *[100-16-3]* M 153.1, m 158°(dec), pK_1^{25} -9.2 (aq H_2SO_4), pK_2^{25} 3.70. Crystd from EtOH.

3-Nitrophenyl isocyanate *[3320-87-4]* M 164.1, m 52-54°. Crystd from pet ether (b 28-38°).

4-Nitrophenyl isocyanate *[100-28-7]* M 164.1, m 53°. Crystd from pet ether (b 28-38°).

2-Nitrophenylpropiolic acid *[530-85-8]* M 191.1, m 157°(dec), pK^{25} 2.83. Crystd from water.

4-Nitrophenyl trifluoroacetate *[658-78-6]* M 235.1, m 37-39°, b 120°/12mm. Recrystd from $CHCl_3$/hexane [Margolis et al. *J Biol Chem* **253** 7891 *1078*].

4-Nitrophenyl urea *[556-10-5]* M 181.2, m 238°. Crystd from EtOH and hot water.

3-Nitrophthalic acid *[603-11-2]* M 211.1, m 216-218°, pK^{25} 3.93. Crystd from hot water (1.5mL/g). Air dried.

4-Nitrophthalic acid *[610-27-5]* M 211.1, m 165°, pK^{25} 4.12. Crystd from ether or ethyl acetate.

3-Nitrophthalic anhydride *[641-70-3]* M 193.1, m 164°. Crystd from *benzene, *benzene/pet ether, acetic actic or acetone. Dried at 100°.

1-Nitropropane *[108-03-2]* M 89.1, b 131.4°, d 1.004, n 1.40161, n^{25} 1.39936, pK^{25} 8.98. Purified as *nitromethane*.

2-Nitropropane *[79-46-9]* M 89.1, b 120.3°, d 0.989, n 1.3949, n^{25} 1.39206, pK^{25} 7.68. Purified as *nitromethane*.

5-Nitro-2-*n*-propoxyaniline *[553-79-7]* M 196.2, m 47.5-48.5°, pK_{Est} ~2.32. Crystd from *n*-propyl alcohol/pet ether.

3-Nitro-2-pyridinesulfenyl chloride *[68206-45-1]* M 190.2, m 217-222°(dec). Crystallises as yellow needles from CH_2Cl_2. When pure it is stable for several weeks at room temperature, and no decomposition was observed after 6 months at <0°. UV (MeCN) has λmax at 231nm (ε 12,988), 264nm (ε 5,784) and 372nm (ε 3,117). [NMR and UV: Matsuda and Aiba *Chem Lett* 951 *1978*; Wagner et al. *Chem Ber* **75** 935 *1942*.]

5-Nitroquinoline *[607-34-1]* M 174.2, m 70°, pK^{20} 2.69. Crystd from pentane, then from *benzene.

8-Nitroquinoline *[706-35-2]* M 174.2, m 88-89°, pK^{20} 2.55. Crystd from hot water, MeOH, EtOH or EtOH/diethyl ether (3:1).

4-Nitroquinoline 1-oxide *[56-57-5]* M 190.2, m 157°. Recrystd from aqueous acetone [Seki et al. *J Phys Chem* **91** 126 *1987*].

2-Nitroresorcinol *[601-89-8]* M 155.1, m 81-81°, pK_1^{20} 6.37, pK_2^{20} 9.46. Crystd from aq EtOH.

4-Nitrosalicylic acid *[619-19-1]* M 183.1, m 277-288°, pK^{25} 2.23. Crystd from water.

5-Nitrosalicylic acid *[96-97-9]* M 183.1, m 233°, pK_1^{25} 2.32, pK_2^{25} 10.34. Crystd from acetone (charcoal), then twice more from acetone alone.

Nitrosobenzene *[586-96-9]* **M 107.1, m 67.5-68°, b 57-59°/18mm.** Steam distd, then cryst from a small volume of EtOH with cooling below 0°, dried over CaCl$_2$ in a dessicator at atm pressure, and stored under N$_2$ at 0°. Alternatively it can be distd onto a cold finger cooled with brine at ~-10° in a vac at 17mm (water pump), while heating in a water bath at 65-70° [Robertson and Vaughan *J Chem Educ* **27** 605 *1950*].

***N*-Nitrosodiethanolamine** *[1116-54-7]* **M 134.4, b 125°/0.01mm, n 1.485.** Purified by dissolving the amine (0.5g) in 1-propanol (10mL) and 5g of anhydrous Na$_2$SO$_4$ added with stirring. After standing for 1-2h, it was filtered and passed through a chromatographic column packed with AG 50W x 8 (H$^+$form, a strongly acidic cation exchanger). The eluent and washings were combined and evapd to dryness at 35°. [Fukuda et al. *Anal Chem* **53** 2000 *1981*.] *Possible* **CARCINOGEN**.

4-Nitroso-*N,N*-dimethylaniline *[138-89-6]* **M 150.2, m 86-87°, 92.5-93.5°, b 191-192°/100mm, pK^{25} 4.54.** Recryst from pet ether or CHCl$_3$-CCl$_4$ and dried in air. Alternatively suspend in H$_2$O, heat to boiling and add HCl until it dissolves. Filter, cool and collect the *hydrochloride [42344-05-8]*, m 177° after recrystn from H$_2$O containing a small amount of HCl. The *hydrochloride* (e.g. 30g) is made into a paste with H$_2$O (100mL) in a separating funnel. Add cold aq 2.5 NaOH or Na$_2$CO$_3$ to a pH of ~ 8.0 (green color due to free base) and extracted with toluene, CHCl$_3$ or Et$_2$O. Dry extract (K$_2$CO$_3$), filter, distil off the solvent, cool residue and collect the crystalline free base. Recryst as above and dried in air.

***N*-Nitrosodiphenylamine** *[156-10-5]* **M 198.2, m 144-145°(dec).** Crystd from *benzene.

1-Nitroso-2-naphthol *[131-91-9]* **M 173.2, m 110.4-110.8°, pK^{25} 7.63.** Crystd from pet ether (b 60-80°, 7.5mL/g).

2-Nitroso-1-naphthol *[132-53-6]* **M 173.2, m 158°(dec), pK^{25} 7.24.** Purified by recrystn from pet ether (b 60-80°) or by dissolving in hot EtOH, followed by successive addition of small volumes of water.

4-Nitroso-1-naphthol *[605-60-7]* **M 173.2, m 198°, pK^{25} 8.18.** Crystd from *benzene.

2-Nitroso-1-naphthol-4-sulfonic acid (3H$_2$O) *[3682-32-4]* **M 316.3, m 142-146°(dec), pK$_{Est}$ ~6.3 (OH).** Crystd from dilute HCl soln. Crystals were dried over CaCl$_2$ in a vacuum desiccator. Also purified by dissolution in aqueous alkali and pptn by addition of water. Reagent for cobalt.

4-Nitrosophenol *[104-91-6]* **M 123.1, m >124°(dec), pK^{25} 6.36.** Crystd from xylene.

***N*-Nitroso-*N*-phenylbenzylamine** *[612-98-6]* **M 212.2, m 58°.** Crystd from absolute EtOH and dried in air.

***trans*-β-Nitrostyrene** *[5153-67-3]* **M 149.2, m 60°.** Crystd from absolute EtOH, or three times from *benzene/pet ether (b 60-80°) (1:1).

4-Nitrostyrene *[100-13-0]* **M 149.2, m 20.5-21°.** Crystd from CHCl$_3$/hexane. Purified by addition of MeOH to ppte the polymer, then crystd at -40° from MeOH. Also crystd from EtOH. [Bernasconi et al. *J Am Chem Soc* **108** 4541 *1986*.]

2-Nitro-4-sulfobenzoic acid *[552-23-8]* **M 247.1, m 111°, pK$_{Est}$ ~1.65.** Crystd from dilute HCl. Hygroscopic.

2-Nitrotoluene *[88-72-2]* **M 137.1, m -9.55° (α-form), -3.85° (β-form), b 118°/16mm, d 1.163, 222.3°/760mm, n 1.545.** Crystd (repeatedly) from absolute EtOH by cooling in a Dry-ice/alcohol mixture, Further purified by passage of an alcoholic soln through a column of alumina.

3-Nitrotoluene *[99-08-1]* **M 137.1, m 16°, b 113-114°/15mm, 232.6°, d 1.156, n 1.544.** Dried with P$_2$O$_5$ for 24h, then fractionally distd under reduced pressure. [*Org. Synth* Coll Vol I 416 *1948*.]

4-Nitrotoluene *[99-99-0]* **M 137.1, m 52°.** Crystd from EtOH, MeOH/water, EtOH/water (1:1) or MeOH. Air dried, then dried in a vac desiccator over H_2SO_4. [Wright and Grilliom *J Am Chem Soc* **108** 2340 *1986*.]

5-Nitrouracil **(2,4-dihydroxy-5-nitropyrimidine)** *[611-08-5]* **M 157.1, m 280-285°, >300°, pK_1^{20} 0.03, pK_2^{20} 5.55, pK_3^{20} 11.3.** Recrystallises as prisms from boiling H_2O as the monohydrate and loses H_2O on drying *in vacuo*. [UV: Brown *J Chem Soc* 3647 *1959*; Brown *J Appl Chem* **2** 239 *1952*; Johnson *J Am Chem Soc* **63** 263 *1941*.]

Nitrourea *[556-89-8]* **M 105.1, m 158.4-158.8°(dec).** Crystd from EtOH/pet ether.

5-Nitrovanillin **(nitroveratric aldehyde)** *[6635-20-7]* **M 197.2, m 172-175°, 176°, 178°.** Forms yellow plates from AcOH, and needles from EtOH [Slotta and Szyszke *Chem Ber* **68** 184 *1935*]. With diazomethane, 5-nitro-3,4-dimethoxyacetophenone is formed [Brady and Manjunath *J Chem Soc* **125** *1067* *1924*]. The *methyl ether* crystallises from EtOAc or AcOH, **m** 88°, 90-91°, and the *phenylhydrazone* has **m** 108-110° (from aqueous EtOH). [Finger and Schott *J Prakt Chem* [2] **115** 288 *1927*.] For *oxime* **m** 216° (from EtOH or AcOH) and the *oxime acetate* has **m** 147° (from aq EtOH) [Vogel *Monatsh Chem* **20** 384 *1899*; Brady and Dunn *J Chem Soc* **107** 1861 *1915*].

n-**Nonane** *[111-84-2]* **M 126.3, b 150.8°, d 0.719, n 1.40542, n^{25} 1.40311.** Fractionally distd, then stirred with successive volumes of conc H_2SO_4 for 12h each until no further colouration was observed in the acid layer. Then washed with water, dried with $MgSO_4$ and fractionally distd. Alternatively, it was purified by azeotropic distn with 2-ethoxyethanol, followed by washing out the alcohol with water, drying and distilling. [Forziati et al. *J Res Nat Bur Stand* **36** 129 *1946*].

2,5-Norbornadiene *[121-46-0]* **M 92.1, b 89°, d 0.854, n 1.4707.** Purified by distn from activated alumina [Landis and Halpern *J Am Chem Soc* **109** 1746 *1987*].

cis-endo-**5-Norbornene-2,3-dicarboxylic** **anhydride (carbic anhydride, 3aα,4,7,7,$\alpha\alpha$-tetrahydro-4α,7α-methanoisobenzofuran-1,3-dione)** *[129-64-6]* **M 164.2, m 164.1°, 164-165°, 164-167°, d 1.417.** Forms crystals from pet ether, hexane or cyclohexane. It is hydrolysed by H_2O to form the acid [Diels and Alder *Justus Liebigs Ann Chem* **460** 98 *1928*; Maitte *Bull Soc Chim Fr* 499 *1959*]. The *exo-exo-isomer* has **m** 142-143° (from *C_6H_6-pet ether) [Alder and Stein *Justus Liebigs Ann Chem* **504** 216 *1933*].

Norbornylene *[498-66-8]* **M 94.2, m 44-46°, b 96°.** Refluxed over Na, and distd [Gilliom and Grubbs *J Am Chem Soc* **108** 733 *1986*]. Also purified by sublimation *in vacuo* onto an ice-cold finger [Woon et al. *J Am Chem Soc* **108** 7990 *1986*].

Norcamphor **(bicyclo[2.2.1]heptan-2-one)** *[497-38-1]* **M 110.2, m 94-95°.** Crystd from water.

Norcholanic acid *[511-18-2]* **M 346.5, m 177°, 186°, $[\alpha]_D^{20}$ +32° (EtOH), pK_{Est} ~4.8.** Crystd from acetic acid.

Norcodeine *[467-15-2]* **M 285.3, m 185°, 186°, pK 9.10.** Crystd from acetone or ethyl acetate.

Nordihydroguaiaretic **[1,4-bis(3,4-dihydroxyphenyl)-2,3-dimethylbutane] acid** *[500-38-9]* **M 302.4, m 184-185°, $pK_{Est(1)}$~9.7, $pK_{Est(2)}$~12.** Crystd from dilute acetic acid.

Norleucine **(α-amino-*n*-caproic acid)** *[R(+)- 327-56-0; S(-)- 327-57-1]* **M 117.2, m 301° $[\alpha]_{546}^{20}$ (+) and (-) 28° (c 5, 5M HCl); *[RS: 616-06-8]* m 297-300° (sublimes partially at ~280°), pK_1 2.39, pK_2 9.76 (for *RS*).** Crystd from water.

Norvaline (**R-α-amino-n-valeric acid**) *[R(+)- 2031-12-9; S(-)- 6600-40-4]* **M 117.2, m 305°(dec), $[\alpha]_{546}^{20}$ (+) and (-) 25°** (c 10, 5M HCl), **pK_1^{25} 2.36, pK_2^{25} 9.87** (9.72). Crystd from aqueous EtOH or water.

Nylon powder. Pellets were dissolved in ethylene glycol under reflux. Then ppted as a white powder on addition of EtOH at room temperature. This was washed with EtOH and dried at 100° under vacuum.

n-Octacosane *[630-02-4]* **M 394.8, m 62.5°**. Purified by forming its adduct with urea, washing and crystallising from acetone/water. [McCubbin *Trans Faraday Soc* **58** 2307 *1962*.] Crystd from hot, filtered isopropyl ether soln (10mL/g).

n-Octacosanol (octacosyl alcohol) *[557-61-9]* **M 410.8, m 83.4°, 84°**. Recryst from large vols of Me₂CO. It sublimes at 200-250°/1mm instead of distilling.

n-Octadecane *[593-45-3]* **M 254.5, m 28.1°, b 173.5°/10mm, 316.1°/760mm, d_4^{20} 0.7768, n 1.4390**. Crystd from acetone and distd under reduced pressure from sodium.

Octadecyl acetate *[822-23-1]* **M 312.5, m 32.6°**. Distd under vac, then crystd from diethyl ether/MeOH.

n-Octadecyl alcohol (stearyl alcohol) *[112-92-5]* **M 270.5, m 61°, b 153-154°/0.3mm**. Crystd from MeOH, or dry diethyl ether and *benzene, then fractionally distd under reduced pressure. Purified by column chromatography. Freed from cetyl alcohol by zone melting.

Octadecyl ether (dioctadecyl ether) *[6297-03-6]* **M 523.0, m 59.4°**. Vacuum distd, then crystd from MeOH/*benzene.

Octadecyltrimethylammonium bromide *[1120-02-1]* **M 392.5, m 250°(dec)**. See entry on p. 446 in Chapter 6.

2,3,7,8,12,13,17,18-Octaethyl-21H,23H-porphine *[2683-82-1]* **M 534.8, m 322°, 326°**. Chromatographed on SiO₂ using CHCl₃ as eluent. It crystallises from CHCl₃ (dark red), MeOH (blue violet), pyridine (**m** 318°) and *C₆H₆ (deep red). [Fischer and Bämler *Justus Liebigs Ann Chem* **468** 58, 85 *1929*.]

Octafluoropropane (profluorane) *[76-19-7]* **M 188.0, m -183°, b -38°**. Purified for pyrolysis studies by passage through a copper vessel containing CoF₃ at about 270°, then fractionally distd. [Steunenberg and Cady *J Am Chem Soc* **74** 4165 *1952*.] Also purified by several trap-to-trap distns at low temperatures [Simons and Block *J Am Chem Soc* **59** 1407 *1937*].

1,2,3,4,6,7,8,9-Octahydroanthracene *[1079-71-6]* **M 186.3, m 78°**. Crystd from EtOH, then purified by zone melting.

Octamethylcyclotetrasiloxane *[556-67-2]* **M 296.6, m 17.3°, b 175-176°, d_4^{20} 0.957, n 1.396**. Purified by zone melting.

Octan-1,8-diol (octamethylene glycol) *[629-41-4]* **M 146.2, m 59-61°, b 172°/20mm**. Recrystd from EtOH and distd in a vac.

n-Octane *[111-65-9]* **M 114.2, b 126.5°, d_4^{20} 0.704, n 1.39743, n^{25} 1.39505**. Extracted repeatedly with conc H₂SO₄ or chlorosulfonic acid, then washed with water, dried and distd. Also purified by azeotropic distn with EtOH, followed by washing with water to remove the EtOH, drying and distilling. For further details, see n-*heptane*. Also purified by zone melting.

1-Octanethiol *[111-88-6]* **M 146.3, b 86°/15mm, 197-200°/760mm, d_4^{20} 0.8433, n 1.4540, pK^{25} 10.72 (dil *t*-BuOH).** Passed through a column of alumina [Battacharyya et al. *J Chem Soc, Faraday Trans 1* **82** 135 *1986;* Fletcher *J Am Chem Soc* **68** 2727 *1946*] .

1-Octene *[111-66-0]* **M 112.2, b 121°/742mm, d_4^{20} 0.716, n 1.4087.** Distd under nitrogen from sodium which removes water and peroxides. Peroxides can also be removed by percolation through dried, acid washed alumina. Stored under nitrogen in the dark. [Strukul and Michelin *J Am Chem Soc* **107** 7563 *1985.*]

(*trans*)-2-Octene *[13389-42-9]* **M 112.2, b 124-124.5°/760mm, d_4^{20} 0.722, n 1.4132.** Purification as for 1-octene above.

***n*-Octyl alcohol** *[111-87-5]* **M 130.2, b 98°/19mm, 195.3°/760mm, d 0.828, n 1.43018.** Fractionally distd under reduced pressure. Dried with sodium and again fractionally distd or refluxed with boric anhydride and distd (**b** 195-205°/5mm), the distillate being neutralised with NaOH and again fractionally distd. Also purified by distn from Raney nickel and by preparative GLC.

***n*-Octylammonium 9-anthranilate** *[88020-99-9]* **M 351.5, m 134-135°, pK^{25} 10.65 (for octylamine).** Recrystd several times from ethyl acetate.

***n*-Octylammonium hexadecanoate** *[88020-97-7]* **M 385.7, m 52-53°.** Purified by several recrystns from *n*-hexane or ethyl acetate. The solid was then washed with cold anhydrous diethyl ether, and dried *in vacuo* over P_2O_5.

***n*-Octylammonium octadecanoate** *[32580-92-0]* **M 413.7, m 56-57°.** Purified as for the *hexadecanoate* above.

***n*-Octylammonium tetradecanoate** *[17463-35-3]* **M 358.6, m 46-48°.** Purified as for the *hexadecanoate* above.

4-Octylbenzoic acid *[3575-31-3]* **M 234.3, m 99-100°, pK^{25} 6.5 (80% aq EtOH), pK_{Est} ~4.5 (H_2O).** Crystd from EtOH has **m** 139°; crystd from aq EtOH has **m** 99-100°. Forms liquid crystals.

***n*-Octyl bromide** *[111-83-1]* **M 193.1, b 201.5°, d_4^{20} 1.118, n^{25} 1.4503.** Shaken with H_2SO_4, washed with water, dried with K_2CO_3 and fractionally distd.

4-(*tert*-Octyl)phenol *[140-66-9]* **M 206.3, m 85-86°, b 166°/20mm, pK_{Est} ~ 10.4.** Crystd from *n*-hexane.

1-Octyne *[629-05-0]* **M 110.2, b 126.2°/760mm, d_4^{20} 0.717, n^{25} 1.4159.** Distd from $NaBH_4$ to remove peroxides.

α-Oestradiol *[57-91-0]* **M 272.4, m 220-230°, $[\alpha]_D^{20}$ +55° (c 1, dioxane).** Crystd from aq EtOH.

β-Oestradiol-3-benzoate *[50-50-0]* **M 376.5, m 194-195°, $[\alpha]_{546}^{20}$ +70° (c 2, dioxane).** Crystd from EtOH

Oleic acid *[112-80-1]* **M 282.5, m 16°, b 360°(dec), d_4^{20} 0.891, n^{30} 1.4571, pK^{25} 6.42 (50% aq EtOH), pK_{Est} ~4.8 (H_2O).** Purified by fractional crystn from its melt, followed by molecular distn at 10^{-3}mm, or by conversion to its methyl ester, the free acid can be crystd from acetone at -40° to -45° (12mL/g). For purification by the use of lead and lithium salts, see Keffler and McLean [*J Soc Chem Ind (London)* **54** 176T *1935*]. Purification based on direct crystn from acetone is described by Brown and Shinowara [*J Am Chem Soc* **59** 6 *1937;* pK White *J Am Chem Soc* **72** 1857 *1950*].

Oleyl alcohol *[143-28-2]*   **M 268.5, b 182-184°/1.5mm, d_4^{20} 0.847, $n^{27.5}$ 1.4582.**   Purified by fractional crystn at -40° from acetone, then distd under vacuum.

Opianic acid (2-formyl-4,5-dimethylbenzoic acid) *[519-05-1]*   **M 210.2, m 150°, pK^{25} 3.07.** Crystd from water.

Orcinol (5-methylresorcinol) *[504-15-4]*   **M 124.2, m 107.5°, m 59-61° (hydrate), pK_1^{20} 9.36 (9.48), pK_2^{20} 11.6 (11.20).**   Crystd from $CHCl_3$/*benzene (2:3).

L-Ornithine *[70-26-8]*   **M 132.2, m 140°, $[\alpha]_D^{25}$ +16° (c 0.5, H_2O), pK_2^{20} 8.75, pK_3^{25} 10.73.** Crystd from water containing 1mM EDTA (to remove metal ions).

L-Ornithine monohydrochloride *[3184-13-2]*   **M 168.6, m 233°(dec), $[\alpha]_D^{25}$ +28.3° (5M HCl).** Likely impurities are citrulline, arginine and D-ornithine. Crystd from water by adding 4 volumes of EtOH and dried in a vacuum desiccator over fused $CaCl_2$.

Orotic acid (H_2O) *[50887-69-9]*   **M 174.1, m 235-346°(dec), pK_1^{25} 1.8, pK_2^{25} 9.55.**   Crystd from water.

Orthanilic acid (2-aminobenzenesulfonic acid) *[88-21-1]*   **M 173.2, m >300°(dec), pK^{25} 2.49.** Crystd from aqueous soln, containing 20mL of conc HCl per L, then crystd from distilled water, and dried in a vacuum desiccator over Sicapent.

Ouabain {3-[(6-deoxy-α-L-mannopyranosyl)oxy]-1,5,11a,14,19-pentahydroxycard-20(22)-enolide} *[630-60-4]*   **M 728.8, m 180°(dec), $[\alpha]_{546}^{20}$ -30° (c 1, H_2O).**   Crystd from water as the octahydrate. Dried at 130°. Stored in the dark.

Oxalic acid ($2H_2O$) *[6153-56-6]*   **M 90.0, m 101.5°;** *[anhydrous 144-62 -7]* **m 189.5°, pK_1^{25} 1.08 (1.37), pK_2^{25} 3.55 (3.80).**   Crystd from distilled water. Dried in vacuum over H_2SO_4. The anhydrous acid can be obtained by drying at 100° overnight.

Oxaloacetic acid *[328-42-7]*   **M 132.1, m 160°(decarboxylates), pK_1^{25} 2.22, pK_2^{25} 3.89, pK_3^{25} 13.0.**   Crystd from boiling ethyl acetate, or from hot acetone by addition of hot *benzene.

2-Oxoglutaric acid (2-oxopentane-1,5-dioic, α-ketoglutaric acid) *[328-50-7]*   **M 146.1, m 114°, 115-117°, (pK_{Est} see oxaloacetic acid above).**   Crystd repeatedly from Me_2CO/*benzene, EtOAc or ethyl propionate.

Oxalylindigo *[2533-00-8]*   **M 316.3.**   Recrystd twice from nitrobenzene and dried by heating *in vacuo* for several hours. [Sehanze et al. *J Am Chem Soc* **108** 2646 *1986*.]

Oxamide *[471-46-5]*   **M 88.1, m >320°(dec).**   Crystd from water, ground and dried in an oven at 150°.

2-Oxazolidinone *[497-25-6]*   **M 87.1, m 89-90°, 91°, b 152°/0.4mm.**   Crystd from *benzene or dichloroethane.

Oxetane (1.3-trimethylene oxide) *[503-30-0]*   **M 58.1, b 45-46°/736mm, 47-49°/atm, 48°/760mm, d_4^{20} 0.892, n_D^{20} 1.395.**   Distd twice from sodium metal and then fractionated through a small column at atmospheric pressure, **b 47.0-47.2°.** Also purified by preparative gas chromatography using a 2m silica gel column. Alternatively add KOH pellets (50g for 100g of oxetane) and distil through an efficient column or a column packed with 1/4in Berl Saddles and the main portion boiling at 45-50° is collected and redistilled over fused KOH. [Noller *Org Synth* Coll Vol III 835 *1955*; Dittmer et al. *J Am Chem Soc* **79** 4431 *1957*.]

Oxine Blue [3-(4-hydroxyphenyl)-3-(8-hydroxy-6-quinilinyl)-1(3*H*)-isobenzofuranone] *[3733-85-5]* M 369.4, m 134-135°. Recrystd from EtOH. Dried over H_2SO_4.

Palmitic acid anhydride (hexadecanoic anhydride) *[623-65-4]* M 494.9, m 63-64°, 64°, d^{82} 0.838, n^{68} 1.436. It is moisture sensitive and hydrolyses in water. Purified by refluxing with acetic anhydride for 1hr, evaporating and freeing the residue of acetic acid and anhydride by drying the residue at high vac and crystallising from pet ether at low temperature.

[2.2]-Paracyclophane (tricyclo[8.2.2.2^{4,7}]hexadeca-4,6,10,12,13,15-hexaene) *[1633-22-3]* M 208.3, m 284°, 285-287°, 286-288°, 288-290°. Purified by recrystn from AcOH. ^{1}H-NMR δ: 1.62 (Ar-H) and -1.71 (CH$_2$) [Waugh and Fessenden *J Am Chem Soc* 79 846 *1957*; IR and UV: Cram et al. *J Am Chem Soc* 76 6132 *1954*, Cram and Steinberg *J Am Chem Soc* 73 5691 *1951*; complex with unsaturated compounds: Cram and Bauer *J Am Chem Soc* 81 5971 *1959*; Syntheses: Brink *Synthesis* 807 *1975*, Givens et al. *J Org Chem* 44 16087 *1979*, Kaplan et al. *Tetrahedron Lett* 3665 *1976*].

Paraffin (oil) *[8012-95-1]* d 0.880, n 1.482. Treated with fuming H_2SO_4, then washed with water and dilute aqueous NaOH, then percolated through activated silica gel.

Paraffin Wax. Melted in the presence of NaOH, washed with water until all of the base had been removed. The paraffin was allowed to solidify after each wash. Finally, 5g of paraffin was melted by heating on a water-bath, then shaken for 20-30min with 100mL of boiling water and fractionally crystd.

Parafuchsin (4,4',4''-triaminotryllium [triphenylmethane] carbonium ion, para-rosaniline, paramagenta) *[467-62-9]* M 305.4, pK 7.57 and free base has pK >13. Dissolve in EtOH (1.16g in 30mL), filter and add aqueous NH$_3$ till neutral and ppte by adding H$_2$O giving 0.8g m 247° dec (sintering at 230°). Dissolve in EtOH neutralise with NH$_3$ add 0.1g of charcoal filter, and repeat, then add H$_2$O (100mL) to ppte the colourless *carbinol* dry, m 257° dec (sintering at 232°). [Weissberger and Theile *J Chem Soc* 148 *1934*.] The *carbinol* (pseudo-base) was said to have m 232° (186° dec), and is slightly sol in H$_2$O but sol in acids and EtOH [pK: Goldacre and Phillips *J Chem Soc* 172 *1949*]. The *perchlorate* (dark red with a green shine) has m 300° and explodes at 317° [Dilthey and Diaklage *J Prakt Chem* [2] 129 *1931*].

Paraldehyde (acetaldehyde trimer, 2,4,6-trimethyl-1,3,5-trioxane) *[123-63-7]* M 132.2, m 12.5°, 124°, d 0.995, n 1.407. Washed with water and fractionally distd.

Patulin {4-hydroxy-4*H*-furo[3.2-*c*]pyran-2(6*H*)-one} *[149-29-1]* M 154.1, m 110°. Crystd from diethyl ether or chloroform. (*Highly* TOXIC).

Pavatrine hydrochloride *[548-65-2]* M 333.7, m 143-144°. Recrystd from isopropanol, and dried over P_2O_5 under vacuum.

Pelargonic acid (nonanoic acid) *[112-05-0]* M 158, m 15°, b 98.9°/1mm, 225°/760mm, pK^{25} 4.96. Esterified with ethylene glycol and distd. (This removes dibasic acids as undistillable residues.) The acid was regenerated by hydrolysing the ester.

Pelargononitrile (octyl cyanide) *[2243-27-8]* M 139.2, m -34°, b 92°/10mm, 224°, d 0.818, n 1.4255. Stirred with P_2O_5 (~5%), distd from it and redistd under vac. IR should have CN but no OH bands.

Pelargonyl chloride (nonanoyl chloride) *[764-85-2]* M 176.7, b 88°/12mm, d 0.941, n 1.436. Refluxed with acetyl chloride (~ 3 vols) for 1h, then distil off the AcCl followed by the nanoyl chloride under reduced pressure. It is moisture sensitive and should be stored in sealed ampules.

Penicillic acid *[90-65-3]* M 158.2, m 58-64° (monohydrate), 83-84° (anhydrous, lactone). Crystd from water as the monohydrate, or from pet ether. Free acid is in equilibrium with the lactone.

Pentabromoacetone   *[79-49-2]*   **M 452.6, m 76°, pK 8.0 (MeOH), pK$_{Est}$ ~4.6 (H$_2$O).** Crystd from diethyl ether or EtOH and sublimes .

Pentabromophenol   *[608-71-9]*   **M 488.7, m 229°, pK$_{Est}$ ~ 4.5.** Purified by crystn (charcoal) from toluene then from CCl$_4$. Dried for 2 weeks at *ca* 75°.

1-Pentacene   *[135-48-8]*   **M 278.4, m 300°** . Crystd from *benzene.

Pentachloroethane (pentalin)   *[76-01-7]*   **M 202.3, b 69°/37mm, 152.2°/64mm, 162.0°, d 1.678, n^{15} 1.50542.** Usual impurities include trichloroethylene. Partially decomposes if distd at atmospheric pressure. Drying with CaO, KOH or sodium is unsatisfactory because of the elimination of the elements of HCl. It can be purified by steam distn, or by washing with conc H$_2$SO$_4$, water, and then aqueous K$_2$CO$_3$, drying with solid K$_2$CO$_3$ or CaSO$_4$, and fractionally distd under reduced pressure.

Pentachloronitrobenzene   *[82-68-8]*   **M 295.3, m 146°.** Crystd from EtOH.

Pentachlorophenol   *[87-86-5]*   **M 266.3, m 190-191°, pK^{25} 4.8.** Twice crystd from toluene/EtOH. Sublimed *in vacuo*.

Pentachloropyridine   *[2176-62-7]*   **M 251.3, m 122-124°, 123°, 124°, 124-125°, 125-126°, b 279-280°/atm, pK -6.02 (aq H$_2$SO$_4$).** Purified by recryst from EtOH or aqueous EtOH. It sublimes at 150°/3mm. [den Hertog et al. *Recl Trav Chim Pays-Bas* **69** 673 *1950*; Schikh et al. *Chem Ber* **69** 2604 *1936*.]

Pentachlorothiophenol   *[133-49-3]*   **M 282.4, m between 228° and 235°, pK$_{Est}$ ~1.1.** Crystd from *benzene.

Pentadecafluoro octanoic acid (perfluorocaprylic acid)   *[335-67-1]*   **M 414.1, m 54.9-55.6°, b 189°/736mm, pK$_{Est}$ <0.** Recrystd from CCl$_4$ and toluene, and can be distd. It forms micelles in H$_2$O and the solubility is 1% in H$_2$O. [Bernett and Zisman *J Phys Chem* **63** 1911 **1959**; IR: Bro and Sperati *J Polym Sci* **38** 289 *1959*.]

Pentadecanoic acid   *[1002-84-2]*   **M 242.4, m 51-53°, 80°, b 158°/1mm, 257°/760mm, d^{80} 0.8424, pK$_{Est}$ ~5.0.** Cryst from Et$_2$O and distd. Very hygroscopic. See purification of palmitic acid.

Pentadecanolide   (1-oxacyclohexadecan-2-one, pentadecanoic-ω-lactone, 15-hydroxypenta-decanoic lactone, exaltolide, Tibetolide)   *[106-02-5]*   **M 240.4, m 34-36°, 37-37.5°, 37-38°, b 102-103°/0.03mm, 112-114°/0.2mm, 137°/2mm, 169°/10-11mm, d$_4^{40}$ 0.9401.** It has been recrystd from MeOH (4 parts) at -15°. [Hundiecker and Erlbach *Chem Ber* **80** 135 *1947*; Galli and Mandolini *Org Synth* **58** 100 *1978*; Demole and Enggist *Helv Chim Acta* **11** 2318 *1978*.]

Penta-1,3-diene   *[cis: 1574-41-0; trans: 2004-70-8]*   **M 68.1, b 42°, d 0.680, n 1.4316.** Distd from NaBH$_4$. Purified by preparative gas chromatography. [Reimann et al. *J Am Chem Soc* **108** 5527 *1986*.]

Penta-1,4-diene   *[591-93-5]*   **M 68.1, b 25.8-26.2°/756mm, d 0.645, n 1.3890.** Distd from NaBH$_4$. Purified by preparative gas chromatography. [Reimann et al. *J Am Chem Soc* **108** 5527 *1986*.]

Pentaerythritol   *[115-77-5]*   **M 136.2, m 260.5°.** Refluxed with an equal volume of MeOH, then cooled and the ppte dried at 90°. Crystd from dil aq HCl. Sublimed under vacuum at 200°.

Pentaerythritol tetraacetate   *[597-71-7]*   **M 304.3, m 78-79°.** Crystd from hot water, then leached with cold water until the odour of acetic acid was no longer detectable.

Pentaerythrityl laurate   *[13057-50-6]*   **M 864.6, m 50°.** Crystd from pet ether.

Pentaerythritol tetranitrate. *[78-11-5]* **M 316.2, m 140.1°.** Crystd from acetone or acetone/EtOH. **EXPLOSIVE.**

Pentaethylenehexamine *[4067-16-7]* **M 232.4, d 0.950, n_D^{20} 1.510, pK$_1$ 1.2, pK$_2$ 2.7, pK$_3$ 4.3, pK$_4$ 7.8, pK$_5$ 9.1, pK$_6$ 9.9 (all estimated).** Fractionally distd twice at 10-20mm, the fraction boiling at 220-250° being collected. Its soln in MeOH (40mL in 250mL) was cooled in an ice-bath and conc HCl was added dropwise with stirring. About 50mL was added, and the ppted hydrochloride was filtered off, washed with acetone and diethyl ether, then dried in a vacuum desiccator. [Jonassen et al. *J Am Chem Soc* **79** 4279 *1957*.]

Pentafluorobenzene *[363-72-4]* **M 168.1, b 85°/atm, 85-86°/atm, 88-89°/atm, d_4^{20} 1.524, n_D^{20} 1.3931.** Purified by distn and by gas chromatography. IR film: 1535 and 1512 cm⁻¹ (*benzene ring). [UV: Stephen and Tatlow *Chem Ind (London)* 821 *1957*; Nield et al. *J Chem Soc* 166 *1959*.] See triethylenepentamine

2,3,4,5,6-Pentafluorobenzoic acid *[602-94-8]* **M 212.1, m 101-103°, 103-104°, 104-105°, 106-107°, pK^{25}1.75.** Dissolve in Et$_2$O, treat with charcoal, filter, dry (CaSO$_4$), filter, evaporate and recrystallise residue from pet ether (b 90-100°) after adding a little toluene to give large colourless plates. UV (H$_2$O): λmax 265nm (ε 761). The *S-benzylisothiuronium salt* has **m** 187° after recrystn from H$_2$O. [McBee and Rapkin *J Am Chem Soc* **73** 1366 *1951*; Nield et al. *J Chem Soc* 166, 170 *1959*.]

O-(2,3,4,5,6-Pentafluorobenzyl)hydroxylamine hydrochloride (PFBHA) *[57981-02-9]* **M 249.6, m 215°, 215-216°, pK$_{Est}$ ~1.1.** Recrystd from EtOH to form colourless leaflets. Drying the compound at high vacuum and elevated temperature will result in losses by sublimation. [Youngdale *J Pharm Sci* **65** 625 *1976*; Wehner and Handke *J Chromatog* **177** 237 *1979*; Nambara et al. give incorrect **m** as 115-116° *J Chromatogr* **114** 81 *1975*.]

2,3,4,5,6-Pentafluorophenol *[771-61-9]* **M 184.1, m 33-35°, 38.5-39.5°, b 72-74°/48mm, 142-144°/atm, 143°/atm, n_D^{20}1.4270 (liquid prep), pK^{25} 5.53.** A *hygroscopic* low melting solid not freely soluble in H$_2$O. Purified by distn, preferably in a vacuum [Forbes et al. *J Chem Soc* 2019 *1959*; IR and pKa: Birchall and Haszeldine *J Chem Soc* 13 *1959*]. IR film: 3600 (OH) and 1575 (fluoroaromatic breathing) cm⁻¹. The *benzoyl* derivative has **m** 74-75°, *3,4-dinitrobenzoyl* derivative has **m** 107°, the *tosylate* has **m** 64-65° (from EtOH) and the *K salt* crystallises from Me$_2$CO, **m** 242° dec, with 1H$_2$O *salt* the **m** is 248° dec and the 2H$_2$O *salt* has **m** 245° dec.

1-(Pentafluorophenyl)ethanol *[R-(+)- 104371-21-3; S-(-)- 104371-20-2]* **M 212.1, m 41-42°,42°, 42.5-43°, $[\alpha]_{546}^{20}$ (+) and (-) 9°, $[\alpha]_D^{20}$ (+) and (-) 7.5° (c 1, *n*-pentane).** Recrystd from *n*-pentane at -40° and vacuum sublimed at room temp at 0.3mm (use ice cooled cold finger). It has also been purified by column chromatography through Kieselgel 60 (0.063-0.2mm mesh, Merck), eluted with EtOAc-*n*-hexane (1:5), then recrystd from *n*-pentane and vacuum sublimed. It has R$_F$ on Kieselgel 60 F$_{254}$ TLC foil and eluting with EtOAc-*n*-hexane (1:5). [Meese *Justus Liebigs Ann Chem* 2004 *1986*.] The *racemate [75853-08-6]* has **m** 32-34°, **b** 77-79°/8mm, 80-82°/37mm, n_D^{20} 1.4426 and the *3,4-dinitrobenzoate* has **m** 83° [Nield et al. *J Chem Soc* 166 *1959*].

2,2,3,3,3-Pentafluoropropan-1-ol *[422-05-9]* **M 150.1, b 80°, d 1.507, n 1.288, pK^{25} 12.74.** Shaken with alumina for 24h, dried with anhydrous K$_2$CO$_3$, and distd, collecting the middle fraction (b 80-81°) and redistilling.

Pentafluoropyridine *[700-16-3]* **M 169.1, m -41.5°, b 83.5°, 84°, 83-85°, d_4^{20} 1.609, n_D^{20} 1.3818, pK$_{Est}$ ~<0.** Distd through a concentric tube column; has λ max in cyclohexane at 256.8nm. [Chambers et al. *J Chem Soc* 3573 *1964*]; ¹⁹F NMR: Bell et al. *J Fluorine Chem* **1** 51 *1971*.] The *hexafluoroantimonate* has **m** 98-102° dec.

Pentamethylbenzene *[700-12-9]* **M 148.3, m 53.5-55.1°.** Successively crystd from absolute EtOH, toluene and MeOH, and dried under vacuum. [Rader and Smith *J Am Chem Soc* **84** 1443 *1962*.] It has also been crystd from *benzene or aqueous EtOH, and sublimed.

n-**Pentane** *[109-66-0]* **M 72.2, b 36.1°, d 0.626, n^{25} 1.35472.** Stirred with successive portions of conc H_2SO_4 until there was no further coloration during 12h, then with 0.5N $KMnO_4$ in 3M H_2SO_4 for 12h, washed with water and aqueous $NaHCO_3$. Dried with $MgSO_4$ or Na_2SO_4, then P_2O_5 and fractionally distd through a column packed with glass helices. It was also purified by passage through a column of silica gel, followed by distn and storage with sodium hydride. An alternative purification is by azeotropic distn with MeOH, which is subsequently washed out from the distillate (using water), followed by drying and distn. For removal of carbonyl-containing impurities, see n-*heptane*. Also purified by fractional freezing (*ca* 40%) on a copper coil through which cold air was passed, then washed with conc H_2SO_4 and fractionally distd.

Pentane-1-thiol *[110-66-7]* **M 104.2, m -76°, b 122.9°/697.5mm, d^{25} 0.8375, pK$_{Est}$ ~10.1.** Dissolved in aqueous 20% NaOH, then extracted with a small amount of diethyl ether. The soln was acidified slightly with 15% H_2SO_4, and the thiol was distd out, dried with $CaSO_4$ or $CaCl_2$, and fractionally distd under nitrogen. [Ellis and Reid *J Am Chem Soc* **54** 1674 *1932*.]

Pentan-2-ol *[6032-29-7]* **M 88.2, b 119.9°, d 0.810, n 1.41787, n^{25} 1.4052.** Refluxed with CaO, distd, refluxed with magnesium and again fractionally distd.

Pentan-3-ol *[584-02-1]* **M 88.2, b 116.2°, d 0.819, n^{25} 1.4072.** Refluxed with CaO, distd, refluxed with magnesium and again fractionally distd.

Pentan-3-one (diethyl ketone) *[96-22-0]* **M 86.1, b 102.1°, d 0.8099, n 1.392.** See diethyl ketone on p. 204.

Pentaquine monophosphate *[5428-64-8]* **M 395.6, m 189-190°, pK^{70} 8.22.** Crystd from 95% EtOH.

Pent-2-ene (mixed isomers) *[109-68-2]* **M 70.1, b 36.4°, d 0.650, n 1.38003, n^{25} 1.3839.** Refluxed with sodium wire, then fractionally distd twice through a Fenske (glass helices packing) column.

cis-**Pent-2-ene** *[627-20-3]* **M 70.1, b 37.1°, d 0.657, n^{25} 1.3798.** Dried with sodium wire and fractionally distd, or purified by azeotropic distn with MeOH, followed by washing out the MeOH with water, drying and distilling. Also purified by chromatography through silica gel and alumina [Klassen and Ross *J Phys Chem* **91** 3668 *1987*].

trans-**Pent-2-ene** *[646-04-8]* **M 70.1, b 36.5°, d 0.6482, n 1.3793.** It was treated as above and washed with water, dried over anhydrous Na_2CO_3, and fractionally distd. The middle cut was purified by two passes of fractional melting.

Pentobarbital (5-ethyl-5-1'-methylbutyl barbituric acid, Nembutal) *[76-74-4]* **M 226.4, m ~127°(dec), pK$_{Est(1)}$~ 8.0, pK$_{Est(2)}$~12.7.** Soln of the sodium salt in 10% HCl was prepared and the acid was extracted by addition of ether. Then purified by repeated crystn from $CHCl_3$. [Bucket and Sandorfy *J Phys Chem* **88** 3274 *1984*.]

Pentyl acetate (*n*-amyl acetate) *[628-63-7]* **M 130.2, b 147-149°/atm, 149.55°, 149.2°/atm, d$_4^{20}$ 0.8753, n$_D^{20}$ 1.4028.** Purified by repeated fractional distn through an efficient column or spinning band column. [Timmermann and Hennant-Roland *J Chim Phys* **52** 223 *1955*; Mumford and Phillips *J Chem Soc* 75 *1950*; ^{1}H NMR: Crawford and Foster *Can J Phys* **34** 653 *1956*.]

Pent-2-yne *[627-21-4]* **M 68.1, b 26°/2.4mm, d 0.710, n^{25} 1.4005.** Stood with, then distd at low pressure from, sodium or $NaBH_4$.

Perbenzoic acid *[93-59-4]* M 138.1, m 41-43°, b 97-110°/13-15mm, pK$_{Est}$ ~7.7. Crystd from *benzene or pet ether. Readily sublimed and is steam volatile. Sol in CHCl$_3$, CCl$_4$ and Et$_2$O. [*Org Synth* Coll Vol I (2nd Edn) 431 *1948*.] **EXPLOSIVE.**

Perchlorobutadiene *[87-68-3]* M 260.8, b 144.1°/100mm, 210-212°/760mm, d 1.683, n 1.5556. Washed with four or five 1/10th volumes of MeOH (or until the yellow colour has been extracted), then stirred for 2h with H$_2$SO$_4$, washed with distilled water until neutral and filtered through a column of P$_2$O$_5$. Distd under reduced pressure through a packed column. [Rytner and Bauer *J Am Chem Soc* **82** 298 *1960*.]

Perfluorobutyric acid *[375-22-4]* M 214.0, m -17.5°, b 120°/735mm, d 1.651, n^{16} 1.295, pK^{25} -0.17. Fractionally distd twice in an Oldershaw column with an automatic vapour-dividing head, the first distn in the presence of conc H$_2$SO$_4$ as a drying agent.

Perfluorocyclobutane *[115-25-3]* M 200.0, m -40°, b -5°, d^{-20} 1.654, d° 1.72. Purified by trap-to-trap distn, retaining the middle portion.

Perfluorocyclohexane *[355-68-0]* M 300.1, m 51° (sublimes), b 52°. Extracted repeatedly with MeOH, then passed through a column of silica gel (previously activated by heating at 250°).

Perfluoro-1,3-dimethylcyclohexane *[335-27-3]* M 400.1, b 101°, d 1.829, n 1.300. Fractionally distd, then 35mL was sealed with about 7g KOH pellets in a borosilicate glass ampoule and heated at 135° for 48h. The ampoule was cooled and opened, and the liquid was resealed with fresh KOH in another ampoule and heated as before. This process was continued until no further decomposition was observed. The substance was then washed with distilled water, dried (CaSO$_4$) and distd. [Grafstein *Anal Chem* **26** 523 *1954*.]

Perfluoroheptane *[335-57-9]* M 388.1, b 99-101°, d^{25} 1.7200. Purified as for *perfluorodimethylhexane*. Other procedures include shaking with H$_2$SO$_4$, washing with water, drying with P$_2$O$_5$ for 48h and fractionally distilling. Alternatively, it has been refluxed for 24h with saturated acid KMnO$_4$ (to oxidise and remove hydrocarbons), then neutralised, steam distd, dried with P$_2$O$_5$, and passed slowly through a column of dry silica gel. It has been purified by fractional crystn, using partial freezing.

Perfluoro-*n*-hexane *[355-42-0]* M 338.1, m -4°, b 58-60°, d 1.684. Purified by fractional freezing. The methods described for *perfluoroheptane* should be applicable here.

Perfluoro(methylcyclohexane) *[355-02-2]* M 350.1, b 76.3°, d^{25} 1.7878. Refluxed for 24h with saturated acid KMnO$_4$ (to oxidise and remove hydrocarbons), then neutralised, steam distd, dried with P$_2$O$_5$ and passed slowly through a column of dry silica gel. [Glew and Reeves *J Phys Chem* **60** 615 *1956*.] Also purified by percolation through a 1m neutral activated alumina column, and ^{1}H-impurities checked by NMR.

Perfluorononane *[375-96-2]* M 488.1, b 126-127°, d 1.80, n 1.275. Purified as for *perfluorodimethylcyclohexane*.

Perfluoropropyl iodide *[754-34-7]* M 295.9, b 41°, d 2.13, n 1.339. Purified by fractional distn.

Perfluorotributylamine (**heptacosafluorotributylamine**) *[311-89-7]* M 671.1, b 177.6°/760mm, d 1.881, n 1.291, pK$_{Est}$ ~5.0. Purified as for perfluorodimethylcyclopropane, see also perfluorotripropylamine [Hazeldine *J Chem Soc* 102 *1951*].

Perfluorotripropylamine (**heneicosafluorotripropylamine**) *[338-83-0]* M 521.1, b 130°/atm, 129.5-130.5°/atm, d 1.822, n 1.279, pK$_{Est}$ ~5.6. Purified as for *perfluorodimethylcyclopropane*. [Hazeldine *J Chem Soc* 102 *1951*, for azeotropes see Simons and Linevsky *J Am Chem Soc* **74** 4750 *1972*.] **IRRITANT.**

Pericyazine [**10-{3-(4-hydroxy-1-piperidinyl)-propyl}-10*H*-phenothiazine-2-carbonitrile**] *[2622-26-6]* M 365.4, m 116-117°. Recrystd from a saturated soln in cyclohexane. Antipsychotic and is a reagent for Pd and Rh.

Perylene *[198-55-0]* **M 252.3, m 273-274°.** Purified by silica-gel chromatography of its recrystd picrate. [Ware *J Am Chem Soc* **83** 4374 *1961*.] Crystd from *benzene, toluene or EtOH and sublimed in a flow of oxygen-free nitrogen. [Gorman et al. *J Am Chem Soc* **107** 4404 *1985;* Johansson et al. *J Am Chem Soc* **109** 7374 *1987.*]

Petroleum ether *[8032-32-4]* **b 35-60°, d 0.640, n 1.363.** Shaken several times with conc H_2SO_4, then 10% H_2SO_4 and conc $KMnO_4$ (to remove unsatd, including aromatic, hydrocarbons) until the permanganate colour persists. Washed with water, aqueous Na_2CO_3 and again with water. Dried with $CaCl_2$ or Na_2SO_4, and distd. It can be dried further using CaH_2 or sodium wire. Passage through a column of activated alumina, or treatment with CaH_2 or sodium, removes peroxides. For the elimination of carbonyl-containing impurities without using permanganate, see *n-heptane.* These procedures could be used for all fractions of pet ethers.
Rapid purification: Pass through an alumina column and fractionally distilling, collecting the desired boiling fraction.

R(-)-α-Phellandrene ⸱ **(*p*-menta-1,5-diene)** *[4221-98-1]* **M 136.2, b 61°/11mm, 175-176°/760mm, d 0.838, n 1.471.** Purified by gas chromatography on an Apiezon column.

Phenacylamine hydrochloride *[5468-37-1]* **M 171.6, m 194°(dec).** See 2-aminoacetophenone hydrochloride on p. 103.

Phenanthrene *[85-01-8]* **M 178.2, m 98°.** Likely contaminants include, anthracene, carbazole, fluorene and other polycyclic hydrocarbons. Purified by distn from sodium, boiling with maleic anhydride in xylene, crystn from acetic acid, sublimation and zone melting. Has also been recrystd repeatedly from EtOH, *benzene or pet ether (b 60-70°), with subsequent drying under vacuum over P_2O_5 in an Abderhalden pistol. Feldman, Pantages and Orchin [*J Am Chem Soc* **73** 4341 *1951*] separated from most of the anthracene impurity by refluxing phenanthrene (671g) with maleic anhydride (194g) in xylene (1.25L) under nitrogen for 22h, then filtered. The filtrate was extracted with aqueous 10% NaOH, the organic phase was separated, and the solvent was evaporated. The residue, after stirring for 2h with 7g of sodium, was vacuum distd, then recrystd twice from 30% *benzene in EtOH, then dissolved in hot glacial acetic acid (2.2mL/g), slowly adding an aqueous soln of CrO_3 (60g in 72mL H_2O added to 2.2L of acetic acid), followed by slow addition of conc H_2SO_4 (30mL). The mixture was refluxed for 15min, diluted with an equal volume of water and cooled. The ppte was filtered off, washed with water, dried and distd, then recrystd twice from EtOH. Further purification is possible by chromatography from $CHCl_3$ soln on activated alumina, with *benzene as eluent, and by zone refining.

Phenanthrene-9-aldehyde *[4707-71-5]* **M 206.3, m 102-103°, pK -6.39 (aq H_2SO_4).** Crystd from EtOH and sublimed at 95-98°/0.07mm.

9,10-Phenanthrenequinone *[84-11-7]* **M 208.2, m 208°, pK -7.1 (aq H_2SO_4).** Crystd from dioxane or 95% EtOH and dried under vacuum.

Phenanthridine *[229-87-8]* **M 179.2, m 106.5°, 108-109°, b 350°, pK^{20} 4.61 (4.48).** Purified *via* the $HgCl_2$ addition compound formed when phenanthridine (20g) in 1:1 HCl (100mL) was added to aq $HgCl_2$ (60g in 3L), and the mixture was heated to boiling. Conc HCl was then added until all of the solid had dissolved. The compound separated on cooling, and was dec with aq NaOH (*ca* 5M). Phenanthridine was extracted with Et_2O and crystd from pet ether (b 80-100°) or EtOAc. [Cumper et al. *J Chem Soc* 45218 *1962*.] Also purified by zone melting; sublimes in vac. [Slough and Ubbelhode *J Chem Soc* 911 *1957*.] See p. 124.

1,10-Phenanthroline **(*o*-phenanthroline)** *[66-71-7 (anhydr); 5144-89-8 (H_2O)]* **M 198.2, m 98-101°, 108-110° (hydr), 118° (anhydr), b >300°, pK_1^{25}-0.7 (aq $HClO_4$), pK_2^{25} 4.86 (4.96).** Crystd as its *picrate* (m 191°) from EtOH, then the free base was liberated, dried at 78°/8mm over P_2O_5 and crystd from pet ether (b 80-100°). [Cumper, Ginman and Vogel *J Chem Soc* 1188 *1962*.] It can be purified by zone melting. Also crystd from hexane, *benzene/pet ether (b 40-60°) or sodium-dried *benzene, dried and stored over H_2SO_4. The monohydrate is obtained by crystn from aqueous EtOH or ethyl acetate. It has been crystd from H_2O (300 parts) to give the *monohydrate* **m** 102-103° and sublimes at 10^{-3}mm [Fielding and LeFevre *J*

Chem Soc 1811 *1951.*] The *anhydrous* compound has **m** 118° (after drying at high vacuum at 80°), also after recrystn from pet ether or *C_6H_6 (70 parts) and drying at 78°/8mm. [UV: Badger et al. *J Chem Soc* 3199 *1951.*] It has a pKa in H_2O of 4.857 (25°) or 5.02 (20°) and 4.27 in 50% aq EtOH (20°) [Albert et al. *J Chem Soc* 2240 *1948*].

1,10-Phenanthroline hydrochloride (*o*-phenanthroline hydrochloride) *[3829-86-5]* **M 243.7, m 212-219°.** It crystallises from 95% EtOH, **m** 212-219° as the *monohydrate*, the *half hydrate* has **m** 217°. The *3HCl* has **m** 143-145° (sinters at 128°) [Thevenet et al. *Acta Cryst Sect B* **33** 2526 *1977*].

4,7-Phenanthroline-5,6-dione *[84-12-8]* **M 210.2, m 295°(dec).** Crystd from MeOH.

Phenazine *[92-82-0]* **M 180.2, m 171°, pK_1^{20} -4.9 (aq H_2SO_4), pK_2^{20} 1.21.** Crystd from EtOH, $CHCl_3$ or ethyl acetate, after pre-treatment with activated charcoal. It can be sublimed *in vacuo*, and zone refined.

Phenazine methosulfate *[299-11-6]* **M 306.3.** See 5-methylphenazinium methyl sulfate on p. 547 in Chapter 6.

Phenethylamine *[64-04-0]* **M 121.2, b 87°/13mm, d 0.962, n 1.535, pK^{25} 9.88.** Distd from CaH_2, under reduced pressure, just before use.

Phenethyl bromide *[103-63-9]* **M 185.1, b 92°/11mm, d 1.368, n 1.557.** Washed with conc H_2SO_4, water, aq 10% Na_2CO_3 and water again, then dried with $CaCl_2$ and fractionally distd just before use.

N-2-Phenethyl urea *[2158-04-5]* **M 164.2, m 173-174°.** Crystd from water.

Phenetole *[103-73-1]* **M 122.2, b 60°/9mm, 77.5°/31mm, 170.0°/760mm, d 0.967, n 1.50735, n^{25} 1.50485.** Small quantities of phenol can be removed by shaking with NaOH, but this is not a very likely contaminant of commercial material. Fractional distn from sodium, at low pressures, probably gives adequate purification. It can be dissolved in diethyl ether and washed with 10% NaOH (to remove phenols), then water. The ethereal soln was evaporated and the phenetole fractionally distd under vacuum.

Phenocoll hydrochloride (4-ethoxyaniline, *p*-phenetidine HCl) *[536-10-6]* **M 230.7, m 234°, pK^{28} 5.20.** Crystd from water. Sublimes *in vacuo*.

Phenol *[108-95-2]* **M 94.1, m 40.9°, b 85.5-86.0°/20mm, 180.8°/760mm, d 1.06, n^{41} 1.54178, n^{46} 1.53957, pK^{25} 9.86 (10.02).** Steam was passed through a boiling soln containing 1mole of phenol and 1.5-2.0moles of NaOH in 5L of H_2O until all non-acidic material had distd. The residue was cooled, acidified with 20% (v/v) H_2SO_4, and the phenol was separated, dried with $CaSO_4$ and fractionally distd under reduced pressure. It was then fractionally crystd several times from its melt [Andon et al. *J Chem Soc* 5246 *1960*]. Purification *via* the benzoate has been used by Berliner, Berliner and Nelidow [*J Am Chem Soc* **76** 507 *1954*]. The benzoate was crystd from 95% EtOH, then hydrolysed to the free phenol by refluxing with two equivalents of KOH in aq EtOH until the soln became homogeneous. It was acidified with HCl and extracted with diethyl ether. The ether layer was freed from benzoic acid by thorough extraction with aqueous $NaHCO_3$, and, after drying and removing the ether, the phenol was distd.
Phenol has also been crystd from a 75% w/w soln in water by cooling to 11° and seeding with a crystal of the hydrate. The crystals were centrifuged off, rinsed with cold water (0-2°) satd with phenol, and dried. It can be crystd from pet ether [Berasconi and Paschalis *J Am Chem Soc* **108** 2969 *1986*].
Draper and Pollard [*Science* **109** 448 *1949*] added 12% water, 0.1% aluminium (can also use zinc), and 0.05% $NaHCO_3$ to phenol, and distd at atmospheric pressure until the azeotrope was removed, The phenol was then distd at 25mm. Phenol has also been dried by distn from the *benzene soln to remove the water-*benzene azeotrope and the excess *benzene, followed by distn of the phenol at reduced pressure under nitrogen. Processes such as this are probably adequate for analytical grade phenol which has as its main impurity water. Phenol has also been crystd from pet ether/*benzene or pet ether (b 40-60°). Purified material is stored in a vacuum desiccator over P_2O_5 or $CaSO_4$.

Phenol-2,4-disulfonic acid   *[96-77-5]*   M 254.2, pK$_1$ <1, pK$_2$ <1, pK$_3$ ~8.3.   Crystd   from EtOH/diethyl ether.

Phenolphthalein   *[77-09-8]*   M 319.2, m 263°, pK$_{Est(1)}$~ 4.2, pK$_{Est(2)}$~ 9.8.   Dissolved in EtOH (7mL/g), then diluted with eight volumes of cold water. Filtered. Heated on a water-bath to remove most of the alcohol and the pptd phenolphthalein was filtered off and dried under vacuum.

Phenolphthalol   *[81-92-5]*   M 306.3, m 201-202°, pK$_{Est}$ ~ 9.8.   Crystd from aqueous EtOH.

Phenosafranine (3,7-diamino-5-phenylphenazinium chloride)   *[81-93-6]*   M 322.8, m >300°, λ$_{max}$ 530nm (H$_2$O).   Crystd from dilute HCl.

Phenothiazine   *[92-84-2]*   M 199.3, m 184-185°.   Crystd from *benzene or toluene (charcoal) after boiling for 10min under reflux. Filtered on a suction filter. Dried in an oven at 100°, then in a vacuum desiccator over paraffin chips. Also twice recrystd from water and dried in an oven at 100° for 8-10h.

Phenoxazine   *[135-67-1]*   M 199.2, m 156°, 156-158°, 158-159°, b 215°/4mm.   Crystd from EtOH and sublimed *in vacuo*. If too impure then extract in a Soxhlet using toluene. Evaporate the solvent and dissolve residue (*ca* 100g) in *C$_6$H$_6$ (1L) **CARCINOGEN**, use an efficient fume cupboard) and chromatograph through an Al$_2$O$_3$ column (50 x 450 mm). The eluent (*ca* 3L) is evaporated to *ca* 150mL and cooled when *ca* 103g of phenoxazine m 149-153° is obtained. Sublimation yields platelets m 158-159°. It forms a green *picrate* m 141.5-142°. [Gilman and Moore *J Am Chem Soc* **79** 3485 *1957*; Müller et al. *J Org Chem* **24** 37 *1959*.]

Phenoxyacetic acid   *[122-59-8]*   M 152.2, m 98-99°, pK^{25} 3.18.   Crystd from water or aq EtOH.

Phenoxyacetyl chloride   *[701-99-5]*   M 170.6, b 112°/10mm, 102°/16mm, 225-226°/atm, d$_4^{20}$ 1.235, n$_D^{20}$ 1.534.   If it has no OH band in the IR then distil in a vacuum, taking precautions for the moisture-sensitive compound. If it contains free acid (due to hydrolysis, OH bands in the IR) then add an equal volume of redistilled SOCl$_2$, reflux for 2-3h, evaporate and distil the residue in a vacuum as before. The *amide* has m 101°. [McElvain and Carney *J Am Chem Soc* **68** 2592 *1946*.]

4-Phenoxyaniline   *[139-59-3]*   M 185.2, m 95°, pK^{20} 4.44 (50% aq EtOH).   Crystd from water.

Phenoxybenzamine [N-(2-chloroethyl)-N-(1-methyl-2-phenoxyethyl)benzylamine]   *[59-96-1]* M 303.5, m 38-40°, hydrochloride *[63-92-3]* M 340.0, m 137.5-140°, pK$_{Est}$ ~4.2.   The free base is crystd from pet ether and the *HCl* is crystd from EtOH/diethyl ether.

2-Phenoxybenzoic acid   *[2243-42-7]*   M 214.2, m 113°, b 355°/760mm, pK^{15} 3.53.   Crystd from aqueous EtOH.

3-Phenoxybenzoic acid   *[3739-38-6]*   M 214.2, m 145°, pK 3.59.   Crystd from aqueous EtOH.

Phenoxybutyric acid   *[6303-58-8]*   M 180.2, m 64°, 65-66°, 82-83°, 99°, b 180-185°/12mm, pK 3.17.   It has been purified by recrystn from pet ether, *C$_6$H$_6$, Et$_2$O-pet ether, EtOH and from H$_2$O. It can be steam distd or distd in a good vac. [UV: Ramart-Lucas and Hoch *Bull Soc Chim Fr* [4] **51** 824 *1932*; Dann and Arndt *Justus Liebigs Ann Chem* **587** 38 *1954*.] The *acid chloride* has b 154-156°/20mm [Hamford and Adams *J Am Chem Soc* **57** 921 *1935*]; and the *amide* crystallises from *C$_6$H$_6$ as needles m 113°.

2-Phenoxypropionic acid (lactic acid O-phenylether)   *[940-31-8]*   M 166.2, m 115-116°, b 105-106°/5mm, 265-266°/758mm, pK 3.11.   Crystd from water.

Phensuximide (N-methyl-2-phenylsuccinimide)   *[86-34-0]*   M 189.2, m 71-73°.   Crystd from hot 95% EtOH.

Phenylacetamide *[103-81-1]* **M 135.2, m 158.5°.** Crystd repeatedly from absolute EtOH. Dried under vacuum over P_2O_5.

Phenyl acetate *[122-79-2]* **M 136.2, b 78°/10mm, d 1.079, n^{22} 1.5039.** Freed from phenol and acetic acid by washing (either directly or as a soln in pentane) with aqueous 5% Na_2CO_3, then with saturated aqueous $CaCl_2$, drying with $CaSO_4$ or Na_2SO_4, and fractional distn at reduced pressure.

Phenylacetic acid *[103-82-2]* **M 136.2, m 76-77°, b 140-150°/20mm, pK_1 -7.59 (aq H_2SO_4), pK_2 4.31.** Crystd from pet ether (b 40-60°), isopropyl alcohol, aq 50% EtOH or hot water. Dried under vac. It can be distd under reduced pressure.

Phenylacetone (1-phenylpropan-2-one) *[103-79-9]* **M 134.2, b 69-71°/3mm, d 1.00, n 1.516.** Converted to the semicarbazone and crystd three times from EtOH (**m** 186-187°). The semicarbazone was hydrolysed with 10% phosphoric acid and the ketone was distd. [Kumler, Strait and Alpen *J Am Chem Soc* **72** 1463 *1950*.]

4'-Phenylacetophenone *[92-91-1]* **M 196.3, m 120.3-121.2°, b 196-210°/18mm, 325-327°/760mm.** Crystd from EtOH or acetone. Can also be distd under reduced or atmospheric pressure.

Phenylacetylene *[536-74-3]* **M 102.1, b 75°/80mm, d 0.930, n^{25} 1.5463, pK ~19.** Distd through a spinning band column. Should be filtered through a short column of alumina before use [Collman et al. *J Am Chem Soc* **108** 2988 *1986*; for pK see Brandsma *Preparative Acetylenic Chemistry,* 1st Edn Elsevier 1971, p. 15, ISBN 0444409475].

dl-**Phenylalanine** *[150-30-1]* **M 165.2, m 162°, pK_1^{25} 2.58, pK_2^{25} 9.24.** Crystd from water and dried under vacuum over P_2O_5.

L-Phenylalanine *[63-91-2]* **M 165.2, m 280°(dec), $[\alpha]_D^{25}$ -34.0° (c 2, H_2O).** Likely impurities are leucine, valine, methionine and tyrosine. Crystd from water by adding 4 volumes of EtOH. Dried under vac over P_2O_5. Also crystd from satd refluxing aq solns at neutral pH, or 1:1 (v/v) EtOH/water soln, or conc HCl.

Phenylalaninol (2-amino-3-phenylpropan-1-ol) *[R-(+)- 5267-64-1; S-(-)- 3182-95-4]* **M 151.2, m 91-92°, 91.5°, 92-94°, b 80°/11mm (Kügelrohr), $[\alpha]_{546}^{20}$ (+) and (-) 28°, $[\alpha]_D^{20-25}$ (+) and (-) 23-28.7° (c 1-5, EtOH), pK_{Est} ~9.3.** It can be recrystd from Et_2O, *C_6H_6-pet ether (b 40-60°) or toluene and distd in a vacuum. Has been purified by dissolving in Et_2O, drying over K_2CO_3, filtering, evaporating to a small volume, cooling in ice and collecting the plates. Store in the presence of KOH (i.e. CO_2—free atm). [Karrer and Ehrhardt *Helv Chim Acta* **34** 3203 *1951*; Oeda *Bull Chem Soc Jpn* **13** 465 *1938*.] The *picrate* has **m** 141-141.5° (from EtOH-pet ether). The *hydrogen oxalate* has **m** 177°, 161-162° [Hunt and McHale *J Chem Soc* 2073 *1957*]. The *racemate* has **m** 87-88° from *C_6H_6-pet ether (75-77° from Et_2O), and the *hydrochloride* has **m** 139-141° [Fodor et al. *J Chem Soc* 1858 *1951*].

3-Phenylallyl chloride (cinnamyl chloride) *[E: 18685-01-3; Z: 39199-93-4]* **M 152.6, b 92-93°/3mm.** Distd under vacuum three times from K_2CO_3.

Phenyl 4-aminosalicylate *[133-11-9]* **M 229.2, m 153°, $pK_{Est(1)}$~2.0 (NH_2), $pK_{Est(2)}$~9.7 (OH).** Crystd from isopropanol.

4-Phenylanisole (4-methoxybiphenyl) *[361-37-6]* **M 184.2, m 89.9-90.1°.** Crystd from *benzene/pet ether. Dried under vacuum in an Abderhalden pistol.

9-Phenylanthracene *[602-55-1]* **M 254.3, m 153-154°, b 417°.** Chromatographed on alumina in *benzene and crystd from acetic acid.

N-**Phenylanthranilic acid** *[91-40-7]* **M 213.2, m 182-183°, pK_1^{25} -1.28 (aq H_2SO_4), pK_2^{25} 3.86 (CO_2H).** Crystd from EtOH (5mL/g) or acetic acid (2mL/g) by adding hot water (1mL/g).

2-Phenyl-1-azaindolizine   *[56983-95-0]*   M 194.2, m 140°, pK$_{Est}$ ~1.9. Crystd from EtOH or *benzene/pet ether.

p-**Phenylazobenzoyl chloride**   *[104-24-5]*   M 244.7, m 93°. Crystd from pet ether (b 60-80°).

4-Phenylazo-1-naphthylamine   *[131-22-6]*   M 247.3, m 125-125.5°. Crystd from cyclohexane or aq EtOH. [Brode et al. *J Am Chem Soc* **74** 4641 *1952.*]

1-Phenylazo-2-naphthylamine   *[85-84-7]*   M 247.3, m 102-104°. See 1-benzeneazo-2-naphthylamine on p. 120.

4-Phenylazophenacyl bromide   *[62625-24-5]*   M 317.3, m 103-104°. Purified on a column of silica gel, using pet ether/diethyl ether (9:1 v/v) as solvent.

4-Phenylazophenol (4-hydroxyazobenzene)   *[1689-82-3]*   M 198.2, m 155°, pK$_1^{25}$ -0.93, pK$_2^{25}$ 8.2. Crystd from *benzene or 95% EtOH.

Phenyl benzenethiosulfonate (diphenyldisulfoxide)   *[1212-08-4]* M 250.3, m 36-37°, 45-46°, 45-47°. Recrystd from EtOH or MeOH. Also purified from phenylsulfide impurities by dissolving in CHCl$_3$, washing with aq satd NaHCO$_3$, drying (Na$_2$SO$_4$) evaporating and the residual oil was passed through a silica gel column (600g) and eluted with hexane-*C$_6$H$_6$ (1L, 4:1, eluting PhSSPh) then *C$_6$H$_6$ (1L) which elutes PhSSO$_2$Ph. [Trost and Massiot *J Am Chem Soc* **99** 4405 *1977*; Knoevenagel and Römer *Chem Ber* **56** 215 *1923.*]

Phenyl benzoate   *[93-99-2]*   M 198.2, m 69.5°, b 198-199°. Crystd from EtOH using *ca* twice the volume needed for complete soln at 69°.

Phenyl-1,4-benzoquinone   *[363-03-1]*   M 184.2, m 114-115°. Crystd from heptane or pet ether (b 60-70°) and sublimed *in vacuo*. [Carlson and Miller *J Am Chem Soc* **107** 479 *1985.*]

1-Phenylbiguanide   *[102-02-3]*   M 177.2, m 144-146°, pK$_1^{32}$ 2.16, pK$_2^{32}$ 10.74. Crystd from water or toluene.

S-(-)-**1-Phenylbutanol**   *[22135-49-5]*   M 150.2, m 46-47°, 46-48°, 49°, b 90-92°/2mm.   [α]$_D^{18}$ -51.4° (c 5, CHCl$_3$), -44.7° (c 5.13, *C$_6$H$_6$). Purified by distn and crystallises on cooling. The *hydrochloride* has [α]$_D^{20}$ +45.1° (c 4.8, *C$_6$H$_6$). The (-)-*hydroperoxide* has b 58°/0.005mm, n$_D^{20}$ 1.5123, α$_D^{18}$ -2.14°, (*l* = 0.5dcm, neat). [Holding and Ross *J Chem Soc* 145 *1954*; Davies and Feld *J Chem Soc* 4637 *1958.*] The (±)-*racemate* has b 73°/0.05mm, and its *4-nitrophenylhydrazone* has m 58°.

Phenylbutazone (4-butyl-1,2-diphenylpyrazolidin-3,5-dione)   *[50-33-9]*   M 308.4, m 105°, 106-108°. Crystd from EtOH.

2-Phenylbutyramide   *[90-26-6]*   M 163.2, m 86°. Crystd from water.

2-Phenylbutyric acid *[R-(-)- 938-79-4; S-(+)- 4286-15-1]* M 164.2, b 102-104°/atm, d$_4^{20}$ 1.056, n$_D^{20}$ 1.521, [α]$_D^{20}$ (-) and (+) 96° (c 2.5, *C$_6$H$_6$), [α]$_D^{23}$ (-) and (+) 5.8° (neat), pK$_{Est}$ ~4.3. Purified by distn at atmospheric pressure using an efficient column. The *acid chlorides* have b 106-107°/20mm, [α]$_D^{18}$ (-) and (+) 108° (c 2, *C$_6$H$_6$). [Levene et al. *J Biol Chem* **100** 589 *1933*, Gold and Aubert *Helv Chim Acta* **41** 1512 *1958*; ORD in heptane: Rothen and Levene *J Chem Phys* **7** 975 *1939.*]

3-Phenylbutyric acid *[R-(-)- 772-14-5; S-(+)- 772-15-6]* M 164.2, b 94-95°/3mm, 134°/4mm, d$_4^{26}$ 1.066, n$_D^{25}$ 1.5167, [α]$_D^{20}$ (-) and (+) 57° (c 1, *C$_6$H$_6$), pK^{25} 4.40. Purified as the 2-isomer above, i.e. by distn, but under a good vacuum. [Prelog and Scherrer *Helv Chim Acta* **42** 2227 *1959*; Levene and Marker *J Biol Chem* **93** 761 *1932*, **100** 685 *1933*; Cram *J Am Chem Soc* **74** 2137 *1952.*] The *R-amide* crystallises from H_2O, **m** 101.5-102°, $[\alpha]_D^{20}$ -16.5° (c 1.2, EtOH). The *racemic acid* has **m** 39-40°, **b** 134-136°/6mm, 158°/12mm [Marvel et al. *J Am Chem Soc* **62** 3499 *1940*].

4-Phenylbutyric acid *[1821-12-1]* **M 164.2, m 50°, pK^{25} 4.76.** Crystd from pet ether (b 40-60°).

o-**(Phenylcarbamoyl)-1-scopolamine methobromide** *[138-10-3]* **M 518.4, m 200.5-201.5°** (dec). Crystd from 95% EtOH.

9-Phenylcarbazole *[1150-62-5]* **M 243.3, m 94-95°.** Crystd from EtOH or isopropanol and sublimed *in vacuo.*

O-**Phenyl chlorothionoformate** *[1005-56-7]* **M 172.6, b 81-83°/6mm, 91°/10mm, d$_4^{20}$ 1.276, n$_D^{20}$ 1.585.** Purified by dissolving in $CHCl_3$, washing with H_2O, drying ($CaCl_3$), filtering, evaporating and distilling twice under vacuum to give a clear yellow liquid. **It is reactive and POISONOUS - work in a fume cupboard.** Store in sealed ampoules under N_2. Possible impurity is *O,O'-diphenyl thiocarbonate* which has **m** 106° which remains behind in the distilling flask. [Bögemann et al. in *Methoden Der Organischen Chemie (Houben-Weyl)* 4th edn (E. Müller Ed.) **Vol 9** *Schwefel-Selen-Tellur Verbindungen* pp. 807-808 *1955*; Rivier and Schalch *Helv Chim Acta* **6** 612 *1932*; Kalson *Chem Ber* **20**, 2384 *1987*; Rivier and Richard *Helv Chim Acta* **8** 490 *1925*; Schönberg and Varga *Justus Liebigs Ann Chem* **483** 176 *1930*; *Chem Ber* **64** 1390 *1931*.]

Phenyl cinnamate *[2757-04-2]* **M 224.3, m 75-76°, b 205-207°/15mm.** Crystd from EtOH (2mL/g). It can also be distd under reduced pressure.

α-**Phenylcinnamic acid** *[91-48-5]* **M 224.3, m 174°**(*cis*), **m 138-139°**(*trans*), **pK 4.8 (60% aq EtOH).** Crystd from ether/pet ether.

o-**Phenylenediamine** *[95-54-5]* **M 108.1, m 100-101°, pK$_1^{25}$ 0.67 (aq H_2SO_4), pK$_2^{25}$ 4.47 (4.85).** Crystd from aqueous 1% sodium hydrosulfite (charcoal), washed with ice-water and dried in a vacuum desiccator, or sublimed *in vacuo*. It has been purified by recrystn from toluene and zone refined [Anson et al. *J Am Chem Soc* **108** 6593 *1986*]. Purification by refluxing a CH_2Cl_2 solution containing charcoal was also carried out followed by evaporation and recrystn [Koola and Kochi *J Org Chem* **52** 4545 *1987*], protect from light.

m-**Phenylenediamine** *[108-45-2]* **M 108.1, m 61-63°, 62-63°, 62.5°, 63-64°, b 146°/22mm, 282-284°/760mm, 284-287°/atm, d$_{10}^{10}$ 1.1422, n$_D^{57.7}$ 1.6340, pK$_1^{25}$ 2.41, pK$_2^{25}$ 4.98.** Purified by distn under vac followed by recryst from EtOH (rhombs) and if necessary redistn. It should be protected from light otherwise it darkens rapidly. [Neilson et al. *J Chem Soc* 371 *1962*; IR: Katritzky and Jones *J Chem Soc* 3674, 2058 *1959*; UV: Forbes and Leckie *Can J Chem* **36** 1371 *1958*.] The *hydrochloride* has **m** 277-278°, and the *bis-4-chlorobenzenesulfonyl* derivative has **m** 220-221° from H_2O (214-215°, from MeOH-H_2O) [Runge and Pfeiffer *Chem Ber* **90** 1737 *1957*].

p-**Phenylenediamine** *[106-50-3]* **M 108.1, m 140°, pK$_1^{25}$ 2.89, pK$_2^{25}$ 6.16.** Crystd from EtOH or *benzene, and sublimed *in vacuo*, protect from light.

o-**Phenylenediamine dihydrochloride** *[615-28-1]* **M 181.1, m 180°.** Crystd from dilute HCl (60mL conc HCl, 40mL water, with 2g stannous chloride), after treatment of the hot soln with charcoal by adding an equal volume of conc HCl and cooling in an ice-salt mixture. The crystals were washed with a small amount of conc HCl and dried in a vacuum desiccator over NaOH.

2-Phenyl-1,3-diazahexahydroazulene *[2161-31-1]* **M 212.3.** Recrystd three times from de-aerated cyclohexane in the dark.

1,4-Phenylene diisothiocyanate (bitoscanate) *[4044-65-9]* **M 192.3, m 129-131°, 130-131°, 132°.** Purified by recrystn from AcOH, pet ether (b 40-60°), Me_2CO or aq Me_2CO. [van der Kerk et al. *Recl Trav Chim Pays-Bas* **74** 1262 *1955*; Leiber and Slutkin *J Org Chem* **27** 2214 *1962*.]

1-Phenyl-1,2-ethanediol *[R-(-)- 16355-00-3; S-(+)- 25779-13-9]* **M 138.2, m 64-67°, 65-66°,** $[\alpha]_D^{24}$ **(-) and (+) 40.5° (c 2.8, H_2O), $[\alpha]_D^{20}$ (-) and (+) 39° (c 3, EtOH).** Purified by recryst from $*C_6H_6$-ligroin and sublimed at 1-2mm. [Arpesella et al. *Gazetta* **85** 1354 *1955;* Prelog et al. *Helv Chim Acta* **37** 221 *1954*.]

***dl*-1-Phenylethanol** *[13323-81-4]* **M 122.2, b 60.5-61.0°/3mm, 106-107°/22-23mm, d 1.01, n^{25} 1.5254.** Purified *via* its hydrogen phthalate. [See Houssa and Kenyon *J Chem Soc* 2260 *1930*.] Shaken with a soln of ferrous sulfate, and the alcohol layer was washed with distilled H_2O, dried ($MgSO_4$) and fractionally distd.

2-Phenylethanol *[60-12-8]* **M 122.2, b 215-217°, d 1.020.** Purified by shaking with a soln of ferrous sulfate, and the alcohol layer was washed with distd water and fractionally distd.

Phenyl ether (diphenyl ether) *[101-84-8]* **M 170.2, m 27.0°, d 1.074, $n^{30.7}$ 1.57596.** Crystd from 90% EtOH. Melted, washed with 3M NaOH and water, dried with $CaCl_2$ and fractionally distd under reduced pressure. Fractionally crystd from its melt and stored over P_2O_5.

1-Phenylethyl isocyanate (α-methylphenyl isocyanate) *[R-(+)- 33375-06-3; S(-)- 14649-03-7]* **M 147.2, b 82-83°/12-14mm, d_4^{20} 1.045, n_D^{20} 1.513, $[\alpha]_D^{24}$ (+) and (-) 2° (c 3.5, $*C_6H_6$), (+) and (-) 10.5° (neat).** Purified by fractional distn under vacuum. With ammonia it gives the *ureido* derivative which crystallises from H_2O, **m** 121-122°, $[\alpha]_D^{25}$ (+) and (-) 48.8°. [Cairns *J Am Chem Soc* **63** 870 *1941*.] The *racemate* has **b** 90-94°/3mm, 96°/18mm [Seiftan *Justus Liebigs Ann Chem* **562** 75 *1949*].

***p*-α-Phenylethylphenol** *[1988-89-2]* **M 198.3, m 56.0-56.3°, pK_{Est} ~10.3.** Crystd from pet ether.

5-(α-Phenylethyl)semioxamazide *[93-95-8]* **M 207.1, m 167-168° (*l*-), 157° (*dl*-).** Crystd from EtOH.

9-Phenyl-3-fluorone *[975-17-7]* **M 320.3, m >300°(dec), λ_{max} 462nm (ε 4.06 x 10^4, in 1M HCl aq EtOH).** Recrystd from warm, acidified EtOH by addition of ammonia. The crude material (1g) can be extracted with EtOH (50mL) in a Soxhlet apparatus for 10h to remove impurities. Impurities can be detected by paper electrophoresis. [Petrova et al. *Anal Lett* **5** 695 *1972*.]

L-α-Phenylglycine *[2935-35-5]* **M 151.2, m 305-310°, $[\alpha]_{546}$ +185° (c 1, M HCl), pK_1^{25} 1.83, pK_2^{25} 4.39 (for *dl*).** Crystd from EtOH.

Phenylglycine-*o*-carboxylic acid *[612-42-0]* **M 195.2, m 208°.** Crystd from hot water (charcoal).

Phenylhydrazine *[100-63-0]* **M 108.1, m 23°, b 137-138°/18mm, 241-242°/760mm, d 1.10, n 1.607, pK_1^{20} -5.2 (aq H_2SO_4), pK_2^{25} 5.27.** Purified by chromatography, then crystd from pet ether (b 60-80°)/*benzene. [Shaw and Stratton *J Chem Soc* 5004 *1962*.]

Phenylhydrazine hydrochloride *[59-88-1]* **M 144.5, m 244°.** One litre of boiling EtOH was added to 100g of phenylhydrazine hydrochloride dissolved during 1-3h (without heating) in 200mL of warm water (60-70°). The soln was filtered off, while still hot, through Whatman No 2 filter paper and cooled in a refrigerator. The ppte was collected on a medium sintered-glass filter and recrystd twice this way, then washed with cold EtOH, dried thoroughly and stored in a stoppered brown bottle. [Peterson, Karrer and Guerra *Anal Chem* **29** 144 *1957*.] Hough, Powell and Woods [*J Chem Soc* 4799 *1956*] boiled the hydrochloride with three times its weight of water, filtered hot (charcoal), added one-third volume of conc HCl and cooled to 0°. The crystals were washed with acetone, and dried over P_2O_5 under vacuum. The salt has also been crystd from 95% EtOH.

Phenylhydroxylamine (N-hydroxyaniline) *[100-65-2]* **M 109.1, m 82°, pK 3.2.** Impure base deteriorates rapidly. Crystd from H_2O, *C_6H_6 or *C_6H_6/pet ether (40-60°). *Picrate* has **m 186°**.

2-Phenyl-1,3-indandione *[83-12-5]* **M 222.2, m 149-151°, pK^{20} 4.12 (1% aq MeOH).** Crystd from EtOH.

2-Phenylindolizine *[25379-20-8]* **M 193.2, m 214°(dec), pK$_{Est}$ ~4.4.** Crystd from EtOH.

Phenylisocyanate *[103-71-9]* **M 119.1, b 45-47°/10mm, d 1.093, n 1.536.** Distd under reduced pressure from P_2O_5.

Phenylisothiocyanate (phenyl mustard oil) *[103-72-0]* **M 135.2, m -21°, b 95°/12mm, 117.1°/33mm, 221°/760mm, d$_4^{25}$ 1.1288, n$_D^{23.4}$ 1.64918.** It is insol in H_2O, but sol in Et_2O and EtOH. If impure (due to formation of thiourea) then steam dist into a receiver containing 5-10mL of N H_2SO_4. Separate the oil, dry over $CaCl_2$ and distil under vacuum. [Dains et al. *Org Synth* Coll Vol I 447 *1941*.]

8-Phenylmenthol *[1R,2S,5R-(-)- 65253-04-5; 1S,2R,5S-(+)- 57707-91-2]* **M 232.4, [α]$_D^{20}$ (-) and (+) 26°** (c 2, EtOH). Dissolve in toluene, dry (Na_2SO_4), evap and chromatograph on a silica gel column and eluting with 5% Et_2O in pet ether to give an oil with the desired rotation. IR has v 3420cm^{-1} (OH) with consistent ^{1}H NMR [Corey and Ensley *J Org Chem* **43** 1610 *1978*; Whitesell et al. *Tetrahedron* **42** 2993 *1986*; Bednarski and Danishefsky *J Am Chem Soc* **108** 7060 *1986*.]

1-Phenyl-5-mercaptotetrazole *[86-93-1]* **M 178.2, m 150° (dec), 155° (dec), 157-158°, pK^{25} 3.65 (5% aq EtOH).** Purified by recryst from EtOH or CHCl$_3$ (m 152°) [Tautomerism: Kauer and Sheppard *J Org Chem* **32** 3580 *1967*; UV: Leiber et al. *Can J Chem* **37** 563 *1959*]. The *ammonium salt* crystallises from EtOH and dec at 176°, and the *sodium salt* crystallises from EtOH-*C_6H_6, melts at 96° and dec at 145° [Stollé *J Prakt Chem* [2] **133** 60 *1932*].

Phenyl methanesulfonate *[16156-59-5]* **M 172.1, m 61-62°.** Crystd from MeOH.

2-Phenylnaphthalene *[612-94-2]* **M 204.3, m 103-104°.** Chromatographed on alumina in *benzene and crystd from aqueous EtOH.

N-Phenyl-1-naphthylamine *[90-30-2]* **M 219.3, m 63.7-64.0°, pK$_{Est}$ ~0.1.** Crystd from EtOH, pet ether or *benzene/EtOH. Dried under vacuum in an Abderhalden pistol.

N-Phenyl-2-naphthylamine *[135-88-6]* **M 219.3, m 107.5-108.5°, 110°, pK$_{Est}$ ~0.5.** Crystd from EtOH, MeOH, glacial acetic acid or *benzene/hexane.

4-Phenylphenacyl bromide *[135-73-9]* **M 275.2, m 126°.** Crystd (charcoal) from EtOH (15mL/g), or ethyl acetate/pet ether (b 90-100°).

2-Phenylpropanal *[93-53-8]* **M 134.2, b 206°/760mm, d 1.001, n 1.5183.** May contain up to 15% of acetophenone. Purified *via* the bisulfite addition compound [Lodge and Heathcock *J Am Chem Soc* **109** 3353 *1987*] and see Chapter 2 for prepn, and decompn, of bisulfite adduct.

Phenylpropiolic acid *[637-44-5]* **M 146.2, m 137.8-138.4°, pK^{25} 2.23.** Crystd from *benzene, CCl$_4$ or aqueous EtOH.

RS-2-Phenylpropionic acid *[492-37-5]* **M 150.2, m 16-16.5°, b 153-155°/20mm, 189°/48mm, 260-262°/atm, d 1.10, n 1.522, pK^{25} 4.3.** Fractionally distd, or recrystd from pet ether (b 40-60°) strong cooling (see references below).

2-Phenylpropionic acid *[R-(-)- 7782-26-5; S-(+)- 7782-24-3]* **M 150.2, m 30.3-31°, 30-32°, b 115°/1-2mm, 142°/12mm, [α]$_D^{20}$ (-) and (+) 99.7°** (l = 1 dcm, neat), **(-) and (+) 89.1°** (c

1.7, EtOH), (-) and (+) 75° (c 1.6, CHCl₃). Purified by vacuum distn and by recrystn from pet ether. The *S*-anilide has **m** 103-104° (from H₂O or CHCl₃/*C₆H₆), $[\alpha]_D^{25}$ +47° (c 9, Me₃CO) [Argus and Kenyon *J Chem Soc* 916 *1939*; Campbell and Kenyon *J Chem Soc* 25 *1946*; Levene et al. *J Biol Chem* **88** 27, 34 *1930*; *Beilstein* **9**, 3rd Suppl p 2417].

3-Phenylpropionic acid (hydrocinnamic acid) *[501-52-0]* **M 150.2, m 48-48.5°, pK²⁵ 4.56.** Crystd from *benzene, CHCl₃ or pet ether (b 40-60°). Dried in a vacuum.

3-Phenylpropyl bromide *[637-59-2]* **M 199.1, b 110°/12mm, 128-129°/29mm, d 1.31.** Washed successively with conc H₂SO₄, water, 10% aqueous Na₂CO₃ and again with water, then dried with CaCl₂ and fractionally distd just before use.

Phenyl 2-pyridyl ketoxime *[1826-28-4]* **M 198.2, m 151-152°.** Crystd from EtOH (charcoal).

Phenylpyruvic acid *[156-06-9]* **M 164.2, m 150-154°, 158-159°, pK_Est ~2.1.** Recrystd from *C₆H₆. The *phenylhydrazone* has **m** 173° [Zeller *Helv Chim Acta* **26** 1614 *1943*; Hopkins and Chisholm *Can J Research* [B] **24** 89 *1946*]. The *2,4-dinitrophenylhydrazone* has **m** 162-164° (189°, 192-194°) [Fones *J Org Chem* **17** *1952*].

6-Phenylquinoline *[612-95-3]* **M 205.3, m 110.5-111.5°, pK_Est ~5.2.** Crystd from EtOH (charcoal).

2-Phenylquinoline-4-carboxylic acid (cinchophen) *[132-60-5]* **M 249.3, m 215°, pK_Est(1)~0.5 (CO₂H), pK_Est(2)~5.1 (N).** Crystd from EtOH (*ca* 20mL/g).

Phenyl salicylate (Salol) *[118-55-8]* **M 214.2, m 41.8-42.6°, pK_Est ~9.9.** Fractionally crystd from its melt, then crystd from *benzene.

3-Phenylsalicylic acid *[304-06-3]* **M 214.3, m 186-187.5°, pK_Est(1)~2.8 (CO₂H), pK_Est(2)~11.0 (OH).** Dissolved in *ca* 1 equivalent of saturated aqueous Na₂CO₃, filtered and ppted by adding 0.8 equivalents of M HCl. Crystd from ethylene dichloride (charcoal), and sublimed at 0.1mm. [Brooks, Eglington and Norman *J Chem Soc* 661 *1961*.]

1-Phenylsemicarbazide *[103-03-7]* **M 151.2, m 172°.** Crystd from water and dried in vac over KOH.

4-Phenylsemicarbazide *[537-47-3]* **M 151.2, m 122°.** Crystd from water and dried in vac over KOH.

Phenylsuccinic acid *[R-(-)- 46292-93-7; S-(+)- 4036-30-1]* **M 194.2, m 173-176°, 178.5-179°, 179-180°, $[\alpha]_D^{25}$ (-) and (+) 171° (c 2, Me₂CO), $[\alpha]_D^{26-30}$ (-) and (+) 148° (c 0.27-5, EtOH), pK₁²⁵ 3.78, pK₂²⁵ 5.55.** Purified by repptn from alkali and recrystn from H₂O. [Naps and Johns *J Am Chem Soc* **62** 2450 *1940*; Fredga and Matell *Bull Soc Chim Belg* **62** 47 *1953*; Wren and Williams *J Chem Soc* **109** 572 *1916*.] The *racemate* *[635-51-8]* has **m** 166-168°, 168° after recrystn from H₂O or MeCN; its *S-benzylthiouronium* salt has **m** 164-165° (from EtOH) [Griediger and Pedersen *Acta Chem Scand* **9** 1425 *1955*].

1-Phenyl-5-sulfanilamidopyrazole *[526-08-9]* **M 314.3, m 179-183°.** Crystd from EtOH.

1-Phenylthiosemicarbazide *[645-48-7]* **M 167.2, m 200-201°(dec).** Crystd from EtOH.

4-Phenylthiosemicarbazide *[5351-69-9]* **M 167.2, m 140°.** Crystd from EtOH.

1-Phenyl-2-thiourea *[103-85-5]* **M 152.1, m 154°.** Crystd from water and dried at 100° in air.

Phenyltoloxamine [2-(2-dimethylaminoethoxy)-diphenylmethane] hydrochloride *[6152-43-8]* **M 291.8, m 119-120°, pK²⁵ 9.3 (free base).** Crystd from isobutyl methyl ketone.

Phenyl 4-toluenesulfonate *[640-60-8]* **M 248.2, m 94.5-95.5°.** Crystd from MeOH or glacial acetic acid.

Phenyl 4-tolylcarbonate *[13183-20-5]* **M 228.2, m 67°.** Purified by preparative GLC with 20% Apiezon on Embacel, and sublimed *in vacuo*.

4-Phenyl-1,2,4-triazole-3,5-diol (4-phenylurazole) *[15988-11-1]* **M 175.2, m 207-209°.** Crystd from water.

4-Phenyl-1,2,4-triazoline-3,5-dione (PTAD) *[4233-33-4]* **M 175.2, m 165-170°(dec), 170-177°(dec).** Carmine red needles obtained by sublimation (ice cold finger) at 100°/0.1mm, and/or by recrystn from EtOH. IR: ν 1760, 1780 cm^{-1}. [Cookson et al. *Org Synth* **51** 121 *1971*; Moore et al. *J Org Chem* **39** 3700 *1974*.]

1-Phenyl-2,2,2-trifluoroethanol *[R-(-)- 10531-50-7; S-(+)- 340-06-7]* **M 176.1, b 74-76°/10mm, 125-127°/760mm, d_4^{20} 1.301, n_D^{20} 1.4632, $[\alpha]_D^{20}$ (-) and (+) 31° (neat).** Purified by fractional distn preferably in a vacuum. [Morrison and Ridgeway *Tetrahedron Lett* 573 *1969*; NMR: Pirkle and Beare *J Am Chem Soc* **90** 6250 *1968*.] The *racemate [340-05-6]* has **b** 52-54°/2mm, 57-59°/2mm, 64-65°/5mm, d_4^{20} 1.293, n_D^{20} 1.457, and the *2-carbobenzoyl* derivative has **m** 137-138° [Mosher et al. *J Am Chem Soc* **78** 4374 *1956*].

Phenylurea *[64-10-8]* **M 136.2, m 148°, pK^{25} -1.45 (aq H$_2$SO$_4$).** Crystd from boiling water (10mL/g). Dried in a steam oven at 100°.

9-Phenyl-9-xanthenol (hydroxypixyl) *[596-38-3]* **M 274.3, m 158-161°, 158.5-159°, 159°.** Dissolve in AcOH and add H$_2$O whereby it separates as colourless prisms. It is slightly soluble in CHCl$_3$, soluble in *C$_6$H$_6$ but insoluble in pet ether. It sublimes on heating. UV in H$_2$SO$_4$: λmax 450nm (ε 5620) and 370nm (ε 24,900) and the *HClO$_4$* salt in CHCl$_3$ has λmax 450 (ε 404) and 375nm (ε 2420). [Sharp *J Chem Soc* 2558 *1958*; Bünzly and Decker *Chem Ber* **37** 2983 *1904*; Chattopadhyaya and Reece *J Chem Soc, Chem Commun* 639 *1978*; Gomberg and Cone *Justus Liebigs Ann Chem* **370** 142 *1909*.]

Phloretin [2',4',6'-trihydroxy-3-(*p*-hydroxyphenyl)propiophenone] *[60-82-2]* **M 274.3, m 264-271°(dec), pK$_{Est(1)}$~7.5, pK$_{Est(2)}$~8.0, pK$_{Est(3)}$~10, pK$_{Est(4)}$~12 (phenolic OH's).** Crystd from aqueous EtOH.

Phloridzin (2H$_2$O) [phloretin 2'-*O*-β-D-glucoside] *[60-81-1]* **M 472.5, m 110°, $[\alpha]_{546}^{20}$ -62° (c 3.2, EtOH).** Crystd as dihydrate from water.

Phloroacetophenone (2H$_2$O) (2',4',6'-trihydroxyacetophenone) *[480-66-0]* **M 186.2, m 218-219°, pK$_{Est(1)}$~7.9, pK$_{Est(2)}$~12.0.** Crystd from hot water (35mL/g).

Phloroglucinol (2H$_2$O) (benzene-1,3,5-triol) *[6099-90-7 (2H$_2$O); 108-73-6 (anhydr)]* **M 126.1, m 217-219°, 117° (anhydrous), pK$_1^{25}$ -7.74 (HClO$_4$), pK$_2^{20}$ 7.97, pK$_3^{20}$ 9.23.** Crystd from water, and stored in the dark under nitrogen.

Phorone (2,6-dimethylhepta-2,5-dien-4-one) *[504-20-1]* **M 138.2, m 28°, b 197°/743mm.** Crystd repeatedly from EtOH.

"Phosphine" [dye CI 793, Chrysaniline mononitrate, 3-amino-9-(4-aminophenyl)-acridinium mononitrate) *[10181-37-0]* **M 348.4, m >250°(dec), pK$_{Est}$ ~8.0.** Crystd from *benzene/EtOH.

Phthalaldehyde *[643-79-8]* **M 134.1, m 54-56°, 55.5-56°, 58°, b 83-84°/0.8mm.** Purified by steam distillation better by using super heated steam (at 175-180°) and efficient cooling. The distillate is saturated with Na$_2$SO$_4$ extracted exhaustively with EtOAc, dried (Na$_2$SO$_4$), filtered and evaporated. The residue is recrystd from pet ether (b 90-100°) [Beill and Tarbell *Org Synth* Coll Vol IV 808 *1963*]. It can be distd under vacuum. The bis-*2,4-dinitrophenylhydrazone* has **m** 278-280° [Hatt and Stephenson *J Chem Soc* 199 *1952*].

Phthalazine *[253-52-1]* **M 130.2, m 90-91°, pK^{20} 3.47.** Crystd from diethyl ether or *benzene, and sublimed under vacuum.

Phthalazine-1,4-dione **(phthalhydrazide)** *[1445-69-8]* **M 162.2, m 330-333°, 336°, 346°, pK$_1^{20}$ -3.29 pK$_2^{20}$ -0.99, pK$_3^{20}$ 5.67, pK$_4^{20}$ 13.0.** Twice recrystd from 0.1M KOH [Merenyi et al. *J Am Chem Soc* 108 7716 *1986*], EtOH or dimethylformamide and sublimes >300°.

Phthalazone **(1-hydroxyphthalazine)** *[119-39-1]* **M 146.2, m 183-184°, 186-188°, b 337°/760mm, pK$_1^{20}$ -2.2, pK$_2^{20}$ -1.4, pK$_3^{20}$ 11.99.** Crystd from H_2O or EtOH and sublimed *in vacuo*.

o-**Phthalic acid** *[88-99-3]* **M 166.1, m 211-211.5°, pK$_1^{25}$ 2.76 (3.05), pK$_2^{25}$ 4.92 (4.73).** Crystd from water.

Phthalic anhydride *[85-44-9]* **M 148.1, m 132°, b 295°.** Distd under reduced pressure. Purified from the acid by extracting with hot $CHCl_3$, filtering and evaporating. The residue was crystd from $CHCl_3$, CCl_4 or *benzene, or sublimed. Fractionally crystd from its melt. Dried under vacuum at 100°. [Saltiel *J Am Chem Soc* 108 2674 *1986*.]

Phthalide *[87-41-2]* **M 134.1, m 72-73°, pK -7.98 (aq H_2SO_4).** Crystd from water (75mL/g) and dried in air on filter paper.

Phthalimide *[85-41-6]* **M 147.1, m 235°, 238°, pK 8.30.** Crystd from EtOH (20mL/g) (charcoal), or by sublimation. For potassium phthalimide see entry in Chapter 5.

Phthalimidoglycine *[4702-13-0]* **M 205.2, m 192-193°, pK$_{Est}$ ~3.** Crystd from water or EtOH.

Phthalonitrile *[91-15-6]* **M 128.1, m 141°.** Crystd from EtOH, toluene or *benzene. Can also be distd under high vacuum.

Phthalylsulfacetamide *[131-69-1]* **M 362.3, m 196°.** Crystd from water.

Phthiocol **(2-hydroxy-3-methylnaphthaquinone)** *[483-55-6]* **M 188.1, m 173-174°, pK$_{Est}$ ~4.2.** Crystd from diethyl ether/pet ether.

Physalien (*all trans* β-carotene-3,3'-diol dipalmitate) *[144-67-2]* **M 1044, m 98.5-99.5°, $A_{1m}^{1\%}$ (λmax) 1410 (449nm), 1255 (478nm) in hexane.** Purified by chromatography on water-deactivated alumina, using hexane/diethyl ether (19:1) to develop the column. Crystd from *benzene/EtOH. Stored in the dark, in inert atmosphere, at 0°.

Physodic acid **[4,4',6'-trihydroxy-6-(2-oxoheptyl)-2'-pentyl-2,3'-oxydibenzoic acid 1,5-lactone]** *[84-24-2]* **M 470.5, m 205°, pK$_{Est(1)}$~3.0, pK$_{Est(2)}$~10, pK$_{Est(3)}$~13.** Crystd from MeOH. The *diacetate* has **m** 155-156° (from Me_2CO/CS_2).

Phytoene **(7,7',8,8',11,11',12,12'-octahydro-ψ,ψ-carotene)** *[540-04-5]* **M 544.9, $A_{1m}^{1\%}$ (λmax) 850 (287nm) in hexane, λ$_{max}$ 275, 287 and 297nm nm.** Purified by chromatography on columns of magnesium oxide-Supercel (a diatomaceous filter aid) or alumina [Rabourn et al. *Arch Biochem Biophys* 48 267 *1954*]. Stored as a solution in pet ether under nitrogen at -20°.

Phytofluene *[540-05-6]* **M 549.0, b 140-185°(bath temp)/0.0001m $A_{1m}^{1\%}$ (λmax) 1350 (348nm) in pet ether, λ$_{max}$ 331, 348, 267.** Purified by chromatography on partially deactivated alumina [Kushwaha et al. *J Biol Chem* 245 4708 *1970*]. Stored as a soln in pet ether under nitrogen at -20°.

Picein (p-acetylphenyl-β-D-glucopyranoside) [530-14-3] M 298.3, m 195-196°, $[\alpha]_D^{20}$ -88° (c 1, H₂O). Crystd from MeOH or (as monohydrate) from water.

Picene [213-14-3] M 278.3, m 364°. Crystd from isopropylbenzene/xylene. Can also be sublimed.

2-Picoline-N-oxide (**2-methylpyridine-1-oxide**) [931-19-1] M 109.1, m 41-45°, b 89-90°/0.8-0.9mm, 90-100°/1mm, 110°/4mm, 135°/5mm, 123°/9mm, 123-124°/15mm, 259-261°/atm, n_D^{25} 1.5854 (supercooled), pK²⁵ 1.10. Purified by fractional distillation and could be recrystd from *C₆H₆-hexane but is *hygroscopic*. [Bullitt and Maynard *J Am Chem Soc* **76** 1370 *1954*; Ross et al. *J Am Chem Soc* **78** 3625 *1956*; IR: Wiley and SlaymAker *J Am Chem Soc* **79** 2233 *1957*.] The *picrate* has m 125-126.5° (from EtOH) [Boekelheide and Linn *J Am Chem Soc* **76** 1286 *1954*]. The *phthalate* has m 115-116° (from EtOH) [den Hertog et al. *Recl Trav Chim Pays-Bas* **70** 591 *1951*.]

3-Picoline-N-oxide (**3-methylpyridine-1-oxide**) [1003-73-2] M 109.1, m 37-39°, 37-38° (evac capillary) , 84-85°/0.3mm, 101-103°/0.7-0.8mm, 114-115°/1.5mm, 118°/2mm, pK²⁵ 1.08. Purified by careful fractionation *in vacuo*. The distillate remains supercooled for several days before solidifying. It is a slightly *hygroscopic* solid which could melt in the hand. The *picrate* has m 149-151° (from EtOH). [Taylor and Corvetti *Org Synth* Coll Vol IV 654 *1963*; IR: Katritzky et al. *J Chem Soc* 3680 *1959*; Jaffé and Doak *J Am Chem Soc* **77** 4441, 4481 *1955*; Boekelheide and Linn *J Am Chem Soc* **76** 1286 *1954*].

4-Picoline-N-oxide (**4-methylpyridine-1-oxide**) [1003-67-4] M 109.1, m 182-184°, 185-186°, 186-188°, pK²⁵ 1.29. Recryst from EtOH-EtOAc, Me₂CO-Et₂O or *C₆H₆. [Bullitt and Maynard *J Am Chem Soc* **76** 1370 *1954*; Boekelheide and Linn *J Am Chem Soc* **76** 1286 *1954*].

Picolinic acid (pyridine-2-carboxylic acid) [98-98-6] M 123.1, m 138°, pK_1^{25} 1.03 (1.36), pK_2^{25} 5.30 (5.80). Crystd from water or *benzene.

α-Picolinium chloride [14401-91-3] M 129.6, m 200°. 1:1 Mixture of α-picoline and HCl, distd at 275°. Then vacuum sublimed at 91-91.5°.

N-Picolinoylbenzimidazole [100312-29-6] M 173.3, m 105-107°. Recrystd three times from hexane [Fife and Przystas *J Am Chem Soc* **108** 4631 *1986*].

Picric acid [88-89-1] M 229.1, m 122-123°, pK²⁵ 0.33 (0.37). Crystd first from acetic acid then acetone, toluene, CHCl₃, aqueous 30% EtOH, 95% EtOH, MeOH or H₂O. Dried in a vacuum oven at 80° for 2h. Alternatively, dried over Mg(ClO₄)₂ or fused and allowed to freeze under vacuum three times. Because it is **EXPLOSIVE**, picric acid should be stored moistened with H₂O, and only small portions should be dried at any one time. The dried acid should **NOT** be heated.

Picrolic acid [3-methyl-4-nitro-1-(4-nitrophenyl)-2-pyrazolin-5-one, picrolonic acid] [550-74-3] M 264.2, m 120°(dec),116.5° (dec at 125°) 125°. Crystd from water or EtOH (Solubility is 0.123% at 15° and 1.203% at 100° in H₂O; and 1.107% at 0° and 11.68% at 81° in EtOH). It forms Ca, Cu Hg, Mg, Na, Sr, and Pb complexes [Maquestian et al *Bull Soc Chim Belg* **82** 233 *1973*; Isaki et al. *Chem Ber* **74** 1420 *1941*].

Picrotoxin [124-87-8] M 602.6, m 203°, $[\alpha]_{546}^{20}$ -40° (c 1, EtOH). Crystd from water.

Picryl chloride [88-88-0] M 226.3, m 83°. Crystd from CHCl₃ or EtOH.

Picryl iodide [4436-27-5] M 340.0, m 164-165°. Crystd from *benzene.

Pimelic acid (heptane-1,7-dioic acid) [111-16-0] M 160.2, m 105-106°, pK_1^{25} 4.46, pK_2^{25} 5.58. Crystd from water or from *benzene containing 5% diethyl ether.

Pinacol (hexahydrate) *[6091-58-3 (6H₂O); 76-09-5 (anhydr)]* **M 194.3, m 46.5°, b 59°/4mm.** Distd then crystd repeatedly from water.

Pinacol (anhydrous) *[76-09-5]* **M 118.1, m 41.1°, b 172°.** The hydrate is rendered anhydrous by azeotropic distn of water with *benzene. Recrystd from *benzene or toluene/pet ether, absolute EtOH or dry diethyl ether. Recrystn from water gives the hexahydrate.

Pinacolone oxime *[2475-93-6]* **M 115.2, m 78°.** Crystd from aqueous EtOH.

Pinacyanol chloride *[2768-90-3]* **M 388.9, m 270°(dec).** Crystd from EtOH/diethyl ether.

R-α-Pinene *[7785-70-8]* **M 136.2, b 61°/30mm, 156.2°/760mm, d 0.858, n¹⁵ 1.4634, n 1.4658, [α]$_D^{25}$ +47.3°.** Isomerised by heat, acids and certain solvents. Should be distd under reduced pressure under nitrogen and stored in the dark. Purified *via* the nitrosochloride [Waterman et al. *Recl Trav Chim Pays-Bas* **48** 1191 *1929*]. For purification of optically active forms see Lynn [*J Am Chem Soc* **91** 361 *1919*]. Small quantities (0.5mL) have been purified by GLC using helium as carrier gas and a column at 90° packed with 20 wt% of polypropylene sebacate on a Chromosorb support. Larger quantities were fractionally distd under reduced pressure in a column packed with stainless steel gauze spirals. Material could be dried with CaH_2 or sodium, and stored in a refrigerator: $CaSO_4$ and silica gel were not satisfactory because they induced spontaneous isomerisation. [Bates, Best and Williams *J Chem Soc* 1521 *1962*.]

S-α-Pinene *[7785-26-4]* **M 136.2, b 155-156°/760mm, d 0.858, n 1.4634, [α]$_D^{20}$ -47.2°.** Purification as for *R*-α-Pinene above.

***dl*-Pipecolinic acid (piperidine-2-carboxylic acid)** *[4043-87-2]* **M 129.1, m 264°, pK$_1^{25}$ 2.29, pK$_2^{25}$ 10.77.** Crystd from water.

Piperazine *[110-85-0]* **M 86.1, m 110-112°, 44° (hexahydrate *142-63-2*) b 125-130°/760mm, pK$_1^{25}$ 5.33, pK$_2^{25}$ 9.73.** Crystd from EtOH or anhydrous *benzene, and dried at 0.01mm. It can be sublimed under vacuum and purified by zone melting.
§ Piperazine on polystyrene support is commercially available.

Piperazine-*N,N'*-bis(2-ethanesulfonic acid) (PIPES) *[5625-37-6]* **M 302.4, pK$_1^{25}$ <3, pK$_2^{25}$ 6.82 (7.82).** Crystd from boiling water (maximum solubility is about 1g/L) or as described for ADA [*N*-(2-acetamido)iminodiacetic acid, see above].

Piperazine dihydrochloride (H₂O) *[142-64-3 (2HCl); 6094-40-2 (xHCl)]* **M 177.1, m 82.5-83.5°.** Crystd from aqueous EtOH. Dried at 110°.

Piperazine phosphate (H₂O) *[18534-18-4]* **M 197.6.** Crystd twice from water, air-dried and stored for several days over Drierite. The salt dehydrates slowly if heated at 70°.

Piperic acid [*trans,trans*-5-(3,4-methylenedioxyphenyl)-2,4-pentadieneoic acid] *[136-72-1]* **M 218.2, m 217°, pK$_{Est}$ ~4.7.** Crystd from EtOH. Turns yellow in light. Sublimes with partial dec.

Piperidine *[110-89-4]* **M 85.2, f -9°, b 35.4°/40mm, 106°/760mm, d 0.862, n 1.4535, n²⁵ 1.4500, pK²⁵ 11.20.** Dried with BaO, KOH, CaH_2, or sodium, and fractionally distd (optionally from sodium, CaH_2, or P_2O_5). Purified from pyridine by zone melting.
§ Piperidine on polystyrene support is commercially available.

Piperidinium chloride *[6091-44-7]* **M 121.6, m 244-245°.** Crystd from EtOH/diethyl ether in the presence of a small amount of HCl.

Piperidinium nitrate *[6091-45-8]* **M 145.2, m 110°.** Crystd from acetone/ethyl acetate.

Piperine (1-piperoylpiperidine) *[94-62-2]* **M 285.4, m 129-129.5°, pK^{15} 1.98.** Crystd from EtOH or *benzene/ligroin.

Piperonal *[120-57-0]* **M 150.1, m 37°, b 140°/15mm, 263°/760mm.** Crystd from aqueous 70% EtOH or EtOH/water.

Piperonylic acid *[94-53-1]* **M 166.1, m 229°, pK^{25} 4.50.** Crystd from EtOH or water.

Pivalic acid (trimethylacetic acid) *[75-98-9]* **M 102.1, m 35.4°, b 71-73°/0.1mm, pK^{25} 5.03.** Fractionally distd under reduced pressure, then fractionally crystd from its melt. Recrystd from *benzene.

Pivaloyl chloride (trimethylacetyl chloride) *[3282-30-2]* **M 120.6, b 57.6°/150mm, 70.5-71/250mm, 104°/754mm, 104-105°/atm, 105-108°/atm, d$_4^{20}$ 1.003, n$_D^{20}$ 1.4142.** First check the IR to see if OH bands are present. If absent, or present in small amounts, then redistil under moderate vac. If present in large amounts then treat with oxalyl chloride or thionyl chloride and reflux for 2-3h, evap and distil residue. **Strongly LACHRYMATORY - work in a fumecupboard.** Store in sealed ampoules under N$_2$. [Traynham and Battiste *J Org Chem* **22** 1551 *1957*; Grignard reactns: Whitmore et al. *J Am Chem Soc* **63** 647 *1941*.]

Plumbagin (5-hydroxy-2-methyl-1,4-naphthaquinone) *[481-42-5]* **M 188.1, m 78-79°, pK$_{Est(1)}$~9.5, pK$_{Est(2)}$~11.0.** Crystd from aqueous EtOH and sublimed in a vac. Steam distils.

Polyacrylonitrile *[25014-41-9]*. Ppted from dimethylformamide by addition of MeOH.

Poly(diallyldimethylammonium) chloride *[26062-79-3]*. Ppted from water in acetone, and dried in vacuum for 24h. [Hardy and Shriner *J Am Chem Soc* **107** 3822 *1985*.]

Polyethylene *[9002-88-4]*. Crystd from thiophen-free *benzene and dried over P$_2$O$_5$ under vacuum.

Polymethyl acrylate *[9003-21-8]*. Ppted from a 2% soln in acetone by addition of water.

Polystyrene *[9003-53-6]*. Ppted repeatedly from CHCl$_3$ or toluene soln by addition of MeOH. Dried *in vacuo* [Miyasaka et al. *J Phys Chem* **92** 249 *1988*].

Polyvinyl acetate *[9003-20-7]*. Ppted from acetone by addition of *n*-hexane.

Poly(N-vinylcarbazole) *[25067-59-8]*. Ppted seven times from tetrahydrofuran with MeOH, with a final freeze-drying from *benzene. Dried under vacuum.

Polyvinyl chloride *[9002-81-2]*. Ppted from cyclohexanone by addition of MeOH.

Poly(4-vinylpyridine) *[25232-41-1]* **M (105.1)$_n$.** Purified by repeated pptn from solns in EtOH and dioxane, and then EtOH and ethyl acetate. Finally, freeze-dried from *tert*-butanol.

Poly(N-vinylpyrrolidone) *[9003-39-8]* **M (111.1)$_n$, crosslinked** *[25249-54-1]* **m >300°.** Purified by dialysis, and freeze-dried. Also by pptn from CHCl$_3$ soln by pouring into ether. Dried in a vacuum over P$_2$O$_5$. For the crosslinked polymer purification is by boiling for 10min in 10% HCl and then washing with glass-distilled water until free from Cl ions. Final Cl ions were removed more readily by neutralising with KOH and continued washing.

Prednisone *[53-03-2]* **M 358.5, m 238°(dec), [α]$_D^{20}$ +168° (c 1, dioxane), λ$_{max}$ 238nm (log ε 4.18) in MeOH.** Crystd from acetone/hexane.

Pregnane *[481-26-5]* **M 300.5, m 83.5°, [α]$_D^{20}$ +21° (CHCl$_3$).** Crystd from MeOH.

5β-Pregnane-3α,20α-diol *[80-92-2]* **M 320.5, m 243-244°, [α]$_{546}^{20}$ +31° (c 1, EtOH).** Crystd from acetone.

5β-Pregnane-3α,20β-diol *[80-91-1]* **M 320.5, m 244-246°, [α]$_{546}^{20}$ +22° (c 1, EtOH).** Crystd from EtOH.

Procaine   [4-(2-diethylaminomethoxycarbonyl)aniline]   *[59-46-1]* **M 236.3, m 51° (dihydrate), 61° (anhydrous), pK$_1^{15}$ 2.45, pK$_2^{15}$ 8.91.** Crystd as the dihydrate from aqueous EtOH and as anhydrous material from pet ether or diethyl ether. The latter is *hygroscopic*.

Proclavine   (3,6-diaminoacridine) *[92-62-6]* **M 209.2, m 284-286°, pK^{25} 9.60.** Crystd from aqueous MeOH. For proflavin see 3,6-diaminoacridine hydrochloride

Progesterone   *[57-83-0]* **M 314.5, m 128.5°, [α]$_{546}^{20}$ +220° (c 2, dioxane).** Crystd from EtOH. When crystd from pet ether **m** is 121°, λ$_{max}$ 240nm, log ε 4.25 (EtOH).

L-Proline   *[147-85-3]* **M 115.1, m 215-220°(dec)(D-isomer),   220-222°(dec) (L-form), 205°(dec)(DL-isomer), [α]$_D$25 (H$_2$O, L-isomer), pK$_1^{25}$ 1.95, pK$_2^{25}$ 10.64.** Likely impurity are hydroxyproline. Purified *via* its picrate which was crystd twice from water, then decomposed with 40% H$_2$SO$_4$. The picric acid was extracted with diethyl ether, the H$_2$SO$_4$ was pptd with Ba(OH)$_2$, and the filtrate evapd. The residue was crystd from hot absolute EtOH [Mellan and Hoover *J Am Chem Soc* **73** 3879 *1951*] or EtOH/ether. *Hygroscopic.* Stored in a desiccator.

Prolycopene   (all Z-ψ,ψ−carotene)   *[2361-24-2]* **M 536.5, m 111°, λ$_{max}$ 443.5, 470nm in pet ether.** Purified by chromatography on deactivated alumina [Kushwaha et al. *J Biol Chem* **245** 4708 *1970*]. Crystd from pet ether. Stored in the dark, in an inert atmosphere at -20°.

L-Prolylglycine   *[2578-57-6]*   **M 172.2, m 236°, [α]$_D^{20}$ +21.1° (c 4, H$_2$O), pK$_1^{25}$ 3.19, pK$_2^{25}$ 8.97.** Crystd from water at 50-60° by addition of EtOH.

Proneurosporene (3,4,7',8'-tetrahydrolycopene)   *[10467-46-6]* **M 538.9, λ$_{max}$ 408, 432, 461 nm, ε$_{1cm}^{1\%}$ 2040 (432nm) in hexane.** Purified by chromatography on deactivated alumina [Kushwaha et al. *J Biol Chem* **245** 4708 *1970*]. Stored in the dark, in an inert atmosphere at 0°.

Propane   *[74-98-6]*   **M 44.1, m -189.7, b -42.1°/760mm, d 0.5005, n 1.2898.** Purified by bromination of the olefinic contaminants. Propane was treated with bromine for 30min at 0°. Unreacted bromine was quenched, and the propane was distd through two -78° traps and collected at -196° [Skell et al. *J Am Chem Soc* **108** 6300 *1986*].

Propane-1,2-diamine   (propylenediamine)   *[78-90-0]* **M 74.1, b 120.5°, d 0.868, n 1.446, pK$_1^{25}$ 6.61, pK$_2^{25}$ 9.82.** Purified by azeotropic distn with toluene. [Horton, Thomason and Kelly *Anal Chem* **27** 269 *1955*.]

Propane-1,2-diol   (propyleneglycol)   *[57-55-6]*   **M 76.1, b 104°/32mm, d 1.040, n 1.433.** Dried with Na$_2$SO$_4$, decanted and distd under reduced pressure.

Propane-1,3-diol   *[504-63-2]*   **M 76.1, b 110-122°/12mm, d 1.053, n$^{18.5}$ 1.4398.** Dried with K$_2$CO$_3$ and distd under reduced pressure. More extensive purification involved conversion with benzaldehyde to *2-phenyl-1,3-dioxane* (m 47-48°) which was subsequently decomposed by shaking with 0.5M HCl (3mL/g) for 15min and standing overnight at room temperature. After neutralisation with K$_2$CO$_3$, the benzaldehyde was removed by distn and the diol was recovered from the remaining aqueous soln by continuous extraction with CHCl$_3$ for 1day. The extract was dried with K$_2$CO$_3$, the CHCl$_3$ was evaporated and the diol was distd. [Foster, Haines and Stacey *Tetrahedron* **6** 177 *1961*.]

Propane-1-thiol *[107-03-9]* **M 76.1, b 65.3°/702mm, d^{25} 0.83598, n^{25} 1.43511, pK^{20} 10.82.** Purified by soln in aqueous 20% NaOH, extraction with a small amount of *benzene and steam distn until clear. After cooling, the soln was acidified slightly with 15% H$_2$SO$_4$, and the thiol was distd out, dried with anhydrous CaSO$_4$ or CaCl$_2$, and fractionally distd under nitrogen. [Mathias and Filho *J Phys Chem* **62** 1427 *1958*.] Also purified by liberation of the mercaptan by adding dilute HCl to the residue remaining after steam distn. After direct distn from the flask, and separation of the water, the mercaptan was dried (Na$_2$SO$_4$) and distd under nitrogen.

Propane-2-thiol (Isopropyl mercaptan) *[75-33-2]* **M 76.1, b 49.8°/696mm, d^{25} 0.80895, n^{25} 1.42154, pK^{25} 10.86.** Purification as for propane-1-thiol above.

Propargyl alcohol (2-propyn-1-ol) *[107-19-7]* **M 56.1, b 54°/57mm, 113.6°/760mm, d 0.947, n 1.432.** Commercial material contains a stabiliser. An aqueous soln of propargyl alcohol can be concentrated by azeotropic distn with butanol or butyl acetate. Dried with K$_2$CO$_3$ and distd under reduced pressure, in the presence of about 1% succinic acid, through a glass helices-packed column.

Propargyl chloride (3-chloropropyne) *[624-65-7]* **M 74.5, b 58°/760mm, 65°/760mm, d 1.03, n 1.435.** Purified by fractional distn at atm press. Note that a possible impurity propargyl alcohol has **b** 114-115°/atm. [Henry *Chem Ber* **8** 398 *1875*.] **HIGHLY TOXIC** and **FLAMMABLE**.

Propene *[115-07-1]* **M 42.1, m -185.2°, b -47.8°/750mm, d 0.519, n^{-71} 1.357.** Purified by freeze-pump-thaw cycles and trap-to-trap distn.

p-**(1-Propenyl)phenol** *[cis/trans 6380-21-8; 539-12-8]* **M 134.2, m 93-94°, pK$_{Est}$ ~10.2.** Crystd from water.

β-**Propiolactone** *[57-57-8]* **M 72.1, b 83°/45mm, d 1.150, n^{25} 1.4117.** Fractionally distd under reduced pressure, from sodium. **CARCINOGEN.**

Propionaldehyde *[123-38-6]* **M 58.1, b 48.5-48.7°, d 0.804, n 1.3733, n^{25} 1.37115.** Dried with CaSO$_4$ or CaCl$_2$, and fractionally distd under nitrogen or in the presence of a trace of hydroquinone (to retard oxidation). Blacet and Pitts [*J Am Chem Soc* **74** 3382 *1952*] repeatedly vacuum distd the middle fraction until no longer gave a solid polymer when cooled to -80°. It was stored with CaSO$_4$.

Propionamide *[79-05-0]* **M 73.1, m 79.8-80.8°, pK^{24} -0.9 (H$_0$ scale, aq H$_2$SO$_4$).** Crystd from acetone, *benzene, CHCl$_3$, water or acetone/water, then dried in a vacuum desiccator over P$_2$O$_5$ or conc H$_2$SO$_4$.

Propionic acid *[79-09-4]* **M 74.1, b 141°, d 0.992, n 1.3865, n^{25} 1.3843, pK$_1^{25}$-6.8 (H$_0$ scale, aq H$_2$SO$_4$) , pK$_2^{25}$4.88.** Dried with Na$_2$SO$_4$ or by fractional distn, then redistd after refluxing with a few crystals of KMnO$_4$. An alternative purification uses the conversion to the ethyl ester, fractional distn and hydrolysis. [Bradbury *J Am Chem Soc* **74** 2709 *1952*.] Propionic acid can also be heated for 0.5h with an amount of benzoic anhydride equivalent to the amount of water present (in the presence of CrO$_3$ as catalyst), followed by fractional distn. [Cham and Israel *J Chem Soc* 196 *1960*.]

Propionic anhydride *[123-62-6]* **M 130.2, b 67°/18mm, 168°/780mm, d 1.407, n 1.012.** Shaken with P$_2$O$_5$ for several minutes, then distd.

Propionitrile *[107-12-0]* **M 55.1, b 97.2°, d 1.407, n^{15} 1.36812, n^{30} 1.36132.** Shaken with dil HCl (20%), or with conc HCl until the odour of isonitrile has gone, then washed with water, and aqueous K$_2$CO$_3$. After a preliminary drying with silica gel or Linde type 4A molecular sieves, it is stirred with CaH$_2$ until hydrogen evolution ceases, then decanted and distd from P$_2$O$_5$ (not more than 5g/L, to minimise gel formation). Finally, it is refluxed with, and slowly distd from CaH$_2$ (5g/L), taking precautions to exclude moisture.

n-**Propyl acetate**   *[109-60-4]*   **M 102.1, b 101.5°, d 0.887, n 1.38442, pK^{25} -7.18 (H$_0$ scale, aq H$_2$SO$_4$).** Washed with satd aqueous NaHCO$_3$ until neutral, then with satd aqueous NaCl. Dried with MgSO$_4$ and fractionally distd.

n-**Propyl alcohol (1-propanol)**   *[71-23-8]*   **M 60.1, b 97.2°, d^{25} 0.79995, n 1.385, pK^{25} 16.1.** The main impurities in *n*-propyl alcohol are usually water and 2-propen-1-ol, reflecting the commercial production by hydration of propene. Water can be removed by azeotropic distn either directly (azeotrope contains 28% water) or by using a ternary system, e.g. by adding *benzene. Alternatively, for gross amounts of water, refluxing over CaO for several hours is suitable, followed by distn and a further drying. To obtain more nearly anhydrous alcohol, suitable drying agents are firstly NaOH, CaSO$_4$ or K$_2$CO$_3$, then CaH$_2$, aluminium amalgam, magnesium activated with iodine, or a small amount of sodium. Alternatively, the alcohol can be refluxed with *n*-propylsuccinate or phthalate in a method similar to the one described under EtOH. Allyl alcohol is removed by adding bromine (15mL/L) and then fractionally distilling from a small amount of K$_2$CO$_3$. Propionaldehyde, also formed in the bromination, is removed as the 2,4-dinitrophenylhydrazone. *n*-Propyl alcohol can be dried down to 20ppm of water by passage through a column of pre-dried molecular sieves (type 3 or 4A, heated for 3h at 300°) in a current of nitrogen. Distn from sulfanilic or tartaric acids removes impurities. Albrecht [*J Am Chem Soc* **82** 3813 *1960*] obtained spectroscopically pure material by heating with charcoal to 50-60°, filtering and adding 2,4-dinitrophenylhydrazine and a few drops of conc H$_2$SO$_4$. After standing for several hours, the mixture was cooled to 0°, filtered and vac distd. Gold and Satchell [*J Chem Soc* 1938 *1963*] heated *n*-propyl alcohol with 3-nitrophthalic anhydride at 76-110° for 15h, then recrystd the resulting ester from H$_2$O, *benzene/pet ether (b 100-120°)(3:1), and *benzene. The ester was hydrolysed under reflux with aq 7.5M NaOH for 45min under nitrogen, followed by distn (also under nitrogen). The fraction (b 87-92°) was dried with K$_2$CO$_3$ and stirred under reduced pressure in the dark over 2,4-dinitrophenylhydrazine, then freshly distilled. Also purified by adding 2g NaBH$_4$ to 1.5L alcohol, gently bubbling with argon and refluxing for 1day at 50°. Then added 2g of freshly cut sodium (washed with propanol) and refluxed for one day. Distd, taking the middle fraction [Jou and Freeman *J Phys Chem* **81** 909 *1977*].

n-**Propylamine**   *[107-10-8]*   **M 59.1, b 48.5°, d 0.716, n 1.38815, pK^{25} 10.69.** Distd from zinc dust, at reduced pressure, in an atmosphere of nitrogen.

n-**Propyl bromide.**   *[106-94-5]*   **M 123.0, b 71.0°, d 1.354., n^{15} 1.43695, n^{25} 1.43123.** Likely contaminants include *n*-propyl alcohol and isopropyl bromide. The simplest purification procedure uses drying with MgSO$_4$ or CaCl$_2$ (with or without a preliminary washed of the bromide with aq NaHCO$_3$, then water), followed by fractional distn away from bright light. Chien and Willard [*J Am Chem Soc* **79** 4872 *1957*] bubbled a stream of oxygen containing 5% ozone through *n*-propyl bromide for 1h, then shook with 3% hydrogen peroxide soln, neutralised with aq Na$_2$CO$_3$, washed with distilled water and dried. Then followed vigorous stirring with 95% H$_2$SO$_4$ until fresh acid did not discolour within 12h. The propyl bromide was separated, neutralised, washed dried with MgSO$_4$ and fractionally distd. The centre cut was stored in the dark. Instead of ozone, Schuler and McCauley [*J Am Chem Soc* **79** 821 *1957*] added bromine and stored for 4 weeks, the bromine then being extracted with aq NaHSO$_3$ before the sulfuric acid treatment was applied. Distd. Further purified by preparative gas chromatography on a column packed with 30% SE-30 (General Electric ethylsilicone rubber) on 42/60 Chromosorb P at 150° and 40psi, using helium. [Chu *J Phys Chem* **41** 226 *1964*.]

n-**Propyl chloride**   *[540-54-5]*   **M 78.5, b 46.6°, d 0.890, n 1.3880.** Dried with MgSO$_4$ and fractionally distd. More extensively purified using extraction with H$_2$SO$_4$ as for *n*-propyl bromide. Alternatively, Chien and Willard [*J Am Chem Soc* **75** 6160 *1953*] passed a stream of oxygen containing about 5% ozone through the *n*-propyl chloride for three times as long as was needed to cause the first coloration of starch iodide paper by the exit gas. After washing with aqueous NaHCO$_3$ to hydrolyse ozonides and remove organic acids, the chloride was dried with MgSO$_4$ and fractionally distd.

1-Propyl-3-(*p*-chlorobenzenesulfonyl) urea   *[94-20-2]*   **M 260.7, m 127-129°.** Crystd from aqueous EtOH.

Propylene carbonate   *[108-32-7]*   **M 102.1, b 110°/0.5-1mm, 238-239°/760mm, d 1.204, n 1.423.** Manufactured by reaction of 1,2-propylene oxide with CO$_2$ in the presence of a catalyst (quaternary ammonium halide). Contaminants include propylene oxide, carbon dioxide, 1,2- and 1,3-propanediols, allyl alcohol and ethylene carbonate. It can be purified by percolation through molecular sieves (Linde 5A, dried at 350° for 14h under a stream of argon), followed by distn under vac. [Jasinski and Kirkland *Anal Chem* **39** 163 *1967*.] It can be stored over molecular sieves under an inert gas atmosphere. When purified in this way it contains less than 2ppm water. Activated alumina and dried CaO have been also used as drying agents prior to fractional distn under reduced pressure. It has been dried with 3A molecular sieves and distd under nitrogen in the presence of *p*-toluenesulfonic acid. Then redistilled and the middle fraction collected.

dl-**Propylene oxide** *[75-56-9]* **M 58.1, b 34.5°, d 0.829, n 1.3664.** Dried with Na_2SO_4 or CaH_2, and fractionally distilled through a packed column (glass helices), after refluxing with Na, CaH_2, or KOH pellets.

n-**Propyl ether (dipropyl ether)** *[111-43-3]* **M 102.2, b 90.1°, d 0.740, n^{15} 1.38296, n 1.3803, pK -4.40 (aq H_2SO_4).** Purified by drying with $CaSO_4$, by passage through an alumina column (to remove peroxides), and by fractional distn.

Propyl formate *[110-74-7]* **M 88.1, b 81.3°, d 0.9058, n 1.3779.** Distd, then washed with satd aq NaCl, and with satd aq $NaHCO_3$ in the presence of solid NaCl, dried with $MgSO_4$ and fractionally distd.

n-**Propyl gallate** *[121-79-9]* **M 212.2, m 150°.** Crystd from aqueous EtOH.

n-**Propyl iodide (1-iodopropane)** *[107-08-4]* **M 170.0, b 102.5°, d 1.745, n 1.5041.** Should be distd at reduced pressure to avoid decomposition. Dried with $MgSO_4$ or silica gel and fractionally distd. Stored under nitrogen with mercury in a brown bottle. Prior to distn, free iodine can be removed by shaking with copper powder or by washing with aq $Na_2S_2O_3$ and drying. Alternatively, the *n*-propyl iodide can be treated with bromine, then washed with aq $Na_2S_2O_3$ and dried. See also *n-butyl iodide*.

n-**Propyl propionate** *[106-36-5]* **M 120.2, b 122°, d 0.881, n 1.393.** Treated with anhydrous $CuSO_4$, then distd under nitrogen.

6-Propyl-2-thiouracil (propacil, propyail) *[51-52-5]* **M 170.2, m 218-220°, 218-220°, pK_1^{21} -6.54 (aq H_2SO_4), pK_2^{21} -4.22 (aq H_2SO_4), pK_3^{21} 8.25 (4% aq EtOH).** Purified by recrystn from H_2O (sol in 900 parts at 20°, and 100 parts at 100°). UV, MeOH: λmax 277nm. [Anderson et al. *J Am Chem Soc* **67** 2197 *1945*; Vanderhaegue *Bull Soc Chim Belg* **59** 689 *1950*.]

Propyne *[74-99-7]* **M 40.1, m -101.5°, b -23.2°/760mm, d^{-50} 0.7062, n^{-40} 1.3863.** Purified by preparative gas chromatography.

Protocatechualdehyde *[139-85-5]* **M 138.1, m 153°.** Crystd from water or toluene and dried in a vacuum desiccator over KOH pellets or shredded wax respectively.

Protopine [fumarine, macleyine, 4,6,7,14-tetrahydro-5-methyl-bis[1,3]-benzodioxolo[4,5-*c*:5',6'-*g*]azecine-13(5*H*)-one] *[130-86-9]* **M 353.4, m 208°, pK 5.99.** Crystd from EtOH/CHCl₃.

1S,2S-**Pseudoephedrine (1-hydroxy-1-phenyl-2-methylaminopropane)** *[90-82-4]* **M 165.2, m 118-119°, $[\alpha]_D^{20}$ +53.0° (EtOH), +40.0° (H_2O), pK^{25} 9.71.** Crystd from dry diethyl ether, or from water and dried in a vacuum desiccator.

1S,2S-**Pseudoephedrine hydrochloride** *[345-78-8]* **M 210.7, m 181-182°, 185-188°, $[\alpha]_D^{20}$ +61° (c 1 H_2O).** Crystd from EtOH.

Pteridine *[91-18-9]* **M 132.2, m 139.5-140°, pK_1^{20} 4.05 (equilibrium, hydrate), pK_2^{20} 11,90 (OH of hydrate).** Crystd from EtOH, *benzene, *n*-hexane, *n*-heptane or pet ether. It sublimes at 120-130°/20mm. Stored at 0°, in the dark; turns green in the presence of light and on long standing in the dark.

2,4-(1H,3H)-Pteridinedione H$_2$O (lumazine) *[487-21-8]* **M 182.1, m >350°, pK$_1^{20}$ <1.0, pK$_2^{20}$ 7.94.** Crystd from water.

Pterin (2-aminopteridin-4(3H)-one) *[2236-60-4]* **M 163.1, m >300°, pK$_1^{20}$ 2.27 (basic), pK$_2^{20}$ 7.96 (acidic).** It was dissolved in hot 1% aqueous ammonia, filtered, and an equal volume of hot 1M aqueous formic acid was added. The soln was allowed to cool at 0-2° overnight. The solid was collected and washed with distilled water several times by centrifugation and dried *in vacuo* over P$_2$O$_5$ overnight, and then at 100° overnight.

Pterocarpin {(6aR-cis)-6a,12a-dihydro-3-methoxy-6H-[1,3]dioxolo[5,6]benzofuro[3,2c][1]-benzopyran} *[524-97-0]* **M 298.3, m 165°, [α]$_{546}^{20}$ -215° (c 0.5,CHCl$_3$).** Crystd from EtOH, or pet ether.

Pteroic acid (2-amino-6-p-carboxyanilinomethylpteridin-4(3H)-one) *[119-24-4]* **M 312.3, m >300°(dec), pK$_{Est(1)}$~ 2.3 (basic, N1), pK$_{Est(2)}$~ 2.6 (basic, CH$_2$NH), pK$_{Est(3)}$~ 4.5 (COOH), pK$_{Est(4)}$ ~ 7.9 (acidic 4-OH) .** Crystd from dilute HCl. Hygroscopic **IRRITANT**

R(+)-Pulegone *[89-82-7]* **M 152.2, b 69.5°/5mm, n 1.4849, d 0.935, [α]$_{546}^{20}$ +23.5°(neat).** Purified *via* the semicarbazone. [Erskine and Waight *J Chem Soc* 3425 *1960*.]

Purine *[120-73-0]* **M 120.1, m 216-217°, pK$_1^{20}$ 2.30, pK$_2^{20}$ 9.86.** Crystd from toluene or EtOH.

Purpurin (1,2,4-trihydroxy-5,10-anthraquinone) *[81-54-9]* **M 256.2, m 253-256°, pK$_{Est(1)}$~7.0 (2-OH), pK$_{Est(2)}$~9.0 (4-OH), pK$_{Est(3)}$~11.1 (1-OH).** Cryst from aq EtOH, dry at 100°.

Purpurogallin (2,3,4,6-tetrahydroxy-5H-benzocyclohepten-5-one) *[569-77-7]* **M 220.2, m 274° (rapid heating) (pK 7—10, phenolic OH).** Crystd from acetic acid.

Pyocyanine (1-hydroxy-5-methylphenazinium zwitterion) *[85-66-5]* **M 210.2, m 133° (sublimes and dec on further heating).** Crystd from H$_2$O as dark blue needles. *Picrate* has **m** 190° dec.

Pyrazine *[290-37-9]* **M 80.1, m 47°, b 115.5-115.8°, pK$_1^{20}$ -6.25 (aq H$_2$SO$_4$), pK$_2^{25}$ 1.1 (0.51 at 20°).** Distd in steam and crystd from water. Purified by zone melting.

Pyrazinecarboxamide *[98-96-4]* **M 123.1, m 189-191° (sublimes slowly at 159°), pK -0.5.** Crystd from water or EtOH.

Pyrazinecarboxylic acid *[98-97-5]* **M 124.1, m 225-229°(dec), pK^{25} 2.92.** Crystd from water.

Pyrazine-2,3-dicarboxylic acid *[89-01-0]* **M 168.1, m 183-185°(dec), pK$_1$ <-2.0, pK$_2$ 0.9, pK$_3$ 2.77 (2.20).** Crystd from water. Dried at 100°.

Pyrazole *[288-13-1]* **M 68.1, m 70°, pK^{25} 2.48.** Crystd from pet ether, cyclohexane, or water. [Barszcz et al. *J Chem Soc, Dalton Trans* 2025 *1986*.]

Pyrazole-3,5-dicarboxylic acid *[3112-31-0]* **M 174.1, m 287-289°(dec), pK$_{Est(1)}$~1.2 (CO$_2$H), pK$_{Est(2)}$~3.7 (CO$_2$H), pK$_{Est(3)}$~12 (NH).** Crystd from water or EtOH.

Pyrene *[129-00-0]* **M 202.3, m 149-150°.** Crystd from EtOH, glacial acetic acid, *benzene or toluene. Purified by chromatography of CCl$_4$ solns on alumina, with *benzene or n-hexane as eluent. [Backer and Whitten *J Phys Chem* **91** 865 *1987*.] Also zone refined, and purified by sublimation. Marvel and Anderson [*J Am Chem Soc* **76** 5434 *1954*] refluxed pyrene (35g) in toluene (400mL) with maleic anhydride (5g) for 4days, then added 150mL of aqueous 5% KOH and refluxed for 5h with occasional shaking. The toluene layer was separated, washed thoroughly with H$_2$O, concentrated to about 100mL and allowed to cool. Crystalline pyrene was filtered off and recrystd three times from EtOH or acetonitrile. [Chu and Thomas *J Am Chem Soc* **108**

6270 *1986*; Russell et al. *Anal Chem* **50** 2961 *1986*.] The material was free from anthracene derivatives. Another purification step involved passage of pyrene in cyclohexane through a column of silica gel. It can be sublimed in a vacuum and zone refined. [Kano et al. *J Phys Chem* **89** 3748 *1985*.]

Pyrene-1-aldehyde *[3029-19-4]* **M 230.3, m 125-126°.** Recrystd three times from aqueous EtOH.

1-Pyrenebutyric acid *[3443-45-6]* **M 288.4, m 184-186°, pK$_{Est}$ ~4.1.** Crystd from *benzene, EtOH, EtOH/water (7:3 v/v) or *C$_6$H$_6$/AcOH. Dried over P$_2$O$_5$. [Chu and Thomas *J Am Chem Soc* **108** 6270 *1986*.]

1-Pyrenecarboxylic acid *[19694-02-1]* **M 230.3, m 126-127°, pK$_{Est}$ ~3,2.** Crystd from *C$_6$H$_6$ or 95% EtOH.

1-Pyrenesulfonic acid *[26651-23-0]* **M 202.2, m >350°, pK$_{Est}$ <0.** Crystd from EtOH/water. The tetra-Na salt cryst from H$_2$O and the sulfonyl chloride has **m** 120°(dec). [Vollmann et al. *Justus Liebigs Ann Chem* **531** 32 *1937* and *Justus Liebigs Ann Chem* **540** 189 *1939*.]

1,3,6,8-Pyrenetetrasulfonic acid *[6528-53-6]* **M 522.2, m >400°, pK$_{Est}$ <0** Crystd from water [Tietz and Bayer *Justus Liebigs Ann Chem* **540** 189 *1939*.]

Pyridine *[110-86-1]* **M 79.1, f -41.8°, b 115.6°, d 0.9831, n 1.51021, pK^{25} 5.23.** Likely impurities are H$_2$O and amines such as the picolines and lutidines. Pyridine is *hygroscopic* and is miscible with H$_2$O and organic solvents. It can be dried with solid KOH, NaOH, CaO, BaO or sodium, followed by fractional distn. Other methods of drying include standing with Linde type 4A molecular sieves, CaH$_2$ or LiAlH$_4$, azeotropic distn of the H$_2$O with toluene or *benzene, or treated with phenylmagnesium bromide in ether, followed by evaporation of the ether and distn of the pyridine. A recommended [Lindauer and Mukherjee *Pure Appl Chem* **27** 267 *1971*] method dries pyridine over solid KOH (20g/Kg) for 2weeks, and fractionally distils the supernatant over Linde type 5A molecular sieves and solid KOH. The product is stored under CO$_2$-free nitrogen. Pyridine can be stored in contact with BaO, CaH$_2$ or molecular sieves. Non-basic materials can be removed by steam distilling a soln containing 1.2 equivalents of 20% H$_2$SO$_4$ or 17% HCl until about 10% of the base has been carried over along with the non-basic impurities. The residue is then made alkaline, and the base is separated, dried with NaOH and fractionally distd.
Alternatively, pyridine can be treated with oxidising agents. Thus pyridine (800mL) has been stirred for 24h with a mixture of ceric sulfate (20g) and anhydrous K$_2$CO$_3$ (15g), then filtered and fractionally distd. Hurd and Simon [*J Am Chem Soc* **84** 4519 *1962*] stirred pyridine (135mL), water (2.5L) and KMnO$_4$ (90g) for 2h at 100°, then stood for 15h before filtering off the ppted manganese oxides. Addition of solid KOH (*ca* 500g) caused pyridine to separate. It was decanted, refluxed with CaO for 3h and distd.
Separation of pyridine from some of its homologues can be achieved by crystn of the oxalates. Pyridine is ppted as its oxalate by adding it to the stirred soln of oxalic acid in acetone. The ppte is filtered, washed with cold acetone, and pyridine is regenerated and isolated. Other methods are based on complex formation with ZnCl$_2$ or HgCl$_2$. Heap, Jones and Speakman [*J Am Chem Soc* **43** 1936 *1921*] added crude pyridine (1L) to a soln of ZnCl$_2$ (848g) in 730mL of water, 346mL of conc HCl and 690mL of 95% EtOH. The crystalline ppte of ZnCl$_2$.(pyridine)$_2$ was filtered off, recrystd twice from absolute EtOH, then treated with a conc NaOH soln, using 26.7g of solid NaOH to 100g of the complex. The ppte was filtered off, and the pyridine was dried with NaOH pellets and distd. Similarly, Kyte, Jeffery and Vogel [*J Chem Soc* 4454 *1960*] added pyridine (60mL) in 300mL of 10% (v/v) HCl to a soln of HgCl$_2$ (405g) in hot water (2.3L). On cooling, crystals of pyridine-HgCl$_2$ (1:1) complex separated and were filtered off, crystd from 1% HCl (to **m** 178.5-179°), washed with a little EtOH and dried at 110°. The free base was liberated by addition of excess aq NaOH and separated by steam distn. The distillate was saturated with solid KOH, and the upper layer was removed, dried further with KOH, then BaO and distd. Another possible purification step is fractional crystn by partial freezing.
Small amounts of pyridine have been purified by vapour-phase chromatography, using a 180-cm column of polyethyleneglycol-400 (Shell 5%) on Embacel (May and Baker) at 100°, with argon as carrier gas. The Karl Fischer titration can be used for determining water content. A colour test for pyrrole as a contaminant is described by Biddiscombe et al. [*J Chem Soc* 1957 *1954*].
§ Polystyrene supported pyridine is commercially available.

Pyridine-2-aldehyde *[1121-60-4]* **M 107.1, b 81.5°/25mm, d 1.121, n 1.535, pK$_1^{25}$ 3.84, pK$_2^{25}$ 12.68.** Sulfur dioxide was bubbled into a soln of 50g in 250mL of boiled water, under nitrogen, at 0°, until pptn was complete. The addition compound was filtered off rapidly and, after washing with a little water, it was refluxed in 17% HCl (200mL) under nitrogen until a clear soln was obtained. Neutralisation with NaHCO$_3$ and extraction with ether separated the aldehyde which was recovered by drying the extract, then distilling twice, under nitrogen. [Kyte, Jeffery and Vogel *J Chem Soc* 4454 *1960*.]

Pyridine-3-aldehyde *[500-22-1]* **M 107.1, b 89.5°/14mm, d 1.141, n 1.549, pK$_1^{20}$ 3.80, pK$_2^{20}$ 13.10.** Purification as for pyridine-2-aldehyde.

Pyridine-4-aldehyde *[872-85-5]* **M 107.1, b 79.5°/12mm, d 1.137, n 1.544, pK$_1^{20}$ 4,77, pK$_2^{20}$ 12.20.** . Purification as for pyridine-2-aldehyde.

Pyridine-2-aldoxime (pyridine-2-carboxaldoxime) *[873-69-8]* **M 122.1, m 111-113°, 114°, pK$_1^{25}$ 3.56, pK$_2^{25}$ 10.17.** Recrystd from Et$_2$O-pet ether or H$_2$O. The *picrate* has **m** 169-171° (from aqueous EtOH). It is used in peptide synthesis. [UV: Grammaticakis *Bull Chem Soc Fr* 109, 116 *1956*; Ginsberg and Wilson *J Am Chem Soc* **79** 481 *1957*; Hanania and Irvine *Nature* **183** 40 *1959*; Green and Saville *J Chem Soc* 3887 *1956*.]

Pyridine-3-aldoxime *[1193-92-6]* **M 122.1, m 150°, pK$_1^{20}$ 4.07, pK$_2^{20}$ 10.39.** Crystd from water.

Pyridine-4-aldoxime *[696-54-8]* **M 122.1, m 129°, pK$_1^{20}$ 4.73, pK$_2^{20}$ 10.03.** Crystd from water.

2,6-Pyridinedialdoxime *[2851-68-5]* **M 165.1, m 212°, pK$_{Est(1)}$~3.0, pK$_{Est(2)}$~10.** Crystd from water.

Pyridine-2,5-dicarboxylic acid *[100-26-5]* **M 167.1, m 254°, pK$_2^{25}$ 2.49, pK$_3^{25}$ 5.12.** Crystd from dil HCl.

Pyridine-3,4-dicarboxylic acid *[490-11-9]* **M 167.1, m 256°, pK$_1^{25}$ 2.43, pK$_2^{25}$ 4.78.** Crystd from dilute aqueous HCl.

Pyridine hydrobromide perbromide (pyridinium bromide perbromide) *[39416-48-3]* **M 319.9, m 130° (dec), 132-134° (dec).** It is a very good brominating agent - liberating one mol. of Br$_2$. Purified by recrystn from glacial acetic acid (33g from 100mL of AcOH). [Fieser and Fieser *Reagents for Organic Chemistry* Vol 1 967 *1967*.]

Pyridine hydrochloride *[628-13-7]* **M 115.6, m 144°, b 218°.** Crystd from CHCl$_3$/ethyl acetate and washed with diethyl ether.

Pyridine *N*-oxide *[694-59-7]* **M 95.1, m 67°, pK^{24} 0.79.** Purified by vacuum sublimation.

Pyridine 3-sulfonic acid *[636-73-7]* **M 159.2, m 365-366° (dec), 357°, pK^{25} 2.89 (12% aq EtOH), 3.22 (H$_2$O)(protonation on N).** Purified by recrystn from H$_2$O or aqueous EtOH as needles or plates. [pKa: Evans and Brown *J Org Chem* **27** 3127 *1962*; IR: Arnett and Chawla *J Am Chem Soc* **100** 214 *1978*.] UV in 50% aqueous EtOH: λmax at 208 and 262nm. The *ammonium salt* has **m** 243° (from H$_2$O), the *sulfonyl chloride* has **m** 133-134° (from pet ether), the *amide* has **m** 110-111° (from H$_2$O), the *hydrochloride* has **m** > 300° (dec), and the *N-methyl betaine* has **m** 130° (from H$_2$O). [Gastel and Wibaut *Recl Trav Chim Pays Bas* **53** 1031 *1934*; McIlvain and Goese *J Am Chem Soc* **65** 2233 *1943*; Machek *Monatsh Chem* **72** 77 *1938*.]

2-Pyridinethiol (2-mercaptopyridine) *[2637-34-5]* **M 111.2, m 127.4°, 127-130°, 130-132°, pK$_1^{20}$ -1.07, pK$_2^{20}$ 9.97,** If impure, dissolve in CHCl$_3$, wash with dil AcOH, H$_2$O, dry (MgSO$_4$), evaporate under reduced press and recryst residue from*C$_6$H$_6$ or H$_2$O. *2-Methylmercaptopyridine* (b 100-104°/33mm) was formed by treatment with MeI/NaOH. [Albert and Barlin *J Chem Soc* 2394 *1959*; Phillips and Shapiro *J Chem Soc* 584 *1942*.]

Pyridoxal hydrochloride, pyridoxamine hydrochloride and pyridoxine hydrochloride (vitamin B_6) see entries in Chapter 6.

1-(2-Pyridylazo)-2-naphthol (PAN) *[85-85-8]* **M 249.3, m 140-142°, pK_1^{30-36} 2.9, pK_2^{30-36} 11.2.** Purified by repeated crystn from MeOH. It can also be purified by sublimation under vacuum. Purity can be checked by TLC using a mixed solvent (pet ether, diethyl ether, EtOH; 10:10:1) on a silica gel plate.

4-(2-Pyridylazo)resorcinol (PAR) *[1141-59-9]* **M 215.2, m >195°(dec), λ_{max} 415nm, ϵ 2.59 x 10^4 (pH 6-12), pK_1^{25} 2.69, pK_2^{25} 5.50.** Purified as the sodium salt by recrystn from 1:1 EtOH/water. Purity can be checked by TLC using a silica gel plate and a mixed solvent (*n*-BuOH:EtOH:2M NH_3; 6:2:2).

Pyridyldiphenyltriazine *[1046-56-6]* **M 310.4, m 191-192°.** Purified by repeated recrystn from EtOH/dimethylformamide.

1-(4-pyridyl)ethanol *[R-(+)- 27854-88-2; S-(-)- 54656-96-1]* **M 123.2, m 63-65°, 67-69°, $[\alpha]_D^{20}$ (+) and (-) 49.8° (c 0.5, EtOH), pK_{Est} ~5.4.** Purified by recrystn from pet ether. The (-)-*di-O-benzoyl tartrate salt* has **m** 146-148° (from EtOH). [UV, ORD: Harelli and Samori *J Chem Soc Perkin Trans 2* 1462 *1974*.] The *racemate* recrystallises from Et_2O **m** 74-76°, **b** 90-94°/1mm [Ferles and Attia *Collect Czech Chem Commun* 38 611 *1973*; UV, NMR: Nielson et al. *J Org Chem* 29 2898 *1964*.]

Pyrogallol *[87-66-1]* **M 126.1, m 136°, pK_1^{20} 9,05, pK_2^{20} 11.19, pK_3^{20} 14.** Crystd from EtOH/*benzene.

R-pyroglutamic acid (5-oxo-D-proline, R-2-Pyrrolidone-5-carboxylic acid) *[4042-36-8]* **M 129.1, m 156-158°, $[\alpha]_{546}^{20}$ +11° (c 5, H_2O).** Crystd from EtOH/pet ether.

S-Pyroglutamic acid (5-oxo-L-proline) *[98-79-3]* **M 129.1, m 156-158°, 162-164°, $[\alpha]_{546}^{20}$ -11° (c 5, H_2O), pK 12.7 (by electron spin resonance).** Crystd from EtOH by addition of pet ether. NH_4 salt has **m** 184-186° (from EtOH).

Pyromellitic acid (benzene-1,2,4,5-tetracarboxylic acid) *[89-05-4]* **M 254.2, m 276°, 281-284°, pK_1^{25} 1.87, pK_2^{25} 2.72, pK_3^{25} 4.30, pK_4^{25} 5.52.** Dissolved in 5.7 parts of hot dimethylformamide, decolorised and filtered. The ppte obtained on cooling was separated and air dried, the solvent being removed by heating in an oven at 150-170° for several hours. Crystd from water.

Pyromellitic dianhydride *[89-32-7]* **M 218.1, m 286°.** Crystd from ethyl methyl ketone or dioxane. Dried, and sublimed *in vacuo*.

α-Pyrone (2H-pyran-2-one) *[504-31-4]* **M 96.1, m 5°, 8-9°, b 103-111°/19-22mm, 110°/26mm, 104°/30mm, 115-118°/37mm, 206-207°/atm, d_4^{20} 1.1972, n_D^{20} 1.5298, pK -1.14 (aq H_2SO_4).** Dissolve in Et_2O, wash with brine, dry (Na_2SO_4), filter, evaporate, distil residue under vacuum and redistil. It is a colourless liquid. [Zimmermann et al. *Org Synth* Coll Vol V 982 *1973*; Nakagawa and Saegusa *Org Synth* 56 49 *1977*; Elderfield *J Org Chem* 6 566 *1941*.] The *picrate* has **m** 106-107° (from EtOH).

γ-Pyrone (4H-pyran-4-one) *[108-97-4]* **M 96.1, m 32.5-32.6°, 33°, 32-34°, b 88.5°/7mm, 91-91.5°/9mm, 95-97°/13mm, 105°/23mm, 215°/atm, pK^{25} 0.10.** Purified by vacuum distn, the distillate crystallises and is *hygroscopic*. It is non-steam volatile. The *hydrochoride* has **m** 139° (from EtOH), and the *picrate* has **m** 130.2-130.3° (from EtOH or H_2O). [Mayer *Chem Ber* 90 2362 *1957*; IR: Jones et al. *Can J Chem* 37 2007 *1959*; Neelakatan *J Org Chem* 22 1584 *1957*.]

Pyronin Y [3,6-bis(dimethylamino)xanthylium chloride] *[92-32-0]* **M 302.8, m 250-260°, CI 45005, λmax 522nm, pK_{Est} ~7.6.** Commercial material contained a large quantity of zinc. Purified by dissolving 1g in 50mL of hot water containing 5g NaEDTA. Cooled to 0°, filtered, evapd to dryness and the residue extracted with EtOH. The soln was evaporated to 5-10mL, filtered, and the dye pptd by addition of excess dry diethyl ether. It was centrifuged and the crystals were washed with dry ether. The procedure was repeated, then the product was dissolved in CHCl$_3$, filtered and evapd. The dye was stored in a vacuum.

Pyrrole *[109-97-7]* **M 67.1, m 23.4°, b 66°/80mm, 129-130°/atm, d 0.966, n 1.5097, pK_1^{25} -4.4 (Protonation on carbon), pK_2^{25} 17.51 (aq KOH, H$_-$ scale).** Dried with NaOH, CaH$_2$ or CaSO$_4$. Fractionally distd under reduced pressure from CaH$_2$. Stored under nitrogen, turns brown in air. Redistd immediately before use.

Pyrrolidine *[123-75-1]* **M 71.1, b 87.5-88.5°, d 0.860, n 1.443, pK^{25} 11.31.** Dried with BaO or sodium, then fractionally distd, under N$_2$, through a Todd column packed with glass helices (see p. 174).

Pyrrolidine-1-carbodithioic acid ammonium salt *[5108-96-3]* **M 164.3, m 128-130°, pK^{25} 3.25 (free acid).** Purified by recryst twice by dissolving in MeOH and adding Et$_2$O. Also by recrystn from EtOH. [Synth and Polarography: Kitagawa and Taku *Bull Chem Soc Jpn* **64** 2151 *1973*; Malissa and Schöffmann *Mikrochim Acta* 187 *1955*.]

Pyruvic acid *[127-17-3]* **M 88.1, m 13°, b 65°/10mm, pK^{25} 2.39 (2.60).** Distd twice, then fractionally crystd by partial freezing.

p-Quaterphenyl *[135-70-6]* M 306.4, m 312-314°. Recrystd from dimethyl sulfoxide at *ca* 50°.

Quercetin (2H$_2$O) (3,3',4',5,6-pentahydroxyflavone) *[6151-25-3 (2H$_2$O); 117-39-3 (anhydr)]* **M 338.3, m *ca* 315°(dec), (phenolic pKs 7—10).** Crystd from aq EtOH and dried at 100°.

Quercitrin (quercetin glycoside) *[522-12-3]* **M 302.2, m 168°, 176-178°.** Crystd from aq EtOH and dried at 135° to give the higher melting form.

Quinaldic (quinoline-2-carboxylic) acid *[93-10-7]* **M 173.2, m 156-157°, pK_1^{25} 1.45, pK_2^{25} 2.49 (2.97).** Crystd from *benzene.

Quinalizarin (1,2,5,8-tetrahydroxy-9,10-anthraquinone) *[81-61-8]* **M 272.2, m 275°, $pK_{Est(1)}$~7.1 (1-OH), $pK_{Est(2)}$~9.9 (8-OH), $pK_{Est(3)}$~11.1 (5-OH), $pK_{Est(4)}$~11.8 (2-OH).** Crystd from acetic acid or nitrobenzene. It can be sublimed *in vacuo*.

Quinazoline *[253-82-7]* **M 130.2, m 48.0-48.5°, b 120-121°/17-18mm, pK_1^{20}-4.51 (aq H$_2$SO$_4$, anhydrous dication), pK_2^{20} 2.01 (anhydrous monocation), pK_3^{20} 4.3 (equilibrium with 3,4-hydrated species), pK_4^{20} 12.1 (hydrated anion).** Purified by passage through an activated alumina column in *benzene or pet ether (b 40-60°). Distd under reduced pressure, sublimed under vacuum and crystd from pet ether. [Armarego *J Appl Chem* **11** 70 *1961*.]

Quinhydrone *[106-34-3]* **M 218.2, m 168°.** Crystd from H$_2$O at 65°, then dried in a vac desiccator.

1*R*,3*R*,4*R*,5*R*-Quinic acid (1,3,4,5-tetrahydroxy-cyclohexane-carboxylic acid) *[77-95-2]* **M 192.3, m 172°(dec), $[\alpha]_{546}^{20}$ -51° (c 20, H$_2$O), pK^{25} 3.58.** Crystd from water.

Quinidine *[56-54-2]* **M 324.4, m 171°, $[\alpha]_{546}^{20}$ +301.1° (CHCl$_3$ contg 2.5% (v/v) EtOH), pK_1^{15} 4.13, pK_2^{15} 8.77.** Crystd from *benzene or dry CHCl$_3$/pet ether (b 40-60°), discarding the initial, oily crop of crystals. Dried under vacuum at 100° over P$_2$O$_5$.

Quinine *[130-95-0]* **M 324.4, m 177°(dec), $[\alpha]^{20}_{546}$ -160° (c 1, $CHCl_3$), pK^{20}_1 4.13 (quinoline N), pK^{20}_2 8.52 (piperidine N).** Crystd from abs EtOH.

Quinine bisulfate *[6183-68-2 (7H2O); 549-56-4 (anhydr)]* **M 422.4, m 160° (anhydrous).** Crystd from 0.1M H_2SO_4, forms heptahydrate when crystd from water

Quinine sulfate (2H2O) *[6119-70-6 (H2O); 804-63-7 (anhydr)]* **M 783.0, m 205°.** Crystd from water, dried at 110°.

Quinizarin (1,4-dihydroxy-9,10-anthraquinone) *[81-64-1]* **M 240.2, m 200-202°, pK^{25}_1 9.90 (9.5), pK^{25}_2 11.18.** Crystd from glacial acetic acid.

Quinoline *[91-22-5]* **M 129.2, m -16°, b 111.5°, 236°/758mm, d 1.0937, n 1.625, pK^{25} 4.80 (4.93).** Dried with Na_2SO_4 and vac distd from zinc dust. Also dried by boiling with acetic anhydride, then fractionally distilling. Calvin and Wilmarth [*J Am Chem Soc* **78** 1301 *1956*] cooled redistd quinoline in ice and added enough HCl to form its hydrochloride. Diazotization removed aniline, the diazo compound being broken down by warming the soln to 60°. Non-basic impurities were removed by ether extraction. Quinoline was liberated by neutralising the hydrochloride with NaOH, then dried with KOH and fractionally distd at low pressure. Addition of cuprous acetate (7g/L of quinoline) and shaking under hydrogen for 12h at 100° removed impurities due to the nitrous acid treatment. Finally the hydrogen was pumped off and the quinoline was distd. Other purification procedures depend on conversion to the phosphate (m 159°, pptd from MeOH soln, filtered, washed with MeOH, then dried at 55°) or the picrate (m 201°) which, after crystn were reconverted to the amine. The method using the picrate [Packer, Vaughan and Wong *J Am Chem Soc* **80** 905 *1958*] is as follows: quinoline is added to picric acid dissolved in the minimum volume of 95% EtOH, giving yellow crystals which were washed with EtOH, air-dried and crystd from acetonitrile. These were dissolved in dimethyl sulfoxide (previously dried over 4A molecular sieves) and passed through basic alumina, on which the picric acid is adsorbed. The free base in the effluent is extracted with *n*-pentane and distd under vacuum. Traces of solvent can be removed by vapour-phase chromatography. [Moonaw and Anton *J Phys Chem* **80** 2243 *1976*.] The $ZnCl_2$ and dichromate complexes have also been used. [Cumper, Redford and Vogel *J Chem Soc* 1176 *1962*.]

2-Quinolinealdehyde *[5470-96-2]* **M 157.2, m 71°, pK_{Est} ~3.3.** Steam distd. Crystd from H_2O. Protected from light.

8-Quinolinecarboxylic acid *[86-59-9]* **M 173.2, m 186-187.5°, pK^{25}_1 1.82, pK^{25}_2 6.87.** Crystd from water.

Quinoline ethiodide (1-ethylquinolinium iodide) *[634-35-5]* **M 285.1, m 158-159°.** Crystd from aqueous EtOH.

Quinoxaline *[91-19-0]* **M 130.2, m 28° (anhydr), 37°(H_2O), b 108-110°/0.1mm, 140°/40mm, pK^{20}_1-5.52 (-5.8, dication), pK^{25}_2 2.08 (monocation).** Crystd from pet ether. Crystallises as the monohydrate on addition of water to a pet ether soln.

Quinoxaline-2,3-dithiol *[1199-03-7]* **M 194.1, m 345°(dec), pK_1 6.9, pK_2 9.9.** Purified by repeated dissolution in alkali and re-pptn by acetic acid.

p-**Quinquephenyl** (*p*-pentaphenyl) *[61537-20-0]* **M 382.5, m 388.5°.** Recrystd from boiling dimethyl sulfoxide (b 189°, lowered to 110°). The solid obtained on cooling was filtered off and washed repeatedly with toluene, then with conc HCl. The final material was washed repeatedly with hot EtOH. It was also recrystd from pyridine, then sublimed *in vacuo*.

Quinuclidine (1-azabicyclo[2.2.2]octane) *[100-76-5]* **M 111.2, m 158°(sublimes), pK^{25} 10.95.** Crystd from diethyl ether.

D-Raffinose (5H$_2$O) *[17629-30-0 (5H$_2$O); 512-69-6 (anhydr)]* M 594.5, m 80°, $[\alpha]_{546}^{20}$ +124° (c 10, H$_2$O). Crystd from aqueous EtOH.

Rauwolscine hydrochloride *[6211-32-1]* M 390.0, m 278-280°. Crystd from water.

Reductic acid (1,2-dihydroxycyclopent-1,2-en-3-one) *[80-72-8]* M 114.1, m 213°, pK^{20} 4.72. Crystd from ethyl acetate.

Rescinnamine *[24815-24-5]* M 634.7, m 238-239°(vac), $[\alpha]_D^{20}$ -97° (c 1, CHCl$_3$), pK$_{Est(1)}$~ <0 (carbazole N), pK$_{Est(2)}$~7.0 (quinolizidine N). Crystd from *benzene or MeOH.

Reserpic acid *[83-60-3]* M 400.5, m 241-243°, pK$_{Est(1)}$~<0 (carbazole N), pK$_{Est(2)}$~4.0 (CO$_2$H), pK$_{Est(3)}$~7.4 (quinolizidine N). Crystd from MeOH. The *hydrochloride 0.5H$_2$O* has m 257-259°, $[\alpha]_D^{23}$ -81° (H$_2$O).

Reserpine *[50-55-5]* M 608.7, m 262-263°, $[\alpha]_{546}^{20}$ -148° (c 1, CHCl$_3$), pK$_{Est(1)}$~<0 (carbazole N), pK$_2$ 6.6 (7.4)(quinolizidine N). Crystd from aq acetone.

Resorcinol *[108-46-3]* M 110.1, m 111.2-111.6°, pK$_1^{25}$ 9.23, pK$_2^{25}$ 13.05. Crystd from *benzene, toluene or *benzene/diethyl ether.

Retene *[483-65-8]* M 234.3, m 99°. Crystd from EtOH.

Retinal (vitamin A aldehyde), Retinoic acid (vitamin A acid), Retinol (vitamin A alcohol) see entries in Chapter 6.

Retinyl acetate *[127-47-9]* M 328.5, m 57°. Separated from retinol by column chromatography, then crystd from MeOH. See Kofler and Rubin [*Vitamins and Hormones (NY)* **18** 315 *1960*] for review of purification methods. Stored in the dark, under N$_2$ or Ar, at 0°. See Vitamin A acetate p. 574 in Chapter 6.

Retinyl palmitate *[79-81-2]* M 524.9, m 28-29°, $\varepsilon_{1cm}^{1\%}$ (*all-trans*) 1000 (325 nm) in EtOH. Separated from retinol by column chromatography on water-deactivated alumina with hexane containing a very small percentage of acetone. Also chromatographed on TLC silica gel G, using pet ether/isopropyl ether/acetic acid/water (180:20:2:5) or pet ether/acetonitrile/acetic acid/water (190:10:1:15) to develop the chromatogram. Then recrystd from propylene at low temperature.

Rhamnetin (3,3'-4',5-tetrahydroxy-7-methoxy flavone, 7-methyl quercitin) *[90-19-7]* M 316.3, m >300°(dec), several phenolic pKs ~7-10.5. Crystd from EtOH.

L-α-Rhamnose (H$_2$O) *[10030-85-0 (H$_2$O); 3615-41-6 (anhydr)]* M 182.2, m 105°, $[\alpha]_D^{15}$ +9.1° (c 5, H$_2$O). Crystd from water or EtOH.

Rhodamine B chloride [3,5-bis-(diethylamino)-9-(2-carboxyphenyl)xanthylium chloride] *[81-88-9]* M 479.0, m 210-211°(dec), CI 45170, λmax 543nm, {Free base *[509-34-2]* CI 749}, pK 5.53. Major impurities are partially dealkylated compounds not removed by crystn. Purified by chromatography, using ethyl acetate/isopropanol/ammonia (conc)(9:7:4, R$_F$ 0.75 on Kieselgel G). Also crystd from conc soln in MeOH by slow addition of dry diethyl ether; or from EtOH containing a drop of conc HCl by slow addition of ten volumes of dry diethyl ether. The solid was washed with ether and air dried. The dried material has also been extracted with *benzene to remove oil-soluble material prior to recrystn. Store in the dark.

Rhodamine 6G [Basic Red 1, 3,5-bis-(ethylamino)-9-(2-ethoxycarbonylphenyl)-2,7-dimethylxanthylium chloride] *[989-38-8]* M 479.3, CI 45160, λmax 524nm, pK 5.58. Crystd from MeOH or EtOH, and dried in a vac oven.

Rhodanine (2-mercaptothiazolidin-4-one) [141-84-4] M 133.2, m 168.5° (capillary), pK[20] 5.18. Crystd from glacial acetic acid or water.

Riboflavin, riboflavin-5'-phosphate (Na salt, 2H$_2$O) and **ribonucleic acid** see entries in Chapter 6.

α-D-Ribose [50-69-1] M 150.1, m 90°, $[\alpha]_{546}^{20}$ -24° (after 24h, c 10, H$_2$O), pK[25] 12.22. Crystd from aqueous 80% EtOH, dried under vacuum at 60° over P$_2$O$_5$ and stored in a vacuum desiccator.

Ricinoleic acid (dl 12-hydroxyoleic acid) [141-22-0] M 298.5, m 7-8° (α-form), 5.0° (γ-form), n 1.4717, pK$_{Est}$ ~4.5. Purified as methyl acetylricinoleate [Rider J Am Chem Soc 53 4130 1931], fractionally distilling at 180-185°/0.3mm, then 87g of this ester was refluxed with KOH (56g), water (25mL), and MeOH (250mL) for 10min. The free acid was separated, crystd from acetone at -50°, and distd in small batches, b 180°/0.005mm. [Bailey et al. J Chem Soc 3027 1957.]

Rosaniline HCl (Magenta I, Fuschin) [632-99-5] M 337.9, m >200°(dec). Purified by dissolving in EtOH, filtering and adding H$_2$O. Filter or centrifuge and wash the ppte with Et$_2$O and dry in air. Could be crystd from H$_2$O. Also recrystd from water and dried in vacuo at 40°. Crystals have a metallic green lustre. UV max in EtOH is at 543nm (ε 93,000). Solubility in H$_2$O is 0.26%. A carmine red colour is produced in EtOH. [Scalan J Am Chem Soc 57 887 1937.]

p-Rosolic acid (4-[bis-{4-hydroxyphenyl}methylene]-2,5-cyclohexadien-one, 4',4"-di-hydroxy-fuschson, aurin, corallin) [603-45-2] M 290.3, m 292°, 295-300° (dec with liberation of phenol), 308-310°(dec), pK$_1$ 3.11, pK$_2$ 8.62. It forms green crystals with a metallic lustre but the colour depends on the solvent used. When recrystd from brine (satd aqueous NaCl) acidified with HCl it forms red needles, but when recrystd from EtOH-AcOH the crystals have a beetle iridescent green colour. It has been recrystd from Me$_2$CO (although it dissolves slowly), methyl ethyl ketone, 80-95% AcOH and from AcOH-*C$_6$H$_6$. An aq KOH soln is golden yellow and a 70% H$_2$SO$_4$ soln is deep red in colour. An alternative purification is to dissolve this triphenylmethane dye in 1.5% of aq NH$_3$, filter, and heat to 70-80°, then acidify with dilute AcOH by adding it slowly with vigorous stirring, whereby the aurin separates as a brick-red powder or as purplish crystals depending on the temperature and period of heating. Filter off the solid, wash with H$_2$O and a little dilute AcOH then H$_2$O again. Stir this solid with Et$_2$O to remove any ketones and allow to stand overnight in the Et$_2$O, then fiter and dry in air then in a vacuum. [Gomberg and Snow J Am Chem Soc 47 202 1925; Baines and Driver J Chem Soc 123 1216 1923; UV: Burawoy Chem Ber 64 462 1941; Neuk and Schmid J Prakt Chem [2] 23 549 1881.]

Rubijervine (slanid-5-ene-3β-12α-diol) [79-58-3] M 413.6, m 240-246°, $[\alpha]_D^{20}$ +19° (EtOH), pK$_{Est}$ ~7.0. Crystd from EtOH. It has solvent of crystn.

Rubrene [517-51-1] M 532.7, m >315°. See 5,6,11,12-tetraphenylnaphthacene on p. 366.

(+)-Rutin (quercetin-3-rubinoside) [153-18-4] M 610.5, m 188-189, $[\alpha]_{546}^{20}$ +13° (c 5, EtOH) **(polyphenolic flavone pKs 7—10).** Crystd from MeOH or water/EtOH, air dried, then dried for several hours at 110°.

Saccharic acid (D-glucaric acid) [87-73-0] M 210.1, m 125-126°, $[\alpha]_D^{20}$ +6.9°→ +20.6° (H$_2$O), pK$_1$ 3.01, pK$_2$ 3.94 (D-isomer). Crystd from 95% EtOH.

Safranine O [477-73-6] M 350.9, λ_{max} 530nm, pK[25] 6.4. Crystd from *benzene/MeOH (1:1) or water. Dried under vacuum over H$_2$SO$_4$.

Safrole **(5-allyl-1,3-benzodioxole, 4-allyl-1,2-methylenedioxybenzene)** *[94-59-7]* **M 162.1, m~ 11°, b 69-70°/1.5mm, 104-105°/6mm, 231.5-232°/atm, 235-237°/atm, d_4^{20} 1.0993, n_D^{20} 1.53738.** It has been purified by fractional distn, although it has also been recrystd from low boiling pet ether at low temperatures. [IR: Briggs et al. *Anal Chem* **29** 904 *1957*; UV: Patterson and Hibbert *J Am Chem Soc* **65** 1962 *1943*.] The *maleic anhydride adduct* forms yellow crystals from toluene **m** 257° [Hickey *J Org Chem* **13** 443 *1948*], and the *picrate* forms orange-red crystals from $CHCl_3$ [Baril and Magrdichian *J Am Chem Soc* **58** 1415 *1936*].

D(-)-Salicin *[138-52-3]* **M 286.3, m 204-208°, $[\alpha]_D^{25}$ -63.5° (c *ca* 3, H_2O).** Crystd from EtOH.

Salicylaldehyde **(*o*-hydroxybenzaldehyde)** *[90-02-8]* **M 122.1, b 93°/25mm, 195-197°/760mm, d 1.167, n 1.574, pK^{25} 8.37.** Ppted as the bisulfite addition compound by pouring the aldehyde slowly and with stirring into a 25% soln of $NaHSO_3$ in 30% EtOH, then standing for 30min. The ppte, after filtering at the pump, and washing with EtOH, was decomposed with aq 10% $NaHCO_3$, and the aldehyde was extracted into diethyl ether, dried with Na_2SO_4 or $MgSO_4$, and distd, under reduced pressure. Alternatively, salicylaldehyde can be pptd as its copper complex by adding it to warm, satd soln of copper acetate, shaking then standing in ice. The ppte was filtered off, washed thoroughly with EtOH, then with diethyl ether, and decomposed with 10% H_2SO_4, the aldehyde was extracted into diethyl ether, dried and distd. It has also been purified by repeated vacuum distn, and by dry column chromatography on Kieselgel G [Nishiya et al. *J Am Chem Soc* **108** 3880 *1986*]. The *acetyl* derivative has **m** 38-39° (from pet ether or EtOH) and **b** 142°/18mm, 253°/atm.

Salicylaldoxime *[94-67-7]* **M 137.1, m 57°, pK_{Est} ~ 8.3.** Crystd from $CHCl_3$/pet ether (b 40-60°).

Salicylamide *[65-45-2]* **M 137.1, m 142-144°, pK^{20} 8.37.** Crystd from water or repeatedly from chloroform [Nishiya et al. *J Am Chem Soc* **108** 3880 *1986*].

Salicylanilide *[87-17-2]* **M 213.2, m 135°, pK_{Est} ~8.3.** Crystd from water.

Salicylhydroxamic acid *[89-73-6]* **M 153.1, m 179-180°(dec), pK_1^{30} 2.15, pK_2^{30} 7.46, pK_3^{30} 9.72.** Crystd from acetic acid.

Salicyclic acid **(2-hydroxybenzoic acid)** *[69-72-7]* **M 138.1, m 157-159°, 158-160°, 159.5°, 159-160°, 162°, b 211°/20mm, pK_1^{25} 3.01, pK_2^{25} 13.43 (13.01).** It has been purified by steam distn, by recrystn from H_2O (solubility is 0.22% at room temp and 6.7% at 100°), absolute MeOH, or cyclohexane and by sublimation in a vacuum at 76°. The *acid chloride* (needles) has **m** 19-19.5°, **b** 92°/15mm, *amide* **m** 133° (yellow needles from H_2O), and *anilide* (prisms fron H_2O) **m** 135°. The *O-acetyl* derivative has **m** 135° (rapid heating and the liquid resolidifies at 118°) and the *o-benzoyl* derivative has **m** 132° (aq EtOH). [IR: Hales et al. *J Chem Soc* 3145 *1954*; UV: Bergmann et al. *J Chem Soc* 2351 *1950*].

Sarcosine *[107-97-1]* **M 89.1, m 212-213°(dec), pK_1^{20} 2.12, pK_2^{20} 10.19.** Crystd from abs EtOH.

Scopoletin **(7-hydroxy-6-methoxycoumarin)** *[92-61-5]* **M 192.2, m 206°, 208-209°, pK 8.96 (70%aq EtOH).** Crystd from water, acetic acid or *C_6H_6/MeOH.

Sebacic acid *[111-20-6]* **M 202.3, m 134.5°, pK_1^{25} 4.58, pK_2^{25} 5.54.** Purified *via* the disodium salt which, after crystn from boiling water (charcoal), was again converted to the free acid. The free acid was crystd repeatedly from hot distd water or from Me_2CO+pet ether and dried under vacuum.

Sebacic acid monomethyl ester *[818-88-2]* **M 216.3, m 42-43°, b 169-171°/4mm.** Recrystd from Me_3CO+pet ether or pet ether at low temperature and distd in a vacuum.

Sebaconitrile **(decanedinitrile)** *[1871-96-1]* **M 164.3, m 8°, b 199-200°.** Mix with P_2O_5 (10% by wt) and distilled from it, then redistilled.

Secobarbital **(5-allyl-5-1'-methylbutylbarbituric acid)** *[76-73-3]* **M 260.3, m 100°, pK$_{Est(1)}$~3.5, pK$_{Est(2)}$~12.0.** A soln of the salt in 10% HCl was ppted and the acid form was extracted by the addition of ether. Then purified by repeated crystn from CHCl$_3$. [Buchet and Sandorfy *J Phys Chem* **88** 3274 *1984*.]

Semicarbazide hydrochloride **(hydrazine carboxamide hydrochloride)** *[563-41-7]* **M 111.5, m 173°(dec), 175°(dec), pK^{24} 3.66.** Crystd from aqueous 75% EtOH and dried under vacuum over CaSO$_4$. Also crystd from a mixture of 3.6 mole % MeOH and 6.4 mole % of water. [Kovach et al. *J Am Chem Soc* **107** 7360 *1985*.] IR: ν 700, 3500 cm^{-1} [*Org Synth* Coll Vol I 485 *1941*; Davison and Christie *J Chem Soc* 3389 *1955*; Thiele and Stange *Chem Ber* **27** 33 *1894*; pK: Bartlett *J Am Chem Soc* **54** 2853 *1923*]. The *free base* crystd as prisms from abs EtOH, m 96° [Curtius and Heidenreich *Chem Ber* **27** 55 *1894*]. **TOXIC ORALLY**, possible **CARCINOGEN** and **TERATOGEN**.

Sennoside A *[81-27-6]* **M 862.7, m 220-240°(dec), [α]$_D^{20}$ -147° (c 5, Me$_2$CO/H$_2$O 7:1).** Crystd from aq acetone or large vols of H$_2$O.

Sennoside B *[128-57-4]* **M 962.7, m 182-190°(dec), [α]$_D^{20}$ -100° (c 2, Me$_2$CO/H$_2$O 7:3).** Crystd from aq acetone or large vols of H$_2$O.

L-Serine *[56-45-1]* **M 105.1, m 228°(dec), [α]$_D^{25}$ +14.5° (1M HCl), [α]$_{546}^{20}$ +16° (c 5, 5M HCl), pK$_1^{25}$ 2.15, pK$_2^{25}$ 9.21.** Likely impurity is glycine. Crystd from water by adding 4 volumes of EtOH. Dried. Stored in a desiccator.

Serotonin creatinine sulfate (H$_2$O) *[971-74-4]* **M 405.4, m 220°(dec), pK$_1$ 10.1, pK$_2$ 11.1, pK$_3$ 18.25 (NH) for serotonin, pK 4.9 for creatinine.** Crystd (as monohydrate) from water.

Shikimic acid *[138-59-0]* **M 174.2, m 183-184.5°, 190°, [α]$_{546}^{20}$ -210° (c 2, H$_2$O), pK^{14} 4.15.** Crystd from water or MeOH/AcOEt and sublimes in a vac.

Sinomenine hydrochloride **(7,8-didehydro-4-hydroxy-3,7-dimethoxy-17-methyl-9α,13α,14α-morphan-6-one HCl)** *[6080-33-7]* **M 365.9, m 231°, [α]$_D^{17}$ -83° (c 4, H$_2$O), pK$_{Est(1)}$~10.0 (N), pK$_{Est(2)}$~10.4 (OH).** Crystd from water.

β-Sitosterol *[83-46-5]* **M 414.7, m 136-137°, [α]$_{546}^{20}$ -42° (c 2, CHCl$_3$).** Crystd from MeOH. Also purified by zone melting.

Skellysolve A is essentially *n*-pentane, **b** 28-30°,
Skellysolve A is essentially *n*-hexane, **b** 60-68°,
Skellysolve C is essentially *n*-heptane, **b** 90-100°,
Skellysolve D is mixed heptanes, **b** 75-115°,
Skellysolve E is mixed octanes, **b** 100-140°,
Skellysolve F is pet ether, **b** 30-60°,
Skellysolve G is pet ether, **b** 40-75°,
Skellysolve H is hexanes and heptanes, **b** 69-96°,
Skellysolve L is essentially octanes, **b** 95-127°. For methods of purification, see **petroleum ether**.

Smilagenin *[126-18-1]* **M 416.6, m 185°, [α]$_D^{25}$ -69° (c 0.5, CHCl$_3$).** Crystd from acetone.

Solanidine *[80-78-4]* **M 397.6, m 218-219°, [α]$_D^{20}$ -29° (c 0.5, CHCl$_3$), pK^{15} 6.66.** Crystd from CHCl$_3$/MeOH.

α-Solanine *[20562-02-1]* **M 868.1, m 286°(dec), [α]$_D^{20}$ -58° (c 0.8, pyridine), pK^{15} 6.66.** See α-solanine on p. 566 in Chapter 6.

Solanone [$S(+)$-*trans*-2-methyl-5-isopropyl-1,3-nonan-8-one] *[1937-54-8]*   **M 194.3, b 60°/1mm, $[\alpha]_D^{20}$ +14° (neat).** Purified by high vacuum distillation and stored in sealed ampules [Kohda and Sato *J Chem Soc, Chem Commun* 951 *1981*]. It has UV (hexane) at λmax 230nm (ε 11,800).

Solasodine *[126-17-0]*   **M 413.6, m 202°, $[\alpha]_D^{25}$ -100° (c 2, MeOH), pK 7.7.** Crystd (as monohydrate) from MeOH or aq 80% EtOH, and sublimes in a vac.

Solasonine (solasodine-3-O-mannoglucoside) *[19121-58-5]*   **M 884.0, m 279°, $[\alpha]_D^{20}$ -75° (c 0.5, MeOH), pK$_{Est}$ ~ 7.7.** Crystd from aq 80% dioxane or MeOH.

Solochrome Violet R [4-hydroxy-3-(2-hydroxynaphthyl-1-ylazo)benzenesulfonic acid] *[2092-55-9]* **M 367.3, CI 15670, λmax 501nm, pK$_2^{25}$ 7.22 (OH), pK$_3^{25}$ 13.39 (OH).** Converted to the monosodium salt by pptn with NaOAc/AcOH buffer of pH 4, then purified by pptn of the free acid from aq soln with conc HCl, washing and extracting with EtOH in a Soxhlet extractor. The acid ppted on evaporating the EtOH and was reconverted to the sodium salt.as described for *Chlorazole Sky Blue FF*. Dried at 110°. It is *hygroscopic*. [Coates and Rigg *Trans Faraday Soc* 57 1088 *1961*.]

Sorbic acid (2,4-hexadienoic acid) *[110-44-1]*   **M 112.1, m 134°, pK^{25} 4.76.** Crystd from water.

Sorbitol *[50-70-4]*   **M 182.2, m 89-93° (hemihydrate), 110-111° (anhydrous), $[\alpha]_{546}^{20}$ -1.8° (c 10, H$_2$O), pK^{60} 13.00.** Crystd (as hemihydrate) several times from EtOH/water (1:1), then dried by fusing and storing over MgSO$_4$.

(-)-Sparteine sulfate pentahydrate *[6160-12-9]* **M 422.5, m loses H$_2$O at 100° and turns brown at 136° (dec), $[\alpha]_D^{20}$ -22° (c 5, H$_2$O), $[\alpha]_D^{21}$ -16° (c 10, EtOH for free base), pK$_1^{20}$ 2.24, pK$_2^{20}$ 9.46.** Recrystd from aq EtOH or H$_2$O although the solubility in the latter is high. The *free (-)-base* has **b** 173°/8mm and is steam volatile but resinifies in air. The *dipicrate* forms yellow needles from EtOH-Me$_2$CO, **m** 205-206° [Clemo et al. *J Chem Soc* 429 *1931*; see also Bolnmann and Schuman *The Alkaloids* (Ed Manske) Vol 9 175 *1967*]. The *free (±)-base* has **m** 71-72.5° [van Tamelen and Foltz *J Am Chem Soc* 82 2400 *1960*].

Spermidine [N-(3-aminopropyl)-1,4-diaminobutane] *[124-20-9]* **M 145.3, m 23-25°, b 128-131°/15mm, d 0.918, n 1.482, pK$_1^{25}$ 8.25, pK$_2^{25}$ 9.64, pK$_3^{25}$ 10.43.** It is a strong base with an alkylamine odour and absorbs CO$_2$ from the atmosphere. It is purified by shaking with solid K$_2$CO$_3$ or NaOH, decanting and distilling from K$_2$CO$_3$ in a vacuum. Store in the dark under N$_2$.

Spermidine trihydrochloride *[334-50-9]* **M 245.3, m ~250°(dec), 256-258°, for pKa see *free base* above.** Recrystd from dry 3% HCl in ethanol adding dry Et$_2$O if necessary. Filter rapidly and dry in a vac desiccator. Alternatively centrifuge the crystals off wash them with dry Et$_2$O and dry in a vacuum.

Spermine 4HCl (N,N-bis(3-aminopropyl)-1,4-butanediamine 4HCl) *[306-67-2]* **M 348.2, m 313-315°.** The pKs are similar to spermidine above. Purification as for spermidine trihydrochloride above.

Squalene (all-*trans*- 2,6,10,15,19,23-hexamethyltetracosahexa-2,6,10,14,18,22-ene, spinacen) *[111-02-4]* **M 410.7, m ~75°, b 203°/0.15mm.** See squalene on p. 567 in Chapter 6.

Squaric acid (3,4-dihydroxy-3-cyclobutene-1,2-dione) *[2892-51-5]* **M 114.1, m 293°(dec), 294°(dec), >300°, pK$_1^{20}$ 1,50, pK$_2^{20}$ 2.93.** Purified by recryst from H$_2$O — this is quite simple because the acid is ~ 7% soluble in boiling H$_2$O and only 2% at room temperature. It is not soluble in Me$_2$CO or Et$_2$O hence it can be rinsed with these solvents and dried in air or a vacuum. It is not hygroscopic and gives an intense purple colour with FeCl$_3$. It has IR ν at 1820 (C=O) and 1640 (C=C) cm^{-1}; and UV λmax at 269.5nm (ε 37K M^{-1}cm^{-1}).) [Cohn et al. *J Am Chem Soc* 81 3480 *1959*; Park et al. *J Am Chem Soc* 84 2919 *1962*] See also **pKa** values of 0.59 ±0.09 and 3.48 ±0.023 [Scwartz and Howard *J Phys Chem* 74 4374 *1970*].

Starch *[9005-84-9]*   **M (162.1)n.** See entry on p. 567 in Chapter 6.

Stearic acid (octadecanoic acid) *[57-11-4]* **M 284.5, m 71.4°, 72°, b 144-145°/27mm, 383°/760mm, d 0.911, n 1.428, pK_{Est} ~5.0.** Crystd from acetone, acetonitrile, EtOH (5 times), aq MeOH, ethyl methyl ketone or pet ether (b 60-90°), or by fractional pptn by dissolving in hot 95% EtOH and pouring into distd water, with stirring. The ppte, after washing with distd water, was dried under vacuum over P_2O_5. It has also been purified by zone melting and partial freezing. [Tamai et al. *J Phys Chem* **91** 541 *1987*.]

Stigmasterol *[83-48-7]* **M 412.7, m 170°, $[\alpha]_D^{22}$ -51° (CHCl$_3$), $[\alpha]_{546}^{20}$ -59° (c 2, CHCl$_3$).** Crystd from hot EtOH. Dried in vacuum over P_2O_5 for 3h at 90°. Purity was checked by NMR.

cis-**Stilbene** *[645-49-8]* **M 180.3, b 145°/12mm.** Purified by chromatography on alumina using hexane and distd under vacuum. (The final product contains *ca* 0.1% of the *trans*-isomer). [Lewis et al. *J Am Chem Soc* **107** 203 *1985*; Saltiel *J Phys Chem* **91** 2755 *1987*.]

trans-**Stilbene** *[103-30-0]* **M 180.3, m 125.9°, b 305-307°/744mm, d 0.970.** Purified by vac distn. (The final product contains about 1% of the *cis* isomer). Crystd from EtOH. Purified by zone melting. [Lewis et al. *J Am Chem Soc* **107** 203 *1985*; Bollucci et al. *J Am Chem Soc* **109** 515 *1987*; Saltiel *J Phys Chem* **91** 2755 *1987*.]

(-)-Strychnine *[57-24-9]* **M 334.4, m 268°, $[\alpha]_{546}^{20}$ -139° (c 1, CHCl$_3$), pK_1^{20} 2,50, pK_2^{20} 8.2.** Crystd as the hydrochloride from water, then neutralised with ammonia.

Styphnic acid (2,4,6-trinitroresorcinol) *[82-71-3]* **M 245.1, m 177-178°, 179-180°, pK_1^{25} 0.06 (1.74), pK_2^{25} 4.23 (4.86).** Crystd from ethyl acetate or water containing HCl [**EXPLODES violently on rapid heating.**]

Styrene *[100-42-5]* **M 104.2, b 41-42°/18mm, 145.2°/760mm, d 0.907, n 1.5469, n^{25} 1.5441.** Styrene is difficult to purify and keep pure. Usually contains added inhibitors (such as a trace of hydroquinone). Washed with aqueous NaOH to remove inhibitors (e.g. *tert*-butanol), then with water, dried for several hours with MgSO$_4$ and distd at 25° under reduced pressure in the presence of an inhibitor (such as 0.005% *p-tert*-butylcatechol). It can be stored at -78°. It can also be stored and kept anhydrous with Linde type 5A molecular sieves, CaH$_2$, CaSO$_4$, BaO or sodium, being fractionally distd, and distd in a vacuum line just before use. Alternatively styrene (and its deuterated derivative) were passed through a neutral alumina column before use [Woon et al. *J Am Chem Soc* **108** 7990 *1986*; Collman *J Am Chem Soc* **108** 2588 *1986*].

(±)-Styrene glycol (±-1-phenyl-1,2-ethanediol) *[93-56-1]* **M 138.2, m 67-68°.** Crystd from pet ether.

Styrene oxide *[96-09-3]* **M 120.2, b 84-86°/16.5mm, d 1.053, n 1.535.** Fractional distn at reduced pressure does not remove phenylacetaldehyde. If this material is present, the styrene oxide is treated with hydrogen under 3 atmospheres pressure in the presence of platinum oxide. The aldehyde, but not the oxide, is reduced to ß-phenylethanol) and separation is now readily achieved by fractional distn. [Schenck and Kaizermen *J Am Chem Soc* **75** 1636 *1953*.]

Suberic acid (hexane-1,6-dicarboxylic acid) *[505-48-6]* **M 174.2, m 141-142°, pK_1^{25} 4.12, pK_2^{25} 5.40.** Crystd from acetone and sublimes at 300° without dec.

Succinamic acid (succinic acid amide) *[638-32-4]* **M 117.1, m 155°, 156-157°, pK^{25} 4.54.** Crystd from Me$_2$CO or H$_2$O and dried in vac. Not v sol in MeOH. Converted to succinimide above 200°.

Succinamide *[110-14-5]* **M 116.1, m 262-265°(dec).** Crystd from hot water.

Succinic acid *[110-15-6]* **M 118.1, m 185-185.5°, pK_1^{25} 4.21, pK_2^{25} 5.72.** Washed with diethyl ether. Crystd from acetone, distd water, or *tert*-butanol. Dried under vacuum over P_2O_5 or conc H$_2$SO$_4$. Also purified by conversion to the disodium salt which, after crystn from boiling water (charcoal), is treated with mineral acid to regenerate the succinic acid. The acid is then recrystd and vacuum dried.

Succinic anhydride *[108-30-5]* **M 100.1, m 119-120°.** Crystd from redistd acetic anhydride or $CHCl_3$, then filtered, washed with diethyl ether and dried under vacuum.

Succinimide *[123-56-8]* **M 99.1, m 124-125°, pK^{25} 9.62.** Crystd from EtOH (1mL/g) or water.

Succinonitrile *[110-61-2]* **M 80.1, m 57.9°, b 108°/1mm, 267°/760mm.** Purified by vacuum sublimation, also crystd from acetone.

D(+)-Sucrose (β-D-fructofuranosyl-α-D-glucopyranoside) *[57-50-1]* **M 342.3, m 160-186°, 186-188°, $[\alpha]^{20}_{546}$ +78° (c 10, H_2O), $[\alpha]^{20}_{D}$ + 66° (c 26, H_2O), pK 12.62.** Crystd from water (solubility: 1g in 0.5mL H_2O at 20°, 1g in 0.2mL in boiling H_2O). Sol in EtOH (0.6%) and MeOH (1%). *Sucrose diacetate hexaisobutyrate* is purified by melting and, while molten, treated with $NaHCO_3$ and charcoal, then filtered.

D-Sucrose octaacetate *[126-14-7]* **M 678.6, m 83-85°, $[\alpha]^{20}_{546}$ +70° (c 1, $CHCl_3$).** Crystd from EtOH.

Sudan I (Solvent Yellow 14, 1-phenylazo-2-naphthol) *[824-07-9]* **M 248.3, m 135°, CI 12055, λmax 476nm, pK_{Est} ~9.0.** Crystd from EtOH.

Sudan III [Solvent Red 23, 1-(p-phenylazo-phenylazo)-2-naphthol] *[85-86-9]* **M 352.4, m 199°(dec), CI 26100, λ_{max} 354, 508 nm, pK_{Est} ~9.0.** Crystd from EtOH, EtOH/water or *benzene/abs EtOH (1:1).

Sudan IV [Solvent Red 24, 1-(4-o-tolylazo-o-tolylazo)-2-naphthol] *[85-83-6]* **M 380.5, m ~184°(dec), CI 26105, λmax 520nm, pK_{Est} ~9.0.** Crystd from EtOH/water or acetone/water.

Sulfaguanidine *[57-67-0]* **M 214.2, m 189-190°, pK_1 0.48, pK_2 2.75.** Crystd from hot water (7mL/g).

Sulfamethazine *[57-68-1]* **M 278.3, m 198-200°, pK_1 2.65, pK_2 7.4.** Crystd from dioxane.

Sulfanilamide (p-aminobenzenesulfonamide) *[63-74-1]* **M 172.2, m 166°, pK^{20}_1 2.30, pK^{20}_2 10.26.** Crystd from water or EtOH.

Sulfanilic acid (4-aminobenzenesulfonic acid) *[121-57-3]* **M 173.2, pK^{25}_1 <1, pK^{25}_2 3.23.** Crystd (as dihydrate) from boiling water. Dried at 105° for 2-3h, then over 90% H_2SO_4 in a vacuum desiccator.

Sulfapyridine *[144-83-2]* **M 349.2, m 193°, pK^{20} 8.64.** Crystd from 90% acetone and dried at 90°.

o-Sulfobenzoic acid (H_2O) *[123333-68-6 (H_2O); 632-25-7]* **M 202.2, m 68-69°, $pK_{Est(1)}$<1, $pK_{Est(2)}$~3.1 (CO_2H).** Crystd from water.

o-Sulfobenzoic acid (monoammonium salt) *[6939-89-5]* **M 219.5.** Crystd from water.

o-Sulfobenzoic anhydride *[81-08-3]* **M 184.2, m 128°, b 184-186°/18mm.** See also 2,1-benzoxathiol-3-one-1,1-dioxide on p. 126.

Sulfolane (tetramethylenesulfone) *[126-33-0]* **M 120.2, m 28.5°, b 153-154°/18mm, 285°/760mm, d 1.263, n^{30} 1.4820.** Prepared commercially by Diels-Alder reaction of 1,3-butadiene and sulfur dioxide, followed by Raney nickel hydrogenation. The principle impurities are water, 3-sulfolene, 2-sulfolene and 2-isopropyl sulfolanyl ether. It is dried by passage through a column of molecular sieves. Distd under reduced pressure through a column packed with stainless steel helices. Again dried with molecular sieves and distd. [Cram et al. *J Am Chem Soc* **83** 3678 *1961*; Coetzee *Pure Appl Chem* **49** 211 *1977*.] Also, it was stirred at 50° and small portions of solid KMnO$_4$ were added until the colour persisted during 1h. Dropwise addition of MeOH then destroyed the excess KMnO$_4$, the soln was filtered, freed from potassium ions by passage through an ion-exchange column and dried under vacuum. It has also been vacuum distd from KOH pellets. It is *hygroscopic*. [See Sacco et al. *J Phys Chem* **80** 749 *1976*; *J Chem Soc, Faraday Trans 1* **73** 1936 *1977*; **74** 2070 *1978*; *Trans Faraday Soc* **62** 2738 *1966*.] Coetzee has reviewed the methods of purification of sulfolane, and also the removal of impurities. [Coetzee in *Recommended Methods of Purification of Solvents and Tests for Impurities*, Coetzee Ed. Pergamon Press, 1982.]

5-Sulfosalicylic acid *[5965-83-3]* **M 254.2, m 108-110°, pK$_1^{25}$ <0, pK$_2^{25}$ 2.67, pK$_3^{25}$ 11.67.** Crystd from water. Alternatively, it was converted to the monosodium salt which was crystd from water and washed with a little water, EtOH and then diethyl ether. The free acid was recovered by acidification.

Syringaldehyde (3,5-dimethoxy-4-hydroxybenzaldehyde) *[134-96-3]* **M 182.2, m 113°, pK$_{Est}$ ~8.** Crystd from pet ether.

Syringic acid (3,5-dimethoxy-4-hydroxybenzoic acid) *[530-57-4]* **M 198.2, m 204-205°, 206.5°, 206-209°, 209-210°, pK$_1^{25}$ 4.34, pK$_2^{25}$ 9.49.** Recrystd from H$_2$O using charcoal [Bogert and Coyne *J Am Chem Soc* **51** 571 *1929*; Anderson and Nabenhauer *J Am Chem Soc* **48** 3001 *1926*.] The *methyl ester* has **m** 107° (from MeOH), the *4-acetyl* derivative has **m** 190° and the *4-benzoyl* derivative has **m** 229-232°. [Hahn and Wassmuth *Chem Ber* **67** 2050 *1934*; UV: Lemon *J Am Chem Soc* **69** 2998 *1947* and Pearl and Beyer *J Am Chem Soc* **72** 1743 *1950*.]

D(-)-Tagatose *[87-81-0]* **M 180.2, m 134-135°, [α]$_{546}$ -6.5° (c 1, H$_2$O).** Crystd from aqueous EtOH.

d- **Tartaric acid** *[147-71-7]* **M 150.1, m 169.5-170° (2S,3S-form, natural) [α]$_{546}^{20}$ -15° (c 10, H$_2$O); m 208° (2RS,3RS-form), pK$_1^{25}$ 3.03, pK$_2^{25}$ 4.46, pK$_3^{25}$ 14.4.** Crystd from distilled H$_2$O or *benzene/diethyl ether containing 5% of pet ether (b 60-80°) (1:1). Soxhlet extraction with diethyl ether has been used to remove an impurity absorbing at 265nm. It has also been crystd from absolute EtOH/hexane, and dried in a vacuum for 18h [Kornblum and Wade *J Org Chem* **52** 5301 *1987*].

*meso-***Tartaric acid** *[147-73-9]* **M 150.1, m 139-141°, pK$_1^{25}$ 3.17, pK$_2^{25}$ 4.91.** Crystd from water, washed with cold MeOH and dried at 60° under vacuum.

Taurocholic acid *[81-24-3]* **M 515.6, m 125°(dec), [α]$_D$ +38.8 (c 2, EtOH), pK 1.4.** Crystd from EtOH/diethyl ether.

Terephalaldehyde *[623-27-8]* **M 134.1, m 116°, b 245-248°/771mm.** Crystd from water.

Terephthalic acid (benzene-1,4-dicarboxylic acid) *[100-21-0]* **M 166.1, m sublimes >300° without melting, pK$_1^{20}$ 3.4, pK$_2^{20}$ 4.34.** Purified *via* the sodium salt which, after crystn from water, was reconverted to the acid by acidification with mineral acid.

Terephthaloyl chloride *[100-20-9]* **M 203.0, m 80-82°.** Crystd from dry hexane.

*o-***Terphenyl** *[84-15-1]* **M 230.3, m 57-58°.** Crystd from EtOH. Purified by chromatography of CCl$_4$ solns on alumina, with pet ether as eluent, followed by crystn from pet ether (b 40-60°) or pet ether/*benzene. They can also be distd under vacuum.

*m-***Terphenyl** *[92-06-8]* **M 230.3, m 88-89°.** Purification as for *o*-terphenyl above.

p-**Terphenyl** *[92-94-4]* **M 230.3, m 212.7°.** Crystd from nitrobenzene or trichlorobenzene. It was purified by chromatography on alumina in a darkened room, using pet ether, and then crystallizing from pet ether (b 40-60°) or pet ether/*benzene.

Terpin hydrate *[2451-01-6]* **M 190.3, m 105.5° (*cis*), 156-158° (*trans*).** Crystd from H_2O or EtOH.

2,2':6',2"-Terpyridyl (2,2':6',2"-terpyridyl) *[1148-79-4]* **M 233.3, m 91-92°, pK$_1^{23}$ 2.64, pK$_2^{23}$ 4.33.** Crystd from diethyl ether, toluene or from pet ether, then aqueous MeOH, followed by vacuum sublimation at 90°.

Terreic acid (2-hydroxy-3-methyl-1,4-benzoquinone-5,6-epoxide) *[121-40-4]* **M 154.1, m 127-127.5°, $[\alpha]_D^{22}$ +74° (pH 4, phosphate buffer), -17° ($CHCl_3$), pK 4.5.** Crystd from *benzene or hexane. Sublimed *in vacuo*.

Terthiophene (2,5-di[thienyl]thiophene; α−terthienyl) *[1081-34-1]* **M 248.4, m 94-95.5°, 94-96°.** Possible impurities are bithienyl and polythienyls. Suspend in H_2O and steam distil to remove bithienyl. The residue is cooled and extracted with $CHCl_3$, dried ($MgSO_4$), filtered, evaporated and the residue chromatographed on Al_2O_3 using pet ether-3% Me_2CO as eluant. The terphenyl zone is then eluted from the Al_2O_3 with Et_2O, the extract is evaporated and the residue is recrystd from MeOH (40mL per g). The platelets are washed with cold MeOH and dried in air. [UV: Sease and Zechmeister *J Am Chem Soc* **69** 270 *1947*; Uhlenbroek and Bijloo *Recl Trav Chim Pays-Bas* **79** 1181 *1960*.] See also entry on p. 568 in Chapter 6.

Testosterone *[58-22-0]* **M 288.4, m 155°, $[\alpha]_{546}^{20}$ +130° (c 1, dioxane).** Crystd from aq acetone.

Testosterone propionate *[57-85-2]* **M 344.5, m 118-122°, $[\alpha]_{546}^{20}$ +100° (c 1, dioxane).** Crystd from aqueous EtOH.

2,4,5,6-Tetraaminopyrimidine sulfate *[5392-28-9]* **M 238.2, m 255° (dec), >300°, >350° (dec), pK^{20} 6.82.** Purified by recrystn from H_2O, 2N H_2SO_4 (20 parts, 67% recovery) or 0.1N H_2SO_4 (40 parts, 62% recovery), and dried in air. [UV: Konrad and Pfleiderer *Chem Ber* **103** 722 *1970*; Malletta et al. *J Am Chem Soc* **69** 1814 *1947*; Cavalieri et al. *J Am Chem Soc* **70** 3875 *1948*.]

Tetra-*n*-amylammonium bromide *[866-97-7]* **M 378.5, m 100-101°.** Crystd from pet ether, *benzene or acetone/ether mixtures and dried in vacuum at 40-50° for 2 days.

Tetra-*n*-amylammonium iodide *[2498-20-6]* **M 425.5, m 135-137°.** Crystd from EtOH and dried at 35° under vac. Also purified by dissolving in acetone and pptd by adding diethyl ether; and dried at 50° for 2 days.

1,4,8,11-Tetraazacyclotetradecane *[295-37-4]* **M 200.33, m 173° (closed capillary and sublimes at 125°), 183-185°, 185°, pK$_{Est(1)}$~3.8, pK$_{Est(2)}$~6.0, pK$_{Est(3)}$~9.0, pK$_{Est(4)}$~9.6.** Purified by recrystn from dioxane (white needles) and sublimes above 120°. It has been distilled, **b 132-140°/4-8mm.** It forms complexes with metals and gives a sparingly soluble nitrate salt, **m 205° (dec),** which crystallises from H_2O and is dried at 150°. [UV: Bosnich et al. *Inorg Chem* **4** 1102 *1963*, van Alphen *Recl Trav Chim Pays-Bas* **56** 343 *1937*.]

Tetrabenazine (2-oxo-3-isobutyl-9,10-dimethoxy-1,2,3,4,6,7,-hexahydro-11b*H*-benzo[a]-quinolizine) *[58-46-8]* **M 317.4, m 127-128°, pK$_{Est}$ ~ 8.** Crystd from MeOH. The *hydrochloride* has **m 208-210°** and the *oxime* has **m 158°** (from EtOH).

1,1,2,2,-Tetrabromoethane *[79-27-6]* **M 345.7, f 0.0°, b 243.5°, d 2.965, n 1.63533.** Washed successively with conc H_2SO_4 (three times) and H_2O (three times), dried with K_2CO_3 and $CaSO_4$ and distd.

3',3'',5',5''-Tetrabromophenolphthalein ethyl ester *[1176-74-5]* **M 662.0, m 212-214°.** Crystd from *benzene, dried at 120° and kept under vacuum.

Tetra-*n*-butylammonium bromide *[1643-19-2]* **M 322.4, m 119.6°.** Crystd from *benzene (5mL/g) at 80° by adding hot *n*-hexane (three volumes) and allowing to cool. Dried over P_2O_5 or $Mg(ClO_4)_2$, under vacuum. The salt is *very hygroscopic.* It can also be crystd from ethyl acetate or dry acetone by adding diethyl ether and dried *in vacuo* at 60° for 2 days. It has been crystd from acetone by addition of diethyl ether. So *hygroscopic* that all manipulations should be carried out in a dry-box. Purified by precipitation of a saturated solution in dry CCl_4 by addition of cyclohexane or by recrystallisation from ethyl acetate, then heating in vacuum to 75° in the presence of P_2O_5. [Symons et al. *J Chem Soc, Faraday Trans 1* **76** 2251 *1908*.] Also recrystallised from CH_2Cl_2/diethyl ether and dried in a vacuum desiccator over P_2O_5. [Blau and Espenson *J Am Chem Soc* **108** 1962 *1986*.]

Tetra-*n*-butylammonium chloride *[1112-67-0]* **M 277.9, m 15.7°.** Crystd from acetone by addition of diethyl ether. Very **hygroscopic** and forms crystals with $34H_2O$.

Tetra-*n*-butylammonium fluoroborate *[429-42-5]* **M 329.3, m 161-163°.** See tetrabutyl-ammonium fluoroborate on p. 480 in Chapter 5.

Tetra-*n*-butylammonium hexafluorophosphate *[3109-63-5]* **M 387.5, m 239-241°.** Recrystd from satd EtOH/water and dried for 10h in vac at 70°. It was also recrystd three times from abs EtOH and dried for 2 days in a drying pistol under vac at boiling toluene temperature [Bedard and Dahl *J Am Chem Soc* **108** 5933 *1986*].

Tetra-*n*-butylammonium hydrogen sulfate *[32503-27-8]* **M 339.5, m 171-172°.** Crystd from acetone.

Tetra-*n*-butylammonium iodide *[311-28-4]* **M 369.4, m 146°.** Crystd from toluene/pet ether (see entry for the corresponding bromide), acetone, ethyl acetate, EtOH/diethyl ether, nitromethane, aq EtOH or water. Dried at room temperature under vac. It has also been dissolved in MeOH/acetone (1:3, 10mL/g), filtered and allowed to stand at room temperature to evaporate to *ca* half its original volume. Distilled water (1mL/g) was then added, and the ppte was filtered off and dried. It was also dissolved in acetone, ppted by adding ether and dried in vac at 90° for 2 days. It has also been recrystallised from CH_2Cl_2/pet ether or hexane, or anhydrous methanol and stored in a vacuum desiccator over H_2SO_4. [Chau and Espenson *J Am Chem Soc* **108** 1962 *1986*.]

Tetra-*n*-butylammonium nitrate *[1941-27-1]* **M 304.5, m 119°.** Crystd from *benzene (7mL/g) or EtOH, dried in a vacuum over P_2O_5 at 60° for 2 days.

Tetra-*n*-butylammonium perchlorate *[1923-70-2]* **M 341.9°, m 210°(dec).** Crystd from EtOH, ethyl acetate, from *n*-hexane or diethyl ether/acetone mixture, ethyl acetate or hot CH_2Cl_2. Dried in vacuum at room temperature over P_2O_5 for 24h. [Anson et al. *J Am Chem Soc* **106** 4460 *1984*; Ohst and Kochi *J Am Chem Soc* **108** 2877 *1986*; Collman et al. *J Am Chem Soc* **108** 2916 *1986*; Blau and Espenson *J Am Chem Soc* **108** 1962 *1986*; Gustowski et al. *J Am Chem Soc* **108** *1986*; Ikezawa and Kutal *J Org Chem* **52** 3299 *1987*.]

Tetra-*n*-butylammonium picrate *[914-45-4]* **M 490.6, m 89°.** Crystd from EtOH. Dried in a vacuum desiccator over P_2O_5.

Tetra-*n*-butylammonium tetrabutylborate (Bu_4N^+ Bu_4B^-) *[23231-91-6]* **M 481.7, m 109.5°.** Dissolved in MeOH or acetone, and crystd by adding distd water. Dried in vacuum at 70°. It has also been successively recrystd from isopropyl ether, isopropyl ether/acetone (50:1) and isopropyl ether/EtOH (50:1) for 10h, then isopropyl ether/acetone for 1h, and dried at 65° under reduced pressure for 1 week. [Kondo et al. *J Chem Soc, Faraday Trans 1* **76** 812 *1980*.]

2,3,4,5-Tetrachloroaniline *[634-83-3]* **M 230.9, m 119-120°, pK$_{Est}$ ~-0.26.** Crystd from EtOH.

2,3,5,6-Tetrachloroaniline *[3481-20-7]* **M 230.9, m 107-108°, pK$_{Est}$ ~-1.8.** Crystd from EtOH.

1,2,3,4-Tetrachlorobenzene *[634-66-2]* **M 215.9, m 45-46°, b 254°/760mm.** Crystd from EtOH.

1,2,3,5-Tetrachlorobenzene *[634-90-2]* **M 215.9, m 51°, b 246°/760mm.** Crystd from EtOH.

1,2,4,5-Tetrachlorobenzene *[95-94-3]* **M 215.9, m 139.5-140.5°, b 240°/760mm.** Crystd from EtOH, ether, *benzene, *benzene/EtOH or carbon disulfide.

3,4,5,6-Tetrachloro-1,2-benzoquinone *[2435-53-2]* **M 245.9, m 130°.** Crystd from AcOH. Dry in vacuum desiccator over KOH.

1,1,2,2-Tetrachloro-1,2-difluoroethane *[72-12-0]* **M 203.8, f 26.0°, b 92.8°/760mm.** Purified as for trichlorotrifluoroethane.

***sym*-Tetrachloroethane** *[79-34-5]* **M 167.9, b 62°/100mm, 146.2°/atm, d 1.588, n^{15} 1.49678.** Stirred, on a steam-bath, with conc H$_2$SO$_4$ until a fresh portion of acid remained colourless. The organic phase was then separated, distd in steam, dried (CaCl$_2$ or K$_2$CO$_3$), and fractionally distd in a vac.

Tetrachloroethylene *[127-18-4]* **M 165.8, b 62°/80mm, 121.2°, d^{15} 1.63109, d 1.623, n^{15} 1.50759, n 1.50566** It decomposes under similar conditions to CHCl$_3$, to give phosgene and trichloroacetic acid. Inhibitors of this reaction include EtOH, diethyl ether and thymol (effective at 2-5ppm). Tetrachloroethylene should be distd under a vac (to avoid phosgene formation), and stored in the dark out of contact with air. It can be purified by washing with 2M HCl until the aq phase no longer becomes coloured, then with water, drying with Na$_2$CO$_3$, Na$_2$SO$_4$, CaCl$_2$ or P$_2$O$_5$, and fractionally distilling just before use. 1,1,2-Trichloroethane and 1,1,1,2-tetrachloroethane can be removed by counter-current extraction with EtOH/water.

Tetrachloro-*N*-methylphthalimide *[14737-80-5]* **M 298.9, m 209.7°.** Crystd from absolute EtOH.

2,3,4,6-Tetrachloronitrobenzene *[879-39-0]* **M 260.9, m 42°.** Crystd from aqueous EtOH.

2,3,5,6-Tetrachloronitrobenzene *[117-18-0]* **M 260.9, m 99-100°.** Crystd from aqueous EtOH.

2,3,4,5-Tetrachlorophenol *[4901-51-3]* **M 231.9, m 116-117°, pK$_{Est}$ ~6.2.** Crystd from pet ether.

2,3,4,6-Tetrachlorophenol *[58-90-2]* **M 231.9, m 70°, b 150°/15mm, pK$_{Est}$ ~5.4.** Crystd from pet ether.

2,3,5,6-Tetrachlorophenol *[935-95-5]* **M 231.9, m 115°, pK$_{Est}$ ~5.0.** Crystd from pet ethers.

Tetrachlorophthalic anhydride *[117-08-8]* **M 285.9, m 255-257°.** Crystd from chloroform or *benzene, then sublimed.

2,3,4,6-Tetrachloropyridine *[14121-36-9]* **M 216.9, m 74-75°, b 130-135°/16-20mm, pK$_{Est}$ ~-5.7.** Crystd from 50% EtOH.

Tetracosane *[646-31-1]* **M 338.7, m 54°, b 243-244°/15mm.** Crystd from diethyl ether.

Tetracosanoic (lignoceric) acid *[557-59-5]* **M 368.7, m 84°, 87.5-88°, pK$_{Est}$ ~5.0.** Crystd from acetic acid, Me$_2$CO, toluene, pet ether/Me$_2$CO, *C$_6$H$_6$/Me$_2$CO.

1,2,4,5-Tetracyanobenzene *[712-74-3]* **M 178.1, m 270-272° (280°).** Crystd from EtOH and sublimed *in vacuo*. [Lawton and McRitchie *J Org Chem* **24** 26 *1959*; Bailey et al. *Tetrahedron* **19** 161 *1963*.]

Tetracyanoethylene *[670-54-2]* **M 128.1, m 199-200° (sealed tube).** Crystd from chlorobenzene, dichloroethane, or dichloromethane [Hall et al. *J Org Chem* **52** 5528 *1987*]. Stored at 0° in a desiccator over NaOH pellets. (It slowly evolves HCN on exposure to moist air.) It can also be sublimed at 120° under vacuum. Also purified by repeated sublimation at 120-130°/0.5mm. [Frey et al. *J Am Chem Soc* **107** 748 *1985*; Traylor and Miksztal *J Am Chem Soc* **109** 2778 *1987*.]

7,7,8,8-Tetracyanoquinodimethane *[1518-16-7]* **M 204.2, m 287-290°(dec).** Recrystd from distd, dried, acetonitrile.

Tetradecane *[629-59-4]* **M 198.4, m 6°, b 122°/10mm, 252-254°, d 0.763, n 1.429.** Washed successively with 4M H_2SO_4 and water. Dried over $MgSO_4$ and distd several times under reduced pressure [Poë et al. *J Am Chem Soc* **108** 5459 *1986*].

1-Tetradecanol *[112-72-1]* **M 214.4, m 39-39.5°, b 160°/10mm, 170-173°/20mm.** Crystd from aq EtOH. Purified by zone melting.

Tetradecyl ether (di-tetradecyl ether) *[5412-98-6]* **M 410.7.** Distd under vac and then crystd repeatedly from MeOH/*benzene.

Tetradecyltrimethylammonium bromide *[1119-97-7]* **M 336.4, m 244-249°.** Crystd from acetone or a mixture of acetone and >5% MeOH. Washed with diethyl ether and dried in a vacuum oven at 60°. [Dearden and Wooley *J Phys Chem* **91** 2404 *1987*.]

Tetraethoxymethane *[78-09-1]* **M 192.3, b 159°.** See tetraethyl orthocarbonate on p. 360.

Tetraethylammonium bromide *[71-91-0]* **M 210.2, m 269°(dec), 284°(dec).** Recrystd from EtOH, CHCl₃ or diethyl ether, or, recrystd from acetonitrile, and dried over P_2O_5 under reduced pressure for several days. Also recrystd from EtOH/diethyl ether (1:2), EtOAc, water or boiling MeOH/acetone (1:3) or by adding equal volume of acetone and allowing to cool. Dried at 100° *in vacuo* for 12 days, and stored over P_2O_5.

Tetraethylammonium chloride hydrate *[56-34-8]* **M 165.7, m dec>200°.** Crystd from EtOH by adding diethyl ether, from warm water by adding EtOH and diethyl ether, from dimethylacetamide or from CH_2Cl_2 by addition of diethyl ether. Dried over P_2O_5 in vacuum for several days. Also crystd from acetone/CH_2Cl_2/hexane (2:2:1) [Blau and Espenson *J Am Chem Soc* **108** 1962 *1986*; White and Murray *J Am Chem Soc* **109** 2576 *1987*].

Tetraethylammonium iodide *[68-05-3]* **M 257.2, m 302°, >300°(dec).** Crystd from acetone/MeOH, EtOH/water, dimethylacetamide or ethyl acetate/EtOH (19:1). Dried under vacuum at 50° and stored over P_2O_5.

Tetraethylammonium perchlorate *[2567-83-1]* **M 229.7, m 345°(dec).** Crystd repeatedly from water, aqueous MeOH, acetonitrile or acetone, and dried at 70° under vacuum for 24h. [Cox et al. *J Am Chem Soc* **106** 5965 *1984*; Liu et al. *J Am Chem Soc* **108** 1740 *1986*; White and Murray *J Am Chem Soc* **109** 2576 *1987*.] Also twice crystd from ethyl acetate/95% EtOH (2:1) [Lexa et al. *J Am Chem Soc* **109** 6464 *1987*].

Tetraethylammonium picrate *[741-03-7]* **M 342.1, m >300°(dec).** Purified by successive crystns from water or 95% EtOH followed by drying in vacuum at 70°.

Tetraethylammonium tetrafluoroborate *[429-06-1]* **M 217.1, m 235°, 275-277°, 289-291°.** Recrystd three times from a mixture of ethyl acetate/hexane (5:1) or MeOH/pet ether, then stored at 95° for 48h under vacuum [Henry and Faulkner *J Am Chem Soc* **107** 3436 *1985*; Huang et al. *Anal Chem* **58** 2889 *1986*]. See entry on p. 481 in Chapter 5.

Tetraethylammonium tetraphenylborate *[12099-10-4]* **M 449.4.** Recrystd from aqueous acetone. Dried in a vacuum oven at 60° for several days. *Similarly for the propyl and butyl homologues.*

Tetraethyl 1,1,2,2-ethanetetracarboxylate *[632-56-4]* **M 318.3, m 73-74°.** Twice recrystd from EtOH by cooling to 0°.

Tetraethylene glycol dimethyl ether *[143-24-8]* **M 222.3, b 105°/1mm, d 1.010, n 1.435.** Stood with CaH_2, $LiAlH_4$ or sodium, and distd when required.

Tetraethylenepentamine *[112-57-2]* **M 189.3, b 169-171°/0.05mm, d 0.999, n 1.506, pK_1^{25} 2.98, pK_2^{25} 4.72, pK_3^{25} 8.08, pK_4^{25} 9.10, pK_5^{25} 9.68.** Distd under vacuum. Purified *via* its pentachloride, nitrate or sulfate. Jonassen, Frey and Schaafsma [*J Phys Chem* **61** 504 *1957*] cooled a soln of 150g of the base in 300mL of 95% EtOH, and added dropwise 180mL of conc HCl, keeping the temperature below 20°. The white ppte was filtered, crystd three times from EtOH/water, then washed with diethyl ether and dried by suction. Reilley and Holloway [*J Am Chem Soc* **80** 2917 *1958*], starting with a similar soln cooled to 0°, added slowly (keeping the temperature below 10°) a soln of 4.5g-moles of HNO_3 in 600mL of aqueous 50% EtOH (also cooled to 0°). The ppte was filtered by suction, recrystd five times from aqueous 5% HNO_3, then washed with acetone and absolute EtOH and dried at 50°. [For purification *via* the sulfate see Reilley and Vavoulis (*Anal Chem* **31** 243 *1959*), and for an additional purification step using the Schiff base with benzaldehyde see Jonassen et al. *J Am Chem Soc* **79** 4279 *1957*].

Tetraethyl orthocarbonate (ethyl orthocarbonate, tetraethoxy ethane) *[78-09-1]* **M 192.3, b 59.6-60°/14mm, 158°/atm, 159°/atm, 160-161°/atm, d_4^{20} 0.9186, n_D^{20} 1.3932.** Likely impurities are hydrolysis products. Shake with brine (satd NaCl; dilute with a little Et_2O if amount of material is small) and dry ($MgSO_4$). The organic layer is filtered off and evaporated, and the residue is distd through a helices packed fractionating column with a total reflux partial take-off head. All distns can be done at atmospheric pressure in an inert atmosphere (e.g. N_2). [Roberts and McMahon *Org Synth* Coll Vol IV 457 *1963*; Connolly and Dyson *J Chem Soc* 828 *1937*; Tieckelmann and Post *J Org Chem* **13** 266 *1948*.]

1,1,2,2-Tetrafluorocyclobutane *[374-12-9]* **M 128.1, b 50-50.7°, d 1.275, n 1.3046.** Purified distn or by preparative gas chromatography using a 2m x 6mm(i.d.) column packed with ß,ß'-oxydipropionitrile on Chromosorb P at 33°. [Conlin and Fey *J Chem Soc, Faraday Trans 1* **76** 322 *1980*; Coffmann et al. *J Am Chem Soc* **71** 490 *1949*.]

Tetrafluoro-1,3-dithietane *[1717-50-6]* **M 164.1, m -6°, b 47-48°/760mm, d^{25} 1.6036, n^{25} 1.3908.** Purified by preparative gas chromatography or by distn through an 18in spinning band column. Also purified by shaking vigorously *ca* 40mL with 25mL of 10% NaOH, 5mL of 30% H_2O_2 until the yellow colour disappeared. The larger layer was separated, dried over silica gel to give a colourless liquid boiling at 48°. It had a singlet at -1.77ppm in the NMR spectrum. [Middleton, Howard and Sharkey, *J Org Chem* **30** 1375 *1965*.]

2,2,3,3-Tetrafluoropropanol *[76-37-9]* **M 132.1, b 106-106.5°, pK^{25} 12.74.** Tetrafluoro-propanol (450mL) was added to a soln of 2.25g of $NaHSO_3$ in 90mL of water, shaken vigorously and stood for 24h. The fraction distilling at or above 99° was refluxed for 4h with 5-6g of KOH and rapidly distd, followed by a final fractional distn. [Kosower and Wu *J Am Chem Soc* **83** 3142 *1961*.] Alternatively, shaken with alumina for 24h, dried overnight with anhydrous K_2CO_3 and distd, taking the middle fraction (**b** 107-108°).

Tetera-*n*-heptylammonium bromide *[4368-51-8]* **M 490.7, m 89-91°.** Crystd from *n*-hexane, then dried in a vacuum oven at 70°.

Tetra-*n*-heptylammonium iodide *[3535-83-9]* **M 537.7.** Crystd from EtOH.

Tetra-*n*-hexylammonium bromide *[4328-13-6]* **M 434.6, m 99-100°.** Washed with ether, and dried in a vacuum at room temperature for 3 days.

Tetra-*n*-hexylammonium chloride *[5922-92-9]* **M 390.1.** Crystd from EtOH.

Tetra-*n*-hexylammonium iodide *[2138-24-1]* **M 481.6, m 99-101°.** Washed with diethyl ether and dried at room temperature *in vacuo* for 3 days.

Tetrahexylammonium perchlorate *[4656-81-9]* **M 454.1, m 104-106°.** Crystd from acetone and dried *in vacuo* at 80° for 24h.

Tetrahydrofuran *[109-99-9]* **M 72.1, b 25°/176mm, 66°/760mm, d_4^{20} 0.889, n_D^{20} 1.4070, pK -2.48 (aq H_2SO_4).** It is obtained commercially by catalytic hydrogenation of furan from pentosan-containing agricultural residues. It was purified by refluxing with, and distilling from $LiAlH_4$ which removes water, peroxides, inhibitors and other impurities [Jaeger et al. *J Am Chem Soc* **101** 717 *1979*]. Peroxides can also be removed by passage through a column of activated alumina, or by treatment with aq ferrous sulfate and sodium bisulfate, followed by solid KOH. In both cases, the solvent is then dried and fractionally distd from sodium. Lithium wire or vigorously stirred molten potassium have also been used for this purpose. CaH_2 has also been used as a drying agent.
Several methods are available for obtaining the solvent almost anhydrous. Ware [*J Am Chem Soc* **83** 1296 *1961*] dried vigorously with sodium-potassium alloy until a characteristic blue colour was evident in the solvent at Dry-ice/cellosolve temperatures. The solvent was kept in contact with the alloy until distd for use. Worsfold and Bywater [*J Chem Soc* 5234 *1960*], after refluxing and distilling from P_2O_5 and KOH, in turn, refluxed the solvent with sodium-potassium alloy and fluorenone until the green colour of the disodium salt of fluorenone was well established. [Alternatively, instead of fluorenone, benzophenone, which forms a blue ketyl, can be used.] The tetrahydrofuran was then fractionally distd, degassed and stored above CaH_2. *p*-Cresol or hydroquinone inhibit peroxide formation. The method described by Coetzee and Chang [*Pure Appl Chem* **57** 633 *1985*] for 1,4-dioxane also applies here. Distns should always be done in the presence of a reducing agent, e.g. $FeSO_4$. **It irritates the skin, eyes and mucous membranes and the vapour should never be inhaled. It is HIGHLY FLAMMABLE and the necessary precautions should be taken.**
Rapid purification: Purification as for diethyl ether.

***l*-Tetrahydropalmatine (2,3,9,10-tetramethoxy-6*H*-dibenzo[a,g]quinolizidine)** *[10097-84-4]* **M 355.4, m 148-149°, $[\alpha]_D^{20}$ -291° (EtOH).** Crystd from MeOH by addition of water [see *J Chem Soc (C)* 530 *1967*].

Tetrahydropyran *[142-68-7]* **M 86.1, b 88.0°, n 1.4202, d 0.885, pK -2.79 (aq H_2SO_4).** Dried with CaH_2, then passed through a column of silica gel to remove olefinic impurities and fractionally distd. Freed from peroxides and moisture by refluxing with sodium, then distilling from $LiAlH_4$. Alternatively, peroxides can be removed by treatment with aqueous ferrous sulfate and sodium bisulfate, followed by solid KOH, and fractional distn from sodium.

Tetrahydro-4*H*-pyran-4-one *[29943-42-8]* **M 100.1, b 57-59°/11mm, 65-66°/15mm, 67-68°/18mm, 73°/20mm, 164.7°/atm, 166-166.5°/atm, d_4^{20} 1.0844, n_D^{20} 1.4551.** Purified by repeated distn preferably in a vacuum. [Baker *J Chem Soc* 296 *1944*; IR: Olsen and Bredoch *Chem Ber* **91** 1589 *1958*.] The *oxime* has **m** 87-88° and **b** 110-111°/13mm [Cornubert et al. *Bull Soc Chim Fr* 36 *1950*]. The *4-nitrophenylhydrazone* forms orange-brown needles from EtOH, **m** 186° [Cawley and Plant *J Chem Soc* 1214 *1938*].

Tetrahydrothiophene *[110-01-0]* **M 88.2, m -96°, b 14.5°/10mm, 120.9°/760mm, d 0.997, n 1.5289.** Crude material was purified by crystn of the mercuric chloride complex to a constant melting point. It was then regenerated, washed, dried, and fractionally distd. [Whitehead et al. *J Am Chem Soc* **73** 3632 *1951*.] It has been dried over Na_2SO_4 and distd in a vacuum [Roberts and Friend *J Am Chem Soc* **108** 7204 *1986*].

Tetrahydro-4*H***-thiopyran-4-one** *[1072-72-6]* **M 116.2, m 60-62°, 61-62°, 64-65°, 65-67°.** Purified by recrystn from diisopropyl ether or pet ether and dried in air. If too impure then dissolve in Et$_2$O, wash with aq NaHCO$_3$, then H$_2$O, dried (MgSO$_4$), filtd, evapd and the residue recrystd as before. [Cardwell *J Chem Soc* 715 *1949*.] The *oxime* can be recrystd from CHCl$_3$-pet ether (at -20°) and has **m** 84-85° [Barkenbus et al. *J Org Chem* **20** 871 *1955*]. The *2,4-dinitrophenylhydrazone* has **m** 186° (from EtOAc) [Barkenbus et al. *J Org Chem* **16** 232 *1951*]. The *S-dioxide* is recrystd from AcOH, **m** 173-174° [Fehnel and Carmack *J Am Chem Soc* **70** 1813 *1948*].

Tetrahydroxy-*p***-benzoquinone (2H$_2$O)** *[5676-48-2]* **M 208.1, pK$_1^{30}$ 4,80, pK$_2^{30}$ 6.8.** Crystd from water.

Tetrakis(dimethylamino)ethylene *[996-70-3]* **M 300.2, b 60°/1mm, d 0.861, n 1.4817, pK$_{Est(1)}$<0, pK$_{Est(2)}$<0, pK$_{Est(3)}$~ 1.5, pK$_{Est(4)}$ 5.1.** Impurities include tetramethylurea, dimethylamine, tetramethylethanediamine and tetramethyloxamide. It was washed with water while being flushed with nitrogen to remove dimethylamine, dried over molecular sieves, then passed through a silica gel column (previously activated at 400°) under nitrogen. Degassed on a vacuum line by distn from a trap at 50° to one at -70°. Finally, it was stirred over sodium-potassium alloy for several days. [Holroyd et al. *J Phys Chem* **89** 4244 *1985*; Wiberg *Angew Chem Int Ed Engl* **7** 766 *1968*.]

Tetralin (1,2,3,4-tetrahydronaphthalene) *[119-64-2]* **M 132.2, b 65-66°/5mm, 207.6°/760 mm, d 0.968, n 1.5413.** It was washed with successive portions of conc H$_2$SO$_4$ until the acid layer no longer became coloured, then washed with aq 10% Na$_2$CO$_3$, and then distd water. Dried (CaSO$_4$ or Na$_2$SO$_4$), filtered, refluxed and fractionally distd at under reduced pressure from sodium or BaO. It can also be purified by repeated fractional freezing.
Bass [*J Chem Soc* 3498 *1964*] freed tetralin, purified as above, from naphthalene and other impurities by conversion to ammonium tetralin-6-sulfonate. Conc H$_2$SO$_4$ (150mL) was added slowly to stirred tetralin (272mL) which was then heated on a water bath for about 2h to give complete soln. The warm mixture, when poured into aq NH$_4$Cl soln (120g in 400mL water), gave a white ppte which, after filtering off, was crystd from boiling water, washed with 50% aq EtOH and dried at 100°. Evapn of its boiling aq soln on a steam bath removed traces of naphthalene. The pure salt (229g) was mixed with conc H$_2$SO$_4$ (266mL) and steam distd from an oil bath at 165-170°. An ether extract of the distillate was washed with aq Na$_2$SO$_4$, and the ether was evapd, prior to distilling the tetralin from sodium. Tetralin has also been purified *via* barium tetralin-6-sulfonate, conversion to the sodium salt and decomposition in 60% H$_2$SO$_4$ using superheated steam.

Tetralin hydroperoxide *[771-29-9]* **M 164.2, m 56°.** Crystd from hexane.

α-Tetralone (1,2,3,4-tetrahydro-1-oxonaphthalene) *[529-34-0]* **M 146.2, m 2-7°, 7.8-8.0°, b 75-85°/0.3mm, 89°/0.5mm, 94-95°/2mm, 132-134°/15mm, 143-145°/20mm, d$_4^{20}$ 1.0695, n$_D^{20}$ 1.5665.** Check the IR first. Purify by dissolving 20mL in Et$_2$O (200mL), washing with H$_2$O (100mL), 5% aq NaOH (100mL), H$_2$O (100mL), 3% aq AcOH (100mL), 5% NaHCO$_3$ (100mL) then H$_2$O (100mL) and dry the ethereal layer over MgSO$_4$. Filter, evap and fractionate the residue through a 6in Vigreux column under reduced pres to give a colourless oil (~17g) with **b** 90-91°/0.5-0.7mm. [Snyder and Werber *Org Synth* Coll Vol III 798 *1955*.] It has also been fractionated through a 0.5metre packed column with a heated jacket under reflux using a partial take-off head. [Olson and Bader *Org Synth* Coll Vol IV 898 *1963*.]

β-Tetralone (1,2,3,4-tetrahydro-2-oxonaphthalene) *[530-93-8]* **M 146.2, m 17-18°, ~18°, b 93-95°/2mm, 104-105°/4mm, 114-115°/4-5mm, 140°/18mm, d$_4^{20}$ 1.0000, n$_D^{20}$ 1.5598.** If reasonably pure then fractionate through an efficient column. Otherwise purify *via* the *bisulfite adduct*. To a soln of NaHSO$_3$ (32.5g, 0.31mol) in H$_2$O (57mL) is added 95% EtOH (18mL) and set aside overnight. Any bisulfite-sulfate that separated is removed by filtration and the filtrate is added to the tetralone (14.6g, 0.1mol) and shaken vigorously. The adduct separates in a few minutes as a white ppte and kept on ice for ~3.5h with occasional shaking. The ppte is collected, washed with 95% EtOH (13mL), then with Et$_2$O (4 x 15mL, by stirring the suspension in the solvent, filtering and repeating the process). The colourless product is dried in air and stored in air tight containers in which it is stable for extended periods (yield is ~17g). This bisulfite (5g) is suspended in H$_2$O (25mL) and Na$_2$CO$_3$.H$_2$O (7.5g) is added (pH of soln is ~10). The mixture is then extracted with Et$_2$O (5 x 10mL, i.e. until the aqueous phase does not test for tetralone — see below). Wash the combined extracts with 10% aqueous HCl (10mL), H$_2$O (10mL, i.e. until the washings are neutral), dry (MgSO$_4$), filter, evaporate and distil the residual oil using Claisen flask under reduced pressure and in a N$_2$ atm. The pure tetralone is a colourless liquid **b** 70-71°/0.25mm (see also above). The yield is ~2g. **Tetralone test:** Dissolve a few drops of the tetralone soln (ethereal or aqueous) in 95% EtOH in a test tube and add 10 drops of 25% NaOH down the side of the tube. A deep blue colour develops at the interface with air. [Soffer *Org Synth* Coll Vol IV 903 *1963*; Cornforth et al. *J Chem Soc* 689 *1942*; UV: Soffer et al. *J Am Chem Soc* 1556 *1952*.] The *phenylhydrazone* has **m** 108° [Crawley and Robinson *J Chem Soc* 2001 *1938*].

Tetramethylammonium bromide *[64-20-0]* **M 154.1, sublimes with dec >230°.** Crystd from EtOH, EtOH/diethyl ether, MeOH/acetone, water or from acetone/MeOH (4:1) by adding an equal volume of acetone. It was dried at 110° under reduced pressure or at 140° for 24h.

Tetramethylammonium chloride *[75-57-0]* **M 109.6, m >230°(dec).** Crystd from EtOH, EtOH/CHCl$_3$, EtOH/diethyl ether, acetone/EtOH (1:1), isopropanol or water. Traces of the free amine can be removed by washing with CHCl$_3$.

Tetramethylammonium hydroxide (5H$_2$O) *[10424-65-4 (5H$_2$O); 75-59-2 (aq soln)]* **M 181.2, m 63°, 65-68°.** Freed from chloride ions by passage through an ion-exchange column (Amberlite IRA-400, prepared in its OH$^-$ form by passing 2M NaOH until the effluent was free from chloride ions, then washed with distilled H$_2$O until neutral). A modification, to obtain carbonate-free hydroxide, uses the method of Davies and Nancollas [*Nature* 165 237 *1950*].

Tetramethylammonium iodide *[75-58-1]* **M 201.1, m >230°(dec).** Crystd from water or 50% EtOH, EtOH/diethyl ether, ethyl acetate, or from acetone/MeOH (4:1) by adding an equal volume of acetone. Dried in a vacuum desiccator.

Tetramethylammonium perchlorate *[2537-36-2]* **M 173.6, m >300 °(dec).** Crystd from acetone and dried *in vacuo* at 60° for several days.

Tetramethylammonium tetraphenylborate *[15525-13-0]* **M 393.3.** Recrystd from acetone, acetone/CCl$_4$ and from acetone/1,2-dichloroethane. Dried over P$_2$O$_5$ in vacuum, or in a vacuum oven at 60° for several days.

1,2,3,4-Teteramethylbenzene (prehnitine) *[488-23-3]* **M 134.2, m -6.3°, b 79.4°/10mm, 204-205°/760mm, d 0.905, n 1.5203.** Dried over sodium and distd under reduced pressure.

1,2,3,5-Tetramethylbenzene (isodurene) *[527-53-7]* **M 134.2, m -23.7°, b 74.4°/10mm, 198°/760mm, d 0.890, n 1.5130.** Refluxed over sodium and distd under reduced pressure.

1,2,4,5-tetramethylbenzene (durene) *[95-93-2]* **M 134.2, m 79.5-80.5°.** Chromatographed on alumina, and recrystd from aqueous EtOH or *benzene. Zone-refining removes duroaldehydes. Dried under vacuum. [Yamauchi et al. *J Phys Chem* 89 4804 *1985*.] It has also been sublimed *in vacuo* [Johnston et al. *J Am Chem Soc* 109 1291 *1987*].

N,N,N',N'-**Tetramethylbenzidine** *[366-29-0]* **M 240.4, m 195.4-195.6°, pK$_{Est(1)}$~3.4, pK$_{Est(2)}$~4.5.** Crystd from EtOH or pet ether, then from pet ether/*benzene, and sublimed in a vacuum. [Guarr et al. *J Am Chem Soc* 107 5104 *1985*.] Dried under vac in an Abderhalden pistol, or carefully on a vacuum line.

2,2,4,4-Tetramethylcyclobutan-1,3-dione *[933-52-8]* **M 140.2, m 114.5-114.9°.** Crystd from *benzene and dried under vacuum over P$_2$O$_5$ in an Abderhalden pistol.

3,3,5,5-Tetramethylcyclohexanone *[14376-79-5]* **M 154.3, m 11-12°, 13.2°, b 59-61°, 80-82°/13mm, 196°/760mm, 203.8-204.8°/760mm, d$_D^{20}$ 0.8954, n$_D^{20}$ 1.4515.** Purified first through a

24in column packed with Raschig rings then a 40cm Vigreux colum under reduced pressure (**b** 69-69.3°/7mm, see above). The *oxime* has **m** 144-145° (from 60% EtOH) and the *semicarbazone* has **m** 196-197°, 197-198° (214.5°, 217-218°) [Karasch and Tawney *J Am Chem Soc* **63** 2308 *1941*; UV: Sandris and Ourisson *Bull Soc Chim Fr* 958 *1956*].

p,p'-**Tetramethyldiaminodiphenylmethane** [bis(*p*-dimethylaminophenyl)methane, **Michler's base**] *[101-61-1]* **M 254.4, m 89-90°, b 155-157°/0.1mm, pK$_{Est(1)}$~5.8, pK$_{Est(2)}$~5.1.** Crystd from EtOH (2mL/g) or 95% EtOH (*ca* 12mL/g). It sublimes on heating.

Tetramethylene sulfoxide (tetrahydrothiophen 1-oxide) *[1600-44-8]* **M 104.2, b 235-237°, d 1.175, n 1.525.** Shaken with BaO for 4 days, then distd from CaH$_2$ under reduced pressure.

N,N,N',N'-**Tetramethylethylenediamine (TMEDA, TEMED)** *[110-18-9]* **M 116.2, b 122°, d 1.175, n^{25} 1.4153, pK$_1^{25}$ 5.90, pK$_2^{25}$ 9.14.** Partially dried with molecular sieves (Linde type 4A), and distd in vacuum from butyl lithium. This treatment removes all traces of primary and secondary amines and water. [Hay, McCabe and Robb *J Chem Soc, Faraday Trans 1* **68** 1 *1972*.] Or, dried with KOH pellets. Refluxed for 2h with one-sixth its weight of *n*-butyric anhydride (to remove primary and secondary amines) and fractionally distd. Refluxed with fresh KOH, and distd under nitrogen. [Cram and Wilson *J Am Chem Soc* **85** 1245 *1963*.] Also distd from sodium.

Tetramethylethylenediamine dihydrochloride *[7677-21-8]* **M 198.2, m ~300°.** Crystd from 98% EtOH/conc HCl. Hygroscopic. [Knorr *Chem Ber* **37** 3510 *1904*.]

1,1,3,3-Tetramethylguanidine *[80-70-6]* **M 115.2, b 159-160°, d 0.917 n 1.470, pK^{25} 13.6.** Refluxed over granulated BaO, then fractionally distd.

N,N,N',N'-**Tetramethyl-1,8-naphthalenediamine** *[20734-58-1]* **M 214.3, m 45-48°, 47-48°, b 144-145°/4mm, pK$_1$ -10.5 (aq H$_2$SO$_4$, diprotonation), pK$_2$ 12.34 (monoprotonation).** It is prepared by methylating 1,8-diaminonaphthalene and likely impurities are methylated products. The tetramethyl compound is a stronger base than the unmethylated, di and trimethylated derivatives. The pKa values are: 1,8-(NH$_2$)$_2$ = 4.61, 1,8-(NHMe)$_2$ = 5.61, 1-NHMe-8-NHMe$_2$ = 6.43 and 1,8-(NMe$_2$)$_2$ = 12.34. The mixture is then treated H$_2$O at pH 8 (where all but the required base are protonated) and extracted with Et$_2$O or CHCl$_3$. The dried extract (K$_2$CO$_3$) yields the tetramethyldiamine on evapn which can be distd. It is a strong base with weak nucleophilic properties, e.g. it could not be alkylated by refluxing with EtI in MeCN for 4 days and on treatment with methyl fluorosulfonate only the fluorosulfonate salt of the base is obtained. [NMR: Adler et al. *J Chem Soc, Chem Commun* 723 *1968* ; *J Am Chem Soc* **63** 358 *1941*.] See Proton sponge p. 134.

Tetramethyl orthocarbonate (methyl orthocarbonate, tetramethoxy methane) *[1850-14-2]* **M 136.2, m -5.6°, -5°, -2°, b 113.5°/760mm, 113.5-114°/755mm, 112-114°/atm, d$_4^{20}$ 1.0202, n$_D^{20}$ 1.3860.** Purified in the same way as for tetraethyl orthocarbonate. [Smith *Acta Chem Scand* **10** 1006 *1956*; Tiekelmann and Post *J Org Chem* **13** 266 *1948*.]

2,6,10,14-Tetramethylpentadecane (pristane, norphytane) *[1921-70-6]* **M 268.5, b 68° (bath temp)/0.004mm, 158°/10mm, 296°/atm, d$_4^{20}$ 0.7827, n$_D^{20}$ 1.4385.** Purified by shaking with conc H$_2$SO$_4$ (care with this acid, if amount of pristane is too small then it should be diluted with pet ether *not* Et$_2$O which is quite sol in the acid), the H$_2$O (care as it may heat up if in contact with conc H$_2$SO$_4$), dried (MgSO$_4$) evaporated and distd over Na metal. [Sörensen and Sörensen *Acta Chem Scand* **3** 939 *1949*.]

N,N,N',N'-**Tetramethyl-1,4-phenylenediamine** *[100-22-1]* **M 164.3, m 51°, b 260°/760mm, pK$_1^{20}$ 2.29, pK$_2^{20}$ 6.35.** Crystd from pet ether or water. It can be sublimed or dried carefully in a vacuum line, and stored in the dark under nitrogen. Also recrystd from its melt.

N,N,N',N'-**Tetramethyl-1,4-phenylenediamine dihydrochloride** *[637-01-4]* **M 237.2, m 222-224°.** Crystd from isopropyl or *n*-butyl alcohols, satd with HCl. Treated with aq NaOH to give the free base which was filtered, dried and sublimed in a vacuum. [Guarr et al. *J Am Chem Soc* **107** 5104 *1985*.]

2,2,6,6-Tetramethylpiperidinyl-1-oxy (TEMPO) *[2564-83-2]* **M 156.3, m 36-38°.** Purified by sublimation (33°, water aspirator) [Hay and Fincke *J Am Chem Soc* **109** 8012 *1987*].

2,2,6,6-Tetramethyl-4-piperidone hydrochloride (triacetoneamine) *[33973-59-0]* **M 191.7, m 190°** (dec), **198-199°** (dec), **pK^{25} 7.90.** Purified by recrystn from EtOH/Et$_2$O, MeCN or Me$_2$CO/MeOH. The *free base* has **m** 37-39° (after sublimation), **b** 102-105°/18mm, and *hydrate* **m** 56-58° (wet Et$_2$O); the *hydrobromide* has **m** 203° (from EtOH-Et$_2$O) and the *picrate* has **m** 196° (from aq EtOH). [Sandris and Ourisson *Bull Soc Chim Fr* 345 *1958*.]

Tetramethylthiuram disulfide [bis-(dimethylthiocarbamyl)-disulfide] *[137-26-8]* **M 240.4, m 146-148°, 155-156°.** Crystd (three times) from boiling CHCl$_3$, then recrystd from boiling CHCl$_3$ by adding EtOH dropwise to initiate pptn, and allowed to cool. Finally it was ppted from cold CHCl$_3$ by adding EtOH (which retained the monosulfide in soln). [Ferington and Tobolsky *J Am Chem Soc* **77** 4510 *1955*.]

1,1,3,3-Tetramethyl urea *[632-22-4]* **M 116.2, f -1.2°, b 175.2°/760mm, d 0.969, n 1.453.** Dried over BaO and distd under nitrogen.

Tetramethyl uric acid *[2309-49-1]* **M 224.2, m 225°, 228°, pK$_{Est}$<0.** Crystd from H$_2$O or MeOH.

1,3,5,5-Tetranitrohexahydropyrimidine *[81360-42-1]* **M 270.1, m 153-154°.** Crystd from EtOH (5x), and sublimed (~65°/0.05mm) [*J Org Chem* **47** 2474 *1982*; *J Labelled Comp Radiopharm* **29** 1197 *1991*].

Tetranitromethane *[509-14-8]* **M 196.0, m 14.2°, b 21-23°/23mm, 126°/760mm, d 1.640, n 1.438.** Shaken with dilute NaOH, washed, steam distd, dried with Na$_2$SO$_4$ and fractionally crystd by partial freezing. The melted crystals were dried with MgSO$_4$ and fractionally distd under reduced pressure. Shaken with a large volume of dilute NaOH until no absorption attributable to the *aci*-nitro anion (from mono- di- and tri-nitromethanes) is observable in the water. Then washed with distilled water, and distilled at room temperature by passing a stream of air or nitrogen through the liquid and condensing in a trap at -80°. It can be dried with MgSO$_4$ or Na$_2$SO$_4$, fractionally crystd from the melt, and fractionally distd under reduced pressure.

Tetra(*p*-nitrophenyl)ethylene *[47797-98-8]* **M 512.4, m 306-307°.** Crystd from dioxane or AcOH and dried at 150°/0.1mm. [Gorvin *J Chem Soc* 678 *1959*.]

4,7,13,18-Tetraoxa-1,10-diazabicyclo[8.5.5]eicosane (Cryptand 211) *[31250-06-3]* **M 288.1, pK$_{Est}$ ~ 7.9.** Redistd, dried under high vacuum over 24h, and stored under nitrogen.

1,7,10,16-Tetraoxa-4,13-diazacyclooctadecane (4,13-diaza-18-crown-6) *[23978-55-4]* **M 262.3, m 118-116°, pK$_{Est}$ ~ 8.8.** Twice recrystd from *benzene/n-heptane, and dried for 24h under high vacuum [E.Weber and F.Vögtle *Top Curr Chem* (Springer Verlag, Berlin) **98** 1 *1981*; D'Aprano and Sesta *J Phys Chem* **91** 2415 *1987*].

Tetrapentylammonium bromide *[866-97-7]* **M 378.5, m 100-101°.** See tetra-*n*-amylammonium bromide on p. 356.

Tetraphenylethylene *[632-51-9]* **M 332.4, m 223-224°, b 415-425°/760mm.** Crystd from dioxane or from EtOH/*benzene. Sublimed under high vacuum.

Tetraphenylhydrazine *[632-52-0]* **M 336.4, m 147°, pK$_{Est}$ ~0.** Crystd from 1:1 CHCl$_3$/toluene or CHCl$_3$/EtOH. Stored in a refrigerator, in the dark.

trans-**1,1,4,4-Tetraphenyl-2-methylbutadiene** *[20411-57-8]* **M 372.5.** Crystd from EtOH.

1,2,3,4-Tetraphenylnaphthalene *[751-38-2]* **M 432.6, m 199-201°, 204-204.5°.** Crystd from MeOH or as EtOH. [Fieser and Haddadin *Org Synth* **46** 107 *1966*.]

5,6,11,12-Tetraphenylnaphthacene (Rubrene) *[517-51-1]* **M 532.7, m>315°, 322°, d 1.255**
Orange crysts by sublimation at 250-260°/3-4mm [UV Badger and Pearce *Spectrochim Acta* **4** 280 *1950*]. Also recrystd from *benzene under red light because it is chemiluminescent and light sensitive.

5,10,15,20-Tetraphenylporphyrin (TPP) *[917-23-7]* **M 614.7, λ_{max} 482nm.** Purified by chromatography on neutral (Grade I) alumina, and recrystd from CH_2Cl_2/MeOH [Yamashita et al. *J Phys Chem* **91** 3055 *1987*].

Tetra-*n*-propylammonium bromide *[1941-30-6]* **M 266.3, m >280°(dec).** Crystd from e thyl acetate/EtOH (9:1), acetone or MeOH. Dried at 110° under reduced pressure.

Tetra-*n*-propylammonium iodide *[631-40-3]* **M 313.3, m >280°(dec).** Purified by crystn from EtOH, EtOH/diethyl ether (1:1), EtOH/water or aqueous acetone. Dried at 50° under vacuum. Stored over P_2O_5 in a vacuum desiccator.

Tetra-*n*-propylammonium perchlorate *[15780-02-6]* **M 285.8, m 239-241°.** See tetrapropyl-ammonium perchlorate on p. 483 in Chapter 5.

5,10,15,20-Tetra-4'-pyridinylporphyrin *[16834-13-2]* **M 618.7, m >300°(dec).** Purified by chromatography on alumina (neutral, Grade I), followed by recrystn from CH_2Cl_2/MeOH [Yamashita et al. *J Phys Chem* **91** 3055 *1987*].

Tetrathiafulvalene *[31366-25-3]* **M 204.4, m 122-124°.** Recrystd from cyclohexane/hexane under an argon atmosphere [Kauzlarich et al. *J Am Chem Soc* **109** 4561 *1987*].

1,2,3,4-Tetrazole *[288-94-8]* **M 70.1, m 156°, pK 4.89 (acidic).** Crystd from EtOH, sublimed under high vacuum at *ca* 120° (*care should be taken due to possible* **EXPLOSION**).

Thebaine *[115-37-7]* **M 311.4, m 193°, $[\alpha]_D^{25}$ -219° (EtOH), pK^{15} 8.15.** Sublimed at 170-180°.

2-Thenoyltrifluoroacetone *[326-91-0]* **M 222.2, m 42-44°, b 96-98°/9mm.** Crystd from hexane or *benzene. (Aqueous solns slowly decompose).

2-Thenylamine *[27757-85-3]* **M 113.1, b 78.5°/15mm, pK^{30} 8.92.** Distd under reduced pressure (nitrogen), from BaO, through a column packed with glass helices.

Theobromine *[83-67-0]* **M 180.2, m 337°, pK$_1^{40}$ -0.16, pK$_2^{25}$ 9.96.** Crystd from water.

Theophylline *[58-55-9]* **M 180.2, m 270-274°, pK$_1^{40}$ -0.24, pK$_2$ 2.5, pK$_3^{40}$ 8.79.** Crystd from H_2O.

Thevetin *[11018-93-2]* **M 858.9, m softens at 194°, m 210°.** Crystd (as trihydrate) from isopropanol. Dried at 100°/10mm to give the hemihydrate (*very hygroscopic*).

Thianthrene *[92-85-3]* **M 216.3, m 158°.** Crystd from Me_2CO (charcoal), AcOH or EtOH. Sublimes in a vacuum.

ε-[2-(4-Thiazolidone)]hexanoic acid *[539-35-5]* **M 215.3, m 140°, pK$_{Est}$ ~4.7.** Crystd from H_2O, Me_2CO or MeOH.

Thiazoline-2-thiol *[96-53-7]* **M 119.2, m 106-107°, 106-108°, pK$_{Est}$ ~13.0.** Purified by dissolution in alkali, pptn by addition of HCl and then recrystd from H_2O as needles. [IR: Flett *J Chem Soc* 347 *1953* and Mecke et al. *Chem Ber* **90** 975; Gabriel and Stelzner *Chem Ber* **28** 2931 *1895*.]

4-(2-Thiazolylazo)resorcinol *[2246-46-0]* **M 221.2, m 200-202°(dec), λ_{max} 500 nm, pK_1^{25} 1.25, pK_2^{25} 6.53, pK_3^{25} 10.76.** Dissolved in alkali, extracted with diethyl ether, and re-ppted with dil HCl. The purity was checked by TLC on silica gel using pet ether/diethyl ether/EtOH (10:10:1) as the mobile phase.

Thietane (trimethylene sulfide) *[287-27-4]* **M 74.1, m -64°, -73.2°, b 93.8-94.2°/752mm, 95°/atm, d_4^{20} 1.0200, n_D^{20} 1.5020.** Purified by preparative gas chromatography on a dinonyl phthalate column. It has also been purified by drying over anhydrous K_2CO_3, and distd through a 25 cm glass helices packed column (for 14g of thietane), then dried over $CaSO_4$ before sealing in a vac. [Haines et al. *J Phys Chem* **58** 270 *1954*.] It is characterised as the *dimethylsulfonium iodide* **m** 97-98° [Bennett and Hock *J Chem Soc* 2496 *1927*]. The *S-oxide* has **b** 102°/25mm, n_D^{21} 1.5075 [Tamres and Searles *J Am Chem Soc* **81** 2100 *1959*].

Thioacetamide *[62-55-5]* **M 75.1, m 112-113°, pK^{25} 13.4.** Crystd from absolute diethyl ether or *benzene. Dried at 70° in vacuum and stored over P_2O_5 at 0° under nitrogen. (*Develops an obnoxious odour on storage*, and absorption at 269nm decreases, hence it should be freshly crystd before use).

Thioacetanilide *[677-53-6]* **M 151.2, m 75-76°, pK_{Est} ~13.1.** Crystd from H_2O and dried *in vacuo*.

Thiobarbituric acid *[504-17-6]* **M 144.2, m 235°(dec), pK_1^{25} 2.25, pK_2^{25} 10.72 (2% aq ETOH).** Crystd from water.

Thiobenzanilide *[636-04-4]* **M 213.2, m 101.5-102°, pK_{Est} ~12.6.** Crystd from MeOH at Dry-ice temperature.

(1R)-(-)-Thiocamphor (1R-bornane-2-thione, 1R-(-)-1,7,7-trimethylbicyclo[2.2.1]heptane-2-thione) *[53402-10-1]* **M 168.3, m 136-138°, 146°, $[\alpha]_D^{22}$ -22° (c 3, EtOAc).** Forms red prisms from EtOH and sublimes under vacuum. It possesses a sulfurous odour and is volatile as camphor. [Sen *J Indian Chem Soc* **12** 647 *1935*; **18** 76 *1941*.] The *racemate* crystallises from $*C_6H_6$ and has **m** 145° [138.6-139°, White and Bishop *J Am Chem Soc* **62** 10 *1940*].

1,1'-Thiocarbonyldiimidazole *[6160-65-2]* **M 178.1, m 100-102°, 105-106°.** It forms yellow crystals by recrystn from tetrahydrofuran or by sublimation at 10^{-3}Torr (bath temp 70-80°). Hydrolysed by H_2O, store dry. [Staab and Walther *Justus Liebigs Ann Chem* **657** 98 *1962*; Pullukat et al. *Tetrahedron Lett* 1953 *1967*.]

Thiochrome {2,7-dimethyl-5H-thiachromine-8-ethanol; 3,8-dimethyl-2-hydroxyethyl-5H-thiazolo[2,3:1',2']pyrimido[4',5'-d]pyrimidine} *[92-35-3]* **M 262.3, m 227-228°, pK_{Est} ~ 5.8 (thiazol-N protonation).** Crystd from chloroform.

Thiodiglycollic acid *[123-93-3]* **M 150.2, m 129°, pK_1^{25} 3.15 (3.24), pK_2^{25} 4.13 (4.56).** Crystd from water.

3,3'-Thiodipropionic acid *[111-17-1]* **M 178.2, pK_1^{25} 3.84, pK_2^{25} 4.66.** Crystd from water.

Thioflavine T *[2390-54-7]* **M 318.9, pK^{25} 2.7.** Crystd from *benzene/EtOH (1:1).

Thioformamide *[115-08-2]* **M 61.0, m 29°, pK_{Est} ~12.4.** Crystd from ethyl acetate or ether/pet ether.

Thioglycollic acid *[68-11-1]* **M 92.1, b 95-96°/8mm, d 1.326, n 1.505, pK_1^{25} 3.42, pK_2^{25} 10.20.** Mixed with an equal volume of *benzene, the *benzene is then distd to dehydrate the acid. After heating to 100° to remove most of the *benzene, the residue was distd under vacuum and stored in sealed ampoules at 3°. [Eshelman et al. *Anal Chem* **22** 844 *1960*.]

Thioguanosine (2-amino-6-mercapto-9-β-D-ribofuranosylpurine) *[85-31-4]* **M 299.3, m 230-231°(dec), $[\alpha]_D^{20}$ -64° (c 1.3, 0.1N NaOH), pK 8.33.** Crystd (as hemihydrate) from water.

Thioindigo *[522-75-8]* **M 296.2, m >280°.** Adsorbed on silica gel from CCl_4/*benzene (3:1), eluted with *benzene, crystd from $CHCl_3$ and dried at 60-65°. [Wyman and Brode *J Am Chem Soc* **73** 1487 *1951*.] This paper also gives details of purification of other thioindigo dyes.

Thiomalic (mercaptosuccinic) acid *[70-49-5]* **M 150.2, m 153-154°, pK_1^{25} 3.64 (3.17), pK_2^{25} 4.64 (4.67), pK_3^{25} 10.37 (10.52).** Extracted from aqueous soln several times with diethyl ether, and the aqueous soln freeze-dried.

Thio-Michler's Ketone *[1226-46-6]* **M 284.6, λ_{max} 457 nm (ε 2.92 x 10^4 in 30% aq n-propanol).** Purified by recrystn from hot EtOH or by triturating with a small volume of $CHCl_3$, followed by filtration and washing with hot EtOH [Terbell and Wystrade *J Phys Chem* **68** 2110 *1964*].

Thionanthone (thioxanthone) *[492-22-8]* **M 212.3, m 212-213°, b 371-373°/712mm.** See 9*H*-thioxanthene-9-one on p. 369.

2-Thionaphthol *[91-60-1]* **M 160.2, m 81°, 82°, b 153.5°/15mm, 286°/760mm, pK_{Est} ~6.1.** Crystd from EtOH.

Thionine (3,7-diaminophenothiazine) *[135-59-1; 581-64-6 (HCl)]* **M 263.7, ε_{590} 6.2 x 10^4 M^{-1} cm^{-1}, pK^{15} 6.9.** The standard biological stain is highly pure. It can be crystd from water or 50% EtOH, then chromatographed on alumina using $CHCl_3$ as eluent [Shepp, Chaberek and McNeil *J Phys Chem* **66** 2563 *1962*]. Dried overnight at 100° and stored in a vacuum. The *hydrochloride* can be crystd from 50% EtOH or dilute HCl and aqueous *n*-butanol. Purified also by column chromatography and washed with $CHCl_3$ and acetone. Dried *in vacuo* at room temperature.

Thiooxine hydrochloride (8-mercaptoquinoline hydrochloride) *[34006-16-1]* **M 197.7, m 170-175° (dec), pK_1^{25} 2.16, pK_2^{25} 8.38.** Crystallises from EtOH and the crystals are yellow in colour. It has pKa^{20} values of 2.05 and 8.29 in H_2O. [UV: Albert and Barlin *J Chem Soc* 2384 *1959*.]

Thiophane (tetrahydrothiophene) *[110-01-0]* **M 88.2, b 40.3°/39.7mm.** See tetrahydrothiophene on p. 361.

Thiophene *[110-02-1]* **M 84.1, f -38.5°, b 84.2°, d 1.525, n 1.52890, n^{30} 1.5223.** The simplest purification procedure is to dry with solid KOH, or reflux with sodium, and fractionally distd through a glass-helices packed column. More extensive treatments include an initial wash with aq HCl, then water, drying with $CaSO_4$ or KOH, and passage through columns of activated silica gel or alumina. Fawcett and Rasmussen [*J Am Chem Soc* **67** 1705 *1945*] washed thiophene successively with 7M HCl, 4M NaOH, and distd water, dried with $CaCl_2$ and fractionally distd. *Benzene was removed by fractional crystn by partial freezing, and the thiophene was degassed and sealed in Pyrex flasks. [Also a method is described for recovering the thiophene from the *benzene-enriched portion.]

Thiophene-2-acetic acid *[1918-77-0]* **M 142.2, m 76°, pK^{25} 3.89.** Crystd from ligroin.

Thiophene-3-acetic acid *[6964-21-2]* **M 142.2, m 79-80°, pK_{Est} ~3.1.** Crystd from ligroin.

2-Thiophenecarboxaldehyde *[98-03-3]* **M 112.2, b 106°/30mm, d 1.593, n 1.222.** Washed with 50% HCl and distd under reduced pressure just before use.

Thiophene-2-carboxylic acid *[527-72-0]* **M 128.2, m 129-130°, pK^{25} 3.89.** Crystd from water.

Thiophene-3-carboxylic acid *[88-31-1]* **M 128.1, m 137-138°, pK^{25} 6.23.** Crystd from water.

Thiophenol (benzenethiol) *[108-98-5]* **M 110.2, f -14.9°, b 46.4°/10mm, 168.0°/760mm, d 1.073, n 1.5897, pK^{25} 6.62.** Dried with $CaCl_2$ or $CaSO_4$, and distd at 10mm pressure or at 100mm (**b** 103.5°) in a stream of nitrogen.

Thiopyronine (2,7-dimethylaminothiaxanthane) *[2412-14-8]* **M 318.9**, λ_{max} **564nm** (ε **78,500**) H_2O, pK_{Est} ~ **7**. Purified as the hydrochloride by recrystn from hydrochloric acid. [Fanghanel et al. *J Phys Chem* **91** 3700 *1987*.]

Thiosalicylic (2-mercaptobenzoic) acid *[147-93-3]* **M 154.2**, **m 164-165°**, pK_1^{25} **3.54**, pK_2^{25} **8.80**. Crystd from hot EtOH (4mL/g), after adding hot distd water (8mL/g) and boiling with charcoal. The hot soln was filtered, cooled, the solid collected and dried *in vacuo* (P_2O_5). Cryst from AcOH and sublimes *in vacuo*.

Thiosemicarbazide *[79-19-6]* **M 91.1**, **m 181-183°**, pK_1^{25} **1.88**, pK_2^{25} **12.81**. Crystd from water.

Thiothienoyltrifluoroacetone *[4552-64-1]* **M 228.2**, **m 61-62°**. Easily oxidised and has to be purified before use. This may be by recrystd from *benzene or by dissolution in pet ether, extraction into 1M NaOH soln, acidification of the aqueous phase with 1-6M HCl soln, back extraction into pet ether and final evapn of the solvent. The purity can be checked by TLC. It was stored in ampoules under nitrogen at 0° in the dark. [Muller and Rother *Anal Chim Acta* **66** 49 *1973*.]

Thiouracil *[141-90-2]* **M 128.2**, **m 240°(dec)**, pK^{25} **7.52**. Crystd from water or EtOH.

Thiourea *[62-56-6]* **M 76.1**, **m 179°**, pK^{20} **-1.19** (aq H_2SO_4). Crystd from absolute EtOH, MeOH, acetonitrile or water. Dried under vacuum over H_2SO_4 at room temperature.

9H-Thioxanthene-9-one (thioxanthone) *[492-22-8]* **M 212.3**, **m 200-202°, 209°, 212-214°, b 371-373°/712mm**. Yellow needles from $CHCl_3$ or EtOH and sublimes *in vacuo*. Sol in CS_2, hot AcOH and soln in concn H_2SO_4 to give a yellow color with green fluorescence in VIS light. The *sulfone* has **m 187°** (from EtOH), and the *hydrazone* has **m 115°** (yellow leaflets from EtOH/*C_6H_6). [Szmant et al. *J Org Chem* **18** 745 *1953*; Ullmann et al. *Chem Ber* **49** 2509 *1916*; NMR: Sharpless et al. *Org Magn Res* **6** 115 *1974*.]

L-Threonine *[72-19-5]* **M 119.1**, **m 251-253°**, $[\alpha]_D^{26}$ **-28.4°** (H_2O), pK_1^{25} **2.17**, pK_2^{25} **9.00**. Likely impurities are *allo*-threonine and glycine. Crystd from water by adding 4 volumes of EtOH. Dried and stored in a desiccator.

Thymidine *[50-89-5]* **M 242.2**, **m 185°**, pK_2^{25} **9.65**. Crystd from ethyl acetate.

Thymine *[65-71-4]* **M 126.1**, **m 326°**, pK^{25} **9.82**. Crystd from ethyl acetate or water. Purified by preparative (2mm thick) TLC plates of silica gel, eluting with ethyl acetate/isopropanol/water (75:16:9, v/v; R_F 0.75). Spot localised by uv lamp, cut from plate, placed in MeOH, shaken and filtered through a millipore filter, then rotary evapd. [Infante et al. *J Chem Soc, Faraday Trans 1* **68** 1586 *1973*.]

Thymolphthalein complexone *[1913-93-5]* **M 720.8**, **m 190°(dec)**, $pK_1^{18.2}$ **7.35**, $pK_2^{18.2}$ **12.25**. Purification as for phthalein complexone except that it was synthesised from thymolphthalein instead of cresolphthalein.

Tiglic acid *[80-59-1]* **M 100.1**, **m 63.5-64°, b 198.5°**, pK^{18} **4.96**. Crystd from water.

Tinuvin P (2-[2H-benzotriazol-2-yl]-p-cresol) *[50936-05-5]* **M 225.3**, **m 131-133°**, $pK_{Est(1)}$~**1.6** (N protonation), $pK_{Est(2)}$~ **8** (phenolic OH). Recrystd from *n*-heptane or Me_2CO/pentane. [Woessner et al. *J Phys Chem* **81** 3629 *1985*.]

o-**Tolidine (3,3'-dimethylbenzidine)** *[119-93-7]* **M 212.3**, **m 131-132°**, pK^{25} **4.45**. Dissolved in *benzene, percolated through a column of activated alumina and crystd from *benzene/pet ether.

p-**Tolualdehyde** *[104-87-0]* **M 120.2**, **b 83-85°/0.1mm, 199-200°/760mm, d 1.018, n 1.548**. Steam distd, dried with $CaSO_4$ and fractionally distd.

o-**Toluamide** *[527-85-5]* **M 135.2, m 141°.** Crystd from hot water (10mL/g) and dried in air.

Toluene   *[108-88-3]*   **M 92.1, b 110.6°, d^{10} 0.87615, d^{25} 0.86231, n 1.49693, n^{25} 1.49413.**
Dried with $CaCl_2$, CaH_2 or $CaSO_4$, and dried further by standing with sodium, P_2O_5 or CaH_2. It can be
fractionally distd from sodium or P_2O_5. Unless specially purified, toluene is likely to be contaminated with
methylthiophenes and other sulfur containing impurities. These can be removed by shaking with conc H_2SO_4,
but the temperature must be kept below 30° if sulfonation of toluene is to be avoided. A typical procedure
consists of shaking toluene twice with cold conc H_2SO_4 (100mL of acid per L), once with water, once with
aqueous 5% $NaHCO_3$ or NaOH, again with H_2O, then drying successively with $CaSO_4$ and P_2O_5, with final
distn from P_2O_5 or over $LiAlH_4$ after refluxing for 30min. Alternatively, the treatment with $NaHCO_3$ can be
replaced by boiling under reflux with 1% sodium amalgam. Sulfur compounds can also be removed by
prolonged shaking of the toluene with mercury, or by two distns from $AlCl_3$, the distillate then being washed
with water, dried with K_2CO_3 and stored with sodium wire. Other purification procedures include refluxing and
distn of sodium dried toluene from diphenylpicrylhydrazyl, and from $SnCl_2$ (to ensure freedom from peroxides).
It has also been co-distd with 10% by volume of ethyl methyl ketone, and again fractionally distd. [Brown and
Pearsall *J Am Chem Soc* **74** 191 *1952*.] For removal of carbonyl impurities see **benzene*. Toluene has been
purified by distn under nitrogen in the presence of sodium benzophenone ketyl. Toluene has also been dried
with $MgSO_4$, after the sulfur impurities have been removed, and then fractionally distd from P_2O_5 and stored in
the dark [Tabushi et al. *J Am Chem Soc* **107** 4465 *1985*]. Toluene can be purified by passage through a
tightly packed column of Fuller's earth.
Rapid purification: Alumina, CaH_2 and 4A molecular sieves (3% w/v) may be used to dry toluene (6h
stirring and standing). Then the toluene is distd, discarding the first 5% of distillate, and is stored over
molecular sieves (3A, 4A) or Na wire.

Toluene-2,4-diamine *[95-80-7]* **M 122.2, m 99°, b 148-150°/8mm, 292°/760mm,
$pK_{Est(1)}$~2.5, $pK_{Est(2)}$~4.4.** Recrystd from water containing a very small amount of sodium dithionite (to
prevent air oxidation), and dried under vacuum. Also cryst from **benzene*.

o-**Toluenesulfonamide**   *[88-19-7]* **M 171.2, m 155.5°.** Crystd from hot water, then from EtOH or
Et_2O-pet ether.

p-**Toluenesulfonamide**   *[70-55-3]* **M 171.2, m 137-137.5°, 138°.** Crystd from hot water, then from
EtOH or Et_2O-pet ether.

p-**Toluenesulfonic acid**   *[6192-52-5]* **M 190.2, m 38° (anhydrous), m 105-107°
(monohydrate), pK 1.55.** Purified by pptn from a satd soln at 0° by introducing HCl gas. Also crystd
from conc HCl, then crystd from dilute HCl (charcoal) to remove benzenesulfonic acid. It has been crystd from
EtOH/water. Dried in a vacuum desiccator over solid KOH and $CaCl_2$. *p*-Toluenesulfonic acid can be dehydrated
by azeotropic distn with **benzene* or by heating at 100° for 4h under water-pump vacuum. The anhydrous acid
can be crystd from **benzene*, $CHCl_3$, ethyl acetate, anhydrous MeOH, or from acetone by adding a large excess
of **benzene*. It can be dried under vacuum at 50°.

Toluenesulfonic acid hydrazide (tosylhydrazide) *[1576-35-8]* **M 186.2, m 108-110°, 109-110°.**
Dissolve in hot MeOH (~1g/4mL), filter through Celite and ppte material by adding 2-2.5 vols of distd H_2O.
Air or vac dry. [Fiedman et al. *Org Synth Coll Vol V* 1055 *1973*.]

p-**Toluenesulfonyl chloride**   **(tosyl chloride)** *[98-59-9]* **M 190.7, m 66-69°, 67.5-68.5°, 69°,
b 138-139°/9mm, 146°/15mm, 167°/36mm.** Material that has been standing for a long time contains
tosic acid and HCl and has **m** *ca* 65-68°. It is purified by dissolving (10g) in the minimum volume of $CHCl_3$
(*ca* 25mL) filtered, and diluted with five volumes (i.e. 125mL) of pet ether (b 30-60°) to precipitate impurities.
The soln is filtered, clarified with charcoal and concentrated to 40mL by evaporation. Further evaporation to a
very small volume gave 7g of white crystals which were analytically pure, **m** 67.5-68.5°. (The insoluble
material was largely tosic acid and had **m** 101-104°). [Pelletier *Chem Ind (London)* 1034 *1953*.]
Also crystd from toluene/pet ether in the cold, from pet ether (b 40-60°) or **benzene*. Its soln in diethyl ether
has been washed with aqueous 10% NaOH until colourless, then dried (Na_2SO_4) and crystd by cooling in

powdered Dry-ice. It has also been purified by dissolving in *benzene, washing with aqueous 5% NaOH , then dried with K_2CO_3 or $MgSO_4$, and distd under reduced pressure and can be sublimed at high vacuum [Ebel *Chem Ber* **60** 2086 *1927*].

p-**Toluenethiol** *[106-45-6]* **M 124.2, m 43.5-44°, pK^{25} 6.82.** Crystd from pet ether (b 40-70°).

Toluhydroquinone (2-methylbenzene-1,4-diol) *[95-71-6]* **M 124.1, m 128-129°, pK$_1^{20}$ 10.15, pK$_2^{20}$ 11.75.** Crystd from EtOH.

o-**Toluic acid** *[118-90-1]* **M 136.2, m 102-103°, pK^{25} 3.91.** Crystd from *benzene (2.5mL/g) and dried in air.

m-**Toluic acid** *[99-04-7]* **M 136.2, m 111-113°, pK^{25} 4.27.** Crystd from water.

p-**Toluic acid** *[99-94-5]* **M 136.2, m 178.5-179.5°, pK^{25} 4.37.** Crystd from water, water/EtOH (1:1), MeOH/water or *benzene.

o-**Toluidine (2-methylaniline)** *[95-53-4]* **M 107.2, f -16.3°, b 80.1°/10mm, 200.3°/760mm, d 0.999, n 1.57246, n^{25} 1.56987, pK^{25} 4.45.** In general, methods similar to those for purifying aniline can be used, e.g. distn from zinc dust, at reduced pressure, under nitrogen. Berliner and May [*J Am Chem Soc* **49** 1007 *1927*] purified *via* the oxalate. Twice-distd *o*-toluidine was dissolved in four times its volume of diethyl ether and the equivalent amount of oxalic acid needed to form the dioxalate was added as its soln in diethyl ether. (If *p*-toluidine is present, its oxalate pptes and can be removed by filtration.) Evapn of the ether soln gave crystals of *o*-toluidine dioxalate. They were filtered off, recrystd five times from water containing a small amount of oxalic acid (to prevent hydrolysis), then treated with dilute aqueous Na_2CO_3 to liberate the amine which was separated, dried ($CaCl_2$) and distd under reduced pressure.

m-**Toluidine (3-methylaniline)** *[108-44-1]* **M 107.2, f -30.4°, b 82.3°/10mm, 203.4°/760mm, d 0.989, n 1.56811, n^{25} 1.56570, pK^{25} 4.71.** It can be purified as for aniline. Twice-distd, *m*-toluidine was converted to the hydrochloride using a slight excess of HCl, and the salt was fractionally crystd from 25% EtOH (five times), and from distd water (twice), rejecting, in each case, the first material that crystd. The amine was regenerated and distd as for *o*-toluidine. [Berliner and May *J Am Chem Soc* **49** 1007 *1927*.]

p-**Toluidine (4-methylaniline)** *[106-49-0]* **M 107.2, m 44.8°, b 79.6°/10mm, 200.5°/760mm, d 0.962, n 1.5636, n $^{59.1}$ 1.5534, pK^{25} 5.08.** In general, methods similar to those for purifying aniline can be used. It can be separated from the *o*- and *m*-isomers by fractional crystn from its melt. *p*-Toluidine has been crystd from hot water (charcoal), EtOH, *benzene, pet ether or EtOH/water (1:4), and dried in a vacuum desiccator. It can also be sublimed at 30° under vacuum. For further purification, use has been made of the oxalate, the sulfate and acetylation. The oxalate, formed as described for *o*-toluidine, was filtered, washed and recrystd three times from hot distd water. The base was regenerated with aq Na_2CO_3 and recrystd three times from distd water. [Berliner and May *J Am Chem Soc* **49** 1007 *1927*.] Alternatively, *p*-toluidine was converted to its acetyl derivative which, after repeated crystn from EtOH, was hydrolysed by refluxing (50g) in a mixture of 500mL of water and 115mL of conc H_2SO_4 until a clear soln was obtained. The amine sulfate was isolated, suspended in water, and NaOH was added. The free base was distd twice from zinc dust under vacuum. The *p*-toluidine was then recrystd from pet ether and dried in a vacuum desiccator or in a vacuum for 6h at 40°. [Berliner and Berliner *J Am Chem Soc* **76** 6179 *1954*; Moore et al. *J Am Chem Soc* **108** 2257 *1986*.]

Toluidine Blue O *[93-31-9]* **M 305.8, CI 52040, λmax 626nm, pK^{25} 7.5.** Crystd from hot water (18mL/g) by adding one and a half volumes of alcohol and chilling on ice. Dried at 100° in an oven for 8-10h.

p-**Toluidine hydrochloride** *[540-23-8]* **M 143.6, m 245.9-246.1°.** Crystd from MeOH containing a few drops of conc HCl. Dried under vacuum over paraffin chips.

2-p-Toluidinylnaphthalene-6-sulfonic acid   *[7724-15-4]*   **M 313.9, pK$_{Est}$ ~ 0.** Crystd twice from 2% aqueous KOH and dried under high vacuum for 4h at room temperature. Crystd from water. Tested for purity by TLC on silica gel with isopropanol as solvent. The free acid was obtained by acidifying a satd aqueous soln.

o-Tolunitrile   *[529-19-1]*   **M 117.2, b 205.2°, d 0.992, n 1.5279.** Fractionally distd, washed with conc HCl or 50% H$_2$SO$_4$ at 60° until the smell of isonitrile had gone (this also removed any amines), then washed with saturated NaHCO$_3$ and dilute NaCl solns, then dried with K$_2$CO$_3$ and redistd.

m-Tolunitrile   *[620-22-4]*   **M 117.2, b 209.5-210°/773mm, d 0.986, n 1.5250.** Dried with MgSO$_4$, fractionally distd, then washed with aqueous acid to remove possible traces of amines, dried and redistd.

p-Tolunitrile   *[104-85-8]*   **M 117.2, m 29.5°, b 104-106°/20mm.** Melted, dried with MgSO$_4$, fractionally crystd from its melt, then fractionally distd under reduced pressure in a 6-in spinning band column. [Brown *J Am Chem Soc* **81** 3232 *1959*.] It can also be crystd from *benzene/pet ether (b 40-60°).

4-Tolyl-2-benzoic acid (4'-methylbiphenyl-2-carboxylic acid)   *[7148-03-0]*   **M 196.2, m 138-139°, pK^{25} 3.64.** Crystd from toluene.

p-Tolylacetic acid   *[622-47-9]*   **M 150.2, m 94°.** See 4-methylphenylacetic acid on p. 297.

p-Tolyl carbinol (4-methylbenzyl alcohol)   *[589-18-4]*   **M 122.2, m 61°, b 116-118°/20mm, 217°/760mm.** Crystd from pet ether (b 80-100°, 1g/mL). It can also be distd under reduced pressure.

p-Tolyl disulfide   *[103-19-5]*   **M 246.4, m 45-46°.** Purified by chromatography on alumina using hexane as eluent, then crystd from MeOH. [Kice and Bowers *J Am Chem Soc* **84** 2384 *1962*.]

p-Tolylsulfonylmethyl isocyanide (tosylmethyl isocyanide, TOSMIC)   *[36635-61-7]*   **M 195.2, m 114-115°(dec), 116-117°(dec).** Use an efficient fume cupboard. Purify by dissolving TOSMIC (50g) in CH$_2$Cl$_2$ (150mL) and pass through a column (40x3cm) containing neutral alumina (100g) in CH$_2$Cl$_2$ and eluting with CH$_2$Cl$_2$. A nearly colorless soln (700mL) is collected, evaporated *in vacuo* and the residue (42-47g) of TOSMIC (m 113-114° dec) is recrystd once from MeOH (m 116-117° dec). [Hoogenboom et al. *Org Synth* **57** 102 *1977*; Lensen *Tetrahedron Lett* 2367 *1972*..] Also recrystd from EtOH (charcoal) [Saito and Itano, *J Chem Soc, Perkin Trans 1* 1 *1986*].

p-Tolyl urea   *[622-51-5]*   **M 150.2, m 181°.** Crystd from EtOH/water (1:1).

trans-Traumatic acid (2-dodecene-1,12-dioic acid)   *[6402-36-4]*   **M 228.3, m 165-166°, 150-160°/0.001mm, pK$_{Est(1)}$~4.2, pK$_{Est(2)}$~4.6.** Crystd from EtOH, acetone or glyme.

α,α'-Trehalose (2H$_2$O)   *[6138-23-4]*   **M 378.3, m 96.5-97.5°, 203° (anhydrous).** Crystd (as the dihydrate) from aqueous EtOH. Dried at 13°.

1,2,3-Triaminopropane trihydrochloride   *[free base 21291-99-6]*   **M 198.7, m 250°, pK$_1^{20}$ 3.72, pK$_2^{20}$ 7.95, pK$_3^{20}$ 9.59.** Cryst from EtOH.

1,5,7-Triazabicyclo[4.4.0]dec-5-ene   (TBD, 1,3,4,6,7,8-hexahydro-2h-pyrimido[1,2-a]-pyrimidine)   *[5807-14-7]*   **M 139.2, m 125-130°, pK ~ 16** Cryst from Et$_2$O but readily forms white crystals of the carbonate. It is a strong base (see pK, i.e. about 100 times more basic than tetramethylguanidine. The *picrate* has **m** 220.5-222° (from EtOH). Forms the *5-nitro deivative* **m** 14.5-160° that gives a *5-nitro nitrate* salt **m** 100-101° (from EtOH-Et$_2$O) and a *5-nitro picrate* **m** 144-145° (from H$_2$O). [McKay and Kreling *Can J Chem* **35** 1438 *1957*; Schwesinger *Chimia* **39** 369 *1985*; Hilpert et al. *J Chem Soc, Chem Commun* 1401 *1983*; Kamfen and Eschenmoser *Helv Chim Acta* **72** 185 *1989*].

1,2,4-Triazole   *[288-88-0]*   **M 69.1, m 121°, 260°, pK$_1^{25}$ 2.27 (basic), pK$_2^{25}$ 10.26 (acidic).** Crystd from EtOH or water [Barszcz et al. *J Chem Soc, Dalton Trans* 2025 *1986*].

Tribenzylamine *[620-40-6]* **M 287.4, m 93-94°, 230°/13mm, pK$_{Est}$ <0.** Crystd from abs EtOH or pet ether. Dried in a vacuum over P$_2$O$_5$ at room temperature. *HCl* has **m** 226-228° (from EtOH) and *picrate* has **m** 191° (from H$_2$O or aq EtOH).

2,4,6-Tribromoacetanilide *[607-93-2]* **M 451.8, m 232°.** Crystd from EtOH.

2,4,6-Tribromoaniline *[147-82-0]* **M 329.8, m 120°, pK$_{Est}$ ~-0.5 (aq H$_2$SO$_4$).** Crystd from MeOH.

***sym*-Tribromobenzene** *[626-39-1]* **M 314.8, m 122°.** Crystd from glacial acetic acid/water (4:1), then washed with chilled EtOH and dried in air.

Tribromochloromethane *[594-15-0]* **M 287.2, m 55°.** Melted, washed with aqueous Na$_2$S$_2$O$_3$, dried with BaO and fractionally crystd from its melt.

2,4,6-Tribromophenol *[118-79-6]* **M 330.8, m 94°, pK^{25} 6.00.** Crystd from EtOH or pet ether. Dried under vacuum over P$_2$O$_5$ at room temperature.

Tri-*n*-butylamine *[102-82-9]* **M 185.4, b 68°/3mm, 120°/44mm, d 0.7788, n 1.4294, pK^{25} 9.93.** Purified by fractional distn from sodium under reduced pressure. Pegolotti and Young [*J Am Chem Soc* 83 3251 *1961*] heated the amine overnight with an equal volume of acetic anhydride, in a steam bath. The amine layer was separated and heated with water for 2h on the steam bath (to hydrolyse any remaining acetic anhydride). The soln was cooled, solid K$_2$CO$_3$ was added to neutralize any acetic acid that had been formed, and the amine was separated, dried (K$_2$CO$_3$) and distd at 44mm pressure. Davis and Nakshbendi [*J Am Chem Soc* 84 2085 *1926*] treated the amine with one-eighth of its weight of benzenesulfonyl chloride in aqueous 15% NaOH at 0-5°. The mixture was shaken intermittently and allowed to warm to room temperature. After a day, the amine layer was washed with aq NaOH, then water and dried with KOH. (This treatment removes primary and secondary amines.) It was further dried with CaH$_2$ and distd under vacuum.

Tri-*n*-butylammonium hydrobromide *[37026-85-0]* **M 308.3, m 75.2-75.9°.** Crystd from ethyl acetate.

Tri-*n*-butylammonium nitrate *[33850-87-2]* **M 304.5.** Crystd from mixtures of *n*-hexane and acetone (95:5). Dried over P$_2$O$_5$.

Tri-*n*-butylammonium perchlorate *[14999-66-7]* **M 285.5.** Recrystd from *n*-hexane.

***sym*-Tri-*tert*-butylbenzene** *[1460-02-2]* **M 246.4, m 73.4-73.9°.** Crystd from EtOH.

2,4,6-Tri-*tert*-butylphenol *[732-26-3]* **M 262.4, m 129-132°, 131°/1mm, 147°/10mm, 278°/760mm, pK^{25} 12.19.** Crystd from *n*-hexane or several times from 95% EtOH until the EtOH soln was colourless [Balasubramanian and Bruice *J Am Chem Soc* 108 5495 *1986*]. It has also been purified by sublimation [Yuan and Bruice *J Am Chem Soc* 108 1643 *1986*; Wong et al. *J Am Chem Soc* 109 3428 *1987*]. Purification has been achieved by passage through a silica gel column followed by recrystn from *n*-hexane [Kajii et al. *J Phys Chem* 91 2791 *1987*].

Tricarballylic acid (propane-1,2,3-tricarboxylic acid) *[99-14-9]* **M 176.1, m 166°, pK$^{25}_1$ 3.47, pK$^{25}_2$ 4.54, pK$^{20}_3$ 5.89.** Crystd from diethyl ether.

Trichloroacetamide *[594-65-0]* **M 162.4, m 139-141°, b 238-240°.** Its xylene soln was dried with P$_2$O$_5$, then fractionally distd.

Trichloroacetanilide *[2563-97-5]* **M 238.5, m 95°.** Crystd from *benzene.

Trichloroacetic acid *[76-03-9]* **M 163.4, m 59.4-59.8°, pK²⁵ 0.51.** Purified by fractional crystn from its melt, then crystd repeatedly from dry *benzene and stored over conc H_2SO_4 in a vac desiccator. It can also be crystd from $CHCl_3$ or cyclohexane, and dried over P_2O_5 or $Mg(ClO_4)_2$ in a vac desiccator. Trichloroacetic acid can be fractionally distd under reduced pressure from $MgSO_4$. Layne, Jaffé and Zimmer [*J Am Chem Soc* **85** 435 *1963*] dried trichloroacetic acid in *benzene by distilling off the *benzene-water azeotrope, then crystd the acid from the remaining *benzene soln. Manipulations were carried out under nitrogen. [*Use a well ventilated fume cupboard*].

2,3,4-Trichloroaniline *[634-67-3]* **M 196.5, m 67.5°, b 292°/774mm, pK_Est ~1.3.** Crystd from ligroin.

2,4,5-Trichloroaniline *[636-30-6]* **M 196.5, m 96.5°, b 270°/760mm, pK 1.09.** Crystd from ligroin.

2,4,6-Trichloroaniline *[634-93-5]* **M 196.5, m 78.5°, b 127°/14mm, 262°/746mm, pK 0.03.** Crystd from ligroin.

1,2,3-Trichlorobenzene *[87-61-6]* **M 181.5, m 52.6°.** Crystd from EtOH.

1,2,4-Trichlorobenzene *[120-82-1]* **M 181.5, m 17°, b 210°.** Separated from a mixture of isomers by washing with fuming H_2SO_4, then water, drying with $CaSO_4$ and slowly fractionally distilling. [Jensen, Marino and Brown *J Am Chem Soc* **81** 3303 *1959*.]

1,3,5-Trichlorobenzene *[108-70-3]* **M 181.5, m 64-65°.** Recrystd from dry *benzene or toluene.

3,4,5-Trichloro-*o*-cresol (3,4,5-trichloro-2-methylphenol) *[608-92-4]* **M 211.5, m 77°, pK_Est ~7.6.** Crystd from pet ether.

2,3,5-Trichloro-*p*-cresol *[608-91-3]* **M 211.5, m 66-67°, pK_Est ~6.9.** Crystd from pet ether.

1,1,1-Trichloroethane *[71-55-6]* **M 133.4, f -32.7°, b 74.0°, d 1.337, n 1.4385.** Washed successively with conc HCl (or conc H_2SO_4), aq 10% K_2CO_3 (Na_2CO_3), aq 10% NaCl, dried with $CaCl_2$ or Na_2SO_4, and fractionally distd. It can contain up to 3% dioxane as preservative. This is removed by washing successively with 10% aq HCl, 10% aq $NaHCO_3$ and 10% aq NaCl; and distd over $CaCl_2$ before use.

1,1,2-Trichloroethane *[79-00-5]* **M 133.4, f -36.3°, b 113.6°, d 1.435, n 1.472.** Purification as for 1,1,1-trichloroethane above.

Trichloroethylene *[79-01-6]* **M 131.4, f -88°, b 87.2°, d 1.463, n²¹ 1.4767.** Undergoes decomposition in a similar way to $CHCl_3$, giving HCl, CO, $COCl_2$ and organic products. It reacts with KOH, NaOH and 90% H_2SO_4, and forms azeotropes with water, MeOH, EtOH, and acetic acid. It is purified by washing successively with 2M HCl, water and 2M K_2CO_3, then dried with K_2CO_3 and $CaCl_2$, and fractionally distd immediately before use. It has also been steam distd from 10% $Ca(OH)_2$ slurry, most of the water being removed from the distillate by cooling to -30° to -50° and filtering off the ice through chamois skin: the trichloroethylene was then fractionally distd at 250mm pressure and collected in a blackened container. [Carlisle and Levine *Ind Eng Chem (Anal Ed)* **24** 1164 *1932*.]

2,4,5-Trichloro-1-nitrobenzene *[89-69-0]* **M 226.5, m 57°.** Crystd from EtOH.

3,4,6-Trichloro-2-nitrophenol *[82-62-2]* **M 242.4, m 92-93°, pK_Est ~4.1.** Crystd from pet ether or EtOH.

2,4,5-Trichlorophenol *[95-95-4]* **M 197.5, m 67°, b 72°/1mm, pK²⁵ 7.0.** Crystd from EtOH or pet ether.

2,4,6-Trichlorophenol *[88-06-2]* **M 197.5, m 67-68°, pK²⁵ 6.23.** Crystd from *benzene, EtOH or EtOH/water.

3,4,5-Trichlorophenol *[609-19-8]* **M 197.5, m 100°, pK²⁵ 7.84.** Crystd from pet ether/*benzene mixture.

2,4,5-Trichlorophenoxyacetic acid (2,4,5-T) *[93-76-5]* **M 255.5, m 153°, 155-158°, pK²⁵ 2.83.** Crystd from *benzene. **(CANCER SUSPECT)**

1,1,2-Trichlorotrifluoroethane *[76-13-1]* **M 187.4, b 47.6°/760mm, d 1.576, n 1.360.** Washed with water, then with weak alkali. Dried with $CaCl_2$ or H_2SO_4 and distd. [Locke et al. *J Am Chem Soc* **56** 1726 *1934*.]

Tricycloquinazoline *[195-84-6]* **M 230.3, m 322-323°.** Crystd repeatedly from toluene, followed by vac sublimation at 210° at a pressure of 0.15-0.3 Torr in subdued light.

Tridecanoic acid *[638-53-9]* **M 214.4, m 41.8°, 44.5-45.5° (several forms), b 199-200°/24mm, pK$_{Est}$ ~5.0.** Crystd from acetone.

7-Tridecanone *[462-18-0]* **M 198.4, m 33°, b 255°/766mm.** Crystd from EtOH.

Tri-*n*-dodecylammonium nitrate *[2305-34-2]* **M 585.0.** Crystd from *n*-hexane/acetone (95:5) and kept in a desiccator over P_2O_5.

Tri-*n*-dodecylammonium perchlorate *[5838-82-4]* **M 622.4.** Recrystd from *n*-hexane or acetone and kept in a desiccator over P_2O_5.

Triethanolamine hydrochloride *[637-39-8]* **M 185.7, m 177°, pK²⁵ 7.92 (free base).** Crystd from EtOH. Dried at 80°.

1,1,2-Triethoxyethane *[4819-77-6]* **M 162.2, b 164°, d 0.897, n 1.401.** Dried with Na_2SO_4, and distd.

Triethylamine *[121-44-8]* **M 101.2, b 89.4°, d 0.7280, n 1.4005, pK²⁵ 10.82.** Dried with $CaSO_4$, $LiAlH_4$, Linde type 4A molecular sieves, CaH_2, KOH, or K_2CO_3, then distd, either alone or from BaO, sodium, P_2O_5 or CaH_2. It has also been distd from zinc dust, under nitrogen. To remove traces of primary and secondary amines, triethylamine has been refluxed with acetic anhydride, benzoic anhydride, phthalic anhydride, then distd, refluxed with CaH_2 (ammonia-free) or KOH (or dried with activated alumina), and again distd. Another purification involved refluxing for 2h with *p*-toluenesulfonyl chloride, then distd. Grovenstein and Williams [*J Am Chem Soc* **83** 412 1961] treated triethylamine (500mL) with benzoyl chloride (30mL), filtered off the ppte, and refluxed the liquid for 1h with a further 30mL of benzoyl chloride. After cooling, the liquid was filtered, distd, and allowed to stand for several hours with KOH pellets. It was then refluxed with, and distd from, stirred molten potassium. Triethylamine has been converted to its hydrochloride, crystd from EtOH (to **m** 254°), then liberated with aq NaOH, dried with solid KOH and distd from sodium under nitrogen.

Triethylammonium hydrobromide *[636-70-4]* **M 229.1, m 248°.** Equimolar portions of triethylamine and aqueous solutions of HBr in acetone were mixed. The ppted salt was washed with anhydrous acetone and dried in vacuum for 1-2h. [Odinekov et al. *J Chem Soc, Faraday Trans 2* **80** 899 *1984*.] Recrystd from $CHCl_3$ or EtOH.

Triethylammonium hydrochloride *[554-68-7]* **M 137.7, m 257-260°(dec).** Purified like the bromide above.

Triethylammonium hydroiodide *[4636-73-1]* **M 229.1, m 181°.** Purified as for triethylammonium bromide, except the soln for pptn was precooled acetone at -10° and the ppte was twice recrystd from a cooled acetone/hexane mixture at -10°.

Triethylammonium trichloroacetate *[4113-06-8]* **M 263.6.** Equimolar solns of triethylamine and trichloroacetic acid in *n*-hexane were mixed at 10°. The solid so obtained was recrystd from $CHCl_3$/*benzene.

Triethylammonium trifluoroacetate *[454-49-9]* **M 196.2.** Purified as for the corresponding trichloroacetate. The salt was a colourless liquid at ambient temperature.

1,2,4-Triethylbenzene *[877-44-1]* **M 162.3, b 96.8-97.1°/12.8mm, d 0.8738, n 1.5015.** For separation from a commercial mixture see Dillingham and Reid [*J Am Chem Soc* **60** 2606 *1938*].

1,3,5-Triethylbenzene *[102-25-0]* **M 162.3, b 102-102.5°, d 0.8631, n 1.4951.** For separation from a commercial mixture see Dillingham and Reid [*J Am Chem Soc* **60** 2606 *1938*].

Triethylene glycol *[112-27-6]* **M 150.2, b 115-117°/0.1mm, 278°/760mm, n^{15} 1.4578, d^{15} 1.1274.** Dried with $CaSO_4$ for 1 week, then repeatedly and very slowly fractionally distd under vacuum. Stored in a vacuum desiccator over P_2O_5. It is very *hygroscopic*.

Triethylene glycol dimethyl ether (triglyme) *[112-49-2]* **M 178.2, b 225°, d 0.987, n 1.425.** Refluxed with, and distd from sodium hydride or $LiAlH_4$.

Triethylenetetramine (TRIEN, TETA, trientine) *[112-24-3]* **M 146.2, m 12°, b 157°/20mm, d 0.971, n 1.497, pK_1^{25} 3.32, pK_2^{25} 6.67, pK_3^{25} 9.20, pK_4^{25} 9.92.** Dried with sodium, then distd under vac. Further purification has been *via* the nitrate or the chloride. For example, Jonassen and Strickland [*J Am Chem Soc* **80** 312 *1958*] separated TRIEN from admixture with TREN (38%) by soln in EtOH, cooling to approximately 5° in an ice-bath and adding conc HCl dropwise from a burette, keeping the temperature below 10°, until all of the white crystalline ppte of TREN.HCl had formed and been removed. Further addition of HCl then ppted thick creamy white TRIEN.HCl which was crystd several times from hot water by adding an excess of cold EtOH. The crystals were finally washed with Me_2CO, then Et_2O and dried in a vacuum desiccator.

Triethylenetetramine tetrahydrochloride (TRIEN.HCl) *[4961-10-4]* **M 292.1, m 266-270°.** Crystd repeatedly from hot water by pptn with cold EtOH or EtOH/HCl. Washed with acetone and abs EtOH and dried in a vacuum oven at 80° (see TRIEN above).

Triethyl orthoformate (1,1,1-triethoxymethane) *[122-51-0]* **M 148.2, m 30°, b 60°/30mm, 144-146°, d 0.891, n 1.392.** Fractionate first at atm press, then in a vac. If impure, then wash with H_2O, dry over anhyd K_2CO_3, filter and fractionate through a Widmer column. [Sah and Ma *J Am Chem Soc* **54** 2964 *1932*; Ohme and Schmitz *Justus Liebigs Ann Chem* **716** 207 *1968*.] **IRRITANT** and **FLAMMABLE**.

Triethyloxonium fluoroborate *[368-39-8]* **M 190.0, m 92-93°(dec).** Crystd from diethyl ether. *Very hygroscopic*, and should be handled in a dry box and stored at 0°. [*Org Synth* **46** 113 *1966*.] Pure material should give a clear and colourless soln in dichloromethane (1 in 50, w/v).

Trifluoroacetic acid *[76-05-1]* **M 114.0, f -15.5°, b 72.4°, d 1.494, n 1.2850, pK^{25} 0.52.** The purification of trifluoroacetic acid, reported in earlier editions of this work, by refluxing over $KMnO_4$ for 24h and slowly distilling has resulted in very **SERIOUS EXPLOSIONS** on various occasions, but not always. This apparently depends on the source and/or age of the acid. The method is NOT RECOMMENDED. Water can be removed by adding trifluoroacetic anhydride (0.05%, to diminish water content) and distd. [Conway and Novak *J Phys Chem* **81** 1459 *1977*]. It can be refluxed and distd from P_2O_5. It is further purified by fractional crystn by partial freezing and again distd. **Highly TOXIC vapour.**

Trifluoroacetic anhydride *[407-25-0]* **M 210.0, b 38-40°/760mm, d 1.508.** Purification by distilling over KMnO$_4$, as for the acid above is **EXTREMELY DANGEROUS** due to the possiblility of **EXPLOSION**. It is best purified by distilling from P$_2$O$_5$ slowly, and collecting the fraction boiling at 39.5°. Store in a dry atmosphere. **Highly TOXIC vapour.**

1,1,1-Trifluoro-2-bromoethane *[421-06-7]* **M 163.0.** See 1-bromo-3,3,3-trifluoroethane on p. 142.

2,2,2-Trifluoroethanol *[75-89-8]* **M 100.0, b 72.4°/738mm, d 1.400, pK^{25} 12.8.** Dried with CaSO$_4$ and a little NaHCO$_3$ (to remove traces of acid). **Highly TOXIC vapour.**

Trifluoromethanesulfonic anhydride (triflic anhydride) *[358-23-6]* **M 282.1, b 82-85°, 84°, d 1.71, n 1.322.** Distil through a short Vigreux column. Could be freshly prepd from the anhydrous acid (11.5g) and P$_2$O$_5$ (11.5g, or half this weight) by setting aside at room temp for 1h, distilling off volatile products then through a short Vigreux column. Readily hydrolysed by H$_2$O and decomposes appreciably after a few days to liberate SO$_2$ and produce a viscous liquid. Store dry at low low temp. [Burdon et al. *J Chem Soc* 2574 *1957*; Beard et al. *J Org Chem* **38** 373 *1973*.] **Highly TOXIC vapour.**

4-(Trifluoromethyl)acetophenone *[709-63-7]* **M 188.2, m 31-33°, b 79-81°/9mm.** Purified by distillation or sublimation *in vacuo*.

3-Trifluoromethyl-4-nitrophenol *[88-30-2]* **M 162.1, m 81°, pK$_{Est}$ ~6.1.** Crystd from *benzene or from pet ether/*benzene mixture.

α,α,α-Trifluorotoluene (benzotrifluoride) *[98-08-8]* **M 144.1, b 102.5°, d 1.190, n^{30} 1.4100.** Purified by repeated treatment with boiling aqueous Na$_2$CO$_3$ (until no test for chloride ion was obtained), dried with K$_2$CO$_3$, then with P$_2$O$_5$, and fractionally distd.

Triglycyl glycine (tetraglycine) *[637-84-3]* **M 246.2, m 270-275°(dec).** Crystd from distilled water (optionally, by the addition of EtOH).

Trigonelline (1-methylnicotinic acid zwitterion) *[535-83-1]* **M 137.1, m 218°(dec).** Crystd (as monohydrate) from aqueous EtOH, then dried at 100°.

2,3,4-Trihydroxybenzoic acid *[610-02-6]* **M 170.1, m 207-208°, pK$_{Est(1)}$~3.4, pK$_{Est(2)}$~7.8, pK$_{Est(3)}$>12.** Crystd from water.

2,4,6-Trihydroxybenzoic acid *[83-30-7]* **M 170.1, m 205-212°(dec), pK$_{Est(1)}$~1.5, pK$_{Est(2)}$~8.0, pK$_{Est(3)}$>12.** Crystd from water.

4',5,7-Trihydroxyflavone (apigenin) *[520-36-5]* **M 270.2, m 296-298°, 300-305°, 345-350° (pK's 7—10, for phenolic OH).** Crystd from aq pyridine or aq EtOH. Dyes wool yellow when with added Cr.

3,4,5-Triiodobenzoic acid *[2338-20-7]* **M 499.8, m 289-290°, 293°, pK^{25} 0.65.** Crystd from aqueous EtOH or water.

3,4,5-Triiodobenzyl chloride *[52273-54-8]* **M 504.3, m 138°.** Crystd from CCl$_4$/pet ether (charcoal).

3,3',5-Triiodo-*S*-thyronine *[6893-02-3]* **M 651.0, m 236-237°(dec), [α]$_D^{29.5}$ +21.5° (EtOH/1M aq HCl, 2:1), pK$_1^{25}$ 6.48, pK$_2^{25}$7.62, pK$_3^{25}$ 7.82.** Likely impurities are as in *thyroxine*. Purified by dissolving in dilute NH$_3$ at room temperature, then crystd by addition of dilute acetic acid to pH 6.

Trimellitic (benzene-1,2,4-tricarboxylic) acid *[528-44-9]* **M 210.1, m 218-220°, pK$_1^{25}$ 2.42, pK$_2^{25}$ 3.71, pK$_3^{25}$ 5.01.** Crystd from acetic acid or aqueous EtOH.

1,2,3-Trimethoxybenzene *[634-36-6]* **M 168.2, m 45-46°.** Sublimed under vacuum.

1,3,5-Trimethoxybenzene *[621-23-8]* **M 168.2, m 53°.** Sublimed under vacuum.

Trimethylamine *[75-50-3]* **M 59.1, b 3.5°, pK^{25} 9.80.** Dried by passage of the gas through a tower filled with solid KOH. Water and impurities containing labile hydrogen were removed by treatment with freshly sublimed, ground, P_2O_5. Has been refluxed with acetic anhydride, and then distd through a tube packed with HgO and BaO. [Comyns *J Chem Soc* 1557 *1955.*] For more extensive purification, trimethylamine has been converted to the hydrochloride, crystd (see below), and regenerated by treating the hydrochloride with excess aq 50% KOH, the gas passing through a $CaSO_4$ column into a steel cylinder containing sodium ribbon. After 1-2 days, the cylinder was cooled at -78° and hydrogen and air were removed by pumping. [Day and Felsing *J Am Chem Soc* **72** 1698 *1950.*] Trimethylamine has also been trap-to-trap distd and then freeze-pump-thaw degassed [Halpern et al. *J Am Chem Soc* **108** 3907 *1986*].

Trimethylamine hydrochloride *[593-81-7]* **M 95.7, m >280°(dec).** Crystd from $CHCl_3$, EtOH or *n*-propanol, and dried under vacuum. It has also been crystd from *benzene/MeOH, MeOH/diethyl ether and dried under vacuum over paraffin wax and H_2SO_4. Stood over P_2O_5. It is *hygroscopic*.

Trimethylamine hydroiodide *[20230-89-1]* **M 186.0, m 263°.** Crystd from MeOH.

1,2,4-Trimethylbenzene (pseudocumene) *[95-63-6]* **M 120.2, m -43.8°, b 51.6°/10mm, 167-168°/760mm, d 0.889, n 1.5048.** Refluxed over sodium and distd under reduced pressure.

2,4,6-Trimethylbenzoic acid (mesitoic acid) *[480-63-7]* **M 164.2, m 155°, pK^{25} 3.45.** Crystd from water, ligroin or carbon tetrachloride [Ohwada et al. *J Am Chem Soc* **108** 3029 *1986*].

Trimethyl-1,4-benzoquinone *[935-92-2]* **M 150.1, m 29-30°, 36°, b 98°/10mm, 108°/18mm.** Distd in a vac or sublimed *in vacuo* before use. [Smith et al. *J Am Chem Soc* **60** 318 *1939.*]

***R*-(-)-2,2,6-Trimethyl-1,4-cyclohexanedione** *[60046-49-3]* **M 154.2, m 88-90°, 91-92°, [α]$_D^{20}$ -270° (c 0.4%, MeOH), [α]$_D^{20}$ -275° (c 1, CHCl$_3$).** Obtained from fermentation and purified by recrystn from diisopropyl ether. [ORD: Leuenberger et al. *Helv Chim Acta* **59** 1832 *1976.*] The *racemate* has **m** 65-67° and the *4-(4-phenyl)semicarbazone* has **m** 218-220° (from CH_2Cl_2-MeOH) [Isler et al. *Helv Chim Acta* **39** 2041 *1956.*]

2,2,5-Trimethylhexane *[3522-94-9]* **M 128.3, m -105.8°, b 124.1°, d 0.716, n 1.39971, n^{25} 1.39727.** Extracted with conc H_2SO_4, washed with H_2O, dried (type 4A molecular sieves), and fractionally distd.

Trimethyl-1,4-hydroquinone (2,3,5-trimethylbenzene-1,4-diol) *[700-13-0]* **M 152.2, m 173-174°, pK$_{Est(1)}$~ 11.1, pK$_{Est(2)}$~ 12.7.** Recrystd from water, under anaerobic conditions.

1',3',3'-Trimethyl-6-nitrospiro[2H-benzopyran-2,2'-indoline] *[1498-88-0]* **M 322.4, m 180°.** Recrystd from absolute EtOH [Hinnen et al. *Bull Soc Chim Fr* 2066 *1968*; Ramesh and Labes *J Am Chem Soc* **109** 3228 *1987*].

Trimethylolpropane *[77-99-6]* **M 134.2, m 57-59°.** Crystd from acetone and ether.

2,2,3-Trimethylpentane *[564-02-3]* **M 114.2, b 109.8°, d 0.7161, n 1.40295, n^{25} 1.40064.** Purified by azeotropic distn with 2-methoxyethanol, which was subsequently washed out with water. The trimethylpentane was then dried and fractionally distd. [Forziati et al. *J Res Nat Bur Stand* **36** 129 *1946*.]

2,2,4-Trimethylpentane (isooctane) *[540-84-1]* **M 114.2, b 99.2°, d 0.693, n 1.39145, n^{25} 1.38898.** Distd from sodium, passed through a column of silica gel or activated alumina (to remove traces of olefins), and again distd from sodium. Extracted repeatedly with conc H_2SO_4, then agitated with aqueous $KMnO_4$, washed with water, dried ($CaSO_4$) and distd. Purified by azeotropic distn with EtOH, which was subsequently washed out with water, and the trimethylpentane was dried and fractionally distd. [Forziati et al. *J Res Nat Bur Stand* **36** 126 *1946*.] Also purified by fractional crystn.

2,3,5-Trimethylphenol *[697-82-5]* **M 136.2, m 95-96°, b 233°/760mm, pK^{25} 10.67.** Crystd from water or pet ether.

2,4,5-Trimethylphenol *[496-78-6]* **M 136.2, m 70.5-71.5°, pK^{25} 10.57.** Crystd from water.

2,4,6-Trimethylphenol *[527-60-6]* **M 136.2, m 69°, b 220°/760mm, pK^{25} 10.86.** Crystd from water and sublimed *in vacuo*.

3,4,5-Trimethylphenol *[527-54-8]* **M 136.2, m 107°, b 248-249°/760mm, pK^{25} 10.25.** Crystd from pet ether.

Trimethylphenylammonium benzenesulfonate *[16093-66-6]* **M 293.3.** Crystd repeatedly from MeOH (charcoal).

2,2,4-Trimethyl-6-phenyl-1,2-dihydroquinoline *[3562-69-4]* **M 249.3, m 102°.** Vacuum distd, then crystd from absolute EtOH.

2,4,6-Trimethylpyridine (sym-collidine) *[108-75-8]* **M 121.2, m -46°, b 10°/2.7mm, 36-37°/2mm, 60.7°/13mm, 65°/31mm, 170.4°/760mm, 175-178°/atm, d^{25} 0.9100, n$_D^{20}$ 1.4939, 1.4981, n^{25} 1.4959, pK^{25} 6.69.** Commercial samples may be grossly impure. Likely contaminants include 3,5-dimethylpyridine, 2,3,6-trimethylpyridine and water. Brown, Johnson and Podall [*J Am Chem Soc* **76** 5556 *1954*] fractionally distd 2,4,6-trimethylpyridine under reduced pressure through a 40-cm Vigreux column and added to 430mL of the distillate slowly, with cooling to 0°, 45g of BF_3-diethyl etherate. The mixture was again distd, and an equal volume of dry *benzene was added to the distillate. Dry HCl was passed into the soln, which was kept cold in an ice-bath, and the hydrochloride was filtered off. It was recrystd from abs EtOH (1.5mL/g) to **m 286-287°**(sealed tube). The free base was regenerated by treatment with aq NaOH, then extracted with *benzene, dried ($MgSO_4$) and distd under reduced pressure. Sisler et al. [*J Am Chem Soc* **75** 446 *1953*] ppted trimethylpyridine as its phosphate from a soln of the base in MeOH by adding 85% H_3PO_4, shaking and cooling. The free base was regenerated as above. Garrett and Smythe [*J Chem Soc* 763 *1903*] purified the trimethylpyridine via the $HgCl_2$ complex. It is more soluble in cold than hot H_2O [sol 20.8% at 6°, 3.5% at 20°, 1.8% at 100°].
Also purified by dissolving in $CHCl_3$, adding solid K_2CO_3 and Drierite, filtering and fractionally distilling through an 8in helix packed column. The *sulfate* has **m 205°**, and the *picrate* (from hot H_2O) has **m 155-156°**. [Frank and Meikle *J Am Chem Soc* **72** 4184 *1950*.]

Trimethylsulfonium iodide *[2181-42-2]* **M 204.1, m 215-220°(dec).** Crystd from EtOH.

1,3,7-Trimethyluric acid *[5415-44-1]* **M 210.2, m 345°(dec), pK 6.0.** Crystd from water.

1,3,9-Trimethyluric acid *[7464-93-9]* **M 210.2, m 340°, 347°, pK^{20} 9.39.** Crystd from water.

1,7,9-Trimethyluric acid *[55441-82-2]* **M 210.2, m 316-318 (dec)°, 345°, pK$_{Est}$ ~9.0.** Crystd from water or EtOH, and sublimed *in vacuo*.

Trimyristin *[555-45-3]* **M 723.2, m 56.5°.** Crystd from diethyl ether.

2,4,6-Trinitroanisole *[606-35-9]* **M 243.1, m 68°.** Crystd from EtOH or MeOH. Dried under vac.

1,3,6-Trinitrobenzene *[99-35-4]* **M 213.1, m 122-123°.** Crystd from glacial acetic acid, $CHCl_3$, CCl_4, EtOH aq EtOH or EtOH/*benzene, after (optionally) heating with dil HNO_3. Air dried. Fused, and crystd under vacuum.

2,4,6-Trinitrobenzenesulfonic acid hydrate (TNBS, picrylsulfonic acid) *[2508-19-2]* **M 293.2, m 180°, λmax 240nm (ε 650 $M^{-1}cm^{-1}$), pK_{Est}~ <0.** It is also available as 0.1M and 5%w/v solns in H_2O. Recrystd from 1M HCl and dried at 100° or a mixt of EtOH (50mL), H_2O (30mL) and conc HCl (70mL) for 65g of acid. The *diethanolamine salt* had **m 182-183°** [Golumbic *J Org Chem* **11** 518 *1946*].

2,4,6-Trinitrobenzoic acid *[129-66-8]* **M 225.1, m 227-228°, pK^{25} 0.65.** Crystd from distilled water. Dried in a vacuum desiccator.

2,4,6-Trinitro-*m*-cresol *[602-99-3]* **M 243.1, m 107.0-107.5°, pK 2,8.** Crystd successively from H_2O, aq EtOH and *benzene/cyclohexane, then dried at 80° for 2h. [Davis and Paabo *J Res Nat Bur Stand* **64A** 533 *1960*.]

2,4,7-Trinitro-9-fluorenone *[129-79-3]* **M 315.2, m 176°.** Crystd from nitric acid/water (3:1), washed with water and dried under vacuum over P_2O_5, or recrystd from dry *benzene.

2,4,6-Trinitrotoluene (TNT) *[118-96-7]* **M 227.1, m 81.0-81.5°.** Crystd from *benzene and EtOH. Then fused and allowed to cryst under vacuum. Gey, Dalbey and Van Dolah [*J Am Chem Soc* **78** 1803 *1956*] dissolved TNT in acetone and added cold water (1:2:15), the ppte was filtered, washed free from solvent and stirred with five parts of aq 8% Na_2SO_3 at 50-60° for 10min. It was filtered, washed with cold water until the effluent was colourless, and air dried. The product was dissolved in five parts of hot CCl_4, washed with warm water until the washings were colourless and TNT was recoverd by cooling and filtering. It was recrystd from 95% EtOH and carefully dried over H_2SO_4. The dry solid should not be heated without taking precautions for a possible **EXPLOSION**.

2,4,6-Trinitro-*m*-xylene *[632-92-8]* **M 241.2, m 182.2°.** Crystd from ethyl methyl ketone.

Tri-*n*-octylamine *[1116-76-3]* **M 353.7, b 164-168°/0.7mm, 365-367°/760mm, d 0.813, n 1.450, pK^{25} 10.65.** It was converted to the amine hydrochloride etherate which was recrystd four times from diethyl ether at -30° (see below). Neutralisation of this salt regenerated the free amine. [Wilson and Wogman *J Phys Chem* **66** 1552 *1962*.] Distd at 1-2mm pressure.

Tri-*n*-octylammonium chloride *[1188-95-0]* **M 384.2, m 78-79°, pK 8.35 (in 70% aq EtOH).** Crystd from Et_2O, then *n*-hexane (see above). [Burrows et al. *J Chem Soc* 200 *1947*.]

Tri-*n*-octylammonium perchlorate *[2861-99-6]* **M 454.2, m >300°(dec).** Crystd from *n*-hexane.

1,3,5-Trioxane *[110-88-3]* **M 90.1, m 64°, b 114.5°/759mm.** Crystd from sodium-dried diethyl ether or water, and dried over $CaCl_2$. Purified by zone refining.

Trioxsalen (2,5,9-trimethyl-7*H*-furo[3,2-*g*]benzopyran-7-one) *[3902-71-4]* **M 228.3, m 233-235°, 234.5-235°.** Purified by recrystn from $CHCl_3$. If too impure it is fractionally crystd from $CHCl_3$-pet ether (b 30-60°) using Norit and finally crystd from $CHCl_3$ alone to give colourless prisms, **m 234.5-235°.** It is a photosensitiser so it should be stored in the dark. [UV: Kaufmann *J Org Chem* **26** 117 *1961*; Baeme et al. *J Chem Soc* 2976 *1949*.]

Tripalmitin *[555-44-2]* **M 807.4, m 66.4°.** Crystd from acetone, diethyl ether or EtOH.

Triphenylamine *[603-34-9]* **M 245.3, m 127.3-127.9°, pK -5.0 (in fluorosulfuric acid).** Crystd from EtOH or from *benzene/abs EtOH, diethyl ether and pet ether. It was sublimed under vacuum and carefully dried in a vacuum line. Stored in the dark under nitrogen.

1,3,5-Triphenylbenzene *[612-71-5]* **M 306.4, m 173-175°.** Purified by chromatography on alumina using *benzene or pet ether as eluents.

Triphenylene *[217-59-4]* **M 228.3, m 198°, b 425°.** Purified by zone refining or crystn from EtOH or $CHCl_3$, and sublimed.

1,2,3-Triphenylguanidine *[101-01-9]* **M 287.3, m 144°, pK 9.10.** Crystd from EtOH or EtOH/water, and dried under vacuum.

Triphenylmethane *[519-73-3]* **M 244.3, m 92-93°.** Crystd from EtOH or *benzene (with one molecule of *benzene of crystallisation which is lost on exposure to air or by heating on a water bath). It can also be sublimed under vacuum. It can also be given a preliminary purification by refluxing with tin and glacial acetic acid, then filtered hot through a glass sinter disc, and ppted by addition of cold water.

Triphenylmethanol (triphenylcarbinol) *[76-84-6]* **M 260.3, m 164°, b 360-380° (without dec), pK^{25} -6.63 (aq H_2SO_4).** Crystd from EtOH,MeOH, CCl_4 (4mL/g), *benzene, hexane or pet ether (b 60-70°). Dried at 90°. [Ohwada et al. *J Am Chem Soc* **108** 3029 *1986*.]

Triphenylmethyl chloride (trityl chloride) *[76-83-5]* **M 278.9, m 111-112°.** Crystd from iso-octane. Also crystd from 5 parts of pet ether (b 90-100°) and 1 part of acetyl chloride using 1.8g of solvent per g of chloride. Dried in a desiccator over soda lime and paraffin wax. [*Org Synth* Coll Vol III 841 *1955*; Thomas and Rochow *J Am Chem Soc* **79** 1843 *1957*; Moisel et al. *J Am Chem Soc* **108** 4706 *1986*.]

2,3,5-Triphenyltetrazolium chloride (TTC) *[298-96-4]* **M 334.8, m 243°(dec).** Crystd from EtOH or $CHCl_3$, and dried at 105°.

Tri-*n*-propylamine *[102-69-2]* **M 143.3, b 156.5°, d 0.757, n 1.419, pK^{25} 10.66.** Dried with KOH and fractionally distd. Also refluxed with toluene-*p*-sulfonyl chloride and with KOH, then fractionally distd. The distillate, after addn of 2% phenyl isocyanate, was redistd and the residue fractionally distd from sodium. [Takahashi et al. *J Org Chem* **52** 2666 *1987*.]

Tripyridyl triazine *[3682-35-7]* **M 312.3, m 245-248°.** Purified by repeated crystn from aq EtOH.

Tris-(2-aminoethyl)amine (TREN) *[4097-89-6]* **M 146.2, b 114°/15mm, 263°/744mm, d 0.977, n 1.498, pK_1^{25} 8.42, pK_2^{25} 9.44, pK_3^{25} 10.13.** For a separation from a mixture containing 62% TRIEN, see entry under triethylenetetramine. Also purified by conversion to the hydrochloride (see below), recrystn and regeneration of the free base [Xie and Hendrickson *J Am Chem Soc* **109** 6981 *1987*].

Tris-(2-aminoethyl)amine trihydrochloride *[14350-52-8]* **M 255.7, m 300°(dec).** Crystd several times by dissolving in a minimum of hot water and precipitating with excess cold EtOH. The ppte was washed with acetone, then diethyl ether and dried in a vacuum desiccator.

Tris(d,d-dicampholylmethanato)europium (III) *[52351-64-1]* **M 108.5, m 220-227.5°, 229-232°, $[\alpha]_D^{25}$ +28.6° (c 5.4, CCl_4; and varies markedly with concentration).** Dissolve in pentane, filter from any insol material, evaporate to dryness and dry the residue (white powder) at 100°/0.1mm for 36h. The IR has ν 1540cm^{-1}. [McCreary et al. *J Am Chem Soc* **96** 1038 *1974*.]

Tris-(dimethylamino)methane *(N,N,N',N',N'',N''*-hexamethylmethanetriamine) *[5762-56-1]* **M 145.3, b 42-43°/12mm, n 1.4349, pK_{Est} ~ 10.** Dry over KOH and dist through a Vigreux column at waterpump vacuum. Store in absence of CO_2. [Bredereck et al. *Chem Ber* **101** 1885 *1968* and *Angew Chem, Int Ed Engl* **5** 132 *1966*.]

Tris-(hydroxymethyl)methylamine (TRIS) *[77-86-1]* **M 121.1, m 172°, pK^{25} 8.07.** Tris can ordinarily be obtained in highly pure form suitable for use as an acidimetric standard. If only impure material is available, it should be crystd from 20% EtOH. Dry in a vacuum desiccator over P_2O_5 or $CaCl_2$.

Alternatively, it is dissolved in twice its weight of water at 55-60°, filtered, concd to half its volume and poured slowly, with stirring, into about twice the volume of EtOH. The crystals which separate on cooling to 3-4° are filtered off, washed with a little MeOH, air dried by suction, then finally ground and dried in a vacuum desiccator over P_2O_5. It has also been crystd from water, MeOH or aq MeOH, and vacuum dried at 80° for 2 days.

Tris-(hydroxymethyl)methylamonium hydrochloride (TRIS-HCl) *[1185-53-1]* **M 157.6, m 149-150°(dec).** Crystd from 50% EtOH, then from 70% EtOH. Tris-hydrochloride is also available commercially in a highly pure state. Otherwise, crystd from 50% EtOH, then 70% EtOH, and dried below 40° to avoid risk of decomposition.

1,1,1-Tris-(hydroxymethyl)ethane (2-hydroxymethyl-2-methyl-1,3-propanediol) *[77-85-0]* **M 120.2, m 200°.** Dissolved in hot tetrahydrofuran, filtered and ppted with hexane. It has also been crystd from acetone/water (1:1). Dried in vacuum.

N-Tris-(hydroxymethyl)methyl-2-aminomethanesulfonic acid (TES) *[7365-44-8]* **M 229.3, m 224-226°(dec), pK^{20} 7.50.** Crystd from hot EtOH containing a little water.

N-Tris-(hydroxymethyl)methylglycine (TRICINE) *[5704-04-1]* **M 179.2, m 186-188°(dec), pK$_1^{20}$ ~2.3, pK$_2^{20}$ 8.15.** Crystd from EtOH and water.

Tris-(hydroxymethyl)nitromethane [2-(hydroxymethyl)-2-nitro-1,3-propanediol] *[126-11-4]* **M 151.1, m 174-175°(dec, tech grade), 214°(pure).** Crystd from $CHCl_3$/ethyl acetate or ethyl acetate/*benzene. It is an acid and a 0.1M sol in H_2O has pH 4.5. **IRRITANT.**

Tris-[(3-trifluoromethylhydroxymethylene)-d-camphorato] europium (III) [Eu(tfc)$_3$] *[34830-11-0]* **M 893.6, m 195-299° (dec), ~220°, $[\alpha]_D^{24}$ +152° (c 2, CCl_4; and varies markedly with concentration).** Purified by extraction with pentane, filtered and filtrate evapd and the residual bright yellow amorphous powder is dried at 100°/0.1mm for 36h. A sample purified by fractional molecular distn at 180-200°/0.004mm gave a liquid which solidified and softened at ~130° and melted at ~180° and was analytically pure. IR (CCl_4) ν: 1630-1680cm^{-1} and NMR (CCl_4) δ broad: -1.3 to 0.5, -0.08 (s), 0.41 (s), 1.6-2.3 and 3.39 (s). [McCreary et al. *J Am Chem Soc* **96** 1038 *1974*; ; Goering et al. *J Am Chem Soc* **93** 5913 *1971*.]

1,3,5-Trithiane *[291-21-4]* **M 138.3, m 220°(dec).** Crystd from acetic acid.

Triuret (1,3-dicarbamoylurea) *[556-99-0]* **M 146.1, m 233°(dec).** Crystd from aq ammonia. Gives mono and dipotassium salts.

Tropaeolin 00, *[554-73-4]* **M 316.3, pK$_{Est(2)}$~ 5.8, pK$_{Est(3)}$~ 10.3.** Recrystd twice from water [Kolthoff and Gus *J Am Chem Soc* **60** 2516 *1938*].

Tropaeolin 000 (see Orange II p. 477 in Chapter 5). Purified by salting out from hot distilled water using sodium acetate, then three times from distilled water and twice from EtOH.

3-Tropanol (Tropine) *[120-29-6]* **M 141.2, m 63°, b 229°/760mm, pK^{15} 3.80.** Distd in steam and crystd from diethyl ether. *Hygroscopic.*

dl-Tropic (3-hydroxy-2-phenylpropionic) acid *[529-64-6]* **M 166.2, m 118°, pK^{25} 4.12.** Crystd from water or *benzene.

Tropolone *[533-75-5]* **M 122.1, m 49-50°, b 81-84°/0.1mm, pK$_1$ -0.53 (protonation of CO, aq H_2SO_4), pK$_2$ 6.67 (acidic OH).** Crystd from hexane or pet ether and sublimed at 40°/4mm.

Tryptamine [(3-2-aminoethyl)indole)] *[61-54-1]* **M 160.1, m 116°, pK$_1^{25}$ -6.31 (aq H_2SO_4, diprotonation), pK$_{Est(2)}$~4.9 , pK$_3^{25}$ 16.60 (acidic indole NH).** Crystd from *benzene.

Tryptamine hydrochloride   *[343-94-2]*   **M 196.7, m 252-253°.** Crystd from EtOH/water.

L-Tryptophan   *[73-22-3]*   **M 204.3, m 278°, $[\alpha]_D^{20}$ -33.4° (EtOH), $[\alpha]_{546}^{20}$ -36° (c 1, H_2O), pK_1^{25} -6.23 (aq H_2SO_4), pK_2^{25} 2.46, pK_3^{25} 9.41, pK_4^{25} 14.82 (acidic NH, in aq NaOH).** Crystd from water/EtOH, washed with anhydrous diethyl ether and dried at room temperature under vac over P_2O_5.

Tryptophol [3-(2-hydroxyethyl)indole]   *[526-55-6]*   **M 161.2, m 59°, b 174°/2mm.** Crystd from diethyl ether/pet ether, *C_6H_6, *C_6H_6/pet ether. The *picrate* has **m 98-100°** (from *C_6H_6).

(+)-Tubocurarine chloride (5H_2O)   *[57-94-3]*   **M 771.7, m 274-275°(dec) (anhydrous), $[\alpha]_{546}^{20}$ +235° (c 0.5, H_2O), $pK_{Est(1)}$~8.5, $pK_{Est(2)}$~8.8.** Crystd from water and forms various hydrates.

D(+)-Turanose   *[547-25-1]*   **M 342.3, m 168-170°, $[\alpha]_D^{20}$ +88° (c 4, H_2O).** Crystd from water by addition of EtOH.

Tyramine (4-hydroxybenzylamine)   *[51-67-2]*   **M 137.2, m 164-165°, pK_1^{25} 9.74 (OH), pK_2^{25} 10.52 (NH_2).** Crystd from *benzene or EtOH.

Tyramine hydrochloride   *[60-19-5]*   **M 173.6, m 274-276°.** Crystd from EtOH by addition of diethyl ether, or from conc HCl.

Tyrocidine A (cyclic decapeptide antibiotic with two D-Phe amino acids)   *[1481-70-5]*   **M 1268.8, m 240°(dec), $[\alpha]_D^{25}$ -115° (c 0.91, MeOH).** Crystd as hydrochloride from MeOH or EtOH and HCl. [Paladin and Craig *J Am Chem Soc* **76** 688 *1954*; King and Craig *J Am Chem Soc* **77** 6624 *1955*; Okamoto et al. *Bull Chem Soc Jpn* **50** 231 *1977*.]

L-Tyrosine   *[60-18-4]*   **M 181.2, m 290-295°(dec), $[\alpha]_D^{25}$ -10.0° (5M HCl), pK_1^{25} 2.18 (CO_2H), pK_2^{25} 9.21 (OH), pK_3^{20} 10.47 (NH_2).** Likely impurities are L-cysteine and the ammonium salt. Dissolved in dilute ammonia, then crystd by adding dilute acetic acid to pH 5. Also crystd from water or EtOH/water, and dried at room temperature under vacuum over P_2O_5.

Umbelliferone   (7-hydroxycoumarin)   *[93-35-6]* **M 162.2, m 225-228°, pK_{Est} ~8.0.** Crystd from water.

Undecan-1-ol   *[112-42-5]*   **M 172.3, m 16.5°.** Purified by repeated fractional crystn from its melt or by distn in a vacuum.

Undec-10-enoic acid   *[112-38-9]*   **M 184.3, m 25-25.5°, b 131°/1mm, 168°/15mm, pK_{Est} ~5.0.** Purified by repeated fractional crystn from its melt or by distn in a vacuum.

Uracil   *[66-22-8]*   **M 122.1, m 335°(dec), pK_1^{25} 9.43, pK_2^{25} 13.3-14.2.** Crystd from water.

Uramil (5-aminobarbituric acid)   *[118-78-5]*   **M 143.1, m 310-312°, 320°, >400°(dec), $pK_{Est(1)}$~3.9, $pK_{Est(2)}$~8.0, $pK_{Est(3)}$~12.5.** Crystd from water.

Urea   *[57-13-6]*   **M 60.1, m 132.7-132.9°, pK^{25} 0.12.** Crystd twice from conductivity water using centrifugal drainage and keeping the temperature below 60°. The crystals were dried under vacuum at 55° for 6h. Levy and Margouls [*J Am Chem Soc* **84** 1345 *1962*] prepared a 9M soln in conductivity water (keeping the temperature below 25°) and, after filtering through a medium-porosity glass sinter, added an equal volume of absolute EtOH. The mixture was set aside at -27° for 2-3 days and filtered cold. The ppte was washed with a small amount of EtOH and dried in air. Crystn from 70% EtOH between 40° and -9° has also been used. Ionic impurities such as ammonium isocyanate have been removed by treating the conc aqueous soln at 50° with

Amberlite MB-1 cation- and anion-exchange resin, and allowing to crystallise. [Benesch, Lardy and Benesch *J Biol Chem* **216** 663 *1955*.] Also crystd from MeOH or EtOH, and dried under vacuum at room temperature.

Urea nitrate *[124-47-0]* **M 123.1, m 152°(dec).** Crystd from dilute HNO_3.

Uric acid *[69-93-2]* **M 168.1, m >300°, pK_1 5.75, pK_2 10.3.** Crystd from hot distilled water.

Uridine *[58-96-8]* **M 244.2, m 165°, $[\alpha]_D^{20}$ +4.0° (H_2O), pK^{25} 9.51 (9.25).** Crystd from aqueous 75% MeOH.

Urocanic acid (4-imidazolylacrylic acid) *[104-98-3]* **M 138.1, m 225°, 226-228°, $pK_{Est(1)}$~2.5, $pK_{Est(2)}$~6, $pK_{Est(3)}$~11.** Crystd from water and dried at 100°.

Ursodeoxycholic acid *[128-13-2]* **M 392.5, m 203°, $[\alpha]_D^{20}$ +60° (c 0.2, EtOH), pK_{Est} ~4.8.** Crystd from EtOH.

(+)-Usnic acid [2,6-diacetyl-3,7,9-trihydroxy-8,9b-dimethyldibenzofuran-1(2H)-one] *[7562-61-0]* **M 344.3, m 204°.** See (+)-usnic acid on p. 573 in Chapter 6.

trans-Vaccenic acid (octadec-11-enoic acid) *[693-72-1]* **M 282.5, m 43-44°, pK_{Est} ~ 4.9.** Crystd from acetone. The *methyl ester* has **b** 174-175°/5mm.

***n*-Valeraldehyde** *[110-62-3]* **M 86.1, b 103°, d 0.811, n^{25} 1.40233.** Purified *via* the bisulfite derivative. [Birrell and Trotman-Dickinson *J Chem Soc* 2059 *1960*.]

***n*-Valeramide (pentanamide)** *[626-97-1]* **M 101.1, m 115-116°.** Crystd from EtOH.

Valeric acid (*n*-pentanoic acid) *[109-52-4]* **M 102.1, b 95°/22mm, 186.4°/atm, d 0.938, n 1.4080, pK^{25} 4.81.** Water was removed from the acid by distn using a Vigreux column, until the boiling point reached 183°. A few crystals of $KMnO_4$ were added, and after refluxing, the distn was continued, [Andrews and Keefer *J Am Chem Soc* **83** 3708 *1961*.]

δ-Valerolactam (2-piperidone) *[675-20-7]* **M 99.1, m 38.5-39.5°, 39-40°, 40°, b 81-82°/0.1mm, 136-137°/15mm, pK 0.75 (in AcOH).** Purified by repeated fractional distn. [Cowley *J Org Chem* **23** 1330 *1958*; Reppe et al. *Justus Liebigs Ann Chem* **596** 198 *1955*; IR: Huisgen et al. *Chem Ber* **90** 1437 *1957*.] The *hydrochloride* has **m** 183-184° (from isoPrOH or EtOH-Et$_2$O) [Hurd et al. *J Org Chem* **17** 865 *1952*], and the *oxime* has **m** 122.5° (from pet ether) [Behringer and Meier *Justus Liebigs Ann Chem* **607** 67 *1957*]. *Picrate* has **m** 92-93°.

γ-Valerolactone (± 4,5-dihydro-5-methyl-2(3H)-furanone) *[108-29-2]* **M 100.1, m -37°, 36°, b 82-85°/10mm, 102-103°/28mm, 125.3°/68mm, 136°/100mm, 205.75-206.25°/754mm, d_4^{20} 1.072, n_D^{20} 1.4322.** Purified by repeated fractional distillation [Boorman and Linstead *J Chem Soc* 577, 580 *1933*]. IR ν: 1790 (CS$_2$), 1775 (CHCl$_3$) cm^{-1} [Jones et al. *Can J Chem* **37** 2007 *1959*]. The *BF$_3$-complex* distils at 110-111°/20mm [Reppe et al. *Justus Liebigs Ann Chem* **596** 179 *1955*]. It is characterised by conversion to γ-hydroxy-*n*-valeramide by treatment with NH$_3$, **m** 51.5-52° (by slow evapn of a CHCl$_3$ soln).

δ-Valerolactone (tetrahydro-2H-pyran-2-one) *[542-28-9]* **M 100.1, m -13°, -12°, b 88°/4mm, 97°/10mm, 124°/24mm, 145-146°/40mm, 229-229.5°/atm, d_4^{20} 1.1081, n_D^{20} 1.4568.** Purified by repeated fractional distn. IR ν: 1750 (in CS$_2$), 1732 (in CHCl$_3$), 1748 (in CCl$_4$) and 1733 (in MeOH) cm^{-1}. [Huisgen and Ott *Tetrahedron* **6** 253 *1959*; Linstead and Rydon *J Chem Soc* 580 *1933*; Jones et al. *Can J Chem* **37** 2007 *1959*.]

Valeronitrile *[110-59-8]* **M 83.1, b 142.3°, d 0.799, n^{15} 1.39913, n^{30} 1.39037.** Washed with half its volume of conc HCl (twice), then with saturated aqueous $NaHCO_3$, dried with $MgSO_4$ and fractionally distd from P_2O_5.

L-Valine *[72-18-4]* **M 117.2, m 315°, $[\alpha]_D^{20}$ +266.7° (6M HCl), pK_1 2.26, pK_2 9.68.** Crystd from water by addition of EtOH.

Vanillin (4-hydroxy-3-methoxybenzaldehyde) *[121-33-5]* **M 152.2, m 83°, b 170°/15mm, pK^{25} 7.40.** Crystd from water or aqueous EtOH, or by distn *in vacuo*.

Veratraldehyde *[120-14-9]* **M 166.2, m 42-43°.** Crystd from diethyl ether, pet ether, CCl_4 or toluene.

Variamine Blue RT [4-(phenylamino)benzenediazonium sulfate (1:1)] *[4477-28-5]* **M 293.3, CI 37240, λ_{max} 377 nm.** Dissolved 10g in 100mL of hot water. Sodium dithionite (0.4g) was added, followed by active carbon (1.5g) and filtered hot. To the colourless or slightly yellow filtrate a soln of saturated NaCl was added and the mixture cooled. The needles were filtered off, washed with cold water, dried at room temperature, and stored in a dark bottle (light sensitive). [Erdey *Chem Analyst* **48** 106 *1959*.]

Vicine (2,4-diamino-5-β-D-glucopyranosidoxy-6-hydroxypyrimidine) *[152-93-2]* **M 304.3, m 243-244°, $[\alpha]_D^{20}$ -12° (c 4, 0.2N NaOH).** Crystd from water or aqueous 85% EtOH, and dried at 135°.

Vinyl acetate *[108-05-4]* **M 86.1, b 72.3°, d 0.938, n 1.396.** Inhibitors such as hydroquinone, and other impurities are removed by drying with $CaCl_2$ and fractionally distilling under nitrogen, then refluxing briefly with a small amount of benzoyl peroxide and redistilling under nitrogen. Stored in the dark at 0°.

9-Vinylanthracene *[2444-68-0]* **M 204.3, m 65-67°, b 61-66°/10mm.** Purified by vacuum sublimation. Also by chromatography on silica gel with cyclohexane as eluent, and recrystd from EtOH [Werst et al. *J Am Chem Soc* **109** 32 *1987*].

Vinyl butoxyethyl ether *[4223-11-4]* **M 144.2.** Washed with aqueous 1% NaOH, dried with CaH_2, then refluxed with and distd from, sodium.

***N*-Vinylcarbazole** *[1484-13-5]* **M 193.3, m 66°.** Crystd repeatedly from MeOH in amber glassware. Vacuum sublimed.

Vinylene carbonate *[872-36-6]* **M 86.1, m 22°.** Purified by zone melting.

Vinyl chloroformate *[5130-24-5]* **M 106.5, b 46.5°/80mm, 67-69°/atm, 109-110°/760mm, d_4^{20} 1.136, n_D^{23} 1.420.** It has been fractionated through a Todd column (Model A with ~60 plates, see p. 174) under atmospheric pressure and purity can be checked by gas chromatography. It has IR with v at 3100 + 2870 (CH_2), 1780 (C=O), 1640 (C=C) and 940 (CH_2 out-of-plane) and 910 (CH_2 wagging) cm^{-1}. [IR: Lee *J Org Chem* **30** 3943 *1965*; Levaillant *Ann Chim (Paris)* **6** 504 *1936*.] Used for protecting NH_2 groups in peptide synthesis [Olofson et al. *Tetrahedron Lett* 1563 *1977*].

1-Vinylnaphthalene *[826-74-4]* **M 154.2, b 124-125°/15mm.** Fractionally distd under reduced pressure on a spinning-band column, dried with CaH_2 and again distd under vacuum. Stored in sealed ampoules in a freezer.

2-Vinylpyridine monomer *[100-69-6]* **M 105.1, b 79-82°/29mm, d 0.974, n 1.550, pK^{25} 4.92.** Steam distd, then dried with $MgSO_4$ and distd under vacuum.

4-Vinylpyridine monomer *[100-43-6]* **M 105.1, b 40-41°/1.4mm, 54°/5mm, 58-61°/12mm, 68°/18mm, 79°/33mm, d_4^{20} 0.9836, n_D^{20} 1.5486, pK^{25} 5.62.** Purified by fractional distillation under a good vacuum and in a N_2 atmosphere and stored in sealed ampoules under N_2, and kept in the dark at -20°.

The *picrate* has **m** 175-176°. [UV: Coleman and Fuoss *J Am Chem Soc* **77** 5472 *1955*; Overberger et al. *J Polymer Sci* **27** 381 *1958*; Petro and Smyth *J Am Chem Soc* **79** 6142 *1957*.] Used for alkylating SH groups in peptides [Anderson and Friedman *Can J Biochem* **49** 1042 *1971* ; Cawins and Friedman *Anal Biochem* **35** 489 *1970*].

Vinyl stearate *[111-63-7]* **M 310.5, m 35°, b 166°/1.5mm.** Vacuum distd under nitrogen, then crystd from acetone (3mL/g) or ethyl acetate at 0°.

Violanthrene (dibenzanthrene, 5,10-dihydroviolanthrene A) *[81-31-2]* **M 428.5.** Purified by vacuum sublimation over Cu in a muffle furnace at 450°/25mm in a CO_2 atmosphere [Scholl and Meyer *Chem Ber* **67** 1229 *1934*]. *Violanthrene A* (anthro[9,1,2-cde]benzo[rst]pentaphene *[188-87-4]* **M** 426.5 has **m** 506°. [Clar *Chem Ber* **76** 458 *1943*.]

Viologen (4,4'-dipyridyl dihydrochloride) *[27926-72-3]* **M 229.1, m >300°.** Purified by pptn on adding excess of acetone to a concentrated solution in aqueous MeOH. It has also been recrystd several times from MeOH and dried at 70° under vacuum for 24h [Prasad et al. *J Am Chem Soc* **108** 5135 *1986*], and recrystd three times from MeOH/isopropanol [Stramel and Thomas *J Chem Soc, Faraday Trans* **82** 799 *1986*].

Visnagin (4-methoxy-7-methyl-5*H*-furo[3,2-g][1]benzopyran-5-one) *[82-57-5]* **M 230.2, m 142-145°.** Crystd from water.

dl-Warfarin (4-hydroxy-3-(3-oxo-1-phenylbutyl)-2*H*-1-benzopyran-2-one) *[81-81-2]* **M 308.3, m 161°.** Crystd from MeOH. The *acetate* has **m** 182-183° and *2,4-dinitrophenylhydrazone* has **m** 215-216°. Effective anticoagulant and rodenticide.

Xanthatin (3-methylene-7-methyl-6-[3-oxo-1-buten-1-yl]cyclohept-5-ene-[10,11-*b*]furan-2-one) *[26791-73-1]* **M 246.3, m 114.5-115°, [α]$_D$ -20° (EtOH).** Crystd from MeOH or EtOH. UV: λ_{max} 213 and 275nm (ε 22800 and 7300).

Xanthene *[92-83-1]* **M 182.2, m 100.5°, b 310-312°/760mm.** Crystd from *benzene or EtOH.

9-Xanthenone (xanthone) *[90-47-1]* **M 196.2, m 175.6-175.4°.** Crystd from EtOH (25mL/g) and dried at 100°. It has also been recrystd from *n*-hexane three times and sublimed *in vacuo*. [Saltiel *J Am Chem Soc* **108** 2674 *1986*].

Xanthopterin (H$_2$O) see entry on p. 576 in Chapter 6.

Xanthorhamnin *[1324-63-6]* **M 770.7, m 195°, [α]$_D^{20}$ +3.75° (EtOH).** Crystd from a mixture of ethyl and isopropyl alcohols, air dried, then dried for several hours at 110°.

Xanthosine (2H$_2$O) [9-(β-D-ribosyl)purin-2,6(1*H*,3*H*)-dione] *[5968-90-1]* **M 320.3, [α]$_D^{20}$ -53° (c 8, 0.3M NaOH), pK$_1^{25}$ <2.5, pK$_2^{25}$ 5.67, pK$_3^{25}$ 12.85.** Crystd from EtOH or water (as dihydrate).

Xanthurenic acid (5,8-dihydroxyquiniline-2-carboxylic acid) *[59-00-7]* **M 205.2, m 286°, 290-295°(dec), pK$_{Est(1)}$~ 1.5, pK$_{Est(2)}$~ 4.9, pK$_{Est(3)}$~ 9.8.** Ppted by the addition of 2N formic acid to its soln in hot 2M ammonia (charcoal). Filter solid off, dry in a vac at ~80° in the dark. UV (H$_2$O) has λmax nm (ε M^{-1}cm^{-1}): 243 (30,000) and 342 (6,500). The *methyl ester* has **m** 262° (from MeOH).

Xanthydrol *[90-46-0]* **M 198.2, m 123-124°.** Crystd from EtOH and dried at 40-50°.

Xylene *[1330-20-7]* **M 106.1 (mixed isomers).** Usual impurites are ethylbenzene, paraffins, traces of sulfur compounds and water. It is not practicable to separate the *m*-, and *p*-isomers of xylene by fractional distn, although, with a sufficiently efficient still, *o*-xylene can be fractionally distd from a mixture of isomers. Purified (and dried) by fractional distn from $LiAlH_4$, P_2O_5, CaH_2 or sodium. This treatment can be preceded by shaking successively with conc H_2SO_4, water, aqueous 10% NaOH, water and mercury, and drying with $CaCl_2$ for several days. Xylene can be purified by azeotropic distn with 2-ethoxyethanol or 2-methoxyethanol, the distillate being washed with water to remove the alcohol, then dried and fractionally distilled.

o-**Xylene** *[95-47-6]* **M 106.2, f -25.2°, b 84°/14mm, 144.4°/760mm, d 0.88020, d^{25} 0.87596, n 1.50543, n^{25} 1.50292.** The general purification methods listed under xylene are applicable [Clarke and Taylor *J Am Chem Soc* **45** 831 *1923*]. *o*-Xylene (4.4Kg) is sulfonated by stirring for 4h with 2.5L of conc H_2SO_4 at 95°. After cooling, and separating the unsulfonated material, the product was diluted with 3L of water and neutralised with 40% NaOH. On cooling, sodium *o*-xylene sulfonate separated and was recrystd from half its weight of water. [A further crop of crystals was obtained by concentrating the mother liquor to one-third of its volume]. The salt was dissolved in the minimum amount of cold water, then mixed with the same amount of cold water, and with the same volume of conc H_2SO_4 and heated to 110°. *o*-Xylene was regenerated and steam distd. It was then dried and redistd.

m-**Xylene** *[108-38-3]* **M 106, f -47.9°, b 139.1°, d 0.86417, d^{25} 0.85990, n 1.49721, n^{25} 1.49464.** The general purification methods listed under *xylene* are applicable. The *o*- and *p*-isomers can be removed by their selective oxidation when a *m*-xylene sample containing them is boiled with dilute HNO_3 (one part conc acid to three parts water). After washing with water and alkali, the product can be steam distd, then distd and purified by sulfonation. [Clarke and Taylor *J Am Chem Soc* **45** 831 *1923*.] *m*-Xylene is selectively sulfonated when a mixture of xylenes is refluxed with the theoretical amount of 50-70% H_2SO_4 at 85-95° under reduced pressure. By using a still resembling a Dean and Stark apparatus, water in the condensate can be progressively withdrawn while the xylene is returned to the reaction vessel. Subsequently, after cooling, then adding water, unreacted xylenes are distd off under reduced pressure. The *m*-xylene sulfonic acid is subsequently hydrolysed by steam distn up to 140°, the free *m*-xylene being washed, dried with silica gel and again distd. Stored over molecular sieves Linde type 4A.

p-**Xylene** *[106-42-3]* **M 106.2, f 13.3, b 138.3°, d 0.86105, d^{25} 0.85669, n 1.49581, n^{25} 1.49325.** The general purification methods listed for *xylene* are applicable. *p*-Xylene can readily be separated from its isomers by crystn from such solvents as MeOH, EtOH, isopropanol, acetone, butanone, toluene, pentane or pentene. It can be further purified by fractional crystn by partial freezing, and stored over sodium wire or molecular sieves Linde type 4A. [Stokes and French *J Chem Soc, Faraday Trans 1* **76** 537 *1980*.]

Xylenol Orange {3*H*-2,1-benzoxathiol-3-ylidene-bis-[(6-hydroxy-5-methyl-*m*-phenylene)-methylnitrilo]tetraacetic acid, S,S-dioxide} *[1611-35-4]* **M 672.6, m 210°(dec), ε_{578} 6.09 x 10^4 (pH 14), ε_{435} 2.62 x 10^4 (pH 3.1), pK_1 -1.74, pK_2 -1.09 (aq H_2SO_4-HNO_3), pK_3 2.58, pK_4 3.23, pK_5 6.46, pK_6 10.46, pK_7 12.28.** Generally contaminated with starting material (cresol red) and semixylenol orange. Purified by ion-exchange chromatography using DEAE-cellulose, eluting with 0.1M NaCl soln which will give the sodium salt. Cresol Red, semixylenol orange and iminodiacetic acid bands elute first. This procedure will give the sodium salt of the dye. To obtain the free acid dissolve the salt in H_2O and acidify with AcOH. Filter off, wash with H_2O and dry first in air and then in a vac deisiccator over P_2O_5 in the dark [Sato, Yokoyama and Momoki *Anal Chim Acta* **94** 317 *1977*].

α-D-**Xylose** *[58-86-6]* **M 150.1, m 146-147°, $[\alpha]_D^{20}$ -18.8° (c 4, H_2O).** Purified by slow crystn from aq 80% EtOH or EtOH, then dried at 60° under vac over P_2O_5. Stored in a vacuum desiccator over $CaSO_4$.

m-**Xylylene diisocyanate** *[3634-83-1]* **M 188.2, b 88-89°/0.02mm, 130°/2mm, d_4^{20} 1.204, n_D^{20} 1.4531.** Purified by repeated distn through a 2 plate column. [Ferstundig and Scherrer *J Am Chem Soc* **81** 4838 *1959*.]

α-Yohimbine [146-48-5] M 354.5, m 278°(dec), $[\alpha]_D^{20}$ +55.6° (c 2, EtOH), pK_1^{22} 3.0, pK_2^{22} 7.45.
Crystd from EtOH, and dried to remove EtOH of crystn. For γ-Yohimbine see ajmalicine on p. 98.

Zeaxanthin [all *trans*-β-carotene-3,3'(R,R')-diol] [144-68-3] M 568.9, m 207°, 215.5°, λ_{max} 275 (log ε 4.34), 453 (log ε 5.12), 480 (log ε 5.07) in EtOH.
Yellow plates (with a blue lustre) from MeOH or EtOH.

CHAPTER 5

PURIFICATION OF INORGANIC AND METAL ORGANIC CHEMICALS

(Including Organic compounds of B, Bi, P, Se, Si, and ammonium and metal salts of organic acids)

The most common method of purification of inorganic species is by recrystallisation, usually from water. However, especially with salts of weak acids or of cations other than the alkaline and alkaline earth metals, care must be taken to minimise the effect of hydrolysis. This can be achieved, for example, by recrystallising acetates in the presence of dilute acetic acid. Nevertheless, there are many inorganic chemicals that are too insoluble or are hydrolysed by water so that no general purification method can be given. It is convenient that many inorganic substances have large temperature coefficients for their solubility in water, but in other cases recrystallisation is still possible by partial solvent evaporation.

Organo-metallic compounds, on the other hand, behave very much like organic compounds, e.g. they can be redistilled and may be soluble in organic solvents. A note of **caution** should be made about handling organo-metallic compounds, e.g. arsines, because of their **potential toxicities**, particularly when they are volatile. Generally the suppliers of such compounds provide details about their safe manipulation. These should be read carefully and adhered to closely. If in any doubt always assume that the materials are lethal and treat them with utmost care. The same **safety precautions** about the handling of substances as stated in Chapter 4 should be followed here (see Chapter 1).

For information on **ionization (pK)** see Chapter 1, p. 7, and Chapter 4, p. 80. In order to avoid repetition, the literature (or predicted) pK values of anionic and/or cationic species are usually reported at least once, and in several cases is entered for the free acid or free base, e.g. Na_2SO_4 will have a pK value for Na^+ at the entry for NaOH and the pK values for SO_4^{2-} at the entries for H_2SO_4. When the pK values of the organic counter-ions are not given in this Chapter, as in case of sodium benzoate, the reader is referred to the value(s) in Chapter 4, e.g. of benzoic acid .

Abbreviations of titles of periodical are defined as in the Chemical Abstracts Service Source Index (CASSI). A note on other abbreviations is in Chapter 1, p. 30.

Benzene, which has been used as a solvent successfully and extensively in the past for reactions and purification by chromatography and crystallisation is now considered a **very dangerous substance** so it has to be used with extreme care. We emphasised that an alternative solvent to benzene (e.g. toluene, toluene-petroleum ether, or a petroleum ether to name a few) should be used first. However, if benzene has to be used then all operations have to be performed in a well ventilated fumehood and precautions taken to avoid inhalation and contact with skin and eyes. Whenever benzene is mentioned in the text and asterisk e.g. $^*C_6H_6$ or *benzene, is inserted to remind the user that special precaution should be adopted.

Organic dyes which are *not* complexed or are salts of metals are included in Chapter 4 (use the CAS Registry Numbers to find them). Commercially available polymer supported reagents are indicated with § under the appropriate reagent.

Acetarsol see *N*-Acetyl-4-hydroxy-*m*-arsanilic acid.

Acetonyltriphenylphosphonium chloride *[1235-21-8]* **M 354.8, m 237-238°, 244-246° (dec).** Recrystd from $CHCl_3$ + $^*C_6H_6$ + pet ether (b 60-80°) and by dissolving in $CHCl_3$ and pouring it into dry Et_2O. λ_{max}^{EtOH} nm(ϵ) 255(3,600), 262(3,700), 268(4,000) and 275(3,100). The *iodide salt* crystallises from H_2O and has **m** 207-209°. [*J Org Chem* **22** 41 *1957*.] **IRRITANT** and *hygroscopic*. When shaken with a 10% aqueous soln of Na_2CO_3 (8h) it gives *acetylmethylene triphenyl phosphorane* which is recrystd from $MeOH-H_2O$ and after drying at 70°/0.1mm has **m** 205-206°. UV: λmax nm(ϵ) 268 (6600), 275 (6500) and 288 (5700); IR:ν (cm^{-1}) 1529 (s), 1470 (m), 1425 (s), 1374 (m), 1105 (s) and 978 (s). [*J Org Chem* **22** 41, 44 *1957*.]

3R,4R,1'R-4-Acetoxy-3-[1-(*tert*-butylmethylsilyloxy)ethyl]-2-azetinone *[76855-69-1]* **M 287.4, m 107-108°, $[\alpha]_D^{20}$+55°** (c 0.5, toluene), **$[\alpha]_D^{20}$+53.7°** (c 1.04, $CHCl_3$). Purified by chromatography on silica gel (3 x 14cm) for 50g of ester using 20% EtOAc in *n*-hexane. The eluate is evaporated and the residue recrystd from hexane as white fluffy crystals. [*Tetrahedron* **39** 2505 *1983*.]

Acetylenedicarboxylic acid monopotassium salt *[928-04-1]* **M 152.2.** Very soluble in H_2O, but can be crystd from small volume of H_2O in small crystals. These are washed with EtOH and dried over H_2SO_4 at 125°. [*Chem Ber* **10** 841 *1877*; *Justus Liebigs Ann Chem* **272** 133 *1893*.]

Acetylferrocene (ferrocenyl methylketone) *[1271-55-2]* **M 228.1, m 86°, 86-87°.** Orange-red crystals, recrystd from isooctane and sublimed at 100°/1mm. The *oxime* has **m** 167-170° (from Et_2O or aq EtOH). The *semicarbazone* has **m** 198-201° (from EtOH). [*J Am Chem Soc* **77** 2022 3009 *1955*; *J Chem Soc* 650 *1958*.]

N-Acetyl-4-hydroxy-*m*-arsanilic acid** *[97-44-9]* **M 275.1, pK$_1$ 3.73, pK$_2$ 7.9, pK$_3$ 9.3.** Crystd from water.

Allyl trimethylsilane (2-propenyltrimethylsilane) *[762-72-1]* **M 114.3, b 83.0-84.5°, 84-88°, 85.5-86.0°, d 0.713, n 1.405.** Fractionate through an efficient column at atm pressure. If impure dissolve in THF, shake with H_2O (2x), dry (Na_2SO_4) and fractionate. [Cudlin and Chvalovský *Collect Czech Chem Commun* **27** 1658 *1962*.]

Allyl tri-*n*-butylstannane (allyl tributyl tin) *[24850-33-7]* **M 331.1, b 88-92°/0.2mm, 115°/17mm, d 1.068, n 1.487.** A possible impurity is tributylchlorostannane — test for Cl as Cl ion after hydrolysing. Dissolve in $^*C_6H_6$ (or toluene), shake with dil aq NaOH, dry ($CaCl_2$), and dist in a vac [Jones et al. *J Chem Soc* 1446 *1947*; *Aldrichimica Acta* **17** 75 *1984* and **20** 45 *1987*].

Alizarin Red S (3,4-dihydroxy-9,10-dioxo-2-anthracene sulfonic acid, Na salt. H_2O) *[130-22-3]* **M 360.4, pK$_1^{25}$ <1, pK$_2^{25}$ 5.49, pK$_3^{25}$ 10.85 (11.01).** Commercial samples contain large amounts of sodium and potassium chlorides and sulfates. It is purified by passing through a Sephadex G-10 column, followed by elution with water, then 50% aq EtOH [King and Pruden *Analyst (London)* **93** 601 *1968*]. Finally dissolve in EtOH and ppte with Et_2O several times [*J Phys Chem* **54** 829 *1950*; polarography *J Am Chem Soc* **70** 3055 *1948*].

Alumina (neutral) *[1344-28-1]* **M 102.0 (anhyd.).** Stirred with hot 2M HNO_3, either on a steam bath for 12h (changing the acid every hour) or three times for 30min, then washed with hot distilled water until the washings had pH 4, followed by three washings with hot MeOH. The product was dried at 270° [Angyal and Young *J Am Chem Soc* **81** 5251 *1959*]. For the preparation of alumina for chromatography see Chapter 1.

Aluminum acetylacetonate *[13963-57-0]* **M 324.3, m 192-194°, 195°.** Crystd several times from *benzene or aqueous MeOH, λ_{max} 216 and 286mn. [*J Phys Chem* **62** 440 *1958*.] It can be purified by sublimation and has the following solubilities in g per cent: $^*C_6H_6$ 35.9 (20°), 47.6 (40°), toluene 15.9 (20°), 22.0 (40°) and acetylacetone 6.6 (20°), 10.4 (40°). [*Inorg Synth* **5** 105 *1957*.]

Aluminium ammonium sulfate (10H$_2$O) *[7784-26-1]* **M 453.3, m 93°, pK$_1^{25}$ 4.89, pK$_2^{25}$ 5.43, pK$_3^{25}$ 5.86 (fAl^{3+} aquo), pK$_4^{25}$ 11.22 [aluminate Al(OH)$_4$].** Crystd from hot H$_2$O and cool in ice.

Aluminium bromide *[7727-15-3]* **M 266.7, m 97°, b 114°/10mm.** Refluxed and then distilled from pure aluminium chips in a stream of nitrogen into a flask containing more of the chips. It was then distd under vacuum into ampoules [Tipper and Walker *J Chem Soc* 1352 *1959*]. Anhydrous conditions are essential, and the white to very light brown solid distillate can be broken into lumps in a dry-box (under nitrogen). Fumes in moist air.

Aluminium caesium sulfate (12H$_2$O) *[7784-17-0 (12H$_2$O); 14284-36-7]* **M 568.2.** Crystd from hot water (3mL/g).

Aluminium chloride (anhydrous) *[7446-70-0]* **M 133.3, m 192.6°.** Sublimed several times in an all glass system under nitrogen at 30-50mm pressure. Has also been sublimed in a stream of dry HCl and has been subjected to a preliminary sublimation through a section of granular aluminium metal [for manipulative details see Jensen *J Am Chem Soc* **79** 1226 *1957*]. Fumes in moist air.

Aluminum ethoxide *[555-75-9]* **M 162.2, m 154-159°, 146-151°, b 187-190°/7mm, 210-214°/13mm.** Crystd from CS$_2$ [m 139°, CS$_2$ complex] and distd in a vacuum. Molecular weight corresponds to [Al(OEt)$_3$]$_4$ [*J Phys Chem* **39** 1127 *1935*; *J Am Chem Soc* **69** 2605 *1947*].

Aluminium fluoride (anhydrous) *[7784-18-4]* **M 84.0, m 250°.** Technical material may contain up to 15% alumina, with minor impurities such as aluminium sulfate, cryolite, silica and iron oxide. Reagent grade AlF$_3$ (hydrated) contains only traces of impurities but its water content is very variable (may be up to 40%). It can be dried by calcining at 600-800° in a stream of dry air (some hydrolysis occurs), followed by vacuum distn at low pressure in a graphite system, heated to approximately 925° (condenser at 900°) [Henry and Dreisbach *J Am Chem Soc* **81** 5274 *1959*].

Aluminium isopropoxide *[555-31-7]* **M 204.3, m 119°, b 94°/0.5mm, 135°/10mm.** Distd under vacuum. *Hygroscopic.*

Aluminium nitrate (9H$_2$O) *[7784-27-2 (9H$_2$O); 13473-90-0]* **M 375.1.** Crystd from dilute HNO$_3$, and dried by passing dry nitrogen through the crystals for several hours at 40°. After 2 recrystns of ACS grade it had S, Na and Fe at 2.2, 0.01 and 0.02 ppm resp.

Aluminium potassium sulfate (12H$_2$O, alum) *[7784-24-9]* **M 474.4, m 92°.** Crystd from weak aqueous H$_2$SO$_4$ (*ca* 0.5mL/g).

Aluminium rubidium sulfate (12H$_2$O) *[7784-29-4]* **M 496.2.** Crystd from aq H$_2$SO$_4$ (*ca* 2.5mL/g).

Aluminium sulfate (anhydrous) *[10043-01-3]* **M 342.2, m 765°(dec); Al$_2$O$_3$ 14-18 H$_2$O** *[17927-65-0]*; **Al$_2$O$_3$ 18 H$_2$O** *[7784-31-8]*. Crystd from hot dilute H$_2$SO$_4$ (1 mL/g) by cooling in ice. When a soln of alumina (Al$_2$O$_3$) in conc H$_2$SO$_4$ is slowly cooled, Al$_2$SO$_4$ 17 or 18H$_2$O deposits as a crystalline mass . Al$_2$SO$_4$ 17H$_2$O is the stable form in equilibrium with its saturated aqueous soln at 25° [Smith *J Am Chem Soc* **64** 41 *1942*]. This is purified by dissolving in a small vol of H$_2$O and adding EtOH until the sulfate readily crystallises from the oily supersaturated soln. It forms Al$_2$O$_3$ 16H$_2$O between 0-112°. On gradual heating the hydrate melts giving the anhydrous salt at *ca* 250°. Several hydrates up to 27H$_2$O have been described. Further heating to red heat (~ 600-800) causes decomposition to Al$_2$O$_3$ + SO$_3$ + SO$_2$ and O$_2$ [Cobb *J Soc Chem Ind* **29** 250 *1910*]. ACS reagent is Al$_2$O$_3$ 18H$_2$O (98+%).

Aluminum triethyl (triethyl aluminum) *[97-93-8]* **M 114.2, b 69°/1.5mm, 76°/2.5mm, 129-131°/55mm, d$_4^{20}$ 0.695, n$_D^{20}$ 1.394.** Purified by fractionation in an inert atmosphere under vacuum in a 50cm column containing a heated nichrome spiral, taking the fraction 112-114°/27mm. It is very sensitive to H$_2$O and should be stored under N$_2$. It should not contain chloride which can be shown by hydrolysis and testing with AgNO$_3$. [*J Am Chem Soc* **75** 4828 5193*1953*; NMR: *J Am Chem Soc* **81** 3826 *1959*.]

Aluminium tri-*tert*-butoxide *[556-91-2]* **M 246.3, m 208-210⁰(dec).** Crystd from *benzene and sublimed at 180⁰.

Aluminium trimethanide (trimethyl aluminium) *[75-24-1]* **M 72.1, m 15.2⁰, b 111.5⁰/488.2mm, 124.5⁰/atm, d_4^{20} 0.725.** Distd through a 10-20 theoretical plates column under 1 atm of N_2 (better with very slow take-off). Attacks grease (use glass joints). Also vac distd over Al in absence of grease, into small glass vials and sealed under N_2. Purity is measured by freezing point. Reacts with H_2O, is non-conducting in *C_6H_6 and is **HIGHLY FLAMMABLE**. [*J Chem Soc* 468 *1946*; *J Am Chem Soc* **68** 2204 *1946*.]

4-Aminophenylmercuric acetate *[6283-24-5]* **M 371.8, m 168⁰, 175⁰(dec), 180⁰(dec).** Recrystd from hot dilute AcOH and dried in air. [*J Indian Chem Soc* **32** 613 *1955; Justus Liebigs Ann Chem* **465** 269 *1928.*]

Ammonia (gas) *[7664-41-7]* **M 17.0, pK²⁵ 9.25.** Major contaminants are water, oil and non-condensible gases. Most of these impurities are removed by passing the ammonia through a trap at -22⁰ and condensing it at -176⁰ under vacuum. Water is removed by distilling the ammonia into a tube containing a small lump of sodium. Also dried by passage through porous BaO, or over alumina followed by glass wool impregnated with sodium (prepared by soaking the glass wool in a solution of sodium in liquid ammonia, and evaporating off the ammonia). It can be rendered oxygen-free by passage through a soln of potassium in liquid ammonia.

Ammonia (liquid) *[7664-41-7]* **M 17.0, m -77.7⁰, b -33.4⁰, n_D 1.325, d 0.597, d⁻⁷⁹ 0.817g/mL.** Dried, and stored, with sodium in a steel cylinder, then distd and condensed by means of liquid air, the non-condensable gases being pumped off. In order to obtain liquid NH_3 from a cylinder turn the cylinder up-side-down (i.e. with the valve at the bottom, use a metal stand to secure it in this position) and lead a plastic tube from the tap to a measuring cylinder placed in an efficient fume cupboard which is kept running. Turn the tap on and allow the ammonia to be released. At first, gas and liquid will splatter out (make sure that the plastic tube is secure) but soon the liquid will drip into the measuring cylinder. The high latent heat of evaporation will cool the ammonia so that the liquid will remain cool and not boil vigorously. If the ammonia is required dry the necessary precautions should be taken, i.e. the gas is allowed to flow through tubes packed with coarse CaO pellets.

Ammonia (aqueous) *[7664-41-7]* **M 17.0 + H_2O, d 0.90 (satd, 27% w/v, 14.3 N), pK²⁵ 9.25.** Obtained metal-free by saturating distilled water, in a cooling bath, with ammonia (from tank) gas. Alternatively, can use isothermal distn by placing a dish of conc aq ammonia and a dish of pure water in an empty desiccator and leaving for several days. **AMMONIA (gas, liquid or aq soln) is very irritating and should not be inhaled in large volumes as it can lead to olfactory paralysis (temporary and partially permanent).**

Ammonium acetate *[631-61-8]* **M 77.1, m 112-114⁰, d 1.04.** Crystd twice from anhydrous acetic acid, dried under vacuum for 24h at 100⁰ [Proll and Sutcliff *Trans Faraday Soc* **57** 1078 *1961*].

Ammonium benzoate *[1863-63-4]* **M 139.2, m 198⁰, 200⁰(dec), d 1.26.** Crystd from EtOH.

Ammonium bisulfate (ammonium hydrogen sulfate) *[7803-63-6]* **M 115.1⁰, m ~147⁰, d 1.79, pK²⁵ 1.96 (HSO_4^-).** Crystd from water at room temperature (1mL/g) by adding EtOH and cooling.

Ammonium bromide *[12124-97-9]* **M 98.0, m 450⁰(sublimes), d 2.43.** Crystd from 95% EtOH.

Ammonium chloride *[12125-02-9]* **M 53.5, m 338⁰(sublime point, without melting), d 1.53.** Crystd several times from conductivity water (1.5mL/g) between 90⁰ and 0⁰. Sublimes. After one crystn, ACS grade had: metal(ppm) As (1.2), K (1), Sb (7.2), V (10.2).

Ammonium chromate *[7788-98-9]* **M 152.1, m 185⁰(dec), d 1.81, pK_1^{25} 0.74, pK_2^{25} 6.49 (for H_2CrO_4).** Crystd from weak aqueous ammonia (*ca* 2.5mL/g) by cooling from room temperature.

Ammonium dichromate *[7789-09-5]* **M 252.1, m 170°(dec), d 1.26.** Crystd from weak aq HCl (*ca* 1mL/g). (Possible **carcinogen**)

Ammonium dihydrogen arsenate *[13462-93-6]* **M 159.0, m 300°(dec).** Crystd from water (1mL/g).

Ammonium dihydrogen orthophosphate *[7722-76-1]* **M 115.0, m 190°, d 1.80.** Crystd from water (0.7mL/g) between 100° and 0°.

Ammonium dodecylsulfate *[2235-54-3]* **M 283.4.** Recrystd first from 90% EtOH and then twice from abs EtOH, finally dried in a vacuum.

Ammonium ferric oxalate (3H$_2$O) *[13268-42-3]* **M 428.1, m ~160°(dec), d 1.77.** Crystd from hot water (0.5mL/g).

Ammonium ferric sulfate (12H$_2$O) *[7783-83-7 (12H$_2$O); 10138-04-2 (anhydr)]* **M 482.2, m ~37°, d 1.71.** Crystd from aqueous ethanol.

Ammonium ferrous sulfate (6H$_2$O) *[7783-85-9 (6H$_2$O); 10045-89-3 (anhydr)]* **M 392.1, m 100°(dec), d 1.86.** A soln in warm water (1.5mL/g) was cooled rapidly to 0°, and the resulting fine crystals were filtered at the pump, washed with cold distilled water and pressed between sheets of filter paper to dry.

Ammonium formate *[540-69-2]* **M 63.1, m 116°, 117.3°, d$_4^{45}$ 1.280.** Heat solid in NH$_3$ vapour and dry in vacuum till NH$_3$ odour is faint. Recryst from abs EtOH and then keep in a desiccator over 99% H$_2$SO$_4$ *in vacuo*. It is very *hygroscopic*. Exists in two forms, stable needles and less stable plates. Also forms acid salts, i.e. HCO$_2$NH$_4$.3HCO$_2$H and HCO$_2$NH$_4$.HCO$_2$H. [*J Am Chem Soc* **43** 1473 *1921*; **63** 3124 *1941* .]

Ammonium hexachloroiridate (IV) *[16940-92-4]* **M 441.0.** Ppted several times from aqueous soln by saturation with ammonium chloride. This removes any palladium and rhodium. Then washed with ice-cold water and dried over conc H$_2$SO$_4$ in a vacuum desiccator. If osmium or ruthenium is present, it can be removed as the tetroxide by heating with conc HNO$_3$, followed by conc HClO$_4$, until most of the acid has been driven off. (This treatment is repeated). The near-dry residue is dissolved in a small amount of water and added to excess NaHCO$_3$ soln and bromine water. On boiling, iridic (but not platinic) hydroxide is ppted. It is dissolved in HCl and ppted several times, then dissolved in HBr and treated with HNO$_3$ and HCl to convert the bromides to chlorides. Saturation with ammonium chloride and cooling precipitates ammonium hexachloroiridate which is filtered off and purified as above [Woo and Yost *J Am Chem Soc* **53** 884 *1931*].

Ammonium hexacyanoferrate II hydrate *[14481-29-9]* **M 284.1, m dec on heating.** The pale yellow trihydrate powder can be washed with 10% aq NH$_3$, filtd, then washed several times with EtOH and Et$_2$O, and dried at room temp. Decomposes in vacuum above 100° and should be stored away from light and under N$_2$. In light and air it decomposes by losing NH$_3$. [*Handbook of Preparative Inorganic Chem (Ed. Brauer)* Vol II 1509 *1965*.]

Ammonium hexafluorophosphate *[16941-11-0]* **M 163.0, d$_4^{18}$ 2.181, pK$_1^{25}$~ 0.5, pK$_2^{25}$ 5.12 (for fluorophosphoric acid H$_2$PO$_3$F).** Crystallises from H$_2$O in square plates. Decomposes on heating before melting. Soluble in H$_2$O at 20° (74.8% w/v), also very soluble in Me$_2$CO, MeOH, EtOH and MeOAc and is decomposed by boiling acids. [*Chem Ber* **63** 1063 *1930*.]

Ammonium hexafluorosilicate *[16919-19-0]* **M 178.1, pK$_2$ 1.92 (for H$_2$SiF$_6$).** Crystd from water (2mL/g). After 3 recrystns of Tech grade it had Li, Na, K and Fe at 0.3, 0.2, 0.1 and 1.0 ppm resp.

Ammonium hypophosphite *[7803-65-8]* **M 83.0.** Crystd from hot EtOH.

Ammonium iodate *[13446-09-8]* **M 192.9, pK²⁵ 0.79 (IO³⁺).** Crystd from water (8mL/g) between 100° and 0°.

Ammonium iodide *[12027-06-4]* **M 144.9, sublimes with dec ~405°, d 2.51.** Crystd from EtOH by addition of ethyl iodide. Very *hygroscopic.* Stored in the dark.

Ammonium ionophore I (Nonactin) *[6833-86-7]* **M 736.9, m 147-148°, $[\alpha]_D^{20}$ 0° (c 1.2, CHCl₃).** Crystd from MeOH in colourless needles and is dried at 20° in high vac. A selectophore with high sensitivity for NH_4^+ ions. [*Helv Chim Acta* **38** 1445 *1955,* **45** 129 *1962,* **55** 1371 *1972; Acta Cryst* **27B** 1680 *1971.*]

Ammonium magnesium chloride (6H₂O) *[60314-43-4]* **M 256.8.** Crystd from water (6mL/g) by partial evapn in a desiccator over KOH (deliquescent).

Ammonium magnesium sulfate (6H₂O) *[20861-69-2]* **M 360.6.** Crystd from water (1mL/g) between 100° and 0°.

Ammonium manganous sulfate (6H₂O) *[13566-22-8]* **M 391.3.** Crystd from water (2mL/g) by partial evapn in a desiccator.

Ammonium metavanadate *[7803-55-6]* **M 117.0, m 200°(dec).** See ammonium (meta) vanadate on p. 395.

Ammonium molybdate *[13106-76-8]* **M 196.0, pK₁²⁵ 0.9 (proton addition), pK₂²⁵ 3.57, pK₃²⁵ 4.08 (for H₂MoO₄).** Crystd from water (2.5mL/g) by partial evapn in a desiccator.

Ammonium nickel sulfate (6H₂O) *[7785-20-8 (6H₂O); 15699-18-0 (anhydr)]* **M 395.0, d 1.923.** Crystd from water (3mL/g) between 90° and 0°.

Ammonium nitrate *[6484-52-2]* **M 80.0, m 210°(dec explosively), d 1.72.** Crystd twice from distilled water (1mL/g) by adding EtOH, or from warm water (0.5mL/g) by cooling in an ice-salt bath. Dried in air, then under vacuum. After 3 recrystns of ACS grade it contained Li and B at 0.03 and 0.74 ppm resp.

Ammonium oxalate (H₂O) *[6009-70-7]* **M 142.1, d 1.50.** Crystd from water (10mL/g) between 50° and 0°.

Ammonium perchlorate *[7790-98-9]* **M 117.5, d 1.95, pK²⁵ -2.4 to -3.1 (for HClO₄).** Crystd twice from distilled water (2.5mL/g) between 80° and 0°, and dried in a vacuum desiccator over P₂O₅. Drying at 110° might lead to slow decomposition to chloride. **POTENTIALLY EXPLOSIVE.**

Ammonium peroxydisulfate *[7727-54-0]* **M 228.2, m dec when heated wet liberating oxygen, d 1.98.** Recrystd at room temperature from EtOH/water.

Ammonium picrate *[131-74-8]* **M 246.1, m EXPLODES above 200°.** Crystd from EtOH and acetone.

Ammonium reineckate (Reineckate salt) *[13573-16-5]* **M 345.5, m 270-273°(dec).** Crystd from water, between 30° and 0°, working by artificial light. Solns of reineckate decompose slowly at room temperature in the dark and more rapidly at higher temperatures or in diffuse sunlight.

Ammonium selenate *[7783-21-3]* **M 179.0, d 2.19, m dec on heating.** Crystd from water at room temperature by adding EtOH and cooling.

Ammonium sulfamate *[7773-06-0]* **M 114.1, m 132-135°, dec at 160°.** Crystd from water at room temperature (1mL/g) by adding EtOH and cooling.

Ammonium sulfate *[7783-20-2]* **M 132.1, m 230°(dec), 280°(dec), d 1.77.** Crystd twice from hot water containing 0.2% EDTA to remove metal ions, then finally from distilled water. Dried in a desiccator for 2 weeks over $Mg(ClO_4)_2$. After 3 recrystns ACS grade had Ti, K, Fe, Na at 11, 4.4, 4.4, 3.2 ppm resp.

Ammonium tetrafluoroborate *[13826-83-0]* **M 104.8, pK^{25} 2.77 (for HBF$_4$).** Crystd from conductivity water (1mL/g) between 100° and 0°.

Ammonium tetraphenylborate *[14637-34-4]* **M 337.3, m ca 220°(dec).** Dissolve in aqueous Me_2CO and allow crystn to proceed slowly otherwise very small crystals are formed. No trace of Me_2CO was left after drying at 120° [*Trans Faraday Soc* **53** 19 *1957*]. The salt was ppted from dilute AcOH soln of sodium tetraphenylborane in the presence of NH_4^+ ions. After standing for 5min, the ppte was filtered off onto a sintered porcelain crucible, washed with very dilute AcOH and dried at room temp for at least 24h [*Anal Chem* **28** 1001 *1956*]. Alternatively a soln of sodium tetraphenylborane (5% excess) in H_2O is added to NH_4Cl soln. After 5min the ppte is collected, washed several times with H_2O and recryst from aqueous Me_2CO. [*Analyt Chim Acta* **19** 342 *1958.*]

Ammonium thiocyanate *[1762-95-4]* **M 76.1, m 138°(dec), 149°(dec), pK^{25} -1.85 (for HSCN), 149.** Crystd three times from dilute $HClO_4$, to give material optically transparent at wavelengths longer than 270nm. Has also been crystd from absolute MeOH and from acetonitrile.

Ammonium tungstate (VI) *[11120-25-5]* **M 283.9, pK$_1^{25}$ 2.20, pK$_2^{25}$ 3.70 (for tungstic acid, H$_2$WO$_4$).** Crystd from warm water by adding EtOH and cooling.

Ammonium (meta) vanadate *[7803-55-6]* **M 117.0, d$_{10}^{20}$ 2.326.** Wash with H_2O until free from Cl$^-$ and dry in air. It is soluble in H_2O (5.18g/100mL at 15°, 10.4g/100mL at 32°) but is more soluble in dilute NH_3. Crystd from conductivity water (20mL/g). When heated at relatively low temperatures it loses H_2O and NH_3 to give vanadium oxide (V_2O_5) and at 210° it forms lower oxides. [*Inorg Synth* **3** 117 *1950.*] After washing Tech grade with H_2O it had Na, Mn and U at 0.06, 0.2 and 0.1 ppm resp.

n-**Amylmercuric chloride** *[544-15-0]* **M 307.2, m 110°.** Crystd from EtOH.

Anthraquinone Blue B (Acid Blue 45, 1,5-diamino-4,8-dihydroxy-9,10-anthraquinone-3,7-disulfonic acid di-Na salt) *[2861-02-1]* **M 474.3, m >300°, CI 63010, λmax 595nm, pK$_{Est(1)}$~<0, pK$_{Est(2)}$~2, pK$_{Est(3)}$~9.** Purified by salting out three times with sodium acetate, followed by repeated extraction with EtOH [McGrew and Schneider *J Am Chem Soc* **72** 2547 *1950*].

Anthraquinone Blue RXO *[4403-89-8]* **M 445.5.** Purified by salting out three times with sodium acetate, followed by repeated extraction with EtOH [McGrew and Schneider *J Am Chem Soc* **72** 2547 *1950*].

Anthraquinone Green G [Acid Green 25, Alizarin Cyanine Green F, 1,4-bis-(4-methyl-2-sulfophenyl-1-amino)-9,10-anthraquinone di-Na salt] *[4403-90-1]* **M 624.6, m 235-238°, CI 61570, λmax 642nm, pK >0.** Purified by salting out three times with sodium acetate, followed by repeated extraction with EtOH [McGrew and Schneider *J Am Chem Soc* **72** 2547 *1950*]. It is a green powder that slightly sol in Me_2CO, EtOH and pyridine. Sol in conc H_2SO_4 to give a blue soln which becomes turquoise on dilution. [Allen et al. *J Org Chem* **7** 63 *1942.*]

9,10-Anthraquinone-2,6-disulfonic acid (disodium salt) *[853-68-9]* **M 412.3, m >325°, pK$_{Est}$ ~<0 (for SO$_3$H).** Crystd three times from water, in the dark [Moore et al. *J Chem Soc. Faraday Trans1* **82** 745 *1986*].

9,10-Anthraquinone-2-sulfonic acid (Na salt, H$_2$O) *[131-08-8]* **M 328.3, pK$_{Est}$ ~<0 (SO$_3$H).** Crystd from H_2O or MeOH (charcoal). [Costa and Bookfield *J Chem Soc, Faraday Trans 1* **82** 991 *1986*].

Antimony (V) pentafluoride *[7783-70-2]* **M 216.7, m 7.0°, 8.3°, b 141°, 150°, 148-150°, d 2.99, pK^{25} 2.55 [for HSb(OH)$_6$ = Sb(OH)$_6^-$ + H$^+$].** Purified by vacuum distillation preferably in a quartz apparatus, and stored in quartz or aluminum bottles. It is a *hygroscopic* viscous liquid which reacts *violently* with H_2O and is hydrolysed by alkalis. *It is* **POISONOUS** *and attacks the skin.* [*J Chem Soc* 2200 *1950*; *Handbook of Preparative Inorganic Chemistry (Ed. Brauer)* Vol I 200 *1965*.]

Antimony trichloride *[10025-91-9]* **M 228.1, m 73°, b 283°, pK_1^{25} 1.4, pK_2^{25} 11.0 (11.8), pK_3^{25} 12.95 (for Sb^{3+} aquo).** Dried over P_2O_5 or by mixing with toluene or xylene and distilling (water is carried off with the organic solvent), then distd twice under dry nitrogen at 50mm, degassed and sublimed twice in a vacuum into ampoules. Can be crystd from CS_2. Deliquescent. Fumes in moist air.

Antimony trifluoride *[7783-56-4]* **M 178.8, m 292°.** Crystd from MeOH to remove oxide and oxyfluoride, then sublimed under vacuum in an aluminium cup on to a water-cooled copper condenser [Woolf *J Chem Soc* 279 *1955*].

Antimony triiodide *[7790-44-5]* **M 502.5, m 167°.** Sublimed under vacuum.

Antimony trioxide *[1309-64-4]* **M 291.5, m 656°.** Dissolved in minimum volume of dilute HCl, filtered, and six volumes of water were added to ppte a basic antimonous chloride (free from Fe and Sb_2O_5). The ppte was redissolved in dilute HCl, and added slowly, with stirring, to a boiling soln (containing a slight excess) of Na_2CO_3. The oxide was filtered off, washed with hot water, then boiled and filtered, the process being repeated until the filtrate gave no test for chloride ions. The product was dried in a vacuum desiccator [Schuhmann *J Am Chem Soc* **46** 52 *1924*]. After on cryst(pptn?), the oxide from a Chinese source had: metal (ppm) Al (8), Ag (0.2), As (56), Cr (6), Ge (0.4), Mn (0.2), Na (16), Ni (2.2) Pb (2.4), Sn (0.4) and V (32).

Aqua regia. This is prepared by adding slowly concentrated HNO_3 (1 vol) to concentrated hydrochloric acid (3 vols) in a glass container. This mixture is used to dissolve metals, including noble metals and alloys, as well as minerals and refractory substances. It is done by suspending the material and boiling (**EFFICIENT FUME CUPBOARD — EYE PROTECTION**] to dryness and repeating the process until the residue dissolves in H_2O. If the aqua regia is to be stored for long periods it is advisable to dilute it with one volume of H_2O which will prevent it from releasing chlorine and other chloro and nitrous compounds which are objectionable. Store cool in a fume cupboard. However, it is good laboratory practice to prepare it freshly and dispose of it down the fume cupboard sink with copious amounts of water.

Argon *[7440-37-1]* **M 39.95, b -185.6°.** Rendered oxygen-free by passage over reduced copper at 450°, or by bubbling through alkaline pyrogallol and H_2SO_4, then dried with $CaSO_4$, $Mg(ClO_4)_2$, or Linde 5A molecular sieves. Other purification steps include passage through Ascarite (asbestos impregnated with sodium hydroxide), through finely divided uranium at about 800° and through a -78° cold trap.
Alternatively the gas is passed over CuO pellets at 300° to remove hydrogen and hydrocarbons, over Ca chips at 600° to remove oxygen and, finally, over titanium chips at 700° to remove nitrogen. Also purified by freeze-pump-thaw cycles and by passage over sputtered sodium [Arnold and Smith *J Chem Soc, Faraday Trans 2* **77** 861 *1981*].

o-**Arsanilic acid** *[2045-00-3]* **M 216.1, m 153°, pK_1^{22} 3.77 (AsO_3H_2), pK_2^{22} 8.66 (AsO_3H^-).** Crystd from water or ethanol/ether. **POISONOUS.**

p-**Arsanilic acid** *[98-50-0]* **M 216.1, m 232°, pK_1^{22} 4.05 (AsO_3H_2), pK_2^{22} 8.66 (AsO_3H^-).** Crystd from water or ethanol/ether. **POISONOUS.**

Arsenazo I [3(2-arsonophenylazo)-4,5-dihydroxy-2,7-naphthalenedisulfonic acid di Na salt] *[66019-20-3]* **M 614.3, ε 2.6 x 10^4 at 500nm, pH 8.0; pK_1 0.6(0.8), pK_2 3.52, pK_3 2.97(AsO_3H_2), pK_4 8.20(AsO_3H^-), pK_5 9.98(OH), pK_6 15.0.** A saturated aqueous soln of the free acid was slowly added to an equal volume of conc HCl. The orange ppte was filtered, washed with acetonitrile and dried for 1-2h at 110° [Fritz and Bradford *Anal Chem* **30** 1021 *1958*].

Arsenazo III [3,6-bis(2-arsonophenylazo)-4,5-dihydroxy-2,7-naphthalenedisulfonic acid di Na salt] *[62337-00-2]* **M 776.4, pK_1 -2.7, pK_2 -2.7, pK_3 0.6, pK_4 0.8, pK_5 1.6, pK_6 3.4,**

pK₇ 6.27, pK₈ 9.05, pK₉ 11.98, pK₁₀ 15.1. Contaminants include monoazo derivatives, starting materials for synthesis and by-products. Partially purified by pptn of the dye from aqueous alkali by addition of HCl. More thorough purification by taking a 2g sample in 15-25mL of 5% aq NH₃ and filter. Add 10mL HCl (1:1) to the filtrate to ppte the dye. Repeat procedure and dissolve solid dye (0.5g) in 7mL of a 1:1:1 mixture of *n*-propanol:conc NH₃:water at 50°. After cooling, filter soln and treat the filtrate on a cellulose column using 3:1:1 mixture of *n*-propanol:conc NH₃:water as eluent. Collect the blue band and evaporate to 10-15mL below 80°, then add 10mL conc HCl to ppte pure Arsenazo III. Wash with EtOH and air-dry [Borak et al. *Talanta* **17** 215 *1970*]. The purity of the dye can be checked by paper chromatography using M HCl as eluent.

Arsenic *[7440-38-2]* **M 74.9, m 816°.** Heated under vacuum at 350° to sublime oxides, then sealed in a Pyrex tube under vacuum and sublimed at 600°, the arsenic condensing in the cooler parts of the tube. Stored under vacuum [Shih and Peretti *J Am Chem Soc* **75** 608 *1953*]. **POISONOUS.**

Arsenic acid (arsenic pentoxide hydrate, arsenic V oxide hydrate, orthoarsenic acid) *[12044-50-7]* **M 229.8 + xH₂O, pK₁²⁵ 2.26, pK₂²⁵ 6.76, pK₃²⁵ 11.29 (H₃AsO₄).** Cryst from conc solns of boiling conc HNO₃ as rhombic crystls. Dried in vac to give hemihydrate (hygroscopic). Heating above 300° yields As₂O₅. [Thaler *Z Anorg Allg Chem* **246** 19 *1941*.] **POISONOUS.**

Arsenic tribromide *[7784-33-0]* **M 314.6, m 31.1°, b 89°/11mm, 221°/760mm.** Distd under vacuum. **POISONOUS.**

Arsenic trichloride (butter of arsenic) *[7784-34-1]* **M 181.3, b 25°/11mm, 130.0°.** Refluxed with arsenic for 4h, then fractionally distd. The middle fraction was stored with sodium wire for two days, then again distd [Lewis and Sowerby *J Chem Soc* 336 *1957*]. Fumes in moist air and readily hydrolysed by H₂O. **POISONOUS.**

Arsenic triiodide *[7784-45-4]* **M 455.6, m 146°, b 400°/atm.** Crystd from acetone, sublimes below 100°. **POISONOUS**

Arsenic III oxide (arsenic trioxide, arsenious oxide) *[1327-53-3]* **M 197.8, three forms: m ~200°(amorphous glass), m 275°(sealed tube, octahedral, common form, sublimes > 125° without fusion but melts under pressure), m ~312°, pK₁²⁰ 9.27, pK₂²⁰ 13.54, pK₃²⁰ 13.99 (for H₃AsO₃).** Crystd in octahedral form from H₂O or from dil HCl (1:2), washed, dried and sublimed (193°/760mm). Analytical reagent grade material is suitable for use as an analytical standard after it has been dried by heating at 105° for 1-2h or has been left in a desiccator for several hours over conc H₂SO₄. **POISONOUS (particulary the vapour, handle in a ventilated fume cupboard).**

Aurothioglucose (gold thioglucose) *[12192-57-3]* **M 392.2.** Purified by dissolving in H₂O (0.05g in 1mL) and ppting by adding EtOH. Yellow cryst with slight mercaptan odour. Decomposes slowly in H₂O, sol in propylene glycol but insol in EtOH and other common organic solvents. [*FEBS Lett* **98** 351 *1970*.]

Barium (metal) *[7440-39-3]* **M 137.3, m 727°.** Cleaned by washing with diethyl ether to remove adhering paraffin, then filed in an argon-filled glove box, washed first with ethanol containing 2% conc HCl, then with dry ethanol. Dried under vacuum and stored under argon [Addison, Coldrey and Halstead *J Chem Soc* 3868 *1962*]. Has also been purified by double distn under 10mm argon pressure.

Barium acetate *[543-80-6]* **M 255.4.** Crystd twice from anhydrous acetic acid and dried under vacuum for 24h at 100°.

Barium bromate *[13967-90-3]* **M 265.3.** Crystd from hot water (20mL/g). The monohydrate melts at 260°(dec).

Barium bromide (2H₂O) *[7791-28-8]* **M 333.2, m at 75° loses first H₂O and at 120° loses the second H₂O.** Crystd from water (1mL/g) by partial evaporation in a desiccator.

Barium chlorate (H₂O) *[10294-38-9 (hydrate); 13477-00-4 (anhydr)]* **M 322.3, m 414°.** Crystd from water (1mL/g) between 100° and 0°.

Barium chloride (2H₂O) *[10326-27-9]* **M 244.3, m ~120°(dec, hydrate), 963° (anhyd).** Twice crystd from water (2mL/g) and oven dried to constant weight.

Barium dithionate (2H₂O) *[13845-17-5]* **M 333.5, m >150° loses SO₂, pK²⁵ 0.49 (for H₂S₂O₆, theory pK₁ -3.4, pK₂ -0.2).** Crystd from water.

Barium ferrocyanide (6H₂O) *[13821-06-2]* **M 594.8, m 80°(dec), pK₃²⁵ 2.57, pK₄²⁵ 4.35 (for ferrocyanide).** Crystd from hot water (100mL/g).

Barium fluoride *[7787-32-8]* **M 175.3, m 1353°, 1368°, b 2260°, d 4.83.** Washed well with distd H₂O and dried in vacuum Sol in H₂O [1.6g (10°), 1.6g (20°) and 1.62g (30°) per L), mineral acids and aq NH₄Cl. May be stored in glass bottles. [*Handbook of Preparative Inorganic Chemistry (Ed. Brauer)* Vol I 234 *1963*.]

Barium formate *[541-43-5]* **M 277.4, pK²⁵ 3.74 (for HCO₂H).** Crystd from warm water (4mL/g) by adding EtOH and cooling.

Barium hydroxide (8H₂O) *[12230-71-6]* **M 315.5, m 78°, pK₁²⁵ 13.13, pK₂²⁵ 13.36.** Crystd from water (1mL/g).

Barium hypophosphite (H₂O) *[14871-79-5]* **M 285.4.** Ppted from aq soln (3mL/g) by adding EtOH.

Barium iodate (H₂O) *[7787-34-0]* **M 487.1, m 130°(loses H₂O), 476°(dec).** Crystd from a large volume of hot water by cooling.

Barium iodide (2H₂O) *[7787-33-9 (2H₂O); 13718-50-8 (anhydr)]* **M 427.2, m 740°(dec).** Crystd from water (0.5mL/g) by partial evapn in a desiccator. **POISONOUS.**

Barium ionophore I *[N,N,N',N'-tetracyclohexyloxy-bis-(o-phenyleneoxy)diacetamide]* *[96476-01-6]* **M 644.9, m 156-158°.** Purified by chromatography on a Kieselgel column and eluted with CH₂Cl₂-EtOAc (5:1), and recryst from EtOH-Me₂CO as colourless crystals. It is an electrically neutral ionophore with high selectivity for Ba²⁺ ions and with high lipophilicity. [*Chem Ber* 118 1071 *1985*.]

Barium manganate (barium permanganate) *[7787-35-1]* **M 256.3, d 3.77.** Wash with conductivity H₂O by decantation until the supernatant gives a faint test for Ba²⁺. Remove excess H₂O in vac (IMPORTANT), then heat at 100° and the last traces of H₂O are removed in a vac desiccator over P₂O₅. Store over KOH. It disproportionates in hot H₂O or dil acid to Ba(MnO₂)₂ and MnO₂, and is a mild oxidant. [*J Am Chem Soc* 44 1965 *1924*; *Inorg Synth* 11 56 *1960*.]

Barium nitrate *[10022-31-8]* **M 261.4, m 593°(dec).** Crystd twice from water (4mL/g) and dried overnight at 110°. **POISONOUS.**

Barium nitrite (H₂O) *[7787-38-4]* **M 247.4, m 217°(dec).** Crystd from water (1mL/g) by cooling in an ice-salt bath. **POISONOUS.**

Barium perchlorate *[13465-95-7]* **M 336.2, m 505°, pK²⁵ -2.4 to -3.1 (for HClO₄).** Crystd twice from water.

Barium propionate (H₂O) *[5908-77-0]* **M 301.5, pK²⁴ 4.88 (for propionic acid).** Crystd from warm water (50mL/g) by adding EtOH and cooling.

Barium sulfate *[7722-43-7]* **M 233.4, m >1580°.** Washed five times by decantation with hot distilled water, dialysed against distd water for one week, then freeze-dried and oven dried at 105° for 12h.

Barium tetrathionate *[82203-66-5]* **M 361.6.** Purified by dissolution in a small volume of water and ppted with EtOH below 5°. After drying the salt was stored in the dark at 0°.

Barium thiocyanate (2 H₂O) *[2092-17-3]* **M 289.6, pK²⁵ -1.85 (for HSCN).** Crystd from water (2.5mL/g) by partial evaporation in a desiccator.

Barium thiosulfate *[35112-53-9]* **M 249.5, m 220°(dec), pK₁²⁵ 0.6, pK₂²⁵ 1.74 (for H₂S₂O₃).** Very slightly soluble in water. Washed repeatedly with chilled water and dried in air at 40°.

Benzaldehyde-2-sulfonic acid sodium salt *[1008-72-6]* **M 208.2, m dec on heating.** Forms prisms or plates by extracting with boiling EtOH, filtering, evaporate to dryness and recrystallise the Na salt from a small volume of H₂O. The *N-phenylhydrazone sodium salt* recrysts from H₂O, **m** 174.5°. [Gnehm and Schüle *Justus Liebigs Ann Chem* **299** 363 *1898*.]

Benzenechromium tricarbonyl *[12082-08-5]* **M 214.1, m 163-166°.** Purified by sublimation *in vacuo*.

Benzeneselenenyl bromide (phenylselenenyl bromide) *[34837-55-3]* **M 236.0, m 58-62°, 60°, 62°, b 107-108°/15mm, 134°/35mm.** Dist in a vac, recryst from pet ether, CHCl₃ (EtOH free), or Et₂O (cooling mixture) as dark red or orange crysts, sublimes at 25°/0.001mm [Behaghel et al. *Chem Ber* **65** 815 *1932*; Pitteloud and Petrzilka *Helv Chim Acta* **62** 1319 *1979*]. **HIGHLY TOXIC**

Benzeneselenenyl chloride (benzeneselenyl chloride, phenylselenenyl chloride) *[5707-04-0]* **M 191.5, m 59-60°, 64-65°, b 92°/5mm, 120°/20mm.** Purified by distn in vac and recrystn (orange needles) from hexane [Foster *J Am Chem Soc* **55** 822 *1933*, Foster et al. *Recl Trav Chim, Pays-Bas* **53** 405, 408 *1934*; Behaghel and Seibert *Chem Ber* **66** 714 *1933*.] **HIGHLY TOXIC.**

Benzeneseleninic acid *[6996-92-5]* **M 189.1, m 122-124°, pK²⁵ 4.70.** Add 10% excess of 15M NH₃ to the solid acid and stir until the solid dissolves, filter, decolorise with charcoal (2x, Norite) and acidify by slow addn of 6M HCl, filter solid and wash with H₂O. Dissolve the acid in the minimum vol of MeOH and this soln is added dropwise to boiling H₂O until cloudiness appears. At this point add 25% more boiling H₂O, filter hot (decolorise if necessary) and cool rapidly with scratching to 0°. After 30min the solid is filtd off and recryst as before but with very slow cooling. The colorless needles are filtered off and dried in a vac desiccator (CaCl₂) before the melting point is measured [McCullough and Gould *J Am Chem Soc* **71** 674 *1949*].

Benzeneseleninic anhydride *[17697-12-0]* **M 360.1, m 124-126°, 164-165°, 170-173°.** When the anhydride is recrystd from *C₆H₆ it has **m** 124-126° but when this is heated at 140°/1h in a vac or at 90°/2h it had **m** 164-165° and gives a solid **m** 124-126° when recrystd from *C₆H₆. Both depress the melting point of the acid PhSeO₂H. If the high melting anhydride is dissolved in *C₆H₆ and seeded with the high melting anhydride, the high melting anhydride crystallises out. It readily absorbs H₂O to form the acid (PhSeO₂H, **m** 122-124°). Because of this the commercial anhydride could contain up to 30% of the acid. Best purified by converting to the HNO₃ complex (**m** 112°) and heating this *in vacuo* at 120°/72h to give the anhydride as a white powder **m** 164-165°. Alternatively heat the anhydride *in vacuo* at 120°/72 h until IR shows no OH band. [Ayvrey et al. *J Chem Soc* 2089 *1962*; Barton et al. *J Chem Soc, Perkin Trans 1* 567 *1977*.] **TOXIC** solid.

Benzeneselenol (phenylselenol, selenophenol) *[645-96-5]* **M 157.1, b 57-59°/8mm, 71-72°/18mm, 84-86°/25mm, d 1.480, n 1.616.** Dissolve in aq N NaOH, acidify with conc HCl and extract with Et₂O, dry over CaCl₂, filter, evap on a steam bath and distil from a Claisen flask or through a short column collecting the middle fraction and seal immediately in a glass vial, otherwise the colourless liquid becomes yellow. The alkali insol materials consist of diphenylselenide (**b** 167°/16mm) and diphenyldiselenide, **m** 63° (from EtOH). **TOXIC**, use rubber gloves. It has a foul odour. [Foster *Org Synth* Coll Vol III 771 *1955*.]

Benzeneselenonic acid (benzeneselenoic acid)   *[39254-48-3]* **M 205.1, m 64°, pK²⁵ 4.79.** Purified by dissolving in H_2O and passing through a strong cation exchange resin (H^+ form). Evap under reduced pressure and dry in a high vac to give colourless hygroscopic crysts [Dostal et al. *Z Chem* **6** 153 *1966* and IR: *Chem Ber* **104** 2044 *1971*].

Benzenestibonic acid *[535-46-6]* **M 248.9, m >250°(dec).** Crystd from acetic acid, or from EtOH-$CHCl_3$ mixture by addition of water.

Benzenesulfinic acid Na salt *[873-55-2]* **M 164.2, m >300°, pK²⁵ 2.16 (2.74; for $PhSO_2H$).** Dissolve in the minimum vol of O_2 free H_2O (prepared by bubbling N_2 through for 2 h) and adding O_2 free EtOH (prepared as for H_2O), set aside at 4° overnight under N_2, filtd, washed with abs EtOH, then Et_2O and dried in a vac. The Na salt is relatively stable to air oxidation, but is best kept under N_2 in the dark. Also recrystd from EtOH and dried at 120° for 4h in a vacuum [Kornblum and Wade *J Org Chem* **52** 5301 *1987*].

Benzopurpurin 4B {3,3'-[(3,3'-dimethyl[1,1'-biphenyl]-4,4'-diyl)bis(azo)]bis[4-amino-1-naphthalenesulfonic acid di-Na salt, Direct red 2} *[992-59-6]* **M 724.7, λmax 500nm, CI 23500, pK<0.** Crystd from H_2O. It is a biological stain that is violet at pH 1.2 and red at pH 4.0 and is used for detecting Al, Mg, Hg, Au and U.

Benzotriazol-1-yloxytris(dimethylamino)phosphonium hexafluorophosphate (BOP reagent) *[56602-33-6]* **M 442.29, m > 130° (dec), 147-149° (dec).** Dissolve in CH_2Cl_2, dry ($MgSO_4$), filter, conc under vac then add dry Et_2O and filter off first crop. Add CH_2Cl_2 to the filtrate and concentrate again to give a second crop. Solid is washed with dry Et_2O and dried in vac. Also recryst from dry Me_2CO/Et_2O and check purity by NMR. Store in the dark. [Castro et al. *Synthesis* 751 *1976*.]

Benzylidene-bis-(tricyclohexylphosphine) dichlororuthenium (Grubbs catalyst) *[172222-30-9]* **M 823.0.** Repeatedly wash with Me_2CO and MeOH and dry in vac. Alternatively dissolve in CH_2Cl_2 concentrate to half vol, filter, add MeOH to ppte it as purple microcrystals. Filter off, wash several times with Me_2CO and MeOH and dry in a vac for several hours. [Scwab, Grubbs and Ziller *J Am Chem Soc* **118** 100 *1996*; Miller, Blackwell and Grubbs *J Am Chem Soc* **118** 9606 *1966*.]
§ A polymer supported version is available.

Benzyl Orange [4-(4-benzylaminophenylazo)benzenesulfonic acid potassium salt] *[589-02-6]* **M 405.5, pK$_{Est(1)}$~<0, pK$_{Est(2)}$~3.8.** Crystd from H_2O.

Benzyltriphenylphosphonium chloride *[1100-88-5]* **M 388.9, m 280° (sintering), 287-288°.** Wash with Et_2O and crystallise from EtOH (six sided plates). *Hygroscopic* and forms crystals with one mol H_2O. [*Justus Liebigs Ann Chem* **229** 320 *1885*; *Chem Ber* **83** 291 *1950*.]

Beryllium acetate (basic) [$Be_4O(OAc)_6$] *[1332-52-1]* **M 406.3, m 285-286°.** Crystd from chloroform.

Beryllium potassium fluoride *[7787-50-0]* **M 105.1.** Crystd from hot water (25mL/g).

Beryllium sulfate (4H_2O) *[7787-56-6]* **M 177.1, m ~100°(dec), pK$_1^{25}$ 3.2, pK$_2^{25}$ ~6.5 (Be²⁺).** Crystd from weak aqueous H_2SO_4.

Bicyclo[2.2.1]hepta-2,5-diene rhodium (I) chloride dimer (norbornadiene rhodium chloride complex dimer) *[12257-42-0]* **M 462, m 240°(dec).** Recrystd from hot $CHCl_3$-pet ether as fine crystals soluble in $CHCl_3$ and *C_6H_6 but almost insoluble in Et_2O or pet ether. [*J Chem Soc* 3178 *1959*.]

R-(-)-1,1'-Binaphthyl-2,2'-diylhydrogen phosphate *[39648-67-4]* M 348.3, m 217°, $[\alpha]_D^{20}$ -608° (c 1, MeOH), pK^{20} 0.74. Recrystallise from EtOH. Reflux for 3h in N NaOH is required to hydrolyse the cyclic phosphate. [*Tetrahedron Lett* 4617 *1971* ; *Tetrahedron Lett* **24**, 343 *1983*.]

S-(+)-1,1'-Binaphthyl-2,2'-diylhydrogen phosphate *[35193-64-7]* M 348.3, $[\alpha]_D^{20}$ +608° (c 1, MeOH), pK^{20} 0.74. Recrystallise from EtOH. Reflux for 3h in N NaOH is required to hydrolyse the cyclic phosphate. [*Tetrahedron Lett* 4617 *1971* ; *Tetrahedron Lett* **24**, 343 *1983*.]

2-Biphenylyl diphenyl phosphate *[132-29-6]* M 302.4, n^{25}1.5925. Vacuum distd, then percolated through an alumina column. Passed through a packed column maintained at 150° to remove residual traces of volatile materials by a counter-current stream of nitrogen at reduced pressure. [Dobry and Keller *J Phys Chem* **61** 1448 *1957*.]

2,2'-Bipyridinium chlorochromate *[76899-34-8]* M 292.6. Washed with cold conc HCl then H_2O (sintered glass funnel) and dried in vacuum ($CaCl_2$) to a free flowing yellow-brown powder. Stored in the dark. [*Synthesis* 691 *1980*; *Synth Commun* **10** 951 *1980*.] **SUSPECTED CARCINOGEN.**

2,2'-Biquinolin-4,4'-dicarboxylic acid dipotassium salt *[63451-34-3]* M 420.51. Recryst from H_2O. The *Cu salt* has λ_{max} at 562nm. [*Anal Biochem* **56** 4409 *1973*.]

Bis-(*p-tert*-butylphenyl)phenyl phosphate *[115-87-7]* M 438.5, b 281°/5mm, n^{25} 1.5412. Same as for 2-biphenylyl diphenyl phosphate (above).

Bis-(2-chlorophenyl) phenyl phosphate *[597-80-8]* M 395, b 254°/4mm, n^{25} 1.5767. Same as for 2-biphenylyl diphenyl phosphate above.

Bis-(1,5-cyclooctadiene)nickel (0) *[1295-35-8]* M 275.0, m 142° (dec). Available in sealed ampoules under N_2. All procedures should be carried out in a dry box and in an atmosphere of N_2 or Argon in subdued light because the complex is light and oxygen sensitive and flammable. The solid is washed with dry Et_2O (under Ar) and separates from toluene as yellow crysts. Filter under Ar gas pressure, place the crysts in a container and dry under a vac of 0.01 mm to remove adhered toluene, flush with Ar and seal under Ar or N_2 in glass ampoules. [Semmelhack *Org Reactions* **19** 115 and 178*1972*; Wilke et al. *Justus Liebigs Ann Chem* **699** 1 *1966*.] **SUSPECTED CARCINOGEN.**

Bis(2,9-dimethyl-1,10-phenanthroline) copper(I) perchlorate *[54816-44-5]* M 579.6, pK^{25} -2.4 to -3.1 (for $HClO_4$). Crystd from acetone.

2,2'-Bis-(diphenylphosphino)-1,1'-binaphthyl (BINAP) *[RS 98327-87-8]* M 622.7, m 283-286°, *[R-(+)-76189-55-4]* m 241-242°, *[S-(-)- 76189-56-5]* m 241-242°, $[\alpha]_D^{20}$ (+) and (-) 233° (c 0.3 toluene). Dissolve the individual enantiomers in toluene, wash with 30% aq NaOH, three times with H_2O, dry (Na_2SO_4), evap to ~15% of its vol and add an equal vol of degassed MeOH. Collect the solid, wash with MeOH and dry at 80°/0.005mm for 6h. Recryst from 1:1 mixt of toluene-EtOH to optical purity (**m** 241-242°)[Takaya et al. *Org Synth* **67** 20 *1989*].
§ A polymer supported version is available.

1,4-Bis-(diphenylphosphino)butane *[7688-25-7]* M 426.5, m 135-136°. Recrystd from EtOH [Trippett *J Chem Soc* 4263 *1961*].

2R,3R-(+)-2,3-Bis(diphenylphosphino)butane (**R,R**-CHIRAPHOS) *[74839-84-2]*, 2S,3S-(-)-2,3-bis(di-phenylphosphino)butane (**S,S**-CHIRAPHOS) *[64896-28-2]* M 426.5, m 108-109°, $[\alpha]_D^{20}$ (+) and (-) 200° (c 1.5 $CHCl_3$). Recrystd from abs EtOH (~6g in 60mL) as colorless plates [Fryzuk and Bosnich *J Am Chem Soc* **99** 6262 *1977* and **101** 3043 *1979*].

1,2-Bis-(diphenylphosphino)ethane (DIPHOS) [1663-45-2] **M 398.4, m 139-140°,140-142°, 143-144°, pK_{Est} ~4.5.** Recrystd from aq EtOH or *C_6H_6. The *dimethiodide* recrystd from MeOH has **m** 305-307° and the *dioxide* recrystd from toluene or DMF (needles), or *C_6H_6 (plates) has **m** 252-254° (276-278°) [Isslieb et al. *Chem Ber* **92** 3175 *1959*; NMR: Aquiar et al. *J Org Chem* **29** 1660 *1964*; Bäckvall et al. *J Org Chem* **52** 5430 *1987*].

1,1'-Bis-(diphenylphosphino)ferrocene [12150-46-8] **M 554.4, m 181-183°, 184-194°.** Wash with distilled H_2O and dry in a vacuum. Dissolve in *ca* 5 parts of hot dioxane and cool to give orange crystals **m** 181-183°. Recrystn from *C_6H_6-heptane (1:2) gives product with **m** 183-184°. [*J Organomet Chem* **27** 241 *1971*.]

Bis-(2-ethylhexyl) 2-ethylhexyl phosphonate [25103-23-5] **M 434.6, n^{25} 1.4473.** Purified by stirring an 0.4M soln in *benzene with an equal volume of 6M HCl at *ca* 60° for 8h. The *benzene layer was then shaken successively with equal volumes of water (twice), aqueous 5% Na_2CO_3 (three times), and water (eight times), followed by evaporation of the *benzene and distilled under reduced pressure at room temperature (using a rotating evacuated flask). Stored in dry, dark conditions [Peppard et al. *J Inorg Nucl Chem* **24** 1387 *1962*]. Vacuum distilled, then percolated through an alumina column before finally passed through a packed column maintained at 150° where residual traces of volatile materials were removed by a counter-current stream of N_2 at reduced pressure [Dobry and Keller *J Phys Chem* **61** 1448 *1957*].

Bis-(2-ethylhexyl) phosphoric acid [298-07-7] **M 322.4.** See di-(2-ethylhexyl) phosphoric acid on p. 418.

Bis(ethyl)titanium(IV) chloride [2247-00-9] **M 177.0.** Crystd from boiling toluene.

Bis(ethyl)zirconium(IV) chloride [92212-70-9] **M 220.3.** Crystd from boiling toluene.

2,4-Bis-(methylthio)-1,3,$2\lambda^5$,$4\lambda^5$-dithiadiphosphetane-2,4-dithione (Davy's reagent) [82737-61-9] **M 284.4, m 160°.** Recrystd from *C_6H_6 in yellow plates or from hot trichlorobenzene. The low **m** observed in the literature (112° with gradual softening at 68-102°) has been attributed to the presence of elemental sulfur in the crystals. [*Tetrahedron* **40** 2663 *1984*; *J Org Chem* **22** 789 *1957*.]

Bismuth [7440-69-9] **M 209.0, m 271-273°.** Melted in an atmosphere of dry helium and filtered through dry Pyrex wool to remove any bismuth oxide present [Mayer, Yosim and Topol *J Phys Chem* **64** 238 *1960*].

Bismuthiol I (2,5-dimercapto-1,3,4-thiadiazole) potassium salt [4628-94-8] **M 226.4, m 275-276°(dec), $pK_{Est(1)}$~4.1.** Usually contaminated with disulfide. Purified by crystn from EtOH. Reagent for detection of Bi,Cu, Pb and Sb.

Bismuth trichloride [7787-60-2] **M 315.3, m 233.6°, pK^{25} 1.58 (Bi_3^+ = $BiOH_2^+$ + H^+).** Sublimed under high vacuum, or dried under a current of HCl gas, followed by fractional distn, once under HCl and once under argon.

N,N'-Bis-(salicylidene)ethylenediamine cobalt (II) [Co(SALEN)$_2$, salcomine] [14167-18-1] **M 325.2.** The powder should have an oxygen capacity of 4.7-4.8% as measured by the increase in wt under O_2 at 100 pounds pressure at *ca* 20°. The O_2 is expelled on heating the material to 65°. Recryst from pyridine, $CHCl_3$ or *C_6H_6, and the solvent may be removed by heating at 120° in a vac. However this heating may mean reduced O_2 capacity. In the dry state it absorbs O_2 turning from maroon colour to black. [Diehl and Hack *Inorg Synth* **3** 196 *1950*.]

Bis-(tetrabutylammonium) dichromate [56660-19-6] **M 700.9, m 139-142°.** Wash with water and dry in a vacuum. Crystallises from hexane (**m** 79-80°). [*Synth Commun* **10** 75 *1980*.] (Possible **CARCINOGEN**).

Bis-[4-(1,1,3,3-tetramethylbutyl)phenyl]phosphate calcium salt (Selectophore) *[40835-97-0]* **M 987.3.** The Ca diester salt is washed with H_2O (x3) and MeOH (x3) alternately and dried in a vacuum oven at 50°. If the Ca salt is contaminated with much Ca salt of the monoester then it (10g) is converted to the free acid by adding 6N HCl (*ca* 10vols) and Et_2O (> 50vols) to it and stirred vigorously to form the free acids. When no white ppte remained (*ca* 5min), the Et_2O is separated, washed with H_2O (2 x > 50 mL) and dried by filtering through a bed of anhydrous Na_2SO_4 (11 x 5 cm) which is then washed with Et_2O (2 x >50 mL). Evapn gives an oil (TLC R_F 0.81 for diester and 0.50 for monoester). The oil is dissolved in *benzene (*ca* 25mL) and extracted with ethane-1,2-diol (25mL, 10x). After ten washings, a small sample of the *benzene layer is washed twice with H_2O to remove the diol and showed that it is pure bis-[4-(1,1,3,3-tetramethylbutyl)-phenyl)phosphoric acid by TLC, i.e. no monophosphate. To form the Ca salt the oil is dissolved in MeOH and to it is added the equivalent amount of $CaCl_2$ together with aq NaOH to keep the pH >10. The resulting white ppte is collected washed alternately with 3 batches of H_2O and MeOH and dried in a vacuum oven at 50°. [*J Inorg Nucl Chem* **40** 1483 *1978*.]

2,4-Bis-(*p*-tolylthio)-1,3,2λ^5,4λ^5-dithiadiphosphetane-2,4-dithione (Heimgartner's reagent) *[114234-09-2]* **M 436.6, m 175-176°.** Recrystallise from toluene (light yellow solid), wash with Et_2O and dry in a vacuum. [*Helv Chim Acta* **70** 1001 *1987*.]

***N,O*-Bis-(trimethylsilyl)acetamide (BSA)** *[10416-59-8]* **M 203.4, b 71-73°/35mm, d 0.836, 1.4150.** Fractionate through a spinning band column and collect liquid **b** 71-73°/35mm, and not higher because the main impurity $MeCONHSiMe_3$ distills at **b** 105-107°/35mm. Used for derivatising alcohols and sugars [Klebe et al. *J Am Chem Soc* **88** 3390 *1966*, see *Carbohydr Res* **241** 209 *1993* and **237** 313 *1992*]. It is **FLAMMABLE** and **TOXIC**.

Bis-(trimethylsilyl)acetylene *[14630-40-1]* **M 170.4, m 26°, b 134-136°/atm.** Dissolve in pet ether, wash with ice-cold dilute HCl. The pet ether extract is dried ($MgSO_4$), evaporated and fractionated at atmospheric pressure. [*J Organomet Chem* **37** 45 *1972*.]

Bis-(trimethylsilyl) sulfide (hexamethyldisilathiane) *[3385-94-2]* **M 178.5, b 65-67°/16mm, 162.5-163.5°/750mm corr, 164°/760mm, d 0.85, n 1.4598.** Dissolve in pet ether (b *ca* 40°), remove solvent and distilled. Redistilled under atmospheric pressure of dry N_2. It is collected as a colourless liquid which solidifies to a white solid in Dry-ice. On standing for several days it turns yellow possibly due to liberation of sulfur. Store below 4° under dry N_2. [*J Chem Soc* 3077 *1950*.]

Bis-(triphenylphosphine)nickel(II) chloride *[14264-16-5]* **M 654.2, m 225°(dec).** Wash with glacial AcOH and dry in vac over H_2SO_4 and KOH until AcOH is removed. [*J Chem Soc* 719 *1958*.]

Boric acid (boracic acid) *[10043-35-3]* **M 61.8, m 171°, pK^{25} 9.23.** Crystd three times from H_2O (3mL/g) between 100° and 0°, after filtering through sintered glass. Dried to constant weight over metaboric acid in a desiccator. It is steam volatile. After 2 recrystns of ACS grade it had Ag at 0.2 ppm.

9-Borabicyclo[3.3.1]nonane (9BBN) *[monomer 280-64-8]* *[dimer 21205-91-4 or 70658-61-6]* *[1:1 coordination compound with tetrahydrofuran 76422-63-4]* **M 122.0 (monomer), 244.0 (dimer), m 141-143° (monomer), 150-152°, 154-155° (dimer), b 195°/12mm.** Available as the solid dimer or in tetrahydrofuran soln. The solid is relatively stable and can be purified by distn in a vacuum (as dimer) and by recrystn from tetrahydrofuran (solubility at room temp is 9.5%, 0.78M), filter solid under N_2 wash with dry pentane and dry *in vacuo* at *ca* 100°. The solid is a dimer (IR 1567cm^{-1}), stable in air (for *ca* 2 months), and can be heated for 24h at 200° in an inert atmosphere without loss of hydride activity. It is a dimer in tetrahydrofuran soln also (IR 1567cm^{-1}). It is sensitive to H_2O and air (O_2) in soln. Concentration in soln can be determined by reaction with MeOH and measuring the vol of H_2 liberated, or it can be oxidised to *cis*-cyclooctane-1,5-diol (**m** 73.5-74.5°). [IR: *J Am Chem Soc* **90** 5280 *1968*, **96** 7765 *1974*; *J Org Chem* **41** 1778 *1976*, **46** 3978 *1981*.]

Borane pyridine complex *[110-51-0]* **M 92.9, m 8-10°, 10-11°, b 86°/7mm, 100-101°/12mm, d$_4^{20}$ 0.785.** Dissolve in Et_2O and wash with H_2O in which it is insol. Evap Et_2O and distil (gives better than 99.8% purity). Its vap pressure is less than 0.1mm at room temp. [*J Am Chem Soc* **77** 1506 *1955*.]

Borane triethylamine complex *[1722-26-5]* **M 115.0, b 76°/4mm, 97.0°/12mm, d_4^{20} 0.78.** Distil in a vacuum using a 60cm glass helices packed column. [*J Am Chem Soc* **64** 325 *1942*, **84** 3407 *1962*; *Tetrahedron Lett* 4703 *1968*.]

Borane trimethylamine complex *[75-22-9]* **M 73.0, m 94-94.5°, b 171°/atm.** Sublimed using equipment described in *J Am Chem Soc* **59** 780 *1937*. Its vapour pressure is 86mm at 100°. Colourless hexagonal crystals varying from needles to short lumps, slightly soluble in H_2O (1.48% at 30°), EtOH (1%), hexane (0.74%) but very soluble in Et_2O, *C_6H_6 and AcOH. Stable at 125°. [*J Am Chem Soc* **59** 780 *1939*, **104** 325 *1942*.]

Boron trichloride (trichloroborane) *[10294-34-5]* **M 117.2, b 0°/476mm.** Purified (from chlorine) by passage through two mercury-filled bubblers, then fractionally distd under vacuum. In a more extensive purification the nitrobenzene addition compound is formed by passage of the gas over nitrobenzene in a vacuum system at 10°. Volatile impurities are removed from the crystalline yellow solid by pumping at -20°, and the BCl_3 is recovered by warming the addition compound at 50°. Passage through a trap at -78° removes entrained nitrobenzene; the BCl_3 finally condensing in a trap at -112° [Brown and Holmes *J Am Chem Soc* **78** 2173 *1956*]. Also purified by condensing into a trap cooled in acetone/Dry-ice, where it was pumped for 15min to remove volatile impurities. It was then warmed, recondensed and again pumped.

Boron trifluoride *[7637-07-2]* **M 67.8, b -101°/760mm.** The usual impurities - bromine, BF_5, HF and non-volatile fluorides - are readily separated by distn. Brown and Johannesen [*J Am Chem Soc* **72** 2934 *1950*] passed BF_3 into benzonitrile at 0° until the latter was satd. Evacuation to 10^{-5}mm then removed all traces of SiF_4 and other gaseous impurities. [A small amount of the BF_3-benzonitrile addition compound sublimed and was collected in a U-tube cooled to -80°]. Pressure was raised to 20mm by admitting dry air, and the flask containing the BF_3 addition compound was warmed with hot water. The BF_3 evolved was passed through a -80° trap (to condense any benzonitrile) into a tube cooled in liquid air. The addition compound with anisole can also be used. For drying, BF_3 can be passed through H_2SO_4 saturated with boric oxide. Fumes in moist air. [Commercially available as a 1.3M soln in MeOH or PrOH.]

Boron trifluoride diethyl etherate *[109-63-7]* **M 141.9, b 67°/43mm, b 126°/760mm, d 1.154, n 1.340.** Treated with a small quantity of diethyl ether (to remove an excess of this component), and then distd under reduced pressure, from CaH_2. Fumes in moist air. **TOXIC.**

Bromine *[7726-95-6]* **M 159.8, b 59°, d 3.102, n 1.661.** Refluxed with solid KBr and distd, dried by shaking with an equal volume of conc H_2SO_4, then distd. The H_2SO_4 treatment can be replaced by direct distn from BaO or P_2O_5 A more extensive purification [Hildenbrand et al. *J Am Chem Soc* **80** 4129 *1958*] is to reflux about 1L of bromine for 1h with a mixture of 16g of CrO_3 in 200mL of conc H_2SO_4 (to remove organic material). The bromine is distd into a clean, dry, glass-stoppered bottle, and chlorine is removed by dissolving *ca* 25g of freshly fused CsBr in 500mL of the bromine and standing overnight. To remove HBr and water, the bromine was then distd back and forth through a train containing alternate tubes of MgO and P_2O_5. **HIGHLY TOXIC.**

Bromine pentafluoride *[7789-30-2]* **M 174.9, m -60.5°, b 41.3°, d^{25} 2.466.** Purified *via* its KF complex, as described for chlorine trifluoride. **HIGHLY TOXIC.**

2-Bromoallyltrimethylsilane *[81790-10-5]* **M 193.2, b 64-66°/10mm, 82-85°/58-60mm, d_4^{20} 1.13.** Fractionally distd through an efficient column. It is **flammable**. [*J Am Chem Soc* **104** 3733 6879 *1982*.]

2-Bromo-1,3,2-benzodioxaborole *[51901-85-0]* **M 198.8, m 47°, 51-53°, b 76°/9mm.** Keep at 20°/15mm for some time and then fractionally distil. [*J Chem Soc* 1529 *1959*.]

1R(endo,anti)-3-**Bromocamphor-8-sulfonic acid ammonium salt** *[55870-50-3]* **M 328.2, m 284-285°(dec), $[\alpha]_D^{25}$ +84.8° (c 4, H$_2$O).** Passage of a hot aqueous soln through an alumina column removed water-soluble coloured impurities which remained on the column when the ammonium salt was eluted with hot water. The salt was crystd from water and dried over CaCl$_2$ [Craddock and Jones *J Am Chem Soc* **84** 1098 *1962*; Kauffmann *J Prakt Chem* **33** 295 *1966*].

Bromopyrogallol Red *[16574-43-9]* **M 576.2, ε 5.45 x 10^4 at 538nm (water pH 5.6-7.5).** See Bromopyrogallol Red (5,5'-dibromopyrogallolsulfonephthalein) on p. 141 in Chapter 4.

Bromosulfalein **(phenoltetrabromophthalein 3',3'-disulfonic acid disodium salt)** *[71-67-0]* **M 838.0.** Purified by TLC on silica Gel G (Merck 250μ particle size) in two solvent systems (BuOH-AcOH-H$_2$O 30:7.5:12.5 v/v; and BuOH-propionic acid-H$_2$O 30:20:7.5 v/v). When the solvent reached a height of 10cm the plate was removed, dried in air and developed with NH$_3$ vapour giving blue coloured spots. Also the dye was chromatographed on MN Silica Gel with *t*-BuOH-H$_2$O-*n*-BuOH (32:10:5 v/v and visualised with a dilute KOH (or NaOH if the Na salt is required) spray. The product corresponding to bromosulfalein was scraped off and eluted with H$_2$O, filtered and evap to dryness in a vacuum. It was dissolved in H$_2$O and filtered through Sephadex G-25 and evaporated to dryness. [UV and IR identification: *J Pharm Sci* **57** 819 *1968*; NMR: *Chem Pharm Bull Jpn* **20** 581 *1972*; *Anal Biochem* **83** 75 *1977*.]

Bromtrimethylsilane **(trimethylbromosilane, trimethylsilyl bromide)** *[2857-97-8]* **M 153.1, m -43.5° to -43.2°; b 40.5°/200mm, 77.3°/735mm, 79°/744mm, 79.8-79.9°/754mm, d_4^{20} 1.1805, d_4^{20} 1.190, n_D^{20} 1.422.** Purified by repeated fractional distillation and stored in sealed ampoules in the dark. [*J Am Chem Soc* **75** 1583 *1953*.] Also fractionally distd through a 15 plate column (0.8 x 32cm packed with 1/16in single turn helices from Pt-Ir wire). [*J Am Chem Soc* **68** 1161 *1946*; **70** 433 *1948*.]

tert-**Butyldiphenylchlorosilane** **(TBDPSCl, *tert*-butylchlorodiphenylsilane)** *[58479-61-1]* **M 274.9, b 90°/0.015mm, d 1.057, n 1.568.** Purified by repeated fractional distn. It is soluble in DMF and pentane [Hanessian and Lavalee *Can J Chem* **53** 2975 *1975*; Robl et al. *J Med Chem* **34** 2804 *1991*].

n-**Butylmercuric chloride** *[543-63-5]* **M 293.1, m 130°.** Crystd from EtOH.

n-**Butylphenyl *n*-butylphosphonate** *[36411-99-1]* **M 270.3.** Crystd three times from hexane as its compound with uranyl nitrate. See *tri-n-butyl phosphate* below.

tert-**Butyldimethylsilyl chloride** **(TBDMSCl)** *[18162-48-6]* **M 150.7, m 87-89°, 92.5°, b 125°/760mm.** Fractionally distd at atmospheric pressure. [*J Am Chem Soc* **76** 1030 *1954*; **94** 6190 *1972*.]

p-tert-**Butylphenyl diphenyl phosphate** *[981-40-8]* **M 382.4, b 261°/6mm, n^{25} 1.5522.** Purified by vacuum distn, and percolation through an alumina column, followed by passage through a packed column maintained at 150° to remove residual traces of volatile materials in a counter-current stream of N$_2$ at reduced pressure [Dobry and Keller *J Phys Chem* **61** 1448 *1957*].

n-**Butylstannoic acid** **[PhSn(OH)$_3$, trihydroxy-*n*-butylstannane]** *[22719-01-3]* **M 208.8.** Purified by adding excess KOH in CHCl$_3$ to remove *n*-BuSn(OH)Cl$_2$ and *n*-BuSn(OH)$_2$Cl, and isolated by acidification [Holmes et al. *J Am Chem Soc* **109** 1408 *1987*].

Cacodylic acid (dimethylarsinic acid) *[75-60-5]* M 138.0, m 195-196°, pK^{25} 6.15

[Me$_2$As(O)OH]. Crystd from warm EtOH (3mL/g) by cooling and filtering. Dried in vacuum desiccator over CaCl$_2$. Has also been twice recrystd from propan-2-ol. [Koller and Hawkridge *J Am Chem Soc* **107** 7412 *1985*.]

Cadion [1-(4-nitrophenyl)-3-(4-phenylazophenyl)-triazene] *[5392-67-6]* **M 346, m 198°.** Commercial cadion is purified by recrystn from 95% EtOH and dried. It is stable in 0.2 N KOH (in 20% aqueous EtOH) at 25°. It is a sensitive reagent for Cd, and the Cd complex has λmax (EtOH) 475nm. [*Aust Chem Inst J Proc* **4** 26 *1937* ; *Anal Chim Acta* **19** 377 *1958*.]

Cadmium *[7440-43-9]* **M 112.4, m 321.1°, b 767°.** Oxide has been removed by filtering the molten metal, under vacuum through quartz wool.

Cadmium acetate (2H$_2$O) *[5743-04-4]* **M 230.5, m 255°(anhydr), d 2.01 (hydr), 2.34 (anhydr), pK$_1^{25}$ 9.7, pK$_2^{25}$ ~11.0 (for Cd^{2+}).** Crystd twice from anhydrous acetic acid and dried under vacuum for 24h at 100°.

Cadmium bromide (4H$_2$O) *[13464-92-1 (4H$_2$O); 7789-42-6 (anhydr)]* **M 344.2, m 566°, b 963, d 5.19°.** Crystd from water (0.6mL/g) between 100° and 0°, and dried at 110°. Forms monohydrate below 36° and the 4H$_2$O above 36°.

Cadmium chloride *[10108-64-2]* **M 183.3, m 568°, b 960°, d 4.06.** Crystd from water (1mL/g) by addition of EtOH and cooling.

Cadmium fluoride *[7790-79-6]* **M 150.4, m >1000°, b 1748°, d 6.35.** Crystd by dissolving in water at room temperature (25mL/g) and heating to 60°.

Cadmium iodide *[7790-80-9]* **M 366.2, m 388°, b 787°, d 5.66.** Crystd from ethanol (2mL/g) by partial evaporation.

Cadmium ionophore I [*N*,*N*,*N'*,*N'*-tetramethyl-3,6-dioxooctanedi-(thioamide)] *[73487-00-0]* **M 432.7, m 35-36°.** Wash well with pet ether, then several times with 2N HCl (if it has a slight odour of pyridine) then H$_2$O and dry in a vacuum over H$_2$SO$_4$. It is a polar selectrophore for Cd. [*Helv Chim Acta* **63** 217 *1980*.]

Cadmium lactate *[16039-55-7]* **M 290.6.** Crystd from water (10mL/g) by partial evapn in a desiccator.

Cadmium nitrate (4H$_2$O) *[10022-68-1]* **M 308.5, m 59.5°.** Crystd from water (0.5mL/g) by cooling in ice-salt.

Cadmium potassium iodide *[13601-63-3]* **M 532.2.** Crystd from ethanol by partial evapn.

Cadmium salicylate *[19010-79-8]* **M 248.5, 242°(dec).** Crystd from distd H$_2$O by evapn in a desiccator.

Cadmium sulfate *[7790-84-3 (for 3CdSO$_4$ 8H$_2$O); 10124-36-4 (anhydr)]* **M 208.4 (anhydr), 769.5 (hydr).** Crystd from distd water by partial evapn in a desiccator. On heating gives monohydrate at 80°.

Calcein sodium salt [2',7'-bis-{*N*,*N*-di(carboxymethyl)aminomethyl}fluorescein Na salt, **Fluorexon, Fluorescein Complexon**] *[1461-15-0]* **M 666.5, pK$_{Est(1)}$~ 1.9, pK$_{Est(2)}$~ 2.5, pK$_{Est(3)}$~ 8.0, pK$_{Est(4)}$ ~ 10.5 (all for N-CH$_2$COOH), and pK$_{Est(5)}$ ~ 3.5 (for benzoic COOH).** Dissolve in distilled H$_2$O and acidify with dilute HCl to pH 3.5. Filter off the solid acid and wash well with H$_2$O. Redissolve *ca* 10g in 300mL H$_2$O containing 12g of NaOAc. Ppte again by adding HCl, filter and wash with H$_2$O. Add the solid to 200mL of EtOH stir for 1h and filter. Repeat the EtOH wash and dry the bright yellow solid in a vacuum. This acid decomposes on heating at *ca* 180°. See below for the prepn of the Na salt. [*Anal Chem* **28** 882 *1956*].
Dissolve in H$_2$O and acidify with 3N HCl to pH 3.5. Collect the solid and wash with H$_2$O. The air-dried ppte is extracted with 70% aqueous EtOH, filtered hot and cooled slowly. Fine yellow needles of the acid crystallise out, are filtered and dissolved in the minimum quantity of 0.01N NaOH and reppted with N HCl to pH 3.5. It is then recrystd from 70% aqueous EtOH (3x). The final product (acid) is dried at 80° in a vacuum for 24h, **m >300°dec.** It contains one mol of water per mol of acid (C$_{30}$H$_{36}$N$_4$O$_{13}$.H$_2$O). The product is pure as revealed

by electrophoresis at pH 5.6 and 8.6, and by TLC in i-BuOH-i-PrOH-AcOH-H_2O (60:60:5:5 by vol) or i-PrOH or pH 8.0 borate buffer. [Wallach et al. *Anal Chem* **31** 456 *1959*.]
The Na salt is prepared by dissolving the in H_2O containing 2 mols of NaOH per mol of acid reagent and lyophilising. It complexes with Ca and Mg ions.

Calcium *[7440-70-2]* **M 40.1, m 845°.** Cleaned by washing with ether to remove adhering paraffin, filed in an argon-filled glove box, and washed with ethanol containing 2% of conc HCl. Then washed with dry ethanol, dried in a vac and stored under pure argon [Addison, Coldrey and Halstead, *J Chem Soc* 3868 *1962*].

Calcium acetate monohydrate *[5743-26-0 (H$_2$O), 62-54-4 (xH$_2$O)]* **M 176.2 (H$_2$O), m 150° (loses H$_2$O), pK^{25} 12.7 (for Ca$_2$$^+$).** Crystd from water (3mL/g) by partial evapn in a desiccator.

Calcium benzoate (3H$_2$O) *[2090-05-3]* **M 336.4.** Crystd from water (10m/g) between 90° and 0°.

Calcium bromide (H$_2$O) *[62648-72-0; 71626-99-8 (xH$_2$O); 7789-41-5 (anhydr)]* **M 217.9, d 3.35.** Crystd from EtOH or Me_2CO. It loses H_2O on heating and is anhydrous at 750° then it loses Br. Deliquescent.

Calcium butyrate *[5743-36-2]* **M 248.2.** Crystd from water (5mL/g) by partial evapn in a desiccator.

Calcium carbamate *[543-88-4]* **M 160.1.** Crystd from aqueous ethanol.

Calcium chloride (anhydrous) *[10043-52-4]* **M 111.0, m 772°, b >1600°, d$_4^{15}$ 2.15.** Available as fused granules or cubic crystals. It is very *hygroscopic*. Very soluble in H_2O (exothermic), and EtOH. Store in a tightly closed container.

Calcium chloride (2H$_2$O) *[10035-04-8]* **M 147.0, m 175°(dehydr), 772°(dec).** Crystd from ethanol, and is hygroscopic. Loses H_2O at 200° so it can be dried at high temperatures to dehydrate. *Hexahydrate [7774-34-7]* has **m** 30° and **d** 1.67.

Calcium dithionite *[13812-88-9]* **M 168.2, m dec on heating.** Crystd from water, or water followed by acetone and dried in air at room temperature.

Calcium D-gluconate monohydrate *[299-28-5]* **M 448.4, m dec on heating, [α] $_{546}^{20}$ +11.0°, [α]$_D^{20}$ +9.0° (c 1.2, H$_2$O).** It is sol in H_2O (3.5g in 100g at 25°). Dissolve in H_2O, filter and ppte by adding MeOH. Filter off solid and dry in a vacuum at 85°. Alternatively, dissolve in H_2O, filter (from insol inorganic Ca) and evaporate to dryness under vacuum at 85°. [*J Am Pharm Assoc* **41** 366 *1952*.]

Calcium D-heptagluconate dihydrate *[17140-60-2]* **M 526.4, [α] $_{546}^{20}$ +5.2°, [α]$_D^{20}$ +4.4° (c 5, H$_2$O).** Purified same as calcium D-gluconate.

Calcium formate *[544-17-2]* **M 130.1, m dec on heating, d 2.01.** Crystd from water (5mL/g) by partial evaporation in a desiccator.

Calcium hexacyanoferrate (II) (11H$_2$O) *[13821-08-4]* **M 490.3.** Recrystd three times from conductivity H_2O and air dried to constant weight over partially dehydrated salt. [*Trans Faraday Soc* **45** 855 *1949*.] Alternatively the Ca salt can be purified by pptn with absolute EtOH in the cold (to avoid oxidation) from an air-free saturated aqueous soln. The pure lemon yellow crystals are centrifuged, dried in a vacuum desiccator first over dry charcoal for 24h, then over partly dehydrated salt and stored in a dark glass stoppered bottle. No deterioration occurred after 18 months. No trace of Na, K or NH_4 ions could be detected in the salt from the residue after decomposition of the salt with conc H_2SO_4. Analyses indicate 11mols of H_2O per mol of salt. The solubility in H_2O is 36.45g (24.9°) and 64.7g (44.7°) per 100g of solution. [*J Chem Soc* 50 *1926*.]

Calcium hydroxide *[1305-62-0]* **M 74.1, m loses H$_2$O on heating, pK^{25} 12.7 (for Ca^{2+}).** Heat analytical grade calcium carbonate at 1000° during 1h. Allow the resulting oxide to cool and add slowly to water. Heat the suspension to boiling, cool and filter through a sintered glass funnel of medium porosity (to remove soluble alkaline impurities). Dry the solid at 110° and crush to a uniformly fine powder.

Calcium iodate *[7789-80-2 (H₂O)]* **M 389.9, m >540°, pK²⁵ 0.79 (for HIO₃).** Crystd from water (100mL/g).

Calcium iodide (xH₂O) *[71626-98-7 (xH₂O); 10102-68-8 (anhydr)]* **M 293.9 (for 4H₂O), m 740°, b 1100°.** Dissolved in acetone, which was then diluted and evaporated. This drying process was repeated twice, then the CaI₂ was crystd from acetone-diethyl ether and stored over P₂O₅. Very *hygroscopic* when anhydrous and is light sensitive [Cremlyn et al. *J Chem Soc* 528 *1958*]. *Hexahydrate* has **m** 42°.

Calcium ionophore I (ETH 1001) *[58801-34-6]* **M 685.0.** This is a neutral Ca selectophore. It can be purified by thick layer (2mm) chromatography (Kieselgel F₂₄₅) and eluted with Me₂CO-CHCl₃ (2:1). [*Helv Chim Acta* **56** 1780 *1973*.]

Calcium ionophore II (ETH 129) *[74267-27-9]* **M 460.7, m 153-154°.** Recrystd from Me₂CO. It forms 1:2 and 1:3 metal/ligand complexes with Mg²⁺ and Ca²⁺ ions respectively, and induces selectivity in membranes for Ca²⁺ over Mg²⁺ by a factor of *ca* 10⁴. [*Helv Chim Acta* **63** 191 *1980*.]

Calcium ionophore III [A23187 calcimycin] *[52665-69-7]* **M 523.6, m 181-182°, [α]²⁵_D -56.0° (c 1, CHCl₃).** Recrystallises from Me₂CO as colourless needles. Protect from light and moisture, store in a refrigerator. Soluble in Me₂SO or EtOH and can be stored for 3 months without loss of activity. Mg and Ca salts are soluble in organic solvents and cross biological membranes. It has a pKa of 6.9 in 90% Me₂SO. The Ca complex cryst from 50% EtOH as colourless prisms. *Highly* **TOXIC** [*Ann Rev Biochem* **45** 501 *1976*; *J Am Chem Soc* **96** 1932 *1974, J Antibiotics* **29** 424 *1976*.]

Calcium isobutyrate *[533-90-4]* **M 248.2.** Crystd from water (3mL/g) by partial evapn in a desiccator.

Calcium lactate (5H₂O) *[814-80-2]* **M 308.3, m anhydr at 120°.** Crystd from warm water (10mL/g) by cooling to 0°.

Calcium nitrate (4H₂O) *[13477-34-4]* **M 236.1, m 45°(dehydr), 560°(anhydr).** Crystd four times from water (0.4mL/g) by cooling in a CaCl₂-ice freezing mixture. The tetrahydrate was dried over conc H₂SO₄ and stored over P₂O₅, to give the anhydrous salt. It is deliquescent. After 3 recrystns of ACS grade it had Co, Fe, Mg, Sr and Zn at 0.2, 1. 0, 0.02, 10 and 0.02 ppm resp.

Calcium nitrite (2H₂O) *[13780-06-8 (30%w/w aq soln)]* **M 150.1(hydr), m dec on heating, d 2.22.** Crystd from hot water (1.4mL/g) by adding ethanol and cooling to give the hydrate. It is deliquescent.

(+)-Calcium pantothenate (H₂O) (D(+)- *137-08-6; 63409-48-3]* **M 476.5.** See *R*(+)-pantothenic acid calcium salt on p. 555 in Chapter 6.

Calcium permanganate (4H₂O) *[10118-76-0 (anhydr)]* **M 350.0 (for 4H₂O).** Crystd from water (3.3mL/g) by partial evapn in a desiccator. It is deliquescent.

Calcium propionate *[4075-81-4]* **M 186.2, m dec on heating.** Crystd from water (2mL/g) by partial evapn in a desiccator.

Calcium salicylate (2H₂O) *[824-35-1]* **M 350.4.** Crystd from water (3mL/g) between 90° and 0°.

Calcium sulfate dihydrate *[10101-41-4]* **M 172.1, m 150(dec), d 2.32.** Loses only part of its H₂O at 100-150° (see below). Soluble in H₂O and very slowly soluble in glycerol. Insoluble in most organic solvents.

Calcium sulfate hemihydrate *[10034-76-1]* **M 145.2.** Sol in H_2O (0.2 parts/100 at 18.75°). Completely dehydrated >650°. Dry below 300° to give a solid with estimated pore size *ca* 38% of vol. Anhydrous $CaSO_4$ has high affinity for H_2O and will absorb 6.6% of its weight of H_2O to form the hemihydrate (gypsum). It sets to a hard mass with H_2O, hence should be kept in a tightly sealed container.

Calcium thiosulfate *[10124-41-1]* **M 152.2, m 43-49°, pK_1^{25} 0.6, pK_2^{25} 1.74 (for $H_2S_2O_3$).** Recrystd from water below 60° in a N_2 atmosphere, followed by drying with EtOH and Et_2O. Stored in a refrigerator. [Pethybridge and Taba *J Chem Soc, Faraday Trans 1* **78** 1331 *1982*.]

(4-Carbamylphenylarsylenedithio)diacetic acid *[531-72-6]* **M 345.1, pK_{Est}~3.5.** Crystd from MeOH or EtOH.

Carbonate ionophore I [ETH 6010] (heptyl 4-trifluoroacetylbenzoate) *[129476-47-7]* **M 316.3, b 170°/0.02 Torr, d 0.909.** Purified by flash chromatography (2g of reagent with 30g of Silica Gel 60) and eluted with EtOAc/hexane (1:19). The fractions that absorbed at 260nm were pooled, evapd and dried at room temp (10.3 Torr). The oily residue was distd in a bubbled-tube apparatus (170°/0.02 Torr). Its IR ($CHCl_3$) had peaks at 1720, 1280, 940cm^{-1} and its sol in tetrahydrofuran is 50mg/0.5mL. It is a lipophilic neutral ionophore selective for carbonate as well as being an optical humidity sensor. [*Anal Chim Acta* **233** 41 *1990*.]

Carbon dioxide *[124-38-9]* **M 44.0, sublimes at -78.5°, pK_1^{25} 6.35, pK_2^{25} 10.33 (for H_2CO_3).** Passed over CuO wire at 800° to oxidise CO and other reducing impurities (such as H_2), then over copper dispersed on Kieselguhr at 180° to remove oxygen. Drying at -78° removed water vapour. Final purification was by vacuum distn at liquid nitrogen temperature to remove non-condensable gases [Anderson, Best and Dominey *J Chem Soc* 3498 *1962*]. Sulfur dioxide can be removed at 450° using silver wool combined with a plug of platinised quartz wool. Halogens are removed by using Mg, Zn or Cu, heated to 450°.

Carbon disulfide, see entry on p. 156 in Chapter 4.

Carbon monoxide *[630-08-0]* **M 28.0, b -191.5°.** Iron carbonyl is a likely impurity in CO stored under pressure in steel tanks. It can be decomposed by passage of the gas through a hot porcelain tube at 350-400°. Passage through alkaline pyrogallol soln removes oxygen (and CO_2). Removal of CO_2 and water are effected by passage through soda-lime followed by $Mg(ClO_4)_2$. Carbon monoxide can be condensed and distd at -195°. **HIGHLY POISONOUS gas.**

Carbonyl bromide *[593-95-3]* **M 187.8, b 64.5°/760mm.** Purified by distn from Hg and from powdered Sb to remove free bromine, then vacuum distd to remove volatile SO_2 (the major impurity) [Carpenter et al. *J Chem Soc, Faraday Trans 2* 384 *1977*]. **TOXIC**

Carbonyl sulfide *[463-58-1]* **M 60.1, m -138°, b -47.5°, -50°.** Purified by scrubbing through three consecutive fritted washing flasks containing conc NaOH at 0° (to remove HCN), and then through conc H_2SO_4 (to remove CS_2) followed by a mixture of NaN_3 and NaOH solution; or passed through traps containing satd aq lead acetate, then through a column of anhydrous $CaSO_4$. Then freeze-pumped repeatedly and distd through a trap packed with glass wool and cooled to -130° (using an *n*-pentane slurry). It liquefies at 0°/12.5mm. The gas is stored over conc H_2SO_4. **TOXIC**

Catecholborane (1,3,2-Benzodioxaborole) *[274-07-7]* **M 119.2, b 50°/50mm, 66°/80mm, 76-77°/100mm, 88°/165mm, d 1.125, n 1.507 (also available as a 1.0M soln in THF).** A moisture sensitive flammable liquid which is purified by distn in a vacuum under a N_2 atmosphere and stored under N_2 at 0-4°. It liberates H_2 when added to H_2O or MeOH. A soln in THF after 25h at 25° has residual hydride of 95%(under N_2) and 80% (under air) [Brown and Gupta *J Am Chem Soc* **97** 5249 *1975*].

Celite 545 (diatomaceous earth) *[12003-10-0]*. Stood overnight in conc HCl after stirring well, then washed with distilled water until neutral and free of chloride ions. Washed with methanol and dried at 50°.

Ceric ammonium nitrate *[16774-21-3]* **M 548.2, pK$_1^{25}$ -1.15, pK$_2^{25}$ -0.72, pK$_3^{25}$ 1.68, pK$_4^{25}$ 2.29 (for aquo Ce^{4+}).** Ceric ammonium nitrate (125g) is warmed with 100mL of dilute HNO$_3$ (1:3 v/v) and 40g of NH$_4$NO$_3$ until dissolved, and filtered off on a sintered-glass funnel. The solid which separates on cooling in ice is filtered off on a sintered funnel (at the pump) and air is sucked through the solid for 1-2 h to remove most of the nitric acid. Finally, the solid is dried at 80-85°.

Cerous acetate *[537-00-8]* **M 317.3, pK$_1^{25}$ 8.1 (9.29), pK$_2^{25}$ 16.3, pK$_3^{25}$ 26.0 (for Ce^{3+}).** Crystd twice from anhydrous acetic acid, then pumped dry under vacuum at 100° for 8h.

Cesium bromide *[7787-69-1]* **M 212.8, m 636°, b *ca* 1300°, d 4.44.** Very soluble in H$_2$O, soluble in EtOH but insoluble in Me$_2$CO. Dissolve in the minimum volume of H$_2$O, filter and ppte by adding Me$_2$CO. Filter solid and dry at 100°. Also recrystd from water (0.8mL/g) by partial evaporation in a desiccator.

Cesium carbonate *[534-17-8]* **M 325.8, m 792°(at red heat).** Crystd from ethanol (10mL/g) by partial evaporation.

Cesium chloride *[7647-17-8]* **M 168.4, m 645°, b 1303°, d 3.99.** Soluble in H$_2$O but can be purified by crystn from H$_2$O [sol in g per cent: 162.3(0.7°), 182.2(16.2°) and 290(at bp 119.4°)] and dried in high vac. Sol in EtOH and is deliquescent, keep in a tightly closed container. [*Handbook of Preparative Inorganic Chemistry (Ed. Brauer)* Vol I 951 *1963*.] For further purification of CsCl, a conc aqueous soln of the practically pure reagent is treated with an equivalent weight if I$_2$ and Cl$_2$ bubbled into the soln until pptn of CsCl$_2$I ceased. Recrystn yields a salt which is free from other alkali metals. It is then decomposed to pure CsCl on heating. [*J Am Chem Soc* 52 3886 *1930*.] Also rerystd from acetone-water, or from water (0.5mL/g) by cooling in CaCl$_2$/ice. Dried at 78° under vacuum.

Cesium chromate *[56320-90-2]* **M 381.8, pK$_1^{25}$ 0.74, pK$_2^{25}$ 6.49 (for H$_2$CrO$_4$).** Crystd from water (1.4mL/g) by partial evapn in a desiccator.

Cesium fluoride *[13400-13-0]* **M 151.9, m 703°.** Crystd from aqueous soln by adding ethanol.

Cesium iodide *[7789-17-5]* **M 259.8, m 621°, b~1280°, d 4.5.** Crystd from warm water (1mL/g) by cooling to -5°.

Cesium nitrate *[7789-18-6]* **M 194.9, m 414°(dec), d 3.65.** Crystd from water (0.6mL/g) between 100° and 0°. After 1 crystn of 99.9% grade it had K, Na and Se at 0.8, 0.4 and 0.2 ppm resp.

Cesium oleate *[31642-12-3]* **M 414.4.** Crystd from EtOAc, dried in an oven at 40° and stored over P$_2$O$_5$.

Cesium perchlorate *[13454-84-7]* **M 232.4, pK^{25} -2.4 to -3.1 (for HClO$_4$).** Crystd from water (4mL/g) between 100° and 0°.

Cesium perfluoro-octanoate *[17125-60-9]* **M 546.0.** Recrystd from a butanol-petroleum ether mixture, dried in an oven at 40° and stored over P$_2$O$_5$ under vacuum.

Cesium sulfate *[10294-54-9]* **M 361.9, m 1005°, d 4.24.** Crystd from water (0.5mL/g) by adding ethanol and cooling.

Chloramine-T (*N*-chloro-*p*-toluenesulfonamide sodium salt) 3H$_2$O *[7080-50-4]* **M 281.7, m 168-170°(dec).** Crystd from hot water (2mL/g). Dried in a desiccator over CaCl$_2$ where it loses water. Protect from sunlight. Used for detection of bromate and halogens, and Co, Cr, Fe, Hg, Mn, Ni and Sb ions.

Chlorazol Sky Blue FF {6,6'-[(3,3'-dimethoxy[1,1'-biphenyl]-4,4'-diyl)bis(azo)bis(4-amino-5-hydroxy-1,3-naphthylenedisulfonic acid) tetra-Na salt *[2610-05-1]* **M 996.9, m**

>300°(dec). Freed from other electrolytes by adding aqueous sodium acetate to a boiling soln of the dye in distd water. After standing, the salted-out dye was filtered on a Büchner funnel, the process being repeated several times. Finally, the ppted dye was boiled several times with absolute EtOH to wash out any sodium acetate, then dried (as the sodium salt) at 105°. [McGregor, Peters and Petropolous *Trans Faraday Soc* **58** 1045 *1962*.]

Chlorine *[7782-50-5]* **M 70.9, m -101.5°, b -34.0°, d 2.898.** Passed in succession through aqueous $KMnO_4$, dilute H_2SO_4, conc H_2SO_4, and a drying tower containing $Mg(ClO_4)_2$. Or, bubbled through with water, dried over P_2O_5 and distd from bulb to bulb in a vac line. **HIGHLY TOXIC.**

Chlorine trifluoride *[7790-91-2]* **M 92.5, b 12.1°.** Impurities include chloryl fluoride, chlorine dioxide and hydrogen fluoride. Passed first through two U-tubes containing NaF to remove HF, then through a series of traps in which the liquid is fractionally distd. Can be purified *via* the KF complex, $KClF_4$, formed by adding excess ClF_3 to solid KF in a stainless steel cylinder in a dry-box and shaking overnight. After pumping out the volatile materials, pure ClF_3 is obtained by heating the bomb to 100-150° and condensing the evolved gas in a -196° trap [Schack, Dubb and Quaglino *Chem Ind (London)* 545 *1967*]. **HIGHLY TOXIC.**

Chlorodiphenylphosphine (diphenylphosphinous chloride) *[1079-66-9]* **M 220.6, m 15-16°, b 124-126°/0.6mm, 174°/5mm, 320°/atm, d 1.229, n 1.636.** Air sensitive, pale yellow lachrymatory liquid which is purified by careful fractional distn and discarding the lower boiling fraction which contains the main impurity $PhPCl_2$ (**b** 48-51°/0.7mm); and checking for tmpurities by NMR. [Weinberg *J Org Chem* **40** 3586 *1975*; Honer et al. *Chem Ber* **94** 2122 *1961*.]

4-(Chloromercuri)benzenesulfonic acid monosodium salt *[14110-97-5]* **M 415.2, dec on heating.** The free acid is obtained by acidifying an aq soln, filtering off the acid, washing it with H_2O and recrystallising from hot H_2O to give a colourless solid which is dried in a vacuum over P_2O_5 and should give negative Cl^- ions. The Na salt is made by dissolving in one equivalent of aqueous NaOH and evaporate to dryness. [*Chem Ber* **67** 130 *1934*; *J Am Chem Soc* **76** 4331 *1954*.] **HIGHLY TOXIC.**

4-Chloromercuribenzoic acid *[59-85-8]* **M 357.2, m >300°.** Its suspension in water is stirred with enough 1M NaOH to dissolve most of it: a small amount of insoluble matter is removed by centrifugation. The chloromercuribenzoic acid is then ppted by adding 1M HCl and centrifuged off. The pptn is repeated twice. Finally, the ppte is washed three times with distilled water (by centrifuging), then dried in a thin layer under vacuum over P_2O_5 [Boyer *J Am Chem Soc* **76** 4331 *1954*].

Chloromethylphosphonic acid dichloride *[1983-26-2]* **M 167.4, b 50°/0.5mm, 52-53(59)°/2mm, 63-65°/3mm, 78-79°/10mm, 87-88°/15mm, 102-103°/30mm, d_4^{20} 1.638, n_D^{20} 1.4971.** Fractionally distd using a short Claisen column and redistd. The *aniline salt* has **m** 199-201°. The ^{31}P NMR has a line at -38±2 ppm from 85% H_3PO_4. [Kinnear and Perren *J Chem Soc* 3437 *1952*; NMR: *J Am Chem Soc* **78** 5715 *1956*; *J Org Chem* **22** 462 *1957*.]

2-Chloro-2-oxo-1,3,2-dioxaphospholane *[6609-64-9]* **M 142.5, m 12-14°, b 89-91°/0.8mm, d_4^{20} 1.549, n_D^{20} 1.448.** Should be distd at high vacuum as some polymerisation occurs on distn. It has IR bands at 3012, 2933, 1477, 1366, 1325, 1040, 924 and 858 cm^{-1}. In H_2O at 100° it is hydrolysed to $HOCH_2CH_2OPO_3H_2$ in 30min [IR: Cox and Westheimer *J Am Chem Soc* **80** 5441 *1958*].

2-Chlorophenyl diphenyl phosphate *[115-85-5]* **M 360.7, b 236°/4mm, n^{25} 1.5707.** Purified by vacuum distn, percolated through a column of alumina, then passed through a packed column maintained by a countercurrent stream of nitrogen at reduced pressure [Dobry and Keller *J Phys Chem* **61** 1448 *1957*].

Chlorosulfonic (chlorosulfuric) acid *[7790-94-5]* **M 116.5, b 151-152°/750mm, d_4 1.753, n 1.4929, pK -5.9 (aq H_2SO_4).** Distd in an all-glass apparatus, taking the fraction boiling at 156-158°. Reacts **EXPLOSIVELY** with water [Cremlyn *Chlorosulfonic acid: A Versatile Reagent*, Royal Society of Chemistry UK, 2002, 308 pp, ISBN 0854044981].

Chloro-(2,2':6',2'-terpyridine)platinum (II) chloride (2H$_2$O) *[60819-00-3]* **M 535.3.** Recrystd from hot dilute HCl and cooling to give the red dihydrate. The trihydrate crysts slowly from a cold aq soln and is air dried. The red dihydrate can be obtained from the trihydrate by desiccation over conc H$_2$SO$_4$, by washing with EtOH or by precipitating from a warm aq soln with HCl. The dihydrate is also formed by decomposing the black trihydrate form by heating in water (slowly), or more rapidly with hot 2N HCl. [*J Chem Soc* 1498 *1934*.]

Chloro-tri-isopropyl titanium *[20717-86-6]* **M 260.6, m 45-50°, b 61-65°/0.1mm.** Distd under vacuum and sets slowly to a solid on standing. Stock reagents are made by dissolving the warm liquid in pentane, toluene, Et$_2$O, THF, CH$_2$Cl$_2$, and can be stored in pure state or in soln under dry N$_2$ for several months. The reagent is *hygroscopic* and is hydrolysed by H$_2$O. [*Chem Ber* **118** 1421 *1985*.]

Chlorotriphenylsilane **(triphenylchlorosilane)** *[76-86-8]* **M 294.9, m 90-92°, 91-93°, 94-95°, 97-99°, b 156°/1mm, 161°/0.6mm.** Likely impurities are tetraphenylsilane, small amounts of hexaphenyldisiloxane and traces of triphenylsilanol. Purified by distn at 2mm, then crystd from EtOH-free CHCl$_3$, and from pet ether (b 30-60°) or hexane by cooling in a Dry-ice/acetone bath. [*J Chem Soc* 3671 *1957*; *J Am Chem Soc* **72** 4471 *1958*, **77** 6395 *1955*, **79** 1843 *1957*.]

Chlorotris(triphenylphosphine) rhodium I (Wilkinson's catalyst) *[14694-95-2]* **M 925.2, m 138°(dec), 140°(dec), 157-158°(dec).** Forms dark burgundy crysts from hot EtOH after refluxing for 30min. When the soln is heated for only 5min orange crystals are formed, Heating the orange crystals in EtOH yields red crystals. Crystn from Me$_2$CO gives the orange crystals. The two forms have similar IR spectra but X-rays are slighly different. [Osborne et al. *J Chem Soc (A)* 1711 *1966*; Osborne and Wilkinson *Inorg Synth* **10** 67 *1967*; Bennett and Donaldson *Inorg Chem* **16** 655 *1977*.] Sol in CH$_2$Cl$_2$ ~2% (25°), in toluene 0.2% (25°), and less sol in Me$_2$CO, MeOH, BuOH and AcOH, but insol in pet ethers and cyclohexane. It reacts with donor solvents such as pyridine, DMSO and MeCN.

Chromeazurol S *[1667-99-8]* **M 539.3,** λ_{max} **540nm,** ϵ **7.80 x 10^4 (10M HCl), CI 43825, pK$_1^{25}$ <0, pK$_2^{25}$ 2.25, pK$_3^{25}$ 4.88, pK$_4^{25}$ 11.75.** Crude *phenolic triphenylmethanecarboxysulfonic acid triNa salt* (40g) is dissolved in water (250mL) and filtered. Then added conc HCl (50mL) to filtrate, with stirring. Ppte is filtered off, washed with HCl (2M) and dried. Redissolved in water (250mL) and pptn repeated twice more in water bath at 70°. Then dried under vacuum over solid KOH (first) then P$_2$O$_5$ [Martynov et al. *Zh Analyt Khim* **32** 519 *1977*]. It has also been purified by paper chromatography using *n*-butanol, acetic acid and water (7:3:1). First and second spots were extracted. It chelates Al and Be. Used for estimating fluoride.

Chromic chloride (anhydrous) *[10025-73-7]* **M 158.4, m 1152°, pK$_1^{25}$ 3.95, pK$_2^{25}$ 5.55, pK$_3^{25}$ 10.5 (for Cr^{3+}).** Sublimed in a stream of dry HCl. Alternatively, the impure chromic chloride (100g) was added to 1L of 10% aq K$_2$Cr$_2$O$_7$ and several millilitres of conc HCl, and the mixture was brought to a gentle boil with constant stirring for 10 min. (This removed a reducing impurity.) The solid was separated, and washed by boiling with successive 1L lots of distilled water until the wash water no longer gave a test for chloride ion, then dried at 110° [Poulsen and Garner *J Am Chem Soc* **81** 2615 *1959*].

Chromium (III) acetylacetonate *[21679-31-2]* **M 349.3, m 212-216°, 216°, pK^{25} 4.0 (see chromic chloride).** Purified by dissolving 6g in hot *C$_6$H$_6$ (20mL) and adding 75mL of pet ether slowly. Cool to room temp then chill on ice, filter off and dry in air to give 2.9g. Also crystallises from EtOH. Sol in heptane, *C$_6$H$_6$, toluene and pentane-2,4-dione at 20-40°. It forms a 1:2 complex with CHCl$_3$. [*Inorg Synth* **5** 130 *1957*; *J Am Chem Soc* **80** 1839 *1958*.]

Chromium ammonium sulfate (12H$_2$O) *[34275-72-4 (hydr); 13548-43-1 (anhydr)]* **M 478.4, m 94° loses 9H$_2$O then dehydr at 300°, d 1.72.** Crystd from a saturated aqueous soln at 55° by cooling slowly with rapid mechanical stirring. The resulting fine crystals were filtered on a Büchner funnel, partly dried on a porous plate, then equilibrated for several months in a vacuum desiccator over crude chromium ammonium sulfate (partially dehydrated by heating at 100° for several hours before use) [Johnson, Hu and Horton *J Am Chem Soc* **75** 3922 *1953*].

Chromium (II) chloride (anhydrous) *[10049-05-5]* **M 122.9, m 824°, d_4^{14} 2.75.** Obtained from the dihydrate by heating *in vacuo* at 180°. It is a very *hygroscopic* white powder which dissolves in H_2O to give a sky blue solution. Stable in dry air but oxidises rapidly in moist air and should be stored in air tight containers. It sublimes at 800° in a current of HCl gas and cooled in the presence of HCl gas. Alternatively it can be washed with air-free Et_2O and dried at 110-120°. [*Inorg Synth* **3** 150 *1950*.]

Chromium hexacarbonyl *[13007-92-6]* **M 220.1, m 130°(dec), d 1.77.** Wash with cold EtOH then Et_2O and allow to dry in air. Alternatively recrystallise from dry Et_2O. This is best accomplished by placing the hexacarbonyl in a Soxhlet extractor and extracting exhaustively with dry Et_2O. Pure $Cr(CO)_6$ is filtered off and dried in air. Completely colourless refracting crystals are obtained by sublimation at 40-50°/<0.5mm in an apparatus where the collecting finger is cooled by Dry Ice and in which there is a wide short bore between the hot and cold sections to prevent clogging by the crystals. Loss of product in the crystn and sublimation is slight. It is important not to overdo the drying as the solid is appreciably volatile and **TOXIC** [vapour pressure is 0.04(8°), 1.0(48°) and 66.5(100°) mm]. Also do not allow the Et_2O solns to stand too long as a brown deposit is formed which is sensitive to light, and to avoid the possibility of violent decomposition. It sinters at 90°, dec at 130°, and **EXPLODES** at 210°. [*Inorg Synth* **3** 156 *1950*; *J Am Chem Soc* **83** 2057 *1961*.]

Chromium potassium sulfate (12H$_2$O) *[7788-99-0]* **M 499.4, pK_1^{25} 0.74, pK_2^{25} 6.49 (for H_2CrO_4, chromic acid).** Crystd from hot water (2mL/g) by cooling.

Chromium trioxide (chromic anhydride) *[1333-82-0]* **M 100.0, m 197°, dec at 250° to Cr_2O_3, d 2.70 (pK_1^{25} 0.74, pK_2^{25} 6.49, for H_2CrO_4, chromic acid).** Red crystals from water (0.5mL/g) between 100° and -5°, or from water/conc HNO_3 (1:5). It separates when potassium or sodium dichromate are dissolved in conc H_2SO_4. Dried in a vacuum desiccator over NaOH pellets; hygroscopic, powerful oxidant, can ignite with organic compounds. It is a skin and pulmonary **IRRITANT**.
§ Commercially available on polymer support. **CANCER SUSPECT.**

Chromium (III) tris-2,4-pentanedionate *[21679-31-2]* **M 349.3, m 216°, pK^{25} 4.0 (see chromic chloride).** See chromium (III) acetylacetonate on p. 412.

Chromoionophore I [ETH 5294] [9-diethylamino-5-octadecanoyl-imino-5-*H*-benzo[a]-phenoxazine] *[125829-24-5]* **M 583.9.** Purified by flash chromatography (Silica Gel) and eluted with EtOAc. The coloured fractions are pooled, evaporated and recrystd from EtOAc. It is a lipophilic chromoionophore and is a selectophore for K and Ca ions. [*Anal Chem* **62** 738 *1990*.]

Chromotropic acid (4,5-dihydroxynaphthalene-2,7-disulfonic acid di-Na salt) *[5808-22-0 (2H$_2$O)]* **M 400.3, m >300°, pK_1 0.61(SO_3^-), pK_2 0.7(SO_3^-), pK_3 5.45(OH), pK_4 15.5(OH).** See disodium 4,5(1,8)-dihydroxynaphthalene-2,7(3,6)-disulfonate (2H$_2$O) on p. 421.

Chromyl chloride *[14977-61-8]* **M 154.9, b 115.7°, d 1.911.** Purified by distn under reduced pressure. **TOXIC.**

Claisen alkali (alkali Claisen). Prepared from KOH (35g) in H_2O (25mL) and diluted to 100mL with MeOH. **STRONGLY CAUSTIC.**

Cobalt (II) *meso*-5.10,15,20-tetraphenylporphine complex *[14172-90-8]* **M 671.7.** Brown crystals from Et_2O or $CHCl_3$-MeOH (*cf iron chloride complex*). Recrystd by extraction (Soxhlet) with *C_6H_6. Sol in most organic solvents except MeOH and pet ether. [UV, IR: *J Am Chem Soc* **70** 1808 *1948*; **81** 5111 *1959*.]

Cobaltous acetate (4H$_2$O) *[6147-53-1]* **M 249.1, pK_1^{25} 9.85 (for Co^{2+}).** Crystd several times as the tetrahydrate from 50% aqueous acetic acid. Converted to the anhydrous salt by drying at 80°/1mm for 60h.

Cobaltous acetylacetonate *[14024-48-7]* **M 257.2, m 172°.** Crystd from acetone.

Cobaltous ammonium sulfate (6H₂O) *[13596-46-8]* **M 395.5, d 1.90.** Crystd from boiling water (2mL/g) by cooling. Washed with ethanol.

Cobaltous bromide (6H₂O) *[85017-77-2 (xH₂O); 7789-43-7 (anhydr)]* **M 326.9 (6H₂O), m 47°(dec), b 100°(dec), d 4.9.** Crystd from water (1mL/g) by partial evaporation in a desiccator.

Cobaltous chloride (6H₂O) *[7791-13-1 (6H₂O); 7646-79-9 (anhydr)]* **M 237.9, m 87°(dec), d 1.92.** A saturated aqueous soln at room temperature was fractionally crystd by standing overnight. The first half of the material that crystd in this way was used in the next crystn. The process was repeated several times, water being removed in a dry-box using air filtered through glass wool and dried over CaCl₂ [Hutchinson *J Am Chem Soc* **76** 1022 *1954*]. Has also been crystd from dilute aq HCl.

Cobaltous nitrate (6H₂O) *[10026-22-9]* **M 291.0, m ~55(6H₂O), 100-105°(dec), d 1.88.** Crystd from water (1mL/g), or ethanol (1mL/g), by partial evapn. After 3 crystns (H₂O) it contains: metal(ppm) As (8), Fe (1.2), K (1), Mg (4), Mn (4), Mo (4), Na (0.6), Ni (18), Zn (1.6).

Cobaltous perchlorate (6H₂O) *[13478-33-6]* **M 365.9, pK²⁵ -2.4 to -3.1 (for HClO₄).** Crystd from warm water (0.7mL/g) by cooling.

Cobaltous potassium sulfate *[13596-22-0]* **M 329.4.** Crystd from water (1mL/g) between 50° and 0°, and dried in a vacuum desiccator over conc H₂SO₄.

Cobaltous sulfate (7H₂O) *[10026-24-1 (7H₂O); 60459-08-7 (xH₂O); 10124-43-3 (anhydr)]* **M 281.1, m 41°(dec), d 2.03.** Crystd three times from conductivity water (1.3mL/g) between 100° and 0°.

Copper (I) thiophenolate *[1192-40-1]* **M 172.7, m ca 280°, pK₁²⁵ 6.62 (for PhS⁻).** The Cu salt can be extracted from a thimble (Soxhlet) with boiling MeOH. It is a green-brown powder which gives a yellow-green soln in pyridine. Wash with EtOH and dry in a vacuum. It can be ppted from a pyridine soln by addition of H₂O, collect ppte, wash with EtOH and dry in a vacuum. [*Synthesis* 662 *1974*; *J Am Chem Soc* **79** 170 *1957*; *Chem Ber* **90** 425 *1957*.]

12-Crown-4 (lithium ionophore V, 1,4,7,10-tetraoxacyclododecane) *[294-93-9]* **M 176.2, m 17°.** The distilled crude product had to be crystd from pentane at -20° to remove acyclic material. It is then dried over P₂O₅. [*Acta Chem Scand* **27** 3395 *1973*.]

Cupferon ammonium salt (*N*-nitroso-*N*-phenylhydroxylamine ammonium salt) *[135-20-6]* **M 155.2, m 150-155°(dec), 162.5-163.5°, 163-164°, pK²⁵ 4.16 (free base).** Recrystd twice from EtOH after treatment with Norite and finally once with EtOH. The crystals are washed with diethyl ether and air dried then stored in the dark over solid ammonium carbonate. A standard soln (ca 0.05M prepared in air-free H₂O) is prepared daily from this material for analytical work and is essentially 100% pure. [*Anal Chem* **26** 1747 *1954*.] It can also be washed with Et₂O, dried and stored as stated. In a sealed, dark container it can be stored for at least 12 months without deterioration. λmax 260nm (CHCl₃). [*Org Synth* Coll Vol I 177 *1948*; *J Am Chem Soc* **78** 4206 *1956*.] *Possible* **CARCINOGEN.**

Cupric acetate (H₂O) *[6046-93-1 (H₂O); 142-71-2 (anhydr)]* **M 199.7, m 115°, 240°(dec), d 1.88, pK₁²⁵ 8.0, pK₂²⁵ 13.1 (for Cu²⁺).** Crystd twice from warm dilute acetic acid solns (5mL/g) by cooling.

Cupric ammonium chloride (2H₂O) *[10534-87-9 (hydr); 15610-76-1 (anhydr)]* **M 277.5, m 110-120°(anhydr) then dec at higher temp, d 2.0.** Crystd from weak aqueous HCl (1mL/g).

Cupric benzoate *[533-01-7]* **M 305.8.** Crystd from hot water.

Cupric bromide *[7789-45-9]* **M 223.4, m 498°, b 900°, d 4.7.** Crystd twice by dissolving in water (140mL/g), filtering to remove any Cu_2Br_2, and concentrating under vac at 30° until crystals appeared. The cupric bromide was then allowed to crystallise by leaving the soln in a vac desiccator containing P_2O_5 [Hope, Otter and Prue *J Chem Soc* 5226 *1960*].

Cupric chloride *[7447-39-4]* **M 134.4, m 498°, 630°(dec).** Crystd from hot dilute aq HCl (0.6mL/g) by cooling in a $CaCl_2$-ice bath. Dehydrated by heating on a steam-bath under vacuum. It is deliquescent in moist air but efflorescent in dry air.

Cupric lactate (H_2O) *[814-81-3]* **M 295.7.** The monohydrate crysts from hot H_2O (3mL/g) on cooling.

Cupric nitrate ($3H_2O$) *[10031-43-3 (3H_2O); 3251-23-8 (anhydr)]* **M 241.6, m 114°, b 170°(dec), d 2.0.** Crystd from weak aqueous HNO_3 (0.5mL/g) by cooling from room temperature. The anhydrous salt can be prepared by dissolving copper metal in a 1:1 mixture of liquid NO_2 and ethyl acetate and purified by sublimation [Evans et al. *J Chem Soc, Faraday Trans 1* **75** 1023 *1979*]. The *hexahydrate* dehydr to trihydrate at 26°, and the *anhydrous salt* sublimes between 150 and 225°, but melts at 255-256° and is deliquescent.

Cupric oleate *[1120-44-1]* **M 626.5.** Crystd from diethyl ether.

Cupric perchlorate ($6H_2O$) *[10294-46-9 (hydr); 13770-18-8]* **M 370.5, m 230-240°, pK^{25} -2.4 to -3.1 (for $HClO_4$).** Crystd from distilled water. The *anhydrous salt* is hygroscopic.

Cupric phthalocyanine *[147-14-8]* **M 576.1.** Precipitated twice from conc H_2SO_4 by slow dilution with water. Also purified by two or three sublimations at 580° in an argon flow at 300-400Pa.

Cupric sulfate *[7758-98-7]* **M 159.6, m >560°.** After adding 0.02g of KOH to a litre of nearly saturated aq soln, it was left for two weeks, then the ppte was filtered on to a fibreglass filter with pore diameter of 5-15 microns. The filtrate was heated to 90° and allowed to evaporate until some $CuSO_4.5H_2O$ had crystd. The soln was then filtered hot and cooled rapidly to give crystals which were freed from mother liquor by filtering under suction [Geballe and Giauque *J Am Chem Soc* **74** 3513 *1952*]. Alternatively crystd from water (0.6mL/g) between 100° and 0°.

Cupric trifluoromethylsulfonate (copper II triflate) *[34946-82-2]* **M 361.7, pK^{25} <-3.0 (for triflic acid).** Dissolve in MeCN, add dry Et_2O until cloudy and cool at -20° in a freezer. The light blue ppte is collected and dried in a vacuum oven at 130°/20mm for 8h. It has λmax 737nm (ϵ 22.4$M^{-1}cm^{-1}$) in AcOH. [*J Am Chem Soc* **95** 330 *1973*]. It has also been dried in a vessel at 0.1Torr by heating with a Fischer burner [*J Org Chem* **43** 3422 *1978*]. It has been dried at 110-120°/5mm for 1h before use and forms a *benzene complex which should be handled in a dry box because it is air sensitive [*Chem Pharm Bull Jpn* **28** 262 *1980*; *J Am Chem Soc* **95** 330 *1973*].

Cuprous bromide *[7787-70-4]* **M 143.4, m 497°, b 1345°, d 4.72.** Purified as for cuprous iodide but using aqueous NaBr.

Cuprous bromide dimethylsulfide complex *[54678-23-8]* **M 205.6, m ca 135°(dec).** Purified by recrystn in the presence of Me_2S. A soln of the complex (1.02g) in Me_2S (5mL) is slowly diluted with hexane (20mL) and the pure colourless prisms of the complex (0.96g) separate and are collected and dried, **m** 124-129°dec. The complex is insoluble in hexane, Et_2O, Me_2CO, $CHCl_3$ and CCl_4. It dissolves in DMF and DMSO but the soln becomes hot and green indicating dec. It dissolves in $^*C_6H_6$, Et_2O, MeOH and $CHCl_3$ if excess of Me_2S is added a colourless soln is obtained. [*J Org Chem* **40** 1460 *1975*.] Prior to use, the complex was dissolved in Me_2S and evaporated to dryness in the weighed reaction flask [*J Organomet Chem* **228** 321 *1983*].

Cuprous chloride *[7758-89-6]* **M 99.0, m 430°, b~1400°.** Dissolved in strong HCl, ppted by dilution with water and filtered off. Washed with ethanol and diethyl ether, then dried and stored in a vacuum desiccator [Österlöf *Acta Chem Scand* **4** 375 *1950*]. Alternatively, to an aq. soln of $CuCl_2.2H_2O$ was added, with stirring, an aqueous soln of anhydrous sodium sulfite. The colourless product was dried at 80° for 30min and stored under N_2. $CuCl_2$ can be purified by zone-refining [Hall et al. *J Chem Soc, Faraday Trans 1* **79** 243 *1983*].

Cuprous cyanide *[544-92-3]* **M 89.6, m 474°.** Wash thoroughly with boiling H_2O, then with EtOH. Dry at 100° to a fine soft powder. [*J Chem Soc* 79 *1943*.]

Cuprous iodide *[7681-65-4]* **M 190.5, m 605°, b 1336°, d_4^{25} 5.63.** It can be freshly prepared by dissolving an appropriate quantity of CuI in boiling saturated aqueous NaI over 30min. Pure CuI is obtained by cooling and diluting the soln with water, followed by filtering and washing sequentially with H_2O, EtOH, EtOAc and Et_2O, pentane, then drying *in vacuo* for 24h [Dieter, *J Am Chem Soc* **107** 4679 *1985*]. Alternatively wash with H_2O then EtOH and finally with Et_2O containing a little iodine. Traces of H_2O are best removed first by heating at 110° and then at 400°. Exess of I_2 is removed completely at 400°. It dissolves in Et_2O if an amine is present to form the amine complex. [*Chem Ind (London)* 1180 *1957*.]

Cuprous iodide trimethylphosphite *[34836-53-8]* **M 314.5, m 175-177°, 192-193°.** Cuprous iodide dissolves in a $^*C_6H_6$ soln containing trimethylphosphite to form the complex. The complex crystallises from $^*C_6H_6$ or pet ether. [*Chem Ber* **38** 1171 *1905*; *Bull Chem Soc Jpn* **34** 1177 *1961*.]

Cuprous thiocyanate *[18223-42-2]* **M 121.6, pK^{25} -1.85 (for HSCN).** Purified as for cuprous iodide but using aq NaSCN.

Cyanamide *[420-04-2]* **M 42.0, m 43°, 45°, 46°, b 85-87°/0.5mm, pK_1^{20} -0.36 (1.1 at 29°), pK_2^{20} 10.27.** Purified by placing *ca* 15g in a Soxhlet thimble and extracting exhaustively (2-3h) with two successive portions of Et_2O (400mL, saturated with H_2O by shaking before use) containing two drops of 1N acetic acid. Two successive portions of Et_2O are used so that the NH_2CN is not heated for too long. Each extract is dried over Na_2SO_4(30g), then combined and evaporated under reduced pressure. The NH_2CN may be stored unchanged at 0° in Et_2O soln in the presence of a trace of AcOH. Extracts from several runs may be combined and evaporated together. The residue from evaporation of an Et_2O soln is a colourless viscous oil which sets to a solid, and can be recrystd from a mixture of 2 parts of $^*C_6H_6$ and 1 part of Et_2O. Concentrating an aqueous soln of NH_2CN at high temps causes **EXPLOSIVE** polymerisation. [*Org Synth Coll Vol IV* 645 *1963*; *Inorg Synth* **3** 39 *1950*; *J Org Chem* **23** 613 *1958*.] *Hygroscopic.*

Cyanogen bromide *[506-68-3]* **M 105.9, m 49-51°, b 60-62°/atm.** *All operations with this substance should be performed in a very efficient fume cupboard - it is very* **POISONOUS** *and should be handled in small amounts. Fresh commercial material is satisfactory for nearly all purposes and does not need to be purified. It is a white crystalline solid with a strong cyanide odour. If it is reddish in colour and partly liquid or paste-like then it is too far gone to be purified, and fresh material should be sought.* It can be purified by distn using small amounts at a time, and using a short wide-bore condenser because it readily solidifies to a crystalline white solid and may clog the condenser. *An appropriate gas mask should be used when transferring the molten solid from one container to another and the operation should be done in an efficient fume cupboard.* The melting point (**m** 49-51°) should be measured in a sealed tube. [*Org Synth* Coll Vol II 150 *1948*.]

Cyanogen iodide *[506-78-5]* **M 152.9, m 146-147°.** *This compound is* **POISONOUS** *and the precautions for cyanogen bromide (above) apply here.* The reagent (*ca* 5.9g) is dissolved in boiling $CHCl_3$ (15mL), filtered through a plug of glass wool into a 25mL Erlenmeyer flask. Cool to room temperature for 15min, then place in an ice-salt bath and cool to -10°. This cooling causes a small aqueous layer to separate as ice. The ice is filtered with the CNI, but melts on the filter and is also removed with the $CHCl_3$ used as washing liquid. The CNI which is collected on a sintered glass funnel is washed 3x with $CHCl_3$ (1.5mL at 0°) and freed from last traces of solvent by being placed on a watch glass and exposed to the atmosphere in a good fume cupboard at room temp for 1h to give colourless needles (*ca* 4.5g), **m** 146-147° (sealed capillary totally immersed in the oil bath). The yield depends slightly on the rapidity of the operation, in this way loss by sublimation can be minimised. If desired, it can be sublimed under reduced pressure at temps at which CNI is only slowly decomposed into I_2 and $(CN)_2$. The vacuum will need to be renewed constantly due to the volatility of CNI. [*Org Synth* **32** 29 *1952*.]

Decaborane [17702-41-9] M 122.2, m 99.7-100°. Purified by vacuum sublimation at 80°/0.1mm, followed by crystn from methylcyclohexane, CH_2Cl_2, or dry olefin-free-n-pentane, the solvent being subsequently removed by storing the crystals in a vacuum desiccator containing $CaCl_2$.

Deuterium [7782-39-0] **M 4.** Passed over activated charcoal at -195° [MacIver and Tobin *J Phys Chem* **64** 451 *1960*]. Purified by diffusion through nickel [Pratt and Rogers, *J Chem Soc, Faraday Trans I* **92** 1589 *1976*]. Always check deuterium for radioactivity to find out the amount of tritium in it (see D_2O below).

Deuterium oxide [7789-20-0] **M 20, f 3.8°/760mm, b 101.4°/760mm, d 1.105.** Distd from alkaline $KMnO_4$ [de Giovanni and Zamenhof *Biochem J* **92** 79 *1963*]. **NOTE that D_2O invariably contains tritiated water and will therefore be RADIOACTIVE; always check the radioactivity of D_2O in a scintillation counter before using.**

cis-**Diamminedichloroplatinum(II) (Cisplatin)** [15663-27-1] **M 300.1, m 270°(dec).** Recrystd from dimethylformamide and the purity checked by IR and UV-VIS spectroscopy. [Raudaschl et al. *Inorg Chim Acta* **78** 143 *1983*.] **HIGHLY TOXIC, SUSPECTED CARCINOGEN.**

Diammonium hydrogen orthophosphate [7783-28-0] **M 132.1.** Crystd from water (1mL/g) between 70° and 0°. After one crystn of ACS grade had Fe, Mo, Na, Se and Ti at 1, 0.2, 1.4, 0.2 and 0.8ppm resp.

Di-n-amyl n-amylphosphonate [6418-56-0] **M 292.4, b 150-151°/2mm, n 1.4378.** Purified by three crystns of its uranyl nitrate complex from hexane (see *tributyl phosphate*). Extracts Zr^{2+} from NaCl solns.

6,6-Dibenzyl-14-crown-4 (lithium ionophore VI; 6,6-dibenzyl-1,4,8,11-tetra-oxa-cyclo-tetradecane) [106868-21-7] **M 384.5, m 102-103°.** Dissolve in $CHCl_3$, wash with saturated aqueous NaCl, dry with $MgSO_4$, evaporate and purify by chromatography on silica gel and gradient elution with $^*C_6H_6$-MeOH followed by preparative reverse phase HPLC on an octadecyl silanised silica (ODS) column and eluting with MeOH. It can be crystd from MeOH (v_{KBr} 1120 cm^{-1}, C-O-C). [*J Chem Soc Perkin Trans 1* 1945 *1986*.]

Di-n-butyl boron triflate (di-n-butylboryl trifluoromethanesulfonate) [60669-69-4] **M 274.1, b 37°/0.12mm, 60°/2mm, pK^{25} <-3.0 (for triflic acid).** Distil in vacuum under argon and store under argon. Should be used within 2 weeks of purchase or after redistn. Use a short path distn system. It has IR bands in CCl_4 at v 1405, 1380, 1320, 1200 and 1550cm^{-1}; and ^{13}C NMR ($CDCl_3$) with δ at 118.1, 25.1, 21.5 and 13.6. [*Org Synth* **68** 83 *1990*; *J Am Chem Soc* **103**, 3099 *1981*.] **TOXIC**

Di-n-butyl cyclohexylphosphonate [1085-92-3] **M 245.4.** The compound with uranyl nitrate was crystd three times from hexane. For method see *tributyl phosphate*.

Di-*tert*-butyl dichlorosilane (DTBCl$_2$) [18395-90-9] **M 213.2, m -15°, b 190°/729mm, 195-197°/atm, d 1.01.** Purified by fractional distn. It is a colourless liquid with a pleasant odour and does not fume in moist air, but does not titrate quantitatively with excess of dil alkali. [*J Am Chem Soc* **70** 2877 *1948*.]

Di-n-butyl n-butylphosphonate [78-46-6] **M 250.3, b 150-151°/10mm, 160-162°/20mm, n^{25} 1.4302.** Purified by three recrystallisations of its compound with uranyl nitrate, from hexane. For method, see *tributyl phosphate*.

Di-*tert*-butyl silyl bis(trifluoromethanesulfonate) [85272-31-7] **M 440.5, b 73.5-74.5°/0.35mm, d 1.36 (see pK for triflic acid).** Purified by fractional distillation. It is a pale yellow liquid which should be stored under argon. It is less reactive than the diisopropyl analogue. The presence of the intermediate monochloro compound can be detected by ^1H NMR, ($CHCl_3$): *tert*-Bu$_2$Si(OTf)$_2$ [δ 1.25s]; but impurities have δ 1.12s for *tert*-Bu$_2$Si(H)OTf and δ 1.19s for *tert*-Bu$_2$HSi(Cl)OTf. [*Tetrahedron Lett* **23** 487 *1982*.] **TOXIC**.

Di-*n*-butyltin oxide *[818-08-6]* **M 248.9, m >300°.** It is prepd by hydrolysis of di-*n*-butyltin dichloride with KOH. Hence wash with a little aq M KOH then H$_2$O and dry at ~80°/10mm until the IR is free from OH bands. [Cummings *Aust J Chem* **18** 98 *1965*.]

Dicarbonyl(cyclopentadienyl)Co (I) *[1207-25-0]* **M 180.1, b 75°/22mm, b 139-140°(dec)/710mm.** Best distd in an atmosphere of CO in a vac. The red brown liquid decomposes slightly on distn even in a vac to liberate some CO. Operations should be performed in an efficient fume cupboard. It is sol in organic solvents and stable in air but decomposes slowly in sunlight and rapidly under UV. [Piper et al. *J Inorg Nucl Chem* **1** 165 *1955*.] **TOXIC**.

Dichlorodimethylsilane   see dimethylchlorosilane p. 419.

2,6-Dichlorophenol-indophenol sodium salt (2H$_2$O) *[620-45-1]* **M 326.1, ε 2.1 x 10^4 at 600nm and pH 8, pK^{30} 5.7 (oxidised form), pK$_1^{30}$ 7.0, pK$_2^{30}$ 10.1 (reduced form).** Dissolved in 0.001M phosphate buffer, pH 7.5 (alternatively, about 2g of the dye was dissolved in 80mL of M HCl), and extracted into diethyl ether. The extract was washed with water, extracted with aqueous 2% NaHCO$_3$, and the sodium salt of the dye was ppted by adding NaCl (30g/100mL of NaHCO$_3$ soln), then filtered off, washed with dilute NaCl soln and dried.

Dicobalt octacarbonyl *[10210-68-1]* **M 341.9, m 51°.** Orange-brown crystals by recrystn from *n*-hexane under a carbon monoxide atmosphere [Ojima et al. *J Am Chem Soc* **109** 7714 *1987*; see also Hileman in *Preparative Inorganic Reactions*, Jolly Ed. Vol 1 101 1987].

Diethyl aluminium chloride *[96-10-6]* **M 120.6, m -75.5°, b 106.5-108°/24.5mm, d 0.96.** Distd from excess dry NaCl (to remove ethyl aluminium dichloride) in a 50-cm column containing a heated nichrome spiral.

O,O-**Diethyl-*S*-2-diethylaminoethyl phosphorothiolate** *[78-53-5]* **M 269.3, m 98-99°.** Crystd from isopropanol/diethyl ether.

Di-(2-ethylhexyl)phosphoric acid ('diisooctyl' phosphate) *[27215-10-7; 298-07-7]* **M 322.4, b 209°/10mm, d 0.965, pK$_{Est}$ ~1.7.** Contaminants of commercial samples include the monoester, polyphosphates, pyrophosphate, 2-ethylhexanol and metal impurities. Dissolved in *n*-hexane to give an 0.8M soln. Washed with an equal volume of M HNO$_3$, then with saturated (NH$_4$)$_2$CO$_3$ soln, with 3M HNO$_3$, and twice with water [Petrow and Allen *Anal Chem* **33** 1303 *1961*]. Similarly, the impure sodium salt, after scrubbing with pet ether, was acidified with HCl and the free organic acid was extracted into pet ether and purified as above. [Peppard et al. *J Inorg Nucl Chem* **7** 231 *1958*] or Stewart and Crandall [*J Am Chem Soc* **73** 1377 *1951*]. Purified also *via* the copper salt [McDowell et al. *J Inorg Nucl Chem* **38** 2127 *1976*].

Diethylmethylsilane *[760-32-7]* **M 102.3, b 78.4°/760mm,   77.2-77.6°/atm, d 0.71.** Fractionally distilled through a *ca* 20 plate column and the fraction boiling within a range of less than 0.5° is collected. [*Izv Akad Nauk SSSR Otd Khim* 1416 *1957*; *J Am Chem Soc* **69** 2600 *1947*.]

Diethyl trimethylsilyl phosphite *[13716-45-5]* **M 210.3, b 61°/10mm, 66°/15mm, d 0.9476, n 1.4113.** Fractionated under reduced pressure and has δ$_P$ -128 ±0.5 relative to H$_3$PO$_4$. [*J Org Chem* **46** 2097 *1981*; *J Gen Chem USSR (Engl Transl)* **45** 231 *1975*.]

N,N-**Diethyltrimethylsilylamine** *[996-50-9]* **M 145.3, b 33°/26mm, 126.8-127.1°/738mm, 126.1-126.4°, d 0.763, n 1.411.** Fractionated through a 2ft vac-jacketed column containing Helipak packing with a reflux ratio of 10:1. [*J Am Chem Soc* **68** 241 *1946*; *J Org Chem* **23** 50 *1958*; *J Prakt Chem* **9** 315 *1959*.]

N,N'-Diheptyl-*N,N'*-5,5-tetramethyl-3,7-dioxanonanediamide [lithium ionophore I (ETH 149)] *[58821-96-8]* M 442.7. Purified by chromatography on Kieselgel using $CHCl_3$ as eluent (IR ν 1640cm^{-1}). [*Helv Chim Acta* **60** 2326 *1977*.]

Dihexadecyl phosphate *[2197-63-9]* **M 546.9, m 75°, pK$_{Est}$ ~1.2.** Crystd from MeOH [Lukac *J Am Chem Soc* **106** 4387 *1984*].

1,2-Dihydroxybenzene-3,5-disulfonic acid, di-Na salt (TIRON) *[149-45-1]* **M 332.2, ε 6.9 x 10^4 at 260nm, pH 10.8, pK$_1$ and pK$_2$ <2 (for SO$_3^-$), pK$_3$ 7.7, pK$_4$ 12.6 (for OHs of disulfonate dianion).** Recrystd from water [Hamaguchi et al. *Anal Chim Acta* **9** 563 *1962*]. Indicator color reagent for Fe, Mn, Ti and Mo ions and complexes with Al, Cd, Co, Co, Fe (III), Mn, Pd, UO_2^{2+}, VO^{2+} and Zn.

Diisooctyl phenylphosphonate *[49637-59-4]* **M 378.5, n^{25} 1.4780.** Vacuum distilled, percolated through a column of alumina, then passed through a packed column maintained at 150° to remove residual traces of volatile materials in a countercurrent stream of N_2 at reduced pressure [Dobry and Keller *J Phys Chem* **61** 1448 *1957*].

Diisopropyl chlorosilane (chlorodiisopropylsilane) *[2227-29-4]* **M 150.7, b 59°/8mm, 80°/10mm, 200°/738mm, d 0.9008, n, 1.4518.** Impurities can be readily detected by ^{1}H NMR. Purified by fractional distn [*J Am Chem Soc* **69** 1499 *1947*; *J Chem Soc* 3668 *1957*; *J Organometal Chem* **282** 175 *1985*].

Dilongifolyl borane *[77882-24-7]* **M 422.6, m 169-172°.** Wash with dry Et_2O and dry in a vacuum under N_2. It has **m** 160-161° in a sealed evacuated capillary. It is sparingly soluble in pentane, tetrahydrofuran, carbon tetrachloride, dichloromethane, and chloroform, but the suspended material is capable of causing asymmetric hydroboration. Disappearance of solid indicates that the reaction has proceeded. [*J Org Chem* **46** 2988 *1981*.]

Dimethyl carbonate *[616-38-6]* **M 90.1, b 89.5°/755mm, 90.2°/atm, d 1.0446, n 1.3687.** If the reagent has broad intense bands at 3300cm^{-1} and above (i.e. OH streching) then it should be purified further. Wash successively with 10% Na_2CO_3 soln, saturated $CaCl_2$, H_2O and dried by shaking mechanically for 1h with anhydrous $CaCl_2$, and fractionated. [*J Chem Soc* **78** *1939*, 1847 *1948*.]

Dimethyl dicarbonate (dimethyl pyrocarbonate) *[4525-33-1]* **M 134.1, m 15.2°, b 45-46°/5mm, d 1.2585, n 1.3950.** Dissolve in Et_2O, shake with a small vol of 0.1N HCl, dry Et_2O with Na_2SO_4 and distil in vac below 100° to give a clear liquid. It dec to CO_2 and dimethyl carbonate on heating at 123-149°. It is readily hydrolysed by H_2O and is an **IRRITANT**. [*J Gen Chem USSR* **22** 1546 *1952*; see also *Chem Ber* **71** 1797 *1938*.]

Dimethyldichlorosilane *[75-78-5]* **M 129.1, m -75.5°, b 68.5-68.7°/750mm, 70.5°/760mm, d 1.0885, n 1.4108.** Other impurities are chlorinated silanes and methylsilanes. Fractionated through a 3/8in diameter 7ft Stedman column rated at 100 theoretical plates at almost total reflux (see p. 441). See purification of $MeSiCl_2$. [*J Am Chem Soc* **70** 3590 *1948*.]

2,6-Dimethyl-1,10-phenanthrolinedisulfonic acid, di-Na salt (H$_2$O) (bathocuproinedisulfonic acid di-Na salt) *[52698-84-7]* **M 564.5, pK$_{Est}$~0 (for free acid).** Inorganic salts and some coloured species can be removed by dissolving the crude material in the minimum volume of water and precipitating by adding EtOH. Purified reagent can be obtained by careful evapn of the filtrate.

Dimethylphenylsilyl chloride (chlorodimethylphenylsilane) *[768-33-2]* **M 170.7, b 85-87°/32mm, 196°/atm, d 1.017, n 1.509.** See phenyl methyl chlorosilane on p. 449.

Dinitrogen tetroxide (nitrogen dioxide, N_2O_4) *[10544-72-6]* **M 92.0 m -11.2°, b 21.1°.** Purified by oxidation at 0° in a stream of oxygen until the blue colour changed to red-brown. Distd from P_2O_5, then solidified on cooling in a deep-freeze (giving nearly colourless crystals). Oxygen can be removed by alternate freezing and melting. **TOXIC VAPOUR.**

Dioctyl phenylphosphonate *[1754-47-8]* **M 378.8, d 1.485, n^{25} 1.4780.** Purified as described under diisooctyl phenylphosphonate.

(1,3-Dioxalan-2-ylmethyl)triphenylphosphonium bromide *[52509-14-5]* **M 429.3, m 191.5-193°, 193-195°.** Wash the crysts with Et_2O, dry in a vac and recryst from CH_2Cl_2-dry Et_2O to give prisms **m** 172-174°, which is raised to 191.5-193°, on drying at 56°/0.5mm. [Cresp et al. *J Chem Soc, Perkin Trans 1* 37 *1974*.]

Diphenyldiselenide *[1666-13-3]* **M 312.1, m 62-64.** Crystd twice from hexane [Kice and Purkiss *J Org Chem* 52 3448 *1987*].

Diphenyl hydrogen phosphate *[838-85-7]* **M 250.2, m 99.5°, pK^{20} 0.26.** Crystd from $CHCl_3$/pet ether.

Diphenylmercury *[587-85-9]* **M 354.8, m 125.5-126°.** Sublimed, then crystd from nitromethane or ethanol. If phenylmercuric halides are present they can be converted to phenylmercuric hydroxide which, being much more soluble, remains in the alcohol or *benzene used for crystn. Thus, crude material (10g) is dissolved in warm ethanol (*ca* 150mL) and shaken with moist Ag_2O (*ca* 10g) for 30min, then heated under reflux for 30min and filtered hot. Concentration of the filtrate by evaporation gives diphenylmercury, which is recrystd from *benzene [Blair, Bryce-Smith and Pengilly *J Chem Soc* 3174 *1959*]. **TOXIC.**

4,7-Diphenyl-1,10-phenanthrolinedisulfonic acid, di-Na salt $3H_2O$ (bathophenanthroline-disulfonic acid di-Na salt) *[52746-49-3]* **M 590.6, m 300°, pK_{Est}~0 (for free acid).** Dissolve crude sample in the minimum volume of water and add EtOH to ppte the contaminants. Carefully evaporate the filtrate to obtain pure material.
It forms a dark red complex with Fe^{2+} with λ_{max} 535nm (ε 2.23 x 10^4mol^{-1} cm^{-1}) [*Anal Chim Acta* 115 407 *1980*]. Prepared by sulfonating bathophenanthroline with $ClSO_3H$: to 100g of bathophenanthroline was added 0.5mL of Fe free $ClSO_3H$ and heated over a flame for 30sec. Cool and carefully add 10mL of pure distd H_2O and warm on a water bath with stirring till all solid dissolved. A stock soln is made by diluting 3mL of this reagent to 100mL with 45% aq NaOAc, filter off the solid and store in a dark bottle. In this way it is stable for several months. [*Am J Clinical Pathology* 29 590 *1958*.]

Diphenylphosphinic acid *[1707-03-5]* **M 218.2, m 194-195°, pK^{20} 1.72.** Recrystd from 95% EtOH and dried under vacuum at room temperature. [see Kosolapoff *Organophosphorus Compounds* J Wiley, NY, *1950*; Kosolapoff and Maier *Organic Phosphorus Compounds* Wiley-Interscience, NY, *1972-1976*.]

Diphenylsilane *[775-12-2]* **M 184.3, b 75-76°/0.5mm, 113-114°/9mm, 124-126°/11mm, 134-135°/16mm, d 1.0027, n 1.5802, 1.5756.** Dissolve in Et_2O, mix slowly with ice-cold 10% AcOH. The Et_2O layer is then shaken with H_2O until the washings are neutral to litmus. Dry over Na_2SO_4, evaporate the Et_2O and distil the residual oil under reduced pressure using a Claisen flask with the take-off head modified into a short column. Ph_2SiH_2 boils at 257°/760mm but it cannot be distd at this temp because exposure to air leads to flashing, decomposition and formation of silica. It is a colourless, odourless oil, miscible with organic solvents but not H_2O. A possible impurity is Ph_3SiH which has **m** 43-45° and would be found in the residue. [*J Org Chem* 18 303 *1953*; *J Am Chem Soc* 74 648*1952*, 81 5925 *1959*.]

Diphenylsilanediol *[947-42-2]* **M 216.3, m 148°(dec).** Crystd from $CHCl_3$-methyl ethyl ketone.

Diphenyl tolyl phosphate *[26444-49-5]* **M 340.3, n^{25} 1.5758.** Vac distd, then percolated through a column of alumina. Finally, passed through a packed column maintained at 150° to remove traces of volatile impurities in a countercurrent stream of nitrogen at reduced pressure. [Dobry and Keller *J Phys Chem* **61** 1448 *1947*.]

Disodium calcium ethylenediaminetetraacetate *[39208-14-5]* **M 374.3, (see pKs for EDTA in entry below).** Dissolved in a small amount of water, filtered and ppted with excess EtOH. Dried at 80°.

Disodium dihydrogen ethylenediaminetetraacetic acid (2H$_2$O) *[6381-92-6]* **M 372.2, m 248°(dec), pK$_1^{25}$ 0.26 pK$_2^{25}$ 0.96, pK$_3^{25}$ 2.60, pK$_4^{25}$ 2.67, pK$_5^{25}$ 6.16, pK$_6^{25}$ 10.26 (see EDTA Cha 4).** Analytical reagent grade material can be used as primary standard after drying at 80°. Commercial grade material can be purified by crystn from water or by preparing a 10% aqueous soln at room temperature, then adding ethanol slowly until a slight permanent ppte is formed, filtering, and adding an equal volume of ethanol. The ppte is filtered off on a sintered-glass funnel, is washed with acetone, followed by diethyl ether, and dried in air overnight to give the dihydrate. Drying at 80° for at least 24h converts it to the anhydrous form.

Disodium 4,5(1,8)-dihydroxynaphthalene-2,7(3,6)-disulfonate (2H$_2$O) *[5808-22-0]* **M 400.3, m >300°, pK$_1$ 0.61(SO$_3^-$), pK$_2$ 0.7(SO$_3^-$), pK$_3$ 5.45(OH), pK$_4$ 15.5(OH).** Crystd from H$_2$O or H$_2$O by addition of EtOH. Complexes with Ag, ClO$_3^-$, Cr, Hg, NO$_2^-$, NO$_3^-$ and Ti. [cf Chromotropic acid p. 413.]

Disodium ethylenebis[dithiocarbamate] *[142-59-6]* **M 436.5, pK$_{Est}$ ~ 3.0.** Crystd (as hexahydrate) from aqueous ethanol.

Disodium-ß-glycerophosphate *[819-83-0 (4H$_2$O)]* **M 216.0, m 102-104°, pK$_2^{25}$ 6.66 (free acid).** Crystd from water.

Disodium hydrogen orthophosphate (anhydrous) *[7558-79-4]* **M 142.0, (see pK of H$_3$PO$_4$).** Crystd twice from warm water, by cooling. Air dried, then oven dried overnight at 130°. *Hygroscopic*: should be dried before use.

Disodium magnesium ethylenediaminetetraacetate *[14402-88-1]* **M 358.5, pK$_1^{25}$ 0.26 pK$_2^{25}$ 0.96, pK$_3^{25}$ 2.60, pK$_4^{25}$ 2.67, pK$_5^{25}$ 6.16, pK$_6^{25}$ 10.26 (see EDTA on p. 237 in Chapter 4).** Dissolved in a small amount of water, filtered and ppted with an excess of MeOH. Dried at 80°.

Disodium naphthalene-1,5-disulfonate *[1655-29-4]* **M 332.3, pK$_{Est}$ ~0.** Recrystd from aqueous acetone [Okahata et al. *J Am Chem Soc* **108** 2863 *1986*].

Disodium 4-nitrophenylphosphate (6H$_2$O) *[4264-83-9]* **M 371.1** Dissolve in hot aqueous MeOH, filter and ppte by adding Me$_2$CO. Wash the solid with Me$_2$CO and repeat the purification. Aq MeOH and Et$_2$O can also be used as solvents. The white fibrous crystals contain less than 1% of free 4-nitrophenol [assay: *J Biol Chem* **167** 57 *1947*].

Disodium phenylphosphate (2H$_2$O) *[3279-54-7]* **M 254.1, pK$_1^{25}$ 1.46, pK$_2^{25}$ 6.29 [for PhPO(OH)$_2$].** Dissolved in a minimum amount of methanol, filtering off an insoluble residue of inorganic phosphate, then ppted by adding an equal volume of diethyl ether. Washed with diethyl ether and dried [Tsuboi *Biochim Biophys Acta* **8** 173 *1952*].

Disodium succinate *[150-90-3]* **M 162.1.** Crystd twice from water (1.2mL/g) and dried at 125°. Freed from other metal ions by passage of a 0.1M soln through a column of Dowex resin A-1 (Na form).

Di-*p*-tolylmercury *[50696-65-6]* **M 382.8, m 244-246°.** Crystd from xylene.

Di-*p*-tolyl phenylphosphonate *[94548-75-1]* **M 388.3, n^{25} 1.5758.** Purified as described under diisooctyl phenylphosphonate.

1,3-Divinyl-1,1,3,3-tetramethyldisiloxane *[2627-95-4]* **M 186.4, m -99.7°; b 128-129°/atm, 139°/760mm, d 0.811, n 1.4122.** Dissolve in Et_2O, wash with H_2O, dry over $CaCl_2$ and distil. [*J Am Chem Soc* **77** 1685 *1955*; *Collect Czech Chem Comm* **24** 3758 *1959*.]

Eosin B (Bluish, Eosin Scarlet, 4',5'-dibromo-2',7'-dinitrofluorescein disodium salt)
[548-24-3] **M 624.1,** λ_{max} **514nm, CI 45400.** Freed from inorganic halides by repeated crystn from butan-1-ol.

Eosin Y (as di-Na salt) *[17372-87-1]* **M 691.9.** Dissolved in water and ppted by addition of dilute HCl. The ppte was washed with water, crystd from ethanol, then dissolved in the minimum amount of dilute NaOH soln and evaporated to dryness on a water-bath. The purified disodium salt was then crystd twice from ethanol [Parker and Hatchard *Trans Faraday Soc* **57** 1894 *1961*].

Eosin YS (Eosin Yellowish, 2',4',5'7'-tetrabromofluorescein di-Na salt) *[17372-87-1]* **M 691.9, CI 45380.** Dissolve in the minimum vol of H_2O (1g/mL), filter and add EtOH until separation of salt is complete. Filter off, wash with abs EtOH, then Et_2O and dry first in air, then at 100°. Used for staining blood cells and for estimating traces of Ag. [Selsted and Becker *Anal Biochem* **155** 270 *1986*; El-Ghamry and Frei *Anal Chem* **40** 1986 *1968*.]

Eriochrome Black T *[1787-61-7]* **M 416.4,** $A_{1cm}^{\%}(\lambda max)$ **656(620nm) at pH 10, using the dimethylammonium salt, pK_2^{25} 5.81, pK_3^{25} 11.55.** The sodium salt (200g) was converted to the free acid by stirring with 500mL of 1.5M HCl, and, after several minutes, the slurry was filtered on a sintered-glass funnel. The process was repeated and the material was air dried after washing with acid. It was extracted with *benzene for 12h in a Soxhlet extractor, then the *benzene was evaptd and the residue was air dried. A further desalting with 1.5M HCl (1L) was followed by crystn from dimethylformamide (in which it is very soluble) by forming a saturated soln at the boiling point, and allowing to cool slowly. The crystalline dimethylammonium salt so obtained was washed with *benzene and treated repeatedly with dilute HCl to give the insoluble free acid which, after air drying, was dissolved in alcohol, filtered and evaporated. The final material was air dried, then dried in a vacuum desiccator over $Mg(ClO_4)_2$ [Diehl and Lindstrom, *Anal Chem* **31** 414 *1959*]. Indicator for complexometry of alkaline earth metals.

Eriochrome Blue Black R (Palatine Chrome Black 6BN, Calcon, 3-hydroxy-4-(2-hydroxy-1-naphthylazo)naphthalene-1-sulfonic acid Na salt] *[2538-85-4]* **M 416.4, pK_2^{25} 7.0, pK_3^{25} 13,5.** Freed from metallic impurities by three pptns from aqueous soln by addn of HCl. The ppted dye was dried at 60° under vacuum. Indicator for complexometry of Al, Fe and Zr.

Ethoxycarbonylmethylene triphenylphosphonium bromide *[1530-45-6]* **M 429.3, m 155-155.5°, 158°(dec).** Wash with pet ether (b 40-50°) and recryst from $CHCl_3/Et_2O$ and dry in high vac at 65°. [Isler et al. *Helv Chim Acta* **40** 1242 *1957*; Wittig and Haag *Chem Ber* **88** 1654, 1664 *1955*.]

(Ethoxycarbonylmethylene)triphenylphosphorane [ethyl (triphenylphosphoranylidene)-acetate] *[1099-45-2]* **M 348.4, m 116-117°, 128-130°.** Cryst by dissolving in AcOH and adding pet ether (b 40-50°) to give colorless plates. UV λmax ($A_{1mm}^{1\%}$): 222nm (865) and 268nm (116) [Isler et al. *Helv Chim Acta* **40** 1242 *1957*].

Ethylarsonic acid *[507-32-4]* **M 154.0, m 99.5°, pK_1 4.72 (As(OH)O⁻), pK_2 8.00 [AsO_2^{2-}].** Crystd from ethanol.

2-Ethyl-1,2-benzisoxazolium tetrafluoroborate *[4611-62-5]* **M 235.0, m 107-109°, 109.5-110.2°.** Recrystd from MeCN-EtOAc to give magnificent crystals. It is not hygroscopic but on long exposure to moisture it etches glass. It is light-sensitive and should be stored in brown glass bottles. UV (H_2O), λmax 258nm (ϵ 13 100) and λmax 297nm (ϵ 2 900); IR (CH_2Cl_2): 1613 (C=N) and 1111-1000 (BF_4^+) [UV, IR, NMR: Kemp and Woodward *Tetrahedron* **21** 3019 *1965*].

Ethylene bis(diphenylphosphine) [1,2-bis(diphenylphosphino)ethane] *[1663-45-2]* **M 398.4, m 139-140°.** See 1,2-bis-(diphenylphosphino)ethane (DIPHOS) on p. 402.

Ethylmercuric chloride *[107-27-7]* **M 265.1, m 193-194°.** Mercuric chloride can be removed by suspending ethylmercuric chloride in hot distilled water, filtering with suction in a sintered-glass crucible and drying. Then crystd from ethanol and sublimed under reduced pressure. It can also be crystd from water.

Ethylmercuric iodide *[2440-42-8]* **M 356.6, m 186°.** Crystd once from water (50mL/g).

Ethyl Orange (sodium 4,4'-diethylaminophenylazobenzenesulfonate) *[62758-12-7]* **M 355.4, $pK_{Est} \sim 3.8$.** Recrystd twice from water.

Ethyl trimethylsilylacetate *[4071-88-9]* **M 160.3, b 74.5°/41mm, 75.5°/42mm, 157°/730mm, d 0.8762, n 1.4149.** Purified by distilling *ca* 10g of reagent through a 15cm, Vigreux column and then redistilling through a 21cm glass helices-packed column [*J Am Chem Soc* **75** 994 *1953*]. Alternatively, dissolve in Et_2O, wash with H_2O, dilute Na_2CO_3, dry over Na_2CO_3, evaporate Et_2O, and distil through a column of 15 theoretical plates [*J Am Chem Soc* **70** 2874 *1948*].

Ethyl 3-(trimethylsilyl)propionate *[17728-88-0]* **M 174.3, b 93°/40mm, 178°-180°/atm, d 0.8763, n 1.4198.** Dissolve in Et_2O, wash with H_2O, dilute Na_2CO_3, dry over Na_2SO_4, evaporate Et_2O and fractionally distil. [*J Am Chem Soc* **72** 1935 *1950*.]

Ethynyl tributylstannane *[994-89-8]* **M 315.1, b 76°/0.2mm, 130-135°/0.7mm, 200°/2mm, d 1.1113, n 1.4770.** Purified by dissolving the reagent (*ca* 50g) in heptane (250mL), washing with H_2O (100mL), drying ($MgSO_4$), evaporating and distilling in a vacuum. It has IR v 3280 ($\equiv C-H$), 2950, 2850, 2005 (C$\equiv$C), 1455, 1065 and 865cm^{-1}. [*J Org Chem* **46** 5221 *1981*; *J Am Chem Soc* **109** 2138 *1987*; *J Gen Chem USSR (Engl Edn)* **37** 1469 *1967*.]

Ethynyl trimethylsilane *[1066-54-2]* **M 98.2, b 53°/atm, 52.5°/atm, d 0.71, n 1.3871.** Distil through an efficient column. The IR has bands at 2041 (C$\equiv$C) and 3289 ($\equiv$C-H) cm^{-1}. [*Chem Ber* **92** 30 *1959*.]

Ethyl triphenylphosphonium bromide *[1530-32-1]* **M 371.3, m 203-205°.** Recrystd from H_2O and dried in high vacuum at 100°. IR has bands at 1449, 1431 and 997cm^{-1}. [*Justus Liebigs Ann Chem* **606** 1 *1957*; *J Org Chem* **23** 1245 *1958*.]

Europium (III) acetate (2H_2O) *[62667-64-5]* **M 383.1, pK_1^{25} 8.31 (for aquo Eu^{3+}).** Recrystd several times from water [Ganapathy et al. *J Am Chem Soc* **108** 3159 *1986*]. For europium shift reagents see lanthanide shift reagents in Chapter 4.

Ferric acetylacetonate *[14024-18-1]* M 353.2, m 181.3-182.3°.
Recrystd twice from *benzene-pet ether **m** 181.3-182.3° corr [*J Chem Soc* 1256 *1938*]. Recrystd from EtOH or Et_2O, **m** 179° [*Justus Liebigs Ann Chem* **323** 13 *1902*]. Recrystd from absolute EtOH, **m** 159.5° [*Chem Ber* **67** 286 *1934*]. Dry for 1hr at 120°.

Ferric Bromide *[10031-26-2]* **M 395.6, m >130°(dec).** Subimed in a sealed tube with Br_2 at 120°-200°. [Lux in *Handbook of Preparative Inorganic Chemistry (Ed. Brauer)* Vol II, p 1494 *1963*.]

Ferric chloride (anhydrous) *[7705-08-0]* **M 162.2, m >300°(dec).** Sublimed at 200° in an atmosphere of chlorine. Stored in a weighing bottle inside a desiccator.

Ferric chloride (6H$_2$O)   *[10025-77-1]* **M 270.3, m 37°(dec), pK$_1^{25}$ 2.83, pK$_2^{25}$ 4.59 (for hydrolysis of Fe^{3+}).** An aqueous soln, saturated at room temperature, was cooled to -20° for several hours. Pptn was slow, even with scratching and seeding, and it was generally necessary to stir overnight. The presence of free HCl retards the pptn [Linke *J Phys Chem* **60** 91 *1956*].

Ferric nitrate (9H$_2$O) *[7782-61-8]* **M 404.0, m 47°(dec).** Cryst from aqueous solutions of moderately strong HNO$_3$ as the violet nonahydrate. With more concentrated aqueous solns (containing some HNO$_3$), the hexahydrate crysts out. The anhydrous salt is slightly deliquescent and decomposes at 47°.

Ferric perchlorate (9H$_2$O) *[13537-24-1]* **M 516.3, pK^{25} -2.4 to -3.1 (for HClO$_4$).** Crystd twice from conc HClO$_4$, the first time in the presence of a small amount of H$_2$O$_2$ to ensure that the iron is fully oxidised [Sullivan *J Am Chem Soc* **84** 4256 *1962*]. Extreme care should be taken with this preparation because it is potentially **DANGEROUS**.

Ferric sulfate (xH$_2$O) *[10028-22-5]* **M 399.9 + xH$_2$O.** Dissolve in the minimum volume of dilute aqueous H$_2$SO$_4$ and allow to evaporate at room temp until crystals start to form. Do not concentrate by boiling off the H$_2$O as basic salts will be formed. Various *hydrates* are formed the—common ones are the *dodeca* and *nona hydrates* which are violet in colour. The anhydrous salt is colourless and very *hygroscopic* but dissolves in H$_2$O slowly unless ferrous sulfate is added.

Ferrocene *[102-54-5]* **M 186.0, m 173-174°.** Purified by crystn from pentane or cyclohexane (also *C$_6$H$_6$ or MeOH can be used). Moderately soluble in Et$_2$O. Sublimes readily above 100°. Crystallisation from EtOH gave m 172.5-173°. [*Org Synth* Coll Vol IV 473 *1963*; *J Chem Soc* 632 *1952*.] Also crystd from methanol and sublimed *in vacuo*. [Saltiel et al. *J Am Chem Soc* **109** 1209 *1987*.]

Ferrocene carboxaldehyde *[12093-10-6]* **M 214.1, m 117-120°, 118-120°, 121°, 124.5°.** Red crystals from EtOH or pet ether and sublimed at 70°/1mm. *Semicarbazone* **m** 217-219°(dec) cryst from aqueous EtOH. *O-Acetyloxime* **m** 80-81° cryst from hexane [*J Org Chem* **22** 355 *1957*]. *2,4-Dinitrophenylhydrazone* **m** 248°(dec). [*Beilstein* **16 4th Suppl** 1798; *J Am Chem Soc* **79** 3416 *1957*; *J Chem Soc* 650 *1958*.]

Ferrocene carboxylic acid *[1271-42-7]* **M 230.1, m 210°(dec), 225-230°(dec).** Yellow crystals from pet ether. Also crystd from aqueous ethanol. [Matsue et al. *J Am Chem Soc* **107** 3411 *1985*.] *Acid chloride* **m** 49° crystallises from pentane, λmax 458nm [*J Org Chem* **24** 280 *1959*]. *Methyl ester* crystallises from aq MeOH **m** 70-71°. *Anhydride* **m** 143-145° from pet ether [*J Org Chem* **24** 1487 *1959*]. *Amide* **m** 168-170° from CHCl$_3$-Et$_2$O or **m** 167-169° from *C$_6$H$_6$-MeOH. [*J Am Chem Soc* **77** 6295 *1955*; **76** 4025 *1954*.]

Ferrocene-1,1'-dicarboxylic acid *[1293-87-4]* **M 274.1, m >250°(dec), >300°.** Orange-yellow crystals from AcOH. Sublimes above 230°. *Monomethyl ester* **m** 147-149° [*Dokl Acad Nauk USSSR* **115**, 518 *1957*]. *Dimethyl ester* **m** 114-115° [*J Am Chem Soc* **74**, 3458 *1958*]. *Diacid chloride* **m** 92-93° from pet ether. [*Dokl Acad Nauk SSSR* **120** 1267 1958; **127** 333 *1959*.]

Ferrocene-1,1,-dimethanol *[1291-48-1]* **M 246.1, m 107-108°.** Obtained from the diacid with LiAlH$_4$ reduction and recrystd from Et$_2$O-pet ether. [*J Am Chem Soc* **82** 4111 *1960*.]

Ferrous bromide *[20049-65-4]* **M 215.7 + xH$_2$O, m 684°, d^{25} 4.63.** Crystn from air-free H$_2$O provides the *hexahydrate* as pale green to bluish-green rhombic prisms. On heating at 49° H$_2$O is lost and the *tetrahydrate* is formed. Further heating at 83° more H$_2$O is lost and the *dihydrate* is formed as a light yellow to dark brown *hygroscopic* powder. The ferrous iron in the aqueous solns of these salts readily oxidises to ferric iron. The salts should be stored over H$_2$SO$_4$ under N$_2$ in tightly closed containiners. They have some solubility in EtOH. [*Chem Ber* **38** 236 *1904*.]

Ferrous chloride (4H$_2$O) *[13478-10-9]* **M 198.8, m 105°(dec), pK$_1^{25}$ 6.7, pK$_2^{25}$ 9.3 (for aquo Fe^{2+}).** A 550mL round-bottomed Pyrex flask was connected, *via* a glass tube fitted with a medium porosity sintered-glass disc, to a similar flask. To 240g of $FeCl_2.4H_2O$ in the first flask was added conductivity water (200mL), 38% HCl (10mL), and pure electrolytic iron (8-10g). A stream of purified N_2 was passed through the assembly, escaping through a mercury trap. The salt was dissolved by heating which was continued until complete reduction had occurred. By inverting the apparatus and filtering (under N_2 pressure) through the sintered glass disc, unreacted iron was removed. After cooling and crystn, the unit was again inverted and the crystals of ferrous chloride were filtered free from mother liquor by applied N_2 pressure. Partial drying by overnight evacuation at room temperature gave a mixed hydrate which, on further evacuation on a water bath at 80°, lost water of hydration and its absorbed HCl (with vigorous effervescence) to give a white powder, $FeCl_2.2H_2O$ [Gayer and Wootner *J Am Chem Soc* **78** 3944 *1956*].

Ferrous chloride *[7758-94-3]* **M 126.8, m 674°, b 1023°, d^{25} 3.16.** Sublimes in a stream of HCl at *ca* 700°, or in H_2 below 300°. Its vapour pressure at 700° is 12mm. Anhydrous $FeBr_2$ can be obtained by carefully dehydrating the *tetrahydrate* in a stream of HBr and N_2, and it can be sublimed under N_2. White *hygroscopic* rhombohedral crystals with a green tint. They oxidise in air to $FeCl_3 + Fe_2O_3$. Sol in H_2O, EtOH Me_2CO but insol in Et_2O. The *tetrahydrate* is pale green to pale blue in colour and loses $2H_2O$ at 105-115°. The *dihydrate* loses H_2O at 120°. The ferrous iron in the aqueous solns of these salts readily oxidises to ferric iron. [*Inorg Synth* **6** 172 *1960*; *Handbook of Preparative Inorganic Chemistry (Ed Brauer)* Vol II 1491 *1965.*]

Ferrous perchlorate (6H$_2$O) *[13933-23-8]* **M 362.9, pK^{25} -2.4 to -3.1 (for HClO$_4$).** Crystd from $HClO_4$.

Ferrous sulfate (7H$_2$O) *[7782-63-0]* **M 278.0, m ~60°(dec).** Crystd from 0.4M H_2SO_4.

Fluorine *[7782-41-4]* **M 38.0, b -129.2°.** Passed through a bed of NaF at 100° to remove HF and SiF_4. [For description of stills used in fractional distn, see Greenberg et al. *J Phys Chem* **65** 1168 *1961*; Stein, Rudzitis and Settle *Purification of Fluorine by Distillation, Argonne National Laboratory,* ANL-6364 1961 (from Office of Technical Services, US Dept of Commerce, Washington 25).] **HIGHLY TOXIC.**

Fluoroboric acid *[16872-11-0]* **M 87.8, pK -4.9.** Crystd several times from conductivity water.

Fluorotrimethylsilane (trimethylsilyl fluoride, TMSF) *[420-56-4]* **M 92.2, m -74°, b 16°/760mm, 19°/730mm, d^0 0.793.** It is a **FLAMMABLE** gas which is purified by fractional distn through a column at low temperature and with the exclusion of air [Booth and Suttle *J Am Chem Soc* **68** 2658 *1946*; Reid and Wilkins *J Chem Soc* 4029 *1955*].

Gallium *[7440-55-3]* **M 69.7, m 29.8°.** Dissolved in dilute HCl and extracted into Et_2O. Pptn with H_2S removed many metals, and a second extraction with Et_2O freed Ga more completely, except for Mo, Th(III) and Fe which were largely removed by pptn with NaOH. The soln was then electrolysed in 10% NaOH with a Pt anode and cathode (2-5A at 4-5V) to deposit Ga, In, Zn and Pb, from which Ga was obtained by fractional crystn of the melt [Hoffman *J Res Nat Bur Stand* **13** 665 *1934*]. Also purified by heating to boiling in 0.5-1M HCl, then heating to 40° in water and pouring the molten Ga with water under vacuum onto a glass filter (30-50 µ pore size), to remove any unmelted metals or oxide film. The Ga was then fractionally crystd from the melt under water.

Gallium (III) Chloride *[13450-90-3]* **M 176.1, m 77.8°, b 133°/100mm, 197.7°/700mm, d 2.47, pK$_1^{25}$ 2.91, pK$_2^{25}$ 3.70, pK$_3^{20}$ 4.42 (for Ga^{3+}).** Pure compound can be obtained by redistn in a stream of Cl_2 or Cl_2/N_2 followed by vacuum sublimation or zone refining. Colourless needles which give *gallium dichloride* [$Ga(GaCl_4)$, m 172.4°] on heating. Dissolves in H_2O with liberation of heat. Soluble in Et_2O. [*Handbook of Preparative Inorganic Chemistry (Ed. Brauer)* Vol I 846 *1963.*]

Gallium (III) nitrate (9H$_2$O) *[63462-65-7]* **M 417.9, m *ca* 65°.** Recrystd from H_2O (sol: 295g/100mL at 20°). White deliquescent colourless powder soluble in H_2O, absolute EtOH and Et_2O. Loses

HNO_3 upon heating at 40°. Addition of Et_2O to a warm ethanolic soln ($40-50^o$) of $Ga(NO_3)_3$ $9H_2O$ precipitates $Ga(OH)_2NO_3.Ga(OH)_3.2H_2O$. If the salt has partly hydrolysed, dissolve in conc HNO_3, reflux, dilute with H_2O and concentrate on a sand bath. Wash several times by adding H_2O and evaporate until there is no odour of acid. Dilute the residue to a Ga concentration of 26g/100mL. At this concentration, spongy $Ga(NO_3)_3.xH_2O$ separates from the viscous soln. After standing for several days the crystals are collected and dried in a stream of dry air first at room temp then at 40°. Dehydration is complete after 2 days. Recrystallise from H_2O and dry on a water pump at room temperature. [*Z Naturforsch* **20B** 71 *1965*; *Handbook of Preparative Inorganic Chemistry (Ed. Brauer)* Vol I 856*1963*.]

Gallium (III) sulfate *[13494-91-2 (anhydr); 13780-42-2 (hydr)]* **M 427.6.** Recrystn from H_2O gives the $16-18H_2O$ hydrate (sol at 20° is 170g/100mL). Alternatively dissolve in 50% H_2SO_4 and evaporate ($60-70^o$), cool and ppte by adding $EtOH/Et_2O$. On heating at 165° it provides the *anhydrous* salt which is a white *hygroscopic* solid. [*Z Naturforsch* **20B** 71 *1965*.]

Germanium *[7440-56-4]* **M 72.6, m 937°, 925-975°, b 2700°, d 5.3.** Copper contamination on the surface and in the bulk of single crystals of Ge can be removed by immersion in molten alkali cyanide under N_2. The Ge was placed in dry cyanide powder in a graphite holder in a quartz or porcelain boat. The boat was then inserted into a heated furnace which, after a suitable time, was left to cool to room temperature. At 750°, a 1mm thickness requires about 1min, whereas 0.5cm needs about half hour. The boat was removed and the samples were taken out with plastic-coated tweezers, carefully rinsed in hot water and dried in air [Wang *J Phys Chem* **60** 45 *1956*].

Germanium (IV) oxide *[1310-53-8]* **M 104.6, m 1080°(soluble form), d^{25} 6.239; m 1116°(insoluble form) d^{25} 4.228, pK_1^{25} 9.02, pK_2^{25} 12.82 (for germanic acid H_2GeO_3).** The oxide (GeO_2) is usually prepared by hydrolysing redistd $GeCl_4$ and igniting in order to remove H_2O and chloride. It can be further purified by dissolving in hot H_2O (sol: 4g/L cold) evaporating and drying the residual crystalline solid. When the *soluble* form (which is produced in H_2O at 355°) is heated for 100h it is converted to the *insoluble* form. This form is stable at temperatures up to 1033°, and fusion at 1080° for 4h causes complete devitrification and it reverts to the *soluble* form. [*J Am Chem Soc* **46** 2358 *1924*, **47** 1945 *1925*, **54** 2303 *1032*.]

Germanium tetrachloride *[10038-98-9]* **M 214.4, m -49.5° (α), -52.0° (β), b 83.1°/760mm, 86.5°/760mm corr, d_4^{20} 1.84.** Traces of Cl_2 and HCl can be removed from the liquid by blowing dry air through it for a few hours at room temperature or shake it with Hg or Hg_2Cl_2 and then fractionally distil in a vacuum. It decomposes on heating at 950°. It has a sharp penetrating odour and fumes in moist air to give a chalky coat of GeO_2. It is slowly hydrolysed by H_2O to give GeO_2. [*J Am Chem Soc* **44** 306 *1922*.]

Germanium tetraethoxide *[14165-55-0]* **M 252.8, m -72°; b 54.5°/5mm, 71-72°/11mm, 188-190°/722mm, d^{25} 1.1288.** Distil through a 10cm Vigreux column under reduced pressure. Alternatively distil through a Fensche glass helices column fitted with a total condensation variable take-off stillhead. Fractionate under reduced pressure using a reflux ratio of 10:1. [*J Am Chem Soc* **75** 718 *1953*; *J Chem Soc* 4916 *1956*.]

Glass powder (100-300 mesh). Washed with 10% HNO_3, water and dried.

Gold (III) bromide (gold tribromide) *[10294-28-7]* **M 436.7, m 150°(dec).** Purified by adding pure Br_2 to the dark powder, securely stopper the container, warm a little and shake while keeping away from light for *ca* 48h. Remove the stopper and place over NaOH until free Br_2 is no longer in the apparatus (48-60h). The bright yellow needles of the tribromide are stable over NaOH in the dark. It is sol in H_2O and in EtOH where it is slowly reduced. Keep in a cooled closed container and protect from light as decomposition causes gold to be formed. *Aurobromic acid* can be obtained by adding the calculated amount of conc HBr to $AuBr_3$ (actually Au_2Br_6) until all dissolves, whereby the acid crystallises out as $HAuBr_4.5H_2O$, deliquescent solid soluble in EtOH with **m** *ca* 27°, and store as above. [*J Chem Soc* 2410 *1931*, 217, 219 *1935*.]

Gold (III) chloride (hydrate) *[16903-35-8]* **M 339.8 + xH₂O, m 229°, b 354°(dec), d 3.9.**
Obtained as a dark red crystalline mass by dissolving Au in aqua regia and evaporating. When sublimed at 180° the crystals are ruby red. The anhydrous salt is *hygroscopic* sol in H_2O but sparingly soluble in EtOH and Et_2O. *Aurochloric acid* is formed when $AuCl_3$ is dissolved in HCl. [*J Am Chem Soc* **35** 553 *1913; Handbook of Preparative Inorganic Chemistry (Ed. Brauer)* Vol II 1056 *1965*.]

Gold (I) cyanide *[506-65-0]* **M 223.0, m dec on heating.** The lemon yellow powder is sparingly soluble in H_2O and EtOH but soluble in aqueous NH_3. It is obtained by heating $H[Au(CN)_2]$ at 110°. Wash well with H_2O and EtOH and dry at 110°. It has an IR band at v $2239 cm^{-1}$ typical fo C≡N stretching vibration. [*Handbook of Preparative Inorg anic Chemistry (Ed. Brauer)* Vol II 1064 *1965*.] CARE: may evolve HCN.

Gold (I) iodide *[10294-31-2]* **M 323.9, m 120°(dec), d 8.25.** It has been prepared by heating gold and iodine in a tube at 120° for 4 months. Since it decomposes to Au and I_2 in the presence of UV light and heat then the main impurity is Au. The salt is therefore purified by heating at 120° with I_2 for several weeks. The crystals should be kept dry and in a cool place in the dark. [*Z Naturforsch* **11B** 604 *1956*.]

Gold (III) oxide hydrate *[1303-58-8]* **M 441.9 + xH₂O, evolves O₂ at 110°, pK_1^{25} <11.7, pK_2^{25} 13.36, pK_3^{25} >15.3 [for Au(OH)₃].** Most probable impurities are Cl^- ions. Dissolve in strong boiling KOH soln (*ca* 5M) and precipitate (**care**) with excess of 3N H_2SO_4. Then shake and centrifuge, resuspend in H_2O and repeat wash several times until free from SO_4 and Cl ions. This gives a *wet* oxide which is dried in air, and dec to Au in sunlight. It is best to keep it wet as it decomposes on drying (analyse wet sample). Store away from light in the presence of H_2O vapour. It evolves O_2 at 110°. It is insoluble in H_2O but soluble in HCl and conc HNO_3. [*J Am Chem Soc* **49** 1221 *1927*.]

Graphite *[7782-42-5]*. Treated with hot 1:1 HCl. Filtered, washed, dried, powdered and heated in an evacuated quartz tube at 1000° until a high vacuum was obtained. Cooled and stored in an atmosphere of helium [Craig, Van Voorhis and Bartell *J Phys Chem* **60** 1225 *1956*].

Haematoporphyrin IX [8,13-bis(1-hydroxyethyl)-3,7,12,17-tetramethyl-21*H*-23*H*-porphin-2,18-dipropionic acid *[14459-29-1]* M 598.7, pK_{Est} ~ 4.8 (-CH₂CH₂COOH).

See hematoporphyrin on p. 541 in Chapter 6.

Helium *[7440-59-7]* **M 4.0.** Dried by passage through a column of Linde 5A molecular sieves and $CaSO_4$, then passed through an activated-charcoal trap cooled in liquid N_2, to adsorb N_2, argon, xenon and krypton. Passed over CuO pellets at 300° to remove hydrogen and hydrocarbons, over Ca chips at 600° to remove oxygen, and then over titanium chips at 700° to remove N_2 [Arnold and Smith *J Chem Soc, Faraday Trans 2* **77** 861 *1981*].

Hexabutyldistannane [hexabutylditin, bis(tributyl)tin] *[813-19-4]* **M 580.4, b 160-162°/0.3mm, d 1.148, n 1.512.** Purified by distn in a vacuum and stored in the dark. [Shirai et al. *Yakugaku Zasshi* **90** 59 *1970, Chem Abstr* **72** 90593 *1970*.]

Hexachlorocyclotriphosphazene *[940-71-6]* **M 347.7, m 113-114°, 113-115°.** See phosphonitrilic chloride trimer on p. 450.

Hexachloroplatinic acid hydrate (H₂PtCl₆, chloroplatinic acid, platinum IV chloride soution) *[16941-12-1]* **M 409.8 + H₂O, m 60° (deliquescent solid).** If it is to be purified, or regenerated from Pt recovered from catalytic hydrogenations, it was dissolved in aqua regia followed by evaporation to dryness and dissolution in the minimum vol of H_2O. Then the aqueous solution was treated with saturated ammonium chloride until all the ammonium hexachloroplatinate separated. The $(NH_4)_2PtCl_6$ was filtered off and dried at 100°. Ignite the salt to give Pt sponge, dissolve the Pt sponge in aqua regia, boil to dryness, dissolve in concentrated HCl, boil to dryness again and repeat the process. Protect from light. [Hickers *J Am Chem Soc* **43** 1268 *1921*; *Org Synth* Coll Vol I 463, 466 *1941*; Bruce *J Am Chem Soc* **58** 687 *1936*.]

Hexaethyldisiloxane *[924-49-0]*   **M 246.5, b 114-115°/16mm, 235.5°/760mm, d 0.8443, n 1.4330.**   Distil in a vacuum, but can be distilled at atmospheric pressure without decomposition. It is characterised by completely dissolving in conc H_2SO_4. [*J Chem Soc* 3077 *1950*.]

2,2,4,4,6,6-Hexamethylcyclotrisiloxane   *[1009-93-4]* **M 219.5, m -10°, b 81-82°/19mm, 111-112°/85mm, 188°/756mm, d 0.9196, n 1.448.** Purified by fractional distillation at atmospheric pressure until the temperature reaches 200°. The residue in the flask is mostly octamethylcyclotetrasilazane. [*J Am Chem Soc* **70** 3888 *1948*.]

Hexamethyldisilane *[1450-14-2]*   **M 164.4, m 9-12°, b 113.1°/750mm, d 0.7272, n 1.4229.** Most likely impurity is trimethylchlorosilane (*cf* boiling point). Wash with H_2O, cold conc H_2SO_4, H_2O again then aqueous $NaHCO_3$, dry over $CaSO_4$ and fractionate at atmospheric pressure. [*J Chem Soc* 2811 *1958*.] Grossly impure sample (25% impurities) was purified by repeated spinning band distn. This lowered the impurity level to 500 ppm. The main impurity was identified as 1-hydroxypentamethyldisilane.

Hexamethyldisilazane (HMDS)   *[999-97-3]* **M 161.4, b 125-125.6°/atm, 126°/760mm, d 0.7747, n 1.407.** Possible impurity is Me_3SiCl. Wash well with pet ether and fractionate through a vacuum jacketed column packed with Helipac using a reflux ratio of 10:1. [*J Org Chem* **23** 50 *1958*.]

Hexamethyldisiloxane   *[107-46-0]*   **M 162.4, b 99.4°/760mm, 100.4°/764mm, d 0.7633, n 1.3777.** Fractionally distilled through a column packed with glass helices with *ca* 15 theoretical plates. [*J Am Chem Soc* **76** 2672 *1954*; *J Gen Chem USSR (Engl ed)* **25** 469 *1955*.]

Hexamethylditin (hexamethyldistannane)   *[661-69-8]*   **M 327.6, m 23.5°, b 85-88°/45mm, 182°/756mm, d^{25} 1.57.** Wash with H_2O and extract with $^*C_6H_6$, dry by filtering through powdered Na_2SO_4, remove $^*C_6H_6$ on a rotary evaporator and fractionally dist the oily residue under vacuum (**b** 85-88°/45mm). *It boils at ca 182° at atmospheric press but it cannot be distilled in air because the hot vapours flash in the condenser.* [*J Am Chem Soc* **47** 2361 *1925*, **63** 2509 *1941*; *Trans Faraday Soc* **53** 1612 *1957*.]

Hexamethylphosphoric triamide (HMPA)   *[680-31-9]*   **M 179.2, f 7.2°, b 68-70°/1mm, 235°/760mm, d 1.024, n 1.460.** The industrial synthesis is usually by treatment of $POCl_3$ with excess of dimethylamine in isopropyl ether. Impurities are water, dimethylamine and its hydrochloride. It is purified by refluxing over BaO or CaO at about 4mm pressure in an atmosphere of nitrogen for several hours, then distd from sodium at the same pressure. The middle fraction (**b** *ca* 90°) is collected, refluxed over sodium under reduced pressure under nitrogen and distd. It is kept in the dark under nitrogen, and stored in solid CO_2. Can also be stored over 4A molecular sieves.
Alternatively, it is distd under vacuum from CaH_2 at 60° and crystd twice in a cold room at 0°, seeding the liquid with crystals obtained by cooling in liquid nitrogen. After about two-thirds frozen, the remaining liquid is drained off [Fujinaga, Izutsu and Sakara *Pure Appl Chem* **44** 117 *1975*]. For tests of purity see Fujinaga et al. in *Purification of Solvents*, Coetzee Ed., Pergamon Press, Oxford, 1982. For efficiency of desiccants in drying HMPT see Burfield and Smithers [*J Org Chem* **43** 3966 *1978*; Sammes et al. *J Chem Soc, Faraday Trans 1* 281 *1986*]. **CARCINOGEN.**

Hexamethylphosphorous triamide (HMPT) *[1608-26-0]* **M 163.2, m 7.2°, b 49-51°/12mm, 162-164°/12mm, d 0.989, n 1.466.** It may contain more than 1% of phosphoric triamide. The yellow oil is first distd at atm press then under reduced press and stored under N_2. It is air sensitive, **TOXIC**, should not be inhaled and is absorbed through the skin. [Mark *Org Synth* Coll Vol V 602 *1973*.]

Hexamminecobalt(III) chloride *[10534-89-1]* **M 267.5.** Crystd from warm water (8mL/g) by cooling. [Bjerrum and McReynolds *Inorg Synth* **2** 217 *1946*.]

Hexammineruthenium(III) chloride *[14282-91-8]* **M 309.6.** Crystd twice from 1M HCl.

Hexarhodium hexadecacarbonyl *[28407-51-4]* **M 1065.6, m 220°(dec, in air), d 2.87.** Slowly loses CO when heated in air; may be regenerated by heating at 80-200° in the presence of CO at 200atm pressure for 15h, preferably in the presence of Cu. Forms black crystals which are insoluble in hexane. It has bands at 2073, 2026 and 1800cm^{-1} in the IR. [*Z Anorg Allg Chem* **251** 96 *1963*; *J Am Chem Soc* **85** 1202 *1963*; *Tetrahedron Lett* **22** 1783 *1981*.]

Hydrazine (anhydrous) *[302-01-2]* **M 32.1, f 1.5-2.0°, b 47°/26mm, 56°/71mm, 113-113.5°/atm, n 1.470, d 1.91, pK$_1^{25}$ -0.88, pK$_2^{25}$ 8.11.** Hydrazine hydrate is dried by refluxing with an equal weight of KOH pellets for 3h, then distilled from fresh NaOH or BaO in a current of dry N_2.

Hydrazine dihydrochloride *[5341-61-7]* **M 105.0, m 198°, d 1.42.** Crystd from aqueous EtOH and dried under vacuum over $CaSO_4$.

Hydrazine monohydrochloride *[2644-70-4]* **M 68.5, m 89°.** Prepared by dropwise addition of cold conc HCl to cold liquid hydrazine in equimolar amounts. The crystals were harvested from water and were twice recrystd from absolute MeOH and dried under vacuum. [Kovack et al. *J Am Chem Soc* **107** 7360 *1985*.]

Hydriodic acid *[10034-85-2]* **M 127.9, b 127°(aq azeotrope), d 1.701, pK^{25} -8.56.** Iodine can be removed from aqueous HI, probably as the amine hydrogen triiodide, by three successive extractions using a 4% soln of Amberlite LA-2 (a long-chain aliphatic amine) in CCl_4, toluene or pet ether (10mL per 100mL of acid). [Davidson and Jameson *Chem Ind (London)* 1686 *1963*.] Extraction with tributyl phosphate in $CHCl_3$ or other organic solvents is also suitable. Alternatively, a De-acidite FF anion-exchange resin column in the OH$^-$-form using 2M NaOH, then into its I$^-$-form by passing dilute KI soln, can be used. Passage of an HI solution under CO_2 through such a column removes polyiodide. The column can be regenerated with NaOH. [Irving and Wilson *Chem Ind (London)* 653 *1964*]. The earlier method was to reflux with red phosphorus and distil in a stream of N_2. The colourless product was stored in ampoules in the dark [Bradbury *J Am Chem Soc* **74** 2709 *1952*; *Inorg Synth* **1** 157 *1939*]. Fumes in moist air. **HARMFUL VAPOURS.**

Hydrobromic acid *[10035-10-6]* **M 80.9, b 125°(aq azeotrope, 47.5% HBr), d 1.38 (34% HBr), pK^{25} -8.69.** A soln of aqueous HBr ca 48% (w/w, constant boiling) was distilled twice with a little red phosphorus, and the middle half of the distillate was taken. (The azeotrope at 760mm contains 47.8% (w/w) HBr.) [Hetzer, Robinson and Bates *J Phys Chem* **66** 1423 *1962*]. Free bromine can be removed by Irvine and Wilson's method for HI (see above), except that the column is regenerated by washing with an ethanolic solution of aniline or styrene. Hydrobromic acid can also be purified by aerating with H_2S, distilling and collecting the fraction boiling at 125-127°. [*Inorg Synth* **1** 155 *1939*.] **HARMFUL VAPOURS.**

Hydrochloric acid *[7647-01-0]* **M 36.5, b 108.6°(aq azeotrope, 20.2% HCl), d 1.09(20%), pK^{25} -6.1.** Readily purified by fractional distillation as constant boiling point acid, following dilution with H_2O. The constant-boiling fraction contains 1 mole of HCl in the following weights of distillate at the stated pressures: 179.555g (730mm), 179.766g (740mm), 179.979 (750mm), 180.193 (760mm), 180.407 (770mm) [Foulk and Hollingsworth *J Am Chem Soc* **45** 1220 *1923*..] **HARMFUL VAPOURS.**

Hydrofluoric acid *[7664-39-3]* **M 20.0, b 112.2°(aq azeotrope, 38.2% HF), d 1.15 (47-53% HF), pK^{25} 3.21.** Freed from lead (Pb ca 0.002ppm) by co-precipitation with SrF_2, by addition of 10mL of 10% $SrCl_2$ soln per kilogram of the conc acid. After the ppte has settled, the supernatant is decanted through a filter in a hard-rubber or paraffin lined-glass vessel [Rosenqvist *Am J Sci* **240** 358 *1942*. Pure aqueous HF solutions (up to 25M) can be prepared by isothermal distn in polyethylene, polypropylene or platinum apparatus [Kwestroo and Visser *Analyst* **90** 297 *1965*]. **HIGHLY TOXIC.**

Hydrogen *[1333-74-0]* **M 2.0, m -259.1°, -252.9°.** Usually purified by passage through suitable absorption train. Carbon dioxide is removed with KOH pellets, soda-lime or NaOH pellets. Oxygen is removed with a "De-oxo" unit or by passage over Cu heated to 450-500°, Cu on Kieselguhr at 250°. Passage over a mixture of MnO_2 and CuO (Hopcalite) oxidises any CO to CO_2 (which is removed as above). Hydrogen can be dried by passage through dried silica-alumina at -195°, through a dry-ice trap followed by a liquid-N_2 trap packed with glass wool, through $CaCl_2$ tubes, or through $Mg(ClO_4)_2$ or P_2O_5. Other purification steps include passage through a hot palladium thimble [Masson *J Am Chem Soc* **74** 4731 *1952*], through an activated-charcoal trap at -195°, and through non-absorbent cotton-wool filter or small glass spheres coated with a thin layer of silicone grease. *Potentially* **VERY EXPLOSIVE** *in air.*

Hydrogen bromide (anhydrous) *[10035-10-6]* **M 80.9.** Dried by passage through $Mg(ClO_4)_2$ towers. This procedure is **hazardous**, see Stoss and Zimmermann [*Ind Eng Chem* **17** 70 *1939*]. Shaken with mercury, distd through a -78° trap and condensed at -195°/10⁻⁵mm. Fumes in moist air. **HARMFUL VAPOURS.**

Hydrogen chloride *[7647-01-0]* **M 36.5.** Passed through conc H_2SO_4, then over activated charcoal and silica gel. Fumes in moist air. Hydrogen chloride in gas cylinder include ethylene, 1,1-dichloroethane and ethyl chloride. The latter two may be removed by fractionating the HCl through a trap cooled to -112°. Ethylene is difficult to remove. Fumes in moist air. **HARMFUL VAPOURS.**

Hydrogen cyanide (anhydrous) *[74-90-8]* **M 27.0, b 25.7°, pK²⁵ 9.21 (aq acid).** Prepared from NaCN and H_2SO_4, and dried by passage through H_2SO_4 and over $CaCl_2$, then distilled in a vacuum system and degassed at 77°K before use [Arnold and Smith *J Chem Soc, Faraday Trans 2* **77** 861 *1981*]. Cylinder HCN may contain stabilisers against explosive polymerisation, together with small amounts of H_3PO_4, H_2SO_4, SO_2, and water. It can be purified by distn over P_2O_5, then frozen in Pyrex bottles at Dry-ice temperature for storage. **EXTREMELY POISONOUS.**

Hydrogen fluoride (anhydrous) *[7664-39-3]* **M 20.0, b 19.4°.** Can be purified by trap-to-trap distn, followed by drying over CoF_2 at room temperature and further distn. Alternatively, it can be absorbed on NaF to form $NaHF_2$ which is then heated under vacuum at 150° to remove volatile impurities. The HF is regenerated by heating at 300° and stored with CoF_3 in a nickel vessel, being distilled as required. (Water content *ca* 0.01%.) To avoid contact with base metal, use can be made of nickel, polychlorotrifluoroethylene and gold-lined fittings [Hyman, Kilpatrick and Katz *J Am Chem Soc* **79** 3668 *1957*]. **HIGHLY TOXIC.**

Hydrogen iodide (anhydrous) *[10034-85-2]* **M 127.9, b -35.5°.** After removal of free iodine from aqueous HI, the solution is frozen, then covered with P_2O_5 and allowed to melt under vacuum. The gas evolved is dried by passage through P_2O_5 on glass wool. It can be freed from iodine contamination by repeated fractional distillation at low temperatures. Fumes in moist air. **HARMFUL VAPOURS.**

Hydrogen ionophore II   (ETH 1907) (4-nonadecylpyridine - Proton ionophore) *[70268-36-9]* **M 345.6, b 180°/0.07mm, pK$_{Est}$~ 6.0.** Dissolve the waxy solid (*ca* 60g) in $CHCl_3$ (200mL), wash with H_2O (3 x 200mL), dry and evaporate to dryness then distil in vacuum. A waxy solid is formed on cooling the distillate. UV: 257nm (ε 1.86 x 10³ M⁻¹cm⁻¹), 308nm (ε 1.7 x 10² M⁻¹cm⁻¹). [IR, NMR UV: *Inorg Chem* **18** 2160 *1979.*]

Hydrogen ionophore III   (*N,N*-dioctadecyl methylamine) *[4088-22-6]* **M 536.0, m 40°, 44-46°, 48-49°, b 252-259°, pK$_{Est}$~ 10.** It can be distd at high vacuum; but dissolving in *C_6H_6, filtering and evaporating gives a waxy solid suitable for electrode use. It recrystallises from Me_2CO. [*Chem Ber* **69** 60 *1936*; *Talanta* **34** 435 *1987.*]

Hydrogen ionophore IV   ETH 1778   (octadecyl isonicotinate) *[103225-02-1]* **M 375.6, m 57.5°, pK$_{Est}$~ 3.5** . Dissolve in Et_2O and wash 3 times with H_2O. Dry, evaporate, and recrystallise the residue from EtOAc/hexane (4:1). [*Anal Chem* **58** 2285 *1986.*]

Hydrogen peroxide *[7722-84-1]* **M 34.0, d 1.110, pK²⁵ 11.65.** The 30% material has been steam distilled using distilled water. Gross and Taylor [*J Am Chem Soc* **72** 2075 *1950*] made 90% H_2O_2 approximately 0.001M in NaOH and then distilled under its own vapour pressure, keeping the temperature below 40°, the receiver being cooled with a Dry-ice/isopropyl alcohol mush. The 98% material has been rendered anhydrous by repeated fractional crystn in all-quartz vessels. **EXPLOSIVE IN CONTACT WITH ORGANIC MATERIAL.**

Hydrogen sulfide *[7783-06-4]* **M 34.1, b -59.6°, pK_1^{25} 7.05, pK_2^{25} 12.89.** Washed, then passed through a train of flasks containing saturated $Ba(OH)_2$ (2x), water (2x), and dilute HCl [Goates et al. *J Am Chem Soc* **73** 707 *1951*]. **HIGHLY POISONOUS.**

Hydroquinone-2-sulfonic acid K salt *[21799-87-1]* **M 228.3, m 250°(dec), $pK_{Est(1)}$~1, $pK_{Est(2)}$~8.5, $pK_{Est(3)}$~11.** Recrystd from water.

Hydroxylamine *[7803-49-8]* **M 33.0, m 33.1°, b 56.5°/22mm, d 1.226, pK^{20} 5.96.** Crystd from *n*-butanol at -10°, collected by vacuum filtration and washed with cold diethyl ether. **Harmful vapours.**

Hydroxylamine hydrochloride *[5470-11-1]* **M 69.5, m 151°.** Crystallised from aqueous 75% ethanol or boiling methanol, and dried under vacuum over $CaSO_4$ or P_2O_5. Has also been dissolved in a minimum of water and saturated with HCl; after three such crystns it was dried under vacuum over $CaCl_2$ and NaOH.

Hydroxylamine sulfate *[10039-54-0]* **M 164.1, m 170°(dec).** Crystallised from boiling water (1.6mL/g) by cooling to 0°.

Hydroxylamine-*O*-sulfonic acid *[2950-43-8]* **M 113.1, m 210-211°, 215°(dec), pK^{45} 1.48.** Stir the solid vigorously with anhydrous Et_2O and filter off using large volumes of dry Et_2O. Drain dry at the pump for 5min and then for 12-14h in a vacuum. Store in a vacuum desiccator/conc H_2SO_4. Determine the purity by oxidation of iodide to I_2. Must be stored in a dry atmosphere at 0-4°. It decompose slowly in H_2O at 25° and more rapidly above this temperature. [*Inorg Synth* **5** 122 *1957*.]

Hydroxynaphthol Blue tri-Na salt *[63451-35-4]* **M 620.5, m dec on heating, pK_{Est} <0.** Crude material was treated with hot EtOH to remove soluble impurities, then dissolved in 20% aqueous MeOH and chromatographed on a cellulose powder column with propanol:EtOH:water (5:5:4) as eluent. The upper of three zones was eluted to give the pure dye which was ppted as the monosodium salt trihydrate by adding conc HCl to the concentrated eluate [Ito and Ueno *Analyst* **95** 583 *1970*].

4-Hydroxy-3-nitrobenzenearsonic acid *[121-19-7]* **M 263.0.** See 2-nitrophenol-4-arsonic acid on p. 446.

Hydroxyurea *[127-07-1]* **M 76.1, m 70-72° (unstable form), m 133-136°, 141° (stable form), pK 10.6.** Recrystallise from absolute EtOH (10g in 150mL). Note that the rate of solution in boiling EtOH is slow (15-30 min). It should be stored in a cool dry place but some decomposition could occur after several weeks. [*Org Synth* Coll Vol V 645 *1973*.] It is very soluble in H_2O and can be crystd from Et_2O. [*Acta Chem Scand* **10** 256 *1956*.]

Hypophosphorous acid (Phosphinic acid) *[6303-21-5]* **M 66.0, m 26.5°, d_4^{30} 1.217, 1.13 and 1.04 for 50, 30-32, and 10% aq solns resp, pK^{25} 1.31 (H_3PO_2).** Phosphorous acid is a common contaminant of commercial 50% hypophosphorous acid. Jenkins and Jones [*J Am Chem Soc* **74** 1353 *1952*] purified this material by evaporating about 600mL in a 1L flask at 40°, under reduced pressure (in N_2), to a volume of about 300mL. After the soln was cooled, it was transferred to a wide-mouthed Erlenmeyer flask which was stoppered and left in a Dry-ice/acetone bath for several hours to freeze (if necessary, with scratching of the wall). When the flask was then left at *ca* 5° for 12h, about 30-40% of it liquefied, and again filtered. This process was repeated, then the solid was stored over $Mg(ClO_4)_2$ in a vacuum desiccator in the cold. Subsequent crystns from *n*-butanol by dissolving it at room temperature and then cooling in an ice-salt bath at -20° did not appear to purify it further. The free acid forms deliquescent crystals **m** 26.5°, and is soluble in H_2O and EtOH. The NaH_2PO_2 salt can be purified through an anion exchange resin [*Z Anorg Allg Chem* **260** 267 *1949*.]

Indigocarmine (2[1,3-dihydro-3-oxo-5-sulfo-2*H*-indol-2-ylidene]-2,3-dihydro-3-oxo-1*H*-indole-5-sulfonic acid di-Na salt) *[860-22-0]* **M 466.4, pK_1^{20} 2.8, p K_2^{20} 12.3.** Its

solubility in H_2O is 1g/100mL at 25°. Could be purified by dissolving in H_2O, filtering and adding EtOH to cause the salt to separate. Wash the solid with EtOH, Et_2O and dry *in vacuo*. [Vörlander and Schubert *Chem Ber* **34** 1860 *1901*; UV: Smit et al. *Anal Chem* **27** 1159 *1955;* Preisler et al. *J Am Chem Soc* **81** 1991 *1959*.]

Indium *[7440-74-6]* **M 114.8, m 156.6°, b 2000°, d 7.31.** Before use, the metal surface can be cleaned with dilute HNO_3, followed by a thorough washing with water and an alcohol rinse.

Indium (III) chloride *[10025-82-8]* **M 211.2, m 586°, d 4.0, pK_1^{25} 3.54, pK_2^{25} 4.28, pK_3^{25} 5.16 (for aquo In^{3+}).** The anhydrous salt forms yellow deliquescent crystals which can be sublimed at 600° in the presence of Cl_2/N_2 (1:1) {does not melt}. It is resublimed in the presence of Cl_2/N_2 (1:10) and finally heated to 150° to expel excess Cl_2. It is soluble in H_2O and should be stored in a tightly closed container. [*J Am Chem Soc* **55** 1943 *1933*.]

Indium (III) oxide *[1312-43-2]* **M 277.6, d 7.18, m sublimes at 850°.** Wash with H_2O and dry below 850°. Volatilises at 850° and dissolves in hot mineral acids to form salts. Store away from light because it darkens due to formation of In.

Indium sulfate *[13464-82-9]* **M 517.8.** Crystd from dilute aqueous H_2SO_4.

Indium (III) sulfate ($5H_2O$) *[17069-79-3]* **M 607.9, d 3.44.** Dissolve in strong H_2SO_4 and slowly evaporate at *ca* 50°. Wash crystals with glacial AcOH and then heat in a furnace at a temperature of 450-500° for 6h. Sol in H_2O is 5%. The pentahydrate is converted to an anhydrous *hygroscopic* powder on heating at 500° for 6h; but heating above this temperature over N_2 yields the oxide sulfate. Evaporation of neutral aqueous solutions provides basic sulfates. [*J Am Chem Soc* **55** 1943 *1933*, **58** 2126 *1936*.]

Iodic acid *[7782-68-5]* **M 175.9, m 118°(dec), d 4.628, pK^{25} 0.79.** Dissolve in the minimum volume of hot dilute HNO_3, filter and evaporate in a vacuum desiccator until crystals are formed. Collect crystals and wash with a little cold H_2O and dry in air in the dark. Soluble in H_2O: 269g/100mL at 20° and 295g/100mL at 40°. Soluble in dilute EtOH and darkens on exposure to light. It is converted to $HIO_3.I_2O_5$ on heating at 70°, but at 220° complete conversion to HIO_3 occurs. [*J Am Chem Soc* **42** 1636 *1920*, **53** 44 *1931*.]

Iodine *[7553-56-2]* **M 253.8, m 113.6°.** Usually purified by vacuum sublimation. Preliminary purifications include grinding with 25% by weight of KI, blending with 10% BaO and subliming; subliming with CaO; grinding to a powder and treating with successive portions of H_2O to remove dissolved salts, then drying; and crystn from *benzene. Barrer and Wasilewski [*Trans Faraday Soc* **57** 1140 *1961*] dissolved I_2 in conc KI and distilled it, then steam distilled three times, washing with distilled H_2O. Organic material was removed by sublimation in a current of O_2 over platinum at about 700°, the iodine being finally sublimed under vacuum. **HARMFUL VAPOURS.**

Iodine monobromide *[7789-33-5]* **M 206.8, m 42°.** Purified by repeated fractional crystallisation from its melt.

Iodine monochloride *[7790-99-0]* **M 162.4, m 27.2°.** Purified by repeated fractional crystallisation from its melt.

Iodine pentafluoride *[7783-66-6]* **M 221.9, m -8.0°, b 97°.** Rogers et al. [*J Am Chem Soc* **76** 4843 *1954*] removed dissolved iodine from IF_5 by agitating with a mixture of dry air and ClF_3 in a Fluorothene beaker using a magnetic stirrer. The mixture was transferred to a still and the more volatile impurities were pumped off as the pressure was reduced below 40mm. The still was gradually heated (kept at 40mm) to remove the ClF_3 before IF_5 distilled. Stevens [*J Org Chem* **26** 3451 *1961*] pumped IF_5 under vacuum from its cylinder, trapping it at -78°, then allowing it to melt in a stream of dry N_2. **HARMFUL VAPOURS.**

Iodine trichloride *[865-44-1]* **M 233.3, m 33°, b 77°(dec).** Purified by sublimation at room temperature.

Iodomethyl trimethylsilane *[4206-67-1]* **M 214.1, b 139.5°/744mm, d 1.44, n_D^{25} 1.4917.** If slightly violet in colour wash with aqueous 1% sodium metabisulfite, H_2O, dry over Na_2SO_4 and fractionally distil at atmospheric pressure. [*J Am Chem Soc* **68** 481 *1946*.]

Iodotrimethylsilane (trimethylsilyl iodide, TMSI) *[16029-98-4]* **M 200.1, b 106.8°/742mm, 107.5°/760mm, d 1.470.** Add a little antimony powder and fractionate with this powder in the still. Stabilise with 1% wt of Cu powder. [*J Chem Soc* 3077 *1950*.]

Iridium *[7439-88-5]* **M 192.2, m 2450°, b ~4500°, d 22.65.** It is a silver white hard solid which oxidises on the surface in air. Scrape the outer tarnished layer until silver clear and store under paraffin. Stable to acids but dissolves in aqua regia. [*Chem Rev* **32** 277 *1943*.]

Iridium (IV) chloride hydrate (hexachloroiridic acid) *[16941-92-7 (6H₂O); 207399-11-9 (xH₂O)]* **M 334.0+H₂O.** If it contains nitrogen then repeatedly concentrate a conc HCl solution until free from nitrogen, and dry free from HCl in a vac over CaO until crystals are formed. The solid is very *hygroscopic*. [*J Am Chem Soc* **53** 884 *1931*; *Handbook of Preparative Inorganic Chemistry (Ed. Brauer)* Vol II 1592 *1965*.]

Iron (wire) *[7439-89-6]* **M 55.9, m 1535°.** Cleaned in conc HCl, rinsed in de-ionised water, then reagent grade acetone and dried under vacuum.

Iron ennecarbonyl (di-iron nonacarbonyl) *[15321-51-4]* **M 363.7, m 100°(dec).** Wash with EtOH and Et₂O and dry in air. Sublimes at 35° at high vacuum. Dark yellow plates stable for several days when kept in small amounts. Large amounts, especially when placed in a desiccator spontaneously *ignite* in a period of one day. It decomposes in moist air. It is insoluble in hydrocarbon solvents but forms complexes with several organic compounds. [*J Am Chem Soc* **72** 1107 *1950*; *Chem Ber* **60** 1424 *1927* .]

Iron (III) *meso*-5,10,15,20-tetraphenylporphine chloride complex *[16456-81-8]* **M 704.0.** Crystallise by extraction from a thimble (Soxhlet) with CHCl₃. Concentrate the extract to *ca* 10mL and add *ca* 80mL of hot MeOH. Dark blue crystals separate on cooling. It can be recrystallised several times from CHCl₃-MeOH. Avoid prolonged heating. It is quite soluble in organic solvents but insoluble in pet ether. [*J Am Chem Soc* **70** 1808 *1948*; UV: **73** 4315 *1951*.]

Iron pentacarbonyl (pentacarbonyl iron) *[13463-40-6]* **M 195.9, m -20°, b 102.8°/749mm, 103°/760mm, n 1.520, d 1.490.** It is a pale yellow viscous liq which is PYROPHORIC and readily absorbed by the skin. **HIGHLY TOXIC (protect from light and air).** It should be purified in a vacuum line by distilling and collecting in a trap at -96° (toluene-Dry ice slush). It has been distd at atm pressure (use a very efficient fume cupboard). At 180°/atm it decomps to give Fe and CO. In UV light in pet ether it forms Fe₂(CO)₉. [Hagen et al. *Inorg Chem* **17** 1369 *1978*; Ewens et al. *Trans Faraday Soc* **35** 6811 *1939*.]

Isopropyldimethyl chlorosilane *[3634-56-8]* **M 140.7, b 109.8-110.0°/738mm, d 0.88, n 1.4158.** Probable impurity is Me₃SiCl (b 56.9°/783mm) which can be removed by efficient fractional distillation. [*J Am Chem Soc* **76** 801 *1954*.]

(2,3-*O*-Isopropylidene)-2,3-dihydroxy-1,4-bis-(diphenylphosphino)butane (DIOP) *[4R,5S-(-)-32305-98-9; 4S,5R-(+)- 37002-48-5]* **M 498.5, m 88-90°, $[\alpha]_D^{19}$ (-) and (+) 26° (c 2.3, CHCl₃), pK_{Est} ~ 4.5.** It has been recrystd from *C₆H₆-pet ether. After 2 recrstns from EtOH it was pure by TLC on silica gel using Me₂CO-hexane as solvent. [Kagan and Dang *J Am Chem Soc* **94** 6429 *1972*.]

Lanthanide shift reagents see p. 277 in Chapter 4.

Lanthanum *[7439-91-0]* **M 138.9, m 920°, b 3470°, d 6.16.** White metal that slowly tarnishes in air due to oxidation. Slowly decomposed by H_2O in the cold and more rapidly on heating to form the hydroxide. The metal is cleaned by scraping off the tarnished areas until the shiny metal is revealed and stored under oil or paraffin. It burns in air at 450°.

Lanthanum triacetate *[917-70-4]* **M 316.0, pK$_1^{25}$ 9.06 (for aquo La^{3+}).** Boil with redistilled Ac$_2$O for 10min (does not dissolve and is a white solid). Cool, filter, wash with Ac$_2$O and keep in a vacuum desiccator (NaOH) till free from solvent. [*J Indian Chem Soc* **33** 877 *1956*.]

N-Lauroyl-N-methyltaurine sodium salt   (sodium N-decanoyl-N-methy-2-aminoethane sulfonate) *[4337-75-1]* **M 343.5, pK$_{Est}$ ~1.5.** Prepared from methyldecanoate (at 180° under N$_2$) or decanoyl chloride and sodium N-methylethane sulfonate and purified by dissolving in H$_2$O and precipitating by addition of Et$_2$O. Decomposes on heating. [Desseigne and Mathian *Mém Services Chim Etat Paris* **31** 359 *1944*, *Chem Abstr* **41** 705 *1947*.]

Lead II acetate *[301-04-2 (anhydr); 6080-56-4 (3H$_2$O)]* **M 325.3, m 280°, pK$_1^{25}$ 7.1 (for Pb^{2+}), pK$_2^{25}$ 10.1 (HPbO$_2^-$), pK$_3^{25}$ 10.8 (PbO$_2^{2-}$).** Crystallised twice from anhydrous acetic acid and dried under vacuum for 24h at 100°.

Lead (bis-cyclopentadienyl) *[1294-74-2]* **M 337.4.** Purified by vacuum sublimation. Handled and stored under N$_2$.

Lead (II) bromide *[10031-22-8]* **M 367.0, m 373°.** Crystallised from water containing a few drops of HBr (25mL of water per gram PbBr$_2$) between 100° and 0°. A neutral solution was evaporated at 110° and the crystals that separated were collected by rapid filtration at 70°, and dried at 105° (to give the *monohydrate*). To prepare the anhydrous bromide, the hydrate is heated for several hours at 170° and then in a Pt boat at 200° in a stream of HBr and H$_2$. Finally fused [Clayton et al. *J Chem Soc, Faraday Trans 1* **76** 2362 *1980*].

Lead (II) chloride *[7758-95-4]* **M 278.1, m 501°.** Crystallised from distilled water at 100° (33mL/g) after filtering through sintered-glass and adding a few drops of HCl, by cooling. After three crystns the solid was dried under vacuum or under anhydrous HCl vapour by heating slowly to 400°.

Lead diethyldithiocarbamate *[17549-30-3]* **M 503.7, pK$_1^{25}$ 3.36 (for N,N-diethyldithio-carbamate).** Wash with H$_2$O and dry at 60-70°, or dissolve in the min vol of CHCl$_3$ and add the same vol of EtOH. Collect the solid that separates and dry as before. Alternatively, recryst by slow evaporation of a CHCl$_3$ soln at 70-80°. Filter the crystals, wash with H$_2$O until all Pb^{2+} ions are eluted (check by adding chromate) and then dry at 60-70° for at least 10h. [*Justus Liebigs Ann Chem* **49** 1146 *1977*.]

Lead (II) formate *[811-54-4]* **M 297.3, m 190°.** Crystd from aqueous formic acid.

Lead (II) iodide *[10101-63-0]* **M 461.0, m 402°.** Crystd from a large volume of water.

Lead monoxide *[1317-36-8]* **M 223.2, m 886°.** Higher oxides were removed by heating under vacuum at 550° with subsequent cooling under vacuum. [Ray and Ogg *J Am Chem Soc* **78** 5994 *1956*.]

Lead nitrate *[10099-74-8]* **M 331.2, m 470°.** Ppted twice from hot (60°) conc aqueous soln by adding HNO$_3$. The ppte was sucked dry in a sintered-glass funnel, then transferred to a crystallising dish which was covered by a clock glass and left in an electric oven at 110° for several hours [Beck, Singh and Wynne-Jones *Trans Faraday Soc* **55** 331 *1959*]. After 2 recrystns of ACS grade no metals above 0.001 ppm were detected.

Lead tetraacetate *[546-67-8]* **M 443.4, m 175-180°, pK$_1$ 1.8, pK$_2$ 3.2, pK$_3$ 5.2, pK$_4$ 6.7.** Dissolved in hot glacial acetic acid, any lead oxide being removed by filtration. White crystals of lead tetraacetate separated on cooling. Stored in a vacuum desiccator over P$_2$O$_5$ and KOH for 24h before use.

Lissamine Green B {1-[bis-(4,4'-dimethylaminophenyl)methyl]-2-hydroxynaphthalene-3,6-disulfonic acid sodium salt, Acid Green 50} *[3087-16-9]* **M 576.6, m >200°(dec), CI 44090, λmax 633nm.** Crystd from EtOH/water (1:1, v/v).

Lissapol C (mainly sodium salt of 9-octadecene-1- sulfate) *[2425-51-6]*. Refluxed with 95% EtOH, then filtered to remove insoluble inorganic electrolytes. The alcohol solution was then concentrated and the residue was poured into dry acetone. The ppte was filtered off, washed in acetone and dried under vacuum. [Biswas and Mukerji *J Phys Chem* **64** 1 *1960*].

Lissapol LS (mainly sodium salt of anisidine sulfate) *[28903-20-0]*. Refluxed with 95% EtOH, then filtered to remove insoluble inorganic electrolytes. The alcohol solution was then concentrated and the residue was poured into dry acetone. The ppte was filtered off, washed in acetone and dried under vacuum. [Biswas and Mukerji *J Phys Chem* **64** 1 *1960*].

Lithium (metal) *[7439-93-2]* **M 6.9, m 180.5°, b 1342°, d 0.534.** After washing with pet ether to remove storage oil, lithium was fused at 400° and then forced through a 10-micron stainless-steel filter with argon pressure. It was again melted in a dry-box, skimmed, and poured into an iron distillation pot. After heating under vacuum to 500°, cooling and returning to the dry-box for a further cleaning of its surface, the lithium was distd at 600° using an all-iron distn apparatus [Gunn and Green *J Am Chem Soc* **80** 4782 *1958*].

Lithium acetate (2H$_2$O) *[546-89-4]* **M 102.0, m 54-56°.** Crystallised from EtOH (5mL/g) by partial evaporation.

Lithium aluminium hydride *[16853-85-3]* **M 37.9, m 125°(dec).** Extracted with Et$_2$O, and, after filtering, the solvent was removed under vacuum. The residue was dried at 60° for 3h, under high vacuum [Ruff *J Am Chem Soc* **83** 1788 *1961*]. **IGNITES in the presence of a small amount of water and reacts EXPLOSIVELY.**

Lithium amide *[7782-89-0]* **M 23.0, m 380-400°, d$^{17.5}$ 1.178.** Purified by heating at 400° while NH$_3$ is passed over it in the upper of two crucibles (the upper crucible is perforated). The LiNH$_2$ will drip into the lower crucible through the holes in the upper crucible. The product is cooled in a stream of NH$_3$. Protect it from air and moisture, store under N$_2$ in a clear glass bottle sealed with paraffin. Store small quantities so that all material is used once the bottle is opened. If the colour of the amide is yellow it should be destroyed as it is likely to have oxidised and to **EXPLODE.** On heating above 450° it is decomposed to Li$_2$NH which is stable up to 750-800°. [*Handbook of Preparative Inorganic Chemistry (Ed. Brauer)* Vol I 463 *1963*; *Inorg Synth* **2** 135 *1953*.]

Lithium benzoate *[553-54-8]* **M 128.1.** from EtOH (13mL/g) by partial evaporation.

Lithium borohydride *[16949-15-8]* **M 21.8, mCrystd 268°, b 380°(dec), d 0.66.** Crystd from Et$_2$O, and pumped free of ether at 90-100° during 2h [Schaeffer, Roscoe and Stewart *J Am Chem Soc* **78** 729 *1956*].

Lithium bromide *[7550-35-8]* **M 86.8, m 550°.** Crystd several times from water or EtOH, then dried under high vacuum for 2 days at room temperature, followed by drying at 100°.

Lithium carbonate *[554-13-2]* **M 73.9, m 552°, 618°.** Crystd from water. Its solubility decreases as the temperature is raised.

Lithium chloride *[7447-41-8]* **M 42.4, m 600°, 723°.** Crystd from water (1mL/g) or MeOH and dried for several hours at 130°. Other metal ions can be removed by preliminary crystallisation from hot aqueous 0.01M disodium EDTA. Has also been crystallised from conc HCl, fused in an atmosphere of dry HCl gas, cooled under dry N$_2$ and pulverised in a dry-box. Kolthoff and Bruckenstein [*J Am Chem Soc* **74** 2529 *1952*] ppted with ammonium carbonate, washed with Li$_2$CO$_3$ five times by decantation and finally with suction, then dissolved in HCl. The LiCl solution was evaporated slowly with continuous stirring in a large evaporating dish, the dry powder being stored (while still hot) in a desiccator over CaCl$_2$.

Lithium diisopropylamide *[4111-54-0]* **M 107.1, b 82-84°/atm, 84°/atm, d^{22} 0.722, flash point -6°.** It is purified by refluxing over Na wire or NaH for 30min and then distilled into a receiver under

N₂. Because of the low boiling point of the amine a dispersion of NaH in mineral oil can be used directly in this purification without prior removal of the oil. It is **HIGHLY FLAMMABLE,** and is decomposed by air and moisture. [*Org Synth* **50** 67 *1970.*]

Lithium dodecylsulfate *[2044-56-6]* **M 272.3.** Recrystd twice from absolute EtOH and dried under vacuum.

Lithium fluoride *[7789-24-4]* **M 25.9, m 842°, 848°, b 1676°, 1681°, d 2.640.** Possible impurities are LiCO₃, H₂O and HF. These can be removed by calcining at red heat, then pulverised with a Pt pestle and stored in a paraffin bottle. Solubility in H₂O is 0.27% at 18°. It volatilises between 1100-1200°. [*Handbook of Preparative Inorganic Chemistry (Ed. Brauer)* Vol I 235 *1963*].

Lithium formate (H₂O) *[6108-23-2 (H₂O); 556-63-8 (anhydr)]* **M 70.0, d 1.46.** Crystd from hot water (0.5mL/g) by chilling.

Lithium hydride. *[7580-67-8]* **M 7.95, m 680°, d 0.76-0.77.** It should be a white powder otherwise replace it. It darkens rapidly on exposure to air and is decomposed by H₂O to give H₂ and LiOH, and reacts with lower alcohols. One gram in H₂O liberates 2.8L of H₂ (Could be explosive).

Lithium hydroxide (H₂O) *[1310-66-3 (H₂O); 1310-65-2 (anhydr)]* **M 42.0, m 471°, d 1.51, pK²⁵ 13.82.** Crystd from hot water (3mL/g) as the monohydrate. Dehydrated at 150° in a stream of CO₂-free air.

Lithium iodate *[13765-03-2]* **M 181.9.** Crystd from water and dried in a vacuum oven at 60°.

Lithium iodide *[10377-51-2]* **M 133.8, m 469°, b 1171°, d 4.06.** Crystd from hot water (0.5mL/g) by cooling in CaCl₂-ice, or from acetone. Dried under vacuum over P₂O₅ for 1h at 60° and then at 120°.

Lithium methylate (lithium methoxide) *[865-34-9]* **M 38.0.** Most probable impurity is LiOH due to hydrolysis by moisture. It is important to keep the sample dry. It can be dried by keeping in a vacuum at 60-80° under dry N₂ using an oil pump for a few hours. Store under N₂ in the cold. It should not have bands above 3000cm⁻¹; IR has v_KBr 1078, 2790, 2840 and 2930cm⁻¹. [*J Org Chem* **21** 156 *1956.*]

Lithium nitrate *[7790-69-4]* **M 68.9, m 253°, d 2.38.** Crystd from water or EtOH. Dried at 180° for several days by repeated melting under vacuum. If it is crystallised from water keeping the temperature above 70°, formation of trihydrate is avoided. The anhydrous salt is dried at 120° and stored in a vac desiccator over CaSO₄. After 99% salt was recrystd 3 times it contained: metal (ppm) Ca (1.6), K (1.1), Mo (0.4), Na (2.2).

Lithium nitrite (H₂O) *[13568-33-7]* **M 71.0.** Crystd from water by cooling from room temperature.

Lithium picrate *[18390-55-1]* **M 221.0.** Recrystd three times from EtOH and dried under vacuum at 45° for 48h [D'Aprano and Sesta *J Phys Chem* **91** 2415 *1987*]. The necessary precautions should be taken in case of **EXPLOSION**.

Lithium perchlorate *[7791-03-9]* **M 106.4, pK²⁵ -2.4 to -3.1 (for HClO₄).** Crystd from water or 50% aq MeOH. Rendered anhydrous by heating the trihydrate at 170-180° in an air oven. It can then be recrystd twice from acetonitrile and again dried under vacuum [Mohammad and Kosower *J Am Chem Soc* **93** 2713 *1971*].

Lithium salicylate *[552-38-5]* **M 144.1.** Crystd from EtOH (2mL/g) by partial evaporation.

Lithium sulfate (anhydrous) *[10377-48-7]* **M 109.9, loses H₂O at 130° and m 859°, d 2.21.** Crystd from H₂O (4mL/g) by partial evaporation.

Lithium tetrafluoroborate *[14283-07-9]* **M 93.7, pK²⁵ 13.82 (Li⁺), pK²⁵ -4.9 (for HBF₄).** Dissolve in THF just below its solubility, filter from insol material and evap to dryness in a vacuum below

50°. Wash the residue with dry Et_2O, and pass dry N_2 gas over the solid and finally heat in an oven at 80-90°. Solubility in Et_2O: 1.9 (1.3)g in 100mL at 25°, in THF: 71g in 100mL at 25°. It is *hygroscopic* and is an **IRRITANT.** [*J Am Chem Soc* **74** 5211 *1952*, **75** 1753 *1953*.]

Lithium thiocyanate (lithium rhodanide) *[556-65-0]* **M 65.0, pK^{25} -1.85 (for HSCN).** It crystallises from H_2O as the dihydrate but on drying at 38-42° it gives the monohydrate. It can be purified by allowing an aqueous soln to crystallise in a vac over P_2O_5. The crystals are collected, dried out in vacuum at 80°/P_2O_5 in a stream of pure N_2 at 110°. [*J Chem Soc* 1245 *1936*.]

Lithium trimethylsilanolate (trimethylsilanol Li salt) *[2004-14-0]* **M 96.1, m 120°(dec in air).** Wash with Et_2O and pet ether. Sublimes at 180°/1mm as fine transparent needles. [*J Org Chem* **17** 1555 *1952*.]

Magnesium   *[7439-95-4]* **M 24.3, m 651°, b 1100°, d 1.739.** Slowly oxidises in moist air and tarnishes. If dark in colour do not use. Shiny solid should be degreased by washing with dry Et_2O, dry and keep in a N_2 atmosphere. It can be activated by adding a crystal of I_2 in the Et_2O before drying and storing.

Magnesium acetate *[142-72-3 (anhydr)*; *16674-78-5 (4H$_2$O)]* **M 214.5, m 80°.** Crystd from anhydrous acetic acid, then dried under vacuum for 24h at 100°

Magnesium benzoate (3H$_2$O) *[553-70-8]* **M 320.6, m ~200°.** Crystd from water (6mL/g) between 100° and 0°.

Magnesium bromide (anhydrous) *[7789-48-2]* **M 184.1, m 711°, d 3.72.** Crystd from EtOH.

Magnesium chloride (6H$_2$O) *[7791-18-6]* **M 203.3, m ~100°(dec), pK$_1^{25}$ 10.3, pK$_2^{25}$ 12.2 (for Mg^{2+} hydrolysis).** Crystd from hot water (0.3mL/g) by cooling.

Magnesium dodecylsulfate *[3097-08-3]* **M 555.1.** Recrystd three times from EtOH and dried in a vacuum.

Magnesium ethylate (magnesium ethoxide) *[2414-98-4]* **M 114.4.** Dissolve *ca* 1g of solid in 12.8mL of absolute EtOH and 20mL of dry xylene and reflux in a dry atmosphere (use $CaCl_2$ in a drying tube at the top of the condenser). Add 10mL of absolute EtOH and cool. Filter solid under dry N_2 and dry in a vacuum. Alternatively dissolve in absolute EtOH and pass through molecular sieves (40 mesh) under N_2, evap under N_2, and store in a tightly stoppered container. [*J Am Chem Soc* **68** 889 *1964*.]

Magnesium D-gluconate *[3632-91-5]* **M 414.6, $[\alpha]_{546}^{20}$ +13.5°, $[\alpha]_D^{20}$ +11.3° (c 1, H$_2$O).** Cryst from dilute EtOH to give *ca* trihydrate, and then dry at 98° in high vacuum. Insol in EtOH and solubility in H_2O is 16% at 25°.

Magnesium iodate (4H$_2$O) *[7790-32-1]* **M 446.2.** Crystd from water (5mL/g) between 100° and 0°.

Magnesium iodide *[10377-58-9]* **M 278.1, m 634°.** Crystd from water (1.2mL/g) by partial evapn in a desiccator.

Magnesium ionophore I (ETH 1117), (N,N'-diheptyl-N,N'-dimethyl-1,4-butanediamide) *[75513-72-3]* **M 340.6.** Purified by flash chromatography (at 40 kPa) on silica and eluting with EtOH-hexane (4:1). IR has ν(CHCl$_3$) 1630cm^{-1}. [*Helv Chim Acta* **63** 2271 *1980*.] It is a good magnesium selectophore compared with Na, K and Ca [*Anal Chem* **52** 2400 *1980*].

Magnesium ionophore II   (ETH 5214), [*N,N''*-octamethylene-bis(*N'*-heptyl-*N''*-methyl methylmalonamide)]   *[119110-37-1]*   **M 538.8.**   Reagent (*ca* 700mg) can be purified by flash chromatography on Silica Gel 60 (30g) and eluting with CH_2Cl_2-Me_2CO (4:1). [*Anal Chem* **61** 574 *1989.*]

Magnesium lactate *[18917-37-1]* **M 113.4.** Crystd from water (6mL/g) between 100° to 0°.

Magnesium nitrate (6H$_2$O) *[13446-18-9]* **M 256.4, m ~95°(dec).** Crystd from water (2.5mL/g) by partial evapn in a desiccator. After 2 recrystns ACS grade has: metal (ppm) Ca (6.2), Fe (8.4), K (2), Mo (0.6), Na (0.8), Se (0.02).

Magnesium perchlorate (Anhydrone, Dehydrite)   *[10034-81-8 (anhydr)]* **M 259.2, m >250°, pK^{25} -2.4 to -3.1 (for HClO$_4$).**   Crystd from water to give the *hexahydrate* M 331.3 *[13346-19-0].* Coll, Nauman and West [*J Am Chem Soc* **81** 1284 *1959*] removed traces of unspecified contaminants by washing with small portions of Et_2O and drying in a vac (**CARE**). The anhydrous salt is commercially available as an ACS reagent, and is as efficient a dehydrating agent as P_2O_5 and is known as "Dehydrite" or "Anhydrone". [Smith et al. *J Am Chem Soc* **44** 2255 *1922* and *Ind Eng Chem* **16** 20 *1924*.] Hygroscopic, Keep in a tightly closed container. **EXPLOSIVE** *in contact with organic materials, and is a* **SKIN IRRITANT**.

Magnesium succinate *[556-32-1]* **M 141.4.** Crystd from water (0.5mL/g) between 100° and 0°.

Magnesium sulfate (anhydrous)   *[7487-88-9]* **M 120.4, m 1127°.** Crystd from warm H_2O (1mL/g) by cooling. Dry heptahydrate at ~250° until it loses 25% of its wt. Store in a sealed container.

Magnesium trifluoromethanesulfonate   *[60871-83-2]* **M 322.4, m >300°.** Wash with CH_2Cl_2 and dry at 125°/2h and 3mmHg. [*Tetrahedron Lett* **24** 169 *1983*.]

Magon [3-hydroxy-4-(hydroxyphenylazo)-2-naphthoyl-2,4-dimethylanilide;   **Xylidyl Blue II**] *[523-67-1]* **M 411.5, m 246-247°.** Suspend in H_2O and add aqueous NaOH until it dissolves, filter and acidify with dil HCl. Collect the dye, dissolve in hot EtOH (sol is 100mg/L at *ca* 25°) concentrate to a small volume and allow to cool. Sol in H_2O of the Na salt is 0.4mg/mL. [*Anal Chim Acta* **16** 155 *1957*; *Anal Chem* **28** 202 *1956*.]

Manganese (III) acetate (2H$_2$O) *[19513-05-4]*   **M 268.1, pK^{25} 0.06 (for Mn^{3+} hydrolysis).** Wash the acetate with AcOH then thoroughly with Et_2O and dry in air to obtain the dihydrate. The *anhydrous* salt can be made by stirring vigorously a mixt of the hydrated acetate (*ca* 6g) and Ac_2O (22.5mL) and heat carefully (if necessary) until the mixture is clear. It is set aside overnight for the material to crystallise. Filter the solid, wash with Ac_2O and dry over P_2O_5. The dihydrate can also be obtained from the *di*- and *tetra*- *hydrate* mixture of the divalent acetate by adding 500mL of Ac_2O and 48g of the hydrated acetate and refluxing for 20min, then add slowly 8.0g of $KMnO_4$. After refluxing for an additional 30min, the mixture was cooled to room temperature and 85mL of H_2O added. It should be noted that larger amounts of H_2O change the yield and nature of the manganese acetate and the yields of reactions that use this reagent, e.g. formation of lactones from olefines. The $Mn(OAc)_3.2H_2O$ is then filtered off after 16h, washed with cold AcOH and air dried. [*J Am Chem Soc* **90** 5903, 5905 *1968*, Heiba et al. **91** 138 *1969*.]
Alternatively dissolve the salt (30g) in glacial acetic acid (200mL) by heating and filter. If crystals do not appear, the glass container should be rubbed with a glass rod to induce crystn which occurs within 1h. If not, allow to stand for a few days. Filter the cinnamon brown crystals which have a sliky lustre and dry over CaO. Keep away from moisture as it is decomposed by cold H_2O. [Lux in *Handbook of Preparative Inorganic Chemistry (Ed. Brauer)* Vol II, p 1469 *1963*; Williams and Hunter *Can J Chem* **54** 3830 *1976*.]

Manganese (II) acetylacetone *[14024-58-9]* **M 253.2, m ~250°.** Purify by stirring 16g of reagent for a few min with 100mL absolute EtOH and filter by suction as rapidly as possible through coarse filter paper. Sufficient EtOH is added to the filtrate to make up for the loss of EtOH and to redissolve any solid that separates. Water (15mL) is added to the filtrate and the solution is evaporated with a stream of N_2 until reduced to half its vol. Cool for a few min and filter off the yellow crystals, dry under a stream of N_2, then in a vacuum at room temp for 6-8h. These conditions are important for obtaining the *dihydrate*. A vacuum to several mm of Hg or much lower pressure for several days produces the anhydrous complex. The degree of hydration can be established by determining the loss in weight of 100g of sample after heating for 4h at 100° and <20mmHg. The theoretical loss in weight for $2H_2O$ is 12.5%. Material sublimes at 200°/2mm. It is soluble in heptane, MeOH, EtOH or $^*C_6H_6$ at 30°. [*Inorg Synth* **6** 164 *1960*, **5** 105 *1957*.]

Manganese decacarbonyl $Mn_2(CO)_{10}$ [*10170-69-1*]   **M 390.0, m 151-152°, 154-155°(sealed tube), d^{25} 1.75.** Golden yellow crystals which in the absence of CO begin to decompose at 110°, and on further heating yield a metallic mirror. In the presence of 3000psi of CO it does not decompose on heating to 250°. It is soluble in common organic solvents, insoluble in H_2O, not very stable in air, to heat or UV light. Dissolves in a lot of $^*C_6H_6$ and can be crystallised from it. It distils with steam at 92-100°. It can be purified by sublimation under reduced pressure (<0.5mm) at room temperature to give well formed golden yellow crystals. If the sample is orange coloured this sublimation leads to a mixture of golden-yellow and dark red crystals of the carbonyl and carbonyl iodide respectively which can be separated by hand picking under a microscope. Separate resublimations provide the pure compounds. **POISONOUS** [*J Am Chem Soc* **76** 3831 *1954*, **80** 6167 *1958*, **82** 1325 *1960*].

Manganous acetate $(4H_2O)$ [*6156-78-1 (4H$_2$O); 638-38-0 (anhydr)*] **M 245.1, m 80°, d 1.59, pK^{25} 10.59 (for Mn^{2+} hydrolysis).** Crystd from water acidified with acetic acid.

Manganous bromide (anhydrous) [*13446-03-2*] **M 214.8, m 695°; $4H_2O$** [*10031-20-6*] **M 286.8, m 64°(dec).** Rose-red deliquescent crystals soluble in EtOH. The H_2O is removed by heating at 100° then in HBr gas at 725° or dry in an atmosphere of N_2 at 200°.

Manganous chloride $(4H_2O)$ [*13446-34-9; 7773-01-5 (anhydr)*] **M 197.9, m 58°, 87.5°, d 2.01.** Crystd from water (0.3mL/g) by cooling.

Manganous ethylenebis(dithiocarbamate) [*12427-38-2*] **M 265.3, pK_{Est} ~ 3.0 (for —NCSSH).** Crystd from EtOH.

Manganous lactate $(3H_2O)$ [*51877-53-3*] **M 287.1.** Crystd from water.

Manganous sulfate (H_2O) [*10034-96-5 (H$_2$O); 15244-36-7 (xH$_2$O)*] **M 169.0, d 2.75.** Crystd from water (0.9mL/g) at 54-55° by evaporating about two-thirds of the water. Dehydr at >400°.

2-Mercaptopyridine N-oxide sodium salt (pyridinethione or pyrithione sodium salt) [*3811-73-2*] **M 149.1, m ~250°(dec), pK_1 -1.95, pK_2 4.65.** When recrystd from water it assayed as 98.7% based on $AgNO_3$ titration [Krivis et al. *Anal Chem* **35** 966 *1963*, see also Krivis et al. *Anal Chem* **48** 1001 *1976*; and Barton and Crich *J Chem Soc. Perkin Trans 1* 1603, 1613 *1986*].

Mercuric acetate [*1600-27-7*] **M 318.7, pK_1^{25} 2.47, pK_2^{25} 3.49 (for Hg^{2+} hydrolysis).** Crystd from glacial acetic acid. **POISONOUS.**

Mercuric bromide [*7789-47-1*] **M 360.4, m 238.1°.** Crystd from hot saturated ethanolic soln, dried and kept at 100° for several hours under vacuum, then sublimed. **POISONOUS.**

Mercuric chloride [*7487-94-7*] **M 271.5, m 276°, b 304°, d 5.6.** Crystd twice from distilled water, dried at 70° and sublimed under high vacuum. **POISONOUS.**

Mercuric cyanide [*592-04-1*] **M 252.6, m 320°(dec), d 4.00.** Crystd from water. **POISONOUS.**

Mercuric iodide (red) [*7774-29-0*] **M 454.4, m 259°(yellow >130°), b ~350°(subl), d 6.3.** Crystd from MeOH or EtOH, and washed repeatedly with distilled water. Has also been mixed thoroughly with excess 0.001M iodine solution, filtered, washed with cold distilled water, rinsed with EtOH and Et_2O, and dried in air. **POISONOUS.**

Mercuric oxide (yellow) *[21908-53-2]* **M 216.6, m 500°(dec).** Dissolved in $HClO_4$ and ppted with NaOH soln.

Mercuric thiocyanate *[592-85-8]* **M 316.8, m 165°(dec), pK^{25} -1.85 (for HSCN).** Recryst from H_2O, and can form various crystal forms depending on conditions. Solubility in H_2O is 0.069% at 25°, but is more soluble at higher temps. Decomposes to Hg above 165°. **Poisonous.** [*J Phys Chem* **35** 1128 *1931*; *Chem Ber* **68** 919 *1935*.]

Mercurous nitrate (2H$_2$O) *[7782-86-7 (2H$_2$O); 7783-34-8 (H$_2$O); 10415-75-5 (anhydr)]* **M 561.2, m 70°(dec), d 4.78, pK^{25} 2.68 (for Hg$_2$$^{2+}$ hydrolysis).** Solubility in H_2O containing 1% HNO_3 is 7.7%. Recrystd from a warm saturated soln of dilute HNO_3 and cool to room temp slowly to give elongated prisms. Rapid cooling gives plates. Colourless crystals to be stored in the dark. **POISONOUS.** [*J Chem Soc* 1312 *1956*.]

Mercurous sulfate *[7783-36-0]* **M 497.3, d 7.56.** Recrystallise from dilute H_2SO_4., and dry in a vacuum under N_2 and store in the dark. Solubility in H_2O is 0.6% at 25°. **POISONOUS.**

Mercury *[7439-97-6]* **M 200.6, m -38.9°, b 126°/1mm, 184°/10mm, 261°/100mm, 356.9°/atm, d 13.534.** After air had been bubbled through mercury for several hours to oxidise metallic impurities, it was filtered to remove coarser particles of oxide and dirt, then sprayed through a 4-ft column containing 10% HNO_3. It was washed with distilled water, dried with filter paper and distilled under vacuum.

Mercury(II) bis(cyclopentadienyl) *[18263-08-6]* **M 330.8.** Purified by low-temp recrystn from Et_2O.

Mercury dibromofluorescein {mercurochrome, merobromin, [2',7'-dibromo-4'-(hydroxy-mercurio)-fluorescein di-Na salt]} *[129-16-8]* **M 804.8, m>300°.** The Na salt is dissolved in the minimum vol of H_2O, or the free acid suspended in H_2O and dilute NaOH added to cause it to dissolve, filter and acidify with dilute HCl. Collect the ppte wash with H_2O by centrifugation and dry in vacuum. The di Na salt can be purified by dissolving in the minimum volume of H_2O and ppted by adding EtOH, filter, wash with EtOH or Me_2CO and dry in a vacuum. Solubility in 95% EtOH is 2% and in MeOH it is 16%. [*J Am Chem Soc* 42 2355 *1920*.]

Mercury orange [1-(4-chloromercuriophenylazo)-2-naphthol] *[3076-91-3]* **M 483.3, m 291.5-293°(corr) with bleaching.** Wash several times with boiling 50% EtOH and recrystallise from 1-butanol (0.9g/L of boiling alcohol). Fine needles insoluble in H_2O but slightly soluble in cold alcohols, $CHCl_3$ and soluble in aqueous alkalis. [*J Am Chem Soc* **70** 3522 *1948*.]

Mercury(II) trifluoroacetate *[13257-51-7]* **M 426.6.** Recrystd from trifluoroacetic anhydride/trifluoroacetic acid [Lan and Kochi *J Am Chem Soc* **108** 6720 *1986*]. Very **TOXIC** and *hygroscopic.*

Metanil Yellow (3[{4-phenylamino}phenylazo]-benzenesulfonic acid) *[587-98-4]* **M 375.4, pK$_{Est}$ <0.** Salted out from water three times with sodium acetate, then repeatedly extracted with EtOH [McGrew and Schneider, *J Am Chem Soc* **72** 2547 *1950*].

(Methoxycarbonylmethyl)triphenylphosphorane [methyl (triphenylphosphoranylidene)-acetate] *[2605-67-6]* **M 334.4, m 162-163°, 169-171°.** Cryst by dissolving in AcOH and adding pet ether (b 40-50°) to give colorless plates. UV λmax ($A_{1mm}^{1\%}$): 222nm (865) and 268nm (116) [Isler et al. *Helv Chim Acta* **40** 1242 *1957*].

Methoxycarbonylmethyltriphenylphosphonium bromide *[1779-58-4]* **M 415.3, m 163°, 165-170°(dec).** Wash with pet ether (b 40-50°) and recryst from $CHCl_3$/Et_2O and dry in high vac at 65°. [Isler et al. *Helv Chim Acta* **40** 1242 *1957*; Wittig and Haag *Chem Ber* **88** 1654, 1664 *1955*.]

Methoxymethyl trimethylsilane (trimethylsilylmethyl methyl ether) *[14704-14-4]* **M 118.3, b 83°/740mm, d_4^{25} 0.758, n_D^{25} 1.3878.** Forms an azeotrope with MeOH (**b** 60°). If it contains MeOH (check IR for bands above 3000cm^{-1}) then wash with H_2O and fractionate. A possible impurity could be chloromethyl trimethylsilane (**b** 97°/740mm). [*J Am Chem Soc* **70** 4142 *1948*.]

1-Methoxy-2-methyl-1-trimethylsiloxypropene (**dimethyl ketene methyl trimethylsilyl acetal**) *[31469-15-5]* **M 174.3, b 121-122°/0.35mm, 125-126°/0.4mm, 148-150°/atm, d 0.86.** Add Et_2O, wash with cold H_2O, dry (Na_2SO_4), filter, evaporate Et_2O, and distil oily residue in a vacuum. [*J Organometal Chem* **46** 59 *1972*.]

***trans*-1-Methoxy-3-(trimethylsilyloxy)-1,3-butadiene (Danishefsky's diene)** *[54125-02-9]* **M 172.3, b 68-69°/14mm, 70-72°/16mm, d 0.885, n 1.4540.** It may contain up to 1% of the precursor 4-methoxybut-4-ene-2-one. Easily distd through a Vigreux column in a vac and taking the middle fraction. [Danishefsky and Kitihara *J Am Chem Soc* **96** 7807 *1974*; Danishefsky *Acc Chem Res* **14** 400 *1981*.]

Methylarsonic acid *[124-58-3]* **M 137.9, m 161°, pK$_1^{25}$ 1.54, pK$_2^{25}$ 6.31** [As(OH)$_2$]. Crystd from abs EtOH.

Methyl dichlorosilane (dichloro methylsilane) *[75-54-7]* **M 115.0, m -92.5°, b 41°/748mm, 40.9°/760mm, 40-45°/atm, d $_{27}^{27}$ 1.105.** Impurities are generally other chloromethyl silanes. Distilled through a conventional Stedman column of 20 theoretical plates or more. [**Stedman column.** A plain tube containing a series of wire-gauze discs stamped into flat, truncated cones and welded together, alternatively base-to-base and edge-to-edge, with a flat disc across each base. Each cone has a hole, alternately arranged, near its base, vapour and liquid being brought into intimate contact on the gauze surfaces (Stedman *Can J Res B* **15** 383 *1937*)]. It should be protected from H_2O by storing over P_2O_5. [*Chem Ber* **52** 695 *1919*; *J Am Chem Soc* **68** 962 *1946*.]

Methylmercuric chloride *[115-09-3]* **M 251.1, m 167°.** Crystd from absolute EtOH (20mL/g).

Methyl Orange (sodium 4,4'-dimethylaminophenylazobenzenesulfonate) *[547-58-0]* **M 327.3, pK 3.47.** Crystd twice from hot water, then washed with a little EtOH followed by diethyl ether. Indicator: pH 3.1 (red) and pH 4.4 (yellow).

Methylphenyl dichlorosilane (dichloro methyl phenylsilane) *[149-74-6]* **M 191.1, b 114-115°/50mm, 202-205°/atm, d 1.17.** Purified by fractionation using an efficient column. It hydrolyses *ca* ten times more slowly than methyltrichlorosilane and *ca* sixty times more slowly than phenyltrichlorosilane. [*J Phys Chem* **61** 1591 *1957*].

Methylphosphonic acid *[993-13-5]* **M 96.0, m 104-106°, 105-107°, 108°, pK$_1^{25}$ 2.12, pK$_2^{25}$ 7.29.** If it tests for Cl$^-$, add H_2O and evaporate to dryness; repeat several times till free from Cl$^-$. The residue solidifies to a wax-like solid. Alternatively, dissolve the acid in the minimum volume of H_2O, add charcoal, warm, filter and evaporate to dryness in a vacuum over P_2O_5. [*J Am Chem Soc* **75** 3379 *1953*.] The *di-Na* *salt* is prepared from 24g of acid in 50mL of dry EtOH and a solution of 23g Na dissolved in 400mL EtOH is added. A white ppte is formed but the mixture is refluxed for 30min to complete the reaction. Filter off and recrystallise from 50% EtOH. Dry crystals in a vacuum desiccator. [*J Chem Soc* 3292 *1952*.]

Methylphosphonic dichloride *[676-97-1]* **M 132.9, m 33°, 33-37°, b 53-54°/10mm, 64-67°/20.5mm, 86°/44mm, 162°/760mm, d_4^{40} 1.4382.** Fractionally redistd until the purity as checked by hydrolysis and acidimetry for Cl$^-$ is correct and should solidify on cooling. [*J Chem Soc* 3437 *1952*; *J Am Chem Soc* **75** 3379 *1952*; for IR see *Can J Chem* **34** 1611 *1956*.]

Methyl Thymol Blue, sodium salt *[1945-77-3]* **M 844.8, ε 1.89 x 10^4 at 435nm, pH 5.5.** Starting material for synthesis is Thymol Blue. Purified as for Xylenol Orange on p. 387 in Chapter 4.

Methyl trichlorosilane *[75-79-6]* **M 149.5, b 13,7°/101mm, 64.3°/710.8mm, 65.5°/745mm, 66.1°/atm, d 1.263, n 1.4110.** If very pure distil before use. Purity checked by ^{29}Si nmr, δ in MeCN is 13.14 with respect to Me$_4$Si. Possible contaminants are other silanes which can be removed by fractional distillation through a Stedman column of >72 theoretical plates with total reflux and 0.35% take-off (see p. 441). The apparatus is under N$_2$ at a rate of 12 bubbles/min fed into the line using an Hg manometer to control the pressure. Sensitive to H$_2$O. [*J Am Chem Soc* **73** 4252 *1951*; *J Org Chem* **48** 3667 *1983*.]

Methyl triethoxysilane *[2031-67-6]* **M 178.31, b 142-144.5°/742mm, 141°/765mm, 141.5°/775mm, d 0.8911, n 1.3820.** Repeated fractionation in a stream of N$_2$ through a 3' Heligrid packed Todd column (see p. 174). Hydrolysed by H$_2$O and yields cyclic polysiloxanes on hydrolysis in the presence of acid in *C$_6$H$_6$. [*J Am Chem Soc* **77** 1292, 3990 *1955*.]

Methyl trimethoxysilane *[1185-55-3]* **M 136.2, b 102°/760mm, d 1.3687, n 1.3711.** Likely impurities are 1,3-dimethyltetramethoxy disiloxane (**b** 31°/1mm) and cyclic polysiloxanes, see methyl triethoxysilane. [*J Org Chem* **16** 1400 *1952*, **20** 250 *1955*.]

N-Methyl-N-trimethylsilylacetamide *[7449-74-3]* **M 145.3, b 48-49°/11mm, 84°/13mm, 105-107°/35mm (solid at room temp), d 0.90, n 1.4379.** Likely impurity is Et$_3$N.HCl which can be detected by its odour. If it is completely soluble in *C$_6$H$_6$, then redistil, otherwise dissolve in this solvent, filter and evaporate first in a vacuum at 12mm then fractionate, all operations should be carried out in a dry N$_2$ atmosphere. [*J Am Chem Soc* **88** 3390 *1966*; *Chem Ber* **96** 1473 *1963*.]

Methyl trimethylsilylacetate *[2916-76-9]* **M 146.3, b 65-68°/50mm, d 0.89.** Dissolved in Et$_2$O, shaken with 1M HCl, washed with H$_2$O, aqueous saturated NaHCO$_3$, H$_2$O again, and dried (a ppte may be formed in the NaHCO$_3$ soln and should be drawn off and discarded). The solvent is distd off and the residue is fractionated through a good column. IR (CHCl$_3$) ν 1728cm^{-1}. [*J Org Chem* **32** 3535 *1967*, **45** 237 *1980*.]

Methyl 2-(trimethylsilyl)propionate *[55453-09-3]* **M 160.3, b 155-157°/atm, d 0.89.** Dissolve in Et$_2$O, wash with aqueous NaHCO$_3$, H$_2$O, 0.1M HCl, H$_2$O again, dry (MgSO$_4$), evaporated and distil. [*J Chem Soc Perkin Trans 1* 541 *1985*; *Tetrahedron* **39** 3695 *1983*.]

Methyl triphenoxyphosphonium iodide *[17579-99-6]* **M 452.2, m 146°.** Gently heat the impure iodide with good grade Me$_2$CO The saturated solution obtained is decanted rapidly from undissolved salt and treated with an equal volume of dry Et$_2$O. The iodide separates as beautiful flat needles which are collected by centrifugation, washed several times with dry Et$_2$O, and dried in a vacuum over P$_2$O$_5$. For this recrystn it is essential to minimise the time of contact with Me$_2$CO and to work rapidly and with rigorous exclusion of moisture. If the crude material is to be used, it should be stored under dry Et$_2$O, and dried and weighed *in vacuo* immediately before use. [*J Chem Soc Perkin Trans 1* 982 *1974*; *J Chem Soc* 224 *1953*.]

Methyl triphenylphosphonium bromide *[1779-49-3]* **M 357.3, m 229-230°(corr), 227-229°, 230-233°.** If the solid is sticky, wash with *C$_6$H$_6$ and dry in a vacuum over P$_2$O$_5$. [Marvel and Gall *J Org Chem* **24** 1494 *1959*; *Chem Ber* **87** 1318 *1954*; Milas and Priesing *J Am Chem Soc* **79** 6295 *1957*; Wittig and Schöllkopf *Org Synth* **40** 66 *1960*..] The *iodide* crystd from H$_2$O has **m** 187.5-188.5° [*J Chem Soc* 1130 *1953*; *Justus Liebigs Ann Chem* **580** 44 *1953*].

N-Methyl-N-trimethylsilyl trifluoroacetamide *[24589-78-4]* **M 199.3, b 78-79°/130mm.** Fractionate through a 40mm Vigreux column. Usually it contains *ca* 1% of methyl trifluoroacetamide and 1% of other impurities which can be removed by gas chromatography or fractionating using a spinning band column. [*J Chromatogr* **42** 103 *1969*, **103** 91 *1975*.]

Methyl vinyl dichlorosilane (dichloro methyl vinyl silane) *[124-70-9]* **M 141.1, b 43-45.5°/11-11.5mm, 91°/742mm, 92.5°/743.2mm, 92.5-93°/atm, d 1.0917, n 1.444.** Likely impurities are dichloromethylsilane, butadienyl-dichloromethylsilane. Fractionate through a column packed with metal filings (20 theoretical plates) at atmospheric pressure. [*Izv Akad Nauk SSSR Ser Khim* 1474 *1957* and 767 *1958*.]

Milling Red SWB {1-[4-[4-[4-toluenesulfonyloxy]phenylazo](3,3'dimethyl-1,1'-biphenyl)-4'-azo]-2-hydroxynaphthalene-6,8-disulfonic acid di-Na salt, Acid Red 114} *[6459-94-5]* **M 830.8, m dec >250°, CI 23635, λmax ~514nm.** Salted out three times with sodium acetate, then repeatedly extracted with EtOH. [McGrew and Schneider *J Am Chem Soc* **72** 2547 *1950*.] See Solochrome Violet R on p. 352 in Chapter 4.

Milling Yellow G *[51569-18-7]*. Salted out three times with sodium acetate, then repeatedly extracted with EtOH. [McGrew and Schneider *J Am Chem Soc* **72** 2547 *1950*.] See Solochrome Violet R on p. 352 in Chapter 4.

Molybdenum hexacarbonyl *[13939-06-5]* **M 264.0, m 150°(dec), b 156°.** Sublimed in a vacuum before use [Connor et al. *J Chem Soc, Dalton Trans* 511 *1986*].

Molybdenum hexafluoride *[7783-77-9]* **M 209.9, b 35°/760mm.** Purified by low-temperature trap-to-trap distillation over predried NaF. [Anderson and Winfield *J Chem Soc, Dalton Trans* 337 *1986*.] **Poisonous vapours.**

Molybdenum trichloride *[13478-18-7]* **M 202.3, m 1027°, d 3.74.** Boiled with 12M HCl, washed with absolute EtOH and dried in a vacuum desiccator.

Molybdenum trioxide (molybdenum IV oxide, MoO$_3$) *[1313-27-5]* **M 143.9, m 795°, d 4.5.** Crystd from water (50mL/g) between 70° and 0°.

Monocalcium phosphate (2H$_2$O) (monobasic) *[7789-77-7 (2H$_2$O); 7757-93-9 (anhydr)]* **M 154.1, m 200°(dec, loses H$_2$O at 100°), d 2.2.** Crystd from a near-saturated soln in 50% aqueous reagent grade phosphoric acid at 100° by filtering through fritted glass and cooling to room temperature. The crystals were filtered off and this process was repeated three times using fresh acid. For the final crystn the solution was cooled slowly with constant stirring to give thin plate crystals that were filtered off on fritted glass, washed free of acid with anhydrous acetone and dried in a vacuum desiccator [Egan, Wakefield and Elmore, *J Am Chem Soc* **78** 1811 *1956*].

Monoperoxyphthalic acid magnesium salt 6H$_2$O. (MMPP) *[84665-66-7]* **M 494.7, m ~93°(dec).** MMPP is a safer reagent than *m*-chloroperbenzoic acid because it is not as explosive and has advantages of solubility. It is sol in H$_2$O, low mol wt alcohols, *i*-PrOH and DMF. The product of reaction, Mg phthalate, is sol in H$_2$O. It has been used in aq phase to oxidise compds in e.g. CHCl$_3$ and using a phase transfer catalyst e.g. methyltrioctylammonium chloride [Brougham *Synthesis* 1015 *1987*]. The oxidising activity can be checked (as for perbenzoic acid in Silbert et al. *Org Synth* Coll Vol V 906 *1973*], and if found to be low it would be best to prepare afresh from phthalic anhydride (1mol), H$_2$O$_2$ (1mol) and MgO at 20-25° to give MMPP. [Hignett, European Pat Appl 27 693 *1981*, *Chem Abstr* **95** 168810 *1981*.]

Naphthalene Scarlet Red 4R [1-(4-sulfonaphthalene-1-azo)-2-hydroxynaphthalene-6,8-disulfonic acid tri-Na salt, New Coccine, Acid Red 18] *[2611-82-7]* M 604.5, m >250°(dec), CI 16255, λmax 506nm.

Dissolved in the minimum quantity of boiling water, filtered and enough EtOH was added to ppte *ca* 80% of the dye. This process was repeated until a soln of the dye in aqueous 20% pyridine had a constant extinction coefficient.

Naphthol Yellow S (citronin A, flavianic acid sodium salt, 8-hydroxy-5,7-dinitro-2-naphthalene sulfonic acid disodium salt) *[846-70-8]* **M 358.2, m dec on heating.** Greenish yellow powder soluble in H$_2$O. The *free sulfonic acid* can be recrystd from dil HCl (**m** 150°) or AcOH-EtOAc (**m** 148-149.5°). The disodium salt is then obtained by dissolving the acid in two equivalents of aqueous NaOH and evaporating to dryness and drying the residue in a vacuum desiccator. The sodium salt can be recrystd from the minimum volume of H_2O or from EtOH [Dermer and Dermer *J Am Chem Soc* **61** 3302 *1939*].

1,2-Naphthoquinone-4-sulfonic acid sodium salt (3,4-dihydro-3,4-dioxo-1-naphthlene sulfonic acid sodium salt) *[521-24-4]* **M 260.2, pK$_{Est}$ <0.** Yellow crystals from aqueous EtOH and dry at 80° *in vacuo*. Solubility in H_2O is 5% [*Org Synth* Coll Vol III 633 *1955*; Danielson *J Biol Chem* **101** 507 *1933*; UV: Rosenblatt et al. *Anal Chem* **27** 1290 *1955*].

1-Naphthyl phosphate disodium salt *[2183-17-7]* **M 268.1, pK$_1^{25}$ 0.97, pK$_2^{25}$ 5.85 (for free acid).** The free acid has **m** 157-158° (from $Me_2CO/^*C_6H_6$). The *free acid* is crystd several times by adding 20 parts of boiling $^*C_6H_6$ to a hot solution of 1 part of *free acid* and 1.2 parts of Me_2CO. It has **m** 157-158°. [*J Am Chem Soc* **77** 4002 *1955*.] The *monosodium salt* was ppted from a soln of the acid phosphate in MeOH by addition of an equivalent of MeONa in MeOH. [*J Am Chem Soc* **72** 624 *1950*.] See entry on p.550 in Chapter 6.

2-Naphthyl phosphate monosodium salt *[14463-68-4]* **M 246.1, m 203-205°, pK$_1^{25}$ 1.25, pK$_2^{25}$ 5.83 (for free acid).** Recrystd from H_2O (10mL) containing NaCl (0.4g). The salt is collected by centrifugation and dried in a vac desiccator, **m** 203-205° (partially resolidifies and melts at 244°). Crystd from MeOH (**m** 222-223°). The free acid is recrystd several times by addn of 2.5 parts of hot $CHCl_3$ to a hot soln of the free acid (1 part) in Me_2CO (1.3 parts), **m** 177-178°. [*J Am Chem Soc.* **73** 5292 *1951*, **77** 4002 *1955*.] See entry on p. 551 in Chapter 6.

Neodynium chloride 6H$_2$O *[13477-89-9]* **M 358.7, m 124°, pK$_1^{25}$ 8.43 (for Nd^{3+} hydrolysis).** Forms large purple prisms from conc solns of dilute HCl. Soluble in H_2O (2.46 parts in 1 part of H_2O) and EtOH.

Neodynium nitrate (6H$_2$O) *[16454-60-7]* **M 438.4, m 70-72°.** Crystallises with 5 and 6 molecules of H_2O from conc solutions in dilute HNO_3 by slow evaporation; 1 part is soluble in 10 parts of H_2O.

Neodymium oxide *[1313-97-9]* **M 336.5, m 2320°.** Dissolved in $HClO_4$, ppted as the oxalate with doubly recrystd oxalic acid, washed free of soluble impurities, dried at room temperature and ignited in a platinum crucible at higher than 850° in a stream of oxygen [Tobias and Garrett *J Am Chem Soc* **80** 3532 *1958*].

Neon *[7440-01-9]* **M 20.2.** Passed through a copper coil packed with 60/80 mesh 13X molecular sieves which is cooled in liquid N_2, or through a column of Ascarite (NaOH-coated silica adsorbent).

Neopentoxy lithium *[3710-27-8]* **M 94.1.** Recrystd from hexane [Kress and Osborn *J Am Chem Soc* **109** 3953 *1987*].

New Methylene Blue N (2,8-dimethyl-3,6-bis(ethylamino)phenothazinium chloride 0.5 ZnCl$_2$) *[6586-05-6]* **M 416.1, m >200°(dec), pK$_1$ 3.54, pK$_2$ 4.82.** Crystd from *benzene/MeOH (3:1).

Nickel (II) acetate (4H$_2$O) *[6018-89-9]* **M 248.9, d 1.744, pK$_1^{25}$ 8.94 (from Ni^{2+} hydrolysis).** Recryst from aqueous AcOH as the green tetrahydrate. Soluble in 6 parts of H_2O. It forms lower hydrates and should be kept in a well closed container. [*Z Anorg Allg Chem* **343** 92 *1966*.]

Nickel (II) acetylacetonate *[3264-82-2]* **M 256.9, m 229-230°, b 220-235°/11mm, d^{17} 1.455.** Wash the green solid with H_2O, dry in a vacuum desiccator and recrystallise from MeOH. [*J Phys Chem* **62** 440 *1958*.] The complex can be conveniently dehydrated by azeotropic distn with toluene and the crystals may be isolated by concentrating the toluene solution. [*J Am Chem Soc* **76** 1970 *1954*.]

Nickel bromide *[13462-88-9]* **M 218.5, m 963°(loses H$_2$O at ~ 200°).** Crystd from dilute HBr (0.5mL/g) by partial evaporation in a desiccator.

Nickel chloride (6H₂O) *[7791-20-0 (6H₂O); 69098-15-3 (xH₂O); 7718-54-9 (anhydr)]* **M 237.7.** Crystd from dilute HCl.

Nickel nitrate (6H₂O) *[13478-00-7]* **M 290.8, m 57°.** Crystd from water (0.3mL/g) by partial evaporation in a desiccator.

Nickelocene [bis-(cyclopentadienyl)nickel II] *[1271-28-9]* **M 188.9, m 173-174°(under N₂).** Dissolve in Et₂O, filter and evaporate in a vacuum. Purify rapidly by recrystn from pet ether using a solid CO₂-Me₂CO bath, m 171-173°(in an evacuated tube). Also purified by vacuum sublimation. [*J Am Chem Soc* **76** 1970 *1954*; *J Inorg Nucl Chem* **2** 95, 110 *1956*.]

Nickel (II) phthalocyanine *[14055-02-8]* **M 571.3, m >300°.** Wash well with H₂O and boiling EtOH and sublime at high vacuum in a slight stream of CO₂. A special apparatus is used (see reference) with the phthallocyanine being heated to red heat. The sublimate is made of needles with an extremely bright red lustre. The powder is dull greenish blue in colour. [*J Chem Soc* 1719 *1936*.]

Nickel sulfate (7H₂O) *[1010-98-1]* **M 280.9, m loses 5H₂O at 100°, anhydr m at ~280°.** Crystd from warm water (0.25mL/g) by cooling.

Nickel 5,10,15,20-tetraphenylporphyrin *[14172-92-0]* **M 671.4, λ_max 414(525)nm.** Purified by chromatography on neutral (Grade I) alumina, followed by recrystn from CH₂Cl₂/MeOH [Yamashita *J Phys Chem* **91** 3055 *1987*].

Niobium (V) chloride *[10026-12-7]* **M 270.2, m 204.7-209.5°, b ~250°(begins to sublime at 125°), d 2.75.** Yellow very deliquescent crystals which decompose in moist air to give HCl. Should be kept in a dry box flushed with N₂ in the presence of P₂O₅. Wash with CCl₄ and dry over P₂O₅. The yellow crystals usually contain a few small dirty white pellets among the yellow needles. These should be easily picked out. Upon grinding in a dry box, however, they turn yellow. NbCl₅ has been sublimed and fractionated in an electric furnace. [*Inorg Synth* **7** 163 *1963*; *J Chem Soc* suppl 233 *1949*.]

Nitric acid *[7697-37-2]* **M 63.0, m -42°, b 83°, d₂₅ 1.5027, [Constant boiling acid has composition 68% HNO₃ + 32% H₂O, b 120.5°, d 1.41], pK²⁵ -1.27 (1.19).** Obtained colourless (approx. 92%) by direct distn of fuming HNO₃ under reduced pressure at 40-50° with an air leak at the head of the fractionating column. Stored in a desiccator kept in a refrigerator. Nitrite-free HNO₃ can be obtained by vac distn from urea.

Nitric oxide *[10102-43-9]* **M 30.0, b -151.8°.** Bubbling through 10M NaOH removes NO₂. It can also be freed from NO₂ by passage through a column of Ascarite followed by a column of silica gel held at -197°K. The gas is dried with solid NaOH pellets or by passing through silica gel cooled at -78°, followed by fractional distillation from a liquid N₂ trap. This purification does not eliminate nitrous oxide. Other gas scrubbers sometimes used include one containing conc H₂SO₄ and another containing mercury. It is freed from traces of N₂ by a freeze and thaw method. **TOXIC.**

p-**Nitrobenzenediazonium fluoroborate** *[456-27-9]* **M 236.9.** Crystd from water. **Can be EXPLOSIVE when dry.**

Nitrogen *[7727-37-9]* **M 28.0, b -195.8°.** Cylinder N₂ can be freed from oxygen by passage through Fieser's soln [which comprises 2g sodium anthraquinone-2-sulfonate and 15g sodium hydrosulfite dissolved in 100mL of 20% KOH [Fieser, *J Am Chem Soc* **46** 2639 *1924*] followed by scrubbing with saturated lead acetate soln (to remove any H₂S generated by the Fieser soln), conc H₂SO₄ (to remove moisture), then soda-lime (to remove any H₂SO₄ and CO₂). Alternatively, after passage through Fieser's solution, N₂ can be dried by washing with a soln of the metal ketyl from benzophenone and Na wire in absolute diethyl ether. [If ether vapour in N₂ is undesirable, the ketyl from liquid Na-K alloy under xylene can be used.]

Another method for removing O₂ is to pass the nitrogen through a long tightly packed column of Cu turnings, the surface of which is constantly renewed by scrubbing it with ammonia (sg 0.880) soln. The gas is then passed through a column packed with glass beads moistened with conc H_2SO_4 (to remove ammonia), through a column of packed KOH pellets (to remove H_2SO_4 and to dry the N_2), and finally through a glass trap packed with chemically clean glass wool immersed in liquid N_2. Nitrogen has also been purified by passage over Cu wool at 723°K and Cu(II) oxide [prepared by heating $Cu(NO_3)_2.6H_2O$ at 903°K for 24h] and then into a cold trap at 77°K.

A typical dry purification method consists of a mercury bubbler (as trap), followed by a small column of silver and gold turnings to remove any mercury vapour, towers containing anhydrous $CaSO_4$, dry molecular sieves or $Mg(ClO_4)_2$, a tube filled with fine Cu turnings and heated to 400° by an electric furnace, a tower containing soda-lime, and finally a plug of glass wool as filter. Variations include tubes of silica gel, traps containing activated charcoal cooled in a Dry-ice bath, copper on Kieselguhr heated to 250°, and Cu and Fe filings at 400°.

2-Nitrophenol-4-arsonic acid (4-hydroxy-3-nitrophenylarsonic acid) *[121-19-7]* **M 263.0, $pK_{Est(1)}$~ 4.4 As(O)-(OH)-(O⁻), $pK_{Est(2)}$~ 7.4 (phenolic OH), $pK_{Est(3)}$~ 7.7 (As(O)-2(O⁻)).** Crystd from water.

1-Nitroso-2-naphthol-3,6-disulfonic acid, di-Na salt, hydrate (Nitroso-R-salt) *[525-05-3]* **M 377.3, m >300°, $pK_{Est(1)}$<0 (SO₃⁻), $pK_{Est(2)}$~7 (OH).** Purified by dissolution in aqueous alkali and precipitation by addition of HCl.

Nitrosyl chloride *[2696-92-6]* **M 65.5, b -5.5°.** Fractionally distilled at atmospheric pressure in an all-glass, low temperature still, taking the fraction boiling at -4° and storing it in sealed tubes.

Nitrous oxide *[10024-97-2]* **M 44.0, b -88.5°.** Washed with conc alkaline pyrogallol solution, to remove O_2, CO_2, and NO_2, then dried by passage through columns of P_2O_5 or Drierite, and collected in a dry trap cooled in liquid N_2. Further purified by freeze-pump-thaw and distn cycles under vacuum [Ryan and Freeman *J Phys Chem* **81** 1455 *1977*].

Nuclear Fast Red (1-amino-2,4-dihydroxy-5,10-anthraquinone-3-sulfonic acid Na Salt) *[6409-77-4]* **M 357.3, m >290°(dec), λ_{max} 518nm.** A soln of 5g of the dye in 250mL of warm 50% EtOH was cooled to 15° for 36h, then filtered on a Büchner funnel, washed with EtOH until the washings were colourless, then with 100mL of diethyl ether and dried over P_2O_5. [Kingsley and Robnett *Anal Chem* **33** 552 *1961*.]

Octadecyl isonicotinate see hydrogen ionophore IV, ETH 1778 on p. 430.

Octadecyl trichlorosilane *[112-04-9]* **M 387.9, b 159-162°/13mm, 185-199°/2-3mm, d_4^{30} 0.98.** Purified by fractional distillation. [*J Am Chem Soc* **69** 2916 *1947*.]

Octadecyl trimethylammonium bromide *[1120-02-1]* **M 392.5, m ~250°dec, 230-240°(dec).** Cryst from EtOH or H_2O (sol 1 in 1000parts). Very soluble in Me_2CO. [*J Am Chem Soc* **68** 714 *1946*.]

Octamethyl cyclotetrasiloxane *[556-67-2]* **M 296.6, m 17-19°, 17.58°, 18.5°, b 74°/20mm, 176.4°/760mm, $d_4^{29.3}$ 0.9451, n_D^{30} 1.3968.** Solid has two forms, m 16.30° and 17.65°. Dry over CaH_2 and distil. Further fractionation can be effected by repeated partial freezing and discarding the liquid phase. [*J Am Chem Soc* **76** 399 *1954*, **75** 6313 *1954*.]

Octamethyl trisiloxane *[107-51-7]* **M 236.5, m -80°, b 151.7°/747mm, 153°/760mm.** Distil twice, the middle fraction from the first distillation is again distilled, and the middle fraction of the second distillation is used. [*J Am Chem Soc* **68** 358, 691 *1946*, *J Chem Soc* 1908 *1953*.]

Octaphenyl cyclotetrasiloxane *[546-56-5]* **M 793.2, m 201-202°, 203-204°, b 330-340°/1mm.** Recryst from AcOH, EtOAc, *C_6H_6 or *C_6H_6/EtOH. It forms two stable polymorphs and both forms as well as the mixture melt at 200-201°. There is a metastable form which melts at 187-189°. [*J Am Chem Soc* **67** 2173 *1945*, **69** 488 *1947*.]

Octyl trichlorosilane *[5283-66-9]* **M 247.7, b 96.5°/10mm, 112°/15mm. 119°/28mm, 229°/760mm, d 1.0744, n 1.4453.** Purified by repeated fractionation using a 15-20 theoretical plates glass column packed with glass helices. This can be done more efficiently using a spinning band column. The purity can be checked by analysing for Cl [*ca* 0.5-1g of sample is dissolved in 25mL of MeOH, diluted with H_2O and titrated with standard alkali. [*J Am Chem Soc* **68** 475 *1946*, **80** 1737 *1958*.]

Orange I [tropaeolin 000 Nr1, 4-(4-hydroxy-1-naphthylazo)benzenesulfonic acid sodium salt] *[523-44-4]* **M 350.3, m >260°(dec).** Purified by dissolving in the minimum volume of H_2O, adding, with stirring, a large excess of EtOH. The salt separates as orange needles. It is collected by centrifugation or filtration, washed with absolute EtOH (3 x) and Et_2O (2x) in the same way and dried in a vacuum desiccator over KOH. The free acid can be recrystallised from EtOH. [*Chem Ber* **64** 86 *1931*.] The purity can be checked by titration with titanium chloride [*J Am Chem Soc* **68** 2299 *1946*].

Orange II [tropaeolin 000 Nr2, 4-(2-hydroxy-1-naphthylazo)benzenesulfonic acid sodium salt] *[633-96-5]* **M 350.3.** Purification is as for Orange I. The solubility in H_2O is 40g/L at 25°. [*Helv Chim Acta* **35** 2579 *1952*.] Also purified be extracting with a small volume of water, then crystd by dissolving in boiling water, cooling to *ca* 80°, adding two volumes of EtOH and cooling. When cold, the ppte is filtered off, washed with a little EtOH and dried in air. It can be salted out from aqueous solution with sodium acetate, then repeatedly extracted with EtOH. Meggy and Sims [*J Chem Soc* 2940 *1956*], after crystallising the sodium salt twice from water, dissolved it in cold water (11mL/g) and conc HCl added to ppte the dye acid which was separated by centrifugation, redissolved and again ppted with acid. After washing the ppte three times with 0.5M acid it was dried over NaOH, recrystd twice from absolute EtOH, washed with a little Et_2O, dried over NaOH and stored over conc H_2SO_4 in the dark.

Orange G (1-phenylazo-2-naphthol-6,8-disulfonic acid di-Na salt) *[1936-15-8]* **M 452.4, pK$_{Est}$~ 9.** Recryst from 75% EtOH, dry for 3h at 110° and keep in a vacuum desiccator over H_2SO_4. The free acid crystallises from EtOH or conc HCl in deep red needles with a green reflex. [*J Am Chem Soc* **48** 2483 *1923*, *J Chem Soc* 292 *1938*.]

Orange RO {acid orange 8, 1,8-[bis(4-n-propyl-3-sulfophenyl-1-amino)]anthra-9,10-quinone di-Na salt} *[5850-86-2]* **M 364.4, CI 15575, λmax 490nm.** Salted out three times with sodium acetate, then repeatedly extracted with EtOH.

Osmium tetroxide (osmic acid) *[20816-12-0]* **M 524.2, m 40.6°, b 59.4°/60mm, 71.5°/100mm, 109.3°/400mm, 130°/760mm, d 5.10, pK$_1^{25}$ 7.2, pK$_2^{25}$ 12.2, pK$_3^{25}$ 13.95, pK$_4^{25}$ 14.17 (H_4OsO_6).** It is **VERY TOXIC** and should be manipulated in a very efficient fume cupboard. It attacks the eyes severely (**use also face protection**) and is a good oxidising agent. It is volatile and has a high vapour pressure (11mm) at room temp. It sublimes and dists well below its boiling point. It is sol in *C_6H_6, H_2O (7.24% at 25°), CCl_4 (375% at 25°), EtOH and Et_2O. It is estimated by dissolving a sample in a glass stoppered flask containing 25mL of a solution of KI (previously saturated with CO_2) and acidified with 0.35M HCl. After gentle shaking in the dark for 30min, the solution is diluted to 200mL with distilled H_2O satd with CO_2 and titrated with standard thiosulfate using Starch indicator. This method is not as good as the gravimetric method. Hydrazine hydrochloride (0.1 to 0.3g) is dissolved in 3M HCl (10mL) in a glass stoppered bottle. After warming to 55-65°, a weighed sample of OsO_4 solution is introduced, and the mixture is digested on a water bath for 1h. The mixture is transferred to a weighed glazed crucible and evaporated to dryness on a hot plate. A stream if H_2 is started through the crucible and the crucible is heated over a burner for 20-30 min. The stream of H_2 is continued until the crucible in cooled to room temperature, and then the H_2 is displaced by CO_2 in order to avoid rapid combustion of H_2. Finally the crucible is weighed. [*Handbook of Preparative Inorganic Chemistry (Ed. Brauer)* Vol II 1603 *1965*; *J Am Chem Soc* **60** 1822 *1938*.]

§ Available commercially on a polymer support.

Oxygen *[7782-44-7]* **M 32.00, m -218.4°, b -182.96°, d^{-183} 1.149, d$^{-252.5}$ 1.426.** Purified by passage over finely divided platinum at 673°K and Cu(II) oxide (see under nitrogen) at 973°, then condensed in liquid N$_2$-cooled trap. **HIGHLY EXPLOSIVE in contact with organic matter.**

Palladium (II) acetate *[3375-31-3]* **M 244.5, m 205°dec, pK$_1^{25}$ 1.0, pK$_2^{25}$ 1.2**
(for Pd$_2^{2+}$). Recrystd from CHCl$_3$ as purple crystals. It can be washed with AcOH and H$_2$O and dried in air. Large crystals can be obtained by dissolving in *C$_6$H$_6$ and allowing to evaporate slowly at room temp. It forms green adducts with nitrogen donors, dissolved in KI soln but is insoluble in aqueous saturated NaCl, and NaOAc. Soluble in HCl to form PdCl$_4^{2-}$. [*Chem Ind (London)* 544 *1964*; *J Chem Soc* 658 *1970*.]

Palladium (II) acetyl acetone *[14024-61-4]* **M 304.6.** Recrystd from *C$_6$H$_6$-pet ether and sublimed *in vacuo*. It is soluble in heptane, *C$_6$H$_6$ (1.2% at 20°, 2.2 at 40°), toluene (0.56% at 20°, 1.4% at 40°) and acetylacetone (1.2% at 20°, 0.05% at 40°). [*J Inorg Nucl Chem* 5 295 *1957/8*; *Inorg Synth* 5 105 *1957*.]

Palladium (II) chloride *[7647-10-1]* **M 177.3, m 678-680°.** The anhydrous salt is insoluble in H$_2$O and dissolves in HCl with difficulty. The dihydrate forms red *hygroscopic* crystals that are readily reduced to Pd. Dissolve in conc HCl through which dry Cl$_2$ was bubbled. Filter this solution which contains H$_2$PdCl$_4$ and H$_2$PdCl$_6$ and on evaporation yields a residue of pure PdCl$_2$. [*Handbook of Preparative Inorganic Chemistry (Ed Brauer)* Vol 2 1582 *1965*; *Org Synth* Coll Vol III 685 *1955*.]

Palladium (II) cyanide *[2035-66-7]* **M 158.1.** A yellow solid, wash well with H$_2$O and dry in air. [*Inorg Chem* 2 245 *1946*]. **POISONOUS.**

Palladium (II) trifluoroacetate *[42196-31-6]* **M 332.4, m 210°(dec).** Suspend in trifluoroacetic acid and evaporate on a steam bath a couple of times. The residue is then dried in vacuum (40-80°) to a brown powder. [*J Chem Soc* 3632 *1965*; *J Am Chem Soc* **102** 3572 *1980*.]

Pentafluorophenyl dimethylchlorosilane (Flophemesyl chloride) *[20082-71-7]* **M 260.7, b 89-90°/10mm, d$_4^{30}$ 1.403, n$_D^{30}$ 1.447.** If goes turbid on cooling due to separation of some LiCl, then dissolve in Et$_2$O, filter and fractionate. [*J Chromatogr* **89** 225 *1974*, **132** 548 *1977*,.]

Perchloric acid *[7601-90-3]* **M 100.5, d 1.665, pK^{25} -2.4 to -3.1 (HClO$_4$).** The 72% acid has been purified by double distn from silver oxide under vacuum: this frees the acid from metal contamination. Anhydrous acid can be obtained by adding gradually 400-500mL of oleum (20% fuming H$_2$SO$_4$) to 100-120mL of 72% HClO$_4$ in a reaction flask cooled in an ice-bath. The pressure is reduced to 1mm (or less), with the reaction mixture at 20-25°. The temperature is gradually raised during 2h to 85°, the distillate being collected in a receiver cooled in Dry-ice. For further details of the distillation apparatus [see Smith *J Am Chem Soc* **75** 184 *1953*]. **HIGHLY EXPLOSIVE, a strong protective screen should be used at all times.**

Phenylarsonic acid (benzenearsonic acid) *[98-05-5]* **M 202.2, m 155-158°(dec), pK$_1^{25}$ 3.65, pK$_2^{25}$ 8.77.** Crystd from H$_2$O (3mL/g) between 90° and 0°.

Phenylboric acid (benzeneboronic acid) *[98-80-6]* **M 121.9, m ~43°, 215-216° (anhydride), 217-220°, pK$_1^{25}$ 8.83.** It recrystallises from H$_2$O, but can convert spontaneously to benzeneboronic anhydride or phenylboroxide on standing in dry air. Possible impurity is dibenzeneborinic acid which can be removed by washing with pet ether. Heating in an oven at 110°/760mm 1h converts it to the *anhydride* m 214-216°. Its solubility in H$_2$O is 1.1% at 0° and 2.5% at 25° and in EtOH it is 10% (w/v). [Gilman and Moore *J Am Chem Soc* **80** 3609 *1958*.] If the acid is required, not the anhydride, the acid (from recrystallisation in H$_2$O) is dried in a slow stream of air saturated with H$_2$O. The anhydride is converted to the acid by recrystallisation from H$_2$O. The acid gradually dehydrates to the anhydride if left in air at room temperature with 30-40% relative humidity. The melting point is usually that of the anhydride because the acid dehydrates before it melts [Washburn et al. *Org Synth* Coll Vol IV 68 *1963*].

Phenyl dimethyl chlorosilane (DMPSCl, chlorodimethylphenylsilane) *[768-33-2]* **M 170.7, b 79°/15mm, 85-87°/32mm, 196°/760mm, d 1.017, n 1.509.** Fractionate through a 1.5 x 18 inch column packed with stainless steel helices; or a spinning band column. [*J Am Chem Soc* **74** 386 *1952*; **70** 1115 *1948*; *J Chem Soc* 494 *1953*.] Used for standardising MeLi or MeMgBr which form Me₃PhSi, estimated by GC [Maienthal et al. *J Am Chem Soc* **76** 6392 *1954*; House and Respess *J Organomet Chem* **4** 95 *1965*.] **TOXIC** and **MOISTURE SENSITIVE**.

1,2-Phenylenephosphorochloridate (2-chloro-1,3,2-benzodioxaphosphole-2-oxide) *[1499-17-8]* **M 190.5, m 52°, 58-59°, 59-61°, b 80-81°/1-2mm, 118°/10mm, 122°/12mm, 125°/16mm, 155°/33mm.** Distil in a vacuum, sets to a colourless solid. It is soluble in pet ether, *benzene and slightly soluble in Et₂O. [*J Chem Soc (C)* 2092 *1970*; *Justus Liebigs Ann Chem* **454** 109 *1927*.]

Phenylmercuric hydroxide *[100-57-2]* **M 294.7, m 195-203°.** Crystd from dilute aqueous NaOH.

Phenylmercuric nitrate *[8003-05-2]* **M 634.4, m 178-188°.** Crystd from water.

Phenylphosphinic acid [benzenephosphinic acid, PhPH(O)(OH)] *[1779-48-2]* **M 142.1, m 70°, 71°, 83-85°, 86°, pK₁²⁵ 1.75.** Cryst from H₂O (sol 7.7% at 25°). Purified by placing the solid in a flask covered with dry Et₂O, and allowed to stand for 1 day with intermittent shaking. Et₂O was decanted off and the process repeated. After filtration, excess Et₂O was removed in vacuum. [*Justus Liebigs Ann Chem* **181** 265 *1876*; *Anal Chem* **29** 109 *1957*; NMR: *J Am Chem Soc* **78** 5715 *1956*.]

Phenylphosphonic acid *[1571-33-1]* **M 158.1, m 164.5-166°, pK₂²⁵ 7.43 (7.07).** Best recryst from H₂O by concentrating an aqueous soln to a small volume and allowing to crystallise. Wash the crystals with ice cold H₂O and dry in a vacuum desiccator over H₂SO₄. [*J Am Chem Soc* **78** 1045 *1954*.] pK²⁵ values in H₂O are 7.07, and in 50% EtOH 8.26. [*J Am Chem Soc* **75** 2209 *1953*.] [IR: *Anal Chem* **23** 853 *1951*.]

Phenylphosphonic dichloride (P,P-dichlorophenyl phosphine oxide) *[824-72-6]* **M 195.0, b 83-84°/1mm, 135-136°/23mm, d₄³⁰ 1.977, n_D³⁰ 1.5578.** Fractionally distilled using a spinning band column. [*J Am Chem Soc* **76** 1045 *1954*; NMR: *J Am Chem Soc* **78** 3557, 5715 *1956*; IR: *Anal Chem* **23** 853 *1951*.]

Phenylphosphonous acid [PhP(OH)₂] *[121-70-0]* **M 141.1, m 71°, pK_Est <0, pK¹⁷ 2.1.** Crystd from hot H₂O.

Phenylphosphonous dichloride (P,P-dichloro phenyl phosphine) *[644-97-3]* **M 179.0, b 68-70°/1mm, 224-226°/atm, d₄³⁰ 1.9317, n_D³⁵ 1.5962.** Vacuum distilled by fractionating through a 20cm column packed with glass helices (better use a spinning band column) [*J Am Chem Soc* **73** 755 *1951*; NMR: *J Am Chem Soc* **78** 3557 *1956*; IR: *Anal Chem* **23** 853 *1951*]. It forms a yellow *Ni complex:* Ni(C₆H₅Cl₂P)₄ (m 91-92°, from H₂O)[*J Am Chem Soc* **79** 3681 *1957*] and a yellow complex with molybdenum carbonyl: Mo(CO)₃.(C₆H₅Cl₂P)₃ (m 106-110°dec) [*J Chem Soc* 2323 *1959*].

Phenyl phosphoro chloridate (diphenyl phosphoryl chloride) *[2524-64-3]* **M 268.6, b 141°/1mm, 194°/13mm, 314-316/272mm, d₄³⁰ 1.2960, n_D³⁵ 1.5490.** Fractionally distd in a good vac, better use a spinning band column. [*J Am Chem Soc* **81** 3023 *1959*; IR: *J Chem Soc* 475, 481 *1952*.]

Phenyl phosphoryl dichloride *[770-12-7]* **M 211.0, m -1°, b 103-104°/2mm, 110-111°/10mm, 130-134°/21mm, 241-243°/atm, d₄³⁰ 1.4160, n_D³⁰ 1.5216.** Fractionally distilled under as good a vacuum as possible using an efficient fractionating column or a spinning band column. It should be redistilled if the IR is not very good [IR: *J Chem Soc* 475, 481 *1952*; *J Am Chem Soc* **60** 750 *1938*, **80** 727 *1958*]. **HARMFUL VAPOURS**.

Phenylthio trimethylsilane (trimethyl phenylthio silane) *[4551-15-9]* **M 182.4, b 95-99°/12mm, d₄³⁰ 0.97.** Purification is as for phenyl trimethyl silylmethyl sulfide on p. 450.

Phenyl trimethoxylsilane (trimethoxysilyl benzene) *[2996-92-1]* **M 198.3, b 103°/20mm, 130.5-131°/45mm, d_4^{35} 1.022, n_D^{35} 1.4698.** Fractionate through an efficient column but note that it forms an azeotrope with MeOH which is a likely impurity. [*J Am Chem Soc* **75** 2712 *1953*; *J Gen Chem USSR (Engl Transl)* **25** 1079 *1955*.]

Phenyl trimethylsilane (trimethylphenylsilane) *[768-32-1]* **M 150.3, b 67.3°/20mm, 98-99°/80mm, 170.6°/738mm, d_4^{25} 0.8646.** See trimethylphenylsilane on p. 489.

Phenyl trimethylsilylmethyl sulfide [(phenylthiomethyl)trimethylsilane] *[17873-08-4]* **M 196.4, b 48°/0.04mm, 113-115°/12mm, 158.5°/52mm, d_4^{30} 0.9671, n_D^{30} 1.5380.** If the sample is suspect then add H_2O, wash with 10% aqueous NaOH, H_2O again, dry (anhydrous $CaCl_2$) and fractionally distil through a 2ft column packed with glass helices. [*J Am Chem Soc* **76** 3713 *1954*.]

Phosgene *[75-44-5]* **M 98.9, b 8.2°/756mm.** Dried with Linde 4A molecular sieves, degassed and distilled under vacuum. This should be done in a closed system such as a vacuum line. **HIGHLY TOXIC, should not be inhaled. If it is inhaled operator should lie still and made to breath ammonia vapour which reacts with phosgene to give urea.**

Phosphine *[7803-51-2]* **M 34.0, m -133°, b -87.7°, pK -14, pK_b 28**. Best purified in a gas line (in a vacuum) in an efficient fume cupboard. It is spontaneously flammable, has a strong odour of decayed fish and is **POISONOUS**. The gas is distd through solid KOH towers (two), through a Dry ice-acetone trap (-78°, to remove H_2O, and P_2H_4 which causes spontaneous ignition with O_2), then through two liquid N_2 traps (-196°), followed by distn into a -126° trap (Dry ice-methylcyclohexane slush), allowed to warm in the gas line and then seal in ampoules preferably under N_2. IR: v 2327 (m), 1121 (m) and 900 (m) cm^{-1}. [Klement in *Handbook of Preparative Inorganic Chemistry (Ed. Brauer)* Vol I, pp. 525-530 *1963*; Gokhale and Jolly *Inorg Synth* **9** 56 *1967*.] PH_3 has also been absorbed into a soln of cuprous chloride in hydrochloric acid (when $CuCl.PH_3$ is formed). PH_3 gas is released when the soln is heated and the gas is purified by passage through KOH pellets and over then P_2O_5. The solubility is 0.26mL/1 mL of H_2O at 20° and a crystalline hydrate is formed on releasing the pressure on an aq soln.

Phosphonitrilic chloride (tetramer) *[1832-07-1]* **M 463.9.** Purified by zone melting, then crystd from pet ether (b 40-60°) or *n*-hexane. [van der Huizen et al. *J Chem Soc, Dalton Trans* 1317 *1986*.]

Phosphonitrilic chloride (trimer) (hexachlorocyclotriphosphazine) *[940-71-6]* **M 347.7, m 112.8°, 113-114°.** Purified by zone melting, by crystallisation from pet.ether, *n*-hexane or *benzene, and by sublimation. [van der Huizen et al. *J Chem Soc, Dalton Trans* 1311 *1986*; Meirovitch et al. *J Phys Chem* **88** 1522 *1984*; Alcock et al. *J Am Chem Soc* **106** 5561 *1984*; Winter and van de Grampel *J Chem Soc, Dalton Trans* 1269 *1986*.]

Phosphoric acid *[7664-38-2]* **M 98.0, m 42.3°, pK_1^{25} 2.15, pK_2^{25} 7.21, pK_3^{25} 12.33.** Pyrophosphate can be removed from phosphoric acid by diluting with distilled H_2O and refluxing overnight. By cooling to 11° and seeding with crystals obtained by cooling a few millilitres in a Dry-ice/acetone bath, 85% orthophosphoric acid crystallises as $H_3PO_4.H_2O$. The crystals are separated using a sintered glass filter. It has pKa^{25} values of 2.15, 7.20 and 12.37 in H_2O.

Phosphorus (red) *[7723-14-0]* **M 31.0, m 590°/43atm, ignites at 200°, d 2.34.** Boiled for 15min with distilled H_2O, allowed to settle and washed several times with boiling H_2O. Transferred to a Büchner funnel, washed with hot H_2O until the washings are neutral, then dried at 100° and stored in a desiccator.

Phosphorus (white) *[7723-14-0]* **M 31.0, m 590, d 1.82.** Purified by melting under dilute $H_2SO_4^-$ dichromate (possible **carcinogen**) mixture and allowed to stand for several days in the dark at room temperature. It remains liquid, and the initial milky appearance due to insoluble, oxidisable material gradually disappears. The phosporus can then be distilled under vacuum in the dark [Holmes *Trans Faraday Soc* **58** 1916

1962]. Other methods include extraction with dry CS_2 followed by evaporation of the solvent, or washing with 6M HNO_3, then H_2O, and drying under vacuum. **POISONOUS.**

Phosphorus oxychloride *[10025-87-3]* **M 153.3, b 105.5°, n 1.461, d 1.675.** Distilled under reduced pressure to separate from the bulk of the HCl and the phosphoric acid, the middle fraction being distilled into ampoules containing a little purified mercury. These ampoules are sealed and stored in the dark for 4-6 weeks with occasional shaking to facilitate reaction of any free chloride with the mercury. The $POCl_3$ is then again fractionally distd and stored in sealed ampoules in the dark until used [Herber *J Am Chem Soc* **82** 792 *1960*]. Lewis and Sowerby [*J Chem Soc* 336 *1957*] refluxed their distilled $POCl_3$ with Na wire for 4h, then removed the Na and again distilled. *Use Na only with almost pure POCl₃ to avoid explosions.* **HARMFUL VAPOURS.**

Phosphorus pentabromide *[7789-69-7]* **M 430.6, m <100°, b 106°(dec).** Dissolved in pure nitrobenzene at 60°, filtering off any insoluble residue on to sintered glass, then crystallised by cooling. Washed with dry Et_2O and removed the ether in a current of dry N_2. (All manipulations should be performed in a dry-box.) [Harris and Payne *J Chem Soc* 3732 *1958*]. Fumes in moist air because of hydrolysis. **HARMFUL VAPOURS.**

Phosphorus pentachloride *[10026-13-8]* **M 208.2, m 179-180°(sublimes).** Sublimed at 160-170° in an atmosphere of chlorine. The excess chlorine was then displaced by dry N_2 gas. All subsequent manipulations were performed in a dry-box [Downs and Johnson *J Am Chem Soc* **77** 2098 *1955*]. Fumes in moist air. **HARMFUL VAPOURS.**

Phosphorus pentasulfide *[1314-80-3]* **M 444.5, m 277-283°.** Purified by extraction and crystallisation with CS_2, using a Soxhlet extractor. Liberates H_2S in moist air. **HARMFUL VAPOURS.**

Phosphorus pentoxide *[1314-56-3]* **M 141.9, m 562°, b 605°.** Sublimed at 250° under vacuum into glass ampoules. Fumes in moist air and reacts violently with water. **HARMFUL VAPOURS and attacks skin.**

Phosphorus sesquisulfide P_4S_3 *[1314-85-8]* **M 220.1, m 172°.** Extracted with CS_2, filtered and evapd to dryness. Placed in H_2O, and steam was passed through for an hour. The H_2O was then removed, the solid was dried, followed by crystallisation from CS_2 [Rogers and Gross *J Am Chem Soc* **74** 5294 *1952*].

Phosphorus sulfochloride (phosphorus thiochloride) *[3982-91-0]* **M 169.4, m -35°, b 122-124°, 125°(corr), d_4^{30} 1.64, n_D^{30} 1.556.** Possible impurities are PCl_5, H_3PO_4, HCl and $AlCl_3$. Gently mix with H_2O to avoid a heavy emulsion, the product decolorises immediately and settles to the bottom layer. **HARMFUL VAPOURS.**

Phosphorus tribromide *[7789-60-8]* **M 270.7, m -41.5°, b 168-170°/725mm, 171-173°/atm, 172.9°/760mm(corr), d_4^{30} 2.852.** It is decomposed by moisture, should be kept dry and is *corrosive*. Purified by distillation through an efficient fractionating column (see Whitmore and Lux *J Am Chem Soc* **54** 3451] in a slow stream of dry N_2, i.e. under strictly dry conditions. [*Inorg Synth* **2** 147 *1946*; *Org Synth* Col Vol II 358 *1943*.] Dissolve in CCl_4, dry over $CaCl_2$, filter and distil. [*Handbook of Preparative Inorganic Chemistry (Ed. Brauer)* vol I 532 *1963*.] Store in sealed ampoules under N_2 and kept away from light. **HARMFUL VAPOURS.**

Phosphorus trichloride *[7719-12-2]* **M 137.3, b 76°, n 1.515, d 1.575.** Heated under reflux to expel dissolved HCl, then distilled. It has been further purified by vacuum fractionation several times through a -45° trap into a receiver at -78°. **HARMFUL VAPOURS.**

Phosphorus triiodide *[13455-01-1]* **M 411.7, m 61°.** Decomposes in moist air and must be kept in a desiccator over $CaCl_2$. It is crystallised from sulfur-free CS_2 otherwise the **m** decreases to *ca* 55°. It is best prepared freshly. [*J Am Chem Soc* **49** 307 *1927*; *Handbook of Preparative Inorganic Chemistry (Ed. Brauer)* vol I 541 *1963*.] **HARMFUL VAPOURS.**

12-Phosphotungstic acid *[12501-23-4]* **M 2880.2, m ~96⁰.** A few drops of conc HNO₃ were added to 100g of phosphotungstic acid dissolved in 75mL of water, in a separating funnel, and the soln was extracted with diethyl ether. The lowest of the three layers, which contained a phosphotungstic acid-ether complex, was separated, washed several times with 2M HCl, then with water and again extracted with ether. Evaporation of the ether, under vacuum with mild heating on a water bath gave crystals which were dried under vacuum and ground [Matijevic and Kerker, *J Am Chem Soc* **81** 1307 *1959*].

Phthalocyanine *[574-93-6]* **M 514.6.** Purified by sublimation (two to three times) in an argon flow at 300-400Pa. Similarly for the Cu(II), Ni(II), Pb(II), VO(II) and Zn(II) phthalocyanine complexes.

Platinum (II) acetylacetonate *[15170-57-7]* **M 393.3, m 249-252⁰.** Recrystd from *C₆H₆ as yellow crystals and dried in air or in a vacuum desiccator. [*Chem Ber* **34** 2584 *1901*.]

Platinum (II) chloride *[10025-65-7]* **M 266.0, d 5.87.** It is purified by heating at 450⁰ in a stream of Cl₂ for 2h. Some sublimation occurs because the PtCl₂ sublimes completely at 560⁰ as red (almost black) needles. This sublimate can be combined to the bulk chloride and while still at *ca* 450⁰ it should be transferred to a container and cooled in a desiccator. A probable impurity is PtCl₄. To test for this add a few drops of H₂O (in which PtCl₄ is soluble) to the salt, filter and add an equal volume of saturated aqueous NH₄Cl to the filtrate. If no ppte is formed within 1 min then the product is pure. If a ppte appears then the whole material should be washed with small volumes of H₂O until the soluble PtCl₄ is removed. The purified PtCl₂ is partly dried by suction and then dried in a vacuum desiccator over P₂O₅. It is insoluble in H₂O but soluble in HCl to form chloroplatinic acid (H₂PtCl₄) by disproportionation. [*Inorg Synth* **6** 209 *1960*.]

Polystyrenesulfonic acid sodium salt **(-CH₂CH(C₆H₄SO₃Na)-)** *[25704-18-1]*. Purified by repeated pptn of the sodium salt from aqueous soln by MeOH, with subsequent conversion to the free acid by passage through an Amberlite IR-120 ion-exchange resin. [Kotin and Nagasawa *J Am Chem Soc* **83** 1026 *1961*.] Recrystd from EtOH. Also purified by passage through cation and anion exchange resins in series (Rexyn 101 cation exchange resin and Rexyn 203 anion exchange resin), then titrated with NaOH to pH 7. The sodium form of polystyrenesulfonic acid ppted by addition of 2-propanol. Dried in a vac oven at 80⁰ for 24h, finally increasing to 120⁰ prior to use. [Kowblansky and Ander *J Phys Chem* **80** 297 *1976*.]

Pontacyl Carmine 2G **(Acid Red 1, Amido Naphthol Red G, Azophloxine, 1-acetamido-8-hydroxy-7-phenylazonaphthalene-3,7-disulfonic acid di-Na salt)** *[3734-67-6]* **M 510.4, CI 18050, λmax 532nm.** Salted out three times with sodium acetate, then repeatedly extracted with EtOH. See Solochrome Violet R on p. 352 in Chapter 4. [McGrew and Schneider *J Am Chem Soc* **72** 2547 *1950*.]

Pontacyl Light Yellow GX **[Acid Yellow 17, 1-(2,5-dichloro-4-sulfophenyl]-3-methyl-4-(4-sulfophenylazo)-5-hydroxypyrazole di-Na Salt]** *[6359-98-4]* **M 551.3, CI 18965, λmax 400nm.** Purification as for Pontacyl Carmine 2G above.

Potassium (metal) *[7440-09-7]* **M 39.1, m 62.3⁰, d 0.89.** Oil was removed from the surface of the metal by immersion in *n*-hexane and pure Et₂O for long periods. The surface oxide was next removed by scraping under ether, and the potassium was melted under vacuum. It was then allowed to flow through metal constrictions into tubes that could be sealed, followed by distillation under vacuum in the absence of mercury vapour (see Sodium). **EXPLOSIVE IN WATER.**

Potassium acetate *[127-08-2]* **M 98.2, m 292⁰, d 1.57, pK²⁵ 16 (for aquo K⁺).** Crystd three times from water-ethanol (1:1) dried to constant weight in a vacuum oven, or crystd from anhydrous acetic acid and pumped dry under vacuum for 30h at 100⁰.

Potassium 4-aminobenzoate *[138-84-1]* **M 175.2.** Crystd from EtOH.

Potassium antimonyltartrate (H₂O) *[28300-74-5]* **M 333.9, [α]_D +141⁰ (c 2, H₂O).** Crystd from water (3mL/g) between 100⁰ and 0⁰. Dried at 100⁰.

Potassium benzoate *[582-25-2]* **M 160.2.** Crystd from water (1mL/g) between 100° and 0°.

Potassium bicarbonate *[298-14-6]* **M 100.1.** Crystd from water at 65-70° (1.25mL/g) by filtering, then cooling to 15°. During all operations, CO_2 is passed through the stirred mixture. The crystals, sucked dry at the pump, are washed with distilled water, dried in air and then over H_2SO_4 in an atmosphere of CO_2.

Potassium biiodate *[13455-24-8]* **M 389.9.** Crystd three times from hot water (3mL/g), stirred continuously during each cooling. After drying at 100° for several hours, the crystals are suitable for use in volumetric analysis.

Potassium bisulfate *[7646-93-7]* **M 136.2, m 214°.** Crystd from H_2O(1mL/g) between 100° and 0°.

Potassium borohydride *[13762-51-1]* **M 53.9, m ~500°(dec).** Crystd from liquid ammonia.

Potassium bromate *[7758-01-2]* **M 167.0, m 350°(dec at 370°), d 3.27.** Crystd from distilled H_2O(2mL/g) between 100° and 0°. To remove bromide contamination, a 5% soln in distilled H_2O, cooled to 10°, has been bubbled with gaseous chlorine for 2h, then filtered and extracted with reagent grade CCl_4 until colourless and odourless. After evaporating the aqueous phase to about half its volume, it was cooled again slowly to about 10°. The crystalline $KBrO_3$ was separated, washed with 95% EtOH and vacuum dried [Boyd, Cobble and Wexler *J Am Chem Soc* **74** 237 *1952*]. Another way to remove Br^- ions was by stirring several times in MeOH and then dried at 150° [Field and Boyd *J Phys Chem* **89** 3767 *1985*].

Potassium bromide *[7758-02-3]* **M 119.0, m 734°, d 2.75.** Crystd from distilled water (1mL/g) between 100° and 0°. Washed with 95% EtOH, followed by Et_2O. Dried in air, then heated at 115° for 1h, pulverised and heated in a vacuum oven at 130° for 4h. Has also been crystd from aqueous 30% EtOH, or EtOH, and dried over P_2O_5 under vacuum before heating in an oven.

Potassium *tert*-butoxide *[865-47-4]* **M 112.2.** It sublimes at 220°/1 Torr. Last traces of *tert*-BuOH are removed by heating at 150-160°/2mm for 1h. It is best prepared fresh; likely impurities are *tert*-BuOH, KOH and K_2CO_3 depending on exposure to air. Its solubility at 25°-26° in hexane, toluene, Et_2O, and THF is 0.27%, 2.27%, 4.34% and 25.0% repectively. [*J Am Chem Soc* **78** 5938, 4364 *1956*.]

Potassium carbonate *[584-08-7]* **M 138.2, m 898°, d 2.3.** Crystd from water between 100° and 0°. After 2 recrystns tech grade had B, Li and Fe at 1.0, 0.04 and 0.01 ppm resp.

Potassium chlorate *[3811-04-9]* **M 122.6, m 368°.** Crystd from water (1.8mL/g) between 100° and 0°, and the crystals are filtered onto sintered glass.

Potassium chloride *[7447-40-7]* **M 74.6, m 771°, d 1.98.** Dissolved in conductivity water, filtered, and saturated with chlorine (generated from conc HCl and $KMnO_4$). Excess chlorine was boiled off, and the KCl was ppted by HCl (generated by dropping conc HCl into conc H_2SO_4). The ppte was washed with water, dissolved in conductivity water at 90-95°, and crystd by cooling to about -5°. The crystals were drained at the centrifuge, dried in a vacuum desiccator at room temperature, then fused in a platinum dish under N_2, cooled and stored in desiccator. Potassium chloride has also been sublimed in a stream of prepurified N_2 gas and collected by electrostatic discharge [Craig and McIntosh *Can J Chem* **30** 448 *1952*].

Potassium chromate *[7789-00-6]* **M 194.2, m 975°, d 2.72, pK_1^{25} 0.74, pK_2^{25} 6.49 (for H_2CrO_4).** Crystd from conductivity water (0.6g/mL at 20°), and dried between 135° and 170°.

Potassium cobalticyanide *[13963-58-1]* **M 332.4, m dec on heating, d 1.91.** Crystd from water to remove traces of HCN.

Potassium cyanate *[590-28-3]* **M 81.1, d 2.05, pK^{25} 3.46 (for HCNO).** Common impurities include ammonia and bicarbonate ion (from hydrolysis). Purified by preparing a saturated aqueous solution at

50°, neutralising with acetic acid, filtering, adding two volumes of EtOH and keeping for 3-4h in an ice bath. (More EtOH can lead to co-precipitation of $KHCO_3$.) Filtered, washed with EtOH and dried rapidly in a vacuum desiccator (P_2O_5). The process is repeated [Vanderzee and Meyers *J Chem Soc* 65 153 *1961*].

Potassium cyanide *[151-50-8]* **M 65.1, m 634°, d 1.52.** A saturated solution in H_2O-ethanol (1:3) at 60° was filtered and cooled to room temperature. Absolute EtOH was added, with stirring, until crystallisation ceased. The solution was again allowed to cool to room temperature (during 2-3h) then the crystals were filtered off, washed with absolute EtOH, and dried, first at 70-80° for 2-3h, then at 105° for 2h [Brown, Adisesh and Taylor *J Phys Chem* 66 2426 *1962*]. Also purified by vacuum melting and zone refining. **HIGHLY POISONOUS.**

Potassium dichromate *[7778-50-9]* **M 294.2, m 398°(dec), d 2.68.** Crystd from water (1mL/g) between 100° and 0° and dried under vacuum at 156°. (Possible **CARCINOGEN**).

Potassium dihydrogen citrate *[866-83-1]* **M 230.2.** Crystd from water. Dried at 80°, or in a vacuum desiccator over Sicapent.

Potassium dihydrogen phosphate *[7778-77-0]* **M 136.1.** Dissolved in boiling distilled water (2mL/g), kept on a boiling water-bath for several hours, then filtered through paper pulp to remove any turbidity. Cooled rapidly with constant stirring, and the crystals were separated on to hardened filter paper, using suction, washed twice with ice-cold water, once with 50% EtOH, and dried at 105°. Alternative crystns are from water, then 50% EtOH, and again water, or from conc aqueous solution by addition of EtOH. Freed from traces of Cu by extracting its aqueous solution with diphenylthiocarbazone in CCl_4, followed by repeated extraction with CCl_4 to remove traces of diphenylthiocarbazone.

Potassium dithionate *[13455-20-4]* **M 238.3, $pK_{Est(1)}$ -3.4, pK_2^{25} 0.49 (for dithionic acid).** Crystd from water (1.5mL/g) between 100° and 0°.

Potassium ethylxanthate *[140-89-6]* **M 160.3, m > 215°(dec).** Crystd from absolute EtOH, ligroin-ethanol or acetone by addition of Et_2O. Washed with ether, then dried in a desiccator.

Potassium ferricyanide *[13746-66-2]* **M 329.3, pK^{25} <1 (for ferricyanide).** Crystd repeatedly from hot water (1.3mL/g). Dried under vacuum in a desiccator.

Potassium ferrocyanide ($3H_2O$) *[14459-95-1]* **M 422.4, pK_3^{25} 2.57, pK_4^{25} 4.35 (for ferrocyanide)** Crystd repeatedly from distilled water, never heating above 60°. Prepared anhydrous by drying at 110° over P_2O_5 in a vacuum desiccator. To obtain the trihydrate, it is necessary to equilibrate in a desiccator over saturated aqueous soln of sucrose and NaCl. Can also be ppted from a saturated solution at 0° by adding an equal volume of cold 95% EtOH, standing for several hours, then centrifuging and washing with cold 95% EtOH. Finally sucked air dry with a water-pump. The anhydrous salt can be obtained by drying in a platinum boat at 90° in a slow stream of N_2 [Loftfield and Swift *J Am Chem Soc* 60 3083 *1938*].

Potassium fluoroborate *[14075-53-7]* **M 125.9, m 530°.** See potassium tetrafluoroborate on p. 458.

Potassium fluorosilicate *[16871-90-2]* **M 220.3, d 2.3, pK 1.92 (for H_2SiF_6).** Crystd several times from conductivity water (100mL/g) between 100° and 0°.

Potassium hexachloroiridate (IV) *[16920-56-2]* **M 483.1.** Crystd from hot aqueous solution containing a few drops of HNO_3.

Potassium hexachloroosmate (IV) *[16871-60-6]* **M 481.1.** Crystd from hot dilute aqueous HCl.

Potassium hexachloroplatinate (IV) *[16921-30-5]* **M 486.0, m 250°(dec).** Crystd from water (20mL/g) between 100° and 0°.

Potassium hexacyanochromate (III) (3H$_2$O) *[13601-11-1]* **M 418.5.** Crystd from water.

Potassium hexafluorophosphate *[17084-13-8]* **M 184.1, pK$_1^{25}$~ 0.5, pK$_2^{25}$5.12 (for fluorophosphoric acid H$_2$PO$_3$F).** Crystd from alkaline aqueous solution, using polyethylene vessels, or from 95% EtOH, and dried in a vacuum desiccator over KOH.

Potassium hexafluorozirconate (K$_2$ZrF$_6$) *[16923-95-8]* **M 283.4, d 3.48.** Recrystd from hot water (solubility is 0.78% at 2° and 25% at 100°).

Potassium hydrogen fluoride *[7789-29-9]* **M 78.1.** Crystd from water.

Potassium hydrogen D-glucarate *[18404-47-2]* **M 248.2, m 188°(dec).** Crystd from water.

Potassium hydrogen malate *[4675-64-3]* **M 172.2.** A saturated aqueous solution at 60° was decolorised with activated charcoal, and filtered. The filtrate was cooled in water-ice bath and the salt was ppted by addition of EtOH. After being crystallised five times from ethanol-water mixtures, it was dried overnight at 130° in air [Eden and Bates *J Res Nat Bur Stand* **62** 161 *1959*].

Potassium hydrogen oxalate (H$_2$O) *[127-95-7]* **M 137.1.** Crystd from water by dissolving 20g in 100mL water at 60° containing 4g of potassium oxalate, filtering and allowing to cool to 25°. The crystals, after washing three or four times with water, are allowed to dry in air.

Potassium hydrogen phthalate *[877-24-7]* **M 204.2.** Crystd first from a dilute aqueous solution of K$_2$CO$_3$, then H$_2$O(3mL/g) between 100° and 0°. Before being used as a standard in volumetric analysis, analytical grade potassium hydrogen phthalate should be dried at 120° for 2h, then allowed to cool in a desiccator.

Potassium hydrogen *d*-tartrate *[868-14-4]* **M 188.2, [α]$_{546}^{20}$ +37.5° (c 10, M NaOH).** Crystd from water (17mL/g) between 100° and 0°. Dried at 110°.

Potassium hydroxide (solution) *[1310-58-3]* **M 56.1, pK^{25} 16 (for aquo K$^+$).** Its carbonate content can be reduced by rinsing KOH sticks rapidly with water prior to dissolving them in boiled out distilled water. Alternatively, a slight excess of saturated BaCl$_2$ or Ba(OH)$_2$ can be added to the soln which, after shaking well, is left so that the BaCO$_3$ ppte will separate out. Davies and Nancollas [Nature **165** 237 *1950*] rendered KOH solutions carbonate free by ion exchange using a column of Amberlite IR-100 in the OH$^-$ form.

Potassium iodate *[7758-05-6]* **M 214.0, pK^{25} 0.80 (for HIO$_3$).** Crystd twice from distilled water (3mL/g) between 100° and 0°, dried for 2h at 140° and cooled in a desiccator. Analytical reagent grade material dried in this way is suitable for use as an analytical standard.

Potassium iodide *[7681-11-0]* **M 166.0, pK^{25} -8.56 (for HI).** Crystd from distilled water (0.5mL/g) by filtering the near-boiling soln and cooling. To minimise oxidation to iodine, the crystn can be carried out under N$_2$ and the salt is dried under vacuum over P$_2$O$_5$ at 70-100°. Before drying, the crystals can be washed with EtOH or with acetone followed by pet ether. Has also been recrystallised from water/ethanol. After 2 recrystns ACS/USP grade had Li and Sb at <0.02 and <0.01 ppm resp.

Potassium ionophore I (valinomycin) *[2001-95-8]* **M 111.3, m 186-187°, 190°, [α]$_D^{20}$ +31.0° (c 1.6, *C$_6$H$_6$).** See valinomycin on p. 573 in Chapter 6.

Potassium isoamyl xanthate *[61792-26-5]* **M 202.4, pK 1.82 (pK^0 2.8 free acid).** Crystd twice from acetone-diethyl ether. Dried in a desiccator for two days and stored under refrigeration.

Potassium laurate *[10124-65-9]* **M 338.4.** Recrystd three times from EtOH [Neto and Helene *J Phys Chem* **91** 1466 *1987*].

Potassium nickel sulfate (6H₂O) *[13842-46-1]* **M 437.1.** Crystd from H_2O(1.7mL/g) between 75° and 0°.

Potassium nitrate *[7757-79-1]* **M 101.1, m 334°.** Crystd from hot H_2O (0.5mL/g) by cooling (*cf* KNO_2 below). Dried for 12h under vacuum at 70°. After 2 recrystns tech grade had < 0.001 ppm of metals.

Potassium nitrite *[7758-09-0]* **M 85.1, m 350°(dec), pK²⁰ 3.20 (for HNO₂).** A saturated solution at 0° can be warmed and partially evaporated under vacuum, the crystals so obtained being filtered from the warm solution. (This procedure is designed to reduce the level of nitrate impurity and is based on the effects of temperature on solubility. The solubility of KNO_3 in water is 13g/100mL at 0°, 247g/100mL at 100°; for KNO_2 the corresponding figures are 280g/100mL and 413g/100mL.)

Potassium nitrosodisulfonate (Fremy's Salt) *[14293-70-0]* **M 268.3.** Yellow needles (dimeric) which dissolve in H_2O to give the violet monomeric free radical. Purified by dissolving (~12g) in 2M KOH (600mL) at 45°, filtering the blue soln and keeping it in a refrigerator overnight. The golden yellow crystals (10g) are filtd off, washed with MeOH (3x), then Et_2O and stored in a glass container in a vac over KOH. It is stable indefinitely when dry. [Cram and Reeves *J Org Chem* **80** 3094 *1958*; Schenk *Handbook of Preparative Inorganic Chemistry (Ed. Brauer)* Vol I p. 505 *1963*.]

Potassium nonafluorobutane sulfonate *[29420-49-3]* **M 338.2.** Wash with H_2O and dry in vacuum. The K salt when distilled with 100% H_2SO_4 gives the free acid which can be distilled (**b** 105°/22mm, 210-212°/760mm) and then converted to the K salt. [*J Chem Soc* 2640 *1957*.]

Potassium oleate *[143-18-0]* **M 320.6.** Crystd from EtOH (1mL/g).

Potassium osmate (VI) dihydrate *[19718-36-6]* **M 368.4.** *Hygroscopic* **POISONOUS** crystals which are soluble in H_2O but insol in EtOH and Et_2O. It decomposes slowly in H_2O to form the tetroxide which attacks the eyes. The solid should be kept dry and in this form it is relatively safe. [*Synthesis* 610 *1972*.]

Potassium oxalate *[6487-48-5]* **M 184.2, m 160°(dec), d 2.13.** Crystd from hot water.

Potassium perchlorate *[7778-74-7]* **M 138.6, m 400(dec), d 2.52, pK²⁵ -2.4 to -3.1 (for HClO₄).** Crystd from boiling water (5mL/g) by cooling. Dried under vacuum at 105°.

Potassium periodate (potassium metaperiodate) *[7790-21-8]* **M 230.0, m 582°, d 3.62.** Crystd from distilled water.

Potassium permanganate *[7722-64-7]* **M 158.0, m 240(dec), d 2.7, pK²⁵ -2.25 (for HMnO₄).** Crystd from hot water (4mL/g at 65°), then dried in a vacuum desiccator over $CaSO_4$. Phillips and Taylor [*J Chem Soc* 4242 *1962*] cooled an aqueous solution of $KMnO_4$, saturated at 60°, to room temperature in the dark, and filtered through a No.4 porosity sintered-glass filter funnel. The solution was allowed to evaporate in air in the dark for 12h, and the supernatant liquid was decanted from the crystals, which were dried as quickly as possible with filter paper.

Potassium peroxydisulfate (potassium persulfate) *[7727-21-1]* **M 270.3.** Crystd twice from distilled water (10mL/g) and dried at 50° in a vacuum desiccator.

Potassium peroxymonosulfate (Oxone, potassium monopersulfate triple salt; 2KHSO₅.KHSO₄.K₂SO₄), *[37222-66-5, triple salt]* *[70693-62-8]* **M 614.8.** This is a stable form of Caro's acid and should contain >4.7% of active oxygen. It can be used in EtOH/H_2O and EtOH/AcOH/H_2O solutions. If active oxygen is too low it is best prepared afresh from 1mole of $KHSO_5$, 0.5moles of $KHSO_4$ and 0.5moles of K_2SO_4. [Kennedy and Stock *J Org Chem* **25** 1901 *1960*; Stephenson US Patent 2,802,722 *1957*.] A rapid prepn of **Caro's acid** is made by stirring finely powdered potassium persulfate (**M** 270.3) into ice cold conc H_2SO_4 (7mL) and when homogeneous add ice (40-50g). It is stable for several days if kept cold.

Keep away from organic matter as it is a **STRONG OXIDANT**. A detailed prepn of **Caro's acid** (*hypersulfuric acid*, H_2SO_5, *[7722-86-3]*) in crystalline form **m** ~45° from H_2O_2 and chlorosulfonic acid was described by Fehér in *Handbook of Preparative Inorganic Chemistry (Ed. Brauer)* Vol I p. 388 *1963*.

Potassium perrhenate ($KReO_4$) *[10466-65-6]* **M 289.3, pK^{25} -1.25 (for $HReO_4$).** Crystd from water (7mL/g), then fused in a platinum crucible in air at 750°.

Potassium phenol-4-sulfonate (4-hydroxybenzene-1-sulfonic acid K salt) *[30145-40-5]* **M 212.3.** Crystd several times from distilled water at 90°, after treatment with charcoal, by cooling to *ca* 10°. Dried at 90-100°.

Potassium phthalimide (phthalimide K salt) *[1074-82-4]* **M 185.2, m >300°.** The solid may contain phthalimide and K_2CO_3 from hydrolysis. If too much hydrolysis has occurred (this can be checked by extraction with cold Me_2CO in which the salt is insoluble, evaporation of the Me_2CO and weighing the residue) it would be better to prepare it afresh. If little hydrolysis had occurred then recryst from a large volume of EtOH, and wash solid with a little Me_eCO and dry in a continuous vacuum to constant weight. [Salzerg and Supriawski *Org Synth* Coll Vol I 119 *1941*; Raman and IR: Hase *J Mol Struct* **48** 33 *1978*; Dykman *Chem Ind (London)* 40 *1972*; IR, NMR: Assef et al. *Bull Soc Chim Fr* II-167 *1979*.]

Potassium picrate *[573-83-1]* **M 267.2.** Crystd from water or 95% EtOH, and dried at room temperature in vacuum. It is soluble in 200 parts of cold water and 4 parts of boiling water. **THE DRY SOLID EXPLODES WHEN STRUCK OR HEATED.**

Potassium propionate *[327-62-8]* **M 112.2.** Crystd from water (30mL/g) or 95% EtOH.

Potassium reineckate *[34430-73-4]* **M 357.5.** Crystd from KNO_3 soln, then from warm water [Adamson *J Am Chem Soc* **80** 3183 *1958*].

Potassium (VI) ruthenate *[31111-21-4]* **M 243.3.** Dissolve in H_2O and evaporate until crystals are formed. The crystals are iridescent green prisms which appear red as thin films. Possible impurity is RuO_4; in this case wash with CCl_4 (which dissolves RuO_4). The concn of an aqueous solution of RuO_4^{2-} (orange colour) can be estimated from the absorbance at 385nm (ε 1030 M^{-1} cm^{-1}), or at 460nm (ε 1820 M^{-1} cm^{-1}). [*Can J Chem* **50** 3741 *1972*; *J Am Chem Soc* **74** 5012 *1952*; *Handbook of Preparative Inorganic Chemistry (Ed. Brauer)* Vol II 1600 *1965*].

Potassium selenocyanate *[3425-46-5]* **M 144.1.** Dissolved in acetone, filtered and ppted by adding Et_2O.

Potassium sodium tartrate ($4H_2O$) *[6381-59-5 ($4H_2O$); 304-59-6 (R,R)]* **M 282.3.** Crystd from distilled water (1.5mL/g) by cooling to 0°.

Potassium sulfate *[7778-80-5]* **M 174.3, m 1069°, d 2.67** Crystd from distilled water (4mL/g at 20°; 8mL/g at 100°) between 100° and 0°.

Potassium *d*-tartrate (H_2O) *[921-53-9 , 6381-59-5]* **M 235.3, m loses H_2O at 150°, d 1.98.** Crystd from distilled water (solubility: 0.4mL/g at 100°; 0.7mL/g at 14°).

Potassium tetrachloroplatinate(II) *[10025-99-7]* **M 415.1, m 500°(dec).** Crystd from aqueous 0.75M HCl (20mL/g) between 100° and 0°. Washed with ice-cold water and dried.

Potassium tetracyanopalladate (II) $3H_2O$ *[10025-98-6]* **M 377.4.** *All operations should be carried out in an efficient fume cupboard -* **Cyanide is very POISONOUS** Dissolve the complex (*ca* 5g) in a solution of KCN (4g) in H_2O (75mL) with warming and stirring and evaporate hot till crystals appear. Cool, filter off the crystals and wash with a few drops of cold H_2O. Further concentration of the mother liquors provides more crystals. The complex is recrystallised from H_2O as the colourless *trihydrate*. It effloresces in dry air and dehydrates at 100° to the *monohydrate*. The *anhydrous salt* is obtained by heating at 200°, but at higher temperatures it decomposes to $(CN)_2$, Pd and KCN. [*Inorg Synth* **2** 245 *1946*.]

Potassium tetrafluoroborate (potassium borofluoride) *[14075-53-7]* **M 125.9, m 530°, d_4^{30} 2.505, pK^{25} -4.9 (for HBF$_4$).** Cryst from H_2O (sol % (temp): 0.3 (3°), 0.45 (20°), 1.4 (40°), 6.27 (100°), and dry under vacuum. Non-hygroscopic salt. A 10% solution is transparent blue at 100°, green at 90° and yellow at 60°. [*Chem Ber* **65** 535 *1932*; *Handbook of Preparative Inorganic Chemistry (Ed. Brauer)* Vol 1 223 *1963*.]

Potassium tetraoxalate (2H$_2$O) [oxalic acid hemipotassium salt] *[6100-20-5 (2H$_2$O); 127-96-8 (anhydr)]* **M 254.2.** Crystd from water below 50°. Dried below 60° at atmospheric pressure.

Potassium tetraphenylborate *[3244-41-5]* **M 358.3.** Ppted from a soln of KCl acidified with dilute HCl, then crystallised twice from acetone, washed thoroughly with water and dried at 110° [Findeis and de Vries *Anal Chem* **28** 1899 *1956*]. It has also been recrystd several times from conductivity water.

Potassium thiocyanate *[333-20-0]* **M 97.2, m 172°, pK^{25} -1.85 (for HSCN).** Crystd from H_2O if much chloride ion is present in the salt, otherwise from EtOH or MeOH (optionally by addition of Et$_2$O). Filtered on a Büchner funnel without paper, and dried in a desiccator at room temperature before being heated for 1h at 150°, with a final 10-20min at 200° to remove the last traces of solvent [Kolthoff and Lingane *J Am Chem Soc* **57** 126 *1935*]. Stored in the dark.

Potassium thiosulfate hydrate *[13446-67-8; 10294-66-3 (75% aq soln)]* **M 190.3, pK_1^{25} 0.6, pK_2^{25} 1.74 (for H$_2$S$_2$O$_3$)** Crystd from warm water (0.5mL/g) by cooling in an ice-salt mixture.

Potassium thiotosylate *[28519-50-8]* **M 226.4.** Recrystallise from absolute EtOH and dry at 130°. In wet EtOH the *monohydrate* can be obtained. [*J Gen Chem USSR (Engl Transl)* **28** 1345 *1958*.]

Potassium trifluoroacetate *[2923-16-2]* **M 152.1, m 140-142°, pK^{25} 0.52 (for CF$_3$CO$_2$H).** To purify dissolve the salt in trifluoroacetic acid with *ca* 2% of trifluoroacetic anhydride, filter and evaporate carefully to dryness (avoid over heating), and finally dry in a vacuum at 100°. It can be recrystallised from trifluoroacetic acid (solubility in the acid is *ca* 50.1%). [*J Am Chem Soc* **74** 4746 *1952*, **76** 4285 *1954*; *J Inorg Nucl Chem* **9** 166 *1959*.]

Potassium trimethylsilanolate (trimethylsilanol K salt) *[10519-96-7]* **M 128.3, m 131-135° (cubic form), d^{25} 1.11, 125°dec (orthorhombic form).** Recryst from H_2O and dried at 100°/1-2mm. [*J Am Chem Soc* **75** 5615 *1953*; IR: *J Org Chem* **17** 1555 *1952*.]

Potassium tungstate (*ortho* 2H$_2$O) *[37349-36-3; 7790-60-5]* **M 362.1, m 921°, d 3.12, pK_1^{25} 2.20, pK_2^{25} 3.70 (for H$_2$WO$_4$).** Crystd from hot water (0.7mL/g).

Praseodymium acetate *[6192-12-7]* **M 318.1.** Recrystd several times from water [Ganapathyl *J Am Chem Soc* **109** 3159 *1986*].

Praseodymium trichloride (6H$_2$O) *[10361-79-2]* **M 355.4, pK_1^{25} 8.55 (for Pr^{3+} hydrol).** Its 1M soln in 6M HCl was passed twice through a Dowex-1 anion-exchange column. The eluate was evaporated in a vac desiccator to about half its vol and allowed to crystallise [Katzin and Gulyas *J Phys Chem* **66** 494 *1962*].

Praseodymium oxide (Pr$_6$O$_{11}$) *[12037-29-5]* **M 1021.4.** Dissolved in acid, ppted as the oxalate and ignited at 650°.

Propargyl triphenyl phosphonium bromide *[2091-46-5]* **M 381.4, m 179°.** Recrystallises from 2-propanol as white plates. Also crystallises from EtOH, **m** 156-158°. IR has ν 1440, 1110cm^{-1} (P-C str). [*Justus Liebigs Ann Chem* **682** 62 *1965*; *J Org Chem* **42** 200 *1977*].

Propenyloxy trimethylsilane *[1833-53-0]* **M 130.3, b 93-95°/atm, d 0.786.** Purified by fractional distillation using a very efficient column at atmospheric pressure. Usually contains 5% of hexamethyldisiloxalane which boils at 99-101°, but is generally non-reactive and need not be removed. [*J Am Chem Soc* **71** 5091 *1952.*] It has been distilled under N_2 through a 15cm column filled with glass helices. Fraction **b** 99-104° is further purified by gas chromatography through a Carbowax column (Autoprep A 700) at a column temperature of 87°, retention time is 9.5min. [*J Organometal Chem* **1** 476 *1963-4.*]

1-Propenyltrimethylsilane (*cis and trans* **mixture**) *[17680-01-2]* **M 114.3, b 85-88°, n $_D^{20}$ 1.4121.** Dissolve (~ 20g) in THF (200mL), shake with H_2O (2x 300 mL), dry (Na_2SO_4) and fractionate. This is a mixture of *cis* and *trans* isomers which can be separated by gas chromatography on an $AgNO_3$ column (for prep: see Seyferth and Vaughan *J Organomet Chem* **1** 138 *1963*) at 25° with He as carrier gas at 9psi. The *cis*-isomer has n_D^{25} 1.4105 and the *trans*-isomer has n_D^{25} 1.4062. [Seyferth et al. *Pure Appl Chem* **13** 159 *1966.*]

Pyridinium chlorochromate *[26299-14-9]* **M 215.6, m 205°(dec).** Dry in a vacuum for 1h. It is not hygroscopic and can be stored for extended periods at room temp without change. If very suspect it can be readily prepared. [*Tetrahedron Lett* 2647 *1975*; *Synthesis* 245 *1982.*]
§ Available commercially on a polymer support.

Pyridinium dichromate *[20039-37-6]* **M 376.2, m 145-148°, 152-153°.** Dissolve in the minimum volume of H_2O and add 5 volumes of cold Me_2CO and cool to -20°. After 3h the orange crystals are collected, washed with a little cold Me_2CO and dried in a vacuum. It is soluble in dimethylformamide (0.9g/mL at 25°), and in H_2O, and has a characteristic IR with ν 930, 875, 765 and 730cm^{-1}. [*Tetrahedron Lett* 399 *1979*; *Chem Ind (London)* 1594 *1969.*] (Possible **carcinogen**).
§ Available commercially on a polymer support.

3-(2-Pyridyl)-5,6-diphenyl-1,2,4-triazine-*p,p'***-disulfonic acid, monosodium salt (H_2O)** *[63451-29-6]* **M 510.5.** Purified by recrystn from water or by dissolving in the minimum volume of water, followed by addition of EtOH to ppte the pure salt.

Pyrocatechol Violet (tetraphenolictriphenylmethanesulfonic acid Na salt) *[115-41-3]* **M 386.4, ε 1.4 x 10^4 at 445nm in acetate buffer pH 5.2-5.4, pK$_{Est(1)}$>0 (SO$_3$H), pK$_{Est(2)}$~ 9.4, pK$_{Est(3)}$~ 13.** It was recrystd from glacial acetic acid. Very *hygroscopic*. Indicator standard for metal complex titrations. [Mustafin et al. *Zh Anal Khim* **22** 1808 *1967.*]

Pyrogallol Red (tetraphenolic xanthyliumphenylsulfonate) *[32638-88-3]* **M 418.4, m >300°(dec), ε 4.3 x 10^4 at 542nm, pH 7.9-8.6, pK as above.** Recrystd from aqueous alkaline solution (Na_2CO_3 or NaOH) by precipitation on acidification [Suk *Collect Czech Chem Commun* **31** 3127 *1966*].

Pyronin B [di-(3,6-bis(diethylamino)xanthylium chloride) diFeCl$_5$ complex] *[2150-48-3]* **M 358.9 (Fe free), m 176-178° (diFe complex), CI 45010, λmax 555nm, pK^{25} 7.7.** Crystd from EtOH. Forms Fe stain.

Quinolinium chlorochromate *[108703-35-1]* **M 265.6, m 127-130°.** A

yellow-brown solid which is stable in air for long periods. If it has deteriorated or been kept for too long, it is best to prepare it freshly. Add freshly distilled quinoline (13mL) to a mixture of chromic acid (CrO_3) (10g) and ~ 5M HCl (11mL of conc HCl and 10mL of H_2O) at 0°. A yellow-brown solid separates, it is filtered off on a sintered glass funnel, dried for 1h in a vacuum, and can be stored for extended periods without serious loss in activity. It is a good oxidant for primary alcohol in CH_2Cl_2. [Singh et al. *Chem Ind (London)* 751 *1986*; method of Corey and Suggs *Tetrahedron Lett* 2647 *1975*.]

Reinecke salt see ammonium reineckate on p. 394.

Resorufin (7-hydroxy-3H-phenoxazine-3-one Na salt) *[635-78-9]* **M 213.2, pK$_1^{30}$ 6.93, pK$_2^{30}$ 9.26, pK$_3^{30}$ 10.0.** Washed with water and recrystd from EtOH several times.

Rhodium (II) acetate dimer (2H$_2$O) *[15956-28-2]* **M 478.0.** Dissolve 5g in boiling MeOH (*ca* 600mL) and filter. Concentrate to 400mL and chill overnight at *ca* 0° to give dark green crystals of the MeOH adduct. Concn of the mother liquors gives a further crop of [Rh(OAc)$_2$]$_2$.2MeOH. The adduct is then heated at 45° in vacuum for 2h (all MeOH is lost) to leave the emerald green crystals of the actetate. [*J Chem Soc (A)* 3322 *1970*.] Alternatively dissolve in glacial AcOH and reflux for a few hs to give an emerald green soln. Evaporate most of the AcOH on a steam bath then heat the residue at 120°/1h. Extract the residue with boiling Me$_2$CO. Filter, concentrate to half its volume and keep at 0°/18h. Collect the crystals, wash with ice cold Me$_2$CO and dry at 110°. It is soluble in most organic solvents with which it forms adducts including NMe$_3$ and Me$_2$S and give solutions with different colours varying from green to orange and red. [UV: *Inorg Synth* **2** 960 *1963*.]

Rhodium (III) chloride *[10049-07-7]* **M 209.3, m >100°(dec), b 717°.** Probable impurities are KCl and HCl. Wash solid well with small volumes of H$_2$O to remove excess KCl and KOH and dissolve in the minimum volume of conc HCl. Evaporate to dryness on a steam bath to give wine-red coloured RhCl$_3$.3H$_2$O. Leave on the steam bath until odour of HCl is lost - do not try to dry further as it begins to decompose above 100° to the oxide and HCL. It is not soluble in H$_2$O but soluble in alkalis or CN solns and forms double salts with alkali chlorides. [*Inorg Synth* **7** 214 *1063*.]

Rhodizonic acid sodium salt (5,6-dihydroxycyclohex-5-ene-1,2,3,4-tetraone di-Na salt) *[523-21-7]* **M 214.0, pK$_1^{30}$ 4.1 (4.25), pK$_2^{30}$ 4.5 (4.72).** The free acid, obtained by acidifiying and extracting with Et$_2$O, drying (MgSO$_4$), filtering, evaporating and distilling in vacuum (**b** 155-160°/14mm). The *free acid* solidifies on cooling and the colourless crystals can be recrystd from tetrahydrofuran-pet ether or *C$_6$H$_6$. It forms a *dihydrate* **m** 130-140°. The pure di Na salt is formed by dissolving in 2 equivs of NaOH and evaporating in a vacuum. It forms violet crystals which give an orange soln in H$_2$O that is unstable for extended periods even at 0°, and should be prepared freshly before use. Salts of rhodizonic acid cannot be purified by recrystn without great loss due to conversion to croconate, so that the original material must be prepared pure. It can be washed with NaOAc soln then EtOH to remove excess NaOAc dried under vacuum and stored in the dark. [UV and tautomerism: Schwarzenbach and Suter *Helv Chim Acta* **24** 617 *1941*; Polarography: Preisler and Berger *J Am Chem Soc* **64** 67 *1942*; Souchay and Taibouet *J Chim Phys* **49** C108 *1952*.]

Rose Bengal [Acid Red 94, 4,5,6,7-tetrachloro-2'.4',5',7'-tetraiodofluorescein di-Na or di-K salt] *[di-Na salt 632-69-9]* **M 1017.7 (di-Na salt)** *[di-K salt 11121-48-5]* **M 1049.8 (di-K salt).** This biological stain can be purified by chromatography on silica TLC using a 35:65 mix of EtOH/acetone as eluent.

Rubidium bromide *[7789-39-1]* **M 165.4, m 682°, b 1340°, d 3.35.** A white crystalline powder which crystallises from H$_2$O (solubility: 50% in cold and 67% in boiling H$_2$O to give a neutral soln). Also crystd from near-boiling water (0.5mL/g) by cooling to 0°.

Rubidium chlorate *[13446-71-4]* **M 168.9, d 3.19.** Crystd from water (1.6mL/g) by cooling from 100°.

Rubidium chloride *[7791-11-9]* **M 120.9, m 715°, d 2.80.** Crystd from water (0.7mL/g) by cooling to 0° from 100°.

Rubidium nitrate *[13126-12-0]* **M 147.5, m 305°, d 3.11.** Crystd from hot water (0.25mL/g) by cooling to room temperature.

Rubidium perchlorate *[13510-42-4]* **M 184.9, d 2.80, pK²⁵ -2.4 to -3.1 (for HClO₄).** Crystd from hot water (1.6mL/g) by cooling to 0°.

Rubidium sulfate *[7488-54-2]* **M 267.0, m 1050°, d 6.31.** Crystd from water (1.2mL/g) between 100° and 0°.

Ruthenium (III) acetylacetonate *[14284-93-6]* **M 398.4, m 240°(dec).** Purified by recrystn from *benzene. [*J Am Chem Soc* **74** 6146 *1952*.]

Ruthenium (III) chloride (2H₂O) (β–form) *[14898-67-0]* **M 207.4 + H₂O, m >500°(dec), d 3.11, pK₁²⁵ 3.40 (for aquo Rh³⁺ hydrolysis).** Dissolve in H₂O, filter and concentrate to crystallise in the absence of air to avoid oxidation. Evaporate the solution in a stream of HCl gas while being heated just below it boiling point until a syrup is formed and finally to dryness at 80-100° and dried in a vacuum over H₂SO₄. When heated at 700° in the presence of Cl₂ the insoluble α-form is obtained [*Handbook of Preparative Inorganic Chemistry (Ed. Brauer)* Vol II 1598 *1965*; *J Org Chem* **46** 3936 *1981*].

Ruthenium (IV) oxide *[12036-10-1]* **M 133.1, d 6.97.** Freed from nitrates by boiling in distilled water and filtering. A more complete purification is based on fusion in a KOH-KNO₃ mix to form the soluble ruthenate and perruthenate salts. The melt is dissolved in water, and filtered, then acetone is added to reduce the ruthenates to the insoluble hydrate oxide which, after making a slurry with paper pulp, is filtered and ignited in air to form the anhydrous oxide [Campbell, Ortner and Anderson *Anal Chem* **33** 58 *1961*].

Ruthenocene   [bis-(cyclopentadienyl)ruthenium] *[1287-13-4]* **M 231.2, m 195.5°, 199-210°.** Sublime in high vacuum at 120°. Yellow crystals which can be recrystallised from CCl₄ as transparent plates. [*J Am Chem Soc* **74** 6146 *1952*].

Samarium (II) iodide *[32248-43-4]* **M 404.2, m 520°, b 1580.** Possible impurity is SmI₃ from which it is made. If present, grind solid to a powder and heat in a stream of pure H₂. The temperature (~ 500-600°) should be below the **m** (~ 628°) of SmI₃, since the molten compounds react very slowly. [Wetzel in *Handbook of Preparative Inorganic Chemistry (Ed. Brauer)* Vol II pp. 1149, 1150 *1965*.]

Selenious acid *[7783-00-8]* **M 129.0, m 70°(dec), d 3.0, pK₁²⁵ 2.62, pK₂²⁵ 8.32 (H₂SeO₃).** Crystd from water. On heating it loses water and SeO₂ sublimes.

Selenium *[7782-49-2]* **M 79.0, m 217.4°, d 4.81.** Dissolved in small portions in hot conc HNO₃ (2mL/g) filtered and evaporated to dryness to give selenious acid which was then dissolved in conc HCl. Passage of SO₂ into the solution ppted selenium (but not tellurium) which was filtered off and washed with conc HCl. This purification process was repeated. The selenium was then converted twice to the selenocyanate by treating with a 10% excess of 3M aqueous KCN, heating for half an hour on a sand-bath and filtering. Addition of an equal weight of crushed ice to the cold solution, followed by an excess of cold, conc HCl, with stirring (in a well ventilated fume hood because HCN is evolved) ppted selenium powder, which, after washing with water until colourless, and then with MeOH, was heated in an oven at 105°, then by fusion for 2h under vacuum. It was cooled, crushed and stored in a desiccator [Tideswell and McCullough *J Am Chem Soc* **78** 3036 *1956*].

Selenium dioxide *[7446-08-4]* **M 111.0, m 340°.** Purified by sublimation, or by solution in HNO₃, pptn of selenium which, after standing for several hours or boiling, is filtered off, then re-oxidised by HNO₃ and cautiously evaporated to dryness below 200°. The dioxide is dissolved in water and again evaporated to dryness.

Selenopyronine *[85051-91-8]* **M 365.8, λmax 571nm (ε 81,000).** Purified as the hydrochloride from hydrochloric acid [Fanghanel et al. *J Phys Chem* **91** 3700 *1987*].

Selenourea *[630-10-4]* **M 123.0, m 214-215°(dec).** Recrystd from water under nitrogen.

Silica *[7631-86-9 (colloidal); 112945-52-5 (fumed)].* Purification of silica for high technology applications uses isopiestic vapour distillation from conc volatile acids and is absorbed in high purity water. The impurities remain behind. Preliminary cleaning to remove surface contaminants uses dip etching in HF or a mixture of HCl, H_2O_2 and deionised water [Phelan and Powell *Analyst* **109** 1299 *1984*].

Silica gel *[63231-67-4; 112926-00-8].* Before use as a drying agent, silica gel is heated in an oven, then cooled in a desiccator. Conditions in the literature range from heating at 110° for 15h to 250° for 2-3h. Silica gel has been purified by washing with hot acid (in one case successively with aqua regia, conc HNO_3, then conc HCl; in another case digested overnight with hot conc H_2SO_4), followed by exhaustive washing with distilled water (one week in a Soxhlet apparatus has also been used), and prolonged oven drying. Alternatively, silica gel has been extracted with acetone until all soluble material was removed, then dried in a current of air, washed with distilled water and oven dried. Silica gel has also been washed successively with water, M HCl, water, and acetone, then activated at 110° for 15h.

Silicon monoxide *[10097-28-6]* **M 44.1, m > 1700°, d 2.18.** Purified by sublimation in a porcelain tube in a furnace at 1250° (4h) in a high vacuum (10^{-4}mm) in a stream of N_2. It is obtained as brownish black scales. [*Handbook of Preparative Inorganic Chemistry (Ed. Brauer)* Vol I 696 *1963*.]

Silicon tetraacetate *[562-90-3]* **M 264.3, m 110-111°, b 148°/5-6mm, pK_1^{25} 9.7, pK_2^{25} 11.9 (for H_4SiO_4 free acid).** It can be crystallised from mixtures of CCl_4 and pet ether or Et_2O, or from acetic anhydride and then dried in a vacuum desiccator over KOH. Ac_2O adheres to the crystals and is removed first by drying at room temp then at 100° for several hours. It is soluble in Me_2CO, is very *hygroscopic* and effervesces with H_2O. It decomposes at 160-170°. [*Z Obshch Khim (Engl Transl)* **27** 985 *1957*; *Handbook of Preparative Inorganic Chemistry (Ed. Brauer)* Vol I 701 *1963*.]

Silicon tetrachloride *[10026-04-7]* **M 169.9, m -70°, b 57.6°, d 1.483.** Distd under vacuum and stored in sealed ampoules under N_2. Very sensitive to moisture.

12-Silicotungstic acid **(tungstosilicic acid; $H_4SiW_{12}O_{40}$)** *[12027-43-9]* **M 2914.5.** Extracted with diethyl ether from a solution acidified with HCl. The diethyl ether was evaporated under vacuum, and the free acid was crystallised twice [Matijevic and Kerker *J Phys Chem* **62** 1271 *1958*].

Silver (metal) *[7440-22-4]* **M 107.9, m 961.9°, b 2212°, d 10.5.** For purification by electrolysis, see Craig et al. [*J Res Nat Bur Stand* **64A** 381 *1960*].

Silver acetate *[563-63-3]* **M 166.9, pK^{25} >11.1 (for aquo Ag^+ hydrolysis).** Shaken with acetic acid for three days, the process being repeated with fresh acid, the solid then being dried in a vacuum oven at 40° for 48h. Has also been recrystallised from water containing a trace of acetic acid, and dried in air.

Silver bromate *[7783-89-3]* **M 235.8, m dec on heating, d 5.21.** Crystd from hot water (80mL/g).

Silver bromide *[7785-23-1]* **M 187.8, m 432°, d 6.47.** Purified from Fe, Mn, Ni and Zn by zone melting in a quartz vessel under vacuum.

Silver chlorate *[7783-92-8]* **M 191.3, m 230°, b 270°(dec), d 4.43.** Recrystd three times from water (10mL/g at 15°; 2mL/g at 80°).

Silver chloride *[7783-90-6]* **M 143.3, m 455°, b 1550°, d 5.56.** Recrystd from conc NH_3 solution.

Silver chromate *[7784-01-2]* **M 331.8, d^{25} 5.625, pK_1^{25} 0.74, pK_2^{25} 6.49 (for H_2CrO_4).** Wash the red-brown powder with H_2O, dry in a vacuum, then powder well and dry again in a vacuum at 90°/5h. Solubility in H_2O is 0.0014% at 10°. [*J Org Chem* **42** 4268 *1977*.]

Silver cyanide *[506-64-9]* **M 133.9, m dec at 320°, d 3.95. POISONOUS** white or grayish white powder. Stir thoroughly with H_2O, filter, wash well with EtOH and dry in air in the dark. It is very insoluble in H_2O (0.000023g in 100mL H_2O) but is soluble in HCN or aqueous KCN to form the soluble $Ag(CN)_2^{2-}$ complex. [*Chem Ber* **72** 299 *1939*; *J Am Chem Soc* **52** 184 *1930*.]

Silver diethyldithiocarbamate *[1470-61-7]* **M 512.3, pK_1^{25} 3.36 (for N,N-diethyldithio-carbamate).** Purified by recrystn from pyridine. Stored in a desiccator in a cool and dark place.

Silver difluoride *[7783-95-1]* **M 145.9, m 690°, d 4.7.** Highly **TOXIC** because it liberates HF and F_2. Very *hygroscopic* and reacts violently with H_2O. It is a powerful oxidising agent and liberates O_3 from dilute acids, and I_2 from I^- soln. Store in quartz or iron ampoules. White when pure, otherwise it is brown-tinged. Thermally stable up to 700°. [*Handbook of Preparative Inorganic Chemistry (Ed. Brauer)* Vol I 241 *1963*.]

Silver fluoride *[7775-41-9]* **M 126.9, m 435°, b *ca* 1150°, d 5.852.** *Hygroscopic* solid with a solubility of 135g/100mL of H_2O at 15°, and forms an insoluble basic fluoride in moist air. Purified by washing with AcOH and dry $*C_6H_6$, then kept in a vacuum desiccator at room temperature to remove $*$benzene and stored in opaque glass bottles. Flaky *hygroscopic* crystals which darken on exposure to light. It *attacks* bone and teeth. [*J Chem Soc* 4538 *1952*; *Handbook of Preparative Inorganic Chemistry (Ed. Brauer)* Vol I 240 *1963*.]

Silver iodate *[7783-97-3]* **M 282.8, m >200°, d 5.53.** Washed with warm dilute HNO_3, then H_2O and dried at 100°, or recrystd from NH_3 soln by adding HNO_3, filtering, washing with H_2O and drying at 100°.

Silver lactate *[128-00-7]* **M 196.9, m ~ 100°.** Recrystd from H_2O by adding EtOH. The solid was collected washed with EtOH then Et_2O and dried at 80° to give the dihydrate. White powder soluble in 15 parts of H_2O but only slightly soluble in EtOH. [*Justus Liebigs Ann Chem* **63** 89 *1847*; *Helv Chim Acta* **2** 251 *1919*.]

Silver nitrate *[7761-88-8]* **M 169.9, m 212°, b 444°(dec), d 4.35.** Purified by recrystn from hot water (solubility of $AgNO_3$ in water is 992g/100mL at 100° and 122g/100mL at 0°). It has also been purified by crystn from hot conductivity water by slow addition of freshly distilled EtOH.
CAUTION: avoid using EtOH for washing the ppte; and avoid concentrating the filtrate to obtain further crops of $AgNO_3$ owing to the risk of EXPLOSION (as has been reported to us) caused by the presence of silver fulminate. When using EtOH in the purification the apparatus should be enveloped in a strong protective shield. [Tully, *News Ed (Am Chem Soc)* **19** 3092 *1941*; Garin and Henderson *J Chem Educ* **47** 741 *1970*; Bretherick, *Handbook of Reactive Chemical Hazards* 4th edn, Butterworths, London, 1985, pp 13-14.] Before being used as a standard in volumetric analysis, analytical reagent grade $AgNO_3$ should be finely powdered, dried at 120° for 2h, then cooled in a desiccator.
Recovery of silver residues as $AgNO_3$ **[use protective shield during the whole of this procedure]** can be achieved by washing with hot water and adding 16M HNO_3 to dissolve the solid. Filter through glass wool and concentrate the filtrate on a steam bath until precipitation commences. Cool the solution in an ice-bath and filter the precipitated $AgNO_3$. Dry at 120° for 2h, then cool in a desiccator in a vacuum. Store over P_2O_5 in a vacuum in the dark. *AVOID contact with hands due to formation of black stains.*

Silver nitrite *[7783-99-5]* **M 153.9, m 141°(dec), d 4.45.** Crystd from hot conductivity water (70mL/g) in the dark. Dried in the dark under vacuum.

Silver(I) oxide *[20667-12-3]* **M 231.7, m ~200°(dec), d 7.13.** Leached with hot water in a Soxhlet apparatus for several hours to remove any entrained electrolytes.

Silver (II) oxide *[1301-96-8]* **M 123.9, m >100°(dec), d^{25} 7.22.** Soluble in 40,000 parts of H_2O, and should be protected from light. Stir with an alkaline solution of potassium peroxysulfate ($K_2S_2O_8$) at 85-

90°. The black AgO is collected, washed free from sulfate with H_2O made slightly alkaline and dried in air in the dark. [*Inorg Synth* **4** 12 *1953*.]

Silver perchlorate (H₂O) *[14242-05-8 (H₂O); 7783-93-9 (anhydr)]* **M 207.3, pK²⁵ -2.4 to -3.1 (for HClO₄).** Refluxed with *benzene (6mL/g) in a flask fitted with a Dean and Stark trap until all the water was removed azeotropically (*ca* 4h). The soln was cooled and diluted with dry pentane (4mL/g of AgClO₄). The ppted AgClO₄ was filtered off and dried in a desiccator over P_2O_5 at 1mm for 24h [Radell, Connolly and Raymond *J Am Chem Soc* **83** 3958 *1961*]. It has also been recrystallised from perchloric acid. [**Caution** *due to* **EXPLOSIVE** *nature in the presence of organic matter.*]

Silver permanganate *[7783-98-4]* **M 226.8, d 4.49.** Violet crystals which can be crystallised from hot H_2O (sol is 9g/L at 20°). Store in the dark. Oxidising agent, decomposed by light.

Silver sulfate *[10294-26-5]* **M 311.8, m 652°, b 1085°(dec), d 5.45.** Crystd form hot conc H_2SO_4 contg a trace of HNO_3, cooled and diluted with H_2O. The ppte was filtd off, washed and dried at 120°.

Silver thiocyanate *[1701-93-5]* **M 165.9, m 265°(dec), d 3.746, pK²⁵ -1.85 (for HSCN).** Digest the solid salt with aqueous NH₄NCS, wash thoroughly with H_2O and dry at 110° in the dark. Soluble in dilute aqueous NH_3. Dissolve in strong aqueous NH₄NCS solution, filter and dilute with large volume of H_2O when the Ag salt separates. The solid is washed with H_2O by decantation until free from NCS⁻ ions, collected, washed with H_2O, EtOH and dried in an air oven at 120°. Alternatively dissolve in dilute aqueous NH_3 and single crystals are formed by free evaporation of the solution in air. [*J Chem Soc* 836, 2405 *1932*; IR and Raman: *Acta Chem Scand* **13** 1607 *1957*; *Acta Cryst* **10** 29 *1957*.]

Silver tosylate *[16836-95-6]* **M 279.1.** The anhydrous salt is obtained by recrystn from H_2O. [*Chem Ber* **12** 1851 *1879*.]

Silver trifluoroacetate *[2966-50-9]* **M 220.9, m 251-255°.** Extract the salt (Soxhlet) with Et_2O. The extract is filtered and evaporated to dryness, then the powdered residue is completely dried in a vacuum desiccator over silica gel. Solubility in Et_2O is 33.5g in 750mL. It can be recrystd from *C_6H_6 (sol: 1.9g in 30mL of *C_6H_6; and 33.5g will dissolve in 750mL of anhydrous Et_2O). [*J Org Chem* **23** 1545 *1958*; *J Chem Soc* 584 *1951*.] It is also soluble in trifluoroacetic acid (15.2% at 30°), toluene, *o*-xylene and dioxane [*J Am Chem Soc* **76** 4285 *1954*].

Silver trifluoromethanesulfonate *[2923-28-6]* **M 256.9.** Recrystd twice from hot CCl_4 [Alo et al. *J Chem Soc, Perkin Trans 1* 805 *1986*].

Sodium (metal) *[7440-23-5]* **M 23.0, m 97.5°, d 0.97.** The metal was placed on a coarse grade of sintered-glass filter, melted under vacuum and forced through the filter using argon. The Pyrex apparatus was then re-evacuated and sealed off below the filter, so that the sodium could be distilled at 460° through a side arm and condenser into a receiver bulb which was then sealed off [Gunn and Green *J Am Chem Soc* **80** 4782 *1958*]. **EXPLODES and IGNITES in water.**

Sodium acetate *[127-09-3]* **M 82.0, m 324°, d 1.53.** Crystd from acetic acid and pumped under vacuum for 10h at 120°. Alternatively, crystd from aqueous EtOH, as the trihydrate. This material can be converted to the anhydrous salt by heating slowly in a porcelain, nickel or iron dish, so that the salt liquefies. Steam is evolved and the mass again solidifies. Heating is now increased so that the salt melts again. (NB: if it is heated too strongly, the salt chars.) After several minutes, the salt is allowed to solidify and cooled to a convenient temperature before being powdered and bottled (water content should now less than 0.02%).

Sodium acetylide *[1066-26-8]* **M 48.0.** It disproportionates at *ca* 180° to sodium carbide. It sometimes contains diluents, e.g. xylene, butyl ether or dioxane which can be removed by filtration followed by a vacuum at 65-60°/5mm. Alternatively the acetylide is purged with HC≡CH at 100-125° to remove diluent. NaC₂H adsorbs 2.2x, 2.0x and 1.6x its wt of xylene, butyl ether and dioxane respectively. Powdered NaC₂H is yellow or yellow-gray in colour and is relatively stable. It can be heated to *ca* 300° in the absence of air. Although no explosion or evolution of gas occurs, it turns brown due to disproportionation. At 170-190° in air it ignites slowly and burns smoothly. At 215-235° in air it flash-ignites and burns quickly. It can be dropped into a *slight* excess of H_2O without flashing or burning but vigorous evolution of $HC{\equiv}CH$ (**HIGHLY FLAMMABLE IN AIR**) occurs. The sample had been stored in the absence of air for one year without deterioration. Due to the high flammability of $HC{\equiv}CH$ the salt should be stored dry, and treated with care. After long storage, $NaC{\equiv}CH$ can be redissolved in liquid NH_3 and used for the same purposes as the fresh material. However it may be slightly turbid due to the presence of moisture. [*J Org Chem* **22** 649 *1957*; *J Am Chem Soc* **77** 5013 *1955*; *Inorg Synth* **2** 76, 81 *1946*; *Org Synth* **30** 15 1950.] See p. 89, Chapter 4 for prepartion.

Sodium alginate *[9005-38-3]*. Freed from heavy metal impurities by treatment with ion-exchange resins (Na^+-form), or with a dilute solution of the sodium salt of EDTA. Also dissolved in 0.1M NaCl, centrifuged and fractionally ppted by gradual addition of EtOH or 4M NaCl. The resulting gels were centrifuged off, washed with aq EtOH or acetone, and dried under vacuum. [Büchner, Cooper and Wassermann *J Chem Soc* 3974 *1961*.]

Sodium *n*-alkylsulfates. Crystd from EtOH/Me_2CO [Hashimoto and Thomas *J Am Chem Soc* **107** 4655 *1985*].

Sodium amide *[7782-92-5]* **M 39.0, m 210°.** It reacts *violently* with H_2O and is soluble in liquid NH_3 (1% at 20°). It should be stored in wax-sealed container is small batches. It is very *hygroscopic* and absorbs CO_2 and H_2O. If the solid is discoloured by being yellow or brown in colour then it should be destroyed as it can be highly **EXPLOSIVE.** It should be replaced if discoloured. It is best destroyed by covering with much toluene and slowly adding dilute EtOH with stirring until all the ammonia is liberated (FUME CUPBOARD). [*Inorg Synth* **1** 74 *1939*; *Handbook of Preparative Inorganic Chemistry (Ed. Brauer)* Vol I 465 *1963*; *Org Synth* Coll Vol III, 778 *1955*.]

Sodium 4-aminobenzoate *[555-06-6]* **M 159.1.** Crystd from water.

Sodium 4-aminosalicylate (2H$_2$O) *[6018-19-5]* **M 175.1.** Crystd from water at room temperature (2mL/g) by adding acetone and cooling.

Sodium ammonium hydrogen phosphate *[13011-54-6]* **M 209.1, m 79°(dec), d 1.55.** Crystd from hot water (1mL/g).

Sodium amylpenicillin *[575-47-3]* **M 350.4.** Crystd from moist acetone or moist ethyl acetate.

Sodium anthraquinone-1,5-disulfonate (H$_2$O) *[853-35-0]* **M 412.3.** Separated from insoluble impurities by continuous extraction with water. Crystd twice from hot water and dried under vacuum.

Sodium anthraquinone-1-sulfonate (H$_2$O) *[107439-61-2]* **M 328.3.** Crystd from hot water (4mL/g) after treatment with active charcoal, or from water by addition of EtOH. Dried under vacuum over $CaCl_2$, or in an oven at 70°. Stored in the dark.

Sodium anthraquinone-2-sulfonate (H$_2$O) *[131-08-8]* **M 328.3.** See 9,10-anthraquinone-2-sulfonic acid disodium salt on p. 395.

Sodium antimonyl tartrate *[34521-09-0]* **M 308.8.** Crystd from water.

Sodium arsenate (7H$_2$O) *[10048-95-0]* **M 312.0, m 50 (loses 5H$_2$O), m 130°, d 1.88 pK_1^{25} 2.22, pK_2^{25} 6.98 (for H$_3$AsO$_4$).** Crystd from water (2mL/g).

Sodium azide *[26628-22-8]* **M 65.0, m 300°(dec, explosive), pK^{25} 4.72 (for HN$_3$).** Crystd from hot water or from water by the addition of absolute EtOH or acetone. Also purified by repeated crystn from an aqueous solution saturated at 90° by cooling it to 10°, and adding an equal volume of EtOH. The crystals were washed with acetone and the azide dried at room temperature under vacuum for several hours in an

Abderhalden pistol. [Das et al. *J Chem Soc, Faraday Trans 1* **78** 3485 *1982*.] **HIGHLY POISONOUS and potentially explosive.**

Sodium barbitone (sodium 5,5-diethylbarbiturate) *[144-02-5]* **M 150.1, pK$_1^{25}$ 3.99, pK$_2^{25}$ 12.5 (barbituric acid).** Crystd from water (3mL/g) by adding an equal volume of EtOH and cooling to 5°. Dried under vacuum over P$_2$O$_5$.

Sodium benzenesulfinate *[873-55-2]* **M 164.2, m >300°.** See benzenesulfinic acid sodium salt on p. 400.

Sodium benzenesulfonate *[515-42-4]* **M 150.1, pK$_1^{25}$ 0.70 (2.55) (for PhSO$_3$H$_2$).** Crystd from EtOH or aqueous 70-100% MeOH, and dried under vacuum at 80-100°.

Sodium benzoate *[532-32-1]* **M 144.1.** Crystd from EtOH (12mL/g).

Sodium benzylpenicillin see *N*-benzylpenicillin sodium salt on p. 514 in Chapter 6.

Sodium bicarbonate *[144-55-8]* **M 84.0, m ~50°(dec, -CO$_2$).** Crystd from hot water (6mL/g). The solid should not be heated above 40° due to the formation of carbonate.

Sodium bis(trimethylsilyl)amide (hexamethyl disilazane sodium salt) *[1070-89-9]* **M 183.4, m 165-167°(sintering at 140°).** It can be sublimed at 170°/2 Torr (bath temp 220-250°) onto a cold finger, and can be recrystd from *C$_6$H$_6$ (sol: 10g in 100mL at 60°). It is slightly soluble in Et$_2$O and is decomposed by H$_2$O. [*Chem Ber* **94** 1540 *1961*.]

Sodium bisulfite *[7631-90-5]* **M 104.1, d 1.48.** Crystd from hot H$_2$O (1mL/g). Dried at 100° under vac for 4h.

Sodium borate (borax) *[1330-43-4]* **M 201.2, m 741°, d 2.37.** Most of the water of hydration was removed from the decahydrate by evacuation at 25° for three days, followed by heating to 100° and evacuation with a high-speed diffusion pump. The dried sample was then heated gradually to fusion (above 966°), allowed to cool gradually to 200°, then tranferred to a desiccator containing P$_2$O$_5$ [Grenier and Westrum *J Am Chem Soc* **78** 6226 *1956*].

Sodium borate (decahydrate, hydrated borax) *[1303-96-4]* **M 381.2, m 75°(loses 5H$_2$O at 60°), d 1.73.** Crystd from water (3.3mL/g) keeping below 55° to avoid formation of the pentahydrate. Filtered at the pump, washed with water and equilibrated for several days in a desiccator containing an aqueous solution saturated with respect to sucrose and NaCl. Borax can be prepared more quickly (but its water content is somewhat variable) by washing the recrystd material at the pump with water, followed by 95% EtOH, then Et$_2$O, and air dried at room temperature for 12-18h on a clock glass.

Sodium borohydride *[16940-66-2]* **M 37.8, m ~400°(dec), d 1.07.** After adding NaBH$_4$ (10g) to freshly distilled diglyme (120mL) in a dry three-necked flask fitted with a stirrer, nitrogen inlet and outlet, the mixture was stirred for 30min at 50° until almost all of the solid had dissolved. Stirring was stopped, and, after the solid had settled, the supernatant liquid was forced under N$_2$ pressure through a sintered-glass filter into a dry flask. [The residue was centrifuged to obtain more of the solution which was added to the bulk.] The solution was cooled slowly to 0° and then decanted from the white needles that separated. The crystals were dried by pumping for 4h to give anhydrous NaBH$_4$. Alternatively, after the filtration at 50° the solution was heated at 80° for 2h to give a white ppte of substantially anhydrous NaBH$_4$ which was collected on a sintered-glass filter under N$_2$, then pumped at 60° for 2h [Brown, Mead and Subba Rao *J Am Chem Soc* **77** 6209 *1955*].
NaBH$_4$ has also been crystd from isopropylamine by dissolving it in the solvent at reflux, cooling, filtering and allowing the solution to stand in a filter flask connected to a Dry-ice/acetone trap. After most of the solvent was passed over into the cold trap, crystals were removed with forceps, washed with dry diethyl ether and dried under vacuum. [Kim and Itoh *J Phys Chem* **91** 126 *1987*.] Somewhat less pure crystals were obtained more rapidly by using Soxhlet extraction with only a small amount of solvent and extracting for about 8h. The

crystals that formed in the flask were filtered off, then washed and dried as before. [Stockmayer, Rice and Stephenson *J Am Chem Soc* **77** 1980 *1955*.] Other solvents used for crystallisation include water and liquid ammonia.

Sodium bromate *[7789-38-0]* **M 150.9, m 381°, d 3.3.** Crystd from hot water (1.1mL/g) to decrease contamination by NaBr, bromine and hypobromite. [Noszticzius et al. *J Am Chem Soc* **107** 2314 *1985*.]

Sodium bromide *[7647-15-6]* **M 102.9, m 747°, b 1390°, d 3.2.** Crystd from water (0.86mL/g) between 50° and 0°, and dried at 140° under vacuum (this purification may not eliminate chloride ion).

Sodium 4-bromobenzenesulfonate *[5015-75-8]* **M 258.7.** Crystd from MeOH, EtOH or distd water.

Sodium *tert*-butoxide *[865-48-5]* **M 96.1.** It sublimes at 180°/1 Torr. Its solubility in *tert*-BuOH is 0.208M at 30.2° and 0.382M at 60°, and is quite soluble in tetrahydrofuran (32g/100g). It should not be used if it has a brown colour. [*J Am Chem Soc* **78** 4364, 3614 *1956, Inorg Synth* **1** 87 1939; IR: *J Org Chem* **21** 156 *1956*.]

Sodium butyrate *[156-54-7]* **M 110.1.** Prepared by neutralisation of the acid and recrystn from EtOH.

Sodium cacodylate (3H$_2$O) *[124-65-2]* **M 214.0, m 60°.** Crystd from aqueous EtOH.

Sodium carbonate *[497-19-8]* **M 106.0, m 858°, d 2.5.** Crystd from water as the decahydrate which was redissolved in water to give a near-saturated soln. By bubbling CO$_2$, NaHCO$_3$ was ppted. It was filtered, washed and ignited for 2h at 280° [MacLaren and Swinehart *J Am Chem Soc* **73** 1822 *1951*]. Before being used as a volumetric standard, analytical grade material should be dried by heating at 260-270° for 0.5h and allowed to cool in a desiccator. For preparation of primary standard sodium carbonate, see *Pure Appl Chem* **25** 459 *1969*. After 3 recrystns tech grade had Cr, Mg, K, P, Al, W, Sc and Ti at 32, 9.4, 6.6, 3.6, 2.4, 0.6, 0.2 and 0.2 ppm resp; another technical source had Cr, Mg, Mo, P, Si, Sn and Ti at 2.6, 0.4, 4.2, 13.4, 32, 0.6, 0.8 ppm resp.

Sodium carboxymethylcellulose *[9004-32-4]*. Dialysed for 48h against distilled water.

Sodium cetyl sulfate *[1120-01-0]* **M 344.5.** See sodium hexadecylsulfate on p. 471.

Sodium chlorate *[7775-09-9]* **M 106.4, m 248°, b >300°(dec), d 2.5.** Crystd from hot water (0.5mL/g).

Sodium chloride *[7647-14-5]* **M 58.4, m 800.7°, b 1413°, d 2.17.** Crystd from saturated aqueous solution (2.7mL/g) by passing in HCl gas, or by adding EtOH or acetone. Can be freed from bromide and iodide impurities by adding chlorine water to an aqueous solution and boiling for some time to expel free bromine and iodine. Traces of iron can be removed by prolonged boiling of solid NaCl in 6M HCl, the crystals then being washed with EtOH and dried at *ca* 100°. Sodium chloride has been purified by sublimation in a stream of pre-purified N$_2$ and collected by electrostatic discharge [Ross and Winkler *J Am Chem Soc* **76** 2637 *1954*]. For use as a primary analytical standard, analytical reagent grade NaCl should be finely ground, dried in an electric furnace at 500-600° in a platinum crucible, and allowed to cool in a desiccator. For most purposes, however, drying at 110-120° is satisfactory.

Sodium chlorite *[7758-19-2]* **M 90.4, m ~180°(dec).** Crystd from hot water and stored in a cool place. Has also been crystd from MeOH by counter-current extraction with liquid ammonia [Curti and Locchi *Anal Chem* **29** 534 *1957*]. Major impurity is chloride ion; can be recrystallised from 0.001M NaOH.

Sodium 4-chlorobenzenesulfonate *[5138-90-9]* **M 214.6, pK$_{Est}$ <0 (for SO$_3$H).** Crystd twice from MeOH and dried under vacuum.

Sodium 3-chloro-5-methylbenzenesulfonate *[5138-92-1]* **M 228.7, pK$_{Est}$ <0 (for SO$_3$H).** Crystd twice from MeOH and dried under vacuum.

Sodium chromate (4H$_2$O) *[10034-82-9]* **M 234.0, m ~20°(for 10H$_2$O), d, 2.7, pK$_1^{25}$ 0.74, pK$_2^{25}$ 6.49 (for H$_2$CrO$_4$).** Crystd from hot water (0.8mL/g).

dl-**Sodium creatine phosphate (4H$_2$O)** *[922-32-7]* **M 327.1.** See creatine phosphate di-Na salt on p. 523 in Chapter 6.

Sodium cyanate *[917-61-3]* **M 65.0, m 550°, d$_4^{20}$ 1.893, pK^{25} 3.47 (for HCNO).** Colourless needles from EtOH. Solubility in EtOH is 0.22g/100g at 0°C. Soluble in H$_2$O but can be recrystallised from small volumes of it.

Sodium cyanoborohydride *[25895-60-7]* **M 62.8, m 240-242°(dec), d^{28} 1.20.** Very *hygroscopic* solid, soluble in H$_2$O (212% at 29°, 121% at 88°), tetrahydrofuran (37% at 28°, 42.2% at 62°), very soluble in EtOH but insoluble in Et$_2$O, *C$_6$H$_6$ and hexane. It is stable to acid up to pH 3 but is hydrolysed in 12N HCl. The rate of hydrolysis at pH 3 is 10^{-8} that of NaBH$_4$. The fresh commercially available material is usually sufficiently pure. If very pure material is required one of the following procedures must be used [*Synthesis* 135 *1975*]: (a) The NaBH$_3$CN is dissolved in tetrahydrofuran (20% w/v), filtered and the filtrate is treated with a fourfold volume of CH$_2$Cl$_2$. The solid is collected and dried in a vacuum [*Inorg Chem.* 9 2146 *1970*]. Dissolve the NaBH$_3$CN in dry MeNO$_2$, filter, and pour the filtrate into a 10-fold volume of CCl$_4$ with vigorous stirring. The white ppte is collected, washed several times with CCl$_4$ and dried in a vacuum [*Inorg Chem* 9 624 *1970*]. (b) When the above procedures fail to give a clean product then dissolve the NaBH$_3$CN (10g) in tetrahydrofuran (80mL) and add N MeOH/HCl until the pH is 9. Pour the solution with stirring into dioxane (250mL). The solution is filtered, and heated to reflux. A further volume of dioxane (150mL) is added slowly with swirling. The solution is cooled slowly to room temp then chilled in ice and the crystalline dioxane complex is collected, dried in a vacuum for 4h at 25°, then 4h at 80° to yield the amorphous dioxane-free powder is 6.7g with purity >98% [*J Am Chem Soc* 93 2897 *1971*]. The purity can be checked by iodometric titration [*Anal Chem* 91 4329 *1969*].

Sodium *p*-cymenesulfonate *[77060-21-0]* **M 236.3.** Dissolved in water, filtered and evaporated to dryness. Crystd twice from absolute EtOH and dried at 110°.

Sodium decanoate (sodium caproate) *[1002-62-6]* **M 194.2.** Neutralised by adding a slight excess of the free acid, recovering the excess acid by Et$_2$O extraction. The salt is crystd from solution by adding pure acetone, repeating the steps several times, then dried in an oven at *ca* 110° [Chaudhury and Awuwallia *Trans Faraday Soc* 77 3119 *1981*].

Sodium 1-decanesulfonate *[13419-61-9]* **M 244.33.** Recrystd from absolute EtOH and dried over silica gel.

Sodium *n*-decylsulfate *[142-87-0]* **M 239.3.** Rigorously purified by continuous Et$_2$O extraction of a 1% aqueous solution for two weeks.

Sodium deoxycholate (H$_2$O) *[302-95-4]* **M 432.6, [α]$_D^{20}$ +48° (c 1, EtOH).** Crystd from EtOH and dried in an oven at 100°. The solution is freed from soluble components by repeated extraction with acid-washed charcoal.

Sodium dibenzyldithiocarbamate *[55310-46-8]* **M 295.4, m 230°(dec), pK^{20} 3.13 (for monobenzyldithiocarbamic acid).** The free acid when recrystd twice from dry Et$_2$O has **m** 80-82°. The Na salt is reppted from aqueous EtOH or EtOH by addition of Et$_2$O or Me$_2$CO [*Anal Chem* 50 896 *1978*]. The *NH$_4$ salt* has **m** 130-133°; *Cu salt* (yellow crystals) has **m** 284-286° and the *Ti salt* has **m** 64-70°.

Sodium 2,5-dichlorobenzenesulfonate *[5138-93-2]* **M 249.0, pK$_{Est}$ <0 (for SO$_3$H).** Crystd from MeOH, and dried under vacuum.

Sodium dichromate *[7789-12-0]* **M 298.0, m 84.6° (2H$_2$O), 356° (anhydr); b 400°(dec), d$_4^{25}$ 2.348.** Crystd from small volumes of H$_2$O by evaporation to crystallisation. Solubility in H$_2$O is 238% at 0° and 508% at boiling. Red dihydrate is slowly dehydrated by heating at 100° for long periods. It is deliquescent, a powerful oxidising agent-*do not place in contact with skin- wash immediately as it is **caustic.*** (Possible **carcinogen**).

Sodium diethyldithiocarbamate (3H$_2$O) *[20624-25-3]* **M 225.3, m 94-96°(anhydr), pK^{20} 3.65 (diethyldithiocarbamic acid).** Recrystd from water.

Sodium di(ethylhexyl)sulfosuccinate (Aerosol-OT) *[577-11-7]* **M 444.6.** Dissolved in MeOH and inorganic salts which ppted were filtered off. Water was added and the solution was extracted several times with hexane. The residue was evaporated to one fifth its original volume, *benzene was added and azeotropic distillation was continued until no water remained. Solvent was then evaporated. The white solid was crushed and dried in vacuum over P$_2$O$_5$ for 48h [El Seoud and Fendler *J Chem Soc, Faraday Trans 1* **71** 452 *1975*].

Sodium diethyloxaloacetate *[63277-17-8]* **M 210.2.** Extracted several times with boiling Et$_2$O (until the solvent remained colourless) and then the residue was dried in air.

Sodium diformylamide *[18197-26-7]* **M 95.0.** Grind under dry tetrahydrofuran (fumehood), filter and wash with this solvent then dry in vacuum. [IR and prepn: *Synthesis* 122 *1990*; *Chem Ber* **100** 355 *1967*, **102** 4089 *1969*.]

Sodium dihydrogen orthophosphate (2H$_2$O) *[13472-35-0 (2H$_2$O); 10049-21-5 (H$_2$O); 7558-80-7 (anhydr)]* **M 156.0, m 60°(dec), d 1.91.** Crystd from warm water (0.5mL/g) by chilling.

Sodium 2,2'-dihydroxy-1-naphthaleneazobenzene-5'-sulfonate *[2092-55-9]* **M 354.3.** See Solochrome Violet R on p. 352 in Chapter 4.

Sodium 2,4-dihydroxyphenylazobenzene-4'-sulfonate *[547-57-9]* **M 304.2.** Crystd from absolute EtOH.

Sodium p-(p-dimethylaminobenzeneazo)-benzenesulfonate *[23398-40-5]* **M 327.3.** Crystd from water.

Sodium p-dimethylaminoazobenzene-o'-carboxylate *[845-10-3]* **M 219.2.** Ppted from aqueous soln as the free acid which was recrystallised from 95% EtOH, then reconverted to the sodium salt.

Sodium p-dimethylaminoazobenzene-p'-carboxylate *[845-46-5]* **M 219.2.** Ppted from aqueous soln as the free acid which was recrystallised from 95% EtOH, then reconverted to the sodium salt.

Sodium 2,4-dimethylbenzenesulfonate *[827-21-4]* **M 208.2.** Crystd from MeOH and dried under vacuum.

Sodium 2,5-dimethylbenzenesulfonate *[827-19-0]* **M 208.2.** Dissolved in distilled water, filtered, then evaporated to dryness. Crystd twice form absolute EtOH or MeOH and dried at 110° under vacuum.

Sodium dimethyldithiocarbamate hydrate *[128-04-1]* **M 143.2, m 106-108°, 120-122°, pK^{25} 3.36 (diethyldithiocarbamic acid).** Crystallise from a small volume of H$_2$O, or dissolve in minimum volume of H$_2$O and add cold Me$_2$CO and dry in air. The solution in Me$_2$CO is ~50g/400mL. The dihydrate loses H$_2$O on heating at 115° to give the hemi hydrate which decomposes on further heating [IR: *Can J Chem.* **34** 1096 *1956*].

Sodium N,N-dimethylsulfanilate *[2244-40-8]* **M 223.2, m >300°.** Crystd from water.

Sodium dithionite (2H$_2$O) *[7631-94-9]* **M 242.1, m 110°(loses 2H$_2$O), 267°(dec), d 2.19, pK$_{Est(1)}$ -3.4, pK$_2^{25}$ 0.49 (for dithionic acid).** Crystd from hot water (1.1mL/g) by cooling.

Sodium dodecanoate (sodium laurate) *[629-25-4]* **M 222.3, pK^{20} 5.3 (-COOH).** Neutralised by adding a slight excess of dodecanoic acid, removing it by ether extraction. The salt is recrystd from aq soln by adding pure Me$_2$CO and repeating the process (see sodium decanoate on p. 468). Also recrystd from MeOH.

Sodium 1-dodecanesulfonate *[2386-53-0]* **M 272.4.** Twice recrystd from EtOH.

Sodium dodecylbenzenesulfonate *[25155-30-0]* **M 348.5.** Recrystd from propan-2-ol.

Sodium dodecylsulfate (SDS, sodium laurylsulfate) *[151-21-3]* **M 288.4, m 204-207°.** Purified by Soxhlet extraction with pet ether for 24h, followed by dissolution in acetone:MeOH:H$_2$O 90:5:5(v/v) and recrystn [Politi et al. *J Phys Chem* **89** 2345 *1985*]. Also purified by two recrystns from absolute EtOH, aqueous 95% EtOH, MeOH, isopropanol or a 1:1 mixture of EtOH:isopropanol to remove dodecanol, and dried under vacuum [Ramesh and Labes *J Am Chem Soc* **109** 3228 *1987*]. Also purified by foaming [see Cockbain and McMullen *Trans Faraday Soc* **47** 322 *1951*] or by liquid-liquid extraction [see Harrold *J Colloid Sci* **15** 280 *1960*]. Dried over silica gel. For DNA work it should be dissolved in excess MeOH passed through an activated charcoal column and evaporated until it crystallises out.
Also purified by dissolving in hot 95% EtOH (14mL/g), filtering and cooling, then drying in a vacuum desiccator. Alternatively, it was crystd from H$_2$O, vacuum dried, washed with anhydrous Et$_2$O, vacuum dried. These operations were repeated five times [Maritato *J Phys Chem* **89** 1341 *1985*; Lennox and McClelland *J Am Chem Soc* **108** 3771 *1986*; Dressik *J Am Chem Soc* **108** 7567 *1986*].

Sodium ethoxide *[141-52-6]* **M 68.1.** *Hygroscopic* powder which should be stored under N$_2$ in a cool place. Likely impurity is EtOH which can be removed by warming at 60-80° under high vacuum. Hydrolysed by H$_2$O to yield NaOH and EtOH. Other impurities, if kept in air for long periods are NaOH and Na$_2$CO$_3$. In this case the powder cannot be used if these impurities affect the reactivity and a fresh sample should be acquired [IR: *J Org Chem* **21** 156 *1956*].

Sodium ethylmercurithiosalicylate *[54-64-8]* **M 404.8.** Crystd from ethanol-diethyl ether

Sodium ethylsulfate *[546-74-7]* **M 166.1.** Recrystd three times from MeOH-Et$_2$O and vacuum dried.

Sodium ferricyanide (H$_2$O) *[14217-21-1; 13601-19-9 (anhydr)]* **M 298.9, pK^{25} <1 (for ferricyanide).** Crystd from hot water (1.5mL/g) or by precipitation from 95% EtOH.

Sodium ferrocyanide (10H$_2$O) *[13601-19-9]* **M 484.1, m 50-80° (loses 10H$_2$O), 435°(dec), d 1.46, pK$_3^{25}$ 2.57, pK$_4^{25}$ 4.35 (for ferrocyanide).** Crystd from hot water (0.7mL/g), until free of ferricyanide as shown by absence of Prussian Blue formation with ferrous sulfate soln.

Sodium fluoride *[7681-49-4]* **M 42.0, m 996°, b 1695°, d 2.56.** Crystd from water by partial evaporation in a vacuum desiccator,. or dissolved in water, and *ca* half of it ppted by addition of EtOH. Ppte was dried in an air oven at 130° for one day, and then stored in a desiccator over KOH.

Sodium fluoroacetate (mono) *[62-74-8]* **M 100.0, m 200-205°(dec).** A free flowing white **TOXIC** powder which is purified by dissolving in *ca* 4 parts of H$_2$O and the pH is checked. If it is alkaline, add a few drops of FCH$_2$CO$_2$H to make the solution just acidic. Evaporate (fumehood) on a steam bath until crystals start to separate, cool and filter the solid off. More solid can be obtained by adding EtOH to the filtrate. Dry at 100° in vacuum. [*J Chem Soc* 1778 *1948*.]

Sodium fluoroborate *[13755-29-8]* **M 109.8, m 384°, d 2.47, pK -4.9 (for fluoroboric acid H$_3$O$^+$BF$_4^-$).** Crystd from hot water (50mL/g) by cooling to 0°. Alternatively, purified from insoluble material by dissolving in a minimum amount of water, then fluoride ion was removed by adding conc lanthanum nitrate in excess. After removing lanthanum fluoride by centrifugation, the supernatant was passed through a cation-exchange column (Dowex 50, Na^+-form) to remove any remaining lanthanum [Anbar and Guttman *J Phys Chem* **64** 1896 *1960*]. Also recrystd from anhydrous MeOH and dried in a vacuum at 70° for 16h. It is affected by moisture. [Delville et al. *J Am Chem Soc* **109** 7293 *1987*.]

Sodium fluorosilicate *[16893-85-9]* **M 188.1.** Crystd from hot water (40mL/g) by cooling.

Sodium formaldehyde sulfoxylate dihydrate (sodium hydroxymethylsulfinate, Rongalite) *[149-44-0]* **M 134.1, m 63-64° (dihydrate).** Crystallises from H_2O as the dihydrate, decomposes at higher temperatures. Store in a closed container in a cool place. It is insoluble in EtOH and Et_2O and is a good reducing agent. [X-ray structure: *J Chem Soc* 3064 *1955*.] Note that this compound $\{HOCH_2SO_2Na\}$ should not be confused with formaldehyde sodium bisulfite adduct $\{HOCH_2SO_3Na\}$ from which it is prepared by reduction with Zn.

Sodium formate (anhydrous) *[141-53-7]* **M 68.0, m 253°, d 1.92.** A saturated aqueous solution at 90° (0.8mL water/g) was filtered and allowed to cool slowly. (The final temperature was above 30° to prevent formation of the hydrate.) After two such crystns the crystals were dried in an oven at 130°, then under high vacuum. [Westrum, Chang and Levitin *J Phys Chem* **64** 1553 *1960*; Roecker and Meyer *J Am Chem Soc* **108** 4066 *1986*.] The salt has also been recrystd twice from 1mM DTPA (diethylenetriaminepentaacetic acid which was recrystd 4x from MilliQ water and dried in a vac), then twice from water [Bielski and Thomas *J Am Chem Soc* **109** 7761 *1987*].

Sodium D-gluconate *[527-07-1]* **M 218.1, m 200-205°dec, $[\alpha]_{546}^{20}$ +14°, $[\alpha]_D^{20}$ +12 (c 20, H_2O).** Crystallise from a small volume of H_2O (sol 59g/100mL at 25°), or dissolve in H_2O and add EtOH since it is sparingly soluble in EtOH. Insoluble in Et_2O. It forms a Cu comples in alkaline soln and a complex with Fe in neutral solution. [*J Am Chem Soc* **81** 5302 *1959*.]

Sodium glycochenodeoxycholate *[16564-43-5]* **M 472.6.** Dissolved in EtOH, filtered and concentrated to crystallisation, and recrystallised from a little EtOH.

Sodium glycocholate *[863-57-0]* **M 488.6.** Dissolved in EtOH, filtered and concentrated to crystallisation, and recrystallised from a little EtOH.

Sodium glycolate (2H_2O) *[2836-32-0]* **M 98.0.** Ppted from aqueous solution by EtOH, and air dried.

Sodium hexadecylsulfate *[1120-01-0]* **M 344.5.** Recrystd from absolute EtOH or MeOH and dried in vac [Abu Hamdiyyah and Rahman *J Phys Chem* **91** 1531 *1987*].

Sodium hexafluorophosphate *[21324-39-0]* **M 167.9, pK_1^{25}~ 0.5, pK_2^{25} 5.12 (for fluorophosphoric acid H_2PO_3F).** Recrystd from acetonitrile and vacuum dried for 2 days at room temperature. It is an **irritant** and is *hygroscopic*. [Delville et al. *J Am Chem Soc* **109** 7293 *1987*.]

Sodium hexanitrocobaltate III ($Na_3[Co(NO)_6]$) *[13600-98-1]* **M 403.9.** Dissolve (*ca* 60g) in H_2O (300mL), filter to obtain a clear solution, add 96% EtOH (250mL) with vigorous stirring. Allow the ppte to settle for 2h, filter, wash with EtOH (4 x 25mL), twice with Et_2O and dry in air [*Handbook of Preparative Inorganic Chemistry (Ed. Brauer)* Vol II 1541 *1965*]. Yellow to brown yellow crystals which are very soluble in H_2O, are decomposed by acid and form an insoluble K salt. Used for estimating K.

Sodium hydrogen diglycollate *[50795-24-9]* **M 156.1.** Crystd from hot water (7.5mL/g) by cooling to 0° with constant stirring, the crystals being filtered off on to a sintered-glass funnel and dried at 110° overnight.

Sodium hydrogen oxalate (2H_2O) *[1186-49-8]* **M 130.0, m 100°(loses 2H_2O), b 200°(dec).** Crystd from hot water (5mL/g) by cooling.

Sodium hydrogen succinate *[2922-54-5]* **M 140.0.** Crystd from water and dried at 110°.

Sodium hydrogen *d*-tartrate *[526-94-3]* **M 190.1, m 100°(loses H$_2$O), b 234°, [α]$_{546}$ +26° (c 1, H$_2$O).** Crystd from warm water (10mL/g) by cooling to 0°.

Sodium hydroxide (anhydrous) *[1310-73-2]* **M 40.0, m 323°, b 1390°, d 2.13.** Common impurities are water and sodium carbonate. Sodium hydroxide can be purified by dissolving 100g in 1L of pure EtOH, filtering the solution under vacuum through a fine sintered-glass disc to remove insoluble carbonates and halides. (This and subsequent operations should be performed in a dry, CO$_2$-free box.) The soln is concentrated under vacuum, using mild heating, to give a thick slurry of the mono-alcoholate which is transferred to a coarse sintered-glass disc and pumped free of mother liquor. After washing the crystals several times with purified alcohol to remove traces of water, they are vacuum dried, with mild heating, for about 30h to decompose the alcoholate, leaving a fine white crystalline powder [Kelly and Snyder *J Am Chem Soc* **73** 4114 *1951*].

Sodium hydroxide solutions (*caustic*), pK^{25} 14.77. Carbonate ion can be removed by passage through an anion-exchange column (such as Amberlite IRA-400; OH$^-$-form). The column should be freshly prepared from the chloride form by slow prior passage of sodium hydroxide soln until the effluent gives no test for chloride ions. After use, the column can be regenerated by washing with dilute HCl, then water. Similarly, other metal ions are removed when a 1M (or more dilute) NaOH soln is passed through a column of Dowex ion-exchange A-1 resin in its Na$^+$-form.

Alternatively, carbonate contamination can be reduced by rinsing sticks of NaOH (analytical reagent quality) rapidly with H$_2$O, then dissolving in distilled H$_2$O, or by preparing a concentrated aqueous soln of NaOH and drawing off the clear supernatant liquid. (Insoluble Na$_2$CO$_3$ is left behind.) Carbonate contamination can be reduced by adding a slight excess of conc BaCl$_2$ or Ba(OH)$_2$ to a NaOH soln, shaking well and allowing the BaCO$_3$ ppte to settle. If the presence of Ba in the soln is unacceptable, an electrolytic purification can be used. For example, sodium amalgam is prepared by the electrolysis of 3L of 30% NaOH with 500mL of pure mercury for cathode, and a platinum anode, passing 15 Faradays at 4Amps, in a thick-walled polyethylene bottle. The bottle is then fitted with inlet and outlet tubes, the spent soln being flushed out by CO$_2$-free N$_2$. The amalgam is then washed thoroughly with a large volume of deionised water (with the electrolysis current switched on to minimize loss of Na). Finally, a clean steel rod is placed in contact in the solution with the amalgam (to facilitate hydrogen evolution), reaction being allowed to proceed until a suitable concentration is reached, before being transferred to a storage vessel and diluted as required [Marsh and Stokes *Aust J Chem* **17** 740 *1964*].

Sodium 2-hydroxy-4-methoxybenzophenone-5-sulfonate *[6628-37-1]* **M 330.3.** Crystd from MeOH and dried under vacuum.

Sodium *p*-hydroxyphenylazobenzene-*p'*-sulfonate *[2623-36-1]* **M 288.2.** Crystd from 95% EtOH.

Sodium hypophosphite monohydrate *[10039-56-2]* **M 106.0 (see pK of hypophosphorous acid).** Dissolve in boiling EtOH, cool and add dry Et$_2$O till all the salt separates. Collect and dry in vacuum. It is soluble in 1 part of H$_2$O. It liberates PH$_3$ on heating and can *ignite* spontaneously when heated. The anhydrous salt is soluble in ethylene glycol (33% w/w) and propylene glycol (9.7%) at 25°.

Sodium iodate *[7681-55-2]* **M 197.9, m dec on heating, d 4.28.** Crystd from water (3mL/g) by cooling.

Sodium iodide *[7681-82-5]* **M 149.9, m 660°, b 1304°, d 3.67.** Crystd from water/ethanol soln and dried for 12h under vacuum, at 70°. Alternatively, dissolved in acetone, filtered and cooled to -20°, the resulting yellow crystals being filtered off and heated in a vacuum oven at 70° for 6h to remove acetone. The NaI was then crystd from very dilute NaOH, dried under vacuum, and stored in a vacuum desiccator [Verdin *Trans Faraday Soc* **57** 484 *1961*].

Sodium ionophore I (ETH 227) **(*N,N',N''*-triheptyl-*N,N',N''*-trimethyl-4,4',4''-propylidyne-tris(3-oxabutyramide)** *[61183-76-4]* **M 642.0.** It is purified (*ca* 200mg) by TLC on Kieselgel F$_{254}$ with CHCl$_3$/Me$_2$CO (1:1) as solvent, followed by HPLC (50mg) with an octadecyltrimethylsilane modified column (Mercksorb SI 100, 10μm) [IR, NMR, MS: *Helv Chim Acta* **59** 2417 *1976*].

Sodium ionophore V (ETH 4120) [4-octadecanoyloxymethyl-*N,N,N',N'*-tetracyclohexyl-1,2-phenylenedioxydiacetamide] *[129880-73-5]* **M 849.3.** Purified by recrystn from EtOAc. [Preparation and properties: *Anal Chim Acta* **233** 295 *1990*].

Sodium ionophore VI {bis[(12-crown-4)methyl]dodecyl methyl malonate} *[80403-59-4]* **M 662.9.** Purified by gel permeation or column chromatography. [Preparation and NMR data: *J Electroanal Chem* **132** 99 *1982*.]

Sodium isopropylxanthate *[140-93-2]* **M 158.2, pK 2.16 (for -S⁻).** Crystd from ligroin/ethanol.

Sodium laurate *[629-25-4]* **M 222.3.** See sodium dodecanoate on p. 470.

Sodium *RS*-mandelate *[114-21-6]* **M 174.1.** Crystd from 95% EtOH.

Sodium 2-mercaptoethanesulfonate (MESNA) *[19767-45-4]* **M 164.2, $pK_1^{20}<0$ (SO_3^-), pK_2^{20} 9.53 (SH).** It can be recrystd from H_2O and does not melt below 250°. It can be purified further by converting to the free acid by passing a 2M soln through an ion exchange (Amberlite IR-120) column in the acid form, evaporating the eluate in a vacuum to give the acid as a viscous oil (readily dec) which can be checked by acid and SH titration. It is then dissolved in H_2O, carefully neutralised with aqueous NaOH, evaporated and recrystd from H_2O [*J Am Chem Soc* **77** 6231 *1955*].

Sodium metanilate *[1126-34-7]* **M 195.2.** Crystd from hot water.

Sodium metaperiodate ($NaIO_4$) *[7790-28-5]* **M 213.9, m ~300°(dec), d 4.17.** Crystd from hot water.

Sodium metasilicate ($5H_2O$) *[6834-92-0]* **M 212.1, m 1088°, d 2.4.** Crystd from aqueous 5% NaOH solution.

Sodium methanethiolate [sodium methylmercaptide] *[5188-07-8]* **M 70.1, pK^{25} 10.33 (MeS⁻).** Dissolve the salt (10g) in EtOH (10mL) and add Et_2O (100mL). Cool and collect the ppte, wash it with Et_2O and dry it in vacuum. It is a white powder which is very soluble in EtOH and H_2O. [*Bull Soc Chim Fr* **3** 2318 *1936*.]

Sodium methoxide *[124-41-4]* **M 54.0.** It behaves the same as sodium ethoxide. It is *hygroscopic* and is hydrolysed by moist air to NaOH and EtOH. Material that has been kept under N_2 should be used. If erratic results are obtained, even with recently purchased NaOMe it should be freshly prepared thus: Clean Na (37g) cut in 1-3g pieces is added in small portions to stirred MeOH (800mL) in a 2L three necked flask equipped with a stirrer and a condenser with a drying tube. After all the Na has dissolved the MeOH is removed by distillation under vacuum and the residual NaOMe is dried by heating at 150° under vacuum and kept under dry N_2 [*Org Synth* **39** 51 *1959*].

Sodium 3-methyl-1-butanesulfonate *[5343-41-9]* **M 174.1.** Crystd from 90% MeOH.

Sodium molybdate ($2H_2O$) *[10102-40-6]* **M 241.9, m 100°(loses $2H_2O$), 687°, d 3.28, pK^{25} 4.08 (for H_2MoO_4).** Crystd from hot water (1mL/g) by cooling to 0°.

Sodium monensin *[22373-78-0]* **M 693.8.** Recrystd from EtOH-H_2O [Cox et al. *J Am Chem Soc* **107** 4297 *1985*].

Sodium 1-naphthalenesulfonate *[130-14-3]* **M 230.2.** Recrystd from water or aqueous acetone [Okadata et al. *J Am Chem Soc* **108** 2863 *1986*].

Sodium 2-naphthalenesulfonate *[532-02-5]* **M 230.2.** Crystd from hot 10% aqueous NaOH or water, and dried in a steam oven.

Sodium 2-naphthylamine-5,7-disulfonate *[79004-97-0]* **M 235.4.** Crystd from water (charcoal) and dried in a steam oven.

Sodium nitrate *[7631-99-4]* **M 85.0, m 307°, b 380°, d 2.26.** Crystd from hot water (0.6mL/g) by cooling to 0°, or from concentrated aqueous solution by addition of MeOH. Dried under vacuum at 140°. After 2 recrystns tech grade had K, Mg, B, Fe Al, and Li at 100, 29, 0.6, 0.4, 0.2 and 0.2 ppm resp.

Sodium nitrite *[7632-00-0]* **M 69.0, m 271°, b 320°, d 2.17.** Crystd from hot water (0.7mL/g) by cooling to 0°, or from its own melt. Dried over P_2O_5.

Sodium 1-octanesulfonate *[5324-84-5]* **M 216.2.** Recrystd from absolute EtOH.
Sodium oleate *[143-19-1]* **M 304.4, m 233-235°.** Crystd from EtOH and dried in an oven at 100°.

Sodium oxalate *[62-76-0]* **M 134.0, m 250-270°(dec), d 2.34.** Crystd from hot water (16mL/g) by cooling to 0°. Before use as a volumetric standard, analytical grade quality sodium oxalate should be dried for 2h at 120° and allowed to cool in a desiccator.

Sodium palmitate *[408-35-5]* **M 278.4, m , 270°, 285-201°.** Crystd from EtOH, dried in an oven.

Sodium perchlorate (anhydrous) *[7601-89-0]* **M 122.4, m 130°(for monohydrate), d 2.02, pK^{25} -2.4 to -3.1 (for $HClO_4$).** Because its solubility in water is high (2.1g/mL at 15°) and it has a rather low temperature coefficient of solubility, sodium perchlorate is usually crystd from acetone, MeOH, water-ethanol or dioxane-water (33g dissolved in 36mL of water and 200mL of dioxane). After filtering and crystallising, the solid is dried under vacuum at 140-150° to remove solvent of crystn. Basic impurities can be removed by crystn from hot acetic acid, followed by heating at 150°. If $NaClO_4$ is ppted from distilled water by adding $HClO_4$ to the chilled solution, the ppte contains some free acid. **EXPLOSIVE**

Sodium phenol-4-sulfonate (2H$_2$O) (4-hydroxybenzenesulfonic acid Na salt) *[825-90-1]* **M 232.2.** Crystd from hot water (1mL/g) by cooling to 0°, or from MeOH, and dried in vacuum.

Sodium phenoxide *[139-02-6]* **M 116.1, m 61-64°.** Washed with Et_2O, then heated under vacuum to 200° to remove any free phenol.

Sodium phenylacetate *[114-70-5]* **M 158.1.** Its aqueous solution was evaporated to crystallisation on a steam bath, the crystals being washed with absolute EtOH and dried under vacuum at 80°.

Sodium o-phenylphenolate (4H$_2$O) *[132-27-4]* **M 264.3.** Crystd from acetone and dried under vacuum at room temperature.

Sodium phosphoamidate *[3076-34-4]* **M 119.0.** Dissolved in water below 10°, and acetic acid added dropwise to pH 4.0 so that the monosodium salt was ppted. The ppte was washed with water and Et_2O, then air dried. Addition of one equivalent of NaOH to the solution gave the sodium salt, the solution being adjusted to pH 6.0 before use [Rose and Heald *Biochem J* 81 339 *1961*].

Sodium phytate (H$_2$O) [*myo*-inositolhexakis(H$_2$PO$_4$) Na salt] *[14306-25-3]* **M 857.9.** Crystd from water.

Sodium piperazine-N,N'-bis(2-ethanesulfonate) H$_2$O (PIPES-Na salt) *[76836-02-7]* **M 364.3.** Crystd from water and EtOH.

Sodium polyacrylate (NaPAA) *[9003-04-7]*. Commercial polyacrylamide was neutralised with an aqueous solution of NaOH and the polymer ppted with acetone. The ppte was redissolved in a small amount of water and freeze-dried. The polymer was repeatedly washed with EtOH and water to remove traces of low molecular weight material, and finally dried in vacuum at 60° [Vink *J Chem Soc, Faraday Trans 1* **75** 1207 *1979*]. Also dialysed overnight against distilled water, then freeze-dried.

Sodium poly(α-L-glutamate). It was washed with acetone, dried, dissolved in water and ppted with isopropanol at 5°. Impurities and low molecular weight fractions were removed by dialysis of the aqueous solution for 50h, followed by ultrafiltration through a filter impermeable to polymers of molecular weights greater the 10^4. The polymer was recovered by freeze-drying. [Mori et al. *J Chem Soc, Faraday Trans 1* 2583 *1978*.]

Sodium propionate *[137-40-6]* **M 96.1, m 287-289°**. Recrystd from H_2O (solubility 10%), and dried by heating at 100° for 4h. Solubility of anhydrous salt in MeOH is 13% at 15° and 13.77% at 68°. It is insoluble in $^*C_6H_6$ and Me_2CO. [*J Chem Soc* 1341 *1934*.]

Sodium pyrophosphate (10H$_2$O) *[13472-36-1]* **M 446.1, d 1.82, pK_1^{25} 1.52, pK_2^{25} 2.36, pK_3^{25} 6.60, pK_4^{25} 9.25 (for pyrophosphoric acid, $H_4P_2O_7$).** Crystd from hot H_2O and air dried at room temp.

Sodium selenate *[13410-01-0]* **M 188.9, pK_1^{25} ~0, pK_2^{25} 1.66 (for selenic acid, H_2SeO_4).** Crystd from water.

Sodium selenite *[10102-18-8]* **M 172.9, m >350°, pK_1^{25} 2.62, pK_2^{25} 8.32 (for H_2SeO_3).** Crystd from water.

Sodium silicate solution *[1344-09-8]* **pK_1^{25} 9.51, pK_2^{25} 11.77 (for silicic acid, H_4SiO_4)** Purified by contact filtration with activated charcoal.

Sodium succinate *[150-90-3]* **M 162.1.** See disodium succinate on p. 421.

Sodium sulfanilate *[515-74-2]* **M 195.2.** Crystd from water.

Sodium sulfate (10H$_2$O) *[7727-73-3 (10H$_2$O); 7757-82-6 (anhydr)]* **M 322.2, m 32°(dec), 884° (anhydr), d 2.68 (anhydr)**. Crystd from water at 30° (1.1mL/g) by cooling to 0°. Sodium sulfate becomes anhydrous at 32°.

Sodium sulfide (9H$_2$O) *[1313-84-4 (9H$_2$O); 1313-82-2 (anhydr)]* **M 240.2, m ~50(loses H$_2$O), 950(anhydr), d 1.43 (10H$_2$O), 1.86 (anhydr).** Some purification of the hydrated salt can be achieved by selecting large crystals and removing the surface layer (contaminated with oxidation products) by washing with distilled water. Other metal ions can be removed from Na_2S solutions by passage through a column of Dowex ion-exchange A-1 resin, Na^+-form. The hydrated salt can be rendered anhydrous by heating in a stream of H_2 or N_2 until water is no longer evolved. (The resulting cake should not be heated to fusion because it is readily oxidised.) Recrystd from distilled water [Anderson and Azowlay *J Chem Soc, Dalton Trans* 469 *1986*].

Sodium sulfite *[7757-83-7]* **M 126.0, d 2.63.** Crystd from warm water (0.5mL/g) by cooling to 0°. Purified by repeated crystns from deoxygenated water inside a glove-box, finally drying under vacuum. [Rhee and Dasgupta *J Phys Chem* **89** 1799 *1985*.]

Sodium R-tartrate (2H$_2$O) *[6106-24-7]* **M 230.1, m 120°(loses H$_2$O), d 1.82.** Crystd from warm dilute aqueous NaOH by cooling.

Sodium taurocholate *[145-42-6]* **M 555.7.** Purified by recrystn and gel chromatography using Sephadex LH-20.

Sodium tetradecylsulfate *[1191-50-0]* **M 316.4.** Recrystd from absolute EtOH [Abu Hamdiyyah and Rahman *J Phys Chem* **91** 1531 *1987*].

Sodium tetrafluoroborate   *[13755-29-8]*   **M 109.8, d 2.47, pK^{25} -4.9 (for HBF$_4$).** See   Sodium fluoroborate on p. 470.

Sodium tetrametaphosphate *[13396-41-3]* **M 429.9, pK$_1^{25}$ 2.60, pK$_2^{25}$ 6.4, pK$_3^{25}$ 8.22, pK$_4^{25}$ 11.4 (tetrametaphosphoric acid, H$_4$P$_4$O$_{12}$).** Crystd twice from water at room temperature by adding EtOH (300g of Na$_4$P$_4$O$_{12}$,H$_2$O, 2L of water, and 1L of EtOH), washed first with 20% EtOH then with 50% EtOH and air dried [Quimby *J Phys Chem* **58** 603 *1954*].

Sodium tetraphenylborate [tetraphenyl boron Na] *[143-66-8]* **M 342.2.** Dissolve in dry MeOH and add dry Et$_2$O. Collect the solid and dry in a vacuum at 80°/2mm for 4h. Also can be extracted (Soxhlet) using CHCl$_3$ and crystallises from CHCl$_3$ as snow white needles. It is freely sol in H$_2$O, Me$_2$CO but insol in pet ether and Et$_2$O. An aqueous soln has pH ~ 5 and can be stored for days at 25° or lower, and for 5 days at 45° without deterioration. Its solubility in polar solvents increases with decrease in temp [*Justus Liebigs Ann Chem* **574** 195 *1950*]. The salt can also be recrystd from acetone-hexane or CHCl$_3$, or from Et$_2$O-cyclohexane (3:2) by warming the soln to ppte the compound. Dried in a vacuum at 80°. Dissolved in Me$_2$CO and added to an excess of toluene. After a slight milkiness developed on standing, the mixture was filtered. The clear filtrate was evaporated at room temperature to a small bulk and again filtered. The filtrate was then warmed to 50-60°, giving clear dissolution of crystals. After standing at this temperature for 10min the mixture was filtered rapidly through a pre-heated Büchner funnel, and the crystals were collected and dried in a vacuum desiccator at room temperature for 3 days [Abraham et al. *J Chem Soc, Faraday Trans 1* **80** 489 *1984*]. If the product gives a turbid aq solution, the turbidity can be removed by treating with freshly prepared alumina gel.

Sodium thioantimonate (Na$_3$SbS$_4$.9H$_2$O, Schlippe's salt) *[13776-84-6]* **M 481.1, m 87°, b 234°, d 1.81.** Crystd from warm water (2mL/g) by cooling to 0°.

Sodium thiocyanate *[540-72-7]* **M 81.1, m 300°, pK^{25} -1.85 (for HSCN).** It is recrystd from EtOH or Me$_2$CO and the mother liquor is removed from the crystals by centrifugation. It is very deliquescent and should be kept in an oven at 130° before use. It can be dried in vacuum at 120°/P$_2$O$_5$ [*Trans Faraday Soc* **30** 1104 *1934*]. Its solubility in H$_2$O is 113% at 10°, 178% at 46°, 225.6% at 101.4°; in MeOH 35% at 15.8°, 51% at 48°, 53.5% at 52.3°; in EtOH 18.4% at 18.8°, 24.4% at 70.9°; and in Me$_2$CO 6.85% at 18.8° and 21.4% at 56° [*J Chem Soc* 2282 *1929*].
Sodium thiocyanate has also been recrystd from water, acetonitrile or from MeOH using Et$_2$O for washing, then dried at 130°, or dried under vacuum at 60° for 2 days. [Strasser et al. *J Am Chem Soc* **107** 789 *1985*; Szezygiel et al. *J Am Chem Soc* **91** 1252 *1987*.] (The latter purification removes material reacting with iodine.) Sodium thiocyanate solns can be freed from traces of iron by repeated batch extractions with Et$_2$O.

Sodium thioglycolate *[367-51-1]* **M 114.1.** Crystd from 60% EtOH (charcoal). Hygroscopic.

Sodium thiosulfate (5H$_2$O) *[10102-17-7 (hydr); 7772-98-7 (anhydr)]* **M 248.2(anhydr), m 48(rapid heat), d 1.69, pK$_1^{25}$ 0.6, pK$_2^{25}$ 1.74 (for H$_2$S$_2$O$_3$)** Crystd from EtOH-H$_2$O solns or from water (0.3mL/g) below 60° by cooling to 0°, and dried at 35° over P$_2$O$_5$ under vacuum.

Sodium p-toluenesulfinate *[824-79-3]* **M 178.2, pK^{25} 2.80 (1.99)(for -SO$_2^-$).** Crystd from water (to constant UV spectrum), and dried under vacuum or extracted with hot *benzene, then dissolved in EtOH-H$_2$O and heated with decolorising charcoal. The solution was filtered and cooled to give crystals of the dihydrate.

Sodium p-toluenesulfonate *[657-84-1]* **M 194.2, pK^{25} -1.34 (for -SO$_3^-$).** Dissolved in distilled water, filtered to remove insoluble impurities and evaporated to dryness. Then crystd from MeOH or EtOH, and dried at 110°. Its solubility in EtOH is not high (maximum 2.5%) so that Soxhlet extraction with EtOH may be preferable. Sodium p-toluenesulfonate has also been crystd from Et$_2$O and dried under vacuum at 50°.

Sodium trifluoroacetate *[2923-18-4]* **M 136.0, m 206-210°(dec), pK^{25} 0.52 (for CF$_3$CO$_2^-$).** A possible contaminant is NaCl. The solid is treated with CF$_3$CO$_2$H and evaporated twice. Its solubility in CF$_3$CO$_2$H is 13.1% at 29.8°. The residue is crystd from dil EtOH and the solid dried in vacuum at 100°. [*J*

Am Chem Soc **76** 4285 *1954*.] It can be ppted from EtOH by adding dioxane, then crystd several times from hot absolute EtOH. Dried at 120-130°/1mm.

Sodium 2,2',4-trihydroxyazobenzene-5'-sulfonate *[3564-26-9]* M 295.3. Purified by precipitating the free acid from aqueous solution using concentrated HCl, then washing and extracting with EtOH in a Soxhlet extractor. Evaporation of the EtOH left the purified acid.

Sodium trimetaphosphate (6H$_2$O) *[7785-84-4]* M 320.2, m 53°, d 1.79, pK$_2^{25}$1.64, pK$_3^{25}$ 2.07 (for trimetaphosphoric acid, H$_3$P$_3$O$_9$). Ppted from an aq soln at 40° by adding EtOH. Air dried.

Sodium 2,4,6-trimethylbenzenesulfonate *[6148-75-0]* M 222.1. Crystd twice from MeOH and dried under vacuum.

Sodium trimethylsilanolate (sodium trimethylsilanol) *[18027-10-6]* M 112.2, m 230°(dec). It is very soluble in Et$_2$O and *C$_6$H$_6$ but moderately soluble in pet ether. It is purified by sublimation at 130-150° in a high vacuum. [IR: *J Am Chem Soc* **75** 5615 *1953*; *J Org Chem* **17** 1555 *1952*.]

Sodium tripolyphosphate *[7758-29-4]* M 367.9, pK$_1^{25}$ ~ 1, pK$_2^{25}$2.0, pK$_3^{25}$ 2.13, pK$_4^{25}$5.78, pK$_5^{25}$8.56 (for tripolyphosphoric acid, H$_5$P$_3$O$_{10}$). Purified by repeated pptn from aqueous solution by slow addition of MeOH and air dried. Also a solution of anhydrous sodium tripolyphosphate (840g) in water (3.8L) was filtered, MeOH (1.4L) was added with vigorous stirring to ppte Na$_5$P$_3$O$_{10}$.6H$_2$O. The ppte was collected on a filter, air dried by suction, then left to dry in air overnight. It was crystd twice more in this way, using a 13% aqueous solution (w/w), and leaching the crystals with 200mL portions of water [Watters, Loughran and Lambert *J Am Chem Soc* **78** 4855 *1956*]. Similarly, EtOH can be added to ppte the salt from a filtered 12-15% aqueous solution, the final solution containing *ca* 25% EtOH (v/v). Air drying should be at a relative humidity of 40-60%. Heat and vac drying should be avoided. [Quimby *J Phys Chem* **58** 603 *1954*.]

Sodium tungstate (2H$_2$O) *[10213-10-2]* M 329.9, m 698°, d 4.18, pK$_1^{25}$2.20, pK$_2^{25}$3.70 (for tungstic acid, H$_2$WO$_4$). Crystd from hot water (0.8mL/g) by cooling to 0°.

Sodium *m*-xylenesulfonate *[30587-85-0]* M 208.2. Dissolved in distilled water, filtered, then evaporated to dryness. Crystd twice form absolute EtOH and dried at 110°.

Sodium *p*-xylenesulfonate *[827-19-0]* M 208.2. See sodium 2,5-dimethylbenzenesulfonate on p. 469.

Stannic chloride (tin IV chloride, stannic tetrachloride) *[7646-78-8]* M 260.5, m -33°, -30.2°, b 114°/760mm, d 2.23, pK^{25} 14.15 (for aquo Sn^{4+} hydrolysis). Fumes in moist air due to hydrate formation. Fractionate in a ground glass still and store in the absence of air. Possible impurities are SO$_2$ and HCl [Baudler *Handbook of Preparative Inorganic Chemistry (Ed. Brauer)* Vol I p. 729 *1963*]. It forms a solid *pentahydrate [10026-06-9]* which smells of HCl and is formed when the anhydrous salt is dissolved in a small vol of H$_2$O. Also refluxed with clean mercury or P$_2$O$_5$ for several hours, then distd under (reduced) N$_2$ pressure into a receiver containing P$_2$O$_5$. Finally redistd. Alternatively, distd from Sn metal under vacuum in an all-glass system and sealed off in large ampoules. Fumes in moist air. SnCl$_4$ is available commercially as 1M solns in CH$_2$Cl$_2$ or hexane. **HARMFUL VAPOURS.**

Stannic iodide (SnI$_4$) *[7790-47-8]* M 626.3, m 144°, b 340, d 4.46. Crystd from anhydrous CHCl$_3$, dried under vacuum and stored in a vacuum desiccator. Sublimes at 180°.

Stannic oxide (SnO$_2$) *[18282-10-5]* M 150.7, m 1630°, d 6.95. Refluxed repeatedly with fresh HCl until the acid showed no tinge of yellow. The oxide was then dried at 110°.

Stannous bis-cyclopentadienyl *[26078-96-6]* M 248.9. Purified by vacuum sublimation. Handled and stored under dry N$_2$. The related thallium and indium compounds are similarly prepared.

Stannous chloride (anhydrous) *[7772-99-8]* **M 189.6, m 247°, b 606°, d 3.95, pK_1^{25} 1.7, pK_2^{25} 3.7 (for aquo Sn^{2+} hydrolysis).** Analytical reagent grade stannous chloride dihydrate is dehydrated by adding slowly to vigorously stirred, redistilled acetic anhydride (120g salt per 100g of anhydride). (In a fume cupboard.) After *ca* an hour, the anhydrous $SnCl_2$ is filtered on to a sintered-glass or Büchner funnel, washed free from acetic acid wth dry Et_2O (2 x 30mL), and dried under vacuum. It is stored in a sealed container. [Stephen *J Chem Soc* 2786 *1930*].

Strontium acetate *[543-94-2]* **M 205.7, d 2.1, pK^{25} 13.0 (for aquo Sr^{2+} hydrolysis).** Crystd from AcOH, then dried under vacuum for 24h at 100°.

Strontium bromide *[10476-81-0]* **M 247.4, m 643°, d 4.22.** Crystd from water (0.5mL/g).

Strontium chloride ($6H_2O$) *[1025-70-4]* **M 266.6, m 61°(rapid heating), 114-150°(loses $5H_2O$), 868°(anhyd).** Crystd from warm water (0.5mL/g) by cooling to 0°.

Strontium chromate *[7789-06-2]* **M 203.6, d 3.9, pK_1^{25} 0.74, pK_2^{25} 6.49 (for H_2CrO_4).** Crystd from water (40mL/g) by cooling.

Strontium hydroxide ($8H_2O$) *[1311-10-0 ($8H_2O$); 18480-07-4 (anhydr)]* **M 265.8, m 100°(loses H_2O), d 1.90, m 375(anhydr), d 3.63 (anhydr).** Crystd from hot water (2.2mL/g) by cooling to 0°.

Strontium lactate ($3H_2O$) *[29870-99-3]* **M 319.8, m 120°(loses $3H_2O$).** Crystd from aq EtOH.

Strontium nitrate *[10042-76-9]* **M 211.6, m 570°, b 645°, d 2.99.** Crystd from hot water (0.5mL/g) by cooling to 0°.

Strontium oxalate (H_2O) *[814-95-9]* **M 193.6, m 150°.** Crystd from hot water (20mL/g) by cooling.

Strontium salicylate *[526-26-1]* **M 224.7.** Crystd from hot water (4mL/g) or EtOH.

Strontium tartrate *[868-19-9]* **M 237.7.** Crystd from hot water.

Strontium thiosalicylate ($5H_2O$) *[15123-90-7]* **M 289.8.** Crystd from hot water (2mL/g) by cooling to 0°.

Sulfamic acid *[5329-14-6]* **M 97.1, m 205°(dec), pK^{25} 0.99 (NH_2SO_3H).** Crystd from water at 70° (300mL per 25g), after filtering, by cooling a little and discarding the first batch of crystals (about 25g) before standing in an ice-salt mixture for 20min. The crystals were filtered by suction, washed with a small quantity of ice water, then twice with cold EtOH and finally with Et_2O. Air dried for 1h, then stored in a desiccator over $Mg(ClO_4)_2$ [Butler, Smith and Audrieth *Ind Eng Chem (Anal Ed)* **10** 690 *1938*]. For preparation of primary standard material see *Pure Appl Chem* **25** 459 *1969*.

Sulfamide *[7803-58-9]* **M 96.1, m 91.5°.** Crystd from absolute EtOH.

Sulfur *[7704-34-9]* **M 32.1, m between 112.8° and 120°, depending on form.** Murphy, Clabaugh and Gilchrist [*J Res Nat Bur Stand* **64A** 355 *1960*] have obtained sulfur of about 99.999% purity by the following procedure: Roll sulfur was melted and filtered through a coarse-porosity glass filter funnel into a 2L round-bottomed Pyrex flask with two necks. Conc H_2SO_4 (300mL) was added to the sulfur (2.5Kg), and the mixture was heated to 150°, stirring continuously for 2h. Over the next 6h, conc HNO_3 was added in about 2mL portions at 10-15min intervals to the heated mixture. It was then allowed to cool to room temperature and the acid was poured off. The sulfur was rinsed several times with distilled water, then remelted, cooled, and rinsed several times with distd water again, this process being repeated four or five times to remove most of the acid entrapped in the sulfur. An air-cooled reflux tube (*ca* 40cm long) was attached to one of the necks of the flask, and a gas delivery tube (the lower end about 1in above the bottom of the flask) was inserted into the other. While the sulfur was boiled under reflux, a stream of helium or N_2 was passed through to remove any water, HNO_3 or H_2SO_4, as vapour. After 4h, the sulfur was cooled so that the reflux tube could be replaced by a bent air-cooled condenser. The sulfur was then distilled, rejecting the first and the final 100mL portions, and transferred in 200mL portions to 400mL glass cylinder ampoules (which were placed on their sides during solidification). After adding about 80mL of water, displacing the air with N_2, and sealing the ampoule was cooled, and the water was titrated with 0.02M NaOH, the process being repeated until the acid content was negligible. Finally, entrapped water was removed by alternate evacuation to 10mm Hg and refilling with N_2 while the sulfur was kept molten. Other purifications include crystn from CS_2 (which is less satisfactory because the sulfur retains appreciable amounts of organic material), *benzene or *benzene/acetone, followed by melting and degassing. Has also been boiled with 1% MgO, then decanted, and dried under vacuum at 40° for 2 days over P_2O_5. [For purification of S_6, "recryst. S_8" and "Bacon-Fanelli sulfur" see Bartlett, Cox and Davis *J Am Chem Soc* **83** 103, 109 *1961*.]

Sulfur dichloride *[10545-99-0]* **M 103.0, m -78°, b 59°/760mm(dec), d 1.621.** Twice distilled in the presence of a small amount of PCl_3 through a 12in Vigreux column, the fraction boiling between 55-61° being redistd (in the presence of PCl_3), and the fraction distilling between 58-61° retained. (The PCl_3 is added to inhibit the decomposition of SCl_2 into S_2Cl_2 and Cl_2). The SCl_2 must be used as quickly as possible after distn, within 1h at room temperature, The sample contains 4% S_2Cl_2. On long standing this reaches 16-18%. **HARMFUL VAPOURS.**

Sulfur dioxide *[7446-09-5]* **M 64.1, b -10°.** Dried by bubbling through concentrated H_2SO_4 and by passage over P_2O_5, then passed through a glass-wool plug. Frozen with liquid air and pumped to a high vacuum to remove dissolved gases. **HARMFUL VAPOURS.**

Sulfuric acid *[7664-93-9]* **M 98.1, d 1.83, $pK_1^{25} \sim$ -8.3, pK_2^{25} 1.99.** Sulfuric acid, and also 30% fuming H_2SO_4, can be distilled in an all-Pyrex system, optionally from potassium persulfate. Also purified by fractional crystn of the monohydrate from the liquid. Dehydrates and attacks skin—wash immediately with H_2O.

Sulfur monochloride (sulfur monochloride) *[10025-67-9]* **M 135.0, m -77°; b 19.1°, 29-30°/12mm, 72°/100mm, 138°/760mm, d^{20} 1.677, n_D^{20} 1.67.** *Pungent, irritating golden yellow liquid.* When impure its colour is orange to red due to SCl_2 formed. It fumes in moist air and liberates HCl, SO_2 and H_2S in the presence of H_2O. Distil and collect the fraction boiling above 137° at atmospheric pressure. Fractionate this fraction over sulfur at *ca* 12mm using ground glass apparatus (b 29-30°). Alternatively purify by distn below 60° from a mixture containing sulfur (2%) and activated charcoal (1%), under reduced pressure (e.g. 50mm). It is soluble in EtOH, *C_6H_6, Et_2O, CS_2 and CCl_4. Store in a closed container in the dark in a refrigerator. [*Handbook of Preparative Inorganic Chemistry (Ed. Brauer)* Vol I 371 *1963*.] **HARMFUL VAPOURS.**

Sulfur trioxide pyridine complex *[26412-87-3]* **M 159.2, m 155-165°, 175°.** Wash the solid with a little CCl_4, then H_2O to remove traces of pyridine sulfate, and dry over P_2O_5 [*Chem Ber* **59** 1166 *1926*; *Synthesis* 59 *1979*].

Sulfuryl chloride *[7791-25-2]* **M 135.0, m -54.1°, b 69.3°/760mm, d_4^{20} 1.67, n_D^{30} 1.44.** *Pungent, irritating colourless liquid.* It becomes yellow with time due to decomposition to SO_2 and HCl. Distil and collect fraction boiling below 75°/atm which is mainly SO_2Cl_2. To remove HSO_3Cl and H_2SO_4 impurities, the distillate is poured into a separating funnel filled with crushed ice and *briefly* shaken. The lower cloudy layer is removed, dried for some time in a desiccator over P_2O_5 and finally fractionated at atmospheric pressure. The middle fraction boils at 69-70° and is pure SO_2Cl_2. It decomposes gradually in H_2O to H_2SO_4 and HCl. Reacts **violently** with EtOH and MeOH and is soluble in *C_6H_6, toluene Et_2O and acetic acid. [*Handbook of Preparative Inorganic Chemistry (Ed Brauer)* Vol I 383 *1963*; *Inorg Synth* **1** 114 *1939*]. **HARMFUL VAPOURS.**

Tantalium (V) chloride (tantalium pentachloride) *[7721-01-9]* M 358.2, m 216.2°, 216.5-220°; b 239°/atm., d 3.68. Purified by sublimation in a stream of Cl_2. Colourless

needles when pure (yellow when contaminated with even less than 1% of $NbCl_5$). Sensitive to H_2O, even in conc HCl it decomposes to tantalic acid. Sol in EtOH. [*J Am Chem Soc* **80** 2952 *1958*; *Handbook of Preparative Inorganic Chemistry (Ed Brauer)* Vol II 1302 *1965*.]

Tantalium pentaethoxide *[6074-84-6]* **M 406.3, b 147º/0.2mm, 202º/10mm, pK_1^{25} 9.6, pK_2^{25} 13.0 (for tantalic acid).** Purified by distillation. It aggregates in $^*C_6H_6$, EtOH, MeCN, pyridine and diisopropyl ether. [*J Chem Soc* 726 *1955*, 5 *1956*.]

Telluric acid *[7803-68-1]* **M 229.6, pK_1^{25} 7.70, pK_2^{25} 11.04 (H_6TaO_6).** Crystd once from nitric acid, then repeatedly from hot water (0.4mL/g).

Tellurium *[13494-80-9]* **M 127.6, m 450º.** Purified by zone refining and repeated sublimation to an impurity of less than 1 part in 10^8 (except for surface contamination by TeO_2). [Machol and Westrum *J Am Chem Soc* **80** 2950 *1958*.] Tellurium is volatile at 500º/0.2mm. Also purified by electrode deposition [Mathers and Turner *Trans Amer Electrochem Soc* **54** 293 *1928*].

Tellurium dioxide *[7446-07-3]* **M 159.6, m 733º, d 6.04.** Dissolved in 5M NaOH, filtered and ppted by adding 10M HNO_3 to the filtrate until the soln was acid to phenolphthalein. After decanting the supernatant, the ppte was washed five times with distilled water, then dried for 24h at 110º [Horner and Leonhard *J Am Chem Soc* **74** 3694 *1952*].

Terbium oxide *[12037-01-3]* **M 747.7, pK^{25} 8.16 (for Tb^{3+} hydrolysis).** Dissolved in acid, ppted as its oxalate and ignited at 650º.

Tetraallyltin (tetraallylstannane) *[7393-43-3]* **M 283.0, b 52º/0.2mm, 69-70º/15mm, d 1.179, n 1.536.** Possible contaminants are allyl chloride and allyltin chloride. Check ^1H NMR and IR [Fishwick and Wallbridge *J Organometal Chem* **25** 69 *1970*], and if impure, dissolve in Et_2O and shake with a 5% aq soln of NaF which ppts allyltin fluoride. Separate the Et_2O layer, dry ($MgSO_4$) and dist at ~ 0.2mm. It decomposes slightly on repeated distn. [O'Brien et al. *Inorg Synth* **13** 75 *1972*; Fishwick et al. *J Chem Soc (A)* 57 *1971*.]

Tetrabutylammonium borohydride *[33725-74-5]* **M 257.3, m 128-129º.** Purified by recrystn from EtOAc followed by careful drying under vacuum at 50-60º. Samples purified in this way showed no signs of loss of *active* H after storage at room temperature for more than 1 year. Nevertheless samples should be stored at *ca* 6º in tightly stoppered bottles if kept for long periods. It is soluble in CH_2Cl_2. [*J Org Chem* **41** 690 *1976*; *Tetrahedron Lett* 3173 *1972*.]

Tetrabutylammonium chlorochromate *[54712-57-1]* **M 377.9, m 184-185º.** Recrystd from EtOAc-hexane. IR v 920cm^{-1} in $CHCl_3$ [*Synthesis* 749 *1983*]. *Powerful oxidant.*

Tetrabutylammonium tetrafluoroborate *[429-42-5]* **M 329.3, m 161.8º, 161-163º, pK^{25} -4.9 (for HBF_4).** Recryst from H_2O, aq EtOH or from EtOAc by cooling in Dry ice. Also recrystd from ethyl acetate/pentane or dry acetonitrile. Dried at 80º under vacuum. [Detty and Jones *J Am Chem Soc* **109** 5666 *1987*; Hartley and Faulkner *J Am Chem Soc* **107** 3436 *1985*.] *Acetate* **m** 118±2º (from BuCl); *bromide* **m** 118º (from EtOAc) and *nitrate* **m** 120º (from $^*C_6H_6$). [*J Am Chem Soc* **69** 2472 *1947*, **77** 2024 *1955*.]

Tetrabutyl orthotitanate monomer (titanium tetrabutoxide) *[5593-70-4]* **M 340.4, b 142º/0.1mm, 134-136º/0.5mm, 160º/0.8mm, 174º/6mm, 189º/13mm, d_4^{35} 0.993, n_D^{25} 1.49.** Dissolve in $^*C_6H_6$, filter if solid is present, evaporate and vacuum fractionate through a Widmer 24inch column. The ester hydrolyses when exposed to air to give hydrated ortho-titanic acid. Titanium content can be determined thus: weigh a sample (*ca* 0.25g) into a weighed crucible and cover with 10mL of H_2O and a few drops of conc HNO_3. Heat (hot plate) carefully till most of the H_2O has evaporated. Cool and add more H_2O (10mL) and conc HNO_3 (2mL) and evaporate carefully (no spillage) to dryness and ignite residue at 600-650º/1h. Weigh the residual TiO_2. [*J Chem Soc* 2773 *1952*; *J Org Chem* **14** 655 *1949*.]

Tetrabutyl tin **(tin tetrabutyl)** *[1461-25-2]* **M 347.2, b 94.5-96°/0.28mm, 145°/11mm, 245-247°/atm, d_4^{20} 1.05, n_D^{24} 1.473.** Dissolve in Et_2O, dry over $MgSO_4$, filter, evaporate and distil under reduced pressure. Although it does not crystallise easily, once the melt has crystallised then it will recrystallise more easily. It is soluble in Et_2O, Me_2CO, EtOAc and EtOH but insoluble in MeOH and H_2O and shows no apparent reaction with H_2O. [*J Org Chem* **19** 74 *1954*, *J Chem Soc* 1992 *1954*.]

Tetraethoxysilane **(tetraethyl orthosilicate)** *[78-10-4]* **M 208.3, m -77°, b 165-166°/atm, d_4^{20} 0.933, n_D^{25} 1.382.** Fractionate through an 80cm Podbielniak type column (see p. 141) with heated jacket and partial take-off head. Slowly decomposed by H_2O, soluble in EtOH. *It is flammable - irritates the eyes and mucous membranes.* [*J Am Chem Soc* **78** 5573 *1956*, cf *J Chem Soc* 5020 *1952*.]

Tetraethylammonium hexafluorophosphate *[429-07-2]* **M 275.2, m >300°, 331°(dec), pK_1^{25}~0.5, pK_2^{25} 5.12 (for fluorophosphoric acid H_2PO_3F).** Dissolve salt (0.8g) in hot H_2O (3.3mL) and cool to crystallise. Yield of prisms is 0.5g. Solubility in H_2O is 8.1g/L at 19° [*Chem Ber* **63** 1067 *1930*].

Tetraethylammonium tetrafluoroborate *[429-06-1]* **M 217.1, m 235°, 356-367°, pK^{25} -4.9 (for HBF_4).** Dissolve in hot MeOH, filter and add Et_2O. It is soluble in ethylene chloride [*J Am Chem Soc* **69** 1016 *1947*, **77** 2025 *1955*]. See entry on p. 359 in Chapter 4.

Tetraethyl lead *[78-00-2]* **M 323.5, b 200°, 227.7°(dec), d 1.653, n 1.5198.** Its more volatile contaminants can be removed by exposure to a low pressure (by continuous pumping) for 1h at 0°. Purified by stirring with an equal volume of H_2SO_4 (d 1.40), keeping the temperature below 30°, repeating this process until the acid layer is colourless. It is then washed with dilute Na_2CO_3 and distilled water, dried with $CaCl_2$ and fractionally distd at low pressure under H_2 or N_2 [Calingaert *Chem Rev* **2** 43 *1926*]. **VERY POISONOUS.**

Tetraethylsilane *[631-36-7]* **M 144.3, b 153.8°/760mm, d_4^{30} 0.77, n_D^{30} 1.427.** Fractionate through a 3ft vacuum jacketted column packed with 1/4" stainless steel saddles. The material is finally percolated through a 2ft column packed with alumina and maintained in an inert atmosphere. [*J Chem Soc* 1992 *1954*; *J Am Chem Soc* **77** 272 *1955*.]

1.1.3.3-Tetraisopropyldisiloxane *[18043-71-5]* **M 246.5, b 129-130°/6mm, d_4^{30} 0.89, n_D^{30} 1.47.** Fractionate under reduced pressure in a N_2 atm. [*J Am Chem Soc* **69** 1500 *1947*.]

Tetraisopropyl orthotitanate **(titanium tetraisopropyl)** *[546-68-9]* **M 284.3, m 18.5°; b 80°/2mm, 78°/12mm, 228-229°/755mm.** Dissolve in dry $^*C_6H_6$, filter if a solid separates, evap and fractionate. It is hydrolysed by H_2O to give solid $Ti_2O(iso\text{-}OPr)_2$ m ca 48°. [*J Chem Soc* 2027, *1952*, 469 *1957*.]

Tetrakis(diethylamino) titanium **[(titanium tetrakis(diethylamide)]** *[4419-47-0]* **M 336.4, b 85-90°/0.1mm, 112°/0.1mm, d_4^{30} 0.93, n_D^{30} 1.54.** Dissolve in $^*C_6H_6$, filter if a solid separates, evaporate under reduced pressure and distil. Orange liquid which reacts violently with alcohols. [*J Chem Soc* 3857 *1960*.]

Tetrakis(hydroxymethyl)phosphonium chloride *[124-64-1]* **M 190.6, m 151°.** Crystd from AcOH and dried at 100° in a vacuum. An 80% w/v aqueous solution has d_4^{20} 1.33 [*J Am Chem Soc* **77** 3923 *1955*].

Tetrakis(triphenylphosphine) palladium *[14221-01-3]* **M 1155.58, m 100-105°(dec).** Yellow crystals from EtOH. It is stable in air only for a short time, and prolonged exposure turns its colour to orange. Store in an inert atmosphere below room temp in the dark. [*J Chem Soc* 1186 *1957*.]

Tetrakis(triphenylphosphine) platinum *[14221-02-4]* **M 1244.3, m 118°.** Recrystd by adding hexane to a cold saturated solution in $^*C_6H_6$. It is soluble in $^*C_6H_6$ and $CHCl_3$ but insoluble in EtOH and hexane. A less pure product is obtained if crystd by adding hexane to a $CHCl_3$ soln. Stable in air for several hours and completely stable under N_2. [*J Am Chem Soc* 2323 *1958*.]

Tetramethoxysilane (tetramethyl orthosilicate) *[681-84-5]* **M 152.2, m 4.5°, b 122°/760mm.** Purification as for tetraethoxysilane. It has a vapour pressure of 2.5mm at 0°. [IR: *J Am Chem Soc* **81** 5109 *1959*.]

Tetramethylammonium borohydride *[16883-45-7]* **M 89.0.** Recrystn from H_2O three times yields *ca* 94% pure compound. Dry in high vacuum at 100° for 3h. The solubility in H_2O is 48% (20°), 61% (40°); and in EtOH 0.5% (25°) and MeCN 0.4% (25°). It decompose slowly in a vacuum at 150°, but rapidly at 250°. The rate of hydrolysis of $Me_4N.BH_4$ (5.8M) in H_2O at 40° is constant over a period of 100h at 0.04% of original wt/h. The rate decreases to 0.02%/h in the presence of Me_4NOH (5% of the wt of $Me_4N.BH_4$). [*J Am Chem Soc* **74** 2346 *1952*.]

Tetramethylammonium hexafluorophosphate *[558-32-7]* **M 219.1, m >300°, d_4^{25} 1.617, pK_1^{25}~ 0.5, pK_2^{25} 5.12 (for fluorophosphoric acid H_2PO_3F).** The salt (0.63g) is recrystd from boiling H_2O (76mL), yielding pure (0.45) $Me_4N.PF_6$ after drying at 100°. It is a good supporting electrolyte. [*Chem Ber* **63** 1067 *1930*.]

Tetramethylammonium perchlorate *[2537-36-2]* **M 123.6, m>300°, pK^{25} -2.4 to -3.1 (for $HClO_4$).** Crystallise twice from H_2O and dry at 100° in an oven. Insol in most organic solvents. [*J Chem Soc* 1210 *1933*.]

Tetramethylammonium triacetoxyborohydride *[109704-53-2]* **M 263.1, m 93-98°, 96.5-98°.** If impure, wash with freshly distd Et_2O and dry overnight in a vac to give a free flowing powder. Check ^{1}H NMR, and if still suspect prepare freshly from Me_4NBH_4 and AcOH in *C_6H_6 and store away from moisture [Banus et al. *J Am Chem Soc* **74** 2346 *1952*; Evans and Chipman *Tetrahedron Lett* **27** 5939 *1986*]. It is an **IRRITANT** and **MOISTURE SENSITIVE**.

Tetramethylammonium triphenylborofluoride *[437-11-6]* **M 392.2.** Crystd from acetone or acetone/ethanol.

2,4,6,8-Tetramethylcyclotetrasiloxane *[2370-88-9]* **M 240.5, m -69°, b 134°/750mm, 134.5-134.9°/755mm, d_4^{20} 0.99, n_D^{20} 1.3872.** It is purified by repeated redistillation, and fractions with the required ^{1}H NMR data are collected. [*J Gen Chem USSR (Engl Transl)* **29** 262 *1959*; *J Am Chem Soc* **68** 962 *1946*].

1,1,3,3-Tetramethyldisiloxane *[3277-26-7]* **M 134.3, b 70.5-71°/731mm, 71-72°/atm, d_4^{30} 0.75, n_D^{25} 11.367.** Possible impurity is 1,1,5,5-tetramethyl-3-trimethylsiloxytrisiloxane b 154-155°/733mm. Fractionate, collect fractions boiling below 80° and refractionate. Purity can be analysed by alkaline hydrolysis and measuring the volume of H_2 liberated followed by gravimetric estimation of silica in the hydrolysate. It is unchanged when stored in glass containers in the absence of moisture for 2-3 weeks. Small amounts of H_2 are liberated on long storage. *Care should be taken when opening a container due to pressure developed.* [*J Am Chem Soc* **79** 974 *1958*; *J Chem Soc* 609 *1958*; IR: *Z Anorg Chem* **299** 78 *1959*.]

N,N,N'N'-Tetramethylphosphonic diamide (methylphosphonic bis-dimethylamide) *[2511-17-3]* **M 150.2, b 60.5°/0.6mm, 138°/32mm, 230-230°/atm, d_4^{30} 1.0157, n_D^{30} 1.4539.** Dissolve in heptane or ethylbenzene shake with 30% aqueous NaOH, stir for 1h, separate the organic layer and fractionate. [*J Org Chem* **21** 413 *1956*]. IR has v 1480, 1460, 1300, 1184, 1065 and 988-970cm[-1] [*Can J Chem* **33** 1552 *1955*].

Tetramethylsilane (TMS) *[75-76-3]* **M 88.2, b 26.3°, n 1.359, d 0.639.** Distilled from conc H_2SO_4 (after shaking with it) or $LiAlH_4$, through a 5ft vacuum-jacketted column packed with glass helices into an ice-cooled condenser, then percolated through silica gel to remove traces of halide.

2,4,6,8-Tetramethyl tetravinyl cyclotetrasiloxane *[2554-06-5]* **M 344.7, m -43.5°, b 111-112°/10mm, 145-146°/13mm, 224-224.5°/758mm, d_4^{30} 0.98, n_D^{30} 1.434.** A 7mL sample was distilled in a small Vigreux column at atmospheric pressure without polymerisation or decomposition. It is soluble in cyclohexane. [*J Am Chem Soc* **77** 1685 *1955*.]

Tetraphenylarsonium chloride hydrate *[507-28-8]* **M 418.8, m 261-263°.** A neutralised aqueous soln was evaporated to dryness. The residue was extracted into absolute EtOH, evaporated to a small volume and ppted by addition of absolute Et_2O. It was again dissolved in a small volume of absolute EtOH or ethyl acetate and reppted with Et_2O. Alternatively purified by adding conc HCl to ppte the chloride dihydrate. Redissolved in water, neutralised with Na_2CO_3 and evaporated to dryness. The residue was extracted with $CHCl_3$ and finally crystallised from CH_2Cl_2 or EtOH by adding Et_2O. If the aqueous layer is somewhat turbid treat with Celite and filter through filter paper. **POISONOUS.**

Tetraphenylarsonium iodide *[7422-32-4]* **M 510.2.** Crystd from MeOH. **POISONOUS.**

Tetraphenylarsonium perchlorate *[3084-10-4]* **M 482.8, pK^{25} -2.4 to -3.1 (for HClO$_4$).** Crystd from MeOH. **POISONOUS.**

Tetraphenylboron potassium *[3244-41-5]* **M 358.2.** See potassium tetraphenylborate on p. 458.

Tetraphenylphosphonium chloride *[2001-45-8]* **M 374.9, m 273-275°.** Crystd from acetone. Dried at 70° under vacuum. Also recrystd from a mixture of 1:1 or 1:2 dichloromethane/pet ether, the solvents having been dried over anhydrous K_2CO_3. The purified salt was dried at room temperature under vasuum for 3 days, and at 170° for a further 3 days. *Extremely hygroscopic.*

Tetraphenylsilane *[1048-08-4]* **M 336.4, m 231-233°.** Crystd from *benzene.

Tetraphenyltin *[595-90-4]* **M 427.1, m 224-225°, 226°.** Yellow crystals from $CHCl_3$, pet ether (b 77-120°), xylene or *benzene/cyclohexane, and dried at 75°/20mm. [*J Am Chem Soc* **74** 531 *1952*.]

Tetrapropylammonium perchlorate *[15780-02-6]* **M 285.8, m 238-240°, pK^{25} -2.4 to -3.1 (for HClO$_4$).** Purified by recrystns from H_2O or MeCN/H_2O (1:4.v/v), and dried in an oven at 60° for several days, or in vacuum over P_2O_5 at 100°. [*Z Phys Chem* **165A** 245 *1933*, **144** 281 *1929*, **140** 97 *1929*.]

Tetra-*n*-propylammonium perruthenate (TPAP, tetrapropyl tetraoxoruthenate) *[114615-82-6]* **M 351.4, m 160°(dec).** It is a strong oxidant and may explode on heating. It can be washed with aq *n*-propanol, then H_2O and dried over KOH in a vac. It is stable at room temp but best stored in a refrigerator. It is sol in CH_2Cl_2 and MeCN. [Dengel et al. *Transition Met Chem* **10** 98 *1985*; Griffith et al. *J Chem Soc, Chem Commun* 1625 *1987*.] § Polymer supported reagent is available commercially.

Tetrasodium pyrene-1,3,6,8-tetrasulfonate *[59572-10-0]* **M 610.5.** Recrystd from aqueous acetone [Okahata et al. *J Am Chem Soc* **108** 2863 *1986*].

Thallium (I) acetate *[563-68-8]* **M 263.4, m 126-128°, 127°, pK^{25} 13.2 (for Tl$^+$).** Likely impurity is H_2O because the white solid is deliquescent. Dry in a vacuum over P_2O_5 or for several days in a desiccator, and store in a well closed container. 7.5g dissolve in 100g of liquid SO_2 at 0°, and *ca* 2mol% in AcOH at 25°. **POISONOUS.** [*Trans Faraday Soc* **32** 1660 *1936*; *J Am Chem Soc* **52** 516.]

Thallous bromide *[7789-40-4]* **M 284.3, m 460°.** Thallous bromide (20g) was refluxed for 2-3h with water (200mL) containing 3mL of 47% HBr. It was then washed until acid-free, heated to 300° for 2-3h and stored in brown bottles. **POISONOUS.**

Thallous carbonate *[6533-73-9]* **M 468.7, m 268-270°.** Crystd from hot water (4mL/g) by cooling. **POISONOUS.**

Thallous chlorate *[13453-30-0]* **M 287.8, d 5.05.** Crystd from hot water (2mL/g) by cooling. **POISONOUS.**

Thallous chloride *[7791-12-0]* **M 239.8, m 429.9°, d 7.0.** Crystd from 1% HCl and washed until acid-free, or crystd from hot water (50mL/g), then dried at 140° and stored in brown bottles. Also purified by subliming in vacuum, followed by treatment with dry HCl gas and filtering while molten. (Soluble in 260 parts of cold water and 70 parts of boiling water). **POISONOUS.**

Thallous hydroxide *[12026-06-1]* **M 221.4, m 139°(dec), pK^{25} 13.2 (for Tl$^+$).** Crystd from hot water (0.6mL/g) by cooling. **POISONOUS.**

Thallous iodide *[7790-30-9]* **M 331.3, m 441.8°, b 824°, d 7.1.** Refluxed for 2-3h with water containing HI, then washed until acid-free, and dried at 120°. Stored in brown bottles. **POISONOUS.**

Thallous nitrate *[10102-45-1]* **M 266.4, m 206°, b 450°(dec), d 5.55.** Crystd from warm water (1mL/g) by cooling to 0°. **POISONOUS.**

Thallous perchlorate *[13453-40-2]* **M 303.8, pK^{25} -2.4 to -3.1 (for HClO$_4$).** Crystd from hot water (0.6mL/g) by cooling. Dried under vacuum for 12h at 100° (protect from possible **EXPLOSION**).

Thallous sulfate *[7446-18-6]* **M 504.8, m 633°, d 6.77.** Crystd from hot water (7mL/g) by cooling, then dried under vacuum over P$_2$O$_5$. **POISONOUS.**

Thexyl dimethyl chlorosilane (dimethyl-[2,3-dimethyl-2-butyl] chlorosilane) *[67373-56-2]* **M 178.8, b 55-56°/10mm, 158-159°/720mm, d$_4^{20}$ 0.970, n$_D^{20}$ 1.428.** Purified by fractional distillation and stored in small aliquots in sealed ampoules. It is very sensitive to moisture and is estimated by dissolving an aliquot in excess of 0.1M NaOH and titrating with 0.1M HCl using methyl red as indicator. [*Helv Chim Acta* **67** 2128 *1984*].

***N*-(Thexyl dimethylsilyl)dimethylamine (*N*-[2,3-dimethyl-2-butyl]dimethylsilyl dimethyl-amine)** *[81484-86-8]* **M 187.4, b 156-160°/720mm.** Dissolve in hexane, filter, evaporate and distil. Colourless oil extremely sensitive to humidity. It is best to store small quatities in sealed ampoules after distillation. For estimation of purity crush an ampoule in excess 0.1N HCl and titrate the excess acid with 0.1M NaOH using methyl red as indicator. [*Helv Chim Acta* **67** 2128 *1984*.]

Thionyl chloride *[7719-09-7]* **M 119.0, b 77°, d 1.636.** Crude SOCl$_2$ can be freed from sulfuryl chloride, sulfur monochloride and sulfur dichloride by refluxing with sulfur and then fractionally distilling twice. [The SOCl$_2$ is converted to SO$_2$ and sulfur chlorides. The S$_2$Cl$_2$ (b 135.6°) is left in the residue, whereas SCl$_2$ (b 59°) passes over in the forerun]. The usual purification is to distil from quinoline (50g SOCl$_2$ to 10g quinoline) to remove acid impurities, followed by distillation from boiled linseed oil (50g SOCl$_2$ to 20g of oil). Precautions must be taken to exclude moisture.
Thionyl chloride for use in organic syntheses can be prepared by distillation of technical SOCl$_2$ in the presence of diterpene (12g/250mL SOCl$_2$), avoiding overheating. Further purification is achieved by redistillation from linseed oil (1-2%) [Rigby *Chem Ind (London)* 1508 *1969*]. Gas chromatographically pure material is obtained by distillation from 10% (w/w) triphenyl phosphite [Friedman and Wetter *J Chem Soc (A)* 36 *1967*; Larsen et al. *J Am Chem Soc* **108** 6950 *1986*]. **Harmful vapours.**

Thorium chloride *[10026-08-1]* **M 373.8, pK$_1^{25}$10.45, pK$_2^{25}$10.62, pK$_3^{25}$ 10.80, pK$_4^{25}$ 11.64 (for aquo Th^{4+}).** Freed from anionic impurities by passing a 2M soln of ThCl$_4$ in 3M HCl through a Dowex-1 anion-resin column. The eluate was partially evaporated to give crystals which were filtered off, washed with Et$_2$O and stored in a desiccator over H$_2$SO$_4$ to dry. Alternatively, a saturated solution of ThCl$_4$ in 6M HCl was filtered through quartz wool and extracted twice with ethyl, or isopropyl, ether (to remove iron), then evaporated to a small volume on a hot plate. (Excess silica ppted, and was filtered off. The filtrate was cooled to 0° and saturated with dry HCl gas.) It was shaken with an equal volume of Et$_2$O, agitating with HCl gas, until the mixture becomes homogeneous. On standing, ThCl$_4$.8H$_2$O ppted and was filtered off, washed with Et$_2$O and dried [Kremer *J Am Chem Soc* **64** 1009 *1942*].

Thorium sulfate (4H$_2$O) *[10381-37-0]* **M 496.2, m 42°(loses H$_2$O), d 2.8.** Crystd from water.

Tin (powder) *[7440-31-5]* **M 118.7.** The powder was added to about twice its weight of 10% aqueous NaOH and shaken vigorously for 10min. (This removed oxide film and stearic acid or similar material sometimes added for pulverisation.) It was then filtered, washed with water until the washings were no longer alkaline to litmus, rinsed with MeOH and air dried. [Sisido, Takeda and Kinugama *J Am Chem Soc* **83** 538 *1961*.]

Tin tetramethyl *[594-27-4]* **M 178.8, m 16.5°, b 78.3°/740mm.** It is purified by fractionation using a Todd column of 35-40 plates at atmospheric pressure (p. 177). The purity of the fractions can be followed by IR [*J Am Chem Soc* **77** 6486 *1955*]. It readily dissolves stopcock silicone greases which give bands in the 8-10µ region. [*J Am Chem Soc* **76** 1169 *1954*.]

Titanium tetrabromide *[7789-68-6]* **M 367.5, m 28.3°, b 233.5°, d 3.3.** Purified by distn. Distillate forms light orange hygroscopic crystals. Store in the dark under N_2 preferably in sealed brown glass ampules. [Olsen and Ryan *J Am Chem Soc* **54** 2215 *1932*.]

Titanium tetrachloride *[7550-45-0]* **M 189.7, b 136.4°, 154°, d 1.730, pK_1^{25} 0.3, pK_2^{25} 1.8, pK_3^{25} 2.1, pK_4^{25} 2.4 (for aquo Ti^{4+} hydrolysis).** Refluxed with mercury or a small amount of pure copper turnings to remove the last traces of light colour [due to $FeCl_3$ and VCl_4], then distilled under N_2 in an all-glass system, taking precautions to exclude moisture. Clabaugh, Leslie and Gilchrist [*J Res Nat Bur Stand* **55** 261 *1955*] removed organic material by adding aluminium chloride hexahydrate as a slurry with an equal amount of water (the slurry being *ca* one-fiftieth the weight of $TiCl_4$), refluxing for 2-6h while bubbling in chlorine, which was subsequently removed by passing a stream of clean dry air. The $TiCl_4$ was then distilled, refluxed with copper and again distilled, taking precautions to exclude moisture. Volatile impurities were then removed using a technique of freezing, pumping and melting. [Baxter and Fertig *J Am Chem Soc* **45** 1228 *1923*; Baxter and Butler *J Am Chem Soc* **48** 3117 *1926*.] **HARMFUL VAPOURS.**

Titanium trichloride *[7705-07-9]* **M 154.3, m >500°, pK_1^{25} 2.55 (for hydrolysis of Ti^{3+} to $TiOH^{2+}$).** Brown purple powder that is very reactive with H_2O and pyrophoric when dry. It should be manipulated in a dry box. It is soluble in CH_2Cl_2 and tetrahydrofuran and is used as a M solution in these solvents in the ratio of 2:1, and stored under N_2. It is a powerful reducing agent. [*Inorg Synth* **6** 52 *1960*; *Synthesis* 833 *1989*.]

Titanocene dichloride *[1271-19-8]* **M 248.9, m 260-280°(dec), 289.2°, 298-291°, d 1.60.** Bright red crystals from toluene or xylene-$CHCl_3$ (1:1) and sublimes at 190°/2mm. It is moderately soluble in EtOH and insoluble in Et_2O, *C_6H_6, CS_2, CCl_4, pet ether and H_2O. [IR: *J Am Chem Soc* **76** 4281 *1954*; NMR and X-ray: *Can J Chem* **51** 2609 *1973*, **53** 1622 *1975*.]

Titanyl sulfate ($TiOSO_4.2H_2O$) *[13825-74-6]* **M 160.0.** Dissolved in water, filtered and crystd three times from boiling 45% H_2SO_4, washing with EtOH to remove excess acid, then with Et_2O. Air dried for several hours, then oven dried at 105-110°. [Hixson and Fredrickson *Ind Eng Chem* **37** 678 *1945*.]

Tribenzyl chlorosilane *[18740-59-5]* **M 336.9, m 139-142°, 141-142°, b 300-360°/100mm.** It is recrystd three times from pet ether; slender colourless needles, **m** 141°, sparingly soluble in pet ether and soluble in Et_2O. Does not fume in air but is decomposed by H_2O to give *tribenzyl silanol* **m** 106° (from pet ether). [*J Chem Soc* **93** 439 *1908*; *J Org Chem* **15** 556 *1950*.]

Tribenzyl phosphine *[76650-89-7]* **M 304.4, m 96-101°, b 203-210°/0.5mm, pK_{Est}~8.8.** Dissolve in Et_2O, dry over Na_2SO_4, evap and distil in an inert atmosphere. Distillate solidifies on cooling and is sublimed at 140°/0.001mm. This has **m** 92-95°(evacuated capillary). When air is bubbled through an Et_2O solution, it is oxidised to *tribenzyl phosphine oxide*, **m** 209-212° (evacuated capillary) (from Me_2CO). [*J Chem Soc* 2835 *1959*.]

Tri-*n*-butyl borate *[688-74-4]* **M 230.2, b 232.4°, n 1.4092, d 0.857.** The chief impurities are *n*-butyl alcohol and boric acid (from hydrolysis). It must be handled in a dry-box, and can readily be purified by fractional distillation, under reduced pressure.

Tri-*n*-butyl chlorosilane *[995-45-9]* **M 234.9**, **b 93-94°/4.5mm**, **134-139°/16mm**, **250-252°/atm**, **142-144°/29mm**, d_4^{20} **0.88**, n_D^{20} **1.447**. Fractionate and store in small aliquots in sealed ampoules. [*J Am Chem Soc* **74** 1361 *1952*; *J Org Chem* **24** 219 *1959*.]

Tri-*n*-butyl phosphate (butyl phosphate) *[126-73-8]*   **M 266.3**, **m -80°**, **b 47°/0.45mm**, **98°/0.1mm**, **121-124°/3mm**, **136-137°/5.5mm**, **166-167°/17mm**, **177-178°/27mm**, **289°/760atm (some dec)**, d_4^{20} **0.980**, n_D^{20} **1.44249**. The main contaminants in commercial samples are organic pyrophosphates, mono- and di- butyl phosphates and butanol. It is purified by washing successively with 0.2M HNO_3 (three times), 0.2M NaOH (three times) and water (three times), then fractionally distilled under vacuum. [Yoshida *J Inorg Nucl Chem* **24** 1257 *1962*.] It has also been purified *via* its uranyl nitrate addition compound, obtained by saturating the crude phosphate with uranyl nitrate. This compound was crystd three times with *n*-hexane by cooling to -40°, and then decomposed by washing with Na_2CO_3 and water. Hexane was removed by steam distn and the water was then evaporated under reduced pressure and the residue was distilled under reduced pressure. [Siddall and Dukes *J Am Chem Soc* **81** 790 *1959*.]
Alternatively wash with water, then with 1% NaOH or 5% Na_2CO_3 for several hours, then finally with water. Dry under reduced pressure and fractionate carefully under vacuum. Stable colourless oil, sparingly soluble in H_2O (1mL dissolves in 165mL of H_2O), but freely miscible in organic solvents. [*J Am Chem Soc* **74** 4953 *1952*, **80** 5441 *1958*; ^{31}P NMR: *J Am Chem Soc* **78** 5715 *1956*; *J Chem Soc* 1488 *1957*.]

Tri-*n*-butyl phosphine *[998-40-3]*   **M 202.3**, **b 109-110°/10mm**, **115-116°/12mm**, **149.5°/50mm**, **240.4-242.2°/atm**, d_4^{20} **0.822**, n_D^{20} **1.4463**, **pK_{Est}~7.6**. Fractionally distilled under reduced pressure in an inert atm (N_2) through an 8" gauze packed column (**b** 110-111°/10mm) and redistilled in a vacuum and sealed in thin glass ampoules. It is easily oxidised by air to *tri-n-butylphosphine oxide*, **b** 293-296°/745mm. It has a characteristic odour, it is soluble in EtOH, Et_2O, and $^*C_6H_6$ but insoluble in H_2O and is less easily oxidised by air than the lower molecular weight phosphines. It forms complexes, e.g. with CS_2 (1:1) **m** 65.5° (from EtOH). [*J Chem Soc* 33 *1929*, 1401 *1956*.]

Tri-*n*-butyl phosphite *[102-85-2]* **M 250.3**, **b 114-115°/5mm**, **122°/12mm**, **130°/17mm**, **137°/26mm**, d_4^{20}**0.926**, n_D^{20} **1.4924**. Fractionate with an efficient column. Stable in air but is slowly hydrolysed by H_2O. [*J Chem Soc* 1464 *1940*, 1488 *1957*; *J Am Chem Soc* **80** 2358, 2999 *1958*.]

Tri-*n*-butyl tin chloride *[1461-22-9]* **M 325.5**, **b 98-100°/0.4mm**, **140-152°/10mm**, **172°/25mm**, d_4^{20} **1.21**, n_D^{20} **1.492**. Fractionate in an inert atmosphere, and seal in small aliquots in glass ampoules. Sensitive to moisture. [*J Chem Soc* 1446 *1947*; *J Appl Chem* **6** 93 *1956*.]

Tributyl tin hydride *[688-73-3]* **M 291.1**, **b 76°/0.7mm**, **81°/0.9mm**, d_4^{20} **1.098**, n_D^{20} **1.473**. Dissolve in Et_2O, add quinol (500mg for 300mL), dry over Na_2SO_4, filter, evaporate and distil under dry N_2. It is a clear liquid if dry and decompose very slowly. In the presence of H_2O traces of tributyl tin hydroxide are formed in a few days. Store in sealed glass ampoules in small aliquots. It is estimated by reaction with aq NaOH when H_2 is liberated. **CARE:** stored samples may be under pressure due to liberated H_2. [*J Appl Chem* **7** 366 *1957*.]

B-Trichloroborazine *[933-18-6]* **M 183.1**, **m 87°**, **b 88-92°/21mm**. Purified by distillation from mineral oil.

Trichloromethyl trimethylsilane (trimethylsilyl trichloromethane) *[5936-98-1]* **M 191.6**, **m 130-132°**, **b 146-156°/749mm**. It distils at atmospheric pressure without decomposition and readily sublimes at 70°/10mm. It has one peak in the ^{1}H NMR spectrum (CH_2Cl_2) δ: 0.38. [*Synthesis* 626 *1980*.]

Tricyclohexylphosphine *[2622-14-2]* **M 280.4**, **m 82-83°**, **pK_{Est}~9.5**. Recrystd from EtOH [Boert et al. *J Am Chem Soc* **109** 7781 *1987*].

Triethoxysilane *[998-30-1]* **M 164.3**, **m -170°**, **b 131.2-131.8°/atm**, **131.5°/760mm**, d_4^{20} **0.98753**, n_D^{20} **1.4377**. Fractionated using a column packed with glass helices of *ca* 15 theoretical plates in an inert atmosphere. Store in aliquots in sealed ampoules because it is sensitive to moisture. [*J Am Chem Soc* **72** 1377, 2032 *1950*; *J Org Chem* **13** 280 *1948*.]

Triethylborane *[97-94-9]* **M 146.0, b 118.6°, n 1.378, d 0.678.** Distilled at 56-57°/220mm.

Triethyl borate *[150-46-9]* **M 146.0, b 118°, n 1.378, d 0.864.** Dried with sodium, then distilled.

Triethyl phosphate *[78-40-0]* **M 182.2, b 40-42°/0.25-0.3mm, 98-98.5°/8-10mm, 90°/10mm, 130°/55mm, 204°/680mm, 215-216°/760mm, d_4^{25} 1.608, n_D^{20} 1.4053.** Dried by refluxing with solid BaO and fractionally distilled under reduced pressure. It is kept with Na and distilled. Stored in the receiver protected from light and moisture. Alternatively it is dried over Na_2SO_4 and distilled under reduced pressure. The middle fraction is stirred for several weeks over anhydrous Na_2SO_4 and again fractionated under reduced pressure until the specific conductance reached a constant low value of κ^{25} 1.19 x 10^8, κ^{40} 1.68 x 10^8, and κ^{55} 2.89 x 10^8 ohm^{-1} cm^{-1}. It has also been fractionated carefully under reduced pressure through a glass helices packed column. It is soluble in EtOH, Et_2O and H_2O (dec). [*J Am Chem Soc* **77** 4767 *1955*, **78** 6413, 3557 (P NMR) *1956*; *J Chem Soc* 3582 *1959*, IR: *J Chem Soc* 475 *1952* and *Can J Chem* **36** 820 *1958*; Kosolapoff *Organophosphorus Compounds*, Wiley p. 258 *1950*.]

Triethylphosphine *[554-70-1]* **M 118.2, b 100°/7mm, 127-128°/744mm, d_4^{15} 0.812, n_D^{18} 1.457, pK^{25} 8.69 (also available as a 1.0M soln in THF). All operations should be carried out in an efficient fume cupboard because it is flammable, toxic and has a foul odour.** Purified by fractional distn at atm pressure in a stream of dry N_2, as it is oxidised by air to the oxide. In 300% excess of CS_2 it forms Et_3PCS_2 (**m** 118-120° cryst from MeOH) which decomposes in CCl_4 to give Et_3PS as a white solid **m** 94° when recryst from EtOH. [Sorettas and Isbell *J Org Chem* **27** 273 *1962*; *J Am Chem Soc* **82** 5791 *1960*; pK: Henderson and Streuli *J Am Chem Soc* **82** 5791 *1960*; see also trimethylphosphine.] Store in a sealed vial under N_2.
Alternatively, dissolve in Et_2O and shake with a solution of AgI and KI to form the insoluble complex. Filter off the complex, dry over P_2O_5 and the Et_3P is regenerated by heating the complex in a tube attached to a vacuum system. [*J Chem Soc* 530 *1953*, 1828 *1937*; *J Org Chem* **27** 2573 *1962*; Kosolapoff *Organophosphorus Compounds*, Wiley p. 31 *1950*.]

Triethyl phosphite *[122-52-1]* **M 166.2, b 48-49°/11mm, b 52°/12mm, 57.5°/19mm, 157.9°/757mm, d_4^{20} 0.9687, n_D^{20} 1.4135.** Treat with Na (to remove water and any dialkyl phosphonate), then decant and distil under reduced pressure, with protection against moisture or distil in vacuum through an efficient Vigreux column or a column packed with Penn State 0.16 x 0.16 in protruded nickel packing and a variable volume take-off head. [*Org Synth* Coll Vol IV 955 *1963*; *J Am Chem Soc* **78** 5817 *1956*, **80** 2999 *1958*; Kosolapoff *Organophosphorus Compounds*, Wiley p. 203 *1950*.]

Triethyl phosphonoacetate **(triethyl carboxymethyl phosphonate)** *[867-13-0]* **M 224.2, b 83-84°/0.5mm, 103°/1.2mm, 143-144°/11mm, 260-262°/atm, d_4^{20} 1.1215, n_D^{20} 1,4310.** Purified by fractional distn, preferably *in vacuo*. ^{31}P NMR has P resonance at 19.5 relative to orthophosphate. [Kosolapoff and Powell *J Am Chem Soc* **68** 1103 *1946*; **72** 4198 *1950*; Speziale and Freeman *J Org Chem* **23** 1586 *1958*.]

Triethyl phosphonoformate *[1474-78-8]* **M 210.2, b 70-72°/0.1mm, 122.5-123°/8mm, 130-131°/10mm, 138.2°/12.5mm, d_4^{20} 1.22, n_D^{20} 1.423.** Dissolve in Et_2O, shake with H_2O (to remove any trace of NaCl impurity), dry (Na_2SO_4), evaporate and distil using an efficient fractionating column. [*Chem Ber* **57** 1035 *1924*.]

Triethyl 2-phosphonopropionate *[3699-66-9]* **M 238.2, b 76-77°/0.2mm, 137-138.5°/17mm, d_0^{20} 1.096, n_D^{20} 1.432.** Purified by fractional distillation with high reflux ratio, preferably using a spinning band column. [*J Am Chem Soc* 4198 *1950*.]

Triethylsilane *[617-86-7]* **M 116.3, b 105-107°, 107-108°, d 0.734. n 1.414.** Refluxed over molecular sieves, then distilled. It was passed through neutral alumina before use [Randolph and Wrighton *J Am Chem Soc* **108** 3366 *1986*].

Triethylsilyl-1,4-pentadiene **(1,4-pentadien-3-yloxy-trimethylsilane)** *[62418-65-9]* **M 198.4, b 72-74°/12mm, d_4^{20} 0.842, n_D^{20} 1.439.** Dissolve in pentane, wash with H_2O, dry (Na_2SO_4), evaporate, and distil under vacuum. R_F values on Kieselgel 60 are 0.15 (pentane) and 0.60 (*C_6H_6). [IR, NMR, MS: *Helv Chim Acta* **64** 2002 *1981* .]

Triethyltin hydroxide *[994-32-1]* **M 222.9.** Treated with HCl, followed by KOH, and filtered to remove diethyltin oxide [Prince *J Chem Soc* 1783 *1959*].

Tri-*n*-hexylborane *[1188-92-7]* **M 265.3.** Treated with hex-1-ene and 10% anhydrous Et_2O for 6h at gentle reflux under N_2, then vacuum distilled through an 18in glass helices-packed column under N_2 taking the fraction **b** 130°/2.1mm to 137°/1.5mm. The distillate still contained some di-*n*-hexylborane [Mirviss *J Am Chem Soc* **83** 3051 *1961*].

Triiron dodecacarbonyl *[17685-52-8]* **M 503.7, m 140°(dec).** It usually contains 10% by weight of MeOH as stabiliser. This can be removed by keeping in a vacuum at 0.5mm for at least 5h. It can be sublimed slowly at high vacuum and is soluble in organic solvents. [*J Org Chem* **37** 930 *1972*, *J Chem Soc* 4632 *1960*; *Inorg Synth* **7** 193 *1963*.]

Triisoamyl phosphate *[919-62-0]* **M 308.4, b 143°/3mm.** Purified by repeated crystallisation, from hexane, of its addition compound with uranyl nitrate. (see *tributyl phosphate*.) [Siddall *J Am Chem Soc* **81** 4176 *1959*].

Triisobutyl phosphate *[126-71-6]* **M 266.3, b 119-129°/8-12mm, 192°/760mm, d 0.962, n 1.421.** Purified by repeated crystallisation, from hexane, of its addition compound with uranyl nitrate. (see *tributyl phosphate*.) [Siddall *J Am Chem Soc* **81** 4176 *1959*.]

Triisooctyl thiophosphate *[30108-39-5]* **M 450.6.** Purified by passage of its solution in CCl_4 through a column of activated alumina.

Triisopropyl phosphite *[116-17-6]* **M 208.2, b 58-59°/7mm, n^{25} 1.4082.** Distilled from sodium, under vacuum, through a column with glass helices. (This removes any dialkyl phosphonate).

Trimesitylphosphine *[23897-15-6]* **M 388.5, m 205-206°, pK_{Est}~8.0.** Recrystd from EtOH [Boert et al. *J Am Chem Soc* **109** 7781 *1987*].

Trimethallyl phosphate *[14019-81-9]* **M 260.3, b 134.5-140°/5mm, n^{25} 1.4454.** Purified as for triisoamyl phosphate.

Trimethoxysilane *[2487-90-3]* **M 122.2, m -114.8°, 81.1°/760mm, 84°/atm, d_4^{20} 0.957, n_D^{20} 1.359.** Likely impurities are $Si(OMe)_4$ and $H_2Si(OMe)_2$. Efficient fractionation is essential for removing these impurities [IR: *J Am Chem Soc* **81** 5109 *1959*].

Trimethyl borate (methylborate, trimethoxyboron) *[121-43-7]* **M 103.9, b 67-68°/742mm, d_4^{20} 0.928, n_D^{20} 1.3610.** Carefully fractionated through a gauze-packed column. Redistil and collect in weighed glass vials and seal. Keep away from moisture. It undergoes alkyl exchange with alcohols and forms azeotropes, e.g. with MeOH the azeotrope consists of 70% $(MeO)_3B$ and 30% MeOH with **b** 52-54°/atm, d 0.87. [*J Chem Soc* 2288 *1952*; *Chem Ind (London)* 53 *1952*; *J Am Chem Soc* **75** 213 *1953*.] Also dried with Na, then distilled.

Trimethyl boroxine *[823-96-1]* **M 125.5, b 80°/742mm, 79.3°/755mm, d_4^{20} 0.902.** Possible impurity is methylboronic acid. If present then add a few drops of conc H_2SO_4 and distil immediately, then fractionate through an efficient column. [*J Am Chem Soc* **79** 5179 *1957*; IR: *Z Anorg Chem* **272** 303 *1953*.]

Trimethyl chlorosilane (chlorotrimethylsilane) *[75-77-4]* **M 108.6, b 56-57°/atm, 58°/760mm, d 0.86, n 1.388.** Likely impurities are other chlorinated methylsilanes, and tetrachlorosilane (**b** 57.6°), some of which can form azeotropes. To avoid the latter very efficient fractional distillation is required. It has been fractionated through a 12 plate glass helices packed column with only the heart-cut material used. It has also been fractionated through a 90cm, 19mm diameter Stedman column (see p. 441). Also purified by redistilling from CaH_2 before use. [*J Am Chem Soc* **70**, 4254, 4258 *1948*; *J Org Chem* **23** 50 *1958*.] **FLAMMABLE and CORROSIVE.**

Trimethyloxonium tetrafluoroborate *[420-37-1]* **M 147.9, m 141-143°(sinters, open capillary), 179.6-180.0°(dec), 210-220°(dec).** The salt must be a white crystalline solid **m** ~ 179.6-180.0° (dec, sealed tube). Under a N_2 atmosphere (e.g. Dry Box), wash twice with CH_2Cl_2 then twice with Na-dried Et_2O, and dry by passing dry N_2 over the salt until free from Et_2O [Curphey *Org Synth* Coll Vol VI 1019 *1988*]. The oxonium salt purified in this way can be handled in air for short periods. The sample kept in a desiccator (Drierite) for 1 month at -20° had the same **m**, and samples stored in this way for >1 year are satisfactory for alkylations. ^1H NMR in liq SO_2 in a sealed tube had a single peak at δ 4.54 (impurities have δ at 3.39). [Meerwein *Org Synth* Coll Vol V 1096 *1973*.] If the sample looks good, dry in a vac desiccator for 2h (25°/1mm) and stored under N_2 at -20°. Melting point depends on heating rate.

Trimethylphenylsilane (phenyltrimethylsilane) *[768-32-1]* **M 150.3, b 67.3°/20mm, 170.6°/738mm, d_4^{25} 0.8646, n 1.491.** Fractionally distd at atm or reduced pressure (Podbielniak column, p.141) and estimated by GC with a column packed with Silicone Fluid No 710 on Chromosorb P support. [Gilman et al. *J Org Chem* **18** 1743 *1953*; Maienthal et al. *J Am Chem Soc* **76** 6392 *1954*; House and Respess *J Organomet Chem* **4** 95 *1965*; *J Am Chem Soc* **71** 2923 *1949*, **73** 4770 *1951*, **75** 2821 *1953*]

Trimethyl phosphate *[512-56-1]* **M 140.1, b 77°/12mm, 94°/22mm, 110°/60mm, 197.2°/atm, d_4^{20} 1.0213, n_D^{20} 1.3961.** Purified by fractionation through and efficient column at high reflux ratio. It is quite soluble in H_2O, solubility is 1:1 at 25°. [*J Am Chem Soc* **74** 2923 *1952*; IR: *J Chem Soc* 847 *1952*; *Can J Chem* **36** 820 *1958*; Kosolapoff *Organophosphorus Compounds*, Wiley p. 258 *1950*.]

Trimethylphosphine *[594-02-2]* **M 76.1, m -86°, b 38-39°/atm, pK^{25} 8.65, (also available as a 1.0M soln in THF or toluene).** All operations should be carried out in an efficient fume cupboard because it is flammable, toxic and has a foul odor. Distd at atm pressure in a stream of dry N_2 (apparatus should be held together with springs to avoid loss of gas from increased pressure in the system) and the distillate run into a soln of AgI in aq KI whereby the complex $[Me_3PAgI]_4$ separates steadily. Filter off the complex, wash with satd aq KI soln, then H_2O and dry in a vac desiccator over P_2O_5. The dry complex is heated in a flask (in a stream of dry N_2) in an oil bath at 140°, when pure Me_3P distils off (bath temp can be raised up to 260°). The vapour pressure of Me_3P at 20° is 466mm and the **b** is 37.8° [Thomas and Eriks *Inorg Synth* **9** 59 *1967*]. Alternatively, freshly distilled Me_3P (6g) is shaken with a solution of AgI (13.2g, 1.1mol) in saturated aqueous KI solution (50mL) for 2h. A white solid, not wetted with H_2O, separates rapidly. It is collected, washed with the KI solution, H_2O, and dried [*J Chem Soc* 1829 *1937*]. The silver complex is stable if kept dry in the dark in which state it can be kept indefinitely. Me_3P can be generated from the complex when required. Store under N_2 in a sealed container. It has been distd in a vacuum line at -78° *in vacuo* and condensed at -96° [IR and NMR: Crosbie and Sheldrick *J Inorg Nucl Chem* **31** 3684 *1969*]. The pK^{22} by NMR was 8.80 [Silver and Lutz *J Am Chem Soc* **83** 786 *1961*; pK 8.65: Henderson and Strueuli *J Am Chem Soc* **82** 5791 *1960*].

[$Me_2PAgI]_4$ *[12389-34-3]* complex is a flammable solid which has **m** 140-142°. It is decomposed by heating gently in one arm of an inverted U tube. The other arm is kept in a freezing mixture. The complex dissociates and pure Me_3P collects in the cold arm and is used at once. It should not be allowed to come in contact with air [*J Chem Soc* 708 *1938*]. The CS_2 complex has **m** 119° (cryst from 95% EtOH) and decomposes in CCl_4 to give Me_3PS **m** 154° (from EtOH) [Sorettas and Isbell *J Org Chem* **27** 273 *1962*].

Triphenylphosphine hydrochloride is unstable and volatilises at 75°/0.4mm (120°/14mm). [*J Am Chem Soc* **67** 503 *1945*; IR: *Trans Faraday Soc* **40** 41 *1944*; Kosolapoff *Organophosphorus Compounds,* Wiley p. 31 *1950*.]

Trimethyl phosphite   *[121-45-9]*   **M 124.1, b 22°/23mm,   86-86.5°/351mm,   111-112°/760mm,   111°/atm, d_4^{20} 1.0495, n_D^{20} 1.408.** Treated with Na (to remove water and any dialkyl phosphonate), then decanted and distilled with protection against moisture. It has also been treated with sodium wire for 24h, then distilled in an inert atmosphere onto activated molecular sieves [Connor et al. *J Chem Soc, Dalton Trans* 511 *1986*]. It has also been fractionally distilled using a spinning band column at high reflux ratio. It is a colourless liquid which is slowly hydrolysed by H_2O. [*J Am Chem Soc* **80** 2999 *1958*; IR: *J Chem Soc* 255 *1950*, P NMR: *J Am Chem Soc* **79** 2719 *1957*; Kosolapoff *Organophosphorus Compounds*, , Wiley p. 203 *1950*.]

Trimethylsilyl acetamide   *[13435-12-6]*   **M 131.3, m 38-43°, 52-54°, b 84°/13mm, 185-186°/atm.** Repeated distillation in an inert atmosphere, all operations to be performed under anhydrous atmosphere. In the presence of moisture trimethylsilanol (**b** 31-34°/26mm) is formed and is a likely impurity (check by NMR). [*Chem Ber* **96** 1473 *1963*.]

Trimethylsilyl acetonitrile (TMSAN) *[18293-53-3]*   **M 113.2, b 49-51°/10mm, 65-70°/20mm, d_4^{20} 0.8729, n_D^{20} 1.4420.** Check if NMR and IR spectra are correct, if not dissolve in *C_6H_6 (10vols), wash with buffer (AcOH-AcONa pH *ca* 7) several times, dry (CaCl$_2$), evaporate and distil. IR: ν (CCl$_4$) 2215 (CN) cm^{-1}; NMR δ (CCl$_4$): 0.23 (s, 9H, SiMe$_3$), and 1.53 (s, 2H, CH$_2$CN). [*J Chem Soc Perkin Trans 1* 26 *1979*.]

Trimethylsilyl azide   *[4648-54-8]*   **M 115.2, b 92-95°/atm, 95-99°/atm, d_4^{20} 0.878, n_D^{20} 1.441.** Distil through a Vigreux column in a N$_2$ atmosphere maintaining the oil bath temperature thermostated at 135-140°. Check the purity by ^1H NMR [CHCl$_3$, δ: single peak at 13cps from Me$_4$Si. Likely impurities are siloxane hydrolysis products. The azide is thermally stable even at 200° when it decomposes slowly without explosive violence. All the same it is advisable to carry out the distillation behind a thick safety screen in a fumehood because unforseen **EXPLOSIVE** azides may be formed on long standing. [Birkofer and Wagner *Org Synth* Coll Vol VI 1030 *1988*.]

Trimethylsilyl chloroacetate   *[18293-71-5]*   **M 166.7, m -20°, b 57-58°/14mm, 70-71°/30mm, 159°/760mm, d_4^{20} 1.057, n_D^{20} 11.4231 .** Purified by repeated fractionation and taking the fractions with clean NMR spectra. [*J Am Chem Soc* 2371 *1952*.]

Trimethylsilyl cyanide   *[7677-24-9]*   **M 99.2, m 8-11°, 10.5-11.5°, 11-12°, 12-12.5°; b 54-55°/87mm, 67-71°/168mm, 114-117°/760mm, 118-119°/760mm, d_4^{20} 0.79 n_D^{20} 1.43916.** Material should have only one sharp signal in the ^1H NMR (in CCl$_4$ with CHCl$_3$ as internal standard) δ: 0.4 and IR with ν at 2210cm^{-1} [*J Am Chem Soc* **74** 5247 *1952*, **77** 3224 *1955*]; otherwise purify by fractionating through an 18 x 1/4 in column. [*J Am Chem Soc* **81** 4493 *1959*.] It has also been carefully distilled using a 60cm vac jacketed column. If volume of sample is small the cyanide can be chased (in the distillation) with xylene that had be previously distilled over P$_2$O$_5$. [*J Org Chem* **39** 914 *1974*.]

2-Trimethylsilyl-1,3-dithiane   *[13411-42-2]*   **M 192.2, b 54.5°/0.17mm, 100°/8mm, d_4^{20} 1.04, n_D^{20} 1.533 .** Fractionally distil through an efficient column and collect the fractions that have the correct NMR and IR spectra. ^1H NMR (CCl$_4$) τ 6.36 (SiMe$_3$), 9.87 (SCHS) and dithiane H at 7 and 8 (ratio 1:9:4:2) from Me$_4$Si; UV λ$_{max}$ 244nm (ε 711); sh 227nm (ε 800). [*J Am Chem Soc* **89** 434 *1967*.]

Trimethylsilyl ethanol *[2916-68-9]*   **M 118.3, b 53-55°/11mm, 75°/41mm, 95°/100mm, d_4^{25} 0.8254, n_D^{25} 1.4220.** If the NMR spectrum is not clean then dissolve in Et$_2$O, wash with aqueous NH$_4$Cl solution, dry (Na$_2$SO$_4$), evaporate and distil. The *3,4-dinitrobenzoyl* deriv has **m** 66° (from EtOH). [NMR: *J Am Chem Soc* **79** 974 *1957*; *Z Naturforsch* **14b** 137 *1959*.]

2-(Trimethylsilyl)ethoxymethyl chloride (SEMCl)   *[76513-69-4]* **M 166.7, b 57-59°/8mm, d 0.942, n 1.4350.** Dissolve in pentane, dry (MgSO$_4$), evaporate and dist residual oil in a vac. Stabilise with 10ppm of diisopropylamine. Store under N$_2$ in a sealed container in a refrigerator. [Lipshutz and Pegram *Tetrahedron Lett* **21** 3343 *1980*.]

2-(Trimethylsilyl)ethoxymethyltrimethylphosphonium chloride *[82495-75-8]* **M 429.0, m 140-142°.** Wash the solid with AcOH and recryst from CH$_2$Cl$_2$-EtOAc. Dry in a vacuum desiccator. *Hygroscopic.* ^{1}H NMR (CDCl$_3$) δ: -0.2 (s, Me$_3$Si), 0.8 (t, 8Hz, CH$_2$Si), 3.83 (t, 8Hz, OCH$_2$), 5.77 (d, J_{PH} 4Hz, P$^+$-CH$_2$O) and 7.70 (m, aromatic H). [*Justus Liebigs Ann Chem* 1031 *1983*.]

Trimethylsilylethyl phenylsulfone **(phenyl-2-trimethylsilylethylsulfone)** *[73476-18-3]* **M 242.4, m 52°.** Dissolve in Et$_2$O, wash with saturated HCO$_3$, saturated NaCl, H$_2$O and dried (MgSO$_4$). Evaporation leaves residual crystals **m** 52°. [*Tetrahedron Lett* **23** 1963 *1982*, *J Org Chem* **53** 2688 *1985*.]

1-Trimethylsilyloxy-1,3-butadiene *[6651-43-0]* **M 142.3, b 131°/760mm (mixt of isomers), 49.5°/25mm (*E*-isomer), d 0.8237, n 1.447.** Purified by fractional distn and collecting the fractions with the required ^{1}H NMR. Store under N$_2$ — it is a flammable and moisture sensitive liquid. [Caseau et al. *Bull Soc Chim Fr* 16658 *1972*; Belge Patent 670,769, *Chem Abstr* **65** 5487d *1966*.]

1-(Trimethylsilyloxy)cyclopentene *[19980-43-9]* **M 156.3, b 45°/11mm, 75-80°/20-21mm, d$_4^{20}$ 0.878, n$_D^{20}$ 1.441.** If too impure as seen by the NMR spectrum then dissolve in 10 vols of pentane, shake with cold NaHCO$_3$(3 x 500mL), then 1.5M HCl (200mL) and aqueous NaHCO$_3$ (200mL) again, dry (Na$_2$SO$_4$), filter, evaporate and distil through a short Vigreux column. ^{1}H NMR: (CDCl$_3$) δ: 0.21 (s, 9H), 1.55 (m, 2H), 1.69 (m, 2H), 2.05 (br d, 4H) and 4.88 (br s, 1H). GLPC in a 6ft x 1/8in with 3% SP2100 on 100-120 mesh Supelcoport column should give one peak. [*Org Synth* Coll Vol VIII 460 *1993*.]

2-(Trimethylsilyloxy)furan *[61550-02-5]* **M 156.3, b 34-35°/9-10mm, 42-50°/17mm, 40-42°/25mm, d$_4^{20}$ 0.950, n$_D^{20}$ 1.436.** Fractionally distilled using a short path column. ^{1}H NMR in CCl$_4$ has δ: 4.90 (dd, *J* 1.3Hz, 3H), 6.00 (t, *J* 3Hz, 4H) and 6.60 (m, 5H). [*Heterocycles* **4** 1663 *1976*.]

4-Trimethylsilyloxy-3-penten-2-one *(cis)* **(acetylacetone enol trimethylsilyl ether)** *[13257-81-3]* **M 172.3, b 66-68°/4mm, 61-63°/5mm, d$_4^{20}$ 0.917, n$_D^{20}$ 1.452.** Fractionally distilled and stored in glass ampoules which are sealed under N$_2$. It hydrolyses readily in contact with moisture giving, as likely impurities, hexamethyldisiloxane and 2,4-pentanedione. [*J Am Chem Soc* **80** 3246 *1958*.]

Trimethylsilyl isocyanate *[1118-02-1]* **M 115.2, b 90-92°/atm, b 91.3-91.6°/atm, d$_4^{20}$ 0.850 n$_D^{20}$ 1.43943.** Purified by repeated fractionation as for the isothiocyanate. [*J Chem Soc* 3077 *1950*.]

Trimethylsilyl isothiocyanate *[2290-65-5]* **M 131.3, m -33°, b 142.6-143.1°/759mm, 143.8°/760mm, n$_D^{20}$ 1.4809.** The ^{1}H NMR spectrum should have only one peak, if not purify by repeated fractionation in an all glass system using a 50cm (4mm internal diameter) column without packing. [*J Am Chem Soc* **69** 3049 *1947*; *Chem Ber* **90** 1934 *1957*; *Synthesis* 51 *1975*.]

(Trimethylsilyl)methanol *[3219-63-4]* **M 104.2, b 120-122°/754mm, 122-123°/768mm, d$_4^{20}$ 0.83 n$_D^{20}$ 1.4176.** If the NMR indicates impurities (should have only two signals) then dissolve in Et$_2$O, shake with aqueous 5N NaOH, M H$_2$SO$_4$, saturated aqueous NaCl, dry (MgSO$_4$) and distil using an efficient column at atmospheric pressure. The *3,5-dinitrobenzoate* has **m** 70-70.5° (from 95% EtOH). [Huang and Wang *Acta Chem Sin* **23** 291 *1957*, *Chem Abs* **52** 19911 *1958*; Speier et al. *J Am Chem Soc* **81** 1844 *1959* and *J Am Chem Soc* **70** 1117 *1949*.]

(Trimethylsilyl)methylamine **(aminomethyl trimethylsilane)** *[18166-02-4]* **M 103.2, b 101.6°/735mm, d$_4^{20}$ 0.77, n$_D^{20}$ 1.416.** A possible contaminant is hexamethyldisiloxane. Should have two ^{1}H NMR signals in CDCl$_3$, if not, dissolve in *C$_6$H$_6$, shake with 15% aq KOH, separate, dry (Na$_2$SO$_4$), filter, evaporate and distil using a still of *ca* 10 theoretical plates. The water azeotrope has **b** 83°/735mm, hence it is important to dry the extract well. The *hydrochloride* has **m** 198/199° (from MeOH or Me$_2$CO). [*J Am Chem Soc* **73** 3867 *1951*; NMR, IR: *J Organometal Chem* **44** 279 *1972*.]

Trimethylsilylmethyl phenylsulfone (phenyltrimethylsilylmethylsulfone) *[17872-92-3]* **M 228.4, m 28-32°, b 121°/0.01mm, 160°/6mm, n$_D^{20}$ 1.5250.** Fractionate at high vacuum and recrystallise from pentane at -80°· If too impure (*cf* IR) dissolve in CH$_2$Cl$_2$ (*ca* 800mL for 100g), wash with 2M aqueous NaOH (2 x 200mL), brine, dry, evaporate and distil. [*J Chem Soc, Perkin Trans 1* 1949 *1985*; IR and NMR: *J Am Chem Soc* **76** 3713 *1954*.]

1-(Trimethylsilyl)-2-phenylacetylene (1-phenyl-2-trimethylsilylacetylene) *[78905-09-6]* **M 174.3, b 45-46°/0.1mm, 67°/5mm, 87.5°/9mm, d 0.8961 n 1.5284.** Dissolve in Et$_2$O, wash with H$_2$O, dry and fractionate through a Todd column (see p. 174). [*J. Am Chem Soc* **80** 5298 *1958*.]

3-(Trimethylsilyl)propyne *[13361-64-3]* **M 112.3, b 99-100°/760mm, d 0.7581, n 1.4091.** Fractionally distilled and 0.5% of 2,6-di-*tert*-butyl-*p*-cresol added to stabilise it. [*Doklady Acad Nauk USSR* **93** 293 *1953*; *Chem Abs* **48** 13616 *1954*.]

1-Trimethylsilyl-1,2,4-triazole *[18293-54-4]* **M 141.3, b 74°/12mm, d$_4^{20}$ 0.99, n$_D^{20}$ 1.4604.** Fractionally distilled at atmospheric pressure in an inert atmosphere because it is moisture sensitive. [*Chem Ber* **93** 2804 *1960*.]

Trimethylsilyl trifluoromethane (trifluoromethyl trimethylsilane) *[81290-20-2]* **M 142.2, b 54-55°, 55-55.5°, d$_4^{20}$ 0.962, n$_D^{20}$ 1.332.** Purified by distilling from trap to trap in a vacuum of 20mm using a bath at 45° and Dry ice-Me$_2$CO bath for the trap. The liquid in the trap is then washed with ice cold H$_2$O (3x), the top layer is collected, dried (Na$_2$SO$_4$), the liquid was decanted and fractionated through a helices packed column at atmospheric pressure. ^{1}H, ^{13}C, ^{19}F, and ^{29}Si NMR can be used for assessing the purity of fractions. [*Tetrahedron Lett* **25** 2195 *1984*; *J Org Chem* **56** 984 *1991*.]

Trimethyl vinyl silane *[754-05-2]* **M 100.2, b 54.4°/744mm, 55.5°/767mm, d\S(25,4) 0.6865, n$_D^{25}$ 1.3880.** If the ^{1}H NMR spectrum shows impurities then dissolve in Et$_2$O, wash with aq NH$_4$Cl soln, dry over CaCl$_2$, filter, evaporate and distil at atmospheric pressure in an inert atmosphere. It is used as a copolymer and may polymerise in the presence of free radicals. It is soluble in CH$_2$Cl$_2$. [*J Org Chem* **17** 1379 *1952*.]

Trineopentyl phosphate *[14540-59-1]* **M 320.4.** Crystd from hexane.

Tri-(4-nitrophenyl)phosphate *[3871-20-3]* **M 461.3, m 155-156°, 156°, 156-158°, 157-159°.** It has been recrystd from AcOH, dioxane, AcOEt and Me$_2$CO and dried in vacuum over P$_2$O$_5$. [*J Am Chem Soc* **72** 5777 *1950*, **79** 3741 *1957*.]

Tri-*n*-octylphosphine oxide *[78-50-2]* **M 386.7, m 59.5-60°, pK$_{Est}$ <0.** Mason, McCarty and Peppard [*J Inorg Nuclear Chem* **24** 967 *1962*] stirred an 0.1M solution in *benzene with an equal volume of 6M HCl at 40° in a sealed flask for 48h, then washed the *benzene solution successively with water (twice), 5% aq Na$_2$CO$_3$ (three times) and water (six times). The *benzene and water were then evaporated under reduced pressure at room temperature. Zingaro and White [*J Inorg Nucl Chem* **12** 315 *1960*] treated a pet ether solution with aqueous KMnO$_4$ (to oxidise any phosphinous acids to phosphinic acids), then with sodium oxalate, H$_2$SO$_4$ and HCl (to remove any manganese compounds). The pet ether solution was slurried with activated alumina (to remove phosphinic acids) and recrystd from pet ether or cyclohexane at -20°. It can also be crystd from EtOH.

Triphenylantimony *[603-36-1]* **M 353.1, m 52-54°.** Recrystd from acetonitrile [Hayes et al. *J Am Chem Soc* **107** 1346 *1985*].

Triphenylarsine *[603-32-7]* **M 306.2, m 60-62°.** Recrystd from EtOH or aqueous EtOH [Dahlinger et al. *J Chem Soc, Dalton Trans* 2145 *1986*; Boert et al. *J Am Chem Soc* **109** 7781 *1987*].

Triphenylbismuth (bismuth triphenyl) *[603-33-8]* M 440.3, m 75-76°, 77-78°, 78.5°, d $^{98.5}_4$ 1.6427(melt). Dissolve in EtOH, ppte with H_2O, extract with Et_2O, dry and evaporate when the residue crystallises. It has been recrystd from EtOH and Et_2O-EtOH and is a stable compound. [*J Chem Soc* suppl p121 *1949*; *Chem Ber* **37** 4620 *1904*; *J Am Chem Soc* **62** 665 *1940*; UV: *J Chem Phys* **22** 1430 *1954*.]

Triphenyl borane (borane triphenyl) *[960-71-4]* M 242.1, m 134-140°, 137°, 139-141°, 142-142.5°, 147.5-148°, 151°, b 203°/15mm. Recryst three times from Et_2O or *C_6H_6 under N_2 and dry at 130°. It can be distilled in a high vacuum at 300-350°, and has been distilled (b 195-215°) in vacuum using a bath temp of 240-330°. N_2 was introduced into the apparatus before dismantling. It forms complexes with amines. [*Chem Ind (London)* 1069 *1957*; *Justus Liebigs Ann Chem* **563** 110 *1949*; *J Am Chem Soc* **57** 1259 *1935*.]

Triphenylchlorostannane (triphenyltin chloride) *[639-58-7]* M 385.5, m 103-106°(dec). See triphenyltin chloride on p. 494.

Triphenyl phosphate *[115-86-6]* M 326.3, m 49.5-50°, b 245°/0.1mm. Crystd from EtOH.

Triphenylphosphine *[603-35-0]* M 262.3, m 77-78°, 79°, 79-81°, 80.5°, 80-81°, b >360°(in inert gas), d^{25}_4 1.194, d^{80}_4 1.075 (liq), pK 2.73. Crystd from hexane, MeOH, diethyl ether, CH_2Cl_2/hexane or 95% EtOH. Dried at 65°/<1mm over $CaSO_4$ or P_2O_5. Chromatographed through alumina using (4:1) *benzene/$CHCl_3$ as eluent. [Blau and Espenson et al. *J Am Chem Soc* **108** 1962 *1986*; Buchanan et al. *J Am Chem Soc* **108** 1537 *1986*; Randolph and Wrighton *J Am Chem Soc* **108** 3366 *1986*; Asali et al. *J Am Chem Soc* **109** 5386 *1987*.] It has also been crystd twice from pet ether and 5 times from Et_2O-EtOH to give m 80.5°. Alternatively, dissolve in conc HCl, and upon dilution with H_2O it separates because it is weakly basic; it is then crystallised from EtOH-Et_2O. It recrystallises unchanged from AcOH. [*J Chem Soc* Suppl. p121 *1949*; *J Am Chem Soc* **78** 3557 *1956*.] $3Ph_3P$ .4HCl crystallises when HCl gas is bubbled through an Et_2O solution; it has m 70-73°, but recrystallises very slowly and is deliquescent. The *HI*, made by adding Ph_3P to hydriodic acid is not hygroscopic and decomps at ~100°. The *chlorate* (1:1) *salt* has m 165-167°, but decomposes slowly at 100°. All salts hydrolyse in H_2O to give Ph_3P [IR, UV: *J Am Chem Soc* **80** 2117 *1958;* pK: *J Am Chem Soc* **82** 5791 *1960*; Kosolapoff, *Organophosphorus Compounds*, Wiley *1950*]. § Available commercially on a polystyrene or polyethyleneglycol support.

Triphenyl phosphine dibromide *[1034-39-5]* M 422.1, m 235°, 245-255°(dec). Recrystd from MeCN-Et_2O. Although it has been recrystd from EtOH, this is not recommended as it converts alcohols to alkyl bromides. It deteriorates on keeping and it is best to prepare it afresh. [Anderson and Freenor *J Am Chem Soc* **86** 5037 *1964*; Horner et al. *Justus Liebigs Ann Chem* **626** 26 *1959*.]

Triphenylphosphine oxide *[791-28-6]* M 278.3, m 152.0°, pK_{Est} ~ -2.10 (aq H_2SO_4). Crystd from absolute EtOH. Dried *in vacuo*.

Triphenyl phosphite *[101-02-0]* M 310.3, b 181-189°/1mm, d 1.183. Its ethereal soln was washed succesively with aqueous 5% NaOH, distilled water and saturated aqueous NaCl, then dried with Na_2SO_4 and distilled under vacuum after evaporating the diethyl ether.

Triphenylphosphorylidene acetaldehyde (formylmethylenetriphenylphosphorane) *[2136-75-6]* M 304.3, m 185-187°, 186-187°(dec). Recryst from Me_2CO, or dissolve in *C_6H_6, wash with N NaOH, dry (MgSO$_4$), evap, and cryst residue from Me_2CO. It can be prepd from its precursor, formylmethyltriphenylphosphonium chloride (crystd from $CHCl_3$/EtOAc), by tratment with Et_3N and extraction with *C_6H_6. [Tripett and Walker *Chem Soc* 1266 *1961*.]

Triphenyl silane *[789-25-3]* M 260.4, m 45°, b 148-151°/1mm. Purified by recrystn from MeOH. [*J Am Chem Soc* **81** 5925 *1959*; *Acta Chem Scand* **9** 947 *1955*; IR: *J Am Chem Soc* **76** 5880 *1954*.]

Triphenylsilanol **(hydroxytriphenylsilane)** *[791-31-1]* **M 276.4, m 150-153°, 151-153°, 154-155°, 156°.** It can be purified by dissolving in pet ether, passing through an Al_2O_3 column, eluting thoroughly with CCl_4 to remove impurities and then eluting the silanol with MeOH. Evaporation gives crystals **m** 153-155°. It can be recrystallised from pet ether, CCl_4 or from *benzene or Et_2O-pet ether (1:1). It has also been recrystallised by partial freezing from the melt to constant melting point. [*J Am Chem Soc* **81** 3288 *1959*; IR: *J Org Chem* **17** 1555 *1952* and *J Chem Soc* 124 *1949*.]

Triphenyltin chloride **(chlorotriphenylstannane)** *[639-58-7]* **M 385.5, m 103-106°(dec), 108°(dec), b 240°/13.5mm.** Purify by distillation, followed by recrystn from MeOH by adding pet ether (b 30-60°), m 105-106° [*Chem Ber* **67** 1348 *1934*], or by crystn from Et_2O [Krause *Chem Ber* **51** 914 *1918*]. It sublimes in a vacuum. **HIGHLY TOXIC.**

Triphenyltin hydroxide *[76-87-9]* **M 367.0, m 122-123.5°, 124-126°.** West, Baney and Powell [*J Am Chem Soc* **82** 6269 *1960*] purified a sample which was grossly contaminated with tetraphenyltin and diphenyltin oxide by dissolving it in EtOH, most of the impurities remaining behind as an insoluble residue. Evaporation of the EtOH gave the crude hydroxide which was converted to triphenyltin chloride by grinding in a mortar under 12M HCl, then evaporating the acid soln. The chloride, after crystallisation from EtOH, had **m** 104-105°. It was dissolved in Et_2O and converted to the hydroxide by stirring with excess aqueous ammonia. The ether layer was separated, dried, and evaporated to give triphenyltin hydroxide which, after crystn from EtOH and drying under vacuum, was in the form of white crystals (**m** 119-120°), which retained some cloudiness in the melt above 120°. The hydroxide retains water (0.1-0.5 moles of water per mole) tenaciously.

Triphenyl vinyl silane *[18666-68-7]* **M 286.5, m 58-59°, 57-59.5°, 67-68°, b 190-210°/3mm.** It has been recrystallised from EtOH, 95% EtOH, EtOH-*C_6H_6, pet ether (b 30-60°) and Et_2O, and has been distilled under reduced pressure. [*J Am Chem Soc* **74** 4582 *1952*; *J Org Chem* **17** 1379 *1952*.]

Tri-*n*-propyl borate *[688-71-1]* **M 188.1, b 175-177°, d 0.857, n 1.395.** Dried with sodium and then distilled.

Triquinol-8-yl phosphate *[52429-99-9]* **M 479.4, m 193-197°, 202-203°.** Purified by recrystn from dimethylformamide. Purity was checked by paper chromatography, R_F 0.90 [*i*-PrOH, saturated $(NH_4)_2SO_4$, H_2O; 2:79:19 as eluent]; IR (KBr) ν 1620–1570 (C=C, C=N) and 1253 (P=O). [*Bull Chem Soc Jpn* **47** 779 *1974*.]

Tri-ruthenium dodecacarbonyl *[15243-33-1]* **M 639.1, m 154-155°.** Recryst from *C_6H_6 or cyclohexane as orange-red crystals, and sublime at 80-100°/0.1mm. It has $ν_{CO}$ 2062 and 2032. [*J Chem Soc, Chem Commun* 684 *1966*; *J Chem Soc (A)* 1238 *1967*; IR,UV: *Angew Chem Int Ed Engl* **7** 427 *1968*.]

Tris-(2-biphenylyl) phosphate *[132-28-5]* **M 554.6, m 115.5-117.5°.** Crystd from MeOH containing a little acetone.

Tris(2,2'-bipyridine)ruthenium(II) dichloride **(6H₂O)** *[50525-27-4]* **M 748.6.** Recrystd from water then from MeOH [Ikezawa et al. *J Am Chem Soc* **108** 1589 *1986*].

Tris-(1,2-dioxyphenyl)cyclotriphosphazine **{trispiro[1,3,5,2,4,6-triazatriphosphorine]-2,2':-2,4":2,6"'-tris(1,3,2)benzodioxaphosphole}** *[311-03-5]* **M 459.0, m 244-245°, 245°, 245-246°.** Recrystd from *C_6H_6 or chlorobenzene, then triple sublimed (175°/0.1mm, 200°/0.1mm, 230°/0.05mm). UV has $λ_{max}$nm (log ε): 276 (3.72), 271 (3.79) 266sh (3.68) and 209 (4.38) in MeCN. IR (ν): 1270 (O-Ph), 1220 (P=N), 835 (P-O-Ph) and 745 (Ph) cm⁻¹. [Alcock *J Am Chem Soc* **86** 2591 *1964*; Alcock et al. *J Am Chem Soc* **98** 5120 *1976*; Meirovitch *J Phys Chem* **88** 1522 *1984*.]

(±)-Tris-(2-ethylhexyl)phosphate **(TEHP, tri-isooctylphosphate, "trioctyl" phosphate,** *[78-42-2; 25103-23-5]* **M 434.6, b 186°/1mm, 219°/5mm, d²⁵ 0.92042, n 1.44464.** TEHP, in an equal volume of diethyl ether, was shaken with aqueous 5% HCl and the organic phase was filtered to remove traces of pyridine (used as a solvent during manufacture) as its hydrochloride. This layer was shaken with aqueous Na_2CO_3, then water, and the ether was distilled off at room temperature. The ester was filtered, dried for 12h at $100^o/15mm$, and again filtered, then shaken intermittently for 2 days with activated alumina (100g/L). It was decanted through a fine sintered-glass disc (with exclusion of moisture), and distd under vacuum. [French and Muggleton *J Chem Soc* 5064 *1957*.] *Benzene can be used as a solvent (to give 0.4M soln) instead of ether. IR: 1702, 1701, 481 and 478cm^{-1} [Bellamy and Becker *J Chem Soc* 475 *1952*]. The *uranyl nitrate* salt was purified by partial crystallisation from hexane [Siddall *J Am Chem Soc* **81** 4176 *1959*].

Trisodium citrate ($2H_2O$) *[68-04-2]* **M 294.1, m 150^o(loses H_2O).** Crystd from warm water by cooling to 0°.

Trisodium 8-hydroxy-1,3,6-pyrenetrisulfonate *[6358-69-6]* **M 488.8, m >300(dec).** Purified by chromatography with an alumina column, and eluted with *n*-propanol-water (3:1, v/v). Recrystd from aqueous acetone (5:95, v/v) using decolorising charcoal.

Trisodium 1,3,6-naphthalenetrisulfonate *[5182-30-9]* **M 434.2.** The free acid was obtained by passage through an ion-exchange column and converted to the lanthanum salt by treatment with La_2O_3. This salt was crystallised twice from hot water. [The much lower solubility of $La_2(SO_4)_3$ and its retrograde temperature dependence allows a good separation from sulfate impurity]. The lanthanum salt was then passed through an appropriate ion-exchange column to obtain the free acid, the sodium or potassium salt. (The sodium salt is *hygroscopic*). [Atkinson, Yokoi and Hallada *J Am Chem Soc* **83** 1570 *1961*.] Also recrystd from aqueous acetone [Okahata et al. *J Am Chem Soc* **108** 2863 *1986*].

Trisodium orthophosphate ($12H_2O$) *[10101-89-0]* **M 380.1, pK_1^{25} 2.15, pK_2^{25} 7.21, pK_3^{25} 12.33 (for H_3PO_4).** Crystd from warm dilute aqueous NaOH (1mL/g) by cooling to 0°.

Tris(2,4-pentandionate)aluminium *[13963-57-0]* **M 324.3.** See aluminum acetylacetonate on p. 390.

Tris-(trimethylsilyl)silane (TTMSS) *[1873-77-4]* **M 248.7, b 73°/5mm, d 0.808, n 1.49.** Purified by fractional distn and taking the middle cut. Store under N_2 or Ar as it is an **IRRITANT** and **PYROPHORIC**. [Chatgilialoglu *Acc Chem Res* **25** 188 *1992*; NMR: Gilman et al. *J Organomet Chem* **4** 163 *1965*.]

Tritium *[10028-17-8]* **M 6.0.** Purified from hydrocarbons and ^{3}He by diffusion through the wall of a hot nickel tube [Landecker and Gray *Rev Sci Instrum* **25** 1151 *1954*]. **RADIOACTIVE.**

Tri-*p*-tolyl phosphate *[20756-92-7; 1330-78-5 (isomeric tritolyl phosphate mixture)]* **M 368.4, b 232-234°, d^{25} 1.16484, n 1.56703.** Dried with $CaCl_2$, then distd under vacuum and percolated through a column of alumina. Passage through a packed column at 150°, with a counter-current stream of nitrogen, under reduced pressure, removed residual traces of volatile impurities.

Tri-*o*-tolylphosphine *[6163-58-2]* **M 304.4, m 129-130°, pK_{Est} <0.** Crystd from EtOH [Boert et al. *J Am Chem Soc* **109** 7781 *1987*].

Tungsten (rod) *[7440-33-7]* **M 183.6. m 3410°, b 5900°, d 19.0.** Cleaned with conc NaOH solution, rubbed with very fine emery paper until its surface was bright, washed with previously boiled and cooled conductivity water and dried with filter paper.

Tungsten hexacarbonyl *[14040-11-0]* **M 351.9, d 2.650.** Sublimed *in vacuo* before use [Connoe et al. *J Chem Soc, Dalton Trans* 511 *1986*].

Tungsten (VI) trichloride *[13283-01-7]* **M 396.6, m 265^o(dec), 275°, b 346°, d$_4^{25}$ 3.520, pK_1^{25} 2.20, pK_2^{25} 3.70 (for tungstic acid, H_2WO_4).** Sublimed in a stream of Cl_2 in a high temperature furnace and collected in a receiver cooled in a Dry Ice-acetone bath in an inert atmosphere because it is sensitive to moisture. It is soluble in CS_2, CCl_4, $CHCl_3$, $POCl_3$, *C_6H_6, pet ether and Me_2CO. Solns decompose on standing. Good crystals can be obtained by heating WCl_6 in CCl_4 to 100° in a sealed tube, followed by slow cooling (tablets of four-sided prisms). Store in a desiccator over H_2SO_4 in the dark. [*Inorg Synth* **3** 163 *1950*, **9** 133*1967*; *Handbook of Preparative Inorganic Chemistry (Ed. Brauer)* Vol II 1417 *1965*.]

Uranium hexafluoride *[7783-81-5]* M 352.0, b 0°/17.4mm, 56.2°/765mm, m 64.8°, pK^{25} 1.68 (for hydrolysis of U^{4+} to UOH^{3+}).
Purified by fractional distillation to remove HF. Also purified by low temperature trap-to-trap distillation over pre-dried NaF [Anderson and Winfield *J Chem Soc, Dalton Trans* 337 *1986*].

Uranium trioxide *[1344-58-7]* M 286.0, d 7.29. The oxide was dissolved in $HClO_4$ (to give a uranium content of 5%), and the solution was adjusted to pH 2 by addition of dilute ammonia. Dropwise addition of 30% H_2O_2, with rapid stirring, ppted U(VI) peroxide, the pH being held constant during the pptn, by addition of small amounts of the ammonia soln. (The H_2O_2 was added until further quantities caused no change in pH.) After stirring for 1h, the slurry was filtered through coarse filter paper in a Büchner funnel, washed with 1% H_2O_2 acidified to pH 2 with $HClO_4$, then heated at 350° for three days in a large platinum dish [Baes *J Phys Chem* **60** 878 *1956*].

Uranyl nitrate (6H₂O) *[13520-83-7]* M 502.1, m 60.2°, b 118°, pK^{25} 5.82 (for aquo UO_2^{2+}). Crystd from water by cooling to -5°, taking only the middle fraction of the solid which separated. Dried as the hexahydrate over 35-40% H_2SO_4 in a vacuum desiccator.

Vanadium (metal) *[7440-62-2]* M 50.9, m 1910°, d 6.0.
Cleaned by rapid exposure consecutively to HNO_3, HCl, HF, de-ionised water and reagent grade acetone, then dried in a vacuum desiccator.

Vanadium (III) acetonylacetonate *[13476-99-8]* M 348.3, m 181-184°, 185-190°, pK_1^{25} 2.92, pK_2^{25} 3.5(for aquo V^{3+} hydrolysis). Crystd from acetylacetone as brown plates. It can be distilled in small quantities without decomposition. It is soluble in $CHCl_3$ and *C_6H_6 and evaporation of a $CHCl_3$ solution yields brown crystals which are washed with cold EtOH and dried in vacuum or at 100° in a CO_2 atmosphere. Under moist conditions it readily oxidises [V(AcAc)₃ to V(AcAc)₂O]. [*J Chem Soc* **103** 78 *1913*, *Inorg Synth* **5** 105 *1957*; *Anal Chem* **30** 526 *1958*; UV: *J Am Chem Soc* **80** 5686 *1958*.]

Vanadyl acetylacetonate *[3153-26-2]* M 265.2, m 256-259°. Crystd from acetone.

Vanadyl trichloride (VOCl₃) *[7727-18-6]* M 173.3, m-79.5°, b 124.5-125.5°/744mm, 127.16°/760mm, d^0 1.854, d^{32} 1.811. Should be lemon yellow in colour. If red it may contain VCl_4 and Cl_2. Fractionally distil and then redistil over metallic Na but be careful to leave some residue because the residue can become **EXPLOSIVE** in the presence of the metal **USE A SAFETY SHIELD** *and avoid contact with moisture*. It readily hydrolyses to vanadic acid and HCl. Store in a tightly closed container or in sealed ampoules under N_2. [*Inorg Synth* **1** 106 *1939*, **4** 80 *1953*.]

Vinyl chlorosilane *[75-94-5]* M 161.5, b 17.7°/46.3mm, 82.9°/599.4mm, 92°/742mm, 91-91.5°/atm, d_4^{20} 0.1.2717, n_D^{20} 1.435. Fractionally distil at atmospheric pressure. It is H_2O sensitive and is stored in the dark and is likely to polymerise. [*Chem Ber* **91** 1805 *1958*, **92** 1012 *1959*; *Anal Chem* **24** 1827 *1952*]

Vinylferrocene (ferroceneylethene) *[1271-51-8]* M 212.1, m 51-52.5°, b 80-85°/0.2mm. Dissolve in Et_2O, wash with H_2O and brine, dry (Na_2SO_4), evap to a small vol. Purify through an Al_2O_3 (Spence grade H) column by eluting the yellow band with pet ether (b 40-60°). The low melting orange crystals which can be sublimed. The *tetracyanoethylene adduct* [49716-63-4] crysts from *C_6H_6-pentane and has **m** 137-

139°(dec). [Horspool and Sutherland *Can J Chem* **46** 3453 *1968*; Berger et al. *J Org Chem* **39** 377 *1974*; Rauch and Siegel *J Organomet Chem* **11** 317 *1968*.]

Vinyltributylstannane (vinyltributyltin) *[7486-35-3]* **M 317.1, b 104-106°/3.5mm, d 1.081, n 1.4751.** Fractionate under reduced pressure and taking the middle fraction to remove impurities such as (*n*-Bu)$_3$SnCl. [Seyferth and Stone *J Am Chem Soc* **79** 515 *1957*.]

Water *[7732-18-5]* M 18.0, m 0°, b 100°, pK^{25} 14.00.

Conductivity water (specific conductance *ca* 10^{-7} mho) can be obtained by distilling water in a steam-heated tin-lined still, then, after adding 0.25% of solid NaOH and 0.05% of KMnO$_4$, distilling once more from an electrically heated Barnstead-type still, taking the middle fraction into a Jena glass bottle. During these operations suitable traps must be used to protect against entry of CO$_2$ and NH$_3$. Water only a little less satisfactory for conductivity measurements (but containing traces of organic material) can be obtained by passing ordinary distilled water through a mixed bed ion-exchange column containing, for example, Amberlite resins IR 120 (cation exchange) and IRA 400 (anion exchange), or Amberlite MB-1. This treatment is also a convenient one for removing traces of heavy metals. (The metals Cu, Zn, Pb, Cd and Hg can be tested for by adding pure concentrated ammonia to 10mL of sample and shaking vigorously with 1.2mL 0.001% dithizone in CCl$_4$. Less than 0.1µg of metal ion will impart a faint colour to the CCl$_4$ layer.) For almost all laboratory purposes, simple distillation yields water of adequate purity, and most of the volatile contaminants such as ammonia and CO$_2$ are removed if the first fraction of distillate is discarded.

Xylenol Orange (sodium salt) *[3618-43-7]*. See entry on p. 387 in Chapter 4.

Zinc (dust) *[7440-66-6]* M 65.4.

Commercial zinc dust (1.2Kg) was stirred with 2% HCl (3L) for 1min, (then the acid was removed by filtration), and washed in a 4L beaker with a 3L portion of 2% HCl, three 1L portions of distilled water, two 2L portions of 95% EtOH, and finally with 2L of absolute Et$_2$O. (The wash solutions were removed each time by filtration.) The material was then dried thoroughly and if necessary, any lumps were broken up in a mortar.

Zinc (metal) *[7440-66-6]* **M 65.4, m 420°, d 7.141.** Fused under vacuum, cooled, then washed with acid to remove the oxide.

Zinc acetate (2H$_2$O) *[5970-45-6]* **M 219.5, m 100°(loses 2H$_2$O), 237°, d 1.74, pK^{25} 8.96 (for hydrolysis of Zn^{2+} to ZnOH$^+$).** Crystd (in poor yield) from hot water or, better, from EtOH.

Zinc acetonylacetate *[14024-63-6]* **M 263.6, m 138°.** Crystd from hot 95% EtOH.

Zinc bromide *[7699-45-8]* **M 225.2, m 384, b 697.** Heated to 300° under vacuum (2 x 10^{-2}mm) for 1h, then sublimed.

Zinc caprylate *[557-09-5]* **M 351.8.** Crystd from EtOH.

Zinc chloride *[7646-85-7]* **M 136.3, m 283°, 290°.** The anhydrous material can be sublimed under a stream of dry HCl, followed by heating to 400° in a stream of dry N$_2$. Also purified by refluxing (50g) in dioxane (400mL) with 5g zinc dust, filtering hot and cooling to ppte ZnCl$_2$. Crystd from dioxane and stored in a desiccator over P$_2$O$_5$. It has also been dried by refluxing in thionyl chloride. [Weberg et al. *J Am Chem Soc* **108** 6242 *1986*.] *Hygroscopic: minimal exposure to the atmosphere is necessary.*

Zinc cyanide *[557-21-1]* **M 117.4, m 800°(dec), d 1.852.** It is a **POISONOUS** white powder which becomes black on standing if $Mg(OH)_2$ and carbonate are not removed in the preparation. Thus wash well with H_2O, then well with EtOH, Et_2O and dry in air at 50°. Analyse by titrating the cyanide with standard $AgNO_3$. Other likely impurities are $ZnCl_2$, $MgCl_2$ and traces of basic zinc cyanide; the first two salts can be washed out. It is soluble in aq KCN solns. However, if purified in this way $Zn(CN)_2$ is not reactive in the Gattermann synthesis. For this the salt should contain at least 0.33 mols of KCl or NaCl which will allow the reaction to proceed faster. [*J Am Chem Soc* **45** 2375 *1923,* **60** 1699 *1938*; *Org Synth* Coll Vol III 549 *1955.*]

Zinc diethyldithiocarbamate *[14324-55-1]* **M 561.7, pK^{25} 3.04 (for Et$_2$NCS$_2^-$).** Crystd several times from hot toluene or from hot $CHCl_3$ by addition of EtOH.

Zinc dimethyldithiocarbamate *[137-30-4]* **M 305.8, m 248-250°, pK^{25} 3.36 (for Me$_2$NCS$_2^-$).** Crystd several times from hot toluene or from hot $CHCl_3$ by addition of EtOH.

Zinc ethylenebis[dithiocarbamate] *[12122-67-7]* **M 249.7.** Crystd several times from hot toluene or from hot $CHCl_3$ by addition of EtOH.

Zinc fluoride *[7783-49-5]* **M 103.4, m 872°, b 1500°, d^{25} 5.00.** Possible impurity is H_2O which can be removed by heating at 100° or by heating to 800° in a dry atmosphere. Heating in the presence of NH_4F produces larger crystals. It is sparingly sol in H_2O (1.51g/100mL) but more sol in HCl, HNO_3 and NH_4OH. It can be stored in glass bottles. [*Handbook of Preparative Inorganic Chemistry (Ed. Brauer)* Vol I 242 *1963.*]

Zinc formate (2H$_2$O) *[557-41-5]* **M 191.4, m 140°(loses H$_2$O), d 2.21.** Crystd from water (3mL/g).

Zinc iodide *[10139-47-6]* **M 319.2, m 446, b 624°(dec), d 4.74.** Heated to 300° under vacuum (2 x 10^{-2}mm) for 1h, then sublimed.

Zinc *RS*-lactate (3H$_2$O) *[554-05-2; 16039-53-5 (L)]* **M 297.5.** Crystd from water (6mL/g).

Zincon (*o*-**[1-(2-hydroxy-5-sulfo)-3-phenyl-5-formazono]-benzoic acid**) *[135-52-4]* **M 459.4.** Main impurities are inorganic salts which can be removed by treatment with dilute acetic acid. Organic contaminants are removed by refluxing with ether. It can be recrystd from dilute H_2SO_4. [Fichter and Schiess *Chem Ber* **33** 751 *1900.*]

Zincon disodium salt (*o*-**[1-(2-hydroxy-5-sulfo)-3-phenyl-5-formazono]-benzoic acid di-Na salt**) *[135-52-4; 56484-13-0]* **M 484.4, m ~250-260° (dec).** Zincon soln is prepared by dissolving 0.13g of the powder in aqueous N NaOH (2mL diluted to 100mL with H_2O). This gives a deep red colour which is stable for one week. It is a good reagent for zinc ions but also forms stable complexes with transition metal ions. [UV-VIS: Bush and Yoe *Anal Chem* **26** 1345 *1954*; Hunter and Roberts *J Chem Soc* 820 *1941*; Platte and Marcy *Anal Chem* **31** 1226 *1959*] The free acid has been recrystd from dilute H_2SO_4. [Fichter and Scheiss *Chem Ber* **33** 751 *1900.*]

Zinc perchlorate (6H$_2$O) *[13637-61-1]* **M 372.4, m 105-107°, pK^{25} -2.4 to -3.1 (for HClO$_4$).** Crystd from water.

Zinc phenol-*o*-sulfonate (8H$_2$O) *[127-82-2]* **M 555.8.** Crystd from warm water by cooling to 0°.

Zinc phthalocyanine *[14320-04-8]* **M 580.9.** Sublimed repeatedly in a flow of oxygen-free N_2.

Zinc sulfate (7H$_2$O) *[7446-20-0]* **M 287.5, m 100°(dec), 280°(loses all 7H$_2$O), >500(anhydr), d 1.97.** Crystd from aqueous EtOH.

Zinc 5,10,15,20-tetraphenylporphyrin *[14074-80-7]* **M 678.1, λ_{max} 418(556)nm.** Purified by chromatography on neutral (Grade I) alumina, followed by recrystallisation from CH_2Cl_2/MeOH [Yamashita et al. *J Phys Chem* **91** 3055 *1987*].

Zinc trifluoromethanesulfonate *[54010-75-2]* **M 363.5, m >300°.** It should be dried at 125° for 2h at 3mm. It is soluble in CH_2Cl_2 but insoluble in pet ether. [*Tetrahedron Lett* **24** 169 *1983*.]

Zirconium (IV) propoxide *[23519-77-9]* **M 327.6, b 198°/0.03mm, 208°/0.1mm, d_4^{20} 1.06, n_D^{20} 1.454.** Although it was stated that it could not be crystallised or sublimed even at 150°/10^{-4}mm [*J Chem Soc* 280 *1951*], the propoxide has, when properly prepared, been purifed by distn in a high vacuum [*J Chem Soc* 2025 *1953*].

Zirconium tetrachloride *[10026-11-6]* **M 233.0, m 300°(sublimes), pK_1^{25} -0.32, pK_2^{25} 0.06, pK_3^{25} 0.35, pK_4^{25} 0.46 (for hydrolysis of aquo Zr^{4-}).** Crystd repeatedly from conc HCl.

Zirconocene chloride hydride **(bis[cyclopentadienyl]zirconium hydride chloride) (Schwartz reagent)** *[37342-97-5]* **M 257.9.** It is a moisture and light sensitive compound. Its purity can be determined by reaction with a slight excess of Me_2CO whereby the active H reacts to produce $Cp_2ZrClOPr^i$ and the integrals of the residual Me_2CO in the ^1H NMR will show how pure the sample is. The presence of Cp_2ZrH_2 can be determined because it forms $Cp_2Zr(OPr^i)_2$. For very active compound it is best to prepare freshly from the dichloride by reduction with Vitride [$LiAl(OCH_2CH_2OH)_2H_2$], the white ppte is filtered off, washed with tetrahydrofuran, *C_6H_6, Et_2O, dried in vacuum and stored under anhydrous conditions and in the dark. [IR: *J Chem Soc, Chem Commun* 1105 *1969*; *J Am Chem Soc* **96** 8115 *1974*, **101** 3521 *1979*; *Synthesis* 1 *1988*.]

Zirconocene dichloride **(bis[cyclopentadienyl]zirconium dichloride)** *[1291-32-3]* **M 292.3, m 242-245°, 248°.** Purified by recrystn from $CHCl_3$ or xylene, and dried in vacuum. ^1H NMR ($CDCl_3$) δ: 6.52 from Me_4Si. Store in the dark under N_2 as it is moisture sensitive. [IR, NMR, MS: *Aust J Chem* **18** 173 *1965*; method of *J Am Chem Soc* **81** 1364 *1959*; and references in the previous entry.]

Zirconyl chloride ($6H_2O$) *[7699-43-6]* **M 286.2, m 150°(loses $6H_2O$).** Crystd repeatedly from 8M HCl as $ZrOCl_2.8H_2O$. On drying $ZrOCl_2.6H_2O$ **m 150°.** The product was not free from hafnium.

Zirconyl chloride ($8H_2O$) *[13520-92-8]* **M 322.3, m 150°(loses $6H_2O$), 210°(loses all H_2O). 400°(anhydr dec), d 1.91.** Recrystd several times from water [Ferragina et al. *J Chem Soc, Dalton Trans* 265 *1986*]. Recrystn from 8M HCl gives the octahydrate as white needles on concentrating. It is also formed by hydrolysing $ZrCl_4$ with water. After one recryst from H_2O, 99+% grade had Ag, Al, As, Cd, Cu, Hf, Mg, Na, Sc and V at 20, 1.8, 0.6, 0.6, 0.4, 8.4, 0.4, 2.4, 80 and 3 ppm resp.

CHAPTER 6

PURIFICATION OF BIOCHEMICALS AND RELATED PRODUCTS

Biochemicals are chemical substances produced by living organisms. They range widely in size, from simple molecules such as formic acid and glucose to macromolecules such as proteins and nucleic acids. Their *in vitro* synthesis is often impossibly difficult and in such cases they are available (if at all) only as commercial tissue extracts which have been subjected to purification procedures of widely varying stringency. The desired chemical may be, initially, only a minor constituent of the source tissue which may vary considerably in its composition and complexity. Recent advances in molecular biology have made it possible to produce substantial amounts of biological materials, which are present in nature in extremely small amounts, by recombinant DNA technology and expression in bacteria, yeast, insect and mammalian cells. The genes for these substances can be engineered such that the gene products, e.g. polypeptides or proteins, can be readily obtained in very high states of purity. However, many such products which are still obtained from the original natural sources are available commercially and may require further purification.

As a preliminary step the tissue might be separated into phases [e.g. whole egg into white and yolk, blood into plasma (or serum) and red cells], and the desired phase may be homogenised. Subsequent treatment usually comprises filtration, solvent extraction, salt fractionation, ultracentrifugation, chromatographic purification, gel filtration and dialysis. Fractional precipitation with ammonium sulfate gives crude protein species. Purification is finally judged by the formation of a single band of macromolecule (e.g. protein) on electrophoresis and/or analytical ultracentrifugation. Although these generally provide good evidence of high purity, none-the-less it does not follow that one band under one set of experimental conditions is an absolute indication of homogeneity.

During the past 20 or 30 years a wide range of methods for purifying substances of biological origin have become available. For small molecules (including many sugars and amino acids) reference should be made to Chapters 1 and 2. The more important methods used for large molecules, polypeptides and proteins in particular, comprise:

1. *Centrifugation.* In addition to centrifugation for sedimenting proteins after ammonium sulfate precipitation in dilute aqueous buffer, this technique has been used for fractionation of large molecules in a denser medium or a medium of varying density. By layering sugar solutions of increasing densities in a centrifuge tube, proteins can be separated in a sugar-density gradient by centrifugation. Smaller DNA molecules (e.g. plasmid DNA) can be separated from RNA or nuclear DNA by centrifugation in aqueous cesium chloride (*ca* 0.975g/mL of buffer) for a long time (e.g. 40h at 40,000 x g). The plasmid DNA band appears at about the middle of the centrifuge tube, and is revealed by the fluorescent pink band formed by the binding of DNA to ethidium bromide which is added to the CsCl buffer. *Microfuges* are routinely used for centrifugation in Eppendorf tubes (1.2-2mL) and can run up to speeds of 12,000 x g. *Analytical centrifugation*, which is performed under specific conditions in an analytical ultracentrifuge is very useful for determining purity, aggregation of protein subunits and the molecular weight of macromolecules. [D.Rickwood, T.C.Ford and J.Steensgaard *Centrifugation: Essential Data Series*, J Wiley & Sons, NY, *1994*].

2. *Gel filtration* with polyacrylamide (mol wt exclusion limit from 3000 to 300,000) and agarose gel (mol wt exclusion limit 0.5 to 150 x 10^6) is useful for separating macromolecules. In this technique high-molecular weight substances are too large to fit into the gel microapertures and pass rapidly through the matrix (with the void volume), whereas low molecular weight species enter these apertures and are held there for longer periods of time, being retarded by the column material in the equilibria, relative to the larger molecules. This method is also used for desalting solutions of macromolecules. *Dry gels* and *crushed beads* are also

useful in the gel filtration process. Selective retention of water and inorganic salts by the gels or beads (e.g. Sephadex G-25) results in increased concentration and purity of the protein fraction which moves with the void volume. (See also Chapter 1, pp 23, 41).

3. *Ion exchange matrices* are microreticular polymers containing carboxylic acid (e.g. Bio-Rad 70) or phosphoric acid (Pharmacia, Amersham Biosciences, Mono-P) exchange functional groups for weak acidic cation exchangers, sulfonic acid groups (Dowex 50W) for strong acidic cation exchangers, diethylaminoethyl (DEAE) groups for weakly basic anion exchangers and quaternary ammonium (QEAE) groups for strong anion exchangers. The old cellulose matrices for ion exchanges have been replaced by Sephadex, Sepharose or Fractogel which have more even particle sizes with faster and more reproducible flow rates. Some can be obtained in fine, medium or coarse grades depending on particle size. These have been used extensively for the fractionation of peptides, proteins and enzymes. The use of *p*H buffers controls the strength with which the large molecules are bound to the support in the chromatographic process. Careful standardisation of experimental conditions and similarly the very uniform size distribution of Mono beads has led to high resolution in the purification of protein solutions. MonoQ (Pharmacia, Amersham Biosciences) is a useful strong anion exchanger, and MonoS (Pharmacia, Amersham Biosciences) is a useful strong cation exchanger whereas MonoP is a weak cation exchanger. These have been successful with medium pressure column chromatography (FPLC, see below in 8). Chelex 100 binds strongly and removes metal ions from macromolecules. [See also Chapter 1, pp. 22-24.]

4. *Hydroxylapatite* is used for the later stages of purification of enzymes. It consists essentially of hydrated calcium phosphate which has been precipitated in a specific manner. It combines the characteristics of gel and ionic chromatography. Crystalline hydroxylapatite is a structurally organised, highly polar material which, in aqueous solution (in buffers) strongly adsorbs macromolecules such as proteins and nucleic acids, permitting their separation by virtue of the interaction with charged phosphate groups and calcium ions, as well as by physical adsorption. The procedure therefore is not entirely ion-exchange in nature. Chromatographic separations of singly and doubly stranded DNA are readily achievable whereas there is negligible adsorption of low molecular weight species.

5. *Affinity chromatography* is a chromatographic technique whereby the adsorbant has a particular and specific affinity for one of the components of the mixture to be purified. For example the adsorbant can be prepared by chemically binding an inhibitor of a specific enzyme (which is present in the crude complex mixture) to a matrix (e.g. Sepharose). When the mixture of impure enzyme is passed through the column containing the adsorbant, only the specific enzyme binds to the column. After adequate washing, the pure enzyme can be released from the column by either increasing the salt concentration (e.g. NaCl) in the eluting buffer or adding the inhibitor to the eluting buffer. The salt or inhibitor can then be removed by dialysis, gel filtration (above) or ultrafiltration (see below). [See W.H.Scouten, *Affinity Chromatography: Bioselective Adsorption on Inert Matrices*, J.Wiley & Sons, NY, *1981*, ISBN 0471026492; H.Schott, *Affinity Chromatography: Template Chromatography of Nucleic Acids and Proteins*, Marcel Dekker, NY, 1984, ISBN 0824771117; P.Matejtschuk ed. *Affinity Separations* Oxford University Press 1997 ISBN 0199635501 (paperback); M.A.Vijayalakshmi, *Biochromatography, Theory and Practice*, Taylot & Francis Publ, 2002, ISBN 0415269032; and Chapter 1, p. 25.]

6. In the *Isoelectric focusing* of large charged molecules on polyacrylamide or agarose gels, slabs of these are prepared in buffer mixtures (e.g. ampholines) which have various pH ranges. When a voltage is applied for some time the buffers arrange themselves on the slabs in respective areas according to their pH ranges (prefocusing). Then the macromolecules are applied near the middle of the slab and allowed to migrate in the electric field until they reach the pH area similar to their isoelectric points and focus at that position. This technique can also be used in a chromatographic mode, *chromatofocusing*, whereby a gel in a column is run (also under HPLC conditions) in the presence of ampholines (narrow or wide pH ranges as required) and the macromolecules are then run through in a buffer. *Capillary electrophoresis* systems in which a current is applied to set the gradient are now available in which the columns are fine capillaries and are used for qualitative and quantitative purposes [See R.Kuhn and S.Hoffstetter-Kuhn, *Capillary Electrophoresis*: Principles and Practice, Springer-Verlag Inc, NY, *1993*; P.Camilleri ed. *Capillary Electrophoresis - Theory and Practice*, CRC Press, Boca Raton, Florida, *1993*; D.R.Baker, *Capillary Electrophoresis*, J Wiley & Sons, NY, *1995*; P.G.Righetti, A.Stoyanov and M.Zhukov, *The Proteome Revisited, Isoelectric Focusing; J.Chromatography Library Vol* **63** 2001, Elsevier, ISBN 0444505261.] The bands are eluted according to their isoelectric points. Isoelectric focusing standards are available which can be used in a preliminary run in order to calibrate the effluent from the column, or alternatively the pH of the effluent is recorded using a glass electrode designed for the purpose. Several efficient commercially available apparatus are available for separating proteins on a preparative and semi-preparative scale.

7. *High performance liquid chromatography* (HPLC) is liquid chromatography in which the eluting liquid is sent through the column containing the packing (materials as in 2-6 above, which can withstand higher than atmospheric pressures) under pressure. On a routine basis this has been found useful for purifying

proteins (including enzymes) and polypeptides after enzymic digestion of proteins or chemical cleavage (e.g. with CNBr) prior to sequencing (using reverse-phase columns such as μ-Bondapak C18). Moderate pressures (50-300psi) have been found most satisfactory for large molecules (FPLC). [See Scopes *Anal Biochem* **114** 8 *1981*; *High Performance Liquid Chromatography and Its Application to Protein Chemistry,* Hearn in *Advances in Chromatography,* **20** 7 *1982*; B. A. Bidlingmeyer *Practical HPLC Methodology and Applications,* J Wiley & Sons, NY *1991*; L.R.Snyder, J.L.GlajCh and J.J.Kirkland *Practical HPLC Method Development,* J Wiley & Sons, NY *1988*; ISBN 0471627828; R.W.A.Oliver, *HPLC of Macromolecules: A Practical Approach,* 2nd Edn, Oxford University Press, *1998*, T.Hanai, *HPLC: A Practical Guide,* Royal Society of Chemistry (UK), *1999*, ISBN 084045155; P.Millner *High Resolution Chromatography,* Oxford University Press, *1999* ISBN 0199636486; see also Chapter 1, bibliography.]

8. *Ultrafiltration* using a filter (e.g. Millipore) can remove water and low-molecular weight substances without the application of heat. Filters with a variety of molecular weight exclusion limits not only allow the concentration of a particular macromolecule to be determined, but also the removal (by washing during filtration) of smaller molecular weight contaminants (e.g. salts, inhibitors or cofactors). This procedure has been useful for changing the buffer in which the macromolecule is present (e.g. from Tris-Cl to ammonium carbonate), and for desalting. Ultrafiltration can be carried out in a stirrer cell (Amicon) in which the buffer containing the macromolecule (particularly protein) is pressed through the filter, with stirring, under argon or nitrogen pressure (e.g. 20-60psi). During this filtration process the buffer can be changed. This is rapid (e.g. 2L of solution can be concentrated to a few mLs in 1 to 2h depending on pressure and filter). A similar application uses a filter in a specially designed tube (Centricon tubes, Amicon) and the filtration occurs under centrifugal force in a centrifuge (4-6000rpm at 0°/40min). The macromolecule (usually DNA) then rests on the filter and can be washed on the filter by centrifugation. The macromolecule is recovered by inverting the filter, placing a conical receiver tube on the same side where the macromolecule rests, filling the other side of the filter tube with eluting solution (usually a very small volume e.g. 100 μL), and during further centrifugation this solution passes through the filter and collects the macromolecule from the underside into the conical receiver tube.

9. *Partial precipitation* of a protein in solution can often be achieved by controlled addition of a strong salt solution, e.g ammonium sulfate. This is commonly the first step in the purification process. Its simplicity is offset by possible denaturation of the desired protein and the (sometimes gross) contamination with other proteins. It should therefore be carried out by careful addition of small aliquots of the powdered salt or concentrated solution (below 4°, with gentle stirring) and allowing the salt to be evenly distributed in the solution before adding another small aliquot. Under carefully controlled conditions and using almost pure protein it is sometimes possible to obtain the protein in crystalline form suitable for X-ray analysis (see below).

10. *Dialysis.* This is a process by which small molecules, e.g. ammonium sulfate, sodium chloride, are removed from a solution containing the protein or DNA using a membrane which is porous to small molecules. The solution (e.g. 10mL) is placed in a dialysis bag or tube tied at both ends, and stirred in a large excess of dialysing solution (e.g. 1.5 to 2 L), usually a weak buffer at *ca* 4°. The dialysing buffer is replaced with fresh buffer several times, e.g. four times in 24h. This procedure is similar to ultrafiltration (above) and allows the replacement of buffer in which the protein, or DNA, is dissolved. It is also possible to concentrate the solutions by placing the dialysis tube or bag in Sephadex G25 which allows the passage of water and salts from the inside of the bag thus concentrating the protein (or DNA) solution. Dialysis tubing is available from various distibutors but "Spectra/por" tubing (from Spectrum Medical Industries, Inc, LA) is particularly effective because it retains macromolecules and allows small molecules to dialyse out very rapidly thus reducing dialysing time considerably. This procedure is used when the buffer has to be changed so as to be compatible with the next purification or storage step, e.g. when the protein (or DNA) needs to be stored frozen in a particular buffer for extended periods.

11. *Gel Electrophoresis.* This is becoming a more commonly used procedure for purifying proteins, nucleic acids, nucleoproteins, polysaccharides and carbohydrates. The gels can be electroblotted onto membranes and the modern procedures of identifying, sequencing (proteins and nucleic acids) and amplifying (nucleic acids) on sub-micro scales have made this technique of separation a very important one. See below for polyacrylamide gel electrophoresis (PAGE), [D.Patel *Gel Electrophoresis,* J.Wiley-Liss, Inc., 1994; P.Jones and D.Rickwood, *Gel Electrophoresis: Nucleic Acids* , J.Wiley and Sons,1999 (paperback) ISBN 0471960438; D.M.Gersten and D.Gersten, *Gel Electrophoresis: Proteins,* J.Wiley and Sons Inc, 1996 ISBN 0471962651; R.Westermeier *Electrophoresis in Practice,* 3rd Edn, Wiley-VCH, NY, 2001, ISBN 3527303006].

12. *Crystallisation.* The ultimate in purification of proteins or nucleic acids is crystallisation. This involves very specialised procedures and techniques and is best left to the experts in the field of X-ray crystallography who provide a complete picture of the structure of these large molecules. [A. Ducruix and R. Giegé eds, *Crystallisation of Nucleic Acids and Proteins: A Practical Approach,* 2nd Edition, 2000,

Oxford University Press, ISBN 0199636788 (paperback); T.L.Blundell and L.N.Johnson *Protein Crystallisation*, Academic Press, NY, 1976; A.McPherson *Preparation and Analysis of Protein Crystals*, J.Wiley & Sons, NY, 1982; A.McPherson, *Crystallisation of Biological Macromolecules*, Cold Spring Harbour Laboratory Press, 2001 ISBN 0879696176.]

Other details of the above will be found in Chapters 1 and 2 which also contain relevant references.

Several illustrations of the usefulness of the above methods are given in the *Methods Enzymol* series (Academic Press) in which 1000-fold purifications or more, have been readily achieved. In applying these sensitive methods to macromolecules, reagent purity is essential. It is disconcerting, therefore, to find that some commercial samples of the widely used affinity chromatography ligand Cibacron Blue F3GA contained this dye only as a minor constituent. The major component appeared to be the dichlorotriazinyl precursor of this dye. Commercial samples of Procion Blue and Procion Blue MX-R were also highly heterogeneous [Hanggi and Cadd *Anal Biochem* **149** 91 *1985*]. Variations in composition of sample dyes can well account for differences in results reported by different workers. The purity of substances of biological origin should therefore be checked by one or more of the methods given above. Water of high purity should be used in all operations. Double glass distilled water or water purified by a MilliQ filtration system (see Chapter 2) is most satisfactory.

Brief general procedures for the purification of polypeptides and proteins. Polypeptides of up to *ca* 1-2000 (10-20 amino acid residues) are best purified by reverse phase HPLC. The desired fractions that are collected are either precipitated from solution with EtOH or lyophilised. The purity can be checked by HPLC and identified by microsequencing (1-30 picomoles) to ascertain that the correct polypeptide was in hand. Polypeptides larger than these are sometimes classified as proteins, and are purified by one or more of the procedures described above. The purification of enzymes and functional proteins which can be identified by specific interactions is generally easier to follow because enzyme activities or specific protein interactions can be checked (by assaying) after each purification step. The commonly used procedures for purifying soluble proteins involve the isolation of an aqueous extract from homogenised tissues or extracts from ruptured cells from microorganisms or specifically cultured cells, for example, by sonication, freeze shocking or passage through a small orifice under pressure. Contaminating nucleic acids are removed by precipitation with a basic protein, e.g. protamine sulfate. The soluble supernatant is then subjected to fractionation with increasing concentrations of ammonium sulfate. The required fractions are then further purified by the procedures described in sections 2-9 above. If an affinity adsorbant has been identified then affinity chromatography can provide an almost pure protein in one step sometimes even from the crude extract. The rule of thumb is that a solution with a protein concentration of 1mg/mL has an absorbance A_{1cm} at 280nm of 1.0 units. Membrane-bound proteins are usually insoluble in water or dilute aqueous buffer and are obtained from the insoluble fractions, e.g. the microsomal fractions from the >100,000 x g ultracentrifugation supernatant. These are solubilised in appropriate detergents, e.g. Mega-10 (nonionic), Triton X-100 (ionic) detergents, and purified by methods 2 to 8 (previous section) in the presence of detergent in the buffer used. They are assayed also in the presence of detergent or membrane lipids.

The purity of proteins is best checked by *polyacrylamide gel electrophoresis* (PAGE). The gels are either made or purchased as pre-cast gels and can be with uniform or gradient gel composition. Proteins are applied onto the gels *via* wells set into the gels or by means of a comb, and travel along the gel surface by means of the current applied to the gel. When the buffer used contains sodium dodecylsulfate (SDS) the proteins are denatured and the denatured proteins (e.g. as protein subunits) separate on the gels mainly according to their molecular sizes. These can be identified by running marker proteins, with a range of molecular weights, simultaneously on a track alongside the proteins under study. The protein bands are visualised by fixing the gel (20% acetic acid) and staining with Coomassie blue followed by silver staining if higher sensitivity is required. An Amersham-Pharmacia "Phast Gel Electrophoresis" apparatus is very useful for rapid analysis of proteins. It uses small pre-cast polyacrylamide gels (two gels can be run simultaneously) with various uniform or gradient polyacrylamide concentrations as well as gels for isoelectric focusing. The gels are usually run for 0.5-1h and can be stained and developed (1-1.5h) in the same apparatus. The equipment can be used to electroblot the protein bands onto a membrane from which the proteins can be isolated and sequenced or subjected to antibody or other identification procedures. It should be noted that all purification procedures are almost always carried out at *ca* 4° in order to avoid denaturation or inactivation of the protein being investigated. Anyone contemplating the purification of a protein is referred to: Professor R.K.Scopes's monograph *Protein Purification*, 3rd edn, Springer-Verlag, New

York, *1994*, ISBN 0387940723; M.L.Ladisch ed. *Protein Purification - from Molecular Mechanisms to Large-scale Processes*, American Chemical Society, Washington DC, *1990*; E.L.V.Harris and S.Angal, *Protein Purification Applications - A Practical Approach*, IRL Press, Oxford, *1990*; J.C.Janson and L.Rydén, *Protein Purification - Principles, High Resolution Methods and Applications*, VCH Publ. Inc., *1989*; ISBN 0895731223 R.Burgess, *Protein Purification - Micro to Macro*, A.R.Liss, Inc., NY, *1987*; S.M.Wheelwright, *Protein Purification: Design and Scale up of Downstream Processing*, J Wiley & Sons, NY, *1994*, references in the bibliography in Chapter 1, and selected volumes of *Methods Enzymol*, e.g. M.P.Deutscher ed. Guide to Protein Purification, *Methods Enzymol*, Academic Press, NY, Vol 182 *1990*, ISBN 0121820831; T.Palzkill, *Proteomics*, Kluwer Academic Publ, *2001*, ISBN 0792375653; M.A.Vijayalakshmi, *Biochromatography, Theory and Practice*, Taylot & Francis Publ, *2002*, ISBN 0415269032; J.S.Davies, *Amino Acids, Peptides and Proteins Vol* **32** *2001*, A Specialist Periodical Report, Royal Society of Chemistry, ISBN 0854042326; S.Roe, *Protein Purification Techniques: A Practical Approach*, 2nd Edn, Oxford University Press, *2001*, ISBN 0199636737; T.Palmer, *Enzymes, Biochemistry, Biotechnology, Clinical Chemistry*, Horwood Publishing, *2001*, ISBN 1898563780.

Brief general procedures for purifying DNA. Oligo-deoxyribonucleotides (up to *ca* 60-mers) are conveniently purified by HPLC (e.g. using a Bio-Rad MA7Q anion exchange column and a Rainin Instrument Co, Madison, Dynamax-300A C_8 matrix column) and used for a variety of molecular biology experiments. Plasmid and chromosomal DNA can be isolated by centrifugation in caesium chloride buffer (see section 1. centrifugation above), and then re-precipitated with 70% ethanol at -70° (18h), collected by centrifugation (microfuge) and dried in air before dissolving in TE (10mM TrisHCl, 1mM EDTA pH 8.0). The DNA is identified on an Agarose gel slab (0.5 to 1.0% DNA grade in 45mM Tris-borate + 1mM EDTA or 40mM Tris-acetate + 1mM EDTA pH 8.0 buffers) containing ethidium bromide which binds to the DNA and under UV light causes it be visualised as pink fluorescent bands. Marker DNA (from λ phage DNA cut with the restriction enzymes Hind III and/or EcoRI) are in a parallel track in order to estimate the size of the unknown DNA. The DNA can be isolated from their band on the gel by transfer onto a nitro-acetate paper (e.g. NA 45) electrophoretically, by binding to silica or an ion exchange resin, extracted from these adsorbents and precipitated with ethanol. The DNA pellet is then dissolved in TE buffer and its concentration determined. A solution of duplex DNA (or RNA) of 50µg/mL gives an absorbance of 1.0units at 260nm/1cm cuvette (single stranded DNA or RNA gives a value of 1.3 absorbance units). DNA obtained in this way is suitable for molecular cloning. For experimental details on the isolation, purification and manipulation of DNA and RNA the reader is referred to: J.Sambrook, E.F.Fritsch and T.Maniatis, *Molecular Cloning - A Laboratory Manual* , 2nd edn, (3 volumes), Cold Spring Harbor Laboratory Press, NY, *1989*, ISBN 0879693096 (paperback); P.D.Darbre, *Basic Molecular Biology: Essential Techniques*, J.Wiley and Sons, *1998*, ISBN 0471977055; J.Sambrook and D.W.Russell, *Molecular Cloning - A Laboratory Manual* , 3rd edn, (3 volumes), Cold Spring Harbor Laboratory Press, NY, *2001*, ISBN 0079695773 (paperback), ISBN 0079695765 (cloth bound), also available on line; M.A.Vijayalakshmi, *Biochromatography, Theory and Practice*, Taylot & Francis Publ, *2002*, ISBN 0415269032; A.Travers and M.Buckle, *DNA-Protein Interactions: A Practical Approach,* Oxford University Press, *2000*, ISBN 0199636915 (paperback); R.Rapley and D.L.Manning eds *RNA: Isolation and Characterisation Protocols*, Humana Press *1998* ISBN 086034941; R.Rapley, *The Nucleic Acid Protocols Handbook*, Humana Press 2000 ISBN 0896038416 (paperback).

This chapter lists some representative examples of biochemicals and their origins, a brief indication of key techniques used in their purification, and literature references where further details may be found. Simpler low molecular weight compounds, particularly those that may have been prepared by chemical syntheses, e.g. acetic acid, glycine, will be found in Chapter 4. Only a small number of enzymes and proteins are included because of space limitations. The purification of some of the ones that have been included has been described only briefly. The reader is referred to comprehensive texts such as the *Methods Enzymol* (Academic Press) series which currently runs to more than 344 volumes and *The Enzymes* (3rd Edn, Academic Press) which runs to 22 volumes for methods of preparation and purification of proteins and enzymes. Leading references on proteins will be found in *Advances in Protein Chemistry* (59 volumes, Academic Press) and on enzymes will be found in *Advances in Enzymology* (72 volumes, then became *Advances in Enzymology and Related Area of Molecular Biology,* J Wiley & Sons). The *Annual Review of Biochemistry* (Annual Review Inc. Patlo Alto California) also is an excellent source of key references to the up-to-date information on known and new natural compounds, from small molecules, e.g. enzyme cofactors to proteins and nucleic acids.

Abbreviations of titles of periodical are defined as in the Chemical Abstracts Service Source Index (CASSI).

Ionisation constants of ionisable compounds are given as **pK** values (published from the literature) and refer to the **pKa** values at room temperature (~ 15°C to 25°C). The values at other temperatures are given as superscripts, e.g. **pK**^{25} for 25°C. Estimated values are entered as **pK**$_{Est(1)}$~ (see Chapter 1, p 6 for further information).

Benzene, which has been used as a solvent successfully and extensively in the past for reactions and purification by chromatography and crystallisation is now considered a **very dangerous substance** so it has to be used with extreme care. We emphasise that an alternative solvent to benzene (e.g. toluene, toluene-petroleum ether, or a petroleum ether to name a few) should be used first. However, if benzene has to be used then all operations have to be performed in a well ventilated fumehood and precautions taken to avoid inhalation and contact with skin and eyes. Whenever benzene is mentioned in the text and asterisk e.g. $^{*}C_6H_6$ or *benzene, is inserted to remind the user that special precaution should be adopted.

Amino acids, **carbohydrates** and **steroids** not found below are in Chapter 4 (see also CAS Registry Numbers Index and General Index).

Abrin A and Abrin B *[1393-62-0]* M_r **63,000-67,000.** Toxic proteins from seeds of
Abras precatorius. Purified by successive chromatography on DEAE-Sephadex A-50, carboxymethylcellulose, and DEAE-cellulose. [Wei et al. *J Biol Chem* **249** 3061 *1974.*]

Acetoacetyl coenzyme A trisodium salt trihydrate *[102029-52-7]* **M 955.6, pK$_1$ 4.0 (NH$_2$), pK$_2$ 6.4 (PO$_4$$^-$).** The pH of solution (0.05g/mL H$_2$O) is adjusted to 5 with 2N NaOH. This solution can be stored frozen for several weeks. Further purification can be carried out on a DEAE-cellulose formate column, then through a Dowex 50 (H$^+$) column to remove Na ions, concentrated by lyophilisation and redissolved in H$_2$O. Available as a soln of 0.05g/mL of H$_2$O. The concn of acetoacetylcoenzyme A is determined by the method of Stern et al. *J Biol Chem* **221** 15 *1956*. It is stable at pH 7-7.5 for several hours at 0° (half life *ca* 1-2h). At room temperature it is hydrolysed in *ca* 1-2h at pH 7-7.5. At pH 1.0/20° it is more stable than at neutrality. It is stable at pH 2-3/-17° for at least 6 months. [*J Biol Chem* **159** 1961 *1964*; **242** 3468 *1967*; Clikenbeard et al. *J Biol Chem* **250** 3108 *1975*; *J Am Chem Soc* **75** 2520 *1953*, **81** 1265 *1959*; see Simon and Shemin *J Am Chem Soc* **75** 2520 *1953*; Salem et al. *Biochem J* **258** 563 *1989.*]

Acetobromo-α-D-galactose *[3068-32-4]* **M 411.2, m 87°, [α]$_{546}^{20}$ +255°, [α]$_D^{20}$ +210° (c 3, CHCl$_3$).** Purified as for the glucose analogue (see next entry). If the compound melts lower than 87° or is highly coloured then dissolve in CHCl$_3$ (*ca* 3 vols) and extract with H$_2$O (2 vols), 5% aqueous NaHCO$_3$, and again with H$_2$O and dry over Na$_2$SO$_4$. Filter and evaporate in a vacuum. The partially crystalline solid or syrup is dissolved in dry Et$_2$O (must be very dry) and recrystd by adding pet ether (b 40-60°) to give a white product. [McKellan and Horecker *Biochem Prep* **11** 111 *1960.*]

Acetobromo-α-D-glucose *[572-09-8]* **M 411.2, m 87-88°, 88-89°, [α] $_{546}^{20}$ +230°, [α]$_D^{20}$ +195° (c 3, CHCl$_3$).** If nicely crystalline recryst from Et$_2$O-pentane. Alternatively dissolve in diisopropyl ether (dried over CaCl$_2$ for 24hours, then over P$_2$O$_5$ for 24hours) by shaking and warming (for as short a period as possible), filter warm. Cool to *ca* 45° then slowly to room temperature and finally at 5° for more than 2hours. Collect the solid, wash with cold dry diisopropyl ether and dry in a vacuum over Ca(OH)$_2$ and NaOH. Store dry in a desiccator in the dark. Solutions can be stabilised with 2% CaCO$_3$. [Redemann and Niemann *Org Synth* **65** 236 *1987*, Coll Vol III 11 *1955.*]

Acetoin dehydrogenase **[from beef liver; acetoin NAD oxidoreductase]** *[9028-49-3]* **M$_r$ 76000, [EC 1.1.1.5].** Purified *via* the acetone cake then Ca-phosphate gel filtration (unabsorbed), lyophilised and then fractionated through a DEAE-22 cellulose column. The Km for diacetyl in 40μM and for

NADH it is 100μM in phosphate buffer at pH 6.1. [Burgos and Martin *Biochim Biophys Acta* **268** 261 *1972*; **289** 13 *1972*.]

(-)-3-β-Acetoxy-5-etienic acid   [3-β-acetoxy-5-etiocholenic acid,   androst-5-ene-17-β-carboxylic acid] *[51424-66-9]* **M 306.5, m 238-240°, 241-242°, 243-245°, 246-247°, [α]$_D^{20}$ -19.9° (c 1, Me$_2$CO), -36° (c 1, Dioxane), -33.5° (CHCl$_3$), pK$_{Est}$ ~ 4.7.** It is purified by recrystn from Me$_2$CO, Et$_2$O-pentane, or AcOH, and dried in a vacuum oven (105°/20mm) and sublimed at high vacuum. [Staunton and Eisenbram *Org Synth* **42** 4 *1962*; Steiger and Reichstein *Helv Chim Acta* **20** 1404 *1937*.]

Acetylcarnitine   chloride   (2-acetoxy-3-carboxy-N,N,N-trimethylpropanamine HCl) *[S(D+)-5080-50-2; R(L-)- 5061-35-8; RS 2504-11-2]* **M 239.7, m 181°, 197°(dec), [α]$_D^{25}$ -28° (c 2, H$_2$O)** for *S*-isomer, **pK^{25} 3.6.** Recrystd from isopropanol. Dried over P$_2$O$_5$ under high vacuum.

Acetylcholine bromide *[66-23-9]* **M 226.1, m 143°, 146°.** *Hygroscopic* solid but less than the hydrochloride salt. It crystd from EtOH as prisms. Some hydrolysis occurs in boiling EtOH particularly if it contains some H$_2$O. It can also be recryst from EtOH or MeOH by adding dry Et$_2$O. [*Acta Chem Scand* **12** 1492, 1497, 1502 *1958*.]

Acetylcholine chloride *[60-31-1]* **M 181.7, m 148-150°, 151°.** It is very sol in H$_2$O (> 10%), and is very *hygroscopic*. If pasty, dry in a vacuum desiccator over H$_2$SO$_4$ until a solid residue is obtained. Dissolve in abs EtOH, filter and add dry Et$_2$O and the hydrochloride separates. Collect by filtration and store under very dry conditions. [*J Am Chem Soc* **52** 310 *1930*.] The *chloroplatinate* crystallises from hot H$_2$O in yellow needles and can be recrystd from 50% EtOH, **m** 242-244° [*Biochem J* **23** 1069 *1929*], other **m** given is 256-257°. The *perchlorate* crystallises from EtOH as prisms **m** 116-117°. [*J Am Pharm Assocn* **36** 272 *1947*.]

N^4-Acetylcytosine *[14631-20-0]* **M 153.1, m >300°, 326-328°, pK$_{Est(1)}$ ~1.7, pK$_{Est(2)}$ ~10.0.** If TLC or paper chromatography show that it contains unacetylated cytosine then reflux in Ac$_2$O for 4h, cool at 3-4° for a few days, collect the crystals, wash with cold H$_2$O, then EtOH and dry at 100°. It is insoluble in EtOH and difficulty soluble in H$_2$O but crystallises in prisms from hot H$_2$O. It is hydrolysed by 80% aq AcOH at 100°/1h. [*Am Chem J* **29** 500 *1903*; UV: *J Chem Soc* 2384 *1956*; *J Am Chem Soc* **80** 5164 *1958*.] It forms an Hg salt [*J Am Chem Soc* **79** 5060 *1957*].

β-D-N-Acetylglucosaminidase [from M sexta insects] *[9012-33-3]* **M$_r$ ~61,000, [EC 3.2.1.52].** Purified by chromatography on DEAD-Biogel, hydroxylapatite chromatography and gel filtration through Sephacryl S200. Two isoforms: a hexosaminidase EI with Km 177μM (V_{max} 328 sec^{-1}) and EII a chitinase with Km 160μM (V_{max} 103 sec^{-1}) with 4-nitrophenyl-β-acetylglucosamine as substrate. [Dziadil-Turner *Arch Biochem Biophys* **212** 546 *1981*.]

β-D-N-Acetylhexosaminidase A and B   (from human placenta) *[9012-33-3]* **M$_r$ ~61,000, [EC 3.2.1.52].** Purified by Sephadex G-200 filtration and DEAE-cellulose column chromatography. Hexosaminidase A was further purified by DEAE-cellulose column chromatography, followed by an ECTEOLA-cellulose column, Sephadex-200 filtration, electrofocusing and Sephadex G-200 filtration. Hexosaminidase B was purified by a CM-cellulose column, electrofocusing and Sephadex G-200 filtration. [Srivastava et al. *J Biol Chem* **249** 2034 *1974*.]

N-Acetyl-D-lactosamine   [2-acetylamino-O-β-D-lactopyranosyl-2-deoxy-D-glucose] *[32181-59-2]* **M 383.4, m 169-171°, 170-171°, [α]$_D^{18}$ +51.5°→ +28.8° (in 3h, c 1, H$_2$O].** Purified by recrystn from MeOH (with 1 mol of MeOH) or from H$_2$O. It is available as a soln of 0.5g /mL of H$_2$O. [Zilliken *J Biol Chem* **271** 181 *1955*.]

O-Acetyl-β-methylcholine chloride   [Methacholine chloride, Amechol, Provocholine, 2-acetoxypropyl-ammonium chloride] *[62-51-1]* **M 195.7, m 170-173°, 172-173°.** It forms white *hygroscopic* needles from Et$_2$O and is soluble in H$_2$O, EtOH and CHCl$_3$. It decomposes readily in alkaline solns and slowly in H$_2$O. It should be handled and stored in a dry atmosphere. The *bromide* is less hygroscopic and the *picrate* has **m** 129.5-131° (from EtOH). [racemate: Annis and Ely *Biochem J* **53** 34 *1953*; IR of iodide: Hansen *Acta Chem Scand* **13** 155 *1959*.]

N-Acetyl muramic acid   [NAMA, *R*-2-(acetylamino)-3-*O*-(1-carboxyethyl)-2-deoxy-D-glucose] *[10597-89-4]*   **M 292.3, m ~125°(dec), $[\alpha]_D^{20}$ +41.2°** (c 1.5, H_2O, after 24h), **pK$_{Est}$ ~ 3.6.** See muramic acid below.

N-Acetyl neuraminic acid (NANA, *O*-Sialic acid, 5-acetamido-3,5-dideoxy-D-*glycero*-D-*glacto*-2-nonulosonic acid, lactaminic acid) *[131-48-6]* M 309.3, m 159°(dec), 181-183°(dec), 185-187°(dec), $[\alpha]_D^{25}$ -33° (c 2, H_2O, l 2), pK 2.6. A Dowex-1x8 (200-400 mesh) in the formate form was used, and was prepd by washing with 0.1M NaOH, then 2N sodium formate, excess formate was removed by washing with H_2O. N-Acetyl neuraminic acid in H_2O is applied to this column, washed with H_2O, then eluted with 2N formic acid at a flow rate of 1mL/min. Fractions (20mL) were collected and tested (Bial's orcinol reagent, *cf Biochem Prep* **7** 1 *1959*). NANA eluted at formic acid molarity of 0.38 and the Bial positive fractions are collected and lyophilised. The residue is recrystd from aqueous AcOH: Suspend 1.35g of residue in AcOH, heat rapidly to boiling, add H_2O dropwise until the suspension dissolves (do not add excess H_2O, filter hot and then keep at +5° for several hours until crystn is complete. Collect and dry in a vacuum over P_2O_5. Alternatively dissolve 1.35g of NANA in 14mL of H_2O, filter, add 160mL of MeOH followed by 360mL of Et_2O. Then add pet ether (b 40-60°) until heavy turbidity. Cool at 20° overnight. Yield of NANA is *ca* 1.3g. Dry over P_2O_5 at 1mm vacuum and 100° to constant weight. It mutarotates in Me_2SO: $[\alpha]_D^{20}$ -115° (after 7min) to -32° (after 24h). It is available as aqueous soln (0.01g/mL). [IR and synthesis: Cornforth et al. *Biochem J* **68** 57 *1958*; Zillikin and O'Brien *Biochem Prep* **7** 1 *1960*; ^{13}C NMR and 1-^{13}C synthesis: Nguyen, Perry *J Org Chem* **43** 551 *1978*; Danishevski, DeNinno *J Org Chem* **51** 2615 *1986*; Gottschalk, *The Chemistry and Biology of Sialic Acids and Related Substances,* Cambridge University Press, London, *1960*.]

N-Acetyl neuraminic acid aldolase **[from** *Clostridium perfringens,* **N-acetylneuraminic acid pyruvate lyase]** *[9027-60-5]* **M$_r$ 32,000 [EC 4.1.3.3].** Purified by extraction with H_2O, protamine pptn, $(NH_4)_2SO_4$ pptn, Me_2CO pptn, acid treatment at pH 5.7 and pptn at pH 4.5. The equilibrium constant for pyruvate + n-acetyl-D-mannosamine ⇌ N-acetylneuraminidate at 37° is 0.64. The Km for N-acetylneuraminic acid is 3.9mM in phosphate at pH 7.2 and 37°. [Comb and Roseman *Methods Enzymol* **5** 391 *1962*.] The enzyme from Hogg kidney (cortex) has been purified 1700 fold by extraction with H_2O, protamine sulfate pptn, $(NH_4)_2SO_4$ pptn, heating between 60-80°, a second $(NH_4)_2SO_4$ pptn and starch gel electrophoresis. The Km for N-acetylneuraminic acid is 1.5mM. [Brunetti et al. *J Biol Chem* **237** 2447 *1962*.]

N-Acetyl penicillamine *[D- 15537-71-0, DL-59-53-0]* **M 191.3, m 183°, 186-187° (DL-form), 189-190° (D-form), D-form $[\alpha]_D^{25}$ +18°** (c 1, 50% EtOH), **pK$_{Est(1)}$~3.0 (CO$_2$H), pK$_{Est(2)}$~ 8.0 (SH).** Both forms are recrystd from hot H_2O. A pure sample of the D-form was obtained after five recrystns. [Crooks in *The Chemistry of Penicillin* Clarke, Johnson and Robinson eds, Princeton University Press, 470 *1949*.]

***p*-Acetylphenyl sulfate potassium salt,** *[38533-41-4]* **M 254.3, m dec on heating, pK$_{Est}$ ~2,1.** Purified by dissolving in the minimum vol of hot water (60°) and adding EtOH, with stirring, then left at 0° for 1h. Crystals were filtd off and recrystd from H_2O until free of Cl⁻ and SO_4^{2-} ions. Dried in a vac over P_2O_5 at room temperature. It is a specific substrate for arylsulfatases which hydrolyse it to *p*-acetylphenol [λmax 327nm (ε 21700 $M^{-1}cm^{-1}$)] [Milsom et al. *Biochem J* **128** 331 *1972*].

S-Acetylthiocholine bromide *[25025-59-6]* **M 242.2, m 217-223°(dec).** It is a *hygroscopic* solid which can be recrystd from ligroin-EtOH (1:1), dried and kept in a vacuum desiccator. Crystn from $^*C_6H_6$-EtOH gave **m** 227° or from propan-1-ol the **m** was 213°. [*Acta Chem Scand* **11** 537 *1957*, **12** 1481 *1958*.]

S-Acetylthiocholine chloride *[6050-81-3]* **M 197.7, m 172-173°** The chloride can be purified in the same way as the bromide, and it can be prepared from the iodide. A few milligrams dissolved in H_2O can be purified by applying onto a Dowex-1 Cl⁻ resin column (prepared by washing with N HCl followed by CO_3^{2-}—free H_2O until the pH is 5.8). After equilibration for 10min elution is started with CO_3^{2-}—free distilled H_2O and

3mL fractions are collected and their OD at 229nm measured. The fractions with appreciable absorption are pooled and lyophilised at 0-5°. Note that at higher temps decomposition of the ester is appreciable; hydrolysis is appreciable at pH >10.5/20°. The residue is dried *in vacuo* over P_2O_5, checked for traces of iodine (conc H_2SO_4 and heat, violet vapours are released), and recrystd from propan-1-ol. [*Clin Chim Acta* **2** 316 *1957*.]

S-Acetylthiocholine iodide *[1866-15-5]* **M 289.2, m 203-204°, 204°, 204-205°.** Recrystd from propan-1-ol (or *iso*-PrOH, or EtOH/Et_2O) until almost colourless and dried in a vacuum desiccator over P_2O_5. Solubility in H_2O is 1% w/v. A 0.075M (21.7mg/mL) solution in 0.1M phosphate buffer pH 8.0 is stable for 10-15 days if kept refrigerated. Store away from light. It is available as a 1% soln in H_2O. [*Biochemical Pharmacology* **7**, 88 *1961*; IR: Hansen *Acta Chem Scand* **13** 151 *1959*, **11** 537 *1957* ; *Clin Chim Acta* **2** 316 *1957*; *Zh Obshch Khim* **22** 267 *1952*.]

Actinomycin C (Cactinomycin) *[8052-16-2]* **M ~1255.** (A commercial mixture of Actinomycin C_1 ~5%, C_2 ~30% and C_3 ~65%). *Actinimycin C_1 (native)* crysts from EtOAc as red crystals, is sol in $CHCl_3$, *C_6H_6 and Me_2CO and has **m** 246-247°(dec), $[\alpha]_D^{20}$ -328° (0.22, MeOH) and λ_{max} 443nm (ε 25,000) and 240nm (ε 34,000). *Actinimycin C_2 (native)* crysts as red needles from EtOAc and has **m** 244-246°(dec), $[\alpha]_D^{20}$ -325° (c 0.2, MeOH), λ_{max} 443nm (ε 25,300) and (ε 33,400). *Actinimycin C_3 (native)* recryst from cyclohexane, or *C_6H_6/MeOH/cyclohexane as red needles **m** 238-241° (dec), $[\alpha]_D^{20}$ -321° (c 0.2, MeOH), λ_{max} 443nm (ε 25,000) and 240nm (ε 33,300). [Brockman and Lackner, *Chem Ber* **101** 1312 *1968*.] It is *light sensitive*.

Actinomycin D (Dactinomycin) *[50-76-0]* **M 1255.5, m 241-243°(dec), $[\alpha]_D^{22}$ -296° (c 0.22, MeOH).** Crystallises as bright red rhombic crystals from absolute EtOH or from MeOH-EtOH (1:3). It will also crystallise from EtOAc-cyclohexane (**m** 246-247° dec), $CHCl_3$-pet ether (**m** 245-246° dec), and EtOAc-MeOH-*C_6H_6 (**m** 241-243° dec). Its solubility in MeCN is 1mg/mL. $[\alpha]_D^{20}$ varies from -296° to -327° (c 0.2, MeOH). λ_{max} (MeOH) 445, 240nm (log ε 4.43, 4.49), λ_{max} (MeOH, 10N HCl, 1:1) 477nm (log ε 4.21) and λ_{max} (MeOH, 0.1N NaOH) 458, 344, 285 (log ε 3.05, 4.28, 4.13). It is *HIGHLY* **TOXIC**, light sensitive and antineoplastic. [Bullock and Johnson, *J Chem Soc* 3280 *1957*.]

Acyl-coenzyme A Synthase **[from beef liver]** *[9013-18-7]* **M$_r$ 57,000, [EC 6.2.1.2].** Purified by extraction with sucrose-HCO_3 buffer, protamine sulfate pptn, $(NH_4)_2SO_4$ (66-65%) pptn at pH 4.35 and a second $(NH_4)_2SO_4$ (35-60%) pptn at pH 4.35. It has Km 0.15mM (V_{rel} 1.0) for octanoate; 0.41mM (V_{rel} 2.37) for heptanoate and 1.59mM (V_{rel} 0.63). Km for ATP is 0.5mM all at pH 9.0 in ethylene glycol buffer at 38°. [Jencks et al. *J Biol Chem* **204** 453 *1953; Methods Enzymol* **5** 467 *1962*.]

Acyl-coenzyme A Synthase (from yeast) *[9012-31-1]* **[EC 6.2.1.1].** This enzyme has been purified by extraction into phosphate buffer pH 6.8-7.0 containing 2-mercaptoethanol and EDTA, protamine sulfate pptn, polyethylene glycol fractionation, Alumina γ gel filtration, concentration by $(NH_4)_2SO_4$ pptn, Bio-Gel A-0.5m chromatography and DEAE-cellulose gradient chromatography. It has M_r ~151,000, Km (apparent) 0.24mM (for acetate) and 0.035mM (for CoA); 1.2 mM (for ATP) and Mg^{2+} 4.0mM. [Frenkel and Kitchens *Methods Enzymol* **71** 317 *1981*.]

Adenosine-5'-diphosphate **[adenosine-5'-pyrophosphate, ADP]** *[58-64-0]* **M 427.2, $[\alpha]_D^{25}$ -25.7° (c 2, H_2O), pK_1^{25} <2 (PO_4H), pK_2^{25} <2 (PO_4H), pK_3^{25} 3.95 (NH_2), pK_4^{25} 6.26 (PO_4H).** Characterised by conversion to the *acridine salt* by addition of alcoholic acridine (1.1g in 50mL), filtering off the yellow salt and recrystallising from H_2O. The salt has **m** 215°(dec), λ_{max} 259nm (ε 15,400) in H_2O. [Baddiley and Todd *J Chem Soc* 648 *1947*, 582 *1949*, *cf* LePage *Biochem Prep* **1** 1 *1949*; Martell and Schwarzenbach *Helv Chim Acta* **39** 653 *1956*].

Adenosine-3'-monophosphoric acid hydrate **[3'-adenylic acid, 3'-AMP]** *[84-21-9]* **M 347.3, m 197°(dec, as 2H_2O), 210°(dec), m 210°(dec), $[\alpha]_{546}$ -50° (c 0.5, 0.5M Na_2HPO_4), pK_1^{25} 3.65, pK_2^{25} 6.05.** It crystallises from large volumes of H_2O in needles as the monohydrate, but is not very soluble in boiling H_2O. Under acidic conditions it forms an equilibrium mixture of 2' and 3' adenylic acids *via* the 2',3'-cyclic phosphate. When heated with 20% HCl it gives a quantitative yield of furfural after 3hours, unlike 5'-adenylic acid which only gives traces of furfural. The yellow *monoacridine salt* has **m** 175°(dec) and the *diacridine salt* has **m** 177° (225°)(dec). [Brown and Todd *J Chem Soc* 44 *1952*; Takaku et al. *Chem Pharm Bull Jpn* **21** 1844 *1973*; NMR: Ts'O et al. *Biochemistry* **8** 997 *1969*.]

Adenosine-5'-monophosphoric acid monohydrate [5'-adenylic acid, 5'-AMP] *[18422-05-4]* **M 365.2, m 178°, 196-200°, 200° (sintering at 181°),** $[\alpha]_D^{20}$ **-47.5°,** $[\alpha]_{546}$ **-56° (c 2, in 2% NaOH), -26.0° (c 2, 10% HCl), -38° (c 1, 0.5M Na$_2$HPO$_4$), pK$_1^{25}$ 3.89, pK$_2^{25}$ 6.14, pK$_3^{25}$ 13.1.** It has been recrystd from H$_2$O (fine needles) and is freely soluble in boiling H$_2$O. Crysts also from H$_2$O by addition of acetone. Purified by chromatography on Dowex 1 (in formate form), eluting with 0.25M formic acid. It was then adsorbed onto charcoal (which had been boiled for 15min with M HCl, washed free of chloride and dried at 100°), and recovered by stirring three times with isoamyl alcohol/H$_2$O (1:9 v/v). The aqueous layer from the combined extracts was evaporated to dryness under reduced pressure, and the product was crystallised twice from hot H$_2$O. [Morrison and Doherty *Biochem J* **79** 433 *1961*]. It has λ_{max} 259nm (ϵ 15,400) in H$_2$O at pH 7.0. [Alberty et al. *J Biol Chem* **193** 425 *1951*; Martell and Schwarzenbach *Helv Chim Acta* **39** 653 *1956*]. The *acridinium salt* has **m** 208° [Baddiley and Todd *J Chem Soc* 648 *1947*; Pettit *Synthetic Nucleotides*, van Nostrand-Reinhold, NY, **Vol 1** 252 *1972*; NMR: Sarma et al. *J Am Chem Soc* **96** 7337 *1974*; Norton et al. *J Am Chem Soc* **98** 1007 *1976*; IR of *diNa salt*: Miles *Biochem Biophys Acta* **27** 324 *1958*].

Adenosine 5"-[β-thio]diphosphate tri-lithium salt *[73536-95-5]* **M 461.1.** Purified by ion-exchange chromatography on DEAE-Sephadex A-25 using gradient elution with 0.1-0.5M triethylammonium bicarbonate. [*Biochem Biophys Acta* **276** 155 *1972*.]

Adenosine 5"-[α-thio]monophosphate di-lithium salt *[19341-57-2]* **M 375.2.** Purified as for the diNa salt [Murray and Atkinson *Biochemistry* **7** 4023 *1968*]. Dissolve 0.3g in dry MeOH (7mL) and M LiI (6mL) in dry Me$_2$CO containing 1% of mercaptoethanol and the Li salt is ppted by adding Me$_2$CO (75mL). The residue is washed with Me$_2$CO (4 x 30mL) and dried at 55°/25mm. λ_{max} (HCl, pH 1.2) 257nm (ϵ 14,800); (0.015M NaOAc, pH 4.8) 259nm (ϵ 14,800); and (0.015M NH$_4$OH, pH 10.1) 259nm (ϵ 15,300).

Adenosine-5'-triphosphate (ATP) *[56-65-5]* **M 507.2,** $[\alpha]_{546}^{25}$ **-35.5 (c 1, 0.5 M Na$_2$HPO$_4$), pK$_1^{25}$ 4.00, pK$_2^{25}$ 6.48.** Ppted as its barium salt when excess barium acetate soln was added to a 5% soln of ATP in water. After filtering off, the ppte was washed with distd water, redissolved in 0.2M HNO$_3$, and again pptd with barium acetate. The ppte, after several washings with distd water, was dissolved in 0.2M HNO$_3$ and slightly more 0.2M H$_2$SO$_4$ than was needed to ppte all the barium as BaSO$_4$, was added. After filtering off the BaSO$_4$, the ATP was ppted by addition of a large excess of 95% ethanol, filtered off, washed several times with 100% EtOH and finally with dry diethyl ether. [Kashiwagi and Rabinovitch *J Phys Chem* **59** 498 *1955*.]

S-(5'-Adenosyl)-L-homosysteine *[979-92-0]* **M 384.4, m 202°(dec), 204°(dec), 205-207°(dec),** $[\alpha]_D^{25}$ **+93° (c 1, 0.2N HCl),** $[\alpha]_D^{23}$ **+44° (c 0.1, 0.05N HCl), (pK see SAM hydrochloride below).** It has been recrystd several times from aqueous EtOH or H$_2$O to give small prisms and has λ_{max} 260nm in H$_2$O. The *picrate* has **m** 170°(dec) from H$_2$O. [Baddiley and Jameison *J Chem Soc* 1085 *1955*; de la Haba and Cantoni *J Biol Chem* **234** 603 *1959*; Borchardt et al. *J Org Chem* **41** 565 *1976*; NMR: Follmann et al. *Eur J Biochem* **47** 187 *1974*.]

(-)-S-Adenosyl-L-methionine chloride (SAM hydrochloride) *[24346-00-7]* **M 439.9, pK$_{Est(1)}$~ 2.13, pK$_{Est(2)}$~ 4.12, pK$_{Est(3)}$~ 9.28.** Purified by ion exchange on Amberlite IRC-150, and eluting with 0.1-4M HCl. [Stolowitz and Minch *J Am Chem Soc* **103** 6015 *1981*.] It has been isolated as the tri-reineckate salt by adding 2 volumes of 1% solution of ammonium reineckate in 2% perchloric acid. The reineckate salt separates at once but is kept at 2° overnight. The salt is collected on a sintered glass funnel, washed with 0.5% of ammonium reineckate, dried (all operations at 2°) and stored at 2°. To obtain adenosylmethionine, the reineckate is dissolved in a small volume of methyl ethyl ketone and centrifuged at room temp to remove a small amount of solid. The clear dark red supernatant is extracted (in a separating funnel) with a slight excess of 0.1 N H$_2$SO$_4$. The aqueous phase is re-extracted with fresh methyl ethyl ketone until it is colourless. [Note that reineckates have UV absorption at 305nm (ϵ 15,000), and the optical density at 305nm is used to detect the presence of reineckate ions.] Methyl ethyl ketone is removed from the aqueous layer containing adenosylmethionine sulfate, the pH is adjusted to 5.6-6.0 and extracted with two volumes of Et$_2$O.

The *sulfate* is obtained by evaporating the aqueous layer in *vacuo*. The *hydrochloride* can be obtained in the same way but using HCl instead of H_2SO_4. SAM-HCl has a solubility of 10% in H_2O. The salts are stable in the cold at pH 4-6 but decompose in alkaline media. [Cantoni *Biochem Prep* **5** 58 *1957*.] The purity of SAM can be determined by paper chromatography [Cantoni *J Biol Chem* **204** 403 *1953*; *Methods Enzymol* **3** 601 *1957*], and electrophoretic methods or enzymic analysis [Cantoni and Vignos *J Biol Chem* **209** 647 *1954*].

L-Adrenaline   [L-epinephrine, L(-)-(3,4-dihydroxyphenyl)-2-methylaminoethanol] *[51-43-4]* **M 183.2, m 210°(dec), 211°(dec), 211-212°(dec), 215°(dec), [α]\s(20,D) -52° (c 2, 5% HCl), pK$_1^{25}$ 8.88, pK$_2^{25}$ 9.90, pK$_3^{25}$ 12.0.** It has been recrystd from EtOH + AcOH + NH$_3$ [Jensen *J Am Chem Soc* **57** 1765 *1935*]. It is sparingly soluble in H_2O, readily in acidic or basic solns but insoluble in aqueous NH$_3$, alkali carbonate solns, EtOH, CHCl$_3$, Et$_2$O or Me$_2$CO. It is readily oxidised in air and turns brown on exposure to light and air. Store in the dark under N$_2$. Its pKa values in H_2O are 8.88 and 9.90 [Lewis *Br J Pharmacol Chemother* **9** 488 *1954*]. The *hydrogen oxalate salt* has **m** 191-192°(dec, evac capillary) after recrystn from H_2O or EtOH [Pickholz *J Chem Soc* 928 *1945*].

Adrenolone hydrochloride [3',4'-dihydroxy-2-methylaminoacetophenone hydrochloride] *[62-13-5]* **M 217.7, m 244-249°(dec), 248°(dec), 256°(dec), pK 5.5.** It was purified by recrystn from EtOH or aqueous EtOH. [Gero *J Org Chem* **16** 1222 *1951*; Kindler and Peschke *Arch Pharm* **269** 581, 603 *1931*.]

ADP-Ribosyl transferase (from human placenta) *[9026-30-6]*. Purified by making an affinity absorbent for ADP-ribosyltransferase by coupling 3-aminobenzamide to Sepharose 4B. [Burtscher et al. *Anal Biochem* **152** 285 *1986*.]

Agglutinin (from peanuts) [*Arachis hypogaea*] *[1393-62-0]* **M$_r$ 134,900 (tetramer).** Purified by affinity chromatography on Sepharose-ζ-aminocaproyl-ß-D-galactopyranosylamine. [Lotan et al. *J Biol Chem* **250** 8518 *1974*.]

Alamethicin (from *Tricoderma viridae*). *[27061-78-5]* **M 1964.3, m 259-260°, 275-270°, [α]$_D^{22}$ -45° (c 1.2, EtOH), pK 6.04 (aq EtOH).** Recrystd from MeOH. [Panday et al. *J Am Chem Soc* **99** 8469 *1977*.] The *acetate [64918-47-4]* has **m** 195-180° from MeOH/Et$_2$O and the *acetate-methyl ester [64936-53-4]* has **m** 145-140° from aq MeOH.

Albumin (bovine and human serum) *[9048-46-8 (bovine)*; *70024-90-7 (human)]* **M$_r$ ~67,000 (bovine), 69 000 (human), UV: A$_{280nm}^{1\%}$ 6.6 (bovine) and 5.3 (human) in H_2O, [α]$_{546}^{25}$ -78.2° (H$_2$O).** Purified by soln in conductivity water and passage at 2-4° through two ion-exchange columns, each containing a 2:1 mixture of anionic and cationic resins (Amberlite IR-120, H-form; Amberlite IRA-400, OH-form). This treatment removed ions and lipid impurities. Care was taken to exclude CO_2, and the soln was stored at -15°. [Möller, van Os and Overbeek *Trans Faraday Soc* **57** 312 *1961*.] More complete lipid removal was achieved by lyophilising the de-ionised soln, covering the dried albumin (human serum) with a mixture of 5% glacial acetic acid (v/v) in iso-octane (previously dried with Na$_2$SO$_4$) and allowing to stand at 0° (without agitation) for upwards of 6h before decanting and discarding the extraction mixture, washing with iso-octane, re-extracting, and finally washing twice with iso-octane. The purified albumin was dried under vacuum for several hours, then dialyzed against water for 12-24h at room temperature, lyophilised, and stored at -10°C [Goodman *Science* **125** 1296 *1957*]. It has be recrystd in high (35%) and in low (22%) EtOH solutions from Cohn's Fraction V.

The **high EtOH recrystn** was as follows: To 1 Kg of Fraction V albumin paste at -5° was added 300mL of 0.4 M pH (pH 5.5) acetate buffer in 35% EtOH pre-cooled to -10° and 430 mL of 0.1 M NaOAc in 25% EtOH also at -10°. Best results were obtained by adding all of the buffer and about half of the NaOAc and stirring slowly for 1hour. The rest of the NaOAc was added when all the lumps had disintegrated. The mixture was set aside at -5° for several days to crystallise. 35% EtOH (1 L) was then added to dilute the crystalline suspension and lower the ionic strength prior to centrifugation at -5° (yield 80%). The crystals were further dissolved in 1.5 volumes of 15% EtOH-0.02M NaCl at -5° and clarified by filtration through washed, calcined diatomaceous earth. This soln may be recrystd by re-adjusting to the conditions in the first crystallisation, or it may be recrystd at 22% EtOH with the aid of a very small amount of decanol (enough to give a final concn of 0.02%).

Note that crystn from lower EtOH gave better purification (i.e. by removing globulins and carbohydrates) and producing a more stable product.

The **low EtOH recrystn** was as follows: To 1 Kg of Fraction V at -10° to -15° was added 500mL of 15% EtOH at -5°, stirred slowly until a uniform suspension was formed. 15% EtOH (500mL) and sufficient 0.2M NaHCO$_3$ soln at 0° to bring the pH (1:10 diln) to 5.3. This required 125- 150mL . Some temp rise occurs and care must be taken to keep the temp < -5°. If the albumin is incompletely dissolved a small amount of H$_2$O was added (100mL at a time at 0°, allowing 15min between additions). Undissolved albumin can be easily distinguished from small amounts of undissolved globulins, or as the last albumin dissolves, the appearance of the soln changes from milky white to hazy grey-green in colour. Keep the soln at -5° for 12h and filter by suspending in 15g of washed fine calcined diatomaceous earth, and thus filtering using a Büchner funnel precoated with coarser diatomaceous earth. The filtrate may require two or more similar filtrations to give a clear soln. To crystallise the filtrate add through a capillary pipette, and with careful stirring, 1/100volume of a soln containing10% decanol and 60% EtOH (at -10°), and seeded with the needle-type albumin crystals. After 2- 3 days crystn is complete. The crystals are centrifuged off. These are suspended with gentle mechanical stirring in one third their weight of 0.005 M NaCl pre-cooled to 0°. With careful stirring, H$_2$O (at 0°) is added slowly in an amount equal to 1.7 times the weight of the crystals. At this stage there is about 7% EtOH and the temp cannot be made lower than -2.5° to -1°. Clarify and collect as above. [Cohn et al. *J Am Chem Soc* **69** 1753 *1947*.]

Human serum albumin has been purified similarly with 25% EtOH and 0.2% decanol. The isoelectric points of bovine and human serum albumins are 5.1 and 4.9.

Amethopterin (**Methotrexate, 4-amino-4-deoxy-N^{10}-methylpteroyl-L-glutamic acid**) *[59-05-2]* M 454.4, m 185-204°(dec), $[\alpha]_D^{20}$ +19° (c 2, 0.1N aq NaOH), pK$_1$ <0.5 (pyrimidine^{2+}), p K$_2$ 2.5 (N5-Me$^+$), pK$_3$ 3.49 (α-CO$_2$H), pK$_4$ 4.99 (γ-CO$_2$H), pK$_5$ 5.50 (pyrimidine$^+$). Commonest impurities are 10-methyl pteroylglutamic acid, 4-amino-10-methylpteroylglutamic acid, aminopterin and pteroylglutamic acid. Purified by chromatography on Dowex-1 acetate, followed by filtration through a mixture of cellulose and charcoal. It has been recrystd from aqueous HCl or by dissolution in the minimum volume of N NaOH and acidified until pptn is complete, filter or collect by centrifugation, wash with H$_2$O (also by centrifugation) and dry at 100°/3mm. It has UV λ$_{max}$ at 244 and 307nm (ε 17300 and 19700) in H$_2$O at pH 1; 257, 302 and 370nm (ε 23000, 22000 and 7100) in H$_2$O at pH 13. [Momle *Biochem Prep* **8** 20 *1961*; Seeger et al. *J Am Chem Soc* **71** 1753 *1949*.] It is a potent inhibitor of dihydrofolate reductase and used in cancer chemotherapy. [Blakley *The Biochemistry of Folic Acid and Related Pteridines* (North-Holland Publ Co., Amsterdam, NY) pp157-163 *1969*.] It is **CARCINOGENIC, HANDLE WITH EXTREME CARE.**

α-Amino acids. All the α-amino acids 'natural' configuration [*S* (L), except for cysteine which is *R*(L)] at the α- carbon atom are available commercially in a very high state of purity. Many of the 'non-natural' α-amino acids with the [*R*(D)] configuration as well as racemic mixtures are also available and generally none require further purification before use unless they are of "Technical Grade'. The *R* or *S* enantiomers are optically active except for glycine which has two hydrogen atoms on the α- carbon atom, but these are *pro*-chiral and enzymes or proteins do distinguish between them, e.g. serine hydroxymethyltransferase successfully replaces the *pro*-α- hydrogen atom of glycine with CH$_2$OH (from formaldehyde) to make *S*-serine. The twenty common natural α-amino acids are: **amino acid**, three letter abbreviation, **one letter abbreviation**, pK (-COOH) and pK (-NH$_3^+$): **Alanine**, Ala, **A**, 2.34, 9.69; **Arginine**, Arg, **R**, 2.17, 9.04; **Asparagine**, Asn, **N**, 2.01, 8.80; **Aspartic acid**, Asp, **D**, 1.89, 9.60; **Cysteine**, Cys, **C**, 1.96, 8.18; **Glutamine**, Gln, **Q**, 2.17, 9.13; **Glutamic acid**, Glu, **E**, 2.19, 9.67; **Glycine**, Gly, **G**, 2.34, 9.60; **Histidine**, His, **H**, 1.8, 9.17; **Isoleucine**, Ile, **I**, 2.35, 9.68; **Leucine**, Leu, **L**, 2.36, 9.60; **Lysine**, Lys, **K**, 2.18, 8.95; **Methionine**, Met, **M**, 2.28, 9.20; **Phenylalanine**, Phe, **F**, 1.83, 9.12; **Proline**, Pro, **P**, 1.99, 10.96; **Serine**, Ser, **S**, 2.21, 9.15; **Threonine**, Thr, **T**, 2.11, 9.62; **Tryptophan**, Trp, **W**, 2.38, 9.39; **Tyrosine**, Tyr, **Y**, 2.2, 9.11, **Valine**, Val, **V**, 2.32, 9.61 repectively. Technical grade amino acids can be purified on ion exchange resins (e.g. Dowex 50W and eluting with a gradient of HCl or AcOH) and the purity is checked by TLC in two dimensions and stained with ninhydrin. (J.P.Greenstein and M.Winitz, *Chemistry of the Amino Acids* (3 Volumes), J.Wiley & Sons, NY, 1961; C.Cooper, N.Packer and K.Williams, *Amino Acid Analysis Protocols*, Humana Press, 2001, ISBN 0896036561). Recently codons for a further two amino acids have been discovered which are involved in ribosome-mediated protein synthesis giving proteins containing these amino acids. The amino acids are R(L)-selenocysteine [Stadtman *Ann Rev Biochem* **65** 83 *1996*] and pyrrolysine [(4R, 5R)-4-substituted (with Me, NH_2 or OH) pyrroline-5-carboxylic acid] [Krzychi and Chan et al. *Science* **296** 1459 and 1462 *2002*.] They are, however, rare at present and only found in a few microorganisms.

9-Aminoacridine hydrochloride monohydrate (Acramine yellow, Monacrin) *[52417-22-8]* **M 248.7, m >355°, pK_1^{20} 4.7, pK_2^{20} 9.99.** Recrystd from boiling H_2O (charcoal; 1g in 300 mL) to give pale yellow crystals with a neutral reaction. It is one of the most fluorescent substances known. At 1:1000 dilution in H_2O it is pale yellow with only a faint fluorescence but at 1:100,000 dilution it is colourless with an intense blue fluorescence. [Albert and Ritchie *Org Synth* Coll Vol III 53 *1955*; Falk and Thomas *Pharm J* **153** 158 *1944*.] See entry in Chapter 4 for the free base.

Aminopterin (4-amino-4-deoxypteroyl-L-glutamic acid) *[54-62-6]* **M 440.4, m 231-235°(dec), $[\alpha]_D^{20}$ +18° (c 2, 0.1N aq NaOH), pK_1 <0.5 (pyrimidine^{2+}), pK_2 2.5 (N5-Me$^+$), pK_3 3.49 (α-CO_2H), pK_4 4.65 (γ-CO_2H), pK_5 5.50 (pyrimidine$^+$).** Purified by recrystn from H_2O, and has properties similar to those of methotrexate. It has UV at λ_{max} 244, 290 and 355nm (ϵ 18600, 21300 and 12000) in H_2O at pH 1; 260, 284 and 370nm (ϵ 28500, 26400 and 8600) in H_2O at pH 13. [Seeger et al. *J Am Chem Soc* **71** 1753 *1949*; Angier and Curran *J Am Chem Soc* **81** 2814 *1959*; Blakley *The Biochemistry of Folic Acid and Related Pteridines* (North-Holland Publ Co., Amsterdam, NY) pp157-163 *1969*.] For small quantities chromatography on DEAE cellulose with a linear gradient of ammonium bicarbonate pH 8 and increasing the molarity from 0.1 to 0.4 and followed by UV is best. For larger quantities a near boiling solution of aminopterin (5g) in H_2O (400mL) was slowly treated with small portions of MgO powder (~0.7g) calcined magnesia) with vigorous stirring until a small amount of MgO remained undissolved and the pH rises from 3-4 to 7-8. Charcoal (1g) is added to the hot solution, filtered at once through a large sintered glass funnel of medium porosity and lined with a hot wet pad of Celite (~2-3 mm thick). The filtrate is cooled in ice and the crystals of the Mg salt are collected by filtration and recrystd form boiling H_2O (200mL) and the crystals washed with EtOH. The Mg salt is redissolved in boiling H_2O (200mL) and carefully acidified with vigorous agitation with AcOH (2mL). Pure aminopterin (3g) separates in fine yellow needles (dihydrate) which are easily filtd. The filtrate is washed with cold H_2O, then Me_2CO and dried in vac. If a trace of impurity is still present as shown by DEAE cellulose chromatography, then repetition of the process will remove it, see UV above. [Loo *J Med Chem* **8** 139 *1965*.] **CARCINOGENIC**

3-Aminopyridine adenine dinucleotide *[21106-96-7]* **M 635.4 (see NAD for pK)** Purified by ion exchange chromatography [Fisher et al. *J Biol Chem* **248** 4293 *1973*; Anderson and Fisher *Methods Enzymol* **66** 81 *1980*].

α–Amino-thiophene-2-acetic acid 2-(2-thienyl)glycine *[R(+)- 65058-23-3; S(-)- 4052-59-9; (-)- 43189-45-3; RS(±)- 21124-40-3]* **M 57.2, m 236-237° (R), 189-191°, 235-236° (S), 208-210°, 223-224° (dec)(RS), $[\alpha]_D^{20}$ (+) and (-) 84° (c 1, 1% aq HCl), $[\alpha]_D^{25}$ (+) and (-) 71° (c 1 H_2O), $pK_{Est(1)}$~ 1.5, $pK_{Est(2)}$~ 8.0.** Recrystd by dissolving in H_2O (1g in 3 mL), adjusting the pH to 5.5 with aq NH_3, diluting with MeOH (20 mL), stirring, adjusting the pH to 5.5 and cooling to 0°. Also recrystd from small vols of H_2O. [*R*-isomer: Nishimura et al. *Nippon Kagaku Zasshi* **82** 1688 *1961; S*-isomer: Johnson and Panetta *Chem Abstr* **63** 14869h *1965*; Johnson and Hardcastle *Chem Abstr* **66** 10930m *1967; RS*-isomer:LiBassi et al. *Gazz Chim Ital* **107** 253 *1977*.] The *(±) N-acetyl* derivative has **m** 191° (from H_2O) [Schouteenten et al. *Bull Soc Chim Fr* II-248, II-252 *1978*].

4(6)-Aminouracil (4-amino-2,6-dihydroxypyrimidine) *[873-83-6]* **M 127.1, m >350°, pK_1^{20} 0.00 (basic), pK_2^{20} 8.69 (acidic), pK_3^{20} 15.32 (acidic).** Purified by dissolving in 3M aq NH_3, filter hot, and add 3M formic acid until pptn is complete. Cool, filter off (or centrifuge), wash well with cold H_2O, then EtOH and dry in air. Dry further in a vac at ~80°. [Barlin and Pfeiderer *J Chem Soc (B)* 1424 *1971*.]

Amylose *[9005-82-7]* **$(C_6H_{10}O_5)_n$ *(for use in iodine complex formation)*.** Amylopectin was removed from impure amylose by dispersing in aqueous 15% pyridine at 80-90° (concn 0.6-0.7%) and leaving the soln stand at 44-45° for 7 days. The ppte was re-dispersed and recrystd during 5 days. After a further dispersion in 15% pyridine, it was cooled to 45°, allowed to stand at this temperature for 12hours, then cooled to 25° and left for a further 10 hours. The combined ppte was dispersed in warm water, ppted with EtOH, washed with absolute EtOH, and vacuum dried [Foster and Paschall *J Am Chem Soc* **75** 1181 *1953*].

Angiotensin (from rat brain) *[70937-97-2]* **M 1524.8.** Purified using extraction, affinity chromatography and HPLC [Hermann et al. *Anal Biochem* **159** 295 *1986*].

Angiotensinogen (from human blood serum) *[64315-16-8]*. Purified by chromatography on Blue Sepharose, Phenyl-Sepharose, hydroxylapatite and immobilised 5-hydroxytryptamine [Campbell et al. *Biochem J* **243** 121 *1987*].

Anion exchange resins. Should be conditioned before use by successive washing with water, EtOH and water, and taken through two OH⁻—H⁺—OH⁺ cycles by successive treatment with N NaOH, water, N HCl, water and N NaOH, then washed with water until neutral to give the OH⁻ form. (See commercial catalogues on ion exchange resins).

ß-Apo-4'-carotenal *[12676-20-9]* **M 414.7, m 139°,** $A_{1cm}^{1\%}$ **2640 at 461nm** Recrystd from CHCl₃/EtOH mixture or *n*-hexane. [Bobrowski and Das *J Org Chem* **91** 1210 *1987*.]

ß-Apo-8'-carotenal *[1107-26-2]* **M 414.7, m 136-139°.** Recrystd from CHCl₃/EtOH mixture or *n*-hexane. [Bobrowski and Das *J Org Chem* **91** 1210 *1987*.]

ß-Apo-8'-carotenoic acid ethyl ester *[1109-11-1]* **M 526.8, m 134-138°,** $A_{1cm}^{1\%}$ **2550 at 449nm.** Crystd from pet ether or pet ether/ethyl acetate. Stored in the dark in an inert atmosphere at -20°.

ß-Apo-8'-carotenoic acid methyl ester *[16266-99-2]* **M 512.7, m 136-137°,** $A_{1cm}^{1\%}$ **2575 at 446nm and 2160 at 471nm, in pet ether.** Crystd from pet ether or pet ether/ethyl acetate. Stored in the dark in an inert atmosphere at -20°.

Apocodeine *[641-36-1]* **M 281.3, m 124°, $pK_{Est(1)}$~ 7.0, $pK_{Est(2)}$~ 8.2.** Crystd from MeOH and dried at 80°/2mm.

Apomorphine *[58-00-4]* **M 267.3, m 195°(dec), pK_1^{15} 7.20 (NH₂), pK_2^{15} 8.91 (phenolic OH).** Crystd from CHCl₃ and pet ether, also from Et₂O with 1 mol of Et₂O which it loses at 100°. It is white but turns green in moist air or in alkaline soln. **NARCOTIC**

Apomorphine hydrochloride *[41372-20-7]* **M 312.8, m 285-287°(dec), $[\alpha]_D^{20}$ -48° (c 1 H₂O).** Cryst from H₂O and EtOH. Crystals turn green on exposure to light. **NARCOTIC**

Aureomycin (7-chlorotetracycline) *[57-62-5]* **M 478.5, m 172-174°(dec), $[\alpha]_D^{23}$ -275° (MeOH), pK₁ 3.3, pK₂ 7.44, pK₃ 9.27.** Dehydrated by azeotropic distn of its soln with toluene. On cooling anhydrous material crystallises out and is recrystd from *C₆H₆, then dried under vacuum at 100° over paraffin wax. (If it is crystd from MeOH, it contains MeOH which is not removed on drying.) [Stephens et al. *J Am Chem Soc* **76** 3568 *1954*; *Biochem Biophys Res Commun* **14** 137 *1964*].

Aureomycin hydrochloride (7-chlorotetracycline hydrochloride) *[64-72-2]* **M 514.0, m 234-236°(dec), $[\alpha]_D^{25}$ -23.5° (H₂O).** Purified by dissolving 1g rapidly in 20mL of hot water, cooling rapidly to 40°, treating with 0.1mL of 2M HCl, and chilling in an ice-bath. The process is repeated twice. Also recrystd from Me₂NCHO + Me₂CO. [Stephens et al. *J Am Chem Soc* **76** 3568 *1954* ; UV: McCormick et al. *J Am Chem Soc* **79** 2849 *1975*.]

Avidin (from egg white) *[1405-69-2]* **M_r ~70,000.** Purified by chromatography of an ammonium acetate soln on CM-cellulose [Green *Biochem J* **101** 774 *1966*]. Also purified by affinity chromatography on 2-iminobiotin-6-aminohexyl-Sepharose 4B [Orr *J Biol Chem* **256** 761 *1981*]. It is a biotin binding protein.

Azurin (from *Pseudomonas aeruginosa*) *[12284-43-4]* M_r **30,000.** Material with $A_{625/A_{280}} = 0.56$ was purified by gel chromatography on G-25 Sephadex with 5mM phosphate pH 7 buffer as eluent [Cho et al. *J Phys Chem* **91** 3690 *1987*]. It is a blue Cu protein used in biological electron transport and its reduced form is obtained by adding a slight excess of $Na_2S_2O_4$. [See *Structure and Bonding* Springer Verlag, Berlin **23** 1 *1975*.]

Bacitracin   (Altracin, Topitracin)   *[1405-87-4]* M **1422.7,** $[\alpha]_D^{23}$ **+5°**
(H_2O). It has been purified by carrier displacement using *n*-heptanol, *n*-octanol and *n*-nonanol as carriers and 50% EtOH in 0.1 N HCl. The pure material gives one spot with R_F ~0.5 on paper chromatography using AcOH:*n*-BuOH: H_2O (4:1:5). [Porath *Acta Chem Scand* **6** 1237 *1952*.] It has also been purified by ion-exchange chromatography. It is a white powder soluble in H_2O and EtOH but insoluble in Et_2O, $CHCl_3$ and Me_2CO. It is stable in acidic soln but unstable in base. (Abraham and Bewton *Biochem J* **47** 257 *1950*; Synthesis: Munekata et al. *Bull Chem Soc Jpn* **46** 3187, 3835 *1973*.]

N^6**-Benzyladenine** *[1214-39-7]* **M 225.3, m 231-232°, 232.5°(dec),** $pK_{Est(1)}$~ **4.2,** $pK_{Est(2)}$~ **10.1.** Purified by recrystn from aqueous EtOH. It has λ_{max} at 207 and 270nm (H_2O), 268 nm (pH 6), 274nm (0.1 N HCl) and 275nm (0.1 N NaOH). [Daly *J Org Chem* **21** 1553 *1956*; Bullock et al. *J Am Chem Soc* **78** 3693 *1956*.]

N^6**-Benzyladenosine** *[4294-16-0]* **M 357.4, m 177-179°, 185-187°,** $[\alpha]_D^{25}$**-68.6°** **(c 0.6, EtOH)(see above entry for pK).** Purified by recrystn from EtOH. It has λ_{max} 266nm (aq EtOH-HCl) and 269 nm (aqueous EtOH-NaOH). [Kissman and Weiss *J Org Chem* **21** 1053 *1956*.]

N-Benzylcinchoninium chloride   (9S-benzyl-9-hydroxycinchoninium chloride) *[69221-14-3]* **M 421.0,** $[\alpha]_D^{20}$ **+169° (c 0.4, H_2O), pK_{Est} ~ 5.** Recrystd from isoPrOH, toluene or small volumes of H_2O. Good chiral phase transfer catalyst [Julia et al. *J Chem Soc Perkin Trans 1* 574 *1981*; Hughes et al. *J Am Chem Soc* **106** 446 *1984*; Hughes et al. *J Org Chem* **52** 4745 *1987*]. See cinchonine below.

R-(-)-N-Benzylcinchonidinium chloride *[69257-04-1]* **M 421.0, m 212-213° (dec),** $[\alpha]_D^{20}$ **-175.4°, -183° (c 5, 0.4, H_2O), pK_{Est} ~5.** Dissolve in minimum volume of H_2O and add absolute Me_2CO. Filter off and dry in a vacuum. Also recrystd from hot EtOH or EtOH-Et_2O. (A good chiral phase transfer catalyst - see above) [Colonna et al. *J Chem Soc Perkin Trans 1* 547 *1981*, Imperali and Fisher *J Org Chem* **57** 757 *1992*]. See cinchonidine below.

N-Benzylpenicillin sodium salt *[69-57-8]* **M 356.37, m 215° (charring and dec), 225° (dec),** $[\alpha]_D^{20}$ **+269° (c 0.7, MeOH),** $[\alpha]_D^{25}$ **+305° (c 1, H_2O), pK^{25} 2.76 (4.84 in 80% aq EtOH)(for free acid).** Purified by dissolving in a small volume of MeOH (in which it is more soluble than EtOH) and treating gradually with ~5 volumes of EtOAc. This gives an almost colourless crystalline solid (rosettes of clear-cut needles) and recrystallising twice more if slightly yellow in colour. The salt has also been conveniently recrystd from the minimum amount of 90% Me_2CO and adding an excess of absolute Me_2CO. A similar procedure can be used with wet *n*-BuOH. If yellow in colour then dissolve (~3.8g) in the minimum volume of H_2O (3mL), add *n*-BuOH and filter through a bed of charcoal. The salt forms long white needles on standing in a refrigerator overnight. More crystals can be obtained on concentrating the mother liquors *in vacuo* at 40°. A further recrystn (without charcoal) yields practically pure salt. A good preparation has ~600 Units/mg. The presence of H_2O in the solvents increases the solubility considerably. The solubility in mg/100mL at 0° is 6.0 (Me_2CO), 15.0 (Me_2CO + 0.5% H_2O), 31.0 (Me_2CO + 1.0% H_2O), 2.4 (methyl ethyl ketone), 81.0 (*n*-butanol) and 15.0 (dioxane at 14°). Alternatively it is dissolved in H_2O (solubility is 10%), filtered if necessary and ppted by addition of EtOH and dried in a vacuum over P_2O_5. A sample can be kept for 24h at 100° without loss of physiological activity. [IR: *Anal Chem* **19** 620 *1947*; *The Chemistry of Penicillin* [Clarke, Johnson and Robinson eds.] Princeton University Press, Princeton NJ, Chapter V 85 *1949*.]
Other salts, e.g. the **potassium salt** can be prepared from the Na salt by dissolving it (147mg) ice-cold in H_2O acidified to pH 2, extracting with Et_2O (~50mL), wash once with H_2O, and extract with 2mL portions of 0.3% $KHCO_3$ until the pH of the extract rose to ~6.5 (~7 extractns). The combined aqueous extracts are

lyophilised and the white residue is dissolved in n-BuOH (1mL, absolute) with the addition of enough H_2O to effect soln. Remove insoluble material by centrifugation and add absolute n-BuOH to the supernatant. Crystals should separate on scratching, and after 2.5h in a refrigerator they are collected, washed with absolute n-BuOH and EtOAc and dried (yield 51.4mg). The *potassium salt* has **m** 214-217° (dec) (block preincubated at 200°; heating rate of 3°/min) and $[\alpha]_D^{22}$ +285° (c 0.748, H_2O). The *free acid* has **m** 186-187° (MeOH-Me_2CO), 190-191° (H_2O) $[\alpha]_D^{25}$ +522°.

(+)-Bicuculine [R-6(5,6,7,8-tetrahydro-6-methyl-1,3-dioxolo[4,5-g]isoquinolon-5-yl)-furo-[3,4-c]-1,3-benzodioxolo-8(6H)-one] *[485-49-4]* **M 367.4, m 177°, 193-195°, 193-197°, 215°, $[\alpha]_D^{20}$ +126° (c 1, $CHCl_3$), $[\alpha]_{546}^{20}$ +159° (c 1, $CHCl_3$), pK 4.84.** Recrystallises from $CHCl_3$-MeOH as plates. Crystals melt at 177° then solidify and re-melt at 193-195° [Manske *Canad J Research* **21B** 13 *1943*]. It is soluble in $CHCl_3$, *C_6H_6, EtOAc but sparingly soluble in EtOH, MeOH and Et_2O. [Stereochem: Blaha et al. *Collect Czech Chem Commun* **29** 2328 *1964*; Snatzke et al. *Tetrahedron* **25** 5059 *1969*; Pharmcol: Curtis et al. *Nature* **266** 1222 *1970*].

L-*erythro*-Biopterin (2-amino-4-hydroxy-6-[{1R,2S}-1,2-dihydroxypropyl]pteridine) *[22150-76-1]* **M 237.2, m >300°(dec), $[\alpha]_{546}^{20}$ -65° (c 2.0, M HCl), pK_1^{25} 2.23(2.45), pK_2^{25} 7.89(8.05).** Purified by chromatography on Florisil washed thoroughly with 2M HCl, and eluted with 2M HCl. The fractions with the UV-fluorescent band are evapd *in vacuo* and the residue recrystd. Biopterin is best recrystd (90% recovery) by dissolving in 1% aq NH_3 (ca 100 parts), and adding this soln dropwise to an equal vol of M aq formic acid at 100° and allowing to cool at 4° overnight. It is dried at 20° to 50°/01mm in the presence of P_2O_5. [Schircks, Bieri and Viscontini *Helv Chim Acta* **60** 211 *1977*; Armarego, Waring and Paal *Aust J Chem* **35** 785 *1982*.] Also crystd from *ca* 50 parts of water or 100 parts of hot 3M aq HCl by adding hot 3M aq NH_3 and cooling. It has UV: λ_{max} at 212, 248 and 321nm (log ε 4.21, 4.09 and 3.94) in H_2O at pH 0.0; 223infl, 235.5, 274.5 and 345nm (log ε 4.07inflexion, 4.10, 4.18 and 3.82) in H_2O at pH 5.0; 221.5, 254.5 and 364nm (log ε 3.92, 4.38 and 3.84) in H_2O at pH 10.0 [Sugimoto and Matsuura *Bull Chem Soc Jpn* **48** 3767 *1875*].

D-(+)-Biotin (vitamin H, hexahydro-2-oxo-1H-thieno[3,4-d]imidazole-4-pentanoic acid) *[58-85-5]* **M 244.3, m 229-231°, 230.2°(dec), 230-231°, 232-234°(dec), $[\alpha]_{546}^{20}$ +108°, $[\alpha]_D^{20}$ +91.3° (c 1, 0.1N NaOH), pK_{Est} ~ 4.8.** Crystd from hot water in fine long needles with a solubility of 22 mg/100mL at 25°. Its solubility in 95% EtOH is 80 mg/100 mL at 25°. Its isoelectric point is at pH 3.5. Store solid and solutions under sterile conditions because it is susceptible to mould growth. [Confalone *J Am Chem Soc* **97** 5936 *1975*; Wolf et al. *J Am Chem Soc* **67** 2100 *1945*; Synthesis: Ohuri and Emoto *Tetrahedron Lett* 2765 *1975*; Harris et al. *J Am Chem Soc* **66** 1756 *1944*.] The *(+)-methyl ester* has **m** 166-167° (from MeOH-Et_2O), $[\alpha]_D^{22}$ +57° (c 1, $CHCl_3$) [du Vigneaud et al. *J Biol Chem* **140** 643, 763 *1941*]; the *(+)-S-oxide* has **m** 200-203°, $[\alpha]_D^{20}$ +130° (c 1.2, 0.1N NaOH) [Melville *J Biol Chem* **208** 495 *1954*]; the *SS-dioxide* has **m** 274-275°(dec, 268-270°) and the *SS-dioxide methyl ester* has **m** 239-241° (from MeOH-Et_2O) [Hofmann et al. *J Biol Chem* **141** 207, 213 *1941*.]

D-(+)-Biotin hydrazide *[66640-86-6]* **M 258.4, m 238-240°, 245-247°, $[\alpha]_D^{20}$ +66° (c 1, Me_2NCHO).** Wash the material with H_2O, dry, wash with MeOH then Et_2O, dry, and recrystallise from hot H_2O (clusters of prisms) [Hofmann et al. *J Biol Chem* **144** 513 *1942*].

D-(+)-Biotin N-hydroxysuccinimide ester (+-biotin N-succinimidyl ester) *[35013-72-0]* **M 342.4, m 210°, 212-214°, $[\alpha]_D^{20}$ +53° (c 1, Me_2NCHO).** Recrystd from refluxing isoPrOH and dried in a vacuum over P_2O_5 + KOH. [Jasiewicz et al. *Exp Cell Biol* **100** 213 *1976*.]

D-(+)-Biotin 4-nitrophenyl ester *[33755-53-2]* **M 365.4, m 160-163°, 163-165°, $[\alpha]_D^{25}$ +47° (c 2, Me_2NCHO containing 1% AcOH).** It has been recrystd by dissolving 2g in 95% EtOH (30mL), heated to dissolve, then cooled in an ice-water bath. The crystals are collected, washed with ice-cold 95% EtOH (5mL) and dried over P_2O_5. The R_F on silica plates ($CHCl_3$:MeOH-19:1) is 0.19 [Bodanszky and Fagan *J Am Chem Soc* **99** 235 *1977*].

N-(+)-Biotinyl-4-aminobenzoic acid   *[6929-40-4]* M 363.4, m 295-297º, 295-300º, [α]$_D^{23}$ +56.55º (c 0.5, 0.1N NaOH), pK$_{Est}$ ~4.0. Dissolve in NaHCO$_3$ soln, cool and ppte by adding N HCl. Collect the solid, dry at 100º and recrystallise from MeOH. Note that it is hydrolysed by aq 3M, 1M and 0.2M HCl at 120º, but can be stored in 5% aq NaHCO$_3$ at -20º without appreciable hydrolysis [Knappe et al. *Biochem Zeitschrift* **338** 599 *1963*; *J Am Chem Soc* **73** 4142 *1951*; Bayer and Wilchek *Methods Enzymol* **26** 1 *1980*]

N-Biotinyl-6-aminocaproic *N*-succinimidyl ester   *[72040-63-2]* M 454.5, m 149-152º. Dissolve ~400mg in dry propan-2-ol (~25mL) with gentle heating. Reduce the volume to ~10mL by gentle boiling and allow the soln to cool. Decant the supernatant carefully from the white crystals, dry the crystals in a vacuum over P$_2$O$_5$ at 60º overnight. Material gives one spot on TLC. [Costello et al. *Clin Chem* **25** 1572 *1979*; Kincaid et al. *Methods Enzymol* **159** 619 *1988*.]

N-(+)-Biotinyl-6-aminocaproyl hydrazide (biotin-6-aminohexanoic hydrazide)   *[109276-34-8]* M 371.5, m 189-191º, 210º, [α]$_D^{20}$ +23º (c 1, Me$_2$NCHO). Suspend in ice-water (100mg/mL), allow to stand overnight at 4º, filter and dry the solid in a vacuum. Recrystd from isoPrOH. R$_F$ 0.26 on SiO$_2$ plate using CHCl$_3$-MeOH (7:3) as eluent. [O'Shannessy et al. *Anal Biochem* **163** 204 *1987*.]

N-(+)-Biotinyl-L-lysine (Biocytin)   *[576-19-2]* M 372.5, m 228.5º, 228-230º (dec), 241-243º, 245-252º (dec, sintering at 227º), [α]$_D^{25}$ +53º (c 1.05, 0.1 N NaOH). Recrystd rapidly from dilute MeOH or Me$_2$CO. Also recrystd from H$_2$O by slow evaporation or by dissolving in the minimum volume of H$_2$O and adding Me$_2$CO until solid separates. It is freely soluble in H$_2$O and AcOH but insoluble in Me$_2$CO. [Wolf et al. *J Am Chem Soc* **76** 2002 *1952*, **72** 1048 *1050*.] It has been purified by chromatography on superfiltrol-Celite, Al$_2$O$_3$ and by countercurrent distribution and then recrystd [IR: Peck et al. *J Am Chem Soc* **74** 1991 *1952*]. The *hydrochloride* can be recrystd from aqueous Me$_2$CO + HCl and has m 227º (dec).

2-(4-Biphenylyl)-5-phenyl-1,3,4-oxadiazole   *[852-38-0]* M 298.4, m 166-167º, 167-170º. Recrystd from toluene. It is a good scintillation material [Brown et al. *Discussion Faraday Soc* **27** 43 *1959*].

2,5-Bis(4-biphenylyl)-1,3,4-oxadiazole (BBOD)   *[2043-06-3]* M 374.5, m 229-230º, 235-238º. Recrystd from heptane or toluene. It is a good scintillant. [Hayes et al. *J Am Chem Soc* **77** 1850 *1955*.]

4,4-Bis(4-hydroxyphenyl)valeric acid [diphenolic acid]   *[126-00-1]* M 286.3, m 168-171º, 171-172º, pK$_{Est(1)}$~ 4.8 (CO$_2$H), pK$_{Est(2)}$~ 7.55 (OH), pK$_{Est(3)}$~9.0 (OH). When recrystd from *C$_6$H$_6$ the crystals have 0.5 mol of *C$_6$H$_6$ (m 120-122º) and when recrystd from toluene the crystals have 0.5 mol of toluene. Purified by recrystn from hot H$_2$O. It is sol in Me$_2$CO, AcOH, EtOH, propan-2-ol, methyl ethyl ketone. It is also recrystallised from AcOH, heptane-Et$_2$O or Me$_2$CO + *C$_6$H$_6$. It has λ$_{max}$ 225 and 279nm in EtOH. The *methyl ester* has m 87-89º (aqueous MeOH to give the trihydrate). [Bader and Kantowicz *J Am Chem Soc* **76** 4465 *1954*.]

Bis(2-mercaptoethyl)sulfone (BMS)   *[145626-87-5]* M 186.3, m 57-58º, pK$_1^{25}$ 7.9, pK$_2^{25}$ 9.0. Recrystd from hexane as white fluffy crystals. Large amounts are best recrystd from de-oxygenated H$_2$O (charcoal). It is a good alternative reducing agent to dithiothreitol. Its IR (film) has ν 2995, 2657, 1306, 1248, 1124 and 729 cm^{-1}. The synthetic intermediate *thioacetate* has m 82-83º (white crystals from CCl$_4$). The *disulfide* was purified by flash chromatography on SiO$_2$ and elution with 50% EtOAc-hexane and recrystd from hexane, m 137-139º. [Lamoureux and Whitesides *J Org Chem* **58** 633 *1993*.]

Bombesin (2-L-glutamin-3-6-L-asparaginealytesin)   *[31362-50-2]* M 619.9. Purified by gel filtration on a small column of Sephadex G-10 and eluted with 0.01 M AcOH. This procedure removes lower molecular weight contaminants which are retarded on the column. The procedure should be repeated twice and the material should now be homogeneous on electrophoresis, and on chromatography gives a single active spot which is negative to ninhydrin but positive to Cl$_2$ and iodoplatinate reagents. R$_F$ on paper chromatography (*n*-BuOH-pyridine-AcOH-H$_2$O (37.5: 25:7.5: 30) is 0.55 for Bombesin and 0.65 for Alytin. [Bernardi *Experientia* **B 27** 872 *1971*; **A 27** 166 *1971*.] The *hydrochloride* has m 185º(dec) (from EtOH) [α]$_D^{24}$ -20.6º [c 0.65, Me$_2$NCHO-(Me$_2$N)$_3$PO (8:2)].

Bradykinin [ArgProProGlyPheSerProPheArg] *[5979-11-3]* **M$_r$ 1,240.4.** Purified by ion-exchange chromatography on CMC (O-carboxymethyl cellulose) and partition chromatography on Sephadex G-25. Purity was checked by paper chromatography using BuOH:AcOH:H$_2$O (4:1:5) as eluent. [Park et al. *Can J Biochem* **56** 92 *1978*; ORD and CD: Bodanszky et al. *Experientia* **26** 948 *1970*; activity: Regoli and Barabé *Pharmacol Rev* **32** 1 *1980*.]

Brefeldin A [1-*R*-2*c*,15*c*-dihydroxy-7*t*-methyl-(1*r*,13*t*)-6-oxa-bicyclo[11.3.0]hexadeca-3*t*,11*t*-dien-5-one, Decumbin] *[20350-15-6]* **M 280.4, m 200-202°, 204°, 204-205°, [α]$_D^{22}$ +95° (c 0.81, MeOH).** Isolated from *Penicillium brefeldianum* and recrystd from aqueous MeOH-EtOAc or MeOH. Solubility in H$_2$O is 0.6mg/mL, 10mg/mL in MeOH and 24.9mg/mL in EtOH. The *O-acetate* recrystallises from Et$_2$O-pentane and has **m** 130-131°, [α]$_D^{22}$ +17° (c 0.95, MeOH). [Sigg *Helv Chim Acta* **47** 1401 *1964*; UV and IR: Härri et al. *Helv Chim Acta* **46** 1235 *1963*; total synthesis: Kitahara et al. *Tetrahedron* 3021 *1979*; X-ray : Weber et al. *Helv Chim Acta* **54** 2763 *1971*.]

Bromelain (anti-inflammatory Ananase from pineapple) *[37189-34-7]* **M$_r$ ~33 000, [EC 3.4.33.4].** This protease has been purified *via* the acetone powder, G-75 Sephadex gel filtration and Bio-Rex 70 ion-exchange chromatography and has A$_{1cm}^{1\%}$ 20.1 at 280nm. The protease from pineapple hydrolyses benzoyl glycine ethyl ester with a Km (app) of 210mM and k$_{cat}$ of 0.36 sec^{-1}. [Murachi *Methods Enzymol* **19** 273 *1970*; Balls et al. *Ind Eng Chem* **33** 950 *1941*.]

5-Bromo-2'-deoxyuridine *[59-14-3]* **M 307.1, m 193-197°(dec), 217-218°, [α]$_D^{25}$ -41° (c 0.1, H$_2$O), pK^{25} ~ 8.1.** Recrystd from EtOH or 96% EtOH. It has λ$_{max}$ 279 nm at pH 7.0, and 279 nm (log ε 3.95) at pH 1.9. Its R$_F$ values are 0.49, 0.46 and 0.53 in *n*-BuOH-AcOH-H$_2$O (4:1:1), *n*-BuOH-EtOH-H$_2$O (40:11:19) and *i*-PrOH-25% aq NH$_3$-H$_2$O (7:1:1) respectively. [*Nature* **209** 230 *1966*; *Collect Czech Chem Comm* **29** 2956 *1964*.]

6-Bromo-2-naphthyl-α-D-galactopyranoside *[25997-59-5]* **M 385.2, m 178-180°, 224-226°, 225°, [α]$_D^{28}$ +60° (c 1.2, pyridine).** It was prepared from penta-*O*-acetyl-D-galactoside and 6-bromo-2-naphthol and ZnCl$_2$. The resulting tetra-acetate (2g) was hydrolysed by dissolving in 0.3N KOH (100mL) and heated until the soln was clear, filtered and cooled to give colourless crystals of the α-isomer which are collected and recrystd twice from hot MeOH. The high specific rotation is characteristic of the α-isomer. The *tetra-acetate* has **m** 155-156° [α]$_D^{20}$ +60° (c 1, CHCl$_3$) [Dey and Pridham *Biochem J* **115** 47 *1969*] [reported **m** 75-85°, [α]$_D^{24}$ +94° (c 1.3, dioxane), Monis et al. *J Histochem Cytochem* **11** 653 *1963*].

5-Bromouridine *[957-75-5]* **M 323.1, m 215-217°, 217-218°, [α]$_D^{25}$ -4.1° (c 0.1, H$_2$O), pK^{25} 8.1.** Recrystd from 96% EtOH. UV λ$_{max}$ 279nm (log ε 3.95) in H$_2$O pH 1.9. R$_F$ in *n*-BuOH:AcOH:H$_2$O (4:4:1) is 0.49; in *n*-BuOH:EtOH:H$_2$O (40:11:9) is 0.46 and in isoPrOH:25%NH$_3$:H$_2$O (7:1:2) is 0.53 using Whatman No 1 paper. [Prystas and Sorm *Collect Czech Chem Commmun* **29** 2956 *1964*.]

Brucine *[357-57-3 (anhydr), 5892-11-5 (4H$_2$O)]* **M 430.5, m 178-179°, [α]$_{546}^{20}$ -149.9° (anhydrous; c 1, in CHCl$_3$), pK$_1^{15}$ 2.50, pK$_2^{15}$ 8.16 (pK$_2^{25}$ 8.28).** Crystd once from water or aq Me$_2$CO, as tetrahydrate, then suspended in CHCl$_3$ and shaken with anhydrous Na$_2$SO$_4$ (to dehydrate the brucine, which then dissolves). Ppted by pouring the soln into a large bulk of dry pet ether (b 40-60°), filtered and heated to 120° in a high vacuum [Turner *J Chem Soc* 842 *1951*]. **VERY POISONOUS**

α-Brucine sulfate (hydrate) *[4845-99-2]* **M 887.0, m 180°(dec).** Crystd from water.

Butyryl choline iodide [(2-butyryloxyethyl)trimethyl ammonium iodide] *[2494-56-6]* **M 301.7, m 85-89°, 87°, 93-94°.** Recrystd from isoPrOH or Et$_2$O. [Tammelin *Acta Chem Scand* **10** 145 *1956*.] The *perchlorate* has **m** 72° (from isoPrOH). [Aldridge *Biochem J* **53** 62 *1953*.]

S-Butyryl thiocholine iodide [(2-butyrylmercaptoethyl)trimethyl ammonium iodide] *[1866-16-6]* **M 317.2, m 173°, 173-176°.** Recrystd from propan-1-ol and dried *in vacuo*; store in the dark under N$_2$. The *bromide* has **m** 150° (from Me$_2$CO) or **m** 140-143° (from butan-1-ol). [Gillis *Chem and Ind (London)* 111 *1957*; Hansen *Acta Chem Scand* **11** 537 *1957*.]

L-Canavanine sulfate (from jackbean, *O*-guanidino-L-homoserine) *[2219-31-0]*
M 274.3, m 160-165°(dec), 172°(dec), $[\alpha]_D^{18.5}$ +19.8° (c 7, H_2O), pK_1^{25} 7.40 (CO_2H), pK_2^{25} 9.25 (α-NH_2), pK_3^{25} 11.5 (guanidinoxy).   Recrystd by dissolving (~1g) in H_2O (10mL), and adding with stirring 0.5 to 1.0 vols of 95% EtOH whereby crystals separate. These are collected, washed with Me_2CO-EtOH (1:1) and dried over P_2O_5 in a vacuum. [Hunt and Thompson *Biochem Prep* **13** 416 *1971*; Feacon and Bell *Biochem J* **59** 221 *1955*.]

Carbonic anhydrase (carbinate hydrolase) *[9001-03-0]* M_r 31,000 [EC 4.2.1.1]. Purified by hydroxylapatite and DEAE-cellulose chromatography [Tiselius et al. *Arch Biochem Biophys* **65** 132 *1956*, *Biochim Biophys Acta* **39** 218 *1960*], and is then dialysed for crystn. A 0.5 to 1% soln of the enzyme in 0.05 M Tris-HCl pH 8.5 was dialysed against 1.75M soln of $(NH_4)_2SO_4$ in the same buffer, and this salt soln was slowly increased in salt concn by periodic removal of small amounts of dialysate and replacement with an equal volume of 3.5M $(NH_4)_2SO_4$. The final salt concn in which the DEAE-cellulose fractions which gave beautiful birefringent suspensions of crystals ranged from 2.4 to 2.7M, and appeared first as fine crystals then underwent transition to thin fragile plates.   Carbonic anhydrase is a Zn enzyme which exists as several isoenzymes of varying degrees of activity [*J Biol Chem* **243** 6474 *1968*; crystal structure: *Nature, New Biology* **235** 131 *1972*; see also P.D. Boyer Ed. *The Enzymes* Academic Press NY, pp 587-665 *1971*].

Carboxypeptidase A   (from bovine pancreas, peptidyl-L-aminoacid lyase) *[11075-17-5]* M_r **34,600** [EC 3.4.17.1]. Purified by DEAE-cellulose chromatography, activation with trypsin and dialysed against 0.1M NaCl, yielding crystals. It is recrystd by dissolving in 20 mL of M NaCl and dialysed for 24hours each against the following salts present in 500mL of 0.02M sodium veronal pH 8.0: ,0.5M NaCl, 0.2M NaCl and 0.15M NaCl. The last dialysate usually induces crystn. If it does not crystallise then dialyse the last soln against 0.02M sodium veronal containing 0.10M NaCl. Only 2 or 3 recrystns are required to attain maximum activity. [Cox et al. *Biochemistry* **3** 44 *1964*.] Enzyme activity is measured by hydrolysing hippuryl-L-phenylalanine (or phenylacetic acid) and observing the rate of change of optical density at 254nm (reaction extinction coefficient is ~0.592 cm^2/μmole at pH 7.5 [Bergmyer *Methods in Enzymatic Analysis* (Academic Press) **1** 436 *1974*].

Carminic acid   (7-α-D-glucopyranosyl-9,10-dihydro-3,5,6,8-tetrahydroxy-1-methyl-9,10-dioxo-2-anthracene carboxylic acid, Neutral Red 4: CI 75470) *[1260-17-9]* M 492.4, m 120°(dec), $[\alpha]_{654}^{15}$ +51.6° (H_2O), (several phenolic pKs). Forms red prisms from EtOH.   It gives a red colour in Ac_2O and yellow to violet in acidic solution. UV: λ_{max} (H_2O) 500nm (ε 6,800); (0.02N HCl) 490-500nm (ε 5,800) and (0.0001N NaOH) 540nm (ε 3,450). IR: ν_{max} (Nujol) 1708s, 1693s, 1677m,1648m, 1632m, 1606s, 1566s, 1509 cm^{-1}. Periodate oxidation is complete after 4h at 0° with the consumption of 6.2 mols. The *tetra-O-methyl carminate* has **m** 186-188° (yellow needles from $^*C_6H_6$ + pet ether). [IR: Ali and Haynes *J Chem Soc* 1033 *1959*; Bhatia and Venkataraman *Indian J Chem* **3** (2) 92 *1965*; Synthesis: Davis and Smith *Biochem Prep* **4** 38 *1955*.]

Carnitine (α-hydroxy-β-N,N,N-trimethylaminopropionic acid) *[R(+)- 541-14-0 ; S(L-) 541-15-1; RS 461-06-3]* M 161.2, m *R* or *S* isomer 197-198°(dec), 210-212°(dec), *RS isomer* 195-197°, $[\alpha]_{546}^{20}$ (+) and (-) 36° (c 10, H_2O), pK^{25} 3.6. The *S*(L) isomer is **levocarnitine, Vitamin** B_7. The *R* or *S* isomers crystallise from EtOH + Me_2CO (hygroscopic). The *R* or *S hydrochlorides* crystallise from hot EtOH and have **m** 142°(dec). The *RS* isomer crystallises from hot EtOH (hygroscopic). The *RS hydrochloride* crystallises in needles from hot EtOH and has **m** 196°(dec).

L-Carnosine (β-alanyl-L-histidine) *[305-84-0]* M 226.2, m 258-260°(dec), 260°(capillary tube), 262°(dec), $[\alpha]_D^{25}$ +20.5° (c 1.5, H_2O), pK_1^{25} 2.64, pK_2^{25} 6.83, pK_3^{25} 9.51.   Likely impurities: histidine, β-alanine. Crystd from water by adding EtOH in excess. Recrystd from aqueous EtOH by slow addition of EtOH to a strong aqueous soln of the dipeptide. Its solubility in H_2O is 33.3% at 25°. [Vinick and Jung *J Org Chem* **48** 392 *1983*; Turner *J Am Chem Soc* **75** 2388 *1953*; Sifford and du Vigneaud *J Biol Chem* **108** 753 *1935*.]

α-**Carotene** *[7488-99-5]* **M 536.9, m 184-188°, [α]$_{643}^{20}$ +385° (c 0.08, $^*C_6H_6$) λ$_{max}$ 422, 446, 474 nm, in hexane, $A_{1cm}^{1\%}$ 2725 (at 446nm), 2490 (at 474nm).** Purified by chromatography on columns of calcium hydroxide, alumina or magnesia. Crystd from CS_2/MeOH, toluene/MeOH, diethyl ether/pet ether, or acetone/pet ether. Stored in the dark, under inert atmosphere at -20°.

all-trans-β-**Carotene** *[7235-40-7]* **M 536.9, m 178-179°, 179-180°, 180°, 181°, 183° (evac capillary), ε$_{1cm}^{1\%}$ 2590 (450nm), 2280 (478nm), in hexane.** It forms purple prisms when crystd from $^*C_6H_6$-MeOH and red rhombs from pet ether. Its solubility in hexane is 0.1% at 0°. It is **oxygen sensitive** and should be stored under N_2 at -20° in the dark. It gives a deep blue colour with $SbCl_3$ in $CHCl_3$. UV: ($^*C_6H_6$) 429infl, λ$_{max}$ 454 and 484nm. The principal peak at 454nm has $A_{1cm}^{1\%}$ 2000. [Synthesis: Surmatis and Ofner *J Org Chem* **26** 1171 *1961*; Milas et al. *J Am Chem Soc* **72** 4844 *1950*.] β-Carotene was also purified by chromatography (Al_2O_3 activity I-II) - it was dissolved in pet ether-$^*C_6H_6$ (10:1), applied to the column and eluted with pet ether-EtOH, the desired fraction was evaporated and the residue recrystd from $^*C_6H_6$-MeOH as violet-red plates. [UV: Inhoffen et al. *Justus Liebigs Ann Chem* **570** 54,68 *1950*; Review: Fleming *Selected Organic Synthesis* (J Wiley, Lond) pp. 70-74 *1973*.] Alternatively it can be purified by chromatography on a magnesia column, thin layer of Kieselguhr or magnesia. Crystd from CS_2/MeOH, Et_2O/pet ether, acetone/pet ether or toluene/MeOH. Stored in the dark, under inert atmosphere, at -20°. Recrystd from 1:1 EtOH/$CHCl_3$ [Bobrowski and Das *J Phys Chem* **89** 5079 *1985*; Johnston and Scaiano *J Am Chem Soc* **108** 2349 *1986*].

γ-**Carotene** *[472-93-5, 10593-83-6]* **M 536.9, $A_{1cm}^{1\%}$ (λmax) 2055 (437nm), 3100 (462nm), 2720 (494nm) in hexane.** Purified by chromatography on alumina or magnesia columns. Crystd from $^*C_6H_6$/MeOH (2:1). Stored in the dark, under inert atmosphere, at 0°.

ξ-**Carotene** *[38894-81-4]* **M 536.9, m 38-42°, λ$_{max}$ 378, 400, 425nm, $A_{1cm}^{1\%}$ (λmax) 2270 (400nm), in pet ether.** Purified by chromatography on 50% magnesia-HyfloSupercel, developing with hexane and eluting with 10% EtOH in hexane. It was crystd from toluene/MeOH. [Gorman et al. *J Am Chem Soc* **107** 4404 *1985*.] Stored in the dark under inert atmosphere at -20°.

λ−**Carrageenan** *[9064-57-7, 9000-07-1 (κ + little of λ)]*. This D-galactose-anhydro-D or L-galactoside polysaccharide is ppted from 4g of Carrageenan in 600mL of water containing 12g of KOAc by addn of EtOH. The fraction taken, ppted between 30 and 45% (v/v) EtOH. [Pal and Schubert *J Am Chem Soc* **84** 4384 *1962*.]

Cation exchange resins. Should be conditioned before use by successive washing with water, EtOH and water, and taken through two H^+-OH^--H^+ cycles by successive treatment with M HCl, water, M NaOH, water and M HCl, then washed with water until neutral to give the H^+ form. (See commercial catalogues on ion exchange resins).

Cathepsin B (from human liver) *[9047-22-7]* **M_r 27,500 [EC 3.4.22.1].** Purified by affinity chromatography on the semicarbazone of Gly-Phe-glycinal-linked to Sepharose 4B, with elution by 2,2'-dipyridyl disulfide [Rich et al. *Biochem J* **235** 731 *1986*; *Methods Enzymol* **80** 551 *1981*].

Cathepsin D (from bovine spleen) *[9025-26-7]* **M_r 56,000, [EC 3.4.23.5].** Purified on a CM column after ammonium sulfate fractionation and dialysis, then starch-gel electrophoresis and by ultracentrifugal analysis. Finally chromatographed on a DEAE column [Press et al. *Biochem J* **74** 501 *1960*].

Cephalosporin C potassium salt *[28240-09-7]* **M 453.5, [α]$_D^{20}$ +103° (H_2O), pK$_1$ <2.6, pK$_2$ 3.1, pK$_3$ 9.8.** Purified by dissolving in the minimum volume of H_2O (filter) and adding EtOH until separation of solid is complete. A soln is stable in the pH range 2.5-8. It has UV λ$_{max}$ is 260nm (log ε 3.95) in H_2O. The Ba salt has [α]$_D^{20}$ +80° (c 0.57, H_2O) [Woodward et al. *J Am Chem Soc* **88** 852 *1966*; Abraham and Newton *Biochem J* **79** 377 *1961*; Hodgkin and Maslen *Biochem J* **79** 402 *1961*; see also *Quart Reviews Chem Soc* London **21** 231 *1967*].

Ceruloplasmin (from human blood plasma) *[9031-37-2]* **M_r 134,000.** This principle Cu transporter (90-90% of circulating Cu) is purified by precipitation with polyethylene glycol 4000, batchwise adsorption and elution from QAE-Sephadex, and gradient elution from DEAE-Sepharose CL-6B. Ceruloplasmin was purified 1640-fold. Homogeneous on anionic polyacrylamide gel electrophoresis (PAGE), SDS-PAGE, isoelectric focusing and low speed equilibrium centrifugation. [Oestnuizen *Anal Biochem* **146** 1 *1985*; Cohn et al. *J Am Chem Soc* **68** 459 *1946*.]

Chemokines. These are small proteins formed from longer precursors and are chemoattractants for lymphocytes and lymphoid organs. They are characterised by having cysteine groups in specific relative positions. The two largest families are the α and β families that have four cysteine residues arranged (C-X-C) and (C-C) respectively. The mature chemokines have ~70 amino acids with internal cys S-S bonds and attract myeloid type cells *in vitro*. The γ-family (Lymphotactin) has only two cys residues. The δ-family (Neurotactin, Fractalkine) has the C-C-X-X-X-C sequence (*ca* 387 amino acids), binds to membrane promoting adhesion of lymphocytes. The soluble domain of human Fractalkine chemoattracts monocytes and T cells. Several chemokines are available commercially (some prepared by recombinant DNA techniques) including 6Ckine/exodus/SLC which belongs to the β-family with 6 cysteines (110 amino acids mature protein), as the name implies (C-C-C-C-X.....X-C-C) and homes lymphocytes to secondary lymphoid organs with lymphocyte adhesion antitumor properties. Other chemokines available are C10 (βCC) and Biotaxin. Several chemokine receptors and antibodies are available commercially and can generally be used without further purification. [Murphy 'Molecular biology of lymphocyte chemoattractant receptors' in *Ann Rev Immunol* **12** 593 *1994*.]

Chirazymes. These are commercially available enzymes e.g. lipases, esterases, that can be used for the preparation of a variety of optically active carboxylic acids, alcohols and amines. They can cause regio and stereospecific hydrolysis and do not require cofactors. Some can be used also for esterification or trans-esterification in neat organic solvents. The proteases, amidases and oxidases are obtained from bacteria or fungi, whereas esterases are from pig liver and thermophilic bacteria. For preparative work the enzymes are covalently bound to a carrier and do not therefore contaminate the reaction products. Chirazymes are available form Roche Molecular Biochemicals and are used without further purification.

Chlorambucil   [4-{bis(2-chloroethyl)amino}benzene)butyric acid]   *[305-03-3]* **M 304.2, m 64-66°, pK$_1$ 5.8 (6.0 at 66°, 50% aq Me$_2$CO), pK$_2$ 8.0.** It is recrystd from pet ether (flat needles) and has a solubility at 20° of 66% in EtOH, 40% in CHCl$_3$, 50% in Me$_2$CO but is insoluble in H$_2$O [Everett et al. *J Chem Soc* 2386 *1953*]. **CARCINOGEN.**

Chloramphenicol   [Amphicol, 1R,2R-(-)-2-{2,2-dichloroacetylamino}-1-{4-nitrophenyl}-propan-1,3-diol]   *[56-75-7]* **M 323.1, m 149-151°, 150-151°, 151-152°, [α]$_D^{20}$ +20.5° (c 3, EtOH), [α]$_D^{25}$ -25.5° (EtOAc).** Purified by recrystn from H$_2$O (sol 2.5mg/mL at 25°) or ethylene dichloride as needles or long plates and by sublimation at high vacuum. It has $A_{1cm}^{1\%}$ 298 at λ$_{max}$278nm and it is slightly soluble in H$_2$O (0.25%) and propylene glycol (1.50%) at 25° but is freely soluble in MeOH, EtOH, BuOH, EtOAc and Me$_2$CO. [Relstock et al. *J Am Chem Soc* **71** 2458 *1949*; Confroulis et al. *J Am Chem Soc* **71** 2463 *1949*; Long and Troutman *J Am Chem Soc* **71** 2469, 2473 *1949*; Ehrhart et al. *Chem Ber* **90** 2088 *1957*.]

Chloramphenicol palmitate *[530-43-8]* **M 561.5, m 90°, [α]$_D^{26}$ +24.6° (c 5, EtOH).** Crystd from *benzene.

2-Chloroadenosine *[146-77-0]* **M 301.7, m 145-146°(dec), 147-149°(dec), pK$_{Est(1)}$~ 0.5, pK$_{Est(2)}$~ 7.6.** Purified by recrystn from H$_2$O (~1% in cold) and has λ$_{max}$ at 264 nm (pH 1 and 7) and 265 nm (pH 13) in H$_2$O. [Brown and Weliky *J Org Chem* **23** 125 *1958*; Schaeffer and Thomas *J Am Chem Soc* **80** 3738 *1958*; IR: Davoll and Lewy *J Am Chem Soc* **74** 1563 *1952*.]

Chlorophylls *a* and *b* see entries on p. 167 in Chapter 4.

6-Chloropurine riboside (6-chloro-9-β-D-ribofuranosyl-9H-purine) *[2004-06-0]* **M 286.7, m 158-162°(dec), 165-166°(sintering At 155°), 168-170°(dec), [α]$_D^{26}$ -45° (c 0.8, H$_2$O).** Purified by suspending the dry solid (~12 g) in hot MeOH (130 mL) and then adding enough hot H$_2$O (~560 mL) to cause solution, filter and set aside at 5° overnight. The colourless crystals of the riboside are filtered off, washed with Me$_2$CO, Et$_2$O and dried at 60°/0.1mm. More material can be obtained from the filtrate by evapn to dryness and recrystn of the residue from MeOH-H_2O (2:1) (15mL/g). It has λ_{max} 264nm (ε 9140) in H_2O. [Robins *Biochem Prep* **10** 145 *1963*; Baker et al. *J Org Chem* **22** 954 *1957*.]

Chromomycin A$_3$ *[7059-24-7]* **M 1183.3, m 185°dec, [α]$_D^{23}$ -57° (c 1, EtOH).** Dissolve reagent (10g) in EtOAc and add to a column of Silica Gel (Merck 0.05-0.2microns, 4x70cm) in EtOAc containing 1% oxalic acid. Elute with EtOAc+1% oxalic acid and check fractions by TLC. Pool fractions, wash with H_2O thoroughly, dry and evaporate. Recryst from EtOAc. The *hepta-acetate* has **m** 214°, [α]$_D^{23}$ -20° (c 1, EtOH). [*Tetrahedron* **23** 421 *1967*; *J Am Chem Soc* **91** 5896 *1969*.]

α-Chymotrypsin *[9004-07-3]* **M$_r$ ~25000 [EC 3.4.21.1].** Crystd twice from four-tenths saturated ammonium sulfate soln, then dissolved in 1mM HCl and dialysed against 1mM HCl at 2-4°. The soln was stored at 2° [Lang, Frieden and Grunwald *J Am Chem Soc* **80** 4923 *1958*].

Cinchonidine *[485-71-2]* **M 294.4, m 210.5°, [α]$_{546}^{20}$ -127.5° (c 0.5, EtOH), pK$_1^{15}$ 4.17, pK$_2^{15}$ 8.4.** Crystd from aqueous EtOH. For *N-benzyl chloride* see entry in Chapter 6.

Cinchonine *[118-10-5]* **M 294.4, m 265°, [α]$_{546}^{20}$ +268° (c 0.5, EtOH), pK$_1^{15}$ 4.28, pK$_2^{15}$ 8.35.** Crystd from EtOH or diethyl ether. For *N-benzyl chloride* see entry in Chapter 6.

Citranaxanthin *[3604-90-8]* **M 456.7, m 155-156°, $A_{1cm}^{1\%}$ (λmax) 410 (349nm), 275 (466nm) in hexane.** Purified by chromatography on a column of 1:1 magnesium oxide and HyfloSupercel (diatomaceous filter aid). Crystd from pet ether. Stored in the dark, under inert atmosphere, at 0°.

Citric acid cycle components (from rat heart mitochondria). Resolved by anion-exchange chromatography [LaNoue et al. *J Biol Chem* **245** 102 *1970*].

Clonidine hydrochloride **[Catapres, 2-(2,6-dichloroanilino)-2-imidazoline hydrochloride]** *[4205-91-8]* **M 266.6, m 305°, pK 5.88 (free base).** It is recrystd from EtOH-Et$_2$O and dried in a vacuum (solubility in H_2O is 5%). It has a pKa of 5.88. The *free base* has **m** 124-125° and is recrystallised from hexane. [Jen et al. *J Med Chem* **18** 90 *1975*; NMR: Jackman and Jen *J Am Chem Soc* **97** 2811 *1975*.]

Clostripain *[9028-00-6]* **[EC 3.4.22.8] M$_r$ ~55,000.** Isolated from *Clostridium histolyticum* callogenase by extraction in pH 6.7 buffer, followed by hydroxylapatite chromatography with a 0.1-0.2 M phosphate gradient, then Sephadex G-75 gel filtration with 0.05M phosphate pH 6.7, dialysis and a second hydroxylapatite chromatography (gradient elution with 0.1M → 0.3M phosphate, pH 6.7). It has proteinase and esterase activity and is assayed by hydrolysing *n*-benzoyl-L-arginine methyl ester. [Mitchell and Harrington *J Biol Chem* **243** 4683 *1968*, *Methods Enzymol* **19** 635 *1970*.]

Cloxacillin sodium salt (sodium 3-o-chlorophenyl-5-methyl-4-isoxazolyl penicillin monohydrate) *[642-78-4]* **M 457.9, m 170°, [α]$_D^{20}$ +163° (H_2O pH 6.0-7.5), pK$_{Est}$ ~ 2.8 (COOH).** Purified by dissolving in isoPrOH containing 20% of H_2O, and diluting with isoPrOH to a water content of 5% and chilled, and recrystd again in this manner. The sodium salt is collected and dried at 40° in air to give the colourless monohydrate. It is soluble in H_2O (5%), MeOH, EtOH, pyridine and ethylene glycol. [Doyle et al. *J Chem Soc* 5838 *1963*; Naylor et al. *Nature* **195** 1264 *1962*.]

β-Cocaine {2β-carbomethoxy-3-β-benzoxytropane, methyl [1R-(exo,exo)]-3-(benzoyloxy)-2-methyl-8-azabicyclo[3.2.1]octane-2-carboxylate} *[50-36-2]* **M 303.4, m 98°, b 187-188°/0.1mm, [α]$_D^{20}$ -15.8° (c 4, CHCl$_3$), pK^{24} 8.39.** Crystallises from EtOH and sublimes below 90° in a vacuum in a non-crystalline form.

Cocarboxylase tetrahydrate (aneurine pyrophosphoric acid tetrahydrate, thiamine pyrophosphoric acid tetrahydrate) *[136-09-4]* **M 496.4, m 220-222°(sinters at 130-140°), 213-214°, pK$_{Est(1)}$~2, pK$_{Est(2)}$~6, pK$_{Est(3)}$~9.** Recrystd from aqueous Me$_2$CO. [Wenz et al. *Justus Liebigs Ann Chem* **618** 210 *1958*; UV: Melnick *J Biol Chem* **131** 615 *1939*; X-ray: Carlisle and Cook *Acta Cryst (B)* **25** 1359 *1969*.] The *hydrochloride salt* has **m** 242-244°(dec), 241-243°(dec) or 239-240°(dec) and is recrystd from aqueous HCl + EtOH, EtOH containing HCl or HCl + Me$_2$CO. [Weijlard *J Am Chem Soc* **63** 1160 *1941*; Synthesis: Weijlard and Tauber *J Am Chem Soc* **60** 2263 *1938*.]

Codeine *[76-57-3]* **M 299.4, m 154-156°, [α]$_D^{20}$ -138° (in EtOH), pK^{25} 8.21.** Crystd from water or aqueous EtOH. Dried at 80°.

Coenzyme A trihydrate *[85-61-0]* **M 821.6, pK$_1$ 4.0 (adenine NH$_2$), pK$_2$ 6.5 (PO$_4$H), pK$_3$ 9.6 (SH).** White powder best stored in an inert atmosphere in the dark in sealed ampoules after drying *in vacuo* over P$_2$O$_5$ at 34°. It has UV: λ$_{max}$ 259 nm (ε 16,800) in H$_2$O. [Buyske et al. *J Am Chem Soc* **76** 3575 *1954*.] It is sol in H$_2$O but insol in EtOH, Et$_2$O and M$_2$CO. It is readily oxidised in air and is best kept as the more stable *trilithium salt* [Moffat and Khorana *J Am Chem Soc* **83** 663 *1961*; see also Beinert et al. *J Biol Chem* **200** 384 *1953*; De Vries et al. *J Am Chem Soc* **72** 4838 *1950*; Gregory et al. *J Am Chem Soc* **74** 854 *1952* and Baddiley *Adv Enzymol* **16** 1 *1955*].

Coenzyme Q$_0$ (2,3-Dimethoxy-5-methyl-1,4-benzoquinone,   3,4-dimethoxy-2,5-tolu-quinone, fumigatin methyl ether) *[605-94-7]* **M   182.2, m 56-58°, 58-60°, 59°.** It crystallises in red needles from pet ether (b 40-60°) and can be sublimed in high vacuum with a bath temperature of 46-48° [Ashley, Anslow and Raistrick *J Chem Soc* 441 *1938*; UV in EtOH: Vischer *J Chem Soc* 815 *1953*; UV in cyclohexane: Morton et al. *Helv Chim Acta* **41** 2343 *1858*; Aghoramurthy et al. *Chem Ind (London)* 1327 *1954*].

Coenzyme Q$_4$ (Ubiquinone-4,   2,3-dimethoxy-5-methyl-6-[3,7,11,15-tetramethyl-hexadeca-2*t*,6*t*,10*t*,14-tetraenyl]-[1,4]benzoquinone *[4370-62-1]* **M 454.7, m   30°, 33-45°, A$_{1cm}^{1\%}$ (275nm) 185.** A red oil purified by TLC chromatography on SiO$_2$ and eluted with Et$_2$O-hexane. Purity can be checked by HPLC (silica column using 7% Et$_2$O-hexane). It has λ$_{max}$ 270 nm (ε 14,800) in pet ether. [NMR and MS: Naruta *J Org Chem* **45** 4097 *1980;* cf Morton *Biochemical Spectroscopy* (Adam Hilger, London, 1975) p 491]. It has also been dissolved in MeOH/EtOH (1:1 v/v) and kept at 5° until crystals appear [Lester and Crane *Biochim Biophys Acta* **32** 497 *1958*].

Coenzyme Q$_9$ (Ubiquinone-9,   2,3-dimethoxy-5-methyl-6-[3,7,11,15,19,23,27,31,35-nonamethyl-hexatriaconta-2*t*,6*t*,10*t*,14*t*,18*t*,22*t*,26*t*,30*t*,34-nonaenyl]-1,4-benzoquinone) *[303-97-9]* **M 795.3, m 40.5-42.5°, 44-45°, 45°.** Yellow crystals purified by recrystn from pet ether and by TLC chromatography on SiO$_2$ and eluted with Et$_2$O-hexane. Purity can be checked by HPLC (silica column using 7% Et$_2$O-hexane). It has λ$_{max}$ 270nm (ε 14,850) in pet ether. [NMR and MS: Naruta *J Org Chem* **45** 4097 *1980;* Le et al. *Biochem Biophys Acta* **32** 497 *1958*; cf Morton *Biochemical Spectroscopy* (Adam Hilger, London, 1975) p 491; IR: Lester et al. *Biochim Biophys Acta* **33** 169 *1959*; UV: Rüegg et al. *Helv Chim Acta* **42** 2616 *1959*; Shunk *J Am Chem Soc* **81** 5000 *1959*.]

Coenzyme Q$_{10}$ (Ubiquinone-10,   2,3-dimethoxy-5-methyl-6-[3,7,11,15,19,23,27,31,35,-39-decamethyl-tetraconta-2*t*,6*t*,10*t*,14*t*,18*t*,22*t*,26*t*,30*t*,34*t*,38-decaenyl]-1,4-benzoquinone) *[303-98-0]* **M 795.3, m 48-49°,49°, 49.5-50.5°, 50°.** Purified by recrystn from EtOH, EtOH + Me$_2$CO or Et$_2$O-EtOH and by chromatography on silica gel using isoPrOH-Et$_2$O (3:1) to give orange crystals. It has λ$_{max}$ 270nm (ε 15,170) in pet ether. [Terao et al. *J Org Chem* **44** 868 *1979*; NMR and MS: Naruta et al. *J Org Chem* **45** 4097 *1980*; IR: Lester et al. *Biochem Biophys Acta* **42** 1278 *1959*, NMR: Planta et al. *Helv Chim Acta* **42** 1278 *1959*; cf Morton *Biochemical Spectroscopy* (Adam Hilger, London, 1975) p 491].

Colcemide (Demecocine) *[477-30-5]* **M 371.4, m 182-185°, 183-185°, 186°, [α]$_D^{20}$ -129° (c 1, CHCl$_3$).** It has been purified by chromatography on silica and eluting with CHCl$_3$-MeOH (9:1) and recrystn from EtOAc-Et$_2$O and forms yellow prisms. UV in EtOH has λ$_{max}$ 243nm (ε 30,200) and 350nm (ε 16,3000). [Synthesis, IR, NMR, MS: Capraro and Brossi *Helv Chim Acta* **62** 965 *1979*.]

Colchicine *[64-86-8]* **M 399.5, m 155-157°(dec), [α]$_{546}^{20}$ -570° (c 1, H$_2$O), pK^{20} 1.85.** Commercial material contains up to 4% desmethylcolchicine. Purified by chromatography on alumina, eluting with CHCl$_3$ [Ashley and Harris *J Chem Soc* 677 *1944*]. Alternatively, an acetone solution on alkali-free alumina has been used, and eluting with acetone [Nicholls and Tarbell *J Am Chem Soc* **75** 1104 *1953*].

Colchicoside *[477-29-2]* **M 547.5, m 216-218°.** Crystd from EtOH.

Colicin E (**from** *E.coli*) *[11032-88-5]*. Purified by salt extraction of extracellular-bound colicin followed by salt fractionation and ion-exchange chromatography on a DEAE-Sephadex column, and then by CM-Sephadex column chromatography [Schwartz and Helinski *J Biol Chem* **246** 6318 *1971*].

Collagenase (**from human polymorphonuclear leukocytes**) *[9001-12-1]* **M$_r$ 68,000-125,000 [EC 3.4.24.3].** Purified by using *N*-ethylmaleimide to activate the enzyme, and wheat germ agglutinin-agarose affinity chromatography [Callaway et al. *Biochemistry* **25** 4757 *1986*].

Compactin *[73573-88-3]* **M 390.5, m 151-153°, 152°, $[\alpha]_D^{22}$ +283° (c 0.48, acetone).** Purified by recrystn from aqueous EtOH. UV: λ_{max} 230, 237 and 246nm (log ε 4.28, 4.30 and 4.11); IR (KBr): ν 3520, 1750 (lactone CO) and 1710 (CO ester) cm^{-1}. [Clive et al. *J Am Chem Soc* **110** 6914 *1988*; Synthesis review: Rosen and Heathcock *Tetrahedron* **42** 4909 *1986*; IR, NMR, MS: Brown et al. *J Chem Soc Perkin Trans 1* 1165 *1976*.]

Convallatoxin (**α cardenolide mannoside**) *[508-75-8]* **M 550.6, m 238-239°, $[\alpha]_D^{25}$ 9.4° (c 0.7, dioxane).** Crystd from EtOAc. *Tetra-acetate* has **m** 238-242° (MeOH/Et$_2$O), $[\alpha]_D^{25}$ -5° (CHCl$_3$).

Copper-zinc-superoxide dismutase (**from blood cell haemolysis**) *[9054-89-1]* **M$_r$ ~32,000 [EC 1.15.1.1].** Purified by DEAE-Sepharose and copper chelate affinity chromatography. The preparation was homogeneous by SDS-PAGE, analytical gel filtration chromatography and by isoelectric focusing [Weselake et al. *Anal Biochem* **155** 193 *1986*; Fridovich *J Biol Chem* **244** 6049 *1969*].

Coproporphyrin I *[531-14-6]* **M 654.7, λ_{max} 591, 548, 401nm in 10% HCl.** Crystd from pyridine/glacial acetic acid. The *dihydrochloride [69477-27-6]* has **M** 727.7 and λmax 395nm in water.

Corticosterone (**11β, 21-dihydroxypregn-4-en-3,20-dione**) *[50-22-6]* **M 346.5, m 180-181°, 180-182°, 181-184°, $[\alpha]_D^{15}$ +223° (c 1.1, EtOH), $[\alpha]_D^{23-25}$ +194° (c 0.1, dioxane).** Purified by recrystn from Me$_2$CO (trigonal plates), EtOH or isoPrOH. UV λ_{max} at 240nm, and gives an orange-yellow soln with strong fluorescence on treatment with concentrated H$_2$SO$_4$. Insoluble in H$_2$O but soluble in organic solvents. [Reichstein and Euw *Helv Chim Acta* **21** 1197 *1938*, **27** 1287 *1944*; Mason et al. *J Biol Chem* **114** 613 *1936*; ORD: Foltz et al. *J Am Chem Soc* **77** 4359 *1955*; NMR: Shoolery and Rogers *J Am Chem Soc* **80** 5121 *1958*.] The *21-O-benzoyl* derivative has **m** 201-202° [Reichstein *Helv Chim Acta* **20** 953 *1937*].

Corticotropin *[92307-52-3]* **polypeptide M$_r$ ~4697.** Extract separated by ion-exchange on CM-cellulose, desalted, evapd and lyophilised. Then run on gel filtration (Sephadex G-50) [Lande et al. *Biochemical Preparations* **13** 45 *1971*; Esch et al. *Biochem Biophys Res Commun* **122** 899 *1984*].

Cortisol see hydrocortisone on p. 541.

Cortisone *[53-06-5]* **M 360.5, m 230-231°, $[\alpha]_{546}^{20}$ +225° (c 1, in EtOH).** Crystd from 95% EtOH or acetone.

Cortisone-21-acetate *[50-04-4]* **M 402.5, m 242-243°, $[\alpha]_{546}^{20}$ +227° (c 1, in CHCl$_3$).** Crystd from acetone.

Creatine (**H$_2$O**) (**N-guanidino-N-methylglycine**) *[6020-87-7]* **M 131.1, m 303°, pK$_1^{25}$ 2.63, pK$_2^{25}$ 14.3.** Likely impurities are creatinine and other guanidino compounds. Crystd from water as monohydrate. Dried under vacuum over P$_2$O$_5$ to give anhydrous material.

Creatine phosphate di Na, 4H$_2$O salt (**phosphocreatine**) *[922-32-7]* **M 327.1, pK$_1^{27}$ 2.7, pK$_2^{27}$ 4.58, pK$_3^{27}$~ 12.** To 3-4g of salt in H$_2$O (220mL) is added 4 vols of EtOH with thorough stirring and allowed to stand at 20° for 12hrs (this temp is critical as crystals did not readily form at 23° or 25°). The salt first appears as oily droplets which slowly settle and crystallise. After 12hrs the supernatant is clear. Stirring and scratching the flask containing the filtrate brings out additional (0.3-1g) crystals if the salt is kept at 20° for 12hrs. Filter at room temp, wash with 3 x 5mL of ice-cold 90% EtOH then 5mL of abs EtOH and dry in a vac desiccator (Drierite or CaCl$_2$) for 16-30hrs. The hexahydrate (plates) is converted to the tetrahydrate salt (needles) in vac at -10°. [Ennor and Stocken *Biochem Prep* **5** 9 *1957*; *Biochem J* **43** 190 *1958*.]

Creatinine (2-imino-1-methyl-4-oxoimidazolidine) *[60-27-5]* **M 113.1, m 260°(dec), pK$_1^{25}$ 4.80, pK$_2^{25}$9.2.** Likely impurities are creatine and ammonium chloride. Dissolved in dilute HCl, then neutralised by adding ammonia. Recrystd from water by adding excess of acetone.

Crotaline (monocrotaline, 12,13-dihydroxy-(13β-14βH)-14,19-dihydro-20-norcrotalanan-11,15-dione) *[315-22-0]* **M 325.4, m 196-197°(dec), 197-198°(dec), 203°(dec), [α]$_D^{20}$ -55° (c 1, EtOH).** It forms prisms from absolute EtOH and recrystallises also from CHCl$_3$. UV in 96% EtOH has λ$_{max}$ 217nm (log ε 3.32). [Adams et al. *J Am Chem Soc* **74** 5612 *1952*; Culvenor and Smith *Aust J Chem* **10** 474 *1957*.] The *hydrochloride* has **m** 212-214° (from MeOH-Et$_2$O) and [α]$_D^{28}$ -38.4° (c 5, H$_2$O) [Adams and Gianturco *J Am Chem Soc* **78** 1922 *1956*]. The *picrate* has **m** 230-231.5°(dec) [Adams et al. *J Am Chem Soc* **74** 5614 *1952*].

α-Cyclodextrin (H$_2$O) *[10016-20-3]* **M 972.9, m >280°(dec), [α]$_{546}$ +175° (c 10, H$_2$O).** Recrystd from 60% aq EtOH, then twice from water, and dried for 12hours in a vacuum at 80°. Also purified by pptn from water with 1,1,2-trichloroethylene. The ppte was isolated, washed and resuspended in water. This was boiled to steam distil the trichloroethylene. The soln was freeze-dried to recover the cyclodextrin. [Armstrong et al. *J Am Chem Soc* **108** 1418 *1986*].

ß-Cyclodextrin (H$_2$O) *[7585-39-9]* **M 1135.0, m >300°(dec), [α]$_{546}$ +170° (c 10, H$_2$O).** Recrystd from water and dried for 12hours in a vacuum at 110°, or 24hours in a vacuum at 70°. The purity was assessed by TLC on cellulose with a fluorescent indicator. [Taguchi, *J Am Chem Soc* **108** 2705 *1986*; Tabushi et al. *J Am Chem Soc* **108** 4514 *1986*; Orstam and Ross *J Phys Chem* **91** 2739 *1987*.]

D-(R-natural) and L-(S-non-natural) Cycloserine (2-amino-3-isoxazolidone) *[R- 68-41-7 and S- 339-72-0]* **M 102.1, m 145-150° (dec), 154-155°, 155-156° (dec), 156° (dec), [α]$_D^{25}$ (+) and (-) 137° (c 5, 2N NaOH), pK$_1^{10}$ 4.5, pK$_2^{10}$ 7.74, pK$_1^{25}$ 4.50, pK$_2^{25}$ 7.43, pK$_1^{50}$ 4.44, pK$_2^{50}$ 7.20.** Purified by recrystn from aqueous EtOH or MeOH or aqueous NH$_3$ + EtOH or isoPrOH. Also recrystd from aqueous ammoniacal soln at pH 10.5 (100mg/mL) by diluting with 5 volumes of isopropanol and then adjusting to pH 6 with acetic acid. An aqueous soln buffered to pH 10 with Na$_2$CO$_3$ can be stored in a refrigerator for 1 week without decomposition. UV: λ$_{max}$ 226nm ($A_{1cm}^{1\%}$ 4.02). The *tartrate salt* has **m** 165-166° (dec), 166-168° (dec), and [α]$_D^{24}$ -41° (c 0.7, H$_2$O). [Stammer et al. *J Am Chem Soc* **79** 3236 *1959*; UV: Kuehl *J Am Chem Soc* **77** 2344 *1955*.]

Cystamine dihydrochloride [2,2'-diaminodiethylene disulfide dihydrochloride, 2,3'-dithio-bis(ethylamine) dihydrochloride] *[56-17-7]* **M 225.2, m 219-220°(dec), pK$_1^{30}$ 8.82, pK$_2^{30}$ 9.58.** Recrystd by dissolving in EtOH containing a few drops of dry EtOH-HCl, filtering and adding dry Et$_2$O. The solid is dried in a vacuum and stored in dry and dark atmosphere. It has been recrystd from EtOH (solubility: 1g in 60mL of boiling EtOH) or MeOH (plates). The *free base* has **b** 90-100°/0.001mm, 106-108°/5mm and 135-136°/atm, d$_4^{20}$ 1.1559, n$_D^{20}$ 1.5720. [Verly and Koch *Biochem J* **58** 663 *1954*; Gonick et al. *J Am Chem Soc* **76** 4671 *1954*; Jackson and Block *J Biol Chem* **113** 137 *1936*.] The *dihydrobromide* has **m** 238-239° (from EtOH-Et$_2$O) [Viscontini *Helv Chim Acta* **36** 835 *1953*].

S,S-(L,L)-Cystathionine (S-2-amino-2-carboxyethyl-L-homocysteine, L-2-amino-4[(2-amino-2-carboxyethyl)thio]butyric acid) *[56-88-2]* **M 222.3, m >300°, dec at 312° with darkening at 270°, [α]$_D^{20}$ +23.9° (c 1, M HCl).** Could be converted to the *HCl* salt by dissolving in 20% HCl and carefully basifying with aqueous NH$_3$ until separation is complete. Filter off and dry in a vacuum. It forms prisms from H$_2$O. The *dibenzoyl* derivative has **m** 229° (from EtOH). [IR: Greenstein and Winitz *Chemistry of the Amino Acids (J Wiley)* **Vol 3** 2690 *1961* and Tallan et al. *J Biol Chem* **230** 707 *1958*; Synthesis: du Vigneaud et al. *J Biol Chem* **143** 59 *1942*; Anslow et al. *J Biol Chem* **166** 39 *1946*.]

[Prepn: Weiss and Stekol *J Am Chem Soc* **73** 2497 *1951*; see also du Vigneaud et al. *J Biol Chem* **143** 60 *1942*; Biological synthesis: Greenberg *Methods Enzymol* **5** 943 *1962*.]

Cysteamine (2-aminoethanethiol, 2-mercaptoethylamine) *[60-23-1]* **M 77.2, m 97-98.5°, 98-99°, 99-100°, pK$_1^0$ 9.15, pK$_2^0$ 11.93, pK$_1^{30}$ 8.42, pK$_2^{30}$ 10.83.** Soluble in H_2O giving an alkaline reaction and it has a disagreeable odour. Likely impurity is the disulfide, cystamine which is not soluble in alkaline solution. Under a N_2 atmosphere dissolve in EtOH, evaporate to dryness and wash the white residue with dry pet ether, then sublime at 0.1mm and store under N_2 (out of contact with air) at 0-10° in the dark. Its *HgCl$_2$ (2:3) complex* has **m** 181-182° (from H_2O), and its *picrate* has **m** 125-126°. [Mills and Bogert *J Am Chem Soc* **57** 2328 *1935*, **62** 1173 *1940*; Baddiley and Thain *J Chem Soc* 800 1952; Shirley *Preparation of Organic Intermediates (J. Wiley)* **Vol 3** 189 *1951*; Barkowski and Hedberg *J Am Chem Soc* **109** 6989 *1987*.]

Cysteamine hydrochloride *[156-57-0]* **M 113.6, m 70.2-70.7°, 70-72°.** Purified by recrystn from EtOH. It is freely soluble in H_2O and should be stored in a dry atmosphere. [Mills and Bogert *J Am Chem Soc* **62** 1177 *1940*.] The *picrate* has **m** 125-126°, see previous entry for *free base*.

(±)-Cysteic acid (3-sulfoalanine, 1-amino-3-sulfopropionic acid) *[13100-82-8, 3024-83-7]* **M 169.2, m 260°(dec).** Likely impurities are cystine and oxides of cysteine. Crystd from water by adding 2 volumes of EtOH. When recrystd from aqueous MeOH it has **m** 264-266°, and the anhydrous acid has **m** ~260°(dec). [Chapeville and Formageot *Biochim Biophys Acta* **26** 538 *1957*; *J Biol Chem* **72** 435 *1927*.]

R(L)-Cysteic acid (H$_2$O) *[23537-25-9]* **M 187.2, m 275-280° (dec), 289°, [α]$_D^{20}$ +8.66° (c 7.4, H$_2$O, pH 1) and +1.54° (H$_2$O, pH 13), pK$_1^{25}$ 1.9 (SO$_3$H), pK$_2^{25}$ 8.7 (CO$_2$H), pK$_3^{25}$ 12.7 (NH$_2$).** Likely impurities are cystine and oxides of cysteine. Crystd from water by adding 2 volumes of EtOH. When recrystd from aqueous MeOH it has **m** 264-266°, and the anhydrous acid has **m** ~260°(dec). [Chapeville and Formageot *Biochim Biophys Acta* **26** 538 *1957*; *J Biol Chem* **72** 435 *1927*.]

D-(S)- and L-(R)- Cysteine (S- and R-2-amino-3-mercaptopropionoic acid) *[S(+)-: 921-01-7 and R(-)-: 52-90-4]* **M 121.2, m 230°, 240° (dec), [α]$_D^{20}$ (+) and (-) 7.6° (c 2, M HCl) and (+) and (-) 10.1° (c 2, H$_2$O, pH 10), pK$_1^{25}$ 1.92 (CO$_2$), pK$_2^{25}$ 8.35 (NH$_2$), pK$_3^{25}$ 10.46 (SH).** Purified by recrystn from H_2O (free from metal ions) and dried in a vacuum. It is soluble in H_2O, EtOH, Me$_2$CO, EtOAc, AcOH, *C$_6$H$_6$ and CS$_2$. Acidic solns can be stored under N_2 for a few days without deterioration. [For synthesis and spectra see Greenstein and Winitz *Chemistry of the Amino Acids (J. Wiley)* **Vol 3** p1879 *1961*.]

L-Cysteine hydrochloride (H$_2$O) *[52-89-1]* **M 175.6, m 175-178° (dec), [α]$_D^{25}$ +6.53° (5M HCl).** Likely impurities are cystine and tyrosine. Crystd from MeOH by adding diethyl ether, or from hot 20% HCl. Dried under vacuum over P_2O_5. *Hygroscopic.*

(±)-Cysteine hydrochloride *[10318-18-0]* **M 157.6.** Crystd from hot 20% HCl; dried under vacuum over P_2O_5.

L-Cystine *[56-89-3]* **M 240.3, [α]$_D^{18.5}$ -229° (c 0.92 in M HCl), pK$_1^{25}$ 1.04 (1.65), pK$_2^{25}$ 2.05 (2.76), pK$_3^{25}$ 8.00 (7.85), pK$_4^{25}$ 10.25 (8.7, 9.85).** Cystine disulfoxide was removed by treating an aqueous suspension with H_2S. The cystine was filtered off, washed with distilled water and dried at 100° under vacuum over P_2O_5. Crystd by dissolving in 1.5M HCl, then adjusting to neutral pH with ammonia. Likely impurities are D-cystine, meso-cystine and tyrosine.

Cytidine *[65-46-3]* **M 243.2, m 210-220°(dec), 230° (dec), 251-252° (dec), [α]$_{546}^{20}$ +37° (c 9, H$_2$O), [α]$_D^{20}$ +29° (c 9, H$_2$O), pK 3.85.** Crystd from 90% aqueous EtOH. Also has been converted to the *sulfate* by dissolving (~200mg) in a soln of EtOH (10mL) containing H_2SO_4 (50mg), whereby the salt crystallises out. It is collected, washed with EtOH and dried for 5hours at 120°/0.1mm. The *sulfate* has **m** 225°. The *free base* can be obtained by shaking with a weak ion-exchange resin, filtering, evaporating and recrystallising the residue from EtOH as before. [Fox and Goodman *J Am Chem Soc* **73** 3256 *1956*; Fox and

Shugar *Biochim Biophys Acta* **9** 369 *1952*; see Prytsas and Sorm in *Synthetic Procedures in Nucleic Acid Chemistry* (Zorbach and Tipson Eds) **Vol 1** 404 *1973.*]

Cytisine see entry in Chapter 4.

Cytochalasin B **(from dehydrated mould matter)** *[14930-96-2]* **M 479.6.** Purified by MeOH extraction, reverse phase C18 silica gel batch extraction, selective elution with 1:1 v/v hexane/tetrahydrofuran, crystn, subjected to TLC and recrystallised [Lipski et al. *Anal Biochem* **161** 332 *1987*].

Cytochrome c_1 **(from horse, beef or fishes' heart, or pigeon breast muscle)** *[9007-43-6]* **M ~ 13,000.** Purified by chromatography on CM-cellulose (CM-52 Whatman) [Brautigan et al. *Methods Enzymol* **53D** 131 *1978*]. It has a high PI (isoelectric point) and has been purified further by adsorption onto an acidic cation exchanger, e.g. Amberlite IRC-50 (polycarboxylic) or in ground form Amberlite XE-40 (100-200 mesh) or Decalso-F (aluminium silicate), where the non-cytochrome protein is not adsorbed and is readily removed. The cytochrome is eluted using a soln containing 0.25g ions/L of a univalent cation at pH 4.7 adsorbed onto the NH_4^+ salt of Amberlite IRC-50 at pH 7, washed with H_2O and then with 0.12M NH_4OAc to remove non-cytochrome protein. When the cytochrome begins to appear in the eluate then the NH_4OAc concn is increased to 0.25 M. The fractions with *ca* Fe = 0.465—0.467 are collected, dialysed against H_2O and adsorbed onto a small IRC-50 column and eluted with 0.5M NH_4OH, then dialysed and lyophilised. (A second fraction (II) can be eluted from the first resin with 0.5M NH_4OH but is discarded). [Keilin and Hartree *Biochem Prep* **1** 1 *1952*; Margoliash *Biochem Prep* **8** 33 *1957.*]

Cytochrome c has been recrystd as follows: The above eluate (*ca* 100mL) is dialysed against H_2O (10 vols) at 4° for 24 h (no more), then passed through an XE-40 column (2 x 1 cm above) which is equilibrated with 0.1M NH_4OAc pH 7.0. The column is washed with 0.1% $(NH_4)_2SO_4$ pH 8.0 and the dark red resin in the upper part of the column is collected and in 0.1% $(NH_4)_2SO_4$ pH 8.0 transferred to another column (7 mm diameter) and the cytochrome c is eluted with 5% $(NH_4)_2SO_4$ pH 8.0. More than 98% of the red colour is collected in a volume of *ca* 4mL in a weighed centrifuge tube. Add a drop of octanol, 0.43g of $(NH_4)_2SO_4$/g of soln. When the salt has dissolved ascorbic acid (5mg), add a few drops of 30% NH_3 and keep the soln at 10° for 10min (turns lighter colour due to reduction). Then add finely powdered $(NH_4)_2SO_4$ in small portions (stir with a glass rod) until the soln becomes turbid. Stopper the tube tightly, and set aside at 15-25° for 2 days while the cytochrome c separates as fine needles or rosettes. Further $(NH_4)_2SO_4$ (20mg) are added per mL of suspension and kept in the cold for a few days to complete the crystallisation. The crystals are collected by centrifugation (5000xg), suspended in saturated $(NH_4)_2SO_4$ (pH 8.0 at 10°) then centrifuged again. For recrystn the crystals are dissolved in the least volume of H_2O, one drop of ammonia and 1 mg of ascorbic acid are added and the above process is repeated. The yield of twice recrystd cytochrome c from 2Kg of muscle is *ca* 200 mg but this varies with the source and freshness of the muscle used. The crystals are stored as a solid after dialysis against 0.08M NaCl or 0.1M sodium buffer and lyophilising, or as a suspension in saturated $(NH_4)_2SO_4$ at 0°. [Hagihara et al. *Biochem Prep* **6** 1 *1958.*] *Purity of cytochrome c:* This is checked by the ratio of the absorbance at 500nm (reduced form) to 280nm (oxidised form), i.e. $\varepsilon_{500}/\varepsilon_{280}$ should be between 1.1 and 1.28, although values of up to 1.4 have been obtained for pure preparations.

For the preparation of the *reduced form* see Margoliash *Biochem Prep* **5** 33 *1957* and Yonetani *Biochem Prep* **11** 19 *1966.*

Cytochrome from *Rhodospirillum rubrum.* ($\varepsilon_{270}/\varepsilon_{551}$ 0.967). Purified by chromatography on a column of CM-Whatman cellulose [Paleus and Tuppy *Acta Chem Scand* **13** 641 *1959*].

Cytochrome c oxidase **(from bovine heart mitochondria).** *[9001-16-5]* **M_r 100,000/haeme, [EC 1.9.3.1].** Purified by selective solubilisation with Triton X-100 and subsequently with lauryl maltoside; finally by sucrose gradient centrifugation [Li et al. *Biochem J* **242** 417 *1978*].

Also purified by extraction in 0.02 M phosphate buffer (pH 7.4) containing 2% of cholic acid (an inhibitor which stabilises as well as solubilises the enzyme) and fractionated with $(NH_4)_2SO_4$ collecting the 26-33% saturation cut and refractionating again and collecting the 26-33% saturation fraction. The pellet collected at 10,000xg appears as an oily paste. The cholate needs to be removed to activate the enzyme as follows: The ppte is dissolved in 10mL of 0.1M phosphate buffer pH 7.4, containing 1% of Tween-80 and dialysed against 1L of 0.01 M PO_4 buffer (pH 7.4) containing 1% of Tween-80 for 10 h at 0° and aliquoted. The enzyme is stable at $0°$ for 2 weeks and at $-15°$ for several months. It is assayed for purity (see reference) by oxidation of reduced cytochrome c (Km 10μM). [Yonetani *Biochem Prep* **11** 14 *1966*; *J Biol Chem* **236** 1680 *1961*.]

Cytokines see chemokines, interferons, interleukins.

Cytosine *[71-30-7]* **M 111.1, m 320-325° (dec), pK_1^{25} 4.6, pK_2^{25} 12.1.** Crystd from water.

Cytosine-1-β-*O*-arabinofuranoside (Cytarabin) *[147-94-4]* **M 243.2, m ~220°(dec), 212-213.5°, $[\alpha]_D^{20}$ +155° (c 1, H_2O), pK^{25} 4.3.** Purified by recrystn from aqueous EtOH. It has λ_{max} 212 and 279nm at pH 2 and 272nm at pH 12. [Walwick et al. *Proc Chem Soc (London)* 84 *1959*.]

N-Decanoyl-*N*-methylglucamine (Mega-10, *N*-D-glucidyl-*N*-methyl deconamide) *[85261-20-7]* **M 349.5, m 91-93°, 92°.** Possible impurities are decanoic acid and *N*-methylglycamine. The former is removed by grinding the solid with Et_2O and then with pet ether and dried over P_2O_5. Twice recrystd from MeOH-Et_2O by dissolving in the minimum volume of MeOH and adding Et_2O and drying in a vacuum. To remove the glycamine the solid (800mg) is dissolved in hot H_2O (10mL) and set aside. Mega-10 crystallises in colourless needles. These are filtered off and dried in a vacuum to constant weight. It is a good non-ionic non-hygroscopic detergent with a critical micelle concentration (CMC) of 7.4mM (0.26%) in 0.1M Tris-HCl pH 7.4 at 25°. [Hildreth *Biochem J* **207** 363 *1982*.]

Demeclocycline hydrochloride **(7-chloro-6-demethyltetracycline hydrochloride, Clortetrin)** *[64-73-3]* **M 501.3, m 174-178°(dec, for sesquihydrate), $[\alpha]_D^{25}$ -258° (c 0.5, 0.1N H_2SO_4), pK 4.45 [H_2O-Me$_2$NCHO (1:1)].** Crystd from EtOH-Et_2O or H_2O and dried in air [McCormick et al. *J Am Chem Soc* **79** 4561 *1957*; Dobrynin et al. *Tetrahedron Lett* 901 *1962*].

2'-Deoxyadenosine (adenine 2'-deoxyriboside) *[16373-93-6]* **M 269.3, m 187-189°, 189-191°, $[\alpha]_D^{20}$ -25° (c 0.5, H_2O), $[\alpha]_{589}^{25}$ -26°, $[\alpha]_{310}^{25}$ -206° (c 0.5, H_2O), pK^{20} 3.79.** Purified by recrystn from H_2O (as hydrated crystals; solubility of mono-hydrate is 1.1% in H_2O at 20°). It has λ_{max} 258nm (pH 1), 260nm (pH 7) and 261nm (pH 13). [Ness and Fletcher *J Am Chem Soc* **81** 4752 *1959*; Walker and Butler *Can J Chem* **34** 1168 *1956*.] The *3',5'-O-diacetyl* derivative has **m** 151-152° (recrystd from EtOAc-pet ether).

3'-Deoxyadenosine (Cordycepin, adenine 3'-deoxyriboside) *[73-03-0]* **M 251.2, m 225-226°, 225-229°, $[\alpha]_D^{20}$ -47° (H_2O), pK_{Est} ~ 4.8.** It forms needles from EtOH, *n*-BuOH and *n*-PrOH, and from H_2O as the mono-hydrate. It has λ_{max} 260nm (ε 14,600) in EtOH. The *picrate* has **m** 195°(dec, yellow crystals from H_2O). [Kaczka et al. *Biochim Biophys Acta* **14** 456 *1964*; Todd and Ulbricht *J Chem Soc* 3275 *1960*; Lee et al. *J Am Chem Soc* **83** 1906 *1961*; Walton et al. *J Am Chem Soc* **86** 2952 *1964*.]

11-Deoxycorticosterone acetate **(21-acetoxy-4-pregnen-3,20-dione)** *[56-47-3]* **M 372.5, m 154-159°, 154-160°, 155-157°, 155-161°, $[\alpha]_D^{20}$ +174° (c 1, dioxane), $[\alpha]_D^{22-24}$ +196° (c 1, $CHCl_3$).** Recrystallises from EtOH as needles or Me$_2$CO-hexane, and sublimes at high vacuum. Partly soluble in MeOH, Me$_2$CO, Et_2O and dioxane but insoluble in H_2O. [Romo et al. *J Am Chem Soc* **79** 5034 *1957*; NMR: Shoolery and Rogers *J Am Chem Soc* **80** 5121 *1959*.]

2'-Deoxycytidine monohydrate *[951-77-9]* **M 245.2, m 119-200°, 207-209°, 213-215°, $[\alpha]_D^{25}$ +78° (c 0.4, N NaOH), $[\alpha]_D^{23}$ +57.6° (c 2, H_2O), pK 4.25.** Purified by recrystn from MeOH-Et_2O or EtOH and dried in air. [NMR: Miles *J Am Chem Soc* **85** 1007 *1963*; UV: Fox and Shugar *Biochim Biophys Acta* **9** 369 *1952*.] The *hydrochloride* crystallises from H_2O-EtOH and has **m** 174°(dec, 169-173°) [Walker and Butler *Can J Chem* **34** 1168 *1956*.] The *picrate* has **m** 208°(dec). [Fox et al. *J Am Chem Soc* **83** 4066 *1961*.]

2'-Deoxycytidine 5'-monophosphoric acid **(deoxycytidylic acid)** *[1032-65-1]* **M 307.2, m 170-172°(dec), 183-184°(dec), 183-187°(dec), $[\alpha]_D^{21}$ +35° (c 0.2, H_2O), pK_1 4.6, pK_2 6.6.**

Recrystd from H_2O or aqueous EtOH and dried in a vacuum. [Volkin et al. *J Am Chem Soc* **73** 1533 *1951*; UV: Fox et al. *J Am Chem Soc* **75** 4315 *1953*; IR: Michelson and Todd *J Chem Soc* 3438 *1954*.]

2'-Deoxyguanosine monohydrate (9-[2-deoxy-β-D-ribofuranosyl]guanidine) *[961-07-9]* **M 285.3**, **m** *ca* **200°(dec)**, $[\alpha]_D^{20}$ **+37.5°** (**c** 2, H_2O), $[\alpha]_D^{14}$ **-47.7°** (**c** 0.9, N NaOH), $pK_{Est(1)}$~ **3.3**, $pK_{Est(2)}$~ **9.2**. Recrystd from H_2O as the monohydrate. [Brown and Lythgoe *J Chem Soc* 1990 *1950*; Levene and London *J Biol Chem* **81** 711 *1929*, **83** 793 *1929*]; UV: Hotchkiss *J Biol Chem* **175** 315 *1948*; ORD: Levendahl and James *Biochim Biophys Acta* **26** 89 *1957*.] The *3',5'-di-O-acetyl derivative* crystd from aqueous EtOH has **m** 222°(dec), $[\alpha]_D^{18}$ -38° (**c** 0.3, 10% aq EtOH) [Hayes et al. *J Chem Soc* 808, 813 *1955*].

2'-Deoxyinosine *[890-38-0]* **M 252.2**, **m 206°(dec)**, **218-220°(dec)**, $[\alpha]_D^{25}$ **-21°** (**c** 2, N NaOH), $[\alpha]_D^{21.5}$ (**c** 1, H_2O), $pK_{Est(1)}$~ **8.9**, $pK_{Est(2)}$~ **12.4**. Purified by recrystn from H_2O. [Brown and Lythgoe *J Chem Soc* 1990 *1950*; UV: : MacNutt *Biochem J* **50** 384 *1952*.]

5-Deoxy-5-(methylthio)adenosine *[2457-80-9]* **M 297.3**, **m 210-213°(dec)**, **211°**, **212°**, **213-214°**, $[\alpha]_D^{20}$ **-23.7°** (**c** 0.02, pyridine), $[\alpha]_D^{20}$ **-8°** (**c** 1, 5% aq NaOH), $[\alpha]_D^{25}$ **+15°** (**c** 0.4-1.0, **0.3N aq AcOH**), pK_{Est} ~**3.5**. It has been recrystd from H_2O and sublimed at 200°/0.004mm. [v.Euler and Myrbäck *Z physiol Chem* **177** 237 *1928*; Weygand and Trauth *Chem Ber* **84** 633 *1951*; Baddiley et al. *J Chem Soc* 2662 *1953*.] The *hydrochloride* has **m** 161-162° [Kuhn and Henkel *Hoppe Seyler's Z Physiol Chem* **269** 41 *1941*]. The *picrate* has **m** 183°(dec) (from H_2O).

Deoxyribonucleic acid (from plasmids). Purified by two buoyant density ultracentrifugations using ethidium bromide-CsCl. The ethidium bromide was extracted with Et_2O and the DNA was dialysed against buffered EDTA and lyophilised. [Marmur and Doty *J Mol Biol* **5** 109 *1962*; Guerry et al. *J Bacteriol* **116** 1064 *1973*.] See p. 504.

3'-Deoxythymidine   {2',3'-dideoxythymidine, 1-[(2r)-5c-hydroxymethyltetrahydro(2r)-furyl]-5-methylpyrimidine-2,4-dione} *[3416-05-5]* **M 226.2**, **m 145°**, **149-151°**, $[\alpha]_D^{26}$ **+18°** (**c** 1, H_2O), pK_{Est} ~ **9.2**. Recrystd from Me_2CO + MeOH. [Michelson and Todd *J Chem Soc* 816 *1955*.]

2'-Deoxyuridine [1-(β-D-*erythro*-2-deoxypentofuranosyl)-1*H*-pyrimidine-2,4-dione] *[951-78-0]* **M 228.2**, **m 163°**, **163-163.5°**, **165-167°** **167°**, $[\alpha]_D^{26}$ **+30°** (**c** 2, H_2O), $[\alpha]_D^{22}$ **+50°** (**c** 1, N NaOH), **pK^{25} 9.3**. Forms needles from absolute EtOH or 95% EtOH. [Dekker and Todd *Nature* **166** 557 *1950*; Brown et al. *J Chem Soc* 3035 *1958*; NMR Jardetzky *J Am Chem Soc* **83** 2919 *1961*; Fox and Shugar *Biochim Biophys Acta* **9** 369 *1952*; UV: MacNutt *Biochem J* **50** 384 *1952*.]

3'-Deoxyuridine   {1-[(2R)-5c-hydroxymethyltetrahydro(2r)furyl]-5-methylpyrimidin-2,4-dione, 2'.3'-dideoxythymidine} *[7057-27-4, 3416-05-5]* **M 226.2**, **m 149-151°**, $[\alpha]_D^{20}$ **+18°** (**c** 1, H_2O), pK_{Est} ~ **9.3**. Recrystd from Me_2CO + MeOH and dried in a vacuum. [Michelson and Todd *J Chem Soc* 816 *1955*.]

Dermatan sulfate (condroitin sulfate B from pig skin) *[54328-33-5 (Na salt)]*. Purified by digestion with papain and hyaluronidase, and fractionation using aqueous EtOH. [Gifonelli and Roden *Biochem Prep* **12** 1 *1968*.]

Desthiobiotin *[533-48-2]* **M 214.3**, **m 156-158°**, $[\alpha]_D^{20}$ **+10.5°** (**c** 2, H_2O), pK_{Est} ~**2.8**. Crystd from H_2O or 95% EtOH.

Dextran *[9004-54-0]* **M_r 6,000-220,000**. Solutions keeps indefinitely at room temperature if 0.2mL of Roccal (10% alkyldimethylbenzylammonium chloride) or 2mg phenyl mercuric acetate are added per 100mL solution. [Scott and Melvin *Anal Biochem* **25** 1656 *1953*.]

Diacetone-D-Glucose (1,2:5,6-di-*O*-isopropylidene-α-D-glucofuranoside) *[582-52-5]* **M 260.3**, **m 107-110°**, **110.5°**, **111-113°**, **112°**, $[\alpha]_D^{15}$ **-18.4°** (**c** 1, H_2O). It crystallises from Et_2O, (needles), pet ether or $^*C_6H_6$ and sublimes *in vacuo*. It is sol in 7 vols of H_2O and 200 vols of pet ether at their boiling points. The solubility in H_2O at 17.5° is 4.3%. It pptes from aq solns on basification with NaOH. [Schmid and Karrer *Helv Chim Acta* **32** 1371 *1949*; Fischer and Rund *Chem Ber* **49** 90, 93 *1916*; IR: Kuhn *Anal Chem* **22** 276 *1950*.]

N,N'-Diacetylchitobiose (2-acetyl-O^4-[2-acetylamino-2-deoxy-β-D-glucopyranosyl]-2-deoxy-D-glucose) *[35061-50-8]* **M 424.4, m 245-247°(dec), 251.5-252.5°, 260-262°, $[\alpha]_D^{25}$ +39.5° (extrapolated) → +18.5° (after 60 min, c 1, H_2O).** Recrystd from aqueous MeOH or aqueous EtOH + 1,2-dimethoxyethane. [Zilliken et al. *J Chem Soc* **77** 1296 *1955*.]

1,8-Diazafluorenone (cyclopenta[1.2-*b*:4,3-*b'*]dipyridin-9-one) *[54078-29-4]* **M 182.2, m 205°, 229-231°, pK_{Est} ~ 2.6.** Recrystd from Me_2CO. The *oxime* has **m** 119-200°. [Druey and Schmid *Helv Chim Acta* **33** 1080 *1950*.]

Di- and tri-carboxylic acids. Resolution by anion-exchange chromatography. [Bengtsson and Samuelson *Anal Chim Acta* **44** 217b *1969*.]

Digitonin *[11024-24-1]* **M 1229.3, m >270°(dec), $[\alpha]_{546}^{20}$ -63° (c 3, MeOH).** Crystd from aqueous 85% EtOH or MeOH/diethyl ether.

Digitoxigenin *[143-62-4]* **M 374.5, m 253°, $[\alpha]_{546}^{20}$ +21° (c 1, MeOH).** Crystd from aqueous 40% EtOH.

D(+)-Digitoxose *[527-52-6]* **M 148.2, m 112°, $[\alpha]_{546}^{20}$ +57° (c 1, H_2O).** Crystd from MeOH/diethyl ether, or ethyl acetate.

Dihydrofolate reductase (from *Mycobacterium phlei*) *[9002-03-3]* **M_r ~18,000 [EC 1.5.1.3].** Purified by ammonium sulfate pptn, then fractionated on Sephadex G-75 column, applied to a Blue Sepharose column and eluted with 1mM dihydrofolate. [Al Rubeai and Dole *Biochem J* **235** 301 *1986*.]

7,8-Dihydrofolic acid (7,8-dihydropteroyl-L-glutamic acid, DHFA) *[4033-27-6]* **M 443.4, pK_1 2.0 (basic 10-NH), pK_2 2.89 (2-NH_2), pK_3 3.45 (α-CO_2H), pK_4 4.0 (basic 5N), pK_5 4.8 (γ-CO_2H), pK_6 9.54 (acidic 3NH).** Best purified by suspending (1g mostly dissolved)) in ice-cold sodium ascorbate (300mL of 10% at pH 6.0 [prepared by adjusting the pH of 30g of sodium ascorbate in 150mL of H_2O by adding 1N NaOH dropwise using a glass electrode till the pH is 6.0]). This gave a clear solution with pH ~5. While stirring at 0° add N HCl dropwise slowly (0.1mL/min) until the pH drops to 2.8 when white birefringent crystals separate. These are collected by centrifugation (1000xg for 5min), washed 3x with 0.001N HCl by centrifugation and decantation. The residue is then dried in a vacuum (0.02mm) over P_2O_5 (change the P_2O_5 frequently at first) and KOH at 25° in the dark. After 24hours the solid reaches constant weight.
For the assay of *dihydrofolate reductase* (see below): suspend ~66.5mg of DHFA in 10mL of 0.001M HCl containing 10mM dithiothreitol (DTT stock made from 154mg in 10mL H_2O making 0.1M), shake well and freeze in 400µL aliquots. Before use mix 400µL of this suspension with 0.1M DTT (200µL, also made in frozen aliquots), and the mixture is diluted with 200µL of 1.5M Tris-HCl pH 7.0 and 1.2mL of H_2O (making a total volume of ~2mL) to give a clear solution. To estimate the concentration of DHFA in this solution, dilute 20µL of this solution to 1mL with 0.1M Tris-HCl pH 7.0 and read the OD at 282nm in a 1cm pathlength cuvette. ε at 282nm is 28,000$M^{-1}cm^{-1}$. [Reyes and Rathod *Methods Enzymol* **122** 360 *1986*.]

Dihydropteridine reductase (from sheep liver) *[9074-11-7]* **M_r 52,000 [EC 1.6.99.7].** Purified by fractionation with ammonium sulfate, dialysed *versus* tris buffer, adsorbed and eluted from hydroxylapatite gel. Then run through a DEAE-cellulose column and also subjected to Sephadex G-100 filtration. [Craine et al. *J Biol Chem* **247** 6082 *1972*.]

Dihydropteridine reductase (from human liver) *[9074-11-7]* **M_r 52,000 [EC 1.6.99.7].** Purified to homogeneity on a naphthoquinone affinity adsorbent, followed by DEAE-Sephadex and CM-Sephadex chromatography. [Firgaira, Cotton and Danks, *Biochem J* **197** 31 *1981*.] [For other dihydropteridine reductases see Armarego et al. *Med Res Rev* **4**(3) 267 *1984*.]

DL-*erythro*-Dihydrosphingosine (dl-*erythro*-2-aminooctadecan-1,3-diol) *[3102-56-5]* **M 301.5, m 85-86°, 85-87°, pK_{Est} ~ 8.8.** Purified by recrystn from pet ether-EtOAc or $CHCl_3$. The (±)-*N-dichloroacetyl* derivative has **m** 142-144° (from MeOH). [Shapiro et al. *J Am Chem Soc* **80** 2170 *1958*; Shapiro and Sheradsky *J Org Chem* **28** 2157 *1963*.] The D-isomer crystallises from pet ether-Et_2O and has **m** 78.5-79°, $[\alpha]_{546}^{28}$ +6° ($CHCl_3$ + MeOH, 10:1). [Grob and Jenny *Helv Chim Acta* **35** 2106 *1953*, Jenny and Grob *Helv Chim Acta* **36** 1454 *1953*.]

Dihydrostreptomycin sesquisulfate *[5490-27-7]* **M 461.4, m 250°(dec), 255-265°(dec), $[\alpha]_D^{20}$ -92.4° (c 1, H_2O), $pK_{Est(1)}$~ 9.5 (NMe), $pK_{Est(2,3)}$~ 13.4 (guanidino).** It crystallises from H_2O with MeOH, *n*-BuOH or methyl ethyl ketone. The crystals are not hygroscopic like the amorphous powder, however both forms are soluble in H_2O but the amorphous solid is about 10 times more soluble than the crystals. The *free base* also crystallises from H_2O-Me_2CO and has $[\alpha]_D^{26}$ -92° (aqueous solution pH 7.0). [Solomons and Regina *Science* **109** 515 *1949*; Wolf et al. *Science* **109** 515 *1949*; McGilveray and Rinehart *J Am Chem Soc* **87** *4003* 1956].

3-(3,4-Dihydroxyphenyl)-L-alanine **(DOPA, EUODOPA)** *[59-92-7, 5796-17-8]* **M 197.2, m 275°(dec), 267-268°(dec), 284-286°(dec), ~295°(dec), $[\alpha]_D^{13}$ -13.1° (c 5.12, N HCl), pK_1^{25} 2.32 (CO_2H), pK_2^{25} 8.72 (NH_2), pK_3^{25} 9.96 (OH), pK_4^{25} 11.79 (OH).** Likely impurities are vanillin, hippuric acid, 3-methoxytyrosine and 3-aminotyrosine. Recryst from large vols of H_2O as colourless white needles; solubility in H_2O is 0.165%, but it is insoluble in EtOH, $^*C_6H_6$, $CHCl_3$, and EtOAc. Also crystd by dissolving in dilute HCl and adding dilute ammonia to give pH 5, under N_2. Alternatively, crystd from dil aqueous EtOH. It is rapidly oxidised in air when moist, and darkens; particularly in alkaline solution. Dry in a vacuum at 70° in the dark, and store in a dark container preferably under N_2. λ_{max} 220.5nm (log ε 3.79) and 280nm (log ε 3.42) in 0.001N HCl. [Yamada et al. *Chem Pharm Bull Jpn* **10** 693 *1962*; Bretschneider et al. *Helv Chim Acta* **56** 2857 *1973*; NMR: Jardetzky and Jardetzky *J Biol Chem* **233** 383 *1958*.]

3,4-Dihydroxyphenylalanine-containing proteins. Boronate affinity chromatography is used in the selective binding of proteins containing 3,4-dihydroxyphenylalanine to a *m*-phenylboronate agarose column and eluting with 1M NH_4OAc at pH 10. [Hankus et al. *Anal Biochem* **150** 187 *1986*.]

3-(3,4-Dihydroxyphenyl)-2-methyl-L-alanine **[methyldopa, 2-amino-3-(3,4-dihydroxy-phenyl)-2-methylpropionic acid]** *[555-30-6]* **M 238.2, m >300°, 300-301°(dec), pK_1^{25} 2.2, pK_2^{25} 9.2, pK_3^{25} 10.6, pK_4^{25} 12.0.** Recrystd from H_2O. [Reinhold et al. *J Org Chem* **33** 1209 *1968*.] The L-*isomer* forms a sesquihydrate from H_2O **m** 302-304° (dec), and the anhydrous crystals are *hygroscopic*, $[\alpha]_D^{23}$ -4.0° (c 1, 0.1N HCl), $[\alpha]_{546}$ +154.5° (c 5, $CuSO_4$ solution). It has λ_{max} 281nm (ε 2780). Solubility in H_2O at 25° is ~10mg/mL and the pH of an aqueous solution is ~5.0. It is almost insoluble in most organic solvents. [Stein et al. *J Am Chem Soc* **77** 700 *1955*.]

(±)-7-(2,3-Dihydroxypropyl)theophylline **(Diprophylline, Dyphylline)** *[479-18-5]* **M 254.3, m 158°, 160-164°, 161°, 161-164°, pK_{Est} ~ 8.7.** Recrystd from EtOH or H_2O. Solubility in H_2O is 33% at 25°, in EtOH it is 2% and in $CHCl_3$ it is 1%. λ_{max} (H_2O) 273nm (ε 8,855). [Roth *Arch Pharm* **292** 234 *1959*.] The *4-nitrobenzoyl* derivative has **m** 178° [Oshay *J Chem Soc* 3975 *1956*].

3,5-Diiodo-L-thyronine (3,5-diiodo-4-[4-hydroxyphenoxy]-1-phenylalanine) *[1041-01-6]* **M 525.1, m 255°(dec), 255-257°(dec), $[\alpha]_D^{22}$ +26° [2N HCl-EtOH (1:2)], pK_1^{20} 3.25, pK_2^{20} 5.32, pK_3^{20} 9.48.** Recrystd from EtOH. [Chambers et al. *J Chem Soc* 3424 *1949*.]

3,5-Diiodo-L-tyrosine dihydrate *[300-39-0]* **M 469.0, m 199-210°, 202°(dec), $[\alpha]_D^{20}$ +2.89° (c 4.9, 4% HCl), pK_1^{25} 2.12, pK_2^{25} 6.48, pK_3^{25} 7.82.** It forms crystals from H_2O [solubility (g/L): 0.204 at 0°, 1.86 at 50°, 5.6 at 75° and 17.0 at 100°]. Also recrysts from 50% or 70% EtOH. When boiled in EtOH the crystals swell and on further boiling a gelatinous ppte is formed [Harrington *Biochem J* **22** 1434 *1928* ; Jurd *J Am Chem Soc* **77** 5747 *1955*]. Also crystd from cold dilute ammonia by adding acetic acid to pH 6.

1,2-Dilauroyl-sn-glycero-3-phosphoethanolamine (±-dilauroyl-α-cephalin, 3-sn-phosphat-idylethanolamine 1,2-didodecanoyl) *[59752-57-7]* M 579.8, m 210°, $pK_{Est(1)}$~ 5.8 (PO₄H), $pK_{Est(2)}$~ 10.5 (NH₂). Recrystd from EtOH or tetrahydrofuran. [Bevan and Malkin *J Chem Soc* 2667 *1951*; IR: Bellamy and Beecher *J Chem Soc* 728 *1953*.]

1-(3-Dimethylaminopropyl)-3-ethylcarbodiimide hydrochloride [1-ethyl-3-(3-dimethyl-aminopropyl) carbodiimide hydrochloride] see entry on p. 212 in Chapter 4.

1,2-Dimyristoyl-sn-glycero-3-phosphocholine monohydrate (dimyristoyl-L-α-lecithin) *[18194-24-6]* M 696.0, $[\alpha]_D^{24}$ +7° (c 8, EtOH-CHCl₃ 1:1 for α₁ form), pK_{Est} ~ 5.8 (PO₄). Three forms α₁, α₂ and β'. Recrystd from aqueous EtOH or EtOH-Et₂O. Solubility at 22-23° in Et₂O is 0.03%, in Me₂CO it is 0.06% and in pyridine it is 1.3%. [Baer and Kates *J Am Chem Soc* 72 942 *1950*; Baer and Maurakas *J Am Chem Soc* 74 158 *1952*; IR: Marinetti and Stotz *J Am Chem Soc* 76 1347 *1954*.] The *S-isomer* with 1 H₂O is recrystd from 2,6-dimethylheptan-4-one and has m 226-227° (sintering at 90-95°), and $[\alpha]_D$ -7° (c 6, MeOH-CHCl₃ 1:1). [Baer and Martin *J Biol Chem* 193 835 *1951*.]

(±)-1,2-Dimyristoyl-sn-glycero-3-phosphoethanolamine (dimyristoyl-α--kephalin) *[998-07-2]* M 635-9, m 207°, $pK_{Est(1)}$~ 5.8 (PO₄H), $pK_{Est(2)}$~ 10.5 (NH₂). Recrystd from EtOH [Bevan and Malkin *J Chem Soc* 2667 *1951*]. The *R-isomer* has m 195-196° (sintering at 130-135°) after recrystn from CHCl₃-MeOH, $[\alpha]_D^{26}$ +6.7° (c 8.5, CHCl₃-AcOH 9:1). [Baer *Can J Biochem Physiol* 35 239 *1957*; Baer et al. *J Am Chem Soc* 74 152 *1952*.]

S-**1,2-Dipalmitin** *[761-35-3]* M 568.9, m 68-69° $[\alpha]_D^{20}$-2.9° (c 8, CHCl₃), $[\alpha]_{546}^{20}$ +1.0° (c 10, CHCl₃/MeOH, 9:1). Crystd from chloroform/pet ether.

R-**Dipalmitoyl-sn-glycero-3-phosphatidic acid** *[7091-44-3]* M 648.9, $[\alpha]_D^{26}$ + 4° (c 10, CHCl₃), $pK_{Est(1)}$~ 1.6, $pK_{Est(2)}$~ 6.1. Recrystd from Me₂CO at low temp. At 21° it is soluble in *C₆H₆ (4.2%), pet ether (0.01%), MeOH (2%), EtOH (2.5%), AcOH (1.3%), Me₂CO (1.76%), and Et₂O (1.5%). [Baer *J Biol Chem* 189 235 *1951*.]

R-**1,2-Dipalmitoyl-sn-glycero-3-phosphocholine monohydrate** (dipalmitoyl-α-L-lecithin) *[63-89-8]* M 752.1, m sinters at 120°, $[\alpha]_D^{25}$ +7.0° (c 5.6, abs CHCl₃), pK_{Est} ~ 5.8 (PO₄). It has three crystn forms α₁, α₂ and β' which change at 60-70° and at 229° respectively. In order to obtain a fine powder, ~2 g are dissolved in CHCl₃ (15mL) and pet ether (b 35-60°) is added and the soln evaporated to dryness *in vacuo* <20°, and then dried at 0.1mm over CaCl₂. [Baer and Maurukas *J Am Chem Soc* 74 158 *1952*; Baer and Kates *J Biol Chem* 185 615 *1950*.]

d,l-**βγ-Dipalmitoylphosphatidyl choline** *[2797-68-4]* M 734.1, m 230-233°, pK_{Est} ~ 5.8 (PO₄). Recrystd from chloroform and dried for 48h at 10⁻⁵ torr [O'Leary and Levine *J Phys Chem* 88 1790 *1984*].

Dipeptidyl aminopeptidase (from rat brain) *[9031-94-1]* [EC 3.4.11.10]. Purified about 2000-fold by column chromatography on CM-cellulose, hydroxylapatite and Gly-Pro AH-Sepharose. [Imai et al. *J Biochem (Tokyo)* 93 431 *1983*.]

1,2-Distearoyl-sn-glycerol *[1429-59-0]* M 625.0. The dl-form recrystallises from CHCl₃-pet ether (b 40-60°), m 59.5° (α form) and 71.5-72.5° (β form). Recrystn from solvents (e.g. EtOH, MeOH, toluene, Et₂O) gives the higher melting form and resolidification gives the lower melting forms. [IR: Chapman *J Chem Soc* 4680 *1958*, 2522 *1956*; .] The *S-isomer* is recrystd from CHCl₃-pet ether and has m 76-77°, $[\alpha]_D^{24}$ -2.8° (c 6, CHCl₃). [Baer and Kates *J Am Chem Soc* 72 942 *1950*.]

1,2-Distearoyl-sn-glycero-3-phosphoethanolamine (distearoyl-α-kephalin) *[1069-79-0]* M 748.1, m 180-182° (*R*-form, sintering at 130-135°), m 196° (± form), $pK_{Est(1)}$~ 5.8 (PO₄H), $pK_{Est(2)}$~ 10.5 (NH₂). The *R*-form is recrystd from CHCl₃-MeOH and the ±-form is recrystd from EtOH. [Bevan and Malkin *J Chem Soc* 2667 *1951*; Baer *Can J Biochem Physiol* 81 1758 *1959*.]

Dolichol (from pig liver) *[11029-02-0]* C_{80}-C_{105} **polyprenols**. Cryst 6 times from pet ether/EtOH at -20°C. Ran as entity on a paper chromatogram on paraffin impregnated paper, with acetone as the mobile phase. [Burgos *Biochem J* **88** 470 *1963*.]

Domoic acid [4-(2-carboxyhexa-3,5-dienyl)-3-carboxymethylproline] *[14277-97-5]* **M 311.3, m 215°, 217°, $[\alpha]_D^{20}$ -108° (c 1,H$_2$O), pK$_1$ 2.20 (2-CO$_2$H), pK$_2$ 3.72 (CO$_2$H), pK$_3$ 4.93 (3-CH$_2$CO$_2$H), pK$_4$ 9.82 (NH).** The acid (~300 mg) is purifed on a Dowex 1 column (3.5 x 40 mm, 200-400 mesh, acetate form), washed with H$_2$O until neutral, then eluted with increasing concentrations of AcOH (8L) from 0 to 0.25M. The fraction containing domoic acid (in 50mL) is collected, evaporated to dryness under reduced pressure and recrystd from aqueous EtOH. Glutamate and Kainate receptor agonist. [Impellizzeri et al. *Phytochemistry* **14** 1549 *1975*; Takemoto and Diago *Arch Pharm* **293** 627 *1960*.]

DNA (deoxyribonucleic acids). The essential structures of chromosomes are DNA and contain the genetic "blue print" in the form of separate genes. They are made up of the four deoxyribonucleic acids (nucleotides): adenylic acid, guanylic acid, cytidylic acid and thymidylic acid (designated A, G, C, T respectively) linked together by their phosphate groups in ester bonds between the 3' and 5' hydroxy groups of the 2'-deoxy-D-ribose moiety of the nucleotides. The chains form a double stranded spiral (helix) in which the two identical nucleotide sequences run antiparallel with the heterocyclic bases hydrogen bonded (A..T, G..C) forming the "ladder" between the strands. Short sequences of DNA are available commercially, are commercially custom made or synthesised in a DNA synthesiser and purified by HPLC. Their purity can be checked by restriction enzyme cleavage followed by gel electophoresis, or directly by gel electrophoresis or analytical HPLC. Commercial DNAs are usually pure enough for direct use but can be further purified using commercially available kits involving binding to silica or other matrices and eluting with tris buffers.

Dopamine-β-hydroxylase (from bovine adrenal medulla) *[9013-38-1]* **M$_r$ ~290,000, [EC 1.14.17.1].** The Cu-containing glycoprotein enzyme has been isolated by two procedures. The first is an elaborate method requiring extraction, two (NH$_4$)$_2$SO$_4$ fractionations, calcium phosphate gel filtration, EtOH fractionation, DEAE-cellulose chromatography followed by two Sephadex-G200 gel filtrations giving enzyme with a specific activity of 65 Units/mg. [Friedman and Kaufman *J Biol Chem* **240** 4763 *1965*; Rush et al. *Biochem Biophys Res Commun* **61** 38 *1974*.] The second procedure is much gentler and provides good quality enzyme. Sedimented chromaffin vesicles were lysed in 10 volumes of 5mM K-phosphate buffer pH 6.5 using a loosely fitting Teflon-glass homogeniser. The mixture is centrifuged at 40,000xg/0.5 h and the supernatant is diluted with an equal volume of 100mM phosphate buffer (pH 6.5) containing 0.4M NaCl. This lysate is applied to a concanavelin A-Sepharose column (4 x 0.7cm) which had been equilibrated with 50 mM of phosphate buffer (pH 6.5 + 0.2M NaCl) with a flow rate of ~ 0.3 mL/min. The column is washed thoroughly with the buffer until OD$_{280nm}$ is 0.005. The enzyme is then eluted with the same buffer containing 10% α-methyl-D-mannoside (flow rate 0.1 mL/min) and the enzyme is collected in twenty column volumes. The pooled eluate is concentrated by ultrafiltration in an Amicon Diaflo stirrer cell using an XM100A membrane. The concentrated enzyme is dialysed against 50 mM phosphate buffer (pH 6.5) containing 0.1% NaCl. The enzyme gives one band (+ two very weak band) on disc gel electrophoresis indicating better than 93% purity (67% fold purification) and has a specific activity of 5.4Units/mg. [Rush et al. *Biochem Biophys Res Commun* **57** 1301 *1974*; Stewart and Klinman *Ann Rev Biochem* **57** 551 *1988*.]

Ellipticine   (5,11-dimethylpyrido[4,3:*b*]carbazole) *[519-23-3]* **M 246.3, m 311-315°(dec), 312-314°(dec), pK 5.78 (80% aq methoxyethanol).** This DNA intercalator is purified by recrystn from CHCl$_3$ or MeOH and dried *in vacuo*. The UV λ_{max} values in aqueous EtOH-HCl are at 241, 249, 307, 335 and 426nm. [Marini-Bettolo and Schmutz *Helv Chim Acta* **42** 2146 *1959*.] The *methiodide* has **m** 360°(dec), with UV λ_{max} (EtOH-KOH) 223, 242, 251, 311, 362 and 432nm. [Goodwin et al. *J Am Chem Soc* **81** 1903 *1959*.]

Enniatin A *[11113-62-5]* **M 681.9, m 122-122.5°, $[\alpha]_D^{18}$ -92° (c 0.9, CHCl$_3$).** A cyclic peptidic ester antibiotic which is recrystd from EtOH/water but is deactivated in alkaline soln. [Ovchinnikov and Ivanov in *The Proteins* (Neurath and Hill eds) Academic Press, NY, Vol V pp. 365 and 516 *1982*.]

(-)-Ephedrine **(1R,2S-2-methylamino-1-phenylpropanol)** *[299-42-3]* **M 165.2, m ~34°, 36°, 38.1°, b 126-129°/7mm, 225-227°/760mm, d^{22} 1.0085, [α]$_D^{26}$ -42° (c 4, 3% HCl), [α]$_D^{22.5}$ +15.1° (c 0.8, H$_2$O), -9.36° (c 3, MeOH), pK^{22} 9.58 (pK^{25} 8.84 in 80% aq methoxyethanol).** Purified by vacuum distn (dehydrates) and forms waxy crystals or granules, and may pick up 0.5 H$_2$O. The presence of H$_2$O raises its **m** to 40°. [Moore and Taber *J Amer Pharm Soc* 24 211 *1935*.] The anhydrous base recrystd from dry ether [Fleming and Saunders *J Chem Soc* 4150 *1955*]. It gradually decomposes on exposure to light and is best stored in an inert atmosphere in the dark (preferably at -20°). Sol in H$_2$O is 5%, in EtOH it is 1% and it is soluble in CHCl$_3$, Et$_2$O and oils. It has pKa values in H$_2$O of 10.25 (0°) and 8.69 (60°) [Everett and Hyne *J Chem Soc* 1136 *1958*; Prelog and Häflinger *Helv Chim Acta* 33 2021 *1950*] and pKa^{25} 8.84 in 80% aqueous methoxyethanol [Simon *Helv Chim Acta* 41 1835 *1958*]. The *hydrochloride* has **m** 220° (from EtOH-Et$_2$O) and [α]$_D^{20}$ -38.8° (c 2, EtOH). [IR: Chatten and Levi *Anal Chem* 31 1581 *1959*.] The anhydrous base crystallises from Et$_2$O [Fleming and Saunders *J Chem Soc* 4150 *1955*].

(+)-Ephedrine hydrochloride **(1S-2R-2-methylamino-1-phenylpropan-1-ol hydrochloride)** *[24221-86-1]* **M 201.7, m 216-219°, [α]$_D^{20}$ +34° (c 11.5, H$_2$O).** Recrystd from EtOH-Et$_2$O. The *free base* recrystallises from *C$_6$H$_6$ with **m** 40-41° (Skita et al. *Chem Ber* 66 974 *1933*].

Erythromycin A *[114-07-8]* **M 733.9, m 133-135°(dec), 135-140°, 137-140°, [α]$_D^{20}$ -75° (c 2, EtOH), pK 8.9.** It recrystallises from H$_2$O to form hydrated crystals which melt at *ca* 135-140°, resolidifies and melts again at 190-193°. The **m** after drying at 56°/8mm is that of the **anhydrous** material at 137-140°. Its solubility in H$_2$O in ~2mg/mL. The *Hydrochloride* has **m** 170°, 173° (from aq EtOH, EtOH-Et$_2$O). [Flynn et al. *J Am Chem Soc* 76 3121 *1954*; constitution : Wiley et al. *J Am Chem Soc* 79 6062 *1957*].

β-Estradiol **(1,3,5-estratrien-3,17β-diol)** *[50-28-2]* **M 272.4, m 173-179°, 176-178°, [α]$_D^{20}$ +76° to +83° (c 1, dioxane).** Purified by chromatography on SiO$_2$ (toluene-EtOAc 4:1) and recrystd from CHCl$_3$-hexane or 80% EtOH. It is stable in air and insoluble in H$_2$O and is ppted by digitonin. UV λ$_{max}$ at 225 and 280 nm. [Oppolzer and Roberts *Helv Chim Acta* 63 1703 *1980*.]

β-Estradiol-6-one **(1,3,5-estratriene-3,17β-diol-6-one)** *[571-92-6]* **M 359.4, m 278-280°, 281-283°, [α]$_D^{20}$ +4.2° (c 0.7, EtOH).** It forms plates from EtOH. The *3,17-diacetate* has **m** 173-175° after recrystn from aqueous EtOH. [Longwell and Wintersteiner *J Biol Chem* 133 219 *1940*.] The UV has λ$_{max}$ 255 and 326nm in EtOH [Slaunwhite et al. *J Biol Chem* 191 627 *1951*].

Ethidium bromide *[1239-45-8]* **M 384.3, m 260-262°.** Crystd from MeOH or EtOH [Lamos et al. *J Am Chem Soc* 108 4278 *1986*]. Sol in H$_2$O is 1%. **POSSIBLE CARCINOGEN.**

Ethoxyquin **(1,2-dihydro-6-ethoxy-2,2,4-trimethylquinoline)** *[91-53-2]* **M 217.3, b 169°/12-13mm, d$_4^{20}$ 1.000, pK$_{Est}$ ~ 5.8.** Purified by fractional distn *in vacuo* and solidifies to a glass. [Knoevenagel *Chem Ber* 54 1723, 1730 *1921*]. The *methiodide* has **m** 179° (from EtOH) and the *1-phenylcarbamoyl* derivative has **m** 146-147° (from EtOH). [Beaver et al. *J Am Chem Soc* 79 1236 *1957*.]

17-α-Ethynylestradiol *[57-63-6]* **M 296.4, m 141-146°, 145-146°, [α]$_D^{20}$ +4° (c 1, CHCl$_3$).** It forms a hemihydrate on recrystn from MeOH-H$_2$O. It dehydrated on melting and re-melts on further heating at **m** 182-184°. UV λ$_{max}$ at 281nm (ε 2040) in EtOH. Solubility is 17% in EtOH, 25% in Et$_2$O, 20% in Me$_2$CO, 25% in dioxan and 5% in CHCl$_3$. [Petit and Muller *Bull Soc Chim Fr* 121 *1951*.] The *diacetyl* derivative has **m** 143-144° (from MeOH) and [α]$_D^{20}$ +1° (c 1, CHCl$_3$) [Mills et al. *J Am Chem Soc* 80 6118 *1958*].

Exonucleases. Like the endonucleases they are restriction enzymes which act at the 3' or 5' ends of linear DNA by hydrolysing off the nucleotides. Although they are highly specific for hydrolysing nucleotides at the 3' or 5' ends of linear DNA, the number of nucleotides cleaved are time dependent and usually have to be estimated from the time allocated for cleavage. Commercially available exonucleases are used without further purification.

Farnesol   (*trans-trans*-3,7,11-trimethyl-2,6,10-dodecatrien-1-ol)   *[106-28-5]* **M 222.4,**
b 111°/0.35mm, 126-127°/0.5mm, 142-143°/2mm, d_4^{20} 0.8871, n_D^{25} 1.4870. Main impurity is
the *cis-trans* isomer. Purified by gas chromatography using a 4ft x 0.125in 3%OV-1 column at 150°. [Corey
et al. *J Am Chem Soc* **92** 6637 *1970*; Popjak et al. *J Biol Chem* **237** 56 *1962*.] Also purifed through a 14-in
Podbielniak column at 11°/0.35mm (see p. 141). Alternatively it has been purified by gas chromatography
using SF96 silicone on Fluoropak columns or Carbowax 20M on Fluoropak or base-washed 30:60 firebrick (to
avoid decomp of alcohol, prepared by treating the firebrick with 5N NaOH in MeOH and washed with MeOH to
pH 8) at 210° with Helium carrier gas at 60 mL/min flow rate. The *diphenylcarbamoyl* derivative has **m** 61-63°
(from MeOH) and has IR band at 3500 cm^{-1}. [Bates et al. *J Org Chem* **28** 1086 *1963*.]

Farnesyl pyrophosphate *[13058-04-3; E,E: 372-97-4]* **M 382.3, $pK_{Est(1)}$~<2, $pK_{Est(2)}$~<2,**
$pK_{Est(3)}$~3.95, $pK_{Est(4)}$~6.26. Purified by chromatography on Whatman No3 MM paper in a system of
isopropanol-isobutanol-ammonia-water (40:20:1:30) (v/v). Stored as the Li or NH$_4$ salt at 0°.

Ferritin (from human placenta) *[9007-73-2]* **M_r ~445,000 (Fe free protein).** The purification of
this major iron binding potein was achieved by homogenisation in water and precipitating with ammonium
sulfate, repeating the cycle of ultracentrifuging and molecular sieve chromatography through Sephadex 4B
column. Isoelectric focusing revealed a broad spectrum of impurities which were separated by ion-exchange
chromatography on Sephadex A-25 and stepwise elution. [Konijn et al. *Anal Biochem* **144** 423 *1985*.]

Fibrinogen (from human plasma) *[9001-32-5]* **M_r 341,000.** A protein made up of 2Aα,2Bβ and
2γ subunits connected by disulfide bridges. Possible impurity is plasminogen. Purified by glycine pptn
[Mosesson and Sherry *Biochemistry* **5** 2829 *1966*] to obtain fractions 1-2, then further purified [Blombäck and
Blombäck *Arkiv Kemi* **10** 415 *1956*] and contaminating plasminogen is removed by passage through a lysine-
Sepharose column. Such preparations were at least 95% clottable as determined by Mosesson and Sherry's
method (above ref.) in which the OD$_{280}$ was measured before and after clotting with 5 Units/mL of thrombin
(> 3000U/mg). All fibrinogen preps were treated with calf intestinal alkaline phosphatase to convert any
fibrinogen peptide-AP to fibrinogen peptide-A by removing serine-bound phosphate. Solutions are then
lyophilised and stored at -20°. [Higgins and Shafer *J Biol Chem* **256** 12013 *1981*.] It is sparingly soluble in
H$_2$O. Aqueous solns are viscous with isoelectric point at pH 5.5. Readily denatured by heating above 56° or by
chemical agents, e.g. salicylaldehyde, naphthoquinone sulfonates, ninhydrin or alloxan. [Edsall et al. *J Am*
Chem Soc **69** 2731 *1947*; Purification: Cama et al. *Naturwissenschaften* **48** 574 *1961*; Lorand and
Middlebrook *Science* **118** 515 *1953*; *cf.* Fuller in *Methods Enzymol* **163** 474 *1988*.]
For plasminogen-deficient fibrinogen from blood plasma, the anticoagulated blood was centrifuged and the
plasma was frozen and washed with saline solution. Treated with charcoal and freeze-thawed. Dialysed *versus*
Tris/NaCl buffer. [Maxwell and Nikel *Biochem Prep* **12** 16 *1968*.]

Fibronectin (from human plasma) *[86088-83-7]* **M_r ~220,000.** This glycoprotein contains 5-12%
of carbohydrate. It has been purified by glycine fractionation and DEAE-cellulose chromatography. Material is
dissolved in 0.25M Tris-phosphate buffer pH 7.0, diluted to 20% and glycine added gradually till 2.1M when the
temperature falls to below 15°. The ppte contains mainly fibrinogen. The supernatant is discarded and the ppte
is treated with an equal volume of H$_2$O, cooled (to 0°) and ppted by adding EtOH to 16% (v/v) at -4°. The ppte
contains some CI (Cold Insoluble) globulin, fibronectin and small quantities of other proteins. To remove these
the ppte is dissolved in 0.25M Tris-phosphate buffer (pH 7.0) *ca* 0.5% and purified by DEAE-cellulose
chromatography after diluting the buffer to 0.05M buffer. [Morrison et al. *J Am Chem Soc* **70** 3103 *1948*;
Mosesson and Umfleet *J Biol Chem* **245** 5728 *1970*; Mosesson and Amrani *Blood* **56** 145 *1980*; Akiyama and
Yamada *Adv Enzymol* **59** 51 *1987*.]

Flavin adenine dinucleotide (di-Na, 2H$_2$O salt, FAD) *[146-14-5]* **M 865.6, $[\alpha]_{546}$ -54° (c 1,**
H$_2$O). Small quantities, purified by paper chromatography using *tert*-butyl alcohols/water, cutting out the
main spot and eluting with water. Larger amounts can be ppted from water as the uranyl complex by adding a
slight excess of uranyl acetate to a soln at pH 6.0, dropwise and with stirring. The soln is set aside overnight in
the cold, and the ppte is centrifuged off, washed with small portions of cold EtOH, then with cold, peroxide-free
diethyl ether. It is dried in the dark under vacuum over P$_2$O$_5$ at 50-60°. The uranyl complex is suspended in
water and, after adding sufficient 0.01M NaOH to adjust the pH to 7, the ppte of uranyl hydroxide is removed by

centrifugation [Huennekens and Felton *Methods Enzymol* **3** 954 *1957*]. It can also be crystd from water. Should be kept in the dark. More recently it was purified by elution from a DEAE-cellulose (Whatman DE 23) column with 0.1M phosphate buffer pH 7, and the purity was checked by TLC. [Holt and Cotton, *J Am Chem Soc* **109** 1841 *1987*.]

Flavin mononucleotide (Na, 2H$_2$O salt, FMN) *[130-40-5]* M **514.4, pK$_1$ 2.1 (PO$_4$H$_2$), pK$_2$ 6.5 (PO$_4$H$^-$), pK$_3$ 10.3 (CONH), fluorescence λmax 530nm (870nm for reduced form).** Purified by paper chromatography using *tert*-butanol-water, cutting out the main spot and eluting with water. Also purified by adsorption onto an apo-flavodoxin column, followed by elution and freeze drying Crystd from acidic aqueous soln. [Mayhew and Strating *Eur J Biochem* **59** 539 *1976*.]

4-Fluoro-7-nitrobenzofurazan   (4-fluoro-7-nitrobenzo-2-oxa-1,3-diazole) *[29270-56-2]* M **183.1, m 52.5-53.5°, 53-56°, 53.5-54.5°.** Purified by repeated recrystn from pet ether (b 40-60°). On treatment with MeONa in MeOH it gave *4-methoxy-7-nitrobenzo-2-oxa-1,3-diazole* m 115-116°. [Nunno et al. *J Chem Soc (C)* 1433 *1970*.] It is a very good fluorophore for amino acids [Imai and Watanabe *Analyt Chim Acta* **130** 377 *1981*], as it reacts with primary and secondary amines to form fluorescent adducts with λ$_{ex}$ 470nm and λ$_{em}$ 530nm. It gives a *glycine derivative* with m 185-187° [Miyano et al. *Anal Chim Acta* **170** 81 *1985*].

4-Fluoro-3-nitrophenylazide *[28166-06-5]* M **182.1, m 53-55°, 54-56°.** Dissolve in Et$_2$O, dry over MgSO$_4$, filter, evaporate and recryst the residue from pet ether (b 20-40°) to give orange needles. Store in a stoppered container at ~0°. The NMR has δ 7.75 (m 1H) and 7.35 (m 2H) in CDCl$_3$. [Hagedorn et al. *J Org Chem* **43** 2070 *1978*.]

2-Fluorophenylalanine *[R(+)- 97731-02-7; S(-)- 19883-78-4]* M **183.2, m 226-232°, 231-234°, [α]$_D^{25}$ (+) and (-) 15° (c 2, H$_2$O pH 5.5), pK$_1^{24}$2.12, pK$_2^{24}$9.01.** Recryst from aqueous EtOH. The *hydrochloride* has m 226-231°(dec), and the *N-acetyl* derivative has m 147-149° (from aqueous EtOH). [Bennett and Nieman *J Am Chem Soc* **72** 1800 *1950*.]

4-Fluorophenylalanine *[R(+)- 18125-46-7; S(-)- 1132-68-9]* M **183.2, m 227-232°, [α]$_D^{25}$ (+) and (-) 24° (c 2, H$_2$O), pK$_1^{24}$2.13, pK$_2^{24}$9.05.** It is recrystd from aqueous EtOH. The R-*N-acetyl* derivative has m 142-145°, [α]$_D^{25}$ -38.6° (c 8, EtOH). [Bennett and Nieman *J Am Chem Soc* **72** 1800 *1950*.]

5-Fluoro-L-tryptophan monohydrate *[16626-02-1]* M **240.2, m >250°(dec), [α]$_D^{20}$ +5.5° (c 1, 0.1N HCl), pK$_{Est(1)}$~ 2.5 (CO$_2$H), pK$_{Est(2)}$~ 9.4 (NH$_2$), pK$_{Est(3)}$~16 (indole-NH).** Recrystd from aqueous EtOH.

5-Fluorouridine   (5-fluoro-1-β-D-ribofuranosyl-1H-pyrimidine-2,4-dione) *[316-46-1]* M **262.2, m 180-182°, 182-184°, [α]$_D^{20}$ +18° (c 1, H$_2$O), pK$_{Est(1)}$~ 8.0, pK$_{Est(2)}$~ 13.** Recrystd from EtOH-Et$_2$O and dried at 100° in a vacuum. UV: λ$_{max}$ 269nm (pH 7.2, H$_2$O), 270nm (pH 14, H$_2$O). [Liang et al. *Mol Pharmacol* **21** 224 *1982*.]

5-Fluorouracil (5-fluoropyrimidinedi-2,4-[1H,3H]-one) *[51-21-8]* M **130.1, m 282-283°(dec), 282-286°(dec), pK$_1^{25}$8.04, pK$_2^{25}$13.0.** Recrystd from H$_2$O or MeOH-Et$_2$O, and sublimed at 190-200°/0.1mm or 210-230°/0.5mm. UV: λ$_{max}$ 265-266nm (ε 7070). [Barton et al. *J Org Chem* **37** 329 *1972*; Duschinsky and Pleven *J Am Chem Soc* **79** 4559 *1957*.]

Fluram   (Fluorescamine, 4-phenyl-spiro[furan-2(3H)-1-phthalan]-3,3'-dione) *[38183-12-9]* M **278.3, m 153-155°, 154-155°.** A non-fluorescent reagent that reacts with primary amines to form highly fluorescent compounds. Purified by dissolving (~1g) in Et$_2$O-*C$_6$H$_6$ (1:1, 180 mL), wash with 1% aq NaHCO$_3$ (50mL), dry (Na$_2$SO$_4$), evaporate in a vacuum. Dissolve the residue in warm CH$_2$Cl$_2$ (5mL), dilute with Et$_2$O (12mL) and refrigerate. Collect the solid and dry in a vacuum. IR (CHCl$_3$): ν 1810, 1745, 1722, 1625 and 1600 cm^{-1}, and NMR (CDCl$_3$): δ 8.71 (s, -OHC=). [Weigele et al. *J Am Chem Soc* **94** 5927 *1972*, *J Org Chem* **41** 388 *1976*; *Methods Enzymol* **47** 236 *1977*.]

Folic acid   (FA, pteroyl-S-glutamic acid)   *[75708-92-8]* **M 441.4, m >250º(dec), $[\alpha]_D^{25}$ +23º (c 0.5, 0.1N NaOH), pK$_1$ 2.35 (protonation N10), pK$_2$ 2.75 (protonation N1), pK$_3$ 3.49 (α-CO$_2$H), pK$_4$ 4.65 (γ-CO$_2$H), pK$_5$ 8.80 (acidic N3).** If paper chromatography indicates impurities then recrystallise from hot H$_2$O or from dilute acid [Walker et al. *J Am Chem Soc* **70** 19 *1948*]. Impurities may be removed by repeated extraction with *n*-BuOH of a neutral aqueous solns of folic acid (by suspending in H$_2$O and adding N NaOH till the solid dissolves then adjusting the pH to ~7.0-7.5) followed by pptn with acid, filtration, and recrystn form hot H$_2$O. [Blakley *Biochem J* **65** 331 *1975*; Kalifa, Furrer, Bieri and Viscontini *Helv Chim Acta* **61** 2739 *1978*.] Chromatography on cellulose followed by filtration through charcoal has also been used to obtain pure acid. [Sakami and Knowles *Science* **129** 274 *1959*.] UV: λ_{max} 247 and 296nm (ε 12800 and 18700) in H$_2$O pH 1.0; 282 and 346nm (ε 27600 and 7200) in H$_2$O pH 7.0; 256, 284 and 366nm (ε 24600, 24500 and 8600) in H$_2$O pH 13 [Rabinowitz in *The Enzymes* (Boyer et al. Eds **2** 185 *1960*].

Follicle Stimulating Hormone   **(FSH, follitropin)**   *[9002-68-0]* **M$_r$ ~36,000.** Purified by Sephadex G100 gel filtration followed by carboxymethyl-cellulose with NH$_4$OAc pH 5.5. The latter separates luteinising hormone from FSH. Solubility in H$_2$O is 0.5%. It has an isoelectric point of 4.5. A soln of 1mg in saline (100mL) can be kept at 60º for 0.5h. Activity is retained in a soln at pH 7-8 for 0.5h at 75º. The activity of a 50% aq EtOH soln is destroyed at 60º in 15 min. [Bloomfield et al. *Biochim Biophys Acta* **533** 371 *1978*; Hartree *Biochem J* **100** 754 *1966*; Pierce and Parsons *Ann Rev Biochem* **50** 465 *1981*.]

Fructose-1,6-diphosphate (trisodium salt) *[38099-82-0]* **M 406.1, pK$_3^{25}$ 6.14, pK$_4^{25}$ 6.93 (free acid).** For purification *via* the acid strychnine salt, see Neuberg, Lustig and Rothenberg [*Arch Biochem* **3** 33 *1943*]. The calcium salt can be partially purified by soln in ice-cold M HCl (1g per 10mL) and repptn by dropwise addition of 2M NaOH: the ppte and supernatant are heated on a boiling water bath for a short time, then filtered and the ppte is washed with hot water. The magnesium salt can be pptd from cold aqueous soln by adding four volumes of EtOH.

Fructose-6-phosphate *[643-13-0]* **M 260.1, $[\alpha]_D^{21}$ +2.5 (c 3, H$_2$O), pK^{25} 5.84.** Crystd as the barium salt from water by adding four volumes of EtOH. The barium can be removed by passage through the H$^+$ form of a cation exchange resin and the free acid collected by freeze-drying.

6-Furfurylaminopurine   **(Kinetin)** *[525-79-1]* **M 215.2, m 266-267º, 269-271º, 270-272º, 272º (sealed capillary), pK$_1$ <1, pK$_2$ 3.8, pK$_3$ 10.** Platelets from EtOH and sublimes at 220º, but is best done at lower temperatures in a good vacuum. It has been extracted from neutral aqueous solns with Et$_2$O. [Miller et al *J Am Chem Soc* **78** 1375 *1956*; Bullock et al. *J Am Chem Soc* **78** 3693 *1956*.]

Fusaric acid   **(5-*n*-butylpyridine-2-carboxylic acid)**   *[536-69-6]* **M 179.2, m 96-98º, 98º, 98-100º, 101-103º, pK$_1$ 5.7, pK$_2$ 6.16 (80% aq methoxyethanol).** Dissolve in CHCl$_3$, dry (Na$_2$SO$_4$), filter, evaporate and recrystallise the residue from 50 parts of pet ether (b 40-60º) or EtOAc, then sublime *in vacuo*. The *copper salt* forms bluish violet crystals from H$_2$O and has **m** 258-259º. [Hardegger and Nikles *Helv Chim Acta* **39** 505 *1956*; Schreiber and Adam *Chem Ber* **93** 1848 *1960*; NMR and MS: Tschesche and Führer *Chem Ber* **111** 3500 *1978*.]

Fuschin   **(Magenta I, rosaniline HCl)**   *[632-99-5]* **M 337.9, m >200º(dec).** See rosaniline hydrochloride on p. 349 in Chapter 4.

D-Galactal   *[21193-75-9]* **M 146.2, m 100º, 100-102º, 104º, 103-106º, $[\alpha]_D^{20}$ -21.3º (c 1, MeOH).** Recryst from EtOAc, EtOH or EtOAc + MeOH. [Overend et al. *J Chem Soc* 675 *1950*; Wood and Fletcher *J Am Chem Soc* **79** 3234 *1957*; Distler and Jourdian *J Biol Chem* **248** 6772 *1973*.]

ß-Galatosidase (from bovine testes) *[9031-11-2]* M_r 510,000, [EC 3.2.1.23]. Purified 600-fold by ammonium sulfate precipitation, acetone fractionation and affinity chromatography on agarose substituted with terminal thio-ß-galactopyranosyl residues. [Distlern and Jourdian *J Biol Chem* **248** 6772 *1973*.]

Gangcyclovir [9-{(1,3-dihydroxy-2-propoxy)methyl}guanine; 2-amino-1,9-{(2-hydroxy-1-hydroxymethyl)-ethoxymethyl}-6H-purin-6-one; Cytovene; Cymeva(e)n(e)] *[82410-32-0]* **M 255.2, m >290°(dec), >300°(dec), monohydrate m 248-249°(dec), $pK_{Est(1)}$~ -1.1, $pK_{Est(2)}$~ 4.1, $pK_{Est(3)}$~ 9.7.** Recryst from MeOH. Alternatively dissolve ~90g of reagent in 700mL of distilled H_2O, filter and cool (*ca* 94% recovery). UV: λ_{max} in MeOH 254nm (ϵ 12,880), 270sh nm (ϵ 9040), solubility in H_2O at 25° is 4.3mg/mL at pH 7.0. **ANTIVIRAL.** [Ogilvie et al. *Can J Chem* **60** 3005 *1982*; Ashton et al. *Biochem Biophys Res Commun* **108** 1716 *1982*; Martin et al. *J Med Chem* **26** 759 *1983*.]

Geranylgeranyl pyrophosphate *[6699-20-3 (NH₄ salt)]* **M 450.5, $pK_{Est(1)}$~<2, $pK_{Est(2)}$~<2, $pK_{Est(3)}$~3.95, $pK_{Est(4)}$~6.26.** Purified by counter-current distribution between two phases of a butanol/isopropyl ether/ammonia /water mixture (15:5:1:19) (v/v), or by chromatography on DEAE-cellulose (linear gradient of 0.02M KCl in 1mM Tris buffer, pH 8.9). Stored as a powder at 0°.

Geranyl pyrophosphate *[763-10-0 (NH₄ salt)]* **M 314.2, $pK_{Est(1)}$~<2, $pK_{Est(2)}$~<2, $pK_{Est(3)}$~3.95, $pK_{Est(4)}$~6.26.** Purified by paper chromatography on Whatman No 3 MM paper in a system of isopropyl alcohol/isobutyl alcohol/ammonia/water (40:20:1:39), R_F 0.77-0.82. Stored in the dark as the ammonium salt at 0°.

Gitoxigenin (3β,14,16β,21-tetrahydroxy-20(22)norcholenic acid lactone) *[545-26-6]* **M 390.5, m 223-226°, 234°, 239-240° (anhydrous by drying at 60°), $[\alpha]_D^{20}$ +30° (c 1, MeOH).** Recrystn from aqueous EtOH produces plates of the sesquihydrate which dehydrate on drying at 100° *in vacuo*. It has also been recrystd from Me_2CO-MeOH and from EtOAc the crystals contain 1 mol of EtOAc with $[\alpha]_D^{21}$ +24.8° (c 1, dioxane). It has UV has λ_{max} at 310, 485 and 520nm in 96% H_2SO_4. On heating with ethanolic HCl it yields *digitaligenin* with loss of H_2O. [Smith *J Chem Soc* 23 *1931*.]

Gliotoxin (3R-6t-hydroxy-3-hydroxymethyl-2-methyl-(5at)-2,3,6,10-tetrahydro-5aH-3,10ac-epidisulfido[1,2-a]-indol-1,4-dione) *[67-99-2]* **M 326.4, m 191-218°(dec), 220°(dec), 221°(dec), $[\alpha]_D^{20}$ -254° (c 0.6, $CHCl_3$), $[\alpha]_D^{25}$ -270° (c 1.7, pyridine).** Purified by recrystn from MeOH. Its solubility in $CHCl_3$ is 1%. The *dibenzoyl* derivative has **m** 202° (from $CHCl_3$-MeOH). [Glister and Williams *Nature* 153 651 *1944*; Elvidge and Spring *J Chem Soc* Suppl 135 *1949*; Johnson et al. *J Am Chem Soc* **65** 2005 *1943*; Bracken and Raistrick *Biochem J* 41 569 *1947*.]

Glucose oxidase (from *Aspergillus niger*) *[9001-37-0]* M_r **186,000, [EC 1.1.3.4].** Purified by dialysis against deionized water at 6° for 48hours, and by molecular exclusion chromatography with Sephadex G-25 at room temperature. [Holt and Cotton *J Am Chem Soc* **109** 1841 *1987*.]

Glucose-1-phosphate *[59-56-3]* **M 260.1, $[\alpha]_D^{25}$ +120° (c 3, H_2O), $[\alpha]_D^{20}$ +78° (c 4, H_2O of di-K salt), pK_1 1.11, pK_2 6.13 [pK^{25} 6.50]** . Two litres of 5% aq soln was brought to pH 3.5 with glacial acetic acid (+ 3g of charcoal, and filtered). An equal volume of EtOH was added, the pH was adjusted to 8.0 (glass electrode) and the soln was stored at 3° overnight. The ppte was filtered off, dissolved in 1.2L of distd water, filtered and an equal volume of EtOH was added. After standing at 0° overnight, the crystals were collected at the centrifuge, and washed with 95% EtOH, then absolute EtOH, ethanol/diethyl ether (1:1), and diethyl ether. [Sutherland and Wosilait, *J Biol Chem* **218** 459 *1956*.] Its barium salt can be crystd from water and EtOH. Heavy metal impurities can be removed by passage of an aqueous soln (*ca* 1%) through an Amberlite IR-120 column (in the appropriate H^+, Na^+ or K^+ forms). *Di-K salt* cryst as $2H_2O$ from EtOH.

Glucose-6-phosphate *[acid 156-73-5; Ba salt 58823-95-3; Na salt 54010-71-8]* **M 260.1, m 205-207°(dec) mono Na salt, $[\alpha]_{546}^{20}$ +41° (c 5, H_2O), pK_1 1.65, pK_2 6.11, pK_3^{25} 11.71 [-$C_1(OH)O^-$].** Can be freed from metal impurities as described for glucose-1-phosphate. Sol of Na Salt is 5% in H_2O at 20°. Its barium salt can be purified by solution in dilute HCl and pptn by neutralising the soln. The ppte is washed with small volumes of cold water and dried in air.

Glucose-6-phosphate dehydrogenase *[9001-40-5]* M_r 128,000 (from Baker's yeast), 63,300 (from rat mammary gland) [EC 1.1.1.49]. The enzyme is useful for measuring pyridine nucleotides in enzyme recycling. The enzyme from Baker's yeast has been purified by $(NH_4)_2SO_4$ fractionation, Me_2CO pptn, a second $(NH_4)_2SO_4$ fractionation, concentration by DEAE-SF chromatography, a third $(NH_4)_2SO_4$ fractionation and recrystn. Crystn is induced by addition of its coenzyme NADP, which in its presence causes rapid separation of crystals at $(NH_4)_2SO_4$ concentration much below than required to ppte the amorphous enzyme. To recryst, the crystals are dissolved in 0.01M NADP (pH 7.3) with $(NH_4)_2SO_4$ at 0.55 saturation and the crystals appear within 10 to 60 min. After standing for 2-3 days (at 4°) the $(NH_4)_2SO_4$ is increased to 0.60 of saturation and more than 80% of the activity in the original crystals is recovered in the fresh crystals. [Noltmann et al. *J Biol Chem* **236** 1255 *1961*]. Large amounts can be obtained from rat livers. The livers are extracted with 0.025M phosphate buffer (pH 7.5), and ppted with 3M $(NH_4)_2SO_4$ (70% of activity). The ppte is dissolved in 3volumes of 0.025M phosphate (pH 7.5), dialysed against this buffer + 0.2mM EDTA at 4° for 5h, then diluted to 1% protein and the nucleic acids ppted by addition of 0.4volumes of 1% protamine sulfate. $(NH_4)_2SO_4$ is added to a concentration of 2M (pH adjusted to 7.0 with NH_3), the ppte is discarded and the supernatant is adjusted to 2.8M $(NH_4)_2SO_4$, dialysed, protein adjusted to 1% and treated with $Ca_3(PO_4)_2$ gel. The gel is added in three steps (1.5mL of 0.4% gel/mL per step) and the gel is removed by centrifugation after each addn. The third gel adsorbed 50% of the activity. The gel is eluted with 0.2M phosphate (pH 7.4, 40mL/g of gel; 60% recovery). The extract is ppted in 3volumes with $(NH_4)_2SO_4$ (adjusted to 4M) to give enzyme with an activity of 30μmoles/mg of protein x hour. [Lowry et al. *J Biol Chem* **236** 2746 *1961*.] Km values for the yeast enzyme are 20μM for G-6P and 2μM for NADP (Tris pH 8.0, 10^{-2} M $MgCl_2$, 38°) [Noltmann and Kuby *The Enzymes* **VII** 223 *1963*].

L-Glutathione (reduced form, γ-L-glutamyl-L-cysteinyl-glycine) *[70-18-8]* M 307.3, m 188-190°(dec), 195°(dec), $[\alpha]_D^{20}$ -20.1° (c 1, H_2O), pK_1^{25} 2.12 (CO_2H), pK_2^{25} 3.59 (CO_2H), pK_3^{25} 8.75 (NH_2), pK_4^{25} 9.65 (10.0, SH). Crystd from 50% aq EtOH, dry in a vac and store below 5°. Alternatively recrystd from aqueous EtOH under N_2, and stored dry in a sealed container below 4°. It is soluble in H_2O. [Weygand and Geiger *Chem Ber* **90** 634 *1957*; Martin and Edsall *Bull Soc Chim Fr* **40** 1763 *1958*; *Biochem Prep* **2** 87 *1952*.]

L-Glutathione (oxidised) *[27025-41-8]* M 612.6, m 175-195°, 195°, $[\alpha]_D^{20}$ -98° (c 2, H_2O), pK_1 3.15, pK_2 4.03, pK_3 8.75. Purified by recrystn from 50% aq EtOH. Its solubility in H_2O is 5%. Store at 4°. [Li et al. *J Am Chem Soc* **76** 225 *1954*; Berse et al. *Can J Chem* **37** 1733 *1959*.]

Glutathione S-transferase (human liver) *[50812-37-8]* M_r 25,000, [EC 2.5.1.18]. Purified by affinity chromatography using a column prepared by coupling glutathione to epoxy-saturated Sepharose. After washing contaminating proteins the pure transferase is eluted with buffer containing reduced glutathione. The solution is then concentrated by ultrafiltration, dialysed against phosphate buffer at pH ~7 and stored in the presence of dithiothreitol (2mM) in aliquots at <-20°. [Simons and Vander Jag *Anal Biochem* **52** 334 *1977*.]

Glyceraldehyde-3-phosphate dehydrogenase *[9001-50-7]* M_r 144,000, [EC 1.2.1.12]. Purified from rabbit muscle by extraction with 0.03N KOH and ppted with $(NH_4)_2SO_4$ (0.52 of saturation). The clear supernatant was adjusted to pH 7.5 and NH_3 was added dropwise to pH 8.2-8.4. Crystals appear sometimes even without seeding. The crystals are dissolved in H_2O, filtered to remove suspended material and 2 volumes of saturated $(NH_4)_2SO_4$ at pH 8.2-8.4 is added. After 1hour the crystals appear. Recrystallise in the same way. [Cori et al. *J Biol Chem* **173** 605 *1948*; Furfine and Velick *J Biol Chem* **240** 844 *1965*, *The Enzymes* **7** 243 *1963*; Lui and Huskey *Biochemistry* **31** 6998 *1992*.] The Km values are: NADH (3.3μM) and 1,3-diphosphoglycerate (8×10^{-7}M) in pH 7.4 imidazole buffer at 26°, NAD (13μM), glyceraldehyde-3-P (90μM), P_i (2.9×10^{-4}M), and arsenate (69μM) in 8.6 M $NaHCO_3$ buffer at 26°. [Orsi and Cleland *Biochemistry* **11** 102 *1972*.]

Glycerol kinase (from *Candida mycoderma, E coli*, rat or pigeon liver glycerokinase) *[9030-66-4]* M_r 251,000, [EC 2.7.1.30]. Commercial enzyme has been dialysed against 2mM Hepes, 5mM dithiothreitol and 0.3mM EDTA, followed by several changes of 20mM Hepes and 5mM dithiothreitol prior to storage under N_2 at -20°. [Knight and Cleland *Biochemistry* **28** 5728 *1989*.] The enzyme from pigeon liver was purified by acid-pptn (acetate buffer at pH 5.1), $(NH_4)_2SO_4$ fractionation, heat treatment (60°/ 1 h), calcium phosphate gel filtration, a second $(NH_4)_2SO_4$ fractionation, dialysis, elution of inert proteins and crystn. This was done by repeatedly extracting the ppte from the last step with 0.05M sodium pyrophosphate (pH 7.5) containing 1mM EDTA and 0.2M $(NH_4)_2SO_4$ was added. Careful addition of solid $(NH_4)_2SO_4$ to this soln lead to crystn of the enzyme. Recrystn was repeated. The enzyme is activated by Mg^{2+} and Mn^{2+} ions and is most stable in solns in the pH 4.5-5.5 range. The stability is greatly increased in the presence of glycerol. It has Km for glycerol of 60μM and for ATP 9μM in glycine buffer pH 9.8 and 25°. [Kennedy *Methods Enzymol* **5** 476 *1962*.]

L-Glycerol-3-phosphate dehydrogenase (GDH, from rabbit muscle) *[9075-65-4]* M_r **78,000 [EC 1.1.1.8]**. Recrystd by adding $(NH_4)_2SO_4$ till 0.45 saturation at pH 5.5 at 4° and the small amount of ppte is removed then satd $(NH_4)_2SO_4$ is added dropwise from time to time over several days in the cold room. The crystals are collected and recrystd until they have maximum activity. The enzyme is stable in half saturated $(NH_4)_2SO_4$ for several weeks at 4°. The equilibrium [dihydroxyacetone][NADH][H^+]/[G-3-P][NAD] is 1.0 x 10^{-12}M in Tris buffer at 25°. It uses NAD ten times more efficiently than NADP. The Km for G-3-P is 1.1 x 10^{-4}M, for NAD it is 3.8 x 10^{-4}M and for dihydroxyacetone it is 4.6 x 10^{-4}M in phosphate buffer pH 7.0 and at 23.3°. Dihydroxyacetone phosphate and fructose-1,6-diphosphate are inhibitors. [Branowski *J Biol Chem* **180** 515 *1949*, *The Enzymes* **7** 85 *1963*; Young and Pace *Arch Biochem Biophys* **75** 125 *1958*; Walsh and Sallach *Biochemistry* **4** 1076 *1965*.]

L-α-Glycerol phosphocholine (Cadmium Chloride)$_x$ complex *[64681-08-9]* **M 257.2 + (183.3)$_x$, pK_{Est} ~ 5.5**. Glycerol phosphocholine is purified *via* the $CdCl_2$ complex which is purified by four recrystns from 99% EtOH by standing at 0° for 1h. The white ppte is collected, washed with EtOH, Et_2O and dried in a vacuum. The amorphous Cd complex can be converted to the crystalline form [$C_8H_{20}O_6NP.CdCl_2.3H_2O$] by dissolving 34.4g in H_2O (410mL) and 99% EtOH (1650mL total) added slowly with stirring and allowing the clear soln to stand at 25° for 12hours, then at 5° for 12h. The crystallised complex is filtered off, washed with cold 80% EtOH and dried in air. Glycerol phosphocholine can be recovered from the complex by dissolving in H_2O (2% soln), passing through an ion-exchange column (4.9 x 100cm, of 1vol IRC-50 and 2vol of IR-45). The effluent is concentrated to a thick syrup at 45°. It is dried further at 50°/P_2O_5/48h. The vitreous product (~8.25g) is dissolved in 99% EtOH (50mL) and the clear soln is cooled at 5°, whereby crystals appear, and then at -15° for 16h. The crystals are filtered off, washed with 99% EtOH, and Et_2O then dried at 50° in a vacuum over P_2O_5. It can be recrystd from 99.5% EtOH, long prisms) which are *hygroscopic* and must be handled in a H_2O-free atmosphere [Tattrie and McArthur *Biochem Prep* **6** 16 *1958*; Baer and Kates *J Am Chem Soc* **70** 1394 *1948* ; *Acta Cryst* **21** 79, 87 *1966*].

Glycine anhydride (2,5-diketopiperazine) *[106-57-0]* **M 114.1, m 309-310°, 311-312°(dec), ~315°(dec), pK_1 -4.45, pK_2 -2.16 (pK_2 -1.94 in AcOH).** Recrystd from H_2O (plates) and can be sublimed (slowly) at 260° or at 140-170°/0.5mm. The *dihydrochloride* has **m** 129-130°, is prepd by dissolving in conc HCl and on adding EtOH to crystallisation point; dried in a vac. The *bis-1-naphthylurethane* has **m** 232°(dec), and the *diperchlorate* has **m** 117° (hygroscopic). [MS: Johnstone *J Chem Soc Perkin Trans 1* 1297 *1975*; NMR: Blaha and Samek *Collect Czech Chem Commun* **32** 3780 *1967*; Sauborn *J Phys Chem* **36** 179 *1932*; Corey *J Am Chem Soc* **60** 1599 *1938*.]

Glycocyamine (N-guanylglycine) *[352-97-6]* **M 117.1, m 280-284°(dec), >300°·, pK^{25} 2.86 (NH_3^+).** Recrystd from 15 parts of hot H_2O, or by dissolving in slightly more than the calculated amount of 2N HCl and ppting by adding an equivalent of 2N NaOH, filtering washing with cold H_2O and drying first *in vacuo* then at 60° *in vacuo*. The *hydrochloride* has **m** 200°(dec) after recrystn from aqueous HCl as plates. The *picrate* forms needles from hot H_2O and has **m** 210°(dec). [Brand and Brand *Org Synth* Coll Vol III 440 *1955*; Failey and Brand *J Biol Chem* **102** 768 *1933*; King *J Chem Soc* 2375 *1930*.]

Glycodeoxycholic acid monohydrate (N-[3α-12α-dihydroxy-5β-cholan-24-oyl]glycine) *[360-65-6]* **M 467.6, m 186-177°(dec), 187-188°, [α]$_D^{23}$ +45.9° (c 1, EtOH), pK_{Est} ~ 4.4.** Recrystallises from H_2O or aqueous EtOH with 1 mol of H_2O and dried at 100° *in vacuo*. Solubility in EtOH is 5%. [UV: Lindstedt and Sjövall *Acta Chem Scand* **11** 421 *1957*.] The *Na salt* is recrystd from EtOH/Et_2O, **m** 245-250°, [α]$_D^{23}$ +41.2° (c 1, H_2O) [Wieland *Hoppe Seyler's Z Physiol Chem* **106** 181 *1919*; Cortese *J Am Chem Soc* **59** 2532 *1937*].

D(+)-Glycogen   *[9005-79-2]* **M 25,000-100,000, m 270-280°(dec), [α]$_{546}$ +216° (c 5, H$_2$O).**
A 5% aqueous soln (charcoal) was filtered and an equal volume of EtOH was added. After standing overnight at 3° the ppte was collected by centrifugation and washed with absolute EtOH, then EtOH/diethyl ether (1:1), and diethyl ether. [Sutherland and Wosilait *J Biol Chem* **218** 459 *1956*.]

Glycogen synthase (from bovine heart) *[9014-56-6]* **M$_r$ 60,000, [EC 2.4.1.11].** Purified by pptn of the enzyme in the presence of added glycogen by polyethylene glycol, chromatography on DEAE-Sephacel and high speed centrifugation through a sucrose-containing buffer. [Dickey-Dunkirk and Kollilea *Anal Biochem* **146** 199 *1985*.]

Gramicidin A   (a pentadecapeptide from *Bacillus brevis*) *[11029-61-1]* **m ~229-230°(dec).**
Purified by countercurrent distribution from *C$_6$H$_6$-CHCl$_3$, MeOH-H$_2$O (15:15:23:7) with 5000 tubes. Fractions were examined by UV (280nm) of small aliquots. Separation from Gramicidin C and other material occurred after 999 transfers. [Gross and Witkop *Biochemistry* **4** 2495 *1965*; Bauer et al. *Biochemistry* **11** 3266 *1972*.] Purified finally by recrystn from EtOH-H$_2$O and dried at 100°/10^{-2} mm over KOH and forms platelets **m** 229-230°. Almost insoluble in H$_2$O (0.6%) but soluble in lower alcohols, dry Me$_2$CO, dioxane, acetic acid and pyridine. The commercial material is more difficult to crystallise than the synthetic compound. [Sarges and Witkop *J Am Chem Soc* **86** 1861, **87** 2011, 2020 *1965*.] It has characteristic [α]$_D^{20}$ +27.3° (c 1.3, MeOH) and UV λ$_{max}$ 282nm (ε 22,100). The *N-carbamoyldeformyl gramicidine A* pptes from EtOAc-pet ether (b 40-60°).

Gramicidin C   (gramicidin S, a pentadecapeptide from *Bacillus brevis*) *[9062-61-7]*. Same as Gramicidin A since they are isolated together and separated. [Sarges and Witkop *Biochemistry* **4** 2491 *1965*; Hunter and Schwartz "Gramicidins" in *Antibotics I* (Gotlieb and Shaw Eds) Springer-Verlag, NY, p.642 *1967*; as well as references above for Gramicidin A.]

Gramicidin S 2HCl (from *Bacillus brevis* Nagano) *[15207-30-4]* **M 1214.4, m 277-278°(dec), [α]$_D^{24}$ -289° (c 0.4, 70% aq EtOH).** Crysts in prisms from EtOH + aq HCl.

Gramicidin S *[113-73-5]* **M 1141.4, m 268-270°, [α]$_D^{25}$ -290° (c 0.5, EtOH + 30mM aq HCl {7:3}].** Crystd from EtOH. *Di-HCl* *[15207-30-4]* cryst from EtOH (+ few drops of HCl) has **m** 277-278°.

N-Guanyltyramine hydrochloride *[60-20-8]* **M 215.7, m 218°, pK$_1$ 10.2 (phenolic OH), pK$_2$ 12.4 (guanidino N).** Purified on a phosphocellulose column and eluted with a gradient of aqueous NH$_3$ (0-10%). The second major peak has the characteristic tryptamine spectrum and is collected, lyphilised to give white crystals of the *dihydrate* which dehydrates at 100°. It has UV λ$_{max}$ at 274.5nm (ε 1310) in 0.1N NaOH and 274.5nm (ε 1330) at pH 7.0. Excitation λ$_{max}$ is at 280nm and emission λ$_{max}$ is at 330nm. [Mekalanos et al. *J Biol Chem* **254** 5849 *1979*.]

Haemoglobin A (from normal human blood) *[9008-02-0]* M$_r$ ~64,500, amorphous.
Purified from blood using CM-32 cellulose column chromatography. [Matsukawa et al. *J Am Chem Soc* **107** 1108 *1985*.] For the purification of the α and β chains see Hill et al. *Biochem Prep* **10** 55 *1963*.

Harmaline   (7-methoxy-1-methyl-4,9-dihydro-3*H*-β-carboline,   4,9-dihydro-7-methoxy-1-methyl-3*H*-pyrido[3,4-*b*]indole) *[304-21-2]* **M 214.3, m 229-230°, 229-231°, 235-237° (after distn at 120-140°/10^{-3}), pK$_1$ 4.2.** Recrystd from MeOH and sublimed at high vacuum. It has UV in MeOH has λ$_{max}$ 218, 260 and 376nm (log ε 4.27, 3.90 and 4.02 respectively); IR (Nujol) ν 1620, 1600, 1570 and 1535cm^{-1} and in CHCl$_3$ ν 1470 and 1629cm^{-1}. [Spenser *Can J Chem* **37** 1851 *1959*; Marion et al. *J Am Chem Soc* **73** 305 *1951*; UV Prukner and Witkop *Justus Liebigs Ann Chem* **554** 127 *1942*.] The *hydrochloride dihydrate* has **m** 234-236°(dec), the *picrate* has **m** 228-229° (sinters at 215°) from aqueous EtOH, and the *N-acetate* forms needles **m** 204-205°.

Hematin (ferrihaeme hydroxide) *[15489-90-4]* **M 633.5, m 200°(dec), pK$_{Est}$ ~ 4.** Crystd from pyridine. Dried at 40° *in vacuo*.

Hematoporphyrin (3,3'-[7,12-bis-(1-hydroxyethyl)-3,8,13,17-tetramethyl-porphyrin-2,18-diyl]-dipropionic acid) *[14459-29-1]* **M 598.7, pK$_{Est}$ ~4.8.** Purified by dissolving in EtOH and adding H$_2$O or Et$_2$O to give deep red crystals. Also recrystd from MeOH. UV has λ_{max} at 615.5, 565, 534.4 and 499.5nm in 0.1 N NaOH, and 597, 619, 634,653, 683 and 701nm in 2 N HCl. [Falk *Porphyrins and Metalloporphyrins* Elsevier, NY, p 175 *1964.*] It is used in the affinity chromatographic purification of Heme proteins [Olsen *Methods Enzymol* 123 324 *1986.*] The *O-methyl-dimethyl ester* has **m** 203-206° (from CHCl$_3$-MeOH) and the *O,O'-dimethyl-dimethyl ester* has **m** 145° (from CHCl$_3$-MeOH). [Paul *Acta Chem Scand* **5** 389 *1951.*]

Hematoporphyrin dimethyl ester *[33070-12-1]* **M 626.7, m 212°.** Crystd from CHCl$_3$/MeOH.

Hematoxylin (±-11*bc*-7,11*b*-dihydroindeno[2,1-*c*]-chromen-3,4-6*ar*-9,10-pentaol) *[517-28-2]* **M 302.3, m 200°(dec), 210-212°(dec).** Recrystd from H$_2$O (as trihydrate) in white-yellow crystals which become red on exposure to light and then melt at 100-120°. It has been recrystd from Me$_2$CO-*C$_6$H$_6$. Crystd also from dil aqueous NaHSO$_3$ until colourless. Soluble in alkali, borax and glycerol. Store in the dark below 0°. [Morsingh and Robinson *Tetrahedron* **26** 182 *1970*; Dann and Hofmann *Chem Ber* **98** 1498 *1955.*]

Hemin (ferriproptoporphyrin IX chloride) *[16009-13-5]* **M 652.0, m sinters at 240°, pK$_{Est}$ ~4.8.** It is purified by recrystn from AcOH. Also heme (5g) is shaken in pyridine (25mL) till it dissolves, then CHCl$_3$ (40mL) is added, the container is stoppered and shaken for 5min (releasing the stopper occasionally). The soln is filtered under slight suction, and the flask and filter washed with a little CHCl$_3$ (15mL). During this period, AcOH (300mL) is heated to boiling and saturated aqueous NaCl (5mL) and conc HCl (4mL) are added. The CHCl$_3$ filtrate is poured in a steady stream, with stirring, into the hot AcOH mixture and set aside for 12hours. The crystals are filtered off, washed with 50% aqueous AcOH (50mL), H$_2$O (100mL), EtOH (25mL), Et$_2$O and dried in air. [Fischer *Org Synth* Coll Vol III 442 *1955.*]

Heparin (from pig intestinal mucosa) *[9005-49-6]* **M$_r$ ~3,000, amorphous, $[\alpha]_D^{20}$ ~+55° (H$_2$O).** Most likely contaminants are mucopolysaccharides including heparin sulfate and dermatan sulfate. Purified by pptn with cetylpyridinium chloride from saturated solutions of high ionic strength. [Cifonelli and Roden *Biochem Prep* **12** 12 *1968.*]

Heparin (sodium salt) *[9041-08-1]* **M$_r$ ~ 3000 (low Mol Wt, Bovine), amorphous, $[\alpha]_D^{20}$ +47° (c 1.5, H$_2$O).** Dissolved in 0.1M NaCl (1g/100mL) and ppted by addition of EtOH (150mL).

Histones (from S4A mouse lymphoma). Purification used a macroprocess column, heptafluorobutyric acid as solubilising and ion-pairing agent and an acetonitrile gradient. [McCroskey et al. *Anal Biochem* **163** 427 *1987.*]

Hyaluronidase *[9001-54-1, 37326-33-3]* **M$_r$ 43,000 (bovine testes), 89,000 (bacterial), [EC 3.2.1.35].** Purified by chromatography on DEAE-cellulose prior to use. [Distler and Jourdain *J Biol Chem* **248** 6772 *1973.*]

Hydrocortisone (11β,17α,21-trihydroxy-pregn-4-ene-3,20-dione) *[50-23-7]* **M 362.5, m 212-213°, 214-217°, 218-221°, 220-222°, $[\alpha]_D^{22}$ +167° (c 1, EtOH).** Recrystd from EtOH or isoPrOH. It is bitter tasting and has UV λ_{max} at 242 nm (log ε 4.20). Its solubility at 25° is: H$_2$O (0.28%), EtOH (1.5%), MeOH (0.62%), Me$_2$CO (0.93%), CHCl$_3$ (0.16%), propylene glycol (1.3%) and Et$_2$O (0.35%). It gives an intense green colour with conc H$_2$SO$_4$. [Wendler et al. *J Am Chem Soc* **72** 5793 *1950.*]

Hydrocortisone acetate (21-acetoxy-11β,17α-trihydroxy-pregn-4-ene-3,20-dione) *[50-03-3]* **M 404.5, m 218-221.5°, 221-223°, 222-225°, $[\alpha]_D^{25}$ +166° (c 0.4, dioxane), +150.7° (c 0.5, Me$_2$CO).** Recrystd from Me$_2$CO-Et$_2$O or aqueous Me$_2$CO as somewhat *hygroscopic* monoclinic crystals. UV has λ_{max} 242 nm ($A_{1cm}^{1\%}$ 390) in MeOH. Its solubility at 25° is: H$_2$O (0.001%), EtOH (0.45%), MeOH (0.04%), Me$_2$CO (1.1%), CHCl$_3$ (0.5%), Et$_2$O (0.15%) and is very soluble in Me$_2$NCHO. [Wendler et al. *J Am Chem Soc* **74** 3630 *1952*; Antonucci et al. *J Org Chem* **18** 7081 *1953.*]

(+)-Hydroquinidine anhydrous   **(9S-6'-methoxy-10,11-dihydrocinchonan-9-ol)** *[1435-55-8]* **M 326.4, m 168-169°, 169°, 169-170°, 171-172°, $[\alpha]_D^{20}$ +231° (c 2, EtOH), +299° (c 0.82, 0.1N H_2SO_4), pK_{Est} ~ 8.8.** Forms needles from EtOH and plates from Et_2O. Slightly soluble in Et_2O and H_2O but readily soluble in hot EtOH. [Heidelberger and Jacobs *J Am Chem Soc* **41** 826 *1919*; King *J Chem Soc* 523 *1946*.] The *hydrochloride* has m 273-274°, $[\alpha]_D^{26}$ +184° (c 1.3, MeOH) and is very soluble in MeOH and $CHCl_3$, but less soluble in H_2O, EtOH and less soluble in dry Me_2CO. [Kyker and Lewis *J Biol Chem* **157** 707 *1945*; Emde *Helv Chim Acta* **15** 557 *1932*.]

Hydroquinine *[522-66-7]* **M 326.4, m 168-171°, 171.5°, $[\alpha]_D^{16}$ +143° (c 1.087, EtOH), pK^{15} 8.87.** Recrystd from EtOH. [Rabe and Schultz *Chem Ber* **66** 120 *1933*.]

19-Hydroxy-4-androsten-3,17-dione   *[510-64-5]* **M 302.4, m 167-169°, 168-170°, 169-170°, 172-173°, $[\alpha]_D^{20}$ +190° (c 1, $CHCl_3$).** Recrystd from Me_2CO-hexane or Et_2O-hexane. It has UV λ_{max} at 242nm in EtOH or MeOH. The *19-acetoxy* derivative has $[\alpha]_D^{26}$ +185° ($CHCl_3$) and λ_{max} 237.5nm in EtOH. [Ehrenstein and Dünnenberger *J Org Chem* **21** 774 *1956*.]

3-Hydroxy butyrate dehydrogenase (from *Rhodopseudomonas spheroides*) *[9028-38-0]* **M_r ~85,000, [EC 1.1.1.30], amorphous.** Purified by two sequential chromatography steps on two triazine dye-Sepharose matrices. [Scavan et al. *Biochem J* **203** 699 *1982*.]

25-Hydroxycholesterol   **(cholest-5-en-3β,25-diol)** *[2140-46-7]* **M 402.7, m 177-179°, 178-180°, 181.5-182.5°, $[\alpha]_D^{25}$ -39° (c 1.05, $CHCl_3$).** Forms colourless needles from MeOH. [Schwartz *Tetrahedron Lett* **22** 4655 *1981*.] The *3β–acetoxy* derivative has m 142-142.8° (from Me_2CO), $[\alpha]_D^{25}$ -40.4° (c 2, $CHCl_3$). The *3β,25-diacetyl* derivative has m 119-120.5° (from MeOH), $[\alpha]_D^{25}$ -35.5° ($CHCl_3$). [Dauben and Bradlow *J Am Chem Soc* **72** 4248 *1950*; Ryer et al. *J Am Chem Soc* **72** 4247 *1950*.]

18-Hydroxy-11-deoxycorticosterone   **(18,21-dihydroxypregn-4-en-3,20-dione tautomeric with 18,20-epoxy-20,21-dihydroxypregn-4-en-3-one)**   *[379-68-0]* **M 346.5, m 168-170°, 171-173°, 191-195°, 200-205°, $[\alpha]_D^{20}$ +151° (c 1, $CHCl_3$).** Recrystn from Et_2O-Me_2CO gave crystals m 200-205°, when recrystd from M_2CO it had m 191-195°. It has UV λ_{max} at 240nm. The *21-O-acetoxy-18-hydroxy* derivative has m 158-159° (from Et_2O-$^*C_6H_6$) and the *21-O-acetoxy-18,20-epoxy* derivative has m 149-154° (from Et_2O). [Kahnt et al. *Helv Chim Acta* **38** 1237 *1955*; Pappo *J Am Chem Soc* **81** 1010 *1959*.]

R-(-)-2-Hydroxy-3,3-dimethyl-γ-butyrolactone   **(3-hydroxy-4,4-dimethyl-4,5-dihydrofuran-2-one, D-pantolactone)** *[599-04-2]* **M 130.1, m 89-91°, 90.5-91.5°, 91°, 92-93°, b 120-122°/15mm, $[\alpha]_D^{20}$ -28° (c 5, MeOH), $[\alpha]_D^{20}$ -51° (c 3, H_2O).** Recrystallise from Et_2O-pet ether, diisopropyl ether or $^*C_6H_6$-pet ether and sublime at 25°/0.0001mm. It hydrolyses readily to the hydroxy-acid and racemises when heated above 145°. The *Brucine salt* has m 211-212° (from EtOH). [Kuhn and Wieland *Chem Ber* **73** 1134 *1940*; and Stiller et al. *J Am Chem Soc* **62** 1779 *1940*; Bental and Tishler *J Am Chem Soc* **68** 1463 *1946*.]

(±)-Ibotenic acid monohydrate (α-[3-hdyroxy-5-isoxazolyl]-glycine, α-amino-3-hydroxy-5-isoxazoleacetic acid) *[2552-55-8]* **M 176.1, m 144-146° (monohydrate), 151-152° (anhydrous), 148-151°, pK_1 2, pK_2 5.1, pK_3 8.2.** It has been converted to the ammonium salt (**m** 121-123° dec) dissolved in H_2O and passed through an Amberlite IR 120 resin (H^+ form) and eluted with H_2O. The acidic fractions were collected, evaporated to dryness and the residue recrystd from H_2O as the monohydrate (**m** 144-146°). The anhydrous acid is obtained by making a slurry with MeOH, decanting and evaporating to dryness and repeating the process twice more to give the anhydrous acid (**m** 151-152°). Recrystn from H_2O gives the monohydrate. [Nakamura *Chem Pharm Bull Jpn* **19** 46 *1971*.] The *ethyl ester* forms needles when crystd from a small volume of Et_2O and has **m** 78-79° and IR ($CHCl_3$) with ν 3500–2300 (OH), 1742 (ester CO), 1628, 1528cm^{-1}, and UV with λ_{max} (EtOH) at 206nm (ε 7080). The *hydrazide* has **m** 174-175° (from MeOH) with IR (KBr) 1656 (C=O)cm^{-1}.

2-Iminothiolane hydrochloride (2-iminotetrahydrothiophene) *[4781-83-3]* **M 137.6, m 187-192°, 190-195°, 193-194°, 202-203°, pK <2 (free base).** Recryst from MeOH-Et$_2$O (**m** 187-192°) but after sublimation at ~180°/0.2mm the melting point rose to 202-203°. It has NMR with δ 2.27 (2H, t), 3.25 (2H, t) and 3.52 (2H, t) in (CD$_3$)$_2$SO. [King et al. *Biochemistry* **17** 1499 *1978.*] The *free base* is purifed by vacuum distn (**b** 71-72°/6mm) with IR (film) with ν 1700 (C=N)cm^{-1} and NMR (CDCl$_3$) with δ at 3.58 (2H, t) and 2.10-2.8 (4H, m). The *free base* is stable on storage but slowly hydrolyses in aqueous solns with half lives at 25° of 390h at pH 9.1, 210h at pH 10 and 18 h at pH 11. [Alagon and King *Biochemistry* **19** 4343 *1980.*]

trans-**Indol-3-ylacrylic acid** *[1204-06-4]* **M 187.2, m 190-195°(dec), 195°(dec), 196°(dec), 195-196°(dec), pK$_{Est}$ ~ 4.2.** Recrystd from AcOH, H$_2$O or EtOAc-cyclohexane. UV in MeOH has λ$_{max}$ at 225, 274 and 325nm. [Shaw et al. *J Org Chem* **23** 1171 *1958*; constitution: Rappe *Acta Chem Scand* **18** 818 *1964*; Moffatt *J Chem Soc* 1442 *1957*; Kimming et al. *Hoppe Seyler's Z Physiol Chem* **371** 234 *1958.*]

3-Indolylbutyric acid *[133-32-4]* **M 203.2, m 120-123°, 123-125°, 124°, pK 4.84.** Recrystd from H$_2$O. It is soluble in EtOH, Et$_2$O and Me$_2$CO but insoluble in CHCl$_3$. [Bowman and Islip *Chem Ind London* 154 *1971*; Jackson and Manske *J Am Chem Soc* **52** 5029 *1930*; Albaum and Kaiser *Am J Bot* **24** 420 *1937.*] UV has λ$_{max}$ 278 and 320nm in isoPrOH [Elvidge *Quart J Pharm Pharmacol* **13** 219 *1940*]. The *methyl ester* has **m** 73-74° (from *C$_6$H$_6$-pet ether) and **b** 230°/6mm [Bullock and Hand *J Am Chem Soc* **78** 5854 *1951*]. Also recrystd from EtOH/water [James and Ware *J Phys Chem* **89** 5450 *1985*].

3-Indolylpyruvic acid *[392-12-1]* **M 203.2, m~210°(dec), 208-210°(dec), 219°(dec), pK$_{Est}$ ~ 2.4.** Recrystd from Me$_2$CO-*C$_6$H$_6$, EtOAc-CHCl$_3$, Me$_2$CO-AcOH (crystals with 1 molecule of AcOH) and dioxane-*C$_6$H$_6$ (with 0.5 molecule of dioxane) [Shaw et al. *J Org Chem* **23** 1171 *1958*; Kaper and Veldstra *Biochim Biophys Acta* **30** 401 *1958*]. The *ethyl ester* has **m** 133° (from Et$_2$O) and its *2,4-dinitrophenylhydrazone* has **m** 255° (from Me$_2$CO). [Baker *J Chem Soc* 461 *1946.*]

myo-**Inositol (cyclohexane[1r,2c,3c,4t,5c,6t]-hexol)** *[87-89-8]* **M 180.2, m 218° (dihydrate), 225-227°, 226-230°.** Recrystd from aq 50% ethanol or H$_2$O forming a dihydrate, or anhydrous crystals from AcOH. The dihydrate is efflorescent and becomes anhydrous when heated at 100°. The anhydrous crystals are not hygroscopic. Solubility in H$_2$O at 25° is 14%, at 60° it is 28%, slightly soluble in EtOH but insoluble in Et$_2$O. [Ballou and Anderson *J Am Chem Soc* **75** 748 *1953*; Anderson and Wallis *J Am Chem Soc* **70** 2931 *1948.*]

Interferons [αIFN, βIFN and γIFN]. Interferons are a family of glycosylated proteins and are cytokines which are produced a few hours after cells have been infected with a virus. Interferons protect cells from viral infections and have antiviral activities at very low concentrations (~3 x 10^{-4} M, less than 50 molecules are apparently sufficient to protect a single cell). Double stranded RNA are very efficient inducers of IFNs. There are three main types of IFNs. The αIFNs are synthesised in lymphocytes and the βIFNs are formed in infected fibroblasts. The α and β families are fairly similar consisting of ca 166 to 169 amino acids. Although γIFNs are also small glycosylated proteins (ca 146 amino acids), they are different because they are not synthesised after viral infections but are produced by lymphocytes when stimulated by **mitogens** (agents that induced cell division).

Several of these IFNs of mouse and human lymphocytes and fibroblasts are available commercially and have been best prepared in quantity by recombinant DNA procedures because they are produced in very small amounts by the cells. The commercial materials do not generally require further purification for their intended purposes. [Pestkas, Interferons and Interferon standards and general abbreviations, *Methods Enzymol*, Wiley & Sons, **119** *1986,* ISBN 012182019X; Lengyel, Biochemistry of interferons and their actions, *Ann Rev Biochem* **51** 251-282 *1982*; De Maeyer and De Maeyer-Guignard, Interferons in *The Cytokine Handbook*, 3rd Edn, Thomson et al. Eds, pp. 491-516 *1998* Academic Press, San Diego, ISBN 0126896623.]

Interleukin (from human source). Purified using lyophilisation and desalting on a Bio-Rad P-6DC desalting gel, then two steps of HPLC, first with hydroxylapatite, followed by a TSK-125 size exclusion column. [Kock and Luger *J Chromatogr* **296** 293 *1984.*]

Interleukin-2 (recombinant human) *[94218-72-1]* **M_r ~15,000, amorphous.** Purified by reverse phase HPLC. [Weir and Sparks *Biochem J* **245** 85 *1987*; Robb et al. *Proc Natl Acad Sci USA* **81** 6486 *1984*.]

Interleukins (IL-1, IL-2 —IlL18]. Interleukins are cytokines which cause a variety of effects including stimulation of cell growth and proliferation of specific cells, e.g. stem cells, mast cells, activated T cells, colony stimulating factors etc, as well as stimulating other ILs, prostaglandins release etc. They are small glycosylated proteins (*ca* 15 kD, 130-180 amino acids produced from longer precursors) and are sometimes referred to by other abbreviations, e.g. IL-2 as TCGF (T cell growth factor), IL-3 as multi-CSF (multilineage colony stimulating factor, also as BPA, HCSF, MCSF and PSF). They are produced in very small amounts and are commercially made by recombinant DNA techniques in bacteria or Sf21 insect cells. Interleukins for human (h-IL), mouse (m-IL) and rat (r-IL) are available and up to IL-18 are available commercially in such purity that they can be used directly without further refinement, particularly those that have been obtained by recombinant DNA procedures which are specific. As well as the interleukins, a variety of antibodies for specific IL reactions are available for research or IL identification. [Symons et al. Lymphokines and Interferons, *A Practical Approach*, Clemens et al. Eds, p. 272 *1987*; IRL Press, Oxford, ISBN 1852210354, 1852210362; Thomson et al. Eds, *The Cytokine Handbook*, 3rd Edn, *1998*; Academic Press, San Diego, ISBN 0126896623.]

Iodonitrotetrazolium chloride (2[4-iodophenyl]-3-[4-nitrophenyl]-5-phenyl-2H-tetrazolium chloride) *[146-68-9]* **M 505.7, m 229°(dec), ~245°(dec).** Recrystd. from H_2O, aqueous EtOH or EtOH-Et$_2$O. Alternatively dissolve in the minimum volume of EtOH and add Et$_2$O; or dissolve in hot H_2O (charcoal), filter and ppte by adding conc HCl. Filter solid off and dry at 100°. Solubility in H_2O at 25° is 0.5%, and in hot MeOH-H_2O (1:1) it is 5%. [Fox and Atkinson *J Am Chem Soc* **72** 3629 *1950*.]

Iodonitrotetrazolium violet-Formazan *[7781-49-9]* **M 471.3, m 185-186°.** Dissolve in boiling dioxane (20g in 300mL), add H_2O (100mL) slowly, cool, filter and dry *in vacuo* at 100°. Its solubility in CHCl$_3$ is ~1%. [UV: Fox and Atkinson *J Am Chem Soc* **72** 3629 *1950*.]

5-Iodouridine (5-iodo-1-[β-D-ribofuranosyl]-pyrimidine-2,4(1H)-dione) *[1024-99-3]* **M 370.1, m 205-208°(dec), 210-215°(dec), $[\alpha]_D^{20}$ -23.5° (c 1, H_2O), pK^{20} 8.5.** Recrystd from H_2O and dried *in vacuo* at 100°. UV has λ_{max} 289nm (0.01N HCl) and 278nm (0.01N NaOH). [Prusoff et al. *Cancer Res* **13** 221 *1953*.]

3-Isobutyl-1-methylxanthine (3-isobutyl-1-methylpurine-2,6-dione) *[28822-58-4]* **M 222.3, m 199-210°, 202-203°, pK$_{Est}$ ~ 6.7 (acidic NH).** Recrystd from aqueous EtOH.

Isopentenyl pyrophosphate *[358-71-4]* **M 366.2, pK$_{Est(1)}$~<2, pK$_{Est(2)}$~<2, pK$_{Est(3)}$~3.95, pK$_{Est(4)}$~6.26.** Purified by chromatography on Whatman No 1 paper using *tert*-butyl alcohol/formic acid/water (20:5:8, R$_F$ 0.60) or 1-propanol/ammonia/water (6:3:1. R$_F$ 0.48). Also purified by chromatography on a DEAE-cellulose column or a Dowex-1 (formate form) ion-exchanger using formic acid and ammonium formate as eluents. A further purification step is to convert it to the monocyclohexylammonium salt by passage through a column of Dowex-50 (cyclohexylammonium form) ion-exchange resin. Can also be converted into its lithium salt.

DL-Isoserine (±-3-amino-2-hydroxypropionic acid) *[632-12-2]* **M 105.1, m 250-252°(dec), 235°(dec), 237°(dec), 245°(dec), pK$_1^{25}$ 2.78 (acidic), pK$_2^{25}$ 9.27 (basic).** Recrystd from H_2O or 50% aqueous EtOH. It has an isoelectric pH of 6.02. [Rinderknocht and Niemann *J Am Chem Soc* **75** 6322 *1953*; Gundermann and Holtmann *Chem Ber* **91** 160 *1958*; Emerson et al. *J Biol Chem* **92** 451 *1931*.] The *hydrobromide* has **m** 128-130° (from aqueous HBr) [Schöberl and Braun *Justus Liebigs Ann Chem* **542** 288 *1939*].

Isoxanthopterin (2-amino-4,7-dihydroxypteridine) *[529-69-1]* **M 179.4, m>300°, pK$_1^{20}$ -0.5 (basic), pK$_2^{20}$ 7.34 (acidic), pK$_3^{20}$ 10.06 (acidic).** Purified by repeated pptn from alkaline solutions by acid (preferably AcOH), filter, wash well with H_2O then EtOH and dried at 100°. Purity is checked by paper chromatography [R$_F$ 0.15 (*n*-BuOH, AcOH, H_2O, 4:1:1); 0.33 (3% aq NH$_4$OH). [Goto et al. *Arch Biochem*

Biophys **111** 8 *1965*.] For biochemistry see Blakley *Biochemistry of Folic Acid and Related Pteridines* North Holland Publ Co, Amsterdam 1969.]

Kanamycin B (Bekanamycin, 4-*O*-[2,6-diamino-2,6-dideoxy-α-D-glucopyranosyl]-6-*O*-[3-amino-3-deoxy-α-D-glucopyranosyl]-2-deoxystreptamine) *[4696-76-8, 29701-07-3 (sulfate salt)]* M 483.5, m 170-179°(dec), 178-182°(dec), $[\alpha]_D^{18}$ +130° (c 0.5, H_2O), pK 7.2.

A small quantity (24mg) can be purified on a small Dowex 1 x 2 column (6 x 50mm), the correct fraction is evapd to dryness and the residue crystd from EtOH containing a small amount of H_2O. [Umezawa et al. *Bull Chem Soc Jpn* **42** 537 *1969*.] It has been crystd from H_2O by dissolving ~1g in H_2O (3mL), adding Me_2NCHO (3mL) setting aside at 4° overnight, The needles are collected and dried to constant weight at 130°. It has also been recrystd from aq EtOH. It is slightly sol in $CHCl_3$ and isoPrOH. [IR: Wakazawa et al. *J Antibiot* **14A** 180, 187 *1961*; Ito et al. *J Antibiot* **17** A 189 *1964*.]

Lactate dehydrogenase (from dogfish, Beef muscle) *[9001-60-9]* M_r 140,000 [EC 1.1.1.27].

40-Fold purification by affinity chromatography using Sepharose 4B coupled to 8-(6-aminohexyl)amino-5'-AMP or -NAD⁺. [Lees et al. *Arch Biochem Biophys* **163** 561 *1974*; Pesce et al. *J Biol Chem* **239** 1753 *1964*.]

Lactoferrin (from human whey). Purified by direct adsorption on cellulose phosphate by batch extraction, then eluted by a stepped salt and pH gradient. [Foley and Bates *Anal Biochem* **162** 296 *1987*.]

Lecithin (1,2-diacylphosphatidylcholine mixture) *[8002-43-5]* M ~600-800, amorphous. From hen egg white. Purified by solvent extraction and chromatography on alumina. Suspended in distilled water and kept frozen until used [Lee and Hunt *J Am Chem Soc* **106** 7411 *1984*, Singleton et al. *J Am Oil Chem Soc* **42** 53 *1965*]. For purification of commercial egg lecithin see Pangborn [*J Biol Chem* **188** 471 *1951*].

Lectins (proteins and/or glycoproteins of non-immune origin that agglutinate cells, from seeds of *Robinia pseudoacacia*), M ~100,000. Purified by pptn with ammonium sulfate and dialysis; then chromatographed on DE-52 DEAE-cellulose anion-exchanger, hydroxylapatite and Sephacryl S-200. [Wantyghem et al. *Biochem J* **237** 483 *1986*.]

Leucopterin (2-amino-5,8-dihydropteridine-4,6,7(1*H*)-trione) *[492-11-5]* M 195.1, m >300° (dec), pK_1^{20} -1.66, pK_2^{20} 7.56, pK_3^{20} 9.78, pK_4^{20} 13.6. Purified by dissolving in aqueous NaOH, stirring with charcoal, filtering and precipitating by adding aqueous HCl, then drying at 100° in a vacuum. It separates with 0.5 moles of H_2O. Its solubility in H_2O is 1g/750 litres [Albert et al. *J Chem Soc* 4219 *1952*; Albert and Wood *J Appl Chem (London)* **2** 591 *1952*; Pfleiderer *Chem Ber* **90** 2631 *1957*].

DL-α-Lipoamide (±-6,8-thioctic acid amide, 5-[1,2]-dithiolan-3-ylvaleric acid amide) *[3206-73-3]* M 205.3, m 124-126°, 126-129°, 130-131°. Recrystd from EtOH and has UV with λ_{max} 331nm in MeOH. [Reed et al. *J Biol Chem* **232** 143 *1958*; IR: Wagner et al. *J Am Chem Soc* **78** 5079 *1956*.]

DL-α-Lipoic acid (±-6,8-thioctic acid, 5-[1,2]-dithiolan-3-ylvaleric acid) *[1077-28-7]* M 206.3, m 59-61°, 60.5-61.5° and 62-63°, b 90°/10⁻⁴mm, 150°/0.1mm, pK^{25} 4.7. It forms yellow needles from cyclohexane or hexane and has been distd at high vacuum, and sublimes at ~90° and very high vacuum. Insoluble in H_2O but dissolves in alkaline soln. [Lewis and Raphael *J Chem Soc* 4263 *1962*; Soper et al. *J Am Chem Soc* **76** 4109; Reed and Niu *J Am Chem Soc* **77** 416 *1955*; Tsuji et al. *J Org Chem* **43** 3606 *1978*; Calvin *Fed Proc USA* **13** 703 *1954*.] The *S-benzylthiouronium salt* has m 153-154° (evacuated capillary; from MeOH), 132-134°, 135-137° (from EtOH). The *d*- and *l*- forms have m 45-47.5° and $[\alpha]_D^{23}$ ±113° (c 1.88, *C_6H_6) and have UV in MeOH with λ_{max} at 330nm (ε 140).

Lipoprotein lipase (from bovine skimmed milk)   *[9004-02-8]* **[EC 3.1.1.34]**.   Purified by affinity chromatography on heparin-Sepharose [Shirai et al. *Biochim Biophys Acta* **665** 504 *1981*].

Lipoproteins (from human plasma).   Individual human plasma lipid peaks were removed from plasma by ultracentrifugation, then separated and purified by agarose-column chromatography. Fractions were characterised immunologically, chemically, electrophoretically and by electron microscopy. [Rudel et al. *Biochem J* **13** 89 *1974*.]

Lipoteichoic acids (from gram-positive bacteria)   *[56411-57-5]*.   Extracted by hot phenol/water from disrupted cells. Nucleic acids that were also extracted were removed by treatment with nucleases. Nucleic resistant acids, proteins, polysaccharides and teichoic acids were separated from lipoteichoic acids by anion-exchange chromatography on DEAE-Sephacel or by hydrophobic interaction on octyl-Sepharose [Fischer et al. *Eur J Biochem* **133** 523 *1983*].

D-Luciferin   (firefly luciferin, *S*-2[6-hydroxybenzothiazol-2-yl]-4,5-dihydrothiazol-4-carboxylic acid),   *[2591-17-5]* **M 280.3**, **m 189.5-190°(dec)**, **196°(dec)**, **201-204°**, **205-210°(dec, browning at 170°)**, $[\alpha]_D^{22}$ **-36° (c 1.2, DMF)**, **p$K_{Est(1)}$~ 1.2 (benzothiazole-N)**, **p$K_{Est(2)}$~ 1.6 (thiazolidine-N)**, **p$K_{Est(3)}$~ 6.0 (CO_2H)**, **p$K_{Est(4)}$ 8.5 (6OH)**. Recrystallises as pale yellow needles from H_2O, or MeOH (83mg from 7mL). It has UV λ_{max} at 263 and 327nm (log ε 3.88 and 4.27) in 95% EtOH. The Na salt has a solubility of 4mg in 1 mL of 0.05M glycine. [White et al. *J Am Chem Soc* **83** 2402 1*1961*, **85** 337 *1963*; UV and IR: Bitler and McElroy *Arch Biochem* **72** 358 *1957*; Review: Cormier et al. *Fortschr Chem Org Naturst* **30** 1 *1973*.]

Lumiflavin   (7,8,10-trimethylbenzo[*g*]pteridine-2,4(3*H*,10*H*)-dione)   *[1088-56-8]* **M 256.3**, **m 330°(dec)**, **340°(dec)**, **pK 10.2**. Forms orange crystals upon recrystn from 12% aqueous AcOH, or from formic acid. It sublimes at high vacuum. It is freely soluble in $CHCl_3$, but not very soluble in H_2O and most organic solvents. In H_2O and $CHCl_3$ soln it has a green fluorescence. UV has λ_{max} at 269, 355 and 445nm (ε 38,800, 11,700 and 11,800 respectively) in 0.1N NaOH and 264, 373 and 440nm (ε 34,700, 11,400 and 10,400 respectively) in 0.1N HCl while UV in $CHCl_3$ has λ_{max} at 270, 312, 341, 360, 420, 445 and 470nm. [Hemmerich et al. *Helv Chim Acta* **39** 1242 *1956*; Holiday and Stern *Chem Ber* **67** 1352 *1834*; Yoneda et al. *Chem Pharm Bull Jpn* **20** 1832 *1972*; Birch and Moye *J Chem Soc* 2622 *1958*; Fluorescence: Kuhn and Moruzzi *Chem Ber* **67** 888 *1934*.]

Magnesium protoporphyrin dimethyl ester   *[14724-63-1]* **M 580.7**.
Crude product dissolved in as little hot dry *C_6H_6 as possible and left overnight at room temperature to cryst. [Fuhrhop and Graniek *Biochem Prep* **13** 55 *1971*.]

α-Melanotropin   *[581-05-5]* **(13 amino acids peptide)**, $[\alpha]_D^{25}$ **-58.5° (c 0.4, 10% aq AcOH)**. Extract separated by ion-exchange on carboxymethyl cellulose, desalted, evapd and lyophilised, then chromatographed on Sephadex G-25. [Lande et al. *Biochem Prep* **13** 45 *1971*.]

ß-Melanotropin.   *[9034-42-8]* **(18-22 amino acids peptide), amorphous**.   Extract separated by ion-exchange on carboxymethyl cellulose, desalted, evapd and lyophilised, then chromatographed on Sephadex G-25. [Lande et al. *Biochem Prep* **13** 45 *1971*.]

6-Mercaptopurine monohydrate   *[6112-76-1]* **M 170.2, m 314-315°(dec), ~315°(dec), 313-315°(dec), pK_1^{20} 7.77, pK_2^{20} 10.84**. Recrystallises from H_2O as yellow crystals of the monohydrate which become anhydrous on drying at 140°. It has UV λ_{max} at 230 and 312nm (ε 14,000 and 19,600) in 0.1N NaOH; 222 and 327nm (ε 9,2400 and 21,300), and 216 and 329nm (ε 8,740 and 19,300) in MeOH. [Albert and Brown *J Chem Soc* 2060 *1954*; IR: Brown and Mason *J Chem Soc* 682 *1957*; UV: Fox et al. *J Am Chem Soc* **80** 1669 *1958*; UV: Mason *J Chem Soc* 2071 *1954*.]

6-Mercaptopurine-9-β-D-ribofuranoside *[574-25-4]* **M 284.3, m 208-210°(dec), 210-211°(dec), 220-223°(dec), 222-224°(dec), $[\alpha]_D^{25}$ -73°** (c 1, 0.1N NaOH), pK 7.56. Recrystd from H_2O or EtOH. It has UV λ_{max} in H_2O at 322nm (pH 1), 320 nm (pH 6.7) and 310nm (pH 13). [IR: Johnson et al. *J Am Chem Soc* **80** 699 *1958*; UV: Fox et al. *J Am Chem Soc* **80** 1669 *1958*.]

Metallothionein (from rabbit liver) *[9038-94-2]*. Purified by precipitation to give Zn- and Cd-containing protein fractions and running on a Sephadex G-75 column, then isoelectric focusing to give two protein peaks [Nordberg et al. *Biochem J* **126** 491 *1972*].

Methadone hydrochloride (2-dimethylamino-4-ethoxycarbonyl-4,4-diphenylbutane HCl) *[1095-90-5]* **M 345.9, m 241-242°, pK_1^{25} 8.94, pK_2^{20} 10.12 (free base).** Crystd from EtOH.

Methoxantin coenzyme (PQQ, pyrrolo quinoline quinone, 2,7,9-tricarboxy-1H-pyrrolo-[2,3-f]-quinoline-4,5-dione, 4,5-dihydro-4,5-dioxo-1H-pyrrolo[2,3-f]quinoline-2,7,9-tricarboxylic acid) *[72909-34-3]* **M 330.2, m 220°(dec).** Efflorescent yellow-orange needles on recrystn from H_2O by addition of Me_2CO, or better from a supersaturated aqueous soln, as it forms an acetone adduct. [Forrest et al. *Nature* **280** 843 *1979*.] It has also been purified by passage through a C-18 reverse phase silica cartridge or a silanised silica gel column in aqueous soln whereby methoxantin remains behind as a red-orange band at the origin. This band is collected and washed thoroughly with dilute aqueous HCl (pH 2) and is then eluted with $MeOH$-H_2O (7:3) and evapd *in vacuo* to give the coenzyme as a red solid. It has also been purified by dissolving in aqueous 0.5M K_2CO_3 and acidified to pH 2.5 whereby PQQ pptes as a deep red solid which is collected and dried *in vacuo*. Methoxantin elutes at 3.55 retention volumes from a C18 µBondapak column using H_2O-$MeOH$ (95:5) + 0.1% AcOH pH 4.5. It has UV λ_{max} at 247 and 330nm (shoulder at 270nm) in H_2O and λ_{max} at 250 and 340nm in H_2O at pH 2.5. With excitation at λ_{ex} 365nm it has a λ_{max} emission at 483nm. The ^{13}C NMR has δ: 113.86, 122.76, 125.97, 127.71, 130.68, 137.60, 144.63, 146.41, 147.62, 161.25, 165.48, 166.45, 173.30 and 180.00.

When a soln in 10% aqueous MeCO is adjusted to pH 9 with aqueous NH_3 and kept at 25° for 30 min, the *acetone adduct* is formed; UV has λ_{max} at 250, 317 and 360nm (H_2O, pH 5.5) and with λ_{ex} at 360nm it has max fluorescence at λ_{max} at 465nm; and the ^{13}C NMR [$(CD_3)_2SO$, TMS] has δ: 29.77, 51.06, 74.82, 111.96, 120.75, 121.13, 125.59, 126.88, 135.21, 139.19, 144.92, 161.01, 161.47, 165.17, 168.61, 190.16 and 207.03. It also forms a *methanol adduct*.

When it is reacted with Me_2SO_4-K_2CO_3 in dry Me_2NCHO at 80° for 4h, it forms the *trimethyl ester* which has **m 265-267°(dec)** [260-263°(dec)] after recrystn from hot MeCN (orange crystals) with UV λ_{max} at 252 and 344nm (H_2O) and 251, 321 and 373nm (in MeOH; MeOH adduct ?). [Duine et al. *Eur J Biochem* **108** 187 *1980*; Duine et al. *Adv Enzymology* **59** 169 *1987*; Corey and Tramontano *J Am Chem Soc* **103** 5599 *1981*; Gainor and Weinreb *J Org Chem* **46** 4319 *1981*; Hendrickson and de Vries *J Org Chem* **17** 1148 *1982*; McKenzie, Moody and Reese *J Chem Soc Chem Commun* 1372 *1983*.]

Methyl benzylpenicillinate *[653-89-4]* **M 348.3, m 97°, $[\alpha]_D^{20}$ +328°** (c 1, MeOH). Crystd from CCl_4.

5-Methylphenazinium methyl sulfate *[299-11-6]* **M 306.3, m 155-157° (198°dec by rapid heating).** It forms yellow prisms from EtOH (charcoal). Solubility in H_2O at 20° is 10%. In the presence of aqueous KI it forms a *semiquinone* which crystallises as blue leaflets from EtOH. [Wieland and Roseen *Chem Ber* **48** 1117 *1913*; Voriskova *Collect Czech Chem Commun* **12** 607 *1947*; Bülow *Chem Ber* **57** 1431 *1924*.]

1-Methyl-4-phenyl-1,2,3,6-tetrahydropyridine hydrochloride (MPTP) *[23007-85-4]* **M 209.7, m 196-198°, pK_{Est} ~ 9.3.** Purified by recrystn from Me_2CO + isoPrOH. The *free base* has **b 137-142°/0.8 mm, n_D^{25} 1.5347.** [Schmidle and Mansfield *J Am Chem Soc* **78** 425 *1956*; Defeudis *Drug Dev Res* **15** 1 *1988*.]

6-α-Methylprednisolone (Medrol, 11β,17-21-trihydroxy-6α-methylpregna-1,4-dien-3,20-dione) *[83-43-2]* **M 347.5, m 226-237°, 228-237°, 240-242°, $[\alpha]_D^{24}$ +91°** (c 0.5, dioxane). Recrystd from EtOAc. UV has λ_{max} in 95% EtOH 243nm (ε 14,875). The *21-acetoxy derivative* has **m 205-**

208° (from EtOAc), $[\alpha]_D^{24}$ +95° (c 1, CHCl$_3$). [Spero et al. *J Am Chem Soc* **78** 6213 *1956*; Fried et al. *J Am Chem Soc* **81** 1235 *1959*; ^{1}H NMR: Slomp and McGarvey *J Am Chem Soc* **81** 2200 *1959*.]

5-Methyltetrahydrofolic acid disodium salt (prefolic A) *[68792-52-9]* M **503.4**, pK$_1$ **2.4** (**N10 protonation**), pK$_2$ **2.7** (**pyrimidine N1 protonation**), pK$_3$ **3.5** (α-CO$_2$H), pK$_4$ **4.9** (γ-CO$_2$H), pK$_5$ **5.6** (**N5-Me**) , pK$_6$ **8.5** (**3NHCO acidic**). Check purity by measuring UV at pH 7.0 (use phosphate buffer) and it should have λ_{max} 290nm and λ_{min} 245nm with a ratio of A$_{290}$/A$_{250}$ of 3.7. This ratio goes down to 1.3 as oxidation to the dihydro derivative occurs. The latter can be reduced back to the tetrahydro compound by reaction with 2-mercaptoethanol at room temp. If oxidation had occurred then the compound should be chromatographed on DEAE-cellulose (~0.9 milliequiv/g, in AcO$^-$ form) in (NH$_4$)$_2$CO$_3$ (1.5 M) and washed with 1M NH$_4$OAc containing 0.01M mercaptoethanol till free from UV absorption and then washed with 0.01M mercaptoethanol. All is done in a nitrogen atmosphere. The reduced folate is then eluted with a gradient between 0.01M mercaptoethanol and 1M NH$_4$OAc containing 0.01M mercaptoethanol and the fractions with absorption at 290nm are collected. These are evapd under reduced pressure at 25° and traces of NH$_4$OAc and H$_2$O are removed at high vacuum/25° (~24-48h). The residue is dissolved in the minimum volume of 0.01M mercaptoethanol and an equivalent of NaOH is added to convert the acid to the diNa salt and evaporated to dryness at high vacuum/25°. The product should have λ_{max} 290nm (ϵ 32,000) in pH 7.0 buffer. [Sakami *Biochem Prep* **10** 103 *1963*.]

5-Methyltryptamine hydrochloride **(3-[2-aminoethyl]-5-methylindole hydrochloride)** *[1010-95-3]* M **210.7**, m **289-291°(dec)**, **290-292°**, pK$_{Est(1)}$~ **-3** (**protonation of ring NH**), pK$_{Est(2)}$~ **9.0** (CH$_2$NH$_2$), pK$_{Est(3)}$~ **10.9** (**acidic indole NH**). Recrystd from H$_2$O. The *free base* has m 93-95° (from **C$_6$H$_6$-cyclohexane), and the *picrate* has m 243°(dec) (from EtOH). [Young *J Chem Soc* 3493 *1958*; Gaddum et al. *Quart J Exp Physiol* **40** 49 *1955*; Röhm *Hoppe Seyler's Z Physiol Chem* **297** 229 *1954*.]

4-Methylumbelliferone(β) hydrate **(7-hydroxy-4-methylcoumarin)** *[90-33-5]* M **194.2, m 185-186°, 185-188°, 194-195°**, pK$_{Est}$ ~ **10.0** (**phenolic OH**). Purified by recrystn from EtOH. It is insoluble in cold H$_2$O, slightly soluble in Et$_2$O and CHCl$_3$, but soluble in MeOH and AcOH. It has blue fluorescence in aqueous EtOH, and has UV λ_{max} 221, 251 and 322.5nm in MeOH. IR has ν 3077 br, 1667, 1592, 1385, 1267, 1156, 1130 and 1066 cm^{-1}. The *acetate* has m 153-154°. [Woods and Sapp *J Org Chem* **27** 3703 *1962*.]

4-Methylumbellifer-7-yl-α-D-glucopyranoside *[17833-43-1]* M **338.3, m 221-222°**, $[\alpha]_D^{20}$ **237° (c 3, H$_2$O)**. Recrystd from hot H$_2$O.

4-Methylumbellifer-7-yl-β-D-glucopyranoside *[18997-57-4]* M **338.3, m 210-212°, 211°**, $[\alpha]_D^{20}$ **-61.5° (c 2, pyridine), -89.5° (c 0.5, H$_2$O for half hydrate)**. Recrystallises as the half hydrate from hot H$_2$O. [Constantzas and Kocourek *Collect Czech Chem Commun* **24** 1099 *1959*; De Re et al. *Ann Chim (Rome)* **49** 2089 *1959*.]

1-Methyluric acid *[708-79-2]* M **182.1, m >350°**, pK$_1$ **5.75** (**basic**), pK$_2$ **10.6** (**acidic**). Recrystd from H$_2$O. [Bergmann and Dikstein *J Am Chem Soc* **77** 691 *1955*.] It has UV λ_{max} at 231 and 283. nm (pH 3) and 217.5 and 292.5nm (pH >12) [Johnson *Biochem J* **5** 133 *1952*].

Mevalonic acid lactone *[674-26-0]* M **130.2, m 28°, b 145-150°/5mm**. Purified *via* the *dibenzylethylenediammonium salt* (**m 124-125°**) [Hofmann et al. *J Am Chem Soc* **79** 2316 *1957*], or by chromatography on paper or on Dowex-1 (formate) column. [Bloch et al. *J Biol Chem* **234** 2595 *1959*.] Stored as *N,N'*-dibenzylethylenediamine (DBED) salt, or as the lactone in a sealed container at 0°.

Mevalonic acid 5-phosphate *[1189-94-2]* M **228.1**, pK$_{Est(1)}$~ **1.5** (PO$_4$H$_2$), pK$_{Est(2)}$~ **4.4** (CO$_2$H), pK$_{Est(3)}$~ **6.31** (PO$_4$H$^-$). Purified by conversion to the *tricyclohexylammonium salt* (**m 154-156°**) by treatment with cyclohexylamine. Crystd from water/acetone at -15°. Alternatively, the phosphate was chromatographed by ion-exchange or paper (Whatman No 1) in a system isobutyric acid/ammonia/water (66:3:30; R$_F$ 0.42). Stored as the cyclohexylammonium salt.

Mevalonic acid 5-pyrophosphate *[1492-08-6]* **M 258.1, $pK_{Est(1)}$~<2, $pK_{Est(2)}$~<2, $pK_{Est(3)}$~3.95 (PO_4), $pK_{Est(4)}$ 4.4 (CO_2H), $pK_{Est(5)}$ ~6.26 (PO_4).** Purified by ion-exchange chromatography on Dowex-1 formate [Bloch et al. *J Biol Chem* **234** 2595 *1959*], DEAE-cellulose [Skilletar and Kekwick, *Anal Biochem* **20** 171 *1967*], on by paper chromatography [Rogers et al. *Biochem J* **99** 381 *1966*]. Likely impurities are ATP and mevalonic acid phosphate. Stored as a dry powder or as a slightly alkaline (pH 7-9) soln at -20°.

Mithramycin A (Aureolic acid, Plicamycin) *[18378-89-7]* **M 1085.2, m 180-183°, $[\alpha]_D^{20}$ -51° (c 0.3, EtOH), pK_{Est} ~ 9.2.** Purified from $CHCl_3$, and is soluble in MeOH, EtOH, Me_2CO, EtOAc, Me_2SO and H_2O, and moderately soluble in $CHCl_3$, but is slightly soluble in $^*C_6H_6$ and Et_2O. Fluorescent antitumour agent used in flow cytometry. [Thiem and Meyer *Tetrahedron* **37** 551 *1981*; NMR: Yu et al. *Nature* **218** 193 *1968*.]

Mitomycin C *[50-07-7]* **M 334.4, m >360°, $pK_{Est(2)}$~ 8.0.** Blue-violet crystals form $^*C_6H_6$-pet ether. It is soluble in Me_2CO, MeOH and H_2O, moderately soluble in $^*C_6H_6$, CCl_4 and Et_2O but insoluble in pet ether. It has UV λ_{max} at 216, 360 and a weak peak at 560nm in MeOH. [Stevens et al. *J Med Chem* **8** 1 *1965*; Shirahata and Hirayama *J Am Chem Soc* **105** 7199 *1983*.]

Muramic acid *[R-2(2-amino-2-deoxy-D-glucose-3-yloxy)-propionic acid]* *[1114-41-6]* **M 251.2, m 145-150°(dec), 152-154°(dec), 155°(dec), $[\alpha]_D^{25}$ +109° (c 2, H_2O), +165.0° (extrapolated to 0 time) → +123° [after 3h (c 3, H_2O)], $pK_{Est(1)}$~ 3.8 (CO_2), $pK_{Est(2)}$~ 7.7 (NH_2).** It has been recrystd from H_2O or aqueous EtOH as monohydrate which loses H_2O at 80° *in vacuo* over P_2O_5. Sometimes contains some NaCl. It has been purified by dissolving 3.2g in MeOH (75mL), filtered from some insoluble material, concentrated to ~10mL and refrigerated. The colourless crystals are washed with absolute MeOH. This process does not remove NaCl; to do so the product is recrystd from a equal weight of H_2O to give a low yield of very pure acid (0.12g). On paper chromatography 0.26µg give one ninhydrin positive spot after development with 75% phenol (R_F 0.51) or with *sec*-BuOH-HCO_2-H_2O (7:1:2) (R_F 0.30). [Matsushima and Park *Biochem Prep* **10** 109 *1963*; *J Org Chem* **27** 3581 *1962*.] The acid has been also purified by dissolving 990mg in 50% aqueous EtOH (2mL), cooling, collecting the colourless needles on a sintered glass funnel and dried over P_2O_5 at 80°/0.1mm to give the anhydrous acid. [Lambert and Zilliken *Chem Ber* **93** 2915 *1960*.] Alternatively the acid is dissolved in a small volume of H_2O, neutralised to pH 7 with ion exchange resin beads (IR4B in OH⁻ form), filtered, evaporated and dried. The residue is recrystd from 90% EtOH (v/v) and dried as above for 24h. [Strange and Kent *Biochem J* **71** 333 *1959*.] The *N-acetyl* derivative has **m** ~125° (dec) and $[\alpha]_D^{20}$ +41.2° after 24h (c 1.5, H_2O). [Watanabe and Saito *J Bacteriol* **144** 428 *1980*.]

Muscimol (pantherine, 5-aminoethyl-3[2h]-isoxazolone) *[2763-96-4]* **M 114.1, m 170-172°(dec), 172-174°(dec), 172-175°, 175°, 176-178°(dec), $pK_{Est(1)}$~ 6 (acidic, ring 2-NH), $pK_{Est(2)}$~ 8 ($CH_2CH_2NH_2$).** Recrystd from MeOH-tetrahydrofuran or EtOH and sublimed at 110-140° (bath) at 10^{-4} mm and gives a yellow spot with ninhydrin which slowly turns purple [NMR: Bowden et al. *J Chem Soc (C)* 172 *1968*]. Also purified by dissolving in the minimum volume of hot H_2O and adding EtOH dropwise until cloudy, cool, and colourless crystals separate; IR: ν 3445w, 3000-2560w br, 2156w, 1635s and 1475s cm⁻¹. [NMR: Jager and Frey *Justus Liebigs Ann Chem* 817 1982.] Alternatively it has been purified by two successive chromatographic treatments on Dowex 1 x 8 with the first elution with 2M AcOH and a second with a linear gradient between 0—2M AcOH and evaporating the desired fractions and recrystallising the residue from MeOH. [McCarry and Savard *Tetrahedron Lett* **22** 5153 *1981*; Nakamura *Chem Pharm Bull Jpn* **19** 46 *1971*.]

Mycophenolic acid (6-[1,3-dihydro-7-hydroxy-5-methoxy-4-methyl-1-oxoisobenzofuran-6-yl]-4-methylhex-4-enoic acid) *[24280-93-1]* **M 320.3, m 141°, 141-143°, $pK_{Est(1)}$~ 2.5 (CO_2H), $pK_{Est(2)}$~ 9.5 (phenolic OH).** Purified by dissolving in the minimum volume of EtOAc, applying to a silica gel column (0.05-0.2 mesh) and eluting with a mixture of EtOAc + $CHCl_3$ + AcOH (45:55:1) followed by recrystn from heptane-EtOAc, from aqueous EtOH or from hot H_2O and drying *in vacuo*. It is a weak dibasic acid moderately soluble in Et_2O, $CHCl_3$ and hot H_2O but weakly soluble in $^*C_6H_6$ and toluene. [Birch and Wright *Aust J Chem* **22** 2635 *1969*; Canonica et al. *J Chem Soc Perkin Trans 1* 2639 *1972*; Birkinshaw, Raistrick and Ross *Biochem J* **50** 630 *1952*.]

Myoglobin (from sperm whale muscle). *[9047-17-0]* M_r ~17,000. Purified by CM-cellulose chromatography and Sephadex G-50 followed by chromatography on Amberlite IRC-50 Type III or BioRex 70 (<400mesh). The crystalline product as a paste in saturated $(NH_4)_2SO_4$ at pH 6.5-7.0 may be stored at 4° for at least 4 years unchanged, but must not be kept in a freezer. [Anres and Atassi *Biochemistry* **12** 942 *1980*; Edmundson *Biochem Prep* **12** 41 *1968*.]

Myricetin (Cannabiscetin, 3,3',4',5,5',7-hexahydroxyflavone) *[529-44-2]* **M 318.2, m >300°, 357°(dec) (polyphenolic pK_{Est}~8-11).** Recrystd from aq EtOH (m 357° dec, as monohydrate) or Me_2CO (m 350° dec, with one mol of Me_2CO) as yellow crystals. Almost insol in $CHCl_3$ and AcOH. The *hexaacetate* has **m** 213°. [Hergert *J Org Chem* **21** 534 *1956*; Spada and Cameroni *Gazzetta* **86** 965, 975 *1956*; Kalff and Robinson *J Chem Soc* **127** 181 *1925*.]

Nalidixic acid (1-ethyl-7-methyl-1,8-naphthyridin-4-one-3-carboxylic acid) *[389-08-2]* M 232.3, m 226.8-230.2°, 228-230°, 229-230°, pK 6.0.
Crystd from H_2O or EtOH as a pale buff powder. It is soluble at 23° in $CHCl_3$ (3.5%), toluene (0.16%), MeOH (0.13%), EtOH (0.09%), H_2O (0.01% and Et_2O (0.01%). It inhibits nucleic acid and protein synthesis in yeast. [Lesher et al. *J Med and Pharm Chem* **5** 1063 *1962*.]

Naloxone hydrochloride hydrate (Narcan, 1-N-propenyl-7,8-dihydro-14-hydroxymorphinan-6-one hydrochloride) *[51481-60-8]* **M 399.9, m 200-205°, $[\alpha]_D^{20}$ -164° (c 2.5, H_2O), $pK_{Est(1)}$~ 6 (N-propenyl), $pK_{Est(2)}$~ 9.6 (phenolic OH).** This opiate antagonist has been recrystd from EtOH + Et_2O or H_2O. It is soluble in H_2O (5%) and EtOH but insoluble in Et_2O. The *free base* has **m** 184° (177-178°) after recrystn from EtOAc, $[\alpha]_D^{20}$ -194.5° (c 0.93, $CHCl_3$). [Olofson et al. *Tetrahedron Lett* 1567 *1977*; Gold et al. *Med Res Rev* **2** 211 *1982*.]

Naltrexone hydrochloride dihydrate (1-N-cyclopropylmethyl-7,8-dihydro-14-hydroxy-morphinan-6-one hydrochloride) *[16676-29-2]* **M 413.9, m 274-276°, $[\alpha]_D^{20}$ -173° (c 1, H_2O), $pK_{Est(1)}$~ 6 (N-cyclopropylmethyl), $pK_{Est(2)}$~ 9.6 (phenolic OH).** This narcotic antagonist has been purified by recrystn from MeOH and dried air. The *free base* has **m** 168-170° after recrystn from Me_2CO. [Cone et al. *J Pharm Sci* **64** 618 *1975*; Gold et al. *Med Res Rev* **2** 211 *1982*.]

α-Naphthoflavone (7,8-benzoflavone) *[604-59-1]* **M 272.3, m 153-155°, 155°, pK 8-9 (phenolic OH).** Recrystd from EtOH or aqueous EtOH. [IR: Cramer and Windel *Chem Ber* **89** 354 *1956*; UV Pillon and Massicot *Bull Soc Chim Fr* 26 *1954*; Smith *J Chem Soc* 542 *1946*; Mahal and Venkataraman *J Chem Soc* 1767 *1934*.] It is a competitive inhibitor of human estrogen synthase. [Kellis and Vickery *Science* **225** 1032 *1984*.]

Naphthol AS-acetate (3-acetoxynaphthoic acid anilide) *[1163-67-3]* **M 305.3, m 152°, 160°.** Recrystd from hot MeOH and dried *in vacuo* over P_2O_5. It is slightly soluble in AcOH, EtOH, $CHCl_3$ or $^*C_6H_6$. It is a fluorogenic substrate for albumin esterase activity. [Chen and Scott *Anal Lett* **17** 857 *1984*.] At λ_{ex} 320nm it had fluorescence at λ_{em} 500nm. [Brass and Sommer *Chem Ber* **61** 1000 *1928*.]

1-Naphthyl phosphate disodium salt *[2183-17-7]* **M 268.1, pK_1^{26} 0.97, pK_2^{26} 5.85 (for free acid).** Purified through an acid ion-exchange column (in H^+ form) to give the *free acid* which is obtained by freeze drying and recrystn from Me_2CO + $^*C_6H_6$, or by adding 2.5 vols of hot $CHCl_3$ to a hot soln of 1 part acid and 1.2 parts Me_2CO and cooling (m 155-157°, 157-158°). The acid is dissolved in the minimum volume of H_2O to which 2 equivalents of NaOH are added and then freeze dried, or by adding the equivalent amount of MeONa in MeOH to a soln of the acid in MeOH and collecting the Na salt, washing with cold MeOH then Et_2O and drying in a vacuum. [Friedman and Seligman *J Am Chem Soc* **72** 624 *1950*; Chanley and Feageson *J*

Am Chem Soc **77** 4002 *1955*.] It is a substrate for alkaline phosphatase [Gomori *Methods Enzymol* **4** 381 *1957*, **128** 212 *1968*], and prostatic phosphatase [Babson *Clin Chem* **30** 1418 *1984*]. See entry on p. 444.

2-Naphthyl phosphate monosodium salt *[14463-68-4]* **M 246.2, m 296° (sintering at 228°), pK$_1^{26}$ 1.28, pK$_2^{26}$ 5.53 (for free acid).** The *free acid* is purified as for the preceding 1-isomer and has **m** 176-177°, 177-178° after recrystn from CHCl$_3$ + Me$_2$CO as the 1-isomer above. It is neutralised with one equivalent of NaOH and freeze dried or prepared as the 1-isomer above. Its solubility in H$_2$O is 5%. It also forms a 0.5 Na.1 H$_2$O salt which has **m** 203-205° (244° ?). [Friedman and Seligman *J Am Chem Soc* **72** 624 *1950*; Chanley and Fegeason *J Am Chem Soc* **77** 4002 *1955*.] See entry on p. 444 in Chapter 5.

D(+)-Neopterin *[2009-64-5]* **M 253.2, m >300°(dec), [α]$_{546}^{20}$ +64.5° (c 0.14, 0.1M HCl), [α]$_D^{25}$ +50.1° (c 0.3, 0.1N HCl), pK$_1$ 2.23 (basic), pK$_2$ 7.89 (acidic).** Purified as biopterin. Also purified on a Dowex 1 x 8 (formate form) column and eluted with 0.03M ammonium formate buffer pH 8.0 then pH 7.2. The fluorescent neopterin fraction is evapd under reduced pressure leaving neopterin and ammonium formate (the latter can be sublimed out at high vacuum). The residue is stirred for 24h with EtOH and the solid is collected and recrystd from H$_2$O [Viscontini et al. *Helv Chim Acta* **53** 1202 *1970*; see Wachter et al. Eds *Neopterin* W de Guyter, Berlin *1992*].

β-Nicotinamide adenine dinucleotide (diphosphopyridine nucleotide, NAD, DPN) *[53-84-9]* **M 663.4, [α]$_D^{23}$ -34.8° (c 1, H$_2$O), pK$_1$ 2.2 (PO$_4$H), pK$_2$ 4.0 (adenine NH$_2$), pK$_3$ 6.1 (PO$_4^-$).** Purified by paper chromatography or better on a Dowex-1 ion-exchange resin. The column was prepared by washing with 3M HCl until free of material absorbing at 260nm, then with water, 2M sodium formate until free of chloride ions and, finally, with water. NAD, as a 0.2% soln in water, adjusted with NaOH to pH 8, was adsorbed on the column, washed with water, and eluted with 0.1M formic acid. Fractions with strong absorption at 360nm were combined, acidified to pH 2.0 with 2M HCl, and cold acetone (*ca* 5L/g of NAD) was added slowly and with constant agitation. It was left overnight in the cold, then the ppte was collected in a centrifuge, washed with pure acetone and dried under vacuum over CaCl$_2$ and paraffin wax shavings [Kornberg *Methods Enzymol* **3** 876 *1957*]. Purified by anion-exchange chromatography [Dalziel and Dickinson *Biochemical Preparations* **11** 84 *1966*.] The purity is checked by reduction to NADH (with EtOH and yeast alcohol dehydrogenase) which has ε$_{340mn}$ 6220 M^{-1}cm^{-1}. [Todd et al. *J Chem Soc* 3727,3733 *1957*.] [pKa, Lamborg et al. *J Biol Chem* **231** 685 *1958*.] The *free acid* crystallises from aq Me$_2$CO with 3H$_2$O and has **m** 140-142°. It is stable in cold neutral aqueous solns in a desiccator (CaCl$_2$) at 25°, but decomposes at strong acid and alkaline pH. Its purity is checked by reduction with yeast alcohol dehydrogenase and EtOH to NADH and noting the OD at 340nm. Pure NADH has ε$_{340}$ 6.2 x 10^{4}M^{-1}cm^{-1}, i.e. 0.1μmole of NADH in 3mL and in a 1cm path length cell has an OD at 340nm of 0.207.

β-Nicotinamide adenine dinucleotide reduced di-Na salt trihydrate (reduced diphosphopyridine nucleotide sodium salt, NADH) *[606-68-8]* **M 763.5, pK as for NAD.** This coenzyme is available in high purity and it is advised to buy a fresh preparation rather than to purify an old sample as purification will invariably lead to a more impure sample contaminated with the oxidised form (NAD). It has UV λ$_{max}$ at 340nm (ε 6,200 M^{-1}cm^{-1}) at which wavelength the oxidised form NAD has no absorption. At 340 nm a 0.161mM solution in a 1cm (pathlength) cell has an absorbance of 1.0 unit. The purity is best checked by the ratio A$_{280nm}$/A$_{340nm}$ ~2.1, a value which increases as oxidation proceeds. The dry powder is stable indefinitely at -20°. Solutions in aqueous buffers at pH ~7 are stable for extended periods at -20° and for at least 8h at 0°, but are oxidised more rapidly at 4° in a cold room (e.g. almost completely oxidised overnight at 4°). [UV: Drabkin *J Biol Chem* **175** 563 *1945*; Fluorescence: Boyer and Thorell *Acta Chem Scand* **10** 447 *1956*; Redox: Rodkey *J Biol Chem* **234** 188 *1959*; Schlenk in *The Enzymes* **2** 250, 268 *1951*; Kaplan in *The Enzymes* **3** 105, 112 *1960*.] Deuterated NADH, i.e. NADD, has been purified through the anion exchange resin AG-1 x 8 (100-200 mesh, formate form) and through a Bio-Gel P-2 column. [Viola, Cook and Cleland *Anal Biochem* **96** 334 *1979*.]

β-Nicotinamide adenine dinucleotide phosphate (NADP, TPN) *[53-59-8]* **M 743.4, pK$_1$ 1.1 (PO$_4$H$_2$), pK$_2$ 4.0 (adenine NH$_2$), pK$_3$ 6.1 (PO$_4^-$).** Purified by anion-exchange chromatography in much the same way as for NAD [Dalziel and Dickinson *Biochem J* **95** 311 *1965*; *Biochemical Preparations* **11** 87 *1966*]. Finally it is purified by dissolving in H$_2$O and precipitating with 4 volumes of Me$_2$CO and dried *in* *vacuo* over P_2O_5. It is unchanged by storing *in vacuo* at 2°. [Hughes et al. *J Chem Soc* 3733 *1957*, Schuster and Kaplan *J Biol Chem* **215** 183 *1955*.] Deuterated NADPH, i.e. NADPD, has been purified through the anion exchange resin AG-1 x 8 (100-200 mesh, formate form) and through a Bio-Gel P-2 column. λ_{min} 259nm (ε 18.000) at pH 7.0. [Viola, Cook and Cleland *Anal Biochem* **96** 334 *1979*.]

β-Nicotinamide adenine dinucleotide phosphate reduced tetrasodium salt (reduced diphosphopyridine nucleotide phosphate sodium salt, NADPH) *[2646-71-1]* **M 833.4, pK as for NADP.** Mostly similar to NADH above.

β-Nicotinamide mononucleotide (NMN) *[1094-61-7]* **M 334.2, $[\alpha]_D^{23}$ -38.3° (c 1, H_2O), pK_{Est} ~ 6.1 (PO_4^-).** Purified by passage through a Dowex 1 (Cl^- form), washed with H_2O until no absorbance at 260 nm. The tubes containing NMN are pooled, adjusted to pH 5.5-6 and evapd *in vacuo* to a small volume. This is adjusted to pH 3 with dilute HNO_3 in an ice bath and treated with 20 volumes of Me_2CO at 0-5°. The heavy white ppte is collected by centrifugation at 0°. It is best stored wet and frozen or can be dried to give a gummy residue. It has λ_{max} 266nm (ε 4600) and λ_{min} 249nm (ε 3600) at pH 7.0 (i.e. no absorption at 340nm). It can be estimated by reaction with CN^- or hydrosulfite which form the 4-adducts equivalent to NADH) which has UV λ_{max} 340nm (ε 6200). Thus after reaction an OD_{340} of one is obtained from a 0.1612mM soln in a 1cm path cuvette. [Plaut and Plaut *Biochem Prep* **5** 56 *1957*; Maplan and Stolzenbach *Methods Enzymol* **3** 899 *1957*; Kaplan et al. *J Am Chem Soc* **77** 815 *1955*.]

(-)-Nicotine (1-methyl-2[3-pyridyl]-pyrrolidine) *[54-11-5]* **M 162.2, b 123-125°/17mm, 246.1°/730.5mm, 243-248°/atm (partial dec), d_4^{20} 1.097, n_D^{20} 1.5280, $[\alpha]_D^{20}$ -169° (c 1, Me_2CO), pK_1^{15} 6.16 (pyridine N^+), pK_2^{15} 10.96 (pyrrolidine N^+).** Very pale yellow *hygroscopic* oil with a characteristic odour (tobacco extract) with browns in air on exposure to light. Purifed by fractional distn under reduced pressure in an inert atmosphere. A freshly distd sample should be stored in dark sealed containers under N_2. It is a strong base, at 0.05 M soln it has a pH of 10.2. Very soluble in organic solvents. It is soluble in H_2O and readily forms salts. [UV: Parvis *J Chem Soc* **97** 1035 *1910*; Dobbie and Fox *J Chem Soc* **103** 1194 *1913*.] The *hydrochlorides* (mono- and di-) form deliquescent crystals soluble in H_2O and EtOH but insoluble in Et_2O. It has also been purified *via* the $ZnCl_2$ double salt. [Ratz *Monatsh Chem* **26** 1241 *1905*; Biosynthesis: Nakan and Hitchinson *J Org Chem* **43** 3922 *1978*.] The *picrate* has **m** 218° (from EtOH). **POISONOUS.**

(±)-Nicotine *[22083-74-5]* **M 162.2, b 242.3°/atm, d_4^{20} 1.082 (pK see above).** Purified by distn. Its solubility in EtOH is 5%. The *picrate* forms yellow needles from hot H_2O and has **m** 218°. The *methiodide* has **m** 219° (from MeOH).

Nisin *[1414-45-5]* **M 3354.2.** Polypeptide from *S. lactis*. Crystd from EtOH. [Berridge et al. *Biochem J* **52** 529 *1952*; synthesis by Fukase et al. *Tetrahedron Lett* **29** 795 *1988*.]

2-Nitrophenyl-β-D-galactopyranoside *[369-07-3]* **M 301.3, m 185-190°, 193°, 193-194°, $[\alpha]_D^{18}$ -51.9° (c 1, H_2O).** Purified by recrystn from EtOH. [Seidman and Link *J Am Chem Soc* **72** 4324 *1950*; Snyder and Link *J Am Chem Soc* **75** 1758 *1953*]. It is a chromogenic substrate for β-galactosidases [Jagota et al. *J Food Sci* **46** 161 *1981*].

4-Nitrophenyl-α-D-galactopyranoside *[7493-95-0]* **M 301.3, m 166-169°, 173°, $[\alpha]_D^{25}$ +248 (c 1, H_2O).** Purified by recrystn from H_2O or aqueous EtOH. The *monohydrate* has **m** 85° which resolidifies and melts at 151-152° (the hemihydrate) which resolidifies and melts again at 173° as the anhydrous form. Drying the monohydrate at 60° yields the hemihydrate and drying at 100° gives the anhydrous compound. The *tetraacetate* has **m** 147° after drying at 100°. [Jermyn *Aust J Chem* **15** 569 *1962*; Helfreich and Jung *Justus Liebigs Ann Chem* **589** 77 *1954*.] It is a substrate for α-galactosidase [Dangelmaier and Holmsen *Anal Biochem* **104** 182 *1980*].

4-Nitrophenyl-β-D-galactopyranoside *[3150-24-1]* **M 301.3, m 178°, 178-181°, 181-182°, $[\alpha]_D^{20}$ -83° (c 1, H_2O).** Purified by recrystn from EtOH. [Horikoshi *J Biochem (Tokyo)* **35** 39 *1042*;

Goebel and Avery *J Exptl Medicine* **50** 521 *1929*; Snyder and Link *J Am Chem Soc* **75** 1758.] It is a chromogenic substrate for β-galactosidases [Buoncore et al. *J Appl Biochem* **2** 390 *1980*].

4-Nitrophenyl-α-D-glucopyranoside *[3767-28-0]* **M 301.3, m 206-212°, 216-217° (sinters at 210°), $[\alpha]_D^{20}$ +215° (c 1, H_2O).** Purified by recrystn from H_2O, MeOH or EtOH. [Jermyn *Aust J Chem* **7** 202 *1954*; Montgomery et al. *J Am Chem Soc* **64** 690 *1942*.] It is a chromogenic substrate from α-glucosidases [Oliviera et al. *Anal Biochem* **113** 188*1981*], and is a substrate for glucansucrases [Binder and Robyt *Carbohydr Res* **124** 287*1983*]. It is a chromogenic substrate for β-glucosidases [Weber and Fink *J Biol Chem* **255** 9030 *1980*].

4-Nitrophenyl-β-D-glucopyranoside *[2492-87-7]* **M 301.2, m 164°, 164-165°, 165°, $[\alpha]_D^{20}$ -107° (c 1, H_2O).** Purified by recrystn from EtOH or H_2O. [Montgomery et al. *J Am Chem Soc* **64** 690 *1942*; Snyder and Link *J Am Chem Soc* **75** 1758 *1953*.]

Nonactin *[6833-84-7]* **M 737.0, m 147-148°, $[\alpha]_D^{20}$ 0° (±2°) (c 1.2, $CHCl_3$).** This macrotetrolide antibiotic was rerystd from MeOH as colourless needles, and dries at 90°/20h/high vacuum. [*Helv Chim Acta* **38** 1445 *1955*, **55** 1371 *1972*; *Tetrahedron Lett* 3391 *1975*.]

***N*-Nonanoyl-*n*-methylglucamine** **(Mega-9)** *[85261-19-4]* **M 335.4, m 87-89°.** A non-ionic detergent purified as *n*-decanoyl-*N*-methylglucamine above. [Hildreth *Biochem J* **207** 363 *1982*.]

Nonyl-β-D-glucopyranoside *[69984-73-2]* **M 306.4, m 67.5-70°, $[\alpha]_D^{20}$ -34.4° (c 5, H_2O), $[\alpha]_D^{25}$ -28.8° (c 1, MeOH).** Purified by recrystn from Me_2CO and stored in well stoppered containers as it is *hygroscopic*. [Pigman and Richtmyer *J Am Chem Soc* **64** 369 *1942*.] It is a UV transparent non-ionic detergent for solubilising membrane proteins [Schwendener et al. *Biochem Biophys Res Commun* **100** 1055 *1981*.]

L-Noradrenaline **(Adrenor, *R*-2-amino-1-[3,4-dihydroxyphenyl]ethan-1-ol, L-norepinephrine)** *[51-41-2, 69815-49-2 (bitartrate salt)]* **M 169.2, m 216.5-218°(dec), ~220-230°(dec), $[\alpha]_D^{20}$ -45° (c 5, N HCl), $[\alpha]_D^{25}$ 37.3° (c 5, 1 equiv aqueous HCl), pK_1^{25} 5.58 (phenolic OH), pK_2^{25} 8.90 (phenolic OH), pK_3^{25} 9.78 (NH_2).** Recrystd from EtOH and stored in the dark under N_2. [pKa, Lewis *Brit J Pharmacol Chemother* **9** 488 *1954*; UV: Bergström et al. *Acta Physiol Scand* **20** 101 *1950*; Fluorescence: Bowman et al. *Science NY* **122** 32 *1955*; Tullar *J Am Chem Soc* **70** 2067 *1948*.] The *L-tartrate salt monohydrate* has **m** 102-104.5°, $[\alpha]_D^{25}$ -11° (c 1.6, H_2O), after recrystn from H_2O or EtOH.

L-Noradrenaline hydrochloride **(Arterenol)** *[329-56-6]* **M 205.6, m 145.2-146.4°, ~150°(dec), $[\alpha]_D^{25}$ -40° (c 6, H_2O), pK see above.** Recrystd from isoPrOH and stored in the dark as it is oxidised in the presence of light (see preceding entry). [Tullar *J Am Chem Soc* **70** 2067 *1948*.]

Novobiocin **(7-[*O^3*-carbamoyl-5-*O^4*-dimethyl-β-L-*lyso*-6-desoxyhexahydropyranosyloxy]-4-hydroxy-3[4-hydroxy-3-{3-methylbut-2-enyl}-benzyl-amino]-8-methylcoumarin)** *[303-81-1]* **M 612.6, two forms m 152-156° and m 172-174°, 174-178°, λ_{max} at 330nm (acid EtOH), 305nm (alk EtOH), $[\alpha]_D^{25}$ -63° (c 1, EtOH), pK_1 4.03 (4.2), pK_2 9.16.** Crystd from EtOH and stored in the dark. It has also been recrystd from Me_2CO-H_2O. [Hoeksema et al. *J Am Chem Soc* **77** 6710 *1955*; Kaczka et al. *J Am Chem Soc* **77** 9404 *1955*.]
The **sodium salt** *[1476-53-5]* **M 634.6, m 210-215°, 215-220°(dec), 222-229°, $[\alpha]_D^{25}$ -38° (c 1, H_2O)** has been recrystd from MeOH, then dried at 60°/0.5mm. [Sensi, Gallo and Chiesa, *Anal Chem* **29** 1611 *1957*; Kaczka et al. *J Am Chem Soc* **78** 4126 *1956*.]

5'-Nucleotidase **(from Electric ray, *Torpedo sp*)** *[9027-73-0]* **[EC 3.1.3.5], amorphous.** Purified by dissolving in Triton X-100 and deoxycholate, and by affinity chromatography on concanavalin A-Sepharose and AMP-Sepharose [Grondal and Zimmerman *Biochem J* **245** 805 *1987*].

Nucleotide thiophosphate analogues. The preparation and purification of [^{3}H]ATPγS, [^{3}H]GTPγS, s^{6}ITPγS (6-thioinosine), cl^{6}ITPγS (6-chloroinosine) and [^{3}H]ATPγS are described and the general purification was achieved by chromatography of the nucleotide thiophosphates in the minimum volume of H_2O placed onto a DEAE-Sephadex A25 column and eluting with a linear gradient of triethylammonium bicarbonate (0.1 to 0.6M for G and I nucleotides and 0.2 to 0.5M for A nucleotides). [*Biochim Biophys Acta* **276** 155 *1972*.]

Nystatin dihydrate (Mycostatin, Fungicidin) *[1400-61-9]* **M 962.1, m dec>160° (without melting by 250°),** $[\alpha]_D^{25}$ **-7° (0.1N HCl in MeOH), -10° (AcOH), +12° (Me_2NCHO), +21° (pyridine).** Light yellow powder with the following solubilities at ~28°: MeOH (1.1%), ethylene glycol (0.9%), H_2O (0.4%), CCl_4 (0.12%), EtOH (0.12%), $CHCl_3$ (0.05%) and $^*C_6H_6$ (0.03%). Could be ppted from MeOH soln by addition of H_2O. Aqueous suspensions of this macrolide antifungal antibiotic are stable at 100°/10min at pH 7.0 but decomposes rapidly at pH <2 and >9, and in the presence of light and O_2. [Birch et al. *Tetrahedron Lett* 1491, 1485 *1964*; Weiss et al. *Antibiot Chemother* **7** 374 *1957*.] It contains a mixture of components A_1, A_2 and A_3.

Octyl-β-D-glucopyranoside   *[29836-26-8]* **M 292.4, m 62-65°, 63.8-65°,** $[\alpha]_D^{20}$
-34° (c 4, H_2O). Purified by recrystn from Me_2CO. It is *hygroscopic* and should be stored in a well stoppered container. [Noller and Rockwell *J Am Chem Soc* **60** 2076 *1938*; Pigman and Richtmyer *J Am Chem Soc* **64** 369 *1942*.] It is a UV transparent non-ionic dialysable detergent for solubilising membrane proteins. The *α–D-isomer* with $[\alpha]_D^{20}$ +118° (c 1, MeOH) has similar solubilising properties. [Lazo and Quinn *Anal Biochem* **102** 68 *1980*; Stubbs et al. *Biochim Biophys Acta* **426** 46 *1976*.]

Orcine monohydrate (3,5-dihydroxytoluene) *[6153-39-5]* **M 142.2, m 56°, 56-58°, 58°, b 147°/5 mm, pK_1^{20} 9.48 (9.26), pK_2^{20} 11.20 (11.66).** Purified by recrystn from H_2O as the monohydrate. It sublimes *in vacuo* and the *anhydrous* compound has **m** 106.5-108° (110°, 108°). Also can be recrystd from $CHCl_3$ (plates) or $^*C_6H_6$ (needles or prisms). [UV: Kiss et al. *Bull Soc Chim Fr* 275 *1949*; Adams et al. *J Am Chem Soc* **62** 732 *1940*.]

Orosomucoid (glycoprotein α_1 acid, from human plasma) *[66455-27-4]* **M_r 42000-44000, amorphous.** Purified by passage through a carboxymethyl cellulose column and through a Sephadex G-25 column. [Aronson et al. *J Biol Chem* **243** 4564 *1968*.]

Orotic acid Li salt H_2O (1-carboxy-4,6-dihydroxypyrimidine Li salt H_2O) *[5266-20-6]* **M 180.0, m >300°, pK_1 2.8 (CO_2H), pK_2 9.4 (OH), pK_3 >13 (OH) (for free acid).** It is soluble in H_2O at 17° and 100°. Best to acidify an aqueous soln, isolating the free acid which is recrystd from H_2O (as monohydrate) **m** 345-347° (345-346°), then dissolving in EtOH, adding an equivalent amount of LiOH in EtOH and evaporating. Its solubility in H_2O is 1.28% (17°) and 2.34% (100°). [Bachstez *Chem Ber* **63** 1000 *1930*; Johnson and Shroeder *J Am Chem Soc* **54** 2941 *1932*; UV: Shugar and Fox *Biochim Biophys Acta* **9** 199 *1952*.]

Oxacillin sodium salt (5-methyl-3-phenyl-4-isoxazolylpenicillin sodium salt) *[1173-88-2]* **M 423.4, m 188°(dec), $[\alpha]_D^{20}$ +29° (c 1, H_2O), pK_{Est} ~ 2.7.** This antibiotic which is stable to penicillinase is purified by recrystn from isoPrOH and dried *in vacuo*. Its solubility in H_2O at 25° is 5%. [Doyle et al. *Nature* **192** 1183 *1961*.]

Oxolinic acid (5-ethyl-5,8-dihydro-8-oxo-1,3-dioxolo[4,5-g]quinoline-3-carboxylic acid) *[14698-29-4]* **M 261.2, m 313-314°(dec), 314-316°(dec), pK_{Est} ~ 2.3.** Purified by recrystn from aqueous Me_2CO or 95% EtOH. It has UV λ_{max} 220, (255.5sh), 259.5, 268, (298sh, 311sh), 321 and 326nm [ϵ 14.8, (36.8sh), 38.4, 38.4, (6.4sh, 9.2sh), 10.8 and 11.2 x 10^3]. [Kaminsky and Mettzer *J Med Chem* **11** 160 *1968*.]

Oxytocin *[50-56-6]* **M 1007.2, m dec on heating, $[\alpha]_D^{22}$ -26.2° (c 0.53, N AcOH).** A cyclic nonapeptide which was purified by countercurrent distribution between solvent and buffer. It is soluble in H_2O, n-BuOH and isoBuOH. [Bodanszky and du Vigneaud *J Am Chem Soc* **81** 2504 *1959*; Cash et al. *J Med Pharm Chem* **5** 413 *1962*; Sakakibara et al. *Bull Chem Soc Jpn* **38** 120 *1965*; solid phase synthesis: Bayer and

Hagenmyer *Tetrahedron Lett* 2037 *1968*.] It was also synthesised on a solid phase matrix and finally purified as follows: A Sephadex G-25 column was equilibrated with the aqueous phase of a mixture of 3.5% AcOH (containing 1.5% of pyridine) + *n*-BuOH + *C_6H_6 (2:1:1) and then the organic phase of this mixture was run through. A soln of oxytocin (100mg) in H_2O (2mL) was applied to the column which was then eluted with the organic layer of the above mixture. The fractions containing the major peak [as determined by the Folin-Lowry protein assay [Fryer et al. *Anal Biochem* **153** 262 *1986*] were pooled, diluted with twice their vol of H_2O, evaporated to a small vol and lyophilised to give oxytocin as a pure white powder (20mg, 508 U/mg). [Ives *Can J Chem* **46** 2318 *1968*.]

Palmitoyl coenzyme A [1763-10-6] M 1005.9.
Possible impurities are palmitic acid, S-palmitoyl thioglycolic acid and S-palmitoyl glutathione. These are removed by placing *ca* 200mg in a centrifuge tube and extracting with Me_2CO (20mL), followed by two successive extractions with Et_2O (15mL) to remove S-palmitoyl thioglycolic acid and palmitic acid. The residue is dissolved in H_2O (4 x 4 mL), adjusted to pH 5 and centrifuged to remove insoluble S-palmitoyl glutathione and other insoluble impurities. To the clear supernatant is added 5% $HClO_4$ (6mL) whereby S-palmitoyl CoA pptes. The ppte is washed with 0.8% $HClO_4$ (10mL) and finally with Me_2CO (3 x 5mL) and dried *in vacuo*. It is stable for at least one year in dry form at 0^0 in a desiccator (dark). Solns are stable for several months at -15^0. Its solubility in H_2O is 4%. The adenine content is used as the basis of purity with λ_{max} at 260 and 232nm (ϵ 6.4 x 10^6 and 9.4 x 10^6 cm^2/mol respectively). Higher absorption at 232nm would indicate other thio ester impurities, e.g. S-palmitoyl glutathione, which absorb highly at this wavelength. Also PO_4 content should be determined and acid phosphate can be titrated potentiometrically. [Seubert *Biochem Prep* **7** 80 *1960*; Srer et al. *Biochim Biophys Acta* **33** 31 *1959*; Kornberg and Pricer *J Biol Chem* **204** 329 , 345 *1953*.]

3-Palmitoyl-sn-glycerol (*R*-glycerol-1-palmitate, L-β-palmitin) [32899-41-5] M 330.5, $d^{27.3}$ 0.9014, m 66.5^0 (α-form), 74^0 (β'-form) and 77^0 (β-form). The stable β-form is obtained by crystn from EtOH or Skellysolve B and recrystn from Et_2O provides the β'-form. The α-form is obtained on cooling the melt. [Malkin and el Sharbagy *J Chem Soc* 1631 *1936*; Chapman *J Chem Soc* 58 *1956*; Luton and Jackson *J Am Chem Soc* **70** 2446 *1948* .]

Pancuronium bromide (2β,16β−dipiperidino-5α-androstan-3α,17β−diol diacetate dimethobromide) [15500-66-0] M 732.7, m 212-215^0, 215^0. Odourless crystals with a bitter taste which are purified through acid-washed Al_2O_3 and eluted with isoPrOH-EtOAc (3:1) to remove impurities (e.g. the monomethobromide) and eluted with isoPrOH to give the pure bromide which can be recrystd from CH_2Cl_2-Me_2CO or isoPrOH-Me_2CO. It is soluble in H_2O (50%) and $CHCl_3$ (3.3%) at 20^0. It is a non-depolarising muscle relaxant. [Buckett et al. *J Med Chem* **16** 1116 *1973*.]

D-Panthenol (Provitamin B, *R*-2,4-dihydroxy-3,3-dimethylbutyric acid 3-hydroxy-propylamide) [81-13-0] M 205.3, b 118-120^0/0.02mm, d_{20}^{20} 1.2, n_D^{20} 1.4935, $[\alpha]_D^{20}$ (c 5, H_2O). Purified by distn *in vacuo*. It is a slightly *hygroscopic* viscous oil. Soluble in H_2O and organic solvent. It is hydrolysed by alkali and strong acid. [Rabin *J Am Pharm Assoc (Sci Ed)* **37** 502 *1948*; Bonati and Pitré *Farmaco Ed Scient* **14** 43 *1959*.]

R-(+)-Pantothenic acid sodium salt (*N*-[2,4-dihydroxy-3,3-dimethylbutyryl] β-alanine Na salt) [867-81-2] M 241.2, $[\alpha]_D^{25}$ +27.1^0 (c 2, H_2O), pK^{25} 4.4 (for free acid). Crystd from EtOH, very hygroscopic (kept in sealed ampoules). The free acid is a viscous hygroscopic oil with $[\alpha]_D^{25}$ +37.5^0 (c 5, H_2O), easily destroyed by acids and bases.

R-(+)-Pantothenic acid Ca salt [(D(+)- 137-08-6; 63409-48-3] M 476.5, m 195-196^0, 200-201^0, $[\alpha]_D^{20}$ +28.2^0 (c 5, H_2O). Crysts in needles from MeOH, EtOH or isoPrOH (with 0.5mol of isoPrOH). Moderately *hygroscopic*. The S-benzylisothiuronium salt has m 151-152^0 (149^0 when crystd from Me_2CO). [Kagan et al. *J Am Chem Soc* **79** 3545 *1957*; Wilson et al. *J Am Chem Soc* **76** 5177 *1954*; Stiller and Wiley *J Am Chem Soc* **63** 1239 *1941*.]

Papain *[9001-73-4]* **M$_r$ ~21,000, [EC 3.4.22.2], amorphous.** A suspension of 50g of papain (freshly ground in a mortar) in 200mL of cold water was stirred at 4° for 4h, then filtered through a Whatman No 1 filter paper. The clear yellow filtrate was cooled in an ice-bath while a rapid stream of H$_2$S was passed through it for 3h, and the suspension was centrifuged at 2000rpm for 20min. Sufficient cold MeOH was added slowly and with stirring to the supernatant to give a final MeOH concn of 70 vol%. The ppte, collected by centrifugation for 20min at 2000rpm, was dissolved in 200mL of cold water, the soln was saturated with H$_2$S, centrifuged, and the enzyme again ppted with MeOH. The process was repeated four times. [Bennett and Niemann *J Am Chem Soc* **72** 1798 *1950.*] Papain has also been purified by affinity chromatography on a column of Gly-Gly-Tyr-Arg-agarose [Stewart et al. *J Am Chem Soc* **109** 3480 *1986*].

Papaverine hydrochloride **(6,7-dimethoxy-1-veratrylisoquinoline hydrochloride)** *[61-25-6]* **M 375.9, m 215-220°, 222.5-223.5°(dec), 231°, pK^{25} 6.41.** Recrystd from H$_2$O and sublimed at 140°/0.1mm. Solubility in H$_2$O is 5%. [Saunders and Srivastava *J Pharm Pharmacol* **3** 78 *1951*; Biggs *Trans Faraday Soc* **50** 800 *1954.*] The *free base* has **m** 148-150° [Bobbitt *J Org Chem* **22** 1729 *1957*].

Pargyline hydrochloride **(Eutonyl, *N*-methyl-*n*-propargylbenzylamine hydrochloride)** *[306-07-0]* **M 195.7, m 154-155°, 155°, pK^{25} 6.9.** Recrystd from EtOH-Et$_2$O and dried *in vacuo*. It is very soluble in H$_2$O, in which it is unstable. The *free base* has **b** 101-103°/11mm. It is a glucuronyl transferase inducer and a monoamine oxidase inhibitor. [von Braun et al. *Justus Liebigs Ann Chem* **445** 205 *1928*; Yeh and Mitchell *Experientia* **28** 298 *1972*; Langstrom et al. *Science* **225** 1480 *1984.*]

Pectic acid *[9046-40-6]* **M$_r$ (C$_6$H$_8$O$_6$)$_n$ ~500,000, amorphous, [α]$_D$ +250° (c 1, 0.1M NaOH).** Citrus pectic acid (500g) was refluxed for 18h with 1.5L of 70% EtOH and the suspension was filtered hot. The residue was washed with hot 70% EtOH and finally with ether. It was dried in a current of air, ground and dried for 18h at 80° under vacuum. [Morell and Link *J Biol Chem* **100** 385 *1933.*] It can be further purified by dispersing in water and adding just enough dilute NaOH to dissolve the pectic acid, then passing the soln through columns of cation- and anion-exchange resins [Williams and Johnson *Ind Eng Chem (Anal Ed)* **16** 23 *1944*], and precipitating with two volumes of 95% EtOH containing 0.01% HCl. The ppte is worked with 95% EtOH, then Et$_2$O, dried and ground.

Pectin *[9000-69-5]* **M$_r$ 25,000-100,000, amorphous.** Dissolved in hot water to give a 1% soln, then cooled, and made about 0.05M in HCl by addition of conc HCl, and ppted by pouring slowly, with vigorous stirring into two volumes of 95% EtOH. After standing for several hours, the pectin is filtered onto nylon cloth, then redispersed in 95% EtOH and stood overnight. The ppte is filtered off, washed with EtOH/Et$_2$O, then Et$_2$O and air dried.

D-(-)-Penicillamine **(*R*-3-mercapto-D-valine, 3,3-dimethyl-D-cysteine, from natural penicillin)** *[52-67-5]* **M 149.2, m 202-206°, 214-217°, [α]$_D^{21}$ -63° (c 1, N NaOH or pyridine), pK$_1^{20}$ 2.4 (CO$_2$H), pK$_2^{20}$ 8.0 (SH), pK$_3^{20}$ 10.68 (NH$_2$).** The melting point depends on the rate of heating (**m** 202-206° is obtained by starting at 195° and heating at 2°/min). It is soluble in H$_2$O and alcohols but insoluble in Et$_2$O, CHCl$_3$, CCl$_4$ and hydrocarbon solvents. Purified by dissolving in MeOH and adding Et$_2$O slowly. Dried *in vacuo* and stored under N$_2$. [Weight et al. *Angew Chem, Int Ed Engl* **14** 330 *1975*; Cornforth in *The Chemistry of Penicillin* (Clarke, Johnson and Robinson Eds) Princeton Univ Press, 455 *1949*; Polymorphism: Vidler *J Pharm Pharmacol* **28** 663 *1976.*] The D-S-*benzyl derivative* has **m** 197-198° (from H$_2$O), [α]$_D^{17}$ -20° (c 1, NaOH), -70° (N HCl).

L-(-)-Penicillamine *[1113-41-3]* **M 149.2, m 190-194°, 202-206°, 214-217°, [α]$_D^{21}$ +63° (c 1, N NaOH or pyridine).** Same as preceding entry for its enantiomer.

D-Penicillamine disulfide hydrate **(*S*,*S'*-di-[D-penicillamine] hydrate)** *[20902-45-8]* **M 296.4 + aq, m 203-204°(dec), 204-205°(dec), [α]$_D^{23}$ +27° (c 1.5, N HCl), -82° (c 0.8, N NaOH), pK$_{Est(1)}$~ 2.4 (CO$_2$), pK$_{Est(2)}$~ 10.7 (NH$_2$).** Purified by recrytn from EtOH or aqueous EtOH. [Crooks in *The Chemistry of Penicillin* (Clarke, Johnson and Robinson Eds) Princeton Univ Press, 469 *1949*; Use as a thiol reagent for proteins: Garel *Eur J Biochem* **123** 513 *1982*; Süs *Justus Liebigs Ann Chem* **561** 31 *1948.*]

Pepsin *[9001-75-6]* M_r 31,500(human), 6000(hog) [EC 3.4.23.1]. Rechromatographed on a column of Amberlite CG-50 using a pH gradient prior to use. Crystd from EtOH. [Richmond et al. *Biochim Biophys Acta* **29** 453 *1958*; Huang and Tang, *J Biol Chem* **244** 1085 *1969*, **245** 2189 *1970*.]

Pertussis toxin (from *Bordetella pertussis*) *[70323-44-3]* M_r 117,000. Purified by stepwise elution from 3 columns comprising Blue Sepharose, Phenyl Sepharose and hydroxylapatite, and SDS-PAGE [Svoboda et al. *Anal Biochem* **159** 402 *1986*; *Biochemistry* **21** 5516 *1982*; *Biochem J* **83** 295 *1978*.]

2-Phenylethyl-β-D-thiogalactoside *[63407-54-5]* M 300.4, m 108°, $[\alpha]_D^{23}$ -32.2° (c 5, MeOH). Recryst from H_2O and dried in air to give the 1.5.H_2O and has **m** 80°. Anhydrous surfactant is obtained by drying at 78° over P_2O_5. [Heilfrich and Türk *Chem Ber* **89** 2215 *1856*.]

Phenyl-β-D-galactopyranoside *[2818-58-8]* M 256.3, m 153-54°, 146-148°, 155-156°(dried at 105°), $[\alpha]_D^{20}$ -42° (c 1, H_2O). Recrystd from H_2O as 0.5H_2O. [Conchie and Hay *Biochem J* **73** 327 *1959*; IR: Whistler and House *Analyt Chem* 25 1463 *1953*.] It is an acceptor substrate for fucosyltransferase [Chester et al. *Eur J Biochem* **69** 583 *1976*].

Phenyl-β-D-glucopyranoside *[1464-44-4]* M 256.3, m 174-175° 174-176°, 176°, $[\alpha]_D^{20}$ -72.2° (c 1 for dihydrate, H_2O). Recrystd from H_2O as 2H_2O and can be dried *in vacuo* at 100°/P_2O_5. Dry preparation has $[\alpha]_D^{25}$ -70.7° (c 2, H_2O). [Robertson and Waters *J Chem Soc* 2729 *1930*; IR: Bunton et al. *J Chem Soc* 4419 *1955*; Takahashi *Yakugaku Zasshi (J Pharm Soc Japan)* **74** 7436 *1954*; Whixtler and House *Anal Chem* **25** 1463; UV: Lewis *J Am Chem Soc* **57** 898 *1935*.] It is a substrate for β−D-glucosidase [deBryne *Eur J Biochem* **102** 257 *1979*].

Phenylmercuric acetate (PhHgOAc) *[62-38-4]* M 336.7, m 148-151°, 149°, 151.8-152.8°. Small colourless lustrous prisms from EtOH. Its solubility in H_2O is 0.17% but it is more soluble in EtOH, Me_2CO and *C_6H_6. [Maynard *J Am Chem Soc* **46** 1510 *1925*; Coleman et al. *J Am Chem Soc* **59** 2703 *1937*; *J Am Pharm Assoc* **25** 752 *1936*.] See PhHgOH and $PhHgNO_3$.PhHgOH on p. 449 in Chapter 5.

Phenylmethanesulfonyl fluoride (PMSF) *[329-98-6]* M 174.2, m 90-91°, 92-93°. Purified by recrystn from *C_6H_6, pet ether or $CHCl_3$-pet ether. [Davies and Dick *J Chem Soc* 483 *1932*; cf Tullock and Coffman *J Org Chem* **23** 2016 *1960*.] It is a general protease inhibitor (specific for trypsin and chymotrypsin) and is a good substitute for diisopropylphosphoro floridate [Fahrney and Gould *J Am Chem Soc* **85** 997 *1963*].

Phosphatase alkaline (alkaline phosphatase) *[9001-78-9]* M_r ~40,000 (bovine liver), ~140,000 (bovine intestinal mucosa), 80,000 (*E.coli*) [EC 3.1.3.1]. The *E.coli* supernatant in sucrose (20%, 33mM) in Tris-HCl pH 8.0 was purified through a DEAE-cellulose column and recrystallised. To the column eluates in 0.125M NaCl is added $MgCl_2$ (to 0.01M) and brought to 50% saturation in $(NH_4)_2SO_4$ by adding the solid (0.20g/mL). The mixture is centrifuged to remove bubbles and is adjusted to pH 8.0 (with 2N NaOH). Saturated $(NH_4)_2SO_4$ at pH 8.0 is added dropwise until the soln becomes faintly turbid (~61% saturation). It is set aside at room temp for 1h (turbidity will increase). The mixture is placed in an ice bath for several minutes when turbidity disappears and a clear soln is obtained. It is then placed in a large ice bath at 0° (~5L) and allowed to warm slowly to room temperature in a dark room whereby crystals are formed appearing as a silky sheen. The crystals are collected by centrifugation at 25° if necessary. The crystalline solns are stable at room temperature for many months. They can be stored at 0°, but are not stable when frozen. Cysteine at 10^{-3}M and thioglycolic acid at 10^{-4}M are inhibitory. Inhibition is reversed on addition of Zn^{2+} ions. Many organic phosphates are good substrates for this phosphatase. [Molamy and Horecker *Methods Enzymol* **9** 639 *1966*; Torriani et al. *Methods Enzymol* **12b** 212 *1968*; Engstrom *Biochim Biophys Acta* **92** 71 *1964*.]

Alkaline phosphatase from rat *osteosarcoma* has been purified by acetone pptn, followed by chromatography on DEAE-cellulose, Sephacryl S-200, and hydroxylapatite. [Nair et al. *Arch Biochem Biophys* **254** 18 *1987*.]

3-sn-Phosphatidylethanolamine (L-α-cephalin, from Soya bean) *[39382-08-6]* M_r ~600-800, amorphous, $pK_{Est(1)}$~ 5.8 (PO_4^-), $pK_{Est(2)}$~ 10.5 (NH_2). Purified by dissolving in EtOH, adding $Pb(OAc)_2.3H_3O$ (30g in 100mL H_2O) until excess Pb^{2+} is present. Filter off the solid. Pass CO_2 gas through the soln until pptn of $PbCO_3$ ceases. Filter the solid off and evaporate (while bubbling CO_2) under vacuum. An equal volume of H_2O is added to the residual oil extracted with hexane. The hexane extract is washed with H_2O until the aqueous phase is free from Pb [test with dithizone (2 mg in 100 mL CCl_4 ; Feigel *Spot Tests* Vol I, Elsevier p. 10 *1954*]. The hexane is dried (Na_2SO_4), filtered and evaporated to give a yellow waxy solid which should be dried to constant weight *in vacuo*. It is practically insoluble in H_2O and Me_2CO, but freely soluble in $CHCl_3$ (5%) and Et_2O, and slightly soluble in EtOH. [Schofield and Dutton *Biochem Prep* **5** 5 *1957*.]

O-**Phosphocolamine 2-aminoethyl dihydrogen phosphate)** *[1071-23-4]* **M 141.1, m 237-240°, 242.3°, 234.5-244.5°, 244-245°(capillary), pK_1^{20} <1.5 (PO_4H_2), pK_2^{20} 5.77 (PO_4H^-), pK_3^{20} 10.26 (NH^+).** Purified by recrystn from aqueous EtOH as a hydrate (**m 140-141°**). Its solubility in H_2O is 17% and 0.003% in MeOH or EtOH at 22°. [Fölisch and Österberg *J Biol Chem* **234** 2298 *1959*; Baer aand Staucer *Can J Chem* **34** 434 *1956*; Christensen *J Biol Chem* **135** 399 *1940*.] It is a potent inhibitor of ornithine decarboxylase [Gilad and Gilad *Biochem Biophys Res Commun* **122** 277 *1984*].

Phosphoenolpyruvic acid monopotassium salt (KPEP) *[4265-07-0]* **M 206.1, pK_1^{25} 3.4 (CO_2), pK_2^{25} 6.35 (PO_4H^-) (for free acid).** It is purified *via* the monocyclohexylamine salt (see next entry). The salt (534mg) in H_2O (10mL) is added to Dowex 50Wx4 H^+ form (200-400 mesh, 2mL, H_2O washed) and stirred gently for 30min and filtered. The resin is washed with H_2O (6mL) and the combined solns are adjusted to pH 7.4 with 3N KOH (~1.4mL) and the volume adjusted to 18.4mL with H_2O to give a soln of 0.1M KPEP which can be lyophilised to a pure powder and is very good for enzyme work. It has been recrystd from MeOH-Et_2O. [Clark and Kirby *Biochem Prep* **11** 103 *1966*; Wold and Ballou *J Biol Chem* **227** 301 *1957*; Cherbuliez and Rabinowitz *Helv Chim Acta* **39** 1461 *1956*.]
The triNa salt *[5541-93-5]* **M 360.0,** is purified as follows: the salt (1g) is dissolved in MeOH (40mL) and dry Et_2O is added in excess. The white crystals are collected and dried over P_2O_5 at 20°. [*Chem Ber* **92** 952 *1959*.]

Phosphoenolpyruvic acid tris(cyclohexylamine) salt *[35556-70-8]* **M 465.6, m 155-180°(dec).** Recrystd from aqueous Me_2CO and dried in a vacuum. At 4° it is stable for >2 years and has IR at 1721cm^{-1} (C=O). [Wold and Ballou *J Biol Chem* **227** 301 *1957*; Clark and Kirby *Biochem Prep* **11** 103 *1966* for the monocyclohexylamine salt.]

D-3-Phosphoglyceric acid disodium salt (D-glycerate 3-phosphate di-Na salt) *[80731-10-8]* **M 230.0, $[\alpha]_D^{25}$ +7.7° (c 5, H_2O), -735° (in aq NH_4^+ molybdate), $pK_{Est(1)}$~1.0 (PO_4H_2), $pK_{Est(2)}$~ 6.66 (PO_4H^-) (for free acid).** Best purified by conversion to the Ba salt by pptn with $BaCl_2$ which is recrystd three times before conversion to the sodium salt. The Ba salt (9.5g) is shaken with 200mL of a 1:1 slurry of Dowex 50 (Na^+ form) for 2h. The mixture is filtered and the resin washed with H_2O (2 x 25mL). The combined filtrates (150mL) are adjusted to pH 7.0 and concentrated *in vacuo* to 30-40mL and filtered if not clear. Absolute EtOH is added to make 100mL and then *n*-hexane is added whereby a white solid and/or a second phase separates. When set aside at room temperature complete pptn of the Na salt as a solid occurs. The salt is removed by centrifugation, washed with Me_2CO, dried in air then in an oven at 55° to give a stable powder (4.5g). It did not lose weight when dried further over P_2O_5 at 78°/8h. The high rotation in the presence of $(NH_4)_6Mo_7O_{24}$ is not very sensitive to the concentration of molybdate or pH as it did not alter appreciably in 1/3 volume between 2.5 to 25% (w/v) of molybdate or at pH values ranging between 4 and 7. [Cowgill *Biochim Biophys Acta* **16** 613 *1955*; Embdan, Deuticke and Kraft *Hoppe Seyler's Z Physiol Chem* **230** 20 *1934*.]

Phospholipids. For the removal of ionic contaminants from raw zwitterionic phospholipids, most lipids were purified twice by mixed-bed ionic exchange (Amberlite AB-2) of methanolic solutions. (About 1g of lipid in 10mL of MeOH). With both runs the first 1mL of the eluate was discarded. The main fraction of the solution was evaporated at 40°C under dry N_2 and recryst three times from *n*-pentane. The resulting white powder was dried for about 4h at 50° under reduced pressure and stored at 3°. Some samples were purified by mixed-bed ion exchange of aqueous suspensions of the crystal/liquid crystal phase. [Kaatze et al. *J Phys Chem* **89** 2565 *1985*.]

Phosphoproteins (various). Purified by adsorbing onto an iminodiacetic acid substituted agarose column to which was bound ferric ions. This chelate complex acted as a selective immobilised metal affinity adsorbent for phosphoproteins. [Muszyfiska et al. *Biochemistry* **25** 6850 *1986*.]

5'-Phosphoribosyl pyrophosphate synthetase (from human erythrocytes, or pigeon or chicken liver) *[9015-83-2]* M_r **60,000, [EC 2.7.6.1].** Purified 5100-fold by elution from DEAE-cellulose, fractionation with ammonium sulfate, filtration on Sepharose 4B and ultrafiltration. [Fox and Kelley *J Biol Chem* **246** 5739 *1971*; Flaks *Methods Enzymol* **6** 158 *1963*; Kornberg et al. *J Biol Chem* **15** 389 *1955*.]

O-**Phospho-L-serine** *[407-41-0]* **M 185.1, m 175-176°, $[\alpha]_D^{20}$ +4.3° (c 3.2, H$_2$O), +16.2° (c 3.2, 2N HCl), pK$_1^{25}$ <1 (PO$_4$H$_2$), pK$_2^{25}$ 2.08 (CO$_2$H), pK$_3^{25}$ 5.65 (PO$_4$H$^-$), pK$_4^{25}$ 9.74 (NH$_3$$^+$).** Recrystd by dissolving 10g in H$_2$O (150mL) at 25°, stirring for up to 20min. Undissolved material is filtered off (Büchner) and 95% EtOH (85mL) is added dropwise during 4min, and set aside at 25° for 3h then at 3° overnight. The crystals are washed with 95% EtOH (100mL) then dry Et$_2$O (50mL) and dried in a vacuum (yield 6.5g). A further quantity (1.5mg) can be obtained by keeping the mother liquors and washings at -10° for 1 week. The *DL-isomer* has **m** 167-170°(dec) after recrystn from H$_2$O + EtOH or MeOH. [Neuhaus and Korkes *Biochem Prep* **6** 75 *1958*; Neuhaus and Byrne *J Biol Chem* **234** 113 *1959*; IR: Fölsch and Mellander *Acta Chem Scand* **11** 1232 *1957*.]

O-**Phospho-L-threonine (L-threonine-*O*-phosphate)** *[1114-81-4]* **M 199.1, m 194°(dec), $[\alpha]_D^{24}$ -7.37° (c 2.8, H$_2$O) (pK as above).** Dissolve in the minimum volume of H$_2$O, add charcoal, stir for a few min, filter and apply onto a Dowex 50W (H$^+$ form) then elute with 2N HCl. Evaporate the eluates under reduced pressure whereby the desired fraction produced crystals of the phosphate which can be recrystd from H$_2$O-MeOH mixtures and the crystals are then dried *in vacuo* over P$_2$O$_5$ at ~80°. [de Verdier *Acta Chem Scand* **7** 196 *1953*.]

O-**Phospho-L-tyrosine (L-tyrosine-*O*-phosphate)** *[21820-51-9]* **M 261.2, m 225°, 227°, 253°, $[\alpha]_D^{20}$ -5.5° (c 1, H$_2$O), -9.2° (c 1, 2N HCl), pK$_{Est(1)}$~ 1.6 (PO$_4$H$_2$), pK$_{Est(2)}$~ 2.02 (CO$_2$H), pK$_{Est(3)}$~ 5.65 (PO$_4$H$^-$), pK$_{Est(4)}$ 9.2 (NH$_3$$^+$).** Purified by recrystn from H$_2$O or H$_2$O + EtOH. [Levene and Schormüller *J Biol Chem* **100** 583 *1933*; Posternak and Graff *Helv Chim Acta* **28** 1258 *1945*.]

Phytol (*d*-3,7*R*,11*R*,15-tetramethylhexadec-2-en-1-ol) *[150-86-7]* **M 296.5, b 145°/0.03mm, 150-151°/0.06mm, 202-204°/10mm, d$_4^{25}$ 0.8497, n$_D^{25}$ 1.437, $[\alpha]_D^{22}$+0.06° (neat).** Purified by distn under high vacuum. It is almost insoluble in H$_2$O but soluble in most organic solvents. It has UV λ_{max} at 212nm (log ε 3.04) in EtOH and IR ν at 3300 and 1670cm^{-1}. [Demole and Lederer *Bull Soc Chim Fr* 1128 *1958*; Burrell *J Chem Soc (C)* 2144 *1966*; Bader *Helv Chim Acta* **34** 1632 *1951*.]

D-Pipecolinic acid (*R*-piperidine-2-carboxylic acid) *[1723-00-8]* **M 129.2, m 264°(dec), 267°(dec), ~280°(dec), $[\alpha]_D^{19}$+26.2° (c 2, H$_2$O), $[\alpha]_D^{25}$+35.7° (H$_2$O), pK$_1^{20}$ 2.29 (CO$_2$H), pK$_2^{20}$ 10.77 (NH$^+$).** Recrystallises as platelets from EtOH and is soluble in H$_2$O. The *hydrochloride* has **m** 256-257°(dec) from H$_2$O and $[\alpha]_D^{25}$ +10.8° (c 2, H$_2$O). [Lukés et al. *Collect Czech Chem Commun* **22** 286 *1957*; Bayerman *Recl Trav Chim Pays-Bas* **78** 134 *1959*; Asher et al. *Tetrahedron Lett* **22** 141 *1981*.]

L-Pipecolinic acid (*S*-piperidine-2-carboxylic acid) *[3105-95-1]* **M 129.2, m 268°(dec), 271°(dec), ~280°(dec), $[\alpha]_D^{20}$ -26° (c 4, H$_2$O), $[\alpha]_D^{25}$ -34.9° (H$_2$O).** Recrystd from aqueous EtOH and sublimes as needles in a vacuum. It is sparingly soluble in absolute EtOH, Me$_2$CO and CHCl$_3$ but insoluble in Et$_2$O. The *hydrochloride* has **m** 258-259°(dec, from MeOH) and $[\alpha]_D^{25}$ -10.8° (c 10, H$_2$O). [Fujii and Myoshi *Bull Chem Soc Jpn* **48** 1241 *1975*.]

Piperidine-4-carboxylic acid (isonipecotic acid) *[498-94-2]* **M 129.2, m 336°(dec, darkens at ~300°), pK$_{Est(1)}$~ 4.3 (CO$_2$H), pK$_{Est(2)}$~ 10.6 (NH$^+$).** Recrystallises from H$_2$O or EtOH as needles. The *hydrochloride* recrystallises from H$_2$O or aqueous HCl and has **m** 293°dec (298°dec, 300°dec). [Wibaut *Recl Trav Chim Pays-Bas* **63** 141 1944; IR: Zacharius et al. *J Am Chem Soc* **76** 2908 *1954*.]

Pituitary Growth Factor **(from human pituitary gland)** *[336096-71-0]*. Purified by heparin and copper affinity chromatography, followed by chromatography on carboxymethyl cellulose (Whatman 52). [Rowe et al. *Biochemistry* **25** 6421 *1986*.]

Plasmids. These are circular lengths of DNA which invade bacteria or other cells e.g. insect cells, yeast cells, and have sequences which are necessary for their replication using enzymes and other ingredients, e.g. nucleotides, present in the cells. They contain engineered, or already have, genes which produce enzymes that provide the cells with specific antibiotic resistance and are thus useful for selecting bacteria containing specific plasmids. Plasmids have been extremely useful in molecular biology since they can be very easily identified (from their size or the sizes of the DNA fragments derived from their restriction enzyme digests) and can be readily engineered *in vitro* (outside the cells). Genes coding for specific enzymes or other functional proteins can be inserted into these plasmids which have DNA sequences that allow the expression of large quantities of bacteria or non-bacterial (e.g. human) proteins. They have also been engineered in such a way as to produce 'fusion proteins' (in which the desired protein is fused with a specific "reporter, marker or carrier protein" which will facilitate the isolation of the desired protein (e.g. by binding strongly to a nickel support) and then the desired protein can be cleaved from the eluted fusion protein and obtained in very pure form. A large number of plasmids with a variety of sequences for specific purposes are commercially available in very pure form. They can be used to infect cells and can be isolated and purified from cell extracts in large amounts using a number of available procedures. These procedures generally involve lysis of the cells (e.g. with alkaline sodium dodecylsulfate, SDS), separation from nuclear DNA, precipitation of plasmid DNA from the cell debris, adsorbing it on columns which specifically bind DNA, and then eluting the DNA from the column (e.g. with specific Tris buffers as recommended by the suppliers) and precipitating it (e.g. with Tris buffer in 70% EtOH at -70°C) The purity is checked in agarose gel (containing ethidium bromide to visualise the DNA) by electrophoresis. A large number of plasmids are now commercially available (see Clontech GmbH, http://www.clontech.com; Invitrogen http://www.invitrogen.com, among other suppliers) used as vectors for bacterial, mammalian, yeast and baculovirus expression.

Podophylotoxin *[518-28-5]* **M 414,4, m 181-181°, 183-184°, 188-189°, $[\alpha]_D^{20}$ -132° (c 1, CHCl₃).** Recrystallises form $^*C_6H_6$ (with $0.5C_6H_6$), EtOH-$^*C_6H_6$, aqueous EtOH (with 1-$1.5H_2O$, m 114-115°) and CH_2Cl_2-pentane. When dried at 100°/10 mm it has **m** 183-184°. [UV: Stoll et al. *Helv Chim Acta* **37** 1747 *1954*; IR: Schecler et al. *J Org Chem* **21** 288 *1956*.] Inhibitor of microtubule assembly [Prasad et al. *Biochemistry* **25** 739 *1986*].

Polyethylene glycol *[25322-68-3]* **M_r various, from ~200 to ~35,000.** May be contaminated with aldehydes and peroxides. Methods are available for removing interfering species. [Ray and Purathingal *Anal Biochem* **146** 307 *1985*.]

Polypeptides. These are a string of α-amino acids usually with the natural S(L) [L-cysteine is an exception and has the R absolute configuration] or sometimes "unnatural" R(D) configuration at the α-carbon atom. They generally have less than ~100 amino acid residues. They can be naturally occurring or, because of their small size, can be synthesised chemically from the desired amino acids. Their properties can be very similar to those of small proteins. Many are commercially available, can be custom made commercially or locally with a peptide synthesiser. They are purified by HPLC and can be used without further purification. Their purity can be checked as described under proteins.

Porphobilinogen (5-amino-4-carboxymethyl-1H-pyrrole-3-propionic acid) *[487-90-1]* **M 226.2, m 172-175°(dec), 175-180°(dec, darkening at 120-130°), pK₁ 3.70 (4-CH₂CO₂H), pK₂ 4.95 (3-CH₂CH₂CO₂H), pK₃ 10.1 (NH⁺).** Recrystallises as the monohydrate (pink crystals) from dil NH_4 OAc solns of pH 4, and is dried *in vacuo*. The *hydrochloride monohydrate* has m 165-170°(dec) (from dilute HCl). [Jackson and MacDonald *Can J Chem* **35** 715 *1957*, Westall *Nature* **170** 614 *1952*; Bogarad *J Am Chem Soc* **75** 3610 *1953*.]

Porphyrin a (from ox heart) *[5162-02-1]* **M 799.0, m dec on heating.** Purified on a cellulose powder column followed by extraction with 17% HCl and fractionation with HCl. [Morell et al. *Biochem J* **78**

793 *1961*.] Recrystd from CHCl$_3$/Pet ether or Et$_2$O/*C$_6$H$_6$ [detailed UV-VIS and NMR date: Caughey et al. *J Biol Chem* **250** 7602 *1975*; Lemberg *Adv Enzymol* **23** 265 *1961*].

Prazosin hydrochloride (2[4-{(2-furoyl)piperazin-1-yl}4-amino-6,7-dimethoxyquinazoline hydrochloride) *[19237-84-4]* M **419.9, m 278-280°, 280-282°, pK 6.5.** It is recrystd by dissolving in hot MeOH adding a small volume of MeOH-HCl (dry MeOH saturated with dry HCl gas) followed by dry Et$_2$O until crystn is complete. Dry *in vacuo* over solid KOH till odour of HCl is absent. It has been recrystd from hot H$_2$O, the crystals were washed with H$_2$O, and the H$_2$O was removed azeotropically with CH$_2$Cl$_2$, and dried in a vacuum. [NMR and IR: Honkanen et al. *J Heterocycl Chem* **17** 797 *1980*; *cf* Armarego and Reece *Aust J Chem* **34** 1561 *1981*.] It is an antihypertensive drug and is an α_1-adrenergic antagonist [Brosman et al. *Proc Natl Acad Sci USA* **82** 5915 *1985*].

Prednisolone acetate (21-acetoxypregna-1,4-diene-11β-17α-diol-3,20-dione) *[52-21-1]* **M 402.5, m 237-239°, 240-242°, 240-243°, 244°, $[\alpha]_D^{20}$ +116° (c 1, dioxane).** Recrystd from EtOH, Me$_2$CO, Me$_2$CO-hexane, and has UV λ_{max} at 243nm in EtOH. [Joly et al. *Bull Soc Chim Fr* 366 *1958*; Herzog et al. *J Am Chem Soc* **77** 4781 *1955*.]

Primaquine diphosphate (*RS*- 8-[4-amino-1-methylbutylamino]-6-methoxyquinoline diphosphate) *[63-45-6]* **M 455.4, m 197-198°, 204-206°, pK$_{Est(1)}$~ 3.38 (ring N$^+$), pK$_{Est(2)}$~ 10.8 NH$_3^+$).** It forms yellow crystals from 90% aq EtOH and is moderately soluble in H$_2$O. The *oxalate salt* has m 182.5-185° (from 80% aq EtOH) and the *free base* is a viscous liquid **b** 165-170°/0.002mm, 175-177°/2mm. [Elderfield et al. *J Am Chem Soc* **68** 1526 *1964*; **77** 4817 *1955*.]

Procaine hydrochloride (Novocain, 2-diethylaminoethyl-4-aminobenzoate) *[51-05-8]* **M 272.8, m 153-156°, 154-156°, 156°, pK$_{Est(1)}$~ 2.52 (NH$_2^+$) pK$_2^{20}$ 9.0 (Et$_2$N$^+$).** Recrystd from aqueous EtOH. It has solubility at 25° in H$_2$O (86.3%), EtOH (2.6%) and Me$_2$CO (1%), it is slightly soluble in CHCl$_3$ but is almost insoluble in Et$_2$O. The anhydrous *free base* is recrystd from ligroin or Et$_2$O and has **m 61°.** [Einhorn *Justus Liebigs Ann Chem* **371** 125 *1909*; IR: Szymanski and Panzica *J Amer Pharm Assoc* **47** 443 *1958*.]

L-Propargylglycine (*S*-2-aminopent-4-ynoic acid) *[23235-01-0]* **M 113.1, m 230°(dec starting at 210°), $[\alpha]_D^{20}$ -35° (c 1, H$_2$O), -4° (c 5, 5N HCl), pK$_{Est(1)}$~ 2.3 (CO$_2$H), pK$_{Est(2)}$~ 9.8 (NH$_2$).** The acid crystallises readily when ~4g in 50mL H$_2$O is treated with abs EtOH at 4°/ 3hrs, and is collected washed with cold abs EtOH and Et$_2$O and dried in vac. Also recrystallises from aqueous Me$_2$CO, R$_F$ on SiO$_2$ TLC plates with *n*-BuOH-H$_2$O-AcOH (4:1:1) is 0.26. The *racemate* has **m 238-240°.** [Leukart et al. *Helv Chim Acta* **59** 2181 *1976*; Eberle and Zeller *Helv Chim Acta* **68** 1880 *1985*; Jansen et al. *Recl Trav Chim Pays-Bas* **88** 819 *1969*.] It is a suicide inhibitor of γ-cystathionase and other enzymes [Washtier and Abeles *Biochemistry* **16** 2485 *1977*; Shinozuka et al. *Eur J Biochem* **124** 377 *1982*].

Propidium iodide (3,8-diamino-5-(3-diethylaminopropyl)-6-phenylphenantridinium iodide methiodide) *[25535-16-4]* **M 668.4, m 210-230°(dec), pK$_{Est(1)}$~ 4 (aniline NH$_2$), pK$_{Est(2)}$~ 8.5 (EtN$_2$).** Recrystd as red crystals from H$_2$O containing a little KI. It fluoresces strongly with nucleic acids. [Eatkins *J Chem Soc* 3059 *1952*.] **TOXIC.**

***R*-Propranalol hydrochloride (*R*-1-isopropylamino-3-(1-naphthyloxy)-2-propanol HCl)** *[13071-11-9]* **M 295.8, m 192°, 193-195°, $[\alpha]_D^{20}$ -25° (c 1, EtOH), pK^{20} 9.5 (for free base).** Recryst from *n*-PrOH or Me$_2$CO. It is soluble in H$_2$O and EtOH but is insoluble in Et$_2$O, *C$_6$H$_6$ or EtOAc. The *racemate* has **m 163-164°,** and the *free base* recryst from cyclohexane has **m 96°.** [Howe and Shanks *Nature* **210** 1336 *1966*.] The *S*-isomer (below) is the physiologically active isomer.

***S*-Propranalol hydrochloride (*S*-1-isopropylamino-3-(1-naphthyloxy)-2-propanol HCl)** *[4199-10-4]* **M 295.8, m 192°, 193-195°, $[\alpha]_D^{20}$ +25° (c 1, EtOH) pK^{20} 9.5.** See preceding entry for physical properties. The is the active isomer which blocks isoprenaline tachycardia and is a β-adrenergic blocker. [Leclerc et al. *Trends Pharmacol Sci* **2** 18 *1981*; Howe and Shanks *Nature* **210** 1336 *1966*.]

Protamine kinase (from rainbow trout testes) *[37278-10-7]* **[EC 2.7.1.70].** Partial purification by hydoxylapatite chromatography followed by biospecific chromatography on nucleotide coupled Sepharose 4B (the nucleotide was 8-(6-aminohexyl)amine coupled cyclic-AMP). [Jergil et al. *Biochem J* **139** 441 *1974.*]

Protamine sulfate (from herring sperm) *[9007-31-2]* $[\alpha]_D^{22}$ **-85.5°** (satd H_2O), pK **7.4-8.0.** A strongly basic protein (white powder, see pK) used to ppte nucleic acids from crude protein extracts. It dissolved to the extent of 1.25% in H_2O. It is freely soluble in hot H_2O but separates as an oil on cooling. It has been purified by chromatography on an IRA-400 ion-exchange resin in the SO_4^{2-} form and washed with dilute H_2SO_4. Eluates are freeze-dried under high vacuum below 20°. This method is used to convert proteamine and protamine hydrochloride to the sulfate. [UV: Rasmussen *Hoppe Seyler's Z Physiol Chem* **224** 97 *1934*; Ando and Sawada *J Biochem (Tokyo)* **49** 252 *1961*; Felix and Hashimoto *Hoppe Seyler's Z Physiol Chem* **330** 205 *1963*]

Protease nexin (From cultured human fibroblasts) *[148263-58-5].* Purified by affinity binding of protease nexin to dextran sulfate-Sepharose. [Farrell et al. *Biochem J* **237** 707 *1986.*]

Proteins. These are usually naturally occurring (or deliberately synthesised in microorganisms, e.g. bacteria, insect cells, or animal tissues), and are composed of a large number of α-*S* (L)amino acids residues (except for L-cysteine which has the *R* absolute configuration), selected from the 20 or so natural amino acids, in specific sequences and in which the α–amino group forms an amide (peptide) bond with the α–carboxyl group of the neighboring amino acid. The number of residues are ususally upwards of 100. Proteins with less than 100 amino acids are better referred to as **polypeptides**. Aqueous soluble proteins generally fold into ball-like structures mainly with hydrophilic residues on the outside of the "balls" and hydrophobic residues on the inside. Proteins can exist singly or can for dimers, trimers, tetramers etc, consisting of similar or different protein subunits. They are produced by cells for a large variety of functions, e.g. enzymology, reaction mediation as in regulation of DNA synthesis or chaperonins for aiding protein folding, formation of pores in membranes for transport of ions or organic molecules, or for intra or inter cellular signalling etc. The purity of proteins can be checked in denaturing (SDS, sodium dodecylsulfate) or non-denaturing polyacrylamide gels using electrophoresis (PAGE), and staining appropriately (e.g. with Coommassie Blue, followed by silver staining for higher sensitivity). If the protein is partly impure then it should be purified further according to the specific literature procedures for the individual protein (see specific proteins in the *Methods Enzymol* , Wiley series).

Proteoglycans (from cultured human muscle cells). Separated by ion-exchange HPLC using a Bio-gel TSK-DEAE 5-PW analytical column. [Harper et al. *Anal Biochem* **159** 150 *1986.*]

Prothrombin (Factor II, from equine blood plasma) *[9001-26-7]* M_r **72,000.** Purified by two absorptions on a barium citrate adsorbent, followed by decomposition of the adsorbents with a weak carboxylic cation-exchanger (Amberlite IRF-97), isoelectric pptn (pH 4.7-4.9) and further purification by chromatography on Sephadex G-200 or IRC-50. Finally recrystd from a 1% soln adjusted to pH 6.0-7.0 and partial lyophilisation to *ca* 1/5 to 1/10th vol and set aside at 2-5° to crystallise. Occasionally seeding is required. [Miller *Biochem Prep* **13** 49 *1971.*]

Protoporphyrin IX (3,18-divinyl-2,7,13,17-tetramethylporphine-8,12-dipropionic acid, ooporphyrin) *[553-12-8]* **M 562.7,** pK_{Est} ~ **4.8.** Purified by dissolving (4g) in 98-100% HCOOH (85mL), diluting with dry Et_2O (700mL) and keeping at 0° overnight. The ppte is collected and washed with Et_2O then H_2O and dried in a vacuum at 50° over P_2O_5. It has been recrystd from aqueous pyridine and from Et_2O as monoclinic, brownish-yellow prisms. UV λ_{max} values in 25% HCl are 557.2, 582.2 and 602.4nm. It is freely soluble in ethanolic HCl, AcOH, $CHCl_3$, and Et_2O containing AcOH. It forms sparingly soluble diNa and diK salts. [Ramsey *Biochem Prep* **3** 39 *1953*; UV: Holden *Aust J. Exptl Biol and Med Sci* **15** 412 *1937*; Garnick *J Biol Chem* **175** 333 *1948*; IR: Falk and Willis *Aust J Sci Res* [A] **4** 579 *1951.*]
The **Dimethyl ester** *[5522-66-7]* **M 590.7, m 228-230°,** is prepared by dissolving (0.4g) in $CHCl_3$ (33mL) by boiling for a few min, then diluting with boiling MeOH (100mL) and refrigerating for 2 days. The crystals are collected, washed with $CHCl_3$-MeOH (1:9) and dried at 50° in a vacuum (yield 0.3g). UV has λ_{max} 631, 576, 541, 506 and 407nm in $CHCl_3$ and 601, 556 and 406nm in 25% HCl. [Ramsey *Biochem Prep* **3** 39 *1953.*]

Prymnesin (**toxic protein from phytoflagellate** *Pyrymnesium parvum*) *[11025-94-8]*. Purified by column chromatography, differential soln and pptn in solvent mixtures and differential partition between diphasic mixtures. The product has at least 6 components as observed by TLC. [Ulitzur and Shilo *Biochim Biophys Acta* **301** 350 *1970*.]

Pterin-6-carboxylic acid (**2-amino-4-oxo-3,4-dihydropteridine-6-carboxylic acid**) *[948-60-7]* **M 207.2, m >360°, pK_1^{20} 1.43, pK_2^{20} 2.88, pK_3^{20} 7.72.** Yellow crystals by repeated dissolution in aqueous NaOH and adding aqueous HCl. It has UV with λ_{max} at 235, 260 and 265nm (ε 11000, 10500 and 9000) in 0.1N HCl and 263 and 365nm (ε 20500 and 9000) in 0.1N NaOH. [UV: Pfleiderer et al. *Justus Liebigs Ann Chem* **741** 64 *1970*; Stockstad et al. *J Am Chem Soc* **70** 5 *1948*; Fluorescence: Kavanagh and Goodwin *Arch Biochem* **20** 315 *1949*.]

Purine-9-β-ribifuranoside (**Nebularin**) *[550-33-4]* **M 252.2, m 178-180°, 181-182°, $[\alpha]_D^{25}$ -48.6° (c 1, H_2O), -22° (c 0.8, 0.1N HCl) and -61° (c 0.8, 0.1N NaOH), pK 2.05.** Recrystd from butanone + MeOH or EtOH and forms a MeOH photo-adduct. It is a strong inhibitor of adenosine deaminase [EC 3.5.4.4]. [Nair and Weichert *Bioorg Chem* **9** 423 *1980*; Löfgren et al. *Acta Chem Scand* **7** 225 *1953*; UV: Brown and Weliky *J Biol Chem* **204** 1019 *1953*.]

Puromycin dihydrochloride (*O*-**methyl-l-tyrosine[N^6,N^6-dimethylaminoadenosin-3'-yl-amide]**) *[58-58-2]* **M 616.5, m 174°, $[\alpha]_D^{25}$ -11° (free base in EtOH), pK_1 6.8, pK_2 7.2.** Purified by recrystn from H_2O. The *free base* has **m** 175.5-177° (172-173°) (from H_2O). The *sulfate* has **m** 180-187° dec (from H_2O), and the *picrate monohydrate* has **m** 146-149° (from H_2O). [Baker et al. *J Am Chem Soc* **77** 1 *1955*; Fryth et al. *J Am Chem Soc* **80** 3736 *1958*.] It is an inhibitor of aminopeptidase and terminates protein synthesis [Reboud et al. *Biochemistry* **20** 5281 *1981*].

Pyridoxal hydrochloride *[65-22-5]* **M 203.6, m 176-180°(dec), pK_1^{20} 4.23 (3-OH), pK_2^{20} 8.7 (Pyridinium$^+$), pK_3^{20} 13.04 (CH$_2$OH?).** Dissolve in water and adjust the pH to 6 with NaOH. Set aside overnight to crystallise. The crystals are washed with cold water, dried in a vacuum desiccator over P_2O_5 and stored in a brown bottle at room temperature. [Fleck and Alberty *J Phys Chem* **66** 1678 *1962*.]

Pyridoxal-5'-phosphate monohydrate (**PLP, codecarboxylase**) *[54-47-7]* **M 265.2, pK_1^{25} <2.5 (PO$_4$), pK_2^{25} 4.14 (3-OH), pK_3^{25} 6.20 (PO$_4$), pK_4^{25} 8.69 (pyridinium$^+$).** It has been purified by dissolving 2g in H_2O (10-15mL, in a dialysis bag a third full) and dialysing with gentle stirring against 1L of H_2O (+ two drops of toluene) for 15h in a cold room. The dialysate is evaporated to 80-100mL then lyophilised. Lemon yellow microscopic needles of the monohydrate remain when all the ice crystals have been removed. The purity is checked by paper chromatography (in EtOH or *n*-PrOH-NH$_3$) and the spot(s) visualised under UV light after reaction with *p*-phenylene diamine, NH$_3$ and molybdate. Solutions stored in a freezer are 2-3% hydrolysed in 3-weeks. At 25°, only 4-6% hydrolysis occurs even in N NaOH or HCl, and 2% is hydrolysed at 37° in 1 day - but is complete at 100° in 4h. Best stored as dry solid at -20°. In aqueous acid the solution is colourless but is yellow in alkaline solutions. It has UV λ_{max} at 305nm (ε 1100) and 380nm (ε 6550) in 0.1 N NaOH; 330nm (ε 2450) and 388nm (ε 4900) in 0.05M phosphate buffer pH 7.0 and 295nm (ε 6700) in 0.1N HCl. [Peterson et al. *Biochemical Preparations* **3** 34, 119 *1953*.] The *oxime* dec at 229-230° and is practically insoluble in H_2O, EtOH and Et_2O. The *O-methyloxime* decomposes at 212-213°. [Heyl et al *J Am Chem Soc* **73** 3430 *1951*.] It has also been purified by column chromatography through Amberlite IRC-50 (H$^+$) [Peterson and Sober *J Am Chem Soc* **76** 169 *1954*].

Pyridoxamine hydrochloride *[5103-96-8, 524-36-7 (free base)]* **M 241.2, m 226-227°(dec), pK_1^{25} 3.54 (3-OH), pK_2^{25} 8.21 (ring N$^+$), pK_3^{25} 10.63 (NH$_2$).** Crystd from hot MeOH. The *free base* crysts from EtOH, has **m** 193-193.5° [Harris et al. *J Biol Chem* **154** 315 *1944*, *J Am Chem Soc* **66** 2088 *1944*].

Pyridoxine hydrochloride see **Vitamin B$_6$.**

Pyruvate kinase isoenzymes (**from** *Salmonella typhimurium*) *[9001-59-6]* **M$_r$ 64,000, [EC 2.7.1.40], amorphous.** Purified by (NH$_4$)$_2$SO$_4$ fractionation and gel filtration, ion-exchange and affinity chromatography. [Garcia-Olalla and Garrido-Pertierra *Biochem J* **241** 573 *1987*.]

Quinacrine [Atebrine, 3-chloro-9(4-diethylamino-1-methyl)butylamino-7-methoxy)acridine] dihydrochloride. *[69-05-6]* M 472.9, m 248-250°(dec), pK_1^{30} -6.49 (aq H_2SO_4), pK_2^{30} 7.73 (ring NH^+), pK_3^{30} 10.18 (Et_2N). Cryst from H_2O (sol 2.8% at room temp) as yellow crystals. Slightly sol in MeOH and EtOH. Antimalarial, antiprotozoal and intercalates DNA. [Wolfe *Antibiot* **3** (Springer-Verlag) 203 *1975*.]

Quisqualic acid (3-[3,5-dioxo-1,2,4-oxadiazolin-2-yl]-L-alanine) *[52809-07-1]* M 189.1, m 190-191°, $[\alpha]_D^{20}$ +17° (c 2, 6M HCl), $pK_{Est(1)}$~ 2.1 (CO_2H), $pK_{Est(2)}$~ 8.9 (NH_2). It has been purified by ion-exchange chromatography on Dowex 50W (x 8, H^+ form), the desired fractions are lyophilised and recrystd from H_2O-EtOH. It has IR (KBr) ν: 3400-2750br, 1830s, 1775s, 1745s and 1605s cm^{-1}; and ^1H NMR (NaOD/D_2O, pH 13) δ: 3.55-3.57 (1H m, X of ABX, H-2), 3.72-3.85 (2H, AB of ABX, H-3), ^{13}C NMR (D_2O) δ: 50.1t, 53.4d, 154.8s, 159.7s and 171.3s. [Baldwin et al. *J Chem Soc, Chem Commun* 256 *1985*.] It is a quasiqualate receptor agonist [Joels et al. *Proc Natl Acad Sci USA* **86** 3404 *1989*].

Renal dipeptidase (from porcine kidney cortex) *[9031-96-3]* M_r 47,000 [EC 3.4.13.11]. Purified by homogenising the tissue, extracting with Triton X-100, elimination of insoluble material, and ion-exchange, size exclusion and affinity chromatography. [Hitchcock et al. *Anal Biochem* **163** 219 *1987*.]

Restriction enzymes (endonucleases). These are enzymes which cleave double stranded DNA (linear or circular) at specific nucleotide sequences within the DNA strands which are then used for cloning (by ligating bits of DNA sequences together) or used for identifying particular DNA materials, e.g. plasmids, genes etc. A very large number of restriction enzymes are now available commercially and are extensively used in molecular biology. They are highly specific for particular nucleotide arrangements and are sensitive to the reaction conditions, e.g. composition of the medium, pH, salt concentration, temperature etc, which have to be strictly adhered to. The enzymes do not require further purification and the reaction conditions are also provided by the suppliers from which the necessary reaction media can also be purchased (see commercial catalogues).

Retinal (Vitamin A aldehyde), Retinoic acid (Vitamin A acid), Retinyl acetate, Retinyl palmitate see entries in Chapter 4.

Reverse transcriptase (from avian or murine RNA tumour viruses) *[9068-38-6]* [EC 2.7.7.49]. Purified by solubilising the virus with non-ionic detergent. Lysed virions were adsorbed on DEAE-cellulose or DEAE-Sephadex columns and the enzyme eluted with a salt gradient, then chromatographed on a phosphocellulose column and enzyme activity eluted in a salt gradient. Purified from other viral proteins by affinity chromatography on a pyran-Sepharose column. [Verna *Biochim Biophys Acta* **473** 1 *1977*; Smith *Methods Enzymol* **65** 560 *1980*; see commercial catalogues for other transcriptases.]

Riboflavin *[83-88-5]* M 376.4, m 295-300°(dec), $[\alpha]_D$ -9.8° (H_2O), -125° (c 5, 0.05N NaOH), pK_1 1.7, pK_2 9.69 (10.2, acidic NH). Crystd from 2M acetic acid, then extracted with $CHCl_3$ to remove lumichrome impurity. [Smith and Metzler *J Am Chem Soc* **85** 3285 *1963*.] Has also been crystd from water. (See also p. 575.)

Riboflavin-5'-phosphate (Na salt, $2H_2O$) *[130-40-5]* M 514.4. See flavin mononucleotide (FMN) on p. 535.

D-(+)-Ribonic acid-γ-lactone *[5336-08-3]* M 148.12, m 80°, 84-86°, $[\alpha]_D^{20}$ +18.3° (c 5, H_2O). Purified by recrystn from EtOAc. The *tribenzoate* has m 54-56° (from AcOH), $[\alpha]_D^{25}$ +27° (c 2.37, Me_2NCHO) and the *3,5-O-benzylidene* derivative has m 230-231.5° (needles from Me_2CO-pet ether), $[\alpha]_D^{25}$ -177° ($CHCl_3$). [Chen and Joulié *J Org Chem* **49** 2168 *1984*; Zinner and Voigt *J Carbohydr Res* **7** 38 *1968*.]

Ribonuclease (from human plasma) *[9001-99-4]* M_r ~13,700, [EC 3.1.27.5], amorphous. Purified by $(NH_4)_2SO_4$ fractionation, followed by PC cellulose chromatography and affinity chromatography (using Sepharose 4B to which $(G)_n$ was covalently bonded). [Schmukler et al. *J Biol Chem* **250** 2206 *1975*.]

RNA (ribonucleic acids). Ribonucleic acids are like DNA except that the 2'-deoxy-D-ribose moiety is replaced by a D-ribose moiety and the fourth nucleotide thymidylic acid (T) is replaced by uridylic acid (U). RNA does not generally form complete douplex molecules like DNA, i.e. it is generally monomeric, except in certain viruses. The two main classes of RNA are **messenger-RNA** (mRNA) and **transfer-RNA** (tRNA). mRNA transcribed from the DNA gene followed by the splicing out of the non-coding nucleotides (of the introns) and codes for a specific gene. There are many different tRNAs, at least one of which links to a specific α-amino acid, that bind to the mRNA *via* the ribosome (a set of proteins) to the RNA triplets (three nucleotides) which code for the particular α-amino acids. An enzyme then joins the α-amino acids of two adjacent tRNA-α-amino acid ribosome complexes bound to the mRNA to form a peptide bond. Thus peptide bonds and consequently polypeptides and proteins coded by the DNA *via* the respective mRNA are produced.
Martin et al. [*Biochem J* **89** 327 *1963*] dissolved RNA (5g) in 90mL of 0.1mM EDTA, then homogenised with 90mL of 90% (w/v) phenol in water using a Teflon pestle. The suspension was stirred vigorously for 1h at room temperature, then centrifuged for 1h at 0° at 25000rpm. The lower (phenol) layer was extracted four times with 0.1mM EDTA and the aqueous layers were combined, then made 2% (w/v) with respect to AcOK and 70% (v/v) with respect to EtOH. After standing overnight at -20°, the ppte was centrifuged down, dissolved in 50mL of 0.1mM EDTA, made 0.3M in NaCl and left 3 days at 0°. The purified RNA was then centrifuged down at 10000xg for 30min, dissolved in 100mL of 0.1mM EDTA, dialysed at 4° against water, and freeze-dried. It was stored at -20° in a desiccator. Michelson [*J Chem Soc* 1371 *1959*] dissolved 10g of RNA in water, added 2M ammonia to adjust the pH to 7, then dialysed in Visking tubing against five volumes of water for 24h. The process was repeated three times, then the material after dialysis was treated with 2M HCl and EtOH to ppte the RNA which was collected, washed with EtOH, ether and dried [see commercial catalogues for further examples].

Ricin (toxin from Castor bean *Ricinus communis*) *[A chain 96638-28-7; B chain 96638-29-8]* M_r ~60,000, amorphous. Crude ricin, obtained by aqueous extraction and $(NH_4)_2SO_4$ pptn, was chromatographed on a galactosyl-Sepharose column with sequential elution of pure ricin. The second peak was due to ricin agglutinin. [Simmons and Russell *Anal Biochem* **146** 206 *1985*.] Inhibitor of protein synthesis. **EXTREMELY DANGEROUS, USE EXTREME CARE [instructions accompany product].**

Rifampicin (Rifampin) *[13292-46-1]* M 823.0, m 183-185°, pK_1 1.7, pK_2 7.9. This macrolide antibiotic crystallises form Me_2CO in red-orange plates. It has UV λ_{max} 237, 255,334, and 475nm (ϵ 33,200, 32,100, 27,000 and 15,400) at pH 7.38. It is stable in Me_2SO and H_2O. Freely soluble in most organic solvents and slightly soluble in H_2O at pH <6. [Binda et al. *Arzneim.-Forsch* **21** 1907 *1971*.] It inhibits cellular RNA synthesis without affecting DNA [Calvori et al. *Nature* **207** 417 *1965*].

Rifamycin B *[13929-35-6]* M 755.8, m 300° (darkening at 160-164°), $[\alpha]_D^{20}$ -11° (MeOH), pK_1 2.60, pK_2 7.76. It forms yellow needles from $^*C_6H_6$. It has solubility in H_2O (0,027%), MeOH (2.62%) and EtOH (0.44%). It has UV λ_{max} 223, 304 and 245nm ($A_{1cm}^{1\%}$ 555, 275 and 220). [Oppolzer and Prelog *Helv Chim Acta* **56** 2287 *1973*; Oppolzer et al. *Experientia* **20** 336 *1964*; X-ray: Brufani et al. *Experientia* **20** 339 *1964*.]

Rifamycin SV sodium salt *[15105-92-7]* M 719.8, m 300°(darkening >140°), $[\alpha]_D^{20}$ -4° (MeOH), pK_{Est} ~ 7.8. Yellow orange crystals from Et_2O-pet ether or aq EtOH, very soluble in MeOH, EtOH, Me_2CO and EtOAc, soluble in Et_2O and HCO_3^-, slightly soluble in H_2O and pet ether. Its UV has λ_{max} at 223, 314 and 445nm ($A_{1cm}^{1\%}$ 586, 322 and 204) in phosphate buffer pH 7. [NMR: Bergamini and Fowst *Arzneim.-Forsch* **15** 951 *1965*.]

Saccharides. Resolved by anion-exchange chromatography. [Walberg and Kando *Anal Biochem* 37 320 *1970*.]

Sarcosine anhydride *[5076-82-4]* **M 142.2, m 146-147°, pK$_{Est(1)}$~-4.2, pK$_{Est(2)}$~-1.9.** Crystd from water, EtOH or ethyl acetate. Dried in vacuum at room temperature.

(-)-Scopolamine hydrobromide 3H$_2$O (6β,7β-epoxy-3α-tropanyl S(-)-tropate HBr, hyoscine HBr) *[114-49-8]* **M 438.3, m 193-194°, 195°, 195-199°, [α]$_D^{25}$ -25°(c 5, H$_2$O), pK^{20} 8.15.** Recrystd from Me$_2$CO, H$_2$O or EtOH-Et$_2$O and dried. Soluble in H$_2$O (60%) and EtOH (5%) but insol in Et$_2$O and slightly in CHCl$_3$. The *hydrochloride* has **m** 300° (from Me$_2$CO). The *free base* is a viscous liquid which forms a crystalline *hydrate* with **m** 59° and [α]$_D^{20}$ -28° (c 2.7, H$_2$O). Readily hydrolysed in dilute acid or base. [Meinwald *J Chem Soc* 712 *1953*; Fodor *Tetrahedron* 1 86 *1957*.]

Seleno-DL-methionine   (±2-amino-4-methylselanylbutyric acid) *[2578-28-1]* **M 196.1, m 265°(dec), 267-269°(dec), 270° (see pKs of methionine).** Crystallises in hexagonal plates from MeOH and H$_2$O. [Klosterman and Painter *J Am Chem Soc* **69** 2009 *1949*.] The L-isomer is purified by dissolving in H$_2$O, adjusting the pH to 5.5 with aqueous NH$_3$, evaporating to near-dryness, and the residue is washed several times with absolute EtOH till solid is formed and then recrystd from Me$_2$CO. It has **m** 266-268°(dec), 275°(dec), [α]$_D^{25}$ +18.1°(c 1, N HCl). [Pande et al. *J Org Chem* **35** 1440 *1970*.]

Serotonin hydrochloride   (5-HT, 3-[2-aminoethyl]-5-hydroxyindole HCl) *[153-98-0]* **M 212.7, m 167-168°, 178-180°, pK$_1^{25}$4.9, pK$_2^{25}$9.8 (10.0, NH$_2$), pK$_3^{25}$ 11.1 (5-OH), pK$_4^{25}$ 18.25 (acidic indole NH).** Purified by recrystn from EtOH-Et$_2$O or Et$_2$O to give the *hygroscopic* salt. Store in the dark as it is light sensitive. The *free base* has **m** 84-86° (from Et$_2$O). The *5-benzyloxy* derivative has **m** 84-86° (from Et$_2$O). [Ek and Witkop *J Am Chem Soc* **76** 5579 *1954*; HamLin and Fischer *J Am Chem Soc* **73** 5007 *1951*.] The *picrate 1H$_2$O* has **m** 196-197.5° (dec with sintering at 160-165°) after recrystn from Et$_2$O. Serotonin is a natural neurotransmitter [Chuang *Life Sci* **41** 1051 *1987*].

Sinigrin monohydrate (Myronate K) *[64550-88-5]* **M 415.5, m 125-127°, 127-129°, [α]$_D^{20}$ -17° (c 0.2, H$_2$O), pK$_{Est}$ <0.** Purified by recryst three times from EtOH and once from MeOH. The *tetraacetate* has **m** 193-195°, [α]$_D^{20}$ -16° (c 0.14, H$_2$O). [Benn et al. *J Chem Soc, Chem Commun* 445 *1965*; Kjaer et al. *Acta Chem Scand* **10** 432 *1956*; Marsh et al. *Acta Cryst (Sect B)* **26** 1030 *1970*.] It is a β-D-thioglucopyranoside substrate for thiogluconidase [MacLeod and Rossiter *Phytochem* **25** 1047 *1986*].

α-Solanine   (solan-5-en-3β-yl-[O^3-β-D-glucopyranosyl-O^2-α-L-rhamnopyranosyl-β-D-galacto -pyranoside]) *[20562-02-1]* **M 868.1, m 285°(dec), 286°(dec) (sintering >190°), [α]$_D^{20}$ -58° (c 0.8, pyridine), pK^{15} 6.66.** Recrystd from EtOH, 85% aqueous EtOH, MeOH or aqueous MeOH as *dihydrate* **m** 276-278°. Solubility in H$_2$O is 25mg/L and 5% in pyridine, but it is very soluble in Et$_2$O and CHCl$_3$. The *hydrochloride* is gummy or amorphous but has been crystd (**m** ~212° dec). It has insecticidal properties. [Kuhn et al. *Chem Ber* **88** 1492 *1955*.]

Somatostatin *[38916-34-6]* **M 1637.9, [α]$_D^{25}$ -36° (c 0.57, 1% AcOH).** A tetradecapeptide which is purified by gel filtration on Sephadex G-25, eluting with 2N AcOH, and then by liquid partition chromatography on Sephadex G-25 using *n*-BuOH-AcOH-H$_2$O (4:1:5) and has R$_F$ = 0.4. It is a brain growth hormone releasing-inhibiting factor which has also been synthesised. [Burgus et al. *Proc Natl Acad Sci USA* **70** 684 *1973*; Sorantakis and McKinley *Biochem Biophys Res Commun* **54** 234 *1973*; Hartridt et al. *Pharmazie* **37** 403 *1982*.]

Spectinomycin dihydrochloride pentahydrate   (Actinospectacin) *[21736-83-4]* **M 495.3, m 205-207°(dec), [α]$_D^{20}$ +14.8° (c 0.4, H$_2$O), pK$_1$ 6.95, pK$_2$ 8.70.** Purified from aqueous Me$_2$CO and is soluble in H$_2$O, MeOH and dilute acid and base but only slightly soluble in Me$_2$CO, EtOH, CHCl$_3$ and *C$_6$H$_6$. The *free base* is an amorphous solid, **m** 184-194° with [α]$_D^{20}$ -20° (H$_2$O). [Wiley et al. *J Am Chem Soc* **93** 2652 *1963*; X-ray: Cochran et al. *J Chem Soc Chem Commun* 494 *1972*.] It is an aminoglycoside antibiotic which interacts with 16S ribosomal RNA [Moazet and Noller *Nature* **327** 389 *1987*]; and is used for the treatment of gonorrhea [Rinehart *J Infect Dis* **119** 345 *1969*].

D-Sphingosine (2S,3S-D-*erythro*-2-aminooctadec-4t-ene-1,3-diol from bovine brain) *[123-78-4]* M 299.5, m 79-82°, 82° 82.5° (softens at ~70°), $[\alpha]_D^{22}$ -3.4° (c 2, $CHCl_3$), pK_{Est} ~ 8.8. Purified by recrystn from EtOAc, Et_2O or pet ether (60-80°) It is insoluble in H_2O but is soluble in Me_2CO, EtOH and MeOH. It has IR bands at 1590 and 875 cm^{-1}, and is characterised as the *tribenzoate* m 122-123° (from 95% EtOH). [Tipton *Biochem Prep* **9** 127 *1962*.]

Spirilloxanthin *[34255-08-8]* M 596.9, m 216-218°, λ_{max} 463, 493, 528 nm, $\varepsilon_{1cm}^{1\%}$ 2680 (493 nm) in pet ether (b 40-70°). Crystd from $CHCl_3$/pet ether, acetone/pet ether, *C_6H_6/pet ether or *C_6H_6. Purified by chromatography on a column of $CaCO_3$/$Ca(OH)_2$ mixture or deactivated alumina. [Polgar et al. *Arch Biochem Biophys* **5** 243 *1944*.] Stored in the dark in an inert atmosphere, at -20°.

Squalane (Cosbiol, 2,6,10,15,19,23-hexamethyltetracosane, perhydrosqualene) *[111-01-3]* M 422.8, m -38°, b 176°/0.05mm, 210-215°/1mm, 274°/10mm, ~350°/760mm, d_4^{20} 0.80785, n_D^{20} 1.416. Purified by fractional distn *in vacuo* or evap distn. Soluble in pet ether, *C_6H_6, Et_2O and $CHCl_3$, slightly sol in alcohols, Me_2CO and AcOH but insol in H_2O [Staudinger and Leupold *Helv Chim Acta* **15** 223 *1932*; Sax and Stross *Anal Chem* **29** 1700 *1951*; Mandai et al. *Tetrahedron Lett* **22** 763 *1981*].

Squalene (*all-trans*-2,6,10,15,19,23-hexamethyl-2,6,10,14,18,22-tetracosahexaene) *[111-02-4]* M 410.7, m ~75°, b 203°/0.1mm, 213°/1mm, 285°/25mm, d^{25} 0.8670, n 1.4905. Crystd repeatedly from Me_2CO (1.4mL/g) using a Dry-ice bath, washing the crystals with cold acetone, then freezing the squalene under vacuum. Squalene was further purified by passage through a column of silica gel or chromatographed on activated alumina, using pet ether as eluent and stored in vac in the dark. Dauben et al. [*J Am Chem Soc* **74** 4321 *1952*] purified squalene *via* its hexachloride and is bactericidal. [Capstack et al. *J Biol Chem* **240** 3258 *1965*; Krishna et al. *Arch Biochem Biophys* **114** 200 *1966*; Heilbron and Thompson *J Chem Soc* 883 *1929*; Karrer et al. *Helv Chim Acta* **13** 1084 *1930*; UV: Farmer et al. *J Chem Soc* 544 *1943*.]

Starch *[9005-84-9]* M (162.1)$_n$. Defatted by Soxhlet extraction with Et_2O or 95% EtOH. For fractionation of starch into "amylose" and "amylopectin" fractions, see Lansky et al. [*J Am Chem Soc* **71** 4066 *1949*].

Sterigmatocystin (3a,12c-dihydro-8-hydroxy-6-methoxy-3H-furo[3',2',:4,5]furo[2,3-c]-xanthen-7-one) *[10048-13-2]* M 324.3, m 246°, 247-248°, $[\alpha]_D^{20}$ -398° (c 0.1, $CHCl_3$), pK_{Est} ~ 8.0. Recrystd from amyl acetate, Me_2CO or EtOH and sublimed *in vacuo*. It has UV λ_{max} at 208, 235, 249 and 329nm (log ε 4.28, 4.39, 4.44 and 4.12). [UV: Bullock et al. *J Chem Soc* 4179, *1962*; UV, IR: Holker and Mulheirn *J Chem Soc Chem Commun* 1576, 1576 *1968*; Birkinshaw and Hammady *Biochem J* **65** 162 *1957*.] This mycotoxin induces bone marrow changes in mice [Curry et al. *Mutation Res* **137** 111 *1984*].

Stigmatellin A (2-[4,6-dimethoxy-3,5,11-trimethyltridecatri-7t,9t,11t-enyl]-8-hydroxy-5,7-dimethoxy-3-methyl-4H-1-benzopyran-4-one) *[91682-96-1]* M 514.6, m 128-130°, $[\alpha]_D^{20}$ +38.5° (c 2.3, MeOH), pK_{Est} ~7 (phenolic OH). It is stable in aqueous soln at neutral pH but decomposes at pH <5. Purified by recrystn from toluene-hexane). It has UV λ_{max}: nm (ε) 248sh (41000), 258 (59500) 267 (65500), 279 (41400) and 335 (5200) in MeOH; 249sh (45600), 258 (60000), 268 (72700), 277 (54100), 320 (2500) and 370 (3000) in MeOH + 1 drop of N KOH; 243sh (29300), 264 (63200), 274 (64100), 283sh (45800), 329 (4800) and 420 (21000) in MeOH + 6N HCl; and IR ($CHCl_3$) ν: 3550m, 1645chs, 1635ss, 1620ss, 1590s, 1510m and 905m cm^{-1}. It gives colour reactions at 110° with vanillin/H_2SO_4 (grey), $Ce(IV)/(NH_4)_2SO_4$ (yellow) and phosphomolybdate (blue-grey). [Höfle et al. *Justus Liebigs Ann Chem* 1882 *1984*.] It inhibits electron transport [Jagow and Link *Methods Enzymol* **126** 253 *1986*; Robertson et al. *Biochemistry* **32** 1310 *1933*], and has antibiotic properties [Kunze et al. *J Antibiot* **37** 454 *1984*]. The 7t,9t,11c-isomer is *Stigmatellin B*.

Streptomycin sulfate *[3810-74-0]* M 1457.4, $[\alpha]_D^{20}$ -84.3° (c 3, H_2O), $pK_{Est(1)}$~ 9.5 (MeNH), $pK_{Est(2,3)}$~ 13.4 (guanidino). Recrystd from H_2O-EtOH, washed with a little EtOH, Et_2O and dried in a vacuum. [UV and IR: Grove and Randall *Antibiotics Monographs* **2** 163 *1855*; Heuser et al. *J Am Chem Soc* **75** 4013 *1953*, Kuehl et al. *J Am Chem Soc* **68** 1460 *1946*; Regna et al. *J Biol Chem* **165** 631 *1946*.] During protein synthesis it inhibits initiation and causes misreading of mRNA [Zierhut et al. *Eur J Biochem* **98** 577 *1979*; Chandra and Gray *Methods Enzymol* **184** 70 *1990*].

Streptonigrin (nigrin,   5-amino-6-[7-amino-5,8-dihydro-6-methoxy-5,8-dioxo-2-quinolinyl]-4-[2-hydroxy-3,4-dimethoxyphenyl]-3-methyl-2-pyridinecarboxylic   acid) *[3930-19-6]* **M 506.5, m 262-263°, 275°(dec), pK 6.3 (1:1 aq dioxane).** Purified by TLC on pH 7-buffered silica gel (made from a slurry of Silica Gel 60 and 400mL of 0.05M phosphate buffer pH 7.0) and eluted with 5% MeOH/CHCl$_3$. The extracted band can then be recrystd from Me$_2$CO or dioxane as almost black plates or needles. It is soluble in pyridine, Me$_2$NCHO, aqueous NaHCO$_3$ (some dec), and slightly soluble in MeOH, EtOH, EtOAc and H$_2$O. It has UV λ_{max} 248, 375-380nm (ε 38400 and 17400). [Weinreb et al. *J Am Chem Soc* **104** 536 *1982*; Rao et al. *J Am Chem Soc* **85** 2532 *1963*.] It is an antineoplastic and causes severe bone marrow depression [Wilson et al. *Antibiot Chemother* **11** 147 *1961*].

Streptozotocin   (N-[methylnitrosocarbamoyl]-α-D-glucosamine, streptozocin) *[18883-66-4]* **M 265.2, m 111-114°(dec), 114-115°(dec), 115°(dec with evolution of gas), $[\alpha]_D^{20}$ ~+39° (H$_2$O, may vary due to mutarotation).** Recrystd from 95% EtOH and is soluble in H$_2$O, MeOH and Me$_2$CO. It has UV λ_{max} 228nm (ε 6360) in EtOH. The *tetraacetate* has **m** 111-114°(dec), $[\alpha]_D^{25}$ +41° (c 0.78, 95% EtOH) after recrystn from EtOAc. [Herr et al. *J Am Chem Soc* **89** 4808 *1967*; NMR: Wiley et al. *J Org Chem* **44** 9 *1979*.] It is a potent methylating agent for DNA [Bennett and Pegg *Cancer Res* **41** 2786 *1981*].

Subtilisin (from *Bacillus subtilis*) *[9014-01-1]* **[EC 3.4.21.62].** Purified by affinity chromatography using 4-(4-aminophenylazo)phenylarsonic acid complex to activated CH-Sepharose 4B. [Chandraskaren and Dhar *Anal Biochem* **150** 141 *1985*].

Succinyl coenzyme A trisodium salt *[108347-97-3]* **M 933.5.** If it should be purified further then it should be dissolved in H$_2$O (0.05g/mL) adjusted to pH 1 with 2M H$_2$SO$_4$ and extracted several times with Et$_2$O. Excess Et$_2$O is removed from the aqueous layer by bubbling N$_2$ through it and stored frozen at pH 1. When required the pH should be adjusted to 7 with dilute NaOH and used within 2 weeks (samples should be frozen). Succinyl coenzyme A is estimated by the hydroxamic acid method [*J Biol Chem* **242** 3468 *1967*]. It is more stable in acidic than in neutral aqueous solutions. [*Methods Enzymol* **128** 435 *1986*.]

2-Sulfobenzoic cyclic anhydride   (2,1-benzoxathiazol-3-one 1,1-dioxide) *[81-08-3]* **M 184.2, m 126-127°, 129.5°, 130°, b 184-186°/18mm.** If the sample has hydrolysed extensively (presence of OH band in the IR) then treat with an equal bulk of SOCl$_2$ reflux for 3h (CaCl$_2$ tube), evaporate and distil residue in a vacuum, then recrystd from *C$_6$H$_6$, Et$_2$O-*C$_6$H$_6$ or CHCl$_3$ (EtOH free by passing through Al$_2$O$_3$, or standing over CaCl$_2$). [Clarke and Breger *Org Synth* Coll Vol I 495 *1948*.] Used for modifying ζ-amino functions of lysyl residues in proteins [Bagree et al. *FEBS Lett* **120** 275 *1980*]. (see entry on p. 126.)

Syrexin (from bovine liver). Purified by (NH$_4$)$_2$SO$_4$ pptn, then by pH step elution from chromatofocusing media in the absence of ampholytes. [Scott et al. *Anal Biochem* **149** 163 *1985*.]

Taurodeoxycholic acid sodium salt monohydrate (n-

[desoxycholyl)taurine Na salt H$_2$O) *[1180-95-6]* **M 539.7, m 171-175°, $[\alpha]_D^{23}$ +37° (c 1, H$_2$O), pK 1.4 (free acid).** The salt is recrystd from EtOH-Et$_2$O. Its solubility in H$_2$O is 10%. The *free acid* has **m** 141-144°. [Norman *Ark Kemi* **8** 331 *1956*.] It forms mixed micelles and solubilises some membrane proteins [Hajjar et al. *J Biol Chem* **258** 192 *1983*].

Terramycin (oxytetracycline) *[79-57-2]* **M 460.4 (anhy), 496.5 (2H$_2$O), sinters at 182°, melts at 184-185°(dec), $[\alpha]_D^{20}$ -196.6° (equilibrium in 0.1M HCl), -2.1° (equilibrium in 0.1M NaOH).** Crystd (as dihydrate) from water or aqueous EtOH.

2,2:5',2''-Terthiophene *[1081-34-1]* **M 248.4, m 92-93°, 94-95°, 94-94.5°, 94-96°.** Recrystd from MeOH, *C$_6$H$_6$, pet ether or MeOH. [UV: Zechmeister and Sease *J Am Chem Soc* **69** 273 *1947*; Steinkopf et al. *Justus Liebigs Ann Chem* **546** 180 *1941*.] Phototoxic nematocide [Cooper and Nitsche *Bioorg Chem* **13** 36 *1985*; Chan et al. *Phytochem* **14** 2295 *1975*]. See also Terthiophene on p. 356 in Chapter 4.

Tetracycline *[60-54-8]* **M 444.4, m 172-174°(dec), $[\alpha]^{20}_{546}$ +270° (c 1, MeOH), pK^{25}_1 3.30, pK^{25}_2 7.68, pK^{25}_3 9.69.** Crystd from toluene.

Tetracycline hydrochloride *[64-75-5]* **M 480.9, m 214°(dec), 215-220°, $[\alpha]^{25}_D$ -258° (c 0.5, 0.1N HCl), $[\alpha]^{20}_D$ -245° (c 1, MeOH), pK_1 1.4 (enolic OH), pK_2 7.8 (phenolic OH), pK_3 9.6 (Me_2N).** Recrystd from MeOH + n-BuOH or n-BuOH + HCl. It is insoluble in Et_2O and pet ether. It has UV λ_{max} at 270 and 366nm in MeOH. [Gottstein et al. *J Am Chem Soc* **81** 1198 *1959*; Conover et al. *J Am Chem Soc* **84** 3222 *1962*.]

$6R$-Tetrahydro-*erythro*-biopterin dihydrochloride (BH$_4$.2HCl, $6R$-2-amino-4-hydroxy-6-[{$1R$,$2S$}-1,2-dihydroxypropyl]-5,6,7,8-tetrahydropteridine 2HCl) *[69056-38-8]* **M 316.2, m 245-246°(dec), $[\alpha]^{25}_D$ -6.8° (c 0.67, 0.1N HCl), pK_1 1.37 (pyrimidine$^+$), pK_2 5.6 (5-NH$^+$), pK_3 10.6 (acidic, 3NH).** Recrystn from HCl enriches BH$_4$ in the natural $6R$ isomer. Dissolve the salt (~6g) in conc HCl (15mL) under gentle warming then add EtOH (30mL) dropwise, chill and collect the colourless needles (67%, up to 99% if mother liquors are concentrated), and dried *in vacuo* immediately over P_2O_5 and KOH. Stores indefinitely at -20° in a dry atmosphere, Better store in sealed ampoules under dry N_2. It can be recrystd from 6N aqueous HCl. It has UV λ_{max} (2N HCl) 264nm (ϵ 16770; pH 3.5 phosphate buffer) 265nm (ϵ 13900); (pH 7.6) 297nm (ϵ 9500) and 260nm sh (ϵ 4690). It has been separated from the $6R$-isomer by HPLC on a Partisil-10SCX column using 30mM ammonium phosphate buffer (pH 3.0) containing 3mM $NaHSO_3$ (2mL/min flow rate; 275nm detector) with retention times of 5.87min ($6R$) and 8.45min ($6S$). It is stable in acidic soln and can be stored for extended periods at -20° in 0.04M HCl. Above pH 7 the neutral species are obtained and these are readily oxidised by oxygen in the solvent to quinonoid species and then further oxidation and degradation occurs at room temperatures. These changes are slower at 0°. The *sulfate salt* can be obtained by recrystn from 2M H_2SO_4 and is less soluble than the hydrochloride salt. The $6R$-2,5,1',2'-*tetraacetylbiopterin* derivative has **m** 292° (dec) after recrystn from MeOH (100 parts) and $[\alpha]^{20}_{589}$ -144° (c 0.5, $CHCl_3$), $[\alpha]^{20}_{589}$ +12.8° (c 0.39, Me_2SO). [NMR, UV: Matsuura et al. *Heterocycles* **23** 3115 *1985*; Viscontini et al. *Helv Chim Acta* **62** 2577 *1979*; Armarego et al. *Aust J Chem* **37** 355 *1984*.]

Tetrahydrofolic acid dihydrochloride 2H$_2$O (THFA, $6S$- or $6RS$- 5,6,7,8-tetrahydrofolic acid 2HCl 2H$_2$O, 5,6,7,8-tetrahydropteroyl-L-glutamic acid 2HCl 2H$_2$O) *[135-16-0]* **M 544.4, m >200°(dec), $[\alpha]^{27}_D$ +16.9° (H$_2$O pH 7.0 + 2-mercaptoethanol), pK_1 1.7 (pyrimidine N$^+$), pK_2 2.4 (10N$^+$), pK_3 3.5 (α-CO$_2$H), pK_4 4.9 (γ-CO$_2$H), pK_5 5.6 (5-NH$^+$), pK_6 10.4 (acidic, 3NH).** Very high quality material is now available commercially and should be a white powder. It can be dried over P_2O_5 in a vacuum desiccator and stored in weighed aliquots in sealed ampoules. It is stable at room temp in sealed ampoules for many months and for much more extended periods at -10°. When moist it is extremely sensitive to moist air whereby it oxidises to the yellow 7,8-dihydro derivative. In soln it turns yellow in colour as it oxidises and then particularly in the presence of acids it turns dark reddish brown in colour. Hence aqueous solutions should be frozen immediately when not in use. It is always advisable to add 2-mercaptoethanol (if it does not interfere with the procedure) which stabilises it by depleting the soln of O_2. The *sulfate salt* is more stable but then it is much less soluble. The best way to prepare standard solns of this acid is to dissolve it in the desired buffer and estimate the concentration by UV absorption in pH 7 buffer at 297nm (ϵ 22,000 M^{-1}cm^{-1}). If a sample is suspect it is not advisable to purify it because it is likely to deteriorate further as "dry box" conditions are necessary. Either a new sample is purchased or one is freshly prepared from folic acid. It has pKa values of -0.1, 4.3 and 9.0. [Hafeti et al. *Biochem Prep* **7** 89 *1960*; UV: Mathews and Huennekens *J Biol Chem* **235** 3304 *1960*; Osborn and Huennekens *J Biol Chem* **233** 969 *1958*; O'Dell et al. *J Am Chem Soc* **69** 250 *1947*; Blakley *Biochem J* **65** 331 *1957*; Asahi *Yakugaku Zasshi (J Pharm Soc Japan)* **79** 1548 *1959*.]

5,6,7,8-Tetrahydropterin sulfate (2-amino-5,6,7,8-tetrahydropteridin-4-one H$_2$SO$_4$) *[20350-44-1]* **M 265, m >200°(dec), pK^{25}_1 1.3 (pyrimidine $^+$), pK^{25}_2 5.6 (5-NH$^+$), pK^{25}_3 10.6 (acidic, 3NH).** If it has become too strongly violet in colour then it may need reducing again. Best to check the UV absorption in N HCl where it has a peak at ~265nm which drops sharply to zero having no absorption at *ca* 340nm. The presence of absorption at 340nm indicated oxidation to quinonoid or 7,8-dihydropterin. If the absorption is weak then dissolve in the minimum volume of anhydrous trifluoroacetic acid (fume hood) add charcoal, filter, then add one or two drops of N H_2SO_4 followed by dry Et_2O at 0°, allow the white tetrahydro salt to settle and collect, and wash with dry Et_2O, by centrifugation. Dry the residue in a vacuum desiccator over P_2O_5 and KOH. Store in aliquots in the dark at $<0°$. It has UV λ_{max} at 265nm (ϵ 16980) at pH -1.0 (dication); 219nm (ϵ 23440) and 266nm (ϵ 12880) at pH 3.5 (monocation); 220nm (ϵ 18620), [260nm (ϵ 4270)sh] and 299nm (ϵ 9330) at pH 8.0 (neutral species); and 218nm (ϵ 10000), [240nm (ϵ 5500)sh] and 287nm (ϵ 5500) at pH 13 (anion). [Blakley *Biochem J* **72** 707 *1959*; Asahi *Yakugaku Zasshi (J Pharm Soc Jpn)* **79** 1557 *1959*; Pfleiderer in *Pterins and Folate* (Benkovic and Blakley Eds) J Wiley Vol 2 p97 *1985*.]

Thiamine monophosphate chloride $1H_2O$ **(Aneurine monophosphate chloride)** *[532-40-1]* **M 416.8, m 193°(dec), 200°(dec), 200-203°(dec), pK_1 2.40, pK_2 4.80, pK_3 6.27, pK_4 9.65, pK_5 10.20.** Purified by recrystn from aqueous HCl, EtOH slightly acidified with HCl, EtOH-Me_2CO, H_2O, or H_2O-EtOH + Et_2O. Dissolve in a small volume of H_2O and mix with EtOH + Me_2CO (1:1) to give the $HCl.H_2O$ as crystals. Filter, wash with Et_2O and dry in a vacuum. The *chloride hydrochloride* , m 215-217°(dec) is obtained when crystd from aqueous HCl. [Wenz et al. *Justus Liebigs Ann Chem* **618** 2280 *1958*, Viscontini et al. *Helv Chim Acta* **34** 1388 *1951*, Leichssenring and Schmidt *Chem Ber* **95** 767 *1962*; McCormick and Wright *Methods Enzymol* **18A** 141, 147 *1970*.]

Thiamphenicol **(1R,2R-2-[2,2-dichloroacetylamino]-1-[4-methanesulfonylphenyl]-propan-1,3-diol)** *[15318-45-3 (D-threo), 90-91-5]* **M 356.2, m 163-166°, 165.2-165.6°, 165-166°, $[\alpha]_D^{25}$ +15.6° (c 2, EtOH), pK 7.2.** Recrystd from H_2O or $CHCl_3$. UV λ_{max} 224, 266 and 274nm (ϵ 13700, 800 and 700) in 95% EtOH. The *1S,2S-isomer [14786-51-7]* has m 164.3-166.3° (from H_2O + EtOAc + pet ether) and $[\alpha]_D^{25}$ -12.6° (c 1, EtOH); and the *racemate 1RS,2RS* **Racefenical** *[15318-45-3]*] has m 181-183° (sinter at 180-183°) from $CHCl_3$-EtOAc-pet ether. [Cutler et al. *J Am Chem Soc* **74** 5475, 5482 *1952*; UV: Nachod and Cutler *J Am Chem Soc* **74** 1291 *1952*; Suter et al. *J Am Chem Soc* **75** 4330 *1953*; Cutler et al. *J Am Pharm Assoc* **43** 687 *1954*.]

Thiazolyl blue tetrazolium bromide **(MTT, 3-[4,5-dimethyl-2-thiazolyl]-2,5-diphenyl-2H-tetrazolium bromide)** *[298-93-1, 2348-71-2]* **M 414.3, m 171°.** It is recrystd by dissolving in MeOH containing a few drops of HBr and then adding dry Et_2O to complete the crystn, wash the needles with Et_2O and dry in a vacuum desiccator over KOH. [Beyer and Pyl *Chem Ber* **87** 1505 *1954*.]

2-Thiocytosine **(4-amino-2-mercaptopyrimidine)** *[333-49-3]* **M 127.2, m 236-237°(dec), 285-290°(dec), pK_1^{20} 3.90 (NH_2), pK_2^{20} 11.10 (SH).** It is recrystd from hot H_2O and dried at 100° to constant weight. [Brown *J Appl Chem (London)* **9** 203 *1959*; Russell et al. *J Am Chem Soc* **71** 2279 *1949*.] It is used in transcription and translation studies [Rachwitz and Scheit *Eur J Biochem* **72** 191 *1977*.]

6-Thioguanine *[154-42-7]* **M 167.2, m >300°, pK_1^{23} 8.2 (SH), pK_2^{23} 11.6 (acidic, 9-NH).** Recrystd from H_2O as needles. It has UV λ_{max} at 258 and 347nm (H_2O, pH 1) and 242, 270 and 322nm (H_2O, pH 11). [Elion and Hitchings *J Am Chem Soc* **77** 1676 *1955*; Fox et al. *J Am Chem Soc* **80** 1669 *1958*.] It is an antineoplastic agent [Kataoka et al. *Cancer Res* **44** 519 *1984*].

Thrombin (from bovine blood plasma) *[9002-04-4]* **M_r 32,600 [EC 3.4.4.13].** Purified by chromatography on a DEAE-cellulose column, while eluting with 0.1M NaCl, pH 7.0, followed by chromatography on Sephadex G-200. Final preparation was free from plasminogen and plasmin. [Yin and Wessler *J Biol Chem* **243** 112 *1968*.]
Thrombin from bovine blood was purified by chromatography using p-chlorobenzylamino-ϵ-aminocaproyl agarose, and gel filtration through Sephadex G-25. [Thompson and Davie *Biochim Biophys Acta* **250** 210 *1971*.]
Thrombin from various species was purified by precipitaion of impurities with rivanol. [Miller *Nature* **184** 450 *1959*.]

L-Thyroxine sodium salt ($5H_2O$) *[6106-07-6]* **M 888.9, $[\alpha]_{546}^{20}$ +20° (c 2, 1M HCl + EtOH, 1:4).** Crystd from absolute EtOH and dried for 8h at 30°/1mm.

D-Thyroxine {O-[3,5-diiodo-4-oxyphenyl]-3,5-diiodo-D-(-)-tyrosine, 3,3',5,5'-tetra-iodo-D-thyronine} *[51-49-0]* **M 776.9, m 235°(dec), 235-236°(dec), 340°(dec), $[\alpha]_D^{20}$ +4.5° (c 3, aq**

0.2N NaOH in 70% EtOH), $[\alpha]_D^{20}$ -17° (c 2, aq N HCl + EtOH 1:4), pK_1^{25} 2.2 (CO_2H), pK_2^{25} 8.40 (OH), pK_3^{25} 10,1 (NH_2). Recrystd from H_2O as needles or from an ammonical soln by dilution with H_2O, MeOH or Me_2CO. Also purified by dissolving ~6.5 g in a mixture of MeOH (200mL) and 2N HCl (20mL), add charcoal, filter then add NaOAc soln to pH 6 and on standing the thyroxine separates, is washed with MeOH then Me_2CO and dried *in vacuo*. The *N-formyl-D-thyroxine* derivative has **m** 210° and $[\alpha]_{546}^{21}$ -26.9° (c 5, EtOH). The racemate ±-*thyroxine* has **m** 256° and is purified in the same way. [Nahm and Siedel *Chem Ber* **96** 1 *1963*; Salter *Biochem J* **24** 471 *1930*.]

L-Thyroxine (*O*-[3,5-diiodo-4-oxyphenyl]-3,5-diiodo-L-(+)-tyrosine, 3,3',5,5'-tetraiodo-D-thyronine) *[51-48-9]* **M 776.9, m 229-230°(dec), ~235°(dec), 237°(dec),** $[\alpha]_D^{22}$ **-5.1°** (**c 2, aq N NaOH + EtOH 1:2**), $[\alpha]_D^{22}$ **+15°** (**c 5, aq N HCl in 95% EtOH 1:2**), $[\alpha]_D^{22}$ **+26°** (**EtOH/1M aq HCl; 1:1**) (**pK 6.6**). Purification is the same as for the D-isomer above. Likely impurities are tyrosine, iodotyrosine, iodothyroxines and iodide. Dissolve in dilute ammonia at room temperature, then crystd by adding dilute acetic acid to pH 6. The *N-formyl-L-thyroxine* has **m** 214°(dec) and $[\alpha]_{546}^{21}$ +27.8° (c 5, EtOH). [Harington et al. *Biochem J* **39** 164 *1945*; Nahm and Siedel *Chem Ber* **96** 1 *1963*; Reineke and Turner *J Biol Chem* **161** 613 *1945*; Chalmers et al. *J Chem Soc* 3424 *1949*.]

Tissue inhibitor of metalloproteins (TIMP, from human blood plasma), M_r ~30,000. Purified by a [anti-human amniotic fluid-TIMP]-Sepharose immuno-affinity column eluted with 50mM glycine/HCl pH 3.0 buffer that is 0.5M in NaCl then by gel filtn [Cawston et al. *Biochem J* **238** 677 *1986*].

dl-α-**Tocopherol** (see **vitamin E**) *[59-02-9]* **M 430.7, $A_{1cm}^{1\%}$ 74.2 at 292 nm in MeOH.** Dissolved in anhydrous MeOH (15mL/g) cooled to -6° for 1h, then chilled in a Dry-ice/acetone bath, crystn being induced by scratching with a glass rod.

γ-**Tocopherol** (3,4-dihydro-2,7,8-trimethyl-2-(4,8,12-trimethyltridecyl)-2*H*-benzopyran-6-ol) *[54-28-4]* **M 416.7, m -30°, b 200-210°/0.1mm, d_4^{20} 0.951, n_D^{20} 1.505, $[\alpha]_D^{20}$ -2.4°** (**EtOH**). Purified by distn at high vacuum and stored in dark ampoules under N_2. UV λ_{max} 298nm ($A_{1cm}^{1\%}$ 92.8). It is insoluble in H_2O but soluble in organic solvents. The *allophanate* (used for separating isomers) has **m** 136-138°, $[\alpha]_D^{18}$ +3.4° ($CHCl_3$). [Baxter et al. *J Am Chem Soc* **65** 918*1943*; Emerson et al. *Science* **83** 421 *1936*, *J Biol Chem* **113** 319 *1936*.]

Tolylene-2,4-diisocyanate (toluene-2,4-diisocyanate). *[584-84-9]* **M 174.2, m 19.5-21.5°, 20-22°, 28°, b 126°/11mm, 124-126°/18mm, 250°/760mm.** It is purified by fractionation in a vacuum and should be stored in a dry atmosphere. It is soluble in organic solvents but reacts with H_2O, alcohols (slowly) and amines all of which could cause explosive polymerisation. It darkens on exposure to light. It has a sharp pungent odour, is **TOXIC** and is **IRRITATING TO THE EYES.** [Siefken *Justus Liebigs Ann Chem* **562** 75, 96, 127 *1949*; Bayer *Angew Chem* **59** 257 *1947*.] It is a reagent for covalent crosslinking of proteins [Wold *Methods Enzymol* **25** 623 *1972*.]

Tomatidine (5α,20β,22α,25β,27-azaspirostan-3β-ol) *[77-59-8]* **M 415.7, m 202-206°,** $[\alpha]_D^{20}$ **+5.9°** (**c 1, MeOH**), $[\alpha]_D^{20}$ **+8°** (**$CHCl_3$**). Forms plates from EtOAc. Also purified by dissolving 80mg in *C_6H_6 and applying to an Al_2O_3 column (3.0g) and eluting with *C_6H_6, evaporating and recrystallising three times from EtOAc. The *hydrochloride* has **m** 265-270° from EtOH and $[\alpha]_D^{25}$ -5° (MeOH). [IR: Uhle *J Am Chem Soc* **83** 1460 *1961*; Kessar et al. *Tetrahedron* **27** 2869 *1971* ; Schreiber and Adams *Experientia* **17** 13 *1961*.]

Tomatine (22*S*,25*S*-3β,β-lycotetraosyloxy-5α-spirosolan) *[17406-45-0]* **M 1034.2, m 263-268°(dec), 290-291°(evac capillary), 283.5-287°(dec), 272-277°(dec), 300-305°(dec),** $[\alpha]_D^{20}$ **-18° to -34°** (**c 0.55, pyridine**). Recrystd from MeOH, EtOH, aqueous EtOH or dioxane + NH_3. It is almost insoluble in pet ether, Et_2O or H_2O. [Reichstein *Angew Chem* **74** 887 *1962*.]

N-**Tosyl-L-lysine chloromethyl ketone** (3*S*-1-chloro-3-tosylamino-7-amino-2-heptanone HCl) *[4272-74-6]* **M 369.3, m 150-153°(dec), 156-158°(dec), ~165°(dec),** $[\alpha]_D^{20}$ **-7.3°** (**c 2, H_2O**), pK_{Est} ~ **10.6** (**7-NH_2**). The hydrochloride slowly crystallises from a conc soln in absolute EtOH, thinned with EtOH-Et$_2$O for collection and dried *in vacuo*. It is a suicide enzyme inhibitor [Matsuda et al. *Chem Pharm Bull Jpn* **30** 2512 *1982*; Shaw et al. *Biochemistry* **4** 2219 *1965*].

Transferrin (from human or bovine serum) *[11096-37-0]* **M$_r$~80,000.** Purified by affinity chromatography on phenyl-boronate agarose followed by DEAE-Sephacel chromatography. The product is free from haemopexin. [Cook et al. *Anal Biochem* **149** 349 *1985*; Aisen and Listowsky *Ann Rev Biochem* **49** 357 *1980*.]

Trehalase (from kidney cortex) *[9025-52-9]* **[EC 3.2.1.28].** Purified by solubilising in Triton X-100 and sodium deoxycholate, and submitting to gel filtration, ion-exchange chromatography, conA-Sepharose chromatography, phenyl-Sepharose CL-4B hydrophobic interaction chromatography, Tris-Sepharose 6B affinity and hydrolyapatite chromatography. Activity was increased 3000-fold. [Yoneyama *Arch Biochem Biophys* **255** 168 *1987*.]

Trifluoperazine dihydrochloride (10-[3-{4-methyl-1-piperazinyl}propyl]-2-trifluoro-methyl-phenothiazine 2HCl) *[440-17-5]* **M 480.4, m 240-243°, 242-243°, pK$_1$ 3.9, pK$_2$ 8.1.** Recrystd from abs EtOH dried *in vacuo* and stored in tightly stoppered bottles because it is *hygroscopic*. It is soluble in H$_2$O but insoluble in *C$_6$H$_6$, Et$_2$O and alkaline aqueous soln. It has UV λ_{max} at 258 and 307.5nm (log ε 4.50 and 3.50) in EtOH (neutral species). [Craig et al. *J Org Chem* **22** 709 *1957*.] It is a calmodulin inhibitor [Levene and Weiss *J Parmacol Exptl Ther* **208** 454 *1978*], and is a psychotropic agent [Fowler *Arzneim.-Forsch* **27** 866 *1977*].

T4-RNA ligase (from bacteriophage-infected *E.coli*) **M$_r$ 43,500 [EC 6.5.1.3 for RNA lyase].** Purified by differential centrifugation and separation on a Sephadex A-25 column, then through hydroxylapatite and DEAE-glycerol using Aff-Gel Blue to remove DNAase activity. (Greater than 90% of the protein in the enzyme preparation migrated as a single band on gradient polyacrylamide gels containing SDS during electrophoresis.) [McCoy et al. *Biochim Biophys Acta* **562** 149 *1979*.]

Tubercidin (7-deazaadenosine) *[69-33-0]* **M 266.3, m 247-248°, [α]$_D^{17}$ -67° (50% aq AcOH), pK^{10} 5.2-5.3.** Forms needles from hot H$_2$O. It is soluble in H$_2$O (0.33%), MeOH (0.5%) and EtOH 0.05%). It has UV λ_{max} 270nm (ε 12100) in 0.001N NaOH. The *picrate* has **m** 229-231°(dec). [Tolman et al. *J Am Chem Soc* **91** 2102 *1969*; Mizuno et al. *J Org Chem* **28** 3329 *1963*, IR: Anzai et al. *J Antibiot (Japan)* **[9] 10** 201 *1957*.]

Tunicamycin *[11089-65-9]* **m 234-235°(dec), [α]$_D^{20}$ +52° (c 0.5, pyridine), pK$_{Est}$ ~ 9.4.** The components are purified by recrystallising 3 times from hot glass-distilled MeOH and the white crystals are dissolved in 25% aqueous MeOH and separated on a Partisil ODS-10μ column (9.4 x 25 cm) [Magnum-9 Whatman] using a 260 nm detector. The column was eluted with MeOH:H$_2$O mixture adjusted to 1:4 (v/v) then to 2:4 (v/v). The individual components are recovered and lyophilised. Ten components were isolated and all were active (to varying extents) depending on the lengths of the aliphatic side-chains. The mixture has UV λ_{max} 205 and 260nm ($A_{1cm}^{1\%}$ 230 and 110). Stable in H$_2$O at neutral pH but unstable in acidic soln. It inhibits protein glycosylation. [Mahoney and Duskin *J Biol Chem* **254** 6572 *1979*; Elnein *Trends Biochem Sci* **6** 219 *1981*; Takatsuki *J Antibiot* **24** 215 *1971*.]

Ubiquinol-cytochrome c reductase (from beef heart mitochondria)

[9027-03-6] **[EC 1.10.2.2].** Purified in Triton X-100 by solubilising the crude enzyme with Triton X-100, followed by hydroxylapatite and gel chromatography. The minimum unit contains nine polypeptide subunits of **M$_r$** 6000 - 49000 kD. [Engel et al. *Biochim Biophys Acta* **592** 211 *1980*.]

Uracil, uridine and uridine nucleotides. Resolved by ion-exchange chromatography AG1 (Cl⁻ form). [Lindsay et al. *Anal Biochem* **24** 506 *1968*.]

Uridine 5'-diphosphoglucose pyrophosphorylase (from rabbit skeletal muscle) *[9029-22-6]* M_r 350,000, **[EC 2.7.7.9].** Purified by two hydrophobic chromatographic steps and gel filtration. [Bergamini et al. *Anal Biochem* **143** 35 *1984*.] Also purified from calf liver by $(NH_4)_2SO_4$ (40-58%) pptn, $Ca_3(PO_4)_2$ gel filtration, DEAE-cellulose chromatography and recrystn by dialysis against increasing concentrations of $(NH_4)_2SO_4$ (from 10%) in 0.02M TEA (at 2.5% increments) until at 20% $(NH_4)_2SO_4$ it crystallises out [Hansen et al. *Methods Enzymol* **8** 248 *1966*].

Uridine 5'-(1-thio) monophosphate *[15548-52-4, 18875-72-4 (Abs Stereochem specified)]* **and Uridine 5'-(α-thio) diphosphate** *[RS(α-P) 27988-67-6; R(α-P) 72120-52-6]* , $pK_{Est(1)}\sim$ **6.4**, $pK_{Est(2)}\sim$ **9.5** The Et_3N salt was purified by dissolving ~4g in 500mL of H_2O (add a drop or two of Et_3N if it does not dissolve) and chromatographed by applying to a column (3 x 30cm) of DEAE-Sephadex A-25 and eluted with a 1.4L linear gradient of $Et_3NH.HCO_3$ from 0.05 to 0.55M, pH 7.8 and 4°. The product eluted between 0.2-0.3M $Et_3N.HCO_3$. Pooled fractions were evaporated and the residue was twice taken up in EtOH and evaporated to dryness to remove the last traces of $Et_3NH.HCO_3$. ^{31}P NMR: P_α is a doublet at -40.81 and -40.33, and P_β at 7.02ppm, $J_{\alpha,\beta}$ 32.96Hz. [*Biochemistry* **18** 5548 *1979*.]

Uridylic acid (di-Na salt) *[27821-45-0]* **M 368.2, m 198.5°, pK_3^{25} 6.63, pK_4^{20} 9.71.** Crystd from MeOH.

Urokinase (from human urine) *[9039-53-6]* M_r 53,000, **[EC 3.4.21.31].** Crystn of this enzyme is induced at pH 5.0 to 5.3 (4°) by careful addition of NaCl with gentle stirring until the soln becomes turbid (silky sheen). The NaCl concentration is increased gradually (over several days) until 98% of saturation is achieved whereby the urokinase crystallises as colourless thin brittle plates. It can be similarly recrystd to maximum specific activity [104K CTA units/mg of protein (Sherry et al. *J Lab Clin Med* **64** 145 *1964*)]. [Lesuk et al. *Science* **147** 880 *1965*; NMR: Bogusky et al. *Biochemistry* **28** 6728 *1989*.] It is a plasminogen activator [Gold et al. *Biochem J* **262** *1989*].

(+)-Usnic acid (2,6-diacetyl-7,9-dihydroxy-8,9b-dimethyldibenzofuran-1,3(2H,9bH)-dione) *[7562-61-0, 125-46-2]* **M 344.3, m 201-204°, 203-206°, $[\alpha]_{546}^{20}$ +630° (c 0.7, $CHCl_3$), pK_1 4.4, pK_2 8.8, pK_3 10.7.** This very weak acid is the natural form which is recrystd from Me_2CO, MeOH or *C_6H_6. At 25° it is soluble in H_2O (<0.01%), Me_2CO (0.77%), EtOAc (0.88%), $MeOCH_2CH_2OH$ (0.22%) and furfural (7.32%). [Curd and Robertson *J Chem Soc* 894 *1937*; Barton and Brunn *J Chem Soc* 603 *1953*; resolution: Dean et al. *J Chem Soc* 1250 *1953*; synthesis: Barton et al. *J Chem Soc* 538 *1956*.]

(-)-Usnic acid (2,6-diacetyl-7,9-dihydroxy-8,9b-dimethyldibenzofuran-1,3(2H,9bH)-dione) *[6159-66-6, 7562-61-0]* **M 344.3, m 201-204°, 204°, $[\alpha]_D^{20}$ -495°** (c 0.9, $CHCl_3$). Properties almost similar to those of the preceding entry.

Ustilagic acid (Ustizeain B, di-D-glucosyldihydroxyhexadecanoic acid) *[8002-36-6]* **M ~780, m 146-147°, $[\alpha]_D^{23}$ +7°** (c 1, pyridine), **pK ~ 4.9.** It is a mixture of partly acetylated di-D-glucosyldihydroxyhexadecanoic acid which crysts from diethyl ether. Also purified from the culture by dissolving in hot MeOH, filtering and concentrating by blowing a current of air until the soln becomes turbid, then heating to 50° and adding 4 vols of H_2O (also at 50°) and allowing to cool very slowly. Filter off the white solid and dry in air. [Lemieux et al. *Can J Chem* **29** 409, 415 *1951*; *Can J Biochem Physiol* **33** 289 *1955*.]

Valinomycin (Potassium ionophore I) *[2001-95-8]* **M 111.3, m 186-187°, 190°,** $[\alpha]_D^{20}$ +31.0° (c 1.6, *C_6H_6). Recryst from dibutyl ether or Et_2O. Dimorphic, modification A crystallises from *n*-octane, and modification B crystallises from $EtOH/H_2O$. Soluble in pet ether, $CHCl_3$, AcOH, BuOAc and Me_2CO. [*J Am Chem Soc* **97** 7242 *1975*; UV, IR and NMR see *Chem Ber* **88** 57 *1955*.]

(±)-Verapramil hydrochloride (5-[N-{3,4-dimethoxyphenylethyl}methylamino]-2-[3,4-dimethoxyphenyl]-2-isopropylvaleronitrile HCl) *[23313-68-0]* **M 491.1, m 138.5-140.5°,**

pK$_{Est}$ ~ 10.6. Purified by dissolving in EtOH, filtering (if insoluble particles are present) and adding Et$_2$O, filtering the salt, washing with Et$_2$O and drying *in vacuo*. It has the following solubilities: hexane (0.001%), CH$_2$Cl$_2$ (~10%), MeOH (~10%) EtOH (20%) and H$_2$O (8.3%). It has UV λ_{max} 232 and 278nm. The *free base* is a viscous yellow oil **b** 243-246°/0.01mm (n$_D^{25}$ 1,5448) and is almost insol in H$_2$O but sol in organic solvents. It is a Ca channel antagonist and is a coronary vasodilator. [Ramuz *Helv Chim Acta* **58** 2050 *1975*; Harvey et al. *Biochem J* **257** 95 *1989*.]

Veratridine (3-veratroylveracevine) *[71-62-5]* **M 673.8, pK$_{Est}$ ~7 (quinolizidine N).** An alkaloid neurotoxin purified from veratrine. [McKinney et al. *Anal Biochem* **153** 33 *1986*.]

Vinblastine sulfate (vincaleucoblastine) *[143-67-9]* **M 909.1, m 284-285°, [α]$_D^{25}$ -28° (c 1, MeOH), pK$_1$ 5.4, pK$_2$ 7.4.** Purified by recrystn from H$_2$O and dried *in vacuo*. [Neuss et al. *J Am Chem Soc* **86** 1440 *1964*.] The *free base* is recrystd from MeOH or EtOH and has **m** 210-212°, 211-216°, [α]$_D^{25}$ +42° (CHCl$_3$); and has UV λ_{max} 214 and 259nm (log ε 4.73 and 4.21). The *dihydrochloride dihydrate* has **m** 244-246°. [Bommer et al. *J Am Chem Soc* **86** 1439 *1964*.] It is a monoamine oxidase inhibitor [Keun Son et al. *J Med Chem* **33** 1845 *1990*].

Vincristine sulfate (22-oxovincaleucoblastine sulfate) *[2068-78-2]* **M 925.1, m 218-220°, [α]$_D^{25}$ +26.2° (CH$_2$Cl$_2$), pK$_1$ 5.0, pK$_2$ 7.4 (in 33% aq Me$_2$NCHO).** Recryst from MeOH. It has UV λ_{max} 220, 255 and 296nm (log ε 4.65, 4.21 and 4.18). It is a monoamine oxidase inhibitor and is used in cancer research [Son et al. *J Med Chem* **33** 1845 *1990*; Horio et al. *Proc Natl Acad Sci USA* **85** 3580 *1988*].

Viomycin sulfate (Viocin, Tuberactinomycin B) *[37883-00-4]* **M 685.7, m 266°(dec), [α]$_D^{17}$ -29.5° (c 1, H$_2$O), pK$_1$ 7.2 (8.2), pK$_2$ 10.3.** Crystd from H$_2$O-EtOH and dried in a vacuum. Dry material is *hygroscopic* and should be stored dry. The UV has λ_{max} at 268 and 285nm (log ε 4.4 and 4.2) in H$_2$O. [Kitigawa et al. *Chem Pharm Bull Jpn* **20** 2176 *1972*.] The *hydrochloride* forms *hygroscopic* plates with **m** 270°(dec), [α]$_D^{18}$ -16.6° (c 1, H$_2$O) with λ_{max} 268nm (log ε 4.5) in H$_2$O; 268nm (log ε 4.4) in 0.1N HCl and 285nm (log ε 4.3) in 0.1N NaOH.

Vitamin A acid [Retinoic acid, 3,7-dimethyl-9-(2,6,6-trimethyl-1-cyclohexenyl)-2,4,6,8-nonatetraen-1-oic acid] *[302-79-4]* **M 300.4, m 180-181°, 180-182°, pK$_{Est}$ ~ 4.2.** Purified by chromatography on silicic acid columns, eluting with a small amount of EtOH in hexane. Dissolve in Et$_2$O, wash with H$_2$O, dry (Na$_2$SO$_4$), evaporate and the solid residue crystd from MeOH (0.53g /3.5mL MeOH to give 0.14g) or EtOH. Also recrystd from *i*-PrOH, or as the *methyl ester* from MeOH. UV in MeOH has λ_{max} 351nm (ε 45,000). 9-*Cis*-acid forms yellow needles from EtOH, **m** 189-190°, UV in MeOH has λ_{max} 343nm (ε 36,500) and 13-*cis*-acid forms red-orange plates from *i*-PrOH, **m** 174-175°, UV has λmax 345nm (ε 39,800). Store in the dark, in an inert atmosphere, at 0° [Robeson et al. *J Am Chem Soc* **77** 4111 *1955*].

Vitamin A alcohol (retinol) *[68-26-8]* **M 286.5, A$_{1cm}^{1\%}$ (λmax)(*all-trans*) 1832 (325 nm), (*13-cis*) 1686 (328nm), (*11-cis*) 1230 (319 nm), (*9-cis*) 1480 (323 nm), (*9,13-di-cis*) 1379 (324 nm), (*11,13-di-cis*) 908 (311 nm) in EtOH.** Purified by chromatography on columns of water-deactivated alumina eluting with 3-5% acetone in hexane. Separation of isomers is by TLC plates on silica gel G, developed with pet ether (low boiling)/methyl heptanone (11:2). Stored in the dark, under nitrogen, at 0°, as in diethyl ether, acetone or ethyl acetate. [See Gunghaly et al. *Arch Biochem Biophys* **38** 75 *1952*.]

Vitamin A aldehyde [all-*trans*-retinal; 3,7-dimethyl-9-(2,6,6-trimethyl-1-cyclohexenyl)-2,4,6,8-nonatetraen-1-al] *[116-31-4]* **M 284.4, m 61-64°.** Separated from retinol by column chromatography on water-deactivated alumina. Eluted with 1-2% acetone in hexane, or on TLC plates of silica gel G development. It crystallises from pet ether or *n*-hexane as yellow-orange crystals, and the UV in hexane has λ_{max} 373nm (A$_{1cm}^{1\%}$ 1,548) [368nm (ε 48000)]. It is an **irritant** and is light sensitive. Store in sealed ampoules under N$_2$. The **semicarbazone** forms yellow crystals from CHCl$_3$-Et$_2$O or EtOH, **m** 199-201°(dec). The 9-*cis*-isomer *[514-85-2]* and the 13-*cis*-isomer *[472-86-6]* [λ_{max} 375nm (ε 1250) in EtOH] are also available commercially.

Vitamin B$_1$ Hydrochloride   [Aneurine hydrochloride, Thiamine hydrochloride, 3{(4-amino-2-methyl-5-pyrimidinyl)methyl}-4-methylthiazolium chloride monohydrochloride] *[67-03-8]* M 337.3, m 248°(dec), 249-250°, monohydrate m 135°(dec), pK$_1^{25}$ 4.8, pK$_2^{25}$ 9.2. Crystallises from 95% EtOH (sol, *ca* 1%). The monohydrate is dehydrated by drying at 100° *in vacuo* over H$_2$SO$_4$, but is *hygroscopic* and picks up one mol. of H$_2$O readily. It can be sterilised at 100° if the pH of the solution is below 5.5. The *nitrate* has **m** 196-200°(dec) and is more stable than the hydrochloride. The *picrolonate* crystallises from H$_2$O and is dimorphic, **m** 164-165° and 228-229°(dec). [Todd and Bergel *J Chem Soc* 364, 367 *1937*; *J Am Chem Soc* **58** 1063, 1504 *1936*, **59** 526 *1937*.]

Vitamin B$_2$ [Riboflavin, Lactoflavin, 6,7-dimethyl-9-(D-1'-ribityl)isoalloxazine] *[83-88-5]* M 376.4, m 278-282°(dec with darkening at 240°), 281-282°, [α]$_D^{25}$ -112° to -122° (c 2.5, 0.02M NaOH), [α]$_D^{20}$ -59° (c 0.23, AcOH), pK$_1$ 1.7, pK$_2$ 9.69 (10.2). It crystallises from H$_2$O as a yellow-orange powder in three different forms with differing amounts of H$_2$O. It melts if placed in an oil bath at 250°, but decomposes at 280° if heated at a rate of 5°/min. Solubility in H$_2$O is 1g in 3000-15000mL depending on the crystal structure. Sol in EtOH at 25° is 4.5mg in 100mL. Store in the dark because it is decomposed to lumichrome by UV light.

Vitamin B$_6$ hydrochloride   (adermine, pyridoxine HCl, 3-hydroxy-4,5-bis[hydroxymethyl]-2-methylpyridine HCl)   *[58-56-0]* M 205.6, m 208-208.5°, 208-209°(dec), 209-210°(dec), 205-212° (sublimes), pK$_1^{25}$ 5.0 (3-OH), pK$_2^{25}$ 8.96 (pyridinium$^+$). Purified by recrystn form EtOH-Me$_2$CO, *n*-BuOH or MeOH-Et$_2$O. Its solubility in H$_2$O is 22% and in EtOH it is 1.1%. It is insoluble in Et$_2$O and CHCl$_3$. Acidic aqueous solns are stable at 120°/30 min. The *free base* has **m** 159-160° after recrystn from Me$_2$CO and sublimation at 140-145°/0.0001mm. It has UV λ$_{max}$ at 290nm (ε 84000) in 0.1N aqueous HCl and 253 and 325nm (ε 3700 and 7100). [Khua and Wendt *Chem Ber* **71** 780 *1938*, **72** 311 *1939*; Harris and Folkers *J Am Chem Soc* **61** 1242 *1939*; Harris et al. *J Am Chem Soc* **62** 3198 *1940*.] See also **Pyridoxal-5'-phosphate H$_2$O** above.

Vitamin B$_{12}$ (cyanocobalamine, α-[5,6-dimethylbenzimidazolyl]cyano cobamide) *[68-19-9]* M 1355.4, m darkens at 210-220° and does not melt below 300°, [α]$_{656}^{23}$ -59° (H$_2$O). Crystd from de-ionized H$_2$O, solubility in H$_2$O is 1g/80g and dried under vacuum over Mg(ClO$_4$)$_2$. The dry red crystals are *hygroscopic* and can absorb ~12% of H$_2$O. A soln at pH 4.5-5 can be autoclaved for 20min at 120° without dec. Aqueous solns are stabilised by addition of (NH$_4$)$_2$SO$_4$. [Golding *Comprehensive Organic Chem* Vol 5 (Ed. Haslam; Pergamon Press, NY, 1979) pp 549-584.]
Alternatively an aqueous soln of the coenzyme was concentrated, if necessary in a vacuum at 25° or less, until the concentration was 0.005 to 0.01M (as estimated by the OD at 522 nm of an aliquot diluted with 0.01M K-phosphate buffer pH 7.0). If crystals begin to form on the walls of the container they should be re-dissolved with a little H$_2$O. The concentrated soln is placed in a glass stoppered flask and diluted with 5vols of Me$_2$CO. After 2-3h at 3° it is centrifuged (10,000xg/10 min) in Me$_2$CO-insol plastic tubes to remove some amorphous ppte. The clear supernatant is inoculated with a small crystal of the vitamin and allowed to crystallise overnight at 3°. Crystals are formed on the walls and the bottom of the container. A further 2vols of Me$_2$CO are added and set aside at 3° to further crystallise. Crystallisation is followed by observing the OD$_{522}$ of the supernatant. When the OD falls to 0.27 then *ca* 94% of the crystals have separated. The supernatant is decanted (saved for obtaining a second crop) and the crystals are washed with a little cold 90% aqueous Me$_2$CO (2 x), 100% Me$_2$CO (2 x), Et$_2$O (2 x) at which time the crystals separated from the glass walls. Allow to settle and remove residual Et$_2$O with a stream of dry N$_2$. The process can be repeated if necessary. The crystals can be dried in air or in a vacuum for 2h over silica gel at 100° with an 8-9% weight loss. [Barker et al. *Biochem Prep* **10** 33 *1963*.]
This material gives a single spot of paper chromatography [see Weissbach et al. *J Biol Chem* **235** 1462 *1960*.] The vitamin is soluble in H$_2$O (16.4mM at 24°, 6.4mM at 1°), in EtOH and PhOH but insol in Me$_2$CO, Et$_2$O, CH$_2$Cl$_2$ and dioxane. UV: λ$_{max}$ 260, 375 and 522nm (ε 34.7 x 10^6, 10.9 x 10^6 and 8.0 x 10^6 / mole) in H$_2$O. The dry crystals are stable for months in the dark, but aqueous solns decompose on exposure to VIS or UV light or alkaline CN$^-$, but stable in the dark at pH 6-7. The vitamin is inactivated by strong acids or alkalies. [Barker et al. *J Biol Chem* **235** 480 *1960*; see also *Vitamin B$_{12}$* (Zagalak and Friedrich Eds) W de Gruyter, Berlin *1979*.]

Vitamin C   see **ascorbic acid** entry on p. 116 in Chapter 4.

Vitamin D₂ *[50-14-6]* M 396.7, m 114-116°, $[\alpha]^{20}_{546}$ +122° (c 4, EtOH) Converted into their 3,5-dinitrobenzoyl esters, and crystd repeatedly from acetone. The esters were then saponified and the free vitamins were isolated. [Laughland and Phillips *Anal Chem* **28** 817 *1956*.]

Vitamin D₃ *[67-97-0]* 384.6, m 83-85°, $[\alpha]^{20}_{546}$ +126° (c 2, EtOH). Converted into their 3,5-dinitrobenzoyl esters, and crystd repeatedly from acetone. The esters were then saponified and the free vitamins were isolated. [Laughland and Phillips *Anal Chem* **28** 817 *1956*.]

Vitamin E (**2R,4'R,8'R-α-tocopherol, natural active isomer**) *[59-02-9]* M 430.7, m 2.5-3.5°, b 200-220°/0.1mm, 200°/0.005mm, d^{25}_4 0.950, n^{25}_D 1.5045, $[\alpha]^{25}_D$ +3.58° (c 1.1, *C₆H₆). Viscous yellow oil which is distd at high vacuum. It has λ_{max} 294nm ($E^{1\%}_{1cm}$ 71). It is oxygen and light sensitive and is best stored as its stable acetate which is purified by evaporative distn at b 180-200°(bath temp)/0.7mm, $[\alpha]^{25}_D$ +3.3° (c 5.1, EtOH). [NMR: Cohen et al. *Helv Chim Acta* **64** 1158 *1981*; Burton and Ingold *Acc Chem Res* **19** 194 *1986*; Karrer et al. *Helv Chim Acta* **21** 520 *1938*.]

Vitamin E acetate (**DL-α-tocopheryl acetate**) *[7695-91-2]* M 472.8, m -27.5°, b 194-196°/0.01mm, 222-224°/0.3mm, d^{20}_4 0.958, n^{20}_D 1.4958. It is a viscous liquid which is purified by distn under high vacuum in an inert atm and stored in sealed ampoules in the dark. It is considerably more stable to light and air than the parent unacetylated vitamin. It is insoluble in H₂O but freely soluble in organic solvents. All eight stereoisomers have been synthesised. The commercially pure *d-α–tocopheryl acetate* (2R,4'R,8'R) has **b** 180-200°/0.7mm and $[\alpha]^{20}_D$ +3.9° (c 5, EtOH). [Cohen et al. *Helv Chim Acta* **64** 1158 *1981*.]

Vitamin K₁ (**2-methyl-3-phytyl-1,4-naphthoquinone**) *[84-80-0]* M 450.7, m -20°, b 141-140/0.001mm, b 140-145°/10⁻³ mm, d^{25}_{25} 0.967, n^{25}_D 1.527, $[\alpha]^{20}_D$ -0.4° (c 57.5, *C₆H₆). Yellow viscous oil, which can be distd at high vacuum practically unchanged. Insoluble in H₂O, but soluble in common organic solvents. Store in the dark under N₂, oxygen sensitive. $A^{1\%}_{1cm}$ 328 at 248nm. [*J Am Chem Soc* **61** 2557 *1939*, **76** 4592 *1954*; *Helv Chim Acta* **27** 225 *1954*.]

Vitamin K₃ (**2-methyl-1,4-naphthoquinone, Menadione, Menaphthone**) *[58-27-5]* M 172.2, m 105-106°, 105-107°. Recrystd from 95% EtOH, or MeOH after filtration. Bright yellow crystals which are decomposed by light. Solubility in EtOH is 1.7% and in *C₆H₆ it is 10%. It **IRRITATES** the mucous membranes and skin. [Fieser *J Biol Chem* **133** 391 *1940*.]

Xanthine (**2,6-dihydroxypurine, purine-2,6(1H,3H)dione**) *[69-89-6]* M 152.1, pK₁ 0.8 [protonation of imidazole 7(9)NH], pK₂ 7.44 [monoanion 1(3)NH], pK₃ 11.12 [dianion 1,3-N²⁻]. The monohydrate separates in a microcryst form on slow acidification with acetic acid of a solution of xanthine in dil NaOH. Also ppted by addition of conc NH₃ to its soln in hot 2N HCl (charcoal). After washing with H₂O and EtOH, it is dehydrated on heating above 125°. Sol in H₂O is 1 in 14,000 at 16° and 1 in 1,500 and separates as plates from boiling H₂O. It has no **m**, but the *perchlorate* has **m** 262-264° [Lister *Heterocyclic Compounds, Fused Pyrimidines—Purines Part II*, Ed. Brown, J.Wiley & Sons, *1971*].

Xanthopterin monohydrate (**2-amino-4,6-dihydroxypteridine, 2-amino-pteridin-4,6(1H,5H)-dione**) *[5979-01-1 (H₂O), 119-48-8 (anhydr)]* M 197.2, m <300°, pK₁ 1.6 (basic), pK₂ 6.59 (acidic), pK₃ 9.31 (acidic)(anhydrous species), and pK₁ 1.6 (basic), pK₂ 8.65 (acidic), pK₃ 9.99 (acidic)(7,8-hydrated species). Purification as for isoxanthopterin. Crystd by acidifying an ammoniacal soln, and collecting by centrifugation followed by washing with EtOH, ether and drying at 100° *in vacuo*. Paper chromatography R_F 0.15 (*n*-PrOH, 1% aq NH₃, 2:1), 0.36 (*n*-BuOH,AcOH, H₂O, 4:1:1) and 0.47 (3% aq NH₃). [Inoue and Perrin *J Chem Soc* 260 *1962*; Inoue *Tetrahedron* **20** 243 *1964*; see also Blakley *Biochemistry of Folic Acid and Related Pteridines* North Holland Publ Co, Amsterdam 1969.]

Xanthotoxin (Methoxalen, 9-methoxyfuro[3,2-g][1]benzopyran-7-one) *[298-81-7]* **M 216.2, m 146-148°, 148°, 148-149°.** Purified by recrystn from $^*C_6H_6$-pet ether (b 60-80°) as silky needles, EtOH-Et$_2$O as rhombic prisms or hot H$_2$O as needles. It is soluble in aqueous alkali due to ring opening of a lactone but recyclises upon acidification. It has UV λ_{max} in EtOH at 219, 249 and 300nm (log ε 4.32, 4.35 and 4.06) and ^{1}H NMR in CDCl$_3$ with δ at 7.76 (d, 1H, J 10 Hz), 7.71 (d, 1H, J 2.5 Hz), 7.38 (s, 1H), 6.84 (d, 1H, J 2.5 Hz), 6.39 (d, 1H, J 10 Hz) and 4.28 (s, 3H). [Nore and Honkanen *J Heterocycl Chem* **17** 985 *1980.*] It is a DNA intercalator and is used in the treatment of dermal diseases [Tessman et al. *Biochemistry* **24** 1669 *1985.*]

Xylanase (from *Streptomyces lividans*) *[37278-89-0]* **M$_r$ 43,000 [EC 3.2.1.8].** Purified by anion-exchange chromatography on an Accell QMA column and finally by HPLC using a ProteinPak DEAE 5PW anion-exchange column. Solutions were stored frozen at -70°. [Morosoli et al. *Biochem J* **239** 587 *1986*; Wong et al. *Microbiol Rev* **52** 305 *1988.*]

Zeatin (*trans-N^6*-[4-hydroxy-3-methylbut-2-en-1-yl]adenine) *[1637-39-4]* M 219.3, m 207-208, 209-209.5°, pK$_1$ 4.4 (basic), pK$_2$ 9.8 (acidic).

Purified by recrystn from EtOH or H$_2$O. The UV has λ_{max} at 207 and 275nm (ε 1400 and 14650) in 0.1N aqueous HCl; 212 and 270nm (ε 17050 and 16150) in aqueous buffer pH 7.2; 220 and 276nm (ε 15900 and 14650) in 0.1N aq NaOH. The *picrate* has **m** 192-194° (from H$_2$O) from which zeatin can be recovered by treatment with Dowex 1 x 8 (200-400 mesh, OH$^-$ form). [Letham et al. *Aust J Chem* **22** 205 *1969*; *Proc Chem Soc (London)* 230 *1964*; Shaw and Wilson *Proc Chem Soc (London)* 231 *1964.*] It is a cell division factor (plant growth regulator) [Letham and Palni *Ann Rev Plant Physiol* **34** 163 *1983*] and inhibits mitochondrial function [Miller *Plant Physiol* **69** 1274 *1982*]. Its *9-riboside* is a cytokine [McDonald and Morris *Methods Enzymol* **100** 347 *1985*].

GENERAL INDEX

For individual organic chemicals, listed alphabetically, see Chapter 4, beginning on Page 80; or inorganic and metal-organic chemicals see Chapter 5, beginning on Page 389, and for biochemicals and related products see Chapter 6, beginning on Page 500. It is much faster to use the CAS Registry Numbers Index on p 585 for locating specific compounds irrespective of which chapter they are in.